GPS SATELLITE SURVEYING

GPS SATELLITE SURVEYING

Fourth Edition

ALFRED LEICK
LEV RAPOPORT
DMITRY TATARNIKOV

WILEY

Cover image: Eric Morrison, UMaine and Wiley
Cover design: Wiley

This book is printed on acid-free paper.

Published by John Wiley & Sons, Inc., Hoboken, New Jersey
Published simultaneously in Canada

Library of Congress Cataloging-in-Publication Data

Leick, Alfred.
GPS satellite surveying / Alfred Leick, Lev Rapoport Dmitry Tatarnikov. – Fourth edition.
pages cm
Includes index.
ISBN 978-1-118-67557-1 (cloth) – 9781119018285 (epdf) – 9781119018261 (epub) 1. Artificial satellites in surveying. 2. Global Positioning System. I. Rapoport, Lev. II. Tatarnikov, Dmitry. III. Title.
TA595.5.L45 2015
526.9'82–dc23 2014040585

Printed in the United States of America

10 9 8 7 6 5 4 3 2 1

CONTENTS

PREFACE

GPS Satellite Surveying has undergone a major revision in order to keep abreast with new developments in GNSS and yet maintain its focus on geodesy and surveying. All chapters have been reorganized in a more logical fashion. Because the GNSS systems have developed significantly since the last edition of the book, we have added new material on the GLONASS, Beidou, and Galileo systems, as well as on the ongoing modernization of GPS. A separate chapter was included on recursive least squares. Another chapter on RTK implementation was added that uses these recursive least-squares algorithms to process across-receiver observation differences and is capable of accepting observations from all GNSS systems. Examples are supported by real data processing. A third new chapter was added on GNSS user antennas. This chapter was prepared by an antenna expert to provide the necessary background information and details to allow practicing engineers to select the right antenna for a project. As to GNSS processing approaches, major new sections were added on PPP-RTK and TCAR. Six new additional appendices were added containing in-depth mathematical supplements for those readers who enjoy the mathematical rigor.

The original author of *GPS Satellite Surveying*, Alfred Leick, appreciates the contributions of Lev Rapoport and Dmitry Tatarnikov and most cordially welcomes these very qualified individuals as co-authors. All three of us wish to thank our families for their outstanding support throughout our professional careers. Lev Rapoport wishes to thank Javad GNSS for permission to use their receivers Triumph-1, Delta TRE-G3T, and Delta Duo-G2D for data collection, and Dr. Javad Ashjaee for the opportunity to get acquainted with GNSS technologies and observe its history through the eyes of a company employee. Dmitry Tatarnikov wishes to thank his colleagues at the Moscow Technology Center of Topcon for their contributions to the research, development, and production of antennas, and the management of Topcon Corporation for support of this work. Alfred Leick expresses his sincere appreciation to anybody contributing to this and any of the previous revisions of *GPS Satellite Surveying*. We appreciate Tamrah Brown's assistance in editing the draft in such a short period of time.

AUTHOR BIOGRAPHIES

Alfred Leick received a Ph.D. from Ohio State University, Department of Geodetic Science, in 1977. He is the Editor-in-Chief of the peer-reviewed journal *GPS Solutions* (Springer), and author of numerous technical publications. His teaching career at the University of Maine in the area of GPS (Global Positioning System), geodesy, and estimation spans 34 years. Other teaching assignments included photogrammetry and remote sensing, digital image processing, linear algebra, and differential equations. He was the creator of the online GPS-GAP (GPS, Geodesy and Application Program) program at the University of Maine, which now continues to be available via Michigan Technological University (www.onlineGNSS.edu) in modified form. Dr. Leick launched his GPS research in 1982 when testing the Macrometer satellite receiver prototype at M.I.T. He continued GPS research throughout the years, including while on sabbatical leave at the Air Force Geophysics Laboratory (Cambridge, MA) in 1984, 3S-Navigation (Irvine, CA) in 1996, Jet Propulsion Laboratory (Pasadena, CA) in 2002, as an Alexander von Humboldt Research Associate at the University of Stuttgart in 1985, a Fulbright Scholar at the University of Sao Paulo during the summers of 1991 and 1992, and a GPS Project Specialist on behalf of World Band and NRC (National Research Council) at Wuhan Technical University of Surveying and Mapping (P.R. China) in the Spring of 1990. He is a Fellow of ACSM (American Congress on Surveying and Mapping).

Dmitry V. Tatarnikov holds a Masters in EE (1983), Ph.D. (1990), and Doctor of Science (the highest scientific degree in Russia, 2009) degrees, all in applied electromagnetics and antenna theory and technique from the Moscow Aviation Institute—Technical University (MAI), Moscow, Russia. He joined the Antenna and Microwave Research Dept. of MAI in 1979, and is currently a professor of Radiophysics, Antennas and Microwave Dept. at MAI. In 1979–1994, he was involved in microstrip-phased array antenna research and development. In 1994, he joined Ashtech R&D Center in Moscow as an Antenna Research Fellow in the high-precision GNSS area. In 1997–2001, he was with Javad Positioning Systems as a senior scientist in the antenna area, and since 2001 he has been leading the Antenna Design with Topcon Technology Center, Moscow, Russia. Prof. Tatarnikov has authored and co-authored more than 70 publications in this area, including a book, research papers, conference presentations, and 12 patents. He has developed student courses in applied electromagnetics, numerical electromagnetics, and receiver GNSS antennas. He is a member of IEEE and the Institute of Navigation (ION), USA.

Lev B. Rapoport received a Master's in Electrical Engineering in 1976 from the radio-technical department of the Ural Polytechnic Institute, Sverdlovsk, and a Ph.D.

in 1982 in automatic control from the Institute of System Analysis of the Russian Academy of Science (RAS), Moscow. In 1995, he received a Doctor of Science degree in automatic control from the Institute of Control Sciences RAS. Since 2003, he has held the head of the "Non-linear Systems Control" laboratory position in this institute. From 1993 to 1998, he worked for Ashtech R&D Center in Moscow part time as a researcher. From 1998 to 2001, he was with Javad Positioning Systems as a Real Time Kinematics (RTK) team leader. From 2001 to 2005, he worked for Topcon Positioning Systems where he was responsible for RTK and Machine Control. From 2005 to 2011, he worked for Javad GNSS as an RTK team leader. Dr. Rapaport has been a consultant at Huawei Technologies R&D Center in Moscow since 2011. Dr. Rapoport is professor of the control problems department of the Moscow Institute of Physics and Technology. He has authored 90 scientific papers, many patents, and conference presentations. He is a member of IEEE and the Institute of Navigation (ION). His research interests include navigation and control.

ACKNOWLEDGMENTS

Financial support from the following company is gratefully acknowledged:

TOPCON

ABBREVIATIONS

COMMONLY USED GNSS ABBREVIATIONS

ARNS	Aeronautical Radio Navigation Service
ARP	Antenna reference point
AS	Antispoofing
ASK	Amplitude shift keying
B1	B1 Beidou carrier (1561.098 MHZ)
B2	B2 Beidou carrier (1207.14 MHz)
B3	B3 Beidou carrier (1268.52MHz)
BOC	Binary offset carrier
BPSK	Binary phase shift keying
C/A-code	Coarse/acquisition code (1.023 MHz)
CDMA	Code division multiple access
CIO	Celestial intermediary origin
CCRF	Conventional celestial reference frame
CEP	Celestial ephemeris pole
CORS	Continuously operating reference stations
CTP	Conventional terrestrial pole
CTRS	Conventional terrestrial reference system
DGPS	Differential GPS
DOD	Department of Defense
DOP	Dilution of precision
DOY	Day of year
E6	E6 Galileo carrier (1278.75 MHz)
ECEF	Earth-centered earth-fixed coordinate system

EOP	Earth orientation parameter
FAA	Federal Aviation Administration
FBSR	Feedback shift register
FDMA	Frequency division multiple access
FSK	Frequency shift keying
FOC	Full operational capability
GAST	Greenwich apparent sidereal time
GDOP	Geometric dilution of precision
GEO	Geostationary earth orbit
GIF	Geometry-free and ionospheric-free solution
GIM	Global ionospheric model
GLONASS	Global'naya Navigatsionnaya Sputnikkovaya Sistema
GNSS	Global navigation satellite system
GML	Gauss midlatitude functions
GMST	Greenwich mean sidereal time
GPS	Global positioning system
GPSIC	GPS Information Center
GPST	GPS time
GRS80	Geodetic reference system of 1980
HDOP	Horizontal dilution of precision
HMW	Hatch/Melbourne/Wübbena function
HOW	Handover word
IAG	International Association of Geodesy
IAU	International Astronomical Union
ICRF	International celestial reference frame
IERS	International Earth Rotation Service
IGDG	Internet-based dual-frequency global differential GPS
IGS	International GNSS Service
IGSO	Inclined geosynchronous satellite orbit
ISC	Intersignal Correction
ITRF	International terrestrial reference frame
IOC	Initial operational capability
ION	Institute of Navigation
IWV	Integrated water vapor
JD	Julian date
JPL	Jet Propulsion Laboratory
L1	L1 carrier (1575.42 MHz)
L2	L2 carrier (1227.6 MHz)
L5	L5 carrier (1176.45 MHz)
LAMBDA	Least-squares ambiguity decorrelation adjustment
LC	Lambert conformal mapping
LEO	Low-earth orbiting satellite
LHCP	Left-hand circular polarization
MEO	Medium earth orbit
NAD83	North American datum of 1983

NAVSTAR	Navigation Satellite Timing and Ranging
NEP	North ecliptic pole
NGS	National Geodetic Survey
NIST	National Institute of Standards and Technology
NOAA	National Oceanic and Atmospheric Administration
PCO	Phase center offset
OPUS	Online processing user service
OTF	On-the-fly ambiguity resolution
P-code	Precision code (10.23 MHz)
PCV	Phase center variation
PDOP	Positional dilution of precision
ppb	parts per billion
ppm	parts per million
PPP	Precise point positioning
PPS	Precise positioning service
PRN	Pseudorandom noise
PSK	Phase shift keying
PWV	Precipitable water vapor
QPSK	Quadature phase shift keying
RHCP	Right-hand circular polarization
RINEX	Receiver independent exchange format
RNSS	Radio navigation satellite services
RTCM	Radio Technical Commission for Maritime Services
RTK	Real-time kinematic positioning
SA	Selective availability
SBAS	Satellite-based augmentation system
SINEX	Solution independent exchange format
SLR	Satellite laser ranging
SNR	Signal-to-noise ratio
SP3	Standard product #3 for ECEF orbital files
SPC	State plane coordinate system
SPS	Standard positioning service
SRP	Solar radiation pressure
SVN	Space vehicle launch number
SWD	Slant wet delay
TAI	International atomic time
TDOP	Time dilution of precision
TEC	Total electron content
TECU	TEC unit
TIO	Terrestrial intermediary origin
TLM	Telemetry word
TM	Transverse Mercator mapping
TOW	Time of week
TRANSIT	Navy navigation satellite system
URE	User range error

USNO	U.S. Naval Observatory
UT1	Universal time corrected for polar motion
UTC	Coordinate universal time
VDOP	Vertical dilution of precision
VLBI	Very long baseline interferometry
VRS	Virtual reference station
VSWR	Voltage standing wave ratio
WAAS	Wide area augmentation service
WADGPS	Wide area differential GPS
WGS84	World Geodetic System of 1984
WVR	Water vapor radiometer
Y-code	Encrypted P-code
ZHD	Zenith hydrostatic delay
ZWD	Zenith wet delay

GPS SATELLITE SURVEYING

CHAPTER 1

INTRODUCTION

Over the last decade, the development and application of GNSS (global navigation satellite system) has been unabatedly progressing. Not only is the modernization of the U.S. GPS (global positioning system) in full swing, the Russian GLONASS (Global'naya Navigatsionnaya Sputnikovaya Sistema) system has undergone a remarkable recovery since its decline in the late 1990s to be now fully operational. The first static and kinematic surveys with the Chinese Beidou system are being published, and the signals of the European Galileo system are being evaluated. While many individuals might look back on the exciting times they were fortunate to experience since the launch of the first GPS satellite in 1978, there are many more enthusiastic individuals gearing up for an even more exciting future of surveying and navigation with GNSS. Yes, it seems like a long time has passed since sunset admirers on top of Mount Wachusett, seeing a GPS antenna with cables connected to a big "machine" in a station wagon were wondering if it would "take off," or if you were "on their side," or regular folks in a parking lot approaching a car with a "GPS" license plate were wondering if you had "such a thing."

Much has been published on the subject of GNSS, primarily about GPS because of its long history. Admirably efficient search engines uncover enormous amounts of resources on the Internet to make an author wonder what else is there to write about. We took the opportunity of updating GPS Satellite Surveying to add strength by including two additional authors, while looking at rearranging the material in a way that reflects the maturity and permanency of the subject and de-emphasizes the news of the day or minor things that may have gotten the early pioneers of GPS excited.

Perhaps the most visible outcome of the rearrangement of the material for this edition is that GNSS in earnest starts only in Chapter 5, which may come as a surprise to the unexpected reader. However, if was determined that first presenting the geodetic and statistical foundations for GPS Satellite Surveying would be more efficient, and then focusing on GNSS, thus taking advantage of having the prerequisites available and not being side-tracked by explaining essential fill-in material. Therefore, there are two chapters devoted to least-squares estimation, followed by a chapter on geodesy. These three chapters clearly identify the traditional clientele this book tries to serve, i.e., those who are interested in using GNSS for high-accuracy applications. The other chapters cover GNSS systems, GNSS positioning, RTK (real-time kinematic), troposphere and ionosphere, and GNSS user antennas. There are nine appendices.

Chapter 2, least-squares adjustment, contains enough material to easily fill a regular 3-credit-hour college course on adjustments. The focus is on estimating parameters that do not depend on time. The material is presented in a very general form independently of specific applications, although the classical adjustment of a geodetic or surveying network comes to mind as an example. The approach to the material is fairly unique as compared to a regular course on least squares because it starts with the mixed model in which the observations and the parameters are implicitly related. This general approach allows for an efficient derivation of various other adjustment models simply by appropriate specifications of certain matrices. Similarly, the general linear hypothesis testing is a natural part of the approach. Of particular interest to surveying applications are the sections on minimal and inner constraints, internal and external reliability, and blunder detection.

Chapter 3, recursive least squares, represents new material that has been added to this fourth revision. In particular in view of RTK application where the position of the rover changes with time, it was deemed appropriate to add a dedicated chapter in which the estimation of time-dependent parameters is the focus. Consequently, we changed the notation using the argument of time consistently to emphasize the time dependency. A strength of this chapter is that it explicitly deals with patterned matrices as they occur in RTK and many other applications. Apart from the term "recursive least squares," other terms might be "first-order partitioning regression" or "Helmert blocking," that express the technique applied to these patterned matrices. Although Chapters 2 and 3 are related since there is only one least-squares method, Chapter 3 stands on its own. It also could serve easily as a text for a regular 3-credit-hour college course.

Chapter 4 is dedicated to geodesy. It provides details on reference frames, such as the ITRF (international terrestrial reference frame), as well as the transformation between such frames. The geodetic datum is a key element in this chapter, which is defined as an ellipsoid of defined location, orientation, and size and an associated set of deflection of the vertical and geoid undulations. Establishing the datum, in particular measuring gravity to compute geoid undulations, is traditionally done by geodesists. The fact that here it is assumed that all this foundational material is given indicates that geodesy is treated not as a science by itself in this book but rather as an enabling element that supports accurate GNSS applications. As the "model for all," we present the three-dimensional (3D) geodetic model, which is applicable to networks of any size and assumes that the geodetic datum is available. In addressing the needs of surveying, the topic of conformal mapping of the ellipsoidal surface is treated in great detail. This includes, as a transitional product encountered along the way, computations on the ellipsoidal surface. It is well known that computing on the conformal mapping plane is limited by the area covered by the network since distortions increase with area. Additionally, the respective computations require the geodesic line, which is mathematically complicated, and the respective expressions are a result of lengthy but unattractive series expansions. Clearly, an attempt is made to point out the preference of the 3D geodetic model when there is the opportunity to do so.

Chapter 5, finally, introduces the various GNSS systems. In order to provide background information on satellite motions, the chapter begins with an elementary discussion of satellite motions, the Kepler elements that describe such motions, and the particularly simple theory of normal orbits, i.e., motion in a central gravity field. The disturbing forces that cause satellites to deviate from normal orbits are discussed as well. However, the material is not presented at the level of detail needed for accurate satellite orbit determination. We assume that orbit determination will continue to be handled by existing expert groups and that respective products will be available either through the broadcast navigation message or the International GNSS Service (IGS) and other agencies in the form of precise and/or ultra-rapid ephemeris and satellite clock data. This chapter includes new material on GPS modernization and on the GLONASS, Galileo, and Beidou systems. In the meantime, interface control documents are available for all these GNSS systems and posted on the Internet. The reader is advised to consult these documents and similar publications that expertly address the space segment.

Chapter 6 discusses in detail the various GNSS positioning approaches conveniently in "one place." It begins with specifying the fundamental pseudorange and carrier phase equations. All relevant functions of these observables are then grouped and listed without much additional explanation. These functions are all well known; exceptions might be the triple-frequency functions. We introduce the "across" terminology in order to more easily identify the specific differencing. As such, we have the across-receiver, across-satellite, and across-time observation (single) differences, and then the traditional double-difference and triple-difference functions. A separate section is dedicated to operational details. That section includes everything one needs to know when carrying out high-accuracy positioning with GNSS. We especially stress the "GNSS infrastructure" that has established itself to support users. By this, we mean the totality of GNSS services provided by government agencies, user groups, universities, and above all the IGS and the (mostly) free online computing services. IGS provides products of interest to the sophisticated high-end GNSS user, while the computation services are of most interest to those responsible for processing field data. This is indeed a marvelous GNSS infrastructure that is of tremendous utility.

As to the actual GNSS positioning approaches, Chapter 6 is concerned with three types of approaches, each having been assigned a separate section. The first section deals with navigation solution, which uses the broadcast ephemeris, and the traditional double-differencing technique with ambiguity fixing for accurate positioning. The double differences are formed on the basis of the base station and base satellite concept to conveniently identify the linear dependent double differences. We note that the reason for the popularity of the double-difference functions is the cancelation of common mode errors, such as receiver and satellite clock errors and hardware delays, as well as the tropospheric and ionospheric impacts on the carrier phases in the case of short baselines. The formation of double-difference functions is briefly contrasted with the equivalent undifferenced approach in which only the nonbase-station and nonbase-satellite observation contains an ambiguity parameter, while each of the others contains an epoch-dependent parameter. The latter approach results in a large system of equations that can be efficiently solved by exploring the pattern of the matrices.

In the second section, we discuss PPP (precise point positioning), CORS (continuous operating reference stations), and the classical differential correction that applies to RTK and PPP-RTK, which has been gaining popularity. In the case of PPP, the user operates one dual-frequency receiver and uses the precise ephemeris and satellite clock corrections to determine accurate position; the known drawback of the technique is long station occupation times. The use of the "classical" differential pseudorange and carrier phase correction is also well established, in particular in RTK. The differential correction essentially represents the discrepancies of the undifferenced observations computed at the reference stations. The user receives the differential correction of one or several reference stations and effectively forms double differences to determine its precise position. In the case of PPP-RTK, biases are transmitted to the user. These biases represent the difference of the satellite biases (clock error and hardware delay) and the base station bias (clock error and hardware delay). The user applies the received biases to the observations and carries out an ambiguity-fixed solution for precise point positioning. The advantage of the PPP-RTK approach is that the biases only primarily depend on the changes of the base station clock. Therefore, if the base station is equipped with an atomic clock, the variability of the transmitted biases can be reduced. Using the classical differential correction, the RTK user needs to estimate $(R - 1)(S - 1)$ ambiguities, where R and S denote the number of receivers (reference plus rover) and satellites involved, whereas the PPP-RTK user only needs to estimate $(S - 1)$ ambiguities. In the case of PPP-RTK, some of the work is shifted to the reference network since it computes the biases relative to the base station, whereas the differential corrections refer to the respective reference station and not a specific base station.

In the third section, we deal with TCAR (three carrier phase ambiguity resolutions). This technique is an extension of the popular dual-frequency technique of computing the wide-lane ambiguity first and independently from the actual position solution. In the case of TCAR, one uses triple-frequency observations to resolve the extra-wide-lane, wide-lane, and narrow-lane ambiguities first.

Additionally, a separate section is dedicated to ambiguity fixing. First, the popular LAMBDA (least-squares ambiguity decorrelation adjustment) technique is discussed in detail. This is followed by material on lattice reduction. It was deemed important to add material to see how other disciplines deal with problems similar to ambiguity fixing in GNSS, and in doing so remaining open-minded as to other possible efficient solutions, in particular as the number of ambiguities increases when eventually all visible satellites of all systems are being observed.

Chapter 7 is dedicated to RTK. Since RTK includes static positioning as a special case, it is considered the most general approach. The technique is applicable to short baselines and long baselines if all effects are appropriately modeled. The chapter refers to a practical implementation of RTK algorithms that uses the formalism of recursive least squares given in Chapter 3, uses across-receiver differences as opposed to double differences, and is designed to include observations from all GNSS systems. Its recipes for software implementations are intended for specialists in geodetic software design. All examples are illustrated by way of real data processing.

Chapter 8 deals with the troposphere and ionosphere. The material is presented in a separate chapter in order to emphasis the major contribution of GPS in sensing the troposphere and ionosphere and, conversely, to understand the major efforts made to correct the observations for ionospheric and tropospheric effects in positioning. In addition to dealing with tropospheric refraction and various models for zenith delays and vertical angle dependencies, some material on tropospheric absorption and water vapor radiometers has been included. The chapter ends with a brief discussion on global ionospheric models.

Chapter 9 represents a major addition to this edition of the book. It is well known that multipath is affecting all GNSS positioning techniques, whether based on carrier phases or pseudoranges, since it is directly related to the ability of the user antenna to block reflected signals. Also realizing that geodesist and surveyors typically are not experts in antenna design, it was thought that a dedicated chapter on GNSS user antennas would provide an important addition to the book. We maintained the terminology and (mostly) also the notion that is found in the antenna expert community in the hope that it would make it easier for GPS *Satellite Surveying* readers to transition to the respective antenna literature if needed. Existing texts are often found to be too simple to be useful or too difficult for nonspecialists to understand. As an example of our approach, the Maxwell equations appear in the first section of the chapter but actually are not used explicitly except as support in the appendices. However, the majority of expressions are thoroughly derived and the respective assumptions are clearly identified. In several instances, however, it was deemed necessary to provide additional references for the in-depth study of the subject.

Chapter 9 is subdivided into seven sections. These sections deal with elements of electromagnetic fields and waves, antenna pattern and gain, phase center variation, signal propagation through a chain of circuits, and various antenna types and manufacturing issues and limitations. The material of this chapter is supplemented by six appendices which contain advanced mathematical material and proofs in compact form for readers who enjoy such mathematical depth. In general, the material is presented with sufficient depth for the reader to appreciate the possibilities and limitations of antenna design, to judge the performance of antennas, and to select the right antenna for the task at hand, in particular for high-accuracy applications.

Depending on one's view, one might consider GPS an old or new positioning and timing technology. Considering that the first GPS satellite was launched in 1978, one certainly can see it as old and well-established technology. However, given that new applications of GPS, and now we need to say GNSS, are continuously being developed, it is certainly also fair to characterize this as new technology. Whatever the reader's view might be, it is impossible to trace back all instances of important developments in GNSS unless, of course, one is willing to write a dedicated book on the history of GNSS. Nevertheless, the "pioneering years" of GPS were extremely uplifting as progress could be measured by leaps and bounds, and results were achieved at a level of quality that one had not expected. We present a brief, and probably subjective, review with a slant toward surveying of the major events up to the year 2000. Today, of course, progress continues to be made, in particular as other GNSS systems become operational; the progress is, however, now smooth and less steep.

Table 1.1 lists some of the noteworthy events up to the year 2000. GPS made its debut in surveying and geodesy with a big bang. During the summer of 1982, the testing of the Macrometer receiver, developed by C. C. Counselman at M.I.T., verified a GPS surveying accuracy of 1 to 2 parts per million (ppm) of the station separation. Baselines were measured repeatedly using several hours of observations to study this new surveying technique and to gain initial experience with GPS. During 1983, a first-order network densification of more than 30 stations in the Eifel region of Germany was observed (Bock et al., 1985). This project was a joint effort by the State Surveying Office of North Rhein-Westfalia, a private U.S. firm, and scientists from M.I.T. In early 1984, the geodetic network densification of Montgomery County (Pennsylvania) was completed. The sole guidance of this project rested with a private GPS surveying firm (Collins and Leick, 1985). Also in 1984, GPS was used at Stanford University for a high-precision GPS engineering survey to support construction for extending the Stanford linear accelerator (SLAC). Terrestrial observations (angles and distances) were combined with GPS vectors. The Stanford project yielded a truly millimeter-accurate GPS network, thus demonstrating, among other things, the high quality of the Macrometer antenna. This accuracy could be verified through comparison with the alignment laser at the accelerator, which reproduces a straight line within one-tenth of a millimeter (Ruland and Leick, 1985). Therefore, by the middle of 1984, 1 to 2 ppm GPS surveying had been demonstrated beyond any doubt. No visibility was required between the stations, and data processing could be done on a microcomputer. Hands-on experience was sufficient to acquire most of the skills needed to process the data—i.e., first-order geodetic network densification suddenly became within the capability of individual surveyors.

President Reagan offered GPS free of charge for civilian aircraft navigation in 1983, once the system became fully operational. This announcement can be viewed as the beginning of sharing arrangements of GPS for military and civilian users.

Engelis et al. (1985) computed accurate geoid undulation differences for the Eifel network, demonstrating how GPS results can be combined with orthometric heights, as well as what it takes to carry out such combinations accurately. New receivers became available—e.g., the dual-frequency P-code receiver TI-4100 from Texas Instruments—which was developed with the support of several federal agencies. Ladd et al. (1985) reported on a survey using codeless dual-frequency receivers and claimed 1 ppm in all three components of a vector in as little as 15 min of observation time. Thus, the move toward rapid static surveying had begun. Around 1985, kinematic GPS became available (Remondi, 1985). Kinematic GPS refers

TABLE 1.1 GPS Development and Performance at a Glance until 2000

1978	Launch of first GPS satellite
1982	Prototype Macrometer testing at M.I.T.
	Hatch's synergism paper
1983	Geodetic network densification (Eifel, Germany)
	President Reagan offers GPS to the world "free of charge"
1984	Geodetic network densification (Montgomery County, Pennsylvania)
	Engineering survey at Stanford
	Remondi's dissertation
1985	Precise geoid undulation differences for Eifel network
	Codeless dual-band observations
	Kinematic GPS surveying
	Antenna swap for ambiguity initialization
	First international symposium on precise positioning with GPS
1986	*Challenger* accident (January 28)
	10 cm aircraft positioning
1987	JPL baseline repeatability tests to 0.2–0.04 ppm
1989	Launch of first Block II satellite
	OTF solution
	Wide area differential GPS (WADGPS) concepts
	U.S. Coast Guard GPS Information Center (GPSIC)
1990	GEOID90 for NAD83 datum
1991	NGS ephemeris service
	GIG91 experiment (January 22–February 13)
1992	IGS campaign (June 21–September 23)
	Initial solutions to deal with antispoofing (AS)
	Narrow correlator spacing C/A-code receiver
	Attitude determination system
1993	Real-time kinematic GPS
	ACSM ad hoc committee on accuracy standards
	Orange County GIS/cadastral densification
	Initial operational capability (IOC) on December 8
	1–2 ppb baseline repeatability
	LAMBDA
1994	IGS service beginning January 1
	Antispoofing implementation (January 31)
	RTCM recommendations on differential GPS (Version 2.1)
	National Spatial Reference System Committee (NGS)
	Multiple (single-frequency) receiver experiments for OTF
	Proposal to monitor the earth's atmosphere with GPS (occultations)
1995	Full operational capability (FOC) on July 17
	Precise point positioning (PPP) at JPL
1996	Presidential Decision Directive, first U.S. GPS policy
1998	Vice president announces second GPS civil signal at 1227.60 MHz
	JPL's automated GPS data analysis service via Internet
1999	Vice president announces GPS modernization initiative and third civil GPS signal at 1176.45 MHz
	IGDG (Internet-based global differential GPS) at JPL
2000	Selective availability set to zero
	GPS Joint Program Office begins modifications to IIR-M and IIF satellites

to ambiguity-fixed solutions that yield centimeter (and better) relative accuracy for a moving antenna. The only constraint on the path of the moving antenna is visibility of the same four (at least) satellites at both receivers. Remondi introduced the antenna swapping technique to accomplish rapid initialization of ambiguities. Antenna swapping made kinematic positioning in surveying more efficient.

The deployment of GPS satellites came to a sudden halt due to the tragic January 28, 1986, *Challenger* accident. Several years passed until the Delta II launch vehicle was modified to carry GPS satellites. However, the theoretical developments continued at full speed. They were certainly facilitated by the publication of Remondi's (1984) dissertation, the very successful First International Symposium on Precise Positioning with the Global Positioning System held at the National Geodetic Survey, and a specialty conference on GPS held by the American Society of Civil Engineers in Nashville in 1988.

Kinematic GPS was used for decimeter positioning of airplanes relative to receivers on the ground (Mader, 1986; Krabill and Martin, 1987). The goal of these tests was to reduce the need for traditional and expensive ground control in photogrammetry. These early successes not only made it clear that precise airplane positioning would play a major role in photogrammetry, but they also highlighted the interest in positioning other remote sensing devices carried in airplanes.

Lichten and Border (1987) reported repeatability of 2–5 parts in 10^8 in all three components for static baselines. Note that 1 part in 10^8 corresponds to 1 mm in 100 km. Such highly accurate solutions require satellite positions of about 1 m and better (we note that today's orbit accuracy is in the range of 5 cm). Because accurate orbits were not yet available at the time, researchers were forced to estimate improved GPS orbits simultaneously with baseline estimation. The need for a precise orbital service became apparent. Other limitations, such as the uncertainty in the tropospheric delay over long baselines, also became apparent and created an interest in exploring water vapor radiometers to measure the wet part of the troposphere along the path of the satellite transmissions. The geophysical community requires high baseline accuracy for obvious reasons, e.g., slow-moving crustal motions can be detected earlier with more accurate baseline observations. However, the GPS positioning capability of a few parts in 10^8 was also noticed by surveyors for its potential to change well-established methods of spatial referencing and geodetic network design.

Perhaps the year 1989 could be labeled the year when "modern GPS" positioning began in earnest. This was the year when the first production satellite, Block II, was launched. Seeber and Wübbena (1989) discussed a kinematic technique that used carrier phases and resolved the ambiguity "on-the-way." This technique used to be called on-the-fly (OTF) ambiguity resolution, meaning there is no static initialization required to resolve the ambiguities, but the technique is now considered part of RTK. The navigation community began in 1989 to take advantage of relative positioning, in order to eliminate errors common to co-observing receivers and make attempts to extend the distance in relative positioning. Brown (1989) referred to it as extended differential GPS, but it is more frequently referred to as wide area differential GPS (WADGPS). Many efforts were made to standardize real-time differential

GPS procedures, resulting in several publications by the Radio Technical Commission for Maritime Services. The U.S. Coast Guard established the GPS Information Center (GPSIC) to serve nonmilitary user needs for GPS information.

The introduction of the geoid model GEOID90 in reference to the NAD83 datum represented a major advancement that helped combine GPS (ellipsoidal) and orthometric height differences and paved the way for replacing much of leveling by GPS-determined heights. More recent geoid models are available.

During 1991 and 1992, the geodetic community embarked on major efforts to explore the limits of GPS on a global scale. The efforts began with the GIG91 [GPS experiment for International Earth Rotation Service (IERS) and Geodynamics] campaign and continued the following year resulting in very accurate polar motion coordinates and earth rotation parameters. Geocentric coordinates were obtained that agreed with those derived from satellite laser ranging within 10 to 15 cm, and ambiguities could be fixed on a global scale providing daily repeatability of about 1 part in 10^9. Such results are possible because of the truly global distribution of the tracking stations. The primary purpose of the IGS campaign was to prove that the scientific community is able to produce high-accuracy orbits on an operational basis. The campaign was successful beyond all expectations, confirming that the concept of IGS is possible. The IGS service formally began January 1, 1994.

For many years, users worried about the impact of antispoofing (AS) on the practical uses of GPS. AS implies switching from the known P-code to the encrypted Y-code, expressed by the notation P(Y). The purpose of AS is to make the P-codes available only to authorized (military) users. The anxiety about AS was considerably relieved when Hatch et al. (1992) reported on the code-aided squaring technique to be used when AS is active. Most manufacturers developed proprietary solutions for dealing with AS. When AS was implemented on January 31, 1994, it presented no insurmountable hindrance to the continued use of GPS. GPS users became even less dependent on AS with the introduction of accurate narrow correlator spacing C/A-code receivers (van Dierendonck et al., 1992), since the C/A-code is not subject to AS measures. By providing a second civil code on L2, eventually a third one on L5, and adding new military codes, GPS modernization will make the P(Y)-code encryption a nonissue for civilian applications, and at the same time, provide enhanced performance to civilian and military users.

A major milestone in the development of GPS was achieved on December 8, 1993, when the initial operational capability (IOC) was declared when 24 satellites (Blocks I, II, IIA) became successfully operational. The implication of IOC was that commercial, national, and international civil users could henceforth rely on the availability of the SPS (Standard Positioning Service). Full operational capability (FOC) would be declared on July 17, 1995, when 24 satellites of the type Blocks II and IIA became operational. Also, Teunissen (1993) introduced the least-squares ambiguity decorrelation adjustment (LAMBDA), which is now widely used.

The determination of attitude/orientation using GPS has drawn attention for quite some time. Qin et al. (1992) report on a commercial product for attitude determination. Talbot (1993) reports on a real-time kinematic centimeter accuracy surveying system. Lachapelle et al. (1994) experiment with multiple (single-frequency) receiver configurations in order to accelerate the on-the-fly ambiguity resolution by means of imposing length constraints and conditions between the ambiguities. While much attention was given to monitoring the ionosphere with dual-frequency and single-frequency code or carrier phase observations, Kursinski (1997) discusses the applicability of radio occultation techniques to use GPS in a general earth's atmospheric monitoring system (which could provide high vertical-resolution profiles of atmospheric temperature across the globe).

The surveying community promptly responded to the opportunities and challenges that came with GPS. The American Congress on Surveying and Mapping (ACSM) tasked an ad hoc committee in 1993 to study the accuracy standards to be used in the era of GPS. The committee addressed questions concerning relative and absolute accuracy standards. The National Geodetic Survey (NGS) enlisted the advice of experts regarding the shape and content of the geodetic reference frame; these efforts eventually resulted in the continuously operating reference stations (CORS). Orange County (California) established 2000 plus stations to support geographic information systems (GIS) and cadastral activities. There are many other examples.

Zumberge et al. (1998a,b) report single-point positioning at the couple-of- centimeters level for static receivers and at the subdecimeter level for moving receivers. This technique became available at the Jet Propulsion Laboratory (JPL) around 1995. The technique that requires dual-frequency observations, a precise ephemeris, and precise clock corrections is referred to as precise point positioning (PPP). These remarkable results were achieved with postprocessed ephemerides at a time when selective availability (SA) was still active. Since 1998, JPL has offered automated data processing and analysis for PPP on the Internet (Zumberge, 1998). Since 1999, JPL has operated an Internet-based dual-frequency global differential GPS system (IGDG). This system determines satellite orbits, satellite clock corrections, and earth orientation parameters in real time and makes corrections available via the Internet for real-time positioning. A website at JPL demonstrates RTK positioning at the subdecimeter for several receiver locations.

Finally, during 1998 and 1999, major decisions were announced regarding the modernization of GPS. In 2000, SA was set to zero as per Presidential Directive. When active, SA entails an intentional falsification of the satellite clock (SA-dither) and the broadcast satellite ephemeris (SA-epsilon); when active it is effectively an intentional denial to civilian users of the full capability of GPS.

CHAPTER 2

LEAST-SQUARES ADJUSTMENTS

Least-squares adjustment is useful for estimating parameters and carrying out objective quality control of measurements by processing observations according to a mathematical model and well-defined rules. The objectivity of least-squares quality control is especially useful in surveying when depositing or exchanging observations or verifying the internal accuracy of a survey. Least-squares solutions require redundant observations, i.e., more observations are required than are necessary to determine a set of unknowns exactly. This chapter contains compact but complete derivations of least-squares algorithms. For additional in-depth study of adjustments we recommend Grafarend (2006).

First, the statistical nature of measurements is analyzed, followed by a discussion of stochastic and mathematical models. The mixed adjustment model is derived in detail, and the observation equation and the condition equation models are deduced from the mixed model through appropriate specification. The cases of observed and weighted parameters are presented as well. A special section is devoted to minimal and inner constraint solutions and to those quantities that remain invariant with respect to a change in minimal constraints. Whenever the goal is to perform quality control on the observations, minimal or inner constraint solutions are especially relevant. Statistical testing is important for judging the quality of observations or the outcome of an adjustment. A separate section deals with statistics in least-squares adjustments. The chapter ends with a presentation of additional quality measures, such as internal and external reliability and blunder detection and a brief exposition of Kalman filtering.

In Chapter 3 the least-squares solution is treated in terms of recursive least squares. While both chapters deal with the "same least-squares" principle, the material in Chapter 3 is given in a form that is more suitable for application when the parameters change with time. Chapter 2 is more geared to applications in surveying and geodesy when the parameters usually are not a function of time, such as a typical survey network, leveling network, or a deformation or photogrammetric survey. We like to stress that the treatment starts with the mixed model in which the observations and parameters are implicitly related. Other models are derived by respective specifications.

2.1 ELEMENTARY CONSIDERATIONS

Objective quality control of observations is necessary when dealing with any kind of measurements such as angles, distances, pseudoranges, carrier phases, and the geopotential. It is best to separate conceptually quality control of observations and precision or accuracy of parameters. It is unfortunate that least-squares adjustment is often associated only with high-precision surveying, although it may be as important to discover a 10 m blunder in a low-precision survey as a 1 cm blunder in a high-precision survey.

Least-squares adjustment allows the combination of different types of observations (such as angles, distances, and height differences) into one solution and permits simultaneous statistical analysis. For example, there is no need to treat traverses, intersections, and resections separately. Since these geometric figures consist of angle and distance measurements, the least-squares rules apply to all of them, regardless of the specific arrangements of the observations or the geometric shape they represent.

Least-squares adjustment simulation is a useful tool to plan a survey and to ensure that accuracy specifications will be met once the actual observations have been made. Simulations allow the observation selection to be optimized when alternatives exist. For example, should one primarily measure angles or rely on distances? Considering the available instrumentation, what is the optimal use of the equipment under the constraints of the project? Experienced surveyors often answer these questions intuitively. Even in these cases, an objective verification using least-squares simulation and the concept of internal and external reliability of networks is a welcome assurance to those who carry responsibility for the project.

2.1.1 Statistical Nature of Surveying Measurements

Assume that a distance of 100 m is measured repeatedly with a tape that has centimeter divisions. A likely outcome of these measurements could be 99.99, 100.02, 100.00, 100.01, etc. Because of the centimeter subdivision of the tape, the surveyor is likely to record the observations to two decimal places. The result therefore is a series of numbers ending with two decimal places. One could wrongly conclude that this measurement process belongs to the realm of discrete statistics yielding discrete outcomes with two decimal places. In reality, however, the series is given two decimal places because of the centimeter division of the tape and the fact that the surveyor did not choose to estimate the millimeters. Imagining a reading device that allows us to read the tape to as many decimal places as desired, we readily see that the process of measuring a distance belongs to the realm of continuous statistics. The same is true for other types of measurements typically used in positioning. A classic textbook case for a discrete statistical process is the throwing of a die in which case the outcome is limited to certain integers.

When measuring the distance, we recognize that any value x_i could be obtained, although experience tells us that values close to 100.00 are most likely. Values such as 99.90 or 100.25 are very unlikely when measured with care. Assume that n measurements have been made and that they have been grouped into bins of length Δx, with bin i containing n_i observations. Graphing the bins in a coordinate system of relative frequency n_i/n versus x_i gives the histogram. For surveying measurements, the smoothed step function of the rectangular bins typically has a bell-like shape. The maximum occurs around the sample mean. The larger the deviation from the mean, the smaller the relative frequency, i.e., the probability that such a measurement will actually be obtained. A goodness-of-fit test would normally confirm the hypothesis that the observations have a normal distribution. Thus, the typical measurement process in surveying follows the statistical law of normal distribution.

2.1.2 Observational Errors

Field observations are not perfect, and neither are the recordings and management of observations. The measurement process suffers from several error sources. Repeated measurements do not yield identical numerical values because of random measurement errors. These errors are usually small, and the probability of a positive or a negative error of a given magnitude is the same (equal frequency of occurrence). Random errors are inherent in the nature of measurements and can never be completely overcome. Random errors are dealt with in least-squares adjustment.

Systematic errors are errors that vary systematically in sign and/or magnitude. Examples are a tape that is 10 cm too short or the failure to correct for vertical or lateral refraction in angular measurement. Systematic errors are particularly dangerous because they tend to accumulate. Adequate instrument calibration, care when observing, such as double centering, and observing under various external conditions help avoid systematic errors. If the errors are known, the observations can be corrected before making the adjustment; otherwise, one might attempt to model and estimate these errors. Discovering and dealing with systematic errors requires a great deal of experience with the data. Success is not at all guaranteed.

Blunders are usually large errors resulting from carelessness. Examples of blunders are counting errors in a whole tape length, transposing digits when recording field observations, continuing measurements after upsetting the tripod, and so on. Blunders can largely be avoided through careful observation, although there can never be absolute certainty that all blunders have been avoided or eliminated. Therefore, an important part of least-squares adjustment is to discover and remove remaining blunders in the observations.

2.1.3 Accuracy and Precision

Accuracy refers to the closeness of the observations (or the quantities derived from the observations) to the true value. Precision refers to the closeness of repeated observations (or quantities derived from repeated sets of observations) to the sample mean.

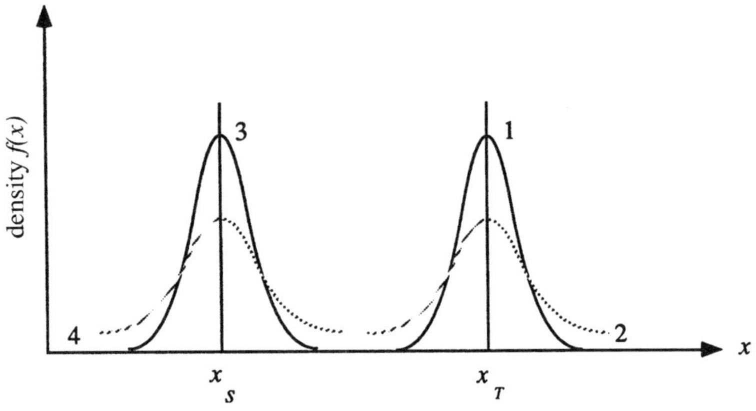

Figure 2.1.1 Accuracy and precision.

Figure 2.1.1 shows four density functions that represent four distinctly different measurement processes of the same quantity. Curves 1 and 2 are symmetric with respect to the true value x_T. These measurements have a high accuracy because the sample mean coincides or is very close to the true value. However, the shapes of both curves are quite different. Curve 1 is tall and narrow, whereas curve 2 is short and broad. Thc observations of process 1 are clustered closely around the mean (true value), whereas the spread of observations around the mean is larger for process 2. Larger deviations from the true value occur more frequently for process 2 than for process 1. Thus, process 1 is more precise than process 2; however, both processes are equally accurate. Curves 3 and 4 are symmetric with respect to the sample mean x_S, which differs from the true value x_T. Both sequences have equally low accuracy, but the precision of process 3 is higher than that of process 4. The difference $x_T - x_S$ is caused by a systematic error. An increase in the number of observations does not reduce this difference.

2.2 STOCHASTIC AND MATHEMATICAL MODELS

This chapter requires some background in statistics. Section A.5 in Appendix A provides selected statistical material that is relevant in what follows. Of particular importance is the law of variance–covariance propagation given in Section A.5.5. It allows computing variances of functions of observations or variances of estimated parameters which are also stochastic quantities.

Least-squares adjustment deals with two equally important components: the stochastic model and the mathematical model. Both components are indispensable and contribute to the adjustment algorithm (Figure 2.2.1). We denote the vector of observation with ℓ_b and the number of observations by n. The observations are random variables, thus the complete notation for the $n \times 1$ vector of observations is $\widetilde{\ell}_b$. To simplify the notation, we do not use the tilde in connection with ℓ_b. The true

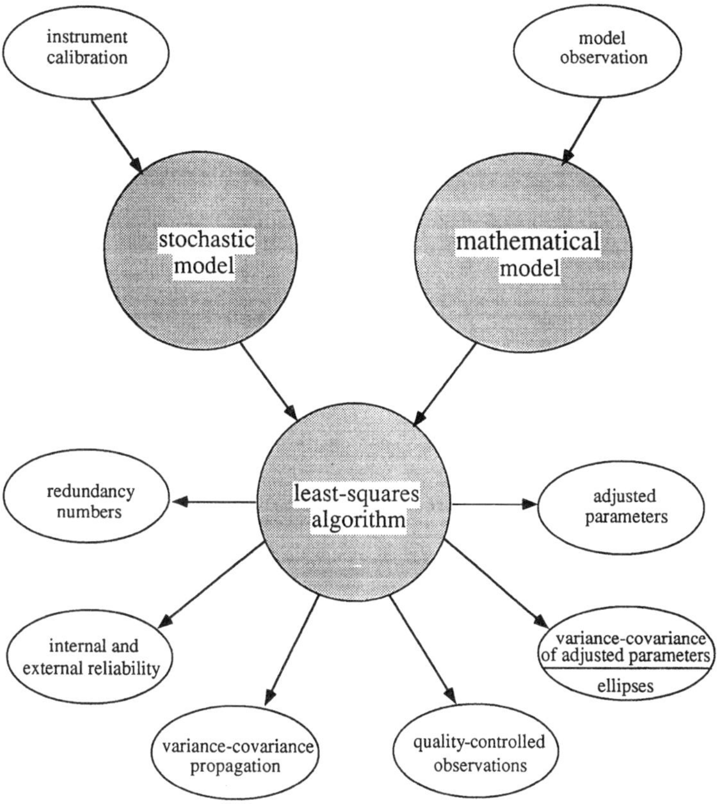

Figure 2.2.1 Elements of least-squares adjustment.

value of the observations, i.e., the mean of the population is estimated from the sample measurements. Since each observation belongs to a different population, the sample size is usually 1. The variances of these distributions comprise the stochastic model. This model introduces information about the precision of the observations (or accuracy if only random errors are present). The variance-covariance matrix $\boldsymbol{\Sigma}_{\ell_b}$ expresses the stochastic model. In many cases, the observations are not correlated and the variance-covariance matrix is diagonal. Occasionally, when so-called derived observations are used which are the outcome of a previous adjustment, or when linear combinations of original observations are adjusted, the variance-covariance matrix contains off-diagonal elements. Because in surveying the observations are normal distributed, the vector of observations has a multivariate normal distribution. We use the notation (A.5.65)

$$\ell_b \sim N\left(\ell_T, \boldsymbol{\Sigma}_{\ell_b}\right) \tag{2.2.1}$$

where ℓ_T is the vector mean of the population, and $\boldsymbol{\Sigma}_{\ell_b}$ is the variance-covariance matrix (A.5.54) or (A.5.63). The cofactor matrix of the observations $\mathbf{Q}_{\ell_b}$ and the weight matrix $\boldsymbol{P}$ are defined by

$$\boldsymbol{Q}_{\ell_b} = \frac{1}{\sigma_0^2} \, \Sigma_{\ell_b} \tag{2.2.2}$$

$$\boldsymbol{P} = \boldsymbol{Q}_{\ell_b}^{-1} = \sigma_0^2 \, \Sigma_{\ell_b}^{-1} \tag{2.2.3}$$

Typically we do not use a subscript to identify $\boldsymbol{P}$ as the weight matrix of the observations. The symbol σ_0^2 denotes the a priori variance of unit weight. It relates the weight matrix and the inverted covariance matrix. An important capability of least-squares adjustment is the estimation of σ_0^2 from observations. We denote that estimate by $\hat{\sigma}_0^2$ and it is the a posteriori variance of unit weight. If the a priori and a posteriori variances of unit weight are statistically equal, the adjustment is said to be correct. More on this fundamental statistical test and its implications will follow in later sections. In general, the a priori variance of unit weight σ_0^2 is set to 1, i.e., the weight matrix is equated with the inverse of the variance-covariance matrix of the observations. The term *variance of unit weight* is derived from the fact that if the variance of an observation equals σ_0^2, then the weight for this observation equals unity. The special cases where $\boldsymbol{P}$ equals the identify matrix, $\boldsymbol{P} = \boldsymbol{I}$, frequently allow a simple and geometrically intuitive interpretation of the minimization.

The mathematical model expresses a simplification of existing physical reality. It attempts to mathematically express the relations between observations and parameters (unknowns) such as coordinates, heights, and refraction coefficients. Least-squares adjustment is a very general tool that can be used whenever a relationship between observations and parameters has been established. Even though the mathematical model is well known for many routine applications, there are always new cases that require a new mathematical model. Finding the right mathematical model can be a challenge.

Much research has gone into establishing a mathematical formulation that is general enough to deal with all types of globally distributed measurement in a unified model. The collection of observations might include distances, angles, heights, gravity anomalies, gravity gradients, geopotential differences, astronomical observations, and GPS observations. The mathematical models become simpler if one does not deal with all types of observations at the same time but instead uses additional external information. See Chapter 4 for a detailed discussion on the 3D geodetic model.

A popular approach is to reduce (modify) the original observations to be compatible with the mathematical model. These are the model observations. For example, if measured vertical angles are used, the mathematical model must include refraction parameters. On the other hand, the original measurements can be corrected for refraction using an atmospheric refraction model. The thus reduced observations refer to a simpler model that does not require refraction parameters. The more reductions are applied to the original observation, the less general the respective mathematical model is. The final form of the model also depends on the purpose of the adjustment. For example, if the objective is to study refraction, one needs refraction parameters in the model. In surveying applications where the objective typically is to determine location, one prefers not to deal with refraction parameters explicitly. The relation between observations and parameterization is central to the success of estimation and at times requires much attention.

In the most general case, the observations and the parameters are related by an implicit nonlinear function:

$$\mathbf{f}(\mathbf{x}_a, \boldsymbol{\ell}_a) = \mathbf{0} \tag{2.2.4}$$

This is the mixed adjustment model. The subscript a is to be read as "adjusted." The symbol $\boldsymbol{\ell}_a$ denotes then $n \times 1$ vector of adjusted observations, and the vector $\mathbf{x}_a$ contains u adjusted parameters. There are r nonlinear mathematical functions in $\mathbf{f}$. Often the observations are explicitly related to the parameters, such as in

$$\boldsymbol{\ell}_a = \mathbf{f}(\mathbf{x}_a) \tag{2.2.5}$$

This is the observation equation model. A further variation is the absence of any parameters as in

$$\mathbf{f}(\boldsymbol{\ell}_a) = \mathbf{0} \tag{2.2.6}$$

This is the condition equation model.

The application usually dictates which model might be preferred. Selecting another model might require a mathematically more involved formulation. In the case of a leveling network, e.g., the observation equation model and the condition equation model can be applied with equal ease.

The observation equation model has the major advantage in that each observation adds one equation. This allows the observation equation model to be implemented relatively easily and generally in software. One does not have to identify independent loop closures, etc.

Figure 2.2.1 indicates some of the outcomes from the adjustment. Statistical tests are available to verify the acceptance of the adjustment or aid in discovering and removing blunders. The adjustment provides probability regions for the estimated parameters and allows variance-covariance propagation to determine functions of the estimated parameters and the respective standard deviations. Of particular interest is the ability of the least-squares adjustment to perform internal and external reliability analysis, in order to quantify marginally detectable blunders and to determine their potential influence on the estimated parameters.

Statistical concepts enter the least-squares adjustment in two distinct ways. The actual least-squares solution merely requires the existence of the variance-covariance matrix; there is no need to specify a particular distribution for the observations. If statistical tests are required, then the distribution of the observations must be known. In most cases, one indeed desires to carry out some statistical testing.

2.3 MIXED MODEL

Observations or functions of observations are always random variables. Typically, a random variable is denoted by a tilde, as is done in Section A.5. In order to simplify the notation, the tilde will not be used in this chapter to identify random variables. A caret is used to identify quantities estimated by least squares, i.e., those quantities that are a solution of a specific minimization. Caret quantities are always random variables because they are functions of observations. To simplify the notation even further, the caret symbol is used consistently only in connection with the parameter $\mathbf{x}$.

In the mixed adjustment model, the observations and the parameters are implicitly related. If $\boldsymbol{\ell}_a$ denotes the vector of n adjusted observations and $\mathbf{x}_a$ denotes u adjusted parameters (unknowns), the nonlinear mathematical model is given by

$$\mathbf{f}(\boldsymbol{\ell}_a, \mathbf{x}_a) = \mathbf{0} \tag{2.3.1}$$

The total number of equations in (2.3.1) is denoted by r. The stochastic model is

$$\mathbf{P} = \sigma_0^2 \, \Sigma_{\ell_b}^{-1} \tag{2.3.2}$$

where $\mathbf{P}$ denotes the $n \times n$ weight matrix, and $\boldsymbol{\Sigma}_{\ell_b}$ denotes the covariance matrix of the observations. The objective is to estimate the parameters. It should be noted that the observations are stochastic (random) variables and that the parameters are deterministic quantities. The parameters exist, but their values are unknown. The estimated parameters, however, will be functions of the observations and therefore random variables.

2.3.1 Linearization

Regular least-squares formulations require that the mathematical model is linear. Nonlinear models, therefore, must be linearized. If we let $\mathbf{x}_0$ denote a vector of known approximate values of the parameters, then the parameter corrections $\mathbf{x}$ are

$$\mathbf{x} = \mathbf{x}_a - \mathbf{x}_0 \tag{2.3.3}$$

If $\boldsymbol{\ell}_b$ denotes the vector of observations, then the residuals are defined by

$$\mathbf{v} = \boldsymbol{\ell}_a - \boldsymbol{\ell}_b \tag{2.3.4}$$

With (2.3.3) and (2.3.4) the mathematical model can be written as

$$\mathbf{f}(\boldsymbol{\ell}_b + \mathbf{v}, \mathbf{x}_0 + \mathbf{x}) = \mathbf{0} \tag{2.3.5}$$

The nonlinear mathematical model is linearized around the known point of expansion $(\boldsymbol{\ell}_b, \mathbf{x}_0)$, giving

$${}_r\mathbf{B}_{n\,n}\mathbf{v}_1 + {}_r\mathbf{A}_{u\,u}\mathbf{x}_1 + {}_r\mathbf{w}_1 = \mathbf{0} \tag{2.3.6}$$

where

$$\mathbf{B} = \left.\frac{\partial \mathbf{f}}{\partial \boldsymbol{\ell}}\right|_{\mathbf{x}_0, \boldsymbol{\ell}_b} \tag{2.3.7}$$

$$\boldsymbol{A} = \left.\frac{\partial \boldsymbol{f}}{\partial \boldsymbol{x}}\right|_{\boldsymbol{x}_0, \ell_b} \tag{2.3.8}$$

$$\boldsymbol{w} = \boldsymbol{f}(\ell_b, \boldsymbol{x}_0) \tag{2.3.9}$$

See Appendix A for linearization of multivariable functions. The coefficient matrices must be evaluated at the point of expansion, which consists of observations and approximate parameters. The discrepancies $\boldsymbol{w}$ must be evaluated for the same point of expansion. The better the approximate values $\boldsymbol{x}_0$, the smaller the parameter corrections $\boldsymbol{x}$.

2.3.2 Minimization and Solution

The least-squares estimate $\hat{\boldsymbol{x}}$ is based on the minimization of the function $\boldsymbol{v}^T\boldsymbol{P}\boldsymbol{v}$. A solution is obtained by introducing a vector of Lagrange multipliers, $\boldsymbol{k}$, and minimizing the function

$$\phi(\boldsymbol{v}, \boldsymbol{k}, \boldsymbol{x}) = \boldsymbol{v}^T\boldsymbol{P}\boldsymbol{v} - 2\boldsymbol{k}^T(\boldsymbol{B}\boldsymbol{v} + \boldsymbol{A}\boldsymbol{x} + \boldsymbol{w}) \tag{2.3.10}$$

Equation (2.3.10) is a function of three variables, namely, $\boldsymbol{v}$, $\boldsymbol{k}$, and $\boldsymbol{x}$. A necessary condition for the minimum is that the partial derivatives must be zero. It can be readily shown that this condition is also sufficient. Differentiating (2.3.10) following the rules of Appendix A and setting the partial derivatives to zero gives

$$\frac{1}{2}\frac{\partial \phi}{\partial \boldsymbol{v}} = \boldsymbol{P}\hat{\boldsymbol{v}} - \boldsymbol{B}^T\hat{\boldsymbol{k}} = \boldsymbol{0} \tag{2.3.11}$$

$$\frac{1}{2}\frac{\partial \phi}{\partial \boldsymbol{k}} = \boldsymbol{B}\hat{\boldsymbol{v}} + \boldsymbol{A}\hat{\boldsymbol{x}} + \boldsymbol{w} = \boldsymbol{0} \tag{2.3.12}$$

$$\frac{1}{2}\frac{\partial \phi}{\partial \boldsymbol{x}} = -\boldsymbol{A}^T\hat{\boldsymbol{k}} = \boldsymbol{0} \tag{2.3.13}$$

The solution of (2.3.11) to (2.3.13) starts with the recognition that $\boldsymbol{P}$ is a square matrix and can be inverted. Thus, the expression for the residuals follows from (2.3.11):

$$\hat{\boldsymbol{v}} = \boldsymbol{P}^{-1} - \boldsymbol{B}^T\hat{\boldsymbol{k}} \tag{2.3.14}$$

Substituting (2.3.14) into (2.3.12), we obtain the solution for the Lagrange multiplier:

$$\hat{\boldsymbol{k}} = -\boldsymbol{M}^{-1}(\boldsymbol{A}\hat{\boldsymbol{x}} + \boldsymbol{w}) \tag{2.3.15}$$

with

$$_{r}\boldsymbol{M}_{r} = {}_{r}\boldsymbol{B}_{n}\, {}_{n}\boldsymbol{P}^{-1}_{n}\, {}_{n}\boldsymbol{B}^{T}_{r} \tag{2.3.16}$$

Finally, the estimate $\hat{\boldsymbol{x}}$ follows from (2.3.13) and (2.3.15)

$$\hat{\boldsymbol{x}} = -\left(\boldsymbol{A}^T\boldsymbol{M}^{-1}\boldsymbol{A}\right)^{-1}\boldsymbol{A}^T\boldsymbol{M}^{-1}\boldsymbol{w} \tag{2.3.17}$$

The estimates $\hat{\mathbf{x}}$ and $\hat{\mathbf{v}}$ are independent of the a priori variance of unit weight. The first step is to compute the parameters $\hat{\mathbf{x}}$ from (2.3.17), then the Lagrange multipliers $\hat{\mathbf{k}}$ from (2.3.15), followed by the residuals $\hat{\mathbf{v}}$ (2.3.14). The adjusted parameters and adjusted observations follow from (2.3.3) and (2.3.4).

The caret symbol in $\hat{\mathbf{v}}$, $\hat{\mathbf{k}}$, and $\hat{\mathbf{x}}$ indicates that all three estimated values follow from minimizing $\mathbf{v}^T\mathbf{P}\mathbf{v}$. However, as stated earlier, the caret is only used consistently for the estimated parameters $\hat{\mathbf{x}}$ in order to simplify the notation.

2.3.3 Cofactor Matrices

Equation (2.3.9) shows that $\mathbf{w}$ is a random variable because it is a function of the observation ℓ_b. With (2.3.2), the law of variance-covariance propagation (A.5.61), and the use of $\mathbf{B}$ in (2.3.7), the cofactor matrix $\mathbf{Q}_w$ becomes

$$\mathbf{Q}_w = \mathbf{B}\mathbf{P}^{-1}\mathbf{B}^T = \mathbf{M} \tag{2.3.18}$$

From (2.3.18) and (2.3.17) it follows that

$$\mathbf{Q}_x = \left(\mathbf{A}^T\mathbf{M}^{-1}\mathbf{A}\right)^{-1} \tag{2.3.19}$$

Combining (2.3.14) through (2.3.17) the expression for the residuals becomes

$$\mathbf{v} = \left[\mathbf{P}^{-1}\mathbf{B}^T\mathbf{M}^{-1}\mathbf{A}\left(\mathbf{A}^T\mathbf{M}^{-1}\mathbf{A}\right)^{-1}\mathbf{A}^T\mathbf{M}^{-1} - \mathbf{P}^{-1}\mathbf{B}^T\mathbf{M}^{-1}\right]\mathbf{w} \tag{2.3.20}$$

It follows from the law of variance propagation (A.5.61) and (2.3.18) that

$$\mathbf{Q}_v = \mathbf{P}^{-1}\mathbf{B}^T\mathbf{M}^{-1}\left[\mathbf{M} - \mathbf{A}\left(\mathbf{A}^T\mathbf{M}^{-1}\mathbf{A}\right)^{-1}\mathbf{A}^T\right]\mathbf{M}^{-1}\mathbf{B}\mathbf{P}^{-1} \tag{2.3.21}$$

The adjusted observations are

$$\begin{aligned}\ell_a &= \ell_b + \mathbf{v}\\ &= \ell_b + \left[\mathbf{P}^{-1}\mathbf{B}^T\mathbf{M}^{-1}\mathbf{A}\left(\mathbf{A}^T\mathbf{M}^{-1}\mathbf{A}\right)^{-1}\mathbf{A}^T\mathbf{M}^{-1} - \mathbf{P}^{-1}\mathbf{B}^T\mathbf{M}^{-1}\right]\mathbf{w}\end{aligned} \tag{2.3.22}$$

Because

$$\frac{\partial \ell_a}{\partial \ell_b} = \mathbf{I} + \mathbf{P}^{-1}\mathbf{B}^T\mathbf{M}^{-1}\mathbf{A}\left(\mathbf{A}^T\mathbf{M}^{-1}\mathbf{A}\right)^{-1}\mathbf{A}^T\mathbf{M}^{-1}\mathbf{B} - \mathbf{P}^{-1}\mathbf{B}^T\mathbf{M}^{-1}\mathbf{B} \tag{2.3.23}$$

it follows that

$$\mathbf{Q}_{\ell_a} = \mathbf{Q}_{\ell_b} - \mathbf{Q}_v \tag{2.3.24}$$

where the inverse of $\mathbf{P}$ has been replaced by $\mathbf{Q}_{\ell_b}$ according to (2.2.3).

2.3.4 A Posteriori Variance of Unit Weight

The minimum of $\boldsymbol{v}^T\boldsymbol{P}\boldsymbol{v}$ follows from (2.3.14), (2.3.15), and (2.3.17) as

$$\boldsymbol{v}^T\boldsymbol{P}\boldsymbol{v} = \boldsymbol{w}^T\left[\boldsymbol{M}^{-1} - \boldsymbol{M}^{-1}\boldsymbol{A}\left(\boldsymbol{A}^T\boldsymbol{M}^{-1}\boldsymbol{A}\right)^{-1}\boldsymbol{A}^T\boldsymbol{M}^{-1}\right]\boldsymbol{w} \tag{2.3.25}$$

The expected value of this random variable is

$$\begin{aligned} E\left(\boldsymbol{v}^T\boldsymbol{P}\boldsymbol{v}\right) &= E\left(\mathrm{Tr}\,\boldsymbol{v}^T\boldsymbol{P}\boldsymbol{v}\right) \\ &= E\left\{\mathrm{Tr}\left[\boldsymbol{w}^T\left(\boldsymbol{M}^{-1} - \boldsymbol{M}^{-1}\boldsymbol{A}\left(\boldsymbol{A}^T\boldsymbol{M}^{-1}\boldsymbol{A}\right)^{-1}\boldsymbol{A}^T\boldsymbol{M}^{-1}\right)\boldsymbol{w}\right]\right\} \\ &= E\left\{\mathrm{Tr}\left[\left(\boldsymbol{M}^{-1} - \boldsymbol{M}^{-1}\boldsymbol{A}\left(\boldsymbol{A}^T\boldsymbol{M}^{-1}\boldsymbol{A}\right)^{-1}\boldsymbol{A}^T\boldsymbol{M}^{-1}\right)\boldsymbol{w}\boldsymbol{w}^T\right]\right\} \\ &= \mathrm{Tr}\left\{\left[\boldsymbol{M}^{-1} - \boldsymbol{M}^{-1}\boldsymbol{A}\left(\boldsymbol{A}^T\boldsymbol{M}^{-1}\boldsymbol{A}\right)^{-1}\boldsymbol{A}^T\boldsymbol{M}^{-1}\right]E\left(\boldsymbol{w}\boldsymbol{w}^T\right)\right\} \end{aligned} \tag{2.3.26}$$

The trace (Tr) of a matrix equals the sum of its diagonal elements. In the first part of (2.3.26), the property that the trace of a 1×1 matrix equals the matrix element itself is used. Next, the matrix products are switched, leaving the trace invariant. In the last part of the equation, the expectation operator and the trace are switched. The expected value $E\left(\boldsymbol{w}\boldsymbol{w}^T\right)$ can be readily computed. Per definition, the expected value of the residuals

$$E(\boldsymbol{v}) = \boldsymbol{0} \tag{2.3.27}$$

is zero because the residuals represent random errors for which positive and negative errors of the same magnitude occur with the same probability. It follows from (2.3.6) that

$$E(\boldsymbol{w}) = -\boldsymbol{A}\boldsymbol{x} \tag{2.3.28}$$

Note that $\boldsymbol{x}$ in (2.3.28) or (2.3.6) is not a random variable. In this expression, $\boldsymbol{x}$ simply denotes the vector of unknown parameters that have fixed values, even though the values are not known. The estimate $\hat{\boldsymbol{x}}$ is a random variable because it is a function of the observations. By using (A.5.53) for the definition of the covariance matrix (2.3.18) and using (2.3.28), it follows that

$$\begin{aligned} E\left(\boldsymbol{w}\boldsymbol{w}^T\right) &= \boldsymbol{\Sigma}_w + E(\boldsymbol{w})E(\boldsymbol{w})^T \\ &= \sigma_0^2\boldsymbol{M} + \boldsymbol{A}\boldsymbol{x}\boldsymbol{x}^T\boldsymbol{A}^T \end{aligned} \tag{2.3.29}$$

Substituting (2.3.29) into (2.3.26) yields the expected value for $\boldsymbol{v}^T\boldsymbol{P}\boldsymbol{v}$:

$$\begin{aligned} E\left(\boldsymbol{v}^T\boldsymbol{P}\boldsymbol{v}\right) &= \sigma_0^2\,\mathrm{Tr}\left\{{}_r\boldsymbol{I}_r - \boldsymbol{M}^{-1}\boldsymbol{A}\left(\boldsymbol{A}^T\boldsymbol{M}^{-1}\boldsymbol{A}\right)^{-1}\boldsymbol{A}^T\right\} \\ &= \sigma_0^2(r-u) \end{aligned} \tag{2.3.30}$$

The difference $r - u$ is called the degree of freedom and equals the number of redundant equations in the model (2.3.1). Strictly, the degree of freedom is $r - R(\boldsymbol{A})$ because the second matrix in (2.3.30) is idempotent. The symbol $R(\boldsymbol{A})$ denotes the rank of the matrix $\boldsymbol{A}$. The a posteriori variance of unit weight is computed from

$$\hat{\sigma}_0^2 = \frac{\hat{\boldsymbol{v}}^T \boldsymbol{P} \hat{\boldsymbol{v}}}{r - u} \tag{2.3.31}$$

Using (2.3.30), we see that

$$E\left(\hat{\sigma}_0^2\right) = \sigma_0^2 \tag{2.3.32}$$

The expected value of the a posteriori variance of unit weight equals the a priori variance of unit weight.

Finally, the estimated covariance matrices are

$$\boldsymbol{\Sigma}_x = \hat{\sigma}_0^2 \boldsymbol{Q}_x \tag{2.3.33}$$

$$\boldsymbol{\Sigma}_v = \hat{\sigma}_0^2 \boldsymbol{Q}_v \tag{2.3.34}$$

$$\boldsymbol{\Sigma}_{\ell_a} = \hat{\sigma}_0^2 \boldsymbol{Q}_{\ell_a} \tag{2.3.35}$$

With equation (2.3.24) it follows that

$$\boldsymbol{\Sigma}_{\ell_a} = \boldsymbol{\Sigma}_{\ell_b} - \boldsymbol{\Sigma}_v \tag{2.3.36}$$

Because the diagonal elements of all three covariance matrices in (2.3.36) are positive, it follows that the variances of the adjusted observations are smaller than those of the original observations. The difference is a function of the geometry of the adjustment, as implied by the covariance matrix $\boldsymbol{\Sigma}_v$.

2.3.5 Iterations

Because the mathematical model is generally nonlinear, the least-squares solution must be iterated. Recall that (2.3.1) is true only for $(\boldsymbol{\ell}_a, \boldsymbol{x}_a)$. Since neither of these quantities is known before the adjustment, the initial point of expansion is chosen as $(\boldsymbol{\ell}_b, \boldsymbol{x}_0)$. For the i th iteration, the linearized model can be written

$$\boldsymbol{B}_{x_{0i}, \ell_{0i}} \overline{\boldsymbol{v}}_i + \boldsymbol{A}_{x_{0i}, \ell_{0i}} \boldsymbol{x}_i + \boldsymbol{w}_{x_{0i}, \ell_{0i}} = \boldsymbol{0} \tag{2.3.37}$$

where the point of expansion $(\boldsymbol{\ell}_{0i}, \boldsymbol{x}_{0i})$ represents the previous solution. The symbols $\boldsymbol{\ell}_{ai}$ and $\boldsymbol{x}_{ai}$ denote the adjusted observations and adjusted parameters for the current (i th) solution. They are computed from

$$\overline{\boldsymbol{v}}_i = \boldsymbol{\ell}_{ai} - \boldsymbol{\ell}_{0i} \tag{2.3.38}$$

$$\boldsymbol{x}_i = \boldsymbol{x}_{ai} - \boldsymbol{x}_{0i} \tag{2.3.39}$$

once the least-squares solution of (2.3.37) has been obtained. The iteration starts with $\ell_{01} = \ell_b$ and $\mathbf{x}_{01} = \mathbf{x}_0$. If the adjustment converges properly, then both $\overline{\mathbf{v}}_i$ and $\mathbf{x}_i$ converge to zero, or, stated differently, ℓ_{ai} and $\mathbf{x}_{ai}$ converge toward ℓ_a and $\mathbf{x}_a$, respectively. The quantity $\overline{\mathbf{v}}_i$ does not equal the residuals. The residuals express the random difference between the adjusted observations and the original observations according to (2.3.4). Defining

$$\mathbf{v}_i = \ell_{ai} - \ell_b \tag{2.3.40}$$

it follows from (2.3.38) that

$$\overline{\mathbf{v}}_i = \mathbf{v}_i + (\ell_b - \ell_{0i}) \tag{2.3.41}$$

Substituting this expression into (2.3.37) gives

$$\mathbf{B}_{x_{0i},\ell_{0i}}\mathbf{v}_i + \mathbf{A}_{x_{0i},\ell_{0i}}\mathbf{x}_i + \mathbf{w}_{x_{0i},\ell_{0i}} + \mathbf{B}_{x_{0i},\ell_{0i}}(\ell_b - \ell_{0i}) = \mathit{0} \tag{2.3.42}$$

The formulation (2.3.42) assures that the vector $\mathbf{v}_i$ converges toward the vector of residuals $\mathbf{v}$. The last term in (2.3.42) will be zero for the first iteration when $\ell_{0i} = \ell_b$. The iteration has converged if

$$\left|\mathbf{v}^T\mathbf{P}\mathbf{v}_i - \mathbf{v}^T\mathbf{P}\mathbf{v}_{i-1}\right| < \varepsilon \tag{2.3.43}$$

where ε is a small positive number.

2.4 SEQUENTIAL MIXED MODEL

Assume that observations are made in two groups, with the second group consisting of one or several observations. Both groups have a common set of parameters. The two mixed adjustment models can be written as

$$\begin{aligned} \mathbf{f}_1(\ell_{1a}, \mathbf{x}_a) &= \mathit{0} \\ \mathbf{f}_2(\ell_{2a}, \mathbf{x}_a) &= \mathit{0} \end{aligned} \tag{2.4.1}$$

Both sets of observations should be uncorrelated, and the a priori variance of unit weight should be the same for both groups, i.e.,

$$\mathbf{P} = \begin{bmatrix} \mathbf{P}_1 & \mathit{0} \\ \mathit{0} & \mathbf{P}_2 \end{bmatrix} = \sigma_0^2 \begin{bmatrix} \Sigma_1^{-1} & \mathit{0} \\ \mathit{0} & \Sigma_2^{-1} \end{bmatrix} \tag{2.4.2}$$

The number of observations in ℓ_{1a} and ℓ_{2a} are n_1 and n_2, respectively; and r_1 and r_2 are the number of equations in the models $\mathbf{f}_1$ and $\mathbf{f}_2$, respectively. The linearization of (2.4.1) yields

$$\mathbf{B}_1\mathbf{v}_1 + \mathbf{A}_1\mathbf{x} + \mathbf{w}_1 = \mathit{0} \tag{2.4.3}$$

$$\mathbf{B}_2\mathbf{v}_2 + \mathbf{A}_2\mathbf{x} + \mathbf{w}_2 = \mathit{0} \tag{2.4.4}$$

where

$$\boldsymbol{B}_1 = \left.\frac{\partial \boldsymbol{f}_1}{\partial \boldsymbol{\ell}_1}\right|_{\ell_{1b},x_0} \quad \boldsymbol{A}_1 = \left.\frac{\partial \boldsymbol{f}_1}{\partial \boldsymbol{x}}\right|_{\ell_{1b},x_0} \quad \boldsymbol{w}_1 = \boldsymbol{f}_1(\ell_{1b}, \boldsymbol{x}_0)$$

$$\boldsymbol{B}_2 = \left.\frac{\partial \boldsymbol{f}_2}{\partial \boldsymbol{\ell}_2}\right|_{\ell_{2b},x_0} \quad \boldsymbol{A}_2 = \left.\frac{\partial \boldsymbol{f}_2}{\partial \boldsymbol{x}}\right|_{\ell_{2b},x_0} \quad \boldsymbol{w}_2 = \boldsymbol{f}_2(\ell_{2b}, \boldsymbol{x}_0) \tag{2.4.5}$$

The function to be minimized is

$$\begin{aligned}\phi(\boldsymbol{v}_1, \boldsymbol{v}_2, \boldsymbol{k}_1, \boldsymbol{k}_2, \boldsymbol{x}) = \boldsymbol{v}_1^T \boldsymbol{P}_1 \boldsymbol{v}_1 + \boldsymbol{v}_2^T \boldsymbol{P}_2 \boldsymbol{v}_2 - 2\boldsymbol{k}_1^T(\boldsymbol{B}_1 \boldsymbol{v}_1 + \boldsymbol{A}_1 \boldsymbol{x} + \boldsymbol{w}_1)\\ -2\boldsymbol{k}_2^T(\boldsymbol{B}_2 \boldsymbol{v}_2 + \boldsymbol{A}_2 \boldsymbol{x} + \boldsymbol{w}_2)\end{aligned} \tag{2.4.6}$$

The solution is obtained by setting the partial derivatives of (2.4.6) to zero,

$$\frac{1}{2}\frac{\partial \phi}{\partial \boldsymbol{v}_1} = \boldsymbol{P}_1 \boldsymbol{v}_1 - \boldsymbol{B}_1^T \boldsymbol{k}_1 = \boldsymbol{0} \tag{2.4.7}$$

$$\frac{1}{2}\frac{\partial \phi}{\partial \boldsymbol{v}_2} = \boldsymbol{P}_2 \boldsymbol{v}_2 - \boldsymbol{B}_2^T \boldsymbol{k}_2 = \boldsymbol{0} \tag{2.4.8}$$

$$\frac{1}{2}\frac{\partial \phi}{\partial \boldsymbol{x}} = -\boldsymbol{A}_1^T \boldsymbol{k}_1 - \boldsymbol{A}_2^T \boldsymbol{k}_2 = \boldsymbol{0} \tag{2.4.9}$$

$$\frac{1}{2}\frac{\partial \phi}{\partial \boldsymbol{k}_1} = \boldsymbol{B}_1 \boldsymbol{v}_1 + \boldsymbol{A}_1 \hat{\boldsymbol{x}} + \boldsymbol{w}_1 = \boldsymbol{0} \tag{2.4.10}$$

$$\frac{1}{2}\frac{\partial \phi}{\partial \boldsymbol{k}_2} = \boldsymbol{B}_2 \boldsymbol{v}_2 + \boldsymbol{A}_2 \hat{\boldsymbol{x}} + \boldsymbol{w}_2 = \boldsymbol{0} \tag{2.4.11}$$

and solving for $\boldsymbol{v}_1$, $\boldsymbol{v}_2$, $\boldsymbol{k}_1$, $\boldsymbol{k}_2$, and $\boldsymbol{x}$. Equations (2.4.7) and (2.4.8) give the residuals

$$\boldsymbol{v}_1 = \boldsymbol{P}_1^{-1} \boldsymbol{B}_1^T \boldsymbol{k}_1 \tag{2.4.12}$$

$$\boldsymbol{v}_2 = \boldsymbol{P}_2^{-1} \boldsymbol{B}_2^T \boldsymbol{k}_2 \tag{2.4.13}$$

Combining (2.4.12) and (2.4.10) yields

$$\boldsymbol{M}_1 \boldsymbol{k}_1 + \boldsymbol{A}_1 \hat{\boldsymbol{x}} + \boldsymbol{w}_1 = \boldsymbol{0} \tag{2.4.14}$$

where

$$\boldsymbol{M}_1 = \boldsymbol{B}_1 \boldsymbol{P}_1^{-1} \boldsymbol{B}_1^T \tag{2.4.15}$$

is an $r_1 \times r_1$ symmetric matrix. The Lagrange multiplier becomes

$$\boldsymbol{k}_1 = -\boldsymbol{M}_1^{-1} \boldsymbol{A}_1 \hat{\boldsymbol{x}} - \boldsymbol{M}_1^{-1} \boldsymbol{w}_1 \tag{2.4.16}$$

Equations (2.4.9) and (2.4.11) become, after combination with (2.4.16) and (2.4.13),

$$\boldsymbol{A}_1^T\boldsymbol{M}_1^{-1}\boldsymbol{A}_1\hat{\boldsymbol{x}} + \boldsymbol{A}_1^T\boldsymbol{M}_1^{-1}\boldsymbol{w}_1 - \boldsymbol{A}_2^T\boldsymbol{k}_2 = \boldsymbol{0} \tag{2.4.17}$$

$$\boldsymbol{B}_2\boldsymbol{P}_2^{-1}\boldsymbol{B}_2^T\boldsymbol{k}_2 + \boldsymbol{A}_2\hat{\boldsymbol{x}} + \boldsymbol{w}_2 = \boldsymbol{0} \tag{2.4.18}$$

By using

$$\boldsymbol{M}_2 = \boldsymbol{B}_2\boldsymbol{P}_2^{-1}\boldsymbol{B}_2^T \tag{2.4.19}$$

we can write equations (2.4.17) and (2.4.18) in matrix form:

$$\begin{bmatrix} \boldsymbol{A}_1^T\boldsymbol{M}_1^{-1}\boldsymbol{A}_1 & \boldsymbol{A}_2^T \\ \boldsymbol{A}_2 & -\boldsymbol{M}_2 \end{bmatrix} \begin{bmatrix} \hat{\boldsymbol{x}} \\ -\boldsymbol{k}_2 \end{bmatrix} = \begin{bmatrix} -\boldsymbol{A}_1^T\boldsymbol{M}_1^{-1}\boldsymbol{w}_1 \\ -\boldsymbol{w}_2 \end{bmatrix} \tag{2.4.20}$$

Equation (2.4.20) shows how the normal matrix of the first group must be augmented in order to find the solution of both groups. The whole matrix can be inverted in one step to give the solution for $\hat{\boldsymbol{x}}$ and $\boldsymbol{k}_2$. Alternatively, one can compute the inverse using the matrix partitioning techniques of Section A.3.5, giving

$$\hat{\boldsymbol{x}} = -\boldsymbol{Q}_{11}\boldsymbol{A}_1^T\boldsymbol{M}_1^{-1}\boldsymbol{w}_1 - \boldsymbol{Q}_{12}\boldsymbol{w}_2 \tag{2.4.21}$$

$$\boldsymbol{k}_2 = \boldsymbol{Q}_{21}\boldsymbol{A}_1^T\boldsymbol{M}_1^{-1}\boldsymbol{w}_1 - \boldsymbol{Q}_{22}\boldsymbol{w}_2 \tag{2.4.22}$$

Setting

$$\boldsymbol{N}_1 = \boldsymbol{A}_1^T\boldsymbol{M}_1^{-1}\boldsymbol{A}_1 \tag{2.4.23}$$

$$\boldsymbol{N}_2 = \boldsymbol{A}_2^T\boldsymbol{M}_2^{-1}\boldsymbol{A}_2 \tag{2.4.24}$$

then using (A.3.53),

$$\boldsymbol{Q}_x \equiv \boldsymbol{Q}_{11} = (\boldsymbol{N}_1 + \boldsymbol{N}_2)^{-1} = \boldsymbol{N}_1^{-1} - \boldsymbol{N}_1^{-1}\boldsymbol{A}_2^T\left[\boldsymbol{M}_2 + \boldsymbol{A}_2\boldsymbol{N}_1^{-1}\boldsymbol{A}_2^T\right]^{-1}\boldsymbol{A}_2\boldsymbol{N}_1^{-1} \tag{2.4.25}$$

$$\boldsymbol{Q}_{12} = \boldsymbol{Q}_{21}^T = \boldsymbol{N}_1^{-1}\boldsymbol{A}_2^T\left[\boldsymbol{M}_2 + \boldsymbol{A}_2\boldsymbol{N}_1^{-1}\boldsymbol{A}_2^T\right]^{-1} \tag{2.4.26}$$

$$\boldsymbol{Q}_{22} = -\left[\boldsymbol{M}_2 + \boldsymbol{A}_2\boldsymbol{N}_1^{-1}\boldsymbol{A}_2^T\right]^{-1} \tag{2.4.27}$$

Substituting $\boldsymbol{Q}_{11}$ and $\boldsymbol{Q}_{12}$ into (2.4.21) gives the sequential solution for the parameters. We denote the solution of the first group by an asterisk and the contribution of the second group by Δ. In that notation, the estimated parameters of the first group are denoted by $\hat{\boldsymbol{x}}^*$, which is simplified to $\boldsymbol{x}^*$. Thus,

$$\hat{\boldsymbol{x}} = \boldsymbol{x}^* + \Delta\boldsymbol{x} \tag{2.4.28}$$

Comparing (2.4.21) and (2.3.17) the sequential solution becomes

$$\boldsymbol{x}^* = -\boldsymbol{N}_1^{-1}\boldsymbol{A}_1^T\boldsymbol{M}_1^{-1}\boldsymbol{w}_1 \tag{2.4.29}$$

and

$$\Delta \boldsymbol{x} = -\boldsymbol{N}_1^{-1}\boldsymbol{A}_2^T\left[\boldsymbol{M}_2 + \boldsymbol{A}_2\boldsymbol{N}_1^{-1}\boldsymbol{A}_2^T\right]^{-1}(\boldsymbol{A}_2\boldsymbol{x}^* + \boldsymbol{w}_2) \tag{2.4.30}$$

Similarly, the expression for the Lagrange multiplier $\boldsymbol{k}_2$ is

$$\boldsymbol{k}_2 = -\left[\boldsymbol{M}_2 + \boldsymbol{A}_2\boldsymbol{N}_1^{-1}\boldsymbol{A}_2^T\right]^{-1}(\boldsymbol{A}_2\boldsymbol{x}^* + \boldsymbol{w}_2) \tag{2.4.31}$$

A different form for the solution of the augmented system (2.4.20) is obtained by using alternative relations of the matrix partitioning inverse expressions (A.3.45) to (A.3.52). It follows readily that

$$\begin{aligned}
\hat{\boldsymbol{x}} &= -(\boldsymbol{N}_1 + \boldsymbol{N}_2)^{-1}\left(\boldsymbol{A}_1^T\boldsymbol{M}_1^{-1}\boldsymbol{w}_1 + \boldsymbol{A}_2^T\boldsymbol{M}_2^{-1}\boldsymbol{w}_2\right)\\
&= -(\boldsymbol{N}_1 + \boldsymbol{N}_2)^{-1}\left(-\boldsymbol{N}_1\boldsymbol{x}^* + \boldsymbol{A}_2^T\boldsymbol{M}_2^{-1}\boldsymbol{w}_2\right)\\
&= \boldsymbol{x}^* - (\boldsymbol{N}_1 + \boldsymbol{N}_2)^{-1}\left(\boldsymbol{N}_2\boldsymbol{x}^* + \boldsymbol{A}_2^T\boldsymbol{M}_2^{-1}\boldsymbol{w}_2\right)
\end{aligned} \tag{2.4.32}$$

The procedure implied by the first line in (2.4.32) is called the method of adding normal equations. The contributions of the new observations are simply added appropriately.

The cofactor matrix $\boldsymbol{Q}_x$ of the parameters can be written in sequential form as

$$\begin{aligned}
\boldsymbol{Q}_x &= \boldsymbol{Q}_{x^*} - \boldsymbol{Q}_{x^*}\boldsymbol{A}_2^T\left[\boldsymbol{M}_2 + \boldsymbol{A}_2\boldsymbol{Q}_{x^*}\boldsymbol{A}_2^T\right]^{-1}\boldsymbol{A}_2\boldsymbol{Q}_{x^*}\\
&= \boldsymbol{Q}_{x^*} + \Delta\boldsymbol{Q}_x
\end{aligned} \tag{2.4.33}$$

where $\boldsymbol{Q}_{x^*}$ is the cofactor matrix of the first group of observations and equals $\boldsymbol{N}_1^{-1}$. The contribution of the second group of observations to the cofactor matrix is

$$\Delta\boldsymbol{Q}_x = -\boldsymbol{Q}_{x^*}\boldsymbol{A}_2^T\left[\boldsymbol{M}_2 + \boldsymbol{A}_2\boldsymbol{Q}_{x^*}\boldsymbol{A}_2^T\right]^{-1}\boldsymbol{A}_2\boldsymbol{Q}_{x^*} \tag{2.4.34}$$

The change $\Delta\boldsymbol{Q}_x$ can be computed without having the actual observations of the second group. This is relevant in simulation studies.

The computation of $\boldsymbol{v}^T\boldsymbol{P}\boldsymbol{v}$ proceeds as usual

$$\begin{aligned}
\boldsymbol{v}^T\boldsymbol{P}\boldsymbol{v} &= \boldsymbol{v}_1^T\boldsymbol{P}_1\boldsymbol{v}_1 + \boldsymbol{v}_2^T\boldsymbol{P}_2\boldsymbol{v}_2\\
&= -\boldsymbol{k}_1^T\boldsymbol{w}_1 - \boldsymbol{k}_2^T\boldsymbol{w}_2
\end{aligned} \tag{2.4.35}$$

The second part of (2.4.35) follows from (2.4.9) to (2.4.13). Using (2.4.16) for $\boldsymbol{k}_1$, (2.4.28) for $\hat{\boldsymbol{x}}$, (2.4.30) for $\Delta\boldsymbol{x}$, and (2.4.31) for $\boldsymbol{k}_2$, then the sequential solution becomes

$$\begin{aligned}
\boldsymbol{v}^T\boldsymbol{P}\boldsymbol{v} &= \boldsymbol{v}^T\boldsymbol{P}\boldsymbol{v}^* + \Delta\boldsymbol{v}^T\boldsymbol{P}\boldsymbol{v}\\
&= \boldsymbol{v}^T\boldsymbol{P}\boldsymbol{v}^* + (\boldsymbol{A}_2\boldsymbol{x}^* + \boldsymbol{w}_2)^T\left[\boldsymbol{M}_2 + \boldsymbol{A}_2\boldsymbol{N}_1^{-1}\boldsymbol{A}_2^T\right]^{-1}(\boldsymbol{A}_2\boldsymbol{x}^* + \boldsymbol{w}_2)
\end{aligned} \tag{2.4.36}$$

with $\boldsymbol{v}^T\boldsymbol{P}\boldsymbol{v}^*$ being obtained from (2.3.25) for the first group only.

The a posteriori variance of unit weight is computed in the usual way:

$$\sigma_0^2 = \frac{\boldsymbol{v}^T \boldsymbol{P} \boldsymbol{v}}{r_1 + r_2 - u} \tag{2.4.37}$$

where r_1 and r_2 are the number of equations in (2.4.1). The letter u denotes, again, the number of parameters.

The second set of observations contributes to all residuals. From (2.4.12), (2.4.16), and (2.4.28) we obtain

$$\begin{aligned} \boldsymbol{v}_1 &= \boldsymbol{v}_1^* + \Delta \boldsymbol{v}_1 \\ &= -\boldsymbol{P}_1^{-1} \boldsymbol{B}_1^T \boldsymbol{M}_1^{-1} (\boldsymbol{A}_1 \boldsymbol{x}^* + \boldsymbol{w}_1) - \boldsymbol{P}_1^{-1} \boldsymbol{B}_1^T \boldsymbol{M}_1^{-1} \boldsymbol{A}_1 \Delta \boldsymbol{x} \end{aligned} \tag{2.4.38}$$

The expression for $\boldsymbol{v}_2$ follows from (2.4.13) and (2.4.31),

$$\boldsymbol{v}_2 = -\boldsymbol{P}_2^{-1} \boldsymbol{B}_2^T \boldsymbol{T} (\boldsymbol{A}_2 \boldsymbol{x}^* + \boldsymbol{w}_2) \tag{2.4.39}$$

where

$$\boldsymbol{T} = \left(\boldsymbol{M}_2 + \boldsymbol{A}_2 \boldsymbol{N}_1^{-1} \boldsymbol{A}_2^T \right)^{-1} \tag{2.4.40}$$

The cofactor matrices for the residuals follow, again, from the law of variance-covariance propagation. The residuals $\boldsymbol{v}_1$ are a function of $\boldsymbol{w}_1$ and $\boldsymbol{w}_2$, according to (2.4.38). Substituting the expressions for $\boldsymbol{x}^*$ and $\Delta \boldsymbol{x}$, we obtain from (2.4.38)

$$\frac{\partial \boldsymbol{v}_1}{\partial \boldsymbol{w}_1} = -\boldsymbol{P}_1^{-1} \boldsymbol{B}_1^T \boldsymbol{M}_1^{-1} \left(\boldsymbol{I} - \boldsymbol{A}_1 \boldsymbol{N}_1^{-1} \boldsymbol{A}_1^T \boldsymbol{M}_1^{-1} + \boldsymbol{A}_1 \boldsymbol{N}_1^{-1} \boldsymbol{A}_2^T \boldsymbol{T} \boldsymbol{A}_2 \boldsymbol{N}_1^{-1} \boldsymbol{A}_1^T \boldsymbol{M}_1^{-1} \right) \tag{2.4.41}$$

$$\frac{\partial \boldsymbol{v}_1}{\partial \boldsymbol{w}_2} = -\boldsymbol{P}_1^{-1} \boldsymbol{B}_1^T \boldsymbol{M}_1^{-1} \boldsymbol{A}_1 \boldsymbol{N}_1^{-1} \boldsymbol{A}_2^T \boldsymbol{T} \tag{2.4.42}$$

Applying the law of covariance propagation to $\boldsymbol{w}_1$ and $\boldsymbol{w}_2$ of (2.4.5) and knowing that the observations are uncorrelated gives

$$\boldsymbol{Q}_{w_1, w_2} = \begin{bmatrix} \boldsymbol{M}_1 & \boldsymbol{0} \\ \boldsymbol{0} & \boldsymbol{M}_2 \end{bmatrix} \tag{2.4.43}$$

By using the partial derivatives (2.4.41) and (2.4.42), expression (2.4.43), and the law of variance-covariance propagation, we obtain, after some algebraic computations, the cofactor matrices:

$$\boldsymbol{Q}_{v_1} = \boldsymbol{Q}_{v_1^*} + \Delta \boldsymbol{Q}_{v_1} \tag{2.4.44}$$

where

$$\boldsymbol{Q}_{v_1^*} = \boldsymbol{P}_1^{-1} \boldsymbol{B}_1^T \boldsymbol{M}_1^{-1} \left(\boldsymbol{P}_1^{-1} \boldsymbol{B}_1^T \right)^T - \left(\boldsymbol{P}_1^{-1} \boldsymbol{B}_1^T \boldsymbol{M}_1^{-1} \boldsymbol{A}_1 \right) \boldsymbol{N}_1^{-1} \left(\boldsymbol{P}_1^{-1} \boldsymbol{B}_1^T \boldsymbol{M}_1^{-1} \boldsymbol{A}_1 \right)^T \tag{2.4.45}$$

$$\Delta \boldsymbol{Q}_{v_1} = \left(\boldsymbol{P}_1^{-1} \boldsymbol{B}_1^T \boldsymbol{M}_1^{-1} \boldsymbol{A}_1 \boldsymbol{N}_1^{-1} \boldsymbol{A}_2^T \right) \boldsymbol{T} \left(\boldsymbol{P}_1^{-1} \boldsymbol{B}_1^T \boldsymbol{M}_1^{-1} \boldsymbol{A}_1 \boldsymbol{N}_1^{-1} \boldsymbol{A}_2^T \right)^T \tag{2.4.46}$$

The partial derivatives of $\mathbf{v}_2$ with respect to $\mathbf{w}_1$ and $\mathbf{w}_2$ follow from (2.4.39),

$$\frac{\partial \mathbf{v}_2}{\partial \mathbf{w}_1} = \mathbf{P}_2^{-1}\mathbf{B}_2^T\mathbf{T}\mathbf{A}_2\mathbf{N}_1^{-1}\mathbf{A}_1^T\mathbf{M}_1^{-1} \tag{2.4.47}$$

$$\frac{\partial \mathbf{v}_2}{\partial \mathbf{w}_2} = -\mathbf{P}_2^{-1}\mathbf{B}_2^T\mathbf{T} \tag{2.4.48}$$

By using, again, the law of variance–covariance propagation and (2.4.43), we obtain the cofactor for $\mathbf{v}_2$:

$$\mathbf{Q}_{v_2} = \mathbf{P}_2^{-1}\mathbf{B}_2^T\mathbf{T}\mathbf{B}_2\mathbf{P}_2^{-1} \tag{2.4.49}$$

The estimated variance–covariance matrix is

$$\hat{\Sigma}_{v_2} = \hat{\sigma}_0^2\mathbf{Q}_{v_2} \tag{2.4.50}$$

The variance–covariance matrix of the adjusted observations is, as usual,

$$\Sigma_{\ell_a} = \Sigma_{\ell_b} - \Sigma_v \tag{2.4.51}$$

As for iterations, one has to make sure that all groups are evaluated for the same approximate parameters. If the first system is iterated, the approximate coordinates for the last iteration must be used as expansion points for the second group. Because there are no observations common to both groups, the iteration with respect to the observations can be done individually for each group.

Occasionally, it is desirable to remove a set of observations from an existing solution. Consider again the uncorrelated case in which the set of observations to be removed is not correlated with the other sets. The procedure is readily seen from (2.4.32), which shows how normal equations are added. When observations are removed, the respective parts of the normal matrix and the right-hand term must be subtracted. Equation (2.4.32) becomes

$$\begin{aligned}\hat{\mathbf{x}} &= -\left(\mathbf{A}_1^T\mathbf{M}_1^{-1}\mathbf{A}_1 - \mathbf{A}_2^T\mathbf{M}_2^{-1}\mathbf{A}_2\right)^{-1}\left(\mathbf{A}_1^T\mathbf{M}_1^{-1}\mathbf{w}_1 - \mathbf{A}_2^T\mathbf{M}_2^{-1}\mathbf{w}_2\right)\\ &= -\left[\mathbf{A}_1^T\mathbf{M}_1^{-1}\mathbf{A}_1 + \mathbf{A}_2^T\left(-\mathbf{M}_2^{-1}\right)\mathbf{A}_2\right]^{-1}\left[\mathbf{A}_1^T\mathbf{M}_1^{-1}\mathbf{w}_1 + \mathbf{A}_2^T\left(-\mathbf{M}_2^{-1}\right)\mathbf{w}_2\right]\end{aligned} \tag{2.4.52}$$

One only has to use a negative weight matrix of the group of observations that is being removed, because

$$-\mathbf{M}_2 = \mathbf{B}_2\left(-\mathbf{P}_2^{-1}\right)\mathbf{B}_2^T \tag{2.4.53}$$

Observations can be removed sequentially following (2.4.30).

The sequential solution can be used in quite a general manner. One can add or remove any number of groups sequentially. A group may consist of a single observation. Given the solution for $i-1$ groups, some of the relevant expressions that include all i groups of observations are

$$\hat{\mathbf{x}}_i = \hat{\mathbf{x}}_{i-1} + \Delta\hat{\mathbf{x}}_i \tag{2.4.54}$$

$$\Delta\hat{\mathbf{x}}_i = -\mathbf{Q}_{i-1}\mathbf{A}_i^T\left(\mathbf{M}_i + \mathbf{A}_i\mathbf{Q}_{i-1}\mathbf{A}_i^T\right)^{-1}\left(\mathbf{A}_i\hat{\mathbf{x}}_{i-1} + \mathbf{w}_i\right) \tag{2.4.55}$$

$$\mathbf{v}^T\mathbf{P}\mathbf{v}_i = \mathbf{v}^T\mathbf{P}\mathbf{v}_{i-1} + \Delta\mathbf{v}^T\mathbf{P}\mathbf{v}_i \tag{2.4.56}$$

$$\Delta\mathbf{v}^T\mathbf{P}\mathbf{v}_i = (\mathbf{A}_i\hat{\mathbf{x}}_{i-1} + \mathbf{w}_i)^T\left(\mathbf{M}_i + \mathbf{A}_i\mathbf{Q}_{i-1}\mathbf{A}_i^T\right)^{-1}(\mathbf{A}_i\hat{\mathbf{x}}_{i-1} + \mathbf{w}_i) \tag{2.4.57}$$

$$\mathbf{Q}_i = \mathbf{Q}_{i-1} - \mathbf{Q}_{i-1}\mathbf{A}_i^T\left(\mathbf{M}_i + \mathbf{A}_i\mathbf{Q}_{i-1}\mathbf{A}_i^T\right)^{-1}\mathbf{A}_i\mathbf{Q}_{i-1} \tag{2.4.58}$$

Every sequential solution is equivalent to a one-step adjustment that contains the same observations. The sequential solution requires the inverse of the normal matrix. Because computing the inverse of the normal matrix requires many more computations than merely solving the system of normal equations, one might sometimes prefer to use the one-step solution instead of the sequential approach.

2.5 MODEL SPECIFICATIONS

The mixed model and the sequential mixed model are the base models from which other solutions can be conveniently derived by appropriate specifications. All of the following models can, of course, be derived separately and independently, i.e., one starts with a minimization of the type (2.3.10), applies partial differentiation, and solves the equations. We first specify the popular observation equation model and then the condition equation model. We then use the sequential solutions and specify a number of very useful specialized cases such as observation equations with observed parameters or observation equations with condition on the parameter.

2.5.1 Observation Equation Model

Often there is an explicit relationship between the observations and the parameters, such as

$$\ell_a = \mathbf{f}(\mathbf{x}_a) \tag{2.5.1}$$

This is the observation equation model. Comparing both mathematical models (2.3.1) and (2.5.1), and taking the definition of the matrix $\mathbf{B}$ (2.3.7) into account, we see that the observation equation model follows from the mixed model using the specification

$$\mathbf{B} \equiv -\mathbf{I} \tag{2.5.2}$$

$$\ell \equiv \mathbf{w} = \mathbf{f}(\mathbf{x}_0) - \ell_b = \ell_0 - \ell_b \tag{2.5.3}$$

It is customary to denote the discrepancy by ℓ instead of $\mathbf{w}$ when dealing with the observation equation model. The symbol ℓ_0 equals the value of the observations as computed from the approximate parameters $\mathbf{x}_0$. The point of expansion for the linearization is $\mathbf{x}_0$; the observation vector is not involved in the iteration because of the explicit form of (2.5.1). The linearized observation equations model is

$${}_n\mathbf{v}_1 = {}_n\mathbf{A}_{u\,u}\mathbf{x}_1 + {}_n\ell_1 \tag{2.5.4}$$

These equations are called the *observation equations*. There is one equation for each observation in (2.5.4).

2.5.2 Condition Equation Model

If the observations are related by a nonlinear function without parameters, we speak of the condition equation model. It is written as

$$\boldsymbol{f}(\ell_a) = \boldsymbol{0} \tag{2.5.5}$$

By comparing this with the mixed model (2.3.1), and applying the definition of the $\boldsymbol{A}$ matrix (2.3.8) we see that the condition equation model follows upon the specification

$$\boldsymbol{A} = \boldsymbol{0} \tag{2.5.6}$$

The linear equations

$$_r\boldsymbol{B}_{n\,n}\boldsymbol{v}_1 + {}_r\boldsymbol{w}_1 = \boldsymbol{0} \tag{2.5.7}$$

are called the condition equations. The iteration for the model (2.5.7) is analogous to a mixed model with the added simplification that there is no $\boldsymbol{A}$ matrix and no parameter vector $\boldsymbol{x}$.

The significance of these three models (observation, condition, and mixed) is that a specific adjustment problem can often be formulated more easily in one of the models. Clearly, that model should be chosen. There are situations in which it is equally easy to use any of the models. A typical example is the adjustment of a level network. Most of the time, however, the observation equation model is preferred, because the simple rule "one observation, one equation" is suitable for setting up general software. Table 2.5.1 lists the important expressions for all three models.

2.5.3 Mixed Model with Observation Equations

The algorithms developed in the previous section can be used to incorporate exterior information about parameters. This includes weighted functions of parameters, weighted individual parameters, and conditions on parameters. These model extensions make it possible to incorporate new types of observations that directly refer to the parameters, to specify parameters in order to avoid singularity of the normal equations, or to incorporate the results of prior adjustments. For example, evaluating conditions between the parameters is the basis for hypothesis testing. These cases are obtained by specifying the coefficient matrices $\boldsymbol{A}$ and $\boldsymbol{B}$ of the mixed model. For example, the mixed models (2.4.1) can be specified as

$$\begin{aligned} \boldsymbol{f}_1(\ell_{1a}, \boldsymbol{x}_a) &= \boldsymbol{0} \\ \ell_{2a} &= \boldsymbol{f}_2(\boldsymbol{x}_a) \end{aligned} \tag{2.5.8}$$

The linearized form is

$$\boldsymbol{B}_1\boldsymbol{v}_1 + \boldsymbol{A}_1\boldsymbol{x} + \boldsymbol{w}_1 = \boldsymbol{0} \tag{2.5.9}$$

$$\boldsymbol{v}_2 = \boldsymbol{A}_2\boldsymbol{x} + \ell_2 \tag{2.5.10}$$

The specifications are $\boldsymbol{B}_2 = -\boldsymbol{I}$ and $\ell_2 = \boldsymbol{w}_2$.

TABLE 2.5.1 Three Adjustment Models

	Mixed Model	Observation Model	Condition Model
Nonlinear model	$\mathbf{f}(\ell_a, \mathbf{x}_a) = 0$	$\ell_a = \mathbf{f}(\mathbf{x}_a)$	$\mathbf{f}(\ell_a) = 0$
Specifications		$\mathbf{B} = -\mathbf{I}, \quad \ell = \mathbf{w}, \ r = n$	$\mathbf{A} = 0$
Linear model	$\mathbf{Bv} + \mathbf{Ax} + \mathbf{w} = 0$	$\mathbf{v} = \mathbf{Ax} + \ell$	$\mathbf{Bv} + \mathbf{w} = 0$
Normal equation elements	$\mathbf{M} = \mathbf{BP}^{-1}\mathbf{B}^T$ $\mathbf{N} = \mathbf{A}^T\mathbf{M}^{-1}\mathbf{A}$ $\mathbf{u} = \mathbf{A}^T\mathbf{M}^{-1}\mathbf{w}$	$\mathbf{M} = \mathbf{P}^{-1}$ $\mathbf{N} = \mathbf{A}^T\mathbf{PA}$ $\mathbf{u} = \mathbf{A}^T\mathbf{P}\ell$	$\mathbf{M} = \mathbf{BP}^{-1}\mathbf{B}^T$
Normal equations	$\mathbf{N}\hat{\mathbf{x}} = -\mathbf{u}$	$\mathbf{N}\hat{\mathbf{x}} = -\mathbf{u}$	
Minimum $\mathbf{v}^T\mathbf{Pv}$	$\mathbf{v}^T\mathbf{Pv} = -\mathbf{u}^T\mathbf{N}^{-1}\mathbf{u} + \mathbf{w}^T\mathbf{M}^{-1}\mathbf{w}$	$\mathbf{v}^T\mathbf{Pv} = -\mathbf{u}^T\mathbf{N}^{-1}\mathbf{u} + \ell^T\mathbf{P}\ell$	$\mathbf{v}^T\mathbf{Pv} = \mathbf{w}^T\mathbf{M}^{-1}\mathbf{w}$
Estimated parameters	$\hat{\mathbf{x}} = -\mathbf{N}^{-1}\mathbf{u}$	$\hat{\mathbf{x}} = -\mathbf{N}^{-1}\mathbf{u}$	—
Estimated residuals	$\hat{\mathbf{v}} = \mathbf{P}^{-1}\mathbf{B}^T\hat{\mathbf{k}}$	$\hat{\mathbf{v}} = \mathbf{A}\hat{\mathbf{x}} + \ell$	$\hat{\mathbf{v}} = \mathbf{P}^{-1}\mathbf{B}^T\hat{\mathbf{k}}$
Estimated variance of unit weight	$\hat{\sigma}_0^2 = \dfrac{\hat{\mathbf{v}}^T\mathbf{P}\hat{\mathbf{v}}}{r-u}$	$\hat{\sigma}_0^2 = \dfrac{\hat{\mathbf{v}}^T\mathbf{P}\hat{\mathbf{v}}}{n-u}$	$\hat{\sigma}_0^2 = \dfrac{\hat{\mathbf{v}}^T\mathbf{P}\hat{\mathbf{v}}}{r}$
Estimated parameter cofactor matrix	$\mathbf{Q}_x = \mathbf{N}^{-1}$	$\mathbf{Q}_x = \mathbf{N}^{-1}$	—
Estimated residual cofactor matrix	$\mathbf{Q}_v = \mathbf{P}^{-1}\mathbf{B}^T\mathbf{M}^{-1}\left(\mathbf{M} - \mathbf{AN}^{-1}\mathbf{A}^T\right)\mathbf{M}^{-1}\mathbf{BP}^{-1}$	$\mathbf{Q}_v = \mathbf{P}^{-1} - \mathbf{AN}^{-1}\mathbf{A}^T$	$\mathbf{Q}_v = \mathbf{P}^{-1}\mathbf{B}^T\mathbf{M}^{-1}\mathbf{BP}^{-1}$
Adjusted observation cofactor matrix	$\mathbf{Q}_{\ell_a} = \mathbf{Q}_{\ell_b} - \mathbf{Q}_v$	$\mathbf{Q}_{\ell_a} = \mathbf{Q}_{\ell_b} - \mathbf{Q}_v$	$\mathbf{Q}_{\ell_a} = \mathbf{Q}_{\ell_b} - \mathbf{Q}_v$

2.5.4 Sequential Observation Equation Model

For the observation equation model we obtain

$$\begin{aligned} \boldsymbol{\ell}_{1a} &= \boldsymbol{f}_1(\boldsymbol{x}_a) \\ \boldsymbol{\ell}_{2a} &= \boldsymbol{f}_2(\boldsymbol{x}_a) \end{aligned} \tag{2.5.11}$$

with the linearized form being

$$\boldsymbol{v}_1 = \boldsymbol{A}_1\boldsymbol{x} + \boldsymbol{\ell}_1 \tag{2.5.12}$$

$$\boldsymbol{v}_2 = \boldsymbol{A}_2\boldsymbol{x} + \boldsymbol{\ell}_2 \tag{2.5.13}$$

The stochastic model is given by the matrices $\boldsymbol{P}_1$ and $\boldsymbol{P}_2$. With proper choice of the elements of $\boldsymbol{A}_2$ and $\boldsymbol{P}_2$, it is possible to introduce a variety of relations about the parameters.

As a first case, consider nonlinear functions of parameters. The design matrix $\boldsymbol{A}_2$ contains the partial derivatives, and $\boldsymbol{\ell}_{2b}$ contains the observed value of the function. This is the case of weighted functions of parameters. Examples are the area or volume of geometric figures as computed from coordinates, angles in geodetic networks, and differences between parameters (coordinates). Each function contributes one equation to (2.5.10) or (2.5.13). The respective expressions are listed in Table 2.5.2 and require no further discussion.

As a second case, consider information about individual parameters. This is a special case of the general method discussed above. Each row of $\boldsymbol{A}_2$ contains a zero with the exception of one position, which contains a 1. The number of rows in the $\boldsymbol{A}_2$ matrix corresponds to the number of weighted parameters. The expressions of Table 2.5.2 are still valid for this case. If information enters into the adjustment in this manner, one speaks of the method of weighted parameters.

2.5.5 Observation Equation Model with Observed Parameters

Consider the case when all parameters are observed and weighted. The specifications for the elements of (2.5.13) are as follows:

$$\boldsymbol{\ell}_{2a} = \boldsymbol{x}_a \tag{2.5.14}$$

$$\boldsymbol{\ell}_{2b} = \boldsymbol{x}_b \tag{2.5.15}$$

$$\boldsymbol{A}_2 = \boldsymbol{I} \tag{2.5.16}$$

$$\boldsymbol{\ell}_2 = \boldsymbol{f}_2(\boldsymbol{x}_0) - \boldsymbol{\ell}_{2b} = \boldsymbol{x}_0 - \boldsymbol{x}_b \tag{2.5.17}$$

The symbols $\boldsymbol{x}_b$ and $\boldsymbol{x}_0$ denote the observed parameters and approximate parameters. During the iterations, $\boldsymbol{x}_0$ converges toward the solution, whereas $\boldsymbol{x}_b$ remains unchanged just as does the vector $\boldsymbol{\ell}_{2b}$. Another special case occurs when the vector $\boldsymbol{\ell}_2$ is zero, which implies that the current values for the approximate parameters

TABLE 2.5.2 Sequential Adjustment Models

	Mixed Model	Observation Model
Nonlinear model	$\boldsymbol{f}_1(\ell_{1a}, \boldsymbol{x}_a) = 0$ $\boldsymbol{f}_2(\ell_{2a}, \boldsymbol{x}_a) = 0$ $\boldsymbol{P} = \begin{bmatrix} \boldsymbol{P}_1 & 0 \\ 0 & \boldsymbol{P}_2 \end{bmatrix}$	$\ell_{1a} = \boldsymbol{f}_1(\boldsymbol{x}_a)$ $\ell_{2a} = \boldsymbol{f}_2(\boldsymbol{x}_a)$ $\boldsymbol{P} = \begin{bmatrix} \boldsymbol{P}_1 & 0 \\ 0 & \boldsymbol{P}_2 \end{bmatrix}$
Linear model	$\boldsymbol{B}_1\boldsymbol{v}_1 + \boldsymbol{A}_1\boldsymbol{x} + \boldsymbol{w}_1 = 0$ $\boldsymbol{B}_2\boldsymbol{v}_2 + \boldsymbol{A}_2\boldsymbol{x} + \boldsymbol{w}_2 = 0$	$\boldsymbol{v}_1 = \boldsymbol{A}_1\boldsymbol{x} + \ell_1$ $\boldsymbol{v}_2 = \boldsymbol{A}_2\boldsymbol{x} + \ell_2$
Normal equation elements	$\boldsymbol{M}_1 = \boldsymbol{B}_1\boldsymbol{P}_1^{-1}\boldsymbol{B}_1^T \quad \boldsymbol{M}_2 = \boldsymbol{B}_2\boldsymbol{P}_2^{-1}\boldsymbol{B}_2^T$ $\boldsymbol{N}_1 = \boldsymbol{A}_1^T\boldsymbol{M}_1^{-1}\boldsymbol{A}_1 \quad \boldsymbol{N}_2 = \boldsymbol{A}_2^T\boldsymbol{M}_2^{-1}\boldsymbol{A}_2$ $\boldsymbol{u}_1 = \boldsymbol{A}_1^T\boldsymbol{M}_1^{-1}\boldsymbol{w}_1 \quad \boldsymbol{u}_2 = \boldsymbol{A}_2^T\boldsymbol{M}_2^{-1}\boldsymbol{w}_2$	$\boldsymbol{M}_1 = \boldsymbol{P}_1^{-1} \quad \boldsymbol{M}_2 = \boldsymbol{P}_2^{-1}$ $\boldsymbol{N}_1 = \boldsymbol{A}_1^T\boldsymbol{P}_1\boldsymbol{A}_1 \quad \boldsymbol{N}_2 = \boldsymbol{A}_2^T\boldsymbol{P}_2\boldsymbol{A}_2$ $\boldsymbol{u}_1 = \boldsymbol{A}_1^T\boldsymbol{P}_1\ell_1 \quad \boldsymbol{u}_2 = \boldsymbol{A}_2^T\boldsymbol{P}_2\ell_2$
Minimum $\boldsymbol{v}^T\boldsymbol{P}\boldsymbol{v}$	$\boldsymbol{v}^T\boldsymbol{P}\boldsymbol{v} = \boldsymbol{v}^T\boldsymbol{P}\boldsymbol{v}^* + \Delta\boldsymbol{v}^T\boldsymbol{P}\boldsymbol{v}$ $\boldsymbol{v}^T\boldsymbol{P}\boldsymbol{v}^* = -\boldsymbol{u}_1^T\boldsymbol{N}_1^{-1}\boldsymbol{u}_1 + \boldsymbol{w}_1^T\boldsymbol{M}_1^{-1}\boldsymbol{w}_1$ $\Delta\boldsymbol{v}^T\boldsymbol{P}\boldsymbol{v} = (\boldsymbol{A}_2\boldsymbol{x}^* + \boldsymbol{w}_2)^T\boldsymbol{T}(\boldsymbol{A}_2\boldsymbol{x}^* + \boldsymbol{w}_2)$	$\boldsymbol{v}^T\boldsymbol{P}\boldsymbol{v} = \boldsymbol{v}^T\boldsymbol{P}\boldsymbol{v}^* + \Delta\boldsymbol{v}^T\boldsymbol{P}\boldsymbol{v}$ $\boldsymbol{v}^T\boldsymbol{P}\boldsymbol{v}^* = -\boldsymbol{u}_1^T\boldsymbol{N}_1^{-1}\boldsymbol{u}_1 + \ell_1^T\boldsymbol{P}_1\ell_1$ $\Delta\boldsymbol{v}^T\boldsymbol{P}\boldsymbol{v} = (\boldsymbol{A}_2\boldsymbol{x}^* + \ell_2)^T\boldsymbol{T}(\boldsymbol{A}_2\boldsymbol{x}^* + \ell_2)$
Estimated parameters	$\hat{\boldsymbol{x}} = \boldsymbol{x}^* + \Delta\boldsymbol{x}$ $\boldsymbol{x}^* = -\boldsymbol{N}_1^{-1}\boldsymbol{u}_1$ $\boldsymbol{T} = \left(\boldsymbol{M}_2 + \boldsymbol{A}_2\boldsymbol{N}_1^{-1}\boldsymbol{A}_2^T\right)^{-1}$ $\Delta\boldsymbol{x} = -\boldsymbol{N}_1^{-1}\boldsymbol{A}_2^T\boldsymbol{T}(\boldsymbol{A}_2\boldsymbol{x}^* + \boldsymbol{w}_2)$	$\hat{\boldsymbol{x}} = \boldsymbol{x}^* + \Delta\boldsymbol{x}$ $\boldsymbol{x}^* = -\boldsymbol{N}_1^{-1}\boldsymbol{u}_1$ $\boldsymbol{T} = \left(\boldsymbol{P}_2^{-1} + \boldsymbol{A}_2\boldsymbol{N}_1^{-1}\boldsymbol{A}_2^T\right)^{-1}$ $\Delta\boldsymbol{x} = -\boldsymbol{N}_1^{-1}\boldsymbol{A}_2^T\boldsymbol{T}(\boldsymbol{A}_2\boldsymbol{x}^* + \ell_2)$
Estimated residuals	$\boldsymbol{v}_1 = \boldsymbol{v}_1^* + \Delta\boldsymbol{v}_1$ $\boldsymbol{v}_1^* = -\boldsymbol{P}_1^{-1}\boldsymbol{B}_1^T\boldsymbol{M}_1^{-1}\left(\boldsymbol{A}_1\boldsymbol{x}^* + \boldsymbol{w}_1\right)$ $\Delta\boldsymbol{v}_1 = -\boldsymbol{P}_1^{-1}\boldsymbol{B}_1^T\boldsymbol{M}_1^{-1}\boldsymbol{A}_1\,\Delta\boldsymbol{x}$	$\boldsymbol{v}_1 = \boldsymbol{v}_1^* + \Delta\boldsymbol{v}_1$ $\boldsymbol{v}_1^* = \boldsymbol{A}_1\boldsymbol{x}^* + \ell_1$ $\Delta\boldsymbol{v}_1 = \boldsymbol{A}_1\,\Delta\boldsymbol{x}$
Estimated variance of unit weight	$\hat{\sigma}_0^2 = \dfrac{\boldsymbol{v}^T\boldsymbol{P}\boldsymbol{v}}{r_1 + r_2 - u}$	$\hat{\sigma}_0^2 = \dfrac{\boldsymbol{v}^T\boldsymbol{P}\boldsymbol{v}}{n_1 + n_2 - u}$
Estimated parameter cofactor matrix	$\boldsymbol{Q}_x = \boldsymbol{Q}_{x^*} + \Delta\boldsymbol{Q}$ $\boldsymbol{Q}_{x^*} = \boldsymbol{N}_1^{-1}$ $\Delta\boldsymbol{Q} = -\boldsymbol{N}_1^{-1}\boldsymbol{A}_2^T\,\boldsymbol{T}\boldsymbol{A}_2\boldsymbol{N}_1^{-1}$	$\boldsymbol{Q}_x = \boldsymbol{Q}_{x^*} + \Delta\boldsymbol{Q}$ $\boldsymbol{Q}_{x^*} = +N_1^{-1}$ $\Delta\boldsymbol{Q} = -\boldsymbol{N}_1^{-1}\boldsymbol{A}_2^T\,\boldsymbol{T}\boldsymbol{A}_2\boldsymbol{N}_1^{-1}$

TABLE 2.5.3 Observed Parameters for the case of Observation Equation Model

$\ell_{1a} = f_1(x_a)$	$\ell_{2a} = x_a$	$P = \begin{bmatrix} P_1 & 0 \\ 0 & P_2 \end{bmatrix}$
$v_1 = A_1 x + \ell_1$	$v_2 = x + \ell_2$	$\ell_2 = x_0 - x_b$
$N_1 = A_1^T P_1 A_1$	$N_2 = P_2$	
$u_1 = A_1^T P_1 \ell_1$	$u_2 = P_2 \ell_2$	
$\hat{x} = -(N_1 + P_2)^{-1}(u_1 + P_2\ell_2)$		
$Q_x = (N_1 + P_2)^{-1}$		

serve as observations of the parameters. This can generally be done if the intent is to define the coordinate system by assigning weights to the current approximate parameters. Table 2.5.3 summarizes the solution for weighted parameters for observation equations. The parameters are weighted simply by adding the respective weights to the diagonal elements of the normal matrix. The parameters not weighted have a zero in the respective diagonal elements of P_2. This is a convenient way of weighting a subset of parameters. Parameters can be fixed by assigning a large weight. Often the specification $P_2 = I$ and $\ell_{2b} = 0$, or $x_b = x_0$, is used as a way to stabilize an ill conditioned system of equations. In the context of least squares this means that the current point of expansion is equally weighted.

It is not necessary that the second group of observations represent the observed parameters. Table 2.5.4 shows the case in which the first group consists of the observed parameters. This approach has the unique feature that all observations can be added to the adjustment in a sequential manner; the first solution is not redundant since it is based solely on the values of the observed parameters. It is important, once again, to distinguish the roles of the observed parameters x_b and the approximations x_0. Because in most cases the P_1 matrix will be diagonal, no matrix inverse computation is required. The size of the matrix T (Table 2.5.2) equals the number of observations in the second group. Thus, if one observation is added at a time, only a 1×1 matrix must be inverted. The residuals can be computed directly from the mathematical model as desired.

2.5.6 Mixed Model with Conditions

A third case pertains to the role of the weight matrix of the parameters. The weight matrix expresses the quality of the information known about the observed parameters. For the adjustment to be meaningful, one must make every attempt to obtain a weight matrix that truly reflects the quality of the additional information. Low weights, or, equivalently, large variances, imply low precision. Even low-weighted parameters can have, occasionally, a positive effect on the quality of the least-squares solution. If the parameters or functions of the parameters are introduced with an infinitely large weight, one speaks of conditions between parameters. The only specifications for implementing conditions are

$$P_2^{-1} = 0 \tag{2.5.18}$$

TABLE 2.5.4 Sequential Solution without Inverting the Normal Matrix. Case: Observation Equation Model

$\ell_{1a} = \mathbf{x}_a$ $\ell_{2a} = \mathbf{f}_{2a}(\mathbf{x}_a)$ $\quad \mathbf{P} = \begin{bmatrix} \mathbf{P}_1 & 0 \\ 0 & \mathbf{P}_2 \end{bmatrix}$
$\mathbf{v}_1 = \mathbf{x} + \ell_1$ $\ell_1 = \mathbf{x}_0 - \mathbf{x}_b$ $\mathbf{v}_2 = \mathbf{A}_2\mathbf{x} + \ell_2$
$\mathbf{N}_1 = \mathbf{P}_1 \quad \mathbf{N}_2 = \mathbf{A}_2^T\mathbf{P}_2\mathbf{A}_2$ $\mathbf{u}_1 = \mathbf{P}_1\ell_1 \quad \mathbf{u}_2 = \mathbf{A}_2^T\mathbf{P}_2\ell_2$
$\hat{\mathbf{x}}_1 = -(\mathbf{x}_0 - \mathbf{x}_b)$ $\mathbf{Q}_1 = \mathbf{P}_1^{-1}$ $\mathbf{v}^T\mathbf{P}\mathbf{v}_1 = 0$
$\hat{\mathbf{x}}_i = \hat{\mathbf{x}}_{i-1} + \Delta\hat{\mathbf{x}}_{i-1}$ $\mathbf{v}^T\mathbf{P}\mathbf{v}_i = \mathbf{v}^T\mathbf{P}\mathbf{v}_{i-1} + \Delta\mathbf{v}^T\mathbf{P}\mathbf{v}_{i-1}$ $\mathbf{Q}_i = \mathbf{Q}_{i-1} + \Delta\mathbf{Q}_{i-1}$
$\mathbf{T} = \left(\mathbf{P}_i^{-1} + \mathbf{A}_i\mathbf{Q}_{i-1}\mathbf{A}_i^T\right)^{-1}$ $\Delta\mathbf{x}_{i-1} = -\mathbf{Q}_{i-1}\mathbf{A}_i^T\mathbf{T}(\mathbf{A}_i\hat{\mathbf{x}}_{i-1} + \ell_i)$ $\Delta\mathbf{v}^T\mathbf{P}\mathbf{v}_{i-1} = (\mathbf{A}_i\hat{\mathbf{x}}_{i-1} + \ell_i)^T\mathbf{T}(\mathbf{A}_i\hat{\mathbf{x}}_{i-1} + \ell_i)$ $\Delta\mathbf{Q}_{i-1} = -\mathbf{Q}_{i-1}\mathbf{A}_i^T\mathbf{T}\mathbf{A}_i\mathbf{Q}_{i-1}$

and

$$\mathbf{P}_2 = \infty \tag{2.5.19}$$

The respective mathematical models are

$$\begin{aligned} \mathbf{f}_1(\ell_{1a}, \mathbf{x}_a) &= 0 \\ \mathbf{g}(\mathbf{x}_a) &= 0 \end{aligned} \tag{2.5.20}$$

with

$$\mathbf{B}_1\mathbf{v}_1 + \mathbf{A}_1\mathbf{x} + \mathbf{w}_1 = 0 \tag{2.5.21}$$

$$\mathbf{A}_2\mathbf{x} + \ell_2 = 0 \tag{2.5.22}$$

2.5.7 Observation Equation Model with Conditions

Similar to the previous case we have for the observation equation model,

$$\begin{aligned} \ell_{1a} &= \mathbf{f}(\mathbf{x}_a) \\ \mathbf{g}(\mathbf{x}_a) &= 0 \end{aligned} \tag{2.5.23}$$

with

$$\mathbf{v}_1 = \mathbf{A}_1\mathbf{x} + \ell_1 \tag{2.5.24}$$

$$\mathbf{A}_2\mathbf{x} + \ell_2 = 0 \tag{2.5.25}$$

TABLE 2.5.5 Conditions on Parameters

	Mixed Model with Conditions	Observation Model with Conditions
Nonlinear model	$\mathbf{f}_1(\ell_{1a}, \mathbf{x}_a) = 0$ $\mathbf{g}(\mathbf{x}_a) = 0,\ \mathbf{P}_1$	$\ell_{1a} = \mathbf{f}_1(\mathbf{x}_a),\ \mathbf{P}_1$ $\mathbf{g}(\mathbf{x}_a) = 0$
Linear model	$\mathbf{B}_1\mathbf{v}_1 + \mathbf{A}_1\mathbf{x} + \mathbf{w}_1 = 0$ $\mathbf{A}_2\mathbf{x} + \ell_2 = 0$	$\mathbf{v}_1 = \mathbf{A}_1\mathbf{x} + \ell_1$ $\mathbf{A}_2\mathbf{x} + \ell_2 = 0$
Normal equation elements	$\mathbf{M}_1 = \mathbf{B}_1\mathbf{P}_1^{-1}\mathbf{B}_1^T$ $\mathbf{N}_1 = \mathbf{A}_1^T\mathbf{M}_1^{-1}\mathbf{A}_1$ $\mathbf{u}_1 = \mathbf{A}_1^T\mathbf{M}_1^{-1}\mathbf{w}_1$	$\mathbf{M}_1 = \mathbf{P}_1^{-1}$ $\mathbf{N}_1 = \mathbf{A}_1^T\mathbf{P}_1\mathbf{A}_1$ $\mathbf{u}_1 = \mathbf{A}_1^T\mathbf{P}_1\ell_1$
Minimum $\mathbf{v}^T\mathbf{P}\mathbf{v}$	$\mathbf{v}^T\mathbf{P}\mathbf{v} = \mathbf{v}^T\mathbf{P}\mathbf{v}^* + \Delta\mathbf{v}^T\mathbf{P}\mathbf{v}$ $\mathbf{v}^T\mathbf{P}\mathbf{v}^* = -\mathbf{u}_1^T\mathbf{N}_1^{-1}\mathbf{u}_1 + \mathbf{w}_1^T\mathbf{M}_1^{-1}\mathbf{w}_1$ $\Delta\mathbf{v}^T\mathbf{P}\mathbf{v} = (\mathbf{A}_2\mathbf{x}^* + \ell_2)^T\mathbf{T}(\mathbf{A}_2\mathbf{x}^* + \ell_2)$	$\mathbf{v}^T\mathbf{P}\mathbf{v} = \mathbf{v}^T\mathbf{P}\mathbf{v}^* + \Delta\mathbf{v}^T\mathbf{P}\mathbf{v}$ $\mathbf{v}^T\mathbf{P}\mathbf{v}^* = -\mathbf{u}_1^T\mathbf{N}_1^{-1}\mathbf{u}_1 + \ell_1^T\mathbf{P}_1\ell_1$ $\Delta\mathbf{v}^T\mathbf{P}\mathbf{v} = (\mathbf{A}_2\mathbf{x}^* + \ell_2)^T\mathbf{T}(\mathbf{A}_2\mathbf{x}^* + \ell_2)$
Estimated parameters	$\hat{\mathbf{x}} = \mathbf{x}^* + \Delta\mathbf{x}$ $\mathbf{x}^* = -\mathbf{N}_1^{-1}\mathbf{u}_1$ $\mathbf{T} = \left(\mathbf{A}_2\mathbf{N}_1^{-1}\mathbf{A}_2^T\right)^{-1}$ $\Delta\mathbf{x} = -\mathbf{N}_1^{-1}\mathbf{A}_2^T\mathbf{T}(\mathbf{A}_2\mathbf{x}^* + \ell_2)$	$\hat{\mathbf{x}} = \mathbf{x}^* + \Delta\mathbf{x}$ $\mathbf{x}^* = -\mathbf{N}_1^{-1}\mathbf{u}_1$ $\mathbf{T} = \left(\mathbf{A}_2\mathbf{N}_1^{-1}\mathbf{A}_2^T\right)^{-1}$ $\Delta\mathbf{x} = -\mathbf{N}_1^{-1}\mathbf{A}_2^T\mathbf{T}(\mathbf{A}_2\mathbf{x}^* + \ell_2)$
Estimated residuals	$\mathbf{v}_1 = \mathbf{v}_1^* + \Delta\mathbf{v}_1$ $\mathbf{v}_1^* = -\mathbf{P}_1^{-1}\mathbf{B}_1^T\mathbf{M}_1^{-1}(\mathbf{A}_1\mathbf{x}^* + \mathbf{w}_1)$ $\Delta\mathbf{v}_1 = -\mathbf{P}_1^{-1}\mathbf{B}_1^T\mathbf{M}_1^{-1}\mathbf{A}_1\,\Delta\mathbf{x}$	$\mathbf{v}_1 = \mathbf{v}_1^* + \Delta\mathbf{v}_1$ $\mathbf{v}_1^* = \mathbf{A}_1\mathbf{x}^* + \ell_1$ $\Delta\mathbf{v}_1 = \mathbf{A}_1\,\Delta\mathbf{x}$
Estimated variance of unit weight	$\hat{\sigma}_0^2 = \dfrac{\mathbf{v}^T\mathbf{P}\mathbf{v}}{r_1 + r_2 - u}$	$\hat{\sigma}_0^2 = \dfrac{\mathbf{v}^T\mathbf{P}\mathbf{v}}{n_1 + n_2 - u}$
Estimated parameter cofactor matrix	$\mathbf{Q}_x = \mathbf{Q}_{x^*} + \Delta\mathbf{Q}$ $\mathbf{Q}_{x^*} = \mathbf{N}_1^{-1}$ $\Delta\mathbf{Q} = -\mathbf{N}_1^{-1}\mathbf{A}_2^T\mathbf{T}\mathbf{A}_2\mathbf{N}_1^{-1}$	$\mathbf{Q}_x = \mathbf{Q}_{x^*} + \Delta\mathbf{Q}$ $\mathbf{Q}_{x^*} = \mathbf{N}_1^{-1}$ $\Delta\mathbf{Q} = -\mathbf{N}_1^{-1}\mathbf{A}_2^T\mathbf{T}\mathbf{A}_2\mathbf{N}_1^{-1}$

Table 2.5.5 contains the expression of the sequential solution with conditions between parameters. If (2.5.19) is used to impose the conditions, the largest numbers that can still be represented in the computer should be used. In most situations, it will be readily clear what constitutes a large weight; the weight must simply be large enough so that the respective observations or parameters do not change during the adjustment. For sequential solution, the solution of the first group must exist. Conditions cannot be imposed sequentially to eliminate a singularity in the first group, e.g., conditions should not be used sequentially to define the coordinate system. A one-step solution is given by (2.4.32).

The a posteriori variance of unit weight is always computed from the final set of residuals. The degree of freedom increases by 1 for every observed parameter function, weighted parameter, or condition. In nonlinear adjustments the linearized condition must always be evaluated for the current point of expansion, i.e., the point of expansion of the last iteration.

The expressions in Table 2.5.2 and Table 2.5.5 are almost identical. The only difference is that the matrix $\boldsymbol{T}$ contains the matrix $\boldsymbol{M}_2$ in Table 2.5.2.

2.6 MINIMAL AND INNER CONSTRAINTS

This section deals with the implementation of minimal and inner constraints to the observation equation model. The symbol r denotes the rank of the design matrix, $R({}_n\boldsymbol{A}_u) = R(\boldsymbol{A}^T\boldsymbol{PA}) = r \leq u$. Note that the use of the symbol r in this context is entirely different from its use in the mixed model, where r denotes the number of equations. The rank deficiency of $u - r$ is generally caused by a lack of coordinate system definition. For example, a network of distances is invariant with respect to translation and rotation, a network of angles is invariant with respect to translation, rotation, and scaling, and a level network (consisting of measured height differences) is invariant with respect to a translation in the vertical. The rank deficiency is dealt with by specifying $u - r$ conditions of the parameters. Much of the theory of inner and minimal constraint solution is discussed by Pope (1971). The main reason for dealing with minimal and inner constraint solutions is that this type of adjustment is important for the quality control of observations. Inner constraint solutions have the additional advantage that the standard ellipses (ellipsoids) represent the geometry as implied by the $\boldsymbol{A}$ and $\boldsymbol{P}$ matrices.

The formulation of the least-squares adjustment for the observation equation model in the presence of a rank deficiency is

$$ {}_n\boldsymbol{v}_1 = {}_n\boldsymbol{A}_u \boldsymbol{x}_B + {}_n\ell_1 \tag{2.6.1} $$

$$ \boldsymbol{P} = \sigma_0^2\, \boldsymbol{\Sigma}_{\ell_b}^{-1} \tag{2.6.2} $$

$$ {}_{u-r}\boldsymbol{B}_u \boldsymbol{x}_B = \boldsymbol{0} \tag{2.6.3} $$

The subscript B indicates that the solution of the parameters $\boldsymbol{x}$ depends on the special condition implied by the $\boldsymbol{B}$ matrix in (2.6.3). This is the observation equation model with conditions between the parameters that was treated in Section 2.5. The one-step solution is given by (2.4.20),

$$\begin{bmatrix} \boldsymbol{A}^T\boldsymbol{PA} & \boldsymbol{B}^T \\ \boldsymbol{B} & \boldsymbol{0} \end{bmatrix} \begin{bmatrix} \hat{\boldsymbol{x}}_B \\ -\hat{\boldsymbol{k}}_2 \end{bmatrix} = \begin{bmatrix} -\boldsymbol{A}^T\boldsymbol{P}\boldsymbol{\ell} \\ \boldsymbol{0} \end{bmatrix} \tag{2.6.4}$$

The matrix on the left side of (2.6.4) is a nonsingular matrix if the conditions (2.6.3) are linearly independent, i.e., the $(u-r) \times u$ matrix $\boldsymbol{B}$ has full row rank, and the rows are linear-independent of the rows of the design matrix $\boldsymbol{A}$. A general expression for the inverse is obtained from

$$\begin{bmatrix} \boldsymbol{A}^T\boldsymbol{PA} & \boldsymbol{B}^T \\ \boldsymbol{B} & \boldsymbol{0} \end{bmatrix} \begin{bmatrix} \boldsymbol{Q}_B & \boldsymbol{S}^T \\ \boldsymbol{S} & \boldsymbol{R} \end{bmatrix} = \begin{bmatrix} \boldsymbol{I} & \boldsymbol{0} \\ \boldsymbol{0} & \boldsymbol{I} \end{bmatrix} \tag{2.6.5}$$

This matrix equation gives the following four equations of submatrices:

$$\boldsymbol{A}^T\boldsymbol{PAQ}_B + \boldsymbol{B}^T\boldsymbol{S} = \boldsymbol{I} \tag{2.6.6}$$

$$\boldsymbol{A}^T\boldsymbol{PAS}^T + \boldsymbol{B}^T\boldsymbol{R} = \boldsymbol{0} \tag{2.6.7}$$

$$\boldsymbol{BQ}_B = \boldsymbol{0} \tag{2.6.8}$$

$$\boldsymbol{BS}^T = \boldsymbol{I} \tag{2.6.9}$$

The solution of these equations requires the introduction of the $(u-r) \times u$ matrix $\boldsymbol{E}$, whose rows span the null space of the design matrix $\boldsymbol{A}$ or the null space of the normal matrix. There is a matrix $\boldsymbol{E}$ such that

$$\left(\boldsymbol{A}^T\boldsymbol{PA}\right)\boldsymbol{E}^T = \boldsymbol{0} \tag{2.6.10}$$

or

$$\boldsymbol{AE}^T = \boldsymbol{0} \quad \text{or} \quad \boldsymbol{EA}^T = \boldsymbol{0} \tag{2.6.11}$$

Because the rows of $\boldsymbol{B}$ are linearly independent of the rows of $\boldsymbol{A}$, the $(u-r) \times (u-r)$ matrix $\boldsymbol{BE}^T$ has full rank and thus can be inverted. Multiplying (2.6.6) by $\boldsymbol{E}$ from the left and using (2.6.11), we get

$$\boldsymbol{S} = \left(\boldsymbol{EB}^T\right)^{-1}\boldsymbol{E} \tag{2.6.12}$$

This expression also satisfies (2.6.9). Substituting $\boldsymbol{S}$ into (2.6.7) gives

$$\boldsymbol{A}^T\boldsymbol{PAE}^T\left(\boldsymbol{BE}^T\right)^{-1} + \boldsymbol{B}^T\boldsymbol{R} = \boldsymbol{0} \tag{2.6.13}$$

Because of (2.6.10), this expression becomes

$$\boldsymbol{B}^T\boldsymbol{R} = \boldsymbol{0} \tag{2.6.14}$$

Because $\boldsymbol{B}$ has full rank, it follows that the matrix $\boldsymbol{R} = \boldsymbol{0}$. Thus,

$$\begin{bmatrix} \boldsymbol{A}^T\boldsymbol{PA} & \boldsymbol{B}^T \\ \boldsymbol{B} & \boldsymbol{0} \end{bmatrix}^{-1} = \begin{bmatrix} \boldsymbol{Q}_B & \boldsymbol{E}^T\left(\boldsymbol{BE}^T\right)^{-1} \\ \left(\boldsymbol{EB}^T\right)^{-1}\boldsymbol{E} & \boldsymbol{0} \end{bmatrix} \tag{2.6.15}$$

Substituting expression (2.6.12) for $\boldsymbol{S}$ into (2.6.6) gives the nonsymmetric matrix

$$\boldsymbol{T}_B \equiv \boldsymbol{A}^T\boldsymbol{PAQ}_B = \boldsymbol{I} - \boldsymbol{B}^T\left(\boldsymbol{EB}^T\right)^{-1}\boldsymbol{E} \tag{2.6.16}$$

This expression is modified with the help of (2.6.8), (2.6.10), and (2.6.16),

$$\left(\boldsymbol{A}^T\boldsymbol{PA} + \boldsymbol{B}^T\boldsymbol{B}\right)\left[\boldsymbol{Q}_B + \boldsymbol{E}^T\left(\boldsymbol{BE}^T\right)^{-1}\left(\boldsymbol{EB}^T\right)^{-1}\boldsymbol{E}\right] = \boldsymbol{I} \tag{2.6.17}$$

It can be solved for $\boldsymbol{Q}_B$:

$$\boldsymbol{Q}_B = \left(\boldsymbol{A}^T\boldsymbol{PA} + \boldsymbol{B}^T\boldsymbol{B}\right)^{-1} - \boldsymbol{E}^T\left(\boldsymbol{EB}^T\boldsymbol{BE}^T\right)^{-1}\boldsymbol{E} \tag{2.6.18}$$

The least-squares solution of $\hat{\boldsymbol{x}}_B$ subject to condition (2.6.3) is, according to (2.6.4), (2.6.5), and (2.6.15),

$$\hat{\boldsymbol{x}}_B = -\boldsymbol{Q}_B\boldsymbol{A}^T\boldsymbol{P}\boldsymbol{\ell} \tag{2.6.19}$$

The cofactor matrix of the parameters follows from the law of variance-covariance propagation

$$\boldsymbol{Q}_{x_B} = \boldsymbol{Q}_B\boldsymbol{A}^T\boldsymbol{PAQ}_B = \boldsymbol{Q}_B \tag{2.6.20}$$

The latter part of (2.6.20) follows from (2.6.16) upon multiplying from the left by $\boldsymbol{Q}_B$ and using (2.6.8). Multiplying (2.6.16) from the right by $\boldsymbol{A}^T\boldsymbol{PA}$ and using (2.6.11) gives

$$\boldsymbol{A}^T\boldsymbol{PA} = \boldsymbol{A}^T\boldsymbol{PAQ}_B\boldsymbol{A}^T\boldsymbol{PA} \tag{2.6.21}$$

The relation implied in (2.6.20) is

$$\boldsymbol{Q}_B\boldsymbol{A}^T\boldsymbol{PAQ}_B = \boldsymbol{Q}_B \tag{2.6.22}$$

There are $u - r$ conditions required to solve the least-squares problem, i.e., the minimal number of conditions is equal to the rank defect of the design (or normal) matrix. Any solution derived in this manner is called a minimal constraint solution. There are obviously many different sets of minimal constraints possible for the same adjustment. The only prerequisite on the $\boldsymbol{B}$ matrix is that it have full row rank and that its rows be linearly independent of $\boldsymbol{A}$. Assume that

$$\boldsymbol{Cx}_C = \boldsymbol{0} \tag{2.6.23}$$

is an alternative set of conditions. The solution $\hat{\boldsymbol{x}}_C$ follows from the expressions given by simply replacing the matrix $\boldsymbol{B}$ by $\boldsymbol{C}$. The pertinent expressions are

$$\hat{\boldsymbol{x}}_C = -\boldsymbol{Q}_C\boldsymbol{A}^T\boldsymbol{P}\boldsymbol{\ell} \tag{2.6.24}$$

$$\boldsymbol{Q}_C = \left(\boldsymbol{A}^T\boldsymbol{PA} + \boldsymbol{C}^T\boldsymbol{C}\right)^{-1} - \boldsymbol{E}^T\left(\boldsymbol{EC}^T\boldsymbol{CE}^T\right)^{-1}\boldsymbol{E} \tag{2.6.25}$$

$$\boldsymbol{T}_C \equiv \boldsymbol{A}^T\boldsymbol{P}\boldsymbol{A}\boldsymbol{Q}_C = \boldsymbol{I} - \boldsymbol{C}^T\left(\boldsymbol{E}\boldsymbol{C}^T\right)^{-1}\boldsymbol{E} \tag{2.6.26}$$

$$\boldsymbol{A}^T\boldsymbol{P}\boldsymbol{A}\boldsymbol{Q}_C\boldsymbol{A}^T\boldsymbol{P}\boldsymbol{A} = \boldsymbol{A}^T\boldsymbol{P}\boldsymbol{A} \tag{2.6.27}$$

$$\boldsymbol{Q}_C\boldsymbol{A}^T\boldsymbol{P}\boldsymbol{A}\boldsymbol{Q}_C = \boldsymbol{Q}_C \tag{2.6.28}$$

The solutions pertaining to the various alternative sets of conditions are all related. In particular,

$$\hat{\boldsymbol{x}}_B = \boldsymbol{T}_B^T\hat{\boldsymbol{x}}_C \tag{2.6.29}$$

$$\boldsymbol{Q}_B = \boldsymbol{T}_B^T\boldsymbol{Q}_C\boldsymbol{T}_B \tag{2.6.30}$$

$$\hat{\boldsymbol{x}}_C = \boldsymbol{T}_C^T\hat{\boldsymbol{x}}_B \tag{2.6.31}$$

$$\boldsymbol{Q}_C = \boldsymbol{T}_C^T\boldsymbol{Q}_B\boldsymbol{T}_C \tag{2.6.32}$$

Equations (2.6.29) to (2.6.32) constitute the transformation of minimal control, i.e., they relate the adjusted parameters and the covariance matrix for different minimal constraints. These transformation expressions are readily proven. For example, by using (2.6.24), (2.6.16), (2.6.26), and (2.6.11), we obtain

$$\begin{aligned}
\boldsymbol{T}_B^T\hat{\boldsymbol{x}}_C &= -\boldsymbol{T}_B^T\boldsymbol{Q}_C\boldsymbol{A}^T\boldsymbol{P}\boldsymbol{\ell} \\
&= -\boldsymbol{Q}_B\boldsymbol{A}^T\boldsymbol{P}\boldsymbol{A}\boldsymbol{Q}_C\boldsymbol{A}^T\boldsymbol{P}\boldsymbol{\ell} \\
&= -\boldsymbol{Q}_B\left[\boldsymbol{I} - \boldsymbol{C}^T\left(\boldsymbol{E}\boldsymbol{C}^T\right)^{-1}\boldsymbol{E}\right]\boldsymbol{A}^T\boldsymbol{P}\boldsymbol{\ell} \\
&= -\boldsymbol{Q}_B\boldsymbol{A}^T\boldsymbol{P}\boldsymbol{\ell} \\
&= \hat{\boldsymbol{x}}_B
\end{aligned} \tag{2.6.33}$$

With (2.6.26), (2.6.21), and (2.6.28), it follows that

$$\begin{aligned}
\boldsymbol{T}_C^T\boldsymbol{Q}_B\boldsymbol{T}_C &= \boldsymbol{Q}_C\boldsymbol{A}^T\boldsymbol{P}\boldsymbol{A}\boldsymbol{Q}_B\boldsymbol{A}^T\boldsymbol{P}\boldsymbol{A}\boldsymbol{Q}_C \\
&= \boldsymbol{Q}_C\boldsymbol{A}^T\boldsymbol{P}\boldsymbol{A}\boldsymbol{Q}_C \\
&= \boldsymbol{Q}_C
\end{aligned} \tag{2.6.34}$$

Instead of using the general condition (2.6.23), we can use the condition

$$\boldsymbol{E}\boldsymbol{x}_P = 0 \tag{2.6.35}$$

The rows of $\boldsymbol{E}$ are linearly independent of $\boldsymbol{A}$ because of (2.6.11). Thus, replacing the matrix $\boldsymbol{C}$ by $\boldsymbol{E}$ in (2.6.24) through (2.6.32) gives this special solution:

$$\hat{\boldsymbol{x}}_P = -\boldsymbol{Q}_P\boldsymbol{A}^T\boldsymbol{P}\boldsymbol{\ell} \tag{2.6.36}$$

$$\boldsymbol{Q}_P = \left(\boldsymbol{A}^T\boldsymbol{P}\boldsymbol{A} + \boldsymbol{E}^T\boldsymbol{E}\right)^{-1} - \boldsymbol{E}^T\left(\boldsymbol{E}\boldsymbol{E}^T\boldsymbol{E}\boldsymbol{E}^T\right)^{-1}\boldsymbol{E} \tag{2.6.37}$$

$$\boldsymbol{T}_P \equiv \boldsymbol{A}^T\boldsymbol{PAQ}_P = \boldsymbol{I} - \boldsymbol{E}^T\left(\boldsymbol{EE}^T\right)^{-1}\boldsymbol{E} \tag{2.6.38}$$

$$\boldsymbol{A}^T\boldsymbol{PAQ}_P\boldsymbol{A}^T\boldsymbol{PA} = \boldsymbol{A}^T\boldsymbol{PA} \tag{2.6.39}$$

$$\boldsymbol{Q}_P\boldsymbol{A}^T\boldsymbol{PAQ}_P = \boldsymbol{Q}_P \tag{2.6.40}$$

$$\hat{\boldsymbol{x}}_B = \boldsymbol{T}_B^T\hat{\boldsymbol{x}}_P \tag{2.6.41}$$

$$\boldsymbol{Q}_B = \boldsymbol{T}_B^T\boldsymbol{Q}_P\boldsymbol{T}_B \tag{2.6.42}$$

$$\hat{\boldsymbol{x}}_P = \boldsymbol{T}_P^T\hat{\boldsymbol{x}}_B \tag{2.6.43}$$

$$\boldsymbol{Q}_P = \boldsymbol{T}_P^T\boldsymbol{Q}_B\boldsymbol{T}_P \tag{2.6.44}$$

The solution (2.6.36) is called the inner constraint solution. The matrix $\boldsymbol{T}_P$ in (2.6.38) is symmetric. The matrix $\boldsymbol{Q}_P$ is a generalized inverse, called the pseudoinverse of the normal matrix; the following notation is used:

$$\boldsymbol{Q}_P = \boldsymbol{N}^+ = \left(\boldsymbol{A}^T\boldsymbol{PA}\right)^+ \tag{2.6.45}$$

The pseudoinverse of the normal matrix is computed from available algorithms of generalized matrix inverses or, equivalently, by finding the $\boldsymbol{E}$ matrix and using equation (2.6.37). For typical applications in surveying, the matrix $\boldsymbol{E}$ can be readily identified. Because of (2.6.11) the solution (2.6.36) can also be written as

$$\hat{\boldsymbol{x}}_P = -\left(\boldsymbol{A}^T\boldsymbol{PA} + \boldsymbol{E}^T\boldsymbol{E}\right)^{-1}\boldsymbol{A}^T\boldsymbol{P}\boldsymbol{\ell} \tag{2.6.46}$$

Note that the covariance matrix of the adjusted parameters is

$$\Sigma_x = \hat{\sigma}_0^2\boldsymbol{Q}_{B,C,P} \tag{2.6.47}$$

depending on whether constraint (2.6.3), (2.6.23), or (2.6.35) is used.

The inner constraint solution is yet another minimal constraint solution, although it has some special features. It can be shown that among all possible minimal constraint solutions, the inner constraint solution also minimizes the sum of the squares of the parameters, i.e.,

$$\boldsymbol{x}^T\boldsymbol{x} = \text{minimum} \tag{2.6.48}$$

This property can be used to obtain a geometric interpretation of the inner constraints. For example, it can be shown that the approximate parameters $\boldsymbol{x}_0$ and the adjusted parameters $\hat{\boldsymbol{x}}_P$ can be related by a similarity transformation whose least-squares estimates of translation and rotation are zero. For inner constraint solutions, the standard ellipses show the geometry of the network and are not affected by the definition of the coordinate system. It can also be shown that the trace of $\boldsymbol{Q}_P$ is the smallest compared to the trace of the other cofactor matrices. All minimal constraint solutions yield the same adjusted observations, a posteriori variance of unit weight, covariance matrices for residuals, and the same values for estimable functions of the parameters and their variances. The next section presents a further explanation of quantities invariant with respect to changes in minimal constraints.

2.7 STATISTICS IN LEAST-SQUARES ADJUSTMENT

If the observations have a multivariate normal distribution as in (2.2.1) and the weight matrix $\boldsymbol{P}$ is the inverse of the variances-covariance matrix as in (2.2.3), we can carry out an adjustment and make objective statements about the data. We first develop the fundamental chi-squared test. Because this test is of such importance, the derivations are given in detail. Next, another test is developed for testing the sequential solution and applied to the testing of a general linear hypothesis. Ellipses of standard deviation, also called error ellipses for short, are very popular in surveying to express positioning accuracy. Therefore, these probability regions are derived and detailed geometric and statistical interpretation is provided.

2.7.1 Fundamental Test

The derivation of the distribution is based on the assumption that the observations have a multivariate normal distribution. The dimension of the distribution equals the number of observations. In the subsequent derivations the observation equation model is used. However, these statistical derivations could just as well have been carried out with the mixed model.

The observation equations are

$$\begin{aligned} \boldsymbol{v} &= \boldsymbol{A}\boldsymbol{x} + \boldsymbol{\ell}_0 - \boldsymbol{\ell}_b \\ &= \boldsymbol{A}\boldsymbol{x} + \boldsymbol{\ell} \end{aligned} \tag{2.7.1}$$

A first assumption is that the residuals are randomly distributed, i.e., the probability for a positive or negative residual of the equal magnitude is the same. From this assumption it follows that

$$E(\boldsymbol{v}) = \boldsymbol{0} \tag{2.7.2}$$

Because $\boldsymbol{x}$ and $\boldsymbol{\ell}_0$ are constant vectors, it further follows that the mean and variance-covariance matrix, respectively, are

$$E(\boldsymbol{\ell}_b) = \boldsymbol{\ell}_0 + \boldsymbol{A}\boldsymbol{x} \tag{2.7.3}$$

$$E\left(\boldsymbol{v}\boldsymbol{v}^T\right) = E\left\{ \left[\boldsymbol{\ell}_b - E(\boldsymbol{\ell}_b)\right]\left[\boldsymbol{\ell}_b - E(\boldsymbol{\ell}_b)\right]^T \right\} = \boldsymbol{\Sigma}_{\ell_b} = \sigma_0^2 \boldsymbol{P}^{-1} \tag{2.7.4}$$

The second basic assumption refers to the type of distribution of the observations. It is assumed that the distribution is multivariate normal. Using the mean (2.7.3) and the covariance matrix (2.7.4), the n-dimensional multivariate normal distribution of $\boldsymbol{\ell}_b$ is written as

$$\boldsymbol{\ell}_b \sim N_n\left(\boldsymbol{\ell}_0 + \boldsymbol{A}\boldsymbol{x}, \boldsymbol{\Sigma}_{\ell_b}\right) \tag{2.7.5}$$

Alternative expressions are

$$\boldsymbol{\ell} \sim N_n\left(-\boldsymbol{A}\boldsymbol{x}, \boldsymbol{\Sigma}_{\ell_b}\right) \tag{2.7.6}$$

$$\boldsymbol{v} \sim N_n\left(\boldsymbol{0}, \boldsymbol{\Sigma}_{\ell_b}\right) = N_n\left(\boldsymbol{0}, \sigma_0^2 \boldsymbol{P}^{-1}\right) \tag{2.7.7}$$

By applying two orthogonal transformations we can conveniently derive $\mathbf{v}^T\mathbf{P}\mathbf{v}$. If $\boldsymbol{\Sigma}_{\ell_b}$ happens to be nondiagonal, one can always find observations that are stochastically independent and have a unit variate normal distribution. As discussed in Appendix A, for a positive definite matrix $\mathbf{P}$ there exists a nonsingular matrix $\mathbf{D}$ such that

$$\mathbf{D} = \mathbf{E}\boldsymbol{\Lambda}^{-1/2} \tag{2.7.8}$$

$$\mathbf{D}^T\mathbf{P}^{-1}\mathbf{D} = \mathbf{I} \tag{2.7.9}$$

$$\mathbf{D}^T\mathbf{v} = \mathbf{D}^T\mathbf{A}\mathbf{x} + \mathbf{D}^T\boldsymbol{\ell} \tag{2.7.10}$$

$$\overline{\mathbf{v}} = \overline{\mathbf{A}}\mathbf{x} + \overline{\boldsymbol{\ell}} \tag{2.7.11}$$

$$\overline{\boldsymbol{\ell}} = \mathbf{D}^T\boldsymbol{\ell}_0 - \mathbf{D}^T\boldsymbol{\ell}_b = \overline{\boldsymbol{\ell}}_0 - \overline{\boldsymbol{\ell}}_b \tag{2.7.12}$$

$$E(\overline{\mathbf{v}}) = \mathbf{D}^T E(\mathbf{v}) = \mathbf{0} \tag{2.7.13}$$

$$\boldsymbol{\Sigma}_{\overline{v}} = \sigma_0^2\mathbf{D}^T\mathbf{P}^{-1}\mathbf{D} = \sigma_0^2\mathbf{I} \tag{2.7.14}$$

$$\overline{\mathbf{v}} \sim N_n\left(\mathbf{0}, \sigma_0^2\mathbf{I}\right) \tag{2.7.15}$$

The columns of the orthogonal matrix $\mathbf{E}$ consist of the normalized eigenvectors of $\mathbf{P}^{-1}$; $\boldsymbol{\Lambda}$ is a diagonal matrix having the eigenvalues of $\mathbf{P}^{-1}$ at the diagonal. The quadratic form $\mathbf{v}^T\mathbf{P}\mathbf{v}$ remains invariant under this transformation because

$$R \equiv \mathbf{v}^T\mathbf{P}\mathbf{v} = \overline{\mathbf{v}}^T\boldsymbol{\Lambda}^{1/2}\mathbf{E}^T\mathbf{P}\mathbf{E}\boldsymbol{\Lambda}^{1/2}\overline{\mathbf{v}} = \overline{\mathbf{v}}^T\boldsymbol{\Lambda}^{1/2}\boldsymbol{\Lambda}^{-1}\boldsymbol{\Lambda}^{1/2}\overline{\mathbf{v}} = \overline{\mathbf{v}}^T\overline{\mathbf{v}} \tag{2.7.16}$$

If the covariance matrix $\boldsymbol{\Sigma}_{\ell_b}$ has a rank defect, then one could use matrix $\mathbf{F}$ of (A.3.17) for the transformation. The dimension of the transformed observations $\overline{\boldsymbol{\ell}}_b$ equals the rank of the covariance matrix.

In the next step, the parameters are transformed to a new set that is stochastically independent. To keep the generality, let the matrix $\overline{\mathbf{A}}$ in (2.7.11) have less than full column rank, i.e., $R(\overline{\mathbf{A}}) = r < u$. Let the matrix $\mathbf{F}$ be an $n \times r$ matrix whose columns constitute an orthonormal basis for the column space of $\overline{\mathbf{A}}$. One such choice for the columns of $\mathbf{F}$ may be to take the normalized eigenvectors of $\overline{\mathbf{A}\mathbf{A}}^T$. Let $\mathbf{G}$ be an $n \times (n-r)$ matrix, such that $[\mathbf{F}\ \mathbf{G}]$ is orthogonal and such that the columns of $\mathbf{G}$ constitute an orthonormal basis to the $n-r$-dimensional null space of $\overline{\mathbf{A}\mathbf{A}}^T$. Such a matrix always exists. There is no need to compute this matrix explicitly. With these specifications we obtain

$$\begin{bmatrix} \mathbf{F}^T \\ \mathbf{G}^T \end{bmatrix} \begin{bmatrix} \mathbf{F} & \mathbf{G} \end{bmatrix} = \begin{bmatrix} \mathbf{F}^T\mathbf{F} & \mathbf{F}^T\mathbf{G} \\ \mathbf{G}^T\mathbf{F} & \mathbf{G}^T\mathbf{G} \end{bmatrix} = \begin{bmatrix} {}_r\mathbf{I}_r & \mathbf{0} \\ \mathbf{0} & {}_{n-r}\mathbf{I}_{n-r} \end{bmatrix} \tag{2.7.17}$$

$$\begin{bmatrix} \mathbf{F} & \mathbf{G} \end{bmatrix} \begin{bmatrix} \mathbf{F} & \mathbf{G} \end{bmatrix}^T = \mathbf{F}\mathbf{F}^T + \mathbf{G}\mathbf{G}^T = \mathbf{I} \tag{2.7.18}$$

$$\overline{\mathbf{A}}^T\mathbf{G} = \mathbf{0} \tag{2.7.19}$$

$$\mathbf{G}^T\overline{\mathbf{A}} = \mathbf{0} \tag{2.7.20}$$

The required transformation is

$$\begin{bmatrix} \mathbf{F}^T \\ \mathbf{G}^T \end{bmatrix} \bar{\mathbf{v}} = \begin{bmatrix} \mathbf{F}^T \\ \mathbf{G}^T \end{bmatrix} \bar{\mathbf{A}}\mathbf{x} + \begin{bmatrix} \mathbf{F}^T \\ \mathbf{G}^T \end{bmatrix} \bar{\ell} \tag{2.7.21}$$

or, equivalently,

$$\begin{bmatrix} \mathbf{F}^T \bar{\mathbf{v}} \\ \mathbf{G}^T \bar{\mathbf{v}} \end{bmatrix} = \begin{bmatrix} \mathbf{F}^T \bar{\mathbf{A}}\mathbf{x} \\ 0 \end{bmatrix} + \begin{bmatrix} \mathbf{F}^T \bar{\ell} \\ \mathbf{G}^T \bar{\ell} \end{bmatrix} \tag{2.7.22}$$

Labeling the newly transformed observations by **z**, i.e.,

$$\mathbf{z} = \begin{bmatrix} \mathbf{z}_1 \\ \mathbf{z}_2 \end{bmatrix} = \begin{bmatrix} \mathbf{F}^T \bar{\ell} \\ \mathbf{G}^T \bar{\ell} \end{bmatrix} \tag{2.7.23}$$

we can write (2.7.22) as

$$\bar{\mathbf{v}}_z = \begin{bmatrix} \bar{\mathbf{v}}_{z1} \\ \bar{\mathbf{v}}_{z2} \end{bmatrix} = \begin{bmatrix} \mathbf{F}^T \bar{\mathbf{A}}\mathbf{x} \\ 0 \end{bmatrix} + \begin{bmatrix} \mathbf{z}_1 \\ \mathbf{z}_2 \end{bmatrix} \tag{2.7.24}$$

There are r random variables in $\mathbf{z}_1$ and $n - r$ random variables in $\mathbf{z}_2$. The quadratic form again remains invariant under the orthogonal transformation, since

$$\begin{aligned} \bar{\mathbf{v}}_z^T \bar{\mathbf{v}}_z &= \bar{\mathbf{v}}^T \left(\mathbf{F}\mathbf{F}^T + \mathbf{G}\mathbf{G}^T \right) \bar{\mathbf{v}} \\ &= \bar{\mathbf{v}}^T \bar{\mathbf{v}} = R \end{aligned} \tag{2.7.25}$$

according to (2.7.18). The actual quadratic form is obtained from (2.7.24),

$$R = \bar{\mathbf{v}}_z^T \bar{\mathbf{v}}_z = \left(\mathbf{F}^T \bar{\mathbf{A}}\mathbf{x} + \mathbf{z}_1 \right)^T \left(\mathbf{F}^T \bar{\mathbf{A}}\mathbf{x} + \mathbf{z}_1 \right) + \mathbf{z}_2^T \mathbf{z}_2 \tag{2.7.26}$$

The least-squares solution requires that R be minimized by variation of the parameters. Generally, equating partial derivatives with respect to $\mathbf{x}$ to zero and solving the resulting equations gives the minimum. The special form of (2.7.26) permits a much simpler approach. The expressions on the right side of equation (2.7.26) consist of the sum of two positive terms (sum of squares). Because only the first term is a function of the parameters $\mathbf{x}$, the minimum is achieved if the first term is zero, i.e.,

$$-{}_r\mathbf{F}_n^T \, {}_n\bar{\mathbf{A}}_u \, {}_u\hat{\mathbf{x}}_1 = \mathbf{z}_1 \tag{2.7.27}$$

Note that the caret identifies the estimated parameters. Consequently, the estimate of the quadratic form is

$$\hat{R} = \mathbf{z}_2^T \mathbf{z}_2 \tag{2.7.28}$$

Because there are $r < u$ equations for the u parameters in (2.7.27), there always exists a solution for $\hat{\mathbf{x}}$. The simplest approach is to equate $u - r$ parameters to zero.

This would be identical to having these $u - r$ parameters treated as constants in the adjustment. They could be left out when setting up the design matrix and, thus, the singularity problem would be avoided altogether. Equation (2.7.27) can be solved subject to $u - r$ general conditions between the parameters. The resulting solution is a minimal constraint solution. If the particular condition (2.6.35) is applied, one obtains the inner constraint solution. If $\overline{\boldsymbol{A}}$ has no rank defect, then the system (2.7.27) consists of u equations for u unknowns.

The estimate for the quadratic form (2.7.28) does not depend on the parameters $\boldsymbol{x}$ and, thus, is invariant with respect to the selection of the minimal constraints for finding the least-square estimate of $\boldsymbol{x}$. Moreover, the residuals themselves are independent of the minimal constraints. Substituting the solution (2.7.27) into (2.7.22) gives

$$\begin{bmatrix} \boldsymbol{F}^T \\ \boldsymbol{G}^T \end{bmatrix} \hat{\overline{\boldsymbol{v}}} = \begin{bmatrix} \boldsymbol{0} \\ \boldsymbol{G}^T \overline{\boldsymbol{\ell}} \end{bmatrix} \tag{2.7.29}$$

Since the matrix $[\boldsymbol{F}\ \boldsymbol{G}]$ is orthonormal, the expression for the residuals becomes

$$\hat{\overline{\boldsymbol{v}}} = [\boldsymbol{F} \quad \boldsymbol{G}] \begin{bmatrix} \boldsymbol{0} \\ \boldsymbol{G}^T \overline{\boldsymbol{\ell}} \end{bmatrix} = \boldsymbol{G}\boldsymbol{G}^T \overline{\boldsymbol{\ell}} \tag{2.7.30}$$

Thus, the residuals are independent of the specific solution for $\hat{\boldsymbol{x}}$. The matrix $\boldsymbol{G}$ depends only on the structure of the design matrix $\overline{\boldsymbol{A}}$. By applying the law of variance-covariance propagation to (2.7.30), we clearly see that the covariance matrix of the adjusted residuals, and thus the covariance matrix of the adjusted observations, does not depend on the specific set of minimal constraints. Note that the transformation (2.7.10) does not invalidate these statements since the $\boldsymbol{D}$ matrix is not related to the parameters.

Returning to the derivation of the distribution of $\boldsymbol{v}^T\boldsymbol{P}\boldsymbol{v}$, we find from (2.7.23) that

$$E(\boldsymbol{z}) = \begin{bmatrix} -\boldsymbol{F}^T \overline{\boldsymbol{A}}\boldsymbol{x} \\ \boldsymbol{0} \end{bmatrix} \tag{2.7.31}$$

using (2.7.20) and the fact that $E(\overline{\boldsymbol{\ell}}) = -\overline{\boldsymbol{A}}\boldsymbol{x}$ according to (2.7.11). Making use of (2.7.15) the covariance matrix is

$$\Sigma_z = \sigma_0^2 \begin{bmatrix} \boldsymbol{F}^T \\ \boldsymbol{G}^T \end{bmatrix} \boldsymbol{I} [\boldsymbol{F} \quad \boldsymbol{G}] = \sigma_0^2 \begin{bmatrix} \boldsymbol{F}^T\boldsymbol{F} & \boldsymbol{F}^T\boldsymbol{G} \\ \boldsymbol{G}^T\boldsymbol{F} & \boldsymbol{G}^T\boldsymbol{G} \end{bmatrix} = \sigma_0^2 \begin{bmatrix} \boldsymbol{I} & \boldsymbol{0} \\ \boldsymbol{0} & \boldsymbol{I} \end{bmatrix} \tag{2.7.32}$$

Since a linear transformation of a random variable with multivariate normal distribution results in another multivariate normal distribution according to (A.5.68), it follows that $\boldsymbol{z}$ is distributed as

$$\boldsymbol{z} \sim N_n \left(\begin{bmatrix} -\boldsymbol{F}^T \overline{\boldsymbol{A}}\boldsymbol{x} \\ \boldsymbol{0} \end{bmatrix}, \sigma_0^2 \begin{bmatrix} {}_r\boldsymbol{I}_r & \boldsymbol{0} \\ \boldsymbol{0} & {}_{n-r}\boldsymbol{I}_{n-r} \end{bmatrix} \right) \tag{2.7.33}$$

The random variables $\mathbf{z}_1$ and $\mathbf{z}_2$ are stochastically independent, as are the individual components. From equation (A.5.71) it follows that

$$\mathbf{z}_2 \sim N_{n-r}\left(\mathbf{0}, \sigma_0^2 \mathbf{I}\right) \tag{2.7.34}$$

Thus

$$z_{2i} \sim n\left(0, \sigma_0^2\right) \tag{2.7.35}$$

$$\frac{z_{2i}}{\sigma_0} \sim n(0, 1) \tag{2.7.36}$$

are unit variate normal distributed. As listed in Appendix A.5.2, the square of a standardized normal distributed variable has a chi-square distribution with one degree of freedom. In addition, the sum of chi-square distributed variables is also a chi-square distribution with a degree of freedom equal to the sum of the individual degrees of freedom. Using these functions of random variables, it follows that $\mathbf{v}^T\mathbf{P}\mathbf{v}$

$$\frac{\hat{R}}{\sigma_0^2} = \frac{\mathbf{z}_2^T\mathbf{z}_2}{\sigma_0^2} = \sum_{i=1}^{n-r} \frac{z_{2i}^2}{\sigma_0^2} \sim \chi^2_{n-r} \tag{2.7.37}$$

has a chi-square distribution with $n - r$ degrees of freedom.

Combining the result of (2.7.37) with the expression for the a posteriori variance of unit weight of Table 2.5.1, we obtain the formulation for a fundamental statistical test in least-squares estimation:

$$\frac{\mathbf{v}^T\mathbf{P}\mathbf{v}}{\sigma_0^2} = \frac{\hat{\sigma}_0^2}{\sigma_0^2}(n - r) \sim \chi^2_{n-r} \tag{2.7.38}$$

Note that $n - r$ is the degree of freedom of the adjustment. If there is no rank deficiency in the design matrix, the degree of freedom is $n - u$. Based on the statistics (2.7.38), the test can be performed to find out whether the adjustment is distorted. The formulation of the hypothesis is as follows:

$$H_0 : \sigma_0^2 = \hat{\sigma}_0^2 \tag{2.7.39}$$

$$H_1 : \sigma_0^2 \neq \hat{\sigma}_0^2 \tag{2.7.40}$$

The zero hypothesis states that the a priori variance of unit weight statistically equals the a posteriori variance of unit weight. Recall that the a posteriori variance of unit weight is a random variable; the adjustment makes a sample value available for this quantity on the basis of the observations (the samples). Both variances of unit weight do not have to be numerically equal but they should be statistically equal in the sense of (2.3.32). If the zero hypothesis is accepted, the adjustment is judged to be correct. If the numerical value

$$\chi^2 = \frac{\hat{\sigma}_0^2}{\sigma_0^2}(n - r) = \frac{\mathbf{v}^T\mathbf{P}\mathbf{v}}{\sigma_0^2} \tag{2.7.41}$$

TABLE 2.7.1 Selected Values for Chi-Square

Degree of Freedom (DF)	Probability α			
	0.975	0.950	0.050	0.025
1	0.00	0.00	3.84	5.02
5	0.83	1.15	11.07	12.83
10	3.25	3.94	18.31	20.48
20	9.59	10.85	31.41	34.17
50	32.36	34.76	67.50	71.42
100	74.22	77.93	124.34	129.56

is such that

$$\chi^2 < \chi^2_{n-r,1-\alpha/2} \tag{2.7.42}$$

$$\chi^2 > \chi^2_{n-r,\alpha/2} \tag{2.7.43}$$

then the zero hypothesis is rejected. The significance level α, i.e., the probability of a type-I error, or the probability of rejecting the zero hypothesis even though it is true, is generally fixed to 0.05. Here the significance level is the sum of the probabilities in both tails. Table 2.7.1 lists selected values from the chi-square distribution $\chi^2_{n-r,\alpha}$. Rejection of the zero hypothesis is taken to indicate that something is wrong with the adjustment. The cause for rejection remains to be clarified. Figure 2.7.1 shows the limits for the a posteriori variance of unit weight as a function of the degree of freedom given the significance level $\alpha = 0.05$.

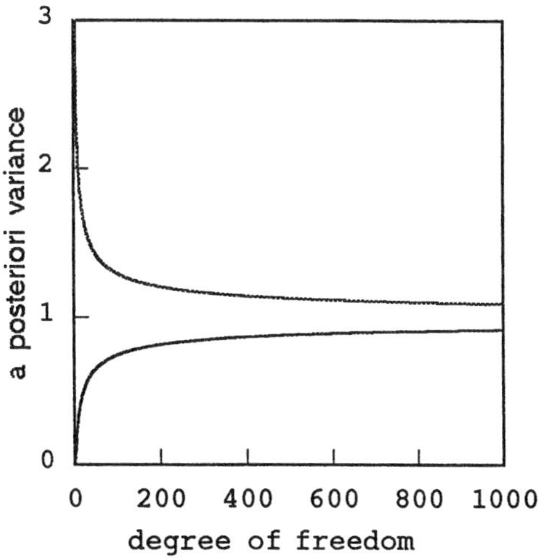

Figure 2.7.1 Limits on the a posteriori variance of unit weight. The figure refers to $\alpha = 0.05$.

The probability β of the type-II error, i.e., the probability of rejecting the alternative hypothesis and accepting the zero hypothesis even though the alternative hypothesis is true, is generally not computed. Type-II errors are considered in Section 2.8.2 in regards to reliability and in Section 6.5.3 in regards to discernibility of estimated ambiguity sets.

2.7.2 Testing Sequential Least Squares

The test statistics for testing groups of observations is based on $\mathbf{v}^T\mathbf{P}\mathbf{v}^*$ and the change $\Delta\mathbf{v}^T\mathbf{P}\mathbf{v}$. According to Table 2.5.2 we have

$$\begin{aligned}\Delta\mathbf{v}^T\mathbf{P}\mathbf{v} &= (\mathbf{A}_2\mathbf{x}^* + \ell_2)^T\mathbf{T}(\mathbf{A}_2\mathbf{x}^* + \ell_2)\\ &= \mathbf{z}_3^T\mathbf{T}\mathbf{z}_3\end{aligned} \tag{2.7.44}$$

The new random variable $\mathbf{z}_3$ is a function of observations ℓ_1 and ℓ_2. Applying the laws of propagation of mean and variance, one finds

$$E(\mathbf{z}_3) = \mathbf{A}_2E(\mathbf{x}^*) + E(\ell_2) = \mathbf{A}_2\mathbf{x} - \mathbf{A}_2\mathbf{x} = \mathbf{0} \tag{2.7.45}$$

$$\Sigma_{z_3} = \mathbf{T}^{-1} \tag{2.7.46}$$

$$\mathbf{z}_3 \sim N\left(\mathbf{0}, \sigma_0^2\mathbf{T}^{-1}\right) \tag{2.7.47}$$

Carrying out the orthonormal transformation yields a random vector whose components are stochastically independent and normally distributed. By standardizing these distributions and summing the squares of these random variables, it follows that

$$\frac{\Delta\mathbf{v}^T\mathbf{P}\mathbf{v}}{\sigma_0^2} = \frac{\mathbf{z}_3^T\mathbf{T}\mathbf{z}_3}{\sigma_0^2} \sim \chi^2_{n_2} \tag{2.7.48}$$

has a chi-square distribution with n_2 degrees of freedom, where n_2 equals the number of observations in the second group. The random variables (2.7.48) and (2.7.38) are stochastically independent. To prove this, consider the new random variable $\mathbf{z} = [\mathbf{z}_1\,\mathbf{z}_2\,\mathbf{z}_3]^T$, which is a linear function of the random variables ℓ (first group) and ℓ_2, according to equations (2.7.10), (2.7.23), and (2.7.44). By using the covariance matrix (2.4.2) and applying variance-covariance propagation, we find that the covariances between the $\mathbf{z}_i$ are zero. Because the distribution of the $\mathbf{z}$ is multivariate normal, it follows that the random variables $\mathbf{z}_i$ are stochastically independent. Since $\Delta\mathbf{v}^T\mathbf{P}\mathbf{v}$ is a function of $\mathbf{z}_3$ only, it follows that $\mathbf{v}^T\mathbf{P}\mathbf{v}$ in (2.7.38), which is only a function of $\mathbf{z}_2$, and $\Delta\mathbf{v}^T\mathbf{P}\mathbf{v}$ in (2.7.48) are stochastically independent. Thus, it is permissible to form the following ratio of random variables:

$$\frac{\Delta\mathbf{v}^T\mathbf{P}\mathbf{v}(n_1 - r)}{\mathbf{v}^T\mathbf{P}\mathbf{v}^*(n_2)} \sim F_{n_2, n_1 - r} \tag{2.7.49}$$

which has an F distribution.

TABLE 2.7.2 Selected Values for F

	n_1			
n_2	1	2	3	4
5	6.61	5.79	5.41	5.19
10	4.96	4.10	3.71	3.48
20	4.35	3.49	3.10	2.87
60	4.00	3.15	2.76	2.53
120	3.92	3.07	2.68	2.45
∞	3.84	3.00	2.60	2.37

Thus the fundamental test in sequential adjustment is based on the F distribution. The zero hypothesis states that the second group of observations does not distort the adjustment, or that there is no indication that something is wrong with the second group of observations. The alternative hypothesis states that there is an indication that the second group of observations contains errors. The zero hypothesis is rejected, and the alternative hypothesis is accepted if

$$F < F_{n_2,\, n_1-r,\, 1-\alpha/2} \tag{2.7.50}$$

$$F > F_{n_2,\, n_1-r,\, \alpha/2} \tag{2.7.51}$$

Table 2.7.2 lists selected values from the F distribution as a function of the degrees of freedom and probability. The tabulation refers to the parameters as specified in $F_{n_1,\, n_2,\, 0.05}$.

2.7.3 General Linear Hypothesis

The general linear hypothesis deals with linear conditions between parameters. Nonlinear conditions are first linearized. The basic idea is to test the change $\Delta \mathbf{v}^T\mathbf{P}\mathbf{v}$ for its statistical significance. Any of the three adjustment models can be used to carry out the general linear hypothesis test. For the observation equation model with additional conditions between the parameters, one has

$$\mathbf{v}_1 = \mathbf{A}_1\mathbf{x} + \boldsymbol{\ell}_1 \tag{2.7.52}$$

$$H_0 : \mathbf{A}_2\mathbf{x} + \boldsymbol{\ell}_2 = \mathbf{0} \tag{2.7.53}$$

Equation (2.7.53) expresses the zero hypothesis H_0. The solution of the combined adjustment is found in Table 2.5.5. Adjusting (2.7.52) alone results in $\mathbf{v}^T\mathbf{P}\mathbf{v}^*$, which has a chi-square distribution with $n - r$ degrees of freedom according to (2.7.48). The change $\Delta \mathbf{v}^T\mathbf{P}\mathbf{v}$ resulting from the condition (2.7.53) is

$$\Delta \mathbf{v}^T\mathbf{P}\mathbf{v} = (\mathbf{A}_2\mathbf{x}^* + \boldsymbol{\ell}_2)^T\, \mathbf{T}(\mathbf{A}_2\mathbf{x}^* + \boldsymbol{\ell}_2) \tag{2.7.54}$$

The expression in (2.7.54) differs from (2.7.44) in two respects. First, the matrix $\boldsymbol{T}$ differs, i.e., the matrix $\boldsymbol{T}$ in (2.7.54) does not contain the $\boldsymbol{P}_2$ matrix. Second, the quantity $\boldsymbol{\ell}_2$ is not a random variable. These differences, however, do not matter in the proof of stochastic independence of $\boldsymbol{v}^T\boldsymbol{P}\boldsymbol{v}^*$ and $\Delta\boldsymbol{v}^T\boldsymbol{P}\boldsymbol{v}$. Analogously to (2.7.44), we can express the change $\Delta\boldsymbol{v}^T\boldsymbol{P}\boldsymbol{v}$ in (2.7.54) as a function of a new random variable $\boldsymbol{z}_3$. The proof for stochastic independence follows the same lines of thought as given before (for the case of additional observations). Thus, just as (2.7.49) is the basis for testing two groups of observations, the basic test for the general linear hypothesis (2.7.53) is

$$\frac{\Delta\boldsymbol{v}^T\boldsymbol{P}\boldsymbol{v}(n_1 - r)}{\boldsymbol{v}^T\boldsymbol{P}\boldsymbol{v}^* \quad n_2} \sim F_{n_2, n_1 - r} \tag{2.7.55}$$

A small $\Delta\boldsymbol{v}^T\boldsymbol{P}\boldsymbol{v}$ implies that the null hypothesis (2.7.53) is acceptable, i.e., the conditions are in agreement with the observations. The conditions do not impose any distortions on the adjustment. The rejection criterion is based on the one-tail test at the upper end of the distribution. Thus, reject H_0 at a $100\alpha\%$ significance level if

$$F > F_{n_2, n_1 - r, \alpha} \tag{2.7.56}$$

The general formulation of the null hypothesis in (2.7.53) makes it possible to test any hypothesis on the parameters, so long as the hypothesis can be expressed in a mathematical equation. Nonlinear hypotheses must first be linearized. Simple hypotheses could be used to test whether an individual parameter has a certain numerical value, whether two parameters are equal, whether the distance between two stations has a certain length, whether an angle has a certain size, etc. For example, consider the hypothesis

$$H_0 \; : \boldsymbol{x} - \boldsymbol{x}_T = \boldsymbol{0} \tag{2.7.57}$$

$$H_1 \; : \boldsymbol{x} - \boldsymbol{x}_T \neq \boldsymbol{0} \tag{2.7.58}$$

The zero hypothesis states that the parameters equal a certain (true) value $\boldsymbol{x}_T$. From (2.7.53) it follows that $\boldsymbol{A}_2 = \boldsymbol{I}$ and $\boldsymbol{\ell}_2 = -\boldsymbol{x}_T$. Using these specifications we can use $\boldsymbol{T} = \boldsymbol{N}$ in (2.7.54), and the statistic (2.7.55) becomes

$$\frac{(\hat{\boldsymbol{x}}^* - \boldsymbol{x}_T)^T\boldsymbol{N}(\hat{\boldsymbol{x}}^* - \boldsymbol{x}_T)}{\hat{\sigma}_0^2 r} \sim F_{r, n_1 - r, \alpha} \tag{2.7.59}$$

where the a posteriori variance of unit weight (first group only) has been substituted for $\boldsymbol{v}^T\boldsymbol{P}\boldsymbol{v}^*$. Once the adjustment of the first group (2.7.52) is completed, the values for the adjusted parameters and the a posteriori variance of unit weight are entered in (2.7.59), and the fraction is computed and compared with the F value (taking the proper degrees of freedom and the desired significance level into account). Rejection or acceptance of the zero hypothesis follows rule (2.7.56).

Note that one of the degrees of freedom in (2.7.59) is $r = R(\boldsymbol{N}) < u$, instead of u, which equals the number of parameters, even though equation (2.7.57) expresses u

conditions. Because of the possible rank defect of the normal matrix $\boldsymbol{N}$, the distribution of $\Delta \boldsymbol{v}^T \boldsymbol{P} \boldsymbol{v}$ in (2.7.54) is a chi-square distribution with r degrees of freedom. Consider the derivation leading to (2.7.48). The u components of $\boldsymbol{z}_3$ are transformed to r stochastically independent unit variate normal distributions that are then squared and summed to yield the distribution of $\Delta \boldsymbol{v}^T \boldsymbol{P} \boldsymbol{v}$. The interpretation is that (2.7.57) represents one hypothesis on all parameters $\boldsymbol{x}$, and not u hypotheses on the u components on $\boldsymbol{x}$.

Expression (2.7.59) can be used to define the r-dimensional confidence region. Replace the particular $\boldsymbol{x}_T$ by the unknown parameter $\boldsymbol{x}$, and drop the asterisk; then

$$P\left[\frac{(\hat{\boldsymbol{x}}-\boldsymbol{x})^T \boldsymbol{N}(\hat{\boldsymbol{x}}-\boldsymbol{x})}{\hat{\sigma}_0^2 r} \leq F_{r,n_1-r,\alpha}\right] = \int_0^{F_{r,n_1-r,\alpha}} F_{r,n_1-r} dF = 1-\alpha \tag{2.7.60}$$

The probability region described by the expression on the left side of equation (2.7.60) is an $R(\boldsymbol{N})$-dimensional ellipsoid. The probability region is an ellipsoid because the normal matrix $\boldsymbol{N}$ is positive definite or, at least, semipositive definite. If one identifies the center of the ellipsoid with $\hat{\boldsymbol{x}}$, then there is $(1-\alpha)$ probability that the unknown point $\boldsymbol{x}$ lies within the ellipsoid. The orientation and the size of this ellipsoid are a function of the eigenvectors and eigenvalues of the normal matrix, the rank of the normal matrix, and the degree of freedom. Consider the orthonormal transformation

$$\boldsymbol{z} = \boldsymbol{F}^T(\boldsymbol{x}-\hat{\boldsymbol{x}}) \tag{2.7.61}$$

with $\boldsymbol{F}$ as specified in (A.3.17) and containing the normalized eigenvectors of $\boldsymbol{N}$, then

$$\boldsymbol{F}^T \boldsymbol{N} \boldsymbol{F} = \boldsymbol{\Lambda} \tag{2.7.62}$$

with $\boldsymbol{\Lambda}$ containing the r eigenvalues of $\boldsymbol{N}$, and

$$(\hat{\boldsymbol{x}}-\boldsymbol{x})^T \boldsymbol{N}(\hat{\boldsymbol{x}}-\boldsymbol{x}) = \boldsymbol{z}^T \boldsymbol{\Lambda} \boldsymbol{z} = \sum_{i=1}^{r} z_i^2 \lambda_i = \sum_{i=1}^{r} \frac{z_i^2}{\left(1/\sqrt{\lambda_i}\right)^2} \tag{2.7.63}$$

Combining equations (2.7.60) and (2.7.63), we can write the r-dimensional ellipsoid, or the r-dimensional confidence region, in the principal axes form:

$$P\left[\frac{z_1^2}{\left(\hat{\sigma}_0\sqrt{rF_{r,n-r,\alpha}/\lambda_1}\right)^2} + \cdots + \frac{z_r^2}{\left(\hat{\sigma}_0\sqrt{rF_{r,n-r,\alpha}/\lambda_r}\right)^2} \leq 1\right] = 1-\alpha \tag{2.7.64}$$

The confidence region is centered at $\hat{\boldsymbol{x}}$. Whenever the zero hypothesis H_0 of (2.7.57) is accepted, the point $\boldsymbol{x}_T$ falls within the confidence region. The probability that the ellipsoid contains the true parameters $\boldsymbol{x}_T$ is $1-\alpha$. For these reasons, one naturally would like the ellipsoid to be small. Equation (2.7.64) shows that the semimajor axes are proportional to the inverse of the eigenvalues of the normal matrix. It is exactly this relationship that makes us choose the eigenvalues of $\boldsymbol{N}$ as large as possible, provided that we have a choice through appropriate network design variation. As an eigenvalue approaches zero, the respective axis of the confidence ellipsoid approaches infinity; this is an undesirable situation, both from a statistical point of view and because of the numerical difficulties encountered during the inversion of the normal matrix.

2.7.4 Ellipses as Confidence Regions

Confidence ellipses are statements of precision. They are frequently used in connection with two-dimensional networks in order to make the directional precision of station location visible. Ellipses of confidence follow by limiting the hypothesis (2.7.57) to two parameters, i.e., the Cartesian coordinates of a station. Of course, in a three-dimensional network one can compute three-dimensional ellipsoids or several ellipses, e.g., one for the horizontal and others for the vertical. Confidence ellipses or ellipsoids are not limited to the specific application of networks. However, in networks the confidence regions can be referenced with respect to the coordinate system of the network and thus can provide an integrated view of the geometry of the confidence regions and the network.

Consider the following hypothesis:

$$H_0 : \boldsymbol{x}_i - \boldsymbol{x}_{i,T} = \boldsymbol{0} \tag{2.7.65}$$

where the notation

$$\boldsymbol{x}_i = \begin{bmatrix} x_1 & x_2 \end{bmatrix}^T \tag{2.7.66}$$

is used. The symbols x_1 and x_2 denote the Cartesian coordinates of a two-dimensional network station P_i. The test of this hypothesis follows the outline given in the previous section. The A_2 matrix is of size $2 \times u$ because there are two separate equations in the hypothesis and u components in $\boldsymbol{x}$. The elements of $\boldsymbol{A}_2$ are zero except those elements of rows 1 and 2, which correspond to the respective positions of x_1 and x_2 in $\boldsymbol{x}$. With these specifications it follows that

$$\boldsymbol{Q}_i = \boldsymbol{A}_2\boldsymbol{N}^{-1}\boldsymbol{A}_2^T = \begin{bmatrix} q_{x_1} & q_{x_1,x_2} \\ q_{x_2,x_1} & q_{x_2} \end{bmatrix} \tag{2.7.67}$$

where $\boldsymbol{Q}_i$ contains the respective elements of the inverse of the normal matrix. With these specifications $\boldsymbol{T} = \boldsymbol{Q}_i^{-1}$ and expression (2.7.55) becomes

$$\frac{1}{2\hat{\sigma}_0^2}(\hat{\boldsymbol{x}}_i - \boldsymbol{x}_{i,T})^T\boldsymbol{Q}_i^{-1}(\hat{\boldsymbol{x}}_i - \boldsymbol{x}_{i,T}) \sim F_{2,n-r} \tag{2.7.68}$$

Given the significance level α, the hypothesis test can be carried out. The two-dimensional confidence region is

$$P\left[\frac{(\hat{\boldsymbol{x}}_i - \boldsymbol{x}_i)^T\boldsymbol{Q}_i^{-1}(\hat{\boldsymbol{x}}_i - \boldsymbol{x}_i)}{2\hat{\sigma}_0^2} \le F_{2,n-r,\alpha}\right] = \int_0^{F_{2,n-r,\alpha}} F_{2,n-r}\,dF = 1 - \alpha \tag{2.7.69}$$

The size of the confidence ellipses defined by (2.7.69) depends on the degree of freedom of the adjustment and the significance level. The ellipses are centered at the adjusted position and delimit the $(1 - \alpha)$ probability area for the true position. The principal axis form of (2.7.69) is obtained through orthogonal transformation. Let $\boldsymbol{R}_i$ denote the matrix whose rows are the orthonormal eigenvectors of $\boldsymbol{Q}_i$, then

$$\boldsymbol{R}_i^T \boldsymbol{Q}_i^{-1} \boldsymbol{R}_i = \boldsymbol{\Lambda}_i^{-1} \tag{2.7.70}$$

according to (A.3.16). The matrix $\boldsymbol{\Lambda}_i$ is diagonal and contains the eigenvalues λ_i^Q and λ_2^Q of $\boldsymbol{Q}_1$. With

$$\boldsymbol{z}_i = \boldsymbol{R}_i^T (\hat{\boldsymbol{x}}_i - \boldsymbol{x}_i) \tag{2.7.71}$$

Expression (2.7.69) becomes

$$P\left\{\left[\frac{z_1^2}{\left(\hat{\sigma}_0\sqrt{\lambda_1^Q 2F_{2,n-r,\alpha}}\right)^2} + \frac{z_2^2}{\left(\hat{\sigma}_0\sqrt{\lambda_2^Q 2F_{2,n-r,\alpha}}\right)^2}\right] \leq 1\right\}$$
$$= \int_0^{F_{2,n-r,\alpha}} F_{2,n-r}\, dF = 1 - \alpha \tag{2.7.72}$$

For $F_{2,n-r,\alpha} = 1/2$, the ellipse is called the *standard ellipse* or the *error ellipse*. Thus, the probability enclosed by the standard ellipse is a function of the degree of freedom $n - r$ and is computed as follows:

$$P(\text{standard ellipse}) = \int_0^{1/2} F_{2,n-r}\, dF \tag{2.7.73}$$

The magnification factor, $\sqrt{2F_{2,\,n-r,\,\alpha}}$, as a function of the probability and the degree of freedom, is shown in Table 2.7.3. The table shows immediately that a small degree of freedom requires a large magnification factor to obtain, e.g., 95% probability. It is seen that in the range of small degrees of freedom, an increase in the degree of freedom rapidly decreases the magnification factor, whereas with a large degree of freedom, any additional observations cause only a minor reduction of the magnification factor. After a degree of freedom of about 8 or 10, the decrease in the magnification factor slows down noticeably. Thus, based on the speed of decreasing magnification factor, a degree of 10 appears optimal, considering the expense of additional observations and the little gain derived from them in the statistical sense. For a degree of freedom of 10, the magnification factor is about 3 to cover 95% probability.

The hypothesis (2.7.65) can readily be generalized to three dimensions encompassing the Cartesian coordinates of a three-dimensional network station. The magnification factor of the respective *standard ellipsoid* is $\sqrt{3F_{3,\,n-r,\,\alpha}}$ for it to contain $(1 - \alpha)$ probability. Similarly, the standard deviation of an individual coordinate is converted to a $(1 - \alpha)$ probability confidence interval by multiplication with

TABLE 2.7.3 Magnification Factor for Standard Ellipses

	Probability $1-\alpha$		
$n-r$	95%	98%	99%
1	20.00	50.00	100.00
2	6.16	9.90	14.10
3	4.37	6.14	7.85
4	3.73	4.93	6.00
5	3.40	4.35	5.15
6	3.21	4.01	4.67
8	2.99	3.64	4.16
10	2.86	3.44	3.89
12	2.79	3.32	3.72
15	2.71	3.20	3.57
20	2.64	3.09	3.42
30	2.58	2.99	3.28
50	2.52	2.91	3.18
100	2.49	2.85	3.11
∞	2.45	2.80	3.03

$\sqrt{F_{1,n-r,\alpha}}$. These magnification factors are shown in Figure 2.7.2 for $\alpha = 0.05$. For higher degrees of freedom, the magnification factors converge toward the respective chi-square values because of the relationship $rF_{r,\infty} = \chi_r^2$.

For drawing the confidence ellipse at station P_i, we need the rotation angle φ between the (x_i) and (z_i) coordinate systems as well as the semimajor and semiminor axis of the ellipse. Let (y_i) denote the translated (x_i) coordinate system through the adjusted point $\hat{\mathbf{x}}_i$; then (2.7.71) becomes

$$\mathbf{z}_i = \mathbf{R}_i^T \mathbf{y}_i \tag{2.7.74}$$

The eigenvectors of $\mathbf{Q}_i$ determine the directions of the semiaxes, and the eigenvalues determine their lengths. Rather than computing the vectors explicitly, we choose to compute the rotation angle φ by comparing coefficients from quadratic forms. Figure 2.7.3 shows the rotational relation

$$\mathbf{z}_i = \begin{bmatrix} \cos\varphi & \sin\varphi \\ -\sin\varphi & \cos\varphi \end{bmatrix} \mathbf{y}_i \tag{2.7.75}$$

and (2.7.70) and (2.7.74) give the two quadratic forms

$$\mathbf{y}_i^T \mathbf{Q}_i \mathbf{y}_i = \mathbf{z}_i^T \Lambda_i \mathbf{z}_i \tag{2.7.76}$$

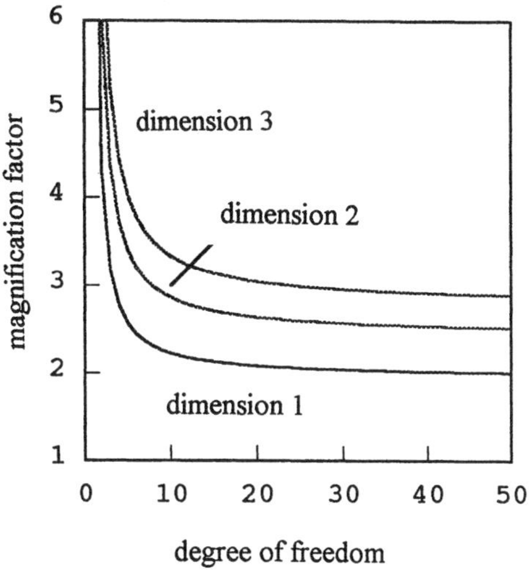

Figure 2.7.2 Magnification factors for confidence regions. The values refer to $\alpha = 0.05$.

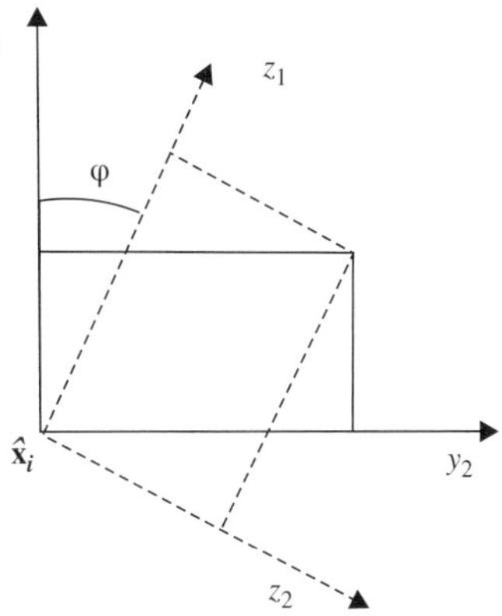

Figure 2.7.3 Rotation of the principal axis coordinate system.

We substitute (2.7.75) into the right-hand side of (2.7.76) and the matrix elements of $\boldsymbol{Q}_i$ of (2.7.67) into the left-hand side and compare the coefficient of $y_1 y_2$ on both sides, giving

$$\sin 2\varphi = \frac{2q_{x_1,x_2}}{\lambda_1^Q - \lambda_2^Q} \tag{2.7.77}$$

The eigenvalues follow directly from the characteristic equation

$$|\boldsymbol{Q}_i - \lambda^Q \boldsymbol{I}| = \begin{vmatrix} q_{x_1} - \lambda^Q & q_{x_1,x_2} \\ q_{x_1,x_2} & q_{x_2} - \lambda^Q \end{vmatrix} = \left(q_{x_1} - \lambda^Q\right)\left(q_{x_2} - \lambda^Q\right) - q_{x_1,x_2}^2 = 0 \tag{2.7.78}$$

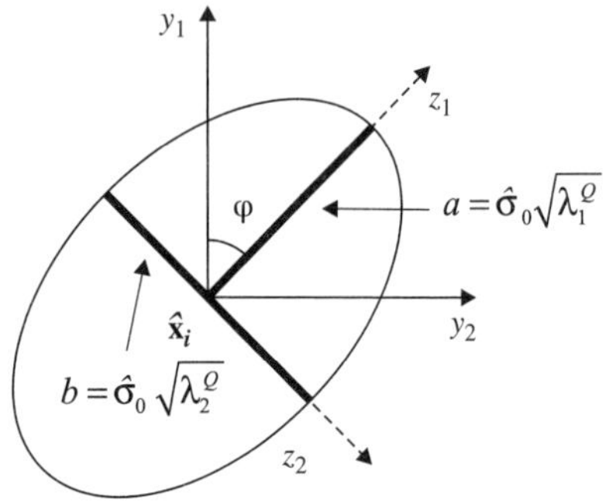

Figure 2.7.4 Defining elements of standard ellipse.

The solution of the quadratic equation is

$$\lambda_1^Q = \frac{q_{x_1} + q_{x_2}}{2} + \frac{1}{2}W \tag{2.7.79}$$

$$\lambda_2^Q = \frac{q_{x_1} + q_{x_2}}{2} - \frac{1}{2}W \tag{2.7.80}$$

$$W = \sqrt{(q_{x_1} - q_{x_2})^2 + 4q_{x_1,x_2}^2} \tag{2.7.81}$$

$$\sin 2\varphi = \frac{2q_{x_1,x_2}}{W} \tag{2.7.82}$$

$$\cos 2\varphi = \frac{q_{x_1} - q_{x_2}}{W} \tag{2.7.83}$$

The terms $\sin 2\varphi$ and $\cos 2\varphi$ determine the quadrant of φ.

Figure 2.7.4 shows the defining elements of the standard ellipse. Recall equation (2.7.72) regarding the interpretation of the standard ellipses as a confidence region. In any adjustment, any two parameters can comprise $\mathbf{x}_i$, regardless of the geometric meaning of the parameters. Examples are the intercept and slope in the fitting of a straight line or ambiguity parameters in the case of GPS carrier phase solutions. The components $\mathbf{x}_i$ can always be interpreted as Cartesian coordinates for drawing the standard ellipse and thus can give a graphical display of the covariance. In surveying networks, the vectors $\mathbf{x}_i$ contain coordinates of stations in a well-defined coordinate system. If $\mathbf{x}_i$ represents latitude and longitude or northing and casting, the horizontal standard ellipse is computed. If $\mathbf{x}_i$ contains the vertical coordinate and easting, then the standard ellipse in the prime vertical is obtained.

Because the shape of the standard ellipses and ellipsoids depends on the geometry of the network through the design matrix and the weight matrix, the geometric interpretation is enhanced if the network and the standard ellipses are displayed together. Occasionally, users prefer to compute coordinate differences and their covariance matrix and plot relative standard ellipses.

2.7.5 Properties of Standard Ellipses

The positional error p of a station is directly related to the standard ellipse, as seen in Figure 2.7.5. The positional error is the standard deviation of a station in a certain direction, say ψ. It is identical with the standard deviation of the distance to a known (fixed) station along the same direction ψ, as computed from the linearized distance equation and variance-covariance propagation. The linear function is

$$r = z_1 \cos \psi + z_2 \sin \psi \tag{2.7.84}$$

Because of equations (2.7.70) and (2.7.71), the distribution of the random variable $\mathbf{z}_i$ is multivariate normal with

$$\begin{bmatrix} z_1 \\ z_2 \end{bmatrix} \sim N\left(\begin{bmatrix} 0 \\ 0 \end{bmatrix}, \hat{\sigma}_0^2 \begin{bmatrix} \lambda_1^Q & 0 \\ 0 & \lambda_2^Q \end{bmatrix}\right) = N\left(\begin{bmatrix} 0 \\ 0 \end{bmatrix}, \begin{bmatrix} a^2 & 0 \\ 0 & b^2 \end{bmatrix}\right) \tag{2.7.85}$$

The variance of the random variable r follows from the law of variance-covariance propagation:

$$\sigma_r^2 = a^2 \cos^2 \psi + b^2 \sin^2 \psi \tag{2.7.86}$$

The variance (2.7.86) is geometrically related to the standard ellipse. Let the ellipse be projected onto the direction ψ. The point of tangency is denoted by P_0. Because the equation of the ellipse is

$$\frac{z_1^2}{a^2} + \frac{z_2^2}{b^2} = 1 \tag{2.7.87}$$

the slope of the tangent is

$$\frac{dz_1}{dz_2} = -\frac{z_2 a^2}{z_1 b^2} = -\tan \psi \tag{2.7.88}$$

See Figure 2.7.5 regarding the relation of the slope of the tangent and the angle ψ. The second part of (2.7.88) yields

$$\frac{z_{01}}{a^2} \sin \psi - \frac{z_{02}}{b^2} \cos \psi = 0 \tag{2.7.89}$$

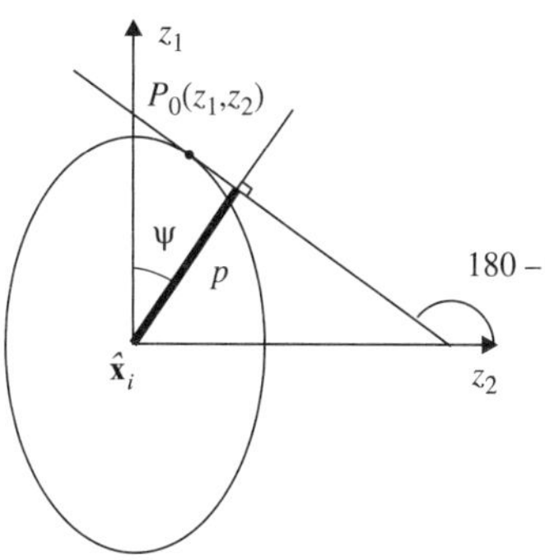

Figure 2.7.5 Position error.

This equation relates the coordinates of the point of tangency P_0 to the slope of the tangent. The length p of the projection of the ellipse is according to the figure,

$$p = z_{01} \cos \psi + z_{02} \sin \psi \tag{2.7.90}$$

Next, equation (2.7.89) is squared and then multiplied with a^2b^2, and the result is added to the square of (2.7.90), giving

$$p^2 = a^2 \cos^2 \psi + b^2 \sin^2 \psi \tag{2.7.91}$$

By comparing this expression with (2.7.86), it follows that $\hat{\sigma}_r = p$, i.e., the standard deviation in a certain direction is equal to the projection of the standard ellipse onto that direction. Therefore, the standard ellipse is not a standard deviation curve. Figure 2.7.6 shows the continuous standard deviation curve. We see that for narrow ellipses there are only small segments of the standard deviations that are close to the length of the semiminor axis. The standard deviation increases rapidly as the direction ψ moves away from the minor axis. Therefore, an extremely narrow ellipse is not desirable if the overall accuracy for the station position is important.

As a by-product of the property discussed, we see that the standard deviations of the parameter x_1 and x_2

$$\hat{\sigma}_{x_1} = \hat{\sigma}_0 \sqrt{q_{x_1}} \tag{2.7.92}$$

$$\hat{\sigma}_{x_2} = \hat{\sigma}_0 \sqrt{q_{x_2}} \tag{2.7.93}$$

are the projections of the ellipse in the directions of the x_1 and x_2 axes. This is shown in Figure 2.7.7. Equations (2.7.92) and (2.7.93) follow from the fact that the diagonal elements of the covariance matrix are the variances of the respective parameters.

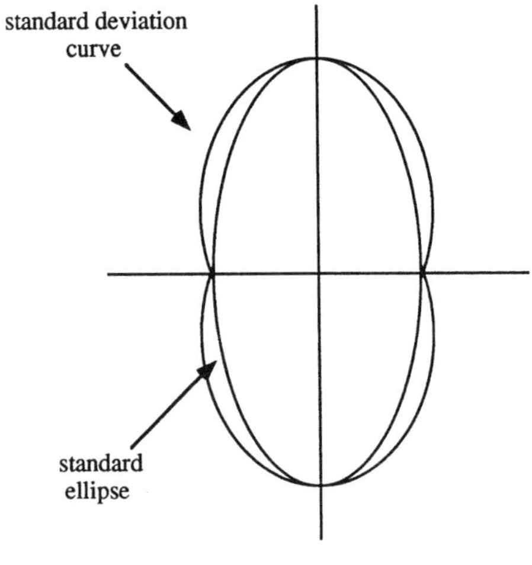

Figure 2.7.6 Standard deviation curve.

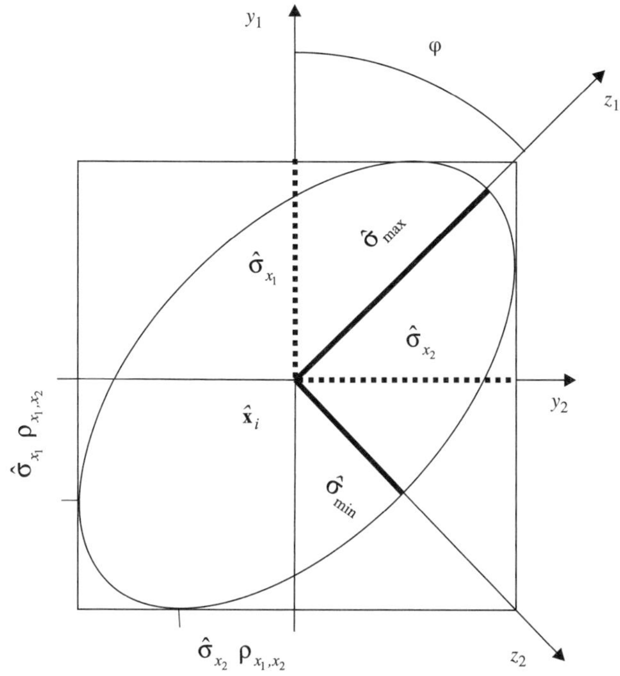

Figure 2.7.7 Characteristics of the standard deviation ellipse.

Equation (2.7.91) confirms for $\psi = 0$ and $\psi = 90°$ that the axes a and b equal the maximum and minimum standard deviations, respectively. The rectangle formed by the semisides $\hat{\sigma}_{x_1}$ and $\hat{\sigma}_{x_2}$ encloses the ellipse. This rectangle can be used as an approximation for the ellipses. The diagonal itself is sometimes referred to as the mean position error $\hat{\sigma}$,

$$\hat{\sigma} = \sqrt{\hat{\sigma}_{x_1}^2 + \hat{\sigma}_{x_2}^2} = \hat{\sigma}_0 \sqrt{q_{x_1} + q_{x_2}} \tag{2.7.94}$$

The points of contact between the ellipse and the rectangle in Figure 2.7.7 are functions of the correlation coefficients. For these points, the tangent on the ellipse is either horizontal or vertical in the (y_i) coordinate system. The equation of the ellipse in the (y) system is, according to (2.7.69),

$$\begin{bmatrix} y_1 & y_2 \end{bmatrix} \begin{bmatrix} q_{x_1} & q_{x_1,x_2} \\ q_{x_1,x_2} & q_{x_2} \end{bmatrix}^{-1} \begin{bmatrix} y_1 \\ y_2 \end{bmatrix} = \hat{\sigma}_0^2 \tag{2.7.95}$$

By replacing the matrix by its inverse, the expression becomes

$$\begin{bmatrix} y_1 & y_2 \end{bmatrix} \begin{bmatrix} q_{x_2} & -q_{x_1,x_2} \\ -q_{x_1,x_2} & q_{x_1} \end{bmatrix} \begin{bmatrix} y_1 \\ y_2 \end{bmatrix} = \left(q_{x_1} q_{x_2} - q_{x_1,x_2}^2\right) \hat{\sigma}_0^2 \tag{2.7.96}$$

Evaluating the left-hand side and dividing both sides by $q_{x_1} q_{x_2}$ gives

$$\frac{y_1^2}{q_{x_1}} + \frac{y_2^2}{q_{x_2}} - \frac{2y_1y_2q_{x_1,x_2}}{q_{x_1}q_{x_2}} = \text{constant} \tag{2.7.97}$$

from which it follows that

$$\frac{dy_1}{dy_2} = \frac{\left(2y_2 / q_{x_2}\right) - \left(2y_1\rho_{x_1,x_2} / \sqrt{q_{x_1} q_{x_2}}\right)}{\left(2y_2\rho_{x_1,x_2} / \sqrt{q_{x_1}q_{x_2}}\right) - \left(2y_1 / q_{x_1}\right)} \tag{2.7.98}$$

Consider the tangent for which the slope is infinity. The equation of this tangent line is

$$y_2 = \hat{\sigma}_0\sqrt{q_{x_2}} \tag{2.7.99}$$

Substituting this expression into the denominator of (2.7.98) and equating it to zero gives

$$\frac{\hat{\sigma}_0\sqrt{q_{x_2}}\rho_{x_1,x_2}}{\sqrt{q_{x_1}q_{x_2}}} = \frac{y_1}{q_{x_1}} \tag{2.7.100}$$

which yields the y_1 coordinate for the point of tangency:

$$y_1 = \hat{\sigma}_0 \sqrt{q_{x_1}}\, \rho_{x_1,x_2} = \hat{\sigma}_{x_1}\, \rho_{x_1,x_2} \tag{2.7.101}$$

The equation for the horizontal tangent is

$$y_1 = \hat{\sigma}_0\sqrt{q_{x_1}} \tag{2.7.102}$$

It follows from the numerator of (2.7.98) that

$$y_2 = \hat{\sigma}_0\sqrt{q_{x_2}}\, \rho_{x_1,x_2} = \hat{\sigma}_{x_2}\, \rho_{x_1,x_2} \tag{2.7.103}$$

Figure 2.7.7 shows that the standard ellipse becomes narrower the higher the correlation. For correlation plus or minus 1 (linear dependence), the ellipse degenerates into the diagonal of the rectangle. The ellipse becomes a circle if $a = b$, or $\sigma_{x_1} = \sigma_{x_2}$, and $\rho_{x_1,x_2} = 0$.

2.7.6 Other Measures of Precision

In surveying and geodesy, the most popular measure of precision is the standard deviation. The confidence regions are usually expressed in terms of ellipses and ellipsoids of standard deviation. These figures are often scaled to contain 95% probability or higher. Because GPS is a popular tool for both surveying and navigation, several of the measures of precision used in navigation are becoming increasingly popular in surveying. Examples include the dilution of precision (DOP) numbers. The DOPs are discussed in detail in Section 6.3.2. Other single-number measures refer to circular or spherical confidence regions for which the eigenvalues of the cofactor matrix have the same magnitude. In these cases, the standard deviations of the coordinates and the semiaxes are of the same size. See equation (2.7.72). When the standard deviations are not equal, these measures become a function of the ratio of the semiaxes. The derivation of the following measures and additional interpretation are given in Greenwalt and Shultz (1962).

The radius of a circle that contains 50% probability is called the circular error probable (CEP). This function is usually approximated by segments of straight lines. The expression

$$\mathrm{CEP} = 0.5887\left(\hat{\sigma}_{x_1} + \hat{\sigma}_{x_2}\right) \tag{2.7.104}$$

is, strictly speaking, valid in the region $\sigma_{\min} / \sigma_{\max} \geq 0.2$, but it is the function used most often. The 90% probability region

$$\mathrm{CMAS} = 1.8227 \times \mathrm{CEP} \tag{2.7.105}$$

is called the circular map accuracy standard. The mean position error (2.7.94) is also called the mean square positional error (MSPE), or the distance root mean square (DRMS), i.e.,

$$\mathrm{DRMS} = \sqrt{\hat{\sigma}^2_{x_1} + \hat{\sigma}^2_{x_2}} \tag{2.7.106}$$

This measure contains 64 to 77% probability. The related measure

$$2\,\mathrm{DRMS} = 2 \times \mathrm{DRMS} \tag{2.7.107}$$

contains about 95 to 98% probability.

The three-dimensional equivalent of CEP is the spherical error probable (SEP), defined as

$$\mathrm{SEP} = 0.5127\ \left(\hat{\sigma}_{x_1} + \hat{\sigma}_{x_2} + \hat{\sigma}_{x_3}\right) \tag{2.7.108}$$

Expression (2.7.108) is, strictly speaking, valid in the region $\sigma_{\min} / \sigma_{\max} \geq 0.35$. The corresponding 90% probability region,

$$\mathrm{SAS} = 1.626 \times \mathrm{SEP} \tag{2.7.109}$$

is called the spherical accuracy standard (SAS). The mean radial spherical error (MRSE) is defined as

$$\mathrm{MRSE} = \sqrt{\hat{\sigma}^2_{x_1} + \hat{\sigma}^2_{x_2} + \hat{\sigma}^2_{x_3}} \tag{2.7.110}$$

and contains about 61% probability.

These measures of precision are sometimes used to capture the achieved or anticipated precision conveniently using single numbers. However, the geometry of the adjustment seldom produces covariance matrices that yield circular distribution. Consequently, the probability levels contained in these measures of precision inevitably are a function of the correlations between the parameters.

2.8 RELIABILITY

Small residuals are not necessarily an indication of a quality adjustment. Equally important is the knowledge that all blunders in the data have been identified and removed and that remaining small blunders in the observations do not adversely impact the adjusted parameters. Reliability refers to the controllability of observations, i.e., the ability to detect blunders and to estimate the effects that undetected blunders may have on a solution. The theory outlined here follows that of Baarda (1967, 1968) and Kok (1984).

2.8.1 Redundancy Numbers

Following the expressions in Table 2.5.1 the residuals for the observation equation model are

$$\bar{\mathbf{v}} = \boldsymbol{Q}_v \boldsymbol{P} \boldsymbol{\ell} \tag{2.8.1}$$

with a cofactor matrix for the residuals

$$\boldsymbol{Q}_v = \boldsymbol{P}^{-1} - \boldsymbol{A}\boldsymbol{N}^{-1}\boldsymbol{A}^T \tag{2.8.2}$$

Compute the trace

$$\begin{aligned} \mathrm{Tr}(\boldsymbol{Q}_v\boldsymbol{P}) &= \mathrm{Tr}\left(\boldsymbol{I} - \boldsymbol{A}\boldsymbol{N}^{-1}\boldsymbol{A}^T\boldsymbol{P}\right) \\ &= n - \mathrm{Tr}\left(\boldsymbol{N}^{-1}\boldsymbol{A}^T\boldsymbol{P}\boldsymbol{A}\right) \\ &= n - u \end{aligned} \tag{2.8.3}$$

A more general expression is obtained by noting that the matrix $\boldsymbol{A}\boldsymbol{N}^{-1}\boldsymbol{A}^T\boldsymbol{P}$ is idempotent. The trace of an idempotent matrix equals the rank of that matrix. Thus,

$$\mathrm{Tr}\left(\boldsymbol{A}\boldsymbol{N}^{-1}\boldsymbol{A}^T\boldsymbol{P}\right) = R\left(\boldsymbol{A}^T\boldsymbol{P}\boldsymbol{A}\right) = R(\boldsymbol{A}) = r \leq u \tag{2.8.4}$$

Thus, from equations (2.8.3) and (2.8.4)

$$\mathrm{Tr}\,(\boldsymbol{Q}_v\boldsymbol{P}) = \mathrm{Tr}\,(\boldsymbol{Q}\boldsymbol{P}_v) = n - R(\boldsymbol{A}) \tag{2.8.5}$$

By denoting the diagonal element of the matrix $\boldsymbol{Q}_v\boldsymbol{P}$ by r_i, we can write

$$\sum_{i=1}^{n} r_i = n - R(\boldsymbol{A}) \tag{2.8.6}$$

The sum of the diagonal elements of $\boldsymbol{Q}_v\boldsymbol{P}$ equals the degree of freedom. The element r_i is called the redundancy number for the observation i. It is the contribution of the

i th observation to the degree of freedom. If the weight matrix $\boldsymbol{P}$ is diagonal, this is usually the case when original observations are adjusted, then

$$r_i = q_i p_i \tag{2.8.7}$$

where q_i is the diagonal element of the cofactor matrix $\boldsymbol{Q}_v$, and p_i denotes the weight of the i th observation. Equation (2.8.2) implies the inequality

$$0 \le q_i \le \frac{1}{p_i} \tag{2.8.8}$$

Multiplying by p_i gives the bounds for the redundancy numbers,

$$0 \le r_i \le 1 \tag{2.8.9}$$

Considering the general relation

$$\boldsymbol{Q}_{\ell_a} = \boldsymbol{Q}_{\ell_b} - \boldsymbol{Q}_v \tag{2.8.10}$$

given in Table 2.5.1 and the specification (2.8.7) for the redundancy number r_i as the diagonal element of $\boldsymbol{Q}_v\boldsymbol{P}$, it follows that if the redundancy number is close to 1, then the variance of the residuals is close to the variance of the observations, and the variance of the adjusted observations is close to zero. If the redundancy number is close to zero, then the variance of the residuals is close to zero, and the variance of the adjusted observations is close to the variance of the observations.

Intuitively, it is expected that the variance of the residuals and the variance of the observations are close; for this case, the noise in the residuals equals that of the observations, and the adjusted observations are determined with high precision. Thus the case of r_i close to 1 is preferred, and it is said that the gain of the adjustment is high. If r_i is close to zero, one expects the noise in the residuals to be small. Thus, small residuals as compared to the expected noise of the observations are not necessarily desirable. Because the inequality (2.8.9) is a result of the geometry as represented by the design matrix $\boldsymbol{A}$, small residuals can be an indication of a weak part of the network.

Because the weight matrix P is considered diagonal, i.e.,

$$p_i = \frac{\sigma_0^2}{\sigma_i^2} \tag{2.8.11}$$

it follows that

$$\hat{\sigma}_{v_i} = \hat{\sigma}_0 \sqrt{q_i} = \hat{\sigma}_0 \sqrt{\frac{r_i}{p_i}} = \hat{\sigma}_0 \sqrt{\frac{r_i \sigma_i^2}{\sigma_0^2}} = \frac{\hat{\sigma}_0}{\sigma_0} \sigma_i \sqrt{r_i} \tag{2.8.12}$$

From (2.8.6) it follows that the average redundancy number is

$$r_{\text{av}} = \frac{n - R(\boldsymbol{A})}{n} \tag{2.8.13}$$

The higher the degree of freedom, the closer the average redundancy number is to 1. However, as seen from Table 2.7.3, the gain, in terms of probability enclosed by the standard ellipses, reduces noticeably after a certain degree of freedom.

2.8.2 Controlling Type-II Error for a Single Blunder

Baarda's (1967) development of the concept of reliability of networks is based on un-Studentized hypothesis tests, which means that the a priori variance of unit weight is assumed to be known. Consequently, the a priori variance of unit weight (not the a posteriori variance of unit weight) is used in this section. The alternative hypothesis H_a specifies that the observations contain one blunder, that the blunder be located at observation *i,* and that its magnitude is ∇_i. Thus the adjusted residuals for the case of the alternative hypothesis are

$$\hat{\boldsymbol{v}}|H_a = \hat{\boldsymbol{v}} - \mathbf{Q}_v\boldsymbol{P}\boldsymbol{e}_i\nabla_i \tag{2.8.14}$$

where

$$\boldsymbol{e}_i = \begin{bmatrix} 0 & \cdots & 0 & 1 & 0 & \cdots & 0 \end{bmatrix}^T \tag{2.8.15}$$

denotes an $n \times 1$ vector containing 1 in position i and zero elsewhere. The expected value and the covariance matrix are

$$E(\hat{\boldsymbol{v}}|H_a) = -\mathbf{Q}_v\boldsymbol{P}\boldsymbol{e}_i\nabla_i \tag{2.8.16}$$

$$\boldsymbol{\Sigma}_{v|H_a} = \hat{\boldsymbol{\Sigma}}_v = \sigma_0^2\mathbf{Q}_v \tag{2.8.17}$$

It follows from (A.5.65) that

$$\hat{\boldsymbol{v}}|H_a \sim N\left(-\mathbf{Q}_v\boldsymbol{P}\boldsymbol{e}_i\nabla_i, \sigma_0^2\mathbf{Q}_v\right) \tag{2.8.18}$$

Since $\boldsymbol{P}$ is a diagonal matrix, the individual residuals are distributed as

$$\hat{v}_i|H_a \sim n\left(-q_ip_i\nabla_i, \sigma_0^2 q_i\right) \tag{2.8.19}$$

according to (A.5.71). Standardizing gives

$$\begin{aligned} w_a|H_a &= \frac{\hat{v}_i|H_a}{\sigma_0\sqrt{q_i}} \sim n\left(\frac{-q_ip_i\nabla_i}{\sigma_0\sqrt{q_i}}, 1\right) \\ &= n\left(\frac{-\sqrt{q_i}p_i\nabla_i}{\sigma_0}, 1\right) \end{aligned} \tag{2.8.20}$$

or

$$H_a : w_a = \frac{\hat{v}_i | H_a}{\sigma_{v_i}} \sim n\left(\frac{-\nabla_i p_i \sqrt{q_i}}{\sigma_0}, 1\right) \tag{2.8.21}$$

The zero hypothesis, which states that there is no blunder, is

$$H_0 : w_0 = \frac{\hat{v}_i \,|\, H_0}{\sigma_{v_i}} \sim n(0, 1) \tag{2.8.22}$$

The noncentrality parameter in (2.8.21), i.e., the mean of the noncentral normal distribution, is denoted by δ_i and is

$$\delta_i = \frac{-\nabla_i p_i \sqrt{q_i}}{\sigma_0} = \frac{-\nabla_i \sqrt{r_i}}{\sigma_i} \tag{2.8.23}$$

The parameter δ_i is a translation parameter of the normal distribution. The situation is shown in Figure 2.8.1. The probability of committing an error of the first kind, i.e., of accepting the alternative hypothesis, equals the significance level α of the test

$$P(|\, w_0 \,| \leq t_{\alpha/2}) = \int_{-t_{\alpha/2}}^{t_{\alpha/2}} n(0, 1)\, dx = 1 - \alpha \tag{2.8.24}$$

or

$$P(|\, w_0 \,| \geq t_{\alpha/2}) = \int_{-\infty}^{t_{1-\alpha/2}} n(0, 1)\, dx + \int_{t_{\alpha/2}}^{\infty} n(0, 1)\, dx = \alpha \tag{2.8.25}$$

In 100 α% of the cases, the observations are rejected and remeasurement or investigations for error sources are performed, even though the observations are correct

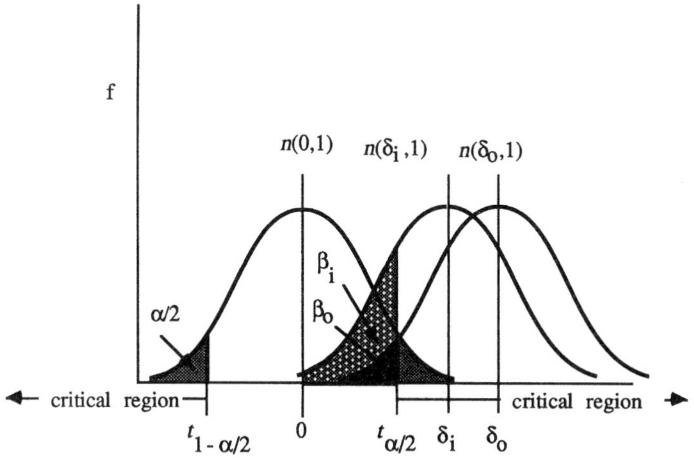

Figure 2.8.1 Defining the noncentrality.

(they do not contain a blunder). From Figure 2.8.1 it is seen that the probability β_i of a type-II error, i.e., the probability of rejecting the alternative hypothesis (and accepting the zero hypothesis) even though the alternative hypothesis is correct, depends on the noncentrality factor δ_i. *Because the blunder* ∇_i *is not known, the noncentrality factor is not known either.* As a practical matter one can proceed in the reverse: one can assume an acceptable probability β_0 for the error of the second kind and compute the respective noncentrality parameter δ_0. This parameter in turn is used to compute the lower limit for the blunder, which can still be detected. The figure shows that

$$P(|w_\alpha| \leq t_{\alpha/2}) = \int_{-t_{a/2}}^{t_{a/2}} n(\delta_i, 1)\, dx \geq \beta_0 \tag{2.8.26}$$

If

$$\delta_\mathrm{i} \leq \delta_0 \tag{2.8.27}$$

Substituting equation (2.8.23) into (2.8.27) gives the limit for the marginally detectable blunder, given the probability levels α and β_0:

$$|\nabla_{0i}| \geq \frac{\delta_0}{\sqrt{r_i}} \sigma_i \tag{2.8.28}$$

Equations (2.8.26) and (2.8.28) state that in $100(1-\beta_0)\%$ of the cases, blunders greater than those given in (2.8.28) are detected. In $100\beta_0\%$ of the cases, blunders greater than those given in (2.8.28) remain undetected. The larger the redundancy number, the smaller is the marginally detectable blunder (for the same δ_0 and σ_i). It is important to recognize that the marginally detectable blunders (2.8.28) are based on adopted probabilities of type-I and type-II errors for the normal distribution. The probability levels α and β_0 refer to the one-dimensional test (2.8.22) of the individual residual v_i, with the noncentrality being δ_0. The assumption is that only one blunder at a time is present. The geometry is shown in Figure 2.8.1. It is readily clear that there is a simple functional relationship $\delta_0 = \delta_n(\alpha, \beta_0)$ between two normal distributions. Table 2.8.1 contains selected probability levels and the respective δ_0 values.

TABLE 2.8.1 Selected Probability Levels in Reliability

α	β_0	δ_0
0.05	0.20	2.80
0.025	0.20	3.1
0.001	0.20	4.12
0.05	0.10	3.24
0.025	0.10	3.52
0.001	0.10	4.57

The chi-square test (2.7.38) of the a posteriori variance of unit weight $\hat{\sigma}_0^2$ is also sensitive to the blunder ∇_i. In fact, the blunder will cause a noncentrality of δ_i for the chi-square distribution of the alternative hypothesis. One can choose the probabilities α_{chi} and β_{chi} for this multidimensional chi-square test such that $\delta_0 = \delta_{\text{chi}}\,(\alpha_{\text{chi}}, \beta_{\text{chi}}, n-u)$. The factor δ_0 depends on the degree of freedom because the chi-square distribution depends on it. Baarda's B method suggests equal traceability of errors through one-dimensional tests of individual residuals, v_i, and the multidimensional test of the a posteriori variance of unit weight $\hat{\sigma}_0^2$. This is achieved by requiring that the one-dimensional test and the multidimensional test have the same type-II error, i.e., $\beta_0 = \beta_{\text{chi}}$. Under this condition there exists a relationship between the probability of type-II error, the significance levels, and the degree of freedom expressed symbolically by $\delta_0 = \delta_n(\alpha, \beta_0) = \delta_{\text{chi}}(\alpha_{\text{chi}}, \beta_0, n-r)$. The B method assures equal traceability but implies different significance levels for the one-dimensional and multidimensional tests. For details see Baarda (1968, p. 25). In practical applications one chooses the factor δ_0 on the basis of a reasonable value for α and δ_0 from Table 2.8.1.

2.8.3 Internal Reliability

Even though the one-dimensional test assumes that only one blunder exists in a set of observations, the limit (2.8.28) is usually computed for all observations. The marginally detectable errors, computed for all observations, are viewed as a measure of the capability of the network to detect blunders with probability $(1-\beta_0)$. They constitute the internal reliability of the network. Because the marginally detectable errors (2.8.28) do not depend on the observations or on the residuals, they can be computed as soon as the configuration of the network and the stochastic model are known. If the limits (2.8.28) are of about the same size, the observations are equally well checked, and the internal reliability is said to be consistent. The emphasis is then on the variability of the marginally detectable blunders rather than on their magnitude. A typical value is $\delta_0 = 4$.

2.8.4 Absorption

According to (2.8.1) the residuals in the presence of one blunder are

$$\mathbf{v} = \mathbf{Q}_v \mathbf{P}(\ell - e_i \nabla_i) \tag{2.8.29}$$

The impact on the residual of observation i is

$$\nabla v_i = -r_i \nabla_i \tag{2.8.30}$$

Equation (2.8.30) is used to estimate the blunders that might cause large residuals. Solving for ∇_i gives

$$\nabla_i = -\frac{\nabla v_i}{r_i} \approx -\frac{v_i^* + \nabla v_i}{r_i} \approx -\frac{v_i}{r_i} \tag{2.8.31}$$

because $v_i^* \ll \nabla v_i$, where v_i^* denotes the residual without the effect of the blunder. The computation (2.8.31) provides only estimates of possible blunders. Because the matrix $\boldsymbol{Q}_v\boldsymbol{P}$ is not a diagonal matrix, a specific blunder has an impact on all residuals. If several blunders are present, their effects overlap and one blunder can mask others; a blunder may cause rejection of a good observation.

Equation (2.8.30) demonstrates that the residuals in least-squares adjustments are not robust with respect to blunders in the sense that the effect of a blunder on the residuals is smaller than the blunder itself, because r varies between 0 and 1. The absorption, i.e., the portion of the blunder that propagates into the estimated parameters and falsifies the solution, is

$$A_i = (1 - r_i)\nabla_i \tag{2.8.32}$$

The factor $(1 - r_i)$ is called the absorption number. The larger the redundancy number, the less is a blunder absorbed, i.e., the less falsification. If $r_i = 1$, the observation is called fully controlled, because the residual completely reflects the blunder. A zero redundancy implies uncontrolled observations in that a blunder enters into the solution with its full size. Observations with small redundancy numbers might have small residuals and instill false security in the analyst. Substituting ∇_i from (2.8.31) expresses the absorption as a function of the residuals:

$$A_i = -\frac{1 - r_i}{r_i} v_i \tag{2.8.33}$$

The residuals can be looked on as the visible parts of errors. The factor in (2.8.33) is required to compute the invisible part from the residuals.

2.8.5 External Reliability

A good and homogeneous internal reliability does not automatically guarantee reliable coordinates. What are the effects of undetectable blunders on the parameters? In deformation analysis, where changes in parameters between adjustments of different epochs indicate existing deformations, it is particularly important that the impact of blunders on the parameters be minimal. The influence of each of the marginally detectable errors on the parameters of the adjustment or on functions of the parameters is called external reliability. The estimated parameters in the presence of a blunder are, for the observation equation model,

$$\hat{\boldsymbol{x}} = -\boldsymbol{N}^{-1}\boldsymbol{A}^T\boldsymbol{P}(\boldsymbol{\ell} - \boldsymbol{e}_i\nabla_i) \tag{2.8.34}$$

The effect of the blunder in observation i is

$$\nabla\boldsymbol{x} = \boldsymbol{N}^{-1}\boldsymbol{A}^T\boldsymbol{P}\boldsymbol{e}_i\nabla_i \tag{2.8.35}$$

The shifts $\nabla \boldsymbol{x}$ are sometimes called local external reliability. The blunder affects all parameters. The impact of the marginally detectable blunder ∇_{0i} is

$$\nabla \boldsymbol{x}_{0i} = \boldsymbol{N}^{-1}\boldsymbol{A}^T\boldsymbol{P}\boldsymbol{e}_i\nabla_{0i} \tag{2.8.36}$$

Because there are n observations, one can compute n vectors (2.8.36), showing the impact of each marginal detectable blunder on the parameters. Graphical representations of these effects can be very helpful in the analysis. The problem with (2.8.36) is that the effect on the coordinates depends on the definition (minimal constraints) of the coordinate system. Baarda (1968) suggested the following alternative expression:

$$\lambda_{0i}^2 = \frac{\nabla \boldsymbol{x}_{0i}^T \boldsymbol{N} \nabla \boldsymbol{x}_{0i}}{\sigma_0^2} \tag{2.8.37}$$

By substituting (2.8.36) and (2.8.28), we can write this equation as

$$\lambda_{0i}^2 = \frac{\nabla_{0i}\boldsymbol{e}_i^T\boldsymbol{P}\boldsymbol{A}\boldsymbol{N}^{-1}\boldsymbol{A}^T\boldsymbol{P}\boldsymbol{e}_i\nabla_{0i}}{\sigma_0^2} = \frac{\nabla_{0i}^2\boldsymbol{e}_i^T\boldsymbol{P}(\boldsymbol{I}-\boldsymbol{Q}_v\boldsymbol{P})\boldsymbol{e}_i}{\sigma_0^2} = \frac{\nabla_{0i}^2 p_i(1-r_i)}{\sigma_0^2} \tag{2.8.38}$$

or

$$\lambda_{0i}^2 = \frac{1-r_i}{r_i}\,\delta_0^2 \tag{2.8.39}$$

The values λ_{0i} are a measure of global external reliability. There is one such value for each observation. If the λ_{0i} are the same order of magnitude, the network is homogeneous with respect to external reliability. If r_i is small, the external reliability factor becomes large and the global falsification caused by a blunder can be significant. It follows that very small redundancy numbers are not desirable. The global external reliability number (2.8.39) and the absorption number (2.8.33) have the same dependency on the redundancy numbers.

2.8.6 Correlated Cases

The derivations for detectable blunders, internal reliability, absorption, and external reliability assume uncorrelated observations for which the covariance matrix $\boldsymbol{\Sigma}_{\ell_b}$ is diagonal. Correlated observations are decorrelated by the transformation (2.7.10). It can be readily verified that the redundancy numbers for the decorrelated observations $\overline{\boldsymbol{\ell}}$ are

$$\overline{r}_i = \left(\overline{\boldsymbol{Q}}_{\overline{v}}\overline{\boldsymbol{P}}\right)_{ii} = \left(\boldsymbol{I} - \boldsymbol{D}^T\boldsymbol{A}\boldsymbol{N}^{-1}\boldsymbol{A}^T\boldsymbol{D}\right)_{ii} \tag{2.8.40}$$

In many applications, the covariance matrix $\boldsymbol{\Sigma}_{\ell_b}$ is of block-diagonal form. For example, for GPS vector observations, this matrix consists of 3×3 full block-diagonal matrices if the correlations between the vectors are neglected. In this case, the matrix $\boldsymbol{D}$ is also block-diagonal and the redundancy numbers can be computed vector by vector from (2.8.40). The sum of the redundancy numbers for the three vector components varies between 0 and 3. Since, in general, the matrix $\boldsymbol{D}$ has a full rank, the degree of freedom $(n - r)$ of the adjustment does not change. Once the redundancy numbers $\bar{r}_i$ are available, the marginal detectable blunders $\overline{\nabla}_{0i}$, the absorption numbers $\overline{A}_i$ and other reliability values can be computed for the decorrelated observations. These quantities, in turn, can be transformed back into the physical observation space by premultiplication with the matrix $\left(\boldsymbol{D}^T\right)^{-1}$.

2.9 BLUNDER DETECTION

Errors (blunders) made during the recording of field observations, data transfer, the computation, etc., can be costly and time-consuming to find and eliminate. Blunder detection can be carried out before the adjustment or as part of the adjustment. Before the adjustment, the discrepancies (angle and/or distance of simple figures such as triangles and traverses) are analyzed. A priori blunder detection is helpful in detecting extra-large blunders caused by, e.g., erroneous station numbering. Blunder detection in conjunction with the adjustment is based on the analysis of the residuals. The problem with using least-squares adjustments when blunders are present is that the adjustments tend to hide (reduce) their impact and distribute their effects more or less throughout the entire network [see (2.8.29) and (2.8.30), noting that the redundancy number varies between zero and 1]. The prerequisite for any blunder-detection procedure is the availability of a set of redundant observations. Only observations with redundancy numbers greater than zero can be controlled.

It is important to understand that if a residual does not pass a statistical test, this does not mean that there is a blunder in that observation. The observation is merely flagged so that it can be examined and a decision about its retention or rejection can be made. Blind rejection is never recommended. A blunder in one observation usually affects the residuals in other observations. Therefore, the tests will often flag other observations in addition to the ones containing blunders. If one or more observations are flagged, the search begins to determine if there is a blunder.

The first step is to check the field notes to confirm that no error occurred during the transfer of the observations to the computer file, and that all observations are reasonable "at face value." If a blunder is not located, the network should be broken down into smaller networks, and each one should be adjusted separately. At the extreme, the entire network may be broken down into triangles or other simple geometric entities, such as traverses, and adjusted separately. Alternatively, the observations can be added sequentially, one at a time, until the blunder is found. This procedure starts with weights assigned to all parameters. The observations are then added sequentially. The sum of the normalized residuals squared is then inspected for unusually large variations. When searching for blunders, the coordinate system should be defined by minimal constraints.

Blunder detection in conjunction with the adjustment takes advantage of the total redundancy and the strength provided by the overall geometry of the network, and thus is more sensitive to smaller blunders. Only if the existence of a blunder is indicated does action need to be taken to locate the blunder. The flagged observations are the best hint where to look for errors and thus avoid unnecessary and disorganized searching of the whole observation data set.

2.9.1 Tau Test

The τ test was introduced by Pope (1976). The test belongs to the group of Studentized tests, which make use of the a posteriori variance of unit weight as estimated from the observations. The test statistic is

$$\tau_i = \frac{v_i}{\hat{\sigma}_{v_i}} = \frac{\sigma_0 v_i}{\hat{\sigma}_0 \sigma_i \sqrt{r_i}} \sim \tau_{n-r} \tag{2.9.1}$$

The symbol τ_{n-r} denotes the τ distribution with $n - r$ degrees of freedom. It is related to Student's t by

$$\tau_{n-r} = \frac{\sqrt{n-r}\, t_{n-r-1}}{\sqrt{n-r-1+t_{n-r-1}^2}} \tag{2.9.2}$$

For an infinite degree of freedom the τ distribution converges toward the Student distribution or the standardized normal distribution, i.e., $\tau_\infty = t_\infty = n(0, 1)$.

Pope's blunder rejection procedure tests the hypothesis $v_i \sim n(0, \hat{\sigma}_{v_i}/\hat{\sigma}_0)$. The hypothesis is rejected, i.e., the observation is flagged for further investigation and possibly rejection, if

$$|\tau_i| \geq c \tag{2.9.3}$$

The critical value c is based on a preselected significance level. For large systems, the redundancy numbers are often replaced by the average value according to equation (2.8.13), in order to reduce computation time; thus

$$\tau_i = \frac{\sigma_0}{\hat{\sigma}_0} \frac{v_i}{\sigma_i \sqrt{(n-r)/n}} \tag{2.9.4}$$

could be used instead of (2.9.1).

2.9.2 Data Snooping

Baarda's data snooping applies to the testing of individual residuals as well. The theory assumes that only one blunder be present in the set of observations. Applying a series of one-dimensional tests, i.e., testing consecutively all residuals, is called a data snooping strategy. Baarda's test belongs to the group of un-Studentized tests which assume that the a priori variance of unit weight is known. The zero hypothesis (2.8.22) is written as

$$n_i = \frac{v_i}{\sigma_0 \sqrt{q_i}} \sim n(0, 1) \tag{2.9.5}$$

At a significant level of 5%, the critical value is 1.96. The critical value for this test is not a function of the number of observations in the adjustment. The statistic (2.9.5) uses the a priori value σ_0 and not the a posteriori estimate $\hat{\sigma}_0$.

Both the τ and the data snooping procedures work best for iterative solutions. At each iteration step, the observation with the largest blunder should be removed. Since least-squares attempts to distribute blunders, several correct observations might receive large residuals and might be flagged mistakenly.

2.9.3 Changing Weights of Observations

This method, although not based on rigorous statistical theory, is an automated method whereby blunders are detected and their effects on the adjustment minimized (or even eliminated). The advantage that this method has, compared to previous methods, is that it locates and potentially eliminates the blunders automatically. The method examines the residuals per iteration. If the magnitude of a residual is outside a defined range, the weight of the corresponding observation is reduced. The process of reweighting and readjusting continues until the solution converges, i.e., no weights are being changed. The criteria for judging the residuals and choice for the reweighting function are somewhat arbitrary. For example, a simple strategy for selection of the new weights at iteration $k+1$ could be

$$p_{k+1,i} = p_{k,i} \begin{cases} e^{-|v_{k,i}|/3\sigma_i} & \text{if} \quad |v_{k,i}| > 3\sigma_i \\ 1 & \text{if} \quad |v_{k,i}| \leq 3\sigma_i \end{cases} \tag{2.9.6}$$

where σ_i denotes the standard deviation of observation i.

The method works efficiently for networks with high redundancy. If the initial approximate parameters are inaccurate, it is possible that correct observations are deweighted after the first iteration because the nonlinearity of the adjustment can cause large residuals. To avoid unnecessary rejection and reweighting, one might not change the weights during the first iteration. Proper use of this method requires some experience. All observations whose weights are changed must be investigated, and the cause for the deweighting must be established.

2.10 EXAMPLES

In the following, we use plane two-dimensional networks to demonstrate the geometry of adjustments. As mentioned above, the geometry of a least-squares adjustment is the result of the combined effects of the stochastic model (weight matrix $\boldsymbol{P}$—representing the quality of the observations) and the mathematical model (design matrix $\boldsymbol{A}$—representing the geometry of the network and the spatial distribution of the observations). For the purpose of these examples, it is not necessary to be concerned about the physical realization of two-dimensional networks. The experienced reader might think of such networks as being located on the conformal mapping plane and that all model observations have been computed accordingly.

We will use the observation equation model summarized in Table 2.5.1. Assume there is a set of n observations, such as distances and angles that determine the points of a network. For a two-dimensional network of s stations, there could be as many as $u = 2s$ unknown coordinates. Let the parameter vector $\boldsymbol{x}_a$ consist of coordinates only, i.e., we do not parameterize refraction, centering errors, etc. To be specific, $\boldsymbol{x}_a$ contains only coordinates that are to be estimated. Coordinates of known stations are constants and not included in $\boldsymbol{x}_a$. The mathematical mode $\boldsymbol{\ell}_a = \boldsymbol{f}(\boldsymbol{x}_a)$ is very simple in this case. The n components of $\boldsymbol{f}$ will contain the functions:

$$d_{ij} = \sqrt{(x_i - x_j)^2 + (y_i - y_j)^2} \tag{2.10.1}$$

$$\alpha_{jik} = \tan^{-1}\frac{x_k - x_i}{y_k - y_i} - \tan^{-1}\frac{x_j - x_i}{y_j - y_i} \tag{2.10.2}$$

In these expressions the subscripts i, j, and k identify the network points. The notation α_{jik} implies that the angle is measured at station i, from j to k in a clockwise sense. The ordering of the components in $\boldsymbol{f}$ does not matter, as long as the same order is maintained with respect to the rows of $\boldsymbol{A}$ and diagonal elements of $\boldsymbol{P}$.

Although the $\boldsymbol{f}(\boldsymbol{x}_a)$ have been expressed in terms of $\boldsymbol{x}_a$, the components typically depend only on a subset of the coordinates. The relevant partial derivatives in a row of $\boldsymbol{A}$ are for distances and angles:

$$\left\{\frac{-\left(y_k - y_i\right)}{d_{ik}}, \frac{-(x_k - x_i)}{d_{ik}}, \frac{y_k - y_i}{d_{ik}}, \frac{x_k - x_i}{d_{ik}}\right\} \tag{2.10.3}$$

$$\left\{\frac{x_i - x_j}{d_{ij}^2}, -\frac{y_i - y_j}{d_{ij}^2}, \frac{x_k - x_j}{d_{kj}^2} - \frac{x_i - x_j}{d_{ij}^2}, -\frac{y_k - y_j}{d_{kj}^2} + \frac{y_i - y_j}{d_{ij}^2}, -\frac{x_k - x_j}{d_{kj}^2}, \frac{y_k - y_j}{d_{kj}^2}\right\} \tag{2.10.4}$$

Other elements are zero. The column location for these partials depends on the sequence in $\boldsymbol{x}_a$. In general, if α is the α th component of $\boldsymbol{\ell}_b$ and β the β th component of $\boldsymbol{x}_a$, then the element $a_{\alpha,\beta}$ of $\boldsymbol{A}$ is

$$a_{\alpha,\beta} = \frac{\partial \ell_\alpha}{\partial x_\beta} \tag{2.10.5}$$

The partial derivatives and the discrepancy $\boldsymbol{\ell}_0$ must be evaluated for the approximate coordinates $\boldsymbol{x}_0$.

Example 1: This example demonstrates the impact of changes in the stochastic model. Figure 2.10.1 shows a traverse connecting two known stations. Three

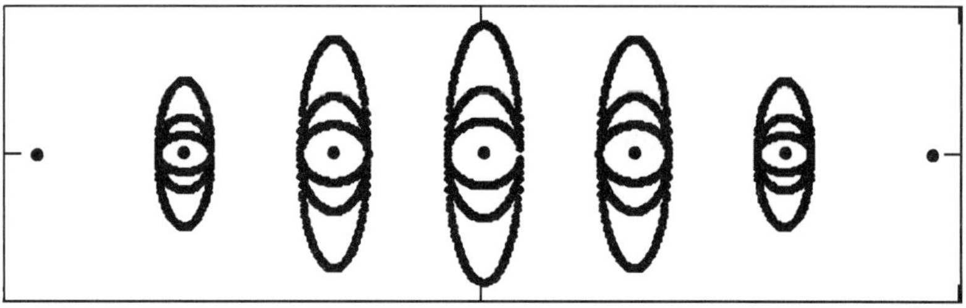

Figure 2.10.1 Impact of changing the stochastic model.

solutions are given. In all cases, the distances are of the same length and observed with the same accuracy. The angle observations are 180° and measured with the same accuracy but are changed by a common factor for each solution. If we declare the solutions with the smallest ellipses in Figure 2.10.1 as the base solutions with observational standard deviation of σ_a, then the other solutions use $2\sigma_a$ and $4\sigma_a$, respectively. The shape of the ellipses elongates as the standard deviation of the angles increases.

Example 2: This example demonstrates the impact of changing network geometry using a resection. Four known stations lie exactly on an imaginary circle with radius r. The coordinates of the new station are determined by angle measurements, i.e., no distances are involved. For the first solution, the unknown station is located at the center of the circle. In subsequent solutions its location moves to $0.5r$, $0.9r$, $1.1r$, and $1.5r$ from the center while retaining the same standard deviation for the angle observations in each case. Figure 2.10.2 shows that the ellipses become more elongated the closer the unknown station moves to the circle. The solution is singular if the new station is located exactly on the circle.

Example 3: Three cases are given that demonstrate how different definitions of the coordinate system affect the ellipses of standard deviation. All cases refer to the same plane network using the same observed angles and distances and the same respective standard deviations of the observations. A plane network that contains angle and distance observations requires three minimal constraints. Simply holding three coordinates fixed imposes such minimal constraints. The particular coordinates are constants and are not included in the parameter vector $\boldsymbol{x}_a$, and, consequently, there are no columns in the $\boldsymbol{A}$ matrix that pertain to these three coordinates. Inner constraints offer another possibility of defining the coordinate system.

Figure 2.10.3 shows the results of two different minimal constraints. The coordinates of station 2 are fixed in both cases. In the first case, we hold one of the coordinates of station 1 fixed. This results in a degenerated ellipse (straight line) at station 1 and a regular ellipse at station 3. In the second case, we hold one of the coordinates of station 3 fixed. The result is a degenerated ellipse at station 3 and a regular ellipse at station 1. The ellipses of standard deviation change significantly due to the

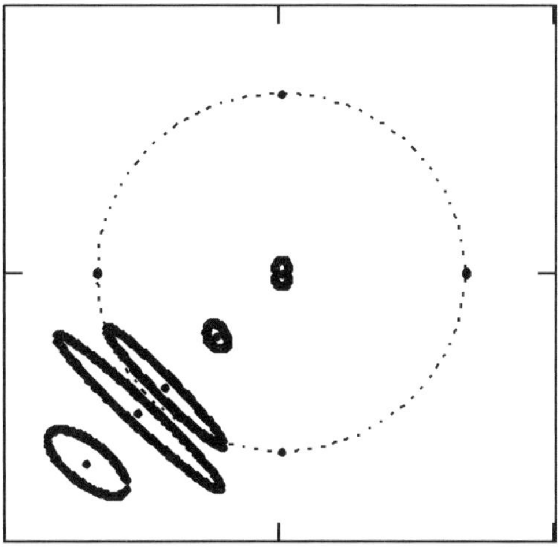

Figure 2.10.2 Impact of changing network geometry.

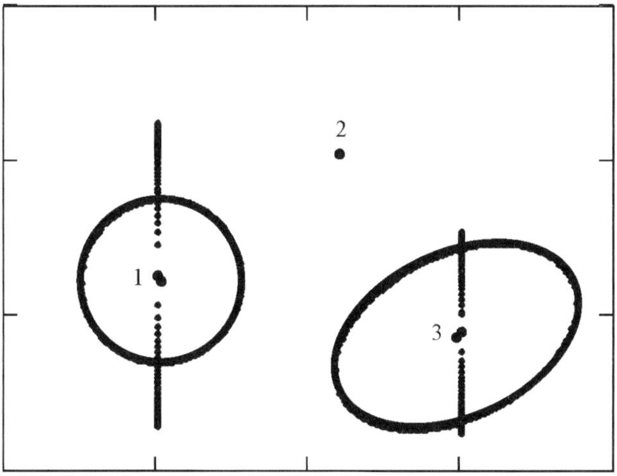

Figure 2.10.3 Changing minimal constraints.

change in minimal constraints. Clearly, if one were to specify the quality of a survey in terms of ellipses of standard deviation, one must also consider the underlying minimal constraints. The figure also shows that the adjusted coordinates for stations 1 and 3 differ in both cases, although the internal shape of the adjusted network 1-2-3 is the same.

The inner constraint solution, which is a special case of the minimal constraint solutions, has the property that no individual coordinates need to be held fixed. All coordinates become adjustable; for s stations of a plane network, the vector $\mathbf{x}_a$

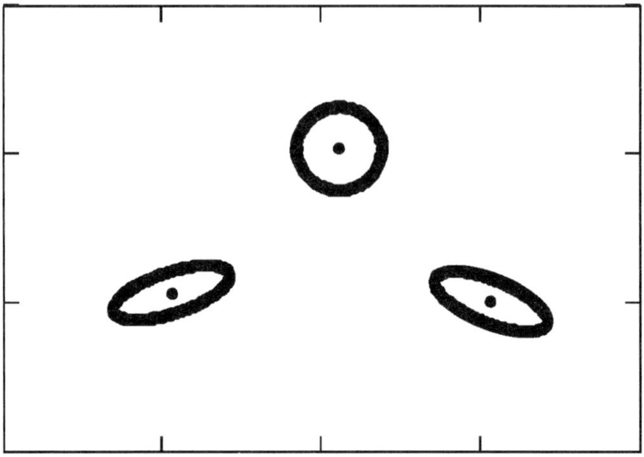

Figure 2.10.4 Inner constraint solution.

contains $2s$ coordinate parameters. The ellipses reflect the geometry of the network, the distribution of the observations, and their standard deviations. Section 2.6 contains the theory of inner constraints. The elements for drawing the ellipses are taken from the cofactor matrix (2.6.37) and equation (2.6.46) gives the adjusted parameters. A first step is to find a matrix $\boldsymbol{E}$ that fulfills $\boldsymbol{AE}^T = \boldsymbol{0}$ according to (2.6.11). The number of rows of $\boldsymbol{E}$ equals the rank defect of $\boldsymbol{A}$. For trilateration networks with distances and angles we have

$$\boldsymbol{E} = \begin{bmatrix} \cdots & 1 & 0 & \cdots & 1 & 0 & \cdots \\ \cdots & 0 & 1 & \cdots & 0 & 1 & \cdots \\ \cdots & -y_i & x_i & \cdots & -y_k & x_k & \cdots \end{bmatrix} \tag{2.10.6}$$

Four constraints are required for triangulation networks that contain only angle observations. In addition to fixing translation and rotation, triangulation networks also require scaling information. The $\boldsymbol{E}$ matrix for such networks is

$$\boldsymbol{E} = \begin{bmatrix} \cdots & 1 & 0 & \cdots & 1 & 0 & \cdots & 1 & 0 & \cdots \\ \cdots & 0 & 1 & \cdots & 0 & 1 & \cdots & 0 & 1 & \cdots \\ \cdots & -y_i & x_i & \cdots & -y_j & x_j & \cdots & -y_k & x_j & \cdots \\ \cdots & x_i & y_i & \cdots & x_j & y_j & \cdots & x_k & y_k & \cdots \end{bmatrix} \tag{2.10.7}$$

The inner constraint solution is shown in Figure 2.10.4. Every station has an ellipse. The minimal constraint solutions and the inner constraint solution give the same estimates for residuals, a posteriori variance of unit weight, and redundancy numbers. While the estimated parameters (station coordinates) and their covariance matrix differ for these solutions, the same result is obtained when using these quantities in covariance propagation to compute other observables and their standard deviations.

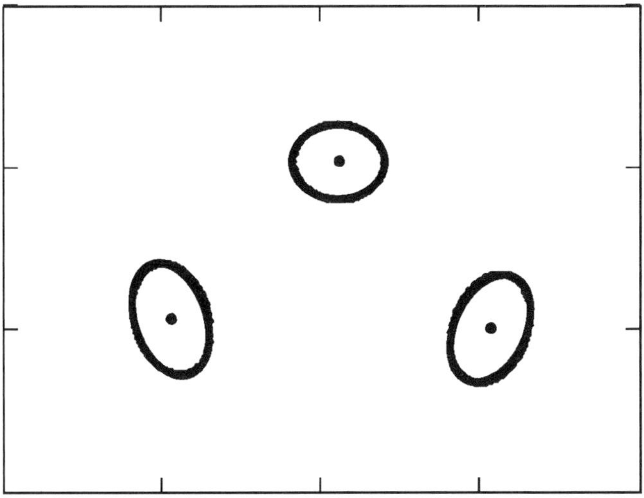

Figure 2.10.5 Weighting approximate coordinates to define the coordinate system.

Example 4: Weighting all approximate coordinates can also provide the coordinate system definition. Table 2.5.3 contains expressions that include a priori weights on the parameters. If the purpose of the adjustment is to control the quality of the observations, it is important that the weights of the approximate coordinate are small enough to allow observations to adjust freely. For example, if the approximate coordinates are accurate to 1 m, one can use a standard deviation of, say, 1 to 2 m, or even larger. Ideally, of course, the weight should reflect our knowledge of the approximate coordinates by using meaningful standard deviation. One may prefer to use large standard deviations just to make sure that the internal geometry of the network solution is not affected.

Figure 2.10.5 shows all ellipses for the case when each approximate station coordinate is assigned a standard deviation of 10 m. The ellipse at each network point is approximately circular. The size of the ellipses is in the range of the a priori coordinate standard deviations. The ellipses in the figure imply a scale factor of 10^6 when compared to those in Figures 2.10.3 and 2.10.4, which roughly corresponds to the ratio of the variances of the approximate coordinates over the average variance of the observations.

The weighted parameter approach is also a convenient way of imposing minimal constraints. Only a subset of three approximate coordinates needs to be weighted in the case of a plane angle and distance network.

2.11 KALMAN FILTERING

Least-squares solutions are often applied to surveying networks whose network points refer to monuments that are fixed to the ground. When using the sequential least-squares approach (2.4.54) to (2.4.58), the parameters $\mathbf{x}$ are typically treated as a time invariant. The subscript i in these expressions identifies the set of additional observations added to the previous solution that contains the sets $1 \leq i \leq i-1$. Each set of observations merely updates $\mathbf{x}$, resulting in a more accurate determination of the fixed monuments.

We generalize the sequential least-squares formulation by allowing the parameter vector $\mathbf{x}$ to change with time. For example, the vector $\mathbf{x}$ might now contain the three-dimensional coordinates of a moving receiver, the coordinates of satellites, tropospheric delay of signals, or other time-varying parameters. We assume that the dynamic model between parameters of adjacent epochs follows the system of linear equations

$$\mathbf{x}_k(-) = \boldsymbol{\Phi}_{k-1}\mathbf{x}_{k-1} + \mathbf{w}_k \tag{2.11.1}$$

We have used the subscript k, instead of i, to emphasize that it now indicates the epoch. The matrix $\boldsymbol{\Phi}_{k-1}$ is called the parameter transition matrix. The random vector $\mathbf{w}_k$ is the system process noise and is distributed as $\mathbf{w}_k \sim N(\mathbf{0}, \mathbf{Q}_{w_k})$. The notation $(-)$ indicates the predicted value. Thus,

$$\hat{\mathbf{x}}_k(-) = \boldsymbol{\Phi}_{k-1}\hat{\mathbf{x}}_{k-1} + \mathbf{w}_k \tag{2.11.2}$$

$\hat{\mathbf{x}}_k(-)$ is the predicted parameter vector at epoch k, based on the estimated parameter $\hat{\mathbf{x}}_{k-1}(-)$ from the previous epoch and the dynamic model. The solution that generated $\hat{\mathbf{x}}_{k-1}$ also generated the respective cofactor matrix $\mathbf{Q}_{k-1}$. The observation equations for epoch k are given in the familiar form

$$\mathbf{v}_k = \mathbf{A}_k\mathbf{x}_k + \boldsymbol{\ell}_k \tag{2.11.3}$$

with $\mathbf{v}_k \sim N\mathbf{0}, \mathbf{Q}_{\ell_k})$.

The first step in arriving at the Kalman filter formulation is to apply variance-covariance propagation to (2.11.1) to predict the parameter cofactor matrix at the next epoch,

$$\mathbf{Q}_k(-) = \boldsymbol{\Phi}_{k-1}\mathbf{Q}_{k-1}\boldsymbol{\Phi}_{k-1}^T + \mathbf{Q}_{w_k} \tag{2.11.4}$$

Expression (2.11.4) assumes that the random variables $\boldsymbol{\ell}_k$ and $\mathbf{w}_k$ are uncorrelated. The various observation sets $\boldsymbol{\ell}_k$ are also uncorrelated, as implied by (2.4.2). The second step involves updating the predicted parameters $\hat{\mathbf{x}}_k(-)$, based on the observations $\boldsymbol{\ell}_k$. Following the sequential least-squares formulation (2.4.54) to (2.4.58), we obtain

$$\mathbf{T}_k = \left[\mathbf{Q}_{\ell_k} + \mathbf{A}_k\mathbf{Q}_k(-)\mathbf{A}_k^T\right]^{-1} \tag{2.11.5}$$

$$\hat{\mathbf{x}}_k = \hat{\mathbf{x}}_k(-) - \mathbf{K}_k\left[\mathbf{A}_k\hat{\mathbf{x}}_k(-) + \boldsymbol{\ell}_k\right] \tag{2.11.6}$$

$$\mathbf{Q}_k = \left[\mathbf{I} - \mathbf{K}_k\mathbf{A}_k\right]\mathbf{Q}_k(-) \tag{2.11.7}$$

$$\mathbf{v}^T\mathbf{P}\mathbf{v}_k = \mathbf{v}^T\mathbf{P}\mathbf{v}_{k-1} + \left[\mathbf{A}_k\hat{\mathbf{x}}_k(-) + \boldsymbol{\ell}_k\right]^T\mathbf{T}_k\left[\mathbf{A}_k\hat{\mathbf{x}}_k(-) + \boldsymbol{\ell}_k\right] \tag{2.11.8}$$

where the matrix

$$\boldsymbol{K}_k = \boldsymbol{Q}_k(-)\boldsymbol{A}_k^T \boldsymbol{T}_k \tag{2.11.9}$$

is called the Kalman gain matrix.

If the parameter x_{k+1} depends only on the past (previous) solution x_k, we speak of a first-order Markov process. If noise w_k has a normal distribution, we talk about a first-order Gauss-Markov process,

$$x_{k+1} = \varphi x_k + w_k \tag{2.11.10}$$

with $w_k \sim n\left(0, q_{w_k}\right)$. In many applications a useful choice for φ is

$$\varphi = e^{-T/\tau} \tag{2.11.11}$$

which implies that the variable x is exponentially correlated, i.e., the autocorrelation function is decreasing exponentially (Gelb, 1974, p. 81). The symbol τ denotes the correlation time, and T denotes the time difference between epochs $k+1$ and k. The variance of the process noise for correlation time τ is

$$q_{w_k} = E(w_k w_k) = \frac{\tau}{2}\left[1 - e^{-2T/\tau}\right] q_k \tag{2.11.12}$$

with q_k being the variance of the process noise (Gelb, 1974, p. 82). The quantities (τ, q_k) could be initially determined from data by fitting a sample mean and sample autocorrelation function.

As τ approaches zero, then $\varphi = 0$. This describes the pure white noise model with no correlation from epoch to epoch. In that case x can be thought of as a random constant, which is a nondynamic quantity.

As τ approaches infinity, we obtain the pure random walk. Applying l'Hôspital rule for computing the limit or using series expansion, we obtain $\varphi = 1$ and $q_{w_k} = Tq_k$. The random noises w_k are uncorrelated.

In general, both the dynamic model (2.11.1) and the observation model (2.11.3) are nonlinear. The extended Kalman filter formulation (Gelb, 1974, p. 187) applies to this general case. The reader is urged to consult that reference or other specialized literature for additional details on Kalman filtering.

CHAPTER 3

RECURSIVE LEAST SQUARES

In many applications of least-squares adjustments the measurements are taken sequentially at discrete epochs in time. Five arrangements are addressed in this chapter: The first case deals with estimation of static parameters. A static parameter represents a time-invariant quantity. In sequential estimation, each new measurement improves the previous estimate of the static parameters. Other cases include parameters that depend on time. Two types of time-dependent parameters are considered. First, we consider time-varying parameters that are not constrained by a dynamic model. They can vary arbitrarily and take independent values at two adjacent epochs. Parameters of the other type represent sequential states of a discrete dynamic process that is subject to a dynamic model. The dynamic model can be of linear or nonlinear functional relationship connecting two sequential states representing parameters at two adjacent time instances. The sequential measurements and estimated parameters are used to update the sequential estimates. For example, in some applications the physical nature of the problem imposes dynamic constraints on the rover coordinates. Another example is across-receiver difference ionospheric delays. Since they do not completely vanish for long baselines, the residual ionospheric delays are slow time-varying parameters that can be constrained by a dynamic model.

The second case discussed in this chapter refers to the mixed problem of estimating both static and arbitrary varying parameters. For example, in real-time kinematics processing of short baselines the carrier phase ambiguities are constant parameters, whereas the time-varying parameters are the rover coordinates which can vary arbitrarily. The third case introduces a dynamic dependence between discrete time epochs of time-varying parameters. The fourth case combines the first and third cases by dealing with a dynamic system that connects sequential states and contains time-invariant parameters. The fifth model is the most general one in that it contains all the features of the fourth model but, in addition, also includes time-varying parameters that are not constrained by a dynamic model. In real-time kinematic processing the most general case includes estimation of static ambiguity parameters, arbitrarily varying across-receiver clock shift, residual ionosphere subject to certain dynamic model, and corrections to the rover position that can be either arbitrarily varying, or subject to a dynamic model.

For each of the five models the batch solutions and real-time sequential solutions are provided. The derivations make use of partial minimization of quadratic forms and of the Cholesky decomposition. Appendix A provides details on both techniques.

The focus of adjustment in Chapter 2 is on static parameters and respective models such as the observation equation model, the mixed model, and the condition equation model. Therefore, the first case relates directly to the observation equation model and the sequential solution discussed in the previous chapter. However, this chapter is exclusively devoted to formulating recursive least squares where time-dependent parameters play the major role. New notation, which shows the time argument explicitly, reflects this focus on time. Additionally, not every matrix or vector is represented by bold letters, as is done in Chapter 2. Only those vectors and matrices that pertain to all sequential time instances are in bold.

3.1 STATIC PARAMETER

Let y be a real-valued parameter to be estimated. It is subject to the linear measurement model

$$W(t)y = b(t) \tag{3.1.1}$$

where t denotes the sequential time instant or the epoch. The matrix $W(t)$ has dimensions $m(t) \times n$, so the real-valued vector y is n-dimensional, i.e., $y \in R^n$, and $b(t) \in R^{m(t)}$. Let y^* be the true value of the parameter and $b^*(t)$ be the true value of the observables, then both obviously satisfying the identity

$$W(t)y^* = b^*(t) \tag{3.1.2}$$

An additive noise with zero mean value and known covariance matrix disturbs the observables vector in such a way that

$$b(t) = b^*(t) + \varepsilon(t) \quad E(\varepsilon(t)) = 0 \quad E(\varepsilon(t)\varepsilon^*(t)) = C(t), \tag{3.1.3}$$

where $E(\cdot)$ denotes the mathematic expectation, and the covariance matrix of the observations $C(t)$ is positive definite and allows the Cholesky decomposition

$$C(t) = L_{C(t)}L_{C(t)}^T \tag{3.1.4}$$

with the lower triangle matrix $L_{C(t)}$. Note that two forms of the Cholesky decompositions are possible: $C = L_C L_C^T$ and $C = \overline{L}_C D \overline{L}_C^T$. In the last case the low triangle matrix $\overline{L}_C$ has unit diagonal entries and the matrix $D = \mathrm{diag}(d_1, d_2, \ldots, d_n)$ is diagonal with positive diagonal entries. Obviously $L_C = \overline{L}_C D^{1/2}$ and the form (3.1.4) of the Cholesky decomposition implies the square root calculations. Either of two forms can be used depending on which one is more convenient.

We assume that the measurements are sequentially accumulated for time instances $t = 1, \ldots, t'$ where t' is the finite time of the accumulation period. The least-squares principle as applied in Chapter 2 minimizes the weighted sum of squares:

$$I(y, t') = \sum_{t=1}^{t'} (W(t)y - b(t))^T C^{-1}(t)(W(t)y - b(t)) \to \min \qquad (3.1.5)$$

The matrix inverse $C^{-1}(t)$ exists since the covariance matrix $C(t)$ is positive definite. Taking (3.1.4) into account, the value $I(y, t')$ in the equivalent form is

$$I(y, t') = \sum_{k=1}^{t'} \left(L_{C(t)}^{-1} W(t)y - L_{C(t)}^{-1} b(t)\right)^T \left(L_{C(t)}^{-1} W(t)y - L_{C(t)}^{-1} b(t)\right) \qquad (3.1.6)$$

After introduction of notations

$$\overline{W}(t) = L_{C(t)}^{-1} W(t) \quad \overline{b}(t) = L_{C(t)}^{-1} b(t) \qquad (3.1.7)$$

rewrite the last expression in the form

$$I(y, t') = \sum_{t=1}^{t'} \left(\overline{W}(t)\, y - \overline{b}(t)\right)^T \left(\overline{W}(t)\, y - \overline{b}(t)\right) \qquad (3.1.8)$$

We now give the problem (3.1.8) a slightly different interpretation. Let $M(t') = m(1) + \cdots + m(t')$ be the total number of all accumulated measurements and let $\overline{\mathbf{W}}(t')$ be the $M(t') \times n$ matrix composed of matrices $\overline{W}(t)$, $t = 1, \ldots, t'$ as shown below,

$$\overline{\mathbf{W}}(t') = \begin{bmatrix} \overline{W}(1) \\ \overline{W}(2) \\ \vdots \\ \overline{W}(t') \end{bmatrix} \qquad (3.1.9)$$

Similarly, the vector $\overline{\mathbf{b}}(t') \in R^{M(t')}$ is composed of the vectors of accumulated measurements,

$$\overline{\mathbf{b}}(t') = \begin{bmatrix} \overline{b}(1) \\ \overline{b}(2) \\ \vdots \\ \overline{b}(t') \end{bmatrix} \qquad (3.1.10)$$

The minimization problem (3.1.8) is now equivalent to the following problem:

$$I(y,t') = \left(\overline{\boldsymbol{W}}(t')y - \overline{\boldsymbol{b}}(t')\right)^T \left(\overline{\boldsymbol{W}}(t')y - \overline{\boldsymbol{b}}(t')\right) \to \min \tag{3.1.11}$$

Assume that the matrix $\overline{\boldsymbol{W}}(t')$ has full column rank,

$$\text{rank}\left(\overline{\boldsymbol{W}}(t')\right) = n \tag{3.1.12}$$

A necessary but not sufficient condition of (3.1.12) is $M(t') \geq n$. Once the condition (3.1.12) holds for a certain time instant t', it will remain valid for larger values $t'' > t'$. We call (3.1.12) the *observability* condition as it guarantees that the parameter y can be estimated from the observables accumulated up to the time instant t'. The best estimate of y is defined as solution to the problem (3.1.11) which is

$$y(t') = \left(\overline{\boldsymbol{W}}^T(t')\overline{\boldsymbol{W}}(t')\right)^{-1}\overline{\boldsymbol{W}}^T(t')\overline{\boldsymbol{b}}(t') \tag{3.1.13}$$

In application of geodesy and real-time navigation, measurements are not all taken at the same time. Instead, they are taken epoch by epoch in an incremental manner. According to this measurement mode, we may want to obtain the best approximation (3.1.13) also epoch by epoch, successively refining the estimate due to better averaging of measurement errors. We have

$$\begin{aligned} y(t'+1) &= \left(\overline{\boldsymbol{W}}^T(t'+1)\overline{\boldsymbol{W}}(t'+1)\right)^{-1}\overline{\boldsymbol{W}}^T(t'+1)\overline{\boldsymbol{b}}(t'+1) \\ &= \left(\overline{\boldsymbol{W}}^T(t')\overline{\boldsymbol{W}}(t') + \overline{W}^T(t'+1)\overline{W}(t'+1)\right)^{-1} \\ &\quad \times \left(\overline{\boldsymbol{W}}^T(t')\overline{\boldsymbol{b}}(t') + \overline{W}^T(t'+1)\overline{b}(t'+1)\right) \end{aligned} \tag{3.1.14}$$

Let us denote

$$\boldsymbol{D}(t') = \overline{\boldsymbol{W}}^T(t')\overline{\boldsymbol{W}}(t') = \sum_{t=1}^{t'} W^T(t)C^{-1}(t)W(t) \tag{3.1.15}$$

Then according to (3.1.14), and taking into account (3.1.15), one obtains the expression

$$\begin{aligned} y(t'+1) &= \boldsymbol{D}^{-1}(t'+1)\left(\overline{\boldsymbol{W}}^T(t')\overline{\boldsymbol{b}}(t') + \overline{W}^T(t'+1)\overline{b}(t'+1)\right) \\ &= \boldsymbol{D}^{-1}(t'+1)\left(\boldsymbol{D}(t')y(t') + \overline{W}^T(t'+1)\overline{b}(t'+1)\right) \\ &= \boldsymbol{D}^{-1}(t'+1)\Big(\boldsymbol{D}(t'+1)y(t') - \overline{W}^T(t'+1)\overline{W}(t'+1)y(t') \\ &\quad + \overline{W}^T(t'+1)\overline{b}(t'+1)\Big) \\ &= y(t') + \boldsymbol{D}^{-1}(t'+1)\overline{W}^T(t'+1)(\overline{b}(t'+1) - \overline{W}(t'+1)y(t')) \end{aligned} \tag{3.1.16}$$

The expression (3.1.16) takes the form of sequentially updating the estimate $y(t')$. It is also called "incremental update" as the estimate $y(t')$ is incremented by a correction in (3.1.16). The first term $y(t')$ is called *projection* to the next time instant. It means that the first guess for the next estimate $y(t' + 1)$ is that it coincides with the previous estimate $y(t')$ until new measurements are available. The second term is called *correction*. It linearly depends on the disagreement of the previous estimate $y(t')$ with new measurement model: $b(t' + 1) - W(t' + 1)y(t') = 0$. This disagreement is called *residual* $r(t' + 1) = b(t' + 1) - W(t' + 1)y(t')$. So, in order to calculate the next estimate of the parameter, we have to calculate the residual vector and update the previous estimate of the parameter with the residual premultiplied by the matrix $\boldsymbol{D}^{-1}(t' + 1)\overline{W}(t' + 1)$:

$$\begin{aligned} r(t' + 1) &= b(t' + 1) - W(t' + 1)y(t') \\ \overline{r}(t' + 1) &= L_{C(t'+1)}^{-1} r(t' + 1) \\ \boldsymbol{D}(t' + 1) &= \boldsymbol{D}(t') + \overline{W}^T(t' + 1)\overline{W}(t' + 1) \\ y(t' + 1) &= y(t') + \boldsymbol{D}^{-1}(t' + 1)\overline{W}(t' + 1)\overline{r}(t' + 1) \end{aligned} \tag{3.1.17}$$

The algorithm (3.1.17) starts with initial data

$$y(0) = 0 \quad \boldsymbol{D}(0) = 0 \tag{3.1.18}$$

Here, and further below, the symbol 0 denotes the zero vector or the zero matrix of the appropriate dimensions.

Let us present the recursive scheme (3.1.17) in a more computationally effective form. The linear systems with a symmetric positive definite matrix

$$Dy = b \quad D = D^T \quad D > 0 \tag{3.1.19}$$

can be rewritten using the Cholesky decomposition $LL^Ty = b$; see expressions (A.3.54) and (A.3.55) of Appendix A. The system with factorized matrix is further equivalent to two systems, (A.3.66) and (A.3.67), presented below for convenience

$$Lz = b \tag{3.1.20}$$

$$L^T y = z \tag{3.1.21}$$

Forward and backward solution runs, explained in Appendix A, solve the linear systems (3.1.20) and (3.1.21), respectively. They are denoted by linear operators,

$$z = \boldsymbol{F}_L b \tag{3.1.22}$$

and

$$y = \boldsymbol{B}_L z \tag{3.1.23}$$

respectively. Forward and backward runs are equivalent to calculations $z = L^{-1}b$ and $y = (L^T)^{-1}z$, respectively. Forward and backward runs can be applied to matrices, assuming that they are applied to all matrix columns sequentially.

Note again, that when writing the expression $z = L^{-1}b$, we usually do not have in mind the explicit calculation of the matrix inverse. Instead, we are interested in

TABLE 3.1.1 Algorithm 1: Estimating Static Parameters

Compute the residual vector	$r(t+1) = b(t+1) - W(t+1)y(t)$
Cholesky decomposition of the matrix $C(t+1)$ and forward substitution calculations	$C(t+1) = L_{C(t+1)}L^T_{C(t+1)}$ $\bar{r}(t+1) = \boldsymbol{F}_{L_{C(t+1)}} r(t+1)$ $\overline{W}(t+1) = \boldsymbol{F}_{L_{C(t+1)}} W(t+1)$
Update the matrix $\boldsymbol{D}(t+1)$ and its Cholesky decomposition	$\boldsymbol{D}(t+1) = \boldsymbol{D}(t) + \overline{W}^T(t+1)\overline{W}(t+1)$ $\boldsymbol{D}(t+1) = L_{\boldsymbol{D}(t+1)}L^T_{\boldsymbol{D}(t+1)}$
Optimal estimate	$y(t+1) = y(t) + \boldsymbol{B}_{L_{\boldsymbol{D}(t+1)}}\left(\boldsymbol{F}_{L_{\boldsymbol{D}(t+1)}}\overline{W}^T(t+1)\bar{r}(t+1)\right)$

the solution of the system $Lz = b$, which is given by an explicit formula $z = \boldsymbol{F}_L b$, provided that the matrix L is a lower triangular.

Thus far we have used the symbol t' to denote the last time instant. Expressions (3.1.17) relate the optimal estimate corresponding to the time instant $t' + 1$ to the previous optimal estimate and the new measurement. In what follows we use the symbol t to denote the last time instant when describing sequential recursive numerical schemes. We complete the description of the incremental least squares by summarizing the steps of the algorithm. Being initialized with $x(0) = 0$, $\boldsymbol{D}(0) = 0$, and $t = 0$, the algorithm proceeds as listed in Table 3.1.1.

Let us present the normal system update step in the more convenient form by describing how to directly calculate the Cholesky decomposition of the matrix $\boldsymbol{D}(t+1)$ given $\boldsymbol{D}(t) = L_{\boldsymbol{D}(t)}L^T_{\boldsymbol{D}(t)}$:

$$\begin{aligned}\boldsymbol{D}(t+1) &= \boldsymbol{D}(t) + \overline{W}^T(t+1)\overline{W}(t+1)\\ &= L_{\boldsymbol{D}(t)}\left(I + L^{-1}_{\boldsymbol{D}(t)}\overline{W}^T(t+1)\overline{W}(t+1)\left(L^T_{\boldsymbol{D}(t)}\right)^{-1}\right)L^T_{\boldsymbol{D}(t)}\\ &= L_{\boldsymbol{D}(t)}\left(I + \hat{W}^T(t+1)\hat{W}(t+1)\right)L^T_{\boldsymbol{D}(t)}\end{aligned} \tag{3.1.24}$$

where

$$\hat{W}^T(t+1) = \boldsymbol{F}_{L_{\boldsymbol{D}(t)}}\overline{W}^T(t+1) \tag{3.1.25}$$

Let $E = I + \hat{W}^T(t+1)\hat{W}(t+1)$ and L_E be its Cholesky factor

$$L_E L_E^T = E \tag{3.1.26}$$

Then it follows from (3.1.25) that

$$L_{\boldsymbol{D}(t+1)} = L_{\boldsymbol{D}(t)}L_E \tag{3.1.27}$$

The calculations (3.1.24) can now be replaced by the following steps:

a. $\hat{W}^T(t+1) = \boldsymbol{F}_{L_{\boldsymbol{D}(t)}}\overline{W}^T(t+1)$
b. $E = I + \hat{W}^T(t+1)\hat{W}(t+1)$

c. $L_E L_E^T = E$

d. $L_{\boldsymbol{D}(t+1)} = L_{\boldsymbol{D}(t)} L_E$

This concludes the description of Algorithm 1. The multiplicative representation (3.1.27) of the Cholesky factor update improves the numerical stability in the case of an ill–conditioned matrix $\boldsymbol{D}(t')$. The observability condition (3.1.12) guarantees nonsingularity of the matrix. The matrix can be either nonsingular or singular, while the conditioning number can be considered a continuous measure of singularity. The greater the conditioning number, the closer the matrix is to singularity. Ill-conditioning can occur at early epochs. Note that the *observability* concept originates from control theory, where it means possibility to recover the system state from the observed measurements.

3.2 STATIC PARAMETERS AND ARBITRARY TIME-VARYING VARIABLES

Let $x(t)$ and y be parameters to be estimated with dimensions $n(t)$ and n, respectively. Let the parameters be subject to the linear measurement model

$$J(t)x(t) + W(t)y = b(t) \tag{3.2.1}$$

for sequential time instances $t = 1, \ldots, t'$. Let the matrices $J(t)$ and $W(t)$ have dimensions $m(t) \times n(t)$ and $m(t) \times n$, respectively. Note that the parameters $x(t)$ are time dependent, while y is time invariant. Each model contains correspondent parameters $x(t)$ and y. The parameter y is common for all models. Define

$$\bar{J}(t) = L_{C(t)}^{-1} J(t) \tag{3.2.2}$$

along with (3.1.7), and define the $M(t') \times (N(t') + n)$ matrix $\bar{\boldsymbol{J}}(t')$ and the $M(t')$-dimensional vector $\bar{\boldsymbol{b}}(t')$

$$\bar{\boldsymbol{J}}(t') = \left[\begin{array}{cccc|c} \bar{J}(1) & 0 & \cdots & 0 & \bar{W}(1) \\ 0 & \bar{J}(2) & \cdots & 0 & \bar{W}(2) \\ \vdots & \vdots & \ddots & \vdots & \vdots \\ 0 & 0 & \cdots & \bar{J}(t') & \bar{W}(t') \end{array}\right] \tag{3.2.3}$$

$$\bar{\boldsymbol{b}}(t') = \begin{bmatrix} \bar{b}(1) \\ \bar{b}(2) \\ \vdots \\ \bar{b}(t') \end{bmatrix} \tag{3.2.4}$$

where $M(t') = \sum_{t=1}^{t'} m(t)$ and $N(t') = \sum_{t=1}^{t'} n(t)$.

Let us consider the least-squares solution,

$$I(x(1),\dots,x(t'),y,t') = \sum_{t=1}^{t'} (J(t)x(t) + W(t)y - b(t))^T C^{-1}(t)(J(t)x(t) + W(t)y - b(t)) \to \min \quad (3.2.5)$$

which is equivalent to the problem

$$I(x(1),\dots,x(t'),y,t') = \sum_{t=1}^{t'} \left(\bar{J}(t)x(t) + \overline{W}(t)y - b(t)\right)^T \left(\bar{J}(t)x(t) + \overline{W}(t)y - b(t)\right) \to \min \quad (3.2.6)$$

for the same reason why (3.1.5) is equivalent to (3.1.8). Let

$$X(t') = \begin{pmatrix} x(1) \\ \vdots \\ x(t') \end{pmatrix} \quad (3.2.7)$$

$$Z(t') = \begin{pmatrix} X(t') \\ y \end{pmatrix} \quad (3.2.8)$$

be the $N(t')$-dimensional vector of the variables $x(1),\dots,x(t')$ and $N(t')+n$-dimensional vector of the variables $x(1),\dots,x(t'),y$, respectively. The problem (3.2.6) is in turn equivalent to the problem

$$I(Z(t'),t') = \left(\bar{\mathbf{J}}(t')Z(t') - \bar{\mathbf{b}}(t')\right)^T \left(\bar{\mathbf{J}}(t')Z(t') - \bar{\mathbf{b}}(t')\right) \to \min \quad (3.2.9)$$

Let us denote

$$\mathbf{D}(t') = \bar{\mathbf{J}}^T(t')\bar{\mathbf{J}}(t') = \left[\begin{array}{c|c|c|c|c} \bar{J}^T(1)\bar{J}(1) & 0 & \cdots & 0 & \bar{J}^T(1)\overline{W}(1) \\ \hline 0 & \bar{J}^T(2)\bar{J}(2) & \cdots & 0 & \bar{J}^T(2)\overline{W}(2) \\ \hline \vdots & \vdots & \ddots & \vdots & \vdots \\ \hline 0 & 0 & \cdots & \bar{J}^T(t')\bar{J}(t') & \bar{J}^T(t')\overline{W}(t') \\ \hline \overline{W}^T(1)\bar{J}(1) & \overline{W}^T(2)\bar{J}(2) & \cdots & \overline{W}^T(t')\bar{J}(t') & \sum_{t=1}^{t'} \overline{W}^T(t)\overline{W}(t) \end{array}\right] \quad (3.2.10)$$

$$\boldsymbol{R}(t') = \bar{\boldsymbol{J}}^T(t')\bar{\boldsymbol{b}}(t') = \begin{bmatrix} \bar{J}^T(1)\bar{b}(1) \\ \bar{J}^T(2)\bar{b}(2) \\ \vdots \\ \bar{J}^T(t')\bar{b}(t') \\ \sum_{t=1}^{t'} \bar{W}^T(t)\bar{b}(t) \end{bmatrix} \tag{3.2.11}$$

then the solution to problem (3.2.9) satisfies the linear equation

$$\boldsymbol{D}(t')Z(t') = \boldsymbol{R}(t') \tag{3.2.12}$$

which has the solution

$$Z(t') = \bar{\boldsymbol{D}}^{-1}(t')\boldsymbol{R}(t') \tag{3.2.13}$$

Let

$$\boldsymbol{D}(t') = L_{\boldsymbol{D}(t')}L^T_{\boldsymbol{D}(t')} \tag{3.2.14}$$

be the Cholesky decomposition, where matrix $L_{\boldsymbol{D}(t')}$ is the $(N(t') + n) \times (N(t') + n)$ low triangle matrix.

Let us present the problem (3.2.9) in the form $I(X(t'), y, t') \to \min$ and note that

$$\min_{X(t'),y} I(X(t'), y, t') = \min_y \left(\min_{X(t')} I(X(t'), y, t') \right) \doteq \min_y \bar{I}(y, t') \tag{3.2.15}$$

See Section A.3.7 of Appendix A for details on partial minimization of quadratic forms. Given fixed y, the internal minimum in (3.2.15) can be split into a sequence of independent minimization problems:

$$I(x(t)) = \left(\bar{J}(t)x(t) + \bar{W}(t)y - \bar{b}(t)\right)^T \left(\bar{J}(t)x(t) + \bar{W}(t)y - \bar{b}(t)\right) \to \min_{x(t)} \tag{3.2.16}$$

each having the solution

$$x(y, t) = \left(\bar{J}^T(t)\bar{J}(t)\right)^{-1}\bar{J}^T(t)\left(\bar{b}(t) - \bar{W}(t)y\right) \tag{3.2.17}$$

which can also be seen from (3.2.10), (3.2.11), and (3.2.12) after substituting the fixed vector y into (3.2.12), taking the term $\bar{J}^T(t)\bar{W}(t)y$ to the right-hand side of (3.2.12), and solving the resulting block-diagonal linear system:

$$\begin{bmatrix} \bar{J}^T(1)\bar{J}(1) & 0 & \cdots & 0 \\ 0 & \bar{J}^T(2)\bar{J}(2) & \cdots & 0 \\ \vdots & \vdots & \ddots & \vdots \\ 0 & 0 & \cdots & \bar{J}^T(t')\bar{J}(t') \end{bmatrix} \begin{bmatrix} x(1) \\ x(2) \\ \vdots \\ x(t') \end{bmatrix} = \begin{bmatrix} \bar{J}^T(1)\bar{b}(1) - \bar{J}^T(1)\overline{W}(1)y \\ \bar{J}^T(2)\bar{b}(2) - \bar{J}^T(2)\overline{W}(2)y \\ \vdots \\ \bar{J}^T(t')\bar{b}(t') - \bar{J}^T(t')\overline{W}(t')y \end{bmatrix} \tag{3.2.18}$$

which is split into t' separated linear systems. Substituting (3.2.17) into (3.2.15) we arrive at the following problem:

$$\begin{aligned} \min_y \bar{I}(y,t') &= \sum_{t=1}^{t'} \left(\bar{J}(t)\left(\bar{J}^T(t)\bar{J}(t)\right)^{-1}\bar{J}^T(t)(\bar{b}(t) - \overline{W}(t)y) + \overline{W}(t)y - \bar{b}(t)\right)^T \\ &\quad \times \left(\bar{J}(t)\left(\bar{J}^T(t)\bar{J}(t)\right)^{-1}\bar{J}^T(t)\left(\bar{b}(t) - \overline{W}(t)y\right) + \overline{W}(t)y - \bar{b}(t)\right) \\ &= \sum_{t=1}^{t'} \left(\bar{b}(t) - \overline{W}(t)y\right)^T \left(I - \bar{J}(t)\left(\bar{J}^T(t)\bar{J}(t)\right)^{-1}\bar{J}^T(t)\right)^T \\ &\quad \times \left(I - \bar{J}(t)\left(\bar{J}^T(t)\bar{J}(t)\right)^{-1}\bar{J}^T(t)\right)\left(\bar{b}(t) - \overline{W}(t)y\right) \end{aligned} \tag{3.2.19}$$

The matrix $\Pi(t) = I - \bar{J}(t)\left(\bar{J}^T(t)\bar{J}(t)\right)^{-1}\bar{J}^T(t)$ is symmetric and idempotent since $\Pi^2(t) = \Pi(t)$. It is the matrix of orthogonal projection on the orthogonal complement to the space spanned on the columns of the matrix $\bar{J}(t)$. Actually, $\Pi(t)\bar{J}(t) = \bar{J}(t) - \bar{J}(t)\left(\bar{J}^T(t)\bar{J}(t)\right)^{-1}\bar{J}^T(t)\bar{J}(t) = 0$ which means the columns of the matrix $\bar{J}(t)$ are mapped to 0. On the other hand, every vector h orthogonal to $\bar{J}(t)$ is mapped to itself: $\Pi(t)h = h - \bar{J}(t)\left(\bar{J}^T(t)\bar{J}(t)\right)^{-1}\bar{J}^T(t)h = h$. The matrix $\Pi(t)$ is singular. Taking the idempotent property into consideration, the last expression for $\bar{I}(y,t')$ takes the form

$$\bar{I}(y,t') = \sum_{t=1}^{t'} \left(\bar{b}(t) - \overline{W}(t)y\right)^T \Pi(t)\left(\bar{b}(t) - \overline{W}(t)y\right) \tag{3.2.20}$$

and the problem $\bar{I}(y,t') \to \min$ has the solution

$$y(t') = \left(\hat{\mathbf{D}}(t')\right)^{-1}\hat{\mathbf{R}}(t') \tag{3.2.21}$$

where

$$\hat{\mathbf{D}}(t') = \sum_{t=1}^{t'} \overline{W}(t)^T \Pi(t)\overline{W}(t) \tag{3.2.22}$$

and

$$\hat{\mathbf{R}}(t') = \sum_{t=1}^{t'} \overline{W}(t)^T \Pi(t)\bar{b}(t) \tag{3.2.23}$$

The matrix $\hat{\boldsymbol{D}}(t')$ is supposed to be nonsingular in (3.2.21). For the case $t' = 1$ we have $\hat{\boldsymbol{D}}(1) = \overline{W}(1)^T \Pi(1)\overline{W}(1)$ and the matrix $\hat{\boldsymbol{D}}(1)$ is singular due to the singularity of the matrix $\Pi(1)$. The necessary conditions for nonsingularity of $\hat{\boldsymbol{D}}(t')$ are $t' > 1$ and

$$\overline{W}(1)^T \Pi(1)\overline{W}(1) \neq \overline{W}(t)^T \Pi(t)\overline{W}(t) \tag{3.2.24}$$

for at least one value t. Conditions (3.2.24) are necessary since their violation leads to the singularity of $\hat{\boldsymbol{D}}(t')$. Actually, $t' = 1$ leads to the singularity of $\hat{\boldsymbol{D}}(t')$ as shown above. Further, if $\overline{W}(1)^T \Pi(1)\overline{W}(1) = \overline{W}(t)^T \Pi(t)\overline{W}(t)$ for all t, then $\hat{\boldsymbol{D}}(t') = t'\overline{W}(1)^T \Pi(1)\overline{W}(1)$ which is singular. On the other hand, conditions (3.2.24) are not sufficient in the general case.

Nonsingularity of the matrix $\hat{\boldsymbol{D}}(t')$ will be called *observability* of the system $\{J(1), W(1), \ldots, J(t'), W(t')\}$. It guarantees that the parameter y can be estimated from the system (3.2.1) for $t = 1, \ldots, t'$. In order to recover parameters $x(t)$ we need also nonsingularity of the matrices $\overline{J}^T(t)\overline{J}(t)$. The observability condition is met if a sufficient number of linearly independent measurements are available. For example, for the problem of carrier phase ambiguity estimation and resolution along with estimation of the time-varying (kinematic) position $x(t)$ using carrier phase observations only, observability is met if the number of satellites is greater or equal to 4 and at least two sets of measurements are received, i.e., $t' \geq 2$. On the other hand, observability is met at the single epoch if pseudorange observations are used along with carrier phases. Conditions (3.2.24) will be met because the movement of satellites ensures that the directional cosine matrix $\overline{J}(t)$ changes in time. On the other hand, if there are only two measurements, which are separated in time by just one second, the matrices $\overline{J}(1)$ and $\overline{J}(2)$ "almost" coincide. This means that the matrix $\hat{\boldsymbol{D}}(t')$ is "nearly" singular or, actually it is ill-conditioned. As time increases the satellite constellation changes and the computed direction cosine matrices will change, eventually leading to an improved conditioning number. Accumulation of a larger number of time-varying independent measurements leads to improvement of observability.

Now let us give a practical way to compute the matrix $\Pi(t)$. First, compute the Cholesky decomposition of the matrix $\overline{J}^T(t)\overline{J}(t)$,

$$\overline{J}^T(t)\overline{J}(t) = L_{\overline{J}(t)} L_{\overline{J}(t)}^T \tag{3.2.25}$$

where $L_{\overline{J}(t)}$ is $n \times n$ lower triangular, then compute $\widetilde{J}^T(t) = \boldsymbol{F}_{\overline{J}(t)}\overline{J}^T(t)$. Finally compute $\Pi(t) = I - \widetilde{J}(t)\widetilde{J}^T(t)$. The updated vector $y(t'+1)$ in the recursive form using (3.2.21) and (3.2.23) is

$$\begin{aligned}
y(t'+1) &= (\hat{\boldsymbol{D}}(t'+1))^{-1}\hat{\boldsymbol{R}}(t'+1) \\
&= (\hat{\boldsymbol{D}}(t'+1))^{-1}\left(\hat{\boldsymbol{R}}(t') + \overline{W}(t'+1)^T \Pi(t'+1)\overline{b}(t'+1)\right) \\
&= (\hat{\boldsymbol{D}}(t'+1))^{-1}\left(\hat{\boldsymbol{D}}(t')y(t') + \overline{W}(t'+1)^T \Pi(t'+1)\overline{b}(t'+1)\right)
\end{aligned}$$

$$= (\hat{\boldsymbol{D}}(t'+1))^{-1}(\hat{\boldsymbol{D}}(t'+1)y(t') - \overline{W}(t'+1)^T \Pi(t'+1)\overline{W}(t'+1)y(t')$$
$$+ \overline{W}(t'+1)^T \Pi(t'+1)\overline{b}(t'+1))$$
$$= y(t') + (\hat{\boldsymbol{D}}(t'+1))^{-1}\overline{W}(t'+1)^T \Pi(t'+1)\left(\overline{b}(t'+1) - \overline{W}(t'+1)y(t')\right) \tag{3.2.26}$$

The algorithm starts with $y(0) = 0$, $\hat{\boldsymbol{D}}(0) = 0$, $t = 0$. A complete description of the incremental least squares is summarized in Table 3.2.1.

Let us now look at the recursive algorithm for estimation of vectors $y(t+1)$, $x(t+1)$ from a slightly different point of view. Note that the data can be processed postmission after accumulating a complete data set. The batch least-squares adjustment leads to a large linear system with a sparse matrix showing a specific pattern. Applying the sparse matrix decomposition technique we will prove that the recursive estimate (3.2.26) can be obtained as Cholesky decomposition of the incrementally updated large-scale matrix of the linear system and incrementally performed forward solution, followed by only a single step backward solution. By performing the full

TABLE 3.2.1 Algorithm 2: Estimating Static Parameters and Arbitrary Time-Varying Variables

Cholesky decomposition of the covariance matrix and forward substitution calculations	$C(t+1) = L_{C(t+1)}L_{C(t+1)}^T$ $\overline{b}(t+1) = \boldsymbol{F}_{L_{C(t+1)}} b(t+1)$ $\overline{W}(t+1) = \boldsymbol{F}_{L_{C(t+1)}} W(t+1)$ $\overline{J}(t+1) = \boldsymbol{F}_{L_{C(t+1)}} J(t+1)$
Cholesky decomposition of the matrix $\overline{J}^T(t+1)\overline{J}(t+1)$ and forward substitution calculations	$\overline{J}^T(t+1)\overline{J}(t+1) = L_{\overline{J}(t+1)}L_{\overline{J}(t+1)}^T$ $\widetilde{J}^T(t+1) = \boldsymbol{F}_{L_{\overline{J}(t+1)}} \overline{J}^T(t+1)$
Projection matrix	$\Pi(t+1) = I - \widetilde{J}(t+1)\widetilde{J}^T(t+1)$
Update the matrix $\hat{\boldsymbol{D}}(t)$	$\hat{\boldsymbol{D}}(t+1) = \hat{\boldsymbol{D}}(t) + \overline{W}(t+1)^T \Pi(t+1)\overline{W}(t+1)$
Cholesky decomposition of $\hat{\boldsymbol{D}}(t)$	$\hat{\boldsymbol{D}}(t+1) = L_{\hat{\boldsymbol{D}}(t+1)}L_{\hat{\boldsymbol{D}}(t+1)}^T$
Residual vector	$\overline{r}(t+1) = \overline{b}(t+1) - \overline{W}(t+1)y(t)$
Update the estimate $y(t+1)$	$y(t+1) = y(t)$ $+ \boldsymbol{B}_{L_{\hat{\boldsymbol{D}}(t+1)}}\left(\boldsymbol{F}_{L_{\hat{\boldsymbol{D}}(t+1)}} \overline{W}(t'+1)^T \Pi(t'+1)\overline{r}(t+1)\right)$
Second residual vector	$r'(t+1) = \overline{b}(t+1) - \overline{W}(t+1)y(t+1)$
Compute estimate $x(t+1)$	$x(t+1) = \boldsymbol{B}_{L_{\overline{J}(t+1)}} \widetilde{J}^T(t+1)r'(t+1)$

backward solution we can improve the estimates $x(t)$ using future observables $b(t')$ corresponding to $t' > t$, which is possible in batch processing. This construction allows for a deeper understanding of the recursive least squares and its connection to batch least squares.

Let us present the matrix $\boldsymbol{D}(t')$ in the expression (3.2.10) in the form

$$\boldsymbol{D}(t') = \left[\begin{array}{c|c|c|c|c} \bar{J}^T(1)\bar{J}(1) & 0 & \cdots & 0 & \bar{J}^T(1)\overline{W}(1) \\ \hline 0 & \bar{J}^T(2)\bar{J}(2) & \cdots & 0 & \bar{J}^T(2)\overline{W}(2) \\ \hline \vdots & \vdots & \ddots & \vdots & \vdots \\ \hline 0 & 0 & \cdots & \bar{J}^T(t')\bar{J}(t') & \bar{J}^T(t')\overline{W}(t') \\ \hline \overline{W}^T(1)\bar{J}(1) & \overline{W}^T(2)\bar{J}(2) & \cdots & \overline{W}^T(t')\bar{J}(t') & \sum_{t=1}^{t'} \overline{W}^T(t)\overline{W}(t) \end{array}\right]$$

$$= \left[\begin{array}{c|cccc} \bar{J}^T(1)\bar{J}(1) & 0 & \cdots & 0 & \bar{J}^T(1)\overline{W}(1) \\ \hline 0 & & & & \\ \vdots & & C & & \\ 0 & & & & \\ \overline{W}^T(1)\bar{J}(1) & & & & \end{array}\right] \tag{3.2.27}$$

and apply formulas (A.3.59) to (A.3.61). We have

$$L(t') = \left[\begin{array}{c|cccc} L_{\bar{J}(1)} & 0 & \cdots & 0 & 0 \\ \hline 0 & & & & \\ \vdots & & M & & \\ 0 & & & & \\ \overline{W}^T(1)\bar{J}(1)\left(L_{\bar{J}(1)}^T\right)^{-1} & & & & \end{array}\right] \tag{3.2.28}$$

where M is the lower triangular matrix of the Cholesky decomposition of the matrix

$$\left[\begin{array}{c|c|c|c|c} \bar{J}^T(2)\bar{J}(2) & 0 & \cdots & 0 & \bar{J}^T(2)\overline{W}(2) \\ \hline 0 & \bar{J}^T(3)\bar{J}(3) & \cdots & 0 & \bar{J}^T(3)\overline{W}(3) \\ \hline \vdots & \vdots & \ddots & \vdots & \vdots \\ \hline 0 & 0 & \cdots & \bar{J}^T(t')\bar{J}(t') & \bar{J}^T(t')\overline{W}(t') \\ \hline \overline{W}^T(2)\bar{J}(2) & \overline{W}^T(3)\bar{J}(3) & \cdots & \overline{W}^T(t')\bar{J}(t') & \begin{array}{c}\sum_{t=1}^{t'} \overline{W}^T(t)\overline{W}(t) - \overline{W}^T(1)\bar{J}(1) \\ \times\left(\bar{J}^T(1)\bar{J}(1)\right)^{-1}\bar{J}^T(1)\overline{W}(1)\end{array} \end{array}\right] \tag{3.2.29}$$

Sequential application of formulas (A.3.59) to (A.3.61) by induction finally gives the following representation:

$$L(t') = \begin{bmatrix} L_{\bar{J}(1)} & 0 & \cdots & 0 & 0 \\ 0 & L_{\bar{J}(2)} & \cdots & 0 & 0 \\ \vdots & \vdots & \ddots & \vdots & \vdots \\ 0 & 0 & \cdots & L_{\bar{J}(t')} & 0 \\ \overline{W}^T(1)\bar{J}(1) \times \left(L^T_{\bar{J}(1)}\right)^{-1} & \overline{W}^T(2)\bar{J}(2) \times \left(L^T_{\bar{J}(2)}\right)^{-1} & \cdots & \overline{W}^T(t') \times \bar{J}(t')\left(L^T_{\bar{J}(t')}\right)^{-1} & \hat{L}(t') \end{bmatrix} \tag{3.2.30}$$

where $\hat{L}(t')$ is a Cholesky factor of the matrix

$$\sum_{t=1}^{t'} \left(\overline{W}^T(t)\overline{W}(t) - \overline{W}^T(t)\bar{J}(t)\left(\bar{J}^T(t)\bar{J}(t)\right)^{-1}\bar{J}^T(t)\overline{W}(t)\right) \tag{3.2.31}$$

which is equal to $\hat{\boldsymbol{D}}(t')$ according to (3.2.22). Now, according to the expressions (3.2.13) and (3.2.14) we have

$$Z(t') = \begin{pmatrix} x(1) \\ \vdots \\ x(t') \\ y(t') \end{pmatrix} = \boldsymbol{B}_{L(t')}(\boldsymbol{F}_{L(t')}\boldsymbol{R}(t')) \tag{3.2.32}$$

Let us first calculate $V(t') = \boldsymbol{F}_{L(t')}\boldsymbol{R}(t')$. We have $V(t') = (v^T(1), \ldots, v^T(t'), w^T(t'))^T$. Taking into account the structure of the matrix (3.2.30), sequentially calculate

$$v(t) = (L_{\bar{J}(t)})^{-1}\bar{J}^T(t)\bar{b}(t) \tag{3.2.33}$$

for $t = 1, \ldots, t'$. The last equation of the system $L(t')V(t') = \boldsymbol{R}(t')$ gives

$$\sum_{t=1}^{t'} \overline{W}^T(t)\bar{J}(t)\left(L^T_{\bar{J}(t)}\right)^{-1} v(t) + \hat{L}(t')w(t') = \sum_{t=1}^{t'} \overline{W}^T(t)\bar{b}(t) \tag{3.2.34}$$

resulting in

$$\begin{aligned} w(t') &= (\hat{L}(t'))^{-1}\left(\sum_{t=1}^{t'} \overline{W}^T(t)\bar{b}(t) - \overline{W}^T(t)\bar{J}(t)\left(L^T_{\bar{J}(t)}\right)^{-1} v(t)\right) \\ &= (\hat{L}(t'))^{-1}\left(\sum_{t=1}^{t'} \overline{W}^T(t)\bar{b}(t) - \overline{W}^T(t)\bar{J}(t)\left(L^T_{\bar{J}(t)}\right)^{-1}\left(L_{\bar{J}(t)}\right)^{-1}\bar{J}^T(t)\bar{b}(t)\right) \end{aligned}$$

$$= (\hat{L}(t'))^{-1} \sum_{t=1}^{t'} \overline{W}^T(t)(I - \overline{J}(t)\left(L^T_{\overline{J}(t)}\right)^{-1}(L_{\overline{J}(t)})^{-1}\overline{J}^T(t))\overline{b}(t)$$

$$= (\hat{L}(t'))^{-1} \sum_{t=1}^{t'} \overline{W}^T(t)\Pi(t)\overline{b}(t) \tag{3.2.35}$$

Then calculate the backward solution $Z(t') = \boldsymbol{B}_{L(t')}V(t')$. Taking into account the structure of the matrix (3.2.30), one obtains

$$\begin{aligned}
y(t') &= (\hat{L}^T(t'))^{-1}w(t') \\
&= (\hat{L}^T(t'))^{-1}(\hat{L}(t'))^{-1} \sum_{t=1}^{t'} \overline{W}^T(t)\Pi(t)\overline{b}(t) \\
&= (\hat{L}^T(t'))^{-1}(\hat{L}(t'))^{-1} \left(\sum_{t=1}^{t'-1} \overline{W}^T(t)\,\Pi(t)\overline{b}(t) + \overline{W}^T(t')\Pi(t')\overline{b}(t') \right) \\
&= (\hat{L}^T(t'))^{-1}(\hat{L}(t'))^{-1}(\hat{\boldsymbol{D}}(t'-1)y(t'-1) + \overline{W}^T(t')\Pi(t')\overline{b}(t')) \\
&= (\hat{L}^T(t'))^{-1}(\hat{L}(t'))^{-1}(\hat{\boldsymbol{D}}(t')y(t'-1) \\
&\quad - \overline{W}^T(t')\Pi(t')\overline{W}(t')y(t'-1) + \overline{W}^T(t')\Pi(t')\overline{b}(t')) \\
&= y(t'-1) + (\hat{L}^T(t'))^{-1}(\hat{L}(t'))^{-1}\overline{W}^T(t')\Pi(t')(\overline{b}(t') - \overline{W}(t')y(t'-1))
\end{aligned} \tag{3.2.36}$$

and

$$\begin{aligned}
x(t) &= \left(L^T_{\overline{J}(t)}\right)^{-1}(v(t) - (L_{\overline{J}(t)})^{-1}\overline{J}^T(t)\overline{W}(t)y(t')) \\
&= \left(L^T_{\overline{J}(t)}\right)^{-1}\left((L_{\overline{J}(t)})^{-1}\overline{J}^T(t)\overline{b}(t) - (L_{\overline{J}(t)})^{-1}\overline{J}^T(t)\overline{W}(t)y(t')\right) \\
&= \boldsymbol{B}_{L_{\overline{J}(t)}}\left(\boldsymbol{F}_{L_{\overline{J}(t)}}\overline{J}^T(t)(\overline{b}(t) - \overline{W}(t)y(t'))\right)
\end{aligned} \tag{3.2.37}$$

with $t = 1, \ldots, t'$. The previous two expressions coincide with (3.2.26) and (3.2.17) obtained earlier.

What is the major difference between (3.2.37) and (3.2.17)? In Equation (3.2.37) we used the estimate $y(t')$ which is available after all t' measurements are received. In other words, we calculate the estimate $x(t)$ for $t < t'$ based on measurements received after the time instant t. That is possible only in the case of offline or postsession processing. In applications to satellite surveying this means that the surveyor collects raw data in the field as long as needed, based on experience of about how long the measurement session should be under certain conditions. Then, during the postsession processing in the office, the surveyor can assume that all data is available simultaneously. This assumption makes it possible to establish an explicit dependence of earlier estimates on the later received data without breaking the causal link. In contrast, when working in real time, we can rely only on measurements received at time instances $t = 1, \dots, t'$ when calculating the estimate $x(t')$. For this reason the estimate $x(t+1)$ in Algorithm 2 depends on the earlier obtained estimate $y(t+1)$. In other words, real-time operation dictates the following calculation order:

$$y(1), x(1), y(2), x(2), \dots, y(t), x(t), \dots \tag{3.2.38}$$

while postsession processing mode suggests both calculation schemes: the scheme (3.2.38) and the scheme

$$y(t'), x(1), x(2), \dots, x(t') \tag{3.2.39}$$

In this section we derived the recursive least-squares algorithm allowing the update of estimates of the static parameter and parameters arbitrarily varying in time. Using the Cholesky decomposition of sparse matrices, we established a connection between recursive processing and batch processing.

3.3 DYNAMIC CONSTRAINTS

In the previous section, the measurement model (3.2.1) contained time-varying variables $x(t)$ which were independent of each other for different values t. In this section we considered the more complex case of dynamic dependency of the variable $x(t)$ on the variable $x(t-1)$.

Consider the discrete dynamic system

$$x(t) = F(t)x(t-1) + \xi(t) \tag{3.3.1}$$

with the n-dimensional state vector $x(t)$ and the $m(t)$-dimensional observation vector $b(t)$, connected to the state by the linear relationship

$$b(t) = H(t)x(t) + \varepsilon(t) \tag{3.3.2}$$

The matrices $F(t)$ and $H(t)$ have dimensions $n \times n$ and $m(t) \times n$, respectively. The stochastic processes $\{\varepsilon(t)\}$ and $\{\xi(t)\}$ are zero centered, stationary, independent of each other, and have covariance matrices

$$E\big(\xi(t)\xi^T(t)\big) = Q, \quad E\big(\varepsilon(t)\varepsilon^T(t)\big) = R \tag{3.3.3}$$

Also, $E(\xi(t)\xi^T(s)) = 0$ and $E(\varepsilon(t)\varepsilon^T(s)) = 0$ for the case when $t \neq s$. In order to complete description, define the initial data

$$x(0) = x_0 + \eta \tag{3.3.4}$$

which is supposed to be known up to the random vector η, which is independent of the vectors $\{\varepsilon(t)\}$ and $\{\xi(t)\}$, and has the covariance matrix

$$E\big(\eta(t)\eta^T(t)\big) = C \tag{3.3.5}$$

Assuming the above specifications, it is assumed that the motion of the system is in accordance with (3.3.1), where the random vector $\xi(t)$ describes the uncertainty of the model, which differs from the real system just for this vector. In other words, a random vector describes the inaccuracy of our knowledge of the real system. Being aware of the limitations of such a description, we continue finding a solution only for applications where this assumption is justified.

Output of the system (3.3.2) is considered as a measurement of a physical quantity $b(t)$, linearly dependent on the state $x(t)$, and measured subject to the random measurement error $\varepsilon(t)$. The estimation problem is to recover the trajectory of the system $\{x(t)\}$ based on the results of successive measurements $\{b(t)\}$ using the description (3.3.1) to (3.3.5).

The least-squares problem is constructed as a minimization of the quadratic function of variables $\{x(t), \ldots, x(t')\}$ as follows:

$$\begin{aligned} I(x(0), x(1), \ldots, x(t'), t') = {} & \sum_{t=1}^{t'} (x(t) - F(t)x(t-1))^T Q^{-1} (x(t) - F(t)x(t-1)) \\ & + \sum_{t=1}^{t'} (b(t) - H(t)x(t))^T R^{-1} (b(t) - H(t)x(t)) \\ & + (x(0) - x_0)^T C^{-1} (x(0) - x_0) \end{aligned} \tag{3.3.6}$$

The measurements are collected at time instants $t = 1, \ldots, t'$. This criterion is the weighted sum of squared residuals of relations (3.3.1), (3.3.2), and (3.3.4). Weighing using the positive definite inverses to the covariance matrices allows taking into account the variance of the entries of the uncertainty vectors as well as correlations between them. This particular weighing uses the inverse of covariance matrices, while other weighing is also possible. Matrices Q, R, and C are supposed to be positive definite. The estimate $\{x(t)\}$, $t = 1, \ldots, t'$ giving the least value to the criterion (3.3.6), is considered the best estimate,

$$I_{\min} = \min_{x(0), x(1), \cdots, x(t')} I(x(0), x(1), \cdots, x(t'), t') \tag{3.3.7}$$

In order to reflect the fact that each of the vectors $\hat{x}(t)$ of the best estimate depends on all measurements vectors $b(1), \ldots, b(t')$, we will be using the notation $\hat{x}(t, b(1), \ldots, b(t'))$. Sometimes the notation expressing dependency on $b(1), \ldots, b(t')$ will be omitted if this does not lead to misunderstandings.

In applications to real-time estimation, it is necessary to obtain the best estimate of the state as soon as measurements become available. In this case, the state $x(t)$ is estimated based on the measurements $b(1), \ldots, b(t)$, while the next state $x(t+1)$ is estimated based on one more measurement $b(1), \ldots, b(t+1)$. Let us denote for the sake of brevity $x^*(t) = \hat{x}(t, b(1), \ldots, b(t))$. Using the next measurement $b(t+1)$ allows for obtaining the next estimate $x^*(t+1)$ and, if necessary, allows for increasing accuracy of earlier obtained estimate $x^*(t)$ because, generally,

$$x^*(t) = \hat{x}(t, b(1), \ldots, b(t)) \neq \hat{x}(t, b(1), \ldots, b(t), b(t+1)) \tag{3.3.8}$$

In order to obtain sequential estimates $x^*(t)$ based on sequential measurements $b(1), \ldots, b(t)$ one can solve problems like (3.3.7) for $t' = t$. However, it is possible to obtain subsequent estimates $x^*(t+1)$ based on previously obtained estimate $x^*(t)$ and a "new" measurement $b(t+1)$.

Further, it is possible to consider the least-squares problem of the form similar to (3.3.6). The difference is that least-squares approach is more general in that it allows a weight matrices that is not necessarily an inverse of a covariance matrix and is arbitrary nonnegative definite with no statistical meaning:

$$\begin{aligned}\widetilde{I}(x(0), x(1), \ldots, x(t'), t') = &\sum_{t=1}^{t'} (x(t) - F(t)x(t-1))^T W^x(t)(x(t) - F(t)x(t-1)) \\ &+ \sum_{t=1}^{t'} (b(t) - H(t)x(t))^T W^y(t) \left(b(t) - H(t)x(t)\right) \\ &+ (x(0) - x_0)^T W^0(t)(x(0) - x_0) \end{aligned} \tag{3.3.9}$$

The only restriction on the choice of the matrices $W^x(t)$, $W^y(t)$, W^0 is that the problem (3.3.9) has a unique solution. The solution of (3.3.9) will be understood as a solution having the lowest Euclidean norm.

In the following we use the dynamic programming approach and derive the recursive relations for Bellman functions (Bellman and Kalaba,1966). In optimal control theory, the Bellman functions describe the dependence of the optimal value of the cost function on the initial state of the dynamic process. Assume that we start the dynamic process with the state $x(0) = x'$ and apply the optimal control strategy. The cost function subject to minimization takes its minimum value denoted by $v(x')$ because this value depends on the initial state. Another choice $x(0) = x''$ gives another optimal value $v(x'')$. Thus, we introduce a conditional minimum since the optimal value depends on the initial state. The Bellman functions explicitly express this dependency. Bellman showed that optimization of a discrete dynamic process can be stated in the recursive form. The relationship that connects values of the Bellman function in two sequential time instances is called the Bellman equation.

Let us denote $\|z\|_W^2 = z^T W z$ and rewrite the problem formulation (3.3.7) in the form

$$\begin{aligned} I_{\min} &= \min_{x(t')} \left[\min_{x(0),x(1),\ldots,x(t'-1)} I(x(0), x(1), \ldots, x(t'), t') \right] \\ &= \min_{x(t')} \left[\min_{x(0),x(1),\ldots,x(t'-1)} \left(\sum_{t=1}^{t'} \|x(t) - F(t)x(t-1)\|_{Q^{-1}}^2 \right.\right. \\ &\qquad \left.\left. + \sum_{t=1}^{t'} \|b(t) - H(t)x(t)\|_{R^{-1}}^2 + \|x(0) - x_0\|_{C^{-1}}^2 \right) \right] \end{aligned} \tag{3.3.10}$$

Denoting the expression in square brackets by

$$v(t', x(t')) = \min_{x(0),x(1),\ldots,x(t'-1)} \left(\sum_{t=1}^{t'} \|x(t) - F(t)x(t-1)\|_{Q^{-1}}^2 + \sum_{t=1}^{t'} \|b(t) - H(t)x(t)\|_{R^{-1}}^2 + \|x(0) - x_0\|_{C^{-1}}^2 \right) \tag{3.3.11}$$

the problem (3.3.10) can be written in the form

$$I_{\min} = \min_{x_{t'}} v(t', x(t')) \tag{3.3.12}$$

Note that the last term $\|b(t') - H(t')x(t')\|_{R^{-1}}^2$ in the second sum of (3.3.11) depends on t' and the variable $x(t')$ and does not depend on the variables $x(0), x(1), \ldots, x(t' - 1)$ for which the minimum is taken. Therefore, the expression (3.3.11) can be rewritten as

$$v(t', x(t')) = \min_{x(0),x(1),\cdots,x(t'-1)} \left(\sum_{t=1}^{t'} \|x(t) - F(t)x(t-1)\|_{Q^{-1}}^2 + \sum_{t=1}^{t'-1} \|b(t) - H(t)x(t)\|_{R^{-1}}^2 + \|x(0) - x_0\|_{C^{-1}}^2 \right) + \|b(t') - H(t')x(t')\|_{R^{-1}}^2 \tag{3.3.13}$$

The problem (3.3.13) is similar to problem (3.3.7) but contains one variable $x(t')$ less in definition of the minimum. Another difference of (3.3.13) compared to (3.3.7) is that the problem (3.3.13) defines the relative minimum, which depends on the variable $x(t')$. Aiming to apply mathematical induction, consider minimization over the variable $x(t' - 1)$ in (3.3.13) as a separate operation,

$$\min_{x(0),x(1),\ldots,x(t'-1)} \left(\sum_{t=1}^{t'} \|x(t) - F(t)x(t-1)\|_{Q^{-1}}^2 + \sum_{t=1}^{t'-1} \|b(t) - H(t)x(t)\|_{R^{-1}}^2 + \|x(0) - x_0\|_{C^{-1}}^2 \right)$$

$$= \min_{x(t'-1)} \left[\min_{x(0),x(1),\ldots,x(t'-2)} \left(\sum_{k=1}^{t'} \|x(t) - F(t)x(t-1)\|_{Q^{-1}}^2 + \sum_{t=1}^{t'-1} \|b(t) - H(t)x(t)\|_{R^{-1}}^2 + \|x(0) - x_0\|_{C^{-1}}^2 \right) \right]$$

$$= \min_{x(t'-1)} \left[\min_{x(0),x(1),\cdots,x(t'-2)} \left(\sum_{t=1}^{t'-1} \|x(t) - F(t)x(t-1)\|^2_{Q^{-1}} + \sum_{t=1}^{t'-1} \|b(t) - H(t)x(t)\|^2_{R^{-1}} \right.\right.$$
$$\left.\left. + \|x(0) - x_0\|^2_{C^{-1}} \right) + \|x(t') - F(t')x(t'-1)\|^2_{Q^{-1}} \right] \tag{3.3.14}$$

Again, denote the internal minimum in the expression (3.3.14) as

$$v(t'-1, x(t'-1)) = \min_{x(0),x(1),\ldots,x(t'-2)} \left(\sum_{t=1}^{t'-1} \|x(t) - F(t)x(t-1)\|^2_{Q^{-1}} \right.$$
$$\left. + \sum_{t=1}^{t'-1} \|b(t) - H(t)x(t)\|^2_{R^{-1}} + \|x(0) - x_0\|^2_{C^{-1}} \right) \tag{3.3.15}$$

then rewrite (3.3.13), taking into account expressions (3.3.14) and (3.3.15), in the form

$$v(t', x(t')) = \min_{x(t'-1)} \left(v(t'-1, x(t'-1)) + \|x(t') - F(t')x(t'-1)\|^2_{Q^{-1}} \right)$$
$$+ \|b(t'-1) - H(t')x(t')\|^2_{R^{-1}} \tag{3.3.16}$$

Continuing the transformation of taking the minimum over the set of variables $x(0), x(1), \ldots, x(t'-2)$ in expression (3.3.15) into the operation of taking the successive minima, we obtain by induction a sequence of functions, called conditional optimum Bellman functions, in accordance with the following recursive relations:

$$v(0, x(0)) = \|x(0) - x_0\|^2_{C^{-1}}$$
$$v(1, x(1)) = \min_{x(0)} \left(v(0, x(0)) + \|x(1) - F(1)x(0)\|^2_{Q^{-1}} \right) + \|b(1) - H(1)x(1)\|^2_{R^{-1}}$$
$$\ldots$$
$$v(t, x(t)) = \min_{x(t-1)} \left(v(t-1, x(t-1)) + \|x(t) - F(t)x(t-1)\|^2_{Q^{-1}} \right)$$
$$+ \|b(t) - H(t)x(t)\|^2_{R^{-1}}$$
$$\ldots$$
$$v(t', x(t')) = \min_{x(t'-1)} \left(v(t'-1, x(t'-1)) + \|x(t') - F(t')x(t'-1)\|^2_{Q^{-1}} \right)$$
$$+ \|b(t') - H(t')x(t')\|^2_{R^{-1}} \tag{3.3.17}$$

Along with the definition of the function $v(t, x(t))$, the value $x(t-1, x(t))$, which minimizes the function $v(t-1, x(t-1)) + \|x(t) - F(t)x(t-1)\|^2_{Q^{-1}}$ over the variable $x(t-1)$ under fixed value of the vector $x(t)$, is defined. These estimates are called *conditionally optimal* since they are defined under condition that $x(t)$ is fixed. In order to find a complete set of optimal estimates $\hat{x}(0), \hat{x}(1), \ldots, \hat{x}(t')$, one needs to find a minimum of $v(t', x(t'))$, achieved at the point $\hat{x}(t')$, and use the recursive relations

$$\hat{x}(t-1) = x(t-1, \hat{x}(t)) \quad t = t', t'-1, \ldots, 1 \tag{3.3.18}$$

These recursive relations allow for sequential definition of the optimal estimate in reverse order, starting with $\hat{x}(t')$. We summarize the above construction in the following statement:

Statement 3.3.1. The solution of the optimal estimation problem (3.3.7) is equivalent to minimizing the function $v(t', x(t'))$ in the variable $x(t')$, resulting in the optimal estimate $\hat{x}(t')$. This function is defined by the recursive relations (3.3.17). The remaining components of the sequence $\hat{x}(t'-1), \hat{x}(t'-2), \ldots, \hat{x}(0)$ of optimal estimates are obtained recursively in accordance with relations (3.3.18).

The above describes batch measurement processing, which is applied after receiving a complete set of measurements $b(1), \ldots, b(t')$. In this case, as has been mentioned, all the components of the sequence $\{\hat{x}(t)\}$ depend on the full set of measurements: $\hat{x}(t) = \hat{x}(t, b(1), \ldots, b(t'))$. The processing is called batch processing as opposed to real-time processing in which case the measurements are obtained sequentially. An estimate of the component of the sequence of state vectors should also be obtained sequentially in real time, starting with $x^*(1)$ (obviously, $x^*(0) = x(0)$) and the total number of measurements (end time instant) is unknown in advance. Therefore, a practical interest consists in obtaining of recursive relations for computing the next estimate $x^*(t)$ using the previous one $x^*(t-1)$, and not vice versa. After receiving a new measurement $b(t)$ and an estimate $x^*(t) = \hat{x}(t, b(1), \ldots, b(t))$, one can specify earlier obtained estimates $x^*(t-1), x^*(t-2), \cdots$ using the relations (3.3.18) obtaining $\hat{x}(t-1), \hat{x}(t-2), \ldots$, but usually this does make much practical sense.

Inductive application of the expression (A.3.38) of Section A.3.4 to expressions (3.3.17) proves the following statement:

Statement 3.3.2. The functions $v(t, x)$ are quadratic in their arguments. Conditionally optimal estimates $x(t-1, x(t))$ are linearly dependent on $x(t)$.

When working in real time, the optimal estimate $x^*(t)$ must be obtained immediately after a measurement $b(t)$ is received. Having a measurement $b(t)$, one can use the tth step of the recursive scheme (3.3.17) for construction of the function $v(t, x(t))$. The measurement $b(t)$ will be the last one among the measurements available at the tth step. The optimal estimate $x^*(t)$ of the vector $x(t)$ is calculated on the base of measurements $b(1), \ldots, b(t)$. Setting $t' = t$ in the expression (3.3.12), one obtains that $x^*(t)$ minimizes the function $v(t, x(t))$ which is quadratic according to Statement 3.3.2. Further, according to the expression (A.3.38) we have

$$v(t, x(t)) = (x(t) - x^*(t))^T \mathbf{D}(t)(x(t) - x^*(t)) \tag{3.3.19}$$

where $\mathbf{D}(t)$ is a positive definite matrix. Specifically, the equation (3.3.19) is correct up to the constant value, which does not affect the argument of the minimum.

Our goal now is construction of the computational scheme for the recursive calculation of matrix $\mathbf{D}(t)$ based on matrix $\mathbf{D}(t-1)$, and the vector of optimal estimate $x^*(t)$ on the basis of vector $x^*(t-1)$ for all $t = 1, 2, \ldots$, starting with

$$\mathbf{D}(0) = C^{-1} \text{ and } x^*(0) = x_0 \tag{3.3.20}$$

In the initial data conditions formulation (3.3.20), we took into account that

$$v(0, x(0)) = (x(0) - x_0)^T C^{-1}(x(0) - x_0) \tag{3.3.21}$$

Therefore, assuming that the function

$$v(t-1, x(t-1)) = (x(t-1) - x^*(t-1))^T \mathbf{D}(t-1)(x(t-1) - x^*(t-1)) \tag{3.3.22}$$

is known, obtain the expressions for function (3.3.19). Consider the function of the variable $x(t-1)$ subject to minimization in (3.3.17), and denote it by $q(x(t-1))$, then

$$\begin{aligned} q(x(t-1)) &= v(t-1, x(t-1)) + \|x(t) - F(t)x(t-1)\|^2_{Q^{-1}} \\ &= (x(t-1) - x^*(t-1))^T \mathbf{D}(t-1)(x(t-1) - x^*(t-1)) \\ &\quad + \|x(t) - F(t)x(t-1)\|^2_{Q^{-1}} \end{aligned} \tag{3.3.23}$$

Expanding the parentheses in the last expression and selecting the quadratic and linear parts, with respect to the variable $x(t-1)$, one obtains

$$\begin{aligned} q(x(t-1)) &= x^T(t-1)(\mathbf{D}(t-1) + F^T(t)Q^{-1}F(t))x(t-1) \\ &\quad - 2x^T(t-1)(\mathbf{D}(t-1)x^*(t-1) + F^T(t)Q^{-1}x(t)) \\ &\quad + x^T(t)Q^{-1}x(t) + c \end{aligned} \tag{3.3.24}$$

where the scalar $c = x^{*T}(t-1)\mathbf{D}(t-1)x^*(t-1)$ does not affect the minimization result in the variable $x(t-1)$. Now given the vector $x(t)$, the argument of the minimum of the function $q(x(t-1))$ is defined by the expression

$$x(t-1, x(t)) = (\mathbf{D}(t-1) + F^T(t)Q^{-1}F(t))^{-1}(\mathbf{D}(t-1)x^*(t-1) + F^T(t)Q^{-1}x(t)) \tag{3.3.25}$$

Substituting (3.3.25) into (3.3.24) and taking into account (3.3.17), we obtain

$$\begin{aligned} v(t,x(t)) &= q(x(t-1,x(t))) + \|b(t) - H(t)x(t)\|^2_{R^{-1}} \\ &= -(\mathbf{D}(t-1)x^*(t-1) + F^T(t)Q^{-1}x(t))^T(\mathbf{D}(t-1) + F^T(t)Q^{-1}F(t))^{-1} \\ &\quad \times (\mathbf{D}(t-1)x^*(t-1) + F^T(t)Q^{-1}x(t)) + \|b(t) - H(t)x(t)\|^2_{R^{-1}} \\ &\quad + x^T(t)Q^{-1}x(t) + c \end{aligned} \tag{3.3.26}$$

Further, selecting quadratic and linear parts with respect to the variable $x(t)$ in the last expression, obtain

$$\begin{aligned} v(t,x(t)) &= x^T(t)\left[Q^{-1} - Q^{-1}F(t)(\mathbf{D}(t-1) + F^T(t)Q^{-1}F(t))^{-1}F^T(t)Q^{-1}\right. \\ &\quad \left.+H^T(t)R^{-1}H(t)\right]x(t) \\ &\quad - 2x^T(t)\left[Q^{-1}F(t)(\mathbf{D}(t-1) + F^T(t)Q^{-1}F(t))^{-1}\mathbf{D}(t-1)x^*(t-1)\right. \\ &\quad \left.+H^T(t)R^{-1}b(t)\right] + c \end{aligned} \tag{3.3.27}$$

where the constant c, which is different from those in the expression (3.3.24), does not affect further calculations and can be omitted. It follows from expression (3.3.27) that

$$\mathbf{D}(t) = Q^{-1} - Q^{-1}F(t)(\mathbf{D}(t-1) + F^T(t)Q^{-1}F(t))^{-1}F^T(t)Q^{-1} + H^T(t)R^{-1}H(t) \tag{3.3.28}$$

The function $v(t,x(t))$ achieves its minimum at the point

$$\begin{aligned} x^*(t) &= \mathbf{D}^{-1}(t)\left[Q^{-1}F(t)(\mathbf{D}(t-1) + F^T(t)Q^{-1}F(t))^{-1}\mathbf{D}(t-1)x^*(t-1)\right. \\ &\quad \left.+H^T(t)R^{-1}b(t)\right] \end{aligned} \tag{3.3.29}$$

Expressions (3.3.28) and (3.3.29) complete a description of the recursive scheme for updating the matrix $\mathbf{D}(t)$ and the vector of the optimal estimate $x^*(t)$, starting with initial data $\mathbf{D}(0) = C^{-1}$, $x(0) = \bar{x}(0)$.

In the following we present these equations in the more convenient form. Let us denote

$$\overline{\mathbf{D}}(t) = Q^{-1} - Q^{-1}F(t)(\mathbf{D}(t-1) + F^T(t)Q^{-1}F(t))^{-1}F^T(t)Q^{-1} \tag{3.3.30}$$

and

$$\bar{x}(t) = F(t)x^*(t-1) \tag{3.3.31}$$

and transform the expression for the matrix

$$Q^{-1}F(t)(\mathbf{D}(t-1) + F^T(t)Q^{-1}F(t))^{-1}\mathbf{D}(t-1)$$

in (3.3.29):

$$
\begin{aligned}
&Q^{-1}F(t)(\mathbf{D}(t-1)+F^T(t)Q^{-1}F(t))^{-1}\mathbf{D}(t-1)\\
&\quad= Q^{-1}F(t)-Q^{-1}F(t)+Q^{-1}F(t)(\mathbf{D}(t-1)+F^T(t)Q^{-1}F(t))^{-1}\mathbf{D}(t-1)\\
&\quad= Q^{-1}F(t)-Q^{-1}F(t)\left[I-(\mathbf{D}(t-1)+F^T(t)Q^{-1}F(t))^{-1}\mathbf{D}(t-1)\right]\\
&\quad= Q^{-1}F(t)-Q^{-1}F(t)(\mathbf{D}(t-1)+F^T(t)Q^{-1}F(t))^{-1}\\
&\qquad\times(\mathbf{D}(t-1)+F^T(t)Q^{-1}F(t)-\mathbf{D}(t-1))\\
&\quad= Q^{-1}F(t)-Q^{-1}F(t)(\mathbf{D}(t-1)+F^T(t)Q^{-1}F(t))^{-1}F^T(t)Q^{-1}F(t)=\overline{\mathbf{D}}(t)F(t)
\end{aligned}
\tag{3.3.32}
$$

Then, as follows from expressions (3.3.28)–(3.3.31), we obtain

$$
\mathbf{D}(t)=\overline{\mathbf{D}}(t)+H^T(t)R^{-1}H(t) \tag{3.3.33}
$$

and

$$
\begin{aligned}
x^*(t)&=\mathbf{D}^{-1}(t)\left[\overline{\mathbf{D}}(t)F(t)x^*(t-1)+H^T(t)R^{-1}b(t)\right]\\
&=\mathbf{D}^{-1}(t)\left[\overline{\mathbf{D}}(t)\overline{x}(t)+H^T(t)R^{-1}b(t)\right]\\
&=\mathbf{D}^{-1}(t)\left[(\mathbf{D}(t)-H^T(t)R^{-1}H(t))\overline{x}(t)+H^T(t)R^{-1}b(t)\right]
\end{aligned}
\tag{3.3.34}
$$

Expanding parentheses in the last expression, we arrive at the final expression

$$
x^*(t)=\overline{x}(t)+\mathbf{D}^{-1}(t)H^T(t)R^{-1}(b(t)-H(t)\overline{x}(t)) \tag{3.3.35}
$$

Expressions (3.3.30), (3.3.31), (3.3.33), and (3.3.35) give a more compact representation of the recursive estimation scheme. Each tth step of the recursive scheme can be presented as projection (or extrapolation) and correction (or update). Expressions (3.3.30) and (3.3.31) define projection; expressions (3.3.33) and (3.3.35) define correction. The last expression (3.3.35) is a sum of projected estimate $\overline{x}(t)$ and correction calculated based on the residual (or disagreement vector) between the new measurement $b(t)$ and the projected or expected measurement $H(t)\overline{x}(t)$.

Now, look at the expression (3.3.28) and recall expressions (A.3.59) to (A.3.61) for block-wise Cholesky decomposition. Let us construct the $2n \times 2n$ matrix

$$
G(t)=\left[\begin{array}{c|c}\mathbf{D}(t-1)+F^T(t)Q^{-1}F(t) & -F^T(t)Q^{-1}\\ \hline -Q^{-1}F(t) & Q^{-1}+H^T(t)R^{-1}H(t)\end{array}\right] \tag{3.3.36}
$$

and apply Cholesky decomposition to it:

$$
G(t)=\left[\begin{array}{c|c}L(t) & 0\\ \hline K(t) & M(t)\end{array}\right]\left[\begin{array}{c|c}L^T(t) & K^T(t)\\ \hline 0 & M^T(t)\end{array}\right] \tag{3.3.37}
$$

where the matrices $L(t)$ and $M(t)$ are lower triangle and the matrix $K(t)$ is dense. From (3.3.37) it directly follows that

$$\begin{aligned} \mathbf{D}(t-1) + F^T(t)Q^{-1}F_k &= L(t)L^T(t) \\ L(t)K^T(t) &= -F^T(t)Q^{-1} \\ Q^{-1} + H^T(t)R^{-1}H(t) - K(t)K^T(t) &= M(t)M^T(t) \end{aligned} \tag{3.3.38}$$

or

$$\begin{aligned} &\mathbf{D}(t-1) + F^T(t)Q^{-1}F(t) = L(t)L^T(t) \\ &K^T(t) = -\mathbf{F}_{L(t)}(F^T(t)Q^{-1}) \\ &Q^{-1} + H^T(t)R^{-1}H(t) - F^T(t)Q^{-1}(\mathbf{D}(t-1) + F^T(t)Q^{-1}F(t))^{-1}Q^{-1}F(t) \\ &\quad = M(t)M^T(t) \end{aligned} \tag{3.3.39}$$

where $\mathbf{F}_{L(t)}$ is the forward run operator defined in (3.1.22). It follows from the last expression that

$$\mathbf{D}(t) = M(t)M^T(t) \tag{3.3.40}$$

which is an alternative form for updating the matrix $\mathbf{D}(t)$. In other words, first the matrix $G(t)$ is calculated and decomposed according to (3.3.37), then the matrix $\mathbf{D}(t)$ is calculated according to the last expression. Now, look at either expression (3.3.29) or (3.3.35), both being equivalent. Taking into account (3.3.39), they can be presented as

$$\begin{aligned} x^*(t) &= (M^T(t))^{-1}(M(t))^{-1} \\ &\quad \times \left[Q^{-1}F(t)(L^T(t))^{-1}(L(t))^{-1}\mathbf{D}(t-1)x^*(t-1) + H^T(t)R^{-1}b(t)\right] \\ &= \mathbf{B}_{M(t)}(\mathbf{F}_{M(t)}(-K(t)\mathbf{F}_{L(t)}(\mathbf{D}(t-1)x^*(t-1)) + H^T(t)R^{-1}b(t))) \end{aligned} \tag{3.3.41}$$

and

$$\begin{aligned} x^*(t) &= \bar{x}(t) + (M^T(t))^{-1}(M(t))^{-1}H^T(t)R^{-1}(b(t) - H(t)\bar{x}(t)) \\ &= \bar{x}(t) + \mathbf{B}_{M(t)}\left(\mathbf{F}_{M(t)}(H^T(t)R^{-1}(b(t) - H(t)\bar{x}(t)))\right) \end{aligned} \tag{3.3.42}$$

respectively.

For the sake of simplicity suppose that the covariance matrices Q and R do not depend on t and are decomposed by Cholesky as

$$Q = L_Q L_Q^T \quad R = L_R L_R^T \tag{3.3.43}$$

Note that $F^T(t)Q^{-1}F(t) = \left(\mathbf{F}_{L_Q}F(t)\right)^T \mathbf{F}_{L_Q}F(t)$. Similar identities hold for other terms including R^{-1}.

TABLE 3.3.1 Algorithm 3: Dynamic Constraints

Extend the $n \times n$ symmetric matrix $\boldsymbol{D}(t)$ to $2n \times 2n$ adding zero matrix blocks	$\hat{\boldsymbol{D}}(t) = \left[\begin{array}{c\|c} \boldsymbol{D}(t) & 0 \\ \hline 0 & 0 \end{array}\right]$
Perform forward substitution calculations with Cholesky factors L_Q and L_R	$\overline{F}(t+1) = \boldsymbol{F}_{L_Q} F(t+1)$ $\overline{H}(t+1) = \boldsymbol{F}_{L_R} H(t+1)$ $\overline{b}(t+1) = \boldsymbol{F}_{L_R} b(t+1)$ $Q^{-1} = \boldsymbol{B}_{L_Q} \boldsymbol{F}_{L_Q}(I)$
Compute the updating matrix	$\Delta(t+1) = \left[\begin{array}{c\|c} \overline{F}^T(t+1)\overline{F}(t+1) & -(\boldsymbol{B}_{L_Q}\overline{F}(t+1))^T \\ \hline -\boldsymbol{B}_{L_Q}\overline{F}(t+1) & Q^{-1} + \overline{H}^T(t+1)\overline{H}(t+1) \end{array}\right]$
Compute the $2n \times 2n$ extended matrix	$G(t+1) = \hat{\boldsymbol{D}}(t) + \Delta(t+1)$ $= \left[\begin{array}{c\|c} \boldsymbol{D}(t) + \overline{F}^T(t+1)\overline{F}(t+1) & -(\boldsymbol{B}_{L_Q}(\overline{F}(t+1)))^T \\ \hline -\boldsymbol{B}_{L_Q}(\overline{F}(t+1)) & Q^{-1} + \overline{H}^T(t+1)\overline{H}(t+1) \end{array}\right]$
Cholesky decomposition of $G(t+1)$	$G(t+1) = \left[\begin{array}{c\|c} L(t+1) & 0 \\ \hline K(t+1) & M(t+1) \end{array}\right] \left[\begin{array}{c\|c} L^T(t+1) & K^T(t+1) \\ \hline 0 & M^T(t+1) \end{array}\right]$
Updated matrix $\boldsymbol{D}(t+1)$	$\boldsymbol{D}(t+1) = M(t+1)M^T(t+1)$
Projected estimate	$\overline{x}(t+1) = F(t+1)x^*(t)$
Residual	$\overline{r}(t+1) = \overline{b}(t+1) - \overline{H}(t+1)\overline{x}(t+1)$
Updated estimate	$x^*(t+1) = \overline{x}(t+1) + \boldsymbol{B}_{M(t+1)}\left(\boldsymbol{F}_{M(t+1)}(\overline{H}^T(t+1)\overline{r}(t+1))\right)$

Let us present the matrix update scheme $\boldsymbol{D}(t) \to \boldsymbol{D}(t+1)$ in a more visually clear form, dividing it into steps presented in Algorithm 3. Starting with $\boldsymbol{D}(0) = C^{-1}$, $x^*(0) = x_0$, and $t = 0$, the algorithm proceeds as described in Table 3.3.1.

When working not in real time, then after the last t'th measurement is received and the last optimal estimated $x^*(t') = \hat{x}(t')$ is obtained, one can make the "backward run" using expression (3.3.25) for $t = t', \ldots, 1$. The factorized representation of that expression is

$$\hat{x}(t) = (L^T(t+1))^{-1}(L(t+1))^{-1}\left[\boldsymbol{D}(t)x^*(t) + F^T(t+1)Q^{-1}\hat{x}(t+1)\right] \tag{3.3.44}$$

It is assumed that vectors $x^*(t+1)$ and matrices $L(t+1)$ and $\boldsymbol{D}(t)$ are calculated and stored in the memory during the "forward run".

Let us make some remarks about solvability of the problem (3.3.6) and (3.3.7). When formulating the problem of optimal estimation, the existence of a unique solution was implicitly assumed. It means that the quadratic part of the functional

$I(x(0), x(1), \ldots, x(t'), t')$ has a positive definite matrix. Positive semidefiniteness obviously follows from expression (3.3.6). The positive definiteness requirement imposes additional conditions on the matrices $F(t), H(t)$, and $D(0)$. Positive definiteness of the quadratic part of $I(x(0), x(1), \ldots, x(t'), t')$ is equivalent to positive definiteness of all matrices $\boldsymbol{D}(t)$ of the quadratic parts of the Bellman functions. Direct verification of positive definiteness of those matrices when executing the recursive estimation scheme also means checking the solvability of the estimation problem. A more careful analysis shows that the matrix $G(1)$ is positive definite if the matrix $D(0)$ is positive definite, which is positive definite due to the condition $D(0) = C^{-1}$ since the matrix C is positive definite. Positive definiteness of $G(1)$ implies positive definiteness of the matrix $D(1)$. Inductively continuing, one comes to the conclusion about positive definiteness of all matrices $D(t)$. Thus, the solvability of problem (3.3.7) for finite t' is a consequence of the positive definite choice of the initial condition $D(0)$. However, the conditionality of the matrix of the quadratic function $I(x(0), x(1), \ldots, x(t'), t')$ could worsen as t' increases. This would mean a "consequent loss" of the property of positive definiteness of the matrix $G(t)$ as t increases. Preservation of positive definiteness of $G(t)$ means so called "observability" of the system (3.3.1) with respect to the output (3.3.2). Observability analysis of the system (3.3.1) with respect to the output (3.3.2) in the time-varying case, when the matrices $F(t)$ and $H(t)$ are dependent on t, can only be numerically performed. Observability analysis is reduced to the estimation of the conditionality of matrices $\boldsymbol{D}(t)$ as they are sequentially calculated.

Now we will derive the recursive estimate $x^*(t+1)$ assuming that the whole set of measurements $b(1), \ldots, b(t')$ is available simultaneously and applying the batch processing. Consider the optimal estimation problem with respect to all variables $x(0), x(1), \ldots, x(t')$ simultaneously. Then, using the Cholesky decomposition of the large sparse matrix, obtain the same recursive estimation equations.

Represent the quadratic function (3.3.6) in the form (A.3.36) and denote

$$X^{t'} = \begin{pmatrix} x(1) \\ \vdots \\ x(t') \end{pmatrix} \tag{3.3.45}$$

Then, the solution of the problem (3.3.7) has the form

$$\hat{X}^{t'} = (\hat{x}(0, b(1), \ldots, b(t'))^T, \hat{x}(1, b(1), \ldots, b(t'))^T, \ldots, \hat{x}(t', b(1), \ldots, b(t'))^T)^T \tag{3.3.46}$$

and satisfies the system of linear equations

$$A^{t'} X = -l^{t'} \equiv r^{t'} \tag{3.3.47}$$

with a matrix $A^{t'}$ and the right-hand side schematically drawn in Figure 3.3.1.

The superscript t' will be omitted if it does not lead to misunderstanding. The matrix A has a block tridiagonal structure. It is called also a "band-like" matrix as the nonzero entries are grouped into a band located around a diagonal, forming a narrow band of a width $3n$ comparing with total dimension $t'n$ of the matrix when $t' \gg 3$. The right hand side vector has form $(C^{-1}\overline{x}(0), H^T(1)R^{-1}b(1), \ldots, H^T(t')R^{-1}b(t'))$.

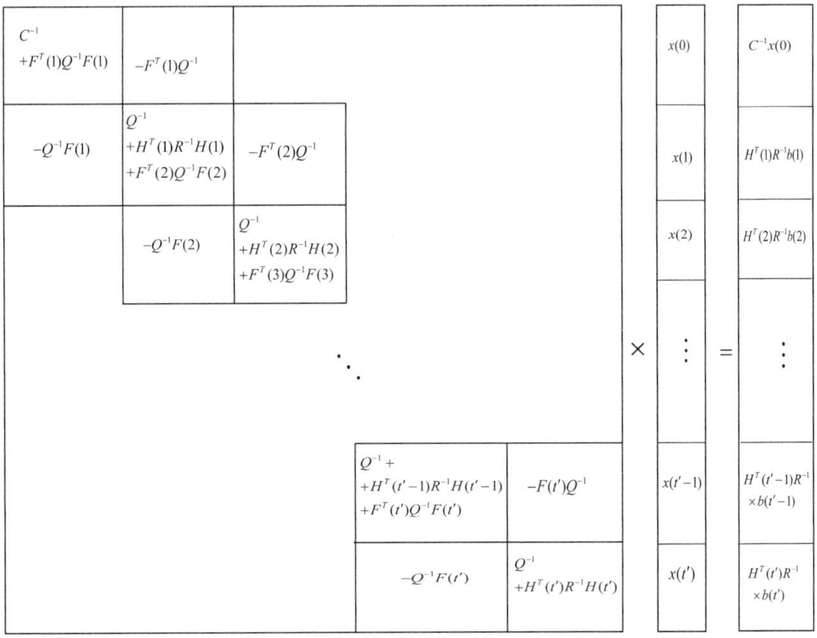

Figure 3.3.1 Schematic structure of the linear system (3.3.47).

The Cholesky decomposition of a band-like matrix preserves the band-like pattern of the lower triangle part of the matrix A. Define

$$A = LL^T \tag{3.3.48}$$

where the lower triangle matrix L has the form shown in Figure 3.3.2. Now sequentially apply formulas (A.3.59) to (A.3.61) to the decomposition (3.3.48). The off-diagonal block entries are expressed as $-Q^{-1}F(t)(L^T(t))^{-1}$ due to the expression (A.3.60) and expressions $-Q^{-1}F(t)$ for off-diagonal blocks of the matrix in Figure 3.3.1.

Let $L(t)$ be the block-diagonal entries of this matrix. In the following we prove that these are exactly the same matrices as described in the lower triangle representation (3.3.37) of matrix $G(t)$ defined in (3.3.36). Furthermore, as it follows from the expression for the first diagonal block of the matrix A, we have $C^{-1} + F^T(1)Q^{-1}F(1) = \mathbf{D}(0) + F^T(1)Q^{-1}F(1) = L(1)L^T(1)$, which corresponds to the expression (3.3.37) for $t = 1$.

For the second diagonal block entry of the matrix A we have

$$\begin{aligned}
&Q^{-1} + H^T(1)R^{-1}H(1) + F^T(2)Q^{-1}F(2) \\
&\quad = Q^{-1}F(1)(L^T(1))^{-1}(L(1))^{-1}F^T(1)Q^{-1} + L(2)L^T(2) \\
&\quad = Q^{-1}F(1)(\mathbf{D}(0) + F^T(1)Q^{-1}F(1))^{-1}F^T(1)Q^{-1} + L(2)L^T(2)
\end{aligned} \tag{3.3.49}$$

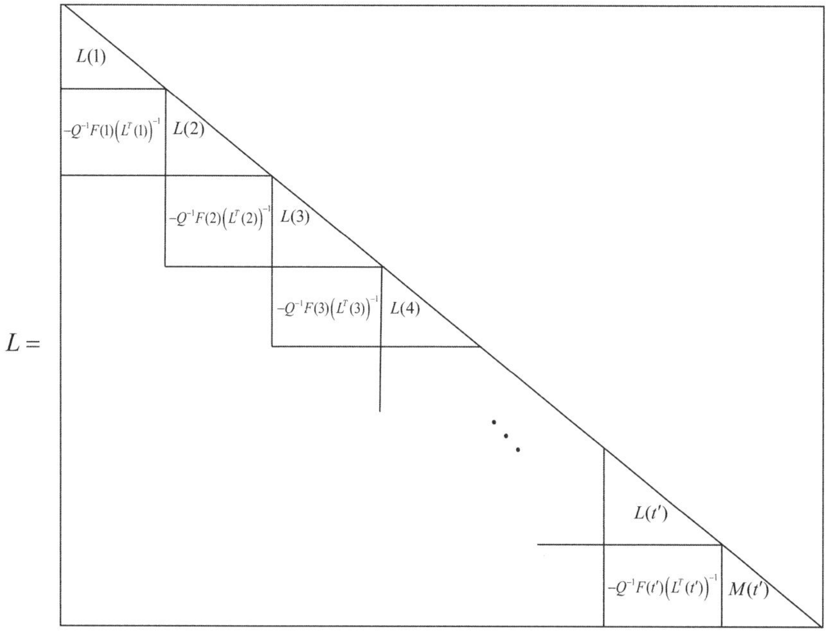

Figure 3.3.2 Band-like structure of the matrix in the Cholesky decomposition (3.3.48).

which together with (3.3.28) gives

$$L(2)L^T(2) = \mathbf{D}(1) + F^T(2)Q^{-1}F(2) \tag{3.3.50}$$

This expression corresponds to (3.3.39) for $t = 2$. Continuing until $t = t'$, we obtain that lower triangle matrices $L(t)$, defined earlier in (3.3.37) and used in the recursive scheme (3.3.39), are exactly diagonal block entries of the lower triangle matrix L in the decomposition (3.3.48). Next we find the expression for the last $t' + 1$ th diagonal block entry of the matrix L. We have

$$\begin{aligned} &Q^{-1} + H^T(t')R^{-1}H(t') \\ &\quad = Q^{-1}F(t')(\mathbf{D}(t'-1) + F^T(t')Q^{-1}F(t'))^{-1}F^T(t')Q^{-1} + M(t')M^T(t') \end{aligned} \tag{3.3.51}$$

which gives

$$M(t')M^T(t') = \mathbf{D}(t') \tag{3.3.52}$$

Therefore, the last diagonal block entry of the matrix L is the lower diagonal block of the Cholesky decomposition (3.3.37). We have proven the following:

Statement 3.3.3. The recursive scheme (3.3.37), (3.3.40) for calculation of matrices $L(t)$, $t = 1, \ldots, t'$, $M(t')$ is equivalent to Cholesky decomposition (3.3.48) of matrix A of the quadratic function (3.3.6).

Let us now apply the forward and backward run formulas (A.3.66), and (A.3.67) to solve the system (3.3.47),

$$\begin{aligned} LZ &= r \\ L^T X &= Z \end{aligned} \tag{3.3.53}$$

Let us take the vector Z as $Z = (z^T(0), z^T(1), \dots, z^T(t'))^T$ and the right-hand side vector r as $r = (r^T(0), r^T(1), \dots, r^T(t'))^T$. Taking into account the structure of the matrix L shown in Figure 3.3.2, and applying block-wise forward run, we can present the solution of the first half of the system (3.3.53) with respect to Z in the form

$$\begin{aligned} z(0) &= (L(1))^{-1} r(0) \\ z(1) &= (L(2))^{-1}(r(1) + Q^{-1}F(1)(L^T(1))^{-1}z(0)) \\ &\dots \\ z(t'-1) &= (L(t'))^{-1}(r(t'-1) + Q^{-1}F(t'-1)(L^T(t'-1))^{-1}z(t'-2)) \\ z(t') &= (M(t'))^{-1}(r(t') + Q^{-1}F(t')(L^T(t'))^{-1}z(t'-1)) \end{aligned} \tag{3.3.54}$$

Then, applying the block-wise backward substitution, the solution of the second half of the system (3.3.53) with respect to X becomes

$$\begin{aligned} \hat{x}(t') &= (M^T(t'))^{-1} z(t') \\ \hat{x}(t'-1) &= (L^T(t'))^{-1}(z(t'-1) + (L(t'))^{-1}F^T(t')Q^{-1}\hat{x}(t')) \\ &\dots \\ \hat{x}(0) &= (L^T(1))^{-1}(z(0) + (L(1))^{-1}F^T(1)Q^{-1}\hat{x}(1)) \end{aligned} \tag{3.3.55}$$

The sequential application of expressions (3.3.54) and (3.3.55) gives the optimal estimate $\hat{x}(t')$, $t = 0, 1, \dots, t'$. Moreover, it was established earlier that $\hat{x}(t') = x^*(t')$. If we are interested only in the estimate $x^*(t')$, then it is sufficient to make only one first step of the backward run after the forward run (3.3.54) has been carried out, i.e., $x^*(t') = (M^T(t'))^{-1}z(t')$. Consider two problems of (3.3.7) formulated with $t'-1$ and t' measurements. Let $A^{t'-1}$ and $A^{t'}$ be correspondent matrices. Let the Cholesky decomposition

$$A^{t'-1} = L^{t'-1}(L^{t'-1})^T \tag{3.3.56}$$

be already calculated. Then, for calculation of the decomposition

$$A^{t'} = L^{t'}(L^{t'})^T \tag{3.3.57}$$

it is sufficient to calculate the block entries $L(t')$, $K(t')$, and $M(t')$ of the decomposition (3.3.48) (see Figure 3.3.2); the superscript t' is omitted because all other block entries of matrices $L \equiv L^{t'}$ and $L^{t'-1}$ coincide. In order to calculate the block $L(t')$ let us use the decomposition

$$\begin{aligned} L(t')L^T(t') &= \mathbf{D}(t'-1) + F^T(t')Q^{-1}F(t') \\ &= M^{t'-1}(t'-1)(M^{t'-1}(t'-1))^T + F^T(t')Q^{-1}F(t') \end{aligned} \tag{3.3.58}$$

Then we obtain

$$K(t') = -Q^{-1}F(t')(L^T(t'))^{-1} \tag{3.3.59}$$

For calculating the matrix $M(t')$ we use the identity for the last lower diagonal block of identity (3.3.48):

$$M(t')M(t')^T + Q^{-1}F(t')\left(L(t')L^T(t')\right)^{-1}F^T(t')Q^{-1} = Q^{-1} + H^T(t')R^{-1}H(t') \tag{3.3.60}$$

The Cholesky decomposition

$$M(t')M^T(t') = Q^{-1} + H^T(t')R^{-1}H(t') - K(t')K^T(t') \tag{3.3.61}$$

completes the calculations.

Now, we establish the connections between estimates $x^*(t')$ and $x^*(t'-1)$ which were calculated at the first step of the recursive schemes (3.3.55) when solving problem (3.3.7) for measurements collected at time instants $1, \ldots, t'$, and solving problem (3.3.7) for measurements collected at time instants $1, \ldots, t'-1$. Also, we need to establish connections with the recursive scheme (3.3.41). Let $z^*(t'-1)$ and $z^*(t')$ be vectors calculated at the last step (step $t'-1$ in the case of $t'-1$ measurements and step t' in the case of t' measurements) of the scheme (3.3.54). It follows from the first expression of (3.3.55) that

$$\begin{aligned} x^*(t') &= (M^{t'}(t')^T)^{-1}z^*(t') \\ x^*(t'-1) &= (M^{t'-1}(t'-1)^T)^{-1}z^*(t'-1) \end{aligned} \tag{3.3.62}$$

We take note of the fact that according to (3.3.54) the vector $z(t'-1)$, obtained at step $t'-1$ of the numerical scheme for the case of t' measurements, is connected with $z^*(t'-1)$ by the relationship

$$z(t'-1) = (L(t'))^{-1}M^{t'-1}(t'-1)z^*(t'-1) \tag{3.3.63}$$

Combination of (3.3.62) and (3.3.63) and the first expression of (3.3.55) gives

$$\begin{aligned} x^*(t') &= (M^T(t'))^{-1}(M(t'))^{-1}(H^T(t')R^{-1}b(t') + Q^{-1}F(t')(L^T(t'))^{-1}z(t'-1)) \\ &= (M^T(t'))^{-1}(M(t'))^{-1}(H^T(t')R^{-1}b(t') \\ &\quad + Q^{-1}F(t')(L^T(t'))^{-1}(L(t'))^{-1}M^{t'-1}(t'-1)z^*(t'-1)) \end{aligned}$$

$$= (M^T(t'))^{-1}(M(t'))^{-1}(H^T(t')R^{-1}b(t')$$
$$+ Q^{-1}F(t')(L^T(t'))^{-1}(L(t'))^{-1}M^{t'-1}(t'-1)M^{t'-1}(t'-1)^T x^*(t'-1))$$
$$= (M^T(t'))^{-1}(M(t'))^{-1}(H^T(t')R^{-1}b(t') - K(t')(L(t'))^{-1}\mathbf{D}(t'-1)x^*(t'-1)) \quad (3.3.64)$$

which coincides with (3.3.41) for $t = t'$. Therefore, we came to the following important statement:

Statement 3.3.4. Sequential computation of block entries of the Cholesky decompositions of the large-scale sparse matrix of the problem (3.3.6) and (3.3.7) gives the same results and the same formulas (3.3.41) (or (3.3.42)) and (3.3.40) as obtained for the recursive optimal estimation when using the Bellman functions approach.

In this section we derived the recursive least-squares algorithm to estimate parameters subject to dynamic constraints. Using the Cholesky decomposition for the band-like matrix, we established a connection between recursive processing and batch processing. We proved that the recursive Algorithm 3 can be obtained as Cholesky decomposition of a continuously increasing band-like matrix and continuously performed forward solution, followed by only a single step backward solution. Performing the full backward solution is equivalent to batch processing.

3.4 STATIC PARAMETERS AND DYNAMIC CONSTRAINTS

The estimation problem is generalized by combining the cases described in Sections 3.2 and 3.3. We consider the dynamic system (3.3.1), which connects two sequential state vectors $x(t) = F(t)x(t-1) + \xi(t)$, and an extension of the measurement model (3.3.2),

$$b(t) = H(t)x(t) + W(t)y + \varepsilon(t) \quad (3.4.1)$$

The extended model incorporates time-varying parameter $x(t)$ and a time-invariant p-dimensional parameter y. The matrix $W(t)$ has dimensions $m(t)$ by p. Actually, we can formally assume a "very slow" dynamics for parameter y, and write dynamics equations as

$$y(t) = y(t-1) + \xi'(t) \quad (3.4.2)$$

where the error $\xi'(t)$ is zero. The last requirement can be taken into account in the form $E(\xi'(t)\xi'(t)^T) \to 0$, i.e., its covariance matrix must take smallest possible value but not destroy the computational stability. However, this approach meets obvious objections when implementing in practice. The numerical scheme described in the previous section assumes the inverse of the covariance matrix, which is not invertible in this case. We can write

$$E\begin{pmatrix} \xi(t)\xi^T(t) & \xi(t)\xi'^T(t) \\ \xi'(t)\xi^T(t) & \xi'(t)\xi'^T(t) \end{pmatrix} = \begin{pmatrix} Q & 0 \\ 0 & 0 \end{pmatrix} \quad (3.4.3)$$

A possible solution would be the introduction of small error $\xi'(t)$ with "near zero" covariance matrix. However, instead of introducing this kind of modification and dealing with the necessity to prove correctness and convergence as diagonal entries tend to zero, we derive a special form of the estimation algorithm for the problems (3.3.1) and (3.4.1). We consider the parameter y as constant, not "slow varying." The uncertainty of the initial data is described by conditions (3.3.4) and (3.3.5). The estimation problem consists of recovering the trajectory $\{x(t)\}$ and the constant vector y using sequential measurements $\{b(t)\}$ and models (3.3.1), and (3.4.1) and specifications (3.3.3) to (3.3.5). Least-squares criterion minimizes the quadratic function of the variables $x(t)$, $t = 0, \ldots, t'$ and y defined by the expression

$$I\big(x(0), x(1), \ldots, x(t'), y, t'\big) = \sum_{t=1}^{t'} \left\|x(t) - F(t)x(t-1)\right\|^2_{Q^{-1}} + \sum_{t=1}^{t'} \left\|b(t) - H(t)x(t) - W(t)y\right\|^2_{R^{-1}} \left\|x(0) - x_0\right\|^2_{C^{-1}} \tag{3.4.4}$$

where t' is the number of accumulated measurements. That is, the weighted sum of squared residuals of the relationships (3.3.1), (3.4.1), and (3.3.4) is minimized. The best estimation $\{\hat{x}(t, b(1), \ldots, b(t'))\}, t = 0, \ldots, t'$, and $\hat{y}(b(1), \ldots, b(t'))$ minimizes the criterion (3.4.4), solving the problem

$$I_{\min} = \min_{x(0),x(1),\ldots,x(t'),y} I(x(0), x(1), \ldots, x(t'), y, t') \tag{3.4.5}$$

which generalizes the early considered problems (3.1.5) and (3.3.7).

Notation identifying dependency of the optimal estimate on the measurements $(b(1), \ldots, b(t'))$ will be omitted if it does not lead to misunderstanding. As done in the previous section, denote $x^*(t) = \hat{x}(t, b(1), \ldots, b(t))$ and $y^*(t) = \hat{y}(b(1), \ldots, b(t))$. Let us derive the recursive relationships for obtaining estimates $x^*(t)$ and $y^*(t)$ on the base of earlier obtained estimates $x^*(t-1)$, $y^*(t-1)$, and the measurement $b(t)$. Rewrite the problem (3.4.5) in the equivalent form

$$I_{\min} = \min_{y}[\min_{x(0),x(1),\ldots,x(t')} I(x(0), x(1), \ldots, x(t'), y, t')] \tag{3.4.6}$$

and consider y fixed, the internal minimization problem in (3.4.6) over the variables $x(0), x(1), \ldots, x(t'), t'$ is

$$\bar{I}_{\min}(y) = \min_{x(0),x(1),\ldots,x(t')} I(x(0), x(1), \ldots, x(t'), y, t') \tag{3.4.7}$$

Changing the order of minimization operations in the last expression, one arrives at the following equivalent formulation of problem (3.4.7):

$$\bar{I}_{\min}(y) = \min_{x(t')} \left[\min_{x(0),x(1),\ldots,x(t'-1)} I(x(0), x(1), \ldots, x(t'), y, t') \right]$$

$$= \min_{x(t')} \left[\min_{x(0),x(1),\ldots,x(t'-1)} \left(\sum_{t=1}^{t'} \left\| x\,(t) - F(t)x(t-1) \right\|_{Q^{-1}}^{2} \right. \right.$$

$$\left. \left. + \sum_{t=1}^{t'} \left\| b\,(t) - H(t)x(t) - W(t)y \right\|_{R^{-1}}^{2} + \left\| x\,(0) - x_0 \right\|_{C^{-1}}^{2} \right) \right] \qquad (3.4.8)$$

Then denote

$$v(t, x(t'), y) = \min_{x(0),x(1),\ldots,x(t'-1)} \left(\sum_{t=1}^{t'} \left\| x\,(t) - F(t)x(t-1) \right\|_{Q^{-1}}^{2} \right.$$

$$\left. + \sum_{t=1}^{t'} \left\| b\,(t) - H(t)x(t) - W(t)y \right\|_{R^{-1}}^{2} + \left\| x\,(0) - x_0 \right\|_{C^{-1}}^{2} \right) \qquad (3.4.9)$$

and problem (3.4.8) takes the following form:

$$\overline{I}_{\min}(y) = \min_{x(t')} v(t', x(t'), y) \qquad (3.4.10)$$

The last term $\|b(t') - H(t')x(t') - W(t')y\|_{R^{-1}}^{2}$ in the second summation operation of expression (3.4.9) does not depend on the variables $x(0), \ldots, x(t'-1)$ and, therefore, the expression can be rewritten as

$$v(t', x(t'), y) = \min_{x(0),x(1),\ldots,x(t'-1)} \left(\sum_{t=1}^{t'} \left\| x\,(t) - F(t)x(t-1) \right\|_{Q^{-1}}^{2} \right.$$

$$\left. + \sum_{t=1}^{t'-1} \left\| b\,(t) - H(t)x(t) - W(t)y \right\|_{R^{-1}}^{2} + \left\| x(0) - x_0 \right\|_{C^{-1}}^{2} \right)$$

$$\left\| b(t') - H(t')x(t') - W(t')y \right\|_{R^{-1}}^{2} \qquad (3.4.11)$$

Repeating the reasoning of the previous section, one arrives at the sequence of the Bellman functions, all of which also depend on the variable y (excluding $v(0, x(0))$, which depending only on $x(0)$),

$$v(0, x(0)) = \|x(0) - x_0\|_{C^{-1}}^{2}$$

$$v(1, x(1), y) = \min_{x(0)} \left(v(0, x(t)) + \|x(1) - F(1)x(0)\|_{Q^{-1}}^{2} \right)$$

$$+ \|b(1) - H(1)x(1) - W(1)y\|_{R^{-1}}^{2}$$

$$v(2, x(2), y) = \min_{x(1)} \left(v(1, x(1), y) + \|x(2) - F(2)x(1)\|_{Q^{-1}}^{2} \right)$$

$$+ \|b(2) - H(2)x(2) - W(2)y\|_{R^{-1}}^{2}$$

$$\cdots$$

$$v(t, x(t), y) = \min_{x(t-1)} \left(v(t-1, x(t-1), y) + \|x(t) - F(t)x(t-1)\|^2_{Q^{-1}} \right) + \|b(t) - H(t)x(t) - W(t)y\|^2_{R^{-1}}$$

$$\ldots$$

$$v(t', x(t'), y) = \min_{x(t'-1)} \left(v(t'-1, x(t'-1), y) + \|x(t') - F(t')x(t'-1)\|^2_{Q^{-1}} \right) + \|b(t') - H(t')x(t') - W(t')y\|^2_{R^{-1}} \quad (3.4.12)$$

The value of the estimate $x(t-1)$, which minimizes the first term of the function $v(t, x(t), y)$, given the variables $x(t)$ and y, is denoted by $x(t-1, x(t), y)$ and called the *conditionally optimal estimate*. In order to find the whole set of optimal estimates $\hat{x}(0), \hat{x}(1), \ldots, \hat{x}(t'), \hat{y}$ one needs to find the minimum of the function $v(t', x(t'), y)$ reached at the point $\hat{x}(t'), \hat{y}$, and exploit the recursive expressions

$$\hat{x}(t-1) = x(t-1, \hat{x}(t), \hat{y}) \quad t = t', t'-1, \cdots, 1 \quad (3.4.13)$$

These relationships allow for sequentially obtaining the components of the optimal estimate in the reverse order. Note again that the first function $v(t', x(t'), y)$ is minimized over two variables $x(t')$ and y, resulting in the estimates $\hat{x}(t')$ and $\hat{y}$. The optimal estimate $\hat{y}$ is then substituted into all expressions (3.4.13). The following statement summarizes the construction described above:

Statement 3.4.1. The solution of the optimal estimation problem (3.4.5) is equivalent to minimization of the function $v(t', x(t'), y)$ over the variables $x(t'), y$ resulting in optimal estimates $\hat{x}(t'), \hat{y}$. The function $v(t', x(t'), y)$ is obtained using recursive relationships (3.4.12). All other components of the sequence $\hat{x}(t'-1), \hat{x}(t'-2), \ldots, \hat{x}(0)$ of optimal estimates are obtained according to relationships (3.4.13).

What is described above is "batch processing" applied to the whole set of measurements $b(1), \ldots, b(t')$ after they have been received. When working in real time, optimal estimate of the sequence of state vectors must be obtained sequentially as the measurements become available, and the total number of measurements t' is unknown beforehand. Therefore, of practical importance are recursive relationships for calculation of the next estimate $x^*(t), y^*(t)$ using the previous estimate $x^*(t'-1), y^*(t'-1)$ and the measurement $b(t)$. After the measurement $b(t)$ has been received and the new estimate $x^*(t')$, $y^*(t'-1)$ has been calculated, earlier obtained estimates $x(t-1), x(t-2), \ldots, x(0)$ can be updated using the "backward run" relationships (3.4.13). However, that cannot be carried out when working in real time. Only some limited backward run depth T is available. In other words, the components $x(t), x(t-1), \ldots, x(t-T+1)$ are updated. This way of recursive processing can be considered as "partial" batch processing of depth T and can be performed in real time.

The depth T is a constant. If $T = 1$, then we have the conventional recursive processing. If T is not constant and $T \equiv t$, then we have a batch processing implemented as a forward run (calculation of Bellman functions (3.4.12)) and a backward run (3.4.13).

Now we can derive the numerical scheme for recursive processing. Inductive application of the expression (A.3.38) to relationships (3.4.12) gives the following result:

Statement 3.4.2. The Bellman functions $v(t, x(t), y)$, defined by relationships (3.4.12) are quadratic functions of their variables. Conditionally optimal estimates $x(t-1, x(t), y)$ are linearly dependent on $x(t)$ and y.

In real-time operations, the optimal estimate $x^*(t)$, y^* must be obtained right after receiving the measurement $b(t)$. Having the measurement $b(t)$, one can use the tth step of the recursive scheme (3.4.12) for constructing the function $v(t, x(t), y)$. The measurement $b(t)$ is the last one received up to time instant t. We are looking for the optimal estimate for $x(t)$ and y at the tth step having available measurements $b(1), \ldots, b(t)$. Setting $t' = t$ in expression (3.4.10), we see that the optimal estimate $x^*(t)$, y^* minimizes the function $v(t, x(t), y)$ which is quadratic due to Statement 3.4.2. According to expression (A.3.38) we have

$$v(t, x(t), y) = \begin{pmatrix} x(t) - x^*(t) \\ y - y^*(t) \end{pmatrix}^T \begin{bmatrix} \mathbf{D}^{xx}(t) & \mathbf{D}^{xy}(t) \\ \mathbf{D}^{yx}(t) & \mathbf{D}^{yy}(t) \end{bmatrix} \begin{pmatrix} x(t) - x^*(t) \\ y - y^*(t) \end{pmatrix} \tag{3.4.14}$$

where the matrix

$$\mathbf{D}(t) = \begin{bmatrix} \mathbf{D}^{xx}(t) & \mathbf{D}^{xy}(t) \\ \mathbf{D}^{yx}(t) & \mathbf{D}^{yy}(t) \end{bmatrix} \tag{3.4.15}$$

is symmetric ($\mathbf{D}^{yx}(t) = \mathbf{D}^{xy}(t)^T$) and positive definite. The equality (3.4.14) is valid up to a constant not affecting the minimum point.

Our goal now is to derive a numerical scheme for calculating the matrix $\mathbf{D}(t)$ and the optimal estimate $x^*(t)$, $y^*(t)$, based on the matrix $\mathbf{D}(t-1)$, the estimates $x^*(t-1)$, $y^*(t-1)$, and the measurement vector $b(t)$. The initial conditions are

$$\mathbf{D}(0) = \begin{bmatrix} C^{-1} & 0 \\ 0 & 0 \end{bmatrix} \tag{3.4.16}$$

$$x^*(0) = x_0 \tag{3.4.17}$$

$$y^*(0) = 0 \tag{3.4.18}$$

The matrix $\mathbf{D}(0)$ is not positive definite which does not contradict our construction since it will not be inversed. Assuming that we know the function

$$v(t-1, x(t-1), y) = \begin{pmatrix} x(t-1) - x^*(t-1) \\ y - y^*(t-1) \end{pmatrix}^T \begin{bmatrix} \mathbf{D}^{xx}(t-1) & \mathbf{D}^{xy}(t-1) \\ \mathbf{D}^{yx}(t-1) & \mathbf{D}^{yy}(t-1) \end{bmatrix} \times \begin{pmatrix} x(t-1) - x^*(t-1) \\ y - y^*(t-1) \end{pmatrix} \tag{3.4.19}$$

we now derive expressions for the function (3.4.14). Consider the function of variables $x(t-1), x(t)$, y, which is subject to minimization over the variable $x(t-1)$ in expression (3.4.12), and denote it by $q(x(t-1), x(t), y)$:

$$q(x(t-1), x(t), y) = v(t-1, x(t-1), y) + \|x(t) - F(t)x(t-1)\|^2_{Q^{-1}}$$
$$= \begin{pmatrix} x(t-1) - x^*(t-1) \\ y - y^*(t-1) \end{pmatrix}^T \begin{bmatrix} \mathbf{D}^{xx}(t-1) & \mathbf{D}^{xy}(t-1) \\ \mathbf{D}^{yx}(t-1) & \mathbf{D}^{yy}(t-1) \end{bmatrix}$$

$$
\times \begin{pmatrix} x(t-1) - x^*(t-1) \\ y - y^*(t-1) \end{pmatrix} + \|x(t) - F(t)x(t-1)\|^2_{Q^{-1}}
$$

$$
= \begin{pmatrix} x(t-1) \\ 0 \\ y \end{pmatrix}^T \begin{bmatrix} \mathbf{D}^{xx}(t-1) & 0 & \mathbf{D}^{xy}(t-1) \\ 0 & 0 & 0 \\ \mathbf{D}^{yx}(t-1) & 0 & \mathbf{D}^{yy}(t-1) \end{bmatrix} \begin{pmatrix} x(t-1) \\ 0 \\ y \end{pmatrix}
$$

$$
- 2 \begin{pmatrix} x(t-1) \\ 0 \\ y \end{pmatrix}^T \begin{bmatrix} \mathbf{D}^{xx}(t-1) & 0 & \mathbf{D}^{xy}(t-1) \\ 0 & 0 & 0 \\ \mathbf{D}^{yx}(t-1) & 0 & \mathbf{D}^{yy}(t-1) \end{bmatrix} \begin{pmatrix} x^*(t-1) \\ 0 \\ y^*(t-1) \end{pmatrix}
$$

$$
+ \begin{pmatrix} x(t-1) \\ x(t) \\ 0 \end{pmatrix}^T \begin{bmatrix} F^T(t)Q^{-1}F(t) & -F^T(t)Q^{-1} & 0 \\ -Q^{-1}F(t) & Q^{-1} & 0 \\ 0 & 0 & 0 \end{bmatrix} \begin{pmatrix} x(t-1) \\ x(t) \\ 0 \end{pmatrix} + c \tag{3.4.20}
$$

Making use of the expression (A.3.83) for the function of partial minimum, we obtain the expression for the function

$$
\begin{aligned}
&\min_{x(t-1)} q(x(t-1), x(t), y) \\
&= \begin{pmatrix} x(t) \\ \hline y \end{pmatrix}^T \left\{ \left[\begin{array}{c|c} Q^{-1} & 0 \\ \hline 0 & \mathbf{D}^{yy}(t-1) \end{array} \right] \right. \\
&\quad - \begin{pmatrix} -Q^{-1}F(t) \\ \hline \mathbf{D}^{xy}(t-1)^T \end{pmatrix} (\mathbf{D}^{xx}(t-1) + F^T(t)Q^{-1}F(t))^{-1} \\
&\quad \times \left. \left(-F^T(t)Q^{-1} \;\middle|\; \mathbf{D}^{xy}(t-1) \right) \right\} \begin{pmatrix} x(t) \\ \hline y \end{pmatrix} \\
&\quad + 2 \begin{pmatrix} x(t) \\ \hline y \end{pmatrix}^T \left\{ - \begin{pmatrix} 0 \\ \mathbf{D}^{xy}(t-1)^T x^*(t-1) + \mathbf{D}^{yy}(t-1) y^*(t-1) \end{pmatrix} \right. \\
&\quad + \begin{pmatrix} -Q^{-1}F(t) \\ \hline \mathbf{D}^{xy}(t-1)^T \end{pmatrix} (\mathbf{D}^{xx}(t-1) + F^T(t)Q^{-1}F(t))^{-1} \\
&\quad \times \left. \left(\mathbf{D}^{xx}(t-1) x^*(t-1) + \mathbf{D}^{xy}(t-1) y^*(t-1) \right) \right\}
\end{aligned} \tag{3.4.21}
$$

and finally, according to expression (3.4.12),

$$
\begin{aligned}
v(t, x(t), y) = &\begin{pmatrix} x(t) \\ \hline y \end{pmatrix}^T \left\{ \left[\begin{array}{c|c} Q^{-1} & 0 \\ \hline 0 & \mathbf{D}^{yy}(t-1) \end{array} \right] \right. \\
&- \begin{pmatrix} -Q^{-1}F(t) \\ \hline \mathbf{D}^{xy}(t-1)^T \end{pmatrix} (\mathbf{D}^{xx}_{k-1} + F^T(t)Q^{-1}F(t))^{-1} \\
&\times \left. \left(-F^T(t)Q^{-1} \;\middle|\; \mathbf{D}^{xy}(t-1) \right) \right\} \begin{pmatrix} x(t) \\ \hline y \end{pmatrix}
\end{aligned}
$$

$$+\begin{pmatrix} x(t) \\ \hline y \end{pmatrix}^T \left[\begin{array}{c|c} H^T(t)R^{-1}H(t) & H^T(t)R^{-1}W(t) \\ \hline W^T(t)R^{-1}H(t) & W^T(t)R^{-1}W(t) \end{array}\right] \begin{pmatrix} x(t) \\ \hline y \end{pmatrix}$$

$$+2\begin{pmatrix} x(t) \\ \hline y \end{pmatrix}^T \left\{ -\begin{pmatrix} 0 \\ \hline \boldsymbol{D}^{xy}(t-1)^T x^*(t-1) + \boldsymbol{D}^{yy}(t-1)y^*(t-1) \end{pmatrix} \right.$$

$$+\begin{pmatrix} -Q^{-1}F(t) \\ \hline \boldsymbol{D}^{yy}(t-1)^T \end{pmatrix} (\boldsymbol{D}^{xx}(t-1) + F^T(t)Q^{-1}F(t))^{-1}$$

$$\left. \times \left(\boldsymbol{D}^{xx}(t-1)x^*(t-1) + \boldsymbol{D}^{xy}(t-1)y^*(t-1)\right) \right\}$$

$$-2\begin{pmatrix} x(t) \\ \hline y \end{pmatrix}^T \begin{pmatrix} H^T(t)R^{-1}b(t) \\ \hline W^T(t)R^{-1}b(t) \end{pmatrix} \tag{3.4.22}$$

In expression (3.4.22) the linear and quadratic parts of the vector variable $(x(t), y)$ are now extracted. According to (3.4.22) and (3.4.14) we have

$$\boldsymbol{D}(t) = \left[\begin{array}{c|c} Q^{-1} - Q^{-1}F(t)\widetilde{\boldsymbol{D}}^{-1}(t-1)F^T(t)Q^{-1} & Q^{-1}F(t)\widetilde{\boldsymbol{D}}^{-1}(t-1)\boldsymbol{D}^{xy}(t-1) \\ \hline \boldsymbol{D}^{xy}(t-1)^T\widetilde{\boldsymbol{D}}^{-1}(t-1)F^T(t)Q^{-1} & \begin{array}{r}\boldsymbol{D}^{yy}(t-1) - \boldsymbol{D}^{xy}(t-1)^T \\ \times\widetilde{\boldsymbol{D}}^{-1}(t-1)\boldsymbol{D}^{xy}(t-1)\end{array} \end{array}\right]$$

$$+\left[\begin{array}{c|c} H^T(t)R^{-1}H(t) & H^T(t)R^{-1}W(t) \\ \hline W^T(t)R^{-1}H(t) & W^T(t)R^{-1}W(t) \end{array}\right] \tag{3.4.23}$$

where

$$\widetilde{\boldsymbol{D}}(t-1) = \boldsymbol{D}^{xx}(t-1) + F^T(t)Q^{-1}F(t) \tag{3.4.24}$$

and

$$\begin{pmatrix} x^*(t) \\ y^*(t) \end{pmatrix} = \boldsymbol{D}^{-1}(t)Y(t) \tag{3.4.25}$$

and

$$Y(t) = \begin{pmatrix} Q^{-1}F(t)\widetilde{\boldsymbol{D}}^{-1}(t-1)(\boldsymbol{D}^{xx}(t-1)x^*(t-1) + \boldsymbol{D}^{xy}(t-1)y^*(t-1)) \\ +H^T(t)R^{-1}b(t) \\ \hline Y^y(t) \end{pmatrix},$$

$$Y^y(t) = \boldsymbol{D}^{xy}(t-1)^T x^*(t-1) + \boldsymbol{D}^{xy}(t-1)y^*(t-1)$$

$$-\boldsymbol{D}^{xy}(t-1)^T\widetilde{\boldsymbol{D}}^{-1}(t-1)(\boldsymbol{D}^{xx}(t-1)x^*(t-1) + \boldsymbol{D}^{xy}(t-1)y^*(t-1))$$

$$+W^T(t)R^{-1}b(t) \tag{3.4.26}$$

Let

$$\overline{\boldsymbol{D}}(t) = \left[\begin{array}{c|c} Q^{-1} - Q^{-1}F(t)\widetilde{\boldsymbol{D}}^{-1}(t-1)F^T(t)Q^{-1} & Q^{-1}F(t)\widetilde{\boldsymbol{D}}^{-1}(t-1)\boldsymbol{D}^{xy}(t-1) \\ \hline \boldsymbol{D}^{xy}(t-1)^T\widetilde{\boldsymbol{D}}^{-1}(t-1)F^T(t)Q^{-1} & \begin{array}{r}\boldsymbol{D}^{yy}(t-1) - \boldsymbol{D}^{xy}(t-1)^T \\ \times\widetilde{\boldsymbol{D}}^{-1}(t-1)\boldsymbol{D}^{xy}(t-1)\end{array} \end{array}\right]$$

$$\equiv \begin{bmatrix} \overline{\boldsymbol{D}}^{xx}(t) & \overline{\boldsymbol{D}}^{xy}(t) \\ \overline{\boldsymbol{D}}^{yx}(t) & \overline{\boldsymbol{D}}^{yy}(t) \end{bmatrix} \tag{3.4.27}$$

and

$$\bar{x}(t) = F(t)x^*(t-1) \tag{3.4.28}$$

then the expression (3.4.23) takes the form

$$\boldsymbol{D}(t) = \bar{\boldsymbol{D}}(t) + \left[\begin{array}{c|c} H^T(t)R^{-1}H(t) & H^T(t)R^{-1}W(t) \\ \hline W^T(t)R^{-1}H(t) & W^T(t)R^{-1}W(t) \end{array}\right] \tag{3.4.29}$$

Transform terms in the expressions (3.4.26) as

$$\begin{aligned}
&Q^{-1}F(t)\tilde{\boldsymbol{D}}^{-1}(t-1)\boldsymbol{D}^{xx}(t-1)x^*(t-1)\\
&\quad = \left[Q^{-1}F(t) - Q^{-1}F(t) + Q^{-1}F(t)(\boldsymbol{D}^{xx}(t-1) + F^T(t)Q^{-1}F(t))^{-1}\boldsymbol{D}^{xx}(t-1)\right]\\
&\qquad \times x^*(t-1)\\
&\quad = \left[Q^{-1}F(t) - Q^{-1}F(t) + Q^{-1}F(t)(\boldsymbol{D}^{xx}(t-1) + F^T(t)Q^{-1}F(t))^{-1}\right.\\
&\qquad \left.\times\, (\boldsymbol{D}^{xx}(t-1) + F^T(t)Q^{-1}F(t) - F^T(t)Q^{-1}F(t))\right] x^*(t-1)\\
&\quad = Q^{-1}F(t) - Q^{-1}F(t)(\boldsymbol{D}^{xx}(t-1) + F^T(t)Q^{-1}F(t))^{-1}F^T(t)Q^{-1}F(t)x^*(t-1)\\
&\quad = \bar{\boldsymbol{D}}^{xx}(t)F(t)x^*(t-1)
\end{aligned} \tag{3.4.30}$$

where expression for $\bar{\boldsymbol{D}}^{xx}(t)$ is taken from (3.4.27), and

$$\begin{aligned}
&\boldsymbol{D}^{xy}(t-1)^T - \boldsymbol{D}^{xy}(t-1)^T\tilde{\boldsymbol{D}}^{-1}(t-1)\boldsymbol{D}^{xx}(t-1)\\
&\quad = \boldsymbol{D}^{xy}(t-1)^T\tilde{\boldsymbol{D}}^{-1}(t-1)(\tilde{\boldsymbol{D}}(t-1) - \boldsymbol{D}^{xx}(t-1))\\
&\quad = \boldsymbol{D}^{xy}(t-1)^T\tilde{\boldsymbol{D}}^{-1}(t-1)F^T(t)Q^{-1}F(t) = \bar{\boldsymbol{D}}^{xy}(t-1)^T
\end{aligned} \tag{3.4.31}$$

Now Equation (3.4.25) can be rewritten as

$$\begin{aligned}
\begin{pmatrix} x^*(t) \\ y^*(t) \end{pmatrix} &= \boldsymbol{D}^{-1}(t)\left[\bar{\boldsymbol{D}}(t)\begin{pmatrix} \bar{x}(t) \\ y^*(t-1) \end{pmatrix} + \begin{pmatrix} H^T(t)R^{-1}b(t) \\ W^T(t)R^{-1}b(t) \end{pmatrix}\right]\\
&= \boldsymbol{D}^{-1}(t)\left\{\boldsymbol{D}(t)\begin{pmatrix} \bar{x}(t) \\ y^*(t-1) \end{pmatrix} - \left[\begin{array}{c|c} H^T(t)R^{-1}H(t) & H^T(t)R^{-1}W(t) \\ \hline W^T(t)R^{-1}H(t) & W^T(t)R^{-1}W(t) \end{array}\right]\right.\\
&\qquad \left.\times \begin{pmatrix} \bar{x}(t) \\ y^*(t-1) \end{pmatrix} + \begin{pmatrix} H^T(t)R^{-1}b(t) \\ W^T(t)R^{-1}b(t) \end{pmatrix}\right\}\\
&= \begin{pmatrix} \bar{x}(t) \\ y^*(t-1) \end{pmatrix} + \boldsymbol{D}^{-1}(t)\left\{-\left[\begin{array}{c|c} H^T(t)R^{-1}H(t) & H^T(t)R^{-1}W(t) \\ \hline W^T(t)R^{-1}H(t) & W^T(t)R^{-1}W(t) \end{array}\right]\right.\\
&\qquad \left.\times \begin{pmatrix} \bar{x}(t) \\ y^*(t-1) \end{pmatrix} + \begin{pmatrix} H^T(t)R^{-1}b(t) \\ W^T(t)R^{-1}b(t) \end{pmatrix}\right\}
\end{aligned} \tag{3.4.32}$$

and finally

$$\begin{pmatrix} x^*(t) \\ y^*(t) \end{pmatrix} = \begin{pmatrix} \bar{x}(t) \\ y^*(t-1) \end{pmatrix} + \boldsymbol{D}^{-1}(t) \begin{pmatrix} H^T(t)R^{-1} \\ W^T(t)R^{-1} \end{pmatrix} (b(t) - H(t)\bar{x}(t) - W(t)y^*(t-1)) \tag{3.4.33}$$

Formulas (3.4.27)–(3.4.29), and (3.4.33) provide the full description of the recursive estimation scheme.

In order to present expressions (3.4.23) [or (3.4.27) and (3.4.29)] in a convenient and numerically stable form we will use the matrix decomposition technique described in the previous subsections. Consider the $(2n+p) \times (2n+p)$ matrix

$$G(t) = \left[\begin{array}{c|c|c} \boldsymbol{D}^{xx}(t-1) + F^T(t)Q^{-1}F(t) & -F^T(t)Q^{-1} & \boldsymbol{D}^{xy}(t-1) \\ \hline -Q^{-1}F(t) & Q^{-1} + H^T(t)R^{-1}H(t) & H^T(t)R^{-1}W(t) \\ \hline \boldsymbol{D}^{xy}(t-1)^T & W^T(t)R^{-1}H(t) & \begin{array}{c} \boldsymbol{D}^{yy}(t-1) \\ +W^T(t)R^{-1}W(t) \end{array} \end{array}\right] \tag{3.4.34}$$

Divide it into blocks according to (A.3.39) where

$$\begin{aligned} N_{11} &= \boldsymbol{D}^{xx}(t-1) + F^T(t)Q^{-1}F(t) \\ N_{12} &= \left(-F^T(t)Q^{-1} \;\middle|\; \boldsymbol{D}^{xy}(t-1)\right) \\ N_{22} &= \left(\begin{array}{c|c} Q^{-1} + H^T(t)R^{-1}H(t) & H^T(t)R^{-1}W(t) \\ \hline W^T(t)R^{-1}H(t) & \boldsymbol{D}^{yy}(t-1) + W^T(t)R^{-1}W(t) \end{array}\right) \end{aligned} \tag{3.4.35}$$

and calculate its Cholesky decomposition

$$G(t) = \begin{bmatrix} L(t) & 0 \\ K(t) & M(t) \end{bmatrix} \begin{bmatrix} L^T(t) & K^T(t) \\ 0 & M^T(t) \end{bmatrix} \tag{3.4.36}$$

It follows from (A.3.59) to (A.3.61) and (3.4.35) that

$$\begin{aligned} L(t)L^T(t) &= \widetilde{\boldsymbol{D}}(t) = \boldsymbol{D}^{xx}(t-1) + F^T(t)Q^{-1}F(t) \\ K(t) &= \begin{pmatrix} -Q^{-1}F(t) \\ \boldsymbol{D}^{xy}(t-1)^T \end{pmatrix} (L^T(t))^{-1} \end{aligned} \tag{3.4.37}$$

and

$$\boldsymbol{D}(t) = M(t)M^T(t) \tag{3.4.38}$$

Let us now look at the expression (3.4.33). Taking into account (3.4.38) we have

$$\begin{aligned} \begin{pmatrix} x^*(t) \\ y^*(t) \end{pmatrix} &= \begin{pmatrix} \bar{x}(t) \\ y^*(t-1) \end{pmatrix} + (M^T(t))^{-1}M^{-1}(t) \begin{pmatrix} H^T(t)R^{-1} \\ W^T(t)R^{-1} \end{pmatrix} \\ &\quad \times (b(t) - H(t)\bar{x}(t) - W(t)y^*(t-1)) \end{aligned} \tag{3.4.39}$$

Expressions (3.4.34), (3.4.36), (3.4.28), and (3.4.39) give full description of the recursive estimation scheme of the parameters $(x^*(t), y^*(t))$ in the decomposed form.

For the sake of simplicity suppose that the covariance matrices Q and R do not depend on t and are decomposed by Cholesky as

$$Q = L_Q L_Q^T \quad R = L_R L_R^T \tag{3.4.40}$$

and also note that $F^T(t)Q^{-1}F(t) = (\mathbf{F}_{L_Q}F(t))^T\mathbf{F}_{L_Q}F(t)$. Similar identities hold for other terms, including R^{-1}. Let us present the estimate update and the matrix update scheme $\mathbf{D}(t-1) \to \mathbf{D}(t)$ in more visual form in Algorithm 4, described in Table 3.4.1. Starting with $\mathbf{D}(0) = C^{-1}$, $x^*(0) = x_0$, and $y^*(0) = 0$, proceed from one row to the next. Then, the recursive update of the optimal estimate is calculated according to the formulas (3.4.28) and (3.4.39).

Now let us return to considering the same problem (3.4.5) but for batch processing. Consider the quadratic function (3.4.4) and present it in the form (A.3.36). Let us present the set of vectors $x(0), x(1), \dots, x(t'), y$ in the form of the $(t'+1)n+p$-dimensional vector

$$X = \begin{pmatrix} x(0) \\ \vdots \\ x(t') \\ y \end{pmatrix} \tag{3.4.41}$$

Then the solution to the problem (3.4.5) satisfies the linear system

$$AX = r \tag{3.4.42}$$

which is schematically shown in Figure 3.4.1. The matrix A has the block band-wise structure with a "bordering." The right-hand side vector r has the form

$$r = \left(C^{-1}x_0, H^T(1)R^{-1}b(1), \dots, H^T(t')R^{-1}b(t'), \sum_{t=1}^{t'} W^T(t)R^{-1}b(t) \right)^T \tag{3.4.43}$$

The location of nonzero block entries of the matrix A and their content is shown in the Figure 3.4.1. The factors of the Cholesky decomposition of matrix A have the pattern that occupies the lower diagonal part of the matrix (Figure 3.4.2). As was done in the Section 3.3, one can show that sequential calculation of the block entries of the Cholesky decomposition of the large matrix A gives the same formulas for the recursive estimation scheme as (3.4.25), (3.4.26), (3.4.34), and (3.4.38).

Statement 3.4.3. The *sequential computation of block entries of the Cholesky decompositions of the large sparse matrix of problems (3.4.4) and (3.4.5) gives the same results and the same formulas (3.4.25), (3.4.26) and (3.4.34), (3.4.38) as were obtained for the recursive optimal estimation using the Bellman functions approach.*

In this section, we derived the recursive least-squares algorithm allowing estimation of parameters subject to dynamic constraint along with static parameters.

TABLE 3.4.1 Algorithm 4: Static Parameters and Dynamic Constraints

Present the $(n+p)\times(n+p)$ symmetric matrix $\boldsymbol{D}(t)$ in the block form	$\boldsymbol{D}(t)=\begin{bmatrix}\boldsymbol{D}^{xx}(t) & \boldsymbol{D}^{xy}(t)\\ \boldsymbol{D}^{xy}(t)^T & \boldsymbol{D}^{yy}(t)\end{bmatrix}$
Extend the matrix $\boldsymbol{D}(t)$ to the $(2n+p)\times(2n+p)$ matrix by adding zero matrix blocks	$\hat{\boldsymbol{D}}(t)=\begin{bmatrix}\boldsymbol{D}^{xx}(t) & 0 & \boldsymbol{D}^{xy}(t)\\ 0 & 0 & 0\\ \boldsymbol{D}^{xy}(t)^T & 0 & \boldsymbol{D}^{yy}(t)\end{bmatrix}$
Perform forward substitution calculations with Cholesky factors L_Q and L_R	$\overline{F}(t+1)=\boldsymbol{F}_{L_Q}F(t+1)$ $\overline{H}(t+1)=\boldsymbol{F}_{L_R}H(t+1)$ $\overline{W}(t+1)=\boldsymbol{F}_{L_R}W(t+1)$ $\overline{b}(t+1)=\boldsymbol{F}_{L_R}b(t+1)$ $Q^{-1}=\boldsymbol{B}_{L_Q}(\boldsymbol{F}_{L_Q}(I))$
Compute the updating matrix	$\Delta(t+1)=\begin{bmatrix}\overline{F}^T(t+1)\overline{F}(t+1) & -(\boldsymbol{B}_{L_Q}\overline{F}(t+1))^T & 0\\ -\boldsymbol{B}_{L_Q}\overline{F}(t+1) & Q^{-1}+\overline{H}^T(t+1)\overline{H}(t+1) & \overline{H}^T(t+1)\overline{W}(t+1)\\ 0 & \overline{W}^T(t+1)\overline{H}(t+1) & \overline{W}^T(t+1)\overline{W}(t+1)\end{bmatrix}$

Compute the $(2n+p)\times(2n+p)$ extended matrix	$G(t+1) = \hat{\boldsymbol{D}}(t) + \Delta(t)$ $= \left[\begin{array}{c:c:c} \boldsymbol{D}^{xx}(t) + \overline{F}^T(t+1)\overline{F}(t+1) & -(\boldsymbol{B}_{L_Q}\overline{F}(t+1))^T & \boldsymbol{D}^{xy}(t) \\ \hdashline -\boldsymbol{B}_{L_Q}\overline{F}(t+1) & Q^{-1} + \overline{H}^T(t+1)\overline{H}(t+1) & \overline{H}^T(t+1)\overline{W}(t+1) \\ \hdashline \boldsymbol{D}^{xy}(t)^T & \overline{W}^T(t+1)\overline{H}(t+1) & \boldsymbol{D}^{yy}(t) + \overline{W}^T(t+1)\overline{W}(t+1) \end{array}\right]$
Calculate Cholesky decomposition of $G(t+1)$	$G(t+1) = \begin{bmatrix} L(t+1) & 0 \\ K(t+1) & M(t+1) \end{bmatrix}\begin{bmatrix} L^T(t+1) & K^T(t+1) \\ 0 & M^T(t+1) \end{bmatrix}$
Updated matrix $\boldsymbol{D}(t+1)$	$\boldsymbol{D}(t+1) = M(t+1)M^T(t+1)$
Compute the projected estimate	$\overline{x}(t+1) = F(t+1)x^*(t)$
Compute the residual vector	$\overline{r}(t+1) = \overline{b}(t+1) - \overline{H}(t+1)\overline{x}(t+1) - \overline{W}(t+1)y^*(t)$
Update the estimate	$\begin{pmatrix} x^*(t+1) \\ y^*(t+1) \end{pmatrix} = \begin{pmatrix} \overline{x}(t+1) \\ y^*(t) \end{pmatrix} + \boldsymbol{B}_{M(t+1)}\left(\boldsymbol{F}_{M(t+1)}\begin{pmatrix} \overline{H}^T(t+1)R^{-1} \\ \overline{W}^T(t+1)R^{-1} \end{pmatrix}\overline{r}(t+1)\right)$

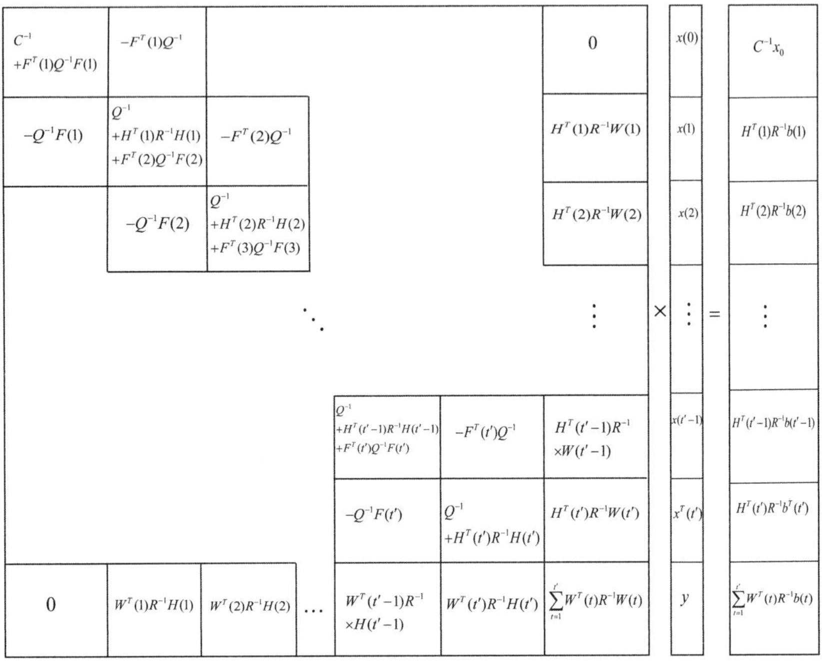

Figure 3.4.1 Schematic structure of the band-like linear system (3.4.42).

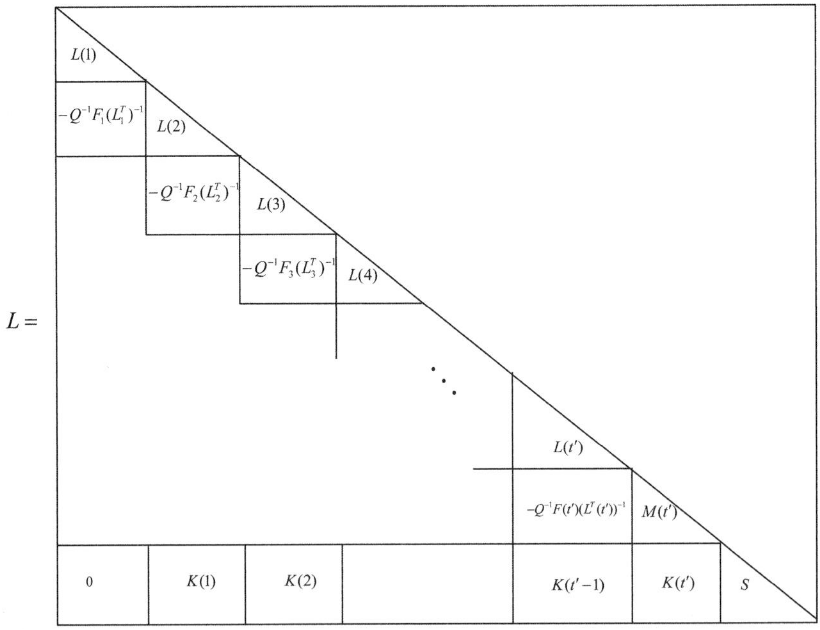

Figure 3.4.2 Band-like structure of the matrix in the Cholesky decomposition.

3.5 STATIC PARAMETER, PARAMETERS SUBJECT TO DYNAMIC CONSTRAINTS, AND ARBITRARY TIME-VARYING PARAMETERS

This section combines the results obtained in Sections 3.3 and 3.4. We consider again the dynamic system (3.3.1) connecting two sequential state vectors: $x(t) = F(t)x(t-1) + \xi(t)$ and extend the measurement model (3.4.1) to include arbitrary time-varying parameter $z(t)$

$$b(t) = H(t)x(t) + J(t)z(t) + W(t)y + \varepsilon(t) \tag{3.5.1}$$

The model (3.5.1) connects the time-varying n-dimensional parameter $x(t)$, which satisfy the dynamic constraints (3.3.1), the arbitrary varying $r(t)$-dimensional parameter $z(t)$, and the constant time-invariant p-dimensional parameter y. The matrix $W(t)$ has dimensions $m(t)$ by p. Let matrices $H(t)$ and $J(t)$ have dimensions $m(t) \times n$ and $m(t) \times r(t)$, respectively. The problem setup is completed by adding initial conditions (3.3.4) and (3.3.5).

We present results omitting intermediate constructions that are similar to those of Sections 3.2 and 3.4. For the sake of simplicity suppose that the covariance matrices Q and R do not depend on k and are factorized by Cholesky as

$$Q = L_Q L_Q^T \qquad R = L_R L_R^T \tag{3.5.2}$$

and note that $F^T(t)Q^{-1}F(t) = \left(\boldsymbol{F}_{L_Q}F(t)\right)^T \boldsymbol{F}_{L_Q}F(t)$. Similar identities hold for other terms, including R^{-1}. Starting with initial conditions

$$\boldsymbol{D}(0) = \begin{bmatrix} C^{-1} & 0 \\ 0 & 0 \end{bmatrix} \tag{3.5.3}$$

$$x^*(0) = x_0 \qquad y^*(0) = 0 \tag{3.5.4}$$

the algorithm proceeds as described in Table 3.5.1.

This completes the description of the algorithms and the chapter. We provided five algorithms for the following estimation problems:

1) Algorithm 1 solves the problem of estimating static parameters. It applies to the processing of across-receiver, across-satellite pseudorange and carrier phase observables in the case of stationary antenna positions. The estimated parameters include corrections to the antenna positions and the double differenced carrier phase ambiguities.

2) Algorithm 2 estimates arbitrary varying parameters and static parameters. It applies to the processing of across-receiver observables for stationary or kinematic processing situations. The across-receiver carrier phase ambiguities are static parameters. Across-receiver clock differences are considered arbitrarily varying parameters taking independent values at adjacent epochs. The position corrections can be considered static parameters or arbitrarily varying parameters for stationary or kinematic cases, respectively.

TABLE 3.5.1 Algorithm 5: Static Parameters, Parameters Subject to Dynamic Constraints, and Arbitrary Time-Varying Parameters

Present the $(n+p)\times(n+p)$ symmetric matrix $\boldsymbol{D}(t)$ in the block form	$\boldsymbol{D}(t) = \begin{bmatrix} \boldsymbol{D}^{xx}(t) & \boldsymbol{D}^{xy}(t) \\ \boldsymbol{D}^{xy}(t)^T & \boldsymbol{D}^{yy}(t) \end{bmatrix}$
Extend the matrix $\boldsymbol{D}(t)$ to $(2n+p)\times(2n+p)$ by adding zero matrix blocks	$\hat{\boldsymbol{D}}(t) = \begin{bmatrix} \boldsymbol{D}^{xx}(t) & 0 & \boldsymbol{D}^{xy}(t) \\ 0 & 0 & 0 \\ \boldsymbol{D}^{xy}(t)^T & 0 & \boldsymbol{D}^{yy}(t) \end{bmatrix}$
Perform forward substitution calculations with Cholesky factors L_Q and L_R	$\overline{F}(t+1) = \boldsymbol{F}_{L_Q} F(t+1)$ $\overline{H}(t+1) = \boldsymbol{F}_{L_R} H(t+1)$ $\overline{W}(t+1) = \boldsymbol{F}_{L_R} W(t+1)$ $\overline{J}(t+1) = \boldsymbol{F}_{L_Q} J(t+1)$ $\overline{b}(t+1) = \boldsymbol{F}_{L_R} b(t+1)$ $Q^{-1} = \boldsymbol{B}_{L_Q}(\boldsymbol{F}_{L_Q}(I))$
Compute Cholesky decomposition of the matrix $\overline{J}^T(t+1)\overline{J}(t+1)$ and forward substitution calculations	$\overline{J}^T(t+1)\overline{J}(t+1) = L_{J(t+1)} L_{J(t+1)}^T$ $\tilde{J}^T(t+1) = \boldsymbol{F}_{L_{J(t+1)}} \overline{J}^T(t+1)$
Projection matrix	$\Pi(t+1) = I - \tilde{J}(t+1)\tilde{J}^T(t+1)$
Compute the updating matrix	$\Delta(t+1) = \begin{bmatrix} \overline{F}^T(t+1)\overline{F}(t+1) & -(\boldsymbol{B}_{L_Q}\overline{F}(t+1))^T & 0 \\ -\boldsymbol{B}_{L_Q}\overline{F}(t+1) & Q^{-1} + \overline{H}^T(t+1)\Pi(t+1)\overline{H}(t+1) & \overline{H}^T(t+1)\Pi(t+1)\overline{W}(t+1) \\ 0 & \overline{W}^T(t+1)\Pi(t+1)\overline{H}(t+1) & \overline{W}^T(t+1)\Pi(t+1)\overline{W}(t+1) \end{bmatrix}$

Compute the $(2n+p)\times(2n+p)$ extended normal matrix	$G(t+1) = \hat{\boldsymbol{D}}(t) + \Delta(t+1)$ $= \left[\begin{array}{c:c:c} \boldsymbol{D}^{xx}(t) + \overline{F}^T(t+1)\overline{F}(t+1) & -(\boldsymbol{B}_{L_Q}\overline{F}(t+1))^T & \boldsymbol{D}^{xy}(t) \\ \hdashline -\boldsymbol{B}_{L_Q}\overline{F}(t+1) & Q^{-1} + \overline{H}^T(t+1)\Pi(t+1)\overline{H}(t+1) & \overline{H}^T(t+1)\Pi(t+1)\overline{W}(t+1) \\ \hdashline \boldsymbol{D}^{xy}(t)^T & \overline{W}^T(t+1)\Pi(t+1)\overline{H}(t+1) & \boldsymbol{D}^{yy}(t) + \overline{W}^T(t+1)\Pi(t+1)\overline{W}(t+1) \end{array}\right]$
Calculate Cholesky decomposition of $G(t+1)$	$G(t+1) = \begin{bmatrix} L(t+1) & 0 \\ K(t+1) & M(t+1) \end{bmatrix}\begin{bmatrix} L^T(t+1) & K^T(t+1) \\ 0 & M^T(t+1) \end{bmatrix}$
Updated matrix $\boldsymbol{D}(t+1)$	$\boldsymbol{D}(t+1) = M(t+1)M^T(t+1)$
Compute the projected estimate	$\overline{x}(t+1) = F(t+1)x^*(t)$
Compute residual	$\overline{r}(t+1) = \overline{b}(t+1) - \overline{H}(t+1)\overline{x}(t+1) - \overline{W}(t+1)y^*(t)$
Update the estimate	$\begin{pmatrix} x^*(t+1) \\ y^*(t+1) \end{pmatrix} = \begin{pmatrix} \overline{x}(t+1) \\ y^*(t) \end{pmatrix} + \boldsymbol{B}_{M(t+1)}\left(\boldsymbol{F}_{M(t+1)}\left(\begin{pmatrix} \overline{H}^T(t+1) \\ \overline{W}(t+1) \end{pmatrix}\Pi(t+1)\overline{r}(t+1)\right)\right)$
Compute second residual vector	$r'(t+1) = \overline{b}(t+1) - \overline{H}(t+1)x^*(t+1) - \overline{W}(t+1)y^*(t+1)$
Compute the estimate	$z^*(t+1) = \boldsymbol{B}_{L_{J(t+1)}}(\widetilde{J}(t+1)r'(t+1))$

3) Dynamic parameters estimation is not explicitly used when processing GNSS observables, but the description of this case gives a basis for further considerations of the fourth and fifth cases. On the other hand, Algorithm 3 pertains to the problem of integration of GNSS measurements and inertial measurements, which is not addressed in this book.

4) Estimation of static parameters and parameters subject to dynamic constraints is handled by Algorithm 4. It applies to processing the across-receiver, across-satellite observables for long baselines. The residual ionosphere is subject to the dynamic model, while ambiguities and corrections to the stationary position are static parameters.

5) Fifth case is the most general one and covers all previously considered cases of static, dynamic, and arbitrarily varying parameters. Algorithm 5 applies to processing the across-receiver or across-receiver, across-satellite observables for stationary or kinematic positioning for short or long base lines, with or without dynamic model applied to the corrections of the kinematic position.

In this chapter we established a connection between the recursive and batch processing. We showed that all recursive algorithms, working in real time can be considered as continuously operating Cholesky decompositions of large matrices involved into batch processing. To be specific, the Cholesky decomposition and the forward solution run continuously. Only a single step backward solution is performed when operating in real time. On the other hand, several steps of the backward solution can be performed if the depth of memory allows it. Finally, by performing the whole backward solution we complete the batch processing. Consideration of these algorithms helps in better understanding the connection between real-time recursive and postmission batch processing.

CHAPTER 4

GEODESY

Geodesy is the theoretical and practical framework for utilizing GNSS vector observations and classical terrestrial observations such as angles and distances. While geodesy has much theoretical and mathematical depth, we limit this chapter to operational aspects of geodesy as needed to process GNSS observations and classical terrestrial observations. Not to exclude fundamentals, we first discuss the international terrestrial and celestial reference frames and then turn to the geodetic datum. The 3D geodetic model plays a pivotal role in the subsequent discussion, followed by the more historical 2D ellipsoidal model, and, last but not least, the popular 3D conformal mapping models.

The precise definition of reference frames becomes more important as the accuracy of geodetic space techniques increases. There are three types of frames we are concerned with—the earth-fixed international terrestrial reference frame (ITRF), the geocentric space-fixed international celestial reference frame (ICRF), and the geodetic datum. Specialized literature considers the ITRF and ICRF as an implementation of theoretical constructions such as the international terrestrial reference system (ITRS) and the geocentric celestial reference system (GCRS). In this chapter, we do not make such a distinction. Given the demand of modern geodetic measurement techniques on precise definitions of reference frames, it is certainly not an understatement to say that the definition and maintenance of such frames has become a science in itself, in particular in connection with properties of the deformable earth. Current solutions have evolved over many years. The literature is rich in contributions that document the interdisciplinary approach and depth needed to arrive at solutions.

The International Earth Rotation Service (IERS) is responsible for establishing and maintaining the ITRF and ICRF frames, whereas typically a national geodetic agency is responsible for establishing and maintaining the nation's geodetic datum.

The IERS relies on the cooperation of many research groups and agencies to accomplish its tasks. Examples of key participants are the U.S. Naval Observatory, the U.S. National Geodetic Survey, the International GNSS Service (IGS), the International Astronomical Union (IAU), and the International Union of Geodesy and Geophysics (IUGG). Our recommended authoritative publications on the broad topic of reference frames and time are Petit and Luzum (2010) and Kaplan (2005). The first addresses IERS Conventions 2010 and is published as IERS Technical Note 36. Its 11 chapters address all aspects of reference frames in great detail. The same wealth of information and depth is found in the other reference, which is the U.S. Naval Observatory Circular No. 179. These publications are available on the Internet. The reader is also encouraged to visit the homepages of the various organizations and groups mentioned above as they are highly recommended resources of additional information about the topics of geodesy, reference frames, and time.

Accurate positioning within the ITRF and ICRF frames requires a number of complex phenomena to be taken into account, such as polar motion, plate tectonic movements, solid earth tides, ocean loading displacements, and precession and nutations. Since there are multiple reference frames, one needs to be able to transform one reference frame to the other. Since much authoritative software is readily available at the homepages of the agencies and organizations mentioned above, we only discuss mathematical expressions to the extent needed for a conceptual presentation of the topic.

The geodetic datum makes the products of space geodesy accessible to practicing surveyors. While most scientists prefer to work with geocentric Cartesian coordinates, it is easier to interpret results in terms of ellipsoidal coordinates such as geodetic latitude, longitude, and height. Consequently, the issues of locating the origin of the ellipsoid and its orientation arise, as well as the need to separate ellipsoidal heights from orthometric heights and geoid undulations. The preferred choice is to use geocentric ellipsoids whose origin and orientation coincides with the ITRF. The location and orientation of the ellipsoid, its size and shape, as well as the respective sets of geoid undulations and deflection of the vertical are all part of the definition of a datum.

In order to understand the fundamental role of the geoid, we need to briefly look at the dependency of observations on gravity. GNSS observations such as pseudoranges and carrier phases depend only indirectly on gravity. For example, once the orbit of the satellites has been computed and the ephemeris is available, there is no need to further consider gravity. To make the GNSS even easier to use, the various GNSS systems broadcast the ephemeris in a well-defined, earth-centered earth-fixed (ECEF) coordinate system, such as the ITRF. Astronomic latitude, longitude, and azimuth determinations with a theodolite using star observations, on the contrary, refer to the instantaneous rotation axis, the instantaneous terrestrial equator of the earth, and the local astronomic horizon (the plane perpendicular to the local plumb line). For applications where accuracy really matters, it is typically the responsibility of the user to apply the necessary reductions or corrections to obtain positions in an ECEF coordinate system. Even vertical and horizontal angles as measured by surveyors with a theodolite or total station refer to the plumb line and the local astronomic horizon. Another type of observation that depends on the plumb line is leveling.

To deal with these types of observations that depend on the direction of gravity, we need to ultimately link the geoid and the ellipsoid. The goal is to reduce observations that depend on the direction of gravity to model observations that refer to the ellipsoid by applying geoid undulation and deflection of the vertical corrections. These "connecting elements" are part of the definition of the datum. For a modern datum these elements are readily available; for example, check out the website of the U.S. National Geodetic Survey for the case of NAD83.

After reducing the observations for the general impact of gravity (and polar motion if applicable), one obtains the model observations of the 3D geodetic model. This model is the simplest, most versatile model for dealing with observations such as angles, distance, and GNSS vectors. The 3D model will be presented in the form of various parameterizations and applied to GNSS vector observations. We start out with the minimal or inner constraint solution for vector networks, and then we generalize the approach by combining GNSS networks and geodetic networks by also estimating a differential scale and three rotation parameters. Several examples are presented. Cases are chosen from the early days of GPS satellite surveying to demonstrate the high accuracy achieved even when the GPS system was under construction.

Many surveyors prefer to work with "plane" coordinates. In order to arrive at model observations to which the laws of plane trigonometry are applicable, two additional reductions must be made. The 3D model observations are further reduced to ellipsoidal surface observations (2D ellipsoidal model observations). The latter observations refer to angles between geodesic lines and the length of geodesic lines. As an intermediary solution we briefly discuss adjustments on the ellipsoidal surface. The ellipsoidal surface observations are then further reduced to the conformal mapping plane (2D conformal model observations). The conformal mapping model observations represent the angle between straight lines on the conformal mapping plane and the straight line distances between mapped points. We discuss the Transverse Mercator (TM) and the Lambert Conformal (LC) mapping in detail, as well as respective adjustments of plane networks on the conformal mapping plane.

4.1 INTERNATIONAL TERRESTRIAL REFERENCE FRAME

A conventional terrestrial reference system (CTRS) must allow the combination of products of space geodesy, such as coordinates and orientation parameters of the deformable earth, into a unified data set. Such a reference system should (a) be geocentric (whole earth, including oceans and atmosphere), (b) incorporate corrections or procedures stemming from the relativistic theory of gravitation, (c) maintain consistency in orientation with earlier definitions, and (d) have no residual global rotation with respect to the crust as viewed over time. The practical realization of such a system is the ITRF. Such a realization is generally referred to as earth-centered earth-fixed (ECEF) coordinate system. To appreciate the demand placed on a modern reference system, consider the fact that geodetic space techniques can provide daily estimates of the center of mass at centimeter-level accuracy and millimeter crustal motion determinations on a global scale. As mentioned above, the authoritative literature on international terrestrial and celestial reference frames are Petit and Luzum (2010, Chapter 4) and Kaplan (2005). The IERS (International Earth Rotation and Reference System Service) is responsible for the realizations of the ITRF (www.iers.org). This section deals only with the major phenomena such as polar motion, plate tectonic motions, solid earth tides, and ocean loading that cause variations of coordinates in the terrestrial reference frame. We also discuss transformations between terrestrial reference systems.

4.1.1 Polar Motion

The intersection of the earth's instantaneous rotation axis and the crust moves in time relative to the crust. This motion is called polar motion. Figure 4.1.1 shows polar motion for the time period 2001–2003. The motion is somewhat periodic. There is a major constituent of about 434 days, called the Chandler period. The amplitude varies but does not seem to exceed 10 m. Several of the polar motion features can be explained satisfactorily from a geophysical model of the earth; however, the fine structures in polar motion are still subject to research.

To avoid variations in latitude and longitude of about 10 m due to polar motion, we want to define a conventional terrestrial pole (CTP) that is fixed relative to the crust. Originally, indeed, such a pole was defined as the center of the figure of polar motion

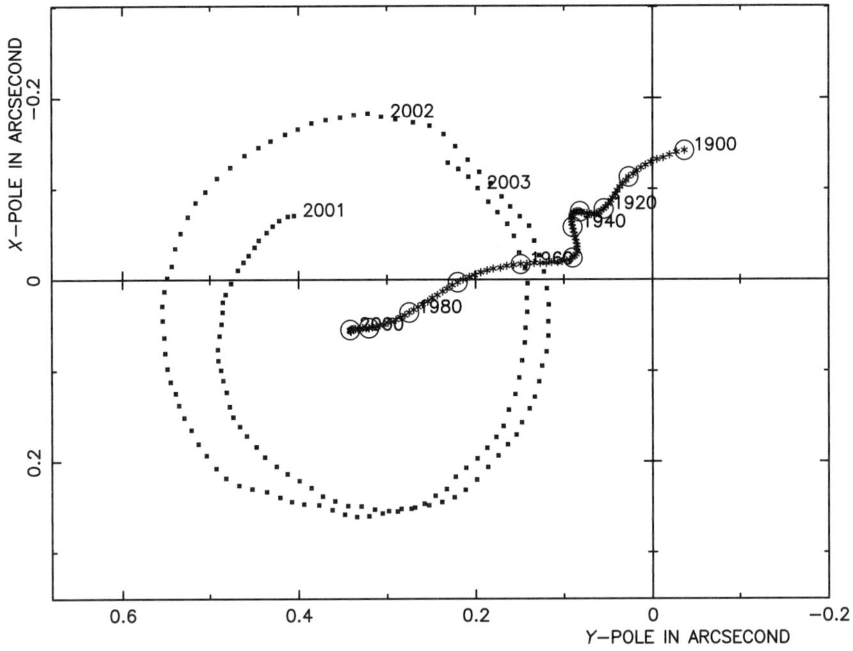

Figure 4.1.1 Polar motion, 2001–2003. The solid line represents the mean pole displacement, 1900–2000 [Courtesy of the International Earth Orientation Service (IERS), Paris Observatory].

for the years 1900–1905. This definition, however, required several refinements as the observation techniques improved over the years. The instantaneous rotation axis is referenced to the CTP by the polar motion coordinates (x_p, y_p). The origin of the polar motion coordinate system is at the CTP, the x axis is along the conventional zero meridian, and the y axis is positive along the 270° meridian. As the figure indicates, there appears to be "polar wander" (gradual shifting of the center of the figure of somewhat periodic motions away from the CTP). The IERS website contains additional graphics on polar motion, plus data files for users.

The CTP is aligned with the direction of the third axis of the ITRF. The definition of an ITRF becomes increasingly complicated because plate tectonic motions cause observation stations to drift, and there are other temporal variations affecting coordinates of a so-called "crust-fixed" coordinate system. As the tectonic plates move, the fixed station coordinates of a global network become inconsistent with each other over time. The solution is to define the reference frame by a consistent set of coordinates and velocities for globally distributed stations at a specific epoch. The center of mass of the earth is the natural choice for the origin of the ITRF because satellite dynamics are sensitive to the center of mass (whole earth plus oceans and atmosphere). As indicated above, the IERS maintains the ITRF using extraterrestrial data from various sources, such as GNSS, very long baseline interferometry (VLBI), and satellite laser ranging (SLR). Because the motions of the deformable earth are complex, there is a need to identify the sites that are part of a particular ITRF definition. Because of continued progress in data reduction techniques, the IERS computes ITRF updates as needed. These solutions are designated by adding the year, e.g., ITRF96, ITRF97, ITRF00, and ITRF2008.

The ITRF-type of reference frame is also called an ECEF frame, as already mentioned above. We denote an ECEF frame by (x) and the coordinate triplet by (x, y, z). The z of the frame is the origin of the polar motion coordinate system. The x and y axes define the terrestrial equatorial plane. In order to maintain continuity with older realizations, the x axis lies in what may be loosely called the Greenwich meridian.

Historically speaking, the International Latitude Service (ILS) was created in 1895, shortly after polar motion had been verified observationally. It was the first international group using globally distributed stations to monitor a reference frame. This service evolved into the International Polar Motion Service (IPMS) in 1962. The IERS was established in 1987 as a single international authority that, henceforth, uses modern geodetic space techniques to establish and maintain reference frames. GNSS systems are major contributors to the definition and maintenance of the terrestrial reference frame, largely a result of strong international cooperation with the IGS (International GNSS Service, see Section 6.2.7.1). The IGS began routine operations in 1994 by providing GPS orbits, tracking data, and offering other data products in support of geodetic and geophysical research.

4.1.2 Tectonic Plate Motion

Figure 4.1.2 shows tectonic plate motions. Even this simple overview of motions over the global map makes clear that they are significant and should be appropriately

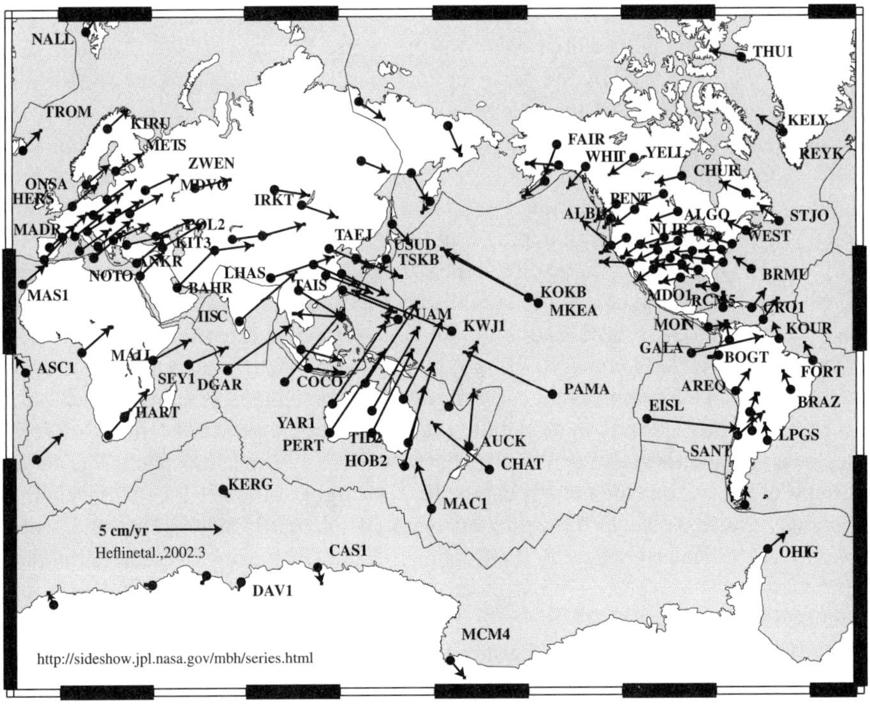

Figure 4.1.2 Observed motions of globally distributed stations.
Velocities for each site were determined from more than 11 years of GPS observations prior to 2000. Results are shown in the ITRF00 reference frame with no net rotation of the crust. The rigid plate motion is clearly visible and describes roughly 80% of observed motion. The remaining 20% is nonrigid motion located in plate boundary zones and is associated with seismic and volcanic activity. The most visible plate boundary zone on the map is southern California (Courtesy of Mike Heflin, JPL).

taken into account when processing GNSS observations and producing coordinates. At the JPL website, http://sideshow.jpl.nasa.gov/post/series.html, the latest update of these motions can be seen. Also available at the site are time series of coordinates for more than 2000 stations. The figure also indicates boundaries of major tectonic plates. The rotations of these plates can be approximated by geophysical models based on geological observations. DeMets et al. (1990) published their basic NUVEL-1 model, which was subjected to a no-net rotation constraint by Argus and Gordon (1991) in their model NNR-NUVEL-1, which in turn was further improved by DeMets et al. (1994) in model NNR-NUVEL-1A. Additional improvements to the model (that also include more plates) are found in DeMets et al. (2010). The smooth motions of these major plates can be combined with actual observed motions at selected stations to create a general station motion function.

4.1.3 Solid Earth Tides

Tides are caused by the temporal variation of the gravitational attraction of the sun and the moon on the earth due to orbital motion. While the ocean tides are very much influenced by the coastal outlines and the shape of the near-coastal ocean floor, the solid earth tides are accurately computable from relatively simple earth models. Their periodicities can be directly derived from the motion of the celestial bodies, similar to nutation. The solid earth tides generate periodic site displacement of stations that depend on latitude. The tidal variation can be as much as 30 cm in the vertical and 5 cm in the horizontal. Petit and Luzum (2010, Chapter 7) list the following expression:

$$\Delta \boldsymbol{x} = \sum_{j=2}^{3} \frac{GM_j}{GM_E} \frac{||\boldsymbol{r}_E||^4}{||\boldsymbol{r}_j||^3} \left\{ h_2 \boldsymbol{e} \left(\frac{3}{2}(\boldsymbol{r}_j \cdot \boldsymbol{e})^2 - \frac{1}{2} \right) + 3 l_2 (\boldsymbol{r}_j \cdot \boldsymbol{e})[\boldsymbol{r}_j - (\boldsymbol{r}_j \cdot \boldsymbol{e})\, \boldsymbol{e}] \right\} \tag{4.1.1}$$

In this expression, GM_E is the gravitational constant of the earth, GM_j is the gravitational constant for the moon ($j = 2$) and the sun ($j = 3$), $\boldsymbol{e}$ is the unit vector of the station in the geocentric coordinate system (x), and $\boldsymbol{r}$ denotes the unit vector of the celestial body; h_2 and l_2 are the nominal degree 2 Love and Shida numbers that describe elastic properties of the earth model. Equation (4.1.1) gives the solid earth tides accurate to at least 5 mm. Additional expressions concerning higher-order terms or expressions for the permanent tide are found in the reference cited above.

4.1.4 Ocean Loading

Ocean loading refers to the deformation of the seafloor and coastal land as a result of redistribution of ocean water during the ocean tides. The earth's crust yields under the weight of the tidal water. Petit and Luzum (2010, Chapter 7) list the following expression for the site displacement components Δc (where the c refers to the radial, west, and south component) at a particular site at time t,

$$\Delta c = \sum_j f_j A_{cj} \cos(\omega_j t + \chi_j + u_j - \Phi_{cj}) \tag{4.1.2}$$

The summation over j represents 11 tidal waves traditionally designated as semidiurnal M_2, S_2, N_2, and K_2, the diurnal K_1, O_1, and P_1, and the long-periodic M_f, M_m, and S_{sa}. The symbols ω_j and χ_j denote the angular velocities and the fundamental astronomic arguments at time $t = 0^h$. The fundamental argument χ_j reflects the position of the sun and the moon, and f_j and u_j depend on the longitude of the lunar node. The station-specific amplitudes A_{cj} and phases Φ_{cj} can be computed using ocean tide models and coastal outline data. The IERS makes these values available for most ITRF reference stations. Typically the M_2 loading deformations are the largest. The total vertical motion can reach 10 cm while the horizontal motion is 2 cm or less. Free ocean tide loading values for individual locations are provided by the Onsala Space Observatory at http://holt.oso.chalmers.se/loading/.

4.1.5 Relating of Nearly Aligned Frames

The transformation of three-dimensional coordinate systems has been given much attention ever since geodetic satellite techniques made it possible to relate local national datums and geocentric datums. Some of the pertinent work from that early era is Veis (1960), Molodenskii et al. (1962), Badekas (1969), Vaniček and Wells (1974), Leick and van Gelder (1975), and Soler and van Gelder (1987). We assume that the Cartesian coordinates of points on the earth's surface are available in two systems. Historically speaking, this was not necessarily the case. Often it was difficult to obtain the Cartesian coordinates in the local geodetic datum because the geoid undulations (see Section 4.3.3) with respect to the local datum were not be accurately known. We first deal with the seven-parameter similarity transformation.

Figure 4.1.3 shows the coordinate system (x) = (x, y, z), which is related to the coordinate system (u) = (u, v, w) by the translation vector $\boldsymbol{t} = [\Delta x \quad \Delta y \quad \Delta z]^T$ between the origins of the two coordinate systems and the small rotations $(\varepsilon, \psi, \omega)$ around the (u, v, w) axes, respectively. The transformation equation expressed in the (x) coordinate system can be seen here:

$$\boldsymbol{t} + (1 + s)\,\boldsymbol{Ru} - \boldsymbol{x} = \boldsymbol{0} \tag{4.1.3}$$

where s denotes the differential scale factor between both systems and $\boldsymbol{R}$ is the product of three consecutive orthogonal rotations around the axes of (u):

$$\boldsymbol{R} = \boldsymbol{R}_3(\omega)\boldsymbol{R}_2(\psi)\boldsymbol{R}_1(\varepsilon) \tag{4.1.4}$$

The symbol $\boldsymbol{R}_i$ denotes the rotation matrix for a rotation around axis i (see Appendix A.2). The angles $(\varepsilon, \psi, \omega)$ are positive for counterclockwise rotations about the respective (u, v, w) axes, as viewed from the end of the positive axis. For nearly aligned coordinate systems these rotation angles are small, allowing the following simplification:

$$\boldsymbol{R} = \boldsymbol{I} + \boldsymbol{Q} = \boldsymbol{I} + \begin{bmatrix} 0 & \omega & -\psi \\ -\omega & 0 & \varepsilon \\ \psi & -\varepsilon & 0 \end{bmatrix} \tag{4.1.5}$$

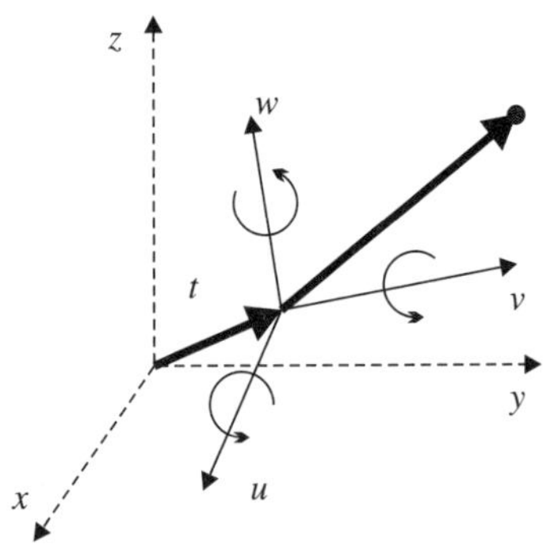

Figure 4.1.3 Differential transformation between Cartesian coordinate systems.

implying $\boldsymbol{R}^{-1} = \boldsymbol{R}^T$ within the same accuracy. Combining (4.1.3) and (4.1.5) gives the linearized form

$$\boldsymbol{t} + \boldsymbol{u} + s\boldsymbol{u} + \boldsymbol{Q}\boldsymbol{u} - \boldsymbol{x} = \boldsymbol{0} \tag{4.1.6}$$

For the purpose of distinguishing various approaches, we call this transformation model 1. The seven transformation parameters $(\Delta x, \Delta y, \Delta z, s, \varepsilon, \psi, \omega)$ can be estimated by least squares. Both sets of Cartesian coordinates $\boldsymbol{u}$ and $\boldsymbol{x}$ are observations. In general, equation (4.1.6) represents a mixed adjustment model $\boldsymbol{f}(\boldsymbol{\ell}_a, \boldsymbol{x}_a) = \boldsymbol{0}$ (see Chapter 2 for an additional explanation of the mixed adjustment model). Each station contributes three equations to (4.1.3).

A variation of (4.1.3), called model 2, is

$$\boldsymbol{t} + \boldsymbol{u}_0 + (1 + s)\,\boldsymbol{R}\,(\boldsymbol{u} - \boldsymbol{u}_0) - \boldsymbol{x} = \boldsymbol{0} \tag{4.1.7}$$

where $\boldsymbol{u}_0$ is a vector in the system (u) to a point located somewhere within the network that is to be transformed. A likely choice for $\boldsymbol{u}_0$ might be the centroid. All other notations are the same as in (4.1.3). If one follows the same procedure as described for the previous model, i.e., omitting second-order terms in scale and rotation and their products, then (4.1.7) becomes

$$\boldsymbol{t} + \boldsymbol{u} + s\,(\boldsymbol{u} - \boldsymbol{u}_0) + \boldsymbol{Q}\,(\boldsymbol{u} - \boldsymbol{u}_0) - \boldsymbol{x} = \boldsymbol{0} \tag{4.1.8}$$

The third model, model 3, uses the same rotation point $\boldsymbol{u}_0$ as model 2, but the rotations are about the axes (n, e, u) of the local geodetic coordinate system at $\boldsymbol{u}_0$. This model thus refers to the local geodetic horizon coordinate system and to ellipsoidal latitude, longitude, and height. Please see Section 4.4 for details on these elements if needed. The n axis is tangent to the geodetic meridian, but the positive direction is toward the south; the e axis is perpendicular to the meridian plane and is positive eastward. The u axis is along the ellipsoidal normal with its positive direction upward, forming a right-handed system with n and e. Similar to (4.1.7), one obtains

$$\boldsymbol{t} + \boldsymbol{u}_0 + (1 + s)\,\boldsymbol{M}\,(\boldsymbol{u} - \boldsymbol{u}_0) - \boldsymbol{x} = \boldsymbol{0} \tag{4.1.9}$$

If (η, ξ, α) denote positive rotations about the (n, e, u) axes and if $(\varphi_0, \lambda_0, h_0)$ are the geodetic coordinates for the point of rotation $\boldsymbol{u}_0$, it can be verified that the $\boldsymbol{M}$ matrix is

$$\boldsymbol{M} = \boldsymbol{R}_3^T(\lambda_0)\boldsymbol{R}_2^T(90 - \varphi_0)\boldsymbol{R}_3(\alpha)\boldsymbol{R}_2(\xi)\boldsymbol{R}_1(\eta)\boldsymbol{R}_2(90 - \varphi_0)\boldsymbol{R}_3(\lambda_0) \tag{4.1.10}$$

Since the rotation angles (η, ξ, α) are differentially small, the matrix $\boldsymbol{M}$ simplifies to

$$\boldsymbol{M}(\lambda_0, \varphi_0, \eta, \xi, \alpha) = \alpha\boldsymbol{M}_\alpha + \xi\boldsymbol{M}_\xi + \eta\boldsymbol{M}_\eta + \boldsymbol{I} \tag{4.1.11}$$

where

$$\boldsymbol{M}_\alpha = \begin{bmatrix} 0 & \sin\varphi_0 & -\cos\varphi_0\sin\lambda_0 \\ -\sin\varphi_0 & 0 & \cos\varphi_0\cos\lambda_0 \\ \cos\varphi_0\sin\lambda_0 & -\cos\varphi_0\cos\lambda_0 & 0 \end{bmatrix} \tag{4.1.12}$$

$$\boldsymbol{M}_\xi = \begin{bmatrix} 0 & 0 & -\cos\lambda_0 \\ 0 & 0 & -\sin\lambda_0 \\ \cos\lambda_0 & \sin\lambda_0 & 0 \end{bmatrix} \tag{4.1.13}$$

$$\boldsymbol{M}_\eta = \begin{bmatrix} 0 & -\cos\varphi_0 & -\sin\varphi_0\sin\lambda_0 \\ \cos\varphi_0 & 0 & \sin\varphi_0\cos\lambda_0 \\ \sin\varphi_0\sin\lambda_0 & -\sin\varphi_0\cos\lambda_0 & 0 \end{bmatrix} \tag{4.1.14}$$

If, again, second-order terms in scale and rotations and their products are neglected, the model (4.1.9) becomes

$$\boldsymbol{t} + \boldsymbol{u} + s\,(\boldsymbol{u} - \boldsymbol{u}_0) + (1+s)(\boldsymbol{M} - \boldsymbol{I})(\boldsymbol{u} - \boldsymbol{u}_0) - \boldsymbol{x} = \boldsymbol{0} \tag{4.1.15}$$

Models 2 and 3 differ in that the rotations in model 3 are around the local geodetic coordinate axes at $\boldsymbol{u}_0$. The rotations $(\eta,\ \xi,\ \alpha)$ are $(\varepsilon,\ \psi,\ \omega)$ are related as follows:

$$\begin{bmatrix} \eta \\ \xi \\ \alpha \end{bmatrix} = \boldsymbol{R}_2(90 - \varphi_0)\boldsymbol{R}_3(\lambda_0)\begin{bmatrix} \varepsilon \\ \psi \\ \omega \end{bmatrix} \tag{4.1.16}$$

Models 1 and 2 use the same rotation angles. The translations for models 1 and 2 are related as

$$\boldsymbol{t}_2 = \boldsymbol{t}_1 - \boldsymbol{u}_0 + (1+s)\boldsymbol{R}\,\boldsymbol{u}_0 \tag{4.1.17}$$

according to (4.1.3) and (4.1.7). Only $\boldsymbol{t}_1$, i.e., the translation vector of the origin as estimated from model 1, corresponds to the geometric vector between the origins of the coordinate systems (x) and (u). The translational component of model 2, $\boldsymbol{t}_2$, is a function of $\boldsymbol{u}_0$, as shown in (4.1.17). Because models 2 and 3 use the same $\boldsymbol{u}_0$, both yield identical translational components. It is not necessary that all seven parameters always be estimated. In small areas it might be sufficient to estimate only the translation components.

4.1.6 ITRF and NAD83

Model 1 discussed above can readily be applied to transforming to ECEF coordinate systems, which we simply call ITRFyy and ITRFzz. Following (4.1.3) and (4.1.5), the transformation is given by

$$\boldsymbol{x}_{t,\,\mathrm{ITRFzz}} = \boldsymbol{t}_t + (1 + s_t)(\boldsymbol{I} - \boldsymbol{Q}(\varepsilon_t))\boldsymbol{x}_{t,\,\mathrm{ITRFyy}} \tag{4.1.18}$$

The vector $\boldsymbol{t}_t$ points to the origin of ITRFyy, i.e., it is the shift between the two frames, where $\boldsymbol{\varepsilon}_t = [\varepsilon_x \quad \varepsilon_y \quad \varepsilon_z]^T$ denotes three differential counterclockwise rotations around the axes of the ITRFyy frame to establish parallelism with the ITRFzz frame. The symbol s_t denotes the differential scale change. Let t_0 denote the epoch of the reference frame, then

$$
\begin{aligned}
\boldsymbol{t}_t &= \boldsymbol{t}_{t_0} + \dot{\boldsymbol{t}}(t - t_0) \\
\varepsilon_t &= \varepsilon_{t_0} + \dot{\varepsilon}(t - t_0)\, m_{\text{masr}} \\
s_t &= s_{t_0} + \dot{s}(t - t_0)
\end{aligned}
\tag{4.1.19}
$$

where $\boldsymbol{t}_{t_0}$, ε_{t_0}, s_{t_0}, $\dot{\boldsymbol{t}}$, $\dot{\varepsilon}$, and $\dot{s}$ are 14. transformation parameters and $m_{\text{masr}} = 4.84813681 \cdot 10^{-9}$ is a factor for converting milliarc seconds (mas) to radians. The time rates of the translations, the rotations, and the differential scale are assumed to be constant. If we further assume that the coordinate velocities are constant for the same frame, then

$$
\boldsymbol{x}_{t,\,\text{IRTFyy}} = \boldsymbol{x}_{t_0,\,\text{IRTFyy}} + (t - t_0)\, \boldsymbol{v}_{t_0,\,\text{IRTFyy}} \tag{4.1.20}
$$

updates the coordinates from reference epoch t_0 to epoch t. Note that all quantities in (4.1.18) refer to the same epoch t. Soler and Marshall (2002) give a more general form that allows the reference epochs on the left and right to be different, respectively.

Equation (4.1.20) readily indicates the difficulties inherent in performing accurate transformation when the coordinates are subject to various changes. For example, the coordinates and their velocities $\boldsymbol{v}$ can abruptly or gradually change due to coseismic motions, and they can also nonlinearly change due to postseismic motions over a time scale from days to decades. Because continuous crustal motion makes station coordinates a function of time, and because of new observations and refinements in processing algorithms and modeling, it is desirable to update the reference frames occasionally. For example, modeling of solar radiation pressure, tropospheric delay, satellite phase center, and ocean loading are the subject of continuous research and refinement.

Table 4.1.1 summarizes transformation parameters between the various ITRF reference systems and the NAD83 (CORS96) system. The latter reference system covers CONUS, is accessed primarily by the U.S. surveying community, and is also the datum of the State Plane Coordinate Systems (Section C.4.4). Details on the methodology and the execution of the major readjustment resulting in the first realization of NAD83 are given in Schwarz and Wade (1990). Two things should be pointed out in this table. The first one is that column 6 is the sum of columns 1 to 5. This is a result of the linearization of the transformation expressions and of neglecting higher order terms of small quantities. The second one is to notice that the origin of NAD83 (CORS96) is offset from the center of mass by (0.99, −1.90, −0.53) meters in Cartesian coordinates. Typically, Department of Defense's (DOD's) publications on the World Geodetic System 1984 (WGS84) contain a complete listing of transformation parameters for all known local and national datums.

Continuing with focus on NAD83, the HTDP (horizontal time-dependent positioning) program, which is available for the National Geodetic Survey and online, is an example of a geodetic-quality transformation program that takes known motions, such as rigid plate tectonic motions and a large number of earthquake-related coseismic and postseismic motions within a certain area of coverage, into account

TABLE 4.1.1 Example of 14 Parameter Transformation between Frames

From To	ITRF2008 ITRF2005	ITRF2005 ITRF2000	ITRF2000 ITRF97	ITRF97 ITRF96	ITRF96 NAD83	ITRF2008 NAD83
$t_x(t_0)$	−0.0029	0.0007	0.0067	−0.00207	0.9910	0.99343
$t_y(t_0)$	−0.0009	−0.0011	0.0061	−0.00021	−1.9072	−1.90331
$t_z(t_0)$	−0.0047	−0.0004	−0.0185	0.00995	−0.5129	−0.52655
$\varepsilon_x(t_0)$	0.000	0.000	0.000	0.12467	25.79	25.91467
$\varepsilon_y(t_0)$	0.000	0.000	0.000	−0.22355	9.65	9.42645
$\varepsilon_z(t_0)$	0.000	0.000	0.000	−0.06065	11.66	11.59935
$s(t_0)$	0.94	0.16	1.55	−0.93496	0.00	1.71504
$\dot{t}_x$	0.0003	−0.0002	0.0000	0.00069	0.0000	0.00079
$\dot{t}_y$	0.0000	0.0001	−0.0006	−0.00010	0.0000	−0.00060
$\dot{t}_z$	0.0000	−0.0018	−0.0014	0.00186	0.0000	−0.00134
$\dot{\varepsilon}_x$	0.000	0.000	0.000	0.01347	0.0532	0.06667
$\dot{\varepsilon}_y$	0.000	0.000	0.000	−0.01514	−0.7423	−0.75744
$\dot{\varepsilon}_z$	0.000	0.000	−0.020	0.00027	−0.0316	−0.05133
$\dot{s}$	0.00	0.08	0.01	−0.19201	0.00	−0.10201

Source: Pearson and Snay (2013).

[a]The units are t[m], $\dot{t}$[m/yr], ε [mas], $\dot{\varepsilon}$ [mas/yr], s[ppb], and $\dot{s}$[ppb/yr].The epoch is $t_0 = 1997.00$. NAD83 is abbreviation for NAD83(CORS96).

(Pearson and Snay, 2013, HTDP User's Guide). This program recognizes currently about 30 different frames. Here are a few examples not listed in Table 4.1.1. The IGS reprocessed observations from the ITRF2008 network stations but used better antenna calibration data, calling it IGS08. While the resulting coordinates slightly differ, the 14 transformation parameters to go from IGS08 to ITFR2008 are zero. NAD83 (2011) results from processing old and more recent GPS observation using the latest models for systematic errors, better refraction modeling, and so on. The processing took place within the IGS08 frame, and then the result was transformed to NAD83 (CORS96) using the ITRF2008 to NAD83 (CORS96) parameters listed in Table 4.1.1. Thus, the transformation between NAD83 (CORS96) and NAD83 (2011) is the identity function. For areas beyond CONUS, there is NAD83 (PA11), the Pacific tectonic plate frame, and NAD83 (MA11), the Mariana tectonic plate frame.

HTDP can carry out several important computations for all frames and any epoch, including:

1. Horizontal crustal velocities. Input: position coordinates and their reference frame. Output: velocity expressed in the same reference frame.

2. Crustal displacement. Input: position coordinates and their reference frame, and times t_1 and t_2. Output: displacement during the time t_1 to t_2, expressed in the same frame.
3. Updating of position coordinates. Input: reference frame, coordinates and their reference epoch t_1, reference epoch t_2 for which the coordinate values are requested. Output: coordinates at epoch t_2.
4. Transforming coordinates from one reference frame to another. Input: starting reference frame, coordinates and their reference epoch t_1, desired reference frame and reference time t_2. Output: coordinates in desired reference frame at epoch t_2.
5. Transforming velocities from one reference frame to another. Input: velocity vector and its reference frame, desired reference frame. Output: velocity vector in desired reference frame.

4.2 INTERNATIONAL CELESTIAL REFERENCE SYSTEM

Historically, the equator, ecliptic, and pole of the rotation of the earth defined the celestial reference frame. The present-day international celestial reference frame (ICRF) is defined by the stable positions of extragalactic radio sources observed by very long baseline interferometry (VLBI), and is maintained by the IERS. Again, historically, we identify the directions of the instantaneous rotation axis as the celestial ephemeris pole (CEP) and the normal of the ecliptic as the north ecliptic pole (NEP). The CEP has recently obtained a companion called the celestial intermediary pole (CIP), as will be explained below.

The angle between directions of both poles, or the obliquity, is about 23.5°, which, by virtue of geometry, is also the angle between the instantaneous equator and the ecliptic. As shown in Figure 4.2.1, the rotation axis can be visualized as moving on a rippled cone whose axis coincides with the NEP. Mathematically, the complete motion is split into a smooth long-periodic motion called lunisolar precession and periodic motions called nutations. Precession and nutation therefore refer to the motion of the earth's instantaneous rotation axis in space. A more differentiated definition in connection with the CIP is provided below. It takes about 26,000 years for the rotation axis to complete one motion around the cone. One may view the nutations as ripples on the circular cone. The longest nutation has a period of 18.6 years and also happens to have the largest amplitude of about 20″. The cause for nutation is the ever-changing gravitational attraction of sun, moon, earth, and planets. Newton's law of gravitation states that the gravitational force between two bodies is proportional to their masses and inversely proportional to the square of their separation. Because of the orbital motions of the earth and the moon, the earth-sun and earth-moon distances change continuously and periodically. As a result, the nutations are periodic in time and reflect the periodic motions of the earth and moon. There are also small planetary precessions stemming from a motion of

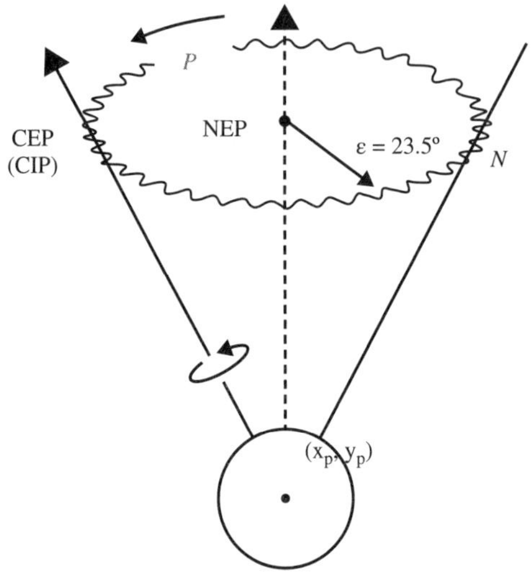

Figure 4.2.1 Lunisolar precession and nutation. The spatial motion of the CEP is parameterized in terms of precession and nutation.

the ecliptic. Nonrigidity effects of the earth on the nutations can be observed with today's high-precision measurement systems. A spherical earth with homogeneous density distribution would neither precess nor nutate.

Because the rotation axis moves in space, the coordinates of stars or extragalactic radio sources change with time due to the motion of the coordinate system. An international celestial reference frame (ICRF) has been defined for the fundamental epoch

$$\text{J2000.0} \equiv \text{January 1, 2000, 12h } TT \tag{4.2.1}$$

The letter "J" in J2000.0 indicates "Julian." We treat the subject of time in greater detail below. Let it suffice to simply state that TT represents terrestrial time, which is realized by the international atomic time (TAI) as

$$TT = TAI + 32^{s}.184 \tag{4.2.2}$$

We denote the respective coordinate system at this initial epoch J2000.0 by $(\overline{X})$. The $\overline{Z}$ axis coincides with the pole. The $\overline{X}$ axis lies in the equatorial plane and points toward the vernal equinox. In reality, in order to maintain consistency the precise definition of the first axis takes earlier definitions into consideration that were based on fundamental star catalogues. Because the ICRF is defined at a specific epoch J2000.0, the directions of the axis of $\overline{X}$ are stable in space per the definition.

Consider two widely separated VLBI antennas on the surface of the earth that are observing signals from a quasar. Because of the great distance to quasars, their direction is the same to any observer regardless where the observer is located on the earth's surface or where the earth happens to be in its orbit around the sun. VLBI observations allow one to relate the orientation of the baseline, and therefore the orientation of the earth, to the inertial directions to the quasars. Any variation in the earth's daily rotation, polar motion, or deficiencies in the adopted mathematical model of nutations, can be detected. The current ICRF solution includes about 600 extragalactic radio sources. The details of VLBI are not discussed here but left to the specialized literature. Let it be mentioned, though, that VLBI and GPS techniques have some similarities. In fact, the early developments in accurate GPS baseline determination very much benefited from existing experiences with VLBI.

4.2.1 Transforming Terrestrial and Celestial Frames

GNSS users typically do not get explicitly involved with transformations between the ITRF and the ICRF described in this section because the satellite ephemeris is generally provided in an ECEF reference frame such as the ITRF, or is provided in the form of the broadcast ephemeris message from which satellite positions can readily be extracted in an ECEF frame. However, those in the field of orbital determination need to know about the ICRF because the motions of the satellites are described in an inertial frame and, therefore, need to appropriately apply earth orientation parameters.

Because of increasing measurement accuracy of all major geodetic measurement systems such as GPS, VLBI, and SLR, more accurate values for several nutation coefficients have been determined, including nutations with periods shorter than two days. Furthermore, it was recognized that GAST (Greenwich apparent sidereal time) does not strictly represent the earth sidereal rotation angle since it depends on the changing nutation in right ascension. However, it was important that a new measure for the earth rotation angle be developed that is independent of nutation, in particular since some of the current nutation coefficients might still be subject to change in the future, and that therefore a new approach for transforming between ITRF and ICRF was needed.

The latest algorithm for transforming ITRF to ICRF and vice versa is given in Petit and Luzum (2010, Chapter 5) and Kaplan (2005). These documents also provide useful background information, including a listing of the various IAU (International Astronomical Union) resolutions that laid the framework for the new procedures. Typically, when using expressions for precession and nutation, the computation load is high because long trigonometric time series need to be evaluated, care must be taken to use a sufficient number of terms, and all parameters must be properly identified and dealt with. The IAU maintains a SOFA (Standards of Fundamental Astronomy) service that maintains an authoritative set of algorithms and procedures and makes respective software available (http://www.iausofa.org). Because any user is very likely to take advantage of such free and authoritative software, and because all mathematical expressions are also available in the above references, we only provide some selected expressions as needed to understand the underlying model concepts.

As a result of implementing the IAU 2000/2006 resolutions, the definition of the new pole, now called CIP (celestial intermediary pole), excludes nutations with a period of less than two days (Figure 4.2.2). In view of including only a subset of nutations, the new pole is called an intermediary pole. The plane perpendicular to the direction of the thus defined CIP is called the intermediary CIP equatorial plane. On

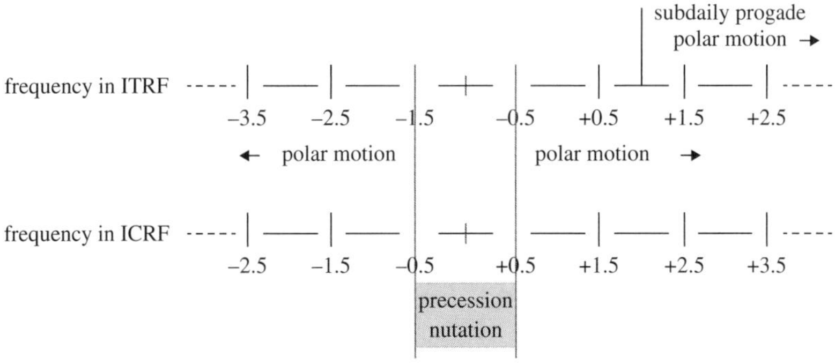

Figure 4.2.2 Conventional separation of nutation and polar motion. The units are cycles per sidereal day.

this equatorial plane/equator, there are two longitude origins, the CIO (celestial intermediary origin) and the TIO (terrestrial intermediary origin). The geocentric angle between the CIO and the TIO at a given time t is the ERA (earth rotation angle). The ERA is a rigorous measure of the sidereal rotation of the earth and is not affected by precession and nutation. The CIO is analogous to the equinox, which is the legacy reference point for sidereal time, but it is not affected by precession and nutation either.

For readers who are familiar with the old procedure and terminology, some clarification might be needed. The acronym CIO was used in the past to identify the conventional international origin of polar motion, which, loosely speaking, is defined as the mean pole position of the instantaneous rotation axis with respect to the curst for the period of about 1900 and 1905. Per definition, the z axis (semiminor axis of the ellipsoid, Section 4.3.2) of the geodetic coordinate systems was aligned to this direction during most of the past century until it was replaced by the BIH (Bureau International de l'Heure) conventional terrestrial pole (CTP) of 1984.0. Today, the z axis of the ITRF approximates this direction and is defined by the set of adopted station coordinates and epoch.

Once again, to benefit readers who are familiar with the old methods of transforming between ITRF and ICRF using Greenwich apparent sidereal time (GAST), the apparent equator, and the equinox, two identical algorithms have been designed. They are referred to as CIO-based transformation and equinox-based transformation.

The CIO-based coordinate transformation from the ITRF to the ICRF at time t can be stated as follows:

$$\overline{\boldsymbol{X}} = \boldsymbol{Q}(t)\, \boldsymbol{R}(t)\, \boldsymbol{W}(t)\boldsymbol{x} \tag{4.2.3}$$

$$\boldsymbol{W}(t) = \boldsymbol{R}_3(-s')\boldsymbol{R}_2(x_p)\boldsymbol{R}_1(y_p) \tag{4.2.4}$$

$$\boldsymbol{R}(t) = \boldsymbol{R}_3(-ERA) \tag{4.2.5}$$

$$\boldsymbol{Q}(t) = \boldsymbol{R}_3(-E)\boldsymbol{R}_2(-d)\boldsymbol{R}_3(E)\boldsymbol{R}_3(s) \tag{4.2.6}$$

The matrix $\boldsymbol{R}_i$ denotes a rotation around axis i (Appendix A). The symbols x_p, y_p denote the polar coordinates of the CIP, i.e., the pole whose definition excludes nutations with a period smaller than 2 days. The quantity s' represents a small additional rotation that becomes necessary because of the new definition. As stated above, the ERA (earth rotation angle) is the angle between CIO and TIO and represents a rigorous measure of the sidereal rotation of the earth. The symbols E and d are functions of precession and nutation, and s represents another rotation similar to s'. Looking at (4.2.3) to (4.2.6), we see that the desired separation has taken place: the matrix $\boldsymbol{Q}(t)$ depends only on precession and nutation, and $\boldsymbol{R}(t)$ depends only on the rotation of the earth.

The complete expression or polar motion, i.e., the coordinates of the CIP with respect to the ITRF z axis, is the sum of three parts (Petit and Luzum, 2010, Chapter 5, equation 5.11):

$$(x_p, y_p) = (x, y)_{IERS} + (\Delta x, \Delta y)_{ocean\ tides} + (\Delta x, \Delta y)_{librations} \tag{4.2.7}$$

The coordinates $(x, y)_{\text{IERS}}$ are the ones published and distributed by IERS. The $(\Delta x, \Delta y)_{ocean\ tides}$ term represents the diurnal and semidiurnal variation of pole coordinates due to ocean tides, and $(\Delta x, \Delta y)_{librations}$ represents the variation of pole coordinates due to those nutations having a period of less than 2 days that are not part of the IAU 2000 nutation model (and the definition of the CIP). Again, the SOFA service provides software to compute the latter two types of pole variations as a function of time.

The equinox-based transformation can be written as

$$\overline{\boldsymbol{X}} = \boldsymbol{B} \cdot \boldsymbol{P}(t)\, \boldsymbol{N}(t) \boldsymbol{R}_3(GAST)\, \boldsymbol{W}(t) \boldsymbol{x} \tag{4.2.8}$$

Here, the matrix $\boldsymbol{B}$ represents a series of small but constant rotations, $\boldsymbol{P}(t)$ and $\boldsymbol{N}(t)$ are the precession and nutation matrices, respectively, and $\boldsymbol{R}_3(GAST)$ is a rotation around the third axis by the GAST angle. Once again, the SOFA service provides software to compute all transformation matrices as a function of time.

It should be noted that the transformations (4.2.3) and (4.2.8) are identical. If the respective transformation matrices in each expression were combined into one matrix, then the elements of the two matrices would be identical. Consistent with these transformations, expression (4.2.7) applies to computing the pole coordinates of the CIP relative to the ITRF. If $\overline{\boldsymbol{X}}$ is used to compute right ascension and declination, one speaks of intermediary right ascension and intermediary declination.

In order to contrast the above new transformation procedures with the old equinox-based transformation in which all nutation terms were used in the definition of the ICRF pole, we summarize the latter briefly and provide at the same time additional insight on some of the parameters used. It starts with (McCarthy, 1996, p. 21; Mueller, 1969, p. 65)

$$\overline{\boldsymbol{X}} = \boldsymbol{P}(t) \boldsymbol{N}(t)\, \boldsymbol{R}_3(-GAST)\, \breve{\boldsymbol{R}}(t) \boldsymbol{x} \tag{4.2.9}$$

where

$$\breve{\boldsymbol{R}}(t) = \boldsymbol{R}_1(y_p)\boldsymbol{R}_2(x_p) \tag{4.2.10}$$

$$\boldsymbol{P}(t) = \boldsymbol{R}_3(\zeta)\boldsymbol{R}_2(-\theta)\boldsymbol{R}_3(z) \tag{4.2.11}$$

$$\boldsymbol{N}(t) = \boldsymbol{R}_1(-\varepsilon)\boldsymbol{R}_3(\Delta\psi)\boldsymbol{R}_1(\varepsilon + \Delta\varepsilon) \tag{4.2.12}$$

with

$$\zeta = 2306''.2181t + 0''.30188t^2 + 0''.017998t^3 \tag{4.2.13}$$

$$z = 2306''.2181t + 1''.09468t^2 + 0''.018203t^3 \tag{4.2.14}$$

$$\theta = 2004''.3109t - 0''.42665t^2 - 0''.041833t^3 \tag{4.2.15}$$

$$\Delta\psi = -17''.1996\sin(\Omega) + 0''.2062\sin(2\Omega) - 1''.3187\sin(2F - 2D + 2\Omega) + \cdots + d\psi \tag{4.2.16}$$

$$\Delta\varepsilon = 9''.2025\cos(\Omega) - 0''.0895\cos(2\Omega) + 0''.5736\cos(2F - 2D + 2\Omega) + \cdots + d\varepsilon \tag{4.2.17}$$

$$\varepsilon = 84381''.448 - 46''.8150t - 0''.00059t^2 + 0''.001813t^3 \tag{4.2.18}$$

where t is the time since J2000.0, expressed in Julian centuries of 36,525 days. The arguments of the trigonometric terms in (4.2.16) and (4.2.17) are simple functions of the fundamental periodic elements l, l', F, D, and Ω, resulting in nutation periods that vary from 18.6 years to about 5 days (recall that this refers to the old set of nutations). Of particular interest is Ω, which appears as a trigonometric argument in the first term of these equations. The largest nutation, which also has the longest period (18.6 years), is a function of Ω only, which represents the rotation of the lunar orbital plane around the ecliptic pole. This old set of nutations contains already more than 100 entries. The amplitudes of the nutations are based on geophysical models of the earth. However, because model imperfections became noticeable as the observation accuracy increased, the so-called celestial pole offsets $d\psi$ and $d\varepsilon$ were added to (4.2.16) and (4.2.17). Eventually these newly determined offsets became part of the IAU nutation model now in use.

The element Ω also describes the 18.6-year tidal period. Since tides and nutations are caused by the same gravitational attraction, it is actually possible to transform the mathematical series of nutations into the corresponding series of tides. Therefore, the solid earth tide expression (4.1.1) could be developed into a series of sine and cosine terms with arguments being simple functions of the fundamental periodic elements. The expressions for the fundamental periodic elements are as follows:

$$l = \text{Mean Anomaly of the Moon}$$
$$= 134^\circ.96340251 + 1717915923''.2178t + 31''.8792t^2 + 0''.051635t^3 + \cdots \tag{4.2.19}$$

$$l' = \text{Mean Anomaly of the Sun}$$
$$= 357^\circ.52910918 + 12596581''.0481t - 0''.5532t^2 - 0''.000136t^3 + \cdots \quad (4.2.20)$$

$$F = L - \Omega$$
$$= 93^\circ.27209062 + 1739527262''.8478t - 12''.7512t^2 - 0''.001037t^3 + \cdots \quad (4.2.21)$$

$$D = \text{Mean Elongation of the Moon from the Sun}$$
$$= 297^\circ.85019547 + 1602961601''.2090t - 6''.3706t^2 + 0''.006593t^3 + \cdots \quad (4.2.22)$$

$$\Omega = \text{Mean Longitude of the Ascending Node of the Moon}$$
$$= 125^\circ.04455501 - 6962890''.2665t + 7''.4722t^2 + 0''.007702t^3 + \cdots \quad (4.2.23)$$

The symbol L denotes the mean longitude of the moon. In these equations, the time t is again measured in Julian centuries of 36,525 days since J2000.0,

$$t = \frac{(TT - \text{J2000.0})_{[\text{days}]}}{36{,}525} \quad (4.2.24)$$

Since the Julian date (JD) of the fundamental epoch is

$$\text{JD(J2000.0)} = 2{,}451{,}545.0\ TT \quad (4.2.25)$$

the time t can be computed as

$$t = \frac{\text{JD} + TT_{[\text{h}]}/24 - 2{,}451{,}545.0}{36{,}525} \quad (4.2.26)$$

The Julian date is a convenient counter for mean solar days. Conversion of any Gregorian calendar date (Y = year, M = month, D = day) to JD is accomplished by the following (van Flandern and Pulkkinen, 1979):

$$\text{JD} = 367 \times Y - 7 \times [Y + (M + 9)/12]/4 + 275 \times M/9 + D + 1{,}721{,}014 \quad (4.2.27)$$

for Greenwich noon. This expression is valid for dates since March 1900. The expression is read as a Fortran-type statement; division by integers implies truncation of the quotients (no decimals are carried). Note that D is an integer.

In order to compute the GAST needed in (4.2.9), we must have universal time (UT1). The latter time is obtained from the UTC (coordinate universal time) of the epoch of observation and the $UT1$-UTC correction. UTC and UT1 will be discussed below. Suffice to say that the correction $UT1$-UTC is a by-product of the observations and is available from IERS publications. GAST can be computed in three steps. First, we compute Greenwich mean sidereal time (GMST) at the epoch $0^{\mathrm{h}}UT1$,

$$\begin{aligned} GMST_{0^{\mathrm{h}}UT1} &= 6^{\mathrm{h}}41^{\mathrm{m}}50^{\mathrm{s}}.54841 + 8640184^{\mathrm{s}}.812866T_u + 0^{\mathrm{s}}.093104T_u^2 \\ &\quad - 6^{\mathrm{s}}.2 \times 10^{-6}T_u^3 \end{aligned} \tag{4.2.28}$$

where $T_u = d_u/36525$ and d_u is the number of days elapsed since January 1, 2000, $12^{\mathrm{h}}UT1$ (taking on values ±0.5, ±1.5, etc.). In the second step, we add the difference in sidereal time that corresponds to UT1 hours of mean time,

$$GMST = GMST_{0^{\mathrm{h}}UT1} + r[(UT1 - UTC) + UTC] \tag{4.2.29}$$

$$r = 1.002737909350795 + 5.9006 \times 10^{-11}T_u - 5.9 \times 10^{-15}T_u^2 \tag{4.2.30}$$

In step three, we apply the nutation to convert the mean sidereal time to apparent sidereal time,

$$GAST = GMST + \Delta\psi \cos\varepsilon + 0''.00264 \sin\Omega + 0''.000063 \sin 2\Omega \tag{4.2.31}$$

Equation (4.2.31) clearly shows that GAST is not a rigorous linear measure of the earth rotation angle. The term $\Delta\psi\cos\varepsilon$ and the last two terms are nonlinear functions of time. Obtaining a measure for the earth's sidereal rotation that does not depend on the nutations (which might still be subject to improvements in the future), was therefore a major objective of the new definition of the CIP and the introduction of the ERA.

The true celestial coordinate system (X), whose third axis coincides with instantaneous rotation axis and the X and Y axes span the true celestial equator, follows from

$$\boldsymbol{X} = \boldsymbol{R}_3(-GAST)\boldsymbol{R}_1(\mathrm{y}_p)\boldsymbol{R}_2(\mathrm{x}_p)\boldsymbol{x} \tag{4.2.32}$$

The intermediary coordinate system ($\breve{\mathrm{x}}$),

$$\breve{\boldsymbol{x}} = \boldsymbol{R}_1(\mathrm{y}_p)\boldsymbol{R}_2(\mathrm{x}_p)\boldsymbol{x} \tag{4.2.33}$$

is not completely crust-fixed, because the third axis moves with polar motion. ($\breve{\mathrm{x}}$) is sometimes referred to as the instantaneous terrestrial coordinate system.

Using (X), the apparent right ascension and declination are computed from the expression

$$\alpha = \tan^{-1}\frac{Y}{X} \tag{4.2.34}$$

$$\delta = \tan^{-1}\frac{Z}{\sqrt{X^2 + Y^2}} \tag{4.2.35}$$

with $0° \leq \alpha < 360°$. Applying (4.2.34) and (4.2.35) to (x) gives the spherical longitude λ and latitude ϕ, respectively. Whereas the zero right ascension is at the vernal equinox and zero longitude is at the reference meridian, both increase counterclockwise when viewed from the third axis.

4.2.2 Time Systems

Twenty-four hours of GAST represents the time for two consecutive transits of the same meridian over the vernal equinox. Unfortunately, these "twenty-four" hours are not suitable to define a constant time interval because of the nonlinear dependencies seen in (4.2.31). The vernal equinox reference direction moves along the apparent celestial equator by the time-varying amount $\Delta\psi \cos\varepsilon$. In addition, the earth's daily rotation varies. This rate variation can affect the length of day by about 1 ms, corresponding to a length of 0.45 m on the equator; therefore, a more constant time scale is needed.

Let us look how one could conceptually compute *UT*1-*UTC* and, as such, UT1 if UTC is known. Assume that a geodetic space technique with a mathematical model relating the observations ℓ and parameters

$$\ell = f(\overline{\mathbf{X}}, \mathbf{x}, GAST, x_p, y_p) \tag{4.2.36}$$

Avoiding the details of such solutions, one can readily imagine different types of solutions, depending on which parameters are unknown and the type of observations available. For simplicity, let $\overline{\mathbf{X}}$ (space object position) and $\mathbf{x}$ (observing station position) be known, and the observations ℓ be taken at known UTC epochs. Then, given sufficient observational strength, it is conceptually possible to solve (4.2.36) for GAST and polar motion x_p, y_p. We could then compute GMST from (4.2.31) and substitute it into (4.2.29). The latter expression can be solved for the correction

$$\Delta UT1 = UT1 - UTC \tag{4.2.37}$$

A brief review on time might be in order. UTC is related to TAI as established by atomic clocks. Briefly, at the 13th General Conference of Weights and Measures (CGPM) in Paris in 1967, the definition of the atomic second, also called the international system (SI) second, was defined as the duration of 9,192,631,770 periods of the radiation corresponding to the state-energy transition between two hyperfine levels of the ground state of the cesium-133 atom. This definition made the atomic second agree with the length of the ephemeris time (ET) second, to the extent that measurement allowed. ET was the most stable time available around 1960 but is no longer in use. ET was derived from orbital positions of the earth around the sun. Its second was defined as a fraction of the year 1900. Because of the complicated gravitational interactions between the earth and the moon, the potential loss of energy due to tidal frictions, etc., the realization of ET was difficult. Its stability eventually did not meet the demands of emerging measurement capabilities. It served as an interim time system. Prior to ET, time was defined in terms of the earth rotation, the so-called earth rotational time scales such as GMST. The rotational time scales were even less constant because of the earth's rotational variations. The rotational time scales and ET were much less stable than atomic time. It takes a good cesium clock 20 to 30 million years to gain or lose one second. Today's modern atomic clocks perform even better. Under the same environmental conditions, atomic transitions are identical from atom to atom and do not change their properties. Clocks based on such transitions should generate the same time. The interested reader is referred to the literature for current atomic clock performance and technology.

TAI is based on the SI second; its epoch is such that $ET - TAI = 32^s.184$ on January 1, 1977. Because TAI is an atomic time scale, its epochs are related to state transitions of atoms and not to the rotation of the earth. Even though atoms are suitable to define an extremely constant time scale, it could, in principle, happen that in the distant future we would have noon, i.e., lunchtime at midnight TAI, just to exaggerate the point that an atomic clock is essentially a machine that is not sensitive to the earth rotation. The hybrid time scale UTC avoids a possible divergence described above and is highly stable at the same time. This is accomplished by using the SI second as scale and changing the epoch labeling such that

$$|\Delta\, UT1| < 0^s.9 \tag{4.2.38}$$

So, UTC follows UT1. One-second adjustments are made on either June 30 or December 31 if a change is warranted. The IERS determines the need for a leap second and announces any forthcoming step adjustment. Figure 4.2.3 shows the history of leap second adjustments. There is an ongoing discussion in the scientific community about possible advantages of discontinuing to make leap second adjustments. $UT1$-UTC shows annual and semiannual variations, as well as variations due to zonal tides.

Simple graphics shows that the mean solar day is longer than the sidereal day by about $24^h/365 \approx 4^m$. The accurate ratio of universal day over sidereal day is given in (4.2.30). UTC is the civilian time system that is broadcast on TV, on radio, and by other time services.

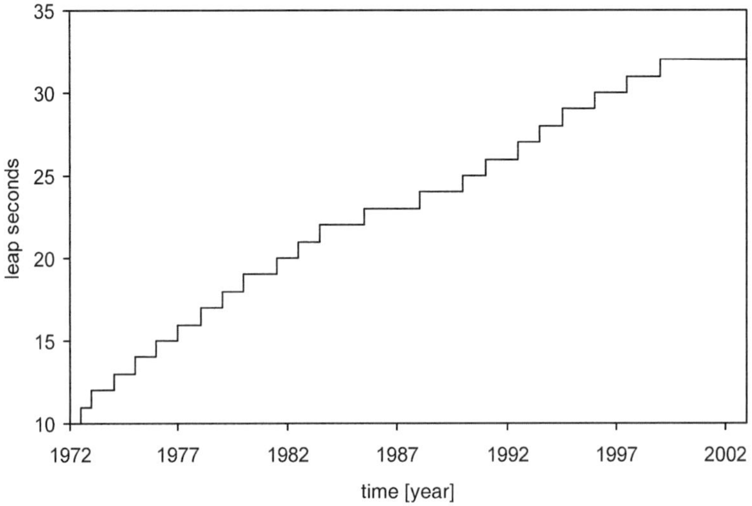

Figure 4.2.3 Leap second adjustments. [Data from IERS (2002)].

The five corrections ($UT1$-UTC, polar motion x_p and y_p, and the celestial pole offsets $d\psi$ and $d\varepsilon$), are called the earth orientation parameters (EOP). The IERS monitors and publishes these values. Modern space techniques allow these parameters to be determined with centimeter accuracy. Visit the IERS homepage at www.IERS.org to see ample graphical displays of the EOP parameters as a function of time.

Various laboratories and agencies operate several atomic clocks and produce their own independent atomic time. For example, the time scale of the U.S. Naval Observatory is called UTC (USNO), and the National Institute of Standards and Technology (NIST) produces the UTC (NIST) scale. The IERS, which uses input from 200 plus clocks and 60 plus different laboratories scattered around the world, computes TAI. UTC and TAI differ only by the integer leap seconds. TAI is not adjusted, but UTC is adjusted for leap seconds as discussed above.

The GPS satellites follow GPS time (GPST). This time scale is steered to be within one microsecond (1 μ sec) of UTC (USNO). The initial epoch of GPST is $0^h UTC$ January 6, 1980. Since that epoch, GPST has not been adjusted to account for leap seconds. It follows that GPST $- TAI = -19^s$, i.e., equal to the offset of TAI and UTC at the initial GPST epoch. The GPS control center synchronizes the clocks of the various space vehicles to GPST.

Finally, the Julian day date (JD) used in (4.2.27) is but a convenient continuous counter of mean solar days from the beginning of the year 4713 B.C. By tradition, the Julian day date begins at Greenwich noon $12^h UT1$. As such, the JD has nothing to do with the Julian calendar that was created by Julius Caesar. It provided for the leap year rule that declared a leap year of 366 days if the year's numerical designation is divisible by 4. This rule was later supplemented in the Gregorian calendar by specifying that the centuries that are not divisible by 400 are not leap years. Accordingly, the year 2000 was a leap year but the year 2100 will not be. The Gregorian calendar reform also included that the day following October 4 (Julian calendar), 1582, was labeled October 15 (Gregorian calendar). The proceedings of the conference to commemorate the 400th anniversary of the Gregorian calendar (Coyne et al., 1983) give background information on the Gregorian calendar. The astronomic justification for the leap year rules stems from the fact that the tropical year consists of $365^d.24219879$ mean solar days. The tropical year equals the time it takes the mean (fictitious) sun to make two consecutive passages over the mean vernal equinox.

4.3 DATUM

The complete definition of a geodetic datum includes the size and shape of the ellipsoid, its location and orientation, and its relation to the geoid by means of geoid undulations and deflection of the vertical. The datum currently used in the United States is NAD83, which was identified above as not being strictly a geocentric datum and is being kept that way for practical reasons. In the discussion below we briefly introduce the geoid and the ellipsoid. A discussion of geoid undulations and deflection of the vertical follows, with emphasis on how to use these elements to reduce observations to the ellipsoidal normal and the geodetic horizon. Finally, the 3D geodetic model is introduced as a general and unified model that deals with observations in all three dimensions and is also, mathematically speaking, the simplest of all.

4.3.1 Geoid

The geoid is a fundamental physical reference surface to which all observations refer if they depend on gravity. Because its shape is a result of the mass distribution inside the earth, the geoid is not only of interest to the measurement specialist but also to scientists who study the interior of the earth. Consider two point masses m_1 and m_2, separated by a distance s. According to Newton's law of gravitation, they attract each other with the force

$$F = \frac{k^2 m_1 m_2}{s^2} \tag{4.3.1}$$

where k^2 is the universal gravitational constant. The attraction between the point masses is symmetric and opposite in direction. As a matter of convenience, we consider one mass to be the "attracting" mass and the other to be the "attracted" mass. Furthermore, we assign to the attracted mass the unit mass ($m_2 = 1$) and denote the attracting mass with m. The force equation then becomes

$$F = \frac{k^2 m}{s^2} \tag{4.3.2}$$

and we speak about the force between an attracting mass and a unit mass as being attracted. Introducing an arbitrary coordinate system as seen in Figure 4.3.1, we decompose the force vector into Cartesian components. Thus,

$$\boldsymbol{F} = \begin{bmatrix} F_x \\ F_y \\ F_z \end{bmatrix} = -F \begin{bmatrix} \cos\alpha \\ \cos\beta \\ \cos\gamma \end{bmatrix} = -\frac{k^2 m}{s^2} \begin{bmatrix} \frac{x-\xi}{s} \\ \frac{y-\eta}{s} \\ \frac{z-\zeta}{s} \end{bmatrix} \tag{4.3.3}$$

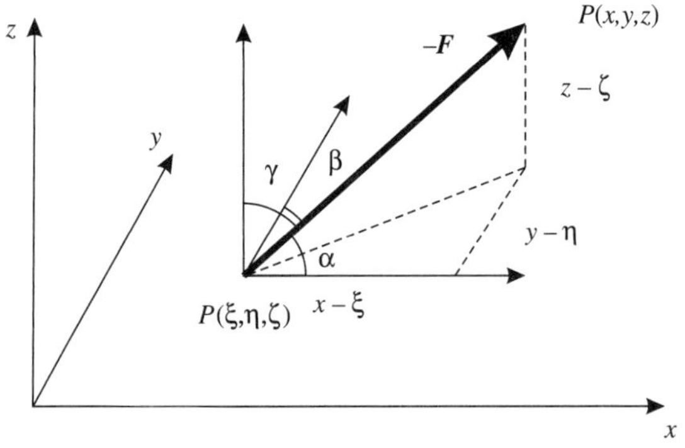

Figure 4.3.1 Components of the gravity vector.

where

$$s = \sqrt{(x - \xi)^2 + (y - \eta)^2 + (z - \zeta)^2} \tag{4.3.4}$$

The negative sign in the decomposition indicates the convention that the force vector points from the attracted mass toward the attracting mass. The coordinates (x, y, z) identify the location of the attracted mass in the specified coordinate system, and (ξ, η, ζ) denotes the location of the attracting mass. The expression

$$V = \frac{k^2 m}{s} \tag{4.3.5}$$

is called the potential of gravitation. It is a measure of the amount of work required to transport the unit mass from its initial position, a distance s from the attracting mass, to infinity. Integrating the force equation (4.3.2) gives

$$V = \int_s^\infty F\, ds = \int_s^\infty \frac{k^2 m}{s^2} ds = -\left.\frac{k^2 m}{s}\right|_s^\infty = \frac{k^2 m}{s} \tag{4.3.6}$$

In vector notation, the potential of gravitation V and the gravitational force vector $\boldsymbol{F}$ are related by

$$F_x = \frac{\partial V}{\partial x} = k^2 m \frac{\partial}{\partial x}\left(\frac{1}{s}\right) = -\frac{k^2 m}{s^2}\frac{\partial s}{\partial x} = -\frac{k^2 m}{s^2}\frac{x - \xi}{s} \tag{4.3.7}$$

Similar expressions can be written for F_y and F_z. Thus, the gradient V is

$$\text{grad } V \equiv \left[\frac{\partial V}{\partial x}\ \frac{\partial V}{\partial y}\ \frac{\partial V}{\partial z}\right]^T = [F_x \quad F_y \quad F_z]^T \tag{4.3.8}$$

From (4.3.5), it is apparent that the gravitational potential is only a function of the separation of the masses and is independent of any coordinate system used to describe the position of the attracting mass and the direction of the force vector $\boldsymbol{F}$. The gravitational potential, however, completely characterizes the gravitational force at any point by means of (4.3.8).

Because the potential is a scalar, the potential at a point is the sum of the individual potentials,

$$V = \sum V_i = \sum \frac{k^2 m_i}{s_i} \tag{4.3.9}$$

Considering a solid body M rather than individual masses, a volume integral replaces the discrete summation over the body,

$$V(x, y, z) = k^2 \iiint_M \frac{dm}{s} = k^2 \iiint_v \frac{\rho\, dv}{s} \tag{4.3.10}$$

where ρ denotes a density that varies throughout the body and v denotes the mass volume.

When deriving (4.3.10), we assumed that the body was at rest. In the case of the earth, we must consider the rotation of the earth. Let the vector $\boldsymbol{f}$ denote the centrifugal force acting on the unit mass. If the angular velocity of the earth's rotation is ω, then the centrifugal force vector can be written

$$\boldsymbol{f} = \omega^2 \boldsymbol{p} = [\omega^2 x \quad \omega^2 y \quad 0]^T \tag{4.3.11}$$

The centrifugal force acts parallel to the equatorial plane and is directed away from the axis of rotation. The vector $\boldsymbol{p}$ is the distance from the rotation axis. Using the definition of the potential and having the z axis coincide with the rotation axis, we obtain the centrifugal potential

$$\Phi = {}^1\!/_2\,\omega^2(x^2 + y^2) \tag{4.3.12}$$

Equation (4.3.12) can be verified by taking the gradient to get (4.3.11). Note again that the centrifugal potential is a function only of the distance from the rotation axis and is not affected by a particular coordinate system definition. The potential of gravity W is the sum of the gravitational and centrifugal potentials

$$W(x,\ y,\ z) = V + \Phi = k^2 \iiint_v \frac{\rho\, dv}{s} + \frac{1}{2}\omega^2(x^2 + y^2) \tag{4.3.13}$$

The gravity force vector $\boldsymbol{g}$ is the gradient of the gravity potential

$$\boldsymbol{g}(x,\ y,\ z) = \text{grad } W = \begin{bmatrix} \dfrac{\partial W}{\partial x} & \dfrac{\partial W}{\partial y} & \dfrac{\partial W}{\partial z} \end{bmatrix}^T \tag{4.3.14}$$

and represents the total force acting at a point as a result of the gravitational and centrifugal forces. The magnitude $||\boldsymbol{g}|| = g$ is called gravity. It is traditionally measured in units of gals where 1 gal = 1 cm/ sec ^{2}. The gravity increases as one moves from the equator to the poles because of the decrease in centrifugal force. Approximate values for gravity are $g_{\text{equator}} \cong 978$ gal and $g_{\text{poles}} \cong 983$ gal. The units of gravity are those of acceleration, implying the equivalence of force per unit mass and acceleration. Because of this, the gravity vector $\boldsymbol{g}$ is often termed gravity acceleration. The direction of $\boldsymbol{g}$ at a point and the direction of the plumb line or the vertical are the same.

Surfaces on which $W(x,\ y,\ z)$ is a constant are called equipotential surfaces, or level surfaces. These surfaces can principally be determined by evaluating (4.3.13) if the density distribution and angular velocity are known. Of course, the density distribution of the earth is not precisely known. Physical geodesy deals with theories that allow estimation of the equipotential surface without explicit knowledge of the density distribution. The geoid is defined to be a specific equipotential surface having gravity potential

$$W(x,\ y,\ z) = W_0 \tag{4.3.15}$$

In practice, this equipotential surface is chosen such that on the average it coincides with the global ocean surface. This is a purely arbitrary specification chosen for ease

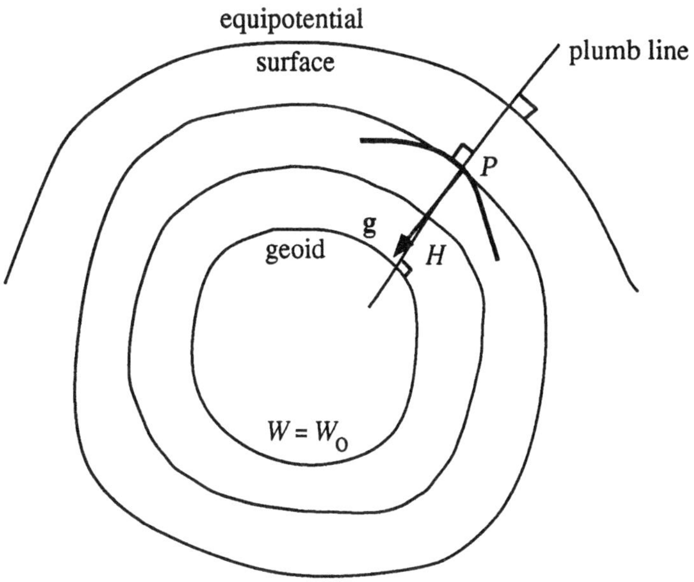

Figure 4.3.2 Equipotential surfaces and the gravity force vector.

of physical interpretation. The geoid is per definition an equipotential surface, not some ideal ocean surface.

There is an important relationship between the direction of the gravity force and equipotential surfaces, demonstrated by Figure 4.3.2. The total differential of the gravity potential at a point is

$$\begin{aligned} dW &= \frac{\partial W}{\partial x}dx + \frac{\partial W}{\partial y}dy + \frac{\partial W}{\partial z}dz \\ &= [\text{grad } W]^T \cdot d\boldsymbol{x} = \boldsymbol{g} \cdot d\boldsymbol{x} \end{aligned} \tag{4.3.16}$$

The quantity dW is the change in potential between two differentially separated points $P(x, y, z)$ and $P'(x + dx, y + dy, z + dz)$. If the vector $d\boldsymbol{x}$ is chosen such that P and P' occupy the same equipotential surface, then $dW = 0$ and

$$\boldsymbol{g} \cdot d\boldsymbol{x} = 0 \tag{4.3.17}$$

Expression (4.3.17) implies that the direction of the gravity force vector at a point is normal or perpendicular to the equipotential surface passing through the point.

The shapes of equipotential surfaces, which are related to the mass distribution within the earth through (4.3.13), have no simple analytic expressions. The plumb lines are normal to the equipotential surfaces and are space curves with finite radii of curvature and torsion. The distance along a plumb line from the geoid to a point is called the orthometric height H. The orthometric height is often misidentified as the "height above sea level." Possibly, confusion stems from the specification that the geoid closely approximates the global ocean surface.

Consider a differential line element $d\boldsymbol{x}$ along the plumb line $\|d\boldsymbol{x}\| = dH$. By noting that H is reckoned positive upward and $\boldsymbol{g}$ points downward, we can rewrite (4.3.16) as

$$\begin{aligned} dW &= \boldsymbol{g} \cdot d\boldsymbol{x} \\ &= g\, dH \cos(\boldsymbol{g},\, d\boldsymbol{x}) = g\, dH \cos(180^\circ) = -g\, dH \end{aligned} \tag{4.3.18}$$

This expression relates the change in potential to a change in the orthometric height. This equation is central in the development of the theory of geometric leveling. Writing (4.3.18) as

$$g = -\frac{dW}{dH} \tag{4.3.19}$$

it is obvious that the gravity g cannot be constant on the same equipotential surface because the equipotential surfaces are neither regular nor concentric with respect to the center of mass of the earth. This is illustrated in Figure 4.3.3, which shows two differentially separate equipotential surfaces. It is observed that

$$g_1 = -\frac{dW}{dH_1} \neq g_2 = -\frac{dW}{dH_2} \tag{4.3.20}$$

The astronomic latitude, longitude, and azimuth refer to the plumb line at the observing station. Figure 4.3.4 shows an equipotential surface through a surface point

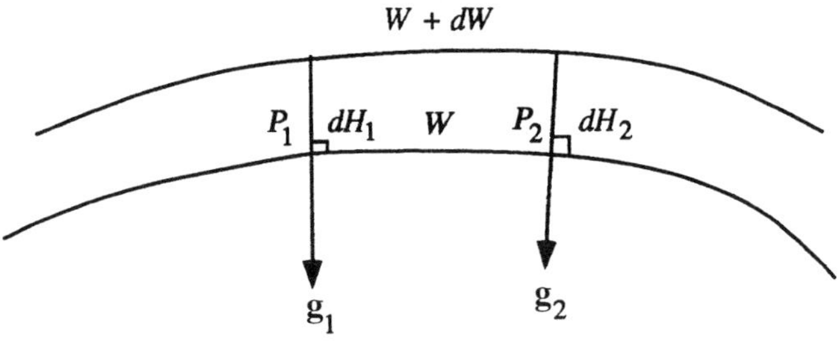

Figure 4.3.3 Gravity on the equipotential surface.

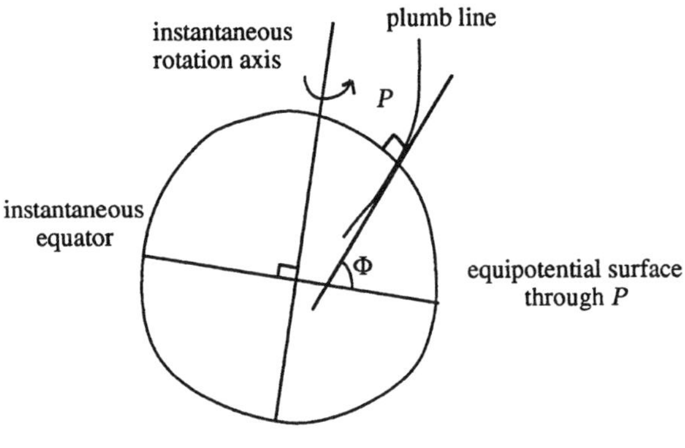

Figure 4.3.4 Astronomic latitude.

P and the instantaneous rotation axis and equator. The astronomic normal at point P, also called the local vertical, is identical to the direction of the gravity force at that point, which in turn is tangent to the plumb line. The astronomic latitude Φ at P is the angle subtended on the instantaneous equator by the astronomic normal. The astronomic normal and the parallel line to the instantaneous rotation axis span the astronomic meridian plane at point P. Note that the instantaneous rotation axis and the astronomic normal may or may not intersect. The astronomic longitude Λ is the angle subtended in the instantaneous equatorial plane between this astronomic meridian and a reference meridian, nominally the Greenwich meridian.

The geopotential number C is simply the algebraic difference between the potentials at the geoid and point P

$$C = W_0 - W \tag{4.3.21}$$

From (4.3.18) it follows that

$$W = W_0 - \int_0^H g \, dH \tag{4.3.22}$$

or

$$C = W_0 - W = \int_0^H g \, dH \tag{4.3.23}$$

or

$$H = -\int_{W_0}^{W} \frac{dW}{g} = \int_0^C \frac{dC}{g} \tag{4.3.24}$$

Equation (4.3.23) shows how combining gravity observations and leveling yields potential differences. The increment dH is obtained from spirit leveling, and the gravity g is measured along the leveling path. Consider a leveling loop as an example. Because one returns to the same point when leveling a loop, i.e., one returns to the same equipotential surface, equation (4.3.23) implies that the integral (or the sum) of the products $g\ dH$ adds up to zero. Because g varies along the loop, the sum over the leveled differences dH does not necessarily add up to zero.

The difference between the orthometric heights and the leveled heights is called the orthometric correction. Expressions for computing the orthometric correction from gravity are available in the specialized geodetic literature. An excellent introduction to height systems is found in Heiskanen and Moritz (1967, Chapter 4). Guidelines for accurate leveling are available from the NGS (Schomaker and Berry, 1981).

4.3.2 Ellipsoid of Rotation

The ellipsoid of rotation, called here simply the ellipsoid, is a relatively simple mathematical figure that closely approximates the actual geoid. When using an ellipsoid for geodetic purposes, we need to specify its shape, location, and orientation with respect to the earth. The size and shape of the ellipsoid are defined by two parameters: the semimajor axis a and the flattening f. The flattening is related to the semiminor axis b by

$$f = \frac{a - b}{a} \tag{4.3.25}$$

Appendix B contains the details of the mathematics of the ellipsoid and common values for a and b. The orientation and location of the ellipsoid often depend on when and how it was established. In the presatellite era, the goal often was to establish a local ellipsoid that best fit the geoid in a well-defined region, i.e., the area of a nation-state. The third axis, of course, always pointed toward the North Pole and the first axis in the direction of the Greenwich meridian. Using local ellipsoids as a reference does have the advantage that some of the reductions (geoid undulation, deflection of the vertical) can possibly be neglected, which is an important consideration when the geoid is not accurately known. With today's advanced geodetic satellite techniques, in particular GPS, and accurate knowledge of the geoid, one prefers so-called global ellipsoids that fit the geoid globally (whose center of figure is at the center of mass, and whose axes coincide with the ITRF). The relationship between the Cartesian coordinates $(x) = (x,\ y,\ z)$ and the geodetic coordinates $(\varphi) = (\varphi,\ \lambda,\ h)$ is given according to (B.1.9 to B.1.11),

$$x = (N + h)\cos\varphi\cos\lambda \tag{4.3.26}$$

$$y = (N + h)\cos\varphi\sin\lambda \tag{4.3.27}$$

$$z = [N(1 - e^2) + h]\sin\varphi \tag{4.3.28}$$

where the auxiliary quantities N and e are

$$N = \frac{a}{\sqrt{1 - e^2 \sin^2\varphi}} \tag{4.3.29}$$

$$e^2 = 2f - f^2 \tag{4.3.30}$$

The transformation from (x) to (φ) is given in Appendix B. It is typically performed iteratively.

4.3.3 Geoid Undulations and Deflections of the Vertical

One approach to estimate the geoid undulation is by measuring gravity or gravity gradients at the surface of the earth. At least in principle, any observable that is a function of the gravity field can contribute to the determination of the geoid. Low-earth orbiting satellites have successfully been used to determine the large structure of the geoid. Satellite-to-satellite tracking is being used to determine the temporal variations of the gravity field, and thus the geoid. The reader may want to check gravity models derived from the Gravity Recovery and Climate Experiment (GRACE) mission which was launched in early 2002. Recent earth gravity solutions show high resolution of geoid features because more observations have become available and the observations have a better global coverage to allow estimation of higher degree spherical harmonic coefficients. Pavlis et al. (2012) discuss one of the latest earth gravity models, the EGM2008, which uses a spherical harmonic expansion up to degree and order 2219 to represent the gravity field.

Actually, the gravity field or functions of the gravity field are typically expressed in terms of a spherical harmonic expansion. For example, the geoid undulation N could be expressed in the form (Lemoine et al., 1998, pp. 5–11),

$$N = \frac{GM}{\gamma r}\sum_{n=2}^{\infty}\left(\frac{a}{r}\right)^n \sum_{m=0}^{n}(\overline{C}_{nm}\cos m\lambda + \overline{S}_{nm}\sin m\lambda)\overline{P}_{nm}(\cos\theta) \tag{4.3.31}$$

In this equation, the following notations are used:

N	Geoid undulation. There should not be cause for confusion using the same symbol for the geoid undulation (4.3.31) and the radius of curvature of the prime vertical (4.3.29); both notations are traditional in the geodetic literature.
φ, λ	Latitude and longitude of station; $\theta = 90 - \varphi$ is the colatitude.
$\overline{C}_{nm}$, $\overline{S}_{nm}$	Normalized spherical harmonic coefficients (geopotential coefficients), of degree n and order m.
$\overline{P}_{nm}$	Associated Legendre functions.
r	Geocentric distance of the station.
GM	Product of the gravitational constant and the mass of the earth. GM is identical to k^2M used elsewhere in this book. Unfortunately, the symbolism is not unique in the literature. We retain the symbols typically used within the respective context.
γ	Normal gravity. Details are given below.
a	Semimajor axis of the ellipsoid.

Figure 4.3.5 shows a map of a global geoid. Although this map is dated, it still represents the global features of the geoid accurately. The geoid undulation and deflections of the vertical are related by differentiation, such as (Heiskanen and Moritz, 1967, p. 112)

$$\xi = -\frac{1}{r}\frac{\partial N}{\partial \theta} \tag{4.3.32}$$

$$\eta = -\frac{1}{r\sin\theta}\frac{\partial N}{\partial \lambda} \tag{4.3.33}$$

Differentiating (4.3.31) gives

$$\xi = -\frac{GM}{\gamma r^2}\sum_{n=2}^{\infty}\left(\frac{a}{r}\right)^n \sum_{m=0}^{n}(\overline{C}_{nm}\cos m\lambda + \overline{S}_{nm}\sin m\lambda)\frac{d\overline{P}_{nm}(\cos\theta)}{d\theta} \tag{4.3.34}$$

$$\eta = -\frac{GM}{\gamma r^2\sin\theta}\sum_{n=2}^{\infty}\left(\frac{a}{r}\right)^n \sum_{m=0}^{n} m(-\overline{C}_{nm}\sin m\lambda + \overline{S}_{nm}\cos m\lambda)\overline{P}_{nm}(\cos\theta) \tag{4.3.35}$$

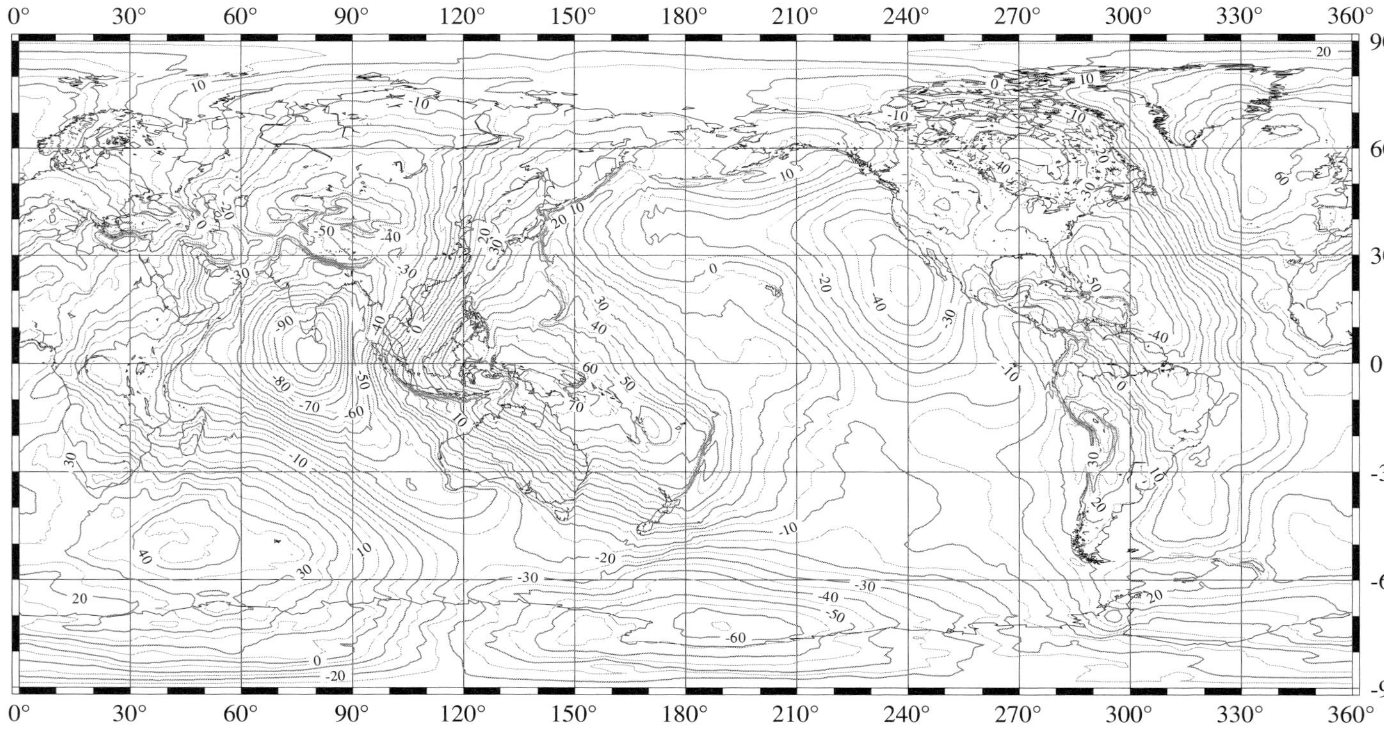

Figure 4.3.5 Geoid undulations of the EGM96 gravity field model computed relative to the GRS80 ellipsoid. The units are in meters [Courtesy German Geodetic Research Institute (DGFI), Munich].

Geoid undulations computed from expressions like (4.3.31) refer to a geocentric ellipsoid with semimajor axis a. In order to obtain the geoid undulations and deflection of the vertical for a nongeocentric ellipsoid, say the NAD83 datum, additional transformations are needed (Soler et al., 2014). Typically, free software is available to compute geoid undulations and deflection of the vertical for a specific datum.

The ellipsoid of rotation provides a simple and adequate model for the geometric shape of the earth. It is the reference for geometric computations in two and three dimensions, as discussed in the next sections. Assigning a gravitational field to the ellipsoid that approximates the actual gravitational field of the earth extends the functionality of the ellipsoid.

Merely a few specifications are needed to fix the gravity and gravitational potential for an ellipsoid, then called a normal ellipsoid. We need to assign an appropriate mass for the ellipsoid and assume that the ellipsoid rotates with the earth. Furthermore, by means of mathematical conditions, the surface of the ellipsoid is defined to be an equipotential surface of its own gravity field. Therefore, the plumb lines of this gravity field intersect the ellipsoid perpendicularly. Because of this property, this gravity field is called the normal gravity field, and the ellipsoid itself is sometimes also referred to as the level ellipsoid.

It can be shown that the normal gravity potential U is completely specified by four defining constants, which are symbolically expressed by

$$U = f(a,\ J_2,\ GM,\ \omega) \tag{4.3.36}$$

In addition to a and GM, which have already been introduced above, we need the dynamical form factor J_2 and the angular velocity of the earth ω. The dynamic form factor is a function of the principal moments of inertia of the earth (polar and equatorial moment of inertia) and is functionally related to the flattening of the ellipsoid. One important definition of the four constants in (4.3.36) comprises the Geodetic Reference System of 1980 (GRS80). The defining constants are listed in Table 4.3.1. A full documentation of this reference system is available in Moritz (1984).

The normal gravitational potential does not depend on the longitude and is given by a series of zonal spherical harmonics:

$$V = \frac{GM}{r}\left[1 - \sum_{n=1}^{\infty} J_{2n}\left(\frac{a}{r}\right)^{2n} P_{2n}(\cos\theta)\right] \tag{4.3.37}$$

TABLE 4.3.1 Constants for GRS80

Defining Constants	Derived Constants
$a = 6378137$ m	$b = 6356752.3141$ m
$GM = 3986005 \times 10^8\ \ \text{m}^3/\text{s}^2$	$1/f = 298.257222101$
$J_2 = 108263 \times 10^{-8}$	$m = 0.00344978600308$
$\omega = 7292115 \times 10^{-11}\ \ \text{rad/s}$	$\gamma_e = 9.7803267715\ \text{m/s}^2$
	$\gamma_p = 9.8321863685\ \text{m/s}^2$

Note that the subscript $2n$ is to be read "2 times n." P_{2n} denotes Legendre polynomials. The coefficients J_{2n} are a function of J_2 that can be readily computed. Several useful expressions can be derived from (4.3.37). For example, the normal gravity, defined as the magnitude of the gradient of the normal gravity field (normal gravitational potential plus centrifugal potential), is given by Somigliana's closed formula (Heiskanen and Moritz, 1967, p. 70),

$$\gamma = \frac{a\gamma_e \cos^2\varphi + b\gamma_p \sin^2\varphi}{\sqrt{a^2\cos^2\varphi + b^2\sin^2\varphi}} \tag{4.3.38}$$

The normal gravity at height h above the ellipsoid is given by (Heiskanen and Moritz, 1967, p. 79)

$$\gamma_h - \gamma = -\frac{2\gamma_e}{a}\left[1 + f + m + \left(-3f + \frac{5}{2}m\right)\sin^2\varphi\right]h + \frac{3\gamma_e}{a^2}h^2 \tag{4.3.39}$$

Equations (4.3.38) and (4.3.39) are often useful approximations of the actual gravity. The value for the auxiliary quantity m in (4.3.39) is given in Table 4.3.1. The normal gravity values for the poles and the equator, γ_p and γ_e are also listed in that table.

4.3.4 Reductions to the Ellipsoid

The primary purpose of this section is to introduce the deflection of the deflection corrections (reduction) and the relation between orthometric and ellipsoidal heights and geoid undulations. The objective is to apply these corrections to convert observed terrestrial angles (azimuths) to angles (azimuths) between normal planes of the ellipsoid, which can then serve as model observations in the three-dimensional geodetic model discussed below. Although precise astronomical latitude, longitude, and azimuth observations are generally no longer a part of the surveyor's tools because of the wide use of GNSS applications, for the sake of completeness some very brief remarks about these "old" techniques are in order.

We already stressed in connection with Figure 4.3.4 that the astronomical latitude refers to the tangent of the instantaneous plumb line and the direction of the instantaneous rotation axis, i.e., the CEP or CIP. By the way, the distinction between the latter two poles is not necessary when it comes to astronomical position determination because of the lack of accuracy of the observation technique. Nevertheless, at least for what used to be called first-order astronomic position determination, the polar motion correction should be considered. The respective correction can certainly be found in old textbooks, e.g., Mueller (1969, p. 87). Applying spherical trigonometry we obtain

$$\begin{aligned} \Phi_{\text{CTP}} &= \Phi + y_p \sin\Lambda - x_p \cos\Lambda \\ \Lambda_{\text{CTP}} &= \Phi - (y_p \cos\Lambda + x_p \sin\Lambda)\tan\Phi \\ A_{\text{CTP}} &= A - (y_p \cos\Lambda + x_p \sin\Lambda)/\cos\Phi \end{aligned} \tag{4.3.40}$$

The observed latitude, longitude, and azimuth are (Φ, Λ, A), and the polar motion coordinates are (x_p, y_p). The reduced astronomic quantities (Φ_{CTP}, Λ_{CTP}, A_{CTP}) are those values which one would have observed if the instantaneous pole CEP (CIP) had coincided with the CTP (ITRF) at the instant of observations.

Now we consider the condition that the semiminor axis of the ellipsoid and the direction of the CTP should be parallel. This condition will show a relationship between the reduced astronomic quantities (Φ_{CTP}, Λ_{CTP}, A_{CTP}) and the corresponding ellipsoidal or geodetic quantities (φ, λ, α), and as such the reductions we are looking for. The geometric relationships are shown in Figures 4.3.6 and 4.3.7. Both figures are not drawn to scale in order to show small angles. The bottom part of Figure 4.3.6 shows the ellipsoid and the ellipsoidal normal passing through a surface point P_1 and intersecting a unit sphere centered at P_1 at point Z_e. The line labeled "equipotential surface" through P_1 should indeed indicate the equipotential surface at P_1; the line $P_1 - Z_a$ is normal to the equipotential surface. The points Z_a, Z_e, CTP,

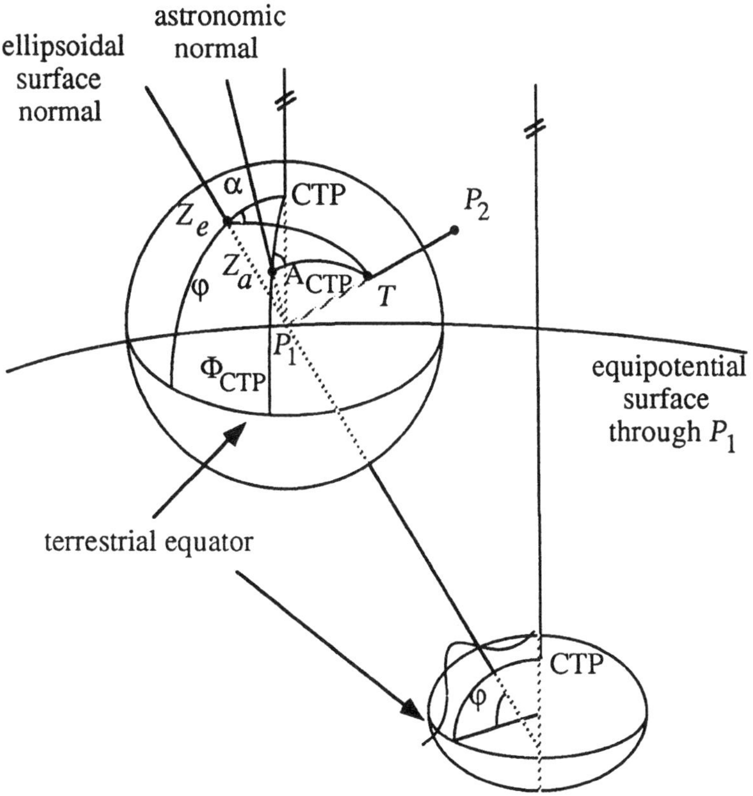

Figure 4.3.6 Astronomic and ellipsoidal normal on a topocentric sphere of direction. The astronomic normal is perpendicular to the equipotential surface at P_1. The ellipsoidal normal passes through P_1.

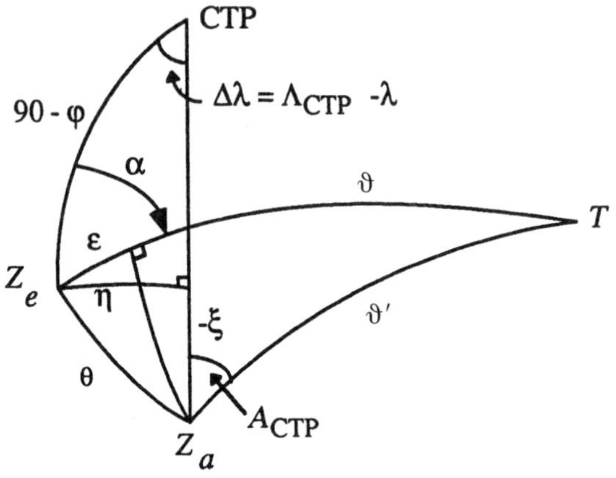

Figure 4.3.7 Deflection of the vertical components.

and T are located on the unit sphere. The line P_1 – CTP is parallel to the semiminor axis of the ellipsoid. The symbols have the following meaning:

Z_a	Astronomic zenith (sensed by instruments)
CTP	Position of the conventional terrestrial pole (z axis of ITRF)
Z_e	Ellipsoidal zenith
T	Target point (intersection $P_1 - P_2$ with sphere)
ϑ'	Observed zenith angle
Φ_{CTP}, Λ_{CTP}	Reduced astronomic latitude and longitude
A_{CTP}	Reduced astronomic azimuth of T and surface point P_2
φ, λ	Ellipsoidal (geodetic) latitude and longitude
α	Ellipsoidal (geodetic) azimuth of T and surface point P_2
ϑ	Ellipsoidal (geodetic) zenith angle
θ	Total deflection of the vertical (not colatitude, same symbol)
ε	Deflection of the vertical in the direction of azimuth
ξ, η	Deflection of the vertical components along the meridian and the prime vertical

The azimuths A_{CTP} and α are angles between normal planes defined by the astronomic and ellipsoidal normal at P_1, respectively. The intersections of these planes with the unit sphere are great circles. By applying spherical trigonometry to the various triangles in Figure 4.3.7, we eventually derive the following relations:

$$A_{\text{CTP}} - \alpha = (\Lambda_{\text{CTP}} - \lambda)\sin\varphi + (\xi\sin\alpha - \eta\cos\alpha)\ \cot\vartheta \tag{4.3.41}$$

$$\xi = \Phi_{\text{CTP}} - \varphi \tag{4.3.42}$$

$$\eta = (\Lambda_{\text{CTP}} - \lambda)\cos\varphi \tag{4.3.43}$$

$$\vartheta = \vartheta' + \xi\cos\alpha + \eta\sin\alpha \tag{4.3.44}$$

These are indeed classical equations whose derivations can be found in most of the geodetic literature, e.g., Heiskanen and Moritz (1967, p. 186). They are also given in Leick (2002). Equation (4.3.41) is called the Laplace equation. It relates the reduced astronomic azimuth and the geodetic azimuths of the target point. The deflection of the vertical, or total deflection of the vertical, is the angle between the directions of the plumb line and the ellipsoidal normal at the same point, i.e., the angle $Z_a - Z_e$. Equations (4.3.42) and (4.3.43) define the deflection of the vertical components. By convention, the deflection of the vertical is decomposed into two components, one along the meridian and one along the prime vertical (orthogonal to the meridian). The deflection components depend directly on the shape of the geoid in the region. Because the deflections of the vertical are merely another manifestation of the irregularity of the gravity field, they are mathematically related to the geoid undulation. See equations (4.3.34) and (4.3.35). Equation (4.3.44) relates the ellipsoidal and the observed zenith angle (refraction not considered).

Several observations are made. First, equations (4.3.41) to (4.3.43) relate reduced astronomic latitude, longitude, and azimuth to the respective ellipsoidal latitude, longitude, and azimuth by means of the deflection of the vertical. Second, the reduction of a horizontal angle due to deflection of the vertical equals the difference of (4.3.41) as applied to both intersecting line segments of the angle. If the zenith angles to the target points are close to 90°, then the corrections are small and can possibly be neglected. This is the reason why deflection of the vertical corrections to angles in surveying can generally be ignored. Third, historically, equation (4.3.41) was used as a condition between the reduced astronomic azimuth and the computed geodetic azimuth to control systematic errors in a network. This can be better accomplished now with GPS. Fourth, if surveyors were to compare the orientation of a GPS vector with the astronomic azimuth derived from solar or Polaris observations, they must expect a discrepancy indicated by (4.3.41). Fifth, if a surveyor were to stake out in the field an azimuth computed from coordinates, the Laplace correction would have to be considered. Sixth, finally, the last term in the Laplace equation (4.3.41) can usually be dropped because of zenith angles close to 90°.

Equations (4.3.42) and (4.3.43) also show how to specify a local ellipsoid that is tangent to the geoid at some centrally located station called the initial point, and whose semiminor axis is still parallel to the CTP. If we specify that at the initial point the reduced astronomic latitude, longitude, and azimuth equal the ellipsoidal latitude, longitude, and azimuth, respectively, then we ensure parallelism of the semimajor axis and the direction of the CTP; the geoid normal and the ellipsoidal normal coincide at that initial point. If, in addition, we set the undulation to zero, then the ellipsoid touches the geoid tangentially at the initial point. Thus the local ellipsoid will have at the initial point

$$\varphi = \Phi_{\mathrm{CTP}} \tag{4.3.45}$$

$$\lambda = \Lambda_{\mathrm{CTP}} \tag{4.3.46}$$

$$\alpha = A_{\mathrm{CTP}} \tag{4.3.47}$$

$$N = 0 \tag{4.3.48}$$

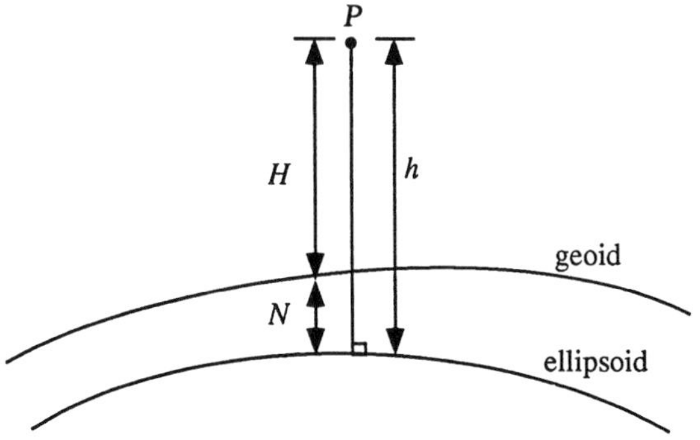

Figure 4.3.8 Geoid undulation, orthometric, and ellipsoidal heights.

Another important quantity linking the ellipsoid and the geoids is the geoid undulation. The relationship between the ellipsoidal height h, the orthometric height H, and the geoid undulation N, is

$$h = H + N \tag{4.3.49}$$

where N is the geoid undulation with respect to the specific ellipsoid. As shown in Figure 4.3.8 the geoid undulation refers to a datum having a well-defined location, orientation, and size of its reference ellipsoid. Once again, the geoid undulation N is computable from expressions like (4.3.31) and is entirely different in meaning from the radius of curvature (4.3.29), so that the traditional use of N for both quantities does not cause confusion.

In regards to quality control of observations, the local ellipsoid can serve as a convenient computation reference in 3-dimensional geodetic adjustments (see next section) in case of small networks, such as local and regional surveys. In these cases, it is not at all necessary to determine the size and shape of a best-fitting local ellipsoid. It is sufficient to adopt the size and shape of any of the current geocentric ellipsoids. Because the deflections of the vertical will be small in the region around the initial point, they can be neglected. Any of the network stations can serve as an initial point and its coordinates do not even have to be accurately known. Similar considerations are valued in regards to the geoid undulations, which are also small because the local ellipsoid is tangent to the geoid at the initial point. For the quality control purpose of observations, the azimuth in (4.3.47) can be freely chosen in such cases, which is yet another convenience. Therefore, the 3D geodetic model is attractive for a quick quality control minimal constraint adjustment to see if the set of observations are consistent, i.e., free of blunders.

4.4 3D GEODETIC MODEL

Once the angular observations have been corrected for the deflection of the vertical, it is a simple matter to develop the mathematics for the 3D geodetic model. The reduced observations, i.e., the observables of the 3D geodetic model, are the geodetic azimuth α, the geodetic horizontal angle δ, the geodetic vertical angle β (or the geodetic zenith angle ϑ), and the slant distance s. Geometrically speaking, these observables refer to the geodetic horizon and the ellipsoidal normal. The reduced horizontal angle is an angle between two normal planes, defined by the target points and the ellipsoidal normal at the observing station. The geodetic vertical angle is the angle between the geodetic horizon and the line of sight to the target.

We assume that the vertical angle has been corrected for atmospheric refraction. The model can be readily extended to include refraction parameters if needed. Thanks to the availability of GNSS systems, we no longer depend on vertical angle observations to support the vertical dimension. The primary purpose of vertical angles in most cases is to support the vertical dimension when adjusting slant distances (because slant distances contribute primarily horizontal information, at least in flat terrain).

Figure 4.4.1 shows the local geodetic coordinate system (w) = (n, e, u), which plays a central role in the development of the mathematical model. The axes n and e span the local geodetic horizon (plane perpendicular to the ellipsoidal normal through the point P_1 on the surface of the earth). The n axis points north, the e axis points east, and the u axis coincides with the ellipsoidal normal (with the positive end outward of the ellipsoid). The spatial orientation of the local geodetic coordinate system is completely specified by the geodetic latitude φ and the geodetic longitude λ. Recall that the z axis coincides with the direction of the CTP.

Figure 4.4.2 shows the geodetic azimuth and vertical angle (or zenith angle) between points P_1 and P_2 in relation to the local geodetic coordinate system. One should keep in mind that the symbol h still denotes the geodetic height of a point above the ellipsoid, whereas the u coordinate refers to the height of the second station P_2 above the local geodetic horizon of P_1. It follows that

$$n = s \cos \beta \cos \alpha \tag{4.4.1}$$

$$e = s \cos \beta \sin \alpha \tag{4.4.2}$$

$$u = s \sin \beta \tag{4.4.3}$$

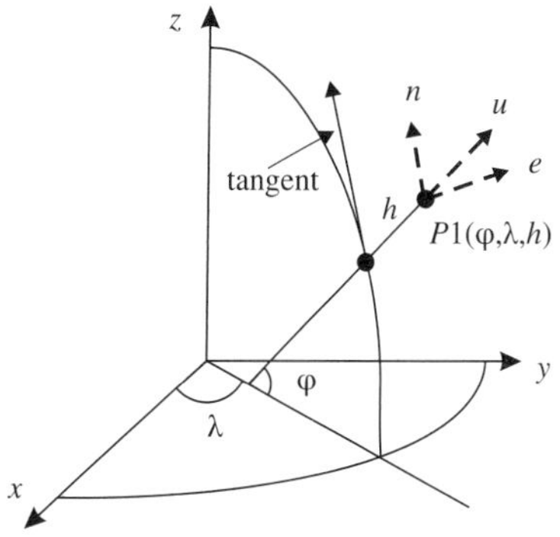

Figure 4.4.1 The local geodetic coordinate system.

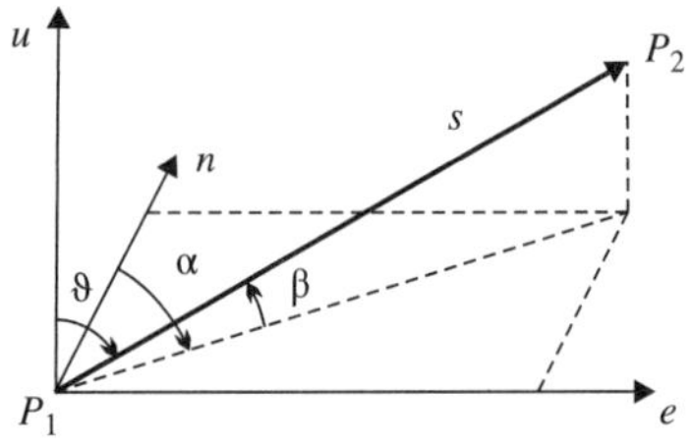

Figure 4.4.2 3D model observations.

The inverses of (4.4.1) to (4.4.3) are

$$\alpha = \tan^{-1}\left(\frac{e}{n}\right) \tag{4.4.4}$$

$$\beta = 90^\circ - \vartheta = \sin^{-1}\left(\frac{u}{s}\right) \tag{4.4.5}$$

$$s = \sqrt{n^2 + e^2 + u^2} \tag{4.4.6}$$

The relationship between the local geodetic coordinate system and the geocentric Cartesian system (x) is illustrated in Figure 4.4.1:

$$\begin{bmatrix} n \\ -e \\ u \end{bmatrix} = \boldsymbol{R}_2\ (\varphi - 90^\circ)\ \boldsymbol{R}_3\ (\lambda - 180^\circ) \begin{bmatrix} \Delta x \\ \Delta y \\ \Delta z \end{bmatrix} \tag{4.4.7}$$

where $\boldsymbol{R}_2$ and $\boldsymbol{R}_3$ denote the rotation matrices given in Appendix A, and

$$\Delta \boldsymbol{X} \equiv \begin{bmatrix} \Delta x \\ \Delta y \\ \Delta z \end{bmatrix} = \begin{bmatrix} x_2 - x_1 \\ y_2 - y_1 \\ z_2 - z_1 \end{bmatrix} \tag{4.4.8}$$

Subscripts will be used when needed to clarify the use of symbols. For example, the differencing operation Δ in (4.4.7) implies $\Delta x \equiv \Delta x_{12} = x_2 - x_1$. The same convention is followed for other differences. A more complete notation for the local geodetic coordinates is (n_1, e_1, u_1) instead of (n, e, u), to emphasize that these components refer to the geodetic horizon at P_1. Similarly, a more unambiguous notation is $(\alpha_{12}, \beta_{12}, \vartheta_{12})$ instead of just $(\alpha, \beta, \vartheta)$ or even $(\alpha_1, \beta_1, \vartheta_1)$, to emphasize that these observables are taken at station P_1 to P_2. For the slant distance, the subscripts do not matter because $s = s_1 = s_{12} = s_{21}$.

Changing the sign of e in (4.4.7) and combining the rotation matrices $\boldsymbol{R}_2$ and $\boldsymbol{R}_3$ one obtains

$$\boldsymbol{w} = \boldsymbol{R}(\varphi, \lambda)\Delta \boldsymbol{x} \tag{4.4.9}$$

with

$$\boldsymbol{R} = \begin{bmatrix} -\sin\varphi\cos\lambda & -\sin\varphi\sin\lambda & \cos\varphi \\ -\sin\lambda & \cos\lambda & 0 \\ \cos\varphi\cos\lambda & \cos\varphi\sin\lambda & \sin\varphi \end{bmatrix} \tag{4.4.10}$$

Substituting (4.4.9) and (4.4.10) into (4.4.4) to (4.4.6) gives expressions for the geodetic observables as functions of the geocentric Cartesian coordinate differences and the geodetic position of P_1:

$$\alpha_1 = \tan^{-1}\left(\frac{-\sin\lambda_1\ \ \Delta x + \cos\lambda_1\ \ \Delta y}{-\sin\varphi_1\cos\lambda_1\ \ \Delta x - \sin\varphi_1\sin\lambda_1\ \ \Delta y + \cos\varphi_1\ \ \Delta z}\right) \tag{4.4.11}$$

$$\beta_1 = \sin^{-1}\left(\frac{\cos\varphi_1\cos\lambda_1\ \ \Delta x + \cos\varphi_1\sin\lambda_1\,\Delta y + \sin\varphi_1\ \ \Delta z}{\sqrt{\Delta x^2 + \Delta y^2 + \Delta z^2}}\right) \tag{4.4.12}$$

$$s = \sqrt{\Delta x^2 + \Delta y^2 + \Delta z^2} \tag{4.4.13}$$

Equations (4.4.11) to (4.4.13) are the backbone of the 3D geodetic model. Other observations such as horizontal angles, heights, and height differences—even GPS vectors—can be readily implemented. Equation (4.4.12) assumes that the vertical angle has been corrected for refraction. One should take note of how little mathematics is required to derive these equations. Differential geometry is not required, and neither is the geodesic line.

4.4.1 Partial Derivatives

Because (4.4.11) to (4.4.13) expressed the geodetic observables explicitly as a function of the coordinates, the observation equation adjustment model $\boldsymbol{\ell}_a = \boldsymbol{f}(\boldsymbol{x}_a)$ can be readily used. The 3D nonlinear model has the general form

$$\alpha_1 = \alpha(x_1,\ y_1,\ z_1,\ x_2,\ y_2,\ z_2) \tag{4.4.14}$$

$$\beta_1 = \beta(x_1,\ y_1,\ z_1,\ x_2,\ y_2,\ z_2) \tag{4.4.15}$$

$$s = s(x_1,\ y_1,\ z_1,\ x_2,\ y_2,\ z_2) \tag{4.4.16}$$

The observables and parameters are $\{\alpha_1,\ \beta_1,\ s\}$ and $\{x_1,\ y_1,\ z_1,\ x_2,\ y_2,\ z_2\}$, respectively. To find the elements of the design matrix, we require the partial derivatives with respect to the parameters. The general form is

$$\begin{bmatrix} d\alpha_1 \\ d\beta_1 \\ ds \end{bmatrix} = \begin{bmatrix} g_{11} & g_{12} & g_{13} & \vdots & g_{14} & g_{15} & g_{16} \\ g_{21} & g_{22} & g_{23} & \vdots & g_{24} & g_{25} & g_{26} \\ g_{31} & g_{32} & g_{33} & & g_{34} & g_{35} & g_{36} \end{bmatrix} \begin{bmatrix} dx_1 \\ dy_1 \\ dz_1 \\ \cdots \\ dx_2 \\ dy_2 \\ dz_2 \end{bmatrix} = [\mathbf{G}_1\ \vdots\ \mathbf{G}_2] \begin{bmatrix} d\boldsymbol{x}_1 \\ \cdots \\ d\boldsymbol{x}_2 \end{bmatrix} \tag{4.4.17}$$

with $d\boldsymbol{x}_i = [dx_i\quad dy_i\quad dz_i]^T$. The partial derivatives are listed in Table 4.4.1. This particular form of the partial derivatives follows from Wolf (1963), after some additional algebraic manipulations.

TABLE 4.4.1 Partial Derivatives with Respect to Cartesian Coordinates

$$g_{11} = \frac{\partial \alpha_1}{\partial x_1} = -g_{14} = \frac{-\sin\varphi_1 \cos\lambda_1 \sin\alpha_1 + \sin\lambda_1 \cos\alpha_1}{s\cos\beta_1} \quad \text{(a)}$$

$$g_{12} = \frac{\partial \alpha_1}{\partial y_1} = -g_{15} = \frac{-\sin\varphi_1 \sin\lambda_1 \sin\alpha_1 - \cos\lambda_1 \cos\alpha_1}{s\cos\beta_1} \quad \text{(b)}$$

$$g_{13} = \frac{\partial \alpha_1}{\partial z_1} = -g_{16} = \frac{\cos\varphi_1 \sin\alpha_1}{s\cos\beta_1} \quad \text{(c)}$$

$$g_{21} = \frac{\partial \beta_1}{\partial x_1} = -g_{24} = \frac{-s\cos\varphi_1 \cos\lambda_1 + \sin\beta_1 \ \Delta x}{s^2\cos\beta_1} \quad \text{(d)}$$

$$g_{22} = \frac{\partial \beta_1}{\partial y_1} = -g_{25} = \frac{-s\cos\varphi_1 \sin\lambda_1 + \sin\beta_1 \ \Delta x}{s^2\cos\beta_1} \quad \text{(e)}$$

$$g_{23} = \frac{\partial \beta_1}{\partial z_1} = -g_{26} = \frac{-s\sin\varphi_1 + \sin\beta_1 \ \Delta z}{s^2\cos\beta_1} \quad \text{(f)}$$

$$g_{31} = \frac{\partial s}{\partial x_1} = -g_{34} = \frac{-\Delta x}{s} \quad \text{(g)}$$

$$g_{32} = \frac{\partial s}{\partial y_1} = -g_{35} = \frac{-\Delta y}{s} \quad \text{(h)}$$

$$g_{33} = \frac{\partial s}{\partial z_1} = -g_{36} = \frac{-\Delta z}{s} \quad \text{(i)}$$

4.4.2 Reparameterization

Often the geodetic latitude, longitude, and height are preferred as parameters instead of the Cartesian components of (x). One reason for such a reparameterization is that humans can visualize changes more readily in latitude, longitude, and height than changes in geocentric coordinates. The required transformation is given by (B.1.16).

$$d\boldsymbol{x} = \begin{bmatrix} -(M+h)\cos\lambda\sin\varphi & -(N+h)\cos\varphi\sin\lambda & \cos\varphi\cos\lambda \\ -(M+h)\sin\lambda\sin\varphi & (N+h)\cos\varphi\cos\lambda & \cos\varphi\sin\lambda \\ (M+h)\cos\varphi & 0 & \sin\varphi \end{bmatrix} \begin{bmatrix} d\varphi \\ d\lambda \\ dh \end{bmatrix}$$

$$= \boldsymbol{J} \begin{bmatrix} d\varphi \\ d\lambda \\ dh \end{bmatrix} \quad (4.4.18)$$

The expressions for the radius of curvatures M and N are given in (B.1.7) and (B.1.6). The matrix $\boldsymbol{J}$ must be evaluated for the geodetic latitude and longitude of the point under consideration; thus, $\boldsymbol{J}_1(\varphi_1, \lambda_1, h_1)$ and $\boldsymbol{J}_2(\varphi_2, \lambda_2, h_2)$ denote the transformation matrices for points P_1 and P_2, respectively. Substituting (4.4.18) into

(4.4.17), we obtain the parameterization in terms of geodetic latitude, longitude, and height:

$$\begin{bmatrix} d\alpha_1 \\ d\beta_1 \\ ds \end{bmatrix} = [\mathbf{G}_1 \mathbf{J}_1 \quad : \quad \mathbf{G}_2 \mathbf{J}_2] \begin{bmatrix} d\varphi_1 \\ d\lambda_1 \\ dh_1 \\ \cdots \\ d\varphi_2 \\ d\lambda_2 \\ dh_2 \end{bmatrix} \tag{4.4.19}$$

To achieve a parameterization that is even easier to interpret, we transform the differential changes in geodetic latitude and longitude parameters $(d\varphi,\ d\lambda)$ into corresponding changes $(dn,\ de)$ in the local geodetic horizon. Keeping the geometric interpretation of the radii of curvatures M and N as detailed in Appendix B, one can further deduce that

$$d\mathbf{w} = \begin{bmatrix} M+h & 0 & 0 \\ 0 & (N+h)\cos\varphi & 0 \\ 0 & 0 & 1 \end{bmatrix} \begin{bmatrix} d\varphi \\ d\lambda \\ dh \end{bmatrix} = \mathbf{H}(\varphi,\ h) \begin{bmatrix} d\varphi \\ d\lambda \\ dh \end{bmatrix} \tag{4.4.20}$$

The components $d\mathbf{w} = [dn \quad de \quad du]^T$ intuitively are related to the "horizontal" and "vertical," and because their units are length, the standard deviations of the parameters can be readily visualized. The matrix $\mathbf{H}$ is evaluated for the station under consideration. The final parameterization becomes

$$\begin{bmatrix} d\alpha_1 \\ d\beta_1 \\ ds \end{bmatrix} = \mathbf{A} \begin{bmatrix} d\mathbf{w}_1 \\ \cdots \\ d\mathbf{w}_2 \end{bmatrix} \tag{4.4.21}$$

with

$$\mathbf{A} = [\mathbf{G}_1 \mathbf{J}_1 \mathbf{H}_1^{-1} \quad : \quad \mathbf{G}_2 \mathbf{J}_2 \mathbf{H}_2^{-1}] = \begin{bmatrix} a_{11} & a_{12} & a_{13} & & a_{14} & a_{15} & a_{16} \\ a_{21} & a_{22} & a_{23} & : & a_{24} & a_{25} & a_{26} \\ a_{31} & a_{32} & a_{33} & & a_{34} & a_{35} & a_{36} \end{bmatrix} \tag{4.4.22}$$

The partial derivatives are listed in Table 4.4.2 (Wolf, 1963; Heiskanen and Moritz, 1967; Vincenty, 1979). Some of the partial derivatives have been expressed in terms of the back azimuth $\alpha_2 \equiv \alpha_{21}$ and the back vertical angle $\beta_2 \equiv \beta_{21}$, meaning azimuth and vertical angle from station 2 to station 1. Early work on the 3D geodetic model is found in Bruns (1878).

4.4.3 Implementation Considerations

The 3D geodetic model is easy to derive since only partial differentiation is required; it is also easy to implement in software. Normally, the observations will be uncorrelated

TABLE 4.4.2 Partial Derivatives with Respect to Local Geodetic Coordinates

$$a_{11} = \frac{\partial \alpha_1}{\partial n_1} = \frac{\sin \alpha_1}{s \cos \beta_{11}} \quad \text{(a)} \qquad a_{12} = \frac{\partial \alpha_1}{\partial e_1} = -\frac{\cos \alpha_1}{s \cos \beta_1} \quad \text{(b)}$$

$$a_{13} = \frac{\partial \alpha_1}{\partial u_1} = 0 \quad \text{(c)}$$

$$a_{14} = \frac{\partial \alpha_1}{\partial n_2} = -\frac{\sin \alpha_1}{s \cos \beta_1}[\cos(\varphi_2 - \varphi_1) + \sin \varphi_2 \sin(\lambda_2 - \lambda_1) \ \cot \alpha_1] \quad \text{(d)}$$

$$a_{15} = \frac{\partial \alpha_1}{\partial e_2} = \frac{\cos \alpha_1}{s \cos \beta_1}[\cos(\lambda_2 - \lambda_1) - \sin \varphi_1 \sin(\lambda_2 - \lambda_1) \tan \alpha_1] \quad \text{(e)}$$

$$a_{16} = \frac{\partial \alpha_1}{\partial u_2} = \frac{\cos \alpha_1 \cos \varphi_2}{s \cos \beta_1}[\sin(\lambda_2 - \lambda_1) + (\sin \varphi_1 \cos(\lambda_2 - \lambda_1) - \cos \varphi_1 \tan \varphi_2) \tan \alpha_1] \quad \text{(f)}$$

$$a_{21} = \frac{\partial \beta_1}{\partial n_1} = \frac{\sin \beta_1 \cos \alpha_1}{s} \quad \text{(g)} \qquad a_{22} = \frac{\partial \beta_1}{\partial e_1} = \frac{\sin \beta_1 \sin \alpha_1}{s} \quad \text{(h)}$$

$$a_{23} = \frac{\partial \beta_1}{\partial u_1} = -\frac{\cos \beta_1}{s} \quad \text{(i)}$$

$$a_{24} = \frac{\partial \beta_1}{\partial n_2} = \frac{-\cos \varphi_1 \sin \varphi_2 \cos(\lambda_2 - \lambda_1) + \sin \varphi_1 \sin \varphi_2 + \sin \beta_1 \cos \beta_2 \cos \alpha_2}{s \cos \beta_1} \quad \text{(j)}$$

$$a_{25} = \frac{\partial \beta_1}{\partial e_2} = \frac{-\cos \varphi_1 \sin(\lambda_2 - \lambda_1) + \sin \beta_1 \cos \beta_2 \sin \alpha_2}{s \cos \beta_1} \quad \text{(k)}$$

$$a_{26} = \frac{\partial \beta_1}{\partial u_2} = \frac{\sin \beta_1 \sin \beta_2 + \sin \varphi_1 \sin \varphi_2 + \cos \varphi_1 \cos \varphi_2 \cos(\lambda_2 - \lambda_1)}{s \cos \beta_1} \quad \text{(l)}$$

$$a_{31} = \frac{\partial s}{\partial n_1} = -\cos \beta_1 \cos \alpha_1 \quad \text{(m)} \qquad a_{32} = \frac{\partial s}{\partial e_1} = -\cos \beta_1 \cos \alpha_1 \quad \text{(n)}$$

$$a_{33} = \frac{\partial s}{\partial u_1} = -\sin \beta_1 \quad \text{(o)} \qquad a_{34} = \frac{\partial s}{\partial n_2} = -\cos \beta_2 \cos \alpha_2 \quad \text{(p)}$$

$$a_{35} = \frac{\partial s}{\partial e_2} = -\cos \beta_2 \sin \alpha_2 \quad \text{(q)} \qquad a_{36} = \frac{\partial s}{\partial u_2} = -\sin \beta_2 \quad \text{(r)}$$

and their contribution to the normal equations can be added one by one. The following are some useful things to keep in mind when using this model:

- **Point of Expansion:** As in any nonlinear adjustment, the partial derivatives must be evaluated at the current point of expansion (adjusted positions of the previous iteration). This applies to coordinates, azimuths, and angles used to express the mathematical functions of the partial derivatives.
- **Reduction to the Mark:** An advantage of the 3D geodetic model is that the observations do not have to be reduced to the marks on the ground. When computing ℓ_0 from (4.4.11) to (4.4.13), use $h + \Delta h$ instead of h for the station heights. The symbol Δh denotes the height of the instrument or that of the target above the mark on the ground. ℓ_b always denotes the measured value, i.e., the geodetic observable that is not further reduced. After completion of the adjustment, the adjusted observations ℓ_a, with respect to the marks on

the ground, can be computed from the adjusted positions using h in (4.4.11) to (4.4.13).

- **Minimal Constraints:** The (φ) or (w) parameterizations are particularly useful for introducing height observations, height difference observations, or minimal constraints by fixing or weighting individual coordinates. The set of minimal constraints depends on the type of observations available within the network. One choice for the minimal constraints might be to fix the coordinates (φ, λ, h) of one station (translation), and the azimuth or the longitude or latitude of another station (rotation in azimuth). One always must make sure that the vertical component of the 3D network is determined by objections observations or, e.g., by height constraints.
- **Transforming Postadjustment Results:** If the adjustment happens to have been carried out with the (x) parameterization, and it is, subsequently, deemed necessary to transform the result into (φ) or (w) coordinates, then the transformations (4.4.18) and (4.4.20) can be used, for example:

$$d\boldsymbol{w} = \boldsymbol{R}\ d\boldsymbol{x} \tag{4.4.23}$$

where

$$\boldsymbol{R} = \boldsymbol{H}\ \boldsymbol{J}^{-1} \tag{4.4.24}$$

according to (4.4.10), (4.4.18), and (4.4.20). The law of variance-covariance propagation provides the 3×3 covariance submatrices,

$$\Sigma_{(\mathrm{w})} = \boldsymbol{R}\ \Sigma_{(x)}\boldsymbol{R}^T \tag{4.4.25}$$

$$\Sigma_{(\varphi,\,\lambda,\,h)} = \boldsymbol{J}^{-1}\,\Sigma_{(\mathrm{x})}\ (\boldsymbol{J}^{-1})^T \tag{4.4.26}$$

- **Leveled Height Differences:** If geoid undulation differences are available, the leveled height differences can be corrected for the undulation differences to yield ellipsoidal height differences. The respective elements of the design matrix are 1 and -1. The accuracy of incorporating leveling data in this manner is limited by our ability to compute accurate undulation differences.
- **Refraction:** If vertical angles are observed for the purpose of providing an accurate vertical dimension, it may be necessary to estimate vertical refraction parameters. If this is done, we must be careful to avoid overparameterization by introducing too many refraction parameters that could potentially absorb other systematic effects not caused by refraction and/or result in an ill-conditioned solution. However, it may be sufficient to correct the observations for refraction using a standard model for the atmosphere.

  In view of GPS capability, the importance of high-precision vertical angle measurement is diminishing. The primary purpose of vertical angles is to give sufficient height information to process the slant distances. Therefore, the types of observations most likely to be used by the modern surveyors are horizontal angles, slant distances, and GPS vectors.

- **Horizontal Angles:** Horizontal angles, of course, are simply the difference of two azimuths. Using the 2-1-3 subscript notation to identify an angle measured at station 1 from station 2 to station 3, in a clockwise sense the mathematical model for the geodetic angle δ_{213} is

$$\delta_{213} = \tan^{-1}\left(\frac{-\sin\lambda_1\ \Delta x_{12} + \cos\lambda_1\ \Delta y_{12}}{-\sin\varphi_1\cos\lambda_1\ \Delta x_{12} - \sin\varphi_1\sin\lambda_1\ \Delta y_{12} + \cos\varphi_1\ \Delta z_{12}}\right) - \tan^{-1}\left(\frac{-\sin\lambda_1\ \Delta x_{13} + \cos\lambda_1\ \Delta y_{13}}{-\sin\varphi_1\cos\lambda_1\ \Delta x_{13} - \sin\varphi_1\sin\lambda_1\ \Delta y_{13} + \cos\varphi_1\ \Delta z_{13}}\right) \tag{4.4.27}$$

  The partial derivatives can be readily obtained from the coefficients a_{2i} listed in Table 4.4.2 by applying them to both line segments of the angles and then subtracting.
- **Height-Controlled 3D Adjustment:** If the observations contain little or no vertical information, i.e., if zenith angles and leveling data are not available, it is still possible to adjust the network in three dimensions. The height parameters h can be weighted using reasonable estimates for their a priori variances. This is the so-called height-controlled three-dimensional adjustment.

A priori weights can also be assigned to the geodetic latitude and longitude or to the local geodetic coordinates n and e. Weighting of parameters is a convenient method for incorporating existing information about control stations into the adjustment.

4.4.4 GPS Vector Networks

Two receivers observing GNSS satellites provide the accurate vector between the stations, expressed in the reference frame of the ephemeris. One can, of course, assume known coordinates for one station and simply add the vector to get the coordinates of the other stations. The drawback of this simplified approach is that there is absolutely no quality control. As surveyors would certainly agree, it is easy to mistakenly set up the instrument over the wrong point; occasionally, a station is marked on the ground by several flags, and each of them might have a distinct but different meaning. Also, the GNSS might have provided an undetected biased solution by fixing the wrong integer ambiguity due to poor satellite visibility condition and too short of a station occupation time. Although processing software has become reliable to flag such biased solutions, problems can occasionally go undetected. Therefore, it is good practice to explore the redundancy of network observations to carry out objective quality control on GNSS vectors, as is done for terrestrial observations such as angles and distances.

The carrier phase processing for two receivers gives not only the vector between the stations but also the 3×3 covariance matrix of the coordinate differences. The covariance matrix of all vector observations is block-diagonal, with 3×3 submatrices along the diagonal. In the case of session solutions where R receivers observe the same satellites simultaneously, the results are $(R-1)$ independent vectors, and a $3(R-1) \times 3(R-1)$ covariance matrix. The covariance matrix is still block-diagonal, but the size of the nonzero diagonal matrices is a function of R.

As mentioned above, a GNSS survey that has determined the relative locations of a cluster of stations should be subjected to a minimal or inner constraint adjustment for purposes of quality control. The network should not contain unconnected vectors whose endpoints are not tied to other parts of the network. At the network level, the quality of the derived vector observations can be assessed, the geometric strength of the overall network can be analyzed, internal and external reliability can be computed, and blunders may be discoverable and removable. For example, a blunder in an antenna height will not be discovered when processing a single baseline, but it will be noticeable in the network solution if stations are reoccupied independently. Covariance propagation for computing distances, angles, or other functions of the coordinates should be done, as usual, with the minimal or inner constraint solution.

The mathematical model is the standard observation equation model,

$$\ell_a = \boldsymbol{f}(\boldsymbol{x}_a) \tag{4.4.28}$$

where $\boldsymbol{\ell}_a$ contains the adjusted observations and $\boldsymbol{x}_a$ denotes the adjusted station coordinates. The mathematical model is linear if the parameterization of receiver positions is in terms of Cartesian coordinates. In this case, the vector observation between stations k and m is modeled simply as

$$\begin{bmatrix} \Delta x_{km} \\ \Delta y_{km} \\ \Delta z_{km} \end{bmatrix} = \begin{bmatrix} x_k - x_m \\ y_k - y_m \\ z_k - z_m \end{bmatrix} \tag{4.4.29}$$

The relevant portion of the design matrix $\boldsymbol{A}$ for the model (4.4.29) is

$$\boldsymbol{A}_{km} = \begin{array}{c} \begin{matrix} x_k & y_k & z_k & x_m & y_m & z_m \end{matrix} \\ \begin{bmatrix} 1 & 0 & 0 & -1 & 0 & 0 \\ 0 & 1 & 0 & 0 & -1 & 0 \\ 0 & 0 & 1 & 0 & 0 & -1 \end{bmatrix} \end{array} \tag{4.4.30}$$

The design matrix looks like one for a leveling network. The coefficients are either 1, −1, or 0. Each vector contributes three rows. Because vector observations contain information about the orientation and scale, one only needs to fix the translational location of the polyhedron. Minimal constraints for fixing the origin can be imposed by simply deleting the three coordinate parameters of one station, holding coordinates of that particular station effectively fixed.

Inner constraints must fulfill the condition

$$\boldsymbol{Ex} = \boldsymbol{0} \tag{4.4.31}$$

according to (2.6.35), or, what amounts to the same condition

$$\boldsymbol{E}^T\boldsymbol{A} = \boldsymbol{0} \tag{4.4.32}$$

It can be readily verified that

$$\boldsymbol{E} = [{}_3\boldsymbol{I}_3 \quad {}_3\boldsymbol{I}_3 \quad {}_3\boldsymbol{I}_3 \quad \cdots] \tag{4.4.33}$$

fulfills these conditions. The matrix $\boldsymbol{E}$ consists of a row of 3×3 identity matrices. There are as many identity matrices as there are stations in the network. The inner constraint solution uses the pseudoinverse (2.6.37)

$$\boldsymbol{N}^+ = (\boldsymbol{A}^T\boldsymbol{P}\boldsymbol{A} + \boldsymbol{E}^T\boldsymbol{E})^{-1} - \boldsymbol{E}^T(\boldsymbol{E}\boldsymbol{E}^T\boldsymbol{E}\boldsymbol{E}^T)^{-1}\boldsymbol{E} \tag{4.4.34}$$

of the normal matrix. If one sets the approximate coordinates to zero, which can be done since the mathematical model is linear, then the origin of the coordinate system is at the centroid of the cluster of stations. For nonzero approximate coordinates, the coordinates of the centroid remain invariant, i.e., the values are the same whether computed from the approximate coordinates or the adjusted coordinates. The standard ellipsoid reflects the true geometry of the network and the satellite constellation. See Chapter 2 for a discussion on which quantities are variant or invariant with respect to different choices of minimal constraints.

The GNSS-determined coordinates refer to the coordinate system of the satellite positions (ephemeris). The broadcast ephemeris coordinate system is given in WGS84, and the precise ephemeris is in ITRF. The latest realizations of these frames agree at the centimeter level.

The primary result of a typical GNSS survey is best viewed as a polyhedron of stations whose relative positions have been accurately determined (to the centimeter or even the millimeter level), but the translational position of the polyhedron is typically known only at the meter level (point positioning with pseudoranges). The orientation of the polyhedron is implied by the vector observations. The Cartesian coordinates (or coordinate differences) of the GNSS survey can, of course, be converted to geodetic latitude, longitude, and height. If geoid undulations are available, the orthometric heights (height differences) can be readily computed. The variance-covariance components of the adjusted parameters can be transformed to the local geodetic system for ease of interpretation using (4.4.25).

4.4.5 Transforming Terrestrial and Vector Networks

We make use of models 2 or 3 of Section 4.1.5 to transform nearly aligned coordinate systems by estimating a scale and three rotation parameters. Assume that a network of terrestrial observations is available that include horizontal angles, slant distances, zenith angles, leveled height differences, and geoid undulations. Assume further that the relative positions of some of these network stations have been determined by GNSS vectors. As a first step one could carry out separate minimal or inner constraint solutions for the terrestrial observations and the GPS vectors, as a matter of quality control. When combining both sets of observations in one adjustment, the definition of the coordinate systems might become important. Let us consider the case when coordinates of some stations are known in the "local datum" (u) and that (u) does not coincide with (x), i.e., the coordinate system of the GNSS vectors. Let it be further required that if the adjusted coordinates should be expressed in (u), i.e., the existing local datum, then the following model

$$\ell_{1a} = \mathbf{f}_1(\mathbf{x}_a) \tag{4.4.35}$$

$$\ell_{2a} = \mathbf{f}_2(s,\ \eta,\ \xi,\ \alpha,\ \mathbf{x}_a) \tag{4.4.36}$$

might be applicable. The model (4.4.35) pertains to the terrestrial observations, denoted here as the $\boldsymbol{\ell}_1$ set. As a special case, these observations could consist of merely the known local station coordinates which would then be treated as observed parameters by the adjustment. Actually, if no terrestrial observations are available and only the coordinates of local stations are known, then the mathematical model (4.4.36) suffices. The GPS vector observations, i.e., the coordinate differences obtained from carrier phase processing, are represented by $\boldsymbol{\ell}_2$. To clarify the notation again, we note that $\mathbf{x}_a$ (adjustment notation) refers to the station coordinates in the geodetic system (u). The respective adjustment models are discussed in Chapter 2.

The additional parameters in (4.4.36) are the differential scale s and three rotation angles $\eta,\ \xi,\ \alpha$. The rotation angles are small since the geodetic coordinate systems (u) and (x) are nearly aligned. Because GNSS yields the coordinate differences, there is no need to include a translation parameter $\mathbf{t}$. Clearly, if $\boldsymbol{\ell}_1$ in (4.4.35) does not contain terrestrial observations at all, the known station coordinates in the (u) system can be treated as observed parameters and thus allow estimation of scale and rotation parameters relative to these known coordinates. This is a simple method to incorporate the GNSS vector observations into the existing local network.

The mathematical model (4.4.36) follows directly from the transformation expression (4.1.9). Applying this expression to the coordinate differences for stations k and m yields

$$(1+s)\mathbf{M}(\lambda_0,\ \varphi_0,\ \eta,\ \xi,\ \alpha)\ (\mathbf{u}_k - \mathbf{u}_m) - (\mathbf{x}_k - \mathbf{x}_m) = \mathbf{0} \tag{4.4.37}$$

The coordinate differences

$$\mathbf{x}_{km} = \mathbf{x}_k - \mathbf{x}_m \tag{4.4.38}$$

represent the observed GPS vector between stations k and m. Thus, the mathematical model (4.4.36) can be written as

$$\mathbf{x}_{km} = (1+s)\mathbf{M}(\lambda_0,\ \varphi_0,\ \eta,\ \xi,\ \alpha)\ (\mathbf{u}_k - \mathbf{u}_m) \tag{4.4.39}$$

After substituting (4.1.11) into (4.4.39), we readily obtain the partial derivatives of the design matrix. Table 4.4.3 lists the partial derivatives with respect to the station coordinates for (a) Cartesian parameterization, (b) parameterization in terms of geodetic latitude, longitude, and height, and (c) parameterization in terms of the local geodetic coordinate systems. The transformation matrices $\mathbf{J}$ and $\mathbf{H}$ referred to in the table are those of (4.4.18) and (4.4.20). Table 4.4.4 contains the partial derivatives of the transformation parameters.

TABLE 4.4.3 Design Submatrix for Stations Occupied with Receivers

Parameterization	Station m	Station k
(u, v, w)	$(1+s)\,\boldsymbol{M}$	$-(1+s)\,\boldsymbol{M}$
(φ, λ, h)	$(1+s)\,\boldsymbol{MJ}(\varphi_m, \lambda_m)$	$-(1+s)\,\boldsymbol{MJ}(\varphi_k, \lambda_k)$
(n, e, u)	$(1+s)\,\boldsymbol{MJ}(\varphi_m, \lambda_m)\boldsymbol{H}^{-1}(\varphi_m)$	$-(1+s)\,\boldsymbol{MJ}(\varphi_k, \lambda_k)\boldsymbol{H}^{-1}(\varphi_k)$

TABLE 4.4.4 Design Submatrix for the Transformation Parameters

s	η	ξ	α
$\boldsymbol{u}_m - \boldsymbol{u}_k$	$\boldsymbol{M}_\eta(\boldsymbol{u}_m - \boldsymbol{u}_k)$	$\boldsymbol{M}_\xi(\boldsymbol{u}_m - \boldsymbol{u}_k)$	$\boldsymbol{M}_\alpha(\boldsymbol{u}_m - \boldsymbol{u}_k)$

4.4.6 GPS Network Examples

The following examples are included because they document some of the first applications of GPS, demonstrating an amazing accuracy that many doubted could be achieved with satellite techniques, possibly because of prior exposure to the earlier TRANSIT (the Navy navigation satellite system). Also, in those early days of GPS satellite surveying there was no "GPS infrastructure" available to support GPS applications, no experience existed for incorporating highly accurate 3-dimensional vectors into existing geodetic networks, and the existing geodetic datums were neither geocentric nor were their axes parallel to the GPS reference system at the time. In many cases there were no geoid undulations available to convert orthometric heights to ellipsoidal heights.

In the following examples, only independent vectors between stations are considered, which means that if three receivers observe simultaneously, only two vectors are used. The stochastic model does not include the mathematical correlation between simultaneously observed vectors, although it should be used if available. The covariance information came directly from baseline processing and does not accommodate small uncertainties in eccentricity, i.e., inaccurate setting up of the antenna over the mark. Only single-frequency carrier phases were available at the time the observations were made.

4.4.6.1 Montgomery County Geodetic Network During the Montgomery County (Pennsylvania) geodetic network densification, the window of satellite visibility was about 5 hours, just long enough to allow two sessions with the then state-of-the-art static technique (Collins and Leick, 1985). The network (Figure 4.4.3) was freely designed, taking advantage of the insensitivity of GPS to the shape of the network (as compared to the many rules of classical triangulation and trilateration). The longest baseline observed was about 42 km. Six horizontal stations with known geodetic latitude and longitude and seven vertical stations with known orthometric height were available for tying the GPS survey to the existing

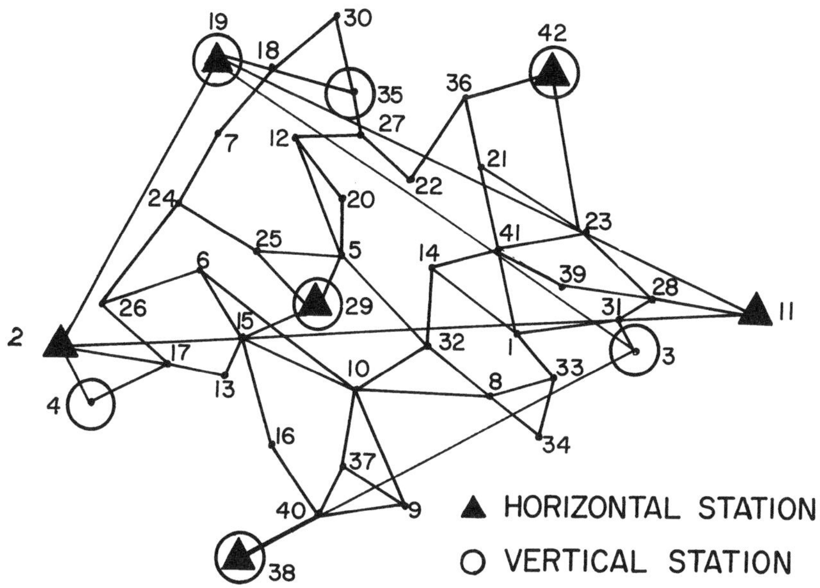

Figure 4.4.3 Existing geodetic control and independent baselines.

geodetic networks. Accurate geoid information was, of course, not available at the time.

Figure 4.4.4 shows two intersections of the ellipsoid of standard deviation for the inner constraint least-squares solution. The top set of ellipses shows the horizontal intersection (i.e., the ellipses of standard deviation in the geodetic horizon), and the bottom set of ellipses shows the vertical intersection in the east-west direction. The figure also shows the daily satellite visibility plot for the time and area of the project. The dots in that figure represent the directions of the semimajor axis of the ellipsoids of standard deviation for each station. These directions tend to be located around the center of the satellite constellation. The standard ellipses show a systematic orientation in both the horizontal and the vertical planes. This dependency of the shape of the ellipses with the satellite constellation enters into the adjustment through the 3×3 correlation matrices. With a better distribution of the satellites over the hemisphere, the alignments seen in Figure 4.4.4 for the horizontal ellipses would not occur. Because satellites are observed above the horizon, the ellipses will still be stretched along the vertical.

The coordinates of the polyhedron of stations are given in the coordinate system of the broadcast ephemeris; at the time of the Montgomery County survey, this was WGS72 (today this would be WGS84 or the latest ITRF). A minimal constraint was specified by equating the geodetic latitude and longitude to the astronomic latitude and longitude of station 29 and equating the ellipsoidal height and the orthometric height. The ellipsoid defined in that manner is tangent to the geoid at station 29. By comparing the resulting ellipsoidal heights with known orthometric heights at the

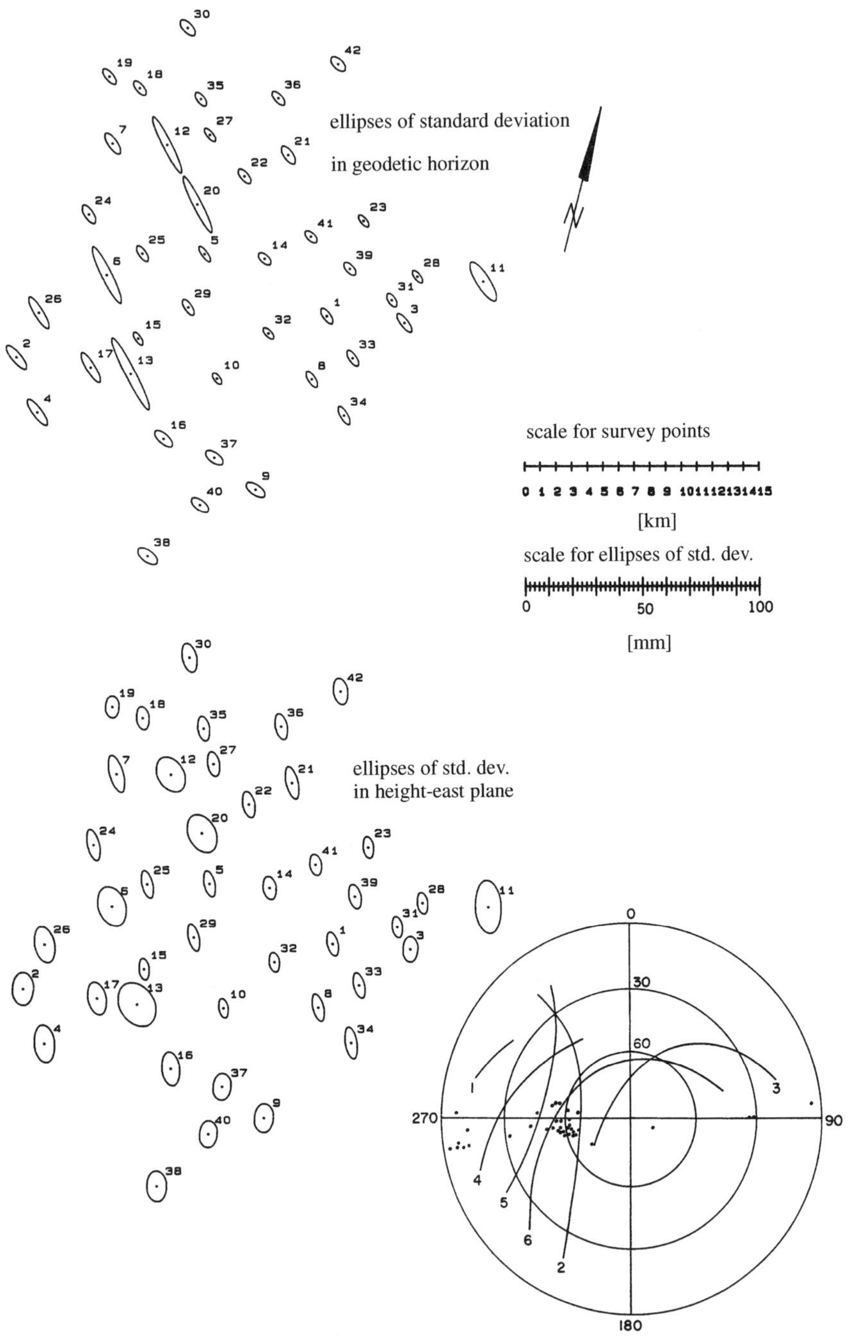

Figure 4.4.4 Inner constraint solution, ellipses of standard deviation and satellite visibility plot.

vertical stations, we can construct a geoid undulation map (with respect to the thus defined ellipsoid). The geoid undulations at other stations can be interpolated to give orthometric height using the basic relation $H = h - N$.

The method described above can be generalized by not using the geodetic positions instead of the astronomic position for station 29 to define minimal constraints. The thus defined local ellipsoid is not tangent to the geoid at station 29. The undulations with respect to such an ellipsoid are shown in Figure 4.4.5.

Alternatively, one can estimate the topocentric rotations (η, ξ, α) and a scale factor implied by model 3 of (4.4.39). There are seven minimal constraints required in this case, e.g., the geodetic latitude and longitude for two stations and the geodetic heights for three stations distributed well over the network. If one were to use orthometric heights for these three stations instead, the angles (ξ, η) would reflect the average deflection of the vertical angles. Using orthometric heights would force the ellipsoid to coincide locally with the geoid (as defined or implied by the orthometric heights at the vertical stations). The rotation in azimuth α is determined by the azimuthal difference between the two stations held fixed and the GPS vector between the same stations. The scale factor is also determined by the two stations held fixed; it contains the possible scale error of the existing geodetic network and the effect of a constant but unknown undulation (i.e., geoid undulations with respect to the ellipsoid of the existing geodetic network).

Simple geometric interpolation of geoid undulations has its limits, of course. For example, any error in a given orthometric height will result inevitably in an erroneous

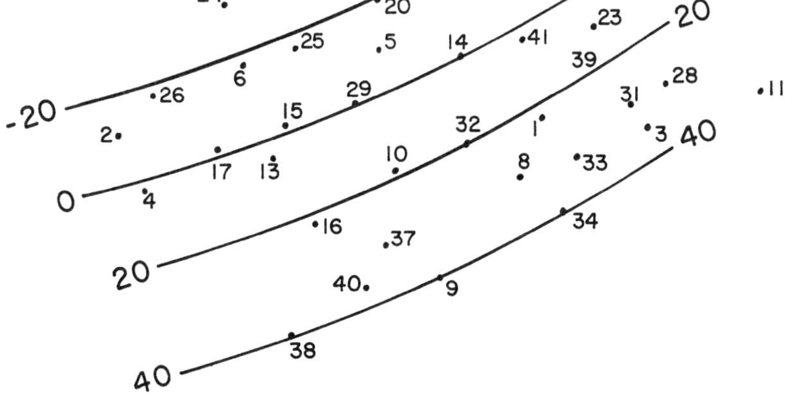

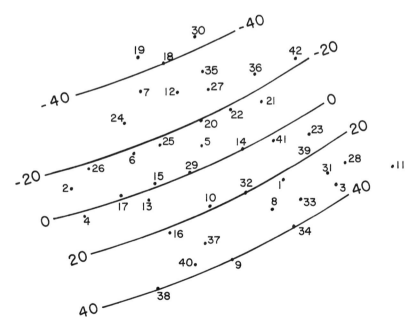

Figure 4.4.5 Geoid undulations with respect to the local ellipsoid. Units are in centimeters.

geoid feature. As a result, the orthometric heights computed from the interpolated geoid undulations will be in error. Depending on the size of the survey area and the "smoothness" of the geoid in that region, such erroneous geoid features might or might not be discovered from data analysis. These difficulties can be avoided if an accurate geoid model is available.

4.4.6.2 SLC Engineering Survey A GPS survey was carried out in 1984 to support construction of the Stanford linear collider (SLC), with the objective of achieving millimeter relative positional accuracy by combining GPS vectors with terrestrial observations (Ruland and Leick, 1985). Because the network was only 4 km long, the broadcast ephemeris errors as well as the impact of the troposphere and ionosphere canceled. The position accuracy in such small networks is limited by the carrier phase measurement accuracy, the phase center variation of the receiver antenna, and the multipath. The Macrometer antenna was used, which was known for its good multipath rejection property and accurate definition of the phase center.

The network is shown in Figure 4.4.6. Stations 1, 10, 19, and 42 are along the 2-mile linear accelerator (linac); the remaining stations of the "loop" were to be determined with respect to these linac stations. The disadvantageous configuration of this network, in regard to terrestrial observations such as angles and distances, is obvious. In order to improve this configuration, one would have to add stations adjacent to the linac; this would have been costly because of local topography and ongoing construction. Such a "degenerate" network configuration is acceptable for GPS positioning because the accuracy of positioning depends primarily on the satellite configuration and not on the shape of the network. Figure 4.4.7 shows the horizontal ellipses of standard deviation and the satellite visibility plot for the inner constraint vector solution. The dark spot on the visibility plot represents the directions of the semimajor axes of the standard ellipsoids.

This project offered an external standard for comparison. For the frequent realignment of the linear accelerator, the linac laser alignment system had been installed. This system is capable of determining positions perpendicular to the axis of the linac to better than ±0.1 mm over the total length of 3050 m. A comparison of the linac stations 1, 10, 19, and 42, as determined from the GPS vector solution with respect to the

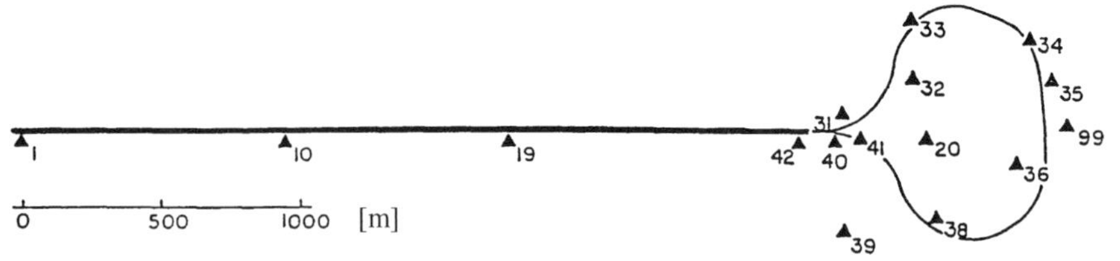

Figure 4.4.6 The SLC network configuration.

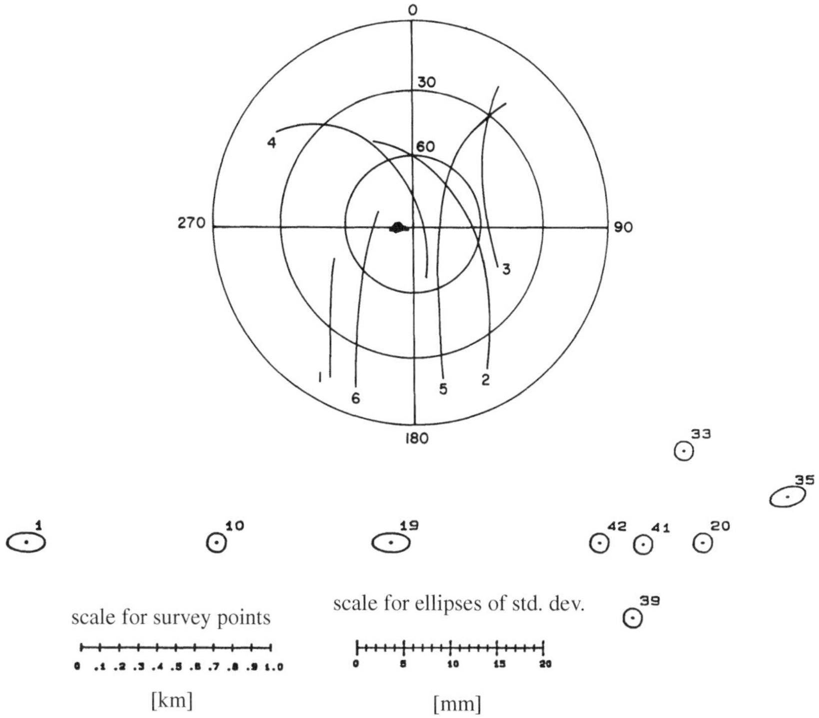

Figure 4.4.7 Horizontal standard ellipses for GPS inner constraint solution and visibility plot.

linac alignment system, was done by means of a transformation. The discrepancies did not exceed ±1 mm for any of the four linac stations.

4.4.6.3 Orange County Densification The Orange County GPS survey consisted of more than 7000 vectors linking 2000 plus stations at about a 0.5-mile spacing. This survey was a major network densification carried out with GPS using several crews operating at the same time. It was considered important to use adjustment techniques to detect and remove blunders that could have resulted from misidentifying stations or from not centering the antenna correctly. As to adjustment quality control capabilities, detected outliers are the prime candidates for in-depth studies and analysis to identify the cause for the outlier and then take corrective action. Redundancy number and internal reliability plots appear useful to identify weak portions of the network (which may result from a deweighting of observations during automated blunder detection). The variance-covariance matrices of the vector observations resulting from individual baseline processing are the determining factor that shapes most of the functions. The analysis begins with graphing the variances of these

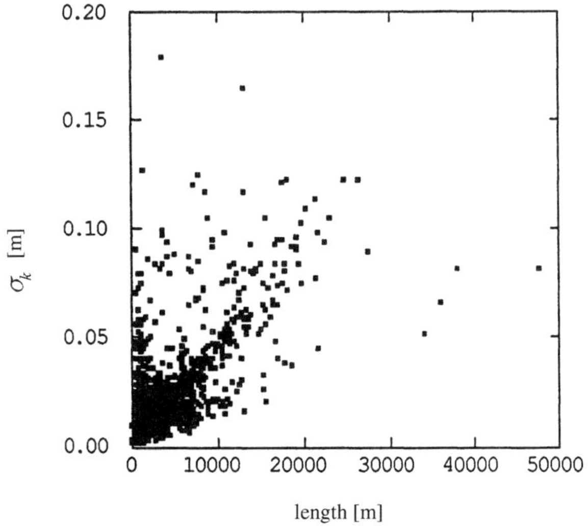

Figure 4.4.8 A priori precision of length of baseline (Permission by ASCE).

baselines, followed by various graphs related to the minimal constraint network solution. Other aspects of the solutions are given in Leick and Emmons (1994).

A priori Stochastic Information: The study begins with using the diagonal elements of the 3 × 3 variance-covariance matrices of the estimated vectors of the phase processing step to compute the simple function

$$\sigma_k = \sqrt{\sigma_{k1}^2 + \sigma_{k2}^2 + \sigma_{k3}^2} \tag{4.4.40}$$

where k identifies the vector. Other simple functions, such as the trace of the variance-covariance matrix, can be used as well. Figure 4.4.8 displays σ_k as a function of the length of the vectors. For longer lines, there appears to be a weak length dependency of about 1:200,000. Several of the shorter baselines show larger-than-expected values. While it is not necessarily detrimental to include vectors with large variances in an adjustment, they are unlikely to contribute to the strength of the network solution. Analyzing the averages of σ_k for all vectors of a particular station is useful in discovering stations that might be connected exclusively by low-precision vector observations.

Variance Factor: As to the minimal constraint network solution, Figures 4.4.9 and 4.4.10 show the square root of the estimated variance factor f_k for each vector k. The factor is computed as

$$f_k = \sqrt{\frac{\overline{\mathbf{v}}_k^T \overline{\mathbf{v}}_k}{R_k}} \tag{4.4.41}$$

with

$$R_k = \overline{r}_{k1} + \overline{r}_{k2} + \overline{r}_{k3} \qquad 0 \le R_k \le 3 \tag{4.4.42}$$

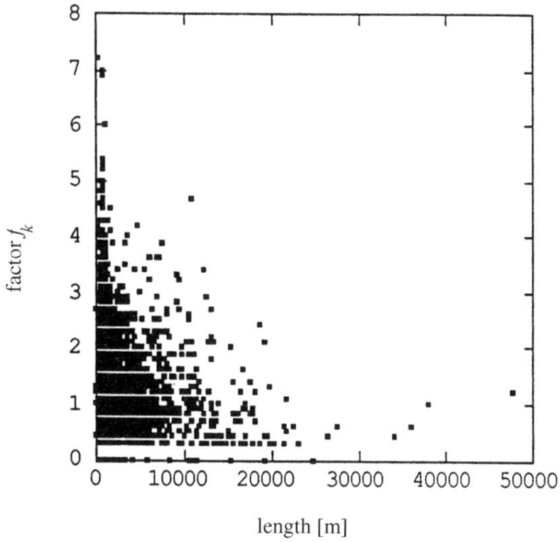

Figure 4.4.9 Variance factor versus length of baseline (Permission by ASCE).

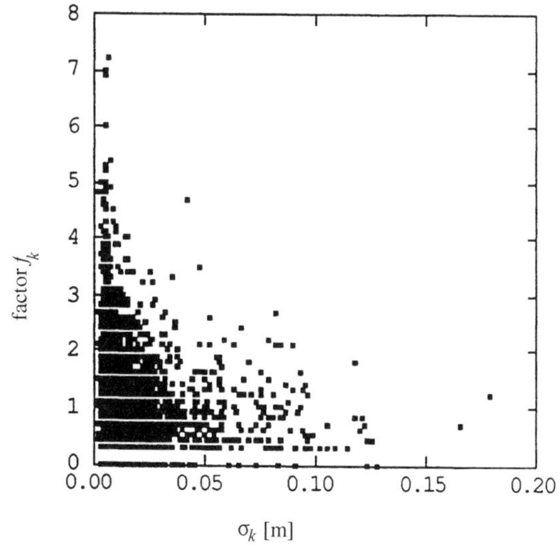

Figure 4.4.10 Variance factor versus precision of baseline (Permission by ASCE).

where $\overline{\boldsymbol{v}}_k$ denote the decorrelated residuals and $\overline{r}_{k1}$, $\overline{r}_{k2}$, and $\overline{r}_{k3}$ are the redundancy numbers of the decorrelated vector components [see equation (2.8.40) regarding the decorrelation of vector observations]. The estimates of f_k are plotted in the Figures 4.4.9 and 4.4.10 as a function of the baseline length and a priori statistics σ_k (4.4.40). The figures shows that the largest factors are associated with the shortest baselines or with lines having small σ_k (which tend to be the shortest baselines). For short

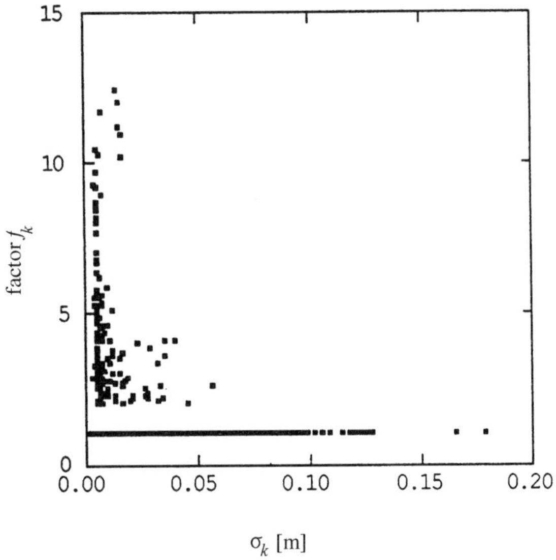

Figure 4.4.11 Applied scale factors (Permission by ASCE).

baselines the centering errors of the antenna and the separation of the electronic and geometric center of the antenna are important; neither is reflected by the stochastic model of the baselines.

The scale factors f_k in Figure 4.4.11 are computed following the procedure of automatically deweighting observations as discussed in Section 2.9.3 (i.e., if the ratio of residual and standard deviation is beyond a threshold value, the scaling factor is computed from an empirical rule and the residuals). All components of the vector are multiplied with the same factor (the largest of three). These scale factors shown in the graph were actually applied.

Redundancy Numbers: The vector redundancy number R_k in (4.4.42) varies between zero and 3. Values close to 3 indicate maximum contribution to the redundancy and minimum contribution to the solution, i.e., the observation is literally redundant. Such observations contribute little, if anything at all, to the adjustment because other usually much more accurate observations determine the solution. A redundancy of zero indicates an uncontrolled observation, which occurs, e.g., if a station is determined by one observation only. A small redundancy number implies little contribution to the redundancy but a big contribution to the solution. Such observations "overpower" other observations and usually have small residuals. As a consequence of their strength, blunders in these observations might not be discovered.

The ordered redundancy numbers in Figure 4.4.12 exhibit a distinctly sharp decrease for the smallest values. Inspection of the data indicates that such very small redundancies occur whenever there is only one good vector observation left to a particular station, while the other vectors to that station have been deweighted by scaling the variance-covariance matrices as part of the automatic blunder detection

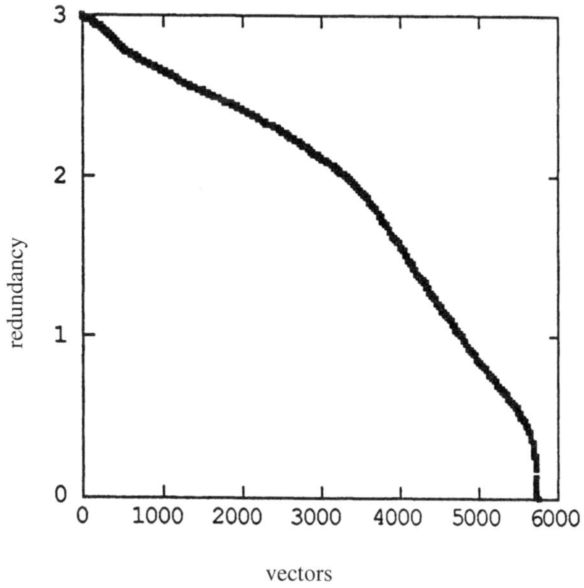

Figure 4.4.12 Ordered vector redundancy (Permission by ASCE).

procedure. Typically, the scaled vectors have a high redundancy number, indicating their diminished contribution. The only remaining unscaled observation contributes the most; therefore, the respective residuals are very small, usually in the millimeter range. Consequently, a danger of automated blunder detection and deweighting is that parts of the network might become uncontrolled.

Figure 4.4.13 indicates that long vectors have large redundancy numbers. The shapes in this figure suggest that it might be possible to identify vectors that can be deleted from the adjustment without affecting the strength of the solution. The steep slope suggests that the assembly of short baselines determines the shape of the network. Mixing short and long baselines is useful only if long baselines have been determined with accuracy comparable to that of shorter lines. This can be accomplished through longer observation times, using dual-frequency observations, and processing with a precise ephemeris.

Internal Reliability: Internal reliability values are shown in Figure 4.4.14. These values are a function of the internal reliability vector components

$$I_k = \sqrt{I_{k1}^2 + I_{k2}^2 + I_{k3}^2} \tag{4.4.43}$$

The internal reliability components are computed according to (2.8.28) for the decorrelated vector observations and are then transformed back to the physical observation space. The values plotted use the factor $\delta_0 = 4.12$. There is essentially a linear relationship between internal reliability and the quality of the observations as expressed by σ_k. The slope essentially equals δ_0. The outliers in this figure are associated with

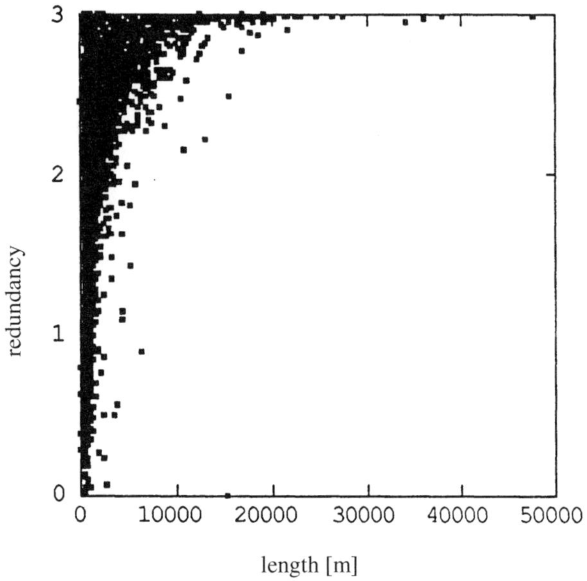

Figure 4.4.13 Vector redundancy versus length of baseline (Permission by ASCE).

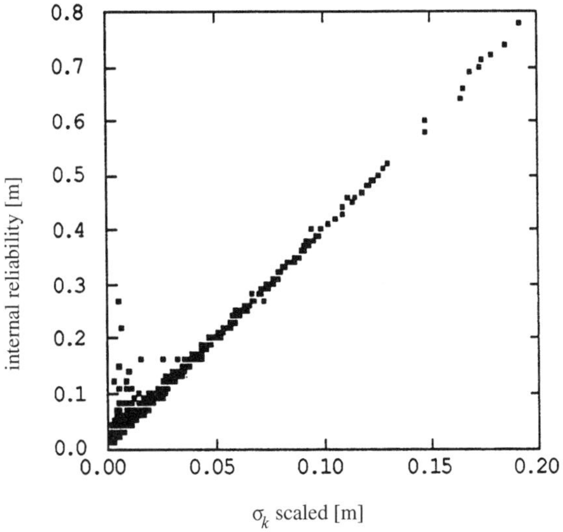

Figure 4.4.14 Internal reliability versus precision of baseline (Permission by ASCE).

small σ_k and pertain to a group of "single vectors" that result when the other vectors to the same station have been deweighted. The linear relationship makes it possible to identify the outliers for further inspection and analysis. Furthermore, this linear relationship nicely confirms that internal reliability is not a function of the shape of the GPS network.

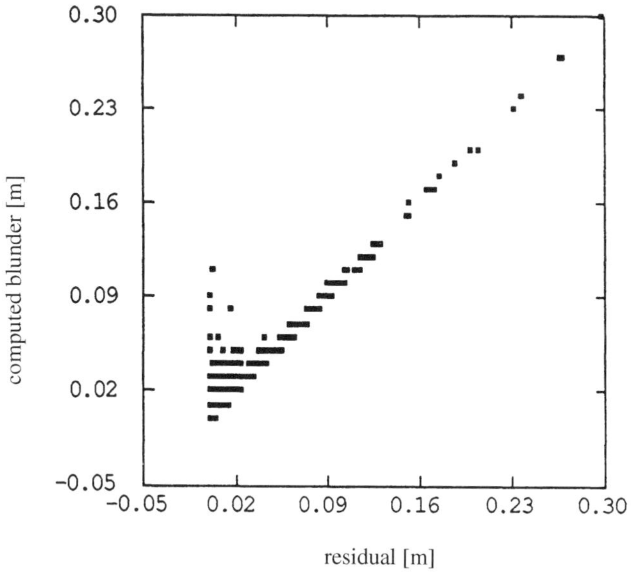

Figure 4.4.15 Computed blunders versus residuals (Permission by ASCE).

Blunders and Absorption: Figure 4.4.15 shows blunders as predicted by the respective residuals. As detailed in (2.8.31), a relationship exists between computed blunders, residuals, and redundancies. The figure shows the blunder function

$$B_k = \sqrt{B_{k1}^2 + B_{k2}^2 + B_{k3}^2} \tag{4.4.44}$$

versus the residual function

$$v_k = \sqrt{v_{k1}^2 + v_{k2}^2 + v_{k3}^2} \tag{4.4.45}$$

The computed blunder and the residuals refer to the physical observation space. This relationship appears to be primarily linear with slope 1:1 (at least for the larger residuals). The outliers seen for small residuals refer to the group of observations with smallest redundancy numbers.

Figure 4.4.16 shows absorption versus redundancy. Absorption specifies that part of a blunder is absorbed in the solution, i.e., absorption indicates falsification of the solution. The values

$$A_k = -v_k + B_k \tag{4.4.46}$$

are plotted. As expected, the observations with lowest redundancy tend to absorb the most. In an extreme case, the absorption is infinite for zero redundancy and zero for a redundancy of 3 (vector observations). Clearly, very small redundancies reflect insensitivity to blunders, which is not desirable.

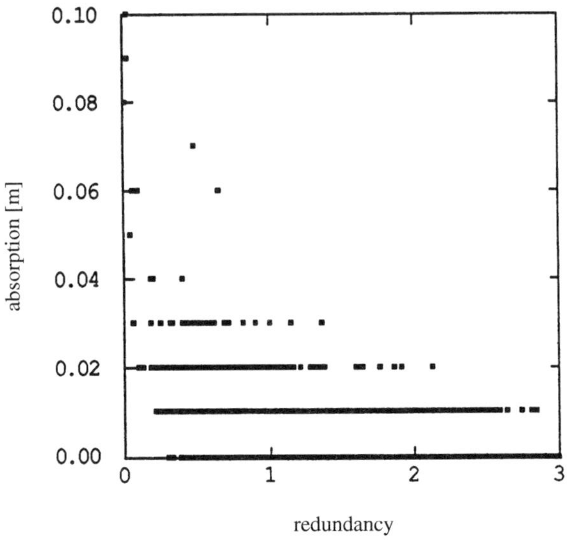

Figure 4.4.16 Absorption versus redundancy (Permission by ASCE).

In summary, as is the case for terrestrial observation, it is not sufficient to limit quality control to residuals and normalized residuals. It is equally important that the quality of the network be presented in terms of redundancy and reliability measures. These functions are, among other things, useful in judging the consequences of deweighting, in particular for large networks when those consequences are not always readily apparent.

4.5 ELLIPSOIDAL MODEL

Computations on either the ellipsoidal surface or the conformal map are inherently two dimensional. The stations are parameterized in terms of geodetic latitude and longitude or conformal mapping coordinates. The third dimension, the height, does not appear explicitly as a parameter but has been "used up" during the reduction of the spatial observations to the ellipsoidal surface. Networks on the ellipsoidal surface or the conformal map have historically been labeled "horizontal networks" and treated separately from a one-dimensional "vertical network." Such a separation was justified at a time when the measurement tools could readily be separated into those that measured primarily "horizontal information" and those that yielded primarily "vertical information." GNSS breaks this separation because it provides accurate three-dimensional positions.

Because two-dimensional geodetic models have a long tradition of having been the backbone of geodetic computations prior to the introduction of geodetic space techniques, the respective solutions belong to the most classical of all geodetic theories and are documented accordingly in the literature. Unfortunately, many of the references on this subject are out of print. We summarize the Gauss midlatitude solution, the transverse Mercator mapping, and Lambert conformal mapping in Appendices B and C. Supporting material from differential geometry is also provided in order to appreciate the "roots and flavor" of the mathematics involved. Additional derivations are available in Leick (2002), which was prepared to support lectures on the subject. The following literature has been found helpful: Dozier (1980), Heck (1987), Kneissl (1959), Grossman (1976), Hristow (1955), Lambert (1772), Lee (1976), Snyder (1982), and Thomas (1952). Publication of many of these "classical" references has been discontinued.

The ellipsoidal and conformal mapping expressions are generally given in the form of mathematical series that are a result of multiple truncations at various steps during the development. These truncations affect the computational accuracy of the expressions and their applicability to areas of a certain size. The expressions given here are sufficiently accurate for typical applications in surveying and geodesy. Some terms may even be negligible when applied over small areas. For unusual applications covering large areas, one might have to use more accurate expressions found in the specialized literature. In all cases, however, given today's powerful computers, one should not be overly concerned about a few unnecessary algebraic operations.

There are only two types of observations that apply to computations on a surface: azimuth (angle) and distance. The reductions, partial derivatives, and other quantities that apply to angles can again be conveniently obtained by differencing the respective expressions for azimuths.

Computations on the ellipsoid and the conformal mapping plane became popular when K. F. Gauss significantly advanced the field of differential geometry and least squares. Gauss used his many talents to develop geodetic computations on the ellipsoidal surface and on the conformal map. The problem presented itself to Gauss when he was asked to observe and compute a geodetic network in northern Germany. Since the curvature of the ellipsoid changes with latitude, the mathematics of computing on the ellipsoidal surface becomes mathematically cumbersome. For the conformal mapping approach, even more mathematical developments are needed. Both approaches require a new element that has not been discussed thus far, the geodesic (the shortest distance between two points on a surface). Developing expressions for the geodesic on the ellipsoidal surface and its image on the map requires advanced mathematical skills, primarily series expansions.

This section contains the mathematical formulations needed to carry out computations on the ellipsoidal surface. We introduce the geodesic line and reduce the 3D geodetic observations to geodesic azimuth and distance. The direct and inverse solutions are based on the Gauss midlatitude expressions. Finally, the partial derivatives are given that allow a network adjustment on the ellipsoid.

4.5.1 Reduction of Observations

The geodetic azimuth α of Section 4.4 is the angle between two normal planes that have the ellipsoidal normal in common; the geodetic horizontal angle δ is defined similarly. These 3D model observations follow from the original observation after a correction is made for the deflection of the vertical. Spatial distances can be used directly in the 3D model presented in Section 4.5. However, angles and distances must be reduced further in order to obtain model observables on the ellipsoidal surface with respect to the geodesic.

4.5.1.1 Angular Reduction to Geodesic Figure 4.5.1 shows the reduction of azimuth. The geodetic azimuth, α, is shown in the figure as the azimuth of the normal plane defined by the ellipsoidal normal of P_1 and the space point P_2. See also Figure 4.3.7. The representatives of these space points are located on their respective ellipsoidal normals on the surface of the ellipsoid and are denoted by P_1' and P_2'. The dotted line P_1' to P_2'' denotes the intersection of the normal plane containing P_2 with the ellipsoid. The azimuth of the normal section defined by the ellipsoidal normal at P_1 and the surface point P_2' is α'. The angular difference $(\alpha' - \alpha)$ is the reduction in azimuth due to the height of P_2; the expression is given in Table 4.5.1. The height of the observing station P_1 does not affect the reduction because α is the angle between planes.

The need for another angular reduction follows from Figure 4.5.2. Assume that two ellipsoidal surface points P_1 and P_2 (labeled P_1' and P_2' in Figure 4.5.1) are located at different latitudes. Line 1 is the normal section from P_1 to P_2 and line 2 indicates the normal section from P_2 to P_1. It can be readily seen that these two normal sections do

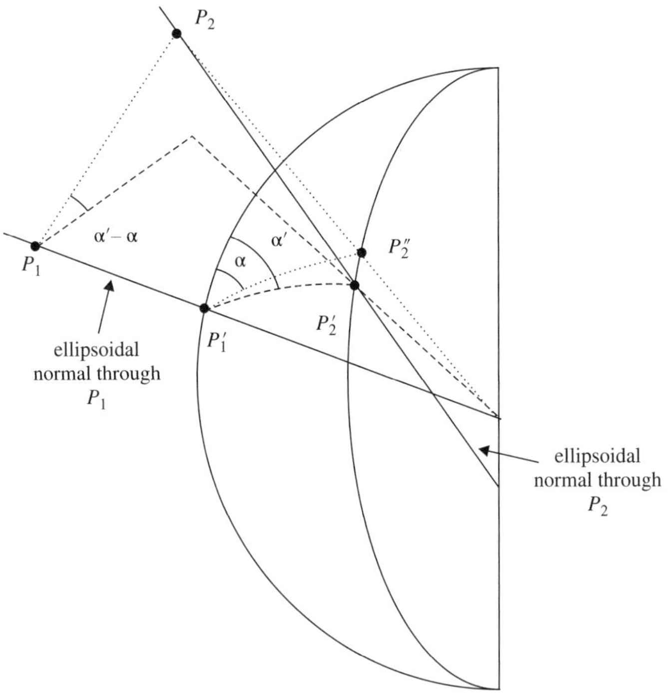

Figure 4.5.1 Normal section azimuth versus height of target.

TABLE 4.5.1 Reducing Geodetic Azimuth to Geodesic Azimuth

$(\alpha_1' - \alpha_1)_{[\text{arcs}]} = 0.108 \cos^2 \varphi_1 \sin 2\alpha_1 \; h_{2\,[\text{km}]}$	(a)
$(\widehat{\alpha}_1 - \alpha_1')_{[\text{arcs}]} = -0.028 \cos^2 \varphi_1 \sin 2\alpha_1 \left(\dfrac{\widehat{s}_{[\text{km}]}}{100}\right)^2$	(b)
$\Delta\alpha_{[\text{arcs}]} = 0.108 \cos^2 \varphi_1 \sin 2\alpha_1 h_{2\,[\text{km}]} - 0.028 \cos^2 \varphi_1 \sin 2\alpha_1 \left(\dfrac{\widehat{s}_{[\text{km}]}}{100}\right)^2$	(c)

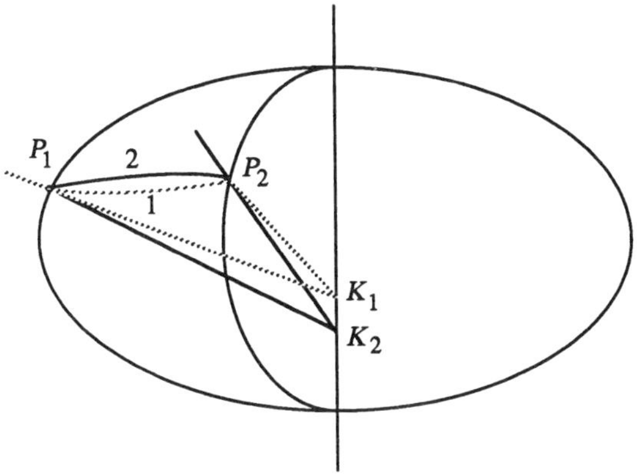

Figure 4.5.2 Normal sections on the ellipsoid.

not coincide because the curvature of the ellipsoidal meridian changes with latitude. The question is, which of these two normal sections should be adopted for the computations? Introducing the geodesic, which connects these two points in a unique way, solves this dilemma. There is only one geodesic from P_1 to P_2. Figure 4.5.3 shows the approximate geometric relationship between the normal sections and the geodesic. The angular reduction $(\widehat{\alpha} - \alpha')$ is required to get the azimuth $\widehat{\alpha}$ of the geodesic. The expression is listed in Table 4.5.1 (note that approximate values for azimuth α and length $\widehat{s}$ of the geodesic are sufficient to evaluate the expressions on the right-hand side of Table 4.5.1).

4.5.1.2 Distance Reduction to Geodesic The slant distance s (not to be confused with the scale correction of Section 4.1.6 which uses the same symbol) must be reduced to the length of a geodesic $\widehat{s}$. Figure 4.5.4 shows an ellipsoidal section along the line of sight. The expression for the lengths $\widehat{s}$ of the geodesic is typically based on a spherical approximation of the ellipsoidal arc. At this level of approximation, there is no need to distinguish between the lengths of the geodesic and the length of the normal section. The radius R, which is evaluated according to Euler's equation (B.1.8) for the center of the line, provides the radius of curvature of the spherical arc. The expressions in Table 4.5.2 relate the slant distance s to the lengths of the geodesic $\widehat{s}$.

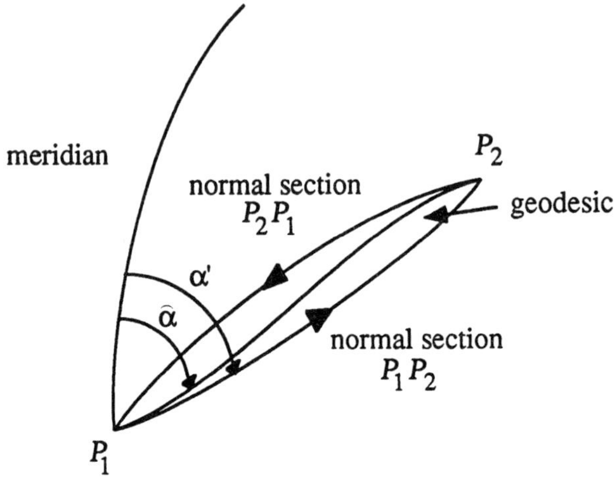

Figure 4.5.3 Normal section azimuth versus geodesic azimuth.

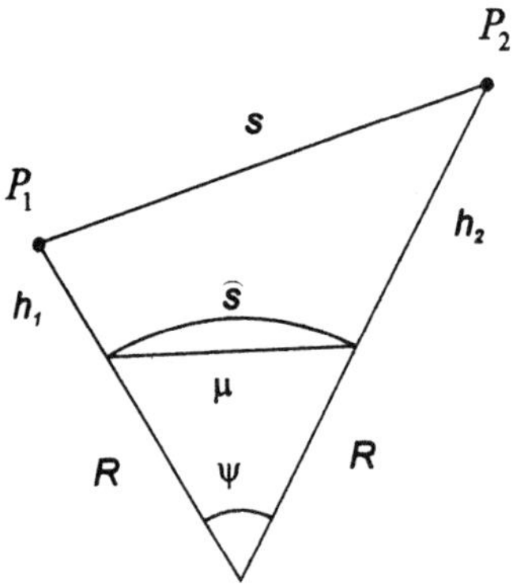

Figure 4.5.4 Slant distance versus geodesic.

TABLE 4.5.2 Reducing Slant Distance to Geodesic

$$\frac{1}{R} = \frac{\cos^2 \alpha}{M} + \frac{\sin^2 \alpha}{N} \quad \text{(d)}$$

$$\mu = \sqrt{\frac{s^2 - \Delta h^2}{\left(1 + \frac{h_1}{R}\right)\left(1 + \frac{h_2}{R}\right)}} \quad \text{(e)}$$

$$\hat{s} = R\psi = 2R \sin^{-1}\left(\frac{\mu}{2R}\right) \quad \text{(f)}$$

One should note that computing the length of the geodesic requires knowledge of the ellipsoidal heights. Using orthometric heights might introduce errors in distance reduction. The height difference $\Delta h = h_2 - h_1$ in expression (e) of Table 4.5.2 must be accurately known for lines with a large slope. Differentiating expression (f) gives the approximate relation

$$d\widehat{s} \approx -\frac{\Delta h}{\widehat{s}} d\Delta h \tag{4.5.1}$$

where $d\Delta h$ represents the error in the height difference. Surveyors often reduce the slant distance in the field to the local geodetic horizon using the elevation angle that is measured together with the slant distance. For observations reduced in such a manner, Δh is small (although not zero), but there is now a corresponding accuracy requirement for the measured elevation angle.

If both stations are located at about the same height $h_1 \approx h_2 \approx h_m$, one obtains from (f)

$$\frac{s - \widehat{s}}{s} = \frac{h_m}{R} \tag{4.5.2}$$

This equation relates the relative error in distance reduction to the mean height of the line. Table 4.5.3 shows that 6 m in height error causes a 1 ppm error in the reduction. This accuracy is routinely achieved with GPS.

Since modern electronic distance measurement instruments are very accurate, it is desirable to apply the height corrections consistently. It is good to remember the rule of thumb that *a 6 m error in height of the line causes a relative change in distance of 1 ppm*. We recognize that geodetic heights are required, not orthometric heights. Since geoid undulations can be as large as 100 m, it is clear that they must be taken into account for high-precision surveying.

4.5.2 Direct and Inverse Solutions on the Ellipsoid

The reductions discussed above produce the geodesic observables, i.e., the geodesic azimuths $\widehat{\alpha}$, the geodesic distance $\widehat{s}$, and the angle between geodesics $\widehat{\delta}$. At the heart of computations on the ellipsoidal surface are the so-called direct and inverse problems, which are summarized in Table 4.5.4. For the direct problem, the geodetic

TABLE 4.5.3 Relative Distance Error

$h_{m[\mathrm{m}]}$	h_m/R
6.37	1:1000000
63.7	1:100000
100	1:64000
500	1:13000
637	1:10000
1000	1:6300

TABLE 4.5.4 Direct and Inverse Solutions on the Ellipsoid

Direct Solution	Inverse Solution
$P_1(\varphi_1,\ \lambda_1),\ \widehat{\alpha}_{12},\ \widehat{s}_{12}$	$P_1(\varphi_1,\ \lambda_1),\ P_2\ (\varphi_2,\ \lambda_2)$
$\downarrow$	$\downarrow$
$(\varphi_2,\ \lambda_2,\ \widehat{\alpha}_{21})$	$(\widehat{\alpha}_{21},\ \widehat{s}_{12},\ \widehat{\alpha}_{21})$

latitude and longitude of one station, say, $P_1(\varphi_1,\ \lambda_1)$, and the geodesic azimuth $\widehat{\alpha}_{12}$ and geodesic distance $\widehat{s}_{12}$ to another point P_2 are given; the geodetic latitude and longitude of station $P_2(\varphi_2,\ \lambda_2)$, and the back azimuth $\widehat{\alpha}_{21}$ must be computed. For the inverse problem, the geodetic latitudes and longitudes of $P_1(\varphi_1,\ \lambda_1)$ and $P_2(\varphi_2,\ \lambda_2)$ are given, and the forward and back azimuth and the length of the geodesic are required. Note that $\widehat{s}_{12} = \widehat{s}_{21}$ but $\widehat{\alpha}_{12} \neq \widehat{a}_{21} \pm 180°$. There are many solutions available in the literature for the direct and inverse problems. Some of these solutions are valid for geodesics that go all around the ellipsoid. We use the Gauss midlatitude (GML) functions given in Table B.2.1 and use Table 4.5.4. Since the GML functions are a result of series developments, they are subject to truncation errors in respective series expansions. The GML solution satisfies typical geodetic applications. In the unlikely case that they are not sufficient because long lines are involved, one can always replace them with other solutions that are valid for long geodesics.

4.5.3 Network Adjustment on the Ellipsoid

The geodesic azimuths, geodesic distances, and the angles between geodesics form a network of stations on the ellipsoidal surface that can be adjusted using standard least-squares techniques. The ellipsoidal network contains no explicit height information, which was used during the transition of the 3D geodetic observables to the geodesic observables on the ellipsoid. Conceptually, this is expressed by $\{\varphi,\ \lambda,\ h\} \rightarrow \{\varphi,\ \lambda\}$ and $\{\alpha,\ \delta,\ \beta,\ s,\ \Delta h,\ \Delta N\} \rightarrow \{\widehat{\alpha},\ \widehat{\delta},\ \widehat{s}\}$. The geodetic height h is no longer a parameter, and geodesic observables do not include quantities that directly correspond to the geodetic vertical angle, the geodetic height difference Δh, or the geoid undulation difference ΔN.

Least-squares techniques are discussed in detail in Chapter 2. For discussion in this section, we use the observation equation model

$$\boldsymbol{v} = \boldsymbol{A}\boldsymbol{x} + (\ell_0 - \ell_b) \tag{4.5.3}$$

In the familiar adjustment notation the symbol $\boldsymbol{v}$ denotes the residuals, $\boldsymbol{A}$ is the design matrix, and $\boldsymbol{x}$ represents the corrections to the approximate parameters $\boldsymbol{x}_0$. The symbol ℓ_b denotes the observations, in this case the geodesic observables, and ℓ_0 represents the observables as computed from the approximate parameters

$$\boldsymbol{x}_0 = [\cdots \quad \varphi_{i,0} \quad \lambda_{i,0} \quad \cdots]^T \tag{4.5.4}$$

using the GML functions. If we further use the (2-1-3) subscript notation to denote the angle measured at station 1 from station 2 to station 3 in a clockwise sense, then the geodesic observables can be expressed as

$$\widehat{\alpha}_{12,b} = \alpha_{12,b} + \Delta\alpha_{12} \tag{4.5.5}$$

$$\widehat{\delta}_{213,b} = \delta_{213,b} + \Delta\alpha_{13} - \Delta\alpha_{12} \tag{4.5.6}$$

$$\widehat{s}_{12} = s\ (s_{12},\ R,\ h_1,\ h_2) \tag{4.5.7}$$

In order to make the interpretation of the coordinate (parameter) shifts easier, it is advantageous to reparameterize the parameters to northing ($dn_i = M_i\ d\varphi_i$) and easting ($de_i = N_i \cos \varphi_i d\lambda_i$). Using the partial derivatives in Table B.2.2, the observation equations for the geodesic observables become

$$v_{\widehat{\alpha}} = \frac{\sin \widehat{\alpha}_{12,0}}{\widehat{s}_{12,0}} dn_1 + \frac{\cos \widehat{\alpha}_{21,0}}{\widehat{s}_{12,0}} de_1 + \frac{\sin \widehat{\alpha}_{21,0}}{\widehat{s}_{12,0}} dn_2 - \frac{\cos \widehat{\alpha}_{21,0}}{\widehat{s}_{12,0}} de_2 + (\widehat{\alpha}_{12,0} - \widehat{\alpha}_{12,b}) \tag{4.5.8}$$

$$\begin{aligned} v_{\widehat{\delta}} = {} & \left(\frac{\sin \widehat{\alpha}_{13,0}}{\widehat{s}_{13,0}} - \frac{\sin \widehat{\alpha}_{12,0}}{\widehat{s}_{12,0}}\right) dn_1 + \left(\frac{\cos \widehat{\alpha}_{31,0}}{\widehat{s}_{13,0}} - \frac{\cos \widehat{\alpha}_{21,0}}{\widehat{s}_{12,0}}\right) de_1 \\ & - \frac{\sin \widehat{\alpha}_{21,0}}{\widehat{s}_{12,0}} dn_2 + \frac{\cos \widehat{\alpha}_{21,0}}{\widehat{s}_{12,0}} de_2 \\ & + \frac{\sin \widehat{\alpha}_{31,0}}{\widehat{s}_{13,0}} dn_3 - \frac{\cos \widehat{\alpha}_{31,0}}{\widehat{s}_{13,0}} de_3 + (\widehat{\delta}_{213,0} - \widehat{\delta}_{213,b}) \end{aligned} \tag{4.5.9}$$

$$v_{\widehat{s}} = -\cos \widehat{\alpha}_{12,0} dn_1 + \sin \widehat{\alpha}_{21,0} de_1 - \cos \widehat{\alpha}_{21,0} dn_2 - \sin \widehat{\alpha}_{21,0} de_2 + (\widehat{s}_{21,0} - \widehat{s}_{12,b}) \tag{4.5.10}$$

The quantities $(\widehat{\alpha}_0,\ \widehat{\beta}_0,\ \widehat{s}_0)$ are computed by the inverse solution. The GLM functions are particularly suitable for this purpose because the inverse solution is noniterative. The results of the adjustment of the ellipsoidal network are the adjusted observations $(\widehat{\alpha}_a,\ \widehat{\beta}_a,\ \widehat{s}_a)$ and the adjusted coordinates

$$\boldsymbol{x}_a = [\cdots \quad \varphi_{i,a} \quad \lambda_{i,a} \quad \cdots]^T \tag{4.5.11}$$

The partial derivatives (4.5.8) to (4.5.10) are also a result of series expansion and are, therefore, approximations and subject to truncation errors. The partial derivatives and the GML functions must have the same level of accuracy.

4.6 CONFORMAL MAPPING MODEL

If the goal is to map the ellipsoid onto a plane in order to display the ellipsoidal surface on the computer screen or to assemble overlays of spatial data, any mapping from the ellipsoid to the plane may be used. In conformal mapping, we map the ellipsoidal surface conformally onto a plane. The conformal property preserves angles. Recall that an angle between two curves, say, two geodesics on the ellipsoid, is defined as the angle between the tangents on these curves. Therefore, conformal mapping preserves the angle between the tangents of curves on the ellipsoid and the respective mapped images. The conformal property makes conformal maps useful for computations in surveying because the directional elements between the ellipsoid and the map have a known relationship.

Users who prefer to work with plane mapping coordinates rather than geodetic latitude and longitude can still use the 3D adjustment procedures developed earlier in this chapter. The given mapping coordinates can be transformed to the ellipsoidal and then used, together with heights, in the 3D geodetic adjustment. The adjusted geodetic positions can subsequently be mapped to the conformal plane.

4.6.1 Reduction of Observations

Let (x, y) denote the Cartesian coordinate system in the mapping plane, and $P_1(x_1, y_1)$ and $P_2(x_2, y_2)$ be the images of corresponding points on the ellipsoid. Reduction of ellipsoidal surface observations to the conformal mapping plane means converting geodesic observations $(\widehat{s}, \widehat{\alpha})$ to the corresponding observables $(\overline{d}, \overline{t})$ on the mapping plane. The symbol $\overline{d}$ denotes the length of the straight line connecting the mapped points P_1 and P_2, and $\overline{t}$ is the grid azimuth of this straight line. This reduction is accomplished by means of the mapping elements $(\gamma, \Delta t, \Delta s)$, which can be identified in Figure 4.6.1.

A couple of notes might be in order. First, consider the geodesic on the ellipsoid between P_1 and P_2 to be mapped point by point; the result is the mapped geodesic as shown in the figure. This image is a smooth but mathematically complicated curve. Second, the length $\widehat{s}$ of the geodesic on the ellipsoid is not equal to the length $\overline{s}$ of the mapped geodesic. The latter does not enter any of the equations below and never needs to be computed explicitly. Third, the straight line between the images P_1 and P_2 is called the rectilinear chord. Fourth, the mapping plane should not be confused with

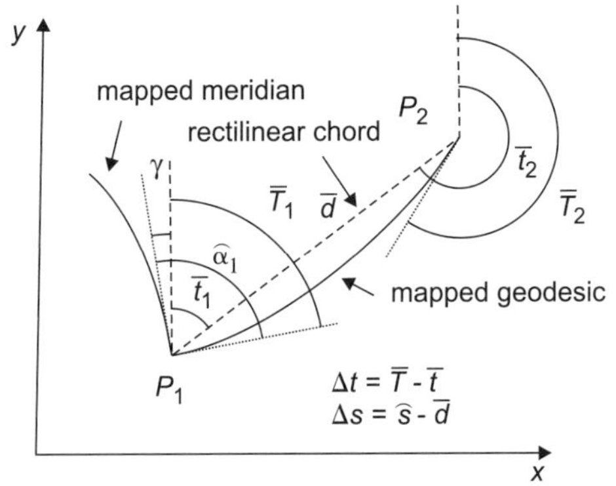

Figure 4.6.1 Mapping elements.

the local astronomic or geodetic horizon. The mapping plane is simply the result of mapping the ellipsoidal surface conformally into a plane. One can generate many such mapping planes for the same ellipsoidal surface area. They all would be conformal mappings.

Like any other curve on the ellipsoid, the ellipsoidal meridian can be mapped. Its image may or may not be a straight line. In order to be general, the figure shows a curved mapped meridian and its tangent. The angle between the y axis and the mapped meridian is the meridian convergence γ. It is one of the three mapping elements and is generally counted positive in the counterclockwise sense. Because of the conformal property, the geodetic azimuth of the geodesic is preserved during the mapping, and it must be equal to the angle between the tangents on the mapped meridian and the mapped geodesic as shown.

The symbols $\overline{T}$ and $\bar{t}$ denote the grid azimuth of the mapped geodesic and the rectilinear chord, respectively. The second mapping element, $\Delta t = \overline{T} - \bar{t}$,

$$\Delta t = \widehat{\alpha} - \gamma - \bar{t} \tag{4.6.1}$$

is called the arc-to-chord correction. It is related to the azimuth $\widehat{\alpha}$ of the geodesic on the ellipsoid, the grid azimuth $\bar{t}$ of the rectilinear chord, and the meridian convergence γ.

The third mapping element, called the map distance reduction $\Delta s = \widehat{s} - \overline{d}$, is the difference in the length of the geodesic on the ellipsoid and the rectilinear chord. Typically, one is not explicitly interested in the length of the projected geodesic $\bar{s}$, but actually needs the length of the rectilinear chord $\overline{d}$. Since there is no specification in conformal mapping as to the preservation of the lengths, the line scale factor

$$k_L = \frac{\overline{d}}{\widehat{s}} \tag{4.6.2}$$

can be arranged to express the third mapping element as

$$\Delta s = \widehat{s}(1 - k_L) \tag{4.6.3}$$

The line scale factor k_L is a ratio of two finite values. It is not unity but is expected to vary with the length of the line and its location within the mapping plane. The line scale factor should not be confused with the point scale factor k, which is the ratio of two differential line elements. See equation (C.2.12).

Let us compute the angle between two chords and two geodesics. The angle between rectilinear chords on the map at station i can be written as

$$\bar{\delta}_i = \bar{t}_{i,i+1} - \bar{t}_{i,i-1} + 2\pi = \overline{T}_{i,i+1} - \Delta t_{i,i+1} - (\overline{T}_{i,i-1} - \Delta t_{i,i-1}) + 2\pi \tag{4.6.4}$$

This relation follows from plane geometry. The angle between the geodesics on either the ellipsoid or their respective mapped images is

$$\widehat{\delta}_i = \widehat{\alpha}_{i,i+1} - \widehat{\alpha}_{i,i-1} + 2\pi = \overline{T}_{i,i+1} + \gamma_i - (\overline{T}_{i,i-1} + \gamma_i) + 2\pi = \overline{T}_{i,i+1} - \overline{T}_{i,i-1} + 2\pi \tag{4.6.5}$$

TABLE 4.6.1 Explicit Functions for Δt and Δs in Terms of Mapping Coordinates

TM: $\Delta t_1 = \dfrac{(x_2 + 2x_1)\ (y_2 - y_1)}{6k_0^2 R_1^2}$	LC: $\Delta t_1 = \dfrac{(2y_1 + y_2)\ (x_1 - x_2)}{6k_0^2 R_0^2}$
$\dfrac{1}{k_L} \equiv \dfrac{\widehat{s}}{d} = \dfrac{1}{6}\left(\dfrac{1}{k_1} + \dfrac{4}{k_m} + \dfrac{1}{k_2}\right)$ $\Delta s = \widehat{s}(1 - k_L)$	

The difference

$$\Delta\delta_i \equiv \widehat{\delta}_i - \overline{\delta}_i = \Delta t_{i,i+1} - \Delta t_{i,i-1} \tag{4.6.6}$$

is the angular arc-to-chord reduction. Equations (4.6.4) to (4.6.6) do not depend on the meridian convergence.

The expressions for the meridian convergence are given in Appendix C for the TM and LC mapping. The expressions for Δt and Δs are listed in Table 4.6.1 as a function of the mapping coordinates. Similar expressions are also available in terms of geodetic latitude and longitude. However, such alternative expressions are really not needed since one can always compute the latitudes and longitudes. The point scale factor k serves merely as an auxiliary quantity to express k_L in a compact form from which Δs can be computed. The subscripts of k indicate the point of evaluation. In the case of m, k is evaluated at the midpoint $[(\varphi_1 + \varphi_2)/2,\ (\lambda_1 + \lambda_2)/2]$. It goes without saying that the expressions in the table are a result of extensive series development and respective truncations. More accurate expressions are available in the literature.

Even though the term "map distortion" has many definitions, one associates a small Δt and Δs with small distortions, meaning that the respective reductions in angle and distance are small and perhaps even negligible. It is important to note that the mapping elements change in size and sign with the location of the line and its orientation. In order to keep Δt and Δs small, we limit the area represented in a single mapping plane in size, thus the need for several mappings to cover large regions of the globe. In addition, the mapping elements are functions of elements specified by the designer of the map, e.g., the factor k_0, the location of the central meridian, or the standard parallel.

4.6.2 Angular Excess

The angular reduction can be readily related to the ellipsoidal angular excess. The sum of the interior angles of a polygon of rectilinear chords on the map (Figure 4.6.2) is

$$\sum_i \overline{\delta}_i = (n - 2) \times 180^\circ \tag{4.6.7}$$

as follows from plane geometry. The sum of the interior angles of the corresponding polygon on the ellipsoid consisting of geodesics is

$$\sum_i \widehat{\delta}_i = (n - 2) \times 180^\circ + \varepsilon \tag{4.6.8}$$

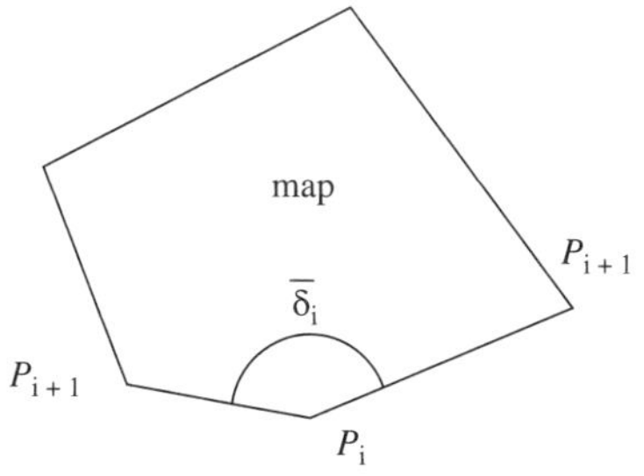

Figure 4.6.2 Angle on the map.

where ε denotes the ellipsoidal angular excess. It follows from (4.6.6) to (4.6.8) that

$$\varepsilon = \sum_i \Delta t_{i,i+1} - \sum_i \Delta t_{i,i-1} \tag{4.6.9}$$

The angular excess can therefore be computed from either the sum of interior angles between geodesics (4.6.8), or from the sum of angular arc-to-chord reductions (4.6.9), or by expression (B.2.44), which uses the Gauss curvature.

4.6.3 Direct and Inverse Solutions on the Map

Having the grid azimuth $\bar{t}$ and the length of the rectilinear chord $\bar{d}$ or the angle $\bar{\delta}$ between rectilinear chords, the rules of plane trigonometry apply in a straightforward manner. In case the geodetic latitude and longitude are given, one can use the mapping equations to compute the map coordinates first. Thus, given $P_1(x_1, y_1)$, the direct solution on the map is

$$\begin{aligned} x_2 &= x_1 + \bar{d}_{12} \sin \bar{t}_{12} \\ y_2 &= y_1 + \bar{d}_{12} \cos \bar{t}_{12} \end{aligned} \tag{4.6.10}$$

and given $P_1(x_1, y_1)$ and $P_2(x_2, y_2)$, the inverse solution on the map is

$$\begin{aligned} \bar{d}_{12} &= \sqrt{(x_2 - x_1)^2 + (y_2 - y_1)^2} \\ \bar{t}_{12} &= \tan \frac{x_2 - x_1}{y_2 - y_1} \end{aligned} \tag{4.6.11}$$

4.6.4 Network Adjustment on the Map

The fact that plane trigonometry can be used makes network adjustments and computations on the conformal plane especially attractive. The observed geodesic azimuth, angle, and distance $(\widehat{\alpha}, \widehat{\delta}, \widehat{s})$ are first corrected by $(\Delta t, \Delta\delta, \Delta s)$ to obtain the respective observables on the map. In regards to adjustments, the current point of expansion (approximate coordinates) should be used for all computations at a specific iteration. At any time during the computations, one may use geodetic latitude and longitude or mapping coordinates as is convenient, since both sets are accurately related by the mapping equations.

Two approaches can be followed. Both require computation of the meridian convergence γ. The scheme shown in (4.6.12) suggests using the GML function to compute the azimuth $\widehat{\alpha}_{12}$ and length $\widehat{s}_{12}$ and then computing the mapping elements Δt_{12} according to (4.6.1) and Δs_{12} by differencing the length of the geodesic line and the rectilinear chord. Alternatively, we may also compute the line scale factor k_L and compute the mapping elements directly from the expressions in Table 4.6.1. Appendix C provides the respective expressions for γ and k_L, either as a function of latitude and longitude or in terms of mapping coordinates. Omitting for simplicity the subscript zero to indicate the point of expansion, we can write

$$\left.\begin{array}{c} \{P_1(\varphi_1, \lambda_1, x_1, y_1), P_2(\varphi_2, \lambda_2, x_2, y_2)\} \\ \downarrow \\ \{\widehat{s}_{12}, \widehat{\alpha}_{12}, \gamma_1, \overline{d}_{12}, \overline{t}_{12}\} \\ \downarrow \\ \Delta t_{12} = \widehat{\alpha}_{12} - \gamma_1 - t_{12} \\ \Delta s_{12} = \widehat{s}_{12} - \overline{d}_{12} \end{array}\right\} \tag{4.6.12}$$

Using again the (2-1-3) subscript notation for angles and standard adjustment notation otherwise, the mapping observables $\overline{t}_{12,b}$, $\overline{\delta}_{213,b}$, and $\overline{d}_{12,b}$ are

$$\overline{t}_{12,b} = \widehat{\alpha}_{12,b} - \gamma_1 - \Delta t_{12} \tag{4.6.13}$$

$$\overline{\delta}_{213,b} = \widehat{\delta}_{213,b} - \Delta t_{13} + \Delta t_{12} \tag{4.6.14}$$

$$\overline{d}_{12,b} = \widehat{s}_{12,b} - \Delta s_{12} \tag{4.6.15}$$

and the observation equations become

$$v_{\overline{t}} = \frac{\sin \overline{t}_{12}}{\overline{d}_{12}} dy_1 - \frac{\cos \overline{t}_{12}}{\overline{d}_{12}} dx_1 - \frac{\sin \overline{t}_{12}}{\overline{d}_{12}} dy_2 + \frac{\cos \overline{t}_{12}}{\overline{d}_{12}} dx_2 + (\overline{t}_{12} - \overline{t}_{12,b}) \tag{4.6.16}$$

$$\begin{aligned} v_{\overline{\delta}} = {} & \left(\frac{\sin \overline{t}_{13}}{\overline{d}_{13}} - \frac{\sin \overline{t}_{12}}{\overline{d}_{12}}\right) dy_1 - \left(\frac{\cos \overline{t}_{13}}{\overline{d}_{13}} - \frac{\cos \overline{t}_{12}}{\overline{d}_{12}}\right) dx_1 + \frac{\sin \overline{t}_{12}}{\overline{d}_{12}} dy_2 \\ & - \frac{\cos \overline{t}_{12}}{d_{12}} dx_2 - \frac{\sin \overline{t}_{13}}{\overline{d}_{13}} dy_3 + \frac{\cos \overline{t}_{13}}{\overline{d}_{13}} dx_3 + (\overline{\delta}_{213} - \overline{\delta}_{213,b}) \end{aligned} \tag{4.6.17}$$

$$v_{\overline{d}} = -\cos \overline{t}_{12} dy_1 - \sin \overline{t}_{12} dx_1 + \cos \overline{t}_{12} dy_2 + (\overline{d}_{12} - \overline{d}_{12,b}) \tag{4.6.18}$$

Just to be sure that there are no misunderstandings about the term *plane,* let us review what created the situation that allows us to use plane trigonometry. The conformal mapping model builds upon the 3D geodetic and 2D ellipsoidal models as visualized by the transition of parameters $\{\varphi, \lambda, h\} \rightarrow \{\varphi, \lambda\} \rightarrow \{x, y\}$ and observables $\{\alpha, \delta, \beta, s, \Delta h, \Delta N\} \rightarrow \{\widehat{\alpha}, \widehat{\delta}, \widehat{s}\} \rightarrow \{\overline{t}, \overline{\delta}, \overline{d}\}$. The height parameter and the vertical observations are not present in the conformal mapping model.

4.6.5 Similarity Revisited

In Appendix C, we state that interpreting the conformal property as a similarity transformation between infinitesimally small figures is permissible. It is difficult to understand such a statement because one typically does not think in terms of infinitesimally small figures. We shed some light on this statement by transforming two clusters of points that were generated with different conformal mappings and look at the discrepancies. For example, if the discrepancies exceed a specified limit, then a similarity transformation cannot be used to transform between both clusters of points.

We construct a simple experiment to demonstrate the similarity transformation. Let there be n equally spaced points on a geodesic circle on the ellipsoid whose center is located at $\varphi_0 = 45°$ and at the central meridian. These points are mapped with the transverse Mercator and Lambert conformal mapping functions using $k_0 = 1$. These two sets of map coordinates are input to a least-squares solution that estimates the parameters of a similarity transformation, i.e., two translations, one scale factor, and one rotation angle. The coordinate residuals v_{x_i} and v_{y_i} for station i are used to compute the station discrepancy $d_i = (v_{x_i}^2 + v_{x_i}^2)^{1/2}$. We use the average of the d_i over all the points as a measure of fit. The radius of the geodesic circle is incremented from 10 to 100 km for the solutions shown in Figure 4.6.3. The figure shows an optimal situation because the circle is centered at the origin of the Lambert conformal mapping and at the central meridian of the Mercator mapping at the same latitude. With $k_0 = 1$, the area around the center of the circle has the least distortion and the similarity model fits relatively well. The 1 m average is reached just beyond a 50 km radius. Both lines overlap in the figure.

Figure 4.6.4 shows discrepancies for different locations of the geodesic circle within the mapping area while the radius remains constant at 10 km. For line 1 (LC), the standard parallel of the Lambert conformal mapping shifts from 45° to 46° while the center of the geodesic circle remains at latitude 45°. In the case of line 2 (TM), the center of the geodesic circle moves from 0° (central meridian) to 1° in longitude, while also maintaining a latitude of 45°. The lines in the figure diverge, indicating

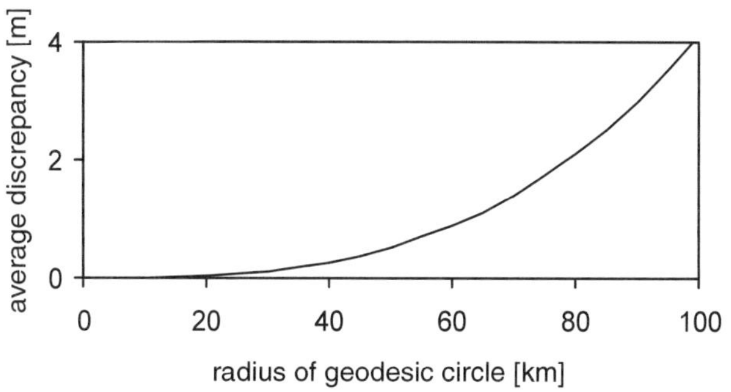

Figure 4.6.3 Similarity transformation of two mapped geodesic circles as a function of radius.

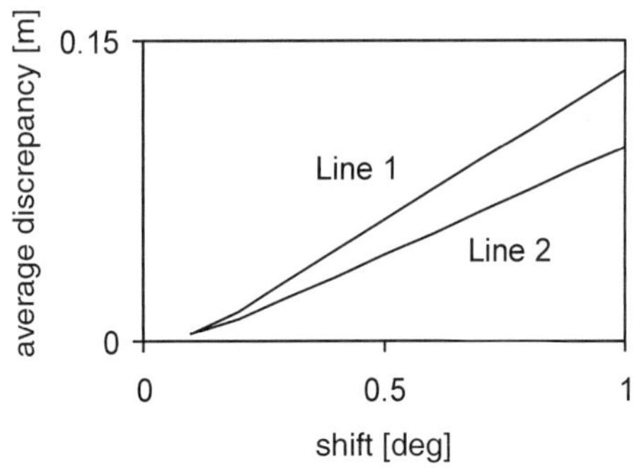

Figure 4.6.4 Similarity transformation of two mapped geodesic circles as a function of location.

that the distortions for both mapping are different and are a function of where the points are located on the mapping plane.

4.7 SUMMARY

This chapter presented a brief summary of geodesy. Some readers might consider it impossible to treat geodesy in just one chapter out of fear for an incomplete or superficial treatment of the subject. In line with the overall objectives of this book, the focus was on operational aspects of geodesy, avoiding long mathematical treatment and instead referring to resources that are publically available on the Internet.

We began with a discussion of the fundamental ITRF and ICRF reference frames. In particular, we explained the definition of the new ICRF pole, called CIP, and provided two ways of transforming between ITRF and ICRF. Our operational view assumes that a geodetic datum is in place, i.e., the locations and orientation of the ellipsoid is known and the geoid undulations and deflection of the vertical are also available for that datum. As an example we referred to the NAD83.

The 3G geodetic model, the ellipsoidal model, and the conformal mapping model are a central part of this chapter. Table 4.7.1 provides a summary of notation used to identify the various model observables. Figure 4.7.1 gives a summary of the type of reductions required to generate the respective model observations. The right column of boxes represents the reductions to be applied to the observations before the adjustment. The left column of boxes are reductions to be added to the adjusted, but now quality-controlled model observations, in order to obtain quality-controlled observations on the surface of earth that can be staked out or otherwise used by the surveyor in physical space. Clearly, the respective corrections on the left and right side of the figure are of the same magnitude but have the opposite sign.

The 3D model was identified to be, mathematically speaking, the simplest and yet most versatile model because it is applicable to a network of any size and can deal with 3-dimensional observations. It was further pointed out that the deflections of the vertical corrections largely cancel for angle observations but need to be applied

TABLE 4.7.1 Summary of Model Notation

3D Model	2D Ellipsoidal	Conformal Map
α geodetic azimuth	$\hat{\alpha}$ geodesic azimuth	
β geodetic vertical angle ϑ geodetic zenith angle		
δ geodetic horizontal angle between normal sections	$\hat{\delta}$ geodesic angle between geodesics	$\overline{\delta}$ map angle between chords
		$\overline{t}$ grid north $\overline{T}$ geodesic north $\overline{\gamma}$ meridian convergence
s slant distance	$\hat{s}$ geodesic length	$\overline{s}$ length of mapped geodesic $\overline{d}$ length of chord

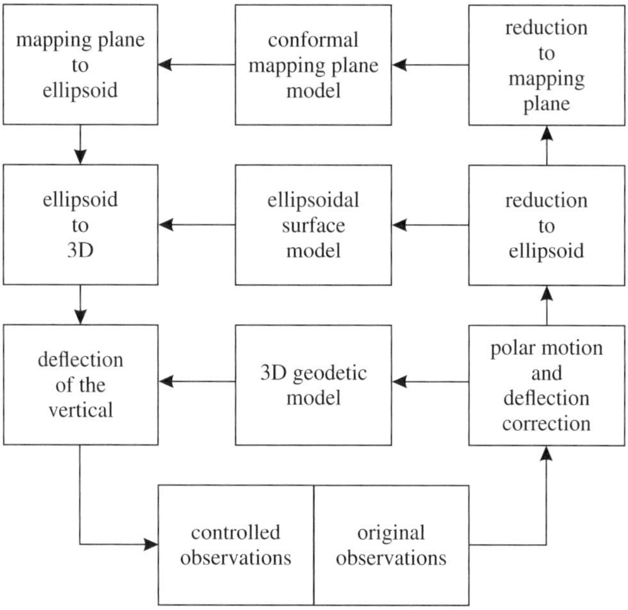

Figure 4.7.1 Three model loops.

when accurate azimuth observations are involved. This model allows the integration of classical terrestrial observations and GNSS vector observations without any further reduction.

The other models are the 2-dimensional ellipsoidal surface and conformal mapping models. It was pointed out that any height information is used to reduce the 3D observations to the ellipsoidal surface and that, consequently, no height parameters are included in 2D network adjustments. The ellipsoidal model is only of historical interest in terms of network adjustments. However, this model is conceptually needed to derive the conformal mapping model. After all, only 2D quantities from the ellipsoidal surface can be mapped onto the 2D conformal mapping plane. Both 2D models are limited according to the size of the network. The ellipsoidal model requires the geodesic line, and the respective expressions are a result of series expansions and therefore suffer from truncation errors. The conformal mapping model not only requires all of the geodesic line formalism but it also needs the conformality condition. The respective expressions also suffer from truncation errors that limit the size and shape of the networks, but most of all can show large-scale distortions in certain parts of the mapped area.

This chapter did not address physical geodesy in any significant detail. Physical geodesy deals with those aspects that involve gravity directly. One could say that the products of physical geodesy are geoid undulations and deflection of the vertical. But, because of our operational approach, we simply assumed that these elements would be available as part of the datum.

Also, leveling was not presented in detail. This technique depends on the direction of the plumb line. For example, much material could be added on the topic of loop closure for orthometric heights. The lack of depth by which the vertical component (leveling) was presented in this chapter is again justified on the basis of the operational approach. It is assumed that sufficiently accurate geoid undulations are available to convert orthometric heights to ellipsoidal heights, and that ellipsoidal height produced by GNSS can be easily converted to orthometric heights. Readers involved in high-accuracy vertical applications that require first-order leveling are advised to contact their national surveying agencies for information on the respective procedures and computation techniques.

CHAPTER 5

SATELLITE SYSTEMS

The satellite motions are introduced by means of normal orbits and the Kepler laws. It follows a summary on the major orbital perturbation of these simple mathematical motions. The first satellite system presented is the global positioning system (GPS). We briefly review the status of the signal transmissions as of the year 2014, including signal structure and navigation message. A section on the modernization of GPS starts with a brief exposure to binary offset carrier modulation, followed by remarks on the new codes L2C, L5, M, and L1C. The GLONASS system is discussed next with emphasis on the broadcast navigation message and brief remarks on GLONASS modernization. The other forthcoming systems, the European Galileo, the Japanese QZSS, and the Chinese Beidou are highlighted next. The details on each satellite system are available in various documents provided by the respective authorities on the Internet. Consider the following references: SPS (2008), IS-GPS-200G (2012), IS-GPS-705C (2012), IS-GPS-800C (2012), GLONASS (2008), Galileo (2010), QZSS (2013), and Beidou (2013).

We do not address signal processing that takes place inside the receiver. The interested reader is referred to specialized texts such as Kaplan (1996), Parkinson et al. (1996), Tsui (2005), Misra and Enge (2006), and Borre et al. (2007).

5.1 MOTION OF SATELLITES

The orbital motion of a satellite is a result of the earth's gravitational attraction, as well as a number of other forces acting on the satellite. The attraction of the sun and the moon and the pressure on the satellite caused by impacting solar radiation particles are examples of such forces. For high-orbiting satellites, the atmospheric drag is negligible. Mathematically, the equations of motion for satellites are differential equations that are solved by numerical integration over time. The integration begins with initial conditions, such as the position and velocity of the satellite at some initial epoch. The computed (predicted) satellite positions can be compared with actual observations. Possible discrepancies are useful to improve the modeled force functions, the accuracy of computed satellite positions, or the estimated positions of the observer.

5.1.1 Kepler Elements

Six Kepler elements are often used to describe the position of satellites in space. To simplify attempts to study satellite motions, we study so-called normal orbits. For normal orbits, the satellites move in an orbital plane that is fixed in space; the actual path of the satellite in the orbital plane is an ellipse in the mathematically strict sense. One focal point of the orbital ellipse is at the center of the earth. The conditions leading to such a simple orbital motion are as follows:

1. The earth is treated as a point mass, or, equivalently, as a sphere with spherically symmetric density distribution. The gravitational field of such a body is radially symmetric; i.e., the plumb lines are all straight lines and point toward the center of the sphere.
2. The mass of the satellite is negligible compared to the mass of the earth.
3. The motion of the satellite takes place in a vacuum; i.e., there is no atmospheric drag acting on the satellite and no solar radiation pressure.
4. No sun, moon, or other celestial body exerts a gravitational attraction on the satellite.

The orbital plane of a satellite moving under such conditions is shown in Figure 5.1.1. The ellipse denotes the path of the satellite. The shape of the ellipse is determined by the semimajor axis a and the semiminor axis b. The symbol e denotes the eccentricity

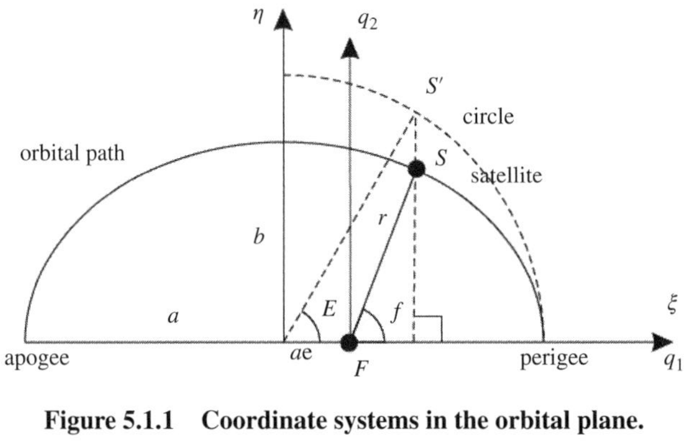

Figure 5.1.1 Coordinate systems in the orbital plane.

of the ellipse. The ellipse is enclosed by an auxiliary circle with radius a. The principal axes of the ellipse form the coordinate system (ξ, η). S denotes the current position of the satellite; the line SS' is in the orbital plane and is parallel to the η axis. The coordinate system (q_1, q_2) is located in the orbital plane, with origin at the focal point F of the ellipse that coincides with the center of the earth. The third axis q_3, not shown in the figure, completes the right-handed coordinate system. The geocentric distance from the center of the earth to the satellite is denoted by r. The orbital locations closest to and farthest from the focal point are called the perigee and apogee, respectively. The true anomaly f and the eccentric anomaly E are measured counterclockwise, as shown in Figure 5.1.1.

The orbital plane is shown in Figure 5.1.2 with respect to the true celestial coordinate system. The center of the sphere of directions is located at the focal point F. The X axis is in the direction of the vernal equinox, the Z axis coincides with the celestial ephemeris pole, and Y is located at the equator, thus completing the right-handed coordinate system. The intersection of the orbital plane with the equator is called the nodal line. The point at which the satellite ascends the equator is the ascending node. The right ascension of the ascending node is denoted by Ω. The line of apsides connects the focal point F and the perigee. The angle subtended by the nodal line and the line of apsides is called the argument of perigee ω. The true anomaly f and the argument of perigee ω lie in the orbital plane. Finally, the angle between the orbital plane and the equator is the inclination i. The figure shows that (Ω, i) determines the position of the orbital plane in the true celestial system, (Ω, ω, i) the orbital ellipse in space, and (a, e, f) the position of the satellite within the orbital plane.

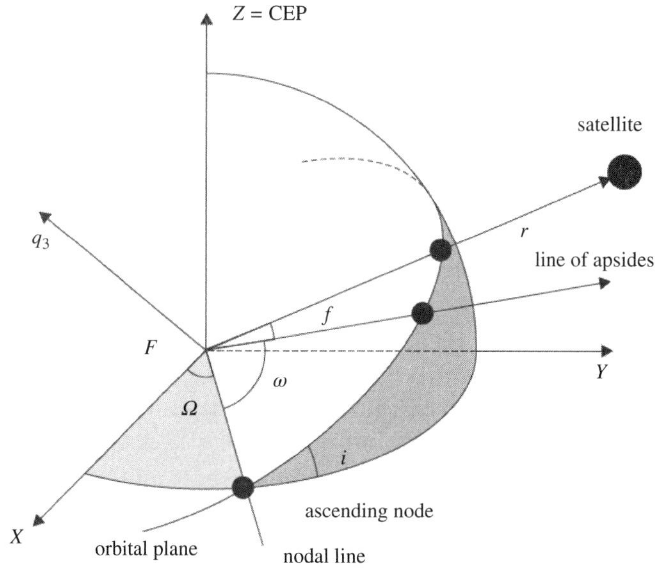

Figure 5.1.2 Orbital plane on the sphere of direction.

The six Kepler elements are $\{\Omega, \omega, i, a, e, f\}$. The true anomaly f is the only Kepler element that is a function of time in the case of normal orbits; the remaining five Kepler elements are constant. For actual satellite orbits, which are not subject to the conditions of normal orbits, all Kepler elements are a function of time. They are called osculating Kepler elements.

5.1.2 Normal Orbital Theory

Normal orbits are particularly useful for understanding and visualizing the spatial motions of satellites. The solutions of the respective equations of motions can be given by simple, analytical expressions. Since normal orbits are a function of the central portion of the earth's gravitational field (which is by far the largest force acting on the satellite), normal orbits are indeed usable for orbital predictions over short periods of time when low accuracy is sufficient. Thus, one of the popular uses of normal orbits is for the construction of satellite visibility charts.

The normal motion of satellites is determined by Newton's law of gravitation,

$$F = \frac{k^2 mM}{r^2} \tag{5.1.1}$$

In (5.1.1), M and m denote the mass of the earth and the satellite, respectively, k^2 is the universal constant of gravitation, r is the geocentric distance to the satellite, and F is the gravitational force between the two bodies. This force can also be written as

$$F = ma \tag{5.1.2}$$

where a in this instance denotes the acceleration experienced by the satellite. Combining (5.1.1) and (5.1.2) gives

$$a = \frac{k^2 M}{r^2} \tag{5.1.3}$$

This equation can be written in vector form as

$$\ddot{\boldsymbol{r}} = -k^2 M \frac{\boldsymbol{r}}{r^3} = -\mu \frac{\boldsymbol{r}}{r^3} \tag{5.1.4}$$

where

$$\mu = k^2 M \tag{5.1.5}$$

is the earth's gravitational constant. Including the earth's atmosphere, it has the value $\mu = 3,986,005 \times 10^8 \mathrm{m}^3\mathrm{s}^{-2}$. The vector $\boldsymbol{r}$ is directed from the central body (earth) to the satellite. The sign has been chosen such that the acceleration is directed toward the earth. The colinearity of the acceleration and the position vector as in (5.1.4) is a characteristic of central gravity fields. A particle released from rest would fall along a straight line toward the earth (straight plumb line).

Equation (5.1.4) is valid for the motion with respect to an inertial origin. In general, one is interested in determining the motion of the satellite with respect to the earth.

The modified equation of motion for accomplishing this is given by Escobal (1965, p. 37) as

$$\ddot{\boldsymbol{r}} = -k^2(M+m)\frac{\boldsymbol{r}}{r^3} \tag{5.1.6}$$

Because $m \ll M$, the second term is often neglected and (5.1.6) becomes (5.1.4).

Figure 5.1.2 gives the position of the satellite in the (q) orbital plane coordinate system $\boldsymbol{q} = [q_1 \quad q_2 \quad q_3]^T$ as

$$\boldsymbol{q} = r\begin{bmatrix} \cos f \\ \sin f \\ 0 \end{bmatrix} \tag{5.1.7}$$

Because the geocentric distance and the true anomaly are functions of time, the derivative with respect to time, denoted by a dot, is

$$\dot{\boldsymbol{q}} = \dot{r}\begin{bmatrix} \cos f \\ \sin f \\ 0 \end{bmatrix} + r\dot{f}\begin{bmatrix} -\sin f \\ \cos f \\ 0 \end{bmatrix} \tag{5.1.8}$$

The second derivatives with respect to time are

$$\ddot{\boldsymbol{q}} = \ddot{r}\begin{bmatrix} \cos f \\ \sin f \\ 0 \end{bmatrix} + 2\dot{r}\dot{f}\begin{bmatrix} -\sin f \\ \cos f \\ 0 \end{bmatrix} + r\ddot{f}\begin{bmatrix} -\sin f \\ \cos f \\ 0 \end{bmatrix} - r(\dot{f})^2\begin{bmatrix} \cos f \\ \sin f \\ 0 \end{bmatrix} \tag{5.1.9}$$

The second derivative is written according to (5.1.4) and (5.1.7) as

$$\ddot{\boldsymbol{r}} = \frac{-\mu}{r^2}\begin{bmatrix} \cos f \\ \sin f \\ 0 \end{bmatrix} \tag{5.1.10}$$

Evaluating (5.1.9) and (5.1.10) at $f = 0$ (perigee) and substituting (5.1.10) for the left-hand side of (5.1.9) gives

$$\ddot{r} - r(\dot{f})^2 = \frac{-\mu}{r^2} \tag{5.1.11}$$

$$r\ddot{f} + 2\dot{r}\dot{f} = 0 \tag{5.1.12}$$

Equation (5.1.12) is developed further by multiplying with r and integrating

$$\int (r^2\ddot{f} + 2r\dot{r}\dot{f})\, dt = C \tag{5.1.13}$$

The result of the integration is

$$r^2\dot{f} + 2r^2\dot{f} = C \tag{5.1.14}$$

as can be readily verified through differentiation. Combining both terms yields

$$r^2 \dot{f} = h \tag{5.1.15}$$

where h is a new constant. Equation (5.1.15) is identified as an angular momentum equation, implying that the angular momentum for the orbiting satellite is conserved.

In order to integrate (5.1.11), we define a new variable:

$$u \equiv \frac{1}{r} \tag{5.1.16}$$

By using equation (5.1.15) for dt/df, the differential of (5.1.16) becomes

$$\frac{du}{df} = \frac{du}{dr}\frac{dr}{dt}\frac{dt}{df} = -\frac{\dot{r}}{h} \tag{5.1.17}$$

Differentiating again gives

$$\frac{d^2u}{df^2} = \frac{d}{dt}\left(-\frac{\dot{r}}{h}\right)\frac{dt}{df} = -\frac{\ddot{r}}{u^2h^2} \tag{5.1.18}$$

or

$$\ddot{r} = -h^2u^2\frac{d^2u}{df^2} \tag{5.1.19}$$

By substituting (5.1.19) in (5.1.11), substituting $\dot{f}$ from (5.1.15) in (5.1.11), and replacing r by u according to (5.1.16), equation (5.1.11) becomes

$$\frac{d^2u}{df^2} + u = \frac{\mu}{h^2} \tag{5.1.20}$$

which can readily be integrated as

$$\frac{1}{r} \equiv u = C\,\cos f + \frac{\mu}{h^2} \tag{5.1.21}$$

where C is a constant.

Equation (5.1.21) is the equation of an ellipse. This is verified by writing the equation for the orbital ellipse in Figure 5.1.1 in the principal axis form:

$$\frac{\xi^2}{a^2} + \frac{\eta^2}{b^2} = 1 \tag{5.1.22}$$

where

$$\xi = ae + r\,\cos f \quad \eta = r\,\sin f \quad b^2 = a^2(1 - e^2) \tag{5.1.23}$$

The expression for b is valid for any ellipse. Substituting (5.1.23) into (5.1.22) and solving the resulting second-order equation for r gives

$$\frac{1}{r} = \frac{1}{a(1-e^2)} + \frac{e}{a(1-e^2)}\,\cos f \tag{5.1.24}$$

with

$$C = \frac{e}{a(1-e^2)} \qquad h = \sqrt{\mu a(1-e^2)} \tag{5.1.25}$$

the identity between the expression for the ellipse (5.1.24) and equation (5.1.21) is established. Thus, the motion of a satellite under the condition of a normal orbit is an ellipse. This is the content of Kepler's first law. The focus of the ellipse is at the center of mass.

Kepler's second law states that the geocentric vector $\boldsymbol{r}$ sweeps equal areas during equal times. Because the area swept for the differential angle df is

$$dA = \frac{1}{2} r^2 \, df \tag{5.1.26}$$

it follows from (5.1.15) and (5.1.25) that

$$\frac{dA}{dt} = \frac{1}{2}\sqrt{\mu a(1-e^2)} \tag{5.1.27}$$

which is a constant.

The derivation of Kepler's third law requires the introduction of the eccentric anomaly E. From Figure 5.1.1 we see that

$$q_1 = \xi - ae = a(\cos E - e) \tag{5.1.28}$$

where

$$\xi = a \cos E \tag{5.1.29}$$

The second coordinate follows from (5.1.22):

$$q_2 \equiv \eta = \sqrt{\left(1 - \frac{\xi^2}{a^2}\right) b^2} \tag{5.1.30}$$

Substitute (5.1.29) in (5.1.30), then

$$q_2 \equiv \eta = b \sin E \tag{5.1.31}$$

With (5.1.28), (5.1.31), and b from (5.1.23), the geocentric satellite distance becomes

$$r = \sqrt{q_1^2 + q_2^2} = a(1 - e \cos E) \tag{5.1.32}$$

Differentiating equations (5.1.32) and (5.1.24) gives

$$dr = ae \sin E \, dE \tag{5.1.33}$$

$$dr = \frac{r^2 e}{a\,(1-e^2)} \sin f \, df \tag{5.1.34}$$

Equating (5.1.34) and (5.1.33), using η and b from (5.1.23), (5.1.31), and (5.1.7), and multiplying the resulting equation by r gives

$$r b \, dE = r^2 \, df \tag{5.1.35}$$

Substituting b from (5.1.23) and (5.1.32) for r, replacing df by dt using (5.1.15), using h from (5.1.25), and then integrating, we obtain

$$\int_{E=0}^{E} (1 - e \cos E) \, dE = \int_{t_0}^{t} \sqrt{\frac{\mu}{a^3}} \, dt \tag{5.1.36}$$

Integrating both sides gives

$$E - e \sin E = M \tag{5.1.37}$$

$$M = n(t - t_0) \tag{5.1.38}$$

$$n = \sqrt{\frac{\mu}{a^3}} \tag{5.1.39}$$

Equation (5.1.39) is Kepler's third law. Equation (5.1.37) is called the Kepler equation. The symbol n denotes the mean motion, M is the mean anomaly, and t_0 denotes the time of perigee passage of the satellite. The mean anomaly M should not be confused with the same symbol used for the mass of the central body in (5.1.1). Let P denote the orbital period, i.e., the time required for one complete revolution, then

$$P = \frac{2\pi}{n} \tag{5.1.40}$$

The mean motion n equals the average angular velocity of the satellite. Equation (5.1.39) shows that the semimajor axis completely determines the mean motion and thus the period of the orbit.

With the Kepler laws in place, one can identify alternative sets of Kepler elements, such as $\{\Omega, \omega, i, a, e, M\}$ or $\{\Omega, \omega, i, a, e, E\}$. Often the orbit is not specified by the Kepler elements but by the vector $\boldsymbol{r} = [X \quad Y \quad Z]^T = \boldsymbol{X}$ and the velocity $\dot{\boldsymbol{r}} = [\dot{X} \quad \dot{Y} \quad \dot{Z}]^T = \dot{\boldsymbol{X}}$, expressed in the true celestial coordinate system (X). Figure 5.1.2 shows that

$$\boldsymbol{q} = \boldsymbol{R}_3(\omega) \, \boldsymbol{R}_1(i) \, \boldsymbol{R}_3(\Omega) \, \boldsymbol{X} = \boldsymbol{R}_{qX}(\Omega, i, \omega) \, \boldsymbol{X} \tag{5.1.41}$$

where $\boldsymbol{R}_i$ denotes a rotation around axis i. The inverse transformation is

$$\boldsymbol{X} = \boldsymbol{R}_{qX}^{-1}(\Omega, i, \omega) \, \boldsymbol{q} \tag{5.1.42}$$

Differentiating (5.1.42) gives

$$\dot{\boldsymbol{X}} = \boldsymbol{R}_{qX}^{-1}(\Omega, i, \omega) \dot{\boldsymbol{q}} \tag{5.1.43}$$

Note that the elements of $\boldsymbol{R}_{qX}$ are constants, because the orbital ellipse does not change its position in space. Using b from (5.1.23), (5.1.28), and (5.1.31), it follows that

$$\boldsymbol{q} = \begin{bmatrix} a(\cos E - e) \\ a\sqrt{1-e^2}\sin E \\ 0 \end{bmatrix} = \begin{bmatrix} r\cos f \\ r\sin f \\ 0 \end{bmatrix} \tag{5.1.44}$$

The velocity becomes

$$\dot{\boldsymbol{q}} = \frac{na}{1 - e\cos E}\begin{bmatrix} -\sin E \\ \sqrt{1-e^2}\cos E \\ 0 \end{bmatrix} = \frac{na}{\sqrt{1-e^2}}\begin{bmatrix} -\sin f \\ e + \cos f \\ 0 \end{bmatrix} \tag{5.1.45}$$

The first part of (5.1.45) follows from (5.1.36), and the second part can be verified using known relations between the anomalies E and f. Equations (5.1.42) to (5.1.45) transform the Kepler elements into Cartesian coordinates and their velocities $(\boldsymbol{X}, \dot{\boldsymbol{X}})$.

The transformation from $(\boldsymbol{X}, \dot{\boldsymbol{X}})$ to Kepler elements starts with the computation of the magnitude and direction of the angular momentum vector

$$\boldsymbol{h} = \boldsymbol{X} \times \dot{\boldsymbol{X}} = [h_X \quad h_Y \quad h_Z]^T \tag{5.1.46}$$

which is the vector form of (5.1.15). The various components of $\boldsymbol{h}$ are shown in Figure 5.1.3. The right ascension of the ascending node and the inclination of the

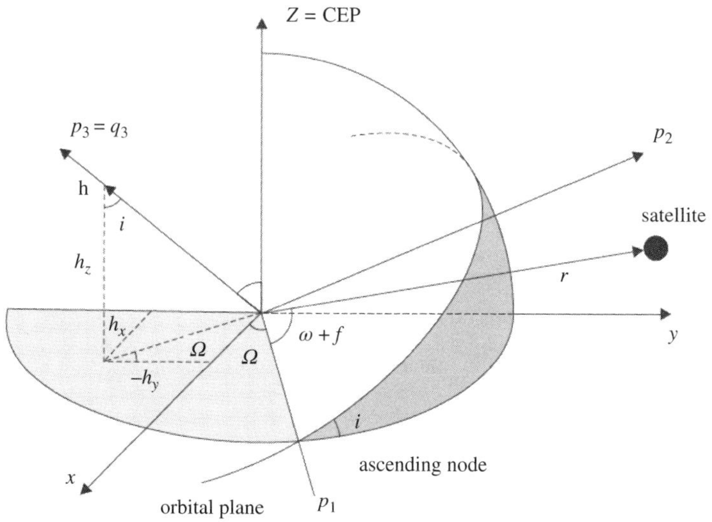

Figure 5.1.3 Angular momentum vector and Kepler elements. The angular momentum vector is orthogonal to the orbital plane.

orbital plane are, according to this figure,

$$\Omega = \tan^{-1}\left(\frac{h_X}{-h_Y}\right) \qquad i = \tan^{-1}\left(\frac{\sqrt{h_X^2 + h_Y^2}}{h_Z}\right) \tag{5.1.47}$$

By defining the auxiliary coordinate system (p) such that the p_1 axis is along the nodal line, p_3 is along the angular momentum vector, and p_2 completes a right-handed coordinate system, we obtain

$$\boldsymbol{p} = \boldsymbol{R}_1(i)\boldsymbol{R}_3(\Omega)\boldsymbol{X} \tag{5.1.48}$$

The sum of the argument of perigee and the true anomaly becomes

$$\omega + f = \tan^{-1}\left(\frac{p_2}{p_1}\right) \tag{5.1.49}$$

Thus far, the orbital plane and the orientation of the orbital ellipse have been determined. The shape and size of the ellipse depend on the velocity of the satellite. The velocity, geocentric distance, and the magnitude of the angular momentum are

$$v = \|\dot{\boldsymbol{X}}\| \quad r = \|\boldsymbol{X}\| \quad h = \|\boldsymbol{h}\| \tag{5.1.50}$$

The velocity expressed in the (q) coordinate system can be written as follows using (5.1.24), (5.1.39), and (5.1.45):

$$\begin{aligned} v^2 &= \dot{q}_1^2 + \dot{q}_2^2 \\ &= \frac{n^2 a^2}{1 - e^2}(\sin^2 f + e^2 + 2e\,\cos f + \cos^2 f) \\ &= \frac{\mu}{a(1 - e^2)}[2 + 2e\,\cos f - (1 - e^2)] \\ &= \mu\left(\frac{2}{r} - \frac{1}{a}\right) \end{aligned} \tag{5.1.51}$$

Equation (5.1.51) yields the expression for the semimajor axis

$$a = \frac{r}{2 - rv^2/\mu} \tag{5.1.52}$$

With h from (5.1.25), it follows that

$$e = \left(1 - \frac{h^2}{\mu a}\right)^{1/2} \tag{5.1.53}$$

and (5.1.32), (5.1.44), and (5.1.45) give an expression for the eccentric anomaly

$$\cos E = \frac{a - r}{a\, e} \qquad \sin E = \frac{\boldsymbol{q} \cdot \dot{\boldsymbol{q}}}{e\sqrt{\mu a}} \tag{5.1.54}$$

These equations determine the quadrant of the eccentric anomaly. Having E, the true anomaly follows from (5.1.44)

$$f = \tan^{-1} \frac{\sqrt{1 - e^2}\, \sin E}{\cos E - e} \tag{5.1.55}$$

Finally, Kepler's equation yields the mean anomaly

$$M = E - e\, \sin E \tag{5.1.56}$$

Equations (5.1.47) to (5.1.56) comprise the transformation from $(\boldsymbol{X}, \dot{\boldsymbol{X}})$ to the Kepler elements.

Table 5.1.1 depicts six examples of trajectories for which the orbital eccentricity is zero, $e = 0$. The satellite position $\boldsymbol{x}$ in the earth-centered earth-fixed (ECEF) coordinate system can be readily computed from $\boldsymbol{X}$ by applying (4.2.32). We can then compute spherical latitude and longitude (ϕ, λ) and the trajectories of the satellites on the sphere. For reasons of convenience, we express the mean motion of the satellites in revolutions per day, $\overline{n} = n / \omega$. The longitude difference between consecutive equatorial crossings can then be computed from

$$\Delta\lambda = \pi \left(1 - \frac{1}{\overline{n}}\right) \tag{5.1.57}$$

Table 5.1.1 also lists the change in longitude of the trajectory over a 24 h period, denoted by $\delta\lambda$. The number in parentheses in the graphs indicates the number of days plotted. In all cases the inclination is $i = 65°$. The maximum and minimum of the trajectories occur at a latitude of i and $-i$, respectively.

Case 1, specified by $\overline{n} = 2$, applies to GPS because the satellite orbits twice per (sidereal) day. Case 2 has been constructed such that the trajectories intersect the equator at 90°. In case 3, the point at which the trajectory touches, having common vertical tangent, and the point of either maximum or minimum have the same longitude. The mean motion must be computed from a nonlinear equation, but $\overline{n} > 1$ is valid. In case 4, the satellite completes one orbital revolution in exactly one (sidereal) day. Case 5 represents a retrograde motion with $\overline{n} < 1$ but with the same properties as case 3. In case 6, the common tangent at the extrema is vertical. The interested reader may verify that

$$\lambda = \tan^{-1}\left(\cos i \frac{\sin\phi}{\sqrt{\sin^2 i - \sin^2\phi}} \right) - \frac{1}{\overline{n}} \sin^{-1}\left(\frac{\sin\phi}{\sin i}\right) \tag{5.1.58}$$

TABLE 5.1.1 Trajectories of Normal Orbits.

$\bar{n} = 2 \qquad \delta\lambda > 2\pi$	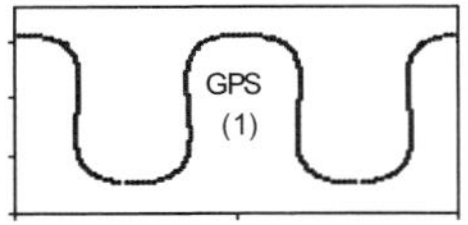
$\bar{n} = \dfrac{1}{\cos i} \qquad \delta\lambda > 2\pi$	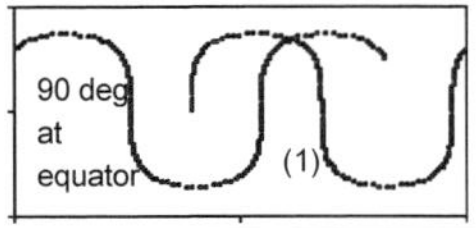
$\dfrac{\pi}{2} + \sin^{-1}\left(\dfrac{\sqrt{1-\bar{n}\cos i}}{\sin i}\right) - \bar{n}\left\{\dfrac{\pi}{2} + \tan^{-1}\left(\dfrac{\sqrt{(1-\bar{n}\cos i)\cos i}}{\bar{n}-\cos i}\right)\right\} = 0 \qquad \delta\lambda > 3\ \Delta\lambda$	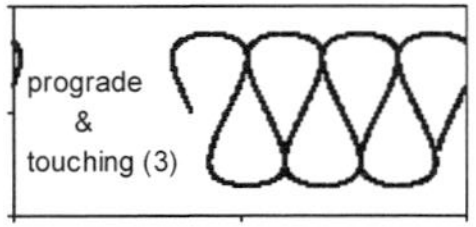
$\bar{n} = 1 \quad \delta\lambda = 2\left\{\tan^{-1}\left(\cos i\sqrt{\dfrac{1-\cos i}{\sin^2 i - 1 + \cos i}}\right) - \sin^{-1}\left(\dfrac{\sqrt{1-\cos i}}{\sin i}\right)\right\}$	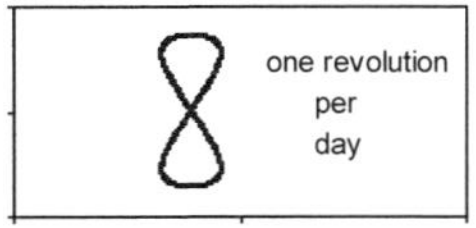
$\dfrac{3\pi}{2} - \sin^{-1}\left(\dfrac{\sqrt{1-\bar{n}\cos i}}{\sin i}\right) - \bar{n}\left\{\dfrac{3\pi}{2} - \tan^{-1}\left(\dfrac{\sqrt{(1-\bar{n}\cos i)\cos i}}{\bar{n}-\cos i}\right)\right\} = 0 \qquad \delta\lambda = 3\ \Delta\lambda$	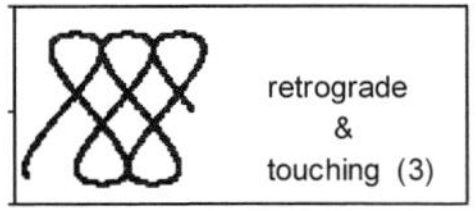
$\bar{n} = \cos i \qquad \delta\lambda = \tan^{-1}\{\cos i\ \tan(2\pi\bar{n})\} - 2\pi$	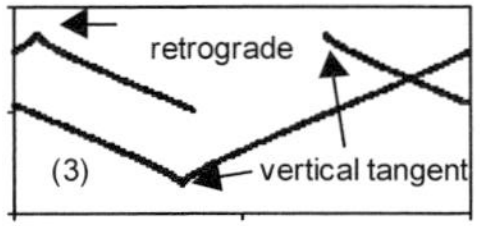

and

$$\frac{d\varphi}{d\lambda} = \frac{\cos\phi\sqrt{\sin^2 i - \sin^2\phi}}{\overline{n}\,\cos i - \cos^2\phi}\,\overline{n} \tag{5.1.59}$$

is valid for all cases.

5.1.3 Satellite Visibility and Topocentric Motion

The topocentric motion of a satellite as seen by an observer on the surface of the earth can be computed from existing expressions. Let $\boldsymbol{X}_S$ denote the geocentric position of the satellite in the celestial coordinate system (X). These positions could have been obtained from (5.1.42), in the case of normal motion or from the integration of perturbed orbits discussed below. The position $\boldsymbol{X}_S$ can then be readily transformed to crust-fixed coordinate system (x), giving $\boldsymbol{x}_S$ by applying (4.2.32). If we further assume that the position of the observer on the ground in the crust-fixed coordinate system is $\boldsymbol{x}_P$, then the topocentric coordinate difference

$$\Delta\boldsymbol{x} = \boldsymbol{x}_S - \boldsymbol{x}_P \tag{5.1.60}$$

can be substituted into Equations (4.4.11) to (4.4.13) to obtain the topocentric geodetic azimuth, elevation, and distance of the satellite. The geodetic latitude and longitude in these expressions can be computed from $\boldsymbol{x}_P$, if necessary. For low-accuracy applications such as the creation of visibility charts it is sufficient to use spherical approximations.

5.1.4 Perturbed Satellite Motion

The accurate determination of satellite positions must consider various disturbing forces. Disturbing forces are all forces causing the satellite to deviate from the simple normal orbit. The disturbances result primarily from the nonsphericity of the gravitational potential, the attraction of the sun and the moon, the solar radiation pressure, and other smaller forces acting on the satellites. For example, albedo is a force due to electromagnetic radiation reflected by the earth. There could be thermal reradiation forces caused by anisotropic radiation from the surface of the spacecraft. Additional forces, such as residual atmospheric drag, affect satellites closer to the earth.

Several of the disturbing forces are computable; others, in particular the smaller forces, require detailed modeling and are still subject to ongoing research. Knowing the accurate location of the satellites and being able to treat satellite position coordinates as known quantities is important in surveying. Most scientific applications of GNSS demand the highest orbital accuracy, almost at the centimeter level. However, even surveying benefits from such accurate orbits, e.g., in precise point positioning with one receiver. See Section 6.6.1 for additional details on this technique. One of the goals of the International GNSS Service (IGS) and its contributing agencies and research groups is to refine orbital computation and modeling in order to make the most accurate satellite ephemeris available to users. In this section, we provide only an introductory exposition of orbital determination. The details are found in the literature, going all the way back to the days of the first artificial satellites.

The equations of motion are expressed in an inertial (celestial) coordinate system, corresponding to the epoch of the initial conditions. The initial conditions are either $(\boldsymbol{X}, \dot{\boldsymbol{X}})$ or the Kepler elements at a specified epoch. Because of the disturbing forces, all Kepler elements are functions of time. The transformation given above can be used to transform the initial conditions from $(\boldsymbol{X}, \dot{\boldsymbol{X}})$ to Kepler elements and vice versa. The equations of motion, as expressed in Cartesian coordinates, are

$$\frac{d\boldsymbol{X}}{dt} = \dot{\boldsymbol{X}} \tag{5.1.61}$$

$$\frac{d\dot{\boldsymbol{X}}}{dt} = -\frac{\mu \boldsymbol{X}}{\|\boldsymbol{X}\|^3} + \ddot{\boldsymbol{X}}_g + \ddot{\boldsymbol{X}}_s + \ddot{\boldsymbol{X}}_m + \ddot{\boldsymbol{X}}_{\text{SRP}} + \cdots \tag{5.1.62}$$

These are six first-order differential equations. The symbol μ denotes the geocentric gravitational constant (5.1.5). The first term in (5.1.62) represents the acceleration of the central gravity field that generates the normal orbits discussed in the previous section. Compare (5.1.62) with (5.1.4). The remaining accelerations are discussed briefly below. The simplest way to solve (5.1.61) and (5.1.62) is to carry out a simultaneous numerical integration. Most of the high-quality engineering or mathematical software packages have such integration routines available. Kaula (1966) expresses the equations of motion and the disturbing potential in terms of Kepler elements. Kaula (1962) gives similar expressions for the disturbing functions of the sun and the moon.

5.1.4.1 *Gravitational Field of the Earth*

The acceleration of the noncentral portion of the gravity field of the earth is given by

$$\ddot{\boldsymbol{X}}_g = \left[\frac{\partial R}{\partial X} \quad \frac{\partial R}{\partial Y} \quad \frac{\partial R}{\partial Z} \right]^T \tag{5.1.63}$$

The disturbing potential R is

$$R = \sum_{n=2}^{\infty} \sum_{m=0}^{n} \frac{\mu a_e^n}{r^{n+1}} \overline{P}_{nm} (\cos\theta) \; [\overline{C}_{nm} \cos m\lambda + \overline{S}_{nm} \sin m\lambda] \tag{5.1.64}$$

with

$$P_{nm} (\cos\theta) = \frac{(1 - \cos^2\theta)^{m/2}}{2^n n!} \frac{d^{(n+m)}}{d(\cos\theta)^{(n+m)}} (\cos^2\theta - 1)^n \tag{5.1.65}$$

$$\overline{P}_n = \sqrt{2n+1}\, P_n \tag{5.1.66}$$

$$\overline{P}_{nm} = \left(\frac{(n+m)!}{2(2n+1)\,(n-m)!} \right)^{-1/2} P_{nm} \tag{5.1.67}$$

Equation (5.1.64) expresses the disturbing potential (as used in satellite orbital computations) in terms of a spherical harmonic expansion. The symbol a_e denotes the mean earth radius, r is the geocentric distance to the satellite, and θ and λ are the spherical co-latitude and longitude of the satellite position in the earth-fixed coordinate system, i.e., $\boldsymbol{x} = \boldsymbol{x}(r, \theta, \lambda)$. The positions in the celestial system (X) follow from (4.2.32). $\overline{P}_{nm}$ denotes the associated Legendre functions, which are known mathematical functions of latitude. $\overline{C}_{nm}$ and $\overline{S}_{nm}$ are the spherical harmonic coefficients of degree n and order m. The bar indicates fully normalized potential coefficients. Note that the summation in (5.1.64) starts at $n = 2$. The term $n = 0$ equals the central component of the gravitational field. It can be shown that the coefficients for $n = 1$ are zero for coordinate systems whose origin is at the center of mass. Equation (5.1.64) shows that the disturbing potential decreases exponentially with the power of n. The high-order coefficients represent the detailed structure of the disturbing potential, and, as such, the fine structure of the gravity field of the earth. Only the coefficients of lower degree and order, say, up to degree and order 36, are significant for satellite orbital computations. The higher the altitude of the satellite, the less the impact of higher-order coefficients on orbital disturbances.

The largest coefficient in (5.1.64) is $\overline{C}_{20}$. This coefficient represents the effect of the flattening of the earth on the gravitational field. Its magnitude is about 1000 times larger than any other spherical harmonic coefficient.

Useful insight into the orbital disturbance of the flattening of the earth is obtained by considering only the effect $\overline{C}_{20}$. An analytical expression is obtained if one expresses the equations of motion (5.1.61) and (5.1.62) in terms of Kepler elements. The actual derivation of such equations is beyond the scope of this book. The reader is referred to Kaula (1966). Mueller (1964) offers the following result:

$$\dot{\omega} = -\left(\frac{\mu}{a^3}\right)^{1/2} \left(\frac{a_e}{a\left(1 - e^2\right)}\right)^2 \frac{3}{2} J_2 \left(1 + \cos^2 i - 1.5 \sin^2 i\right) \tag{5.1.68}$$

$$\dot{\Omega} = -\left(\frac{\mu}{a_3}\right)^{1/2} \left(\frac{a_e}{a\left(1 - e^2\right)}\right)^2 \frac{3}{2} J_2 \cos i \tag{5.1.69}$$

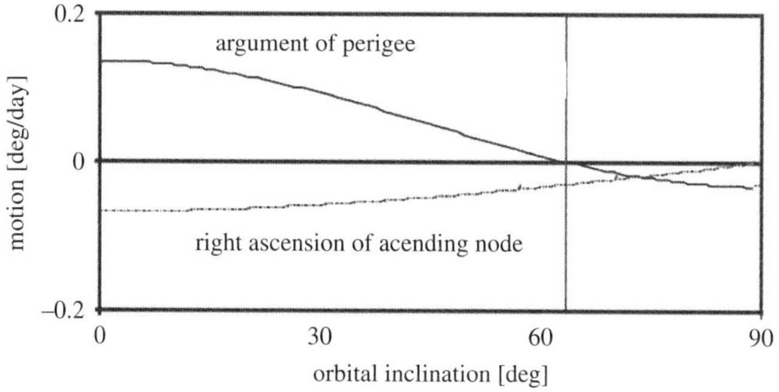

Figure 5.1.4 Impact of the earth's flattening on the motion of the perigee and the nodal line. The data refer to $a = 26{,}600$ km.

In these equations we have made the substitution $\overline{C}_{20} = -J_2\sqrt{5}$. The variations of the argument of perigee and the right ascension of the ascending node are shown in Figure 5.1.4 as a function of the inclination. At the critical inclination of approximately 63.5° the perigee motion is stationary. The perigee and the node regress if $i > 63.5°$. This orbital plane rotation is zero for polar orbits $i = 90°$. Equation (5.1.69) is also useful for understanding the connection between the earth flattening and precession and the 18.6-year nutation/tidal period.

5.1.4.2 Acceleration due to the Sun and the Moon The lunar and solar accelerations on the satellites are (Escobal, 1965, p. 37)

$$\ddot{\boldsymbol{X}}_m = \frac{\mu m_m}{m_e}\left(\frac{\boldsymbol{X}_m - \boldsymbol{X}}{\|\boldsymbol{X}_m - \boldsymbol{X}\|^3} - \frac{\boldsymbol{X}_m}{\|\boldsymbol{X}_m\|^3}\right) \tag{5.1.70}$$

$$\ddot{\boldsymbol{X}}_s = \frac{\mu m_s}{m_e}\left(\frac{\boldsymbol{X}_s - \boldsymbol{X}}{\|\boldsymbol{X}_s - \boldsymbol{X}\|^3} - \frac{\boldsymbol{X}_s}{\|\boldsymbol{X}_s\|^3}\right) \tag{5.1.71}$$

The commonly used values for the mass ratios are $m_m / m_e = 0.0123002$ and $m_s / m_e = 332,946$. Mathematical expressions for the geocentric positions of the moon $\boldsymbol{X}_m$ and the sun $\boldsymbol{X}_s$ are given, for example, in van Flandern and Pulkkinen (1979).

5.1.4.3 Solar Radiation Pressure Solar radiation pressure (SRP) is a result of the impact of light photons emitted from the sun on the satellite's surface. The basic parameters of the SRP are the effective area (surface normal to the incident radiation), the surface reflectivity, the thermal state of the surface, the luminosity of the sun, and the distance to the sun. Computing SRP requires the evaluation of surface integrals over the illuminated regions, taking shadow into account. Even if these regions are known, the evaluation of the surface integrals can still be difficult because of the complex shape of the satellite. The ROCK4 and ROCK42 models represent early attempts to take most of these complex relations and properties into consideration for GPS Block I, Block II, and Block IIa satellites, respectively (Fliegel et al., 1985; Fliegel and Gallini, 1989). Fliegel et al. (1992) describe an SRP force model for geodetic applications. Springer et al. (1999) report on SRP model parameter estimation on a satellite-by-satellite basis, as part of orbital determinations from heavily overdetermined global networks. Ziebart et al. (2002) discuss a pixel array method in connection with finite analysis, in order to even better delineate the illuminated satellite surfaces and surface temperature distribution.

One of the earliest and simplest SRP models uses merely two parameters. Consider the body-fixed coordinate system of Figure 5.1.5. The z' axis is aligned with the antenna and points toward the center of the earth. The satellite finds this direction and remains locked to it with the help of an earth limb sensor. The x' axis is positive toward the half plane that contains the sun. The y' axis completes the right-handed coordinate system and points along the solar panel axis. The satellites are always oriented such that the y' axis remains perpendicular to the earth-satellite-sun plane.

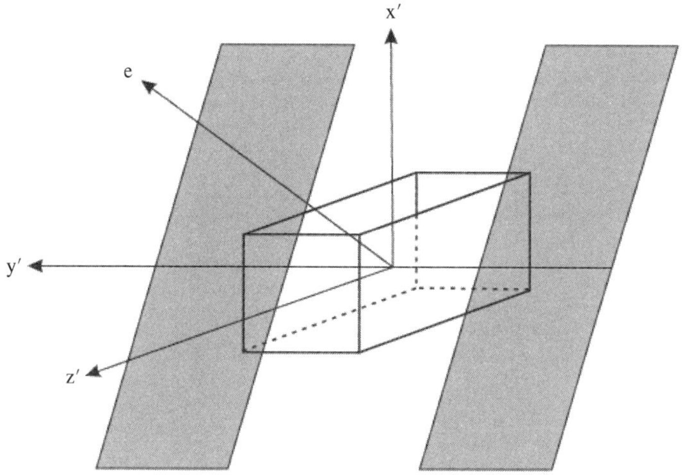

Figure 5.1.5 The satellite body-fixed coordinate system.

The only motion of the spacecraft in this body-fixed frame is the rotation of the solar panels around the y' axis to make the surface of the solar panels perpendicular to the direction of the sun. The direction of the sun is denoted by $\boldsymbol{e}$ in the figure.

In reference to this body-fixed coordinate system, a simple SRP model formulation is

$$\ddot{\boldsymbol{X}}_{SRP} = -p\frac{\boldsymbol{X}_{sun} - \boldsymbol{X}}{\|\boldsymbol{X}_{sun} - \boldsymbol{X}\|} + Y\frac{\boldsymbol{X}_{sun} \times \boldsymbol{X}}{\|\boldsymbol{X}_{sun} \times \boldsymbol{X}\|} \tag{5.1.72}$$

The symbol p denotes the SRP in the direction of the sun. With the sign convention of (5.1.72), p should be positive. The other parameter is called the Y bias. The reasons for its existence could be structural misalignments, thermal phenomena, or possibly misalignment of the solar panels with the direction of the solar photon flux. The fact that a Y bias exists demonstrates the complexity of accurate solar radiation pressure modeling.

Table 5.1.2 shows the effects of the various perturbations over the period of one day. The table shows the difference between two integrations, one containing the specific orbital perturbation and the others turned off. It is found that SRP orbital disturbance reaches close to 100 m in a day. This is very significant, considering that the goal is centimeter orbital accuracy. Over a period of 1 to 2 weeks, the SRP disturbance can grow to over 1 km.

5.1.4.4 Eclipse Transits and Yaw Maneuvers Orbital determination is further complicated when satellites travel through the earth shadow region (eclipse), which occurs twice per year when the sun is in or near the orbital plane. See Figures 5.1.6 and 5.1.7 for a graphical presentation. The umbra is that portion of the shadow cone that no light from the sun can reach. The penumbra is the region of partial shadowing; it surrounds the umbra cone. While the satellite transits through the shadow regions, the solar radiation force acting on the satellite is either

TABLE 5.1.2 Effect of Perturbations on GPS Satellites over One Day[a]

Perturbation	Radial	Along	Cross	Total
Earth flattening	1335	12902	6101	14334
Moon	191	1317	361	1379
Sun	83	649	145	670
$\overline{C}_{2,2}, \overline{S}_{2,2}$	32	175	9	178
SRP	29	87	3	92
$\overline{C}_{n,m}, \overline{S}_{n,m}$ $(n, m = 3 \ldots 8)$	6	46	4	46

Source: Springer et al. (1999).
[a]The units are in meters.

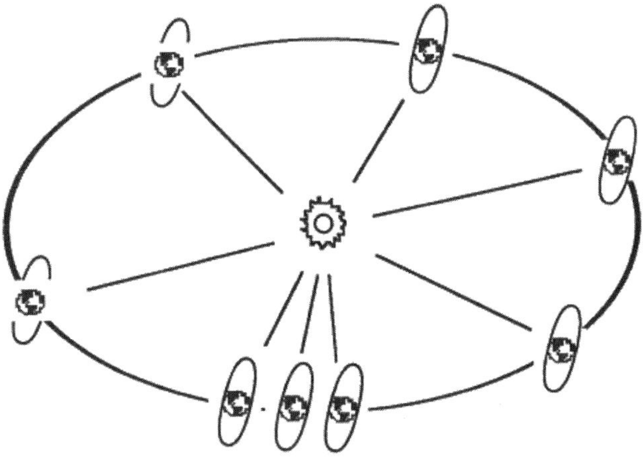

Figure 5.1.6 Biannual eclipse periods.

zero (umbra) or changing (penumbra). These changes in force must be taken into consideration in precise orbital computations. In addition, the thermal reradiation forces change as the temperature of the satellite drops. GPS satellites move through the shadow regions in less than 60 min, twice per day.

The shadow regions cause an additional problem for precise orbit determination. The solar panels are orientated toward the sun by the attitude control system (ACS) solar sensors that are mounted on the solar panels. The condition that the z' axis continuously points toward the center of the earth and the solar panels are continuously normal to the satellite-sun direction, the satellite must yaw, i.e., rotate around the z' axis, in addition to rotating the antennas around the y' axis. While the satellite passes through the shadow region, the ACS solar sensors do not receive sunlight and, therefore, cannot maintain the exact alignment of the solar panels. The satellite starts yawing in a somewhat unpredictable way. Errors in yaw cause errors in GPS observations in two ways. First, the range correction from the center of the satellite

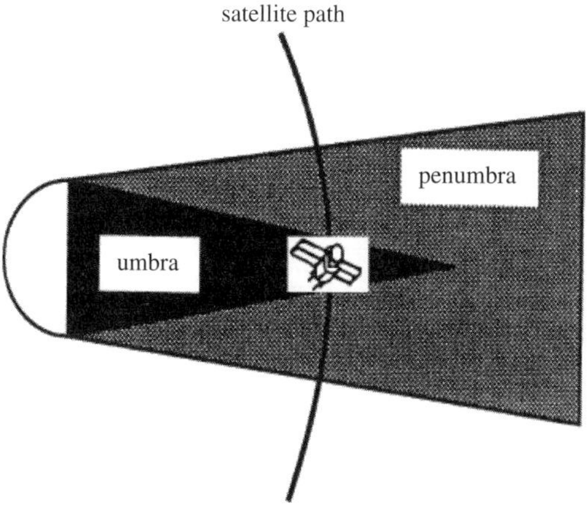

Figure 5.1.7 Earth shadow regions.

antenna to the satellite's center of mass becomes uncertain. Second, there is an additional but unknown windup error. See Section 6.2.4 for more information on the windup error.

Bar-Sever (1996) has investigated the GPS yaw attitude problem and the compensation method in detail. During shadow, the output of the solar sensors is essentially zero and the ACS is driven by the noise of the system. Even a small amount of noise can trigger a significant yaw change. As a corrective action, a small bias signal is added to the signals of the solar sensors, which amounts to a yaw of about 0.5°. As a result, during the time when the sun can be observed, the yaw will be in error by that amount. During eclipse times, the biased noise will yaw the satellite in the direction of the bias, thus avoiding larger and erratic yaw motions. When the satellite leaves the shadow region, the solar sensors provide the information to determine the correct yaw angle. The yaw maneuvers carried out by the satellite from the time it enters the shadow region to the time it leaves it are collectively called "the midnight maneuvers." When the satellite is on the sun-earth axis and between the sun and the earth, the ACS encounters a singularity because any yaw angle represents an optimal orientation of the solar panels for this particular geometry. Any maneuvers that deal with this situation are called "the noon maneuver."

5.2 GLOBAL POSITIONING SYSTEM

Satellite-based positioning has been pursued since the 1960s. An early and very successful satellite positioning system was the Navy navigation satellite system (TRANSIT). Since its release for commercial use in 1967, the TRANSIT positioning system was often used to determine widely spaced networks covering large regions—even the globe. It was instrumental in establishing modern geocentric datums and in connecting various national datums to a geocentric reference frame. The TRANSIT satellites were orbiting in a polar plane at about 1100 km altitude. The TRANSIT satellites were affected more by gravity field variations than the much higher-orbiting GPS satellites. In addition, their transmissions at 150 and 400 MHz were more susceptible to ionospheric delays and disturbances than the higher GPS frequencies. The TRANSIT system was discontinued at the end of 1996.

5.2.1 General Description

The Navigation Satellite Timing and Ranging (NAVSTAR) GPS provides positioning and timing 24 hours per day, anywhere in the world, and under any weather conditions. The U.S. government operates GPS. It is a dual-use system, with its primary purpose being to meet military needs for positioning and timing. Over the past decades, the number of civilian applications has increased seemingly endlessly, and this trend is continuing.

In short, the buildup of the satellite constellation began with the series Block I satellites. These were concept validation satellites that did not have selective availability (SA) or antispoofing (AS) capability. They were launched into three 63° inclined orbital planes. Their positions within the planes were such that optimal observing geometry was achieved over certain military proving grounds in the continental United States. Eleven Block I satellites were launched between 1978 and 1985 (with one launch failure). Their average lifetime was 8 to 9 years. They were designed to provide 3 to 4 days of positioning service without contact with the ground control center. The launch of the second generation of GPS satellites, called Block II, began in February 1989. In addition to radiation-hardened electronics, these operational satellites had full SA/AS capability and carried a navigation data message that was valid for 14 days. Additional modifications resulted in the satellite called Block IIA. These satellites can provide about 6 weeks of positioning service without contact from the control segment. Twenty-eight Block II/IIA satellites were launched between 1989 and 1997 into six planes, 55° inclined. The first third-generation GPS satellite, called Block IIR (R for replenishment), was successfully launched in 1997. These satellites have the capability to determine their orbits autonomously through UHF cross-link ranging and to generate their own navigation message by onboard processing. They are able to measure ranges between themselves and transmit observations to other satellites as well as to ground control. In recent years, GPS has undergone a major modernization. Most importantly, the GPS satellites are transmitting more signals that allow a better delineation of military and civilian uses, and thus increase the performance of GPS even more. Table 5.2.1 shows the current and expected progression of the modernization.

The U.S. government's current policy is to make GPS available in two services. The precise positioning service (PPS) is available to the military and other authorized users. The standard positioning service (SPS) is available to anyone. See SPS (2008) for a detailed documentation of this service. Without going into detail, let it suffice to say that PPS users have access to the encrypted P(Y)-codes (and M-codes starting with Block II R-M) on the L1 and L2 carriers, while SPS users can observe the public codes L1 C/A, L1C, L2C, L5. The encryption of the P-codes began January 31, 1994. SPS positioning capability was degraded by SA measures, which entailed an intentional dither of the satellite clocks and falsification of the navigation

TABLE 5.2.1 Legacy and Modernization of GPS Signals.

Signal	I, IIA, IIR 1978–2004	IIR-M 2005–2009	IIF Since 2010	GPS III Expected in 2015
L1 C/A	X	X	X	X
L1 P(Y)	X	X	X	X
L1M		X	X	X
L2C		X	X	X
L2 P(Y)	X	X	X	X
L2 M		X	X	X
L5			X	X
L1C				X

message. In keeping with the policy, SA was implemented on March 25, 1990, on all Block II satellites. The level of degradation was reduced in September 1990 during the Gulf conflict but was reactivated to its full level on July 1, 1991, until it was discontinued on May 1, 2000. Starting with the Block II R-M, new military signals, L1M and L2M are available. Providing comparable or better performance than L1P(Y) and L2P(Y), the new military signals coexist with them and do not interfere with old user equipment. Careful shaping of the spectrum, based on the specially designed M-code, prevents leakage of new signal power into the spectra of old military signals.

Over time, both satellite and receiver technologies have improved significantly. Whereas older receivers could observe the P(Y)-code more accurately than the C/A-codes, this distinction has all but disappeared with modem receiver technology. Dual-frequency P(Y)-code users have the advantage of being able to correct the effect of the ionosphere on the signals. However, this simple classification of PPS and SPS by no means characterizes how GPS is used today. Researchers have devised various patented procedures that make it possible to observe or utilize the encrypted P(Y)-codes effectively, and in doing so, make dual-frequency observations available, at least to high-end receivers. In certain surveying applications where the primary quantity of interest is the vector between nearby stations, intentional degradation of SA could be overcome by differencing the observations between stations and satellites. However, positioning with GPS works much better without SA. Starting with the Block II R-M, dual-frequency observables are available due to the new civil L2C code. The Block II F satellites (total number is four at the beginning of 2014) start transmission of the third civil L5 signal, allowing the development of triple carrier techniques for standalone and differential positioning.

The six orbital planes of GPS are spaced evenly in right ascension and are inclined by 55° with respect to the equator. Because of the flattening of the earth, the nodal regression is about −0.04187° per day; an annual orbital adjustment keeps the orbits close to their nominal location. Each orbital plane contains four satellites; however, to optimize global satellite visibility, the satellites are not evenly spaced within the orbital plane. The orbits are nominally circular, with a semimajor axis of about 26,660 km. Using Kepler's third law (5.1.39), one obtains an orbital period of slightly less than 12 h. The satellites will complete two orbital revolutions in one sidereal day. This means the satellites will rise about 4 min earlier each day. Because the orbital period is an exact multiple of the period of the earth's rotation, the satellite trajectory on the earth (i.e., the trace of the geocentric satellite vector on the earth's surface) repeats itself each sidereal day.

Because of their high altitude, the GPS satellites can be viewed simultaneously over large portions of the earth. Usually the satellites are observed only above a certain altitude angle, called the mask angle. Typical values for the mask angle are 10 to 15°. At a low elevation angle, the tropospheric effects on the signal can be especially severe and difficult to model accurately. Let ε denote the mask angle, and let α denote the geocentric angle of visibility for a spherical earth, then one can find the relation ($\varepsilon = 0°, \alpha = 152°$), ($\varepsilon = 5°, \alpha = 142°$), ($\varepsilon = 10°, \alpha = 132°$). The viewing angle from the satellite to the limb of the earth is about 27°.

5.2.2 Satellite Transmissions at 2014

The IS-GPS-200G (2012) is the authoritative source for details on the GPS signal structures, usage of these signals, and other information broadcasts by the satellites. All satellite transmissions are coherently derived from the fundamental frequency of 10.23 MHz, made available by onboard atomic clocks. This is also true for the new signals discussed further below. Multiplying the fundamental frequency by 154 gives the frequency for the L1 carrier, $f_1 = 1575.42$ MHz, multiplying by 120 gives the frequency of the L2 carrier, $f_2 = 1227.60$ MHz, and multiplying by 115 gives the frequency $f_5 = 1176.45$ MHz. The chipping (code) rate of the P(Y)-code is that of the fundamental frequency, i.e., 10.23 MHz, whereas the chipping rate of the C/A-code is 1.023 MHz (one-tenth of the fundamental frequency). The navigation message (telemetry) is modulated on both the L1 and the L2 carriers at a chipping rate of 50 bps. It is different for modern signals as will be discussed later. It contains information on the ephemerides of the satellites, GPS time, clock behavior, and system status messages.

Onboard atomic clocks define the space vehicle time. Each satellite operates on its own time system, i.e., all satellite transmissions such as the C/A-code, the P(Y)-codes, and the navigation message are initiated by satellite time. The data in the navigation message, however, are relative to GPS time. Time is maintained by the control segment and follows UTC(USNO) within specified limits. GPS time is a continuous time scale and is not adjusted for leap seconds. The last common epoch between GPS time and UTC(USNO) was midnight January 5–6, 1980. The navigation message contains the necessary corrections to convert space vehicle time to GPS time. The largest unit of GPS time is one week, defined as 604,800 sec. Additional details on the satellite clock correction are given in Section 6.2.2.1.

The atomic clocks in the satellites are affected by both special relativity (the satellite's velocity) and general relativity (the difference in the gravitational potential at the satellite's position relative to the potential at the earth's surface). Jorgensen (1986) gives a discussion in lay terms of these effects and identifies two distinct parts in the relativity correction. The predominant portion is common to all satellites and is independent of the orbital eccentricity. The respective relative frequency offset is $\Delta f/f = -4.4647 \times 10^{-10}$. This offset corresponds to an increase in time of 38.3 μs per day; the clocks in orbit appear to run faster. The apparent change in frequency is $\Delta f = 0.0045674$ Hz at the fundamental frequency of 10.23 MHz. The frequency is corrected by adjusting the frequency of the satellite clocks in the factory before launch to 10.22999999543 MHz. The second portion of the relativistic effect is proportional to the eccentricity of the satellite's orbit. For exact circular orbits, this correction is zero. For GPS orbits with an eccentricity of 0.02 this effect can be as large as 45 ns, corresponding to a ranging error of about 14 m. This relativistic effect can be computed from a simple mathematical expression that is a function of the semimajor axis, the eccentricity, and the eccentric anomaly (see Section 6.2.2). In relative positioning as typically carried out in surveying, the relativistic effects cancel for all practical purposes.

The precision P(Y)-code is used for military navigation. It is a pseudorandom noise (PRN) code which itself is the modulo-2 sum of two other pseudorandom codes. The P(Y)-code does not repeat itself for 37 weeks. Thus, it is possible to assign weekly portions of this code to the various satellites. As a result, all satellites can transmit on the same carrier frequency and yet can be distinguished because of the mutually exclusive code sequences being transmitted. All codes are initialized once per GPS week at midnight from Saturday to Sunday, thus creating, in effect, the GPS week as a major unit of time. The L1 and L2 carriers are both modulated with the same P(Y)-code.

The period of the coarse/acquisition (C/A) code is merely 1 ms and consists of 1023 bits. Each satellite transmits a different set of C/A-codes. These codes are currently transmitted only on L1. The C/A-codes belong to the family of Gold codes, which characteristically have low cross-correlation between all members. This property makes it possible to rapidly distinguish among the signals received simultaneously from different satellites.

One of the satellite identification systems makes use of the PRN weekly number. For example, if one refers to satellite PRN 13, one refers to the satellite that transmits the thirteenth weekly portion of the PRN-code. The short version of PRN 13 is SV 13 (SV=space vehicle). Another identification system uses the space vehicle launch number (SVN). For example, the identification of PRN 13 in terms of launch number is NAVSTAR 9, or SVN 9.

5.2.2.1 *Signal Structure* In electronics, modulation is used for transferring a low-frequency information signal to the radio frequency harmonic wave, which is capable of traveling through space. The wave is called "carrier" because it is used as a media for carrying information. The carrier is modulated by several codes and the navigation (data) message. There are several commonly used digital modulation methods: amplitude shift keying (ASK), frequency shift keying (FSK), and phase shift keying (PSK). The PSK is distinguished further as binary PSK (BPSK), quadrature PSK (QPSK or 4-PSK), 8-PSK, and binary offset carrier modulation (BOC).

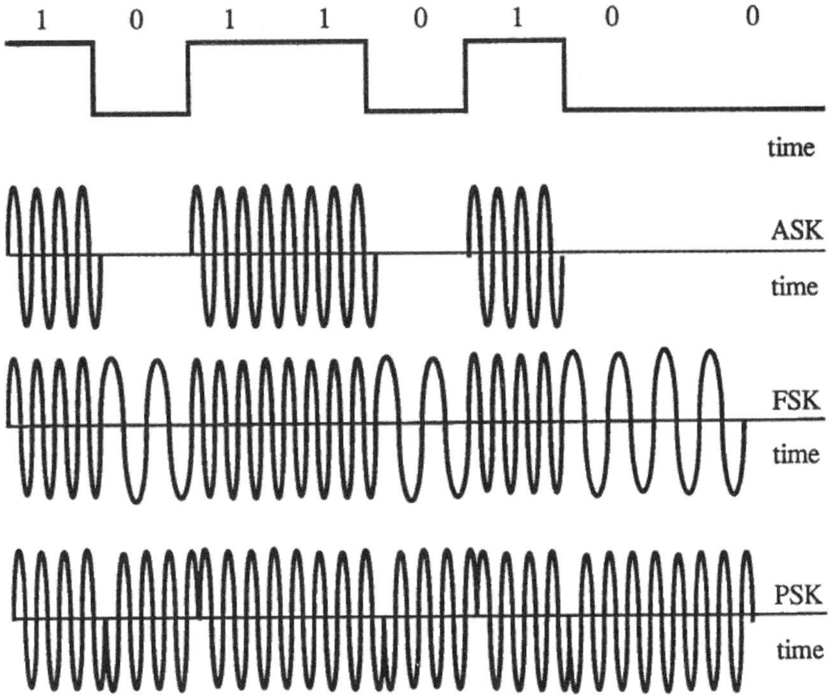

Figure 5.2.1 Digital modulation methods.

More complex modulation schemes referred to as quadrature amplitude modulation (QAM), which combine ASK and PSK, are used in telecommunication. For example 16-QAM is represented by the constellation of 16 points on the complex plain, which means different amplitudes and different phase shifts.

GPS uses BPSK, QPSK, and BOC. Figure 5.2.1 briefly demonstrates some of the BPSK principles involved. The figure shows an arbitrary digital data stream consisting of binary digits 0 and 1. These binary digits are also called chips, bits, codes, or pulses. In the case of GPS, the digital data stream contains the navigation message or the pseudorandom sequences of the codes. The code sequences look random but actually follow a mathematical formula. ASK corresponds to an on/off operation. The digit 1 might represent turning the carrier on and 0 might mean turning it off. FSK implies transmission on one or the other frequency. The transmitting oscillator is required to switch back and forth between two distinct frequencies. In the case of PSK, the same carrier frequency is used, but the phase changes abruptly. With BPSK, the phase shifts 0° and 180°.

Figure 5.2.2 shows two data streams. The sequence (a) could represent the navigation data chipped rate of 50 bits per seconds (bps), and (b) could be the C/A-code or the P(Y)-code chipped at the 1.023 MHz or 10.23 MHz, respectively. The times of bit transition are aligned. The navigation message and the code streams have significantly different chipping rates. A chipping rate of 50 bps implies 50 opportunities

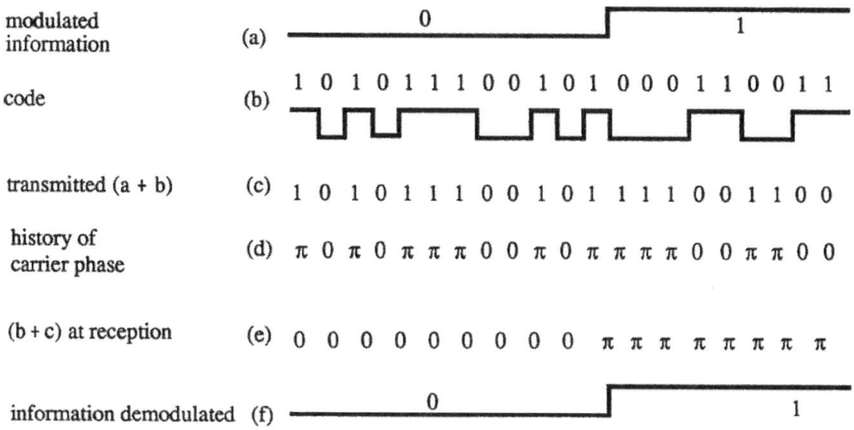

Figure 5.2.2 Modulo-2 addition of binary data streams.

per second for the digital stream to change from 1 to 0 and vice versa. Within the time of a telemetry chip there are 31,508,400 L1 cycles, 20,460 C/A-code chips, and 204,600 P(Y)-code chips. Looking at this in the distance domain, one telemetry chip is 5950 km long, whereas the lengths of the C/A and P(Y)-codes are 293 and 29.3 m, respectively. Thus, the P(Y)-code can change the carrier by 180° every 29.3 m, the C/A-code every 293 m, and the telemetry every 5950 km.

One of the tasks to be accomplished is reading the navigation message at the receiver. We need this information to compute the positions of the satellites. To accomplish this, the data streams (a) and (b) in Figure 5.2.2 are modulo-2 added before transmission at the satellite. Modulo-2 addition follows the rules

$$0 \oplus 0 = 0, \qquad 0 \oplus 1 = 1, \qquad 1 \oplus 0 = 1, \qquad 1 \oplus 1 = 0 \tag{5.2.1}$$

The result is labeled (c). The figure also shows the phase history of the transmitted carrier. Whenever a binary 1 occurs in the 50 bps navigation data stream, the modulo-2 addition inverts 20,460 adjacent digits of the C/A-code. A binary 1 becomes 0 and vice versa. A binary 0 leaves the next 20,460 C/A-codes unchanged. Let the receiver reproduce the original code sequence that is shifted in time to match the transmitted code. We can then modulo-2 add the receiver-generated code with the received phase-modulated carrier. The sum is the demodulated 50 bps telemetry data stream.

The BPSK modulation of the harmonic wave is more conveniently expressed as multiplication. If we represent the binary signal by 1 and -1 instead of 0 and 1, the BPSK modulation converts into a product with the harmonic wave. Conversion of the binary $\{0, 1\}$ signal b, following the modulo-2 addition rule (5.2.1), into the $\{1, -1\}$ signal s, following the multiplication rule $1 \cdot 1 = 1, 1 \cdot (-1) = -1, (-1) \cdot 1 = -1$, and $(-1) \cdot (-1) = 1$, is expressed as

$$s = e^{i\pi b} \tag{5.2.2}$$

Here i is an imaginary unit and the complex exponent is defined as $e^{ix} = \cos x + i \sin x$. The last expression results in $e^{i\pi 0} = 1$, $e^{i\pi 1} = -1$, and

$$e^{i\pi(b_1 \oplus b_2)} = e^{i\pi b_1} e^{i\pi b_2} \tag{5.2.3}$$

The last identity states that sequential application of BPSK modulation is equivalent to the product of the modulating codes. In what follows we assume that the conversion (5.2.2) has already been done and two sequential BPSK modulations are expressed as multiplication.

The modulo-2 addition method must be generalized because the L1 carrier is modulated by three data streams: the navigation data, the C/A-codes, and P(Y)-codes. Thus, the task of superimposing both code streams and the navigation data stream arises. Two sequential superimpositions are not unique, because the C/A-code and the P(Y)-code have identical bit transition epochs (although their length is different). The solution is called quadrature phase shift keying (QPSK). The carrier is split into two parts, the inphase component (I) and the quadrature component (Q). The latter is shifted by 90°. Each component is then binary phase modulated, the inphase component is modulated by the P(Y)-code, and the quadrature component is modulated by the C/A-code. Therefore, the C/A-code signal carrier lags the P(Y)-code carrier by 90°. For the L1 and L2 carriers we have

$$S_1^p(t) = A_P P^p(t) D^p(t) \cos(2\pi f_1 t) + A_C G^p(t) D^p(t) \sin(2\pi f_1 t) \tag{5.2.4}$$

$$S_2^p(t) = B_P P^p(t) D^p(t) \cos(2\pi f_2 t) \tag{5.2.5}$$

In these equations, the superscript p identifies the PRN number of the satellite, A_P, A_C, and B_P are the amplitudes (power) of P(Y)-codes and C/A-code, $P^p(t)$ is the pseudorandom P(Y)-code, $G^p(t)$ is the C/A-code (Gold code), and $D^p(t)$ is the telemetry or navigation data stream. The products $P^p(t)D^p(t)$ and $G^p(t)D^p(t)$ imply modulo-2 addition as suggested by (5.2.3).

In order to explain in greater detail how QPSK relates to (5.2.4), recall the complex expression for the harmonic signal

$$A \cos 2\pi f t = \mathrm{Re}(A e^{i2\pi f t}) \tag{5.2.6}$$

The BPSK modulation of (5.2.6) by the signal (5.2.2) can be expressed as

$$As(t) \cos 2\pi f t = A\mathrm{Re}(e^{i\pi b(t)} e^{i2\pi f t}) = A\mathrm{Re}(e^{i2\pi f t + i\pi b(t)}) \tag{5.2.7}$$

Now rewrite (5.2.4) as

$$S_1^p(t) = A_P \mathrm{Re}(e^{i2\pi f_1 t + i\pi b_P^p(t)}) + A_C \mathrm{Re}(e^{i2\pi f_1 t + i\pi/2 + i\pi b_G^p(t)}) \tag{5.2.8}$$

where $P^p(t)D^p(t) = e^{i\pi b_P^p(t)}$ and $G^p(t)D^p(t) = e^{i\pi b_G^p(t)}$. Further,

$$S_1^p(t) = \mathrm{Re}\left[e^{i2\pi f_1 t}\left(A_P e^{i\pi b_P^p(t)} + A_C e^{i\pi b_G^p(t) + i\pi/2}\right)\right] \tag{5.2.9}$$

Assume that $A_C = A_P = A$ for the sake of simplicity, and consider the expression $Ae^{i\pi b_P^p(t)} + Ae^{i\pi b_G^p(t)+i\pi/2}$ in the internal parentheses of (5.2.9). Denote

$$Q\left(b_P^p(t), b_G^p(t)\right) = e^{i\pi b_P^p(t)} + e^{i\pi b_G^p(t)+i\pi/2} \tag{5.2.10}$$

taking values as shown in Table 5.2.2. The complex valued multiplier $Q(b_P^p(t), b_G^p(t))$ can be considered as a QPSK modulation operator. Application of two $\pi/2$-shifted BPSK modulations is equivalent to one QPSK operator multiplying (modulating) the carrier wave

$$S_1^p(t) = A\mathrm{Re}\left\{Q\left(b_P^p(t), b_G^p(t)\right)e^{i2\pi f_1 t}\right\} \tag{5.2.11}$$

Figure 5.2.3 shows the symbols constellation, which the BPSK operator $B(b) = e^{i\pi b}$ and QPSK operator $Q(b_1, b_2)$ occupy on the complex plane.

The P(Y)-code by itself is a modulo-2 sum of two pseudorandom data streams $X_1(t)$ and $X_2(t - pT)$ as follows:

$$P^p(t) = X_1(t)X_2(t - pT) \tag{5.2.12}$$

$$0 \le p \le 36 \tag{5.2.13}$$

$$\frac{1}{T} = 10.23 \text{ MHz} \tag{5.2.14}$$

Expression (5.2.12) defines the code according to the PRN number p. Using (5.2.13), one can define 37 mutually exclusive P(Y)-code sequences. At the beginning of the GPS week, the P(Y)-codes are reset. Similarly, the C/A-codes are the modulo-2 sum of two 1023 pseudorandom bit codes as follows:

$$G^p(t) = G_1(t)G_2\left[t - N^p(10T)\right] \tag{5.2.15}$$

TABLE 5.2.2 QPSK Modulation Operator $Q\left(b_P^p, b_G^p\right)$.

	$b_P^p = 0$	$b_P^p = 1$
$b_G^p = 0$	$1 + i$	$-1 + i$
$b_G^p = 1$	$1 - i$	$-1 - i$

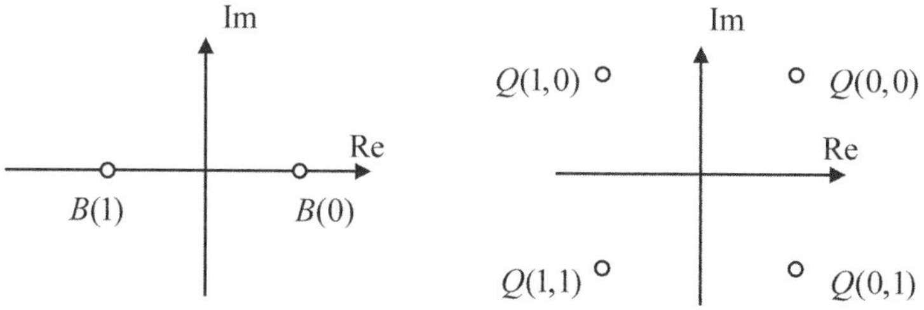

Figure 5.2.3 $B(b)$ and $Q(b_1, b_2)$ constellations on the complex plane.

$G^p(t)$ is 1023 bits long or has a 1 ms duration at a 1.023 Mbps bit rate. The $G^p(t)$ chip is 10 times as long as the X_1 chip. The G_2-code is selectively delayed by an integer number of chips, expressed by the integer N^p, to produce 36 unique Gold codes, one for each of the 36 different P(Y)-codes.

The actual generation of the codes X_1, X_2, G_1, and G_2 is accomplished by a feedback shift register (FBSR). Such devices can generate a large variety of pseudorandom codes. These codes look random over a certain interval, but the feedback mechanism causes the codes to repeat after some time. Figure 5.2.4 shows a very simple register. A block represents a stage register whose content is in either a one or zero state. When the clock pulse is input to the register, each block has its state shifted one block to the right. In this particular example, the output of the last two stages is modulo-2 added, and the result is fed back into the first stage and modulo-2 added to the old state to create the new state. The successive states of the individual blocks, as the FBSR is stepped through a complete cycle, are shown in Table 5.2.3. The elements of the column represent the state of each block, and the successive columns represent the behavior of the shift register as the succession of timing pulses cause it to shift from state to state. In this example, the initial state is (0001). For n blocks, $2^n - 1$ states are possible before repetition occurs. The output corresponds to the state of the last block and would represent the PRN code, if it were generated by such a four-stage FBSR.

The shift registers used in GPS code generation are much more complex. They have many more feedback loops and many more blocks in the sequence. The P(Y)-code is derived from two 12-stage shift registers, $X_1(t)$ and $X_2(t)$, having 15,345,000 and 15,345,037 stages (chips), respectively. Both registers continuously

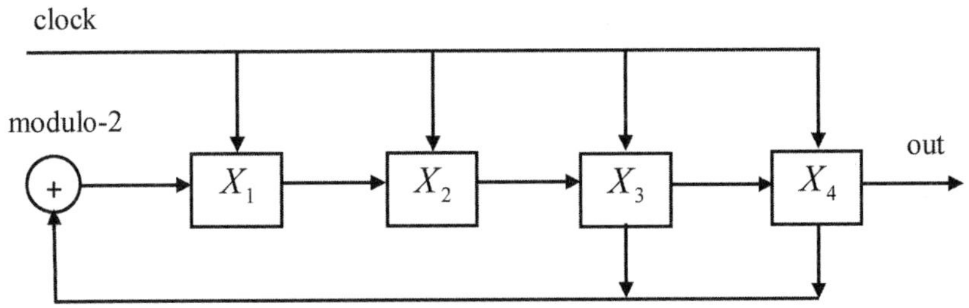

Figure 5.2.4 Simple FBSR.

TABLE 5.2.3 Output of FBSR.

x_1	0	1	0	0	...	1	0	0	0
x_2	0	0	1	0	...	1	1	0	0
x_3	0	0	0	1	...	1	1	1	0
x_4	1	0	0	0	...	1	1	1	1
Output	1	0	0	0	...	1	1	1	1

recycle. The modulo-2 sum of both registers has the length of 15,345,000 times 15,345,037 chips. At the chipping rate of 10.23 MHz, it takes 266.4 days to complete the whole P(Y)-code cycle. It takes 1.5 s for the X_1 register to go through one cycle. The X_1 cycles (epochs) are known as the Z count.

The bandwidth terminology is often used in connection with pseudorandom noise modulation. Let T denote the duration of the chip (rectangular pulse), so then the bandwidth is inverse proportional to T. Therefore, shorter chips (pulses) require greater bandwidth and vice versa. If we subject the rectangular pulse function to a Fourier transform, we obtain the well-known sinc (sine-cardinal) function

$$S(\Delta f, f_c) = \frac{1}{f_c}\left(\frac{\sin\left(\pi\ \Delta f / f_c\right)}{\pi\ \Delta f / f_c}\right)^2 \tag{5.2.16}$$

The symbol Δf is the difference with respect to the carrier frequency L1 or L2. The code frequency 10.23 MHz or 1.023 MHz, respectively, is denoted by f_c. The factor $1/f_c$ serves as a normalizing (unit area) scalar. The top panel of Figure 5.2.5 shows the power spectral density (5.2.16) for the C/A- and P(Y)-codes expressed in watts per hertz (W/Hz). This symmetric function is zero at multiples of the code rate f_c. The first lobe stretches over the bandwidth, covering the range of $\pm f_c$ with respect to the center frequency. The spectral portion signal beyond one bandwidth is filtered out at the satellite and is not transmitted. Power ratios in electronics and in connection with signals and antennas are expressed in terms of decibels (dB) on a logarithmic scale. See Section 9.1.7 for additional detail on the dB scale. The power ratio in terms of decibel units is defined as

$$g_{[\mathrm{dB}]} = 10\ \log_{10}\frac{P_2}{P_1} \tag{5.2.17}$$

Absolute power can be expressed with respect to a unit power P_1. For example, the units dBW or dBm imply $P_1 = 1$ W or $P_1 = 1$ mW, respectively. Frequently, the relation

$$g_{[\mathrm{dB}]} = 20\ \log_{10}\frac{V_2}{V_1} \tag{5.2.18}$$

is seen. In (5.2.18), the symbols V_1 and V_2 denote voltages. Both decibel expressions are related by the fact that the square of voltage divided by resistance equals power. The bottom panel of Figure 5.2.5 shows the power spectral density (5.2.16) for the C/A- and P(Y)-codes, expressed in decibels relative to watt per hertz (dBW/Hz).

The power of the received GPS signals on the ground is lower than the background noise (thermal noise). The specifications call for a minimum power at the user on the earth of −160 dBW for the C/A-code, −163 dBW for the P(Y)-code on L1, and −166 dBW for the P(Y)-code on L2. To track the signal, the receiver correlates the incoming signal by a locally generated replica of the code and accumulates results over certain time. This correlation and accumulation process results in a signal that is well above the noise level.

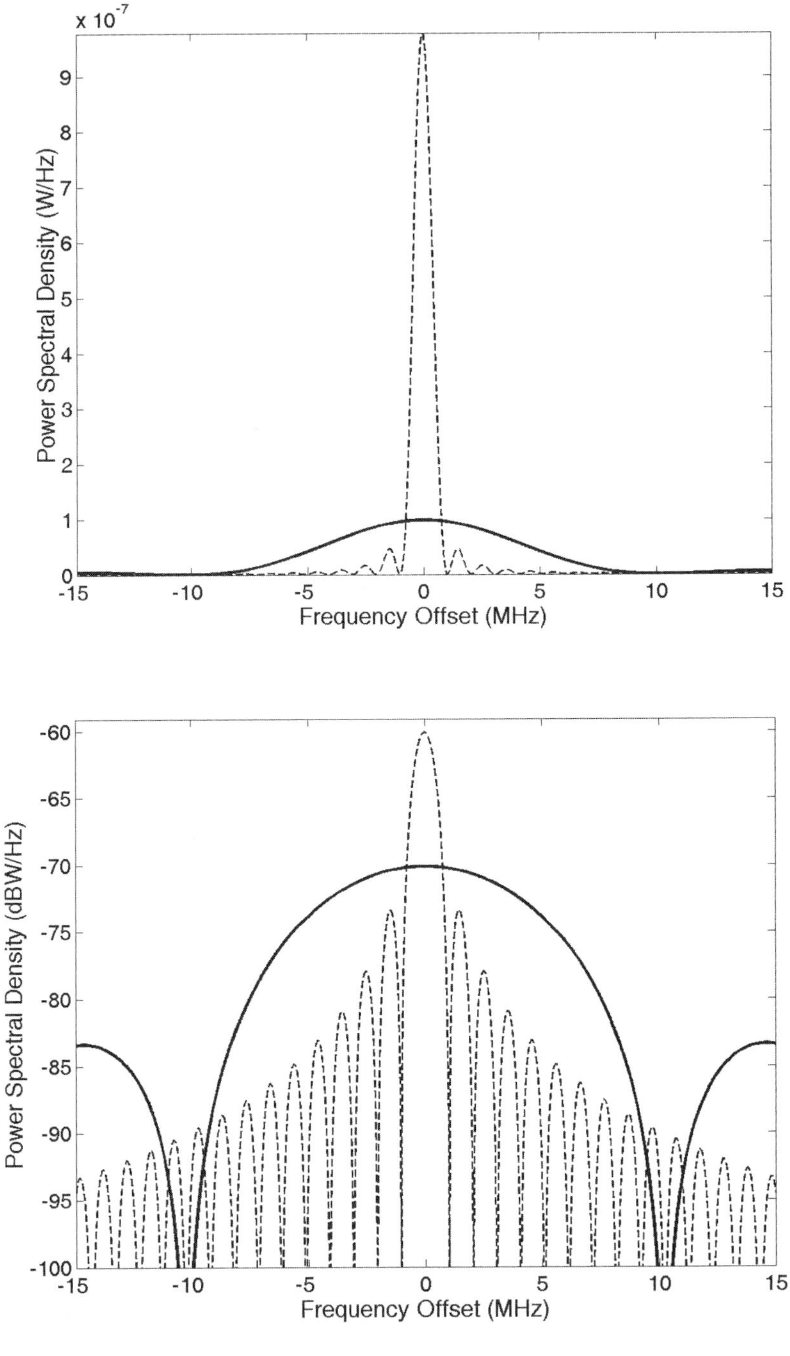

Figure 5.2.5 Power spectral densities of C/A (dashed line) and P(Y) codes (solid line). Top panel is in W/Hz and bottom panel is in dBW/Hz.

5.2.2.2 *Navigation Message* The Master Control Station, located near Colorado Springs, uses data from a network of monitoring stations around the world to monitor the satellite transmissions continuously, compute the broadcast ephemerides, calibrate the satellite clocks, and periodically update the navigation message. This "control segment" ensures that the SPS and PPS are available as specified in SPS (2008).

The satellites transmit a navigation message that contains, among other things, orbital data for computing the positions of all satellites. A complete message consists of 25 frames, each containing 1500 bits. Each frame is subdivided into five 300-bit subframes, and each subframe consists of 10 words of 30 bits each. At the 50 bps rate it takes 6 seconds to transmit a subframe, 30 sec to complete a frame, and 12.5 min for one complete transmission of the navigation message. The subframes 1, 2, and 3 are transmitted with each frame. Subframes 4 and 5 are each subcommutated 25 times. The 25 versions of subframes 4 and 5 are referred to as pages 1 through 25. Thus, each of these pages repeats every 12.5 min.

Each subframe begins with the telemetry word (TLM) and the handover word (HOW). The TLM begins with a preamble and otherwise contains only information that is needed by the authorized user. The HOW is a truncation of the GPS time of week (TOW). HOW, when multiplied by 4, gives the X_1 count at the start of the following subframe. As soon as a receiver has locked to the C/A-code, the HOW word is extracted and is used to identify the X_1 count at the start of the following subframe. In this way, the receiver knows exactly which part of the long P(Y)-code is being transmitted. P(Y)-code tracking can then readily begin, thus the term *handover word*. To lock rapidly to the P(Y)-code, the HOW is included on each subframe (see Figure 5.2.6).

Since military missions might require the jamming of L1, there is a need for equipment capable of acquiring the P(Y)-code directly without the C/A-code by authorized

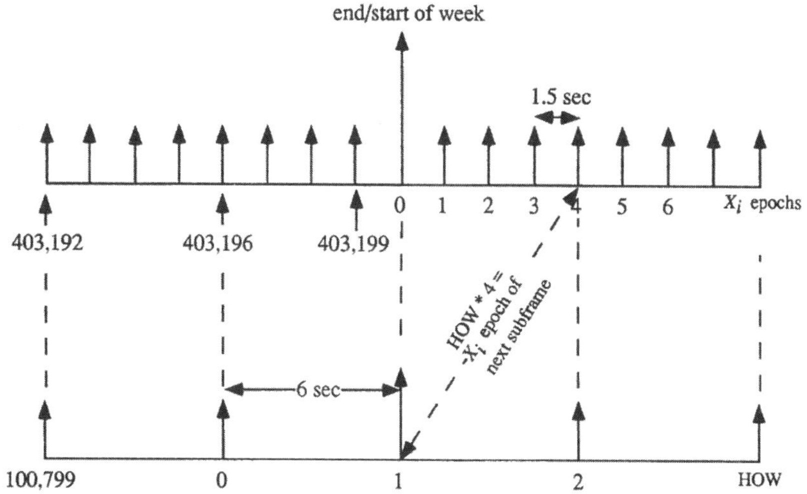

Figure 5.2.6 **HOW versus X_1 epochs.**

users. Such functionality is provided to authorized users of receivers equipped with selective availability antispoofing modules (SAASM). All new military receivers deployed after the end of September 2006 must use SAASM. SAASM does not provide any additional antijam capability, because it uses exactly the same signal in space as current GPS signals (power and modulation). The antijam capabilities will be provided by the M-code, which is a part of GPS modernization available in the Block II R-M.

GPS time is directly related to the X_1 counts of the P(Y)-code. The Z count is a 29-bit number that contains several pieces of timing information. It can be used to extract the HOW, which relates to the X_1 count as discussed above, and the TOW, which represents the number of seconds since the beginning of the GPS week. A full week has 403,199 X_1 counts. The Z count gives the current GPS week number (modulo-1024). The beginning of the GPS week is offset from midnight UTC by the accumulated number of leap seconds since January 5–6, 1980, the beginning of GPS time.

Subframe 1 contains the GPS week number, space vehicle accuracy and health status, satellite clock correction terms a_{f0}, a_{f1}, a_{f2} and the clock reference time t_{oc} (Section 5.3.1), the differential group delay, T_{GD}, and the issue of date clock (IODC) term. The latter term is the issue number of the clock data set and can be used to detect any change in the correction parameters. The messages are updated usually every 4 h.

Subframes 2 and 3 contain the ephemeris parameters for the transmitting satellite. The various elements are listed in Table 5.2.4. These elements are a result of least-squares fitting of the predicted ephemeris over a well-specified interval of time. The issue of the data ephemeris (IODE) term allows users to detect changes in the ephemeris parameters. For each upload, the control center assigns a new number. The IODE is given in both subframes. During the time of an upload, both

TABLE 5.2.4 Elements of Subframes 2 and 3.

M_0	Mean anomaly at reference time
Δn	Mean motion difference from computed value
e	Eccentricity
$\sqrt{a}$	Square root of the semimajor axis
Ω_0	Longitude of ascending node of orbit plane at beginning of week
i_0	Inclination angle at reference time
ω	Argument of perigee
$\dot{\Omega}$	Rate of right ascension
IDOT	Rate of inclination angle
$C_{uc}, C_{us}, C_{rc}, C_{rs}, C_{ic}, C_{is}$	Amplitude of second-order harmonic perturbations
t_{oe}	Ephemeris reference time
IODE	Issue of data (ephemeris)

IODEs will have different values. Users should download ephemeris data only when both IODEs have the same value. The broadcast elements are used with the algorithm of Table 5.2.5. The results are coordinates of the phase center of the space vehicle's antennas in the World Geodetic System of 1984 (WGS84). The latter is an ECEF coordinate system that is very closely aligned with the international terrestrial reference frame (ITRF). There is no need for an explicit polar motion rotation, since the respective rotations are incorporated in the representation parameters. However, when computing the topocentric distance, the user must account for the rotation of the earth during the signal travel time from satellite to receiver.

Subframes 4 and 5 contain special messages, ionospheric correction terms, coefficients to convert GPS time to universal time coordinated (UTC), and almanac data on pages 2–5 and 7–10 (subframe 4) and 1–24 (subframe 5). The ionospheric terms are the eight coefficients $\{\alpha_n, \beta_n\}$ referenced in Table 8.4.3. For accurate computation of UTC from GPS time, the message provides a constant offset term, a linear polynomial term, the reference time t_{ot}, and the current value of the leap second. The almanac provides data to compute the positions of satellites other than the transmitting satellite. It is a reduced-precision subset of the clock and ephemeris parameters of subframes 1 to 3. For each satellite, the almanac contains the following: $t_{\mathrm{oa}}, \delta_i, a_{f0}, a_{f1}, e, \dot{\Omega}, a^{1/2}, \Omega_0, \omega$ and M_0. The almanac reference time is t_{oa}. The correction to the inclination δ_i is given with respect to the fixed value $i_0 = 0.30$ semicircles (= 54°). The clock polynomial coefficients a_{f0} and a_{f1} are used to convert space vehicle (SV) time to GPS time, following equation (6.2.4). The remaining elements of the almanac are identical to those listed in Table 5.2.4. The algorithm of Table 5.2.4 applies, using zero for all elements that are not included in the almanac and replacing the reference time t_{oe} by t_{oa}.

The mean anomaly, the longitude of the ascending node, the inclination, and UTC (if desired) are formulated as polynomials in time; the time argument is GPS time. The polynomial coefficients are, of course, a function of the epoch of expansion. The respective epochs are t_{oc}, t_{oe}, t_{oa}, and t_{ot}.

The navigation message contains other information, such as the user range error (URE). This measure equals the projection of the ephemeris curve fit errors onto the user range and includes effects of satellite timing errors.

5.2.3 GPS Modernization Comprising Block IIM, Block IIF, and Block III

GPS modernization becomes possible because of advances in technology in the satellite and the receiver. The additional signals transmitted by modernized satellites improve the antijamming capability, increase protection against antispoofing, shorten the time to first fix, and provide a civilian "safety of life" signal (L5) within the protected Aeronautical Radio Navigation Service (ARNS) frequency band. The new L2C signals increase signal robustness and resistance to interference and

TABLE 5.2.5 GPS Broadcast Ephemeris Algorithm.

$\mu = 3.986005 \times 10^{14}\ \mathrm{m}^3/\mathrm{s}^2$	Gravitational constant for WGS84
$\dot{\Omega}_e = 7.2921151467 \times 10^{-5}\ \mathrm{rad/s}$	Earth's rotation rate for WGS84
$a = \left(\sqrt{a}\right)^2$	Semimajor axis
$n_0 = \sqrt{\dfrac{\mu}{a^3}}$	Computed mean motion—rad/s
$t_k = t - t^*_{\mathrm{oe}}$	Time from ephemeris reference epoch
$n = n_0 + \Delta n$	Corrected mean motion
$M_k = M_0 + nt_k$	Mean anomaly
$M_k = E_k - e\ \sin E_k$	Kepler's equation for eccentric anomaly
$f_k = \tan^{-1}\left(\dfrac{\sqrt{1-e^2}\sin E_k}{\cos E_k - e}\right)$	True anomaly
$E_k = \cos^{-1}\left(\dfrac{e + \cos f_k}{1 + e\ \cos f_k}\right)$	Eccentricity anomaly
$\phi_k = f_k + \omega$	Argument of latitude
$\left.\begin{aligned}\delta u_k &= C_{us}\ \sin 2\phi_k + C_{uc}\ \cos 2\phi_k \\ \delta r_k &= C_{rc}\ \cos 2\phi_k + C_{rs}\ \sin 2\phi_k \\ \delta i_k &= C_{ic}\ \cos 2\phi_k + C_{is}\ \sin 2\phi_k\end{aligned}\right\}$	Second harmonic perturbations
$u_k = \phi_k + \delta u_k$	Corrected argument of latitude
$r_k = a\,(1 - e\ \cos E_k) + \delta r_k$	Corrected radius
$i_k = i_0 + \delta i_k + (\mathrm{IDOT})\ t_k$	Corrected inclination
$\left.\begin{aligned}r'_k &= r_k\ \cos\ u_k \\ y'_k &= rk\ \sin u_k\end{aligned}\right\}$	Positions in orbital plane
$\Omega_k = \Omega_0 + (\dot{\Omega} - \dot{\Omega}_e)t_k - \dot{\Omega}_e t_{\mathrm{oe}}$	Corrected longitude of ascending node
$\left.\begin{aligned}x_k &= x'_k\ \cos \Omega_k - y'_k\ \cos i_k\ \sin \Omega_k \\ y_k &= x'_k\ \sin \Omega_k + y'_k\ \cos i_k\ \cos \Omega_k \\ z_k &= y'_k\ \sin i_k\end{aligned}\right\}$	Earth-fixed coordinates

Note: t is GPS system time at time of transmission, i.e., GPS time corrected for transit time (range/speed of light). Furthermore, t_k shall be the actual total time difference between the time t and the epoch time t_{oe}, and must account for beginning or end of week crossovers. That is, if t_k is greater than 302,400, subtract 604,800 from t_k. If t_k is less than $-302,400$ seconds, add 604,800 seconds to t_k.

allow longer integration times in the receiver to reduce tracking noise and increase accuracy. The second civil frequency will eliminate the need to use inefficient squaring, cross correlation, or other patented techniques currently used by civilians in connection with L2. Once the GPS modernization is completed, the dual-frequency or triple-frequency receivers are expected to be in common use and affordable to the mass market.

At the same time, L1 and L2 are modulated with new military codes called the M-codes. Although added to L1 and L2, they are spectrally separated from the civilian codes and the old P(Y)-codes because they use more sophisticated binary modulation called binary offset carrier (BOC) modulation. There is no military code planned on L5.

The new L2C signal is described in IS-GPS-200G (2012), details on L5 signal are found in IS-GPS-705C (2012), and a description of L1C is in IS-GPS-800C (2012). For additional material, see Fontana et al. (2001a,b), Barker et al. (2000), and Pozzbon et al. (2011).

5.2.3.1 *Introducing Binary Offset Carrier (BOC) Modulation* For conventional rectangular spreading codes [which are the basis of the P(Y)-codes], the C/A-code heritage signals, and the new L2C and L5 codes, the frequency bandwidth is inversely proportional to the length of the chip. Modulating with faster chipping rates to improve or add additional signals might be impractical because of frequency bandwidth limitations. More advanced modulations have been studied recently that better share existing frequency allocations with each other and with heritage signals by increasing spectral separation, and thus preserve the spectrum. Betz (2002) describes binary-valued modulations, also referred to as binary offset carrier (BOC). Block IIR-M and IIF satellites will transmit a new military M-code signal on L1 and L2 that uses BOC. It is also used in Galileo and QZSS.

Definition of BOC is based on two frequencies, f_c denoting the chipping (code) rate and f_s denoting the subcarrier frequency. Both carriers are multiples of 1.023 MHz, $f_s = \alpha \times 1.023$ MHz, $f_c = \beta \times 1.023$ MHz, and the designation BOC(α, β) is used as abbreviation. The complex envelope (i.e., the complex signal modulating the radio frequency carrier) of BOC(α, β) is expressed in Betz (2002) as

$$s(t) = e^{i\varphi_0} \sum_k a_k \mu_{nT_s}(t - knT_s - t_0) c_{T_s}(t - t_0) \tag{5.2.19}$$

where $\{a_k\}$ is the data-modulated spreading code, which is binary for the binary modulation case, $c_{T_s}(t)$ is the subcarrier — a periodic function with period $2T_s$, and $\mu_{nT_s}(t)$ is a "spreading symbol" — a rectangular pulse lasting from 0 to nT_s. Then n is the number of half-periods of the subcarrier during which the spreading code value remains the same, and the following relationships hold:

$$f_c = \frac{1}{nT_s} = \frac{2}{n} f_s \qquad n = \frac{2\alpha}{\beta} \tag{5.2.20}$$

The normalized power spectral density of the BOC modulation is written as (Betz, 2002)

$$g(f_s,f_c,\Delta f) = \begin{cases} f_c\left(\dfrac{\tan(\pi\ \Delta f/2f_s)\ \cos(\pi\ \Delta f/f_c)}{\pi\ \Delta f}\right)^2 & \text{if } n \text{ is odd} \\ f_c\left(\dfrac{\tan(\pi\ \Delta f/2f_s)\ \sin(\pi\ \Delta f/f_c)}{\pi\ \Delta f}\right)^2 & \text{if } n \text{ is even} \end{cases} \tag{5.2.21}$$

For example, the modulation BOC(10, 5) uses the subcarrier frequency and the spreading code rate of 10.23 and 5.115 MHz, respectively. Furthermore, the value $n = 1$ corresponds to the case of BPSK. For example, BOC(5,10) is BPSK(10), i.e., the BPSK modulation with 10.23 MHz spreading rate which is used for P(Y)-code. The modulation BOC(0.5, 1) is a BPSK(1) used for L1 C/A, having 1.023 MHz spreading rate.

A characteristic difference between the BOC and the conventional rectangular spreading code modulation is seen in the power spectral densities of Figure 5.2.7. The densities for BOC, in this case BOC(10,5), are maximum at the nulls of the P(Y)-codes. Such a property is important for increasing the spectral separation of modulations. The sum of the number of mainlobes and sidelobes between the mainlobes is equal to n, i.e., twice the ratio of the subcarrier frequency to the code rate (5.2.20). As in conventional BPSK the zero crossings of each mainlobe are spaced by twice the code

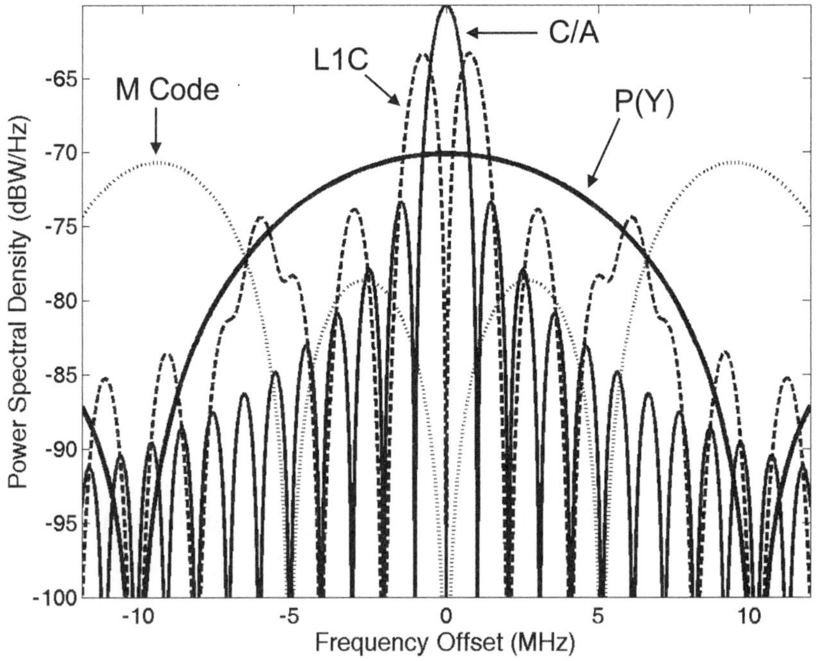

Figure 5.2.7 Spectra of signals, available in the L1 frequency band.

rate, while the zero crossings of each sidelobe are spaced at the code rate. For example, with $n = 5$ the BOC(5,2) modulations have three sidelobes between two mainlobes; with $n = 10$ the BOC(5,1) modulations have eight sidelobes between two mainlobes. In the case of $n = 1$ we have BOC($f_c/2, f_c$) and equations (5.2.21) and (5.2.16) giving the same power spectral density, as already noted.

5.2.3.2 Civil L2C Codes It is the first civilian-use signal to be transmitted on a frequency other than L1, used for the C/A signal. The new L2 is shared between civil and military signals. To increase GPS performance for civilian users, the new space vehicles IIR-M and IIF have two additional civil ranging codes, L2CM (civil moderate length) and L2CL (civil long). As is the case for L1, the new L2 carrier consists of two BPSK modulated carrier components that are inphase quadrature with each other. The inphase carrier continues to be BPSK modulated by the bit train that is the modulo-2 sum of the military P(Y)-code and the legacy navigation data $D^p(t)$. There are three options available for BPSK modulating the quadrature carrier (also called the L2C carrier or the new L2 civil signal):

1. Chip-by-chip time multiplex combinations of bit trains consisting of the modulo-2 sum of the L2CM code and a new navigation message structure $D_C(t)$. The resultant bit trains are then combined with the L2CL code and used to modulate the L2 quadrature carrier. The IIR-M space vehicles will have the option of using the old navigation message $D^p(t)$ instead of $D_C(t)$.
2. Modulo-2 sum of the legacy C/A-code and legacy navigation data $D^p(t)$.
3. C/A-code with no navigation data.

The options are selectable by a ground command. The chipping rate for L2CM and L2CL is 511.5 kbps. L2CM is 10,230 chips long and lasts 20 ms, whereas L2CL has 767,250 chips and lasts 1500 ms. L2CL is 75 times longer than L2CM. $D_C(t)$ is the new navigation data message and has the same structure as the one adopted for the new L5 civil signal. It is both more compact and more flexible than the legacy message.

The spectra of signals available on the L2 band [L2C, L2 P(Y), and L2 M-code] look the same as shown in Figure 5.2.7, excluding the TMBOC signal available only on L1 as L1C. Note that the spectra of signals are shown zero centered which correspond to the base-band representation, only reflecting the modulation. The actual spectra reflecting their allocation in the radio-frequency (RF) signal are shifted from zero to L1 or L2, depending on signals.

5.2.3.3 Civil L5 Code The carrier frequency of L5 is 1176.45 MHz, which is the new third frequency. It is a civilian safety of life signal, and the frequency band is protected by the International Telecommunication Union (ITU) for aeronautical radionavigation service.

As is the case for L1, two L5 carriers are inphase quadrature and each is BPSK(10) modulated separately by bit trains. The bit train of the inphase component is a modulo-2 sum of PRN codes and navigation data. The quadraphase code is a separate PRN code but has no navigation data. The chipping rate of the codes is 10.23 MHz. Each code is a modulo-2 sum of two subsequences, whose lengths are 8190 and 8191 chips that recycle to generate 10,230 chip codes. The navigation data is encoded by the error correcting code, which improves availability of the navigation data. The bandwidth of the L5 code is 24 MHz. Wider bandwidth provides a higher accuracy of ranging. It also has higher transmitting power than L1 and L2, approximately 3 dB.

5.2.3.4 ***M-Code*** One of the underlying objectives behind the GPS modernization is the development of a new military signal, protecting military use of GPS by the United States and its allies and preventing unauthorized use of GPS. On the other hand, the peaceful use of the civil radionavigation service must be preserved.

The new military M-codes uses BOC(10,5), which means the subcarrier frequency and the spreading code rate will be 10.23 and 5.115 MHz, respectively, as well as quadrature phase modulated, i.e., they share the same carrier with the civilian signals. The idea of spectrum separation preserving the legacy signals [civilian L1 C/A and militaryL1/L2 P(Y)] and new signals (L1C and L2C), is illustrated in Figure 5.2.7.

5.2.3.5 ***Civil L1C Code*** The prospective L1C code occupying the L1 band will be transmitted by Block III satellites. It is designed in such a way that it has very little impact on the military M-code. The time-multiplexed BOC (TMBOC) combining BOC(1,1) and BOC(6,1) used for L1C signal, has the spectral density estimated as

$$g_{L1C}(\Delta f) = \frac{10}{11} g_{BOC(1,1)}(\Delta f) + \frac{1}{11} g_{BOC(6,1)}(\Delta f) \tag{5.2.22}$$

where $g_{BOC(1,1)}(\Delta f)$ and $g_{BOC(6,1)}(\Delta f)$ are defined by (5.2.21) for $f_s = 1.023$ MHz, $f_c = 1.023$ MHz, $f_s = 6.138$ MHz, $f_c = 1.023$ MHz, respectively. It is shown in Figure 5.2.7 by the dashed line.

There are many sources of detailed description of this signal, see Macchi-Gernot et al. (2010), for example, while the full description can be found in the IS-GPS-800C (2012) document. The signal is composed of two channels: a pilot channel (denoted by $L1C_p$) and a data channel (denoted $L1C_D$) transmitted inphase quadrature. The pilot channel combines a spreading code and an overlay, or secondary code, denoted by $L1C_O$. The overlay code is generated using FBSR. It is transmitted at 100 bits/s and contains 1800 bits, thus lasting 18 seconds. The overlay code is unique for each PRN. The data channel includes a spreading code and a navigation message. The spreading codes of the pilot and data channels are time synchronized. The spreading codes are broadcasted at the same chipping rate as L1 C/A, that is, 1.023 Mchips/s. On both the pilot and data channels, the spreading codes have a period of 10 ms, and therefore contain 10,230 chips. The L1C spreading codes are generated using a modified Weil code. We refer the reader to the Chapter 3.2.2.1.1 of the IS-GPS-800C (2012) for more detail about the Weil code.

The modulation on $L1C_D$ (data channel) is a BOC(1,1). Its two main lobes are centered at ±1.023 MHz relative to the central frequency (see dashed line plot in Figure 5.2.7) Therefore, a bandwidth of 4.092 MHz is needed to transmit most of its power. The modulation on $L1C_p$ is a TMBOC, which consists of 29/33 chips modulated using a BOC(1,1) and others using a BOC(6,1). When BOC(6,1) is used, the number of chips (and correspondingly the chipping rate) is increased by a factor of six compared to the BOC(1,1), for a time interval equivalent to one chip of the original ranging code. This increased chipping rate has the effect of requiring an increased sampling frequency to extract all the information, because the minimum necessary bandwidth is 14.322 MHz; please look at the third side lobes of the dashed line in Figure 5.2.7.

The L1C navigation message is transmitted on the data channel at 100 symbols per second. The low-density parity-check code (LDPC) with the $\frac{1}{2}$ code rate is used to encode the navigation data. LDPC is one of the error correction codes, see Gallager (1963). Code rate is defined as the ratio between the number of bits necessary to transmit the information and the total number of bits. Encoding aims to improve reliability and antierror protection. Therefore, despite the fact that the message is broadcasted at twice the chipping rate than on L1 C/A, effectively the navigation message rate is the same 50 symbols per second due to the $\frac{1}{2}$ code rate.

5.3 GLONASS

The Russian GLONASS (Global'naya Navigatsionnaya Sputnikovaya Sistema) global navigation satellite system traces its beginnings to 1982, when its first satellite was launched. The time line of the space segment is shown in Figure 5.3.1 by a

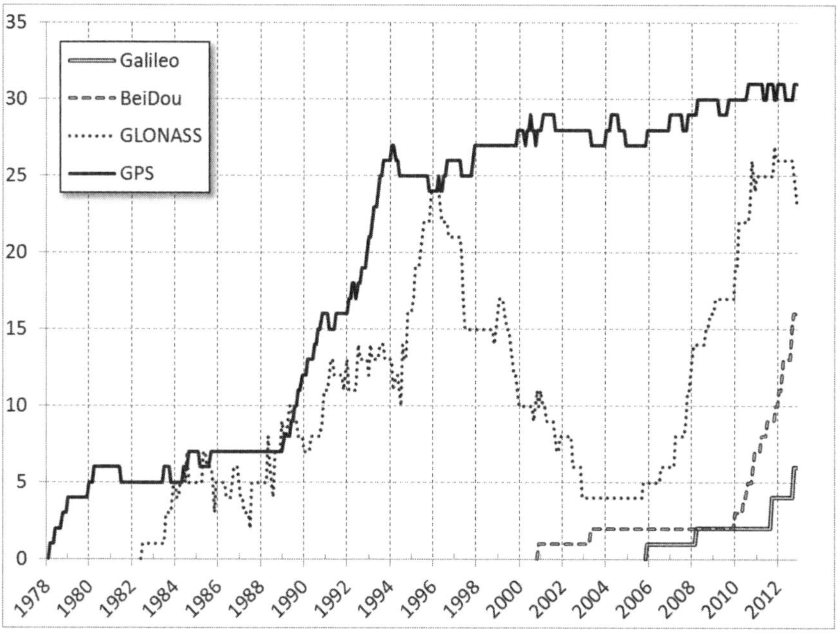

Figure 5.3.1 Operational satellites for GPS, GLONASS, Galileo, and Beidou systems by years. Data from various Internet documents.

dotted line. The number of GPS satellites is shown by a black line for comparison. For technical information about GLONASS, see the interface control document GLONASS (2008). Additional details on the system and its use, plus many references to relevant publications on the subject, are available in Roßbach (2001), Zinoviev (2005), and other papers published in proceedings at various scientific meetings.

Like GPS, GLONASS was intended to encompass at least 24 satellites. The nominal orbits of the satellites are in 3 orbital planes separated by 120°; the satellites are equally spaced within each plane with nominal inclination of 64.8°. The nominal orbits are circular with each radius being about 25,500 km. This translates into an orbital period of about 11 h and 15 min.

A major difference between GLONASS and GPS is that each GLONASS satellite transmits at its own carrier frequency. Let p denote the channel number that is specific to the satellite, then

$$f_1^p = 1602 + 0.5625p \text{ MHz} \tag{5.3.1}$$

$$f_2^p = 1246 + 0.4375p \text{ MHz} \tag{5.3.2}$$

with

$$\frac{f_1^p}{f_2^p} = \frac{9}{7} \tag{5.3.3}$$

The original GLONASS signal structure used $1 \leq p \leq 24$, covering a frequency range in L1 from 1602.5625 to 1615.5 MHz. However, receivers have an interference problem in the presence of mobile-satellite terminals that operate at the 1610 to 1621 MHz range. To avoid such interference, it has been suggested that the channel numbers will be limited to $-7 \leq p \leq 6$ and that satellites located in antipodal slots of the same orbital plane may transmit at the same frequency (GLONASS, 2008). Currently, the L1 frequency covers the range from 1598.0625 to 1605.375 MHz and L2 frequency covers the range from 1242.9375 to 1248.625 MHz.

The L1 and L2 frequencies are coherently derived from common onboard frequency standard running at 5.0 MHz. In order to account for relativistic effects, this value is adjusted to 4.99999999782 MHz. As is the case with GPS, there are C/A-codes on L1 and P-codes on L1 and L2, although the code structures differ. The satellite clocks are steered according to UTC(SU), where SU stands for Russia (former Soviet Union). The GLONASS satellite clocks, therefore, are adjusted for leap seconds.

Two different types of signals are transmitted by GLONASS satellites: Standard precision (ST) and high precision (W) in both the L1 and L2 bands. The GLONASS standard accuracy signal, also known as C/A-code, has a clock rate of 0.511 MHz and is designed for use by civil users. The high accuracy signal (P-code) has a clock rate of 5.11 MHz and is modulated by a special code that is only available to authorized users.

The GLONASS broadcast navigation message contains satellite positions and velocities in the PZ90 ECEF geodetic system and accelerations due to luni-solar attraction at epoch t_0. These data are updated every 30 min and serve as initial conditions for orbital integration. The satellite ephemeris at the epoch t_b with

$|t_b - t_0| \leq 15$ min is calculated by numerical integration of the differential equations of motion (5.1.62). Because the integration time is short, it is sufficient to consider a simplified force model for the acceleration of the gravitational field of the earth. Since the gravitational potential of the earth is in first approximation rotationally symmetric, the contributions of the tesseral harmonics $m \neq 0$ are neglected in (5.1.64). Since $\overline{C}_{20} \gg \overline{C}_{n0}$ for $n > 2$, we neglect the higher-order zonal harmonics. With these simplifications the disturbing potential (5.1.64) becomes

$$\begin{aligned} R &= \frac{\mu a_e^2}{r^3} \overline{C}_{20} \overline{P}_2(\cos\theta) = \frac{\mu a_e^2}{r^3} J_2 P_2(\cos\theta) \\ &= \frac{\mu a_e^2}{r^3} J_2 \left(\frac{3}{2} \cos^2\theta - \frac{1}{2} \right) \end{aligned} \tag{5.3.4}$$

In expression (5.3.4), we switched from the fully normalized spherical harmonic coefficients to regular ones and substituted the expression for the Legendre polynomial $P_2(\cos\theta)$. Since $Z = r \cos\theta$, equation (5.3.4) can be rewritten as

$$R = \frac{\mu a_e^2}{r^3} J_2 \left(\frac{3}{2} \frac{Z^2}{r^2} - \frac{1}{2} \right) \tag{5.3.5}$$

Recognizing that $r = (X^2 + Y^2 + Z^2)^{1/2}$, we can readily differentiate and compute the acceleration $\ddot{\mathbf{X}}_g$ as per (5.1.63),

$$\ddot{X} = -\frac{\mu}{r^3} X - \frac{3}{2} J_2 \frac{\mu a_e^2}{r^5} X \left(1 - 5\frac{Z^2}{r^2} \right) + \ddot{X}_{s+m} \tag{5.3.6}$$

$$\ddot{Y} = -\frac{\mu}{r^3} Y - \frac{3}{2} J_2 \frac{\mu a_e^2}{r^5} Y \left(1 - 5\frac{Z^2}{r^2} \right) + \ddot{Y}_{s+m} \tag{5.3.7}$$

$$\ddot{Z} = -\frac{\mu}{r^3} Z - \frac{3}{2} J_2 \frac{\mu a_e^2}{r^5} Z \left(1 - 5\frac{Z^2}{r^2} \right) + \ddot{Z}_{s+m} \tag{5.3.8}$$

These equations are valid in the inertial system (X) and could be integrated. The PZ90 reference system, however, is ECEF and rotates with the earth. It is possible to rewrite these equations in the ECEF system (x). Since the integration interval is only ± 15 min, we can neglect the change in precession, nutation, and polar motion and only take the rotation of the earth around the z axis into consideration. The final form of the satellite equation of motion then becomes

$$\ddot{x} = -\frac{\mu}{r^3} x - \frac{3}{2} J_2 \frac{\mu a_e^2}{r^5} x \left(1 - 5\frac{z^2}{r^2} \right) + \omega_3^2 x + 2\omega_3 \dot{y} + \ddot{x}_{s+m} \tag{5.3.9}$$

$$\ddot{y} = -\frac{\mu}{r^3} y - \frac{3}{2} J_2 \frac{\mu a_e^2}{r^5} y \left(1 - 5\frac{z^2}{r^2} \right) + \omega_3^2 y + 2\omega_3 \dot{x} + \ddot{y}_{s+m} \tag{5.3.10}$$

$$\ddot{z} = -\frac{\mu}{r^3} z - \frac{3}{2} J_2 \frac{\mu a_e^2}{r^5} z \left(1 - 5\frac{z^2}{r^2} \right) + \ddot{z}_{s+m} \tag{5.3.11}$$

Note that $(\ddot{x}, \ddot{y}, \ddot{z})_{s+m}$ are the accelerations of the sun and the moon given in the PZ90 frame. These values are assumed constant when integrating over the ±15 min interval. In order to maintain consistency, the values for μ, a_e, J_2, and ω_3 should be adopted from GLONASS (2008). This document recommends a four-step Runge-Kutta method for integration.

Various international observation campaigns have been conducted to establish accurate transformation parameters between WGS84 and PZ90, with respect to the ITRF. Efforts are continuing to include the precise GLONASS ephemeris into the IGS products.

The GLONASS program is also undergoing modernization. The new series of satellites are called GLONASS-M. GLONASS-M satellites have better onboard clock stability and a civil code (also called L2C) available at the L2 frequency band. Starting with GLONASS-K1, the first code division multiple access (CDMA) signal becomes available at the frequency band L3 = 1201 MHz. GLONASS-K2, planned for launch starting in 2014, will provide CDMA codes on L1, L2, and L3 bands.

GLONASS satellites have been used successfully for accurate baseline determination since the mid-1990s (Leick et al., 1995). The additional difficulties encountered in baseline processing because of the GLONASS satellites transmitting on different carrier frequencies will be discussed in Chapter 7.

5.4 GALILEO

Galileo, the European global navigation satellite system, is designed to provide a highly accurate, guaranteed, and global positioning service under civilian control that is funded by civilian European institutions. It is interoperable with other global satellite navigation systems. The full constellation is expected to consist of 27 operational and 3 spare space vehicles, located in 3 orbital planes with inclination of 56°, and nominal circular orbits with a semimajor axis of about 29,600 km.

On March 26, 2002, the European Council agreed on the launch of the European Civil Satellite Navigation Program, called Galileo. Basic approaches and critical algorithms were tested by 2003. Two initial satellites, called GIOVE (Galileo In-Orbit Validation Elements) were launched in 2005. They were built for the European Space Agency (ESA) for testing Galileo technology in orbit and eventually to become two satellites of the full Galileo constellation. Four additional satellites were launched by 2012. The full constellation of 30 satellites is planned for completion by 2020. The number of Galileo satellites by years is shown in Figure 5.3.1 by a dual solid line.

It can be seen from Figure 5.4.1 that the Galileo E5A signals share the frequency band with GPS L5. The adjacent region is reserved for Galileo E5B. At the World Radio Conference (WRC) 2000 in Istanbul, Turkey, several decisions were made that deal with the increasing demand for frequency space. For example, the WRC expanded the bottom end of one of the radio navigation satellite services (RNSS) bands to between 1164 and 1260 MHz, putting E5A, E5B, and L5 under RNSS protection. Galileo has also been assigned the range 1260 to 1300 MHz, labeled E6, at the lower L-band region. At the upper L-band, the band labeled El has been

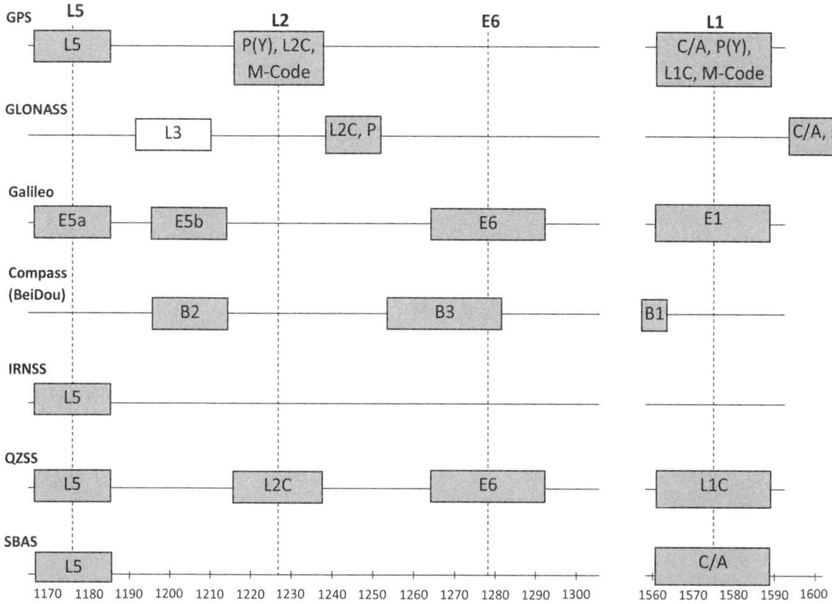

Figure 5.4.1 Allocation of GPS, GLONASS, Galileo, Compass (Beidou), QZSS, IRNSS, and SBAS frequency bands. The S-band (2492.028 MHz) IRNSS signal is not presented.

reserved for Galileo and is centered at the GPS L1 band. Using BOC modulation techniques, the Galileo signal has been constructed to have maximum spectral density at both sides of the E1 band, but covering the whole El band.

In order to make Galileo and GPS compatible, i.e., allow for the use of common receiver components, the carrier frequency for Galileo E1 is 1575.42 MHz, which is the same as GPS L1. Similarly, E5A and L5 use 1176.45 MHz as the common carrier frequency. The modulation (inphase and quadraphase) codes and chipping rate for the various carriers are summarized in Table 5.4.1

TABLE 5.4.1 Galileo Signal Parameters.

Signal	Carrier Frequency (MHz)	Receiver Reference Bandwidth (MHz)	Ranging Code Chip rate (Mchip/s)	Modulation
E1	1575.420	24.552	2.046	BOC(2,2)
E6	1278.750	40.920	5.115	BOC(10,5), BPSK(5)
E5	1191.795	51.150		AltBOC(15,10)
E5a	1176.450	20.460	10.23	
E5b	1207.140	20.460	10.23	

The Galileo E5 signal employs a constant envelope alternate binary offset carrier (AltBOC) modulation. The subcarrier waveforms are chosen in such a way that a constant envelope at the transmitter is obtained. The result of this AltBOC modulation is a split spectrum around the center frequency, as shown in Figure 5.4.1. Each sideband comprises two pseudorandom codes modulated onto the orthogonal components. The inphase components E5AI and E5BI carry the data modulation. The quadrature components E5AQ and E5BQ are pilot signals. While being two parts of the E5 signal, the E5A and E5B signals can be processed independently by the user receiver as though they were two separate QPSK signals with a carrier frequency of 1176.45 and 1207.14 MHz, respectively.

Galileo employs the same ephemerides structure as used by GPS. The user algorithm for ephemeris determination is identical to one used for GPS (see Table 5.2.5). The convolutional code with the code rate $\frac{1}{2}$ is used to increase reliability of the navigation data transmission.

5.5 QZSS

The Japanese quasi zenith satellite system (QZSS) constellation consists of quasi-zenith satellites orbiting the earth. The constellation buildup started in September 2010 with the launch of the first satellite *Michibiki*. It achieved full functionality in late 2011. During the initial phase of technical verification and application demonstration, the goal was to demonstrate that combining GPS and QZSS would significantly improve positioning availability in urban canyon areas of Tokyo.

The government of Japan decided in 2011 to accelerate the QZSS deployment to reach a four-satellite constellation by 2020, while aiming at a final seven-satellite constellation. In March 2013, the Japanese Cabinet Office formally announced a large contract award to Mitsubishi to build one geostationary satellite and two additional quasi-zenith satellites. The three satellites are scheduled to be launched before 2018. In addition, another contract was also signed with a special-purpose company (led by NEC and supported by Mitsubishi UFJ Lease & Finance and Mitsubishi Electric Corporation) to fund the design and construction of the ground control system, as well as its verification and maintenance for a period of 15 years.

Three satellites are planned to be placed in highly elliptical orbit (HEO). The perigee altitude is about 32,000 km and the apogee altitude is about 40,000 km, and all of them will pass over the same "figure-8" ground as shown in Figure 5.5.1. The system is designed so that at least one satellite out of three is always near zenith over Japan. Given its orbit, each satellite appears almost overhead most of the time (i.e., more than 12 h a day with an elevation above 70°). This gives rise to the term "quasi-zenith" for which the system is named. The design life of the quasi-zenith satellites is 10 years. Table 5.5.1 lists the signals planned for QZSS.

The signals L1-C/A, L1C, L2C, and L5 are designed to be compatible with existing GNSS receivers, in order to increase availability of both standalone and high-precision carrier phase differential position services. For example, they allow forming across-satellite, across-receiver mixed GPS-QZSS and Galileo-QZSS

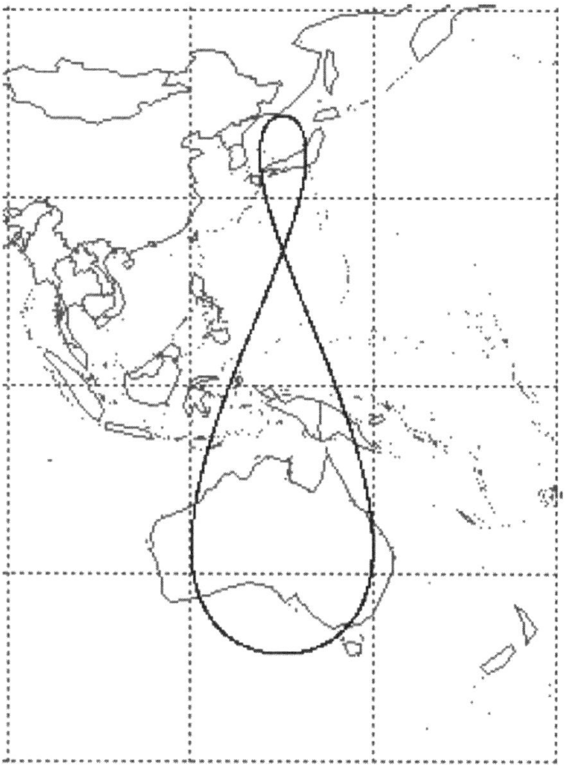

Figure 5.5.1 The QZSS ground track.

TABLE 5.5.1 QZSS Signal Parameters.

Signal	Carrier Frequency (MHz)	Receiver Reference Bandwidth (MHz) (approximate)	Ranging Code Chip rate (Mchip/s)	Modulation
L1 C/A	1575.42	24	1.023	BPSK(1)
L1C	1575.42	24	1.023	BOC(1,1)
L1-SAIF	1575.42	24	1.023	BPSK(1)
L2C	1227.60	24	1.023	BPSK(1)
LEX	1278.75	42	5.115	BPSK(5)
L5	1176.45	24	10.23	BPSK(10)

differences. Experiments with receivers by Javad GNSS show successful resolution of corresponding mixed double difference ambiguities (Rapoport, private communication). In other words, QZSS can be considered an extension of GPS for point and high-precision services. The signal L1-SAIF (submeter-class augmentation with integrity function) is intended to provide submeter augmentation and interoperability with GPS and SBAS (satellite based augmentation system). The LEX signal is the QZSS L-band experimental signal for high-precision service at the 3 cm level. It is compatible with the Galileo E6 signal.

The multi-constellation GNSS interoperable signals L1 C/A, L1C, L2C, and L5 are to be provided free of charge. Compatibility is a mandatory requirement for the QZSS system (working in the same frequency bands as other GNSS systems without harmful interference). For GPS performance enhancement signals, a charging policy for L1-SAIF and LEX signals is under consideration.

Compared to standalone GPS, the system combining GPS and QZSS will improve positioning performance via correction data provided through submeter enhancement signals L1-SAIF and LEX. The ephemeris algorithm is basically the same as used for GPS and described in Table 5.2.4. Detailed descriptions can be found in QZSS (2013). To increase reliability of the navigation data transmission, the LDPC code with the $\frac{1}{2}$ code rate is used.

5.6 BEIDOU

The Chinese Compass Navigation Satellite System (CNSS), also called Beidou-2, is a satellite navigation system that will also be capable of providing positioning, navigation, and timing services to users on a continuous worldwide basis. Since beginning the upgrade in 1997 from a regional navigation system to a global system and achieving formal approval by the government to develop and deploy Beidou-2 in 2004, the system is expected to provide global navigation services by 2020, similar to GPS, GLONASS, or Galileo. The Beidou satellite constellation consists of geostationary earth orbit (GEO), inclined geosynchronous satellite orbit (IGSO), and medium-earth orbit (MEO) satellites.

When fully deployed, the space constellation will include 5 GEOs, 27 MEOs, and 3 IGSOs. The GEO satellites are operating at an altitude of 35,786 km and positioned at 58.75°E, 80°E, 110.5°E, 140°E, and 160°E. The MEO satellites are operating at orbital altitude of 21,528 km with an orbital inclination of 55°. The IGSO satellites are at orbital altitude of 35,786 km with an inclination of 55° to the equatorial plane.

As of December 2011, the Beidou system provides initial operational service for positioning navigation and timing services for the Asia-Pacific region with a constellation of 10 satellites (5 GEOs and 5 IGSOs). During 2012, 5 additional satellites (1 GEO satellite and 4 MEO satellites) were launched, increasing the number of satellites of the constellation to 14. The number of launched Beidou satellites by years is shown in Figure 5.3.1 by a double dashed line.

The Beidou signals are transmitted in three frequency bands: B1, B2 (which equals to E5B of Galileo), and B3. Parameters of the signals are summarized in Table 5.6.1. For MEO and IGSO satellites, the ephemeris calculation algorithm is

TABLE 5.6.1 Beidou Signal Parameters.

Signal	Carrier Frequency (MHz)	Receiver Reference Bandwidth (MHz) (approximate)	Ranging Code Chip rate (Mchip/s)	Modulation
B1	1561.098	4	2.046	BPSK(2)
B2	1207.14	24	10.23	BPSK(10)
B3	1268.52	24	10.23	BPSK(10)

the same as is used for GPS and is described in Table 5.2.4, and the only difference is in using the CGCS2000 constants: $\mu = 3.986004418 \times 10^{14}\ \mathrm{m}^3/\mathrm{s}^2$ for the earth universal gravitational constant and $\dot{\Omega}_e = 7.2921150 \times 10^{-5}$ rad / s for the earth rotation rate.

For the GEO satellites, the two last steps of calculation differ from those described in Table 5.2.4 as follows (Beidou, 2013, p. 34):

$$\Omega_k = \Omega_0 + \dot{\Omega} t_k - \dot{\Omega}_e t_{oe} \tag{5.6.1}$$

$$\begin{aligned} x_{Gk} &= x'_k \cos \Omega_k - y'_k \cos i_k \sin \Omega_k \\ y_{Gk} &= x'_k \sin \Omega_k - y'_k \cos i_k \cos \Omega_k \\ z_{Gk} &= y'_k \sin i_k \end{aligned} \tag{5.6.2}$$

$$\begin{pmatrix} x_k \\ y_k \\ z_k \end{pmatrix} = \boldsymbol{R}_3(\dot{\Omega}_e t_k)\boldsymbol{R}_1\left(-\frac{\pi}{36}\right)\begin{pmatrix} x_{Gk} \\ y_{Gk} \\ z_{Gk} \end{pmatrix} \tag{5.6.3}$$

where matrices $\boldsymbol{R}_1(\varphi)$ and $\boldsymbol{R}_3(\varphi)$ are described in Appendix A.2.

The Beidou reference system is the China Geodetic Coordinate System 2000 (CGCS2000), which is aligned to the ITRS. Beidou has already gained much attention from the geodetic and surveying community as a source for additional observations, but also because it is a mixed constellation consisting of geostationary, inclined geosynchronous, and medium orbits. Shi et al. (2013) processed dual-frequency phase observation of June 2011, at the time the constellation included 3 GEOs and 3 IGSOs, for precise relative positioning. Montenbruck et al. (2013) used a March 2012 triple-frequency data set to provide an initial assessment of the satellite system regarding satellite clock performance, precise absolute positioning, and baseline determination. Tang et al. (2014) evaluated epoch-by-epoch processing of triple-frequency observations to investigate previously suggested optimal triple-frequency processing strategies, such as TCAR (three-frequency carrier ambiguity resolution), for baselines ranging from 2.5 m to 43 km. The implications of using a GPS-type of broadcast ephemeris format for the geostationary and geosynchronous orbits was investigated by Du et al. (2014), who suggest an 18-element GEO broadcast.

5.7 IRNSS

In May 2006, the Indian government approved development of the Indian Regional Navigation Satellite System. The major objective was to have complete Indian control over the space segment, ground segment, and user receivers. Seven satellites will eventually complete the space segment of the IRNSS. Three of them will be GEOs located at 32.5°, 83°, and 131.5° east longitude. Four of them will be IGSO in an orbital plane with 29° inclination. Two IGSO satellites will cross the equator at 55° east and two at 111.75° east. The IGSO satellites will have a “figure-8” ground track. Because of low latitudes, a coverage with low-inclination satellites is optimal. The constellation provides continuous radio visibility with the Indian control stations.

Seven satellites with the prefix “IRNSS-1” form the space segment of the IRNSS. The first satellite, IRNSS-1A, was launched on July 1, 2013. It operates in L5 band (1176.45 MHz) and S band (2492.028 MHz). The second satellite IRNSS-1B was launched on April 4, 2014, and placed in IGSO. In 2014, two more satellites, IRNSS-1C and IRNSS-1D, will be launched. Three more satellites are planned for launch at the beginning of 2015. Thus, by the middle of 2015, India is expected to have the full IRNSS system installed.

IRNSS signals will provide two types of services: standard positioning service (SPS) and restricted service (RS). Both SPS and RS will be carried on L5 (1176.45 MHz) and S band ($2492.08 = 243.6 \times 10.23$ MHz). The SPS signal will be modulated by a BPSK(1) signal. The RS signal will use BOC(5,2) modulation. An additional BOC pilot signal will be provided for the RS service (called an RS-P signal), in order to improve acquisition and performance. Therefore, each of the L5 and S bands will carry three signals: SPS, RS, and RS-P. The user receiver can operate in single- and/or dual-frequency mode. Since an IRNSS interface control document (ICD) has not yet been publicly released, we refer to the research paper by Majithiya et al. (2011) for more detail. Thoelert et al. (2014) recorded and analyzed the spectrum and modulated chip sequences of the signals transmitted by the first IRNSS satellite using a 30 m high-gain antenna.

The IRNSS system is expected to provide absolute position accuracy of better than 10 m over India and better than 20 m in the Indian Ocean and territories surrounding India, up to 1500 km beyond its boundary.

5.8 SBAS: WAAS, EGNOS, GAGAN, MSAS, AND SDCM

The U.S. Federal Aviation Administration (FAA) has developed the wide area augmentation system (WAAS) to improve accuracy, integrity, and availability of GPS, mainly for air navigation applications. A ground segment consists of a network of wide area reference stations (WRS) located in the United States (including Alaska and Hawaii), Canada, Mexico, and Puerto Rico. These ground stations monitor the GPS signals and send information to wide area master stations (WMS) using a communications network. The WMS generate sets of fast and slow corrections. The fast corrections are useful for compensation of rapidly changing errors that affect the positions and clocks of GPS satellites. The slow corrections deal with long-term ephemeris and clock errors, as well as ionospheric delay parameters. Once these corrections have been generated, the WMSs send them to ground uplink stations for transmission to the satellites of the WAAS space segment. The latter modulate the correction on a GPS L1 carrier together with a C/A-code and then transmit to the user. The corrections, therefore, are used in a one-way (satellite to user) manner and can be applied instantly by any receiver inside the WAAS broadcast coverage area. The space segment consists of multiple, multipurpose geostationary communication satellites. Because the transmissions of the SBAS space segment contain a ranging code on L1, they effectively increase the number of satellites available for positioning.

WAAS is one of several satellite-based augmentation systems (SBAS). Europe and Asia are operating or developing their own SBAS: the European Geostationary Navigation Overlay Service (EGNOS), the Indian GPS Aided Geo Augmented Navigation (GAGAN), the Japanese Multifunctional Satellite Augmentation System (MSAS), and the Russian System for Differential Correction and Monitoring (SDCM).

The SBAS satellites enable the users to produce code and carrier phase observations. Eventually, SBAS will be a dual-frequency system. An SBAS L5 interface control document is under development and already circulating in the form of a draft. The signal structure of L1 and L5 SBAS is very similar to those of GPS, but there are differences. The GPS L5 signal is designed with two channels, I and Q. The modulation method is BPSK(10) in both channels with 10,230 code length. The I channel is the data channel and the Q channel is data free. The WAAS L5 signal has only one data channel. Another difference, of course, is the space segment. The SBAS signals are broadcasted from multi functional geostationary satellites and not dedicated navigation satellites. As a result, the space-borne clock errors and orbital errors are greater than those of GPS. However, combining GPS L1/L5, SBAS L1/L5 (providing integrity data for GPS), and Galileo E1/E5 observations in dual-band receivers will enable robust high-precision positioning.

The SDCM system consists of three GEO satellites, Luch-5A, Luch-5B, and Luch-5V, which transmit at GPS L1 frequency band, while serving to improve GLONASS operation; see SDCM (2012).

CHAPTER 6

GNSS POSITIONING APPROACHES

New GNSS positioning and timing techniques continue to be developed and refined. Whereas during the formative years the progress of GPS positioning was measured by leaps and bounds at a breathtaking speed, the current developments are more incremental in nature but, nevertheless, lead to noticeable refinements that result in shortening of observation time, increase in accuracy, and increase in reliability. The incremental improvements are due to the modernization of GPS, the repopulation of the GLONASS satellite system, and the new Beidou, Galileo, and QZSS satellite systems. Equally important to continued improvement of GNSS positioning is the dedicated efforts by many scientists and engineers to complete a GNSS infrastructure for data collection, evaluation, and user service. Examples include the IGS (International GNSS Service), which provides various intermediary products, antenna calibration services, services that provide accurate information about the ionosphere and troposphere, and also processing services that accept user observations and deliver final coordinates in any desired coordinate system.

The first section summarizes undifferenced pseudorange and carrier phase functions between a single station and single satellite. This includes the triple-frequency functions. This is followed by functions that include two stations. These are the single differences, consisting of the across-receiver functions, across-satellite functions, and across-time functions, followed by the traditional double-difference and triple-difference functions.

The second section addresses operational details such as satellite clock corrections, timing group delay, and intersignal corrections, cycle slips, phase windup correction, multipath, and phase center offset and phase center variation. The section concludes with a discussion on the various services available to the user, in particular the IGS.

Section 6.3 deals with the navigation solution. It produces position and time for a single user using pseudorange observations and the broadcast ephemeris. We discuss the DOP (dilution of precision) factors and Bancroft's closed-form navigation solution is presented.

Section 6.4 deals with well-established techniques in surveying and geodesy, i.e., baseline determination using carrier phase observations to determine relative positioning between stations. The focus is on double differencing, although we briefly talk about the equivalent undifferenced formulation. Although the ambiguity function technique does not seem to command the popularity of double differencing, it will be discussed next. The section concludes with initial remarks about GLONASS carrier phase processing.

Section 6.5 is dedicated to double-difference ambiguity fixing. It begins with looking at the problem as a classical least-squares solution with conditions. The popular LAMBDA (least-squares ambiguity decorrelation adjustment) approach for decorrelating the ambiguity parameters is presented in detail, although key references are cited to document various statistical properties of the technique. Because ambiguity fixing plays such a central role in accurate baseline determination, we widen the scope of techniques and talk about lattice reduction in general.

Section 6.6 focuses on the support of networks for precise positioning. First the traditional PPP (precise point positioning) is presented, which makes implicit use of a global network, followed by the traditional use of CORS (continuing operating reference stations) networks, which are typically at the national level. We then present the differential correction, which contains ancillary data from a reference station and is transmitted to the user. The main part of this section is on PPP-RTK and the content of the various network corrections it uses. PPP-RTK has received a lot of attention in recent years.

The last section deals with triple-frequency solutions. These types of solutions are becoming relevant as more satellites of the various GNSS systems transmit on three or more frequencies. The focus is on the extra capability provided by the third frequency as compared to "classical" dual-frequency approaches.

6.1 OBSERVABLES

Pseudoranges and carrier phases are the basic GNSS observations (observables) used for positioning and timing. Carrier phases are always required for accurate surveying at the centimeter level. Obtaining these measurements involves advanced techniques in electronics and digital signal processing. We discuss the pseudorange and carrier phase equations in the form needed to process the observations as downloaded from the receiver. The internal processing of the receiver that produces the pseudorange or carrier phase observables starting satellite signals registered at the antenna is not discussed. The reader is, instead, asked to consult the respective specialized literature on internal receiver processing.

We begin with the derivation of the pseudorange and carrier phase equations and express them in terms of various parameters. These basic observables are then combined into various linear functions. For example, there are functions that do not contain first-order ionospheric terms, functions that are independent of the receiver location, and functions that do not depend on either ionosphere or receiver location. Other functions difference simultaneous observations across receivers, across satellites, or across receivers and satellites. The latter are popularly known as double differences. The triple difference, which is an across-time double difference, is also given. In view of modern GNSS systems, we include, of course, relevant triple-frequency functions.

All functions are provided essentially in the form of a list, thus allowing us to bundle all relevant functions into one location in the book. Only minimal explanations are provided for the functions since they are well known. However, each function is given explicitly in terms of the original observables, making verification of the function easy by simply substituting the original observable expressions.

We begin with explaining a consistent notation that makes it easy to follow the text. The notational elements are summarized first. Consider the following special functions:

$$\gamma_{ij} = \frac{f_i^2}{f_j^2} \tag{6.1.1}$$

$$\lambda_{ij} = \frac{c}{f_i - f_j} \tag{6.1.2}$$

The subscripts are integers used to identify the satellite frequency f. Note that in this particular case, there is no comma between the subscripts i and j on the left side of the equations, and the sequence of the subscripts in γ_{ij} indicates the particular ratio of respective frequencies squared. Similarly, λ_{ij} denotes a wavelength of frequency $f_i - f_j$.

In contrast to the use of double subscripts without comma for the special cases above, it can also indicate a differencing operation as in

$$(\bullet)_{ij} = (\bullet)_{(i)} - (\bullet)_j \tag{6.1.3}$$

An example is $P_{ij} = P_i - P_j$, which can generally be read as the difference of two pseudorange observations. Only for the specific cases of (6.1.1) and (6.1.2) do the double subscripts without comma not indicate differencing.

Occasionally it is advantageous to indicate the specific station and satellite to which an observation refers. This is done with the pk notation, as in $P^p_{k,1}$. Here, the subscript k identifies the station and superscript p identifies the satellite. The numerical value after the comma indicates the frequency. Thus, $P^p_{k,1}$ is the f_1 pseudorange at station k to satellite p. This notation can be generalized by adding another comma and subscript. For example, the notation $I^p_{k,1,P}$ identifies the ionospheric delay of the f_1 pseudorange form receiver k to satellite p. Applying the pk notation to (6.1.3) gives the *across-receiver difference*

$$(\bullet)^p_{km} = (\bullet)^p_k - (\bullet)^p_m \tag{6.1.4}$$

which is the difference of simultaneous observations taken at station k and m to satellite p. Other examples that include the frequency identifier are $P^p_{km,1} = P^p_{k,1} - P^p_{m,1}$

and $P^p_{km,2} = P^p_{k,2} - P^p_{m,2}$. Applying the same differencing notation to superscripts, we obtain the *across-satellite difference*,

$$(\bullet)^{pq}_k = (\bullet)^p_k - (\bullet)^q_k \tag{6.1.5}$$

in which case simultaneous observations taken at station k to satellites p and q are differenced. An example is $P^{pq}_{k,1} = P^p_{k,1} - P^q_{k,1}$. The *across-time difference* is

$$\Delta(\bullet)^p_k = (\bullet)^p_k(t_2) - (\bullet)^p_k(t_1) \tag{6.1.6}$$

The differencing operator Δ indicates differencing over time, and t_1 and t_2 indicate the specific time epochs. Examples are $\Delta P^p_k = P^p_k(t_2) - P^p_k(t_1)$ and $\Delta P^{pq}_{k,1} = P^{pq}_{k,1}(t_2) - P^{pq}_{k,1}(t_1)$. The latter example indicates the across-time and the across-satellite difference of the f_1 pseudorange. The popular *double difference* is

$$(\bullet)^{pq}_{km} = (\bullet)^p_{km} - (\bullet)^q_{km} = (\bullet)^{pq}_k - (\bullet)^{pq}_m \tag{6.1.7}$$

It is an across-receiver and across-satellite difference, or an across-satellite and across-receiver difference. An example is $P^{pq}_{km,1} = P^q_{km,1} - P^q_{km,1} = P^{pq}_{k,1} - P^{pq}_{m,1}$, or the double-differenced ionospheric delay of the f_1 pseudoranges, $I^{pq}_{km,1,P}$. Finally, the *triple difference* is

$$\Delta(\bullet)^{pq}_{km} = (\bullet)^{pq}_{km}(t_2) - (\bullet)^{pq}_{km}(t_1) \tag{6.1.8}$$

which is the across-time double difference.

A complete description of the pseudoranges and carrier phase observables and their functions requires relations that relate the ionospheric delay of the pseudorange and ionospheric carrier phase advance for signals traveling through the ionosphere. The details of the ionosphere and troposphere are presented in Chapter 8. It is sufficient to consider the following relations:

$$I_{j,P} = \gamma_{1j} I_{1,P} = \frac{f_1^2}{f_j^2} I_{1,P} \tag{6.1.9}$$

$$I_{j,\varphi} = \sqrt{\gamma_{1j}}\, I_{1,\varphi} = \frac{f_1}{f_j} I_{1,\varphi} \tag{6.1.10}$$

$$I_{i,\Phi} = \lambda_i I_{i,\varphi} = \frac{c}{f_i} I_{i,\varphi} \tag{6.1.11}$$

$$I_{i,P} = -I_{i,\Phi} \tag{6.1.12}$$

Equation (6.1.9) relates the ionospheric delays for pseudorange of frequencies f_1 and f_j. The factor γ_{1j} is given in (6.1.1). Equations (6.1.10) and (6.1.11) show the respective relations for the ionospheric advance of the carrier phase. The meaning of the subscripts φ and Φ is given below when the carrier phase equation is derived. It suffices to know that φ expresses the carrier phase in radians and Φ represents the scaled carrier phase in units of meters. The factor λ_i denotes the carrier wavelength of frequency f_i. Equation (6.1.12) states that the ionospheric effect on the pseudorange and scaled carrier phase have the same magnitude but have opposite signs.

The basic equations for pseudoranges and carrier phases are given next. We briefly mention how to correct the pseudorange observations for the timing information such as satellite clock errors, timing group delay, and intersignal correction, all of which are available from the broadcast navigation message.

6.1.1 Undifferenced Functions

The basic equations for pseudoranges and carrier phases are derived first. As was mentioned above, the basic observables are developed from the user's point of view. This means that any internal receiver software processing to convert satellite signals registered by the receiver antenna to usable outcomes, i.e., pseudoranges and carrier phases, is not discussed.

6.1.1.1 Pseudoranges Let t denote the system time, such as GPST (GPS time). GPS time is steered by the satellite operators to remain within one microsecond or better of UTC(USNO) time, except for a leap second offset. Temporarily, the nominal receiver time is denoted by $\underline{t}$ and the atomic clock time of a satellite by $\bar{t}$. These are the time values the "hands" of a receiver or space vehicle clock would show. The nominal times equal the true times plus small corrections. At any instant in time, we have $\underline{t}(t) = t + d\underline{t}$ and $\bar{t}(t) = t + d\bar{t}$. The implied sign convention is that the receiver clock error equals the amount by which the receiver clock advances the true time and, similarly, the satellite clock error equals the amount by which the satellite clock time advances relative to the system time. Further, let τ be the transit time of a signal, or the travel time in a vacuum for a specific code to travel from the instant of transmission at the satellite to the instant of reception by the receiver. The signal is recorded at the receiver at the nominal time $\underline{t}(t)$, and transmitted at the nominal satellite time $\bar{t}(t - \tau)$. The pseudorange is then

$$P(t) = c[\underline{t}(t) - \bar{t}(t - \tau)] \tag{6.1.13}$$

where c denotes the velocity of light. Replacing the nominal times by the system time and the respective clock corrections, we obtain

$$\begin{aligned} P(t) &= c[t + d\underline{t} - (t - \tau + d\bar{t}(t - \tau)] = c\tau + c\,d\underline{t} - c\,d\bar{t}(t - \tau) \\ &= \rho(t, t - \tau) + c\,d\underline{t} - c\,d\bar{t} \end{aligned} \tag{6.1.14}$$

In this expression, we have replaced the satellite clock error $d\bar{t}(t - \tau)$ at the instant $t - \tau$ by the satellite clock error $d\bar{t}(t)$ at instant t. This approximation is sufficiently accurate because of the high stability of the satellite clock and considering that τ is about 0.075 sec for GPS-like orbits. The vacuum distance $c\tau$ is denoted by $\rho(t, t - \tau)$ and is henceforth called the topocentric satellite distance.

The derivation of (6.1.14) applies to a vacuum. This equation must be supplemented with additional terms and further specified in order to arrive at a usable pseudorange equation. Since the signal travels through the ionosphere, which acts as a dispersive medium at GPS frequency causing a signal delay, it is necessary to introduce a frequency identifier. We use a subscript to identify the frequency. The troposphere acts as a nondispersive medium in this particular frequency range and also causes a signal delay. Because it is a nondispersive medium, a subscript to identify the frequency is not needed for the tropospheric delay. Other effects to be considered are delays caused by receiver antenna and internal receiver electronic/hardware components, and multipath. Using a numerical subscript to identify the carrier frequency, a more complete expression for the pseudorange observation can be written as

$$\begin{aligned} P_1(t) = {} & \rho(t, t-\tau) + c\,d\underline{t} - c\,d\bar{t} \\ & - c(\Delta t_{SV} - T_{GD} + ISC_{1,P}) + I_{1,P} + T - (d_{1,P} - D_{1,P}) \\ & - (a_{1,P} + A_{1,P}) + M_{1,P} + \varepsilon_{1,P} \end{aligned} \tag{6.1.15}$$

The term Δt_{SV} represents a satellite clock correction as determined by the satellite control center that accounts for the difference of space vehicle time and GPS system time, T_{GD} is the time group delay, and $ISC_{1,P}$ is called the intersignal correction. The latter two corrections, of course, refer to signal delays within the satellite and satellite antenna. All three corrections are available to the user via the navigation message, thus making it possible for the user to correct the observed pseudoranges. These terms are presented in the notation as used in the basic reference IS-GPS-200G (2012). The values Δt_{SV} and T_{GD} are available per satellite, and $ISC_{1,P}$ is given per satellite frequency and code. Additional detail about these terms is found in Section 6.2.2.2. In view of correcting the observations by Δt_{SV} the satellite clock correction $d\bar{t}$ assumes conceptually the role of a residual correction, i.e., anything not taken care of by Δt_{SV} will go into $d\bar{t}$. The ionospheric and tropospheric delays are identified by $I_{1,P}$ and T, respectively. Other relevant terms are the receiver and satellite hardware code delays $d_{1,P}$ and $D_{1,P}$, the receiver and satellite antenna code center offsets $a_{1,P}$ and $A_{1,P}$, the multipath $M_{1,P}$, and the random measurement noise $\varepsilon_{1,P}$. These additional terms and their implications, as well as the degree of cancellation when differencing, are discussed in detail below and in other chapters.

For subsequent discussions, it is convenient to assume that the pseudorange observations have been corrected for the known values Δt_{SV}, T_{GD}, and $ISC_{1,P}$, and therefore, these terms do not have to be explicitly listed any longer on the right side of the equation. Similarly, we assume that the receiver antenna and satellite antenna code phase center offsets are either negligible or known from antenna calibrations and the observed pseudorange can, therefore, be corrected. Also, in order keep the notation to a minimum, we do not introduce new symbols for the corrected pseudorange observations. Thus, we obtain the pseudorange equation in the form commonly given

$$P_1(t) = \rho(t, t-\tau) + c\,d\underline{t} - c\,d\bar{t} + I_{1,P} + T + \delta_{1,P} + \varepsilon_{1,P} \tag{6.1.16}$$

$$\delta_{1,P} = -d_{1,P} + D_{1,P} + M_{1,P} \tag{6.1.17}$$

The term $\delta_{1,P}$ combines the receiver and satellite hardware code delays and the multipath.

The pseudorange equation for the second frequency follows from (6.1.15) or (6.1.16) by changing the subscripts to 2,

$$P_2(t) = \rho(t, t-\tau) + c\,d\underline{t} + c\,d\bar{t} + I_{2,P} + T + \delta_{2,P} + \varepsilon_{2,P} \tag{6.1.18}$$

$$\delta_{2,P} = -d_{2,P} + D_{2,P} + M_{2,P} \tag{6.1.19}$$

Two important differences should be noted. First, whereas the time group delay for the first frequency is T_{GD} in (6.1.15), the $P_2(t)$ pseudorange observation is corrected for $\gamma_{12} T_{GD}$, whereby the ratio γ_{12} is given in (6.1.1). The details for this change are discussed in Section 6.2.2.1. Of course, the intersignal correction $ISC_{2,P}$ is applied instead of $ISC_{1,P}$. The other important difference is the ionospheric term $I_{2,P}$. It is related to the first frequency ionospheric term $I_{1,P}$ in (6.1.9). The tropospheric delay term does not change because the troposphere is a nondispersive medium. Finally, the lumped parameter $\delta_{2,P}$ combines the receiver and satellite hardware code delays and multipath for the second frequency.

Recall that $d\bar{t}$ represents the remaining satellite clock error not accounted for by the Δt_{SV} correction. Similarly, $D_{1,P}$ and $D_{2,P}$ denote the remaining satellite hardware code delays not accounted for by T_{GD}, $ISC_{1,P}$, and $ISC_{2,P}$.

6.1.1.2 Carrier Phases

The carrier phase observation, measured at the receiver at nominal time $\underline{t}$, is the fractional carrier phase which was transmitted τ seconds earlier at the satellite and has traveled the topocentric geometric distance $\rho(t, t-\tau)$. Considering the carrier phase observation of the first frequency, its expression for a vacuum is

$$\varphi_1(t) = \varphi(\underline{t}) - \varphi(\bar{t} - \tau) + N_1 \tag{6.1.20}$$

A major difference compared to the respective pseudorange expression is the presence of the integer ambiguity term N_1. It can be viewed as an initial integer constant that does not change with time unless a cycle slip occurs. If a cycle slip occurs, the observed carrier phase series continues with a different integer ambiguity value. We say that the measurement $\varphi(t)$ is ambiguous with respect to the integer constant. Occasionally it is seen that N_1 is added to the left side of the equation. Since it is an arbitrary constant, it does not matter whether the ambiguity parameter is placed on the left or right side, only the sign changes, which is immaterial.

Equation (6.1.20) is developed further by recognizing that the derivative of a phase with respect to time equals the frequency. Since the satellite clocks are very stable, this derivative can be assumed constant for a short period of time, and we can write

$$\varphi(t - \tau + d\bar{t} + \Delta t_{SV}) = \varphi(t) - f_1 \tau + f_1\, d\bar{t} + f_1\, \Delta t_{SV} \tag{6.1.21}$$

giving the carrier phase equation in the form

$$\begin{aligned} \varphi_1(t) &= \varphi(t + d\underline{t}) - \varphi(t - \tau + d\bar{t} + \Delta t_{SV}) + N_1 \\ &= \varphi(t) + f_1\, d\underline{t} - \varphi(t) + f_1\, \tau - f_1\, d\bar{t} - f_1\, \Delta t_{SV} + N_1 \\ &= \frac{\rho(t, t-\tau)}{\lambda_1} + f_1\, d\underline{t} - f_1\, d\bar{t} - f_1\, \Delta t_{SV} + N_1 \end{aligned} \tag{6.1.22}$$

Note the use of Δt_{SV} in this expression. Applying this known satellite clock correction causes $d\bar{t}$ to again assume the role of a residual satellite clock correction.

The topocentric distance is introduced because $f_1 = c/\lambda_1$, where λ_1 is the wavelength of the first frequency. We now add the hardware delays, the antenna offset terms, and the multipath. Equation (6.1.22) becomes

$$\varphi_1(t) = \frac{\rho(t, t-\tau)}{\lambda_1} + f_1\, d\underline{t} - f_1(d\bar{t} + \Delta t_{SV}) + N_1 + I_{1,\varphi} + \frac{T}{\lambda_1} - (d_{1,\varphi} - D_{1,\varphi}) - (w_{1,\varphi} - W_{1,\varphi}) - (a_{1,\varphi} - A_{1,\varphi}) + M_{1,\varphi} + \varepsilon_{1,\varphi} \tag{6.1.23}$$

The subscript φ is used to indicate that the terms refer to the carrier phase and are expressed units of radians. The ionospheric term is denoted by $I_{1,\varphi}$. Because the troposphere is a nondispersive medium, the tropospheric delay is the same for carrier phases and pseudoranges. The receiver and satellite hardware phase delays are $d_{1,\varphi}$ and $D_{1,\varphi}$. The symbols $w_{k,\varphi}$ and $W_{k,\varphi}$ denote the antenna phase windup angle at the receiver and satellite, respectively. The phase windup angles are a consequence of the circular polarization of the transmissions. Details on the windup angle can be found in Section 6.2.4 and Chapter 9. It is shown there that the phase windup angle tends to cancel in baseline determination using double differences; this technique is discussed further in this chapter. In certain applications the windup angles might be allowed to be absorbed by the hardware delay terms. Finally, $a_{1,\varphi}$ and $A_{1,\varphi}$ denote the phase center offsets at the receiver antenna and satellite antenna, respectively. In the following, we assume again that these quantities are known from antenna calibration and that the observations have been corrected accordingly. The standard form of the carrier phase observation equation becomes

$$\varphi_1(t) = \lambda_1^{-1}\rho(t, t-\tau) + N_1 + f_1\, d\underline{t} - f_1\, d\bar{t} - \lambda_1^{-1} I_{1,P} + \lambda_1^{-1} T + \delta_{1,\varphi} + \varepsilon_{1,\varphi} \tag{6.1.24}$$

$$\delta_{1,\varphi} = -(d_{1,\varphi} + w_{1,\varphi}) + (D_{1,\varphi} + W_{1,\varphi}) + M_{1,\varphi} \tag{6.1.25}$$

Note that the impact of the ionosphere on the carrier phase observation has been parameterized in terms of the pseudorange ionospheric delay $I_{1,p}$. As a matter of standardization, we generally prefer to express the ionospheric delay or advance using $I_{1,P}$. Both ionospheric terms are related by (6.1.11). Both ionospheric terms have opposite signs but the same magnitude when scaled to units of meters. Therefore, the carrier phase advances as the signals travel through the ionosphere. More details on the impact of the ionosphere on GPS signals is found in Chapter 8. Finally, the equation for the carrier phase of the second frequency follows readily by changing subscript 1 to 2.

We notice that both standard forms (6.1.16) and (6.1.24) make use of a lumped parameter that combines the hardware delay and the multipath terms and, in the case of the carrier phase equation, also the windup terms. This specific lumping of terms is for convenience. Later, we will see that the hardware delays cancel in certain observation differences, whereas the multipath will never completely cancel. The windup corrections will also be dealt with in more detail later. The time t serves as common time reference for code and phase measurements. The time error varies with each epoch, whereas the hardware delays typically show little variation over time. As to terminology, the hardware delays $d_{1,\varphi}$ and $D_{1,\varphi}$ are often referred to as uncalibrated phase delays (UPDs), consisting of receiver UPD and satellite UPD. Similarly, one speaks of uncalibrated code delays (UCDs) when referring to $d_{1,\mathrm{P}}$ and $D_{1,\mathrm{P}}$.

Scaled carrier phase function: In many situations it is convenient to refer to the scaled carrier phase equation by multiplying (6.1.24) by the wavelength,

$$\Phi_1(t) \equiv \lambda_1 \varphi = \rho(t, t-\tau) + \lambda_1 N_1 + c\,d\underline{t} - c\,d\bar{t} - I_{1,P} + T + \delta_{1,\Phi} + \varepsilon_{1,\Phi} \tag{6.1.26}$$

$$\delta_{1,\Phi} \equiv \lambda_1^{-1} \delta_{1,\varphi} = -(d_{1,\Phi} + w_{1,\Phi}) + (D_{1,\Phi} + W_{1,\Phi}) + M_{1,\Phi} \tag{6.1.27}$$

The scaled phase $\Phi_1(t)$ is in units of meters. Similarly, the lumped parameter is scaled. The scaling is indicated by the subscript Φ.

Observables with station and satellite subscript and superscript notation: In the above derivations, we used underbar and overbar to indicate the receiver and satellite clock errors, respectively. No station or satellite identifier was used with the other terms. In many situations, it might be desirable or even necessary to identify the specific station and satellite to which an observation refers. We use a subscript letter to indicate the receiver and a superscript letter to identify the satellite as mentioned above when explaining the general notation. Recall that the subscript is separated by a comma from the frequency identification number, and the latter is separated again by a comma from the observation-type identifier (if present). Because this notation is used extensively throughout the book, we summarize the basic observables in this expanded notation:

$$P_{k,1}^{p}(t) = \rho_k^p + c\,dt_k - c\,dt^p + I_{k,1,P}^p + T_k^p + \delta_{k,1,P}^p + \varepsilon_{k,1,P}^p \tag{6.1.28}$$

$$\delta_{k,1,P}^p = -d_{k,1,P} + D_{1,P}^p + M_{k,1,P}^p \tag{6.1.29}$$

$$\varphi_{k,1}^p(t) = \lambda_1^{-1} \rho_k^p + N_{k,1}^p + f_1\,dt_k - f_1\,dt^p - \lambda_1^{-1} I_{k,1,P}^p + \lambda_1^{-1} T_k^p + \delta_{k,1,\varphi}^p + \varepsilon_{k,1,\varphi}^p \tag{6.1.30}$$

$$\delta_{k,1,\varphi}^p = -(d_{k,1,\varphi} + w_{k,1,\varphi}) + \left(D_{1,\varphi}^p + W_{1,\varphi}^p\right) + M_{k,1,\varphi}^p \tag{6.1.31}$$

$$\Phi_{k,1}^p(t) \equiv \lambda_1 \varphi_k^p = \rho_k^p + \lambda_1 N_{k,1}^p + c\,dt_k - c\,dt^p - I_{k,1,P}^p + T_k^p + \delta_{k,1,\Phi}^p + \varepsilon_{k,1,\Phi}^p \tag{6.1.32}$$

$$\delta_{k,1,\Phi}^p \equiv \lambda_1 \delta_{k,1,\varphi}^p = -(d_{k,1,\Phi} + w_{k,1,\Phi}) + \left(D_{1,\Phi}^p + W_{\Phi}^p\right) + M_{k,1,\Phi}^p \tag{6.1.33}$$

Note that the underbar and overbar are no longer needed in this notation. Certainly, this subscript notation may appear as a distraction when the identification of specific stations and satellites is not required. In such a case, we may use the simpler underbar and overbar notation.

The next subsections contain a summary of popular functions of the basic observables. The expressions are grouped according to the terms present, starting with range plus ionosphere, ionospheric free, ionosphere, and multipath. Another group has been added for convenience. It contains several expressions in which the ambiguity terms have been moved to the left side. These so-called ambiguity-corrected functions are convenient when ambiguities have been resolved in prior computations. The last subsection contains new notations for triple-frequency observations. All expressions given below can readily be verified by substituting the basic observables (6.1.16), (6.1.24), or (6.1.26). The expressions are listed as a summary without additional explanations. Whenever an expression is used later, the necessary explanations will then be provided. Each expression is given a name for easy referencing.

6.1.1.3 Range plus Ionosphere

$$\begin{aligned} RI2(\varphi_1, \varphi_2) &\equiv \varphi_1 - \varphi_2 \\ &= \lambda_{12}^{-1}\rho + N_{12} + (f_1 - f_2)\,d\underline{t} - (f_1 - f_2)\,d\bar{t} \\ &\quad - \left(1 - \sqrt{\gamma_{12}}\right) I_{1,\varphi} + \lambda_{12}^{-1} T + \delta_{RI2} + \varepsilon_{RI2} \end{aligned} \tag{6.1.34}$$

$$\begin{aligned} RI3(\varphi_1, \varphi_2) &\equiv \varphi_1 + \varphi_2 \\ &= \frac{f_1 + f_2}{c}\rho + N_1 + N_2 + (f_1 + f_2)(d\underline{t} - d\bar{t}) \\ &\quad - \left(1 - \sqrt{\gamma_{12}}\right) I_{1,\varphi} + \frac{f_1 + f_2}{c} T + \delta_{RI3} + \varepsilon_{RI3} \end{aligned} \tag{6.1.35}$$

$$\begin{aligned} RI4(\Phi_1, \Phi_2) &\equiv \frac{f_1}{f_1 - f_2}\Phi_1 - \frac{f_2}{f_1 - f_2}\Phi_2 \\ &= \rho + \lambda_{12} N_{12} + c\,d\underline{t} - c\,d\bar{t} + \sqrt{\gamma_{12}} I_{1,P} + T + \delta_{RI4} + \varepsilon_{RI4} \end{aligned} \tag{6.1.36}$$

$$\begin{aligned} RI5(P_1, P_2) &\equiv \frac{f_1}{f_1 + f_2} P_1 + \frac{f_2}{f_1 + f_2} P_2 \\ &= \rho + c\,d\underline{t} - c\,d\bar{t} + \sqrt{\gamma_{12}} I_{1,P} + T + \delta_{RI5} + \varepsilon_{RI5} \end{aligned} \tag{6.1.37}$$

6.1.1.4 Ionospheric-Free Functions

$$\begin{aligned} R1(P_1, P_2) \equiv PIF12 &\equiv \frac{f_1^2}{f_1^2 - f_2^2} P_1 - \frac{f_2^2}{f_1^2 - f_2^2} P_2 \\ &= \rho + c\,d\underline{t} - c\,d\bar{t} + T + \delta_{R1} + \varepsilon_{R1} \end{aligned} \tag{6.1.38}$$

$$\begin{aligned} R2(\Phi_1, \Phi_2) \equiv \Phi IF12 &\equiv \frac{f_1^2}{f_1^2 - f_2^2}\Phi_1 - \frac{f_2^2}{f_1^2 - f_2^2}\Phi_2 \\ &= \rho + c\,d\underline{t} - c\,d\bar{t} + c\frac{f_1 - f_2}{f_1^2 - f_2^2} N_1 + c\frac{f_2}{f_1^2 - f_2^2} N_{12} + T + \delta_{R2} + \varepsilon_{R2} \end{aligned} \tag{6.1.39}$$

$$\begin{aligned} R2(\Phi_1, \Phi_2, \text{GPS}) &\equiv \rho + c\,d\underline{t} - c\,d\bar{t} + \frac{2cf_0}{f_1^2 - f_2^2}(17N_1 + 60N_{12}) + T + \delta_{R2} + \varepsilon_{R2} \\ &= \rho + c\,d\underline{t} - c\,d\bar{t} + \lambda_{\Phi IF12} N_{\Phi IF12} + T + \delta_{R2} + \varepsilon_{R2} \end{aligned} \tag{6.1.40}$$

For GPS we have $f_1 = 154f_0, f_2 = 120f_0$, and $f_0 = 10.23$ MHz.

$$R3(\varphi_1, \varphi_2) \equiv \frac{f_1^2}{f_1^2 - f_2^2}\varphi_1 - \frac{f_1 f_2}{f_1^2 - f_2^2}\varphi_2$$

$$= \lambda_1^{-1}\rho + f_1\, d\underline{t} - f_1\, d\bar{t} + \frac{f_1^2 - f_1 f_2}{f_1^2 - f_2^2}N_1 + \frac{f_1 f_2}{f_1^2 - f_2^2}N_{12} + \lambda_1^{-1}T + \delta_{R3} + \varepsilon_{R3} \tag{6.1.41}$$

6.1.1.5 Ionospheric Functions

$$I1(\Phi_1, P_1) \equiv P_1 - \Phi_1 = 2I_{1,P} - \lambda_1 N_1 + \delta_{I1} + \varepsilon_{I1} \tag{6.1.42}$$

$$I2(\varphi_1, \varphi_2, P_1) \equiv \varphi_1 - \varphi_2 - \lambda_{12}^{-1}P_1 = N_{12} - \left(1 - \sqrt{\gamma_{12}}\right)\lambda_{12}^{-1}I_{1,P} + \delta_{I2} + \varepsilon_{I2} \tag{6.1.43}$$

$$I3(\varphi_1, \varphi_2) \equiv \varphi_1 - \sqrt{\gamma_{12}}\varphi_2 = N_1 - \sqrt{\gamma_{12}}N_2 - (1 - \gamma_{12})I_{1,\varphi} + \delta_{I3} + \varepsilon_{I3} \tag{6.1.44}$$

$$I4(\Phi_1, \Phi_2) \equiv \Phi_1 - \Phi_2 = \lambda_1 N_1 - \lambda_2 N_2 - (1 - \gamma_{12})I_{1,P} + \delta_{I4} + \varepsilon_{I4} \tag{6.1.45}$$

$$I5(P_1, P_2) \equiv P_1 - P_2 = (1 - \gamma_{12})I_{1,P} + \delta_{I5} + \varepsilon_{I5} \tag{6.1.46}$$

$$I6(\Phi_1, \Phi_2, \Phi_3) \equiv RI4(\Phi_1, \Phi_3) - RI4(\Phi_1, \Phi_2)$$

$$\equiv \left(\frac{f_1}{f_1 - f_3} - \frac{f_1}{f_1 - f_2}\right)\Phi_1 + \frac{f_2}{f_1 - f_2}\Phi_2 - \frac{f_3}{f_1 - f_3}\Phi_3$$

$$= \lambda_{13}N_{13} - \lambda_{12}N_{12} + \frac{f_1(f_2 - f_3)}{f_2 f_3}I_{1,P} + \delta_{I6} + \varepsilon_{I6} \tag{6.1.47}$$

6.1.1.6 Multipath Functions

$$M1(\Phi_1, \Phi_2, P_1, P_2) \equiv HMW12 \equiv RI4 - RI5$$

$$\equiv \frac{f_1}{f_1 - f_2}\Phi_1 - \frac{f_2}{f_1 - f_2}\Phi_2 - \frac{f_1}{f_1 + f_2}P_1 - \frac{f_2}{f_1 + f_2}P_2$$

$$= \lambda_{12}N_{12} + \delta_{M1} + \varepsilon_{M1} \tag{6.1.48}$$

$$M2(\Phi_1, \Phi_2, P_1, P_2) \equiv AIF12 \equiv R2 - R1$$

$$\equiv \frac{f_1^2}{f_1^2 - f_2^2}\Phi_1 - \frac{f_2^2}{f_1^2 - f_2^2}\Phi_2 - \frac{f_1^2}{f_1^2 - f_2^2}P_1 + \frac{f_2^2}{f_1^2 - f_2^2}P_2$$

$$= c\frac{f_1 - f_2}{f_1^2 - f_2^2}N_1 + c\frac{f_2}{f_1^2 - f_2^2}N_{12} + \delta_{M2} + \varepsilon_{M2} \tag{6.1.49}$$

$$M3(\Phi_1,\Phi_2,P_1) \equiv P_1 + \left(\frac{2}{1-\gamma_{12}} - 1\right)\Phi_1 - \frac{2}{1-\gamma_{12}}\Phi_2$$
$$= -\lambda_1 N_1 + \frac{2}{1-\gamma_{12}}(\lambda_1 N_1 - \lambda_2 N_2) + \delta_{M3} + \varepsilon_{M3} \quad (6.1.50)$$

$$M4(\Phi_1,\Phi_2,P_2) \equiv P_2 - \left(\frac{2\gamma_{12}}{1-\gamma_{12}} + 1\right)\Phi_2 + \frac{2\gamma_{12}}{1-\gamma_{12}}\Phi_1$$
$$= -\lambda_2 N_2 + \frac{2\gamma_{12}}{1-\gamma_{12}}(\lambda_1 N_1 - \lambda_2 N_2) + \delta_{M4} + \varepsilon_{M4} \quad (6.1.51)$$

$$M5(\Phi_1,\Phi_2,\Phi_3) \equiv \left(\lambda_3^2 - \lambda_2^2\right)\Phi_1 + \left(\lambda_1^2 - \lambda_3^2\right)\Phi_2 + \left(\lambda_2^2 - \lambda_1^2\right)\Phi_3$$
$$= \left(\lambda_3^2 - \lambda_2^2\right)\lambda_1 N_1 + \left(\lambda_1^2 - \lambda_3^2\right)\lambda_2 N_2 + \left(\lambda_2^2 - \lambda_1^2\right)\lambda_3 N_3$$
$$+ \delta_{M5} + \varepsilon_{M5} \quad (6.1.52)$$

$$M6(\Phi_1,\Phi_2,\Phi_3) \equiv R2(\Phi_1,\Phi_2) - R2(\Phi_1,\Phi_3)$$
$$\equiv \left(\frac{f_1^2}{f_1^2 - f_2^2} - \frac{f_1^2}{f_1^2 - f_3^2}\right)\Phi_1 - \frac{f_2^2}{f_1^2 - f_2^2}\Phi_2 + \frac{f_3^2}{f_1^2 - f_3^2}\Phi_3$$
$$= c\left(\frac{f_1 - f_2}{f_1^2 - f_2^2} - \frac{f_1 - f_3}{f_1^2 - f_3^2}\right)N_1 + c\frac{f_2}{f_1^2 - f_2^2}N_{12}$$
$$- c\frac{f_3}{f_1^2 - f_3^2}N_{13} + \delta_{M6} + \varepsilon_{M6} \quad (6.1.53)$$

$$M7(P_1,P_2,P_3) \equiv \left(\lambda_3^2 - \lambda_2^2\right)P_1 + \left(\lambda_1^2 - \lambda_3^2\right)P_2 + \left(\lambda_2^2 - \lambda_1^2\right)P_3 = \delta_{M7} + \varepsilon_{M7} \quad (6.1.54)$$

6.1.1.7 Ambiguity-Corrected Functions

$$AC1(\varphi_1,\varphi_2) \equiv (\varphi_{12} - N_{12})\lambda_{12} = \rho + c\,d\underline{t} - c\,d\bar{t} + \sqrt{\lambda_{12}}I_{1,P}$$
$$+ T + \delta_{AC1} + \varepsilon_{AC1} \quad (6.1.55)$$

$$AC2(\varphi_1,\varphi_2) \equiv \left(1 - \frac{\lambda_{12}}{\lambda_1}\right)\varphi_1 + \frac{\lambda_{12}}{\lambda_1}\varphi_2 + N_{12}\frac{\lambda_{12}}{\lambda_1} = N_1 + \left(\sqrt{\gamma_{12}} - 1\right)\frac{I_{1,P}}{\lambda_1}$$
$$+ \delta_{AC2} + \varepsilon_{AC2} \quad (6.1.56)$$

$$AC3(\varphi_1,\varphi_2,\varphi_3) \equiv AC1(\varphi_1,\varphi_2) - AC1(\varphi_1,\varphi_3)$$
$$\equiv (\lambda_{12} - \lambda_{13})\varphi_1 - \lambda_{12}\varphi_2 + \lambda_{13}\varphi_3 - N_{12}\lambda_{12} + N_{13}\lambda_{13}$$
$$= \left(\sqrt{\gamma_{12}} - \sqrt{\gamma_{13}}\right)I_{1,P} + \delta_{AC3} + \varepsilon_{AC3} \quad (6.1.57)$$

$$AC4(\varphi_1, \varphi_2, \varphi_3) \equiv \frac{f_1 AC1(\varphi_1, \varphi_2) - f_3 AC1(\varphi_2, \varphi_3)}{f_1 - f_3}$$

$$\equiv \lambda_{13} \left[\frac{\lambda_{12}}{\lambda_1}\varphi_1 - \left(\frac{\lambda_{12}}{\lambda_1} + \frac{\lambda_{23}}{\lambda_3} \right) \varphi_2 + \frac{\lambda_{23}}{\lambda_3}\varphi_3 - \frac{\lambda_{12}}{\lambda_1}N_{12} + \frac{\lambda_{23}}{\lambda_3}N_{23} \right]$$

$$= \rho + c\,d\underline{t} - c\,d\bar{t} + T + \delta_{AC4} + \varepsilon_{AC4} \tag{6.1.58}$$

The functions summarized above are expressed explicitly in terms of original pseudorange and carrier phase observations. This makes it convenient to determine the lumped parameter for a specific function or even apply variance propagation. For a function $(\bullet)$, the lumped term $\delta_{(\bullet)}$ and the function measurement noise $\varepsilon_{(\bullet)}$ are obtained by applying the respective functions. For example, if $(\bullet) = a\Phi_1 + b\Phi_2$, then $\delta_\bullet = a\delta_{1,\Phi} + b\delta_{2,\Phi}$ and $\varepsilon_\bullet = a\varepsilon_{1,\Phi} + b\varepsilon_{2,\Phi}$. Similarly, one obtains the hardware delays $d_\bullet = ad_{1,\Phi} + bd_{2,\Phi}$, $D_\bullet = aD_{1,\Phi} + bD_{2,\Phi}$, and the multipath $M_\bullet = aM_{1,\Phi} + bM_{2,\Phi}$. If one assumes for the sake of approximate estimation that the standard deviations for all carrier phases and pseudoranges across the frequencies are, respectively, the same, and that they are uncorrelated, the law of variance propagation (A.5.61) provides the standard deviation of the function as $\sigma_\bullet = \sqrt{a^2 + b^2}\sigma_\Phi$ and $\sigma_\bullet = \sqrt{a^2 + b^2}\sigma_P$ if $(\bullet) = aP_1 + bP_2$. Similar expressions are obtained for mixed functions of pseudoranges and carrier phases or when the function contains more than two observables.

The various dual-frequency functions listed above trace their origins back to the beginning of GPS. They were introduced during the time of rapid development of GPS in the late 1970s and early 1980, with apparently no authorship attributed to them in the literature. An exception is function (6.1.48), whose origin is generally acknowledged to go back to Hatch (1982), Melbourne (1985), and Wübbena (1985). We will refer to this function simply as the *HMW* function. This function combines pseudoranges and carrier phases of two frequencies. If it is necessary to clarify to which specific frequencies an application refers, we identify the frequencies by numbers. For example, *HMW*12 would imply the first and second frequency observations as used in (6.1.48). We are not aware of a triple-frequency function that is attributed to a specific author. However, a nice summary of triple-frequency functions is given in Simsky (2006), which we recommend for additional reading.

6.1.1.8 Triple-Frequency Subscript Notation

When dealing with triple-frequency observations, it is often convenient to use a more general notation. Let i, j, and k be constants; then the triple-frequency carrier phase and pseudorange functions can be written as

$$\varphi_{(i,j,k)} \equiv i\varphi_1 + j\varphi_2 + k\varphi_3 \tag{6.1.59}$$

$$\Phi_{(i,j,k)} \equiv \frac{if_1\Phi_1 + jf_2\Phi_2 + kf_3\Phi_3}{if_1 + jf_2 + kf_3} \equiv \frac{c}{if_1 + jf_2 + kf_3}\varphi_{(i,jk)} \tag{6.1.60}$$

$$P_{(i,j,k)} \equiv \frac{i f_1 P_1 + j f_2 P_2 + k f_3 P_3}{i f_1 + j f_2 + k f_3} \tag{6.1.61}$$

where the numerical subscripts identify the frequencies. In addition, we identify the following primary

$$\left.\begin{aligned} f_{(i,j,k)} &= i f_1 + j f_2 + k f_3 \\ \lambda_{(i,j,k)} &= \frac{c}{f_{(i,j,k)}} \\ N_{(i,j,k)} &= i N_1 + j N_2 + k N_3 \end{aligned}\right\} \tag{6.1.62}$$

and secondary

$$\left.\begin{aligned} \beta_{(i,j,k)} &= \frac{f_1^2 (i/f_1 + j/f_2 + k/f_3)}{f_{(i,j,k)}} \\ \mu^2_{(i,j,k)} &= \frac{(i f_1)^2 + (j f_2)^2 + (k f_3)^2}{f^2_{(i,j,k)}} \\ \nu_{(i,j,k)} &= \frac{|i| f_1 + |j| f_2 + |k| f_3}{f_{(i,j,k)}} \end{aligned}\right\} \tag{6.1.63}$$

functions. An even more general linear function can be formed that would include the observation of four frequencies or even combine (6.1.60) and (6.1.61) into one general linear function of pseudoranges and carrier phases. Here we prefer the notation that includes only three frequencies and keeps the carrier phase and pseudorange functions separate. In this notation, the expressions for the carrier phase and pseudorange functions become

$$\begin{aligned} \varphi_{(i,j,k)} = {} & \frac{\rho}{\lambda_{(i,j,k)}} + N_{(i,j,k)} + f_{(i,j,k)} \, d\underline{t} - f_{(i,j,k)} \, d\bar{t} - \frac{\beta_{(i,j,k)} I_{1,p}}{\lambda_{(i,j,k)}} + \frac{T}{\lambda_{(i,j,k)}} \\ & + \delta_{(i,j,k),\varphi} + \varepsilon_{(i,j,k),\varphi} \end{aligned} \tag{6.1.64}$$

$$\begin{aligned} \Phi_{(i,j,k)} &= a_1 \Phi_1 + a_2 \Phi_2 + a_3 \Phi_3 \\ &= \rho + \lambda_{(i,j,k)} N_{(i,j,k)} + c \, d\underline{t} - c \, d\bar{t} - \beta_{(i,j,k)} I_{1,P} + T + \delta_{(i,j,k),\Phi} + \varepsilon_{(i,j,k),\Phi} \end{aligned} \tag{6.1.65}$$

$$\begin{aligned} P_{(i,j,k)} &= b_1 P_1 + b_2 P_2 + b_3 P_3 \\ &= \rho + c \, d\underline{t} - c \, d\bar{t} + \beta_{(i,j,k)} I_{1,P} + T + \delta_{(i,j,k),P} + \varepsilon_{(i,j,k),P} \end{aligned} \tag{6.1.66}$$

The constants a_m and b_m, with $m = 1, \cdots, 3$, can be those of (6.1.60) and (6.1.61). If the factors i, j, and k are integers, then the linear phase combinations preserve the integer nature of the ambiguity as seen from (6.1.62). If $a_1 + a_2 + a_3 = 1$ and $b_1 + b_2 + b_3 = 1$, then the geometric terms and clock error terms are not scaled, i.e., remain unchanged compared to the equations for the original observations. This is the case in (6.1.65) and (6.1.66).

In regard to the secondary functions (6.1.63), $\beta_{(i,j,k)}$ is called the ionospheric scale factor. For the special case of $\beta_{(i,j,k)} = 0$, the ionospheric-free functions are obtained as seen from (6.1.65) and (6.1.66). For the special case that $\sigma_{\Phi_m} = \sigma_\Phi$ and $\sigma_{P_m} = \sigma_P$, with $m = 1, \cdots, 3$ and uncorrelated observations, $\mu^2_{(i,j,k)}$ is called the variance factor because

$$\begin{aligned} \sigma^2_{\Phi_{(i,j,k)}} &= \mu^2_{(i,j,k)} \sigma^2_\Phi \\ \sigma^2_{P_{(i,j,k)}} &= \mu^2_{(i,j,k)} \sigma^2_P \end{aligned} \tag{6.1.67}$$

Assuming that each scaled carrier phase observation has approximately the same multipath $M_{\Phi_m} = M_\Phi$ with $m = 1, \cdots, 3$, and making a similar assumption for the pseudorange multipath, then $\nu_{(i,j,k)}$ is the multipath factor such that

$$\begin{aligned} M_{\Phi_{\max}} &\le \nu_{(i,j,k)} M_\Phi \\ M_{P_{\max}} &\le \nu_{(i,j,k)} M_P \end{aligned} \tag{6.1.68}$$

represents the upper limit for the multipath of the respective function. This superposition of multipath assumes not only that the multipath for each of the three observations is the same but also that the absolute values of the components are added.

As an example of the application of the triple-frequency notation, consider the HMW function (6.1.48), which can be written as $HMW12 = \Phi_{(1,-1,0)} - P_{(1,1,0)}$ or $HMW13 = \Phi_{(1,0,-1)} - P_{(1,0,1)}$.

6.1.2 Single Differences

Let two receivers observe the same satellites at the same nominal times. One can then compute three types of differences among the observations. One difference is the across-receiver difference, obtained when the observations of two stations and the same satellites are differenced. Another difference, called the across-satellite difference, results from differencing observations from the same station and different satellites. The third difference, called across-time difference, is the difference of observations from the same station and the same satellite at different epochs. Even if the observations to a particular satellite are taken at the receivers at exactly the same time, thus being truly simultaneous observations, the respective signals have left the satellite at slightly different times because the respective topocentric satellite distances differ.

6.1.2.1 Across-Receiver Functions The notation (6.1.4) is used to identify the stations and the satellite to form the across-receiver difference. Note that the double subscripts not separated by a comma indicate differencing across the stations. Applying (6.1.16), (6.1.24), and (6.1.26), the across-receiver functions are

$$P^p_{km,1} = \rho^p_{km} + c\,dt_{km} + I^p_{km,1,P} + T^p_{km} - d_{km,1,P} + M^p_{km,1,P} + \varepsilon^p_{km,1,P} \tag{6.1.69}$$

$$\varphi^p_{km,1} = \frac{f_1}{c}\rho^p_{km} + N^p_{km,1} + f_1\,dt_{km} - \lambda_1^{-1} I^p_{km,1,P} + \frac{f_1}{c} T^p_{km} - d_{km,1,\varphi} + M^p_{km,1,\varphi} + \varepsilon^p_{km,1,\varphi} \tag{6.1.70}$$

$$\Phi^{p}_{km,1} \equiv \lambda_1 \varphi^{p}_{km,1} = \rho^{p}_{km} + \lambda_1 N^{p}_{km,1} + c\,dt_{km} - I^{p}_{km,1,P} + T^{p}_{km} - d_{km,1,\Phi} + M^{p}_{km,1,\Phi} + \varepsilon_{km,1,\Phi} \tag{6.1.71}$$

$$\sigma_{(\bullet^{p}_{km})} = \sqrt{2}\sigma_{(\bullet)} \tag{6.1.72}$$

An important feature of this difference is that the satellite clock error dt^p and the satellite hardware delay $D^{p}_{1,\varphi}$ cancel. This cancellation occurs because the satellite clock is very stable, making the satellite clock errors the same for these near-simultaneous transmissions. Similarly, the satellite hardware delays can readily be viewed as constant over such a short time period. The windup angles listed in (6.1.25) and (6.1.27) have been omitted in these expressions and will also be omitted in subsequent expressions for reasons of simplicity. The variance propagation expressed by (6.1.72) assumes that the variances of the respective observations are the same at both stations. This variance propagation expression is not always given explicitly below because it can readily be obtained by applying the respective function of the original observations as needed.

Equally important is noting the terms that do not cancel in the differencing. These are the across-receiver differences of station clock errors, the ionospheric and tropospheric delays, the receiver hardware delays, and multipath. It is noted that in relative positioning over short distances, the tropospheric and ionospheric effects are, respectively, almost the same at each station due to high spatial correlation and, therefore, largely cancel in the differencing. This cancelation is important as it makes relative positioning over short distances very efficient and practicable in surveying.

6.1.2.2 *Across-Satellite Functions* Using the notation in (6.1.5), with superscripts indicating the differencing, the across-satellite differences become

$$P^{pq}_{k,1} = \frac{f_1}{c}\rho^{pq}_{k} + c\,dt^{pq} + I^{pq}_{k,1,P} + T^{pq}_{k} + D^{pq}_{1,P} + M^{pq}_{k,1,P} + \varepsilon^{pq}_{k,1,P} \tag{6.1.73}$$

$$\varphi^{pq}_{k,1} = \frac{f_1}{c}\rho^{pq}_{k} + N^{pq}_{k} + f_1\,dt^{pq} + I^{pq}_{k,1,\varphi} + \frac{f_1}{c}T^{pq}_{k} + D^{pq}_{1,\varphi} + M^{pq}_{k,1,\varphi} + \varepsilon^{pq}_{k,1,\varphi} \tag{6.1.74}$$

$$\Phi^{pq}_{k,1} = \rho^{pq}_{k} + \lambda_1 N^{pq}_{k} + c\,dt^{pq} - I^{pq}_{k,1,P} + T^{pq}_{k} + {}^{pq}_{k,1,\Phi} + \varepsilon^{pq}_{k,1,\Phi} \tag{6.1.75}$$

It is readily seen that the single difference receiver clock error dt_k and receiver hardware delay d_k cancel.

6.1.2.3 *Across-Time Functions* Differencing across time is indicated by the Δ symbol, following the notation in (6.1.6). The relevant functions are

$$\Delta P^{p}_{k,1} = \Delta\rho^{p}_{k} + c\,\Delta\,dt_k - c\,\Delta\,dt^{p} + \Delta I^{p}_{k,1,P} + \Delta T^{p}_{k} + \Delta M^{p}_{k,1,P} + \Delta\varepsilon^{p}_{k,1,P} \tag{6.1.76}$$

$$\Delta\varphi^{p}_{k,1} = \frac{f_1}{c}\Delta\rho^{p}_{k} + f_1\Delta\,dt_k - f_1\Delta\,dt^{p} - \frac{f_1}{c}\Delta I^{p}_{k,1,P} + \frac{f_1}{c}\Delta T^{p}_{k} + \Delta M^{p}_{k,1,\varphi} + \Delta\varepsilon^{p}_{k,1,\varphi} \tag{6.1.77}$$

$$\Delta\Phi^{p}_{k,1} = \Delta\rho^{p}_{k} + c\Delta\,dt_k - c\Delta\,dt^{p} - \Delta I^{p}_{k,1,P} + \Delta T^{p}_{k} + \Delta M^{p}_{k,1,\Phi} + \Delta\varepsilon^{p}_{k,1,\Phi} \tag{6.1.78}$$

$$\Delta I1_{k,1}^{p} = 2\Delta I_{k,1,P}^{p} + \Delta M_{k,1,P}^{p} + \Delta M_{k,1,\Phi}^{p} + \Delta \varepsilon_{k,1,I1}^{p} \tag{6.1.79}$$

$$\Delta \varphi_3 = \frac{\lambda_1}{\lambda_3}\left(\frac{\lambda_3^2 - \lambda_2^2}{\lambda_1^2 - \lambda_2^2}\right)\Delta \varphi_1 - \frac{\lambda_2}{\lambda_3}\left(\frac{\lambda_3^2 - \lambda_1^2}{\lambda_1^2 - \lambda_2^2}\right)\Delta \varphi_2 + \Delta M_{3,\varphi} + \Delta \varepsilon_{3,\varphi} \tag{6.1.80}$$

For differencing over long time intervals, the clock errors might not cancel. In the expressions above, however, we have assumed that the hardware delays are sufficiently stable to cancel. We have further assumed that no cycle slip occurred between the two epochs, therefore the ambiguity cancels. These functions primarily reflect the change of the topocentric satellite distance over time and are often referred to as delta ranges. Equation (6.1.79) represents the time difference of the ionospheric function (6.1.42). One can readily imagine the benefits to be derived from possibly being able to model the ionospheric change $\Delta I1_{k,1}^{p}$ when the time difference is small. An example might be cycle slip fixing. Likewise, some combinations of single-difference functions have certain benefits. For example, applying across-receiver and across-time differencing yields a model that is free of satellite clock errors and depends on changes of receiver clocks, ionosphere, and troposphere.

Function (6.1.80) is a triple-frequency function, expressing the phase difference of the third frequency as a function of the phase differences of the first and second frequencies. In order to simplify the notation, we omitted the subscript and superscripts to identify the station and satellite, respectively.

6.1.3 Double Differences

A double difference can be formed when two receivers observe two satellites simultaneously, or at least near simultaneously. One can either difference two across-receiver differences or two across-satellite differences. In the notation of (6.1.7), the double differences of the basic observables are

$$P_{km,1}^{pq} = \rho_{km}^{pq} + \frac{f_1}{c} I_{km,1,P}^{pq} + T_{km}^{pq} + M_{km,1,P}^{pq} + \varepsilon_{km,1,P}^{pq} \tag{6.1.81}$$

$$\varphi_{km,1}^{pq} = \frac{f_1}{c}\rho_{km}^{pq} + N_{km,1}^{pq} - \frac{f_1}{c} I_{km,1,P}^{pq} + \frac{f_1}{c} T_{km}^{pq} + M_{km,1,\varphi}^{pq} + \varepsilon_{km,1,\varphi}^{pq} \tag{6.1.82}$$

$$\Phi_{km,1}^{pq} = \rho_{km}^{pq} + \lambda_1 N_{km,1}^{pq} - I_{km,1,P}^{pq} + T_{km}^{pq} + M_{km,1,\Phi}^{pq} + \varepsilon_{km,1,\Phi}^{pq} \tag{6.1.83}$$

$$\sigma_{(\bullet_{km}^{pq})} = 2\sigma_{(\bullet)} \tag{6.1.84}$$

The most important feature of the double-difference observation is the cancellation of receiver clock errors, satellite clock errors, receiver hardware delays, and satellite hardware delays. This almost "perfect" cancellation of unwanted errors and delays has made the double-difference observation so popular among users. In addition, since double differencing implies across-receiver differencing, the ionospheric and tropospheric effects on the observations largely cancel in relative positioning over short distances. Unfortunately, since the multipath is a function of the geometry between receiver, satellite, and reflector surface, it does not cancel.

The double-difference integer ambiguity $N^{pq}_{km,1}$ plays an important role in accurate relative positioning. Estimating the ambiguity together with other parameters as a real number is called the float solution. If thus estimated double-differenced ambiguities can be successfully constrained to integers, one obtains the fixed solution. Because of residual model errors such as residual ionosphere and troposphere, the estimated ambiguities will, at best, be close to integers. Successfully imposing integer constraints adds strength to the solution because the number of parameters is reduced and the correlations between parameters reduce as well. The art of accurate relative positioning is inextricably related to successfully fixing the ambiguities. Much effort has gone into extending the baseline length over which ambiguities can be fixed. At the same time, much research has been carried out to develop algorithms that allow the ambiguities to be fixed for short observation spans and short baselines. Being able to relatively easily impose integer constraints on the estimated ambiguities is a major strength of the double differencing approach.

All of the above functions can readily be applied to observations of the second frequency by replacing the subscript 1 by 2. For example, in the across-frequency double difference functions

$$\Phi^{pq}_{km,12} \equiv \Phi^{pq}_{km,1} - \Phi^{pq}_{km,2} = \lambda_1 N^{pq}_{km,1} - \lambda_2 N^{pq}_{km,2} - (1-\gamma_{12}) I^{pq}_{km,1,P} + M^{pq}_{km,12,\Phi} + \varepsilon^{pq}_{km,12,\Phi} \tag{6.1.85}$$

$$\Phi^{pq}_{km,13} \equiv \Phi^{pq}_{km,1} - \Phi^{pq}_{km,3} = \lambda_1 N^{pq}_{km,1} - \lambda_3 N^{pq}_{km,3} - (1-\gamma_{13}) I^{pq}_{km,1,P} + M^{pq}_{km,13,\Phi} + \varepsilon^{pq}_{km,13,\Phi} \tag{6.1.86}$$

the topocentric satellite distances and tropospheric delays cancel. The triple-frequency observations provide three more double-differenced functions. Applying the double-difference operation to (6.1.52), (6.1.57), and (6.1.58) gives

$$\begin{aligned} M5^{pq}_{km} &\equiv (\lambda_3^2-\lambda_2^2)\Phi^{pq}_{km,1} + (\lambda_1^2-\lambda_3^2)\Phi^{pq}_{km,2} + (\lambda_2^2-\lambda_1^2)\Phi^{pq}_{km,3} \\ &= (\lambda_3^2-\lambda_2^2)\lambda_1 N^{pq}_{km,1} + (\lambda_1^2-\lambda_3^2)\lambda_2 N^{pq}_{km,2} + (\lambda_2^2-\lambda_1^2)\lambda_3 N^{pq}_{km,3} \\ &\quad + M^{pq}_{km,M5} + \varepsilon^{pq}_{km,M5} \end{aligned} \tag{6.1.87}$$

$$\begin{aligned} AC3^{pq}_{km} &\equiv (\lambda_{12}-\lambda_{13})\varphi^{pq}_{km,1} - \lambda_{12}\varphi^{pq}_{km,2} + \lambda_{13}\varphi^{pq}_{km,3} - N^{pq}_{km,12}\lambda_{12} + N^{pq}_{km,13}\lambda_{13} \\ &= \left(\sqrt{\gamma_{12}} - \sqrt{\gamma_{13}}\right) I^{pq}_{km,1,P} + M^{pq}_{km,AC3} + \varepsilon^{pq}_{km,AC3} \end{aligned} \tag{6.1.88}$$

$$\begin{aligned} AC4^{pq}_{km} &\equiv \lambda_{13}\left[\frac{\lambda_{12}}{\lambda_1}\varphi^{pq}_{km,1} - \left(\frac{\lambda_{12}}{\lambda_1} + \frac{\lambda_{23}}{\lambda_3}\right)\varphi^{pq}_{km,2} + \frac{\lambda_{23}}{\lambda_3}\varphi^{pq}_{km,3} \right. \\ &\quad \left. - \frac{\lambda_{12}}{\lambda_1} N^{pq}_{km,12} + \frac{\lambda_{23}}{\lambda_3} N^{pq}_{km,23}\right] \\ &= \rho^{pq}_{km,} + M^{pq}_{km,AC4} + \varepsilon^{pq}_{km,AC4} \end{aligned} \tag{6.1.89}$$

Apart from multipath, these functions depend on the ambiguities only, the ionosphere and ambiguities, or the topocentric satellite distance and ambiguities.

6.1.4 Triple Differences

The triple difference (6.1.8) is the difference of two double differences over time:

$$\Delta P_{km,1}^{pq} \equiv \Delta\rho_{km}^{pq} + \Delta I_{km,1,P}^{pq} + \Delta T_{km}^{pq} + \Delta M_{km,1,P}^{pq} + \Delta\varepsilon_{km,1,P}^{pq} \tag{6.1.90}$$

$$\Delta\varphi_{km,1}^{pq} = \frac{f_1}{c}\Delta\rho_{km}^{pq} + \Delta I_{km,1,\varphi}^{pq} + \frac{f}{c}\Delta T_{km}^{pq} + \Delta M_{km,1,\varphi}^{pq} + \Delta\varepsilon_{km,1,\varphi}^{pq} \tag{6.1.91}$$

$$\Delta\Phi_{km,1}^{pq} = \Delta\rho_{km}^{pq} - \Delta I_{km,1,P}^{pq} + \Delta T_{km}^{pq} + \Delta M_{km,1,\Phi}^{pq} + \Delta\varepsilon_{km,1,\Phi}^{pq} \tag{6.1.92}$$

$$\sigma_{(\Delta\bullet_{km}^{pq})} = \sqrt{8}\sigma_{(\bullet)} \tag{6.1.93}$$

The initial integer ambiguity cancels in triple differencing. Because of this cancelation property, the triple-difference observable is probably the easiest observable to process. Often, the triple-difference solution serves as a preprocessor to get good initial positions for a subsequent double-difference solution. The triple differences have another advantage in that cycle slips are mapped as individual outliers in the residuals. Individual outliers can usually be detected and removed.

6.2 OPERATIONAL DETAILS

Whether one develops and improves positioning algorithms, uses GNSS to support research activities, or runs commercially available receivers in engineering applications, there are a number of operational details that one should know. It also helps to know that a lot of what might be called the GNSS infrastructure is in place and ready to be tapped. We begin with some topics of interest to developers of processing techniques and then talk about services that are mostly free of charge and available to the common user.

We briefly address the issue of computing the topocentric satellite distance the GNSS signal travels from the time of emission at the satellite to reception at the user receiver antenna. Detailed information is given on the timing group delay, satellite clock correction, and intersignal corrections, all three of which are transmitted by the navigation message. We then briefly discuss cycle slips in the carrier phase observable, the phase windup correction resulting from the right-circular polarized nature of the signals, and the "ever-present" multipath. Our discussion on service begins with relative and absolute antenna calibration provided specifically by the National Geodetic Service, and continues with a discussion of the International GNSS Service, its products, and online computing series.

6.2.1 Computing the Topocentric Range

The pseudorange equation (6.1.28) and the carrier phase equation (6.1.30) require computation of the topocentric distance ρ_k^p. There are two equivalent solutions available. In Section 7.3, the change of the topocentric distance during the travel time of the signal is computed explicitly. Here we present an iterative solution. In the inertial coordinate system (X), the topocentric distance is expressed by

$$\rho_k^p = \|\boldsymbol{X}_k(t_k) - \boldsymbol{X}^p(t^p)\| \tag{6.2.1}$$

In this coordinate system, the receiver coordinates are a function of time due to the earth's rotation. If the receiver antenna and satellite ephemeris are given in the terrestrial coordinate system, we must take the earth's rotation into account when computing the topocentric satellite distance. If τ denotes the travel time for the signal, then the earth rotates during that time by

$$\theta = \dot{\Omega}_e(t_k - t^p) = \dot{\Omega}_e\tau \tag{6.2.2}$$

where $\dot{\Omega}_e$, is the earth rotation rate. Neglecting polar motion, the topocentric distance becomes

$$\rho_k^p = \|\boldsymbol{x}_k - \boldsymbol{R}_3(\theta)\boldsymbol{x}^p(t^p)\| \tag{6.2.3}$$

where $\boldsymbol{R}_3$ is the orthonormal rotation matrix. Since θ is a function of τ, equation (6.2.3) must be iterated. An initial estimate of the travel time is $\tau_1 = 0.075$ sec. Then compute θ_1 from (6.2.2) and use this value in (6.2.3) to obtain the initial value ρ_1 for the distance. For the second iteration, use $\tau_2 = \rho_1/c$ in (6.2.2) to get θ_2. Continue the iteration until convergence is achieved. Typically, a couple of iterations are sufficient.

6.2.2 Satellite Timing Considerations

There are three timing elements that are or will be transmitted with the GPS broadcast navigation message. They are the satellite clock correction Δt_{SV}, the timing group delay T_{GD}, and the intersignal correction (ISC). The ISC was introduced as part of the modernization of GPS signals and is related to the legacy timing group delay. Therefore, all three timing elements are discussed in detail. Recommended references are Hegarty et al. (2005), Tetewsky et al. (2009), and Feess et al. (2013).

The control segment maintains GPS time (GPST) to within 1 μsec of UTC (USNO) according to the Interface Control Document (IS-GPS-200G, 2012), excluding the occasional UTC leap-second jumps. The current full second offset is readily available from various data services, if needed, allowing the user to convert between GPST and UTC (USNO). Since the satellite transmissions are steered by the nominal time of the individual satellite (satellite time), one needs to know the differences between GPS time and the individual satellite time. In the notation and sign convention as used by the interface control document, the time correction to the nominal space vehicle time t_{SV} is

$$\Delta t_{\mathrm{SV}} = a_{f0} + a_{f1}(t_{\mathrm{SV}} - t_{\mathrm{oc}}) + a_{f2}(t_{\mathrm{SV}} - t_{\mathrm{oc}})^2 + \Delta t_R \tag{6.2.4}$$

with

$$t_{GPS} = t_{SV} - \Delta t_{SV} \tag{6.2.5}$$

and

$$\Delta t_R = -\frac{2}{c^2}\sqrt{a\mu}\, e \sin E = -\frac{2}{c^2}\boldsymbol{X} \cdot \dot{\boldsymbol{X}} \tag{6.2.6}$$

$$\Delta t_{R[\mu\,\text{sec}]} \approx -2e \sin E \tag{6.2.7}$$

The polynomial coefficients are transmitted in units of sec, sec/sec, and sec /sec^2; the clock data reference time t_{oc} is also broadcast in seconds in subframe 1 of the navigation message. The value of t_{SV} must account for the beginning or end-of week crossovers. That is, if $(t_{SV} - t_{oc})$ is greater than 302,400, subtract 604,800 from t_{SV}. If $(t_{SV} - t_{oc})$ is less than $-302,400$, add 604,800 to t_{SV}. The symbol Δt_R is a small relativistic clock correction caused by the orbital eccentricity e. The symbol μ denotes the gravitational constant, a is the semimajor axis of the orbit, and E is the eccentric anomaly. See Chapter 5, equation (5.1.54), for details on these elements. The approximation (6.2.7) follows by taking $a \approx 26,600\,\text{km}$.

A major topic of this subsection is the intersignal corrections (ISC). Such corrections will be available for the modernized GPS signals L1CA, L1P(Y), L1M, L2C, L2P(Y), L2M, L5I, and L5Q, and transmitted with the new navigation message to allow users to correct the observation. In order to provide full flexibility when dealing with the modernized signal, we will adhere in this subsection to the notation used in Tetewsky et al. (2009). This means that the third civil frequency will be referred to as L5. It also means that the pseudorange on L1 is denoted by $P_{1,PY}$. A numerical subscript is again used before the comma to identify the frequency, and the code identification is given after the comma. Omitting the windup terms, the pseudoranges (6.1.16) and (6.1.18) can be written in this new notation as

$$P'_{1,PY} = \rho + c\,d\underline{t} - c\,d\bar{t} - c(\Delta t_{SV} - T_{1,PY}) + I_{1,PY} + T - d_{1,PY} + D_{1,PY} + M_{1,PY} \tag{6.2.8}$$

$$P'_{2,PY} = \rho + c\,d\underline{t} - c\,d\bar{t} - c(\Delta t_{SV} - T_{2,PY}) + \gamma_{12} I_{1,PY} + T - d_{2,PY} + D_{2,PY} + M_{2,PY} \tag{6.2.9}$$

We only need to focus on some of these terms in the context of this section. The satellite clock correction is denoted by Δt_{SV}; $d\bar{t}$ is viewed again as the residual satellite clock error and can be omitted. The satellite hardware code delay of the L1 P-code is now denoted by $T_{1,PY}$. This delay is the difference in time from the instant of signal generation by the satellite clock to the signal departure at the satellite antenna. The delay, therefore, includes the time it needs to pass through the various electronic components of the satellite including the path through the antenna. The delay is a function of frequency and code type. Having introduced the hardware delay $T_{1,PY}$, the term $D_{1,PY}$ is viewed as residual hardware delay and can be omitted. The same is true for the L2 P-code parameters $T_{2,PY}$ versus $D_{2,PY}$. A prime is added to the pseudorange symbols on the left side to indicate that observations are raw, i.e., they have not been corrected for Δt_{SV} and the hardware delays $T_{1,PY}$ and $T_{2,PY}$. Any pseudorange equations for any frequency and code can be written in the form (6.2.8) or (6.2.9).

6.2.2.1 *Satellite Clock Correction and Timing Group Delay* The satellite clock correction Δt_{SV} has traditionally been computed by the GPS operator on the basis of the ionospheric-free L1P(Y) and L2P(Y) pseudorange function. For this purpose, we combine the linear dependent terms $d\bar{t}$ and Δt_{SV}, i.e., we simply omit the former. Recall that from the user's perspective, the satellite clock error Δt_{SV} is considered known and $d\bar{t}$ represents a residual satellite clock error. Furthermore, the user is expected to correct the observations for Δt_{SV}. The GPS operators, on the other hand, need to determine the actual clock correction Δt_{SV} and there is, consequently, no place for another clock term such as $d\bar{t}$. Also, the GPS operator uses the raw observations when attempting to determine Δt_{SV}. The mathematical model for the clock correction is the second-order polynomial

$$\Delta t_{SV} = a_0 + a_1(t - t_0) + a_2(t - t_0)^2 \tag{6.2.10}$$

where t_0 is the reference time and a_0, a_1, and a_2 are the parameters to be determined for a specific satellite, indicated by the subscript SV.

In this new notation, the ionospheric-free function (6.1.38) readily follows form (6.2.8) and (6.2.9):

$$\frac{P'_{2,PY} - \gamma_{12}P'_{1,PY}}{1 - \gamma_{12}} - \rho = c\,d\underline{t} - c\left[\Delta t_{SV} + \frac{1}{1 - \gamma_{12}}\left(T_{2,PY} - \gamma_{12}T_{1,PY}\right)\right] + T - d_{1,PY,2,PY} + M_{1,PY,2,PY} \tag{6.2.11}$$

The topocentric satellite distance ρ has been moved to the left side, assuming that the receivers are located at known stations. The subscript of the receiver hardware delay d and multipath M refer to the specific choice of the ionospheric-free function. The unknown satellite hardware delay term $T_{2,PY} - \gamma_{12}T_{1,PY}$ is considered constant over the time span $t - t_0$. As a result, the constant a_0 in (6.2.10) and the hardware delay term are linear dependent. We combine both terms into a new parameter but label it again a_0 for simplicity to avoid introducing another temporary symbol. This reparameterization can alternatively be accomplished by imposing the condition

$$T_{2,PY} = \gamma_{12}T_{1,PY} \tag{6.2.12}$$

The ionospheric-free function can thus be written as

$$\frac{P'_{2,PY} - \gamma_{12}P'_{1,PY}}{1 - \gamma_{12}} - \rho = c\,d\underline{t} - c\left[a_0 + a_1(t - t_0) + a_2(t - t_0)^2\right] + T - d_{1,PY,2,PY} + M_{1,PY,2,PY} \tag{6.2.13}$$

The receiver clock error, possibly the vertical tropospheric delay, and the polynomial coefficients a_0, a_1, and a_2 per satellite can now be estimated.

Condition (6.2.12) can be arranged as $T_{1,PY} = (T_{1,PY} - T_{2,PY})/(1 - \gamma_{12})$. The satellite manufacturer initially measures the difference $T_{1,PY} - T_{2,PY}$ in the laboratory for each satellite. These measured values may be updated to reflect the actual in-orbit delay difference for each satellite (IS-GPS-200G, 2012, Section 20.3.3.3.3.2). The scaled measured difference is traditionally denoted by T_{GD} and simply referred to as the timing group delay, i.e., $T_{\text{GD}} \equiv (T_{1,PY} - T_{2,PY})/(1 - \gamma_{12})$. With this understanding we can define

$$T_{\text{GD}} \equiv T_{1,PY} \tag{6.2.14}$$

and write (6.2.12) as

$$T_{2,PY} = \gamma_{12} T_{\text{GD}} \tag{6.2.15}$$

Using this legacy T_{GD} notation leads to the familiar form of the pseudorange equations:

$$P'_{1,PY} = \rho + c\,d\underline{t} - c\,d\bar{t} - c(\Delta t_{\text{SV}} - T_{\text{GD}}) + I_{1,PY} + T - d_{1,PY} + M_{1,PY} \tag{6.2.16}$$

$$P'_{2,PY} = \rho + c\,d\underline{t} - c\,d\bar{t} - c(\Delta t_{\text{SV}} - \gamma_{12} T_{\text{GD}}) + \gamma_{12} I_{1,PY} + T - d_{2,PY} + M_{2,PY} \tag{6.2.17}$$

The results of this traditional L1/L2 P-code dual-frequency calibration as discussed briefly above are the common clock error, one per satellite and common to all signals of that satellite, and one T_{GD} per satellite. Implicitly, the satellite hardware delay for L2, $T_{2,PY} = \gamma_{12} T_{\text{GD}}$, is expressed as a scaled value of the timing group delay.

6.2.2.2 Intersignal Correction

The modern version of the pseudorange equation incorporates the ISC, which is the difference of the satellite hardware code delays of the respective codes used in the ionospheric-free functions. Using (6.2.12) and (6.2.14), the ISC for the pseudoranges involved in (6.2.11) becomes

$$ISC_{2,PY} = T_{1,PY} - T_{2,PY} = (1 - \gamma_{12}) T_{1,PY} = (1 - \gamma_{12}) T_{\text{GD}} \tag{6.2.18}$$

The subscript of the intersignal function indicates that the ISC refers to the hardware code delay of L2P(Y) relative to L1P(Y). Substituting (6.2.18) into (6.2.17), the modernized form becomes

$$P'_{2,PY} = \rho + c\,d\underline{t} - c\,d\bar{t} - c(\Delta t_{\text{SV}} - T_{\text{GD}} + ISC_{2,PY}) + \gamma_{12} I_{1,PY} + T - d_{2,PY} + M_{2,PY} \tag{6.2.19}$$

The hardware delays are expressed in terms of the unscaled legacy group delay and the respective ISC. This form can readily be generalized to other pseudoranges. Consider, for example, the L5 inphase pseudorange L5I:

$$\begin{aligned} P'_{5,I} &= \rho + c\,d\underline{t} - c\,d\bar{t} - c(\Delta t_{\text{SV}} - T_{5,I}) + \gamma_{15} I_{1,PY} + T - d_{5,I} + M_{5,I} \\ &= \rho + c\,d\underline{t} - c\,d\bar{t} - c(\Delta t_{\text{SV}} - T_{\text{GD}} + ISC_{5,I}) + \gamma_{15} I_{1,P} + T - d_{5,I} + M_{5,I} \end{aligned} \tag{6.2.20}$$

with $ISC_{5,I} = T_{\text{GD}} - T_{5,I}$. The general form is

$$P'_{i,x} = \rho + c\,d\underline{t} - c\,d\bar{t} - c(\Delta t_{\text{SV}} - T_{\text{GD}} + ISC_{i,x}) + \gamma_{1i} I_{1,P} + T - d_{(i,x),p} + M_{(i,x),p} \tag{6.2.21}$$

with

$$ISC_{i,x} = T_{\mathrm{GD}} - T_{i,x} \tag{6.2.22}$$

Equation (6.2.21) applies to all pseudoranges and codes, even to the L1 pseudorange (6.2.16), since $ISC_{1,PY} = 0$.

The information for computing the satellite clock correction Δt_{SV} and T_{GD} has traditionally been included in the navigation message and transmitted by the user. In the modernized arrangement, the ISC will also be transmitted so that the user can correct the observations for the common satellite clock correction and the known delays:

$$P_{i,x} \equiv P'_{i,x} + c(\Delta t_{\mathrm{SV}} - T_{\mathrm{GD}} + ISC_{i,x}) = \rho + c\,d\underline{t} - c\,d\bar{t} + \gamma_{1i} I_{1,P} + T - d_{i,x} + M_{i,x} \tag{6.2.23}$$

$$\Delta t_{i,x} = \Delta t_{\mathrm{SV}} - T_{\mathrm{GD}} + ISC_{i,x} \tag{6.2.24}$$

$$P''_{i,x} \equiv P'_{i,x} + c\,\Delta t_{\mathrm{SV}} = \rho + c\,d\underline{t} - c\,d\bar{t} - cT_{\mathrm{GD}} - c \cdot ISC_{i,x} + \gamma_{1i} I_{1,P} + T - d_{i,x} + M_{i,x} \tag{6.2.25}$$

In (6.2.23) all three corrections were applied to the observation. The symbol $\Delta t_{i,x}$ in (6.2.24) denotes the known total clock error that is specific to frequency and code. In (6.2.25), only a partial correction is carried out that includes the common clock correction.

Ionospheric-Free and Ionospheric Functions: The ionospheric-free function for the fully corrected pseudoranges is obtained in the familiar form

$$\frac{P_{j,y} - \gamma_{ij} P_{i,x}}{1 - \gamma_{ij}} = \rho + c\,d\underline{t} - c\,d\bar{t} + T - d_{i,x,j,y} + M_{i,x,j,y} \tag{6.2.26}$$

It can readily be verified that the ionospheric delay cancels since $\gamma_{ij}\gamma_{1i} = \gamma_{1j}$. Using partially corrected observations as defined in (6.2.25), the ionospheric-free function looks like this:

$$\begin{aligned}\frac{P''_{j,y} - \gamma_{ij} P''_{i,x} + c \cdot ISC_{j,y} - \gamma_{ij} \cdot c \cdot ISC_{i,x}}{1 - \gamma_{ij}} &- cT_{\mathrm{GD}} = \rho + c\,d\underline{t} - c\,d\bar{t} \\ &+ T - d_{i,x,j,y} + M_{i,x,j,y}\end{aligned} \tag{6.2.27}$$

These general expressions are valid for all frequencies and code types. In the event that L1P(Y) is used, note that $ISC_{1,PY} = 0$.

Using the partially corrected function (6.2.25), and omitting for simplicity the multipath terms, the ionospheric function can be written as

$$I_{1,PY} = \frac{P''_{j,x} - P''_{j,y}}{\gamma_{1i} - \gamma_{1j}} + c \cdot \frac{ISC_{i,x} - ISC_{j,y}}{\gamma_{1i} - \gamma_{1j}} + \frac{d_{i,x} - d_{j,y}}{\gamma_{1i} - \gamma_{1j}} \tag{6.2.28}$$

$$I_{1,PY} = \frac{P''_{1,PY} - P''_{2,PY}}{1-\gamma_{12}} - c \cdot \frac{ISC_{2,PY}}{1-\gamma_{12}} = \frac{P''_{1,PY} - P''_{2,PY}}{1-\gamma_{12}} - T_{GD} + \frac{d_{1,PY} - d_{2,PY}}{1-\gamma_{12}}$$
$$= \frac{P''_{1,PY} - P'_{2,PY}}{1-\gamma_{12}} - \frac{(T_{1,PY} - T_{2,PY}) - (d_{1,PY} - d_{2,PY})}{1-\gamma_{12}} \tag{6.2.29}$$

Equation (6.2.29) represents the popular case that uses $P''_{1,PY}$ and $P''_{2,PY}$ pseudoranges. It follows from the general form by using $ISC_{1,PY} = 0$ and $\gamma_{11} = 1$. The last term represents the receiver and satellite hardware delays for P1Y and P2Y codes. When estimating the ionosphere or the TEC (total electronic content) a Kalman filter usually also estimates the difference in hardware delays. Also, ultimately, one might need to consider variations in the hardware delays due to large diurnal and seasonal temperature changes.

Estimating the ISCs: The intersignal delay for the L1 CA-code follows directly by applying (6.2.23) and (6.2.25) to the P1-code and L1CA pseudoranges, respectively, and differencing both equations, and knowing that $ISC_{1,PY} = 0$,

$$ISC_{2,CA} = P''_{1,PY} - P''_{2,CA} + (d_{1,PY} - d_{2,CA}) \tag{6.2.30}$$

Similarly, applying (6.2.25) to L2C and using (6.2.15) yields

$$ISC_{2,C} = P''_{1,PY} - P''_{2,C} + (1 - \lambda_{12})T_{\text{GD}} \tag{6.2.31}$$

Applying the ionospheric function (6.2.28) to $P_{1,PY}$ and $P_{5,Q}$ to compute $I_{1,PY}$, then substituting (6.2.29) for $I_{1,PY}$, and then making use of (6.2.18), gives

$$ISC_{5,Q} = \left(P''_{1,PY} - P''_{5,Q}\right) - \frac{1-\gamma_{15}}{1-\gamma_{12}}\left(P''_{1,PY} - P''_{2,PY}\right) + (1-\gamma_{15})T_{\text{GD}}$$
$$+ (d_{1,PY} - d_{5,Q}) - \frac{1-\gamma_{15}}{1-\gamma_{12}}(d_{1,PY} - d_{2,PY}) \tag{6.2.32}$$

Similar operations lead to the ISC for L5I,

$$ISC_{5,I} = P''_{1,PY} - P''_{5,I} - \frac{1-\gamma_{15}}{1-\gamma_{12}}\left(P''_{1,PY} - P''_{2,PY}\right) + (1-\gamma_{15})T_{\text{GD}}$$
$$+ (d_{1,PY} - d_{5,I}) - \frac{1-\gamma_{15}}{1-\gamma_{12}}(d_{1,PY} - d_{2,PY}) \tag{6.2.33}$$

All ISCs have been expressed as a function of known timing group delay T_{GD} and are all relative to the common reference L1P(Y). See Feess et al. (2013) for a data example.

6.2.3 Cycle Slips

A cycle slip is a sudden jump in the carrier phase observable by an integer number of cycles. The fractional portion of the phase is not affected by this discontinuity in the observation sequence. Cycle slips are caused by the loss of lock of the phase lock loops. Loss of lock may occur briefly between two epochs or may last several minutes or more if the satellite signals cannot reach the antenna. If receiver software would not attempt to correct for cycle slips, it would be a characteristic of a cycle slip that all observations after the cycle slip would be shifted by the same integer. This situation is demonstrated in Table 6.2.1, where a cycle slip is assumed to have occurred at receiver k while observing satellite q between the epochs $i-1$ and i. The cycle slip is denoted by Δ. Because the double differences are a function of observations at one epoch, all double differences starting with epoch i are offset by the amount Δ. Only one of the triple differences is affected by the cycle slip, because triple differences are differences over time. For each additional slip there is one additional triple-difference outlier and one additional step in the double-difference sequence. A cycle slip may be limited to just one cycle or could be millions of cycles.

This simple relation can break down if the receiver software attempts to fix the slips internally. Assume the receiver successfully corrects for a slip immediately following the epoch of occurrence. The result is one outlier (not a step function) for double differences and two outliers for the triple differences.

There is probably no best method for cycle slip removal, leaving lots of space for optimization and innovation. For example, in the case of simple static applications, one could fit polynomials, generate and analyze higher-order differences, visually inspect the observation sequences using graphical tools, or introduce new ambiguity parameters to be estimated whenever a slip might have occurred. The latter option is very attractive in kinematic positioning.

It is best to inspect the discrepancies rather than the actual observations. The observed double and triple differences show a large time variation that depends on the length of the baseline and the satellites selected. These variations can mask small slips. The discrepancies are the difference between the computed observations and the actual observed values. If good approximate station coordinates are used, the discrepancies are rather flat and allow even small slips to be detected.

For static positioning, one could begin with the triple-difference solution. The affected triple-difference observations can be treated as observations with blunders

TABLE 6.2.1 Effect of Cycle Slips on Carrier Phase Differences.

Carrier Phase				Double Difference	Triple Difference
$\varphi_k^p(i-2)$	$\varphi_m^p(i-2)$	$\varphi_k^q(i-2)$	$\varphi_m^q(i-2)$	$\varphi_{km}^{pq}(i-2)$	$\Delta\varphi_{km}^{pq}(i-1, i-2)$
$\varphi_k^p(i-1)$	$\varphi_m^p(i-1)$	$\varphi_k^q(i-1)$	$\varphi_m^q(i-1)$	$\varphi_{km}^{pq}(i-1)$	$\Delta\varphi_{km}^{pq}(i, i-1)-\Delta$
$\varphi_k^p(i)$	$\varphi_m^p(i)$	$\varphi_k^q(i)+\Delta$	$\varphi_m^q(i)$	$\varphi_{km}^{pq}(i)-\Delta$	$\Delta\varphi_{km}^{pq}(i+1, i)$
$\varphi_k^p(i+1)$	$\varphi_m^p(i+1)$	$\varphi_k^q(i+1)+\Delta$	$\varphi_m^q(i+1)$	$\varphi_{km}^{pq}(i+1)-\Delta$	$\Delta\varphi_{km}^{pq}(i+2, i+1)$
$\varphi_k^p(i+2)$	$\varphi_m^p(i+2)$	$\varphi_k^q(i+2)+\Delta$	$\varphi_m^q(i+2)$	$\varphi_{km}^{pq}(i+2)-\Delta$	

and dealt with using the blunder detection techniques provided in Chapter 2. A simple method is to change the weights of those triple-difference observations that have particularly large residuals. Once the least-squares solution has converged, the residuals will indicate the size of the cycle slips. Not only is triple-difference processing a robust technique for cycle slip detection, it also provides good station coordinates, which, in turn, can be used as approximations in a subsequent double-difference solution.

Before computing the double-difference solution, the double-difference observations should be corrected for cycle slips identified from the triple-difference solution. If only two receivers observe, it is not possible to identify the specific undifferenced phase sequence where the cycle slip occurred from analysis of the double difference. Consider the double differences

$$\varphi_{12}^{1p} = \left(\varphi_1^1 - \varphi_2^1\right) - \left(\varphi_1^p - \varphi_2^p\right) \tag{6.2.34}$$

for stations 1 and 2 and satellites 1 and p. The superscript p denoting the satellites varies from 2 to S, the total number of satellites. Equation (6.2.34) shows that a cycle slips in φ_1^1 or φ_2^1 will affect all double differences for all satellites and cannot be separately identified. The slips Δ_1^1 and $-\Delta_2^1$ cause the same jump in the double-difference observation. The same is true for slips in the phase from station 1 to satellite p and station 2 to satellite p. However, a slip in the latter phase sequences affects only the double differences containing satellite p. Other double-difference sequences are not affected.

For a session network, the double-difference observation is

$$\varphi_{1m}^{1p} = \left(\varphi_1^1 - \varphi_m^1\right) - \left(\varphi_1^p - \varphi_m^p\right) \tag{6.2.35}$$

The superscript p goes from 2 to S, and the subscript m runs from 2 to R. It is readily seen that a cycle slip in φ_1^1 affects all double-difference observations, an error in φ_m^1 affects all double differences pertaining to the baseline 1 to m, an error in φ_1^p affects all double differences containing satellite p, and an error in φ_m^p affects only one series of double differences, namely, the one that contains station m and satellite p. Thus, by analyzing the distribution of a blunder in all double differences at the same epoch, we can identify the undifferenced phase observation sequence that contains the blunder. This identification gets more complicated if several slips occur at the same epoch. In session network processing, it is always necessary to carry out cross checks. The same cycle slip must be verified in all relevant double differences before it can be declared an actual cycle slip. Whenever a cycle slip occurs in the undifferenced phase observations from the base station or to the base satellite, the cycle slip enters several double-difference sequences. In classical double-difference processing, it is not necessary that the undifferenced phase observations be corrected; it is sufficient to limit the correction to the double-difference phase observations if the final position computation is based on double differences. It is also possible to use the geometry-free functions of the observables to detect cycle slips.

6.2.4 Phase Windup Correction

One must go back to the electromagnetic nature of GPS transmissions in order to understand this correction, as has been done in Chapter 9. In short, the GPS

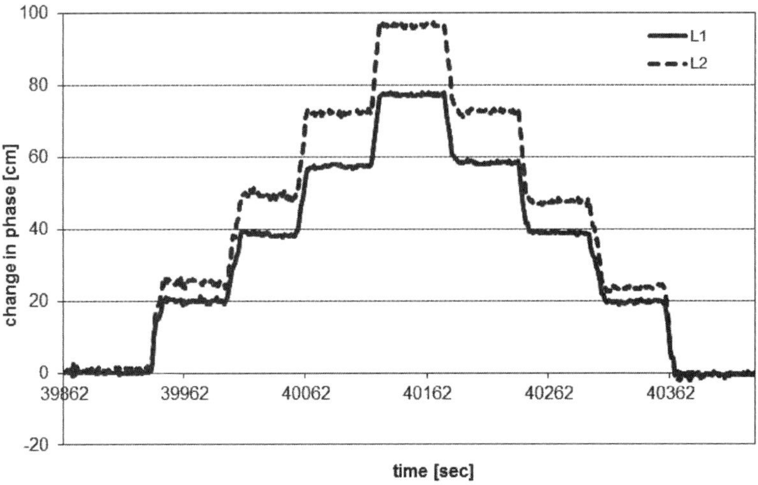

Figure 6.2.1 Antenna rotation test with Javad dual-frequency receiver having two antennae and single oscillator. Data source: Rapoport.

carrier waves are right circularly polarized (RCP). The electromagnetic wave may be visualized as a rotating electric vector field that propagates from the satellite antenna to the receiver antenna. The vector rotates 360° every spatial wavelength or every temporal cycle of the wave. The observed carrier phase can be viewed as the geometric angle between the instantaneous electric field vector at the receiving antenna and some reference direction on the antenna. As the receiving antenna rotates in azimuth, this measured phase changes. The same is true if the transmitting antenna changes its orientation with respect to the receiver antenna. Since the phase is measured in the plane of the receiving antenna, its value depends on the direction of the line of sight to the satellite, in addition to the orientation of the antenna.

Figure 6.2.1 shows the results of a simple test to demonstrate RCP of GPS signals. Two antennas, about 5 m apart, were connected to the same receiver and oscillator and observations were recorded once per second. One of the antennas was rotated 360° in azimuth four times clockwise (as viewed looking down on the antenna), with 1 minute between the rotations, and then four times rotated counterclockwise, again with 1 minute between the rotations. The carrier phase observations were differenced and a linear trend was removed to account for the phase biases and a differential rate (caused by the separation of the antennas). The figure shows the change in the single differences for both L1 and L2. Each complete antenna rotation in azimuth causes a change of one wavelength.

An introductory discussion of the carrier phase windup correction for rotating GPS antennas is found in Tetewsky and Mullen (1997). Wu et al. (1993) derived the phase windup correction expressions for a crossed dipole antenna, but their results are applicable to cases that are more general. Following their derivations, at a given instant the windup correction is expressed as a function of the directions of the dipoles and of the line of sight to the satellite.

Let $\hat{\boldsymbol{x}}$ and $\hat{\boldsymbol{y}}$ denote the unit vectors in the direction of the two-dipole elements in the receiving antenna in which the signal from the y-dipole element is delayed by

90° relative to that from the x-dipole element. $\boldsymbol{k}$ is the unit vector pointing from the satellite to the receiver. We consider a similar definition for $\hat{\boldsymbol{x}}'$ and $\hat{\boldsymbol{y}}'$ at the satellite, i.e., the current in the y'-dipole lags that in the x'- dipole by 90°. They define the effective dipole that represents the resultant of a crossed dipole antenna for the receiver and the transmitter, respectively,

$$\boldsymbol{d} = \hat{\boldsymbol{x}} - \boldsymbol{k}(\boldsymbol{k} \cdot \hat{\boldsymbol{x}}) + \boldsymbol{k} \times \hat{\boldsymbol{y}} \tag{6.2.36}$$

$$\boldsymbol{d}' = \hat{\boldsymbol{x}}' - \boldsymbol{k}(\boldsymbol{k} \cdot \hat{\boldsymbol{x}}') - \boldsymbol{k} \times \hat{\boldsymbol{y}}' \tag{6.2.37}$$

The windup correction is (Wu et al., 1993, p. 95)

$$\delta\varphi = \text{sign}\,[\boldsymbol{k} \cdot (\boldsymbol{d}' \times \boldsymbol{d})]\cos^{-1}\left(\frac{\boldsymbol{d}' \cdot \boldsymbol{d}}{\|\boldsymbol{d}'\|\ \|\boldsymbol{d}\|}\right) \tag{6.2.38}$$

At a given instant in time, the windup correction $\delta\varphi$ cannot be separated from the undifferenced ambiguities, nor is it absorbed by the receiver clock error because it is a function of the receiver and the satellite. In practical applications, it is therefore sufficient to interpret $\hat{\boldsymbol{x}}$ and $\hat{\boldsymbol{y}}$ as unit vectors along northing and easting and $\hat{\boldsymbol{x}}'$ and $\hat{\boldsymbol{y}}'$ as unit vectors in the satellite body coordinate system. Any additional windup error resulting from this redefinition of the coordinate system will also be absorbed by the undifferenced ambiguities. Taken over time, however, the values of $\delta\varphi$ reflect the change in orientation of the receiver and satellite antennas.

The value of the windup correction for across-receiver and double differences has an interesting connection to spherical trigonometry. Consider a spherical triangle whose vertices are given by the latitudes and longitudes of the receivers k and m, and the satellite. In addition, we assume that GPS transmitting antennas are pointing toward the center of the earth and that the ground receiver antennas are pointing upward. This assumption is usually met in the real world. It can be shown that the across-receiver difference windup correction $\delta\varphi^p_{km} = \delta\varphi^p_k - \delta\varphi^p_m$ is equal to the spherical excess if the satellite appears on the left as viewed from station k to station m, and it equals the negative spherical excess if the satellite appears to the right. The double-differencing windup correction $\delta\varphi^{pq}_{km}$ equals the spherical excess of the respective quadrilateral. The sign of the correction depends on the orientation of the satellite with respect to the baseline. For details, refer to Wu et al. (1993).

The windup correction is negligible for short baselines because the spherical excess of the respective triangles is small. Neglecting the windup correction might cause problems when fixing the double-difference ambiguities, in particular for longer lines. The float ambiguities absorb the constant part of the windup correction. The variation of the windup correction over time might not be negligible in float solutions of long baselines. Additional remarks about dealing with the windup corrections are provided in Chapter 7.

There is no windup-type correction for the pseudoranges. Consider the simple case of a rotating antenna that is at a constant distance from the transmitting source and the antenna plane perpendicular to the direction of the transmitting source. Although the measured phase would change due to the rotation of the antenna, the pseudorange will not change because the distance is constant.

6.2.5 Multipath

Once the satellite signals reach the earth's surface, ideally they enter the antenna directly. However, objects in the receiver's vicinity may reflect some signals before they enter the antenna, causing unwanted signatures in pseudorange and carrier phase observations. Although the direct and reflected signals have a common emission time at the satellite, the reflected signals are always delayed relative to the line-of-sight signals because they travel longer paths. The amplitude (voltage) of the reflected signal is always reduced because of attenuation. The attenuation depends on the properties of the reflector material, the incident angle of the reflection, and the polarization. In general, reflections with a very low incident angle have little attenuation. In addition, the impact of multipath on the GPS observables depends on the sensitivity of the antenna in terms of sensing signals from different directions, and the receiver's internal processing to mitigate multipath effects. Multipath is still one of the dominating, if not the dominant, sources of error in GPS positioning. Chapter 9 provides an in-depth treatment of the relationship of antenna properties and multipath effects.

Signals can be reflected at the satellite (satellite multipath) or in the surroundings of the receiver (receiver multipath). Satellite multipath is likely to cancel in the single-difference observables for short baselines. Reflective objects for receivers on the ground can be the earth's surface itself (ground and water), buildings, trees, hills, etc. Rooftops are known to be bad multipath environments because there are often many vents and other reflective objects within the antenna's field of view.

The impact of multipath on the carrier phases can be demonstrated using a planar vertical reflection surface at distance d from the antenna (Georgiadou and Kleusberg, 1988; Bishop et al., 1985). The geometry is shown in Figure 6.2.2. We write the direct line-of-sight carrier phase observable for receiver k and satellite p as

$$S_D = A \cos \varphi \tag{6.2.39}$$

In (6.2.39) we do not use the subscript k and superscript p in order to simplify the notation. The symbols A and φ denote the amplitude (signal voltage) and the phase, respectively. The reflected signal is written as

$$S_R = \alpha A \cos(\varphi + \theta), \qquad 0 \le \alpha \le 1 \tag{6.2.40}$$

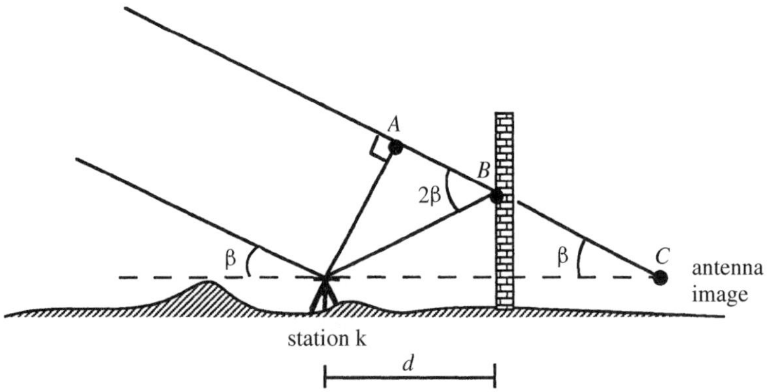

Figure 6.2.2 Geometry for reflection on a vertical planar plane.

The amplitude reduction factor (attenuation) is $\alpha = A'/A$, where A' is the amplitude of the reflected signal. The total multipath phase shift is

$$\theta = 2\pi f \, \Delta\tau + \phi \tag{6.2.41}$$

where f is the frequency, $\Delta\tau$ is the time delay, and ϕ is the fractional shift. The multipath delay shown in Figure 6.2.2 is the sum of the distances AB and BC, which equals $2d \cos \beta$. Converting this distance into cycles and then to radians gives

$$\theta = \frac{4\pi d}{\lambda} \cos \beta + \phi \tag{6.2.42}$$

where λ is the carrier wavelength. The composite signal at the antenna is the sum of the direct and reflected signal,

$$S = S_D + S_R = R \cos(\varphi + \psi) \tag{6.2.43}$$

It can be verified that the resultant carrier phase voltage $R(A, \alpha, \theta)$ and the carrier phase multipath delay $\psi(\alpha, \theta)$ are

$$R(A, \alpha, \theta) = A(1 + 2\alpha \cos \theta + \alpha^2)^{1/2} \tag{6.2.44}$$

$$\psi(\alpha, \theta) = \tan^{-1} \left(\frac{\alpha \sin \theta}{1 + \alpha \cos \theta} \right) \tag{6.2.45}$$

Regarding the notation, we used the symbols $M^p_{k,1}$ and $M^p_{k,2}$ in previous sections to denote the total multipath, i.e., the multipath effect of all reflections on L1 and L2, respectively. If we consider the case of constant reflectivity, i.e., α is constant, the maximum path delay is found when $\partial\psi/\partial\theta = 0$. This occurs at $\theta(\psi_{\max}) = \pm\cos^{-1}(-\alpha)$, the maximum value being $\psi_{\max} = \pm\sin^{-1}\alpha$. The maximum multipath carrier phase error is only a function of the amplitude attenuation in this particular case. The largest value is $\pm 90^\circ$ and occurs when $\alpha = 1$. This maximum corresponds to $\lambda/4$. If $\alpha \ll 1$, then ψ can be approximated by $\alpha \sin \theta$.

The multipath effect on pseudoranges depends among other things on the chipping rate T of the codes and the receiver's internal sampling interval S. A necessary step for each receiver is to correlate the received signal with an internally generated code replica. The offset in time that maximizes the correlation is a measure of the pseudorange. Avoiding the technical details, suffice it to say that time shifting the internal code replica and determining the correlation for early, prompt, and late delays eventually determines the offset. The early and late delays differ from the prompt delay by $-S$ and S, respectively. When the early minus late correlation is zero, i.e., they have the same amplitude, the prompt delay is used as a measure of the pseudorange. Consult Kaplan (1996, p. 148) for additional details on the topic of code tracking loops and correlation. For a single multipath signal, the correlation function consists of the sum of two triangles, one for the direct signal and one for the multipath signal. This is conceptually demonstrated in Figure 6.2.3. The solid thin line and the dashed line represent the correlation functions of the direct and multipath signals, respectively. The thick solid line indicates the combined correlation function, i.e., the sum of the thin line and dashed line. The left figure refers to destructive reflection when the reflected signal arrives out of phase with respect to the direct signal. The right figure

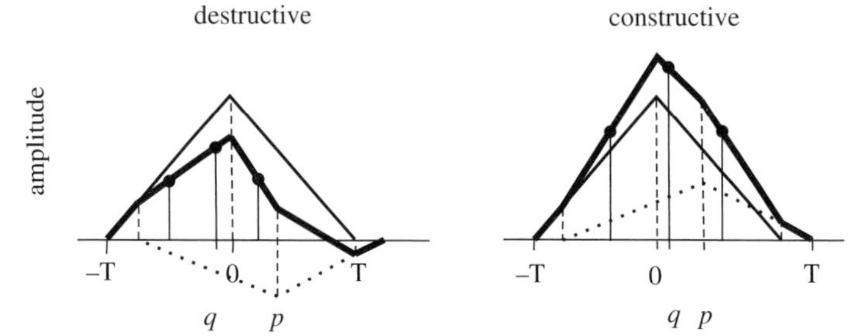

Figure 6.2.3 Correlation function in the presence of multipath. p denotes the time delay of the multipath signal and q is the multipath induced pseudorange error.

refers to constructive reflection when the reflected and direct signals are in phase. Let the combined signal be sampled at the early and late delays. The figure shows that the prompt delay would coincide with the maximum correlation for the direct signal and indicate the correct pseudorange but will be in error by the multipath-induced range error q for the combined signal. The resulting pseudorange measurement errors are negative for destructive reflection and positive for constructive reflection, even though the reflected signal always arrives later than the direct one.

The pseudorange multipath error further depends on whether the sampling interval is greater or smaller than half the chipping period. Byun et al. (2002) provide the following expressions. If $S > T/2$ (wide sampling), then

$$\Delta\tau_P = \begin{cases} \dfrac{\Delta\tau\alpha\cos(2\pi f\Delta\tau + \phi)}{1 + \alpha\cos(2\pi f\Delta\tau + \phi)} & \text{if} \quad \Delta\tau < T - S + \Delta\tau_P \\ \dfrac{(T - S + \Delta\tau)\alpha\cos(2\pi f\Delta\tau + \phi)}{2 + \alpha\cos(2\pi f\Delta\tau + \phi)} & \text{if} \quad T - S + \Delta\tau_P < \Delta\tau < S + \Delta\tau_P \\ \dfrac{(T + S + \Delta\tau)\alpha\cos(2\pi f\Delta\tau + \phi)}{2 - \alpha\cos(2\pi f\Delta\tau + \phi)} & \text{if} \quad s + \Delta\tau_P < \Delta\tau < T + S + \Delta\tau_P \\ 0 & \text{if} \quad \Delta\tau > T + S + \Delta\tau_P \end{cases} \tag{6.2.46}$$

and if $S < T/2$ (narrow sampling), then

$$\Delta\tau_P = \begin{cases} \dfrac{\Delta\tau\alpha\cos(2\pi f\Delta\tau + \phi)}{1 + \alpha\cos(2\pi f\Delta\tau + \phi)} & \text{if} \quad \Delta\tau < S + \Delta\tau_P \\ s\alpha\cos(2\pi f\Delta\tau + \phi) & \text{if} \quad S + \Delta\tau_P < \Delta\tau < T - S + \Delta\tau_P \\ \dfrac{(T + S - \Delta\tau)\alpha\cos(2\pi f\Delta\tau + \phi)}{2 - \alpha\cos(2\pi f\Delta\tau + \phi)} & \text{if} \quad T - S + \Delta\tau_P < \Delta\tau < T + S \\ 0 & \text{if} \quad \Delta\tau > T + S \end{cases} \tag{6.2.47}$$

The pseudorange multipath error is $d_P = c\,\Delta\tau_P$, and $\Delta\tau$ denotes the time delay of the multipath signal. The expressions are valid for the P-codes and the C/A-code as long as the appropriate chipping period T is used.

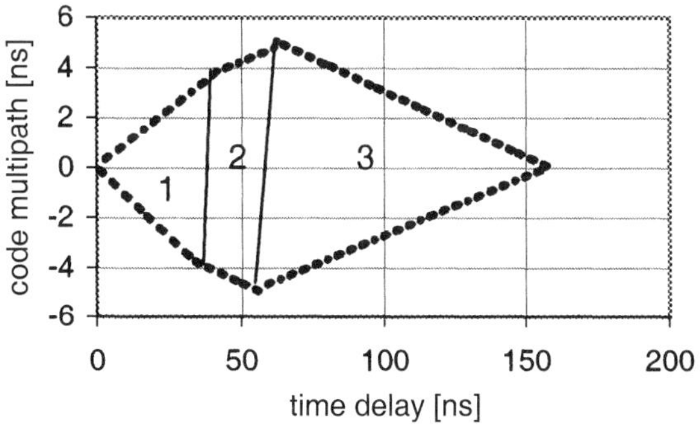

Figure 6.2.4 P1-code pseudorange multipath delay envelope in the case of wide sampling. $T = 98\,\text{nsec}$, $S = 60\,\text{nsec}$, $\alpha_1 = 0.1$, $\phi_1 = 0$.

Figure 6.2.4 shows an example of the envelope for the P1-code multipath range error $\Delta\tau_{P1}$ oscillations versus time delay $\Delta\tau$ for the wide-sampling case $S > T/2$. As the phase varies by π, the multipath error changes from upper to lower bounds and vice versa. The distinct regions of (6.2.46) are readily visible in the figure. Figure 6.2.5 shows an example of the C/A-code multipath range error for the narrow-sampling case $S < T/2$. The main difference between the wide and narrow sampling interval is that the latter has a constant peak at region 2. In fact, shortening the sampling interval S has long been recognized as a means to reduce the pseudorange multipath error. See the second component of (6.2.47), where S appears as a factor. Comparing (6.2.46) and (6.2.47), we find that in region 1 the slopes of the envelopes are the same for wide and narrow correlating. Narrow correlation causes the bounds in region 2 to be smaller. Region 4, for which the multipath error is zero, is reached earlier the narrower the sampling (given the same chipping rate). The lower envelope in these figures corresponds to destructive reflection, while the upper envelope refers to constructive reflection.

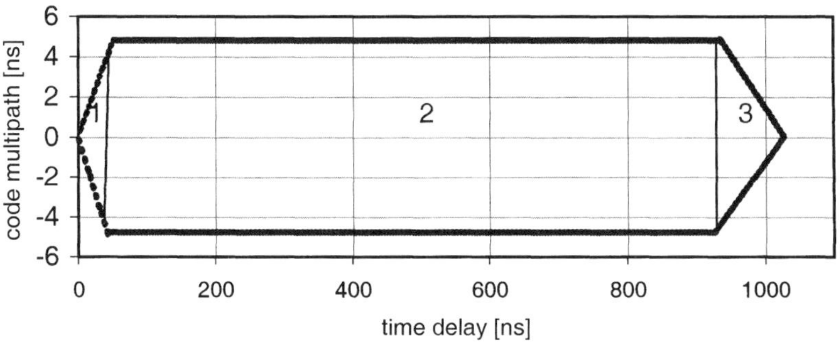

Figure 6.2.5 C/A-code pseudorange multipath delay envelope in the case of narrow sampling. $T = 980\,\text{nsec}$, $S = 48\,\text{nsec}$, $\alpha_1 = 0.1$, $\phi_1 = 0$.

The multipath frequency f_ψ depends on the variation of the phase delay θ, as can be seen from (6.2.40), (6.2.45), (6.2.46), or (6.2.47). Differentiating (6.2.42) gives the expression for the multipath frequency

$$f_\psi = \frac{1}{2\pi}\frac{d\theta}{dt} = \frac{2d}{\lambda}\sin\beta|\dot{\beta}| \tag{6.2.48}$$

The multipath frequency is a function of the elevation angle and is proportional to the distance d and the carrier frequency. For example, if we take $\dot{\beta} = 0.07\,\text{mrad/s}$ (= one-half of the satellite's mean motion) and $\beta = 45°$, then the multipath period is about 5 minutes if $d = 10\,\text{m}$ and about 50 minutes if $d = 1\,\text{m}$. The variation in the satellite elevation angle causes the multipath frequency to become a function of time. According to (6.2.48), the ratio of the multipath frequencies for L1 and L2 equals that of the carrier frequencies, $f_{\psi,1}/f_{\psi,2} = f_1/f_2$.

As an example of a carrier phase multipath, consider a single multipath signal and the ionospheric phase observable (6.1.44). The effect of the multipath for this function is

$$\begin{aligned}\varphi_{MP} &\equiv \psi_1 - \frac{f_1}{f_2}\psi_2 \\ &= \tan^{-1}\left(\frac{\alpha\sin\theta_1}{1+\alpha\cos\theta_1}\right) - \frac{f_1}{f_2}\tan^{-1}\left(\frac{\alpha\sin\theta_2}{1+\alpha\cos\theta_2}\right)\end{aligned} \tag{6.2.49}$$

Figure 6.2.6 shows that the multipath φ_{MP} impacts the ionospheric observable in a complicated manner. The amplitude of the cyclic phase variations is nearly proportional to α. When analyzing the ionospheric observable in order to map the temporal

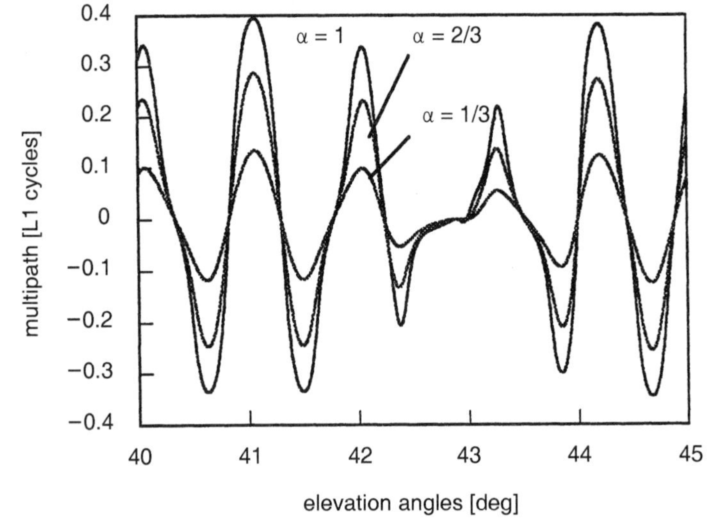

Figure 6.2.6 Example of multipath on the ionospheric carrier phase observable from a vertical planar surface. $d = 10\,\text{m}$, $\phi_1 = \phi_2 = 0$.

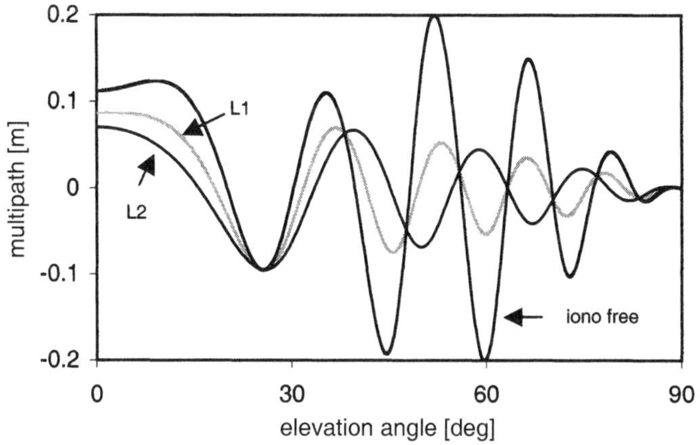

Figure 6.2.7 Pseudorange multipath from a single reflection on a vertical planar surface. $\alpha = 0.1$, $d = 5\,\lambda_1$, $\phi_1 = \phi_2 = 0$.

variation of the ionospheric delay, the multipath signature (6.2.49) cannot be ignored. In fact, the multipath variation of (6.2.45) might occasionally impact our ability to fix the integer ambiguities, even for short baselines.

Figure 6.2.7 shows the effects of multipath on the pseudoranges P1 and P2, and the ionospheric free function (6.1.38). We are using the expression for region 1 in (6.2.46) or (6.2.47), since we consider the case of a nearby reflection. The time delay $\Delta\tau$ is a function of the satellite elevation angle and can be computed from (6.2.42). The figures show the multipath for a satellite that rises ($\beta = 0°$) until it passes overhead ($\beta = 90°$). The multipath is largest for a satellite in the horizon (reflection on vertical surface). In the case of reflection from a horizontal surface, the multipath has a reverse dependency, i.e., it is largest for satellites at the zenith, as can readily be verified.

Fenton et al. (1991) discuss one of the early implementations of narrow correlation in C/A-code receivers. Narrow correlator technology and on-receiver processing methods to reduce carrier phase and pseudorange multipath effects are extensively documented in the literature, e.g., van Dierendonck et al. (1992), Meehan and Young (1992), Veitsel et al. (1998), and Zhdanov et al. (2001). If the phase shift θ changes rapidly, one might even attempt to average the pseudorange measurements. In addition to sophisticated on-receiver signal processing, there are several external ways to mitigate multipath.

1. Since multipath can also arrive from below the antenna (due to edge diffraction), a ground plate is helpful. The ground plate is usually a metallic surface of circular or rectangular form.
2. Partial multipath rejection can be achieved by shaping the gain pattern of the antenna. Since a lot of multipath arrives from reflections near the horizon, multipath may be sharply reduced by using antennas having low gain in these directions.

3. Improved multipath resistance is achieved with choke rings. These are metallic circular grooves with quarter-wavelength depth.
4. Highly reflective surfaces change the polarization from right-hand circular (signal received directly from the GPS satellite) to left-hand circular. GPS antennas that are designed to receive right-hand polarized signals will attenuate signals of opposite polarization.
5. Arrays of antennas can also be used to mitigate multipath. Due to a different multipath geometry, each antenna sees the multipath effect differently. Combined processing of signals from all antennas allows multipath mitigation (Fu et al., 2003). In a design proposed by Counselman, the antenna elements are arranged along the vertical rather than the horizontal (Counselman, 1999).
6. Since the geometry between a GPS satellite and a receiver-reflector repeats every sidereal day, multipath shows the same pattern between consecutive days. Such repetition is useful to verify the presence of multipath by analyzing the repeatability patterns and eventually model the multipath at the station. In relative positioning, the double-difference observable is affected by multipath at both stations.

In practical applications, of course, the various satellite signals are reflected at different objects. The attenuation properties of these objects generally vary; in some cases attenuation might even depend on time. Since the angle of incident also affects attenuation, it can readily be appreciated that the multipath is a difficult error source with which to deal. It is common practice not to observe satellites close to the horizon in order reduce multipath.

Equations (6.1.50) and (6.1.51) are useful to gauge the multipath, in particular the multipath effect on the pseudoranges, if dual-frequency observations are available.

6.2.6 Phase Center Offset and Variation

It is important that the satellite signals are modeled correctly at the satellite and at the receiver. At the satellite one must take the separation of satellite antenna phase and satellite center of mass into consideration. The user antenna phase center offset and variation is generally dealt with in terms of relative and absolute antenna calibration (see Chapter 9 for a more detailed treatment of phase center definition and its variation). The data on both the satellite antenna and the most important user antenna phase center offsets are available from the IGS in the form of ANTEX (antenna exchange format) files. This format was especially designed to be able to handle multiple satellite systems, multiple frequencies per satellite system, and azimuth dependencies of the phase center variations.

6.2.6.1 Satellite Phase Center Offset The satellite antenna phase center offsets are usually given in the satellite-fixed coordinate system (x') that is also used to express solar radiation pressure (see Section 5.1.4.3). The origin of this coordinate system is at the satellite's center of mass. If $\boldsymbol{e}$ denotes the unit vector pointing to the sun, expressed in the ECEF coordinate system (x), then the axes of (x') are defined by the unit vector $\boldsymbol{k}$ (pointing from the satellite toward the earth's center, expressed in (x)), the vector $\boldsymbol{j} = (\boldsymbol{k} \times \boldsymbol{e})/|\boldsymbol{k} \times \boldsymbol{e}\,|$ (pointing along the solar panel axis), and the unit vector $\boldsymbol{i} = \boldsymbol{j} \times \boldsymbol{k}$ that completes the right-handed coordinate system (also located in the sun-satellite-earth plane). It can readily be verified that

$$\boldsymbol{x}_{\mathrm{sa}} = \boldsymbol{x}_{\mathrm{sc}} + [\boldsymbol{i} \quad \boldsymbol{j} \quad \boldsymbol{k}]\ \boldsymbol{x}' \tag{6.2.50}$$

where $\boldsymbol{x}_{\mathrm{sa}}$ is the position of the satellite antenna and $\boldsymbol{x}_{\mathrm{sc}}$ denotes the position of the satellite's center of mass.

The satellite phase center offsets must be determined for each satellite type. When estimating the offsets from observations while the satellite is in orbit, the effect of the offsets might be absorbed, at least in part, by other parameters. This might be the case for the offset in direction $\boldsymbol{k}$ and the receiver clock error. See Mader and Czopek (2001) as an example for calibrating the phase center of the satellite antenna for a Block IIA antenna using ground measurements. The satellite antenna phase center calibration data are available in the ANTEX files from IGS.

6.2.6.2 User Antenna Calibration In the past, the phase center offset and variations of most user antennas were calibrated relative to a reference antenna. This procedure is called relative antenna calibration. Absolute antenna calibrations, where the phase offset and variations are determined independently of a reference antenna, were conducted only for those antennas used at reference stations for which per definition, the best accuracy is needed. However, as in recent years more absolute antenna calibration facilities became available, the trend is moving toward using absolute calibration.

The immediate reference point in positioning with GPS is the phase center of the receiver antenna. Since the phase center cannot be accessed directly with tape, we need to know the relationship between the phase center and an external antenna reference point (ARP) in order to relate the GPS-determined positions to a surveying monument. Unfortunately, the phase center is not well defined. Its location varies with the elevation angle of the arriving signal, and to a lesser extend it also depends on the azimuth. The relationship between the ARP and the phase center, which is the object of antenna calibration, is usually parameterized in terms of phase center offset (PCO) and phase center variation (PCV). The largest offset is in height, which can be as much as 10 cm. The PCO and the PCV also depend on the frequency.

For simplicity, imagine a perfect antenna that has an ARP and a phase center offset that is well known. Imagine further that you connect a "phase meter" to the antenna and that you move the transmitter along the surface of a sphere that is centered on the phase center. In this ideal case, since the distance from the transmitter to the phase center never changes, the output phase will always read a same constant amount. In actuality, there is no perfect antenna, and that situation can never be realized. Instead, one effectively moves a source along a sphere centered on a point that one selects as an average phase center. Now instead of recording a constant phase, one detects phase variations, primarily as a function of satellite elevation. Since the distance from source to antenna is constant, these phase variations must be removed so that constant geometric distance is represented by constant phase measurements. Had one picked another phase center, we would get another set of phase variations. It follows that in general the PCO and PCVs must be used together and why different PCOs and PCV sets will lead one back to the same ARP.

For a long observation series one might hope that the average location of the PCV is well defined and that the position refers to the average phase center. For RTK applications there is certainly no such averaging possible. For short baselines where the antennas at the end of the line see a satellite at approximately the same elevation angle, orienting both antennas in the same direction can largely eliminate the PCO and PCV. This elimination procedure works only for the same antenna types, however. For large baselines or when mixing antenna types, an antenna calibration is necessary and respective corrections must be applied. Antenna calibration is also important when estimating tropospheric parameters, since both the PCV and the tropospheric delay depend on the elevation angle.

Relative antenna calibration using field observations was, e.g., developed at the NGS (Mader, 1999), which also made this service available to users. All test antennas are calibrated with respect to the same reference antenna, which happens to be an AOAD/M_T choke ring antenna. The basic idea is that if the same reference antenna is always used for all calibrations, the PCO and PCV of the reference antenna cancel when double-differencing observations of a new baseline and applying the calibrated PCO and PCV to both antennas. This technique is accurate as long as the elevation difference of a satellite, as seen from both antennas, is negligible since the PCV is parameterized as a function of the elevation angle. Since the PCV amounts to about only 1 to 2 cm and varies only slightly and smoothly with elevation angle, relative phase calibration is applicable to even reasonably long baselines. NGS uses a calibration baseline of 5 m. The reference antenna and the test antenna are connected to the same type of receiver, and both receivers use the same rubidium oscillator as an external frequency standard. Since the test baseline is known, a common frequency standard is used, and because the tropospheric and ionospheric effects cancel over such a short baseline, the single-difference discrepancies over time are very flat and can be modeled as

$$\left(\varphi_{12,b}^{p} - \varphi_{12,0}^{p}\right)_i = \tau_i + \alpha_1 \beta_i^p + \alpha_2 \left(\beta_i^p\right)^2 + \alpha_3 \left(\beta_i^p\right)^3 + \alpha_4 \left(\beta_i^p\right)^4 \tag{6.2.51}$$

The subscript i denotes the epoch, the superscript p identifies the satellite having elevation angle β_i, and τ_i is the remaining relative time delay (receiver clock error). The coefficients α_1 to α_4 and τ_i are estimated by observing all satellites from rising to setting. The result of the relative calibration of the test antenna is then given by

$$\hat{\varphi}_{antenna,PCV}(\beta) = \hat{\alpha}_1 \beta + \hat{\alpha}_2 \beta^2 + \hat{\alpha}_3 \beta^3 + \hat{\alpha}_4 \beta^4 + \xi \tag{6.2.52}$$

The symbol ξ denotes a translation such that $\hat{\varphi}_{antenna,\,PCV}(90^\circ) = 0$. The remaining clock difference estimate $\hat{\tau}$ is not included in (6.2.52). Both $\hat{\tau}$ and ξ cancel in double differencing. Recall that this calibration procedure is relative and therefore (6.2.52) must be applied in the double-differencing mode. We further notice that the model (6.2.52) does not include an azimuthal parameter. The calibration data is available in the ANTINFO (antenna information format) files, which were formatted especially for relative antenna calibration.

Automated absolute and site-independent field calibration of GPS antennas in real time is reported in Wübbena et al. (2000), Schmitz et al. (2002), and references listed therein. They use a precisely controlled three-axes robotic arm to determine the absolute PCO and PCV as a function of elevation and azimuth. This real-time calibration uses undifferenced observations from the test antenna that are differenced over very short time intervals. Rapid changes of orientation of the calibration robot allow the separation of PCV and any residual multipath effects. Several thousand observations are taken at different robot positions. The calibration takes only a few hours.

In order to better serve the high-accurate GNSS community, NGS has also developed an absolute calibration technique (Bilich and Mader, 2010). They move the antenna to be tested on a two-axes robotic arm to view the satellite from different angles. The antenna motion is relatively fast to allow separation of the antenna pattern of the test antenna and the reference antenna and to eliminate errors such as multipath, thus effectively producing absolute calibration. The procedure uses across-receiver, across-time differencing to estimate the antenna calibration parameters. The calibration baseline is about 5 m long, and current procedure requires both receivers to be connected to a common clock. The calibration results are also reported in the ANTEX format.

There are other approaches available for absolute antenna calibration. For example, the antenna can be placed in an anechoic chamber. The interior of such a chamber is lined with radiofrequency absorbent material that reduces signal reflections or "echoes" to a minimum. A signal source antenna generates the signals. Since the source antenna can transmit at different frequencies, these anechoic chamber techniques are suitable for general antenna calibration.

The interested reader is requested to surf the Internet for examples of antenna calibration and additional resources.

6.2.7 GNSS Services

There are numerous services available to help users get the best out of GNSS. One such service is the antenna calibration at NGS mentioned above. Others include the gridded hydrostatic and wet zenith delays available from the TU Vienna and the University of New Brunswick (Chapter 8), ocean loading coefficients from the Onsala Space Observatory (Chapter 4), polar motion and earth rotation parameters from the IERS (Chapter 4), geoid undulations from various geodetic agencies to convert ellipsoidal eights to orthometric heights, and freeware such as LAMBDA (Section 6.5), made available by the UT Delft. Here we briefly focus on two additional services that have a major impact on the use of GNSS. The first service refers to the products provided by the IGS (International GNSS Service), and by the second service we mean the various online services that process field observations to produce final positions and related information.

6.2.7.1 IGS The International GNSS Service (IGS) is a response to a call by international users for an organizational structure that helps maximize the potential of GNSS systems. It is a globally decentralized organization that is self-governed by its members and is without a central source of funding. The support comes from various member organizations and agencies around the world called contributing organizations. Established by the International Association of Geodesy (IAG) in 1993, it officially began its operations on January 1, 1994, under the name International GPS Service for Geodynamics. The current name has been in use since 2005 to convey a stated goal of providing integration and service for all GNSS systems. Details about this important open service, which is available to any GNSS user, including formal statements of goals and objectives, are available at its website http://www.igs.org.

A governing board sets the IGS policies and exercises broad oversight of all IGS functions. The executive arm of the board is the central bureau, which is located at the JPL. There are over 400 globally distributed permanent GPS tracking sites. Figure 6.2.8 shows a subset of participating sites. These stations operate continuously and deliver data almost in real time to the data centers. There are currently 28 data centers—4 global, 6 regional, 17 operational data centers, and 1 project data center. These data centers provide efficient access and storage of data, data redundancy, and data security at the same time. There are 12 analysis centers. These centers use the global data sets to produce products of the highest quality. The analysis centers cooperate with an analysis center coordinator, whose main task is to combine the products of the centers into a single product, which becomes the official IGS product. In addition, there are 28 associate analysis centers that produce information for regional subnetworks, such as ionospheric information and station coordinate velocities.

Table 6.2.2 summarizes the various IGS products. The orbital accuracy in section (1) of the table is RMS values computed from three geocentric coordinates as compared with independently determined laser ranging results. The first accuracy identifier given for the clocks is the RMS computed relative to the IGS time scale; the latter is adjusted to GPS time in daily segments. The second accuracy identifier for the clocks is the standard deviation computed by removing biases for each satellite, which causes the standard deviation to be smaller than the RMS value. The real-time

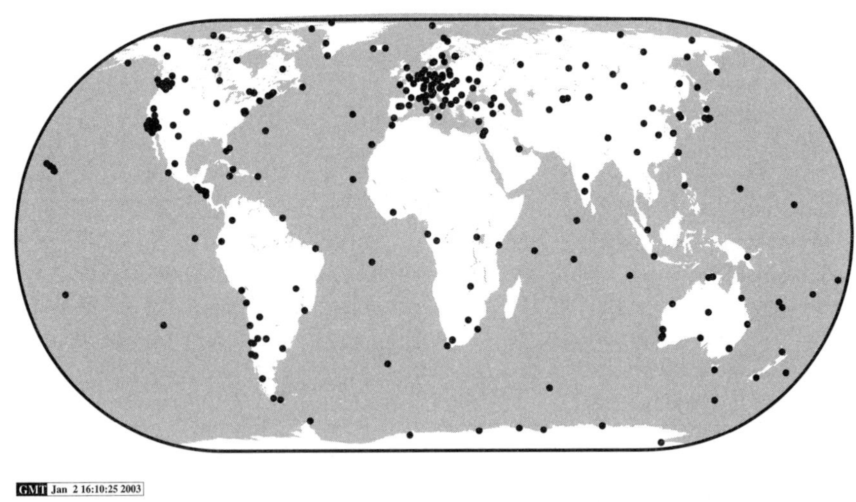

Figure 6.2.8 IGS permanent tracking network in 2002. (Courtesy NASA/JPL/Caltech)

TABLE 6.2.2 IGS Products Product Availability Standards and Quality of Service.

Product	Component	Accuracy	Latency	Updates
GPS Satellite Ephemeris and Satellite Clocks (1)				
Ultra-rapid (predicted half)	Orbits Sat. clocks	~5 cm ~3 ns;1500 ps	Predicted	4x daily
Ultra-rapid (observed half)	Orbits Sat. clocks	~3 cm ~150 ps; ~50 ps	3–9 hours	4x daily
Rapid	Orbits Sat. & sta. clocks	~2.5 cm ~75 ps; 25 ps	17–41 hours	Daily
Final	Orbits Sat. clocks	~2 cm ~75 ps; 20 ps	12–18 days	Weekly
Real time	Orbits Sat. clocks	~5 cm 300 ps; 120 ps	25 sec	Continuous
Geocentric Coordinates and Velocities of IGS Tracking Stations				
Final positions	Horizontal Vertical	3 mm 6 mm	11–17 days	Weekly
Final velocities	Horizontal Vertical	2 mm/yr 3 mm/yr	11–17 days	Weekly
Earth Rotation Parameters (2)				
Ultra-rapid (predicted half)	Polar motion Polar motion rate Length of day	~200 μas ~300 μas/day ~50 μas	Real time	4× daily
Ultra-rapid (observed half)	Polar motion Polar motion rate Length of day	~50 μas ~250 μas/day ~10 μas	3–9 hours	4× daily
Rapid	Polar motion Polar motion rate Length of day	~40 μas ~200 μas/day ~10 μas	17–41 hours	Daily
Final	Polar motion Polar motion rate Length of day	~30 μas ~150 μas/day ~0.01 μas	11–17 days	Weekly
Atmospheric Parameters (3)				
Final troposphere		~4 mm for ZPD	~3 weeks	Daily
Ionosphere TEC grid		2–8 TECU	<11 days	Weekly
Rapid iono TEC grid		2–9 TECU	<24 hours	Daily

Source: Strategic Plan 2013–2016, www.igs.org.

service (IGS-TRS) is the latest addition to the list of products. The service reached full operating capability in 2013 and provides orbit estimates every 5 or 60 seconds and satellite clock estimates every 5 seconds. It uses the Internet protocol NTRIP (Network Transport of RTCM via Internet Protocol) to deliver data to users. Users must run an NTRIP client application, which is available as open source software. Because IGS focuses on all GNSS systems, similar products will be available for other GNSS systems or will be available when these become operational. The first addition will be GLONASS products.

Understanding section (2) of the table, it helps to interpret the angular measurement unit. In units of radians we have 100 μas corresponding to 3.1 mm of equatorial rotation, and in angular units 10 μsec correspond to 4.6 mm of equatorial rotation. In section (3), a TEC unit (TECU) corresponds to 10^{16} electrons per 1 m^2 column.

The IGS is also very instrumental in creating specialized standards for data formats and promoting their universal acceptance. Examples include the series of receiver independent exchange formats (RINEX), standard formats for orbital files (SP3), the solution independent exchange format (SINEX), and the IONspheric Exchange format (IONEX). IGS, being what it is—a federation of voluntary participating agencies, universities, and enthusiastic individual scientists—has made a truly outstanding contribution to the development of GNSS applications. It is a vivid demonstration that the high-accuracy focus on GNSS is understood and valued globally.

6.2.7.2 Online Computing Online GNSS positioning computing services are probably the ones of most immediate interest to users who collect data in the field. These computing services accept input data in common format such as RINEX and use supplementary observations from existing CORS or IGS stations to produce the best solution in a given geodetic frame. Since these services are still evolving and adopt their services to ever-changing GNSS system constellations, it is best to obtain the most up-to-date information from the respective websites. In order to become familiar with these services and their products, it is best to submit test data sets. Testing and verifying is the best way of finding out which of them best fits one's needs. Most of them render the service free of charge.

APPS (Automatic Precise Positioning Service, http://apps.gdgps.net) is operated at the Jet Propulsion Laboratory, California, and is probably the oldest operating online service. A popular service with the U.S. surveying community is OPUS (Online Positioning User Service, http://www.ngs.noaa.gov/OPUS), operated by the National Geodetic Survey. The SCOUT (Scripts Coordinate Update Tool, http://sopac.ucsd.edu/cgi-bin/SCOUT.cgi) service is offered by Scripts Orbit and Permanent Array Center (SOPAC), University of California, San Diego. It traces its origin to the very significant geodetic activities in California in connection with earthquake monitoring. A recent addition to online processing is CenterPoint RTX Post-Processing by Trimble Navigation Limited (http://www.trimblertx.com). This service uses the company's proprietary worldwide CORS network. Other important services are CSRS-PPP (Canadian Spatial Reference System Precise Point Positioning, http://webapp.geod.nrcan.gc.ca/geod/tools-outils/ppp.php), GAPS (GPS Analysis and Positioning Software; http://gaps.gge.unb.ca/), and the Australian AUSPOS (http://www.ga.gov.au/bin/gps.pl), among others.

6.3 NAVIGATION SOLUTION

The navigation solution, also frequently referred to simply as point positioning, is the type of solution the GPS system was originally designed for, achieving position accuracy of about 1m. The solution is available at any time, depending of course on satellite visibility, anywhere on earth. This solution is frequently implemented in nonsurveying products, such as general consumer products or low-accuracy hand-held receivers, or is executed in the background as part of more elaborate solutions.

The navigation solution estimates the receiver coordinates, of course understood to be the receiver antenna coordinates and the receiver clock error using pseudorange observables. Carrier phases can be used to smooth the pseudoranges. There are several simplifying assumptions. The satellite positions at signal transmission times are assumed known and available from the broadcast ephemeris. The satellite clock corrections are also assumed to be available from the navigation message and must be applied to the observations. As discussed in Section 6.2.2, the satellite clocks are monitored by the control center, which models the clock offsets in terms of polynomials in time, and provides an estimate for the time group delay and the intersignal correction. The navigation solution does not estimate a separate satellite clock error. The ionospheric and tropospheric delays are also computed from models as explained in Chapter 8 and applied to the observations, and the hardware delays and multipath are neglected.

6.3.1 Linearized Solution

The navigation solution is based on the pseudorange equation (6.1.28). Applying the simplifying assumptions, we can write equations of the type

$$P_k^p = \|\boldsymbol{x}^p - \boldsymbol{x}_k\| - c\, dt_k + \varepsilon_{k,p}^p = \rho_k^p - \xi_k + \varepsilon_{k,p}^p \tag{6.3.1}$$

The position of the satellite at signal transmission is $\boldsymbol{x}^p$, the receiver location is $\boldsymbol{x}_k$, and $\xi_k = c\, dt_k$ is the receiver clock error expressed in units of meters. We use the notation where the superscript denotes the satellite and the subscript denotes the receiver. The four unknowns $\boldsymbol{x}_k$ and ξ_k can be computed using four pseudoranges measured simultaneously to four satellites. In case more satellites are observed at the same time, the parameters are estimated by the least-squares method. The effect of the earth's rotation during the signal travel time must be taken into consideration when computing the topocentric satellite distance ρ_k^p following Section 6.2.1. Since the receiver clock error ξ_k is solved together with the position coordinates at each epoch, a relatively inexpensive quartz crystal clock in the receiver is sufficient rather than an expensive atomic clock.

As we can see, the basic requirement for a solution to exist is that four satellites are visible at a given epoch. This visibility requirement is a key factor in the design of constellations that aspire to provide global coverage at any time. Modifications of the basic point positioning solution can be readily envisioned. For example, for applications on the ocean it might be possible to determine the ellipsoidal height sufficiently accurately from the height above the water and geoid undulation. Equation (6.3.1) could be parameterized in terms of ellipsoidal latitude, longitude, and height using transformations (4.3.26) through (4.3.30). Therefore, at least in principle, pseudoranges of three satellites are sufficient to determine horizontal position at sea. Other variations, such as connecting the receiver to an accurate atomic clock, could make the receiver clock parameter superfluous or permit a simple modeling of the receiver clock error.

The point positioning accuracy depends on the accuracy of the data provided by the navigation message, the receiver-satellite constellation geometry at the time of observation, the quality of the available ionospheric and tropospheric delays, and the actual measurement error. In practice, one prefers to observe all satellites in view in order to achieve redundancy and the best geometry. Dual-frequency users can use the ionospheric-free pseudorange function (6.1.38) to eliminate the effect of the ionosphere.

If the ordered set of parameters is

$$\boldsymbol{x}^T = [dx_k \quad dy_k \quad dz_k \quad \xi_k] \tag{6.3.2}$$

then the design matrix follows from (6.3.1) after linearization around the nominal station location $\boldsymbol{x}_{k,0}$

$$\boldsymbol{A} = \begin{bmatrix} \mathbf{e}_k^1 & 1 \\ \mathbf{e}_k^2 & 1 \\ \mathbf{e}_k^3 & 1 \\ \vdots & \vdots \end{bmatrix} \qquad \mathbf{e}_k^i = \left[\frac{x^i - x_k}{\rho_k^i} \quad \frac{y^i - y_k}{\rho_k^i} \quad \frac{z^i - z_k}{\rho_k^i} \right] \Bigg|_{\boldsymbol{x}_{k,0}} \tag{6.3.3}$$

The $\boldsymbol{A}$ matrix has as many rows as there are satellites observed, which typically includes all satellites in view. The horizontal 1×3 vector $\mathbf{e}_k^i$ contains the direction cosines for the line from the nominal station location to the satellite. The expression for the least-squares estimate $\boldsymbol{x} = -(\boldsymbol{A}^T\boldsymbol{PA})^{-1}\boldsymbol{A}^T\boldsymbol{P}\ell$ is given in Chapter 2. The weight matrix $\boldsymbol{P}$ is typically diagonal with the diagonal elements reflecting a weighting scheme that is a function of the satellite elevation angle.

We take note that the receiver clock estimate absorbs common mode errors of tropospheric and ionospheric delays and hardware delays. In general, the propagation media delays are a function of azimuth and elevation angle. For example, in the case of the ionosphere we consider splitting the total delay into an average station component $I_{k,P}$ and a component $\delta I^p_{k,P}$ that is a function of the direction of the satellite, giving $I^p_{k,P} = I_{k,P} + \delta I^p_{k,P}$. The tropospheric delay can conceptually be split in a similar manner. The receiver hardware delay is also a common error since it is the same for every satellite observation. These common components can be combined with the receiver clock error into a new epoch parameter ξ_k as

$$\xi_k = c\,dt_k + I_{k,P} + T_k + d_{k,P} \tag{6.3.4}$$

The symbols for ionosphere and the troposphere have no superscript p in this equation in order to identify them as common components at station k. It follows that unmodeled errors that are common to all observations at a particular station do not affect the estimated epoch position. Thus, modeling of the ionosphere and troposphere is useful only if it reduces the variability with respect to the common portion.

6.3.2 DOPs and Singularities

It has become common practice to use DOP (dilution of precision) factors to describe the effect of the receiver-satellite geometry on the accuracy of point positioning. The DOP factors are simple functions of the diagonal elements of the covariance matrix of the adjusted parameters, derived from the linearized model. In general,

$$\sigma = \sigma_0 \text{ DOP} \tag{6.3.5}$$

where σ_0 denotes the standard deviation of the observed pseudoranges, and σ is a one-number representation of the standard deviation of position and/or time. When computing DOPs, the pseudorange observations are considered uncorrelated and of the same accuracy, i.e., the weight matrix is $\boldsymbol{P} = \boldsymbol{I}$. The cofactor matrix of the adjusted receiver position and receiver clock is

$$\boldsymbol{Q}_x = (\boldsymbol{A}^T\boldsymbol{A})^{-1} = \begin{bmatrix} q_x & q_{xy} & q_{xz} & q_{x\xi} \\ & q_y & q_{yz} & q_{y\xi} \\ & & q_z & q_{z\xi} \\ \text{sym} & & & q_\xi \end{bmatrix} \tag{6.3.6}$$

It is often desirable to interpret results in the local geodetic coordinate system, which consists of the coordinates northing n, easting e, and up u. We transform the cofactor matrix (6.3.6) using (4.4.25). The result is

$$\boldsymbol{Q}_w = \begin{bmatrix} q_n & q_{ne} & q_{nu} & q_{n\xi} \\ & q_e & q_{eu} & q_{e\xi} \\ & & q_u & q_{u\xi} \\ \text{sym} & & & q_\xi \end{bmatrix} \tag{6.3.7}$$

The DOP factors are functions of the diagonal elements of (6.3.6) or (6.3.7). Table 6.3.1 shows the various dilution factors: vertical dilution of precision (VDOP) for the height, horizontal dilution of precision (HDOP) for horizontal positions,

TABLE 6.3.1 DOP Expressions.

$\text{VDOP} = \sqrt{q_u}$
$\text{HDOP} = \sqrt{q_n + q_e}$
$\text{PDOP} = \sqrt{q_n + q_e + q_u} = \sqrt{q_x + q_y + q_z}$
$\text{TDOP} = \sqrt{q_\xi}$
$\text{GDOP} = \sqrt{q_n + q_e + q_u + q_\xi}$

positional dilution of precision (PDOP), time dilution of precision (TDOP), and geometric dilution of precision (GDOP). The GDOP is a composite measure reflecting the geometry of the position and the time estimate. The DOPs can be computed in advance, given the approximate receiver location and a predicted satellite ephemeris.

The DOPs were useful for finding the best subset of satellites at the time when a receiver had only four or five channels. They are still useful in identifying a temporal weakness in geometry in kinematic applications, in particular in the presence of signal obstruction. As the constellation observed and the satellites approach a critical configuration, the columns of the design matrix become increasingly linearly dependent, the DOP values increase, and the resulting positioning solution becomes ill conditioned. We consider the case when all satellites, as viewed from the receiver location, appear to be located on the surface of a circular cone (Figure 6.3.1) or in a plane. The vertex of the cone in the figure is located at the receiver. The unit vector $\mathbf{e}_{axis}$ denotes the axis of the cone. The relevant portion of the linearized pseudorange equation is

$$dP_k^p = -\mathbf{e}_k^p \cdot d\mathbf{x}_k \tag{6.3.8}$$

where $\mathbf{e}_k^p$ is the unit vector given in (6.3.3). For all satellites that are located on the cone, the dot product

$$\mathbf{e}_k^i \cdot \mathbf{e}_{\text{axis}}^T = \cos\theta \tag{6.3.9}$$

is constant. The unit vector $\mathbf{e}_k^i$ represents the first three elements of row i of the design matrix. Therefore, (6.3.9) expresses a perfect linear dependency of these three

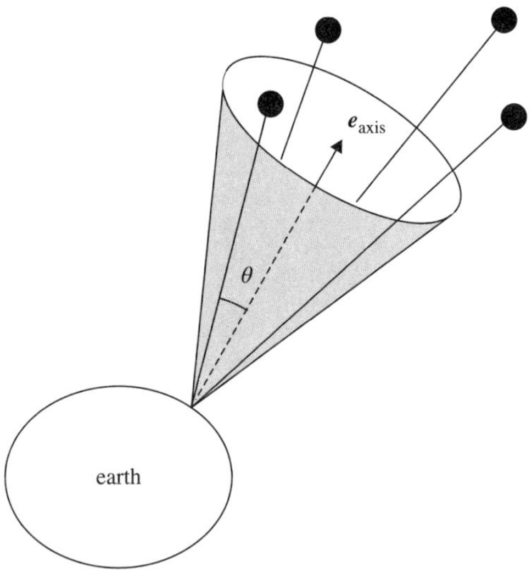

Figure 6.3.1 Critical configuration on a circular cone.

columns. The other critical configuration occurs when the satellites and the receiver are located in the same plane. In this case, the first three columns of the design matrix fulfill the cross-product vector function

$$\mathbf{e}_k^i \times \mathbf{e}_k^j = \mathbf{n} \tag{6.3.10}$$

where $\mathbf{n}$ is perpendicular to the plane. This degenerate solution can readily be visualized since the out-of-plane receiver location is not determined by the linearized model.

Critical configurations usually do not last long because of the continuous motion of the satellites. They present a problem only in continuous kinematic applications or very short rapid static positioning. The more satellites available with unobstructed line of sight, the less likely it is that a critical configuration will ever occur.

6.3.3 Nonlinear Closed Solution

The closed-form point positioning solution has been treated in detail in Grafarend and Shan (2002) and Awange and Grafarend (2002a,b). The reader might consult these publications for an in-depth study of closed expressions, for derivations, and additional references on the topic. Bancroft's (1985) solution is a very early, if not the first, closed-form solution. We merely summarize the solution using the notation of Goad (1998). In order to achieve compact expressions, we define the following product of two arbitrary vectors $\mathbf{g}$ and $\mathbf{h}$ as

$$\langle \mathbf{g}, \mathbf{h} \rangle \equiv \mathbf{g}^T \mathbf{M} \mathbf{h} \tag{6.3.11}$$

where $\mathbf{M}$ is the matrix

$$\mathbf{M} = \begin{bmatrix} {}_3\mathbf{I}_3 & \mathbf{0} \\ \mathbf{0} & -1 \end{bmatrix} \tag{6.3.12}$$

The relevant terms of the pseudorange (6.3.1) are

$$P_k^i + c\,dt_k = \|\mathbf{x}^i - \mathbf{x}_k\| \qquad 1 \le i \le 4 \tag{6.3.13}$$

Squaring both sides gives

$$\left(\mathbf{x}^i \cdot \mathbf{x}^i - P_k^{i2}\right) - 2\left(\mathbf{x}^i \cdot \mathbf{x}_k + P_k^i\, c\,dt_k\right) = -\left(\mathbf{x}_k \cdot \mathbf{x}_k - c^2 dt_k^2\right) \tag{6.3.14}$$

As can be verified, the four pseudorange equations can be written in the compact form

$$\alpha - \mathbf{BM} \begin{bmatrix} \mathbf{x}_k \\ c\,dt_k \end{bmatrix} + \Lambda \tau = 0 \tag{6.3.15}$$

where

$$\Lambda = \frac{1}{2}\left\langle \begin{bmatrix} \mathbf{x}_k \\ c\,dt_k \end{bmatrix}, \begin{bmatrix} \mathbf{x}_k \\ c\,dt_k \end{bmatrix} \right\rangle \tag{6.3.16}$$

$$\alpha^i = \frac{1}{2}\left\langle \begin{bmatrix} \mathbf{x}^i \\ P_k^i \end{bmatrix}, \begin{bmatrix} \mathbf{x}^i \\ P_k^i \end{bmatrix} \right\rangle \tag{6.3.17}$$

$$\boldsymbol{\alpha}^T = [\alpha^1 \quad \alpha^2 \quad \alpha^3 \quad \alpha^4] \tag{6.3.18}$$

$$\boldsymbol{\tau}^T = [1 \quad 1 \quad 1 \quad 1] \tag{6.3.19}$$

$$\mathbf{B} = \begin{bmatrix} x^1 & y^1 & z^1 & -P_k^1 \\ x^2 & y^2 & z^2 & -P_k^2 \\ x^3 & y^3 & z^3 & -P_k^3 \\ x^4 & y^4 & z^4 & -P_k^4 \end{bmatrix} \tag{6.3.20}$$

The solution of (6.3.15) is

$$\begin{bmatrix} \mathbf{x}_k \\ c\,dt_k \end{bmatrix} = \mathbf{MB}^{-1}(\Lambda\boldsymbol{\tau} + \boldsymbol{\alpha}) \tag{6.3.21}$$

We note, however, that Λ is also a function of the unknowns $\mathbf{x}_k$ and dt_k. We substitute (6.3.21) into (6.3.16), giving

$$\langle \mathbf{B}^{-1}\boldsymbol{\tau}, \mathbf{B}^{-1}\boldsymbol{\tau}\rangle \Lambda^2 + 2\left\{\langle \mathbf{B}^{-1}\boldsymbol{\tau}, \mathbf{B}^{-1}\boldsymbol{\alpha}\rangle - 1\right\}\Lambda + \langle \mathbf{B}^{-1}\boldsymbol{\alpha}, \mathbf{B}^{-1}\boldsymbol{\alpha}\rangle = 0 \tag{6.3.22}$$

This is a quadratic equation of Λ. Substituting its roots into (6.3.21) gives two solutions for the station coordinates $\mathbf{x}_k$. Converting the solution to geodetic coordinates and inspecting the respective ellipsoidal heights readily identifies the valid solution.

6.4 RELATIVE POSITIONING

In relative positioning, the vector between two stations is determined when two receivers observe simultaneously. If more than two receivers observe at the same time, we speak of a session network consisting of all the co-observing stations. Session solutions result in a set of correlated vectors between the stations. Our focus will be on short baselines, in which case certain small terms can be neglected.

In relative positioning, one tends to use across-receiver observations, or double- or triple-difference observations. In this subsection, we will deal with double differencing for positioning static receivers as was developed when GPS became available. In Section 6.6, the focus is on network-supported positioning using various forms of differential corrections. Since across-receiver differencing is becoming more popular, Chapter 7 is dedicated to using across-receiver differencing and kinematic applications.

We begin by providing the closed-form solution for double-differenced pseudoranges by modifying the solution presented in the previous section. This is followed by the linearized double-difference and triple-difference solutions. Several aspects of relative positioning are discussed, in particular the impact of the accuracy of the fixed station on the baseline length, the question of what constitutes independent baselines will be addressed, and we will have a look at innovative, but these days less important, antenna swap methods to get a kinematic survey started.

Although double differencing is certainly a popular method for baseline determination, we briefly review the merits of undifferenced processing as proposed by Goad (1985) and then discuss the ambiguity function technique as an alternative to double differencing. The subsection closes with reviewing some peculiarities encountered when processing GLONASS observations.

6.4.1 Nonlinear Double-Difference Pseudorange Solution

This solution is a modification of Bancroft's solution presented in Section 6.3.3 but applied to relative positioning. It assumes that the coordinates of one of the baseline stations, station $\mathbf{x}_k$, and the positions of the satellite are known. The coordinates of the other station, station $\mathbf{x}_m$, are to be determined using double-difference pseudoranges.

The double difference pseudorange equation (6.1.81) can be written as

$$
\begin{aligned}
P_{km,1}^{pq} &= \|\mathbf{x}^p - \mathbf{x}_k\| - \|\mathbf{x}^p - \mathbf{x}_m\| - \left\{ \|\mathbf{x}^q - \mathbf{x}_k\| - \|\mathbf{x}^q - \mathbf{x}_m\| \right\} + M_{km,1,P}^{pq} + \varepsilon_{km,1,P}^{pq} \\
&= \rho_{km}^{pq} + M_{km,1,P}^{pq} + \varepsilon_{km,1,P}^{pq} \qquad (6.4.1)
\end{aligned}
$$

For short baselines, we neglect the double-difference ionospheric and tropospheric terms, and also ignore the multipath. The hardware delays cancel as part of the double differencing. Consider the three independent double differences that can be formed from the observations of four satellites

$$
P_{km}^{pi} = \|\mathbf{x}^p - \mathbf{x}_k\| - \|\mathbf{x}^p - \mathbf{x}_m\| - \left\{ \|\mathbf{x}^i - \mathbf{x}_k\| - \|\mathbf{x}^i - \mathbf{x}_m\| \right\} \quad 1 \le i \le 3 \qquad (6.4.2)
$$

Let p denote the base satellite, in this case we have taken $p = 4$. Since the satellite coordinates and the station coordinates $\mathbf{x}_k$ are known, we can compute the auxiliary quantity Q

$$
Q_m^{pi} = P_{km}^{pi} - \|\mathbf{x}^p - \mathbf{x}_k\| + \|\mathbf{x}^i - \mathbf{x}_k\| \qquad (6.4.3)
$$

Comparing (6.4.2) and (6.4.3), we find that Q relates to the unknown $\mathbf{x}_m$ as

$$
Q_m^{pi} = -\|\mathbf{x}^p - \mathbf{x}_m\| + \|\mathbf{x}^i - \mathbf{x}_m\| \qquad (6.4.4)
$$

Following Chaffee and Abel (1994), we translate the origin of the coordinate system to satellite p

$$
\widetilde{\mathbf{x}}^i = \mathbf{x}^i - \mathbf{x}^p \qquad (6.4.5)
$$

Noting that in the translated coordinate system $\widetilde{\mathbf{x}}^p = 0$, we obtain from (6.4.4)

$$
Q_m^{pi} + \|\widetilde{\mathbf{x}}_m\| = \|\widetilde{\mathbf{x}}^i - \widetilde{\mathbf{x}}_m\| \qquad (6.4.6)
$$

Equations (6.4.6) and (6.3.13) are of the same form. Squaring (6.4.6) gives

$$
\left(\widetilde{\mathbf{x}}^i \cdot \widetilde{\mathbf{x}}^i - Q_m^{pi2}\right) - 2\left(\widetilde{\mathbf{x}}^i \cdot \widetilde{\mathbf{x}}_m + \|\widetilde{\mathbf{x}}_m\|\, Q_m^{pi}\right) = 0 \qquad (6.4.7)
$$

This equation can be verified using

$$\Lambda^2 = \widetilde{\boldsymbol{x}}_m \cdot \widetilde{\boldsymbol{x}}_m \tag{6.4.8}$$

$$\alpha^i = \frac{1}{2}\left\langle \begin{bmatrix} \widetilde{\boldsymbol{x}}^i \\ Q_m^{pi} \end{bmatrix}, \begin{bmatrix} \widetilde{\boldsymbol{x}}^i \\ Q_m^{pi} \end{bmatrix} \right\rangle \tag{6.4.9}$$

$$\boldsymbol{B} = \begin{bmatrix} \widetilde{x}^1 & \widetilde{y}^1 & \widetilde{z}^1 \\ \widetilde{x}^2 & \widetilde{y}^2 & \widetilde{z}^2 \\ \widetilde{x}^3 & \widetilde{y}^3 & \widetilde{z}^3 \end{bmatrix} \tag{6.4.10}$$

$$\boldsymbol{\tau}^T = \begin{bmatrix} -Q_m^{p1} & -Q_m^{p2} & -Q_m^{p3} \end{bmatrix} \tag{6.4.11}$$

$$\widetilde{\boldsymbol{x}}^m = \boldsymbol{B}^{-1}(\Lambda\boldsymbol{\tau} + \boldsymbol{\alpha}) \tag{6.4.12}$$

Substituting (6.4.12) in (6.4.8) gives the quadratic equation for Λ,

$$\left(\langle \boldsymbol{B}^{-1}\boldsymbol{\tau}, \boldsymbol{B}^{-1}\boldsymbol{\tau}\rangle - 1\right)\Lambda^2 + 2\langle \boldsymbol{B}^{-1}\boldsymbol{\tau}, \boldsymbol{B}^{-1}\boldsymbol{\alpha}\rangle\Lambda + \langle \boldsymbol{B}^{-1}\boldsymbol{\alpha}, \boldsymbol{B}^{-1}\boldsymbol{\alpha}\rangle = 0 \tag{6.4.13}$$

The two solutions for Λ are substituted in (6.4.12) to obtain two positions for $\widetilde{\boldsymbol{x}}_m$. The ellipsoidal height can be used to decide which of the positions is correct. Once $\widetilde{\boldsymbol{x}}_m$ is computed, the coordinates can be translated to $\boldsymbol{x}_m$ using (6.4.5).

The closed formulas can be generalized for more than four satellites. In this case, the number of rows in $\boldsymbol{B}$ equals the number of satellites or the number of double differences. We multiply (6.3.15) from the left with $\boldsymbol{B}^T$ and set $\overline{\boldsymbol{\alpha}} = \boldsymbol{B}^T\boldsymbol{\alpha}, \overline{\boldsymbol{B}} = \boldsymbol{B}^T\boldsymbol{B}$, and $\overline{\boldsymbol{\tau}} = \boldsymbol{B}^T\boldsymbol{\tau}$. Equations (6.3.22) or (6.4.13) can then be rewritten in the bar notation and solved for Λ.

6.4.2 Linearized Double- and Triple-Differenced Solutions

Relative positioning with carrier phases of short baselines is presented, assuming again that one baseline station, in this case $\boldsymbol{x}_k$, is known. The key element in this solution is dealing with the ambiguity parameters. Let there be R receivers observing S satellites at T epochs to generate RST carrier phase observations. In many cases, the data set might not be complete due to cycle slips and signal blockage. Let the undifferenced phase observations $\boldsymbol{\psi}$ be ordered first by epoch, then by receiver, and then by satellite. For epoch i, we have

$$\boldsymbol{\psi}_i = \begin{bmatrix} \varphi_1^1(i) & \cdots & \varphi_1^S(i) & \cdots & \varphi_R^1(i) & \cdots & \varphi_R^S(i) \end{bmatrix}^T \tag{6.4.14}$$

$$\boldsymbol{\psi} = \begin{bmatrix} \boldsymbol{\psi}_1 \\ \vdots \\ \boldsymbol{\psi}_T \end{bmatrix} \tag{6.4.15}$$

Regarding the stochastic model, we make the simplifying assumption that all carrier phase observations are uncorrelated and are of the same accuracy. Thus, the complete $RST \times RST$ cofactor matrix of the undifferenced phase observations is

$$\boldsymbol{Q}_{\varphi} = \sigma_{\varphi}^{2}\boldsymbol{I} \tag{6.4.16}$$

with σ_{φ} denoting the standard deviation of the phase measurement expressed in cycles.

The next task is to find the complete set of independent double-difference observations. We designate one station as the base station and one satellite as the base satellite. Without loss of generality, let station 1 be the base station and satellite 1 be the base satellite. The session network of R stations is now thought of as consisting of $R-1$ baselines emanating from the base station. There are $S-1$ independent double differences for each baseline. Thus, a total of $(R-1)(S-1)$ independent double differences can be computed for the session network. On the basis of the ordered observation vector (6.4.14) and the base station and base satellite ordering scheme, an independent set of double differences for epoch i is

$$\boldsymbol{\Delta}_i = \left[\varphi_{12}^{12}(i) \quad \cdots \quad \varphi_{12}^{1S}(i) \quad \cdots \quad \varphi_{1R}^{12}(i) \quad \cdots \quad \varphi_{1R}^{1S}(i)\right]^T \tag{6.4.17}$$

$$\boldsymbol{\Delta} = \begin{bmatrix} \boldsymbol{\Delta}_1 \\ \vdots \\ \boldsymbol{\Delta}_T \end{bmatrix} \tag{6.4.18}$$

The transformation from RST undifferenced observations to $(R-1)(S-1)T$ double-differenced observations is

$$\boldsymbol{\Delta} = \boldsymbol{D}\boldsymbol{\psi} \tag{6.4.19}$$

where $\boldsymbol{D}$ is the $(R-1)(S-1)T \times RST$ transformation matrix having elements -1, 1, and 0 arranged in a well-defined pattern that reflects the number of stations, satellites, and epochs.

For the ordered vector of triple-difference observations,

$$\boldsymbol{\nabla}_i = \left[\varphi_{12}^{12}(i+1,i) \quad \cdots \quad \varphi_{12}^{1S}(i+1,i) \quad \cdots \quad \varphi_{1R}^{12}(i+1,i) \quad \cdots \quad \varphi_{1R}^{1S}(i+1,i)\right]^T \tag{6.4.20}$$

$$\boldsymbol{\nabla} = \begin{bmatrix} \boldsymbol{\nabla}_1 \\ \vdots \\ \boldsymbol{\nabla}_{T-1} \end{bmatrix} \tag{6.4.21}$$

we have

$$\boldsymbol{\nabla} = \boldsymbol{T}\boldsymbol{\Delta} = \boldsymbol{T}\boldsymbol{D}\boldsymbol{\psi} \tag{6.4.22}$$

The matrix $\boldsymbol{T}$ also has elements -1, 1, and 0 arranged in a well-defined pattern.

The double- and triple-difference observations are linear functions of the undifferencedcarrier phases. By applying covariance propagation and taking the cofactor matrix (6.4.16) into account, the respective cofactor matrices are

$$\mathbf{Q}_{\Delta} = \sigma_{\varphi}^2 \mathbf{D}\mathbf{D}^T \tag{6.4.23}$$

$$\mathbf{Q}_{\nabla} = \mathbf{T}\mathbf{Q}_{\Delta}\mathbf{T}^T \tag{6.4.24}$$

The double-difference cofactor matrix $\mathbf{Q}_{\Delta}$ is block-diagonal.The triple-difference cofactor matrix $\mathbf{Q}_{\nabla}$ is band-diagonal for $T > 3$. The triple-difference observations between consecutive (adjacent) epochs are correlated. The inverse of the triple-difference cofactor matrix, which is required in the least-squares solution, is a full matrix.

The relevant terms of the double-difference carrier phase equation (6.1.82) are

$$\begin{aligned}\varphi_{km}^{pq} &= \frac{f}{c}\left\{\|\mathbf{x}^p - \mathbf{x}_k\| - \|\mathbf{x}^p - \mathbf{x}_m\| - \|\mathbf{x}^q - \mathbf{x}_k\| + \|\mathbf{x}^q - \mathbf{x}_m\|\right\} \\ &\quad + N_{km}^{pq} + M_{km,\varphi}^{pq} + \varepsilon_{km,\varphi}^{pq} \\ &= \frac{f}{c}\rho_{km}^{pq} + N_{km}^{pq} + M_{km,\varphi}^{pq} + \varepsilon_{km,\varphi}^{pq}\end{aligned} \tag{6.4.25}$$

The residual ionospheric and tropospheric terms are not explicitly listed in (6.4.25) since they are expected to cancel over short baselines. Notice the presence of the ambiguity term N_{km}^{pq} in (6.4.25) as compared to the expression (6.4.1) for pseudoranges. Assuming that the station coordinates $\mathbf{x}_k$ are known, the parameters to be estimated are $\mathbf{x}_m$ and the double-difference ambiguities. There are $(R-1)(S-1)$ double-difference ambiguities if there are no cycle slips. The multipath $M_{km,\varphi}^{pq}$ is typically treated as a model error and ignored. A row of the design matrix consists of the partial derivatives with respect to the coordinates of station m

$$\frac{\partial \varphi_{km}^{pq}}{\partial \mathbf{x}_m} = \frac{f}{c}\left(\mathbf{e}_m^p - \mathbf{e}_m^q\right) \tag{6.4.26}$$

and contains a 1 in the column of the respective double-difference ambiguity parameter, and zero elsewhere. The least-squares solution that estimates the parameters

$$\mathbf{x} = \begin{bmatrix}\mathbf{x}_m \\ \mathbf{b}\end{bmatrix} \tag{6.4.27}$$

$$\mathbf{b}^T = \begin{bmatrix}N_{12}^{12} & \cdots & N_{12}^{1S} & \cdots & N_{1R}^{12} & \cdots & N_{1R}^{1S}\end{bmatrix}^T \tag{6.4.28}$$

is called the double-difference float solution. If it is possible to also constrain the estimated ambiguities to integers, then we speak of the fixed solution. See Section 6.4.5 for details on ambiguity fixing.

The partial derivatives of triple differences follow from those of double differences by differencing

$$\frac{\partial \varphi_{km}^{pq}(j,i)}{\partial \boldsymbol{x}_m} = \frac{\partial \varphi_{km}^{pq}(j)}{\partial \boldsymbol{x}_m} - \frac{\partial \varphi_{km}^{pq}(i)}{\partial \boldsymbol{x}_m} \tag{6.4.29}$$

since the triple difference is the difference of two double differences. The design matrix of the triple difference contains no columns for the ambiguities because the ambiguities cancel during the differencing across time.

As to software implementation, it is important to avoid repetitious computation when computing the $\boldsymbol{Q}_\Delta$ and $\boldsymbol{Q}_\nabla$ matrices and fully explore the pattern of the $\boldsymbol{D}$ and $\boldsymbol{T}$ matrices to avoid unnecessary zero multiplications. The respective approaches for time- and space-saving software implementations are well known and not discussed in detail here.

The above processing scheme also applies to dual-frequency or multifrequency observations. For each frequency, there is a separate set of ambiguities to be estimated. Similarly, one could transform dual-frequency observations to wide lanes and narrow lanes and process these, taking advantage of being able to fix the wide-lane ambiguities first. Also, relative positioning with pseudoranges and linearized model is very similar to the one for carrier phases. Comparing (6.4.1) and (6.4.25), the major difference is the lack of an ambiguity term in the pseudorange expression. Consequently, there are no ambiguity parameters. The partial derivatives of the coordinates station coordinates are

$$\frac{\partial P_{km}^{pq}}{\partial \boldsymbol{x}_m} = \boldsymbol{e}_m^p - \boldsymbol{e}_m^q \tag{6.4.30}$$

Therefore, in the case of short baselines where the ionospheric and tropospheric effects can be neglected and the multipath is omitted or suitably considered by weighting the observations as a function of the satellite elevation angle, there are only three parameters to be estimated in the double-difference pseudorange solutions.

A general remark is in order regarding cancelation of unmodeled errors in double differencing. For short baselines, the errors common to both stations tend to cancel during differencing. Because the ionospheric and the tropospheric corrections are highly correlated over short distances, most of their delays are common to both stations. An exception might be a tropospheric correction of nearby stations with significantly different elevations. It is useful to apply tropospheric and ionospheric corrections from external sources if these provide accurate differential corrections between the stations. If this is not the case, because, say, the assumed meteorological data are not representative of the actual tropospheric conditions, it might be better to apply no corrections and rely on common-mode elimination. Because of the cancellation of most of the effects of the propagation media, the clock errors, and hardware delays, the technique of relative positioning has become especially popular in surveying. Although the double-difference ambiguity parameters might initially be perceived as a nuisance, they provide a unique vehicle to improving the solution if they can be successfully constrained to integers.

6.4.3 Aspects of Relative Positioning

Although relative posting is well established, this subsection looks at some of the things that one needs to be aware of. Even in relative positioning over short baselines, near singularities may occur when some satellite signals are blocked by obstacles. Also, the user should be aware of the implications of holding one baseline station fixed, i.e., be able to separate absolute positioning from relative positioning. In session solutions, only independent baselines should be used. This subsection also includes material on the antenna-swapping technique which, historically speaking, helped jumpstart kinematic applications.

6.4.3.1 Singularities Similar to point positioning, there are also critical satellite configurations to be concerned with in relative positioning. Whereas the satellites cannot be located simultaneously on a perfectly circular cone as viewed from each of the stations, however, the satellites could be located approximately on circular cones in the case of short baselines, resulting in near singularities. Consider the relevant portion of the double-difference pseudorange or scaled carrier phase equation

$$P_{km}^{pq} = \rho_k^p - \rho_m^p - \left[\rho_k^q - \rho_m^q\right] + \cdots \tag{6.4.31}$$

Let us take station m to be the known location. Then the relevant part of the linearization is

$$dP_{km}^{pq} = -\mathbf{e}_k^p \cdot d\mathbf{x}_k + \mathbf{e}_k^q \cdot d\mathbf{x}_k = \left[\mathbf{e}_k^q - \mathbf{e}_k^p\right] \cdot d\mathbf{x}_k + \cdots \tag{6.4.32}$$

It can readily be verified that the direction vectors $\mathbf{e}_k^i$ are related to the vectors of direction $\mathbf{e}_c^i$ from the center of the baseline as

$$\mathbf{e}_k^i = \mathbf{e}_c^i + \boldsymbol{\varepsilon}_k^i \tag{6.4.33}$$

where the components of $\boldsymbol{\varepsilon}_k^i$ are of the order $O\left(b/\rho_k^p\right)$. The symbol b denotes the length of the baseline. Using (6.4.33), equation (6.4.32) becomes

$$dP_{km}^{pq} = \left[\mathbf{e}_c^q - \mathbf{e}_c^p + \boldsymbol{\varepsilon}_k^q - \boldsymbol{\varepsilon}_k^p\right] \cdot d\mathbf{x}_k + \cdots \tag{6.4.34}$$

For the special case where the vertex of the circular cone (see Figure 6.3.1) is at the center of the baseline, the condition

$$\mathbf{e}_c^i \cdot \mathbf{e}_{axis} = \cos\theta \tag{6.4.35}$$

is valid for all satellites on the cone. This means that the dot products

$$\left[\mathbf{e}_c^q - \mathbf{e}_c^p + \boldsymbol{\varepsilon}_k^q - \boldsymbol{\varepsilon}_k^p\right] \cdot \mathbf{e}_{axis} = \left[\boldsymbol{\varepsilon}_k^q - \boldsymbol{\varepsilon}_k^p\right] \cdot \mathbf{e}_{axis} \tag{6.4.36}$$

are of the order $O\left(b/\rho_k^p\right)$. Such a product applies to each double-difference observation. Therefore, we are dealing with a near-singular situation since the columns of the double-difference design matrix are nearly dependent. The shorter the baseline, the more likely it is that a near singularity is noticeable. As stated for the navigation solution, if all satellites in view are observed and there are no line-of-sight obstructions, such near singularity does not occur.

6.4.3.2 *Impact of a Priori Position Error* At least in the early days of GPS satellite surveying, a frequent concern was the need for a priori knowledge of the geocentric station position of the fixed station, as well as the impact of ephemeris errors on relative positioning. Of course, in today's situation, if one starts the survey at a known location with centimeter accuracy and uses the precise ephemeris, these concerns are no longer valid. However, looking at the geometry of this problem helps us understand why GPS provides accurate relative positions and less accurate geocentric positions.

The answer to these concerns lies again in the linearized double-difference equations. Without loss of generality, it is sufficient to investigate the difference between one satellite and two ground stations. Scaled to distances, the relevant portion of the double-difference equation is

$$P_{km}^{pq}(t) = \rho_k^p(t) - \rho_m^p(t) + \cdots \tag{6.4.37}$$

The linearized form is

$$dP_{km}^{pq} = -\mathbf{e}_k^p \cdot d\mathbf{x}_k + \mathbf{e}_m^p \cdot d\mathbf{x}_m + [\mathbf{e}_k^p - \mathbf{e}_m^p] \cdot d\mathbf{x}^p \tag{6.4.38}$$

Next, we transform the coordinate corrections into their differences and sums. This is accomplished by

$$d\mathbf{x}_k - d\mathbf{x}_m = d(\mathbf{x}_k - \mathbf{x}_m) = d\mathbf{b} \tag{6.4.39}$$

$$\frac{d\mathbf{x}_k + d\mathbf{x}_m}{2} = d\left(\frac{\mathbf{x}_k + \mathbf{x}_m}{2}\right) = d\mathbf{x}_c \tag{6.4.40}$$

The difference (6.4.39) represents the change in the baseline vector, i.e., the change in length and orientation of the baseline, and (6.4.40) represents the change in the geocentric location of the baseline center. The latter can be interpreted as the translatory uncertainty of the baseline, or the uncertainty of the fixed baseline station. Transforming (6.4.38) to the difference and sum gives

$$dP_{km}^{pq} = -\frac{1}{2}\left[\mathbf{e}_k^p + \mathbf{e}_m^p\right] \cdot d\mathbf{b} - \left[\mathbf{e}_k^p - \mathbf{e}_m^p\right] \cdot d\mathbf{x}_c + \left[\mathbf{e}_k^p - \mathbf{e}_m^p\right] \cdot d\mathbf{x}^p \tag{6.4.41}$$

There is a characteristic difference in magnitude between the first bracket and the others. Allowing an error of the order $O(b/\rho_k^p)$, the first bracket simplifies to $2\mathbf{e}_m^p$ or $2\mathbf{e}_k^p$. The second and third brackets are of opposite signs but of the same magnitude. It is readily verified that the terms in the latter two brackets are of the order $O(b/\rho_k^p)$. When the baseline vector is defined by

$$\mathbf{b} \equiv \rho_m^p - \rho_k^p \tag{6.4.42}$$

Equation (6.4.41) becomes, after neglecting the usual small terms,

$$dP_{km}^{pq} = -\mathbf{e}_m^p \cdot d\mathbf{b} + \frac{\mathbf{b}}{\rho_m^p} \cdot d\mathbf{x}_c - \frac{\mathbf{b}}{\rho_m^p} \cdot d\mathbf{x}^p \tag{6.4.43}$$

Equating the first two terms in (6.4.43), we get the relative impact of changes in the baseline and the translatory position of the baseline from

$$\rho_m^p \cdot d\mathbf{b} = \mathbf{b} \cdot d\mathbf{x}_c \tag{6.4.44}$$

Similarly, changes in the baseline vector and ephemeris position are related by

$$\rho_m^p \cdot d\mathbf{b} = \mathbf{b} \cdot d\mathbf{x}^p \tag{6.4.45}$$

These relations are usually quoted in terms of absolute values, thereby neglecting the cosine terms of the dot product. In this sense, a rule of thumb for relating baseline accuracy, a priori geocentric position accuracy, and ephemeris accuracy is

$$\frac{\|d\mathbf{b}\|}{b} = \frac{\|d\mathbf{x}_c\|}{\rho_m^p} = \frac{\|d\mathbf{x}^p\|}{\rho_m^p} \tag{6.4.46}$$

Equation (6.4.46) shows that the accuracy requirements for the a priori geocentric station coordinates and the satellite orbital positions are the same. The accuracy requirement is a function of the baseline length. This means that for short baselines, an accurate position of the reference station might not be required and that the simple point positioning might be sufficient. A 1000 km line can be measured to 1 cm if the ephemeris errors and the geocentric location error can be reduced to 0.2m, according to this rule of thumb. Another interpretation is that the ratio of relative positioning capability $d\mathbf{b}$ to absolute positioning capability $d\mathbf{x}_c$ is about baseline length over the topocentric satellite distance. Equation (6.4.46) explains that GPS observations from closely spaced receivers, as is the case for short baselines, do not provide accurate geocentric locations but provide accurate relative locations, thus the practice of holding one station fixed in relative location determinations. Of course, it is understood that any error in the coordinates of the fixed station propagates directly into the coordinates of the newly determined stations, i.e., speaking in terms of geocentric coordinates, the new position can only be as good as the fixed position. Inner constraint solutions of vector networks allow an objective assessment of the accuracy of the relative positioning achieved. See Section 4.4.4.

6.4.3.3 Independent Baselines The ordering scheme of base station and base satellite used for identifying the set of independent double-difference observations is not the only scheme available; it has been used here because of its simplicity. An example where the base station and base satellite scheme requires a slight modification occurs when the base station does not observe at a certain epoch due to temporary signal blockage. If station 1 does not observe, then the double difference φ_{23}^{pq} can be computed for this particular epoch. Because of the relationship

$$\varphi_{23}^{pq} = \varphi_{13}^{pq} - \varphi_{12}^{pq} \tag{6.4.47}$$

the ambiguity N_{23}^{pq} is related to the base station ambiguities as

$$N_{23}^{pq} = N_{13}^{pq} - N_{12}^{pq} \tag{6.4.48}$$

Introduction of N_{23}^{pq} as an additional parameter would create a singularity of the normal matrix because of the dependency expressed in (6.4.48). Instead of adding this new ambiguity, the base station ambiguities N_{12}^{pq} and N_{13}^{pq} are given the coefficients 1 and −1, respectively, in the design matrix. The partial derivatives with respect to the station coordinates can be computed as required by (6.4.47) and entered directly into the design matrix, because the respective columns are already there. A similar situation arises when the base satellite changes. The linear functions in this case are

$$\varphi_{km}^{23} = \varphi_{km}^{13} - \varphi_{km}^{12} \tag{6.4.49}$$

$$N_{km}^{23} = N_{km}^{13} - N_{km}^{12} \tag{6.4.50}$$

The respective elements for the base satellite ambiguities in the design matrix are, again, 1 and −1.

One must identify $(R-1)(S-1)$ independent double-difference functions in network solutions. In session networks that contain a mixture of long and short baselines, it might be important to take advantage of short baselines because the respective unmodeled errors (troposphere, ionosphere, and possibly orbit) are expected to be small. Fixing the ambiguities to integers adds strength to the solution. This additional strength gained by fixing the ambiguities of a short baseline may also make it possible to fix the ambiguities for the next longer baseline, even though the ambiguity search algorithms might not have been successful without that constraint. The technique is sometimes referred to as "boot-strapping" from shorter to longer baselines. A suitable procedure would be to take baselines in all combinations and order them by increasing length and identify the set of independent baselines, starting with the shortest.

There are several schemes available to identify independent baselines and observations. Hilla and Jackson (2000) report using a tree structure and edges. Here we follow the suggestion of Goad and Mueller (1988) because it highlights yet another useful application of the Cholesky decomposition. Assume that matrix $\boldsymbol{D}$ of (6.4.19) reflects the ordering suggested here, i.e., the first rows of $\boldsymbol{D}$ refer to the double differences of the shortest baseline, the next set of rows refer to the second shortest baseline, and so on. We write the cofactor matrix (6.4.23) as

$$\boldsymbol{Q}_\Delta = \sigma_0^2 \boldsymbol{D}\boldsymbol{D}^T = \sigma_0^2 \boldsymbol{L}\boldsymbol{L}^T \tag{6.4.51}$$

where $\boldsymbol{L}$ denotes the Cholesky factor (A.3.54). The elements of the cofactor matrix $\boldsymbol{Q}_\Delta$ are

$$q_{ij} = \sum_k d_i(k) d_j(k) \tag{6.4.52}$$

where $d_i(k)$ denotes the ith row of the matrix $\boldsymbol{D}$. It is readily verified that the ith and jth columns of $\boldsymbol{Q}_\Delta$ are linearly dependent if the ith and jth rows of $\boldsymbol{D}$ are linearly dependent. In such a case, $\boldsymbol{Q}_\Delta$ is singular. This situation exists when two double differences are linearly dependent. The diagonal element j of the Cholesky factor $\boldsymbol{L}$ will be zero. Thus, one procedure for eliminating the dependent observations is to carry out the computation of $\boldsymbol{L}$ and discard those double differences that cause a zero on the diagonal. The matrix $\boldsymbol{Q}_\Delta$ can be computed row by row starting at the top, i.e., the double differences can be processed sequentially one at a time, from the top to the bottom. For each double difference, the respective row of $\boldsymbol{L}$ can be computed. In this way, the dependent observations can be immediately discovered and removed. Only the independent observations remain. The process ends as soon as the $(R-1)(S-1)$ double differences have been found.

If all receivers observe all satellites for all epochs, this identification process needs to be carried out only once. The matrix $\boldsymbol{L}$, since it is now available, can be used to decorrelate the double differences. The corresponding residuals might be difficult to interpret but could be transformed to the original observational space using $\boldsymbol{L}$ again.

6.4.3.4 Antenna Swap Technique In view of modern processing of multifrequency observations, the antenna swapping technique may today be perceived as impractical, although when introduced by Remondi (1985) it was an innovative and major step forward in making kinematic surveying practical at the time. Although today multifrequency observations are processed recursively as explained in Chapter 7 and the transition to the kinematic survey is automatic as soon as the ambiguities are fixed, let us step back for a moment to see how it used to be (at the time surveyors mostly operated single-frequency receivers).

Basically, a kinematic survey requires an initialization. This means the double-difference ambiguities are resolved first and then held fixed while other points are being surveyed, assuming of course that no cycle slips occurred while the rover moves or that cycle slips are repaired appropriately. A simple way for initial determination of ambiguities is to occupy two known stations. The procedure works best for short baselines where the ionospheric and tropospheric disturbances are negligible. The double-difference equation (6.4.25) can be readily solved for the ambiguity

$$N_{km}^{pq} = \varphi_{km}^{pq} - \lambda^{-1}\rho_{km}^{pq} \tag{6.4.53}$$

when both receiver locations $\boldsymbol{x}_k$ and $\boldsymbol{x}_m$ are known. Usually, simple rounding of the computed values is sufficient to obtain the integers. Once the initial ambiguities are known, the kinematic survey can begin. Let the subscripts k and m now denote the fixed and the moving receiver, then

$$\rho_{m}^{pq} = \rho_{k}^{pq} - \lambda\left[\varphi_{km}^{pq} - N_{km}^{pq}\right] \tag{6.4.54}$$

If four satellites are observed simultaneously, there are three equations like (6.4.54) available to compute the coordinates of the moving receiver $\boldsymbol{x}_m$. If more than four satellites are available, the usual least-squares approach is applicable and cycle slips can be repaired from phase observations. In principle, if five satellites are observed we can repair one slip per epoch; if six satellites are observed, two slips can occur at the same time, etc.

Remondi (1985) introduced the antenna swap procedure in order to initialize the ambiguities for kinematic surveying, requiring only one known station. Assume that four or more satellites were observed at least for one epoch while receiver R_1 and its antenna were located at station k and receiver R_2 and its antenna were at station m.

This is followed by the antenna swap, meaning that antenna R_1 moves to station m and antenna R_2 moves to station k, followed by at least one epoch of observations to the same satellites. The antennas remain connected to their respective receivers. During data processing, it is assumed that the antennas never moved. Using an expanded form of notation to identify the receiver and the respective observation, a double difference at epoch 1 when R_1 was at k and at epoch t when R_1 was at m can be written, respectively, as

$$\varphi_{km}^{pq}(R_2 - R_1, 1) = \lambda^{-1}\left[\rho_k^p(R_1, 1) - \rho_k^q(R_1, 1) - \rho_m^p(R_2, 1) + \rho_m^q(R_2, 1)\right] + N_{km}^{pq} \tag{6.4.55}$$

$$\varphi_{km}^{pq}(R_2 - R_1, t) = \lambda^{-1}\left[\rho_m^p(R_1, t) - \rho_m^q(R_1, t) - \rho_k^p(R_2, t) + \rho_k^q(R_2, t)\right] + N_{km}^{pq} \tag{6.4.56}$$

Notice the sequence of subscripts on the right-hand side of (6.4.56). Differencing both observations gives

$$\begin{aligned}\varphi_{km}^{pq}(R_2 - R_1, 1) - \varphi_{km}^{pq}(R_2 - R_1, t) &= \lambda^{-1}\left[\rho_k^{pq}(t) - \rho_m^{pq}(t) + \rho_k^{pq}(1) - \rho_m^{pq}(1)\right] \\ &\approx 2\lambda^{-1}\left[\rho_k^{pq}(t) - \rho_m^{pq}(t)\right]\end{aligned} \tag{6.4.57}$$

Equation (6.4.57) can be solved for $\boldsymbol{x}_m$, given $\boldsymbol{x}_k$ and observations to at least four satellites (three double differences). Once the position of m is known, the ambiguities can be computed from (6.4.53).

If the topocentric satellite distances would not change during the antenna swapping due to motion of the satellites, the antenna swap technique would yield a baseline vector of twice the actual length. The geometry of antenna swap can be readily visualized in a simplified one-dimensional situation. Consider a horizontal baseline and a satellite located somewhere along the extension of that baseline. As one antenna moves from one end of the baseline to the other, it will register, let's say, a positive accumulated carrier phase change equal to the length of the baseline. As the other antenna switches location, it will also register a carrier phase change equal to the negative of the length of the baseline. Both receivers together will register a motion of twice the length of the baseline.

Initialization by antenna swap on the ground is conveniently done for a very short baseline of a couple of meters. A typical point positioning solution for $\boldsymbol{x}_k$ is sufficient for such short baselines.

6.4.4 Equivalent Undifferenced Formulation

The double-difference algorithm can readily be changed to an equivalent one of single-difference processing. Following Goad (1985) we write the undifferenced phase equation (6.1.30) as

$$\varphi_k^p(t) = \lambda_1^{-1}\rho_k^p + \xi_k^p + \varepsilon_{k,\varphi}^p \tag{6.4.58}$$

where ξ_k^p includes the ambiguity parameter, the receiver and satellite clock terms, ionospheric and tropospheric effects, hardware delays, and multipath. Considering again station 1 as base station and satellite 1 as base satellite, then the undifferenced equations, comprising a double-difference observation containing satellite 2, can be written as

$$\begin{aligned}
\varphi_1^1(t) &= \lambda^{-1}\rho_1^1 + \xi_1^1 + \varepsilon_{1,\varphi}^1 \\
\varphi_2^1(t) &= \lambda^{-1}\rho_2^1 + \xi_2^1 + \varepsilon_{2,\varphi}^1 \\
\varphi_1^2(t) &= \lambda^{-1}\rho_1^2 + \xi_1^2 + \varepsilon_{1,\varphi}^2 \\
\varphi_2^2(t) &= \lambda^{-1}\rho_2^2 + \xi_2^2 + \varepsilon_{2,\varphi}^2
\end{aligned} \tag{6.4.59}$$

Next we compute the double-difference term ξ_{12}^{12},

$$\xi_{12}^{12} = \left(\xi_1^1 - \xi_2^1\right) - \left(\xi_1^2 - \xi_2^2\right) = N_{12}^{12} + M_{12,\varphi}^{12} + \varepsilon_{12,\varphi}^{12} \tag{6.4.60}$$

in which we have neglected the double-difference ionospheric and tropospheric terms. Solving (6.4.60) for ξ_2^2 and substituting into (6.4.59) gives

$$\begin{aligned}
\varphi_1^1(t) &= \lambda^{-1}\rho_1^1 + \xi_1^1 + \varepsilon_{1,\varphi}^1 \\
\varphi_2^1(t) &= \lambda^{-1}\rho_2^1 + \xi_2^1 + \varepsilon_{2,\varphi}^1 \\
\varphi_1^2(t) &= \lambda^{-1}\rho_1^2 + \xi_1^2 + \varepsilon_{1,\varphi}^2 \\
\varphi_2^2(t) &= \lambda^{-1}\rho_2^2 + N_{12}^{12} + \xi_1^1 + \xi_2^1 + \xi_1^2 + \varepsilon_{2,\varphi}^2
\end{aligned} \tag{6.4.61}$$

This is the required reformulation. The undifferenced observations are parameterized in terms of epoch parameters by ξ_1^1, ξ_2^1, and ξ_1^2, which refer to either the base station or the base satellite, and the double difference ambiguity N_{12}^{12}. Note that only the nonbase station nonbase satellite observation contains the ambiguity term.

Given that the ξ -parameters must be estimated every epoch and the stations coordinates and ambiguity parameters are common to all epochs, the resulting normal matrix has a well-known pattern. Although the size matrix increases quickly with time, it can be efficiently stored in computer memory and the normal equation can be solved quickly using either matrix partitioning techniques or recursive least squares. The advantage of the undifferenced formulation is that no variance-covariance propagation is needed for the observations, i.e., the variance-covariance matrix of the undifferenced observations is diagonal.

6.4.5 Ambiguity Function

The least-squares techniques discussed above require partial derivatives and the minimization of $\boldsymbol{v}^T\boldsymbol{P}\boldsymbol{v}$, with $\boldsymbol{v}$ and $\boldsymbol{P}$ being the double-difference residuals and double-difference weight matrix. The derivatives and the discrepancy terms depend on the assumed approximate coordinates of the stations. The least-squares solution is iterated until the solution converges. In the case of the ambiguity function technique, we search for station coordinates that maximize the cosine of the residuals. Consider again the double-difference observation equation

$$v_{km}^{pq} = \varphi_{km,a}^{pq} - \varphi_{km,b}^{pq} = \frac{f}{c}\rho_{km,a}^{pq} + N_{km,a}^{pq} - \varphi_{km,b}^{pq} \tag{6.4.62}$$

In usual adjustment notation, the subscripts a and b denote the adjusted and the observed values, respectively. In (6.4.62), we have neglected again the residual double-difference ionospheric and tropospheric terms, as well as the signal multipath term. The residuals in units of radians are

$$\psi_{km}^{pq} = 2\pi\ v_{km}^{pq} \tag{6.4.63}$$

The key idea of the ambiguity function technique is to realize that a change in the integer N_{km}^{pq} changes the function ψ_{km}^{pq} by a multiple of 2π and that the cosine of this function is not affected by such a change because

$$\cos\left(\psi_{km,L}^{pq}\right) = \cos\left(2\pi v_{km,L}^{pq}\right) = \cos\left[2\pi\left(v_{km,L}^{pq} + \Delta N_{km,L}^{pq}\right)\right] \tag{6.4.64}$$

where $\Delta N_{km,L}^{pq}$ denotes the arbitrary integer. The subscript L, denoting the frequency identifier, has been added for the purpose of generality.

There are $2(R - S)(S - 1)$ double differences available for dual-frequency observations. If we further assume that all observations are equally weighted, then the sum of the squared residuals becomes, with the help of (6.4.63),

$$\mathbf{v}^T\mathbf{P}\mathbf{v}\left(\mathbf{x}_m, N_{km,L}^{pq}\right) = \sum_{L=1}^{2}\sum_{m=1}^{R-1}\sum_{q=1}^{S-1}\left(v_{km,L}^{pq}\right)^2 = \frac{1}{4\pi^2}\sum_{L=1}^{2}\sum_{m=1}^{R-1}\sum_{q=1}^{S-1}\left(\psi_{km,L}^{pq}\right)^2 \tag{6.4.65}$$

If the station coordinates $\mathbf{x}_k$ are known, the function could be minimized by varying the coordinates $\mathbf{x}_m$ and the ambiguities using least-squares estimation. The ambiguity function is defined as

$$\begin{aligned} AF(\mathbf{x}_m) &= \sum_{L=1}^{2}\sum_{m=1}^{R-1}\sum_{q=1}^{S-1}\cos\left(\psi_{km,L}^{pq}\right) \\ &= \sum_{L=1}^{2}\sum_{m=1}^{R-1}\sum_{q=1}^{S-1}\cos\left\{2\pi\left[\frac{f_L}{c}\rho_{km,a}^{pq} + N_{km,L,a}^{pq} - \varphi_{km,L,b}^{pq}\right]\right\} \\ &= \sum_{L=1}^{2}\sum_{m=1}^{R-1}\sum_{q=1}^{S-1}\cos\left\{2\pi\left[\frac{f_L}{c}\rho_{km,a}^{pq} - \varphi_{km,L,b}^{pq}\right]\right\} \end{aligned} \tag{6.4.66}$$

The small double-difference ionospheric, tropospheric, and multipath terms are not listed explicitly in this equation, although they are present and will affect the ambiguity function technique just as they affect the other solution methods. Nevertheless, if we assume that these terms are negligible, and that the receiver positions are perfectly known, then the maximum value of the ambiguity function (6.4.66) is $2(R-1)(S-1)$ because the cosine of each term is 1. Observational noise will cause the value of the ambiguity function to be slightly below the theoretical maximum. Since the ambiguity function does not depend on the ambiguities, it is also independent of cycle slips. This invariant property is the most attractive feature of the ambiguity function and is unique among all the other solution methods.

Because the values $\varphi^{pq}_{km,L}$ in (6.4.63) are small when good approximate coordinates are available (typically corresponding to several hundredths of a cycle), we can expand the cosine function in a series and neglect higher-order terms. Thus,

$$\begin{aligned}
AF(\boldsymbol{x}_m) &= \sum_{L=1}^{2}\sum_{m=1}^{R-1}\sum_{q=1}^{S-1} \cos\left(\Psi^{pq}_{km,L}\right) = \sum_{L=1}^{2}\sum_{m=1}^{R-1}\sum_{q=1}^{S-1}\left[1-\frac{\left(\psi^{pq}_{km,L}\right)^2}{2!}+\cdots\right] \\
&= 2(R-1)(S-1) - \frac{1}{2}\sum_{L=1}^{2}\sum_{m=1}^{R-1}\sum_{q=1}^{S-1}\left(\psi^{pq}_{km,L}\right)^2 \\
&= 2(R-1)(S-1) = 2\pi^2 \boldsymbol{v}^T\boldsymbol{P}\boldsymbol{v}
\end{aligned} \tag{6.4.67}$$

The last part of this equation follows from (6.4.65). The ambiguity function and the least-squares solution are equivalent in the sense that the ambiguity function reaches maximum and $\boldsymbol{v}^T\boldsymbol{P}\boldsymbol{v}$ minimum at the point of convergence, i.e., at the correct $\boldsymbol{x}_m$.

There are several ways to initialize an ambiguity function solution. The simplest procedure is to use a search volume centered at some initial estimate of the station coordinates $\boldsymbol{x}_m$. Such an estimate could be computed from point positioning with pseudoranges; the size of the search volume would be chosen as a function of the accuracy of the coordinate estimates. This physical search volume is subdivided into a narrow grid of points with equal spacing. Each grid point is considered a candidate for the solution and used to compute the ambiguity function (6.4.66). The double-difference ranges $\rho^{pq}_{km,a}$, which are required in (6.4.66), are evaluated for the trial position. As the ambiguity function is computed by adding the individual cosine terms one double difference at a time, an early exit strategy can be implemented to reduce the computational effort. For example, if the trial position differs significantly from the true position, the residuals are likely to be bigger than one would expect just from measurement noise, unmodeled ionospheric and tropospheric effects, and multipath. An appropriate strategy could be to abandon the current trial position, i.e., stop accumulating the ambiguity function, and to begin with the next trial position. This would occur as soon as one term is below the cutoff criteria, e.g.,

$$\cos\left\{2\pi\left[\varphi^{pq}_{km,L,a}(t) - \varphi^{pq}_{km,L,b}(t)\right]\right\}_i < \varepsilon \tag{6.4.68}$$

The choice of the cutoff criteria ε is critical not only for accelerating solutions but also for assuring that the correct solution is not missed. This early exit strategy is unforgiving in the sense that once the correct (trial) position is rejected, the scanning of the remaining trial positions cannot yield the correct solution.

A matter of concern is that the grid of trial positions is close enough to assure that the true solution is not missed. Of course, a very narrow spacing of the trial positions increases the computational load, despite the early exit strategy. The optimal spacing is somewhat related to the wavelength and the number of satellites. On the other hand, the ambiguity function technique can be modified in several ways in order to increase its speed, such as using the double-differenced wide-lanes first. In this case, the trial positions can initially be widely spaced to reflect the wide-lane wavelength of 86 cm. These solutions could serve to identify a smaller physical search space, which can then be scanned using narrowly spaced trial positions.

The ambiguity function technique offers no opportunity to take the correlation between the double-difference observables into account. There is no direct accuracy measure for the final position that maximizes the ambiguity function, such as standard deviations of the coordinates. The quality of the solution is related to the spacing of the trial positions. If the trial positions, e.g., having a 1 cm spacing and a maximum of the ambiguity function is uniquely identified, then one could speak of centimeter-accurate positioning. In order to arrive at a conventional accuracy measure, one can take the position that maximizes the ambiguity function and carry out a regular double-difference least-squares solution. Because the initial positions for this least-squares solution are already very accurate, a single iteration is sufficient and it should be possible to fix the integer. The fixed solution would give the desired statistical information.

The ambiguity function values of all trial positions are ordered by size and normalized (dividing by the number of observations). Often, peaks of lesser value surround the highest peak and it might be impossible to identify the maximum reliably. This situation typically happens when the observational strength is lacking. The solution can be improved by observing for a longer period of time, selecting a better satellite configuration, using dual-frequency observations, etc.

The strength of the ambiguity function approach lies in the fact that the correct solution is obtained even if the data contain cycle slips. Remondi (1984) discusses the application of the ambiguity function technique to single differences. The geodetic use of the ambiguity function technique seems to be traceable to very long baseline interferometry (VLBI) observation processing. Counselman and Gourevitch (1981) present a very general ambiguity function technique and discuss in detail the patterns to be expected for various trial solutions.

6.4.6 GLONASS Carrier Phase

The current GLONASS system implements frequency division multiple access (FDMA) signal modulation as discussed in Chapter 5, whereas other satellite systems utilize CDMA modulation. A consequence of this arrangement is that each GLONASS satellite transmits at a slightly different carrier frequency within its bands while all GPS satellites transmit at the same carrier frequency within the L1 and L2 bands. The GPS carrier phase equation (6.1.30), therefore, must be slightly generalized to allow frequency-dependent hardware delays. We write

$$\varphi^r_{k,1}(t) = \frac{f^r_1}{c}\rho^r_k(t, t-\tau) + N^r_{k,1} + f^r_1 dt_k - f^r_1 dt^r - \frac{f^r_1}{c} I^r_{k,1,P} + \frac{f^r_1}{c} T^r_k + \delta^r_{k,1,\varphi} + \varepsilon^r_{k,1,\varphi} \tag{6.4.69}$$

$$\delta^r_{k,1,\varphi} = d^r_{k,1,\varphi} + D^r_{1,\varphi} + M^r_{k,1,\varphi} \tag{6.4.70}$$

where superscript r denotes the GLONASS channel number that identifies the frequency within the L1 band at which the satellite is transmitting.Note that the hardware delay terms have been given a superscript r. The receiver hardware delay in (6.1.31) has no superscript since in the case of GPS, all satellites transmit on the same L1 carrier frequency. The satellite hardware delay $D^r_{1,\varphi}$ also uses a superscript r to identify the frequency of the GLONASS satellite, whereas in (6.1.31), once again, the superscript p identifies the GPS satellite.

As an introductory example to GLONASS carrier phase processing, we discuss the experimental test of a 10 m baseline collected in 1998 on the roof of 3S Navigation Company, Irvine, California. The receivers were connected to a rubidium clock and recorded single-frequency pseudorange and carrier phases of $S_G = 5$ GPS satellites and $S_R = 4$ GLONASS satellites every second. As Figure 5.3.1 shows, during the mid-1990s, the GLONASS satellite population was sufficiently robust, thus allowing the development of mixed system positioning techniques. This baseline solution was reported in Leick et al. (1998), who used a Kalman filtering program and a least-squares batch program for independent computational verification (both include the LAMBDA ambiguity fixer).

Let the superscripts p and r identify any of the S_G GPS or S_R GLONASS satellites, respectively. Following this notation, the single-difference observations can then be written as

$$\varphi^p_{km,1,G} = \frac{f_1}{c}\rho^p_{km} + N^p_{km,1,G} + d_{km,1,G} - f_1 dt_{km} \tag{6.4.71}$$

$$\varphi^r_{km,1,R} = \frac{f^r_1}{c}\rho^r_{km} + N^r_{km,1,R} + d_{km,1,R} - f^r_1 dt_{km} \tag{6.4.72}$$

These equations utilize a common receiver clock error dt_{km} for GPS and GLONASS observations. The across-receiver hardware delay differences $d_{km,1,G}$ and $d_{km,1,R}$ are dealt with separately. Note, however, that the GLONASS hardware delay term $d_{km,1,R}$ does not have a superscript, which may seem contrary to what one would expect from the undifferenced hardware delay in (6.4.70). Since both receivers were of the same type, produced by the same manufacturer, and were running the same software, the implicit assumption is that the frequency-dependency contribution within the same band is negligible in the across-receiver difference. Studying the validity of this assumption was one of the purposes for collecting this experimental data set.

When processing GPS observations only, one would set $d_{km,1,G} = 0$ since GPS satellites transmit on the same L1 frequency and identical receivers were used, and then estimate the time-dependent clock errors and the constant ambiguities and station coordinates. A suitable ambiguity fixing technique would be applied to fix the across-difference ambiguities to integer.

For the combined processing of GPS and GLONASS across-receiver observations, we used the satellite-dependent parameterization

$$\xi^p_{km,1,G} = N^p_{km,1,G} + d_{km,1,G} \tag{6.4.73}$$

$$\xi^r_{km,1,R} = N^r_{km,1,R} + d_{km,1,R} \tag{6.4.74}$$

According to the model assumptions, ξ parameters are constants in time but not integers because of the receiver hardware delays. Letting the superscripts q and s denote the respective GPS and GLONASS base satellites, we can then write the following set of equations:

$$\varphi^q_{km,1,G} = \frac{f_1}{c}\rho^q_{km} + \xi^q_{km,1,G} - f_1\, dt_{km} \tag{6.4.75}$$

$$\varphi^p_{km,1,G} = \frac{f_1}{c}\rho^p_{km} + \xi^q_{km,1,G} + N^{pq}_{km,1,G} - f_1\, dt_{km} \tag{6.4.76}$$

$$\varphi^s_{km,1,R} = \frac{f^s_1}{c}\rho^s_{km} + \xi^s_{km,1,R} - f^s_1 dt_{km} \tag{6.4.77}$$

$$\varphi^r_{km,1,R} = \frac{f^r_1}{c}\rho^r_{km} + \xi^s_{km,1,R} + N^{rs}_{km,1,R} - f^r_1 dt_{km} \tag{6.4.78}$$

where $N^{pq}_{km,1,G} = \xi^p_{km,1,G} - \xi^q_{km,1,G}$ and $N^{rs}_{km,1,R} = \xi^r_{km,1,R} - \xi^s_{km,1,R}$ are the GPS and GLONASS double-difference ambiguities, respectively. Note that the across-receiver observations (6.4.75) and (6.4.77) refer to the base satellites, and that there are $S_G - 1$ equations (6.4.76) and $S_R - 1$ equations (6.4.78) that refer to nonbase satellites. A Kalman filter, or equivalently recursive least squares, provides the estimated real-valued ambiguity $N^{pq}_{km,1,G}$ and $N^{rs}_{km,1,R}$, the station coordinates, and the receiver clock and base satellite epoch parameters $\hat{\xi}^q_{km,1,G}$ and $\hat{\xi}^s_{km,1,R}$.

The float solution allows a first look at the variation of the hardware delays. First, compute the nonbase parameters

$$\hat{\xi}^p_{km,1,G} = \hat{N}^{pq}_{km,1,G} + \hat{\xi}^q_{km,1,G} \tag{6.4.79}$$

$$\hat{\xi}^r_{km,1,R} = \hat{N}^{rs}_{km,1,R} + \hat{\xi}^s_{km,1,R} \tag{6.4.80}$$

and then analyze the differences $\Delta\hat{\xi}^{qp}_{km,G} = \hat{\xi}^q_{km,1,G} - \hat{\xi}^p_{km,1,G}$ and $\Delta\hat{\xi}^{qr}_{km,G,R} = \hat{\xi}^q_{km,1,G} - \hat{\xi}^r_{km,1,R}$. Note that these differences are taken relative to the estimated GPS base station parameter. The fractional parts of $\Delta\hat{\xi}^{qp}_{km,G}$ estimates the difference $d^q_{km,1,G}$ and $d^p_{km,1,G}$ (we have added the superscripts for clarity). These fractional parts are expected to be zero. Indeed, the computed values are located around zero within a couple of hundredths of a cycle, and the fractional values of $\Delta\hat{\xi}^{qr}_{km,G,R}$, which estimate the difference of $d_{km,1,G}$ and $d^r_{km,1,R}$, are clustered at 0.35 cycles and also vary by a couple of hundredths of a cycle (Leick et al. 1998). Since $\Delta\hat{\xi}^{qp}_{km,G}$ and $\Delta\hat{\xi}^{qr}_{km,G,R}$, respectively, vary only a couple of hundredths of a cycle over time, one can draw two conclusions: first, the offset of $\Delta\hat{\xi}^{qr}_{km,G,R}$ by about 0.35 cycles is significant and second, there is no

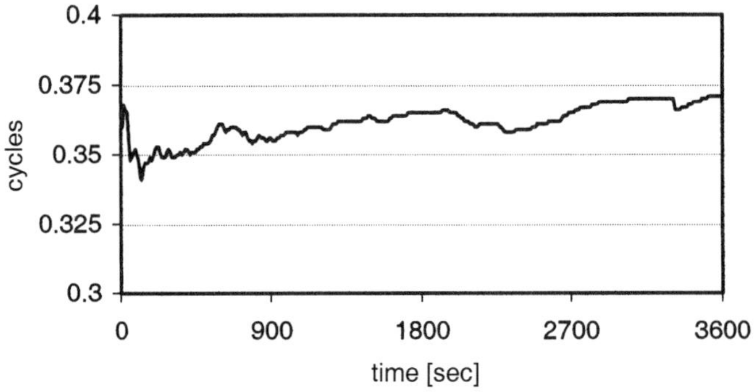

Figure 6.4.1 Fractional parts of across-system across-receiver hardware delays after fixing double-difference system ambiguities.

evidence in the data that the across-receiver hardware delays have a dependency on the GLONASS channel number that exceeds a couple of hundredths of a cycle.

The above conclusions are reconfirmed by the fixed ambiguity solution. The float solution is subjected to an ambiguity fixing routine yielding integer ambiguities, and then the other parameters are updated accordingly. Figure 6.4.1 shows the updated $\Delta\hat{\xi}^{qr}_{km,G,R}$ differences for a period of one hour, in which all double-difference ambiguities could be fixed. The figure shows identical graphs for each GLONASS satellite, i.e., the lines are plotted on top of each other. The fixed solution confirms the offset described above, which is due to the frequency offset of GPS L1 and the bundle of GLONASS frequencies in the L1 band. The remaining minor variation is due to multipath and possibly temperature change and could be modeled as a constant in practical applications.

Having verified the insensitivity of the across-receiver hardware delays to the GLONASS channel number within the same band, the conventional carrier phase double differences have the form

$$\varphi^{pq}_{km,1,G} = \frac{f_1}{c}\rho^{pq}_{km} + N^{pq}_{km,1,G} \tag{6.4.81}$$

$$\varphi^{rs}_{km,1,G} = \frac{f^r_1}{c}\rho^r_{km} - \frac{f^s_1}{c}\rho^s_{km} + N^{rs}_{km,1,R} - \left(f^r_1 - f^s_1\right)dt_{km} \tag{6.4.82}$$

In contrast to GPS double differences, the GLONASS double differences depend on the receiver clock error scaled by the respective frequency difference. This dependency is demonstrated in Figure 6.4.2, which shows the functions

$$\varphi^{rs} = \varphi^{rs}_{km,1,R} - \frac{f^r_1}{c}\rho^r_{km,a} + \frac{f^s_1}{c}\rho^s_{km,a} + \Delta^{rs} \tag{6.4.83}$$

where the observations have been corrected for the adjusted topocentric satellite distances and translated by Δ^{rs} in order to zero the function at the first epoch. The graph shows essentially straight lines because the receivers were connected to a stable

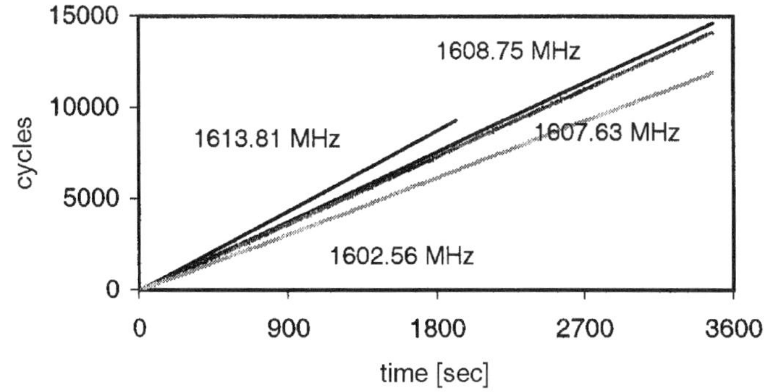

Figure 6.4.2 Impact of receiver clock errors on GLONASS double-differenced observations.

rubidium clock. The slope of the lines is a function of the frequency difference $f_1^r - f_1^s$. Four lines are shown corresponding to the five GLONASS satellites in the data.

Equations (6.4.81) and (6.4.82) could be used in principle to estimate the double-difference integers, as long as the receiver clock differences are also estimated at each epoch. However, caution is required because the coefficients of the clock parameter in (6.4.82) are relatively small compared to, for example, the respective coefficient in the single difference (6.4.78). Another way of looking at this situation is to scale the carrier phases in such a way that the clock term cancels. Consider this example:

$$\varphi_{km,1,R}^r - \frac{f_1^r}{f_1^s}\varphi_{km,1,R}^s = \frac{f_1^r}{c}\rho_{km}^{rs} + N_{km,1,R}^r - \frac{f_1^r}{f_1^s}N_{km,1,R}^s \tag{6.4.84}$$

One could attempt to compute an approximation $N_{km,1,R,0}^s$ of the across-receiver ambiguity using (6.4.72), with station coordinates and receiver clock estimated from pseudoranges and assuming that $d_{km,1,R}$ is negligible. The function (6.4.84) can then be written as

$$\varphi_{km,\,1,\,R}^r - \frac{f_1^r}{f_1^s}\varphi_{km,\,1,\,R}^s + \frac{f_1^r}{f_1^s}N_{km,\,1,\,R,\,0}^s = \frac{f_1^r}{c}\rho_{km}^{rs} + \widetilde{N}_{km,\,1,\,R}^{rs} + \eta^{rs} \tag{6.4.85}$$

with

$$\widetilde{N}_{km,\,1,\,R}^{rs} = N_{km,\,1,\,R}^r - \Delta N^s \qquad \eta^{rs} = \frac{f_1^s - f_1^r}{f_1^s}\Delta N_{km,\,1,\,R}^s \leq 0.01\ \Delta N_{km,\,1,\,R}^s \tag{6.4.86}$$

If $\Delta N_{km,\,1,\,R}^s$ is sufficiently small, i.e., $N_{km,\,1,\,R,\,0}^s$ can be computed sufficiently accurately, it might be possible to neglect η^{rs} and estimate and fix the ambiguity $\widetilde{N}_{km,\,1,\,R}^{rs}$ as an integer.

Other elements of the GLONASS system, such as the form and contents of the broadcast navigation message, the coordinates system, and the system time, are described elsewhere. GLONASS attracted a lot of interest during the mid-1990s because more usable satellites became available to the user, the GLONASS dual-frequency pseudoranges were not encrypted, and the carrier frequencies were different from those of GPS. The following is a sample of relevant literature from the so-called first GLONASS period: Raby and Daly (1993), Leick et al. (1995,

1998), Gourevitch et al. (1996), Povalyaev (1997), Pratt et al. (1997), Rapoport (1997), Kozlov and Tkachenko (1998), Roßach (2001), and Wang et al. (2001). Today, GLONASS once again has a fully deployed constellation and GLONASS observations are routinely combined with GPS observation. For additional details on generalized processing of observations from different satellite systems that allow even nonidentical receivers, see Chapter 7.

6.5 AMBIGUITY FIXING

Ambiguity fixing is essential for achieving centimeter-level accuracy in relative positioning. We first discuss ambiguity fixing in the context of a constrained adjustment, provide a brief background on the various approaches proposed to solve the ambiguity fixing problem, and then discuss in detail the popular LAMBDA method. In the second part of this section, the view is broadened by looking at practices in related disciplines to solve similar problems that might also be of benefit in certain circumstances when applied to GNSS applications.

6.5.1 The Constraint Solution

Fixing ambiguities implies converting real-valued ambiguity estimates to integers. The procedures follow the general linear hypothesis testing as described in Section 2.7.3. The objective is to constrain the estimated ambiguities of the float solution to integers. Let's assume that the parameters are grouped as

$$\boldsymbol{x}^* = \begin{bmatrix} \boldsymbol{a}^* \\ \boldsymbol{b}^* \end{bmatrix} \tag{6.5.1}$$

The symbol $\boldsymbol{a}^*$ denotes the estimated station coordinates and possibly other parameters, such as tropospheric refraction or receiver clock errors. The symbol $\boldsymbol{b}^*$ denotes the estimated float ambiguities. Using the same partitioning, other relevant matrices from the float solution are

$$\boldsymbol{N} = \begin{bmatrix} \boldsymbol{N}_{11} & \boldsymbol{N}_{21} \\ \boldsymbol{N}_{21} & \boldsymbol{N}_{22} \end{bmatrix} = \begin{bmatrix} \boldsymbol{L}_{11} & \boldsymbol{0} \\ \boldsymbol{L}_{12} & \boldsymbol{L}_{22} \end{bmatrix} \begin{bmatrix} \boldsymbol{L}_{11} & \boldsymbol{0} \\ \boldsymbol{L}_{12} & \boldsymbol{L}_{22} \end{bmatrix}^T \tag{6.5.2}$$

$$\boldsymbol{Q}_{x^*} = \boldsymbol{N}^{-1} = \begin{bmatrix} \boldsymbol{Q}_{a^*} & \boldsymbol{Q}_{a^*b^*} \\ \boldsymbol{Q}^T_{a^*b^*} & \boldsymbol{Q}_{b^*} \end{bmatrix} \tag{6.5.3}$$

$$\boldsymbol{Q}_{b^*}^{-1} = \boldsymbol{L}_{22}\boldsymbol{L}_{22}^T \tag{6.5.4}$$

The submatrices $\boldsymbol{L}_{ij}$ are part of the Cholesky factor $\boldsymbol{L}$. The relation (6.5.4) can be readily verified. In the notation of Section 2.7.3, we state the zero hypothesis H_0 as

$$H_0 : \boldsymbol{A}_2\boldsymbol{x}^* + \ell_2 = \boldsymbol{0} \tag{6.5.5}$$

These are n conditions, one for each ambiguity. The hypothesis states that a particular integer set is statistically compatible with the estimated ambiguities from the float solution. When constraining ambiguities, the coefficient matrix $\boldsymbol{A}_2$ takes on the simple form $\boldsymbol{A}_2 = \begin{bmatrix} 0 & \boldsymbol{I} \end{bmatrix}$, where the identity matrix $\boldsymbol{I}$ is of size n. The misclosure is $\ell_2 = -\boldsymbol{b}$, where $\boldsymbol{b}$ is the set of integer ambiguity values that are to be tested. The change in $\boldsymbol{v}^T\boldsymbol{P}\boldsymbol{v}$ due to the n constraints can be written according to (2.7.54)

$$\Delta \boldsymbol{v}^T\boldsymbol{P}\boldsymbol{v} = [\boldsymbol{b}^* - \boldsymbol{b}]^T \boldsymbol{Q}_b^{-1} [\boldsymbol{b}^* - \boldsymbol{b}] \tag{6.5.6}$$

which can he used in the F test (2.7.55)

$$\frac{\Delta \boldsymbol{v}^T\boldsymbol{P}\boldsymbol{v}}{\boldsymbol{v}^T\boldsymbol{P}\boldsymbol{v}^*} \frac{df}{n} \sim F_{n,df} \tag{6.5.7}$$

to test the acceptance of H_0. The value $\boldsymbol{v}^T\boldsymbol{P}\boldsymbol{v}^*$ comes from the float solution and df denotes the degree of freedom of the latter.

Once the hypothesis H_0 has been accepted, thus the best ambiguity candidate $\boldsymbol{b}$ has been identified, the change in the float solution due to the constraints can be computed using expressions from Table 2.5.5. One obtains for the station coordinates, given the integer-constrained ambiguities

$$\hat{\boldsymbol{a}}|\boldsymbol{b} = \boldsymbol{a}^* - \boldsymbol{Q}_{a^*b^*} \boldsymbol{Q}_{b^*}^{-1} (\boldsymbol{b}^* - \boldsymbol{b}) \tag{6.5.8}$$

The respective cofactor matrix after constraining is

$$\boldsymbol{Q}_{\hat{a}|b} = \boldsymbol{Q}_{a^*} - \boldsymbol{Q}_{a^*b^*} \boldsymbol{Q}_{b^*} \boldsymbol{Q}_{a^*b^*}^T \tag{6.5.9}$$

It follows from the positive definiteness properties of the diagonal submatrices of $\boldsymbol{N}$ or $\boldsymbol{Q}$ that the diagonal elements of $\boldsymbol{Q}_{a|b}$ are smaller than the diagonal elements of $\boldsymbol{Q}_{a^*}$, thus expressing a reduction in the variances of the coordinates due to imposing the constraints.

In the early days of GPS surveying, a test set $\boldsymbol{b}$ of integer values was obtained by simply rounding the estimated float ambiguities to the nearest integer. This approach works well for long observation times where many satellites can be observed, and the change in satellite geometry over time significantly improves the float solution. In such cases, the estimated real-valued ambiguities are already close to integers and their estimated variances are small. The situation changes drastically when one attempts to shorten the time of observation, possibly down to the extreme of just one epoch. It is only the distribution of the satellites in the sky and the availability of observations at multiple frequencies that adds strength to the geometry in such a case. The estimated float ambiguities will not necessarily be close to integer, and the estimates will have large variances and be highly correlated in general. A possible solution is to find candidate sets $\boldsymbol{b}_i$ of integers and compute $\Delta \boldsymbol{v}^T\boldsymbol{P}\boldsymbol{v}_i$ according to (6.5.6) for each member of the set. Those with the smallest contribution are subjected to the test (6.5.7).

There are two potential problems with this approach, however. The first one is that we might have many sets $\boldsymbol{b}$ that need to be tested if the variances of the real-valued ambiguities are large. An efficient algorithm, therefore, is needed to shorten the computation time for ambiguity fixing. The second problem is that several candidate sets might pass the test (6.5.7). Naturally, one would like to identify the correct candidate as soon as possible in order to avoid collecting additional observations. The discernibility of the candidate sets will be addressed in Section 6.5.3.

Frei and Beutler (1990) suggest a specific ordering scheme for the candidate ambiguity sets based on the float solution and the covariance matrix. The efficiency of their algorithms relies on the fact that if a certain ambiguity set is rejected, then a whole group of sets is identifiable that can also be rejected and consequently need not be computed explicitly.

Euler and Landau (1992) and Blomenhofer et al. (1993) point out that the matrix $\mathbf{L}_{22}$ in (6.5.4) remains the same for all candidate sets. They further recommend computing (6.5.6) in two steps. First, compute $\mathbf{g} = \mathbf{L}_{22}^T(\mathbf{b}^* - \mathbf{b}_i)$ and then $\Delta\mathbf{v}^T\mathbf{P}\mathbf{v} = \sum g_i^2$, $i = 1 \cdots n$. As soon as the first element g_1 has been computed, it can be squared and taken as the first estimate of the quadratic form. Note that $\Delta\mathbf{v}^T\mathbf{P}\mathbf{v} \geq g_1^2$. The value $\Delta\mathbf{v}^T\mathbf{P}\mathbf{v} = g_1^2$ is substituted in (6.5.7) to compute the test statistic. If that test fails, the trial ambiguity set $\mathbf{b}_i$ can immediately be rejected. There is no need to compute the remaining g_i values. If the test passes, then the next value, g_2, is computed and the test statistic is computed based on $\Delta\mathbf{v}^T\mathbf{P}\mathbf{v} = g_1^2 + g_2^2$. If this test fails, the ambiguity set is rejected; otherwise, g_3 is computed, etc. This procedure continues until either the zero hypothesis has been rejected or all g_i have been computed and the complete sum of n g-squared terms is known. This strategy can be combined with the ordering scheme mentioned above.

Chen and Lachapelle (1995) take advantage of the fact that integer ambiguity resolution accelerates if the range of candidates for a specific ambiguity is small. The smaller these search ranges, the fewer ambiguity sets need to be tested. Their method leads to a sequential reduction in the range of candidates for ambiguities not yet fixed. The procedure is an application of sequential conditional adjustment. When an ambiguity has been fixed and the covariance matrix of the parameters has been propagated, the standard deviations of the remaining ambiguity parameters become smaller. See the explanation given in regard to (6.5.9). The procedure starts with determining the range of the ambiguity which has the smallest variance. There is a strong resemblance between this method and LAMBDA, which will be discussed below in detail. The latter technique first reduced correlation between ambiguities and then applies sequential conditional adjustment.

Melbourne (1985) discusses an approach in which station coordinates are eliminated from the observation equation prior to the search for the ambiguities. The $S - 1$ double-difference epoch observation equations $\mathbf{v} = \mathbf{A}\mathbf{a} + \mathbf{b} + \ell$ are multiplied by $\mathbf{G}^T$, with $\mathbf{G}^T\mathbf{A} = \mathit{0}$, giving $\mathbf{G}^T(\mathbf{b}\text{-}\mathbf{v} + \ell) = \mathit{0}$. The columns of the matrix $\mathbf{G}$ span the null space of $\mathbf{A}$ or $\mathbf{A}\mathbf{A}^T$. Taking $\mathbf{v} = \mathit{0}$ one could attempt to identify by trial and error the correct set of ambiguities that fulfills the condition. Observing five satellites generates one condition; each additional satellites adds another condition. Since the elements of $\mathbf{G}$ change with time enough epochs will eventually be available to allow a unique identification of the ambiguity. Only the correct set of ambiguities will always fulfill the condition. As an alternative to the trial-and-error method, one could use the mixed adjustment model to estimate $\hat{\mathbf{b}}$.

Hatch (1990) suggests a scheme that divides satellites into primary and secondary ones. Consider four satellites, called the primary satellites. The respective three double-difference equations contain the station coordinates and three double-difference ambiguities. When the satellite geometry changes over time, it is possible to estimate all of these parameters. Any secondary satellites in addition to these four primary satellites are, strictly speaking, redundant and are used to develop a procedure for rapidly identifying integer ambiguities. The procedure starts by computing trial sets for the three primary ambiguities using an initial position estimate from a point positioning solution, or from the float solution if several epochs of observations are available and the receivers do not move.

For details on the procedures mentioned above, please check the references cited. Over the years, another method has become the most popular one of all. This is LAMBDA, which we will discuss in some detail in the next section.

Finally, we have pointed out above that there might be several sets of integer ambiguities that pass the test (6.5.7). Since the adjustment has already passed the basic chi-square test, the adjustment as such is correct, i.e., there are no model errors, the observational weights have been chosen correctly, and blunders have been eliminated. In that case, it is natural to look for the smallest $\Delta \mathbf{v}^T \mathbf{P} \mathbf{v}$. From this point of view, the integer fixing problem is called integer least squares and (6.5.8) is the integer least-squares estimator. In short, one seeks the integer vector $\mathbf{b}$ that minimizes $\Delta \mathbf{v}^T \mathbf{P} \mathbf{v}$.

6.5.2 LAMBDA

Teunissen (1993) introduced the least-squares ambiguity decorrelation adjustment (LAMBDA) method. This technique has the highest probability of correct integer estimation among a certain group of estimators (Teunissen, 1999). This probabilistic property and its speed of resolving the ambiguities have resulted in high popularity and general acceptance of the technique. The reader is referred to de Jonge and Tiberius (1996) for details about implementation. The software is available from the TU Delft. This section merely highlights some features of the LAMBDA algorithm.

At the core of the LAMBDA decorrelation is the Z transformation:

$$\mathbf{z} = \mathbf{Z}^T \mathbf{b} \tag{6.5.10}$$

$$\hat{\mathbf{z}} = \mathbf{Z}^T \hat{\mathbf{b}} \tag{6.5.11}$$

$$\mathbf{Q}_z = \mathbf{Z}^T \mathbf{Q}_b \mathbf{Z} \tag{6.5.12}$$

In (6.5.11) we used the symbol $\hat{\mathbf{b}}$ instead of b^* to denote the float ambiguity estimate. The matrix $\mathbf{Z}$ is a regular and square. In order for integers to be preserved, i.e., the integers $\mathbf{b}$ should be mapped into integers $\mathbf{z}$ and vice versa, it is necessary that the elements of both matrices $\mathbf{Z}$ and $\mathbf{Z}^{-1}$ are integers. The condition $|\mathbf{Z}| = \pm 1$ assures that the inverse contains only integer elements if $\mathbf{Z}$ contains integers. Simply consider this: if all elements of $\mathbf{Z}$ are integers, then this is also true for the cofactor matrix $\mathbf{C}$. Therefore, the inverse

$$\mathbf{Z}^{-1} = \frac{\mathbf{C}^T}{|\mathbf{Z}|} \tag{6.5.13}$$

has integer elements because $|\mathbf{Z}| = \pm 1$. The latter condition also implies that

$$|\mathbf{Q}_z| = |\mathbf{Z}^T Q_b \mathbf{Z}| = |\mathbf{Z}^T||\mathbf{Q}_b||\mathbf{Z}| = |\mathbf{Q}_b| \tag{6.5.14}$$

The quadratic form also remains invariant with respect to the Z transformation. Substituting (6.5.10) and (6.5.11) into (6.5.6) and using the inverse of (6.5.12) gives

$$\begin{aligned}\Delta \mathbf{v}^T \mathbf{P} \mathbf{v} &= [\hat{\mathbf{b}}\text{-}\mathbf{b}]^T \mathbf{Q}_b^{-1} [\hat{\mathbf{b}}\text{-}\mathbf{b}] \\ &= [\mathbf{z} - \mathbf{z}]^T \mathbf{Z}^{-1} \mathbf{Q}_b^{-1} (\mathbf{Z}^{-1})^T [\hat{\mathbf{z}}\text{-}\mathbf{z}] \\ &= [\hat{\mathbf{z}}\text{-}\mathbf{z}]^T \mathbf{Q}_z^{-1} [\hat{\mathbf{z}}\text{-}\mathbf{z}]\end{aligned} \tag{6.5.15}$$

Note again that in (6.5.15) we used the symbol $\hat{\mathbf{b}}$ instead of b^* to denote the float ambiguity estimate.

Consider the following example with two random integer variables $\hat{\mathbf{b}} = \left[\hat{b}_1\ \hat{b}_2\right]^T$. Let the respective covariance matrix be

$$\Sigma_b = \begin{bmatrix} \sigma_{b_1}^2 & \sigma_{b_1 b_2} \\ \sigma_{b_2 b_1} & \alpha_{b_2}^2 \end{bmatrix} \tag{6.5.16}$$

where we have omitted the hat notation for simplicity. Let the transformation $\mathbf{z} = \mathbf{Z}^T \mathbf{b}$ utilize a transformation matrix of the special form

$$\mathbf{Z}^T = \begin{bmatrix} 1 & \beta \\ 0 & 1 \end{bmatrix} \tag{6.5.17}$$

where $\hat{\mathbf{z}} = [\hat{z}_1\ \ \hat{z}_2]^T$. We note that $|\mathbf{Z}| = 1$. The element β is obtained by rounding $-\sigma_{b_1 b_2}/\sigma_{b_2}^2$ to the nearest integer $\beta = \text{int}\left(-\sigma_{b_1 b_2}/\sigma_{b_2}^2\right)$. Because β is an integer, the transformed $\mathbf{z}$ variables will also be integers. Applying variance-covariance propagation gives

$$\Sigma_z = \mathbf{Z}^T \Sigma_b \mathbf{Z} = \begin{bmatrix} \beta^2 \sigma_{b_2}^2 + 2\beta\sigma_{b_1 b_2} + \sigma_{b_1}^2 & \beta\sigma_{b_2}^2 + \sigma_{b_1 b_2} \\ \beta\sigma_{b_2}^2 + \sigma_{b_1 b_2} & \sigma_{b_2}^2 \end{bmatrix} \tag{6.5.18}$$

Let ε denote the change due to the rounding, i.e., $\varepsilon = \sigma_{b_1 b_2}/\sigma_{b_2}^2 + \beta$. Using (6.5.18), the variance $\sigma_{z_1}^2$ of the transformed variable can be written as

$$\sigma_{z_1}^2 = \sigma_{b_1}^2 - \left(\frac{\sigma_{b_1 b_2}^2}{\sigma_{b_2}^4} - \varepsilon^2 \right) \sigma_{b_2}^2 \tag{6.5.19}$$

This expression shows that the variance of the transformed variable decreases compared to the original one, i.e., $\sigma_{z_1}^2 < \sigma_{b_1}^2$ whenever

$$\left|\sigma_{b_1 b_2}/\sigma_{b_2}^2\right| > 0.5 \tag{6.5.20}$$

and that both are equal when $\sigma_{b_1 b_2}/\sigma_{b_2}^2 = |\varepsilon| = 0.5$. The property of decreasing the variance while preserving the integer makes the transformation (6.5.17) a favorite for resolving ambiguities because it reduces the search range of the transformed variable. It is interesting to note that z_1 and z_2 would be uncorrelated if one were to choose $\beta = -\sigma_{b_1 b_2}/\sigma_{b_2}^2$. However, such a selection is not permissible because it would not preserve the integer property of the transformed variables.

When implementing LAMBDA, the $\boldsymbol{Z}$ matrix is constructed from the $n \times n$ submatrix $\boldsymbol{Q}_b$ given in (6.5.3). There are n variables $\hat{\boldsymbol{b}}$ that must be transformed. Using the Cholesky decomposition, we find

$$\boldsymbol{Q}_b = \boldsymbol{H}^T \boldsymbol{K} \boldsymbol{H} \tag{6.5.21}$$

The matrix $\boldsymbol{H}$ is the modified Cholesky factor that contains 1 at the diagonal positions and follows from (6.5.4). The diagonal matrix $\boldsymbol{K}$ contains the diagonal squared terms of the Cholesky factor. Assume that we are dealing with ambiguities i and $i+1$ and partition these two matrices as

$$\boldsymbol{H} = \left[\begin{array}{ccc|cc|cc} 1 & & & & & & \\ \vdots & \ddots & & & & & \\ \hline h_{i,1} & & \cdots & 1 & & & \\ h_{i+1,1} & & \cdots & h_{i+1,i} & 1 & & \\ \hline \vdots & & \cdots & \vdots & \vdots & \ddots & \\ h_{n,1} & & \cdots & h_{n,i} & h_{n,i+1} & \cdots & 1 \end{array}\right] = \begin{bmatrix} \boldsymbol{H}_{11} & \boldsymbol{0} & \boldsymbol{0} \\ \boldsymbol{H}_{21} & \boldsymbol{H}_{22} & \boldsymbol{0} \\ \boldsymbol{H}_{31} & \boldsymbol{H}_{32} & \boldsymbol{H}_{33} \end{bmatrix} \tag{6.5.22}$$

$$\boldsymbol{K} = \left[\begin{array}{cc|cc|cc} k_{1,1} & & & & & \\ & \ddots & & & & \\ \hline & & k_{i,i} & & & \\ & & & k_{i+1,i+1} & & \\ \hline & & & & \ddots & \\ & & & & & k_{n,n} \end{array}\right] = \begin{bmatrix} \boldsymbol{K}_{11} & \boldsymbol{0} & \boldsymbol{0} \\ \boldsymbol{O} & \boldsymbol{K}_{22} & \boldsymbol{0} \\ \boldsymbol{O} & \boldsymbol{O} & \boldsymbol{K}_{33} \end{bmatrix} \tag{6.5.23}$$

The transformation matrix $\boldsymbol{Z}$ is partitioned similarly

$$\boldsymbol{Z}_1 = \left[\begin{array}{c|cc|c} \boldsymbol{I} & & & \\ \hline & 1 & 0 & \\ & \beta & 1 & \\ \hline & & & \boldsymbol{I} \end{array}\right] = \begin{bmatrix} \boldsymbol{I}_{11} & \boldsymbol{0} & \boldsymbol{0} \\ \boldsymbol{0} & \boldsymbol{Z}_{22} & \boldsymbol{0} \\ \boldsymbol{0} & \boldsymbol{0} & \boldsymbol{I}_{33} \end{bmatrix} \tag{6.5.24}$$

where $\beta = -\mathrm{int}(h_{i+1,i})$ represents the negative of the rounded value of $h_{i+1,i}$, and

$$\hat{\boldsymbol{z}}_1 = \boldsymbol{Z}_1^T \hat{\boldsymbol{b}} \tag{6.5.25}$$

$$\boldsymbol{Q}_{z,1} = \boldsymbol{Z}_1^T \boldsymbol{Q}_b \boldsymbol{Z}_1 = \boldsymbol{Z}_1^T \boldsymbol{H}^T \boldsymbol{K} \boldsymbol{H} \boldsymbol{Z}_1 = \boldsymbol{H}_1^T \boldsymbol{K}_1 \boldsymbol{H}_1 \tag{6.5.26}$$

It can be shown that the specific form of $\boldsymbol{Z}_1$ and choice for $\boldsymbol{Z}_{22}$ imply the following updates:

$$\boldsymbol{Q}_{z,1} = \begin{bmatrix} \boldsymbol{Q}_{11} & & \text{sym} \\ \boldsymbol{Z}_{22}^T \boldsymbol{Q}_{21} & \boldsymbol{Z}_{22}^T \boldsymbol{Q}_{22} \boldsymbol{Z}_{22} & \\ \boldsymbol{Q}_{31} & \boldsymbol{Q}_{32} \boldsymbol{Z}_{22} & \boldsymbol{Q}_{33} \end{bmatrix} \tag{6.5.27}$$

$$\boldsymbol{H}_1 = \boldsymbol{HZ}_1 = \begin{bmatrix} \boldsymbol{H}_{11} & \boldsymbol{0} & \boldsymbol{0} \\ \boldsymbol{H}_{21} & \overline{\boldsymbol{H}}_{22} & \boldsymbol{0} \\ \boldsymbol{H}_{31} & \overline{\boldsymbol{H}}_{32} & \boldsymbol{H}_{33} \end{bmatrix} \tag{6.5.28}$$

$$\overline{\boldsymbol{H}}_{22} = \begin{bmatrix} 1 & 0 \\ h_{i+1,i} + \beta & 1 \end{bmatrix} \tag{6.5.29}$$

$$\overline{\boldsymbol{H}}_{32} = \begin{bmatrix} h_{i+2,i} + \beta h_{i+2,i+1} & h_{i+2,i+1} \\ h_{i+3,i} + \beta h_{i+3,i+1} & h_{i+3,i+1} \\ \vdots & \vdots \\ h_{n,i} + \beta h_{n,i+1} & h_{n,i+1} \end{bmatrix} \tag{6.5.30}$$

$$\boldsymbol{K}_1 = \boldsymbol{K} \tag{6.5.31}$$

The matrix $\boldsymbol{K}$ does not change as a result of this decorrelation transformation.

If $\beta = 0$, the transformation (6.5.25) is not necessary. However, it is necessary to check whether or not the ambiguities i and $i+1$ should be permuted to achieve further decorrelation. Consider the permutation transformation

$$\boldsymbol{Z}_2 = \left[\begin{array}{c|cc|c} \boldsymbol{I} & & & \\ \hline & 0 & 1 & \\ & 1 & 0 & \\ \hline & & & \boldsymbol{I} \end{array}\right] = \begin{bmatrix} \boldsymbol{I}_{11} & \boldsymbol{0} & \boldsymbol{0} \\ \boldsymbol{0} & \boldsymbol{P} & \boldsymbol{0} \\ \boldsymbol{0} & \boldsymbol{0} & \boldsymbol{I}_{33} \end{bmatrix} \tag{6.5.32}$$

This specific choice for $\boldsymbol{Z}_2$ leads to

$$\overline{\boldsymbol{H}}_{22} = \begin{bmatrix} 1 & 0 \\ h'_{i+1,i} & 1 \end{bmatrix} \begin{bmatrix} 1 & 0 \\ \dfrac{h_{i+1,i}k_{i+1,i+1}}{k_{i,i} + h^2_{i+1,i}k_{i+1,i+1}} & 1 \end{bmatrix} \tag{6.5.33}$$

$$\overline{\boldsymbol{H}}_{21} = \begin{bmatrix} \dfrac{-h_{i+1,i} \atop k_{i,i}}{k_{i,i} + h^2_{i+1,i}k_{i+1,i+1}} & 1 \\ & h'_{i+1,j} \end{bmatrix} \boldsymbol{H}_{21} \tag{6.5.34}$$

$$\overline{\boldsymbol{H}}_{32} = \begin{bmatrix} h_{i+2,i+1} & h_{i+2,i} \\ h_{i+3,i+1} & h_{i+3,i} \\ \vdots & \vdots \\ h_{n,i+1} & h_{n,i} \end{bmatrix} \tag{6.5.35}$$

$$\overline{\boldsymbol{K}}_{22} = \begin{bmatrix} k'_{i,i} & 0 \\ 0 & k'_{i+1,i+1} \end{bmatrix} \begin{bmatrix} k_{i+1,i+1} - \dfrac{h^2_{i+1,i}k^2_{i+1,i+1}}{k_{i,i} + h^2_{i+1,i}k_{i+1,i+1}} & 0 \\ 0 & k_{i,i} + h^2_{i+1,i}k_{i+1,i+1} \end{bmatrix} \tag{6.5.36}$$

Permutation changes the matrix $\boldsymbol{K}$ at $\overline{\boldsymbol{K}}_{22}$. To achieve full decorrelation, the terms $k'_{i+1,i+1}$ and $k_{i+1,i+1}$ must be inspected while the ith and $(i+1)$th ambiguity are considered. Permutation is required if $k'_{i+1,i+1} < k_{i+1,i+1}$. If permutation occurs, the procedure again starts with the last pair of the $(n-1)$th and nth ambiguities and tries to reach the first and second ambiguities. A new $\boldsymbol{Z}$ transformation matrix is constructed whenever decorrelation takes place or the order of two ambiguities is permuted. This procedure is completed when no diagonal elements are interchanged.

The result of the transformations can be written as

$$\hat{\boldsymbol{z}} = \boldsymbol{Z}_q^T \ldots \boldsymbol{Z}_2^T \boldsymbol{Z}_1^T \hat{\boldsymbol{b}} \tag{6.5.37}$$

$$\boldsymbol{Q}_{z,q} = \boldsymbol{Z}_q^T \ldots \boldsymbol{Z}_2^T \boldsymbol{Z}_1^T \boldsymbol{Q}_b \boldsymbol{Z}_1 \boldsymbol{Z}_2 \ldots \boldsymbol{Z}_q = \boldsymbol{H}_q^T \boldsymbol{K}_q \boldsymbol{H}_q \tag{6.5.38}$$

The matrices $\boldsymbol{H}_q$ and $\boldsymbol{K}_q$ are obtained as part of the consecutive transformations. The permuting steps assure that $\boldsymbol{K}_q$ contains decreasing diagonal elements, the smallest element being located at the lower right corner. As a measure of decorrelation between the ambiguities, we might consider the scalar (Teunissen, 1994)

$$r = |\boldsymbol{R}|^{1/2} \quad 0 \le r \le 1 \tag{6.5.39}$$

where $\boldsymbol{R}$ represents a correlation matrix. Applying (6.5.39) to $\boldsymbol{Q}_b$ and $\boldsymbol{Q}_{z,q}$ will give a relative sense of the decorrelation achieved. A value of r close to 1 implies a high decorrelation. Therefore, we expect $r_b < r_{z,q}$. The scalar r is called the ambiguity decorrelation number.

The search step entails finding candidate sets of $\hat{\boldsymbol{z}}_i$ given $(\hat{\boldsymbol{z}}, \boldsymbol{Q}_{z,q})$, which minimize

$$\Delta \boldsymbol{v}^T \boldsymbol{P} \boldsymbol{v} = [\hat{\boldsymbol{z}}\text{-}\boldsymbol{z}]^T \boldsymbol{Q}_{z,q}^{-1} [\hat{\boldsymbol{z}}\text{-}\boldsymbol{z}] \tag{6.5.40}$$

A possible procedure would be to use the diagonal elements of $\boldsymbol{Q}_{z,q}$, construct a range for each ambiguity centered around $\hat{\boldsymbol{z}}_l$, form all possible sets $\boldsymbol{z}_i$, evaluate the quadratic form for each set, and keep track of those sets that produce the smallest $\Delta \boldsymbol{v}^T \boldsymbol{P} \boldsymbol{v}$. A more organized and efficient approach is achieved by transforming the $\hat{\boldsymbol{z}}$ variables into variables $\hat{\boldsymbol{w}}$ that are stochastically independent. First, we decompose the inverse of $\boldsymbol{Q}_{z,q}$ as

$$\boldsymbol{Q}_{z,q}^{-1} = \boldsymbol{M} \boldsymbol{S} \boldsymbol{M}^T \tag{6.5.41}$$

where $\boldsymbol{M}$ denotes the lower triangular matrix with 1 along the diagonal, and $\boldsymbol{S}$ is a diagonal matrix containing positive values that increase toward the lower right corner. The latter property follows from the fact that $\boldsymbol{S}$ is the inverse of $\boldsymbol{K}_q$. The transformed variables $\hat{\boldsymbol{w}}$

$$\hat{\boldsymbol{w}} = \boldsymbol{M}^T [\hat{\boldsymbol{z}} - \boldsymbol{z}] \tag{6.5.42}$$

are distributed as $\hat{\boldsymbol{w}} \sim N(\boldsymbol{0}, \boldsymbol{S}^{-1})$. Because $\boldsymbol{S}$ is a diagonal matrix, the variables $\hat{\boldsymbol{w}}$ are stochastically independent. Using (6.5.42) and (6.5.41), the quadratic form (6.5.40) can he written as

$$\Delta \boldsymbol{v}^T \boldsymbol{P} \boldsymbol{v} = \hat{\boldsymbol{w}}^T \boldsymbol{S} \hat{\boldsymbol{w}} = \sum_{i=1}^{n} \hat{\boldsymbol{w}}_i^2 s_{i,i} \le \chi^2 \tag{6.5.43}$$

The symbol χ^2 acts as a scalar; additional explanations will be given below. Finally, we introduce the auxiliary quantity, also called the conditional estimate

$$\hat{w}_{i|I} = \sum_{j=i+1}^{n} m_{j,i}(\hat{z}_j\text{-}z_j) \tag{6.5.44}$$

The symbol $|I$ indicates that the values for z_j have already been selected, i.e., are known. Note that the subscript j goes from $i = 1$ to n. Since $m_{i,i} = 1$ and using (6.5.44) and (6.5.42), we can write the ith component as

$$\hat{\boldsymbol{w}}_i = \hat{\boldsymbol{z}}_i - \boldsymbol{z}_i + \hat{w}_{i|I} \quad i = 1, n-1 \tag{6.5.45}$$

The bounds of the $\boldsymbol{z}$ parameters follow from (6.5.43). We begin with the nth level to determine the bounds for the nth ambiguity and then proceed to level 1, establishing the bound for the other ambiguities. Using the term with $\hat{w}_n\, s_{n,n}$ in (6.5.43), and knowing that the matrix element $m_{n,n} = 1$ in (6.5.42), we find

$$\hat{w}_n^2 s_{n,n} = (z_n - \hat{z}_n)^2 s_{n,n} \leq \chi^2 \tag{6.5.46}$$

The bounds are

$$\hat{z}_n - \left(\frac{\chi^2}{s_{n,n}}\right)^{1/2} \leq z_n \leq \hat{z}_n + \left(\frac{\chi^2}{s_{n,n}}\right)^{1/2} \tag{6.5.47}$$

Using the terms from i to n in (6.5.43) and (6.5.45), we obtain for level i

$$\hat{w}_i^2 s_{i,i} = (\hat{z}_i - z_i + \hat{w}_{i,I})^2 s_{i,i} \leq \left[\chi^2 - \sum_{j=i+1}^{n} \hat{w}_j^2 s_{j,j}\right] \tag{6.5.48}$$

$$\hat{z}_i + \hat{w}_{i|I} - \frac{1}{\sqrt{s_{i,i}}}\left[\chi^2 - \sum_{j=i+1}^{n} \hat{w}_j^2 s_{j,j}\right]^{1/2} \leq z_1$$

$$\leq \hat{z}_i + \hat{w}_{i|I} + \frac{1}{\sqrt{s_{i,i}}}\left[\chi^2 - \sum_{j=i+1}^{n} \hat{w}_j^2 s_{j,j}\right]^{1/2} \tag{6.5.49}$$

The bounds (6.5.47) and (6.5.49) can contain one or several integer values z_n or z_i. All values must be used when locating the bounds and integer values at the next lower level. The process stops when level 1 is reached. For certain combinations, the process stops earlier if the square root in (6.5.49) becomes negative.

Figure 6.5.1 demonstrates how one can proceed systematically, trying to reach the first level. At a given level, one proceeds from the left to the right while reaching a lower level. This example deals with $n = 4$ ambiguities – z_1, z_2, z_3, and z_4. The fourth level produced the qualifying values $z_4 = \{-1, 0, 1\}$. Using $z_4 = -1$ or $z_4 = 1$ does not produce a solution at level 3 and the branch terminates. Using $z_4 = 0$ gives $z_3 = \{-1, 0\}$ at level 3. Using $z_3 = -1$ and $z_4 = 0$, or in short notation $z = (-1, 0)$, one gets $z_2 = 0$ at level 2. The combination $z = (0, -1, 0)$ does not produce a solution at level 1 so the branch terminates. Returning to level 3, we try the combination $z = (0, 0)$, giving $z_2 = \{-1, 0, 1\}$ at level 2. Trying the left branch with $z = (-1, 0, 0)$ gives

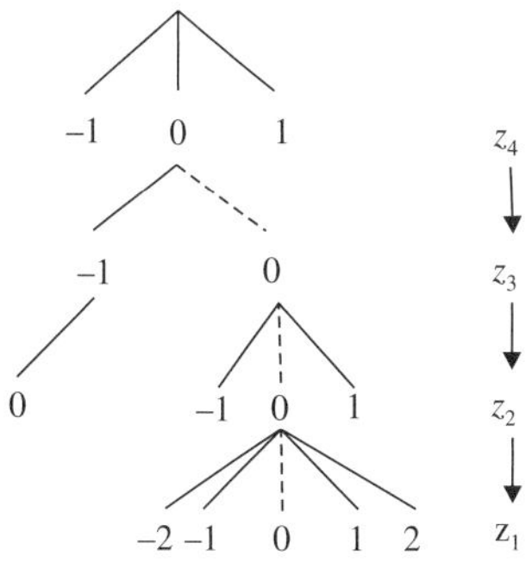

Figure 6.5.1 Candidate ambiguities encountered during the search procedure with decorrelation.

no solution and the branch terminates. Using $z = (0, 0, 0)$ gives $z_1 = \{-2, -1, 0, 1, 2\}$ at the first level. The last possibility, using $z = (1, 0, 0)$, gives no solution. We conclude that five ambiguity sets $\boldsymbol{z}_i = (z_1, 0, 0, 0)$ satisfy the condition (6.5.43). In general, several branches can reach the first level. Because $s_{n,n}$ is the largest value in $\boldsymbol{S}$, the number of z_n candidates is correspondingly small, thus lowering the number of branches that originate from level n and assuring that not many branches reach level 1.

The change $\Delta \boldsymbol{v}^T \boldsymbol{P} \boldsymbol{v}_i$ can be computed efficiently from (6.5.43) because all $\hat{\boldsymbol{w}}_i$ sets become available as part of computing the candidate ambiguity sets. The matrix $\boldsymbol{S}$ does not change. The qualifying candidates $\boldsymbol{z}_i$ are converted back to $\boldsymbol{b}_i$ using the inverse of (6.5.37).

If the constant χ^2 for ambiguity search is chosen improperly, it is possible that the search procedure may not find any candidate vector or that too many candidate vectors are obtained. The latter case results in time-consuming searches. This dilemma can be avoided if the constant is set close to the $\Delta \boldsymbol{v}^{\mathrm{T}} \boldsymbol{P} \boldsymbol{v}$ value of the best candidate ambiguity vector. To do so, the real-valued ambiguities of the float solution are rounded to the nearest integer and then substituted into (6.5.40). The constant is then taken to be equal to $\Delta \boldsymbol{v}^{\mathrm{T}} \boldsymbol{P} \boldsymbol{v}$. This approach guarantees obtaining at least one candidate vector, which probably is the best one because the decorrelated ambiguities are generally of high precision. One can compute a new constant χ^2 by adding or subtracting an increment to one of the nearest integer entries. Using this procedure results in only a few candidate integer ambiguity vectors and guarantees that at least two vectors are obtained.

LAMBDA is a general procedure that requires only the covariance submatrix and the float ambiguity estimates. Therefore, the LAMBDA is applicable even if other parameters are estimated at the same time, such as station coordinates, tropospheric parameters, and clock errors. LAMBDA readily applies to dual-frequency observations, or even future multifrequency situations. Even more generally, LAMBDA applies to any least-squares integer estimation, regardless of the physical meaning of the integer parameters.

LAMBDA can also be used to estimate a subset of ambiguities. For example, in the case of dual-frequency ambiguities one might parameterize in terms of wide-lane and L1 ambiguities. LAMBDA could operate first on the wide-lane covariance submatrix and fix the wide-lane ambiguities and then attempt to fix the L1 ambiguities.

Teunissen (1997) shows that the Z transformation always includes the wide-lane transformation.

6.5.3 Discernibility

The ambiguity testing outlined above is a repeated application of null hypotheses testing for each ambiguity set. The procedure tests the changes $\Delta \mathbf{v}^T \mathbf{P} \mathbf{v}$ due to the constraints. The decision to accept or reject the null hypothesis is based on the probability of the type-I error, which is usually taken to be $\alpha = 0.05$. In many cases, several of the null hypotheses will pass, thus identifying several qualifying ambiguity sets. This happens if there is not enough information in the observations to determine the integers uniquely and reliably. Additional observations might help resolve the situation. The ambiguity set that generates the smallest $\Delta \mathbf{v}^T \mathbf{P} \mathbf{v}$ fits the float solution best and, consequently, is considered the most favored fixed solution. The goal of additional statistical considerations is to provide conditions that make it possible to discard all but one of the ambiguity sets that passed the null hypotheses test.

The alternative hypothesis H_a is always relative to the null hypothesis H_0. The formalism for the null hypothesis is given in Section 2.7.3. In general, the null and alternative hypotheses are

$$H_0\colon\ \mathbf{A}_2 \mathbf{x}^* + \boldsymbol{\ell}_2 = \mathbf{0} \tag{6.5.50}$$

$$H_a\colon\ \mathbf{A}_2 \mathbf{x}^* + \boldsymbol{\ell}_2 + \mathbf{w}_2 = \mathbf{0} \tag{6.5.51}$$

Under the null hypothesis, the expected value of the constraint is zero. See also equation (2.7.45). Thus,

$$E\left(\mathbf{z}_{H_0}\right) \equiv E(\mathbf{A}_2 \mathbf{x}^* + \boldsymbol{\ell}_2) = \mathbf{0} \tag{6.5.52}$$

Because $\mathbf{w}_2$ is a constant, it follows that

$$E\left(\mathbf{z}_{H_a}\right) \equiv E(\mathbf{A}_2 \mathbf{x}^* + \boldsymbol{\ell}_2 + \mathbf{w}_2) = \mathbf{w}_2 \tag{6.5.53}$$

The random variable $\mathbf{z}_{H_a}$ is multivariate normal distributed with mean $\mathbf{w}_2$, i.e.,

$$\mathbf{z}_{H_a} \sim N_{n-r}\left(\mathbf{w}_2, \sigma_0^2 \mathbf{T}^{-1}\right) \tag{6.5.54}$$

See equation (2.7.47) for the corresponding expression for the zero hypothesis. The matrix $\mathbf{T}$ has the same meaning as in Section 2.7.3, i.e.,

$$\mathbf{T} = \left(\mathbf{A}_2 \mathbf{N}_1^{-1} \mathbf{A}_2^T\right)^{-1} \tag{6.5.55}$$

The next step is to diagonalize the covariance matrix of $\mathbf{z}_{H_a}$ and to compute the sum of the squares of the transformed random variables. These newly formed random variables have a unit variate normal distribution with a nonzero mean. According to Section A.5.2, the sum of the squares has a noncentral chi-square distribution. Thus,

$$\frac{\Delta \mathbf{v}^{\mathrm{T}} \mathbf{P} \mathbf{v}}{\sigma_0^2} = \frac{\mathbf{z}_{H_a}^T \mathbf{T} \mathbf{z}_{H_a}}{\sigma_0^2} \sim \chi^2_{n_2,\lambda} \tag{6.5.56}$$

where the noncentrality parameter is

$$\lambda = \frac{\mathbf{w}_2^T \mathbf{T} \mathbf{w}_2}{\sigma_0^2} \tag{6.5.57}$$

The reader is referred to the statistical literature, such as Koch (1988), for additional details on noncentral distributions and their respective derivations. Finally, the ratio

$$\frac{\Delta \mathbf{v}^T \mathbf{P} \mathbf{v}}{\mathbf{v}^T \mathbf{P} \mathbf{v}^*} \frac{n_1 - r}{n_2} \sim F_{n_2, n_1 - r, \lambda} \tag{6.5.58}$$

has a noncentral F distribution with noncentrality λ. If the test statistic computed under the specifications of H_0 fulfills $F \le F_{n_2, n_1 - r, \alpha}$, then H_0 is accepted with a type-I error of α. The alternative hypothesis H_a can be separated from H_0 with the power $1 - \beta(\alpha, \lambda)$. The type-II error is

$$\beta(\alpha, \lambda) = \int_0^{F_{n_2, n_1 - r, \alpha}} F_{n_2, n_1 - r, \lambda} \, dx \tag{6.5.59}$$

The integration is taken over the noncentral F-distribution function from zero to the value $F_{n_2, n_1 - r, \alpha}$, which is specified by the significance level α.

Because the noncentrality is different for each alternative hypothesis according to (6.5.57), the type-II error $\beta(\alpha, \lambda)$ also varies with H_a. Rather than using the individual type-II errors to make decisions, Euler and Schaffrin (1990) propose using the ratio of noncentrality parameters. They designate the float solution as the common alternative hypothesis H_a for all null hypotheses. In this case, the value $\mathbf{w}_2$ in (6.5.51) is

$$\mathbf{w}_2 = -(\mathbf{A}_2 \mathbf{x}^* + \boldsymbol{\ell}_2) \tag{6.5.60}$$

and the noncentrality parameter becomes

$$\lambda \equiv \frac{\mathbf{w}_2^T \mathbf{T} \mathbf{w}_2}{\sigma_0^2} = \frac{\Delta \mathbf{v}^T \mathbf{P} \mathbf{v}}{\sigma_0^2} \tag{6.5.61}$$

where $\Delta \mathbf{v}^T \mathbf{P} \mathbf{v}$ is the change of the sum of squares due to the constraint of the null hypothesis.

Let the null hypothesis that causes the smallest change $\Delta \mathbf{v}^T \mathbf{P} \mathbf{v}$ be denoted by H_{sm}. The changes in the sum of the squares and the noncentrality are $\Delta \mathbf{v}^T \mathbf{P} \mathbf{v}_{\mathrm{sm}}$ and λ_{sm}, respectively. For any other null hypothesis we have $\lambda_j > \lambda_{\mathrm{sm}}$. If

$$\frac{\Delta \mathbf{v}^T \mathbf{P} \mathbf{v}_j}{\Delta \mathbf{v}^T \mathbf{P} \mathbf{v}_{\mathrm{sm}}} = \frac{\lambda_j}{\lambda_{\mathrm{sm}}} \ge \lambda_0(\alpha, \beta_{\mathrm{sm}}, \beta_j) \tag{6.5.62}$$

then the two ambiguity sets comprising the null hypotheses H_{sm} and H_j are sufficiently discernible. Both hypotheses are sufficiently different and are distinguishable by means of their type-II errors. Because of its better compatibility with the float

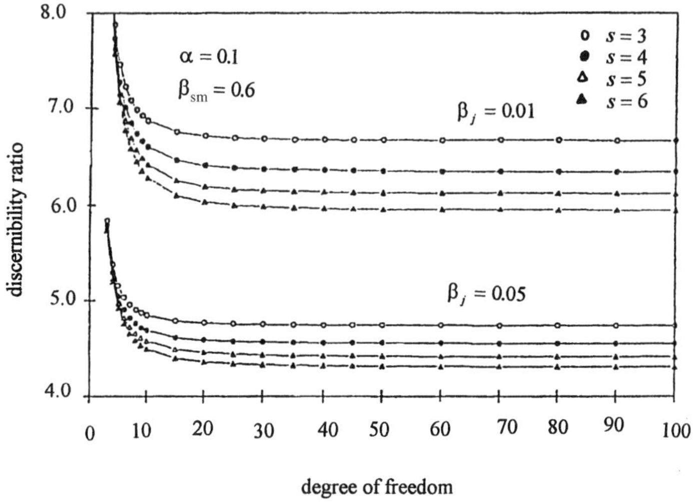

Figure 6.5.2 Discernibility ratio. (Permission by Springer Verlag).

solution, the ambiguity set of the H_{sm} hypothesis is kept, and the set comprising H_j is discarded.

Figure 6.5.2 shows the ratio $\lambda_0(\alpha, \beta_{\text{sm}}, \beta_j)$ as a function of the degree of freedom and the number of conditions. Euler and Schaffrin (1990) recommend a ratio between 5 and 10, which reflects a relatively large β_{sm} and a smaller β_j. Since H_{sm} is the hypothesis with the least impact on the adjustment, i.e., the most compatible with the float solution, it is desirable to have $\beta_{\text{sm}} > \beta_j$ (recall that the type-II error equals the probability of accepting the wrong null hypothesis). Observing more satellites reduces the ratio for given type-II errors.

Many software packages implement a fixed value for the ratio of the best and the second-best solutions, e.g.,

$$\frac{\Delta \boldsymbol{v}^T \boldsymbol{P} \boldsymbol{v}_{\text{2nd smallest}}}{\Delta \boldsymbol{v}^T \boldsymbol{P} \boldsymbol{v}_{\text{sm}}} > 3 \tag{6.5.63}$$

to decide on discernibility. The explanations given above lend some theoretical justification to this commonly used practice, at least for a high degree of freedom.

A lot of work has been done to investigate the theoretical foundations of the ratio test, to suggest better tests for the particular case of integer fixing, and to refine the respective statistics. Examples are Wang et al. (1998), who constructed a test based on the distance between the minimum and the second minimum of $\Delta \boldsymbol{v}^T \boldsymbol{P} \boldsymbol{v}$ instead of the ratio. Teunissen (1998) looked at the success rate of ambiguity fixing for the rounding and bootstrapping techniques. Teunissen (2003) introduced the integer aperture theory and showed that the ratio test is a member of a class of tests provided by the aperture theory. The probability density function of GNSS ambiguity residuals, defined as the difference of float and integer ambiguity, and optimal testing is addressed in Verhagen and Teunissen (2006a,b). There is a lot of literature available on testing and validation of integer estimation. As a first reading, we recommend Verhagen (2004) and Teunissen and Verhagen (2007).

6.5.4 Lattice Reduction and Integer Least Squares

Though the LAMBDA method described above is sufficient to process the GNSS data, another look is essential for understanding how to possibly improve the performance of processing engines as the number of signals increases. At the time when L1, L2, and L5 GPS signals are available along with L1, L2 GLONASS, L1, E5a, E5b, E6 Galileo, L1, L2, L5, E6 QZSS, L1, L5 WAAS, and B1, B2, B3 Beidou signals, the ambiguity resolution problem can encounter 40 and more variables. The need to solve such a large number of ambiguities in real time when performing RTK positioning makes it necessary to revisit the ambiguity resolution problem in view of computational experience accumulated in computer science since the early eighties of the last century.

Usually all integer least-squares methods, including LAMBDA, consist of two parts. The first part transforms the variables in such a way that the new covariance matrix (or its inverse) is closer to the well-conditioned diagonal matrix with diagonal entries sorted in ascending or descending order. This is called the *lattice reduction*. The second part is the integer least squares. Both parts can be performed in many different ways and combined together to form new algorithms.

There are many areas in which the minimization of quadratic functions over a set of integer points is important. Several methods to solve such problems have been developed and are described in the literature—see, for example, Pohst (1981), Fincke and Pohst (1985), Babai (1986), and Agrell et al. (2002). These or similar problems appear, for example, in the implementation of maximum likelihood decoders (MLD) of signals over the finite alphabet performing the search over a certain lattice for a vector closest to a given vector. In order to optimize the search over the lattice, the lattice reduction algorithms have been proposed. A systematic study of this subject starts with Lenstra at al. (1982), giving rise to many applications in mathematics and computer science, as well as data transmission and cryptology. The paper by Korkine and Zolotareff (1873) should also be mentioned. It shows that the conditioning problem for integer lattices has drawn the attention of mathematicians for a long time. The independent statistical study by Teunissen (1993) on the decorrelation method was a response to the need to deal with integer ambiguities in GPS applications.

As a brief introduction to the current state of integer quadratic programming, we start with a description of the branch-and-bound algorithm. It produces a number of candidate solutions, whereas whole subsets of unpromising candidates are discarded by using lower estimated bounds of a cost function. The branch-and-bound algorithm was proposed in Land and Doig (1960) and is still receiving attention in the literature; see, for example, Buchheim et al. (2010) for effective and fast computer implementations.

Then we describe the Finke and Pohst algorithm of Pohst (1981) and Fincke and Pohst (1985). Then the lattice reduction problem will be addressed. In addition, three other algorithms will be briefly identified. Note again that the lattice reduction and integer search can be combined together in different combinations, allowing the derivation of new algorithms that are numerically efficient.

6.5.4.1 Branch-and-Bound Approach

Consider the minimization of

$$q(\hat{\mathbf{z}}) = (\hat{\mathbf{z}} - \mathbf{z})^T \mathbf{D}(\hat{\mathbf{z}} - \mathbf{z}) \tag{6.5.64}$$

over the integer vector $\hat{\mathbf{z}} \in \mathbf{Z}^n$, where $\mathbf{Z}^n$ is the integer-valued lattice in the n-dimensional Euclidean space. In Figure 6.5.3, the contour line ellipse illustrates the constant level set for the positive definite quadratic function. The ellipse is centered at the real-valued vector $\mathbf{z}$. The problem consists of searching for a vector $\mathbf{z}^* \in \mathbf{Z}^n$ which minimizes (6.5.64), i.e., the vector which is closest to $\mathbf{z}$ with respect to the norm $\|\cdot\|^2_{\mathbf{D}}$.

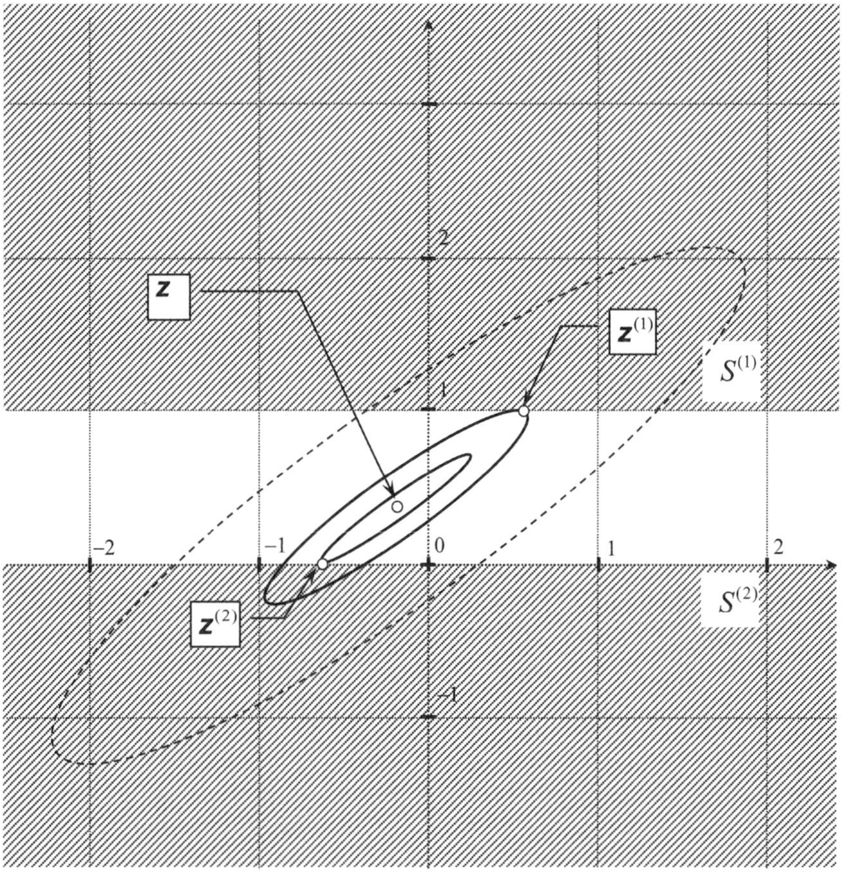

Figure 6.5.3 Contour line of the quadratic function (dashed ellipse centered at the point $\mathbf{z}$) in the two-dimensional case. The vector $\mathbf{z}^*$ closest to $\mathbf{z}$ in the norm $\|\cdot\|^2_{\mathbf{D}}$ must be found.

The branch-and-bound method sequentially reduces the uncertainty of vector $\mathbf{z}^*$ by constructing subsets of the space $\mathbf{R}^n(\mathbf{Z}^n \subset \mathbf{R}^n)$. Each subset corresponds to certain hypothesis about the location of the solution. Each subset is accompanied by the lower estimate of the cost function (6.5.64). The initial hypothesis corresponds to set $S = \mathbf{R}^2$. The lower estimate of the function (6.5.64) is

$$\min_{\mathbf{y} \in S} q(\mathbf{y}) = q(\mathbf{z}) = 0 \tag{6.5.65}$$

We describe the branch-and-bound method using the example shown in Figure 6.5.3. Then we will give its formal description. The point $\mathbf{z}$ has noninteger values of entries. Consider the second entry. As follows from the figure, it belongs to the segment $0 < \mathbf{z}_2 < 1$. Construct two subsets $S^{(1)} \subset S$ and $S^{(2)} \subset S$ in such a way that

$$S^{(1)} \cap S^{(2)} = \varnothing \tag{6.5.66}$$

and

$$\mathbf{Z}^2 \subset S^{(1)} \cup S^{(2)} \tag{6.5.67}$$

More specifically, $S^{(1)} = \{\mathbf{z} : z_2 \geq 1\}$ and $S^{(2)} = \{\mathbf{z} : z_2 \leq 0\}$. The subsets are dashed in the figure. The white (not dashed) strip $\{\mathbf{z} : 0 < z_2 < 1\}$ does not contain integer-valued vectors and can be taken out from further consideration because of (6.5.67). The conditions (6.5.66) and (6.5.67) mean that either $\mathbf{z}^* \in S_1$ or $\mathbf{z}^* \in S_2$. We described the first branching shown in Figure 6.5.4. Calculate the lower bounds of the function $q(\mathbf{y})$ over the subsets $S^{(1)}$ and $S^{(2)}$:

$$v^{(1)} = \min_{\mathbf{y} \in S^{(1)}} q(\mathbf{y}) \leq \min_{\mathbf{y} \in \mathbf{Z}^2 \cap S^{(1)}} q(\mathbf{y}) \tag{6.5.68}$$

$$v^{(2)} = \min_{\mathbf{y} \in S^{(2)}} q(\mathbf{y}) \leq \min_{\mathbf{y} \in \mathbf{Z}^2 \cap S^{(2)}} q(\mathbf{y}) \tag{6.5.69}$$

Note that $v^{(1)}$ and $v^{(2)}$ take their values at the points $\mathbf{z}^{(1)}$ and $\mathbf{z}^{(2)}$, respectively (see Figure 6.5.3). The values $v^{(1)}$ and $v^{(2)}$ estimate the lower bound of the minimum over the integer lattice in the sets $S^{(1)}$ and $S^{(2)}$ because

$$\begin{aligned} v^* &= q(\mathbf{z}^*) = \min_{\hat{\mathbf{z}} \in \mathbf{Z}^2} q(\hat{\mathbf{z}}) \\ &= \min\left\{ \min_{\hat{\mathbf{z}} \in S^{(1)} \cap \mathbf{Z}^2} q(\hat{\mathbf{z}}), \min_{\hat{\mathbf{z}} \in S^{(2)} \cap \mathbf{Z}^2} q(\hat{\mathbf{z}}) \right\} \\ &\geq \min\left\{ \min_{\mathbf{y} \in S^{(1)}} q(\mathbf{y}), \min_{\mathbf{y} \in S^{(2)}} q(\mathbf{y}) \right\} = \min\left\{v^{(1)}, v^{(2)}\right\} \end{aligned} \tag{6.5.70}$$

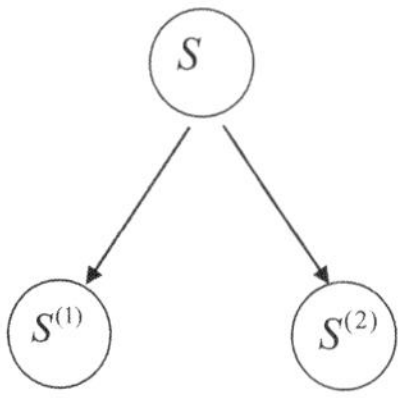

Figure 6.5.4 Branching of the set $S = \mathbf{R}^2$ into two parts $S^{(1)} \cup S^{(2)} \subset S$ in such a way that no one integer point is lost: $\mathbf{Z}^2 \subset S^{(1)} \cup S^{(2)}$.

In other words, $v^* \geq v^{(1)} = q(\mathbf{z}^{(1)})$ and $v^* \geq v^{(2)} = q(\mathbf{z}^{(2)})$. Proceed with the branching to decrease the uncertainty of the location of the optimal point $\mathbf{z}^*$. Look at the tree in Figure 6.5.4 (consisting of root S and two leaves $S^{(1)}$ and $S^{(2)}$) and choose the leaf having the least estimate, which is $v^{(2)}$ as seen from the figure because the ellipse passing through point $\mathbf{z}^{(2)}$ lies inside the ellipse passing through point $\mathbf{z}^{(1)}$. The second entry of the vector $\mathbf{z}^{(2)}$ is an integer, whereas the first one is not and can, therefore, be used for branching. Let us divide the set $S^{(2)}$ (for which $\mathbf{z}^{(2)}$ is the optimum point) into two subsets and exclude the strip $\{\mathbf{z} : -1, < z_1 < 0\}$: $S^{(21)} = \{\mathbf{z} \in S^{(2)} : z_1 \leq -1\}$ and $S^{(22)} = \{\mathbf{z} \in S^{(2)} : z_1 \geq 0\}$. Figures 6.5.5 and 6.5.6 show the partition of the plane corresponding to sets $S^{(1)} \cup S^{(21)} \cup S^{(22)} \subset S$ and the corresponding branching tree, respectively. Points $\mathbf{z}^{(1)}$, $\mathbf{z}^{(21)}$, and $\mathbf{z}^{(22)}$, which are minimizers of $q(\mathbf{y})$ over the sets $S^{(1)}$, $S^{(21)}$, and $S^{(22)}$, respectively, are shown in Figure 6.5.5. The same way as it was proven in (6.5.70), the following conditions are established: $v^* \geq v^{(21)} = q(\mathbf{z}^{(21)})$ and $v^* \geq v^{(22)} = q(\mathbf{z}^{(22)})$.

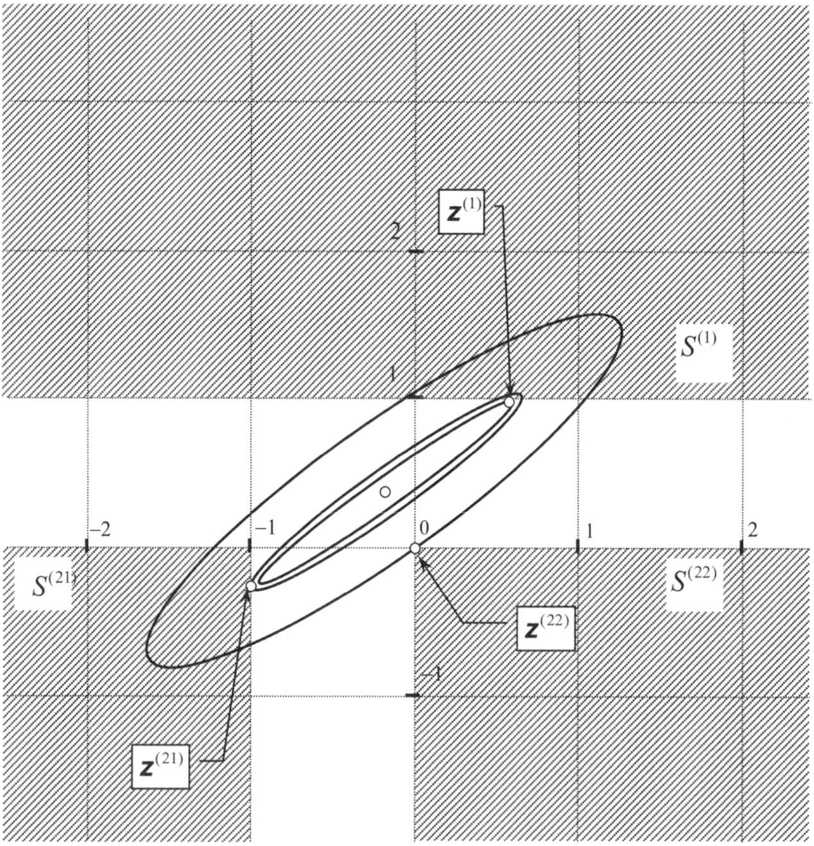

Figure 6.5.5 Partition of the plane corresponding to the sets $S^{(1)} \cup S^{(21)} \cup S^{(22)} \subset S$.

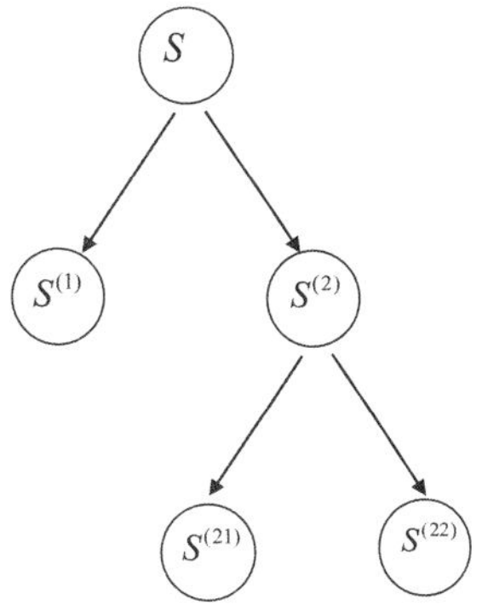

Figure 6.5.6 The branching tree with leaves $S^{(1)}, S^{(21)}, S^{(22)}$. Not one of the integer points is lost in such a partition.

More generally, at each step of branching, the estimates $v^{(i_1 \ldots i_k)}$ corresponding to the leaves of the branching tree do not exceed the optimal value v^*

$$v^* \geq v^{(i_1 \ldots i_k)} \tag{6.5.71}$$

Again, choose the leaf of the tree shown in Figure 6.5.6 corresponding to the least estimate. There are three leaves: $S^{(1)}$, $S^{(21)}$, and $S^{(22)}$. As it follows from Figure 6.5.5, the least estimate is $v^{(1)}$ because the ellipse passing through the point $\boldsymbol{z}^{(1)}$ lies inside two other ellipses. The first entry for point $\boldsymbol{z}^{(1)}$ is not an integer and it, therefore, will be used for branching the leaf node $S^{(1)}$ into two subsets excluding the strip $\{\boldsymbol{z} : 0 < z_1 < 1\}$: $S^{(11)} = \{\boldsymbol{z} \in S^{(1)} : z_1 \leq 0\}$ and $S^{(12)} = \{\boldsymbol{z} \in S^{(1)} : z_1 \geq 1\}$. Figures 6.5.7 and 6.5.8 show the partition of the plane corresponding to the sets $S^{(11)} \cup S^{(12)} \cup S^{(21)} \cup S^{(22)} \subset S$ and the corresponding branching tree, respectively.

Choose the least estimate among $v^{(11)}$, $v^{(12)}$, $v^{(21)}$, and $v^{(22)}$. It is $v^{(21)}$ according to Figure 6.5.7. The estimate $v^{(21)}$ achieves at the point $\boldsymbol{z}^{(21)}$ which has a first integer entry -1 and a second entry satisfying constraints $-1 < z_2 < 0$. The leaf node $S^{(21)}$ is then split into two subsets excluding the strip $\{\boldsymbol{z} : -1 < z_2 < 0\}$: $S^{(211)} = \{\boldsymbol{z} \in S^{(21)} : z_2 \geq 0\}$ and $S^{(212)} = \{\boldsymbol{z} \in S^{(21)} : z_2 \leq -1\}$. Obviously, $S^{(211)} = \{\boldsymbol{z} : z_1 \leq -1, z_2 = 0\}$. Figures 6.5.9 and 6.5.10 show the partition of the plane $S^{(11)} \cup S^{(12)} \cup S^{(211)} \cup S^{(212)} \cup S^{(22)} \subset S$ and the corresponding tree. The least estimate among $v^{(11)}$, $v^{(12)}$, $v^{(211)}$, $v^{(212)}$, and $v^{(22)}$ is $v^{(12)}$ (see Figure 6.5.9). The estimate $v^{(12)}$ satisfies condition $v^{(12)} = q(\boldsymbol{z}^{(12)})$ because the ellipse passing through it lies inside all other ellipses. The set $S^{(12)}$ is split into two sets, $S^{(121)}$ and $S^{(122)}$, according to conditions $S^{(121)} = \{\boldsymbol{z} \in S^{(12)} : z_2 \geq 2\}$ and $S^{(122)} = \{\boldsymbol{z} \in S^{(12)} : z_2 \leq 1\} = \{\boldsymbol{z} : z_1 \geq 1, z_2 = 1\}$. Figures 6.5.11 and 6.5.12 show the partition of the plane $S^{(11)} \cup S^{(121)} \cup S^{(122)} \cup S^{(211)} \cup S^{(212)} \cup S^{(22)} \subset S$ and the corresponding tree.

As it follows from Figure 6.5.11, the least estimate is $v^{(211)}$ which satisfies condition $v^{(211)} = q(\boldsymbol{z}^{(211)})$. Also, as it can be observed from the figure, it is integer valued.

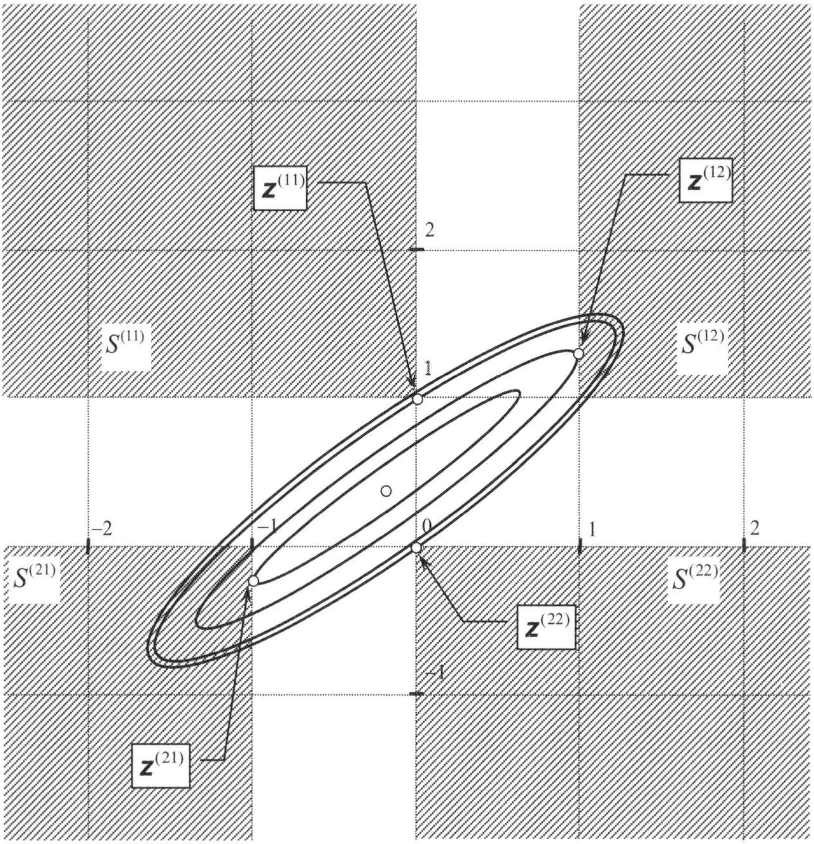

Figure 6.5.7 Partition of the plane corresponding to the sets $S^{(11)} \cup S^{(12)} \cup S^{(21)} \cup S^{(22)} \subset S$.

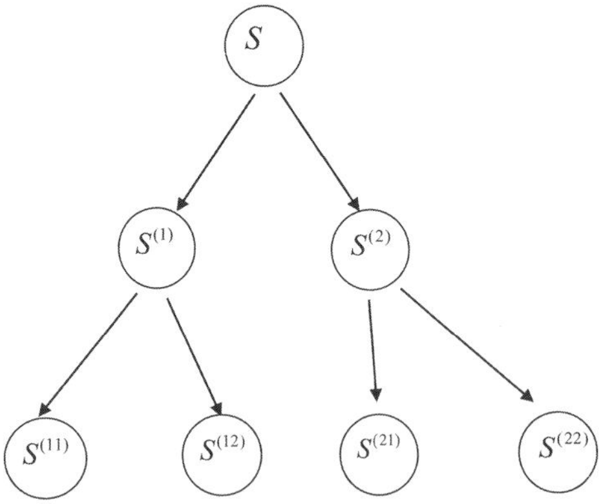

Figure 6.5.8 Branching tree with leaves $S^{(11)}$, $S^{(12)}$, $S^{(21)}$, $S^{(22)}$.

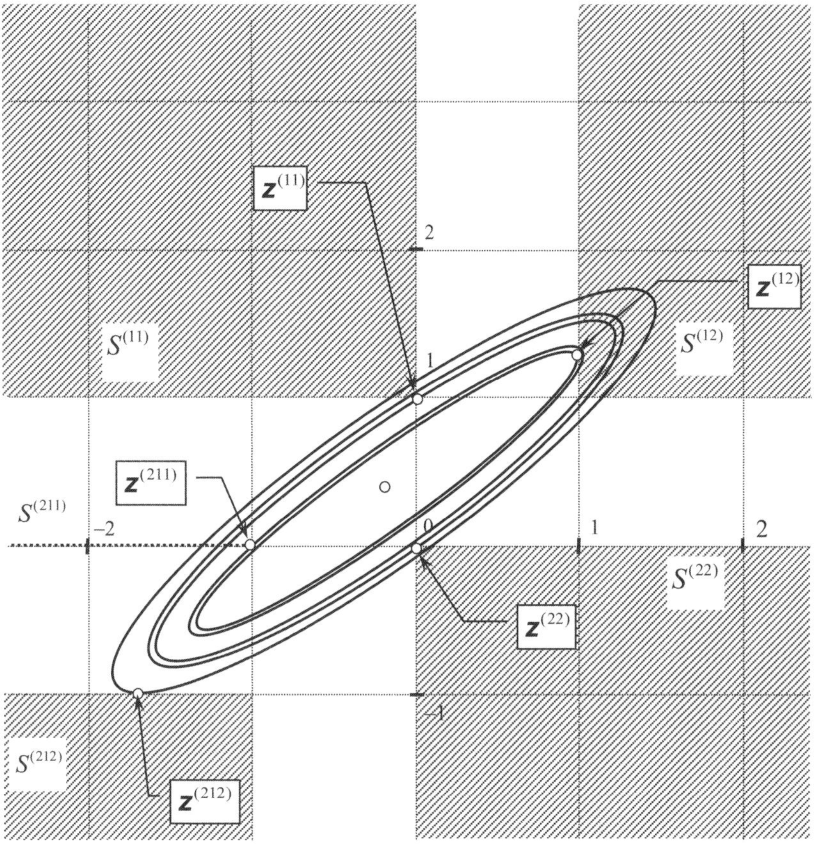

Figure 6.5.9 Partition of the plane corresponding to the sets $S^{(11)} \cup S^{(12)} \cup S^{(211)} \cup S^{(212)} \cup S^{(22)} \subset S$.

Now we conclude that

$$\begin{aligned} v^{(211)} &= \min\left\{v^{(11)}, v^{(121)}, v^{(122)}, v^{(211)}, v^{(212)}, v^{(22)}\right\} \\ &= \min\left\{\min_{\boldsymbol{y}\in S^{(11)}} q(\boldsymbol{y}), \min_{\boldsymbol{y}\in S^{(121)}} q(\boldsymbol{y}), \min_{\boldsymbol{y}\in S^{(122)}} q(\boldsymbol{y}),\right. \\ &\qquad \left.\min_{\boldsymbol{y}\in S^{(211)}} q(\boldsymbol{y}), \min_{\boldsymbol{y}\in S^{(212)}} q(\boldsymbol{y}), \min_{\boldsymbol{y}\in S^{(22)}} q(\boldsymbol{y})\right\} \le \min_{\hat{\boldsymbol{z}}\in \boldsymbol{Z}^2} q(\hat{\boldsymbol{z}}) = q(\boldsymbol{z}^*) = v^* \end{aligned} \tag{6.5.72}$$

On the other hand, $v^{(211)} = q(\boldsymbol{z}^{(211)})$ and taking into account that $\boldsymbol{z}^{(211)}$ is integer valued, we conclude that $v^{(211)} = q(\boldsymbol{z}^{(211)}) \ge v^*$. Together with (6.5.72) this means

$$v^{(211)} = v^* \tag{6.5.73}$$

or, in other words, $\boldsymbol{z}^{(211)}$ is the optimal integer-valued point.

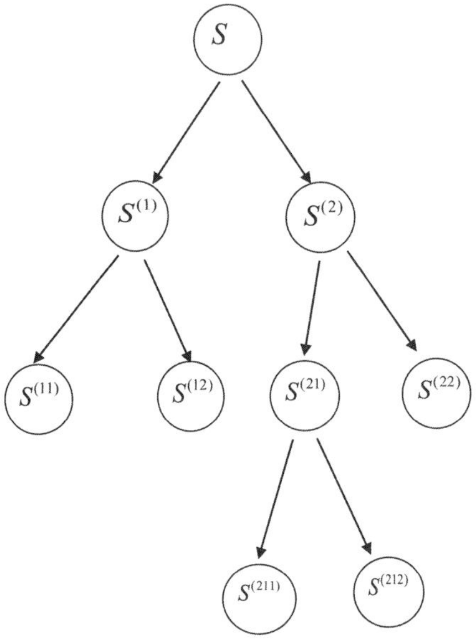

Figure 6.5.10 Branching tree with leaves $S^{(11)}$, $S^{(12)}$, $S^{(211)}$, $S^{(212)}$, $S^{(22)}$.

Now we provide a more formal description of the branch-and-bound method. The binary tree is subject to a transformation at each step k. Each node of the tree corresponds to the subset of space $\boldsymbol{R}^n$. In all figures, the branching starts from the node marked by the symbol S. It is called *the root* and corresponds to $\boldsymbol{R}^n$. At the step $k = 0$, the tree consists of only root. Let $\{S^{(\alpha_1)}, S^{(\alpha_2)}, \ldots, S^{(\alpha_{m_k})}\}$ be subsets corresponding to the leaves of the current tree, α_i being a multi-index. If the leaf node is subjected to branching into another two leaves, the index α_i is transformed into two indices $\alpha_i 1$ and $\alpha_i 2$. The total number of leaves at the step k is m_k, with $m_0 = 1$. Note that a leaf of a tree is a node that has not been subjected to branching (see Figure 6.5.13).

The subset corresponding to leaves has the following property:

$$\boldsymbol{Z}^n \cap \left(\bigcup_{i=1}^{m_k} S^{(\alpha_i)} \right) = \boldsymbol{Z}^n \tag{6.5.74}$$

which means that every point of the integer lattice $\boldsymbol{Z}^n$ belongs to one of the leaf subsets. There is an estimate $v^{(\alpha_i)}$ and a point $\boldsymbol{z}^{(\alpha_i)}$ assigned to a leaf subset as follows:

$$v^{(\alpha_i)} = \min_{y \in S^{(\alpha_i)}} q(\boldsymbol{y}) = q(\boldsymbol{z}^{(\alpha_i)}) \tag{6.5.75}$$

At step k, the leaf subjected to branching is chosen. That is, the leaf corresponding to the minimum value of the estimate

$$v^{(\alpha_*)} = \min_{i=1,\ldots,m_k} v^{(\alpha_i)} \tag{6.5.76}$$

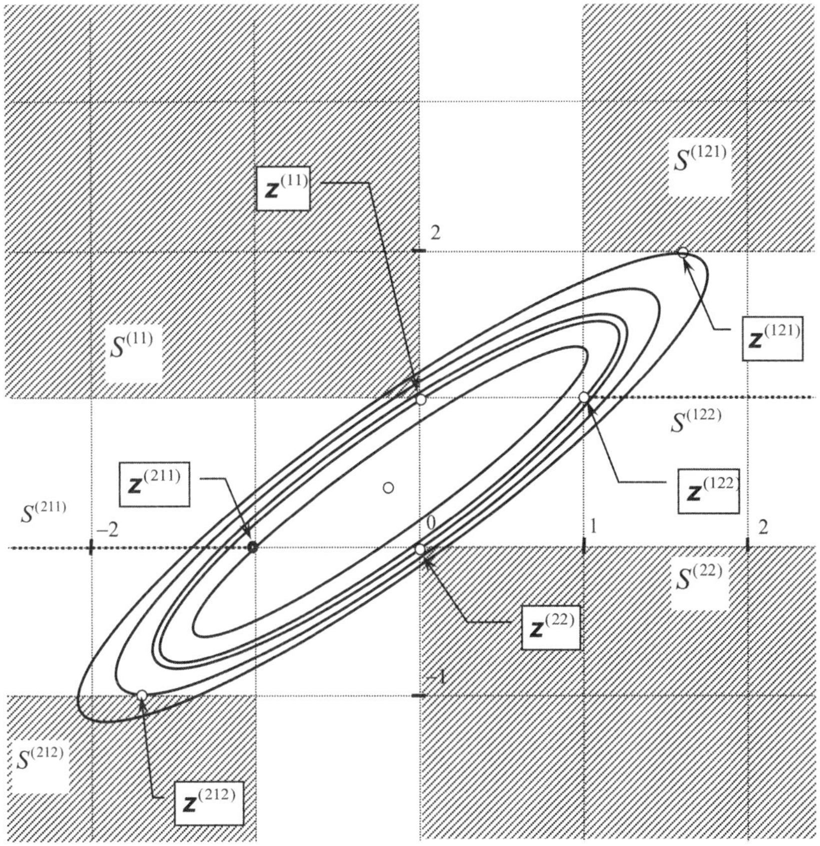

Figure 6.5.11 Partition of the plane corresponding to the sets $S^{(11)} \cup S^{(121)} \cup S^{(122)} \cup S^{(211)} \cup S^{(212)} \cup S^{(22)} \subset S$.

If the vector $\boldsymbol{z}^{(\alpha_*)}$ is integer valued, then it is a solution to the problem (6.5.64). Actually, according to conditions (6.5.75), (6.5.76), and (6.5.74), we have

$$\begin{aligned} v^{(\alpha_*)} &= \min_{i=1,\ldots,m_k} v^{(\alpha_i)} = \min_{i=1,\ldots,m_k} \left\{ \min_{\boldsymbol{y}\in S^{(\alpha_i)}} q(\boldsymbol{y}) \right\} \\ &\le \min_{i=1,\ldots,m_k} \left\{ \min_{\boldsymbol{y}\in \boldsymbol{Z}^n \cap S^{(\alpha_i)}} q(\boldsymbol{y}) \right\} = \min_{\boldsymbol{y}\in \boldsymbol{Z}^n \cap \left(\bigcup_{i=1}^{m_k} S^{(\alpha_i)} \right)} q(\boldsymbol{y}) \\ &= \min_{\boldsymbol{y}\in \boldsymbol{Z}^n} q(\boldsymbol{y}) = v^* \end{aligned} \tag{6.5.77}$$

On the other hand, for any integer point including $\boldsymbol{z}^{(\alpha_*)}$, the following condition holds:

$$v^* \le v^{(\alpha_*)} \tag{6.5.78}$$

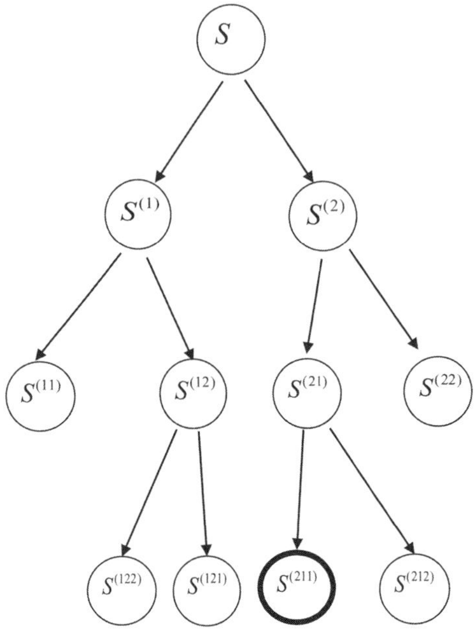

Figure 6.5.12 Branching tree with leaves $S^{(11)} \cup S^{(121)} \cup S^{(122)} \cup S^{(211)} \cup S^{(212)} \cup S^{(22)} \subset S$.

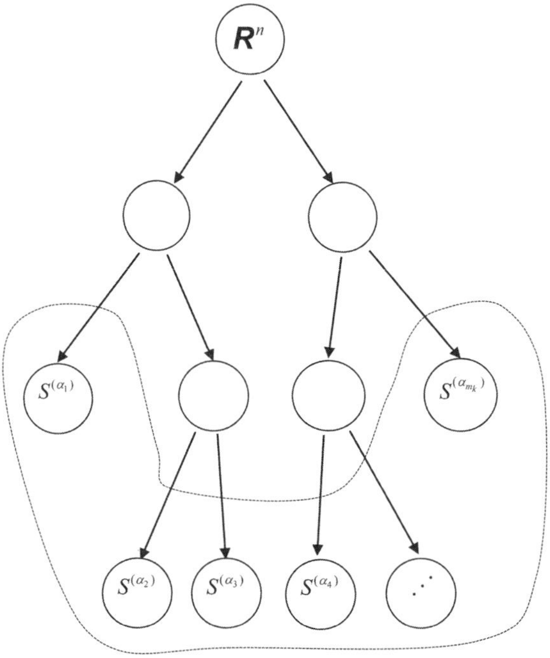

Figure 6.5.13 Binary tree and set of leaves (surrounded by the dashed line).

The combination of (6.5.77) and (6.5.78) proves that $v^* = v^{(\alpha_*)}$, which means optimality of $\boldsymbol{z}^{(\alpha_*)}$.

If the vector $\boldsymbol{z}^{(\alpha_*)}$ is not integer valued, then at least one of its entries, say the lth entry, satisfies the condition

$$\left[z_l^{(\alpha_*)}\right] < z_l^{(\alpha_*)} < \left[z_l^{(\alpha_*)}\right] + 1 \tag{6.5.79}$$

where $[\cdot]$ is the integer part of the value. Then the set $S^{(\alpha_*)}$ is split into two parts

$$S^{(\alpha_* 1)} = S^{(\alpha_*)} \cap \left\{\boldsymbol{z} : z_l \leq \left[z_l^{(\alpha_*)}\right]\right\} \quad S^{(\alpha_* 2)} = S^{(\alpha_*)} \cap \left\{\boldsymbol{z} : z_l \geq \left[z_l^{(\alpha_*)}\right] + 1\right\} \tag{6.5.80}$$

Excluding the slice, we have $\left[z_l^{(\alpha_*)}\right] < z_l < \left[z_l^{(\alpha_*)}\right] + 1$, which means that the leaf node $S^{(\alpha*)}$ is subjected to the branching as shown in Figure 6.5.14. The multi-indices $\alpha_* 1$ and $\alpha_* 2$ are constructed by adding the symbol 1 or 2 to the end of the multi index α_*. This completes the description of the branch-and-bound algorithm.

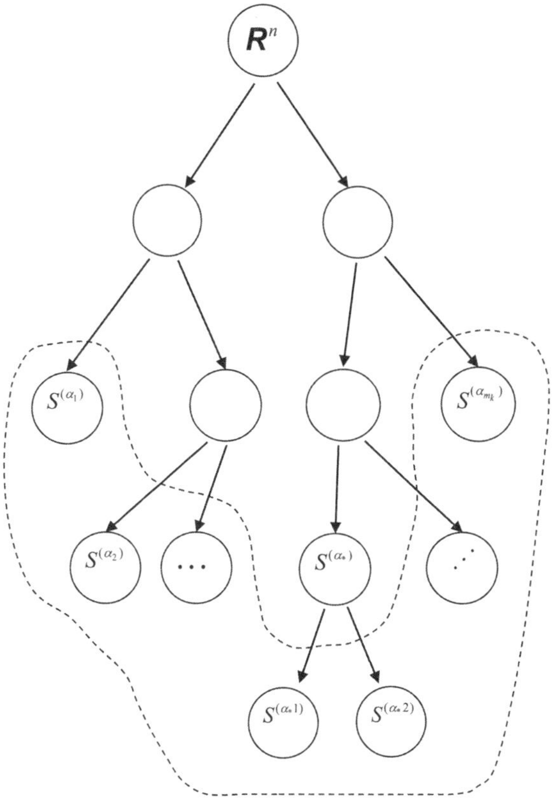

Figure 6.5.14 Binary tree and the set of leaves after the branching is performed.

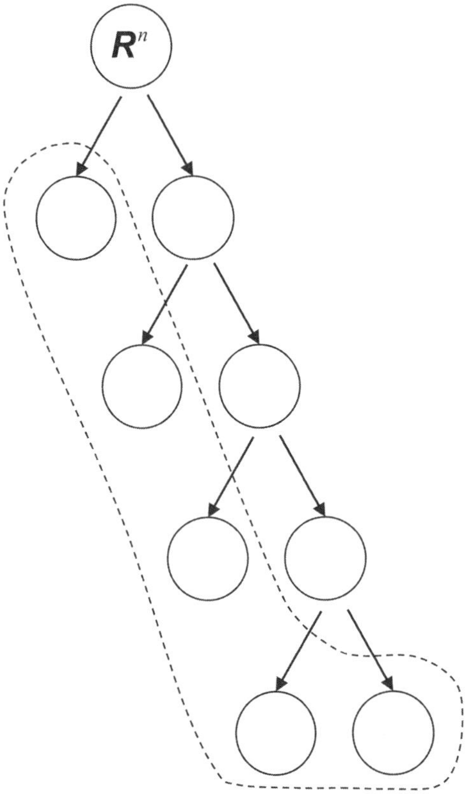

Figure 6.5.15 Single-branch tree.

The estimation problem (6.5.75) is a quadratic programming problem with component-wise box constraints. Algorithms for such problems are considered in Gill et al. (1982). The practical numerical complexity of the branch-and-bound method depends on how much branching has been made. The closer the branching tree is to a single branch (see Figure 6.5.15), the faster the algorithm. On the other hand, the full binary tree supposes exponential computational complexity.

It has been noted that the closer the matrix $\boldsymbol{D}$ in (6.5.64) is to a diagonal matrix (the closer the problem is to a minimization of squares of independent variables), the less branching (fewer leaves) is involved in the resulting tree when performing the algorithm. A significant task, therefore, is to transform problem (6.5.64) into a form that makes the matrix $\boldsymbol{D}$ close to being a diagonal. The integer-valued nature of the problem must be preserved, of course. The unimodular transformations are used for that purpose because they are integer valued together with their inverse. The reduction of the off-diagonal entries of the matrix is tightly connected with the reduction of the lattice, generated by its Cholesky factor. Among all bases generating the same lattice, one looks for the bases having the shortest possible vectors, that is, the subject of lattice reduction theory pioneered by Korkine and Zolotareff (1873). The work of Lenstra et al. (1982) has drawn attention to the lattice reduction problem in modern literature in the area of linear and quadratic integer optimization. We will describe the LLL (Lensta- Lenstra- Lovász) algorithm after a short introduction to another approach to integer minimization known as the Finke-Pohst algorithm [see Pohst (1981) and Fincke and Pohst (1985)].

6.5.4.2 Finke-Pohst Algorithm The algorithm described in this subsection overlaps with the LAMBDA method, while being proposed by other authors independently. The problem (6.5.64) can be equivalently formulated as calculating the vector of the given lattice that is closest to a given vector. Recall that for linearly independent vectors $\boldsymbol{b}_1, \ldots, \boldsymbol{b}_m \in \boldsymbol{R}^n$, the lattice Λ is the set of their linear combinations with integer coefficients

$$\Lambda = \left\{ \hat{\boldsymbol{b}} = \sum_{i=1}^{n} \hat{z}_i \boldsymbol{b}_i \; : \; \hat{z}_i \in \boldsymbol{Z} \right\} \tag{6.5.81}$$

Calculating the Cholesky decomposition of the matrix $\boldsymbol{D} = \boldsymbol{L}\boldsymbol{L}^T$ and $\boldsymbol{b} = \boldsymbol{L}^T\boldsymbol{z}$, we can formulate (6.5.64) as a problem of calculating a vector of the lattice Λ that is closest to $\boldsymbol{b}$ (closest vector problem, or CVP). Note also that there can be different bases $\boldsymbol{b}_1, \ldots, \boldsymbol{b}_n \in \boldsymbol{R}^n$ generating the same lattice Λ. Let $\boldsymbol{B}$ and $\boldsymbol{B}^*$ be matrices composed of columns $\boldsymbol{b}_1, \ldots, \boldsymbol{b}_n$ and $\boldsymbol{b}_1^*, \ldots, \boldsymbol{b}_n^*$, respectively. Two bases generate the same lattice Λ if there exists an integer-valued unimodular (integer-valued invertible) matrix $\boldsymbol{U}$ such that $\boldsymbol{B}^* = \boldsymbol{B}\boldsymbol{U}$. Pohst (1981) and Fincke and Pohst (1985) suggest the following algorithm for CVP:

Let l_{ij} be entries of the matrix $\boldsymbol{L}$ and $q_{ij} = l_{ij}/l_{jj}$ for $1 \leq j \leq i \leq n$. Then (6.5.64) can be written as

$$q(\hat{\boldsymbol{z}}) = \sum_{i=1}^{n} l_{jj}^2 \left(\hat{z}_j - z_j + \sum_{i=j+1}^{n} q_{ij} \left(\hat{z}_i - z_i \right) \right)^2 \tag{6.5.82}$$

The following sequential search over entries of the vector $\hat{\boldsymbol{z}}$ is induced by the triangular structure of the matrix $\boldsymbol{L}$ and (6.5.82). Let C be the upper bound of the minimum in (6.5.64). For example, $C = q([\boldsymbol{z}])$. The value $|\hat{z}_n - z_n|$ is clearly bounded by the value $C^{1/2}/l_{nm}$. More specifically,

$$\left] z_n - \frac{C^{1/2}}{l_{nn}} \right[\leq \hat{z}_n \leq \left[z_n + \frac{C^{1/2}}{l_{nn}} \right] \tag{6.5.83}$$

where $[x]$ is the least integer greater or equal to x and $]x[$ is the largest integer less or equal to x. When introducing LAMBDA we already obtained similar bounds (6.5.47). For each possible value of $\hat{z}_n$ satisfying (6.5.83), we obtain

$$l^2{}_{n-1,n-1}(\hat{z}_{n-1} - z_{n-1} + q_{n,n-1}(\hat{z}_{n-1} - z_{n-1}))^2 \leq C - l_{nn}^2(\hat{z}_n - z_n)^2 \tag{6.5.84}$$

The last inequality implies lower and upper bounds

$$L_{n-1} \leq \hat{z}_{n-1} \leq U_{n-1} \tag{6.5.85}$$

$$L_{n-1} = \left] z_{n-1} - q_{n,n-1}\left(\hat{z}_{n-1} - z_{n-1}\right) - \frac{T_{n-1}^{1/2}}{l_{n-1,n-1}} \right[\tag{6.5.86}$$

$$U_{n-1} = \left[z_{n-1} - q_{n,n-1}\left(\hat{z}_{n-1} - z_{n-1}\right) + \frac{T_{n-1}^{1/2}}{l_{n-1,n-1}} \right] \tag{6.5.87}$$

$$T_{n-1} = C - l_{nn}^2(\hat{z}_n - z_n)^2 \tag{6.5.88}$$

Proceeding with other entries $\hat{z}_{n-2}, \hat{z}_{n-3}, \ldots$, we obtain for each fixed set of values $\hat{z}_n, \hat{z}_{n-1}, \ldots, \hat{z}_{k+1}$

$$l_{kk}^2\left(\hat{z}_k - z_k + \sum_{i=k+1}^{n} q_{ik}\left(\hat{z}_i - z_i\right)\right)^2 \leq T_k \tag{6.5.89}$$

$$T_k = T_{k+1} - l_{k+1,k+1}^2\left(\hat{z}_{k+1} - z_{k+1} + \sum_{i=k+2}^{n} q_{i,k+1}\left(\hat{z}_i - z_i\right)\right)^2 \tag{6.5.90}$$

with $T_n = C$ and k taking values in decreasing order $n-1, n-2, \ldots, 1$. Each time the vector satisfying condition $q(\hat{\mathbf{z}}) < C$ is obtained, C is decreased suitably. Again, similarity with earlier described estimates (6.5.48) and (6.5.49) should be noted. These considerations are summarized in the following algorithm. Denote by $\mathbf{z}^*$ the current record vector and f as the binary flag taking the unit value if the record has been updated at the most outer iteration of the algorithm, and zero otherwise.

1. $\mathbf{z}^* = [\mathbf{z}], C = q([\mathbf{z}]), f = 1$.
2. If $f = 0$, the algorithm terminates with $\mathbf{z}^*$ being a solution.
3. Set $k = n$, $T_n = C$, $S_n = 0, f = 0$.
4. Set $U_k = \left[\frac{T_k^{1/2}}{l_{k,k}} + z_k - S_k\right]$, $L_k = \left]-\frac{T_k^{1/2}}{l_{k,k}} + z_k - S_k\right[$, $\hat{z} = L_k - 1$.
5. Set $\hat{z}_k := \hat{z}_k + 1$. If $\hat{z}_k \leq U_k$, go to step 7; else go step 6.
6. If $k = n$, go to step 2; else set $k := k + 1$ and go to step 5.
7. If $k = 1$, go to step 8; else set $k := k - 1$, $S_k = \sum_{i=k+1}^{n} q_{i,k}(\hat{z}_i - z_i)$,
   $T_k = T_{k+1} - q_{k+1,k+1}(\hat{z}_{k+1} - z_{k+1} + S_{k+1})^2$. Go to step 4.
8. If $q(\hat{\mathbf{z}}) < C$, set $C = q(\hat{\mathbf{z}}), \mathbf{z}^* = \hat{\mathbf{z}}$, and $f = 1$. Go to step 5.

This is the Fincke-Pohst algorithm [Pohst (1981) and Fincke and Pohst (1985)]. Its various modifications differ in the strategy of how the values $\hat{z}_k$ are updated at step 5.

For example, they can be sorted from left to right, or from the center incrementally, i.e., $0, -1, 1, -2, 2, \ldots$.

As noted in Fincke and Pohst (1985), using lattice reduction can significantly decrease the computation complexity of the algorithm. Let r_i denote the columns of the matrix $(\boldsymbol{L}^T)^{-1}$. Then

$$(\hat{z}_i - z_i)^2 = \left(r_i^T \boldsymbol{L}^T(\hat{\boldsymbol{z}} - \boldsymbol{z})\right)^2 \leq \|r_i\|^2 (\hat{\boldsymbol{z}} - \boldsymbol{z})^T \boldsymbol{D} (\hat{\boldsymbol{z}} - \boldsymbol{z}) \leq \|r_i\|^2 C \tag{6.5.91}$$

for all $i = 1, \cdots, n$. This means that by reducing the length of rows of the matrix $\boldsymbol{L}^{-1}$, we reduce the search range of the integer variables. Applying any of the reduction methods to the matrix $(\boldsymbol{L}^T)^{-1}$, we obtain the reduced matrix by multiplying it by appropriately chosen unimodular matrix $\boldsymbol{U}^{-1}$, thus obtaining $(\boldsymbol{M}^T)^{-1} = \boldsymbol{U}^{-1}(\boldsymbol{L}^T)^{-1}$. Then, instead of solving CVP (6.5.64), we solve CVP:

$$q(\hat{\boldsymbol{y}}) = (\hat{\boldsymbol{y}} - \boldsymbol{y})^T \boldsymbol{M}\boldsymbol{M}^T (\hat{\boldsymbol{y}} - \boldsymbol{y}) \tag{6.5.92}$$

using the algorithm described above, $\boldsymbol{y} = \boldsymbol{U}^{-1}\boldsymbol{z}$ and recover the original integer-valued vector

$$\hat{\boldsymbol{z}} = \boldsymbol{U}\hat{\boldsymbol{y}} \tag{6.5.93}$$

The resulting algorithm is as follows:

1′. Compute Cholesky decomposition $\boldsymbol{D} = \boldsymbol{L}\boldsymbol{L}^T$ and $\boldsymbol{L}^{-1}$ (it is computed as a solution of the system $\boldsymbol{L}\boldsymbol{X} = \boldsymbol{I}$ (see Section A.3.6 of Appendix A).

2′. Perform the lattice reduction, computing the row-reduced matrix $(\boldsymbol{M}^T)^{-1}$ together with the unimodular matrix $\boldsymbol{U}^{-1}$ subject to $(\boldsymbol{M}^T)^{-1} = \boldsymbol{U}^{-1}(\boldsymbol{L}^T)^{-1}$. Compute $\boldsymbol{M}^T = \boldsymbol{L}^T\boldsymbol{U}$.

3′. Compute Cholesky decomposition $\overline{\boldsymbol{L}}\,\overline{\boldsymbol{L}}^T = \boldsymbol{M}\boldsymbol{M}^T$ and $q_{ij} = \overline{l}_{ij}/\overline{l}_{jj}$.

4′. Compute $\boldsymbol{z} := \boldsymbol{U}^{-1}\boldsymbol{z}$ and perform the algorithm step 1 to 8 described above.

6.5.4.3 Lattice Reduction Algorithms Now we describe the algorithms of lattice reduction aimed to reduce the rows of the matrix $\boldsymbol{L}$ (columns of the matrix $\boldsymbol{L}^T$). Let $\boldsymbol{b}_1, \ldots, \boldsymbol{b}_n$ be columns of $\boldsymbol{L}^T$ generating the lattice (6.5.81). Starting with the LLL algorithm, we apply the Gram-Schmidt orthogonalization process to vectors $\boldsymbol{b}_1, \ldots, \boldsymbol{b}_n$. The orthogonal vectors $\boldsymbol{b}_1^*, \cdots, \boldsymbol{b}_n^*$ and numbers μ_{ij}, $1 \leq j \leq i \leq n$ are inductively defined by

$$\boldsymbol{b}_1^* = \boldsymbol{b}_1, \; \boldsymbol{b}_i^* = \boldsymbol{b}_i - \sum_{j=1}^{i-1} \mu_{ij}\boldsymbol{b}_j^* \tag{6.5.94}$$

$$\mu_{ij} = \boldsymbol{b}_i^T \boldsymbol{b}_j^* / \boldsymbol{b}_j^{*T} \boldsymbol{b}_j^* \tag{6.5.95}$$

Note that $\boldsymbol{b}_i^*$ with $i > 1$ is the projection of $\boldsymbol{b}_i$ on the orthogonal complement of the linear subspace spanned on the vectors $\boldsymbol{b}_j^*, j = 1, \ldots, i-1$ and $\boldsymbol{b}_1^*, \ldots, \boldsymbol{b}_n^*$. Vectors $\boldsymbol{b}_i^*$ form the orthogonal basis of $\boldsymbol{R}^n$. The smaller the absolute values of μ_{ij}, the closer the original basis for the lattice $\boldsymbol{b}_1, \ldots, \boldsymbol{b}_n$ to the orthogonal basis. The basis for the lattice is called *LLL reduced* if

$$|\mu_{ij}| \leq 1/2 \text{ for } 1 \leq j < i \leq n \tag{6.5.96}$$

and

$$\|\boldsymbol{b}_i^*\|^2 + \mu_{i,i-1}^2 \|\boldsymbol{b}_{i-1}^*\|^2 \geq \delta \|\boldsymbol{b}_{i-1}^*\|^2 \text{ for } 1 < i \leq n \tag{6.5.97}$$

where $\frac{1}{4} < \delta \leq 1$. In the original Lenstra et al. (1982) paper, the case $\delta = 3/4$ was considered. Condition (6.5.96) is called *the size reduction condition*, and condition (6.5.97) is called *the Lovász condition*. The following transformation will be referred to as the *size reduction transformation* and denoted by $T(k, l)$ for $l < k$:

$$\text{If } |\mu_{kl}| > \frac{1}{2} \text{ then} \begin{cases} r = \text{ integer nearest to } \mu_{kl}, \ b_k = b_k - r b_l, \\ \mu_{kj} := \mu_{kj} - r\mu_{lj} \text{ for } j = 1, 2, \cdots, i-1, \\ \mu_{kl} := \mu_{kl} - r \end{cases} \tag{6.5.98}$$

The LLL algorithm for $\delta = 3/4$ consists of the following steps:

1. Perform the Gram-Schmidt orthogonalization according to (6.5.94) and (6.5.95), and denote $B_i = \|\boldsymbol{b}_i^*\|^2$. Set $k = 2$.
2. Perform $T(k, k-1)$. If $B_k < \left(\frac{3}{4} - \mu_{k,k-1}^2\right) B_{k-1}$, then go to step 3.
   Perform $T(k, l)$ for $l = k-2, \ldots, 1$. If $k = n$, terminate $k := k+1$, go to step 2.
3. Set $\mu := \mu_{k,k-1}$, $B := B_k + \mu^2 B_{k-1}$, $\mu_{k,k-1} := \mu B_{k-1} / B$,
   $B_k := B_{k-1} B_k / B$, $B_{k-1} := B$,
   swap vectors $(b_{k-1}, b_k) := (b_k, b_{k-1})$,
   swap values $\begin{pmatrix} \mu_{k-1,j} \\ \mu_{k,j} \end{pmatrix} := \begin{pmatrix} \mu_{k,j} \\ \mu_{k-1,j} \end{pmatrix}$ for $j = 1, 2, \ldots, k-2$
   $\begin{pmatrix} \mu_{i,k-1} \\ \mu_{i,k} \end{pmatrix} := \begin{pmatrix} 1 & \mu_{k,k-1} \\ 0 & 1 \end{pmatrix} \begin{pmatrix} 0 & 1 \\ 1 & -\mu \end{pmatrix} \begin{pmatrix} \mu_{i,k-1} \\ \mu_{i,k} \end{pmatrix}$ for $j = k+1, k+2, \ldots, n$
   if $k > 2$, then $k := k-1$, go to step 2.

The unimodular matrix $\boldsymbol{U}$ is constructed along with the construction of the reduced basis. The transformation $T(k, l)$ is equivalent to the multiplication of the matrix $\boldsymbol{B} = [\boldsymbol{b}_1, \boldsymbol{b}_2, \ldots, \boldsymbol{b}_n]$ by the matrix

$$\begin{bmatrix} 1 & & & \\ & 1 & & \\ & -r & 1 & \\ & & & 1 \end{bmatrix} \begin{matrix} \\ \leftarrow k \\ \leftarrow l \\ \\ \end{matrix} \tag{6.5.99}$$

Step 2 is equivalent to the permutation matrix

$$\begin{bmatrix} 1 & & & \\ & 0 & 1 & \\ & 1 & 0 & \\ & & & 1 \end{bmatrix} \begin{matrix} \\ \leftarrow (k-1) \\ \leftarrow k \\ \\ \end{matrix} \tag{6.5.100}$$

The product of sequentially generated matrices (6.5.99) and (6.5.100) results in the matrix $\boldsymbol{U}$.

There are other definitions of lattice reduction and other reduction algorithms. They can be applied at step 2′ of the algorithm 1′ to 4′.

Note that the LLL algorithm is not the first reduction algorithm proposed in history. Another type of basis reduction is the *Korkine-Zolotareff (KZ)* reduction. To define it, given the basis $\boldsymbol{b}_1, \boldsymbol{b}_2, \ldots, \boldsymbol{b}_n$ we construct the upper triangular matrix, obtained via Gram-Schmidt decomposition (6.5.94) and (6.5.95),

$$\boldsymbol{G} = \begin{bmatrix} \|\boldsymbol{b}_1^*\| & \mu_{21}\|\boldsymbol{b}_1^*\| & \mu_{31}\|\boldsymbol{b}_1^*\| & \cdots & \mu_{n1}\|\boldsymbol{b}_1^*\| \\ 0 & \|\boldsymbol{b}_2^*\| & \mu_{32}\|\boldsymbol{b}_2^*\| & \cdots & \mu_{n2}\|\boldsymbol{b}_2^*\| \\ 0 & 0 & \|b_3^*\| & \cdots & \mu_{n3}\|\boldsymbol{b}_3^*\| \\ \vdots & \vdots & \vdots & \ddots & \vdots \\ 0 & 0 & 0 & \cdots & \|\boldsymbol{b}_n^*\| \end{bmatrix} \tag{6.5.101}$$

The basis $\boldsymbol{b}_1, \boldsymbol{b}_2, \ldots, \boldsymbol{b}_n$ is KZ reduced if its upper triangular representation (6.5.101) is KZ reduced. The matrix (6.5.101) is defined, recursively, to be KZ reduced if either $n = 1$ or each of the following conditions holds: (6.5.96), the vector $(\|\boldsymbol{b}_1^*\|, 0, \ldots, 0)^T$ is shortest in the lattice generated by columns of the matrix (6.5.101), and the submatrix

$$\begin{bmatrix} \|\boldsymbol{b}_2^*\| & \mu_{32}\|\boldsymbol{b}_2^*\| & \cdots & \mu_{n2}\|\boldsymbol{b}_2^*\| \\ 0 & \|\boldsymbol{b}_3^*\| & \cdots & \mu_{n3}\|\boldsymbol{b}_3^*\| \\ \vdots & \vdots & \ddots & \vdots \\ 0 & 0 & \cdots & \|\boldsymbol{b}_n^*\| \end{bmatrix} \tag{6.5.102}$$

is KZ reduced. The KZ-reduced basis is also LLL reduced, but for LLL there exists the LLL algorithm described above, having polynomial complexity, while KZ reduction requires more extensive calculations. Hybrids between KZ and LLL reductions have been proposed in Schnorr (1987). It is noted in Agrell et al. (2002) that the KZ reduction is recommended for applications where the same lattice is to be searched many times for different vectors $\boldsymbol{z}$ in (6.5.64); otherwise, LLL reduction is recommended (the latter is the case for applications to RTK).

Wübben et al. (2011) introduce Seysen's reduction and Brun's reduction algorithms. Both methods use unimodular transformations, differing in definition of the orthogonality measure of the resulting basis.

Another class of methods, called *inverse integer Cholesky decorrelation,* is introduced in Wang et al. (2010) and Zhou and He (2013). In order to make the matrix $\boldsymbol{D}$ or its inverse closer to diagonal, different decorrelation techniques have been developed. The construction of the unimodular transformation starts with Cholesky decomposition in the form

$$\boldsymbol{D} = \boldsymbol{L}\boldsymbol{\Delta}\boldsymbol{L}^T \tag{6.5.103}$$

where $\boldsymbol{L}$ is the lower triangular matrix with unit diagonal and $\boldsymbol{\Delta}$ is a diagonal matrix with positive elements. The unimodular transformation $\boldsymbol{U}_1$ is constructed as inverse $\boldsymbol{L}$ rounded

$$\boldsymbol{U}_1 = [\boldsymbol{L}^{-1}] \tag{6.5.104}$$

The unimodular transformation can be applied to $\boldsymbol{D}$ or $\boldsymbol{D}^{-1}$. In the first case, we have

$$\boldsymbol{D}_1 = \boldsymbol{U}_1 \boldsymbol{D} \boldsymbol{U}_1^T \tag{6.5.105}$$

The matrix $\boldsymbol{D}_1$ is not diagonal since the rounding operation has been applied to $\boldsymbol{L}^{-1}$ in (6.5.104). Repeating calculations (6.5.103) to (6.5.105) construct the unimodular transformation $\boldsymbol{U}^T = \boldsymbol{U}_1^T \boldsymbol{U}_2^T \cdots$. Calculations repeat until either convergence or a predetermined condition number is reached.

6.5.4.4 Other Searching Strategies

The second stage of the integer least-squares algorithm was presented in this section by the branch-and-bound algorithm and sphere decoding (Fincke-Pohst) algorithm. In order to describe other approaches to the searching strategies, we will use the conceptual description presented in Agrell et al. (2002), Section IIIA.

Let the problem (6.5.64) be rewritten in the form

$$||\boldsymbol{G}\hat{\boldsymbol{z}} - \boldsymbol{x}||^2 \to \min_{\hat{\boldsymbol{z}} \in \boldsymbol{Z}^n} \tag{6.5.106}$$

The recursive characterization of the lattice spanned on columns of the matrix $\boldsymbol{G}$ follows from the representation

$$\boldsymbol{G} = \left[\boldsymbol{G}_{n-1} \;\vdots\; \boldsymbol{g}_n\right] \tag{6.5.107}$$

with $\boldsymbol{G}_{n-1}$ being the $n \times (n-1)$ matrix and $\boldsymbol{g}_n$ being the last column of $\boldsymbol{G}$. Moreover, $\boldsymbol{g}_n$ can be written as $\boldsymbol{g}_n = \boldsymbol{g}_{||} + \boldsymbol{g}_{\perp}$ where $\boldsymbol{g}_{||} \in span(\boldsymbol{G}_{n-1})$ belongs to the column space of $\boldsymbol{G}_{n-1}$ and $\boldsymbol{G}_{n-1}\boldsymbol{g}_{\perp} = 0$. If the matrix $\boldsymbol{G}$ is upper triangular as in (6.5.101), then obviously $\boldsymbol{g}_{||} = (g_{n1}, \ldots, g_{n,n-1}, 0)^T$ and $\boldsymbol{g}_{\perp} = (0, \ldots, 0, g_{nn})^T$. Then the lattice $\Lambda(\boldsymbol{G})$ can be factorized as a stack of $(n-1)$-dimensional translated sublattices

$$\Lambda(\boldsymbol{G}) = \bigcup_{z_n=-\infty}^{+\infty} \{\boldsymbol{c} + z_n \boldsymbol{g}_{||} + z_n \boldsymbol{g}_{\perp} \,:\, \boldsymbol{c} \in \Lambda(\boldsymbol{G}^*)\} \tag{6.5.108}$$

The hyperplanes $\{\boldsymbol{c} + z_n \boldsymbol{g}_{||} + z_n \boldsymbol{g}_{\perp} \,:\, \boldsymbol{c} \in \Lambda(\boldsymbol{G}^*)\}$ containing these sublattices are called *layers*. Thus, the number z_n indexes the layers. It denotes to which layer a certain lattice point belongs. The vector $\boldsymbol{g}_{||}$ is the offset by which the sublattice is translated within its layer, relative to the adjacent layers. The distance between two adjacent layers is $||\boldsymbol{g}_{\perp}||$. For the upper triangular case (6.5.101), we have $||\boldsymbol{g}_{\perp}|| = |g_{nn}| = g_{nn}$ because $g_{nn} > 0$.

Now, a large class of search algorithms can be described recursively as a finite number of $(n-1)$-dimensional search operations.

The distance from the vector $\boldsymbol{x}$ in (6.5.106) to the layer indexed by the number z_n is

$$y_n = |z_n - \bar{z}_n| \cdot ||\boldsymbol{g}_{\perp}|| \tag{6.5.109}$$

with $\bar{z}_n$ being defined as

$$\bar{z}_n = \frac{\boldsymbol{x}^T \boldsymbol{g}_{\perp}}{\boldsymbol{g}_{\perp}^T \boldsymbol{g}_{\perp}} \tag{6.5.110}$$

For the upper triangular case (6.5.101), we have $y_n = |z_n g_{nn} - x_n|$ (because $g_{nn} > 0$). Let $\boldsymbol{z}^*$ be a solution to (6.5.106) and ρ_n be the upper bound on $\|\boldsymbol{G}\hat{\boldsymbol{z}} - \boldsymbol{x}\|$. Then only the finite number of layers in (6.5.108) must be searched, indexed by numbers

$$z_n = \left]\bar{z}_n - \frac{\rho_n}{\|\boldsymbol{g}_\perp\|}\right[, \ldots, \left[\bar{z}_n + \frac{\rho_n}{\|\boldsymbol{g}_\perp\|}\right] \tag{6.5.111}$$

The layer with z_n = integer nearest $(\bar{z}_n)$ has the shortest orthogonal distance to $\boldsymbol{x}$. In addition to the two search methods already described above, another three methods will be identified. Each is indexed in search layer segments (6.5.111), but they differ in the order the layers are tried and in the way the upper bound ρ_n is treated and updated. Note that in the branch-and-bound method we dealt with the lower bounds.

Let us denote $[\![z]\!] \equiv$ integer nearest(z). If only $z_n^* = [\![\bar{z}_n]\!]$ is considered in (6.5.111), the problem is immediately reduced to one $(n-1)$-dimensional problem. Sequential application of this strategy yields the Babai nearest plane algorithm (Babai, 1986). Note that the lattice reduction can be repeated or updated for each reduction of dimension. The Babai nearest plane algorithm can be performed without the reduction to the upper triangle from (6.5.101). It is a fast time-polynomial method giving an approximate solution to (6.5.106). In other words, its computational cost has polynomial dependence on the dimension n. The result depends not only on the vector $\boldsymbol{x}$ and the lattice $\Lambda(\boldsymbol{G})$ but also on the lattice reduction lattice basis. Effectively, this means dependence on the lattice reduction method used. The solution $\boldsymbol{z}^B$ found by this algorithm is called the *Babai solution* and the lattice point $\boldsymbol{x}^B = \boldsymbol{G}\boldsymbol{z}^B$ is called the *Babai lattice point*.

Other methods find the strict solution to (6.5.106). Running through all layers and searching each layer with the same value of ρ_{n-1} regardless of z_n yields the Kannan strategy (Kannan 1983, 1987).

The error vector $\boldsymbol{G}\boldsymbol{z}^* - \boldsymbol{x}$ consists of two orthogonal components. The first one belongs to the column space of $\boldsymbol{G}_{n-1}$ $[\boldsymbol{g}_{||} \in \text{span}(\boldsymbol{G}_{n-1})]$ and it represents the $(n-1)$-dimensional error vector. The second one is collinear to $\boldsymbol{g}_\perp$ and its length is y_n as defined in (6.5.109). Since the distance y_n depends on the layer index z_n, the upper bound ρ_{n-1} can be chosen as

$$\rho_{n-1} = \sqrt{\rho_n^2 - y_n^2} \tag{6.5.112}$$

The idea to let the bound be dependent on the layer index represents the Pohst strategy (Pohst,1981; Fincke and Pohst,1985). A detailed description of the algorithm based on this strategy for the case of the upper triangular matrix $\boldsymbol{G}$ has already been given above (see the algorithm steps 1 to 8). The points lying inside the sphere are searched. That is why the method is called "the sphere decoder," since decoding of the vector $\boldsymbol{x}$ is the goal in the communication applications. When the lattice point inside the sphere is found, the bound ρ_n is immediately updated (see step 8 with $C = \rho_n^2$).

The Schnorr-Euchner strategy (Schnorr and Euchner, 1994) combines ideas of the Babai nearest plane algorithm and the Fincke-Pohst decoder. Let $\bar{z}_n \leq [\![\bar{z}_n]\!]$. Then the sequence

$$z_n = [\![\bar{z}_n]\!], [\![\bar{z}_n]\!] - 1, [\![\bar{z}_n]\!] + 1, [\![\bar{z}_n]\!] - 2, \ldots \tag{6.5.113}$$

orders layers in (6.5.111) in non-decreasing distance from $\boldsymbol{x}$. Similarly, they are ordered as

$$z_n = [\![\bar{z}_n]\!], [\![\bar{z}_n]\!] + 1, [\![\bar{z}_n]\!] - 1, [\![\bar{z}_n]\!] + 2, \ldots \tag{6.5.114}$$

if $\bar{z}_n > [\![\bar{z}_n]\!]$. Since the volume of the layer decreases with increasing distance y_n, the chance to find the correct layer earlier maximizes. Another advantage to have the order of layers according to nondecreasing distance to $\boldsymbol{x}$ is that the search can be safely terminated as soon as y_n exceeds the distance to the best lattice point $\boldsymbol{G\hat{z}}$ has found so far. The very first lattice point generated by the algorithm will be the Babai lattice point. Since the ordering (6.5.113) or (6.5.114) does not depend on the bound ρ_n, no initial guess about this bound is needed. The bound is updated each time the record value is found. The first value of the bound is the distance from $\boldsymbol{x}$ to the Babai lattice point.

6.5.4.5 Connection Between LAMBDA and LLL Methods The LLL algorithm plays a significant role in different fields of discrete optimization and communication theory. It is used as a preconditioning step in integer programming algorithms (Schrijver, 1986). In 1993, Teunissen published the LAMBDA method for solving integer least-squares problems for GPS ambiguity resolution [see Teunissen (1993) and Subsection 6.5.2]. At its first stage, the preconditioning unimodular transformation is constructed and applied to the covariance matrix. Historically, LAMBDA appeared later than LLL, but theoretically LAMBDA is independent of LLL and based on different statistical constructions. Overlaps and differences between the decorrelation LAMBDA algorithm and LLL are pointed out in many papers (Lannes, 2013; Grafarend, 2000). Also, there are similarities between the integer search part of LAMBDA and the Fincke-Pohst algorithm. Currently these approaches are developing simultaneously. All developments made on the LLL algorithm can be applied to the LAMBDA algorithm and vice versa.

The statistical proof of optimality of LAMBDA was published in Teunissen (1999). A class of integer estimators is introduced that includes integer rounding, integer bootstrapping; see Blewitt (1989) and Dong and Bock (1989) for details, and the integer least squares. The integer least-squares estimator was proven to be best in the sense of maximizing the probability of correct integer estimation. For the case of ambiguity resolution, this implies that the success rate of any other estimator of integer carrier phase ambiguities will be smaller than, or at most equal to, the ambiguity resolution rate of the integer least-squares estimator. This useful conclusion can be extended on any of the algorithms considered in this subsection but excluding the Babai algorithm, since it provides a fast but approximate solution. It would be interesting to extend the analysis of Teunissen (1999) to the case of the Babai algorithm to see how its success rate relates to the success rate of the approximate algorithms.

With these remarks on the connection between LAMBDA and other integer least-squares estimators, we conclude this subsection. These existing connections may serve as an illustration of the fact that similarly effective results can be obtained by scientists solving different technology problems having similar mathematical meaning.

It should be noted that currently there is no best algorithm showing superior performance among all others in terms of the success rate and the computational cost simultaneously, including computational cost of the preconditioning and search. This leaves a huge opportunity for the creativity of engineers working on efficient geodetic software.

6.6 NETWORK-SUPPORTED POSITIONING

Positioning is always supported in one way or another by a network of reference stations. This is even the case for PPP (precise point positioning), where the "network in the background" is the global IGS network whose observations are used to compute the precise ephemeris and the satellite clock error. We discuss three types of positioning techniques. The first one is the PPP model, which uses lumped parameters that combine the ambiguities and the receiver and satellite hardware delays and clock errors that are part of the receiver code bias and satellite code bias. The second technique is CORS-based relative positioning, which uses double-difference observations to eliminate clock errors and hardware delay terms. RTK (real-time kinematic) is part of this solution group and applies the classical differential correction to the user observations. RTK, with focus on across-receiver differencing, is discussed in Chapter 7. The third technique is PPP-RTK. We reparameterize the unknowns to eliminate the singularities of the system, compute bias parameters instead of the classical differential corrections, use these to correct the user observations, and fix reparameterized undifferenced ambiguities. Several PPP-RTK models will be discussed. The first model is for single-frequency observation. The development starts with the basic carrier phase and pseudorange equations and formulates a network solution to compute the biases. All terms are carried through the development up to the user solution, in order to better understand how various terms are combined as part of the reparameterization. Next, the dual-frequency model is given for network solutions, as well as a line-by-line approach. The last model discussed used dual-frequency across-satellite differences. All dual-frequency PPP-RTK approaches are equivalent in the sense they use the same observational content, although performance differences may occur in practice due to implementation considerations.

6.6.1 PPP

In traditional network-supported positioning which includes traditional RTK, the carrier phase and code (pseudorange) differential corrections of a base station are transmitted to the user. At the user station the observations and differential corrections are transformed into equivalent double differences or across-station single differences in order to carry out the ambiguity-fixed position determination. This technique is discussed in Section 6.6.2. In PPP or PPP-RTK the focus is on transmitting satellite phase and code biases, which consist of clock errors and hardware delays. These biases become estimable after all linear- dependent parameters have been eliminated through reparameterization. In the case of PPP, reparameterization consists of lumping together the integer ambiguities and receiver and satellite hardware delays and estimating the new parameter as a real number, whereas in PPP-RTK (Section 6.2.3) the ambiguities are isolated and thus become accessible to integer constraining.

The estimation of the tropospheric delay is briefly addressed in the PPP section, with the understanding that such an estimation also applies to PPP-RTK.

As to PPP, Zumberge et al. (1998a) introduced precise point positioning utilizing the ionospheric-free carrier phase and pseudorange functions. The ionospheric-free carrier phase equation requires a minor modification to deal with the clock errors, the hardware delays, and the ambiguities. Equation (6.1.39) can be written and modified as

$$\begin{aligned} \Phi IF12_k^p &= \rho_k^p + (c\,dt_k - d_{k,\Phi IF12}) - \left(c\,dt^p - D^p_{\Phi IF12}\right) + \lambda_{\Phi IF12} N^p_{k,\Phi IF12} + T_k^p \\ &\quad + M^p_{k,\Phi IF12} + \varepsilon_{\Phi IF12} \\ &= \rho_k^p + (c\,dt_k - d_{k,PIF12}) - \left(c\,dt^p - D^p_{PIF12}\right) + R_k^p + T_k^p + M^p_{k,\Phi IF12} + \varepsilon_{\Phi IF12} \\ &= \rho_k^p + \xi_{k,PIF12} - \xi^p_{PIF12} + R_k^p + T_k^p + M^p_{k,\Phi IF12} + \varepsilon_{\Phi IF12} \end{aligned} \tag{6.6.1}$$

where $d_{k,\Phi IF12}$ and $D^p_{\Phi IF12}$ are the respective receiver and satellite L1 and L2 hardware phase delays of the function $\Phi IF12$. These delays follow from applying the ionospheric-free function to the hardware delay terms listed in (6.1.33), in particular the receiver delay is a function of $(d_{k,1,\Phi}, d_{k,2,\Phi})$ and the satellite delay is a function of $\left(D^p_{1,\Phi}, D^p_{2,\Phi}\right)$. The terms $\lambda_{\Phi IF12}$ and $N^p_{k,\Phi IF12}$ are the ionospheric-free wavelength and ambiguity, T_k^p is the tropospheric delay, $M^p_{k,\Phi IF12}$ is the multipath, and $\varepsilon_{\Phi IF12}$ is the random measurement noise of the ionospheric-free function. In line two, we add and subtract the receiver and satellite ionospheric-free hardware code delays $d_{k,PIF12}$ and D^p_{PIF12} of the $PIF12$ function. These delays follow from applying the ionospheric-free function to the hardware delay terms listed in (6.1.29), in particular the receiver delay is a function of $(d_{k,1,P}, d_{k,2,P})$ and the satellite delay is a function of $\left(D^p_{1,P}, D^p_{2,P}\right)$. The combined terms are

$$\begin{aligned} R_k^p &= \lambda_{\Phi IF12} N^p_{k,\Phi IF12} + \left(d_{k,PIF12} - D^p_{PIF12}\right) - \left(d_{k,\Phi IF12} - D^p_{\Phi IF12}\right) \\ \xi_{k,PIF12} &= c\,dt_k - d_{k,PIF12} \\ \xi^p_{PIF12} &= c\,dt^p - D^p_{PIF12} \end{aligned} \tag{6.6.2}$$

The lumped parameters R_k^p bundle the ambiguity parameters and the receiver and satellite hardware code and phase delays. In traditional PPP, the hardware delay terms are considered constant, even for long observation sessions. The lumped parameters, one per satellite and station pair, are therefore also constants unless there are cycle slips. The new parameters $\xi_{k,PIF12}$ and ξ^p_{PIF12} are called the ionospheric-free receiver and satellite code biases. They consist of the respective clock errors and hardware code delays of the $PIF12$ function. These delays are functions of the original hardware code delays $d_{k,1,P}$, $d_{k,2,P}$, $D^p_{1,P}$ and $D^p_{2,P}$ listed in (6.1.29), as mentioned above. Similarly, we have an ionospheric-free receiver phase bias $\xi_{k,\Phi IF12} = c\,dt_k - d_{k,\Phi IF12}$ and a satellite phase bias $\xi^p_{\Phi IF12} = c\,dt^p - D^p_{\Phi IF12}$. These biases do not explicitly appear in (6.6.1) because of the introduction of the lumping parameter.

The ionospheric-free pseudorange observations (6.1.38) are the second type of observation in the PPP model. Using the parameterization in terms of ionospheric-free receiver and satellite code biases, this equation becomes

$$PIF12^p_k = \rho^p_k + \xi_{k,PIF12} - \xi^p_{PIF12} + T^p_k + M^p_{k,PIF12} + \varepsilon_{PIF12} \tag{6.6.3}$$

Equations (6.6.1) and (6.6.3) comprise the PPP model. We first discuss the network solution and the user solution, retaining all terms of the equations, and then provide brief remarks on how to deal with the tropospheric and ionospheric terms.

Reparameterization and Network Solution: The network consists of R known stations that observe S satellites in common view. Even a first glance at equations (6.6.1) and (6.6.3) reveals the linear dependency of the code biases. Any change in the receiver code bias can be compensated by a respective change in the satellite code biases. This linear dependency is conveniently eliminated by selecting a base station and estimating the code bias parameters relative to the receiver code bias of that station. This is another type of reparameterization.

The topocentric satellite range term ρ^p_k should be moved to the left side of the equations since the network receivers are located at known stations, and station coordinates are consequently not estimated. Consider the following formulation and solution of a small network consisting of three stations that observe three satellites (omitting the multipath terms):

$$\begin{bmatrix} \boldsymbol{\Phi IF12}_1 - \boldsymbol{\rho}_1 \\ \boldsymbol{PIF12}_1 - \boldsymbol{\rho}_1 \\ \boldsymbol{\Phi IF12}_2 - \boldsymbol{\rho}_2 \\ \boldsymbol{PIF12}_2 - \boldsymbol{\rho}_2 \\ \boldsymbol{\Phi IF12}_3 - \boldsymbol{\rho}_3 \\ \boldsymbol{PIF12}_3 - \boldsymbol{\rho}_3 \end{bmatrix} = \begin{bmatrix} \boldsymbol{I} & \boldsymbol{I} & \boldsymbol{0} & \boldsymbol{0} & \boldsymbol{0} & \boldsymbol{0} \\ \boldsymbol{I} & \boldsymbol{0} & \boldsymbol{0} & \boldsymbol{0} & \boldsymbol{0} & \boldsymbol{0} \\ \boldsymbol{I} & \boldsymbol{0} & \boldsymbol{I} & \boldsymbol{I} & \boldsymbol{0} & \boldsymbol{0} \\ \boldsymbol{I} & \boldsymbol{0} & \boldsymbol{0} & \boldsymbol{I} & \boldsymbol{0} & \boldsymbol{0} \\ \boldsymbol{I} & \boldsymbol{0} & \boldsymbol{0} & \boldsymbol{0} & \boldsymbol{I} & \boldsymbol{I} \\ \boldsymbol{I} & \boldsymbol{0} & \boldsymbol{0} & \boldsymbol{0} & \boldsymbol{0} & \boldsymbol{I} \end{bmatrix} \begin{bmatrix} \widehat{\boldsymbol{\xi}}_{PIF12} \\ \widehat{\boldsymbol{R}}_1 \\ \widehat{\boldsymbol{R}}_2 \\ \widehat{\xi}_{2,PIF12} \\ \widehat{\boldsymbol{R}}_3 \\ \widehat{\xi}_{3,PIF12} \end{bmatrix} \tag{6.6.4}$$

$$\boldsymbol{\Phi IF12}_k - \boldsymbol{\rho}_k = \begin{bmatrix} \Phi IF12^1_k - \rho^1_k \\ \Phi IF12^2_k - \rho^2_k \\ \Phi IF12^3_k - \rho^3_k \end{bmatrix} \qquad \boldsymbol{PIF12}_k - \boldsymbol{\rho}_k = \begin{bmatrix} PIF12^1_k - \rho^1_k \\ PIF12^2_k - \rho^2_k \\ PIF12^3_k - \rho^3_k \end{bmatrix} \tag{6.6.5}$$

$$\widehat{\boldsymbol{\xi}}_{PIF12} = \begin{bmatrix} \widehat{\xi}^1_{PIF12} \\ \widehat{\xi}^2_{PIF12} \\ \widehat{\xi}^3_{PIF12} \end{bmatrix} = \begin{bmatrix} -\xi^1_{PIF12} + \xi_{1,PIF12} + T^1_1 \\ -\xi^2_{PIF12} + \xi_{1,PIF12} + T^2_1 \\ -\xi^3_{PIF12} + \xi_{1,PIF12} + T^3_1 \end{bmatrix} \tag{6.6.6}$$

$$\widehat{\boldsymbol{R}}_1 = \begin{bmatrix} \widehat{R}^1_1 \\ \widehat{R}^2_1 \\ \widehat{R}^3_1 \end{bmatrix} \qquad \widehat{\boldsymbol{R}}_2 = \begin{bmatrix} \widehat{R}^1_2 \\ \widehat{R}^2_2 \\ \widehat{R}^3_2 \end{bmatrix} \qquad \widehat{\boldsymbol{R}}_3 = \begin{bmatrix} \widehat{R}^1_3 \\ \widehat{R}^2_3 \\ \widehat{R}^3_3 \end{bmatrix} \tag{6.6.7}$$

$$\left.\begin{aligned}\widehat{\xi}_{2,PIF12} &= \begin{bmatrix}\xi_{2,PIF12} - \xi_{1,PIF12} - T_{12}^{1}\\ \xi_{2,PIF12} - \xi_{1,PIF12} - T_{12}^{2}\\ \xi_{2,PIF12} - \xi_{1,PIF12} - T_{12}^{3}\end{bmatrix}\\ \widehat{\xi}_{3,PIF12} &= \begin{bmatrix}\xi_{3,PIF12} - \xi_{1,PIF12} - T_{13}^{1}\\ \xi_{3,PIF12} - \xi_{1,PIF12} - T_{13}^{2}\\ \xi_{3,PIF12} - \xi_{1,PIF12} - T_{13}^{3}\end{bmatrix}\end{aligned}\right. \quad (6.6.8)$$

It can readily be verified by direct substitution that (6.6.4) to (6.6.8) indeed represent the system of equations (6.6.1) and (6.6.3) correctly (ignoring the multipath terms). The reparameterizations resulted in estimable parameters that have been denoted by an overhead arc. Station 1 has been selected as the base station in this example. As a consequence of this arbitrary selection, the reparameterized ionospheric-free satellite code biases $\widehat{\xi}^{p}_{PIF12}$, $p = 1, \cdots, S$, are relative to the ionospheric-free receiver code bias $\xi_{1,PIF12,}$ of the base station, i.e., the receiver delay term in (6.6.6) refers to station 1. The reparameterized ionospheric-free receiver code biases $\widehat{\xi}_{2,PIF12,}$ and $\widehat{\xi}_{3,PIF12,}$ are also relative to the receiver code bias of the base station. The latter two estimable biases contain cross-receiver tropospheric delay with respect to the base station. The lumped parameters $\widehat{R}^{p}_{k}$ are defined in (6.6.2).

This example can readily be generalized for a larger network. As the number of satellites increases, so does the number of components in vectors (6.6.5) to (6.6.8). Each additional station adds two rows to the matrix in (6.6.4). These rows are identical to the bottom two rows, but the submatrix in the lower right corner shifts to the right accordingly. For R receivers and S satellites the system consists of $2RS$ equations and as many parameters; the matrix has full rank. These are S reparameterized satellite code biases (6.6.6), RS lumped parameters (6.6.7), and $(R-1)S$ reparameterized receiver code biases (6.6.8).

The estimated reparameterized ionospheric-free satellite code biases $\widehat{\xi}^{p}_{PIF12}$ are transmitted to the user at the unknown station u.

User Solution: The user solution also begins with equations (6.6.1) and (6.6.3). We subtract the received bias corrections (6.6.8) and then balance the equations, recognizing that the received corrections includes a tropospheric term, giving

$$\left.\begin{aligned}\Phi IF12^{p}_{u} - \widehat{\xi}^{p}_{PIF12} &= \rho^{p}_{u} + \widehat{\xi}_{u,PIF12} + R^{p}_{u} + T^{p}_{u1} + M_{\Phi IF12} + \varepsilon_{\Phi IF12}\\ PIF12^{p}_{u} - \widehat{\xi}^{p}_{PIF12} &= \rho^{p}_{u} + \widehat{\xi}_{u,PIF12} + T^{p}_{u1} + M_{PIF12} + \varepsilon_{PIF12}\end{aligned}\right\} \quad (6.6.9)$$

$$\widehat{\xi}_{u,PIF12} = \xi_{u,PIF12} - \xi_{1,PIF12} \quad (6.6.10)$$

The user receiver code bias estimate $\widehat{\xi}_{u,PIF12}$ is relative to the base station code bias. The solution presented contains the unaltered tropospheric term of the original equations. Clearly the multipath is omnipresent in both the estimated code biases and in the user solution. In order to simply the expressions, the multipath terms will only be listed in the user solution. The transmitted satellite bias corrections depend on the tropospheric delay at the base station, as can be seen from (6.6.6). This tropospheric delay appears also on the right side of (6.6.9). Note that in the standard subscript notation the double subscript implies differencing, thus $T^p_{ul} = T^p_u - T^p_1$.

Tropospheric Considerations: The tropospheric delay typically varies with temperature, pressure, and humidity. If the tropospheric model corrections with sufficient accuracy are available at the network stations, a network solution of two stations observing three satellites becomes

$$\begin{bmatrix} \boldsymbol{\Phi IF12}_1 - \rho_1 - \boldsymbol{T}_1 \\ \boldsymbol{PIF12}_1 - \rho_1 - \boldsymbol{T}_1 \\ \boldsymbol{\Phi IF12}_2 - \rho_2 - \boldsymbol{T}_2 \\ \boldsymbol{PIF12}_2 - \rho_2 - \boldsymbol{T}_2 \end{bmatrix} = \begin{bmatrix} \boldsymbol{I} & \boldsymbol{I} & 0 & 0 \\ \boldsymbol{I} & 0 & 0 & 0 \\ \boldsymbol{I} & 0 & \boldsymbol{b} & \boldsymbol{I} \\ \boldsymbol{I} & 0 & \boldsymbol{b} & 0 \end{bmatrix} \begin{bmatrix} \widehat{\xi}_{PIF12} \\ \widehat{\boldsymbol{R}}_1 \\ \widehat{\boldsymbol{R}}_2 \\ \widehat{\xi}_{2,PIF12} \end{bmatrix} \quad \boldsymbol{T}_k = \begin{bmatrix} T^1_k \\ T^2_k \\ T^3_k \end{bmatrix} \tag{6.6.11}$$

$$\boldsymbol{b}^T = \begin{bmatrix} 1 & 1 & 1 \end{bmatrix} \tag{6.6.12}$$

$$\widehat{\xi}_{PIF12} = \begin{bmatrix} \widehat{\xi}^1_{PIF12} \\ \widehat{\xi}^2_{PIF12} \\ \widehat{\xi}^3_{PIF12} \end{bmatrix} = \begin{bmatrix} -\xi^1_{PIF12} + \xi_{1,PIF12} \\ -\xi^2_{PIF12} + \xi_{1,PIF12} \\ -\xi^3_{PIF12} + \xi_{1,PIF12} \end{bmatrix} \quad \widehat{\boldsymbol{R}}^T_k = \begin{bmatrix} \widehat{R}^1_k \\ \widehat{R}^2_k \\ \widehat{R}^3_k \end{bmatrix} \tag{6.6.13}$$

$$\widehat{\xi}_{2,PIF12} = \xi_{2,PIF12} - \xi_{1,PIF12} \tag{6.6.14}$$

The absence of the tropospheric terms in the estimated code biases causes the dimension of $\widehat{\xi}_{2,PIF12}$ to reduce to one. Consequently, the code bias estimates at the nonbase stations do not contain any terms that depend on the satellites. There are $S + RS + (R - 1)$ parameters, i.e., S satellite code biases, RS lumped parameters, and $R - 1$ nonbase station receiver code biases. There are now more equations than parameters. The user solution for this case is

$$\left.\begin{aligned} \Phi IF12^p_u - \widehat{\xi}^p_{PIF12} &= \rho^p_u + \widehat{\xi}_{u,PIF12} + R^p_u + T^p_u + M_{\Phi IF12} + \varepsilon_{\Phi IF12} \\ PIF12^p_u - \widehat{\xi}^p_{PIF12} &= \rho^p_u + \widehat{\xi}_{u,PIF12} + T^p_u + M_{PIF12} + \varepsilon_{PIF12} \end{aligned}\right\} \tag{6.6.15}$$

This user solution differs from the previous one in that it contains only the tropospheric term for the user station.

In practical applications the tropospheric delay is estimated or modeled at the network stations and the user station. The tropospheric slant total delay T^p_k is typically decomposed into the hydrostatic and wet delay components. Following (8.2.18), we write

$$\begin{aligned} T^p_k &= ZHD_k\, m_h(\vartheta^p) + ZWD_k\, m_{wv}(\vartheta^p) \\ &= T^p_{k,0} + dT_k\, m_{wv}(\vartheta^p) \end{aligned} \tag{6.6.16}$$

Examples for the zenith hydrostatic delay (ZHD) and the zenith wet delay (ZWD) models are given in (8.2.14) and (8.2.15). These models use meteorological data as input. The mapping functions m_h and m_{wv} follow from (8.2.19), with ϑ being the zenith angle of the satellite. The term $T^p_{k,0}$ represents an approximation of the slant total tropospheric delay as computed by temperature, pressure, and relative humidity observations using the ZHD and ZWD models, and dT_k is the unknown vertical tropospheric correction at the station. The latter is multiplied by the wet mapping function, assuming that the tropospheric correction becomes necessary because of inaccurate knowledge of the wet delay. As to the mapping function, one can use the well-known Niell mapping function discussed in Section 8.2.2 or other functions developed more recently.

In order to incorporate tropospheric estimation, the system (6.6.4) is expanded to include the new parameters dT_k. The mathematical model for the network now is

$$\begin{bmatrix} \boldsymbol{\Phi IF}12_1 - \rho_1 - \boldsymbol{T}_{1,0} \\ \boldsymbol{PIF}12_1 - \rho_1 - \boldsymbol{T}_{1,0} \\ \hdashline \boldsymbol{\Phi IF}12_2 - \rho_2 - \boldsymbol{T}_{2,0} \\ \boldsymbol{PIF}12_2 - \rho_2 - \boldsymbol{T}_{2,0} \end{bmatrix} = \left[\begin{array}{ccc:ccc} \boldsymbol{I} & \boldsymbol{I} & \boldsymbol{m}_1 & 0 & 0 & 0 \\ \boldsymbol{I} & 0 & \boldsymbol{m}_1 & 0 & 0 & 0 \\ \hdashline \boldsymbol{I} & 0 & 0 & \boldsymbol{I} & \boldsymbol{m}_2 & \boldsymbol{b} \\ \boldsymbol{I} & 0 & 0 & 0 & \boldsymbol{m}_2 & \boldsymbol{b} \end{array}\right] \begin{bmatrix} \widehat{\xi}_{PIF12} \\ \widehat{\boldsymbol{R}}_1 \\ dT_1 \\ \hdashline \widehat{\boldsymbol{R}}_2 \\ dT_2 \\ \widehat{\xi}_{2,PIF12} \end{bmatrix} \tag{6.6.17}$$

$$\boldsymbol{m}_k = \begin{bmatrix} m_{wv}\left(\vartheta_k^1\right) \\ m_{wv}\left(\vartheta_k^2\right) \\ m_{wv}\left(\vartheta_k^3\right) \end{bmatrix} \tag{6.6.18}$$

The observations must be corrected by the model value $T_{k,0}^p$. The subscript zero is borrowed from adjustment notation and means approximate value, i.e., point of linearization. The matrix needs extra columns to accommodate the new tropospheric parameters. In addition to these new parameters, the estimable parameters (6.6.13) and (6.6.14) apply. There are $S + RS + R + R - 1$ parameters, i.e., S satellite code biases, RS lumped parameters, R tropospheric parameters, and $R - 1$ nonbase station receiver code biases. The user solution is

$$\left.\begin{aligned} \Phi IF12_u^p - \widehat{\xi}_{PIF12}^p - T_{u,0}^p &= \rho_u^p + \widehat{\xi}_{u,PIF12} + R_u^p + dT_u m_{wv}\left(\vartheta_u^p\right) + M_{\Phi IF12} + \varepsilon_{\Phi IF12} \\ PIF12_u^p - \widehat{\xi}_{PIF12}^p - T_{u,0}^p &= \rho_u^p + \widehat{\xi}_{u,PIF12} + dT_u m_{wv}\left(\vartheta_u^p\right) + M_{PIF12} + \varepsilon_{PIF12} \end{aligned}\right\} \tag{6.6.19}$$

There are, of course, more refined ways of modeling and estimating the tropospheric delays at the network and at the user, in particular when observing over a longer period of time. This more elaborate modeling is not discussed here.

Let us note that some linear dependencies were eliminated by parameterizing all code bias parameters relative to the base station receiver code bias. Since this bias term includes the receiver clock, it varies accordingly. Also, the corrections (6.6.8) include the across-receiver tropospheric difference which adds additional variability. If the tropospheric corrections are estimated at the network, then the biases do not depend on the troposphere. See equation (6.6.13).

Let us note that, apart from using carrier phase observations, the formulation presented above includes code observations on L1 and code observations on L2. Since the code hardware delays depend on the type of codes, i.e. the P1Y and C/A-code delays differ, the satellite code bias ξ_{PIF12}^p estimated with (6.6.1) and (6.6.3) also depends on the choice of the codes. The IGS estimates its "satellite clock correction", which corresponds to ξ^p_{PIF12}, based in P1Y and P2Y-code observations. If one wishes to remain compatible with the "IGS clock" but uses other code observations, one needs to correct the pseudoranges by what has traditionally been called the differential code bias (DCB). For example, if one observes the C/A-code and the P2Y code, one needs the correction $DCB_{\mathrm{P1Y-CIA}}$, which is the difference of the respective hardware code delays. These can then be transformed to the respective ionospheric-free code delays to correct $PIF12$. In practical applications one needs to be aware that not all receivers observe the same codes. Also, since GPS modernization and other GNSS systems provide new frequencies and codes, such compatibility issues need special attention. Refer to Section 6.2.2.2 for a general approach and notation in regards to intersignal corrections.

6.6.2 CORS

Continuously operating reference stations (CORS) transmit their carrier phase and pseudorange observations in real time to a processing center. The center computes corrections, such as ionospheric and tropospheric corrections and possibly orbital corrections, and transmits these and possibly the original observations of a master reference station to users. The user combines this information with observations collected by the user receiver to determine its position. This conceptual model applies to one CORS station or a network of such stations and to one user or several users.

6.6.2.1 *Differential Phase and Pseudorange Corrections* Let us look at a simple way for computing the differential corrections to the observations. For every satellite p observed at station k, we determine an integer number K^p_k:

$$K^p_k = \left[\frac{P^p_k(1) - \Phi^p_k(1)}{\lambda}\right] = \left[\frac{1}{\lambda}\left(2I^p_{k,P} - \lambda N^p_k + \delta^p_{k,P} - \delta^p_{k,\Phi}\right)\right] \tag{6.6.20}$$

using the observed pseudoranges and carrier phases at some initial epoch. The symbol $[\cdot]$ denotes rounding. The modified carrier phase $\Theta^p_k(t)$ at subsequent epochs is

$$\Theta^p_k(t) = \Phi^p_k(t) + \lambda K^p_k \tag{6.6.21}$$

The numerical value of the carrier phase range is close to that of the pseudorange, differing primarily because of the ionosphere, as can be seen from the right side of (6.6.20); K^p_k is not equal to the ambiguity. The discrepancy of the carrier phase range at epoch t is

$$\ell^p_k = \Theta^p_k - \rho^p_k = \left(\Phi^p_k + \lambda K^p_k\right) - \rho^p_k \tag{6.6.22}$$

where ρ^p_k is the topocentric satellite distance from the known station. The mean discrepancy μ_k of all satellites observed at the site and epoch t is

$$\mu_k(t) = \frac{1}{S}\sum_{p=1}^{S} \ell^p_k(t) \tag{6.6.23}$$

where S denotes the number of satellites. This mean discrepancy is driven primarily by the receiver clock error. The carrier phase correction at epoch t is

$$\Delta\Phi_k^p = \Theta_k^p - \rho_k^p = \left(\Phi_k^p + \lambda K_k^p\right) - \rho_k^p - \mu_k \tag{6.6.24}$$

The second part of this equation follows by substituting (6.6.21) for the carrier phase range. The phase correction (6.6.24) is transmitted to the user receiver u.

The user's carrier phase Φ_u^p is corrected by subtracting the carrier phase correction that was computed at receiver k:

$$\overline{\Phi}_u^p = \Phi_u^p - \Delta\Phi_k^p \tag{6.6.25}$$

Let us recall the across-receiver phase difference

$$\Phi_u^p - \Phi_k^p = \rho_{uk}^p + \lambda N_{uk}^p + c\,dt_{uk} + I_{uk,\Phi}^p + T_{uk}^p + \delta_{uk,\Phi}^p \tag{6.6.26}$$

Substituting (6.6.24) in (6.6.25) gives the expression for the corrected carrier phase at receiver u,

$$\overline{\Phi}_u^p = \rho_u^p + \lambda\left(N_{uk}^p - K_k^p\right) + c\,dt_{uk} + \mu_k + I_{uk,\Phi}^p + T_{uk}^p + \delta_{uk,\Phi}^p \tag{6.6.27}$$

Differencing (6.6.27) between two satellites gives

$$\overline{\Phi}_u^{pq} = \rho_u^{pq} + \lambda\left(N_{uk}^{pq} - K_k^{pq}\right) + I_{uk,\Phi}^{pq} + T_{uk}^{pq} + M_{uk,\Phi}^{pq} + \varepsilon_{uk,\Phi}^{pq} \tag{6.6.28}$$

The position of station u can now be computed at site u using the corrected observation $\overline{\Phi}_u^p$ to at least four satellites and forming three equations like (6.6.28). These equations differ from the conventional double-difference following (6.6.26) by the fact that the modified ambiguity

$$\overline{N}_{uk}^{pq} = N_{uk}^{pq} - K_k^{pq} \tag{6.6.29}$$

is estimated instead of N_{uk}^{pq}.

The telemetry load can be reduced if it is possible to increase the time between transmissions of the carrier phase corrections. For example, if the change in the discrepancy from one epoch to the next is smaller than the measurement accuracy at the moving receiver, or if the variations in the discrepancy are too small to adversely affect the required minimal accuracy for the moving receiver's position, it is possible to average carrier phase corrections over time and transmit the averages. Also, it might be sufficient to transmit the rate of correction $\partial\Delta\Phi/\partial t$. If t_0 denotes the reference epoch, the user can interpolate the correctors over time as

$$\Delta\Phi_k^p(t) = \Delta\Phi_k^p(t_0) + \frac{\partial\Delta\Phi_k^p}{\partial t}(t - t_0) \tag{6.6.30}$$

One way to reduce the size and the slope of the discrepancy is to use the best available coordinates for the fixed receiver and a good satellite ephemeris. Clock errors affect the discrepancies directly. Connecting a rubidium clock to the fixed receiver can effectively control the variations of the receiver clock error. Prior to its termination, selective availability was the primary cause of satellite clock error and was a determining factor that limited modeling like (6.6.30).

In the case of pseudorange corrections, we obtain similarly

$$\ell_k^p = \rho_k^p - P_k^p \tag{6.6.31}$$

$$\Delta P_k^p = \rho_k^p - P_k^p - \mu_k \tag{6.6.32}$$

$$\overline{P}_u^p(t) = P_u^p(t) + \Delta P_k^p(t) \tag{6.6.33}$$

$$\overline{P}_u^{pq}(t) = \rho_u^{pq}(t) + I_{uk,P}^{pq} + T_{uk}^{pq} + M_{uk,P}^{pq} + \varepsilon_{uk,P}^{pq} \tag{6.6.34}$$

The approach described here is applicable to carrier phases of any frequency and to all codes. As seen from (6.6.22) to (6.6.24), the carrier phase and pseudorange corrections contain the ionospheric and tropospheric terms. As suggested in the previous section, the tropospheric delay could be estimated at the network, the ionosphere effects could be eliminated by using dual-frequency observations, and in doing so one would obtain less variability in the corrections. The receiver and satellite clock errors have canceled as part of the implicit double differencing, as have the receiver and satellite hardware delays.

6.6.2.2 RTK In real-time positioning (RTK), the users receive the differential correction from one or several CORS stations (simply referred to as the reference stations) and determine their positions relative to these stations, preferably with an ambiguity-fixed solution. As mentioned above, for short baselines one neglects the tropospheric, ionospheric, and orbital errors. In practical applications it is desirable to extend the reach of RTK over longer baselines. Because of the high spatial correlation of troposphere, ionosphere, and orbital errors, these errors exhibit to some extent a function of distance between the receivers. Wübbena et al. (1996a) took advantage of this dependency and suggested the use of reference station networks to extend the reach of RTK.

There are two requirements at the heart of multiple reference station RTK. First, the positions of the reference stations must be accurately known. This can be readily accomplished using postprocessing and long observation times. The second requirement is that the across-receiver or double-difference integer ambiguities for baselines between reference stations can be computed. It is then possible to compute tropospheric and ionospheric corrections (and possibly orbital corrections) and transmit them to the RTK user.

Let k denote the master reference station and m the other reference stations of the network. Let the master reference station record its own observations and receive observations from the other reference stations in real time. The processor at the master reference station can then generate the corrections T_{km}^p and $I_{km,1,P}^p$ at every epoch for all reference stations and all satellites. These corrections are used to predict the respective corrections at a user location. Various models are in use or have been proposed for computing these corrections and making them available to the user.

Wübbena et al. (1996a) proposed to parameterize the corrections in terms of coordinates. One of the simplest location-dependent models is a plane

$$T^p_{km}(t) = a^p_1(t) + a^p_2(t)\,n_m + a^p_3(t)\,e_m + a^p_4(t)\,u_m \tag{6.6.35}$$

$$I^p_{km,P}(t) = b^p_1(t) + b^p_2(t)\,n_m + b^p_3(t)\,e_m + b^p_4(t)\,u_m \tag{6.6.36}$$

where the symbols n_m, e_m, and u_m denote northing, easting, and up coordinates in the geodetic horizon at the master reference station k. A set of coefficients $a^p_i(t)$ and $b^p_i(t)$, also called the network coefficients, are estimated for every satellite p and network station m as a function of time. Because of the high temporal correlation of the troposphere and ionosphere, simple models in time are sufficient to reduce the amount of data to be transmitted. The master reference station k transmits its own carrier phase observations and the network coefficients a_i, b_i, or alternatively the carrier phase corrections (6.6.24), over the network. A rover u interpolates these corrections for its approximate position and determines its precise location from the set of double-difference observations. This modeling scheme (6.6.35) and (6.6.36) is also referred to as the FKP (flächen korrektur parameter) technique.

Rather than transmitting network coefficients a_i, b_i, one might consider transmitting corrections computed specifically for points on a grid at known locations within the network. The user would interpolate the corrections for the rover's approximate location and apply them to the observations. Wanninger (1997) and Vollath et al. (2000) suggest the use of virtual reference stations (VRS) to avoid changing existing software that double differences the original observations directly. The VRS concept requires that the rover transmit its approximate location to the master reference station, which computes the corrections for the user approximate location. In addition, the master reference station computes virtual observations for the approximate rover location using its own observations and then corrects them for troposphere and ionosphere. The rover merely has to double difference its own observations with those received from the master reference station. No additional tropospheric or ionospheric corrections or interpolations are required at the rover because the effective virtual baseline is very short, typically in the range of meters corresponding to the rover's initial determination of its own location from pseudoranges.

Euler et al. (2001) and Zebhauser et al. (2002) suggest transmitting the observation of the master reference station and the correction differences between pairs of reference stations. The latter would be corrected for location, receiver clock, and ambiguities. The approach is called MAC (master auxiliary concept).

The message formats for data exchange between the reference station, the master station, and the user generally follow standardized formats set by the Radio Technical Commission for Maritime Services (RTCM). This is a nonprofit scientific, professional, and educational organization consisting of international member organizations that include manufacturers, marketing, service providers, and maritime user entities. Special committees address in-depth concerns in radionavigation. The reports prepared by these committees are usually published as RTCM recommendations. The RTCM Special Committee 104 deals with global navigation satellite systems.

As the network area increases, the tropospheric and ionospheric corrections and the orbit corrections require a more elaborate parameterization and are typically transmitted to the user via geostationary satellites. Such networks are called wide area differential GPS (WADGPS) networks. Examples of such systems are WAAS (wide area augmentation system) and EGNOS (European Geostationary Navigation Overlay Service). A more complete listing of systems is found in Chapter 5. RTK solutions are discussed in detail in Chapter 7.

6.6.3 PPP-RTK

The goal of PPP-RTK algorithmic development is to find estimable quantities for undifferenced phase and pseudorange observations that allow the fixing of undifferenced ambiguities to integers. In contrast to the PPP solution, the ambiguity parameters and the receiver and satellite hardware delay terms are not lumped together. Examples of PPP-RTK implementations are reported in Ge et al. (2008) and Loyer et al. (2012).

There are at least two equivalent approaches for finding estimable quantities. One is reparameterization and the other is imposing minimal constraints. The advantage of reparameterization is that all terms remain visible in the expression and thus might make it easier to interpret the impact of any residual errors on the estimable quantities.

Three models are presented. The first model is the one-step single-frequency solution in which observations from all stations are processed in one batch solution, providing a single solution for the estimated parameters and a full variance-covariance matrix. The second model deals with dual-frequency observations. A one-step solution is given and then a sequential solution in which the wide-lane ambiguities are estimated first, followed by a model variation that estimates the parameters and biases by baseline. Only the one-step solutions can take advantage of the full variance-covariance matrix, while the others ignore some correlations between parameters. The third model utilizes across-satellite differences of dual-frequency observations. The network solution provides the PPP-RTK biases to be transmitted to users.

6.6.3.1 *Single-Frequency Solution* The case of single-frequency carrier phase and pseudorange equations (6.1.31) and (6.1.27) are

$$\begin{aligned} \Phi_k^p - \rho_k^p &= \xi_{k,\Phi} - \xi_\Phi^p + \lambda N_k^p + T_k^p - I_{k,p}^p + M_{k,\Phi}^p + \varepsilon_{k,\Phi}^p \\ P_k^p - \rho_k^p &= \xi_{k,P} - \xi_P^p + T_k^p + I_{k,P}^p + M_{k,P}^p + \varepsilon_{k,P}^p \end{aligned} \tag{6.6.37}$$

$$\begin{aligned} \xi_{k,\Phi} &= c\,dt_k - d_{k,1,\Phi} \qquad & \xi_{k,P} &= c\,dt_k - d_{k,1,P} \\ \xi_\Phi^p &= c\,dt^p + D_{1,\Phi}^p \qquad & \xi_P^p &= c\,dt^p + D_{1,P}^p \end{aligned} \tag{6.6.38}$$

The topocentric range term ρ_k^p has been moved to the left side of the equation since the network station coordinates are assumed to be known. Other symbols denote the receiver phase and code biases $\xi_{k,\Phi}$ and $\xi_{k,P}$, the satellite phase and code biases ξ_Φ^p and ξ_P^p, the slant tropospheric delay T_k^p, the slant ionosphere $I_{k,P}^p$, the wavelength λ, the integer ambiguity N_k^p, the multipath terms $M_{k,\Phi}^p$ and $M_{k,P}^p$, and ε denotes the respective observational noise. We note that each ξ-term combines a clock error and a hardware delay term. As to terminology, in Collins (2008) the these terms are referred to as decoupled clock parameters, with $\xi_{k,\Phi}$ and ξ_Φ^p respectively called the receiver and satellite "phase clocks", and $\xi_{k,P}$ and ξ_P^p termed the receiver and satellite "code clocks."

Reparameterization and Network Solution: The system (6.6.37) is singular because a number of linear dependencies exist between the various parameters. Assuming R receivers observing S satellites, there are $2RS$ equations and $2R + 2S + 2RS + 2RS + RS$ unknowns, i.e., $2R$ receiver phase and code biases, $2S$ satellite phase and code biases, $2RS$ tropospheric terms, $2RS$ ionospheric terms, and RS ambiguities. Traditional double differencing removes linear dependencies by introducing the base station and base satellite concept and differencing the observations to eliminate the receiver and satellite biases and creating double-differenced ambiguities. Contrary to the popular double differencing, in PPP-RTK the original observations are kept in undifferenced form and the linear dependencies are eliminated by means of reparameterization (Teunissen et al., 2010). For the current development, all terms are initially retained (except multipath). Eventually the tropospheric delay terms can be omitted because the tropospheric delay will be modeled or estimated at the network. The ionospheric delay terms will not be present when using ionospheric-free dual-frequency carrier phase and pseudorange functions.

Considering the carrier phase observation (6.6.37), we observe that an arbitrary constant added to the satellite phase bias ξ_Φ^p can be offset by adding the same constant to each of the receiver phase biases $\xi_{k,\Phi}$, keeping the observable Φ_k^p unchanged. Similarly, it further shows that any arbitrary constant change in either $\xi_{k,\Phi}$ or ξ_Φ^p can be offset by a corresponding change in the ambiguity N_k^p. The result of reparameterization to eliminate linear dependencies is demonstrated by the following example consisting of three stations observing three satellites:

$$\begin{bmatrix} \boldsymbol{\Phi}_1^p - \boldsymbol{\rho}_1^p \\ \boldsymbol{\Phi}_2^p - \boldsymbol{\rho}_2^p \\ \boldsymbol{\Phi}_3^p - \boldsymbol{\rho}_3^p \end{bmatrix} = \begin{bmatrix} \boldsymbol{I} & \boldsymbol{0} & \boldsymbol{0} & \boldsymbol{0} & \boldsymbol{0} \\ \boldsymbol{I} & \boldsymbol{b} & \boldsymbol{0} & \boldsymbol{A} & \boldsymbol{0} \\ \boldsymbol{I} & \boldsymbol{0} & \boldsymbol{b} & \boldsymbol{0} & \boldsymbol{A} \end{bmatrix} \begin{bmatrix} \widehat{\boldsymbol{\xi}}_\Phi \\ \widehat{\xi}_{2,\Phi} \\ \widehat{\xi}_{3,\Phi} \\ \widehat{\boldsymbol{N}}_2 \\ \widehat{\boldsymbol{N}}_3 \end{bmatrix} \tag{6.6.39}$$

$$\boldsymbol{\Phi}_k - \boldsymbol{\rho}_k = \begin{bmatrix} \Phi_k^1 - \rho_k^1 \\ \Phi_k^2 - \rho_k^2 \\ \Phi_k^3 - \rho_k^3 \end{bmatrix} \qquad \boldsymbol{A} = \begin{bmatrix} 0 & 0 \\ 1 & 0 \\ 0 & 1 \end{bmatrix} \tag{6.6.40}$$

$$\widehat{\boldsymbol{\xi}}_{\Phi} = \begin{bmatrix} \widehat{\xi}^1_{\Phi} \\ \widehat{\xi}^2_{\Phi} \\ \widehat{\xi}^3_{\Phi} \end{bmatrix} = \begin{bmatrix} -\xi^1_{\Phi} & +\xi_{1,\Phi} & +\lambda N^1_1 & +T^1_1 & -I^1_{1,P} \\ -\xi^2_{\Phi} & +\xi_{1,\Phi} & +\lambda N^2_1 & +T^2_1 & -I^2_{1,P} \\ -\xi^3_{\Phi} & +\xi_{1,\Phi} & +\lambda N^3_1 & +T^3_1 & -I^3_{1,P} \end{bmatrix} \tag{6.6.41}$$

$$\begin{aligned} \widehat{\xi}_{2,\Phi} &= \left[\xi_{2,\Phi} - \xi_{1,\Phi} + \lambda N^1_{21} + T^1_{21} - I^1_{21,P}\right] \\ \widehat{\xi}_{3,\Phi} &= \left[\xi_{3,\Phi} - \xi_{1,\Phi} + \lambda N^1_{31} + T^1_{31} - I^1_{31,P}\right] \end{aligned} \tag{6.6.42}$$

$$\begin{aligned} \widehat{\boldsymbol{N}}_2 &= \begin{bmatrix} \widehat{N}^{\,2}_2 \\ \widehat{N}^{\,3}_2 \end{bmatrix} = \begin{bmatrix} \lambda N^{21}_{21} + T^{21}_{21} - I^{21}_{21} \\ \lambda N^{31}_{21} + T^{31}_{21} - I^{31}_{21} \end{bmatrix} \\ \widehat{\boldsymbol{N}}_3 &= \begin{bmatrix} \widehat{N}^{\,2}_3 \\ \widehat{N}^{\,3}_3 \end{bmatrix} = \begin{bmatrix} \lambda N^{21}_{31} + T^{21}_{31} - I^{21}_{31} \\ \lambda N^{31}_{31} + T^{31}_{31} - I^{31}_{31} \end{bmatrix} \end{aligned} \tag{6.6.43}$$

The correctness of the solution (6.6.41) to (6.6.43) can be verified by substituting it into (6.6.39) and comparing the result to (6.6.37). Again, in accordance with the traditional notation, the double subscripts or superscripts indicate a differencing operation. See equation (6.1.7) for a definition of the differencing operation. There are RS observations and as many reparameterized unknowns, i.e., S satellite biases, $R-1$ nonbase receiver phase biases, and $(R-1)(S-1)$ ambiguities. Therefore, the matrix in (6.6.39) has full rank and the system has a unique solution. The estimable quantities are identified by an overhead arc. The base station and base satellite are station 1 and satellite 1, respectively. Similar to the case of PPP, the phase bias estimates are relative to the phase bias $\xi_{1,\Phi}$ of the base station.

Let us consider the view of constraining parameters to eliminate linear dependencies. We could impose the constraint $\xi_{1,\Phi} = 0$, delete the term $\xi_{1,\Phi}$ from (6.6.41) and (6.6.42), and call it definition of the *clock datum* at the base station. This step eliminates one parameter. Second, we realize that (6.6.41) contains only base station ambiguities. Imposing the constraints $N^p_1 = 0$ establishes the *ambiguity datum for the base station* and eliminates S parameters. This step allows us to remove the base station ambiguities in (6.6.41) to (6.6.43). Third, we look at the nonbase station ambiguities contained in (6.6.42) and (6.6.43). For every nonbase station, we constrain its ambiguities to the base satellite to zero, i.e., $N^1_l = 0$, $l = 2 \cdots R$. For example, looking at (6.6.43), we see the ambiguity $N^{21}_{21} = N^{21}_2 - N^{21}_1$. The across-satellite difference N^{21}_1 is already zero because of constraints of the second step. The third step results in $N^1_2 = 0$, and thus N^{21}_{21} becomes N^2_2. One can continue in a similar fashion with nonbase station 3 and other nonbase stations. The third step establishes the *ambiguity datum for each nonbase station* and eliminates $R-1$ additional linear dependencies. All three steps combined generate $S+R$ minimal constraints. Considering there are RS observations and $S+R+RS$ original parameters, i.e., S satellite phase biases, R receiver phase biases, and RS undifferenced ambiguities, imposing that many minimal constraints results in a zero degree of freedom solution, identically to what has been obtained with the reparameterization approach.

Imposing minimal constraints or reparameterization leads to the same set of estimable quantities. The important thing is that in (6.6.43) there are eventually only integers left. As already mentioned, at the network stations the tropospheric delays will either be corrected by a model value or estimated. The tropospheric term will not be present or appear as a separate term to be estimated. The ionospheric delay term will also not be present when dual-frequency observations are used. As a result of the reparameterization and the stipulations regarding tropospheric and ionospheric delays, the ambiguities have been isolated as separate parameters and can be estimated as integers. Yet another view is that of short baselines, for which the double-differenced troposphere and ionosphere are negligible per definition. This leaves only integers in (6.6.43). There is one ambiguity for each nonbase station and nonbase satellite pair. Finally, the system (6.6.39) can readily be generalized to include more satellites and stations.

The reparameterization of the pseudoranges in (6.6.37) requires only the elimination of the base receiver code bias $\xi_{1,P}$. For the case of three receivers observing three satellites and base station 1 and base satellite 1, we have

$$\begin{bmatrix} \boldsymbol{P}_1^p - \rho_1^p \\ \boldsymbol{P}_2^p - \rho_2^p \\ \boldsymbol{P}_3^p - \rho_3^p \end{bmatrix} = \begin{bmatrix} \boldsymbol{I} & \boldsymbol{0} & \boldsymbol{0} \\ \boldsymbol{I} & \boldsymbol{I} & \boldsymbol{0} \\ \boldsymbol{I} & \boldsymbol{0} & \boldsymbol{I} \end{bmatrix} \begin{bmatrix} \widehat{\xi}_P \\ \widehat{\xi}_{2,P} \\ \widehat{\xi}_{3,P} \end{bmatrix} \tag{6.6.44}$$

$$\boldsymbol{P}_k - \rho_k = \begin{bmatrix} P_k^1 - \rho_k^1 \\ P_k^2 - \rho_k^2 \\ P_k^3 - \rho_k^3 \end{bmatrix} \tag{6.6.45}$$

$$\widehat{\xi}_P = \begin{bmatrix} \widehat{\xi}_P^1 \\ \widehat{\xi}_P^2 \\ \widehat{\xi}_P^3 \end{bmatrix} = \begin{bmatrix} -\widehat{\xi}_P^1 + \widehat{\xi}_{1,P} + T_1^1 + I_{1,P}^1 \\ -\widehat{\xi}_P^2 + \widehat{\xi}_{1,P} + T_1^2 + I_{1,P}^2 \\ -\widehat{\xi}_P^3 + \widehat{\xi}_{1,P} + T_1^3 + I_{1,P}^3 \end{bmatrix} \tag{6.6.46}$$

$$\widehat{\xi}_{2,P} = \begin{bmatrix} \xi_{2,P} - \xi_{1,P} + T_{21}^1 + I_{21,P}^1 \\ \xi_{2,P} - \xi_{1,P} + T_{21}^2 + I_{21,P}^2 \\ \xi_{2,P} - \xi_{1,P} + T_{21}^3 + I_{21,P}^3 \end{bmatrix} \quad \widehat{\xi}_{3,P} = \begin{bmatrix} \xi_{3,P} - \xi_{1,P} + T_{31}^1 + I_{31,P}^1 \\ \xi_{3,P} - \xi_{1,P} + T_{31}^2 + I_{31,P}^2 \\ \xi_{3,P} - \xi_{1,P} + T_{31}^3 + I_{31,P}^3 \end{bmatrix} \tag{6.6.47}$$

The combined network solution of the carrier phases (6.6.39) and pseudoranges (6.6.44) provides S satellite phase bias estimates $\widehat{\xi}_{\Phi}^p$ and S satellite code biases $\widehat{\xi}_P^p, p = 1, \cdots, S$, which are transmitted to the user. The estimated ambiguities $\widehat{N}_l^q$, $l = 2, \cdots, R$, and $q = 2, \cdots, S$ may at first glance appear as a by-product in PPP-RTK, but their resolution to integers is important to achieve maximal accuracy for the satellite phase and code biases.

Instead of transmitting the full bias values, it is sufficient to only transmit the fractional parts. Consider the following:

$$n_1^p = \left[\frac{\widehat{\xi}_{\Phi}^p}{\lambda}\right] \qquad \widehat{\xi}_{\Phi,FCB}^p = \frac{\widehat{\xi}_{\Phi}^p}{\lambda} - n_1^p \tag{6.6.48}$$

with $p = 1, \cdots, S$. The symbol $[\bullet]$ denotes the rounding operation to the nearest integer and should not be confused with a matrix bracket, and the subscript FCB denotes the fractional cycle bias. The symbol n_1^p denotes the integer number of wavelengths λ that go into the satellite phase bias $\widehat{\xi}_{\Phi}^p$. In case the tropospheric and ionospheric terms are not present the bias simply consists of $-\xi_{\Phi}^p + \xi_{1,\Phi} + \lambda N_1^p$. Therefore, more precisely and in tune with subsequent sections, n_1^p is the integer number of wavelengths in $-\xi_{\Phi}^p + \xi_{1,\Phi}$ plus N_1^p.The second equation in (6.6.48) provides the computed fractional satellite phase bias $\widehat{\xi}_{\Phi,FCB}^p$. This value is transmitted to the user.

For convenience, this fractional bias is parameterized in terms of Δn^p, which is the number of integer cycles in $-\xi_{\Phi}^p + \xi_{1,\Phi}$:

$$n_1^p = \Delta n^p + N_1^p \tag{6.6.49}$$

Multiplying $\widehat{\xi}_{\Phi,FCB}^p$ in (6.6.48) by λ, substituting (6.6.41) for the satellite phase bias $\widehat{\xi}_{\Phi}^p$ and then substituting (6.6.49), the desired form for the fractional cycle bias becomes

$$\lambda \widehat{\xi}_{\Phi,FCB}^p = -\xi_{\Phi}^p + \xi_{1,\Phi} - \lambda \Delta n^p + T_1^p - I_{1,P}^p \tag{6.6.50}$$

This equation expresses the computed fractional satellite phase bias as a function of satellite and base station phase biases and an unknown integer Δn^p.

User Solution: The user begins with the phase and pseudorange equations (6.6.37). Subtracting the fractional cycle bias of the base satellite from the observation gives

$$\Phi_u^1 - \lambda \widehat{\xi}_{\Phi,FCB}^1 = \rho_u^1 + \xi_{u,\Phi} - \xi_{1,\Phi} + \lambda\left(N_u^1 + \Delta n_1^1\right) + T_{u1}^1 - I_{u1,P}^1 + \varepsilon_{\Phi} \tag{6.6.51}$$

The reparameterized receiver phase bias at the user station, $\widehat{\xi}_{u,\Phi}$, is defined as

$$\widehat{\xi}_{u,\Phi} = \xi_{u,\Phi} - \xi_{1,\Phi} + \lambda\left(N_u^1 + \Delta n^1\right) \tag{6.6.52}$$

This lumped parameter contains the receiver phase difference of station u and base station, the unknown ambiguity of the base satellite, and the unknown number of integer wavelengths defined in (6.6.49). The equation for a nonbase satellite, $q = 2, \cdots, S$, is

$$\Phi_u^q - \lambda \widehat{\xi}_{\Phi,FCB}^q = \rho_u^q + \widehat{\xi}_{u,\Phi} + \lambda\left(N_u^{q1} + \Delta n^{q1}\right) + T_{u1}^q - I_{u1,P}^q + \varepsilon_{\Phi} \tag{6.6.53}$$

The expression has been algebraically rearranged such that the estimable receiver phase parameter is the same as in (6.6.52). In the process of this rearrangement, the ambiguity became an across-satellite ambiguity. The lumped integer

$$\widehat{N}_u^q = N_u^{q1} + \Delta n^{q1} \tag{6.6.54}$$

becomes the estimable ambiguity for station u and satellite q. The corrected pseudorange follows from (6.6.37) by subtracting transmitted code bias (6.6.46)

$$P_u^p - \widehat{\xi}_P^p = \rho_u^p + (\xi_{u,P} - \xi_{1,P}) + T_{u1}^p + I_{u1,P}^p + \varepsilon_P \tag{6.6.55}$$

The code phase bias difference becomes the new estimable code bias at station u

$$\widehat{\xi}_{u,P} = \xi_{u,P} - \xi_{1,P} \tag{6.6.56}$$

Equations (6.6.51), (6.6.53), and (6.6.56) comprise the complete set for the user solution. In summary, they are

$$\left.\begin{aligned} \Phi_u^1 - \lambda\widehat{\xi}^1_{\Phi,FCB} &= \rho_u^1 + \widehat{\xi}_{u,\Phi} + T_{u1}^1 - I_{u1,P}^1 + M_\Phi + \varepsilon_\Phi \\ \Phi_u^q - \lambda\widehat{\xi}^q_{\Phi,FCB} &= \rho_u^q + \widehat{\xi}_{u,\Phi} + \lambda\widehat{N}_u^q + T_{u1}^q - I_{u1,P}^q + M_\Phi + \varepsilon_\Phi \\ P_u^p - \widehat{\xi}_P^p &= \rho_u^p + \widehat{\xi}_{u,P} + T_{u1}^p + I_{u1,P}^p + M_p + \varepsilon_P \end{aligned}\right\} \tag{6.6.57}$$

The superscript q runs from 2 to S and p runs from 1 to S. There are a total of $2S$ observation and $3 + 2 + (S - 1)$ parameters; they are the three baseline components, the receiver phase bias and receiver code bias terms, and the $S - 1$ ambiguities.

In the solution (6.6.57) only the nonbase satellite phase equation contains an ambiguity parameter; all phase equations contain the same receiver phase bias parameter. Keeping these important characteristics in mind, it is clear that the user can select the base satellite independently of which base satellite might have been used in the network solution. Therefore, no information about the identification of the network base satellite needs to be transmitted to the user. The user is free to select any satellite as the base satellite. With tropospheric delays modeled at the network, the user equation will only contain T_u^q. Similar handling could be argued for the ionosphere; however, the dual-frequency solutions discussed below will eliminate the ionospheric term anyway.

Equation (6.6.57) represents the essence of PPP-RTK. The satellite phase and code biases are generated by a network of stations at known locations and transmitted to the user. In the user solution these biases are treated as known quantities and applied to the observations. The network could in principle consist of just a single station. However, with more network stations, the strength of the solution increases by virtue of fixing the ambiguities to integers since a full variance-covariance matrix becomes available. The estimated satellite phase and code biases depend on the base station receiver clock. Unless the base station is equipped with an atomic clock, epoch-wise estimation is required and one cannot readily take advantage of the long-term stability of the satellite clocks and reduce the transmission load for the phase and code biases.

6.6.3.2 Dual-Frequency Solutions All dual-frequency solutions make use of the Hatch-Melbourne-Wübbena (HMW) function for computing the wide-lane ambiguity. Furthermore, the ionospheric-free functions are used in order to eliminate the first-order ionospheric delays. The tropospheric delays are assumed to be modeled or estimated using a mapping function that depends on the satellite elevation angle, and therefore are not relevant to linear independence considerations for parameters. The tropospheric delay and multipath terms will be omitted below. The one-step solution given resolves the integer ambiguities as part of a network solution, and the satellite biases and HMW satellite hardware delays are computed for transmission to the user. The line-by-line method resolves the ambiguities through simple rounding of averaged observation from a receiver-satellite pair. The fractional satellite biases and HMW satellite hardware delays are computed and transmitted. In all cases we assume that the tropospheric delays at the network station have been corrected using a tropospheric model. An initial comparison of various techniques to fix integers in precise point positioning can be found in Geng et al. (2010).

One-Step Network Solution: Collins (2008) proposed a one-step solution using the dual-frequency ionospheric-free phase function, the HMW function, and the ionospheric-free pseudorange function. Although these functions are correlated since the HMW function depends on carrier phase and pseudorange observations, he showed that the correlation is small and suggested that it be neglected. The respective model equations are (6.1.39), (6.1.48), and (6.1.38) and listed for easy reference as:

$$\left.\begin{aligned} \Phi IF12_k^p &= \rho_k^p + \xi_{k,\Phi IF12} - \xi^p_{\Phi IF12} + \lambda_{\Phi IF12} N^p_{k,\Phi IF12} + T_k^p + M^p_{k,\Phi IF12} + \varepsilon_{\Phi IF12} \\ HMW12_k^p &= -d_{k,HMW12} + D^p_{HMW12} + \lambda_{12} N^p_{k,12} + M^p_{k,HMW12} + \varepsilon_{HMW12} \\ PIF12_k^p &= \rho_k^p + \xi_{k,PIF12} - \xi^p_{PIF12} + T_k^p + M^p_{k,HMW12} + \varepsilon_{PIF12} \end{aligned}\right\} \tag{6.6.58}$$

The ionospheric-free receiver phase biases $\xi_{k,\Phi IF12}$ and satellite phase biases $\xi^p_{\Phi IF12}$ contain the receiver and satellite clock errors and receiver and the satellite hardware delays obtained from (6.1.39) by lumping the respective clock and hardware terms contained in δ_{R2}, or by applying the ionospheric-free function to the phase function of (6.6.38). The product of ionospheric-free wavelength and ambiguity is according to (6.1.39),

$$\lambda_{\Phi IF12} N^p_{k,\Phi IF12} = c\frac{f_1 - f_2}{f_1^2 - f_2^2} N^p_{k,1} + c\frac{f_2}{f_1^2 - f_2^2} N^p_{k,12} \tag{6.6.59}$$

where c is the velocity of light. In the case of GPS, the integer $N^p_{k,1}$ is the L1 ambiguity and $N^p_{k,12} = N^p_{k,1} - N^p_{k,2}$ is the wide-lane ambiguity. Using the GPS frequencies $f_1 = 154 f_0$, $f_2 = 120 f_0$, and $f_0 = 10.23$ MHz, the scaled ionospheric-free ambiguity becomes numerically

$$\lambda_{\Phi IF12} N^p_{k,\Phi IF12} = \frac{2cf_0}{f_1^2 - f_2^2}\left(17N^p_{k,1} + 60N^p_{k,12}\right) = 0.107N^p_{k,1} + 0.378N^p_{k,12} \tag{6.6.60}$$

For other frequencies or satellite systems, the numerical values in (6.6.60) change accordingly. We further note that the HMW function does not depend on the receiver and satellite clock errors and the tropospheric delay. The terms $d_{k,HMW12}$ and D^p_{HMW12} are the receiver and satellite hardware delays of the $HMW12$ function. The ionospheric-free receiver code biases $\xi_{k,PIF12}$ and satellite biases ξ^p_{PIF12} contain the receiver and satellite clock errors and respective hardware code delays.

For the network solution the linear dependencies in the mathematical model (6.6.58) are removed by reparameterization, as was done in the previous section. In fact the solution steps applied for the single-frequency case to achieve the reparameterization also apply to this dual-frequency case. The example presented again includes three satellites and two stations. The extension to more network stations observing more common satellites can readily be implemented. As was the case above, station 1 is the base station and satellite 1 is the base satellite. With these specifications the reparameterized solution can be written as

$$\begin{bmatrix} \boldsymbol{\Phi IF12}_1 - \boldsymbol{\rho}_1 \\ \boldsymbol{HMW12}_1 \\ \boldsymbol{\Phi IF12}_2 - \boldsymbol{\rho}_2 \\ \boldsymbol{HMW12}_2 \end{bmatrix} = \begin{bmatrix} \boldsymbol{I} & \boldsymbol{0} & \boldsymbol{0} & \boldsymbol{0} & \boldsymbol{0} \\ \boldsymbol{0} & \boldsymbol{I} & \boldsymbol{0} & \boldsymbol{0} & \boldsymbol{0} \\ \boldsymbol{I} & \boldsymbol{0} & \mathbf{b} & \boldsymbol{0} & \boldsymbol{C} \\ \boldsymbol{0} & \boldsymbol{I} & \boldsymbol{0} & \mathbf{b} & \boldsymbol{D} \end{bmatrix} \begin{bmatrix} \hat{\boldsymbol{\xi}}_{\Phi IF12} \\ \hat{\boldsymbol{D}}_{HMW12} \\ \hat{\xi}_{2,\Phi IF12} \\ \hat{d}^{p}_{HMW12} \\ \hat{\boldsymbol{N}} \end{bmatrix} \tag{6.6.61}$$

$$\boldsymbol{\Phi IF12}_k - \boldsymbol{\rho}_k = \begin{bmatrix} \Phi IF12_k^1 - \rho_k^1 \\ \Phi IF12_k^2 - \rho_k^2 \\ \Phi IF12_k^3 - \rho_k^3 \end{bmatrix} \qquad \boldsymbol{HMW12}_k = \begin{bmatrix} HMW12_k^1 \\ HMW12_k^2 \\ HMW12_k^3 \end{bmatrix} \tag{6.6.62}$$

$$\boldsymbol{C} = \lambda_{\Phi IF12} \begin{bmatrix} 0 & 0 & 0 & 0 \\ 17 & 60 & 0 & 0 \\ 0 & 0 & 17 & 60 \end{bmatrix} \qquad \boldsymbol{D} = \lambda_{12} \begin{bmatrix} 0 & 0 & 0 & 0 \\ 0 & 1 & 0 & 0 \\ 0 & 0 & 0 & 1 \end{bmatrix} \tag{6.6.63}$$

$$\begin{gathered} \hat{\boldsymbol{\xi}}_{\Phi IF12} = \begin{bmatrix} \hat{\xi}^1_{\Phi IF12} \\ \hat{\xi}^2_{\Phi IF12} \\ \hat{\xi}^3_{\Phi IF12} \end{bmatrix} = \begin{bmatrix} -\hat{\xi}^1_{\Phi IF12} + \xi_{1,\Phi IF12} + \lambda_{\Phi IF12} N^1_{1,\Phi IF12} \\ -\hat{\xi}^2_{\Phi IF12} + \xi_{1,\Phi IF12} + \lambda_{\Phi IF12} N^2_{1,\Phi IF12} \\ -\hat{\xi}^3_{\Phi IF12} + \xi_{1,\Phi IF12} + \lambda_{\Phi IF12} N^3_{1,\Phi IF12} \end{bmatrix} \\ \hat{\boldsymbol{D}}_{HMW12} = \begin{bmatrix} \hat{D}^1_{HMW12} \\ \hat{D}^2_{HMW12} \\ \hat{D}^3_{HMW12} \end{bmatrix} = \begin{bmatrix} D^1_{HMW12} - d_{1,HMW12} + \lambda_{12} N^1_{1,12} \\ D^2_{HMW12} - d_{1,HMW12} + \lambda_{12} N^2_{1,12} \\ D^3_{HMW12} + d_{1,HMW12} + \lambda_{12} N^3_{1,12} \end{bmatrix} \end{gathered} \tag{6.6.64}$$

$$\begin{gathered} \hat{\xi}_{2,\Phi IF12} = \xi_{2,\Phi IF12} - \xi_{1,\Phi IF12} + \lambda_{\Phi IF12} N^1_{21,\Phi IF12} \\ \hat{d}_{2,HMW12} = -d_{2,HMW12} + d_{1,HMW12} + \lambda_{12} N^1_{21,12} \end{gathered} \tag{6.6.65}$$

$$\hat{\boldsymbol{N}} = \begin{bmatrix} \hat{N}^2_{2,1} \\ \hat{N}^2_{2,12} \\ \hat{N}^3_{2,1} \\ \hat{N}^3_{2,12} \end{bmatrix} = \begin{bmatrix} N^{21}_{21,1} \\ N^{21}_{21,12} \\ N^{31}_{21,1} \\ N^{31}_{21,12} \end{bmatrix} \tag{6.6.66}$$

For verification purposes, substitute (6.6.64) to (6.6.66) into (6.6.61) and compare with (6.6.58). The equation system is of full rank. It consists of $2RS$ equations and as many parameters, i.e., S satellite phase biases $\hat{\xi}^p_{\Phi IF12}$, S satellite HMW hardware biases $\hat{D}^p_{HMW12}$, $p = 1 \cdots S$, $R-1$ receiver phase biases $\hat{\xi}_{k,\Phi IF12}$, $R-1$ HMW receiver hardware biases $\widehat{d}_{k,HMW12}, k = 2 \cdots R$, and $2(R-1)(S-1)$ ambiguity parameters $N^{q1}_{m1,1}$ and $N^{q1}_{m1,12}$ with $m = 2, \cdots, R$ and $q = 2 \cdots S$. In order to aid in reading the notation, let us recall the notational use in the case of dual frequencies: the subscripts preceding the comma identify stations and the differencing operation, the superscripts identify the satellites and the differencing operation, the numerical 1 after the comma refers to the L1 ambiguity, and the 12 indicates the L1 and L2 wide lane. For example, in this notation $N^{21}_{21,12} = N^{21}_{21,1} - N^{21}_{21,2}$ is the difference of the double-differenced L1 and L2 ambiguities.

The pseudorange solution is identical to the one given in (6.6.44) to (6.6.47), except replacing subscripts P by $PIF12$ and omitting the tropospheric term. It is repeated here for easy referencing. The two-station and three-satellite solution is

$$\begin{bmatrix} \boldsymbol{PIF12}_1 - \boldsymbol{\rho}_1 \\ \boldsymbol{PIF12}_2 - \boldsymbol{\rho}_2 \end{bmatrix} = \begin{bmatrix} \boldsymbol{I} & \boldsymbol{0} \\ \boldsymbol{I} & \boldsymbol{b} \end{bmatrix} \begin{bmatrix} \widehat{\boldsymbol{\xi}}_{PIF12} \\ \widehat{\xi}_{2,PIF12} \end{bmatrix} \tag{6.6.67}$$

$$\boldsymbol{PIF12}_k - \boldsymbol{\rho}_k = \begin{bmatrix} PIF12^1_k - \rho^1_k \\ PIF12^1_k - \rho^1_k \\ PIF12^1_k - \rho^1_k \end{bmatrix} \tag{6.6.68}$$

$$\widehat{\boldsymbol{\xi}}_{PIF12} = \begin{bmatrix} \widehat{\xi}^1_{PIF12} \\ \widehat{\xi}^2_{PIF12} \\ \widehat{\xi}^3_{PIF12} \end{bmatrix} = \begin{bmatrix} -\xi^1_{PIF12} + \xi_{1,PIF12} \\ -\xi^2_{PIF12} + \xi_{1,PIF12} \\ -\xi^3_{PIF12} + \xi_{1,PIF12} \end{bmatrix} \quad \overline{\xi}_{2,PIF12} = \xi_{2,PIF12} - \xi_{1,PIF12} \tag{6.6.69}$$

The vector $\boldsymbol{b}$ is given in (6.6.12). If one combines the pseudorange and carrier phase observations into one solution, there are in total $3RS$ observations and $2RS + S + R - 1$ parameters, giving a degree of freedom of $RS - R - S + 1$.

User Solution: For the user solution, the original equations (6.6.58) are corrected for the transmitted ionospheric-free satellite phase biases, HMW hardware delays, and satellite code biases are applied to the observations. Assuming again without loss of generality that the user selects satellite 1 as the base satellite, the user solution is

$$\begin{aligned} \Phi IF12^1_u - \widehat{\xi}^1_{\Phi IF12} &= \rho^1_u + \xi_{u,\Phi IF12} - \xi_{1,\Phi IF12} + \lambda_{\Phi IF12} + N^1_{u1,\Phi IF12} \\ &\quad + T^1_u + \varepsilon_{\Phi IF12} \\ &= \rho^1_u + \widehat{\xi}_{u,\Phi IF12} + T^1_u + \varepsilon_{\Phi IF12} \end{aligned} \tag{6.6.70}$$

$$\begin{aligned} \Phi IF12^q_u - \widehat{\xi}^q_{\Phi IF12} &= \rho^q_u + \widehat{\xi}_{u,\Phi IF12} + \lambda_{\Phi IF12} N^{q1}_{u1,\Phi IF12} + T^q_u + \varepsilon_{\Phi IF12} \\ &= \rho^q_u + \widehat{\xi}_{u,\Phi IF12} + \lambda_{\Phi IF12} \widehat{N}^q_{u,\Phi IF12} + T^q_u + \varepsilon_{\Phi IF12} \end{aligned} \tag{6.6.71}$$

$$\begin{aligned} HMW12^1_u - \widehat{D}^1_{HMW12} &= -d_{u,HMW12} + d_{1,HMW12} + \lambda_{12} N^1_{u1,12} + \varepsilon_{HMW12} \\ &= \widehat{d}_{u,HMW12} + \varepsilon_{HMW12} \end{aligned} \tag{6.6.72}$$

$$\begin{aligned} HMW12_u^q - \widehat{D}_{HMW12}^q &= \widehat{d}_{u,HMW12} + \lambda_{12} N_{u1,12}^{q1} + \varepsilon_{HMW12} \\ &= \widehat{d}_{u,HMW12} + \lambda_{12} \widehat{N}_{u,12}^{q} + \varepsilon_{HMW12} \end{aligned} \tag{6.6.73}$$

$$\begin{aligned} PIF12_u^p - \widehat{\xi}_{PIF12}^p &= \rho_u^p + \xi_{u,PIF12} - \xi_{1,PIF12}^p + T_u^p + \varepsilon_{PIF12} \\ &= \rho_u^p + \widehat{\xi}_{u,PIF12} + T_u^p + \varepsilon_{PIF12} \end{aligned} \tag{6.6.74}$$

with $q = 2, \cdots, S$. We note again that, as is the case with the single-frequency user solution, the base satellite phase equation and the HMW equation do not contain ambiguity parameters. As long as this property is recognized, the user is free to adopt any satellite as the base satellite, regardless of the choice made during the network solution. The final form of the user solution for all three functions is

$$\left.\begin{aligned} \Phi IF12_u^1 - \widehat{\xi}_{\Phi IF12}^1 - T_{u,0}^1 &= \rho_u^1 + \widehat{\xi}_{u,\Phi IF12} + dT_u m_{wv}(\vartheta^1) + M_{\Phi IF12} + \varepsilon_{\Phi IF12} \\ \Phi IF12_u^q - \widehat{\xi}_{\Phi IF12}^q - T_{u,0}^1 &= \rho_u^q + \widehat{\xi}_{u,\Phi IF12} + \lambda_{\Phi IF12}\left(17\widehat{N}_{u,1}^q + 60\widehat{N}_{u,12}^q\right) \\ &\quad + dT_u m_{wv}(\vartheta^q) + M_{\Phi IF12} + \varepsilon_{\Phi IF12} \\ HMW12_u^1 - \widehat{D}_{HMW12}^1 &= \widehat{d}_{u,HMW12} + M_{HMW12} + \varepsilon_{HMW12} \\ HMW12_u^q - \widehat{D}_{HMW12}^q &= \widehat{d}_{u,HMW12} + \lambda_{12}\widehat{N}_{u,12}^q + M_{HMW12} + \varepsilon_{HMW12} \\ PIF12_u^p - \widehat{\xi}_{PIF12}^p - T_{u,0}^1 &= \rho_u^p + \widehat{\xi}_{u,PIF12} + dT_u m_{wv}(\vartheta^p) + M_{PIF12} + \varepsilon_{PIF12} \end{aligned}\right\} \tag{6.6.75}$$

where $q = 2 \cdots S$ and $p = 1 \cdots S$. We have added one vertical tropospheric parameter. There are $3S$ observations and $3 + 3 + 2(S - 1)$ parameters to be estimated, i.e., three baseline components, three receiver biases, $2(S - 1)$ ambiguities, and one tropospheric parameter. In (6.6.75) the satellite phase and code biases and HMW satellite hardware delays are subtracted.

Network Wide-laning First: The solution of (6.6.61) can be carried in two steps by first estimating the wide-lane ambiguities from the HMW function and then estimating the L1 ambiguities from the ionospheric-free phase function using the wide-lane ambiguities as known quantities. This approach ignores the correlation between both functions. Extracting the HMW equations from (6.6.61), the solution is

$$\begin{bmatrix} \boldsymbol{HMW12}_1 \\ \boldsymbol{HMW12}_2 \end{bmatrix} = \begin{bmatrix} \boldsymbol{I} & \boldsymbol{0} & \boldsymbol{0} \\ \boldsymbol{I} & \boldsymbol{b} & \lambda_{12}\boldsymbol{A} \end{bmatrix} \begin{bmatrix} \widehat{\boldsymbol{D}}_{HMW12} \\ \widehat{d}_{2,HMW12} \\ \widehat{\boldsymbol{N}}_{2,HMW12} \end{bmatrix} \tag{6.6.76}$$

$$\widehat{\boldsymbol{N}}_{2,HMW12} = \begin{bmatrix} \widehat{N}_{2,12}^2 \\ \widehat{N}_{2,12}^3 \end{bmatrix} = \begin{bmatrix} \widehat{N}_{21,12}^{21} \\ \widehat{N}_{21,12}^{31} \end{bmatrix} \tag{6.6.77}$$

The estimates $\widehat{\boldsymbol{D}}_{HMW12}$ and $\widehat{d}_{2,HMW12}$ are identical to those in (6.6.64) and (6.6.65). There are RS equations and as many unknowns, i.e., S satellite hardware biases, $R - 1$ receiver hardware delays, and $(R - 1)(S - 1)$ double-difference ambiguities. Given the wide-lane ambiguities $\widehat{\boldsymbol{N}}_{2,HMW12}$, one can now compute the ionospheric-free phase solution components as

$$\begin{bmatrix} \boldsymbol{\Phi IF12}_1 - \rho_1 \\ \boldsymbol{\Phi IF12}_2 - \rho_2 - 60\lambda_{\Phi IF12}\widehat{\boldsymbol{N}}_{2,HMW12} \end{bmatrix} = \begin{bmatrix} \boldsymbol{I} & \boldsymbol{0} & \boldsymbol{0} \\ \boldsymbol{I} & \boldsymbol{b} & 17\lambda_{\Phi IF12}\boldsymbol{A} \end{bmatrix} \begin{bmatrix} \widehat{\xi}_{\Phi IF12} \\ \widehat{\xi}_{2,\Phi IF12} \\ \widehat{\boldsymbol{N}}_{2,1} \end{bmatrix} \tag{6.6.78}$$

$$\widehat{\boldsymbol{N}}_{2,1} = \begin{bmatrix} \widehat{N}^{2}_{2,1} \\ \widehat{N}^{3}_{2,1} \end{bmatrix} = \begin{bmatrix} N^{21}_{21,1} \\ N^{31}_{21,1} \end{bmatrix} \tag{6.6.79}$$

The estimates $\widehat{\xi}_{\Phi IF12}$ and $\widehat{\xi}_{2,\Phi IF12}$ are identical to those in (6.6.64) and (6.6.65). The system (6.6.78) contains again *RS* equations for as many unknowns. The ionospheric-free phase observations are corrected for the known wide-lane ambiguities $\widehat{\boldsymbol{N}}_{2,HMW12}$ obtained from the first step. Next we can compute the fractional cycle biases $\widehat{\xi}_{\Phi IF12,FCB}$ and $\widehat{\boldsymbol{D}}_{HMW12,FCB}$, which can be transmitted to the user. The fractional biases are explicitly computed in the next approach.

Line-by-Line Approach: This procedure was proposed in Laurichesse and Mercier (2007). We again select station 1 as the base station. The approach first calls for the HMW functions in (6.6.58) to be averaged individually over time and then rounded to determine the integer number of wide lanes of the hardware delays. Next, the fractional satellite hardware delays are computed. The solution of this first step is

$$\overline{HMW12^p_1} = -d_{1,HMW12} + D^p_{HMW12} + \lambda_{12}N^p_{1,12} \tag{6.6.80}$$

$$n^p_{1,HMW12} = \left[\frac{\overline{HMW12^p_1}}{\lambda_{12}}\right] \qquad D^p_{HMW12,FCB} = \frac{\overline{HMW12^p_1}}{\lambda_{12}} - n^p_{1,HMW12} \tag{6.6.81}$$

$$n^p_{1,HMW12} = \Delta n^p_{HMW12} + N^p_{1,12} \tag{6.6.82}$$

$$\lambda_{12}D^p_{HMW12,FCB} = D^p_{HMW12} - d_{1,HMW12} - \lambda_{12}\Delta n^p_{HMW12} \tag{6.6.83}$$

with $p = 1, \cdots, S$. The overbar of $HMW12$ indicates averaging over time. The integer unknown Δn^p_{HMW12} represents, the integer number of wide-lane wavelengths that go into the hardware difference $-d_{1,HMW12} + D^p_{HMW12}$. Equation (6.6.83) follows from the second equation in (6.6.81) multiplied by the wide-lane wavelength, and then substituting (6.6.80) and (6.6.82). The integer $n^p_{1,HMW12}$ can usually be identified reliably after only a short period of observations.

The second step requires a similar treatment of the ionospheric-free phase function (6.6.58). The observation is averaged again over time, then the known integer $n^p_{1,HMW12}$ of (6.6.81) is subtracted from the averaged observation, and finally (6.6.82) is used on the right side. The result is

$$\begin{aligned} a^p_1 \equiv \overline{\Phi IF12^p_1} - \rho^p_1 - \lambda_{\Phi IF12}60n^p_{1,HMW12} &= \xi_{1,\Phi IF12} - \xi^p_{\Phi IF12} \\ &+ \lambda_{\Phi IF12}\left(17N^p_{1,1} - 60\Delta n^p_{HMW12}\right) \end{aligned} \tag{6.6.84}$$

$$n^p_{1,a} = \left[\frac{d^p_1}{\lambda_c}\right] \qquad \xi^p_{a,FCB} = \frac{d^p_1}{\lambda_c} - n^p_{1,a} \tag{6.6.85}$$

$$n^p_{1,a} = \Delta n^p_a + N^p_{1,1} \tag{6.6.86}$$

$$\lambda_c \xi^p_{a,FCB} = \xi_{1,\Phi IF12} - \xi^p_{\Phi IF12} - \lambda_{\Phi IF12}\left(17\Delta n^p_a + 60\Delta n^p_{HMW12}\right) \tag{6.6.87}$$

where $\lambda_c = 17\lambda_{\Phi IF12} \approx 10.7$. Thus, the quantity $n^p_{1,a}$ is the integer number of λ_c units in d^p_1. Since λ_c is small and the function (6.6.84) depends on the receiver clock errors, a longer observation series is required to determine the correct integer $n^p_{1,a}$. This concludes the required computations at the network. The fractional $HMW12$ satellite hardware delays $D^p_{HMW12,FCB}$ and satellite phase biases $\xi^p_{a,FCB}$ are transmitted to the user to correct user observations.

However, the nonbase stations have thus far not been used. The fractional satellite hardware delays $D^p_{HMW12,FCB}$ and satellite phase biases $\xi^p_{a,FCB}$ were computed above without the benefit of observations from nonbase stations. The observations from these stations can serve as quality control. In the case of the $HMW12$ observations, the nonbase station observations are also first averaged over time, then corrected for the $HMW12$ fractional satellite hardware bias, and then rounded to determine the fraction receiver hardware delays. Using (6.6.83), one obtains

$$\overline{HMW12^p_k} - \lambda_{12} D^p_{HMW12,FCB} = -d_{k,HMW12} + d_{1,HMW12} + \lambda_{12}\left(N^p_{k,12} + \Delta n^p_{HMW12}\right) \tag{6.6.88}$$

$$\begin{aligned} n^p_{k,HMW12} &= \left[\frac{\overline{HMW12^p_k} - \lambda_{12} D^p_{HMW12,FCB}}{\lambda_{12}}\right] \\ d_{k,HMW12,FCB} &= \frac{\overline{HMW12^p_k} - \lambda_{12} D^p_{HMW12,FCB}}{\lambda_{12}} - n^p_{k,HMW12} \end{aligned} \tag{6.6.89}$$

with $p = 1, \cdots, S$ and $k = 2, \cdots, R$. The S values for a specific receiver hardware delay $d_{k,HMW12,FCB}$ should agree within random noise. Similarly, the averaged ionospheric-free phase observations are corrected for (6.6.87), giving

$$\begin{aligned} b^p_k &\equiv \overline{\Phi IF12^p_k} - \rho^p_k - \lambda_c \xi^p_{a,FCB} \\ &= \xi_{k,\Phi IF12} - \xi_{1,\Phi IF12} + \lambda_{\Phi IF12}\left(17N^p_{k,1} + 60N^p_{k,12} + 17\Delta n^p_a + 60\Delta n^p_{HMW12}\right) \end{aligned} \tag{6.6.90}$$

$$n^p_{k,b} = \left[\frac{b^p_k}{\lambda_c}\right] \qquad \xi_{k,FCB} = \frac{b^p_k}{\lambda_c} - n^p_{k,b} \tag{6.6.91}$$

with $p = 1, \cdots, S$ and $k = 2, \cdots, R$. The S values for a specific nonbase station receiver phase bias $\xi_{k,FCB}$ should be consistent.

User Solutions: The solution that can readily be built having the fractional cycle delays and biases $D^p_{HMW12,FCB}$ and $\xi^p_{a,FCB}$ available is

$$HMW12^1_u - \lambda_{12} D^1_{HMW12,FCB} = -d_{u,HMW12} + d_{1,HMW12} + \lambda_{12}\left(N^1_{u,12} + \Delta n^1_{HMW12}\right) = \widehat{d}_{u,HMW12} \tag{6.6.92}$$

$$HMW12^p_u - \lambda_{12} D^p_{HMW12,FCB} = \widehat{d}_{u,HMW12} + \lambda_{12}\left(N^{p1}_{u,12} + \Delta n^{p1}_{HMW12}\right) \tag{6.6.93}$$

$$\begin{aligned}\Phi IF12^1_u - \lambda_c \xi^1_{a,FCB} &= \xi_{u,\Phi IF12} - \xi_{1,\Phi IF12} \\ &\quad + \lambda_{\Phi IF12}\left[17\left(N^1_{u,1} + \Delta n^1_a\right) + 60\left(N^1_{u,12} + \Delta n^1_{HMW12}\right)\right] \\ &= \widehat{\xi}_{u,\Phi IF12}\end{aligned} \tag{6.6.94}$$

$$\Phi IF12^p_u - \lambda_c \xi^p_{a,FCB} = \xi_{u,\Phi IF12} + \lambda_{\Phi IF12}\left[17\left(N^{p1}_{u,1} + \Delta n^{p1}_a\right) + 60(N^{p1}_{u,12} + \Delta n^{p1}_{HMW12})\right] \tag{6.6.95}$$

Parameterizing and adding pseudorange observations:

$$\left.\begin{aligned} &HMW12^1_u - \lambda_{12} D^1_{HMW12,FCB} = \widehat{d}_{u,HMW12} + M_{HMW12} + \varepsilon_{HMW12} \\ &HMW12^q_u - \lambda_{12} D^q_{HMW12,FCB} = \widehat{d}_{u,HMW12} + \lambda_{12}\widehat{N}^q_{u,12} + M_{HMW12} + \varepsilon_{HMW12} \\ &\Phi IF12^1_u - \lambda_c \xi^1_{a,FCB} = \rho^1_u + \widehat{\xi}_{u,\Phi IF12} + M_{\Phi IF12} + \varepsilon_{\Phi IF12} \\ &\Phi IF12^q_u - \lambda_c \xi^q_{a,FCB} = \rho^q_u + \xi_{u,\Phi IF12} + \lambda_{\Phi IF12}\left(17\widehat{N}^q_{u,1} + 60\widehat{N}^q_{u,12}\right) \\ &\qquad\qquad + M_{\Phi IF12} + \varepsilon_{\Phi IF12} \\ &PIF12^p_u - \widehat{\xi}^p_{PIF12} = \rho^p_u + \widehat{\xi}_{u,PIF12} + M_{PIF12} + \varepsilon_{PIF12} \end{aligned}\right\} \tag{6.6.96}$$

$q = 2, \cdots, S,\ p = 1, \cdots, S$. The parameterized ambiguities in (6.6.93) and (6.6.95) contain the unknowns Δn^p_{HMW12}, which equal the number of full λ_{12} wavlengths in the satellite hardware delays $D^p_{HMW12} - d_{1,HMW12}$, and the unknown Δn^p_a, which equals the number of λ_c distances in $\xi_{1,\Phi IF12} - \xi^p_{\Phi IF12} - 60\lambda_{\Phi IF12}\Delta n^p_{HMW12}$. The reparameterized wide-lane ambiguities in (6.6.96) are the same.

6.6.3.3 Across-Satellite Differencing

The attraction of across-satellite differencing relates to the cancelation of receiver clock errors and receiver hardware delays. The technique was applied in Garbor and Nerem (1999) and later refined in Ge et al. (2008). The model functions are again the ionospheric-free carrier base and pseudorange functions, the $HMW12$ function, and, for convenience, we add the $AIF12$ function (6.1.49). Assuming satellite 1 as the base satellite, the difference functions are for a general station k:

$$\left.\begin{aligned} &HMW12^{1q}_k = D^{1q}_{HMW12} + \lambda_{12} N^{1q}_{k,12} + \varepsilon_{HMW12} \\ &AIF12^{1q}_k = D^{1q}_{AIF12} + \lambda_{\Phi IF12}\left(17N^{1q}_{k,1} + 60N^{1q}_{k,12}\right) + \varepsilon_{AIF12} \\ &\Phi IF12^{1q}_k = \rho^{1q}_k + \xi^{1q}_{\Phi IF12} + \lambda_{\Phi IF12}\left(17N^{1q}_k + 60N^{1q}_{k,12}\right) + T^{1q}_k + \varepsilon_{\Phi IF12} \\ &PIF12^{1q}_k = \rho^{1q}_k + \xi^{1q}_{PIF12} + T^{1q}_k + \varepsilon_{PIF12} \end{aligned}\right\} \tag{6.6.97}$$

The superscripts indicate the across-differencing operation. The receiver terms $d_{k,HMW12}$, $\xi_{k,\Phi IF12}$, and $\xi_{k,PIF12}$ cancel due to the differencing, i.e., the receiver clock error and the receiver hardware delays cancel. Since $AIF12$ is the difference of $\Phi IF12$ and $PIF12$ according to (6.1.49), the satellite hardware delays of function $AIF12$ are the differences of the $\Phi IF12$ satellite phase bias and $PIF12$ satellite code bias:

$$D^p_{AIF12} = \xi^p_{\Phi IF12} - \xi^p_{PIF12} = D^p_{PIF12} - D^p_{\Phi IF12} \tag{6.6.98}$$

In the difference (6.6.98) the satellite clock error cancels. The receiver clock error and receiver hardware delay in $AIF12$ have canceled due to the cross-satellite differencing, as mentioned above. It follows that D^{1q}_{HMW12}, D^{1q}_{AIF12}, $\xi^{1q}_{\Phi IF12}$, and ξ^{1q}_{PIF12} only contain satellite hardware phase and code delays.

The approach is to determine the fractional cycle biases of the across-satellite hardware delays $D^{1q}_{HMW12,FCB}$ and $D^{1q}_{AIF12,FCB}$ from the network and transmit these values to the user who will utilize (6.6.98) to convert $D^{1q}_{AIF12,FCB}$ to $\xi^{1q}_{\Phi IF12}$ and correct the phase observations. No base station needs to be specified.

Network Solution: For the network solution, the $HMW12$ function of (6.6.97) is averaged over time and then its fractional cycle bias is computed following the regular procedure:

$$\overline{HMW12^{1q}_k} = D^{1q}_{HMW12} + \lambda_{12} N^{1q}_{k,12} \tag{6.6.99}$$

$$n^{1q}_{k,HMW12} = \left[\frac{\overline{HMW12^{1q}_k}}{\lambda_{12}} \right] \qquad D^{1q}_{HMW12,FCB} = \frac{\overline{HMW12^{1q}_k}}{\lambda_{12}} - n^{1q}_{k,HMW12} \tag{6.6.100}$$

$$n^{1q}_{k,HMW12} = \Delta n^{1q}_{HMW12} + N^{1q}_{k,12} \tag{6.6.101}$$

$$\lambda_{12} D^{1q}_{HMW12,FCB} = D^{1q}_{HMW12} - \lambda_{12} \Delta n^{1q}_{HMW12} \tag{6.6.102}$$

The average is again indicated by the overbar. The unknown integer Δn^{1q}_{HMW12} is the number of wide-lane cycles in D^{1q}_{HMW12}. The fractional cycle hardware delay (6.6.100) is averaged over all stations, $k = 1 \cdots R$ and denoted by $\overline{D^{1q}_{HMW12}}$.

Second, the $AIF12$ function of (6.6.97) is averaged over time and corrected for the known integer $n^{1q}_{k,HMW12}$ of (6.6.100), and then the fractional cycle bias is computed. The result is

$$A^{1q}_k \equiv \overline{AIF12^{1q}_k} - 60\lambda_{\Phi IF12} n^{1q}_{k,HMW12} = \lambda_{\Phi IF12}\left(17N^{1q}_{k,1} - 60\Delta n^{1q}_{HMW12}\right) + D^{1q}_{AIF12} \tag{6.6.103}$$

$$n^{1q}_{k,A} = \left[\frac{A^{1q}_k}{\lambda_c} \right] \qquad D^{1q}_{A,FCB} = \frac{A^{1q}_k}{\lambda_c} - n^{1q}_{k,A} \tag{6.6.104}$$

$$n_{k,A}^{1q} = \Delta n_A^{1q} + N_{k,1}^{1q} \tag{6.6.105}$$

$$\lambda_c D_{A,FCB}^{1q} = D_{AIF12}^{1q} - \lambda_{\Phi IF12}\left(17\Delta n_A^{1q} + 60\Delta n_{HMW12}^{1q}\right) \tag{6.6.106}$$

where $\lambda_c = 17\lambda_{\Phi IF12}$. Average the fractional cycle bias over all stations, $k = 1 \cdots R$ and denote it by $\overline{D_{A,FCB}^{1q}}$.

The fractional cycle biases $\overline{D_{HMW12,FCB}^{1q}}$ and $\overline{D_{A,FCB}^{1q}}$ are transmitted to the user, $q = 2, \cdots, S$. The ionospheric-free satellite code bias ξ_{PIF12}^{1q} must also be made available to the user for computing $\xi_{\Phi IF12}^{1q}$ via the relation (6.6.98) and for correcting the pseudorange observations. Ideally, the biases for all across-satellite difference combinations should be available to the users to enable them to select any base satellite.

User Solution: For the user solution, the $HMW12$ and $PIF12$ functions in (6.6.97) can readily be corrected for $\overline{D_{HMW12,FCB}^{1q}}$ and ξ_{PIF12}^{1q}, respectively. The correction $\xi_{\Phi IF12}^{1q}$ follows immediately from (6.6.98) and (6.6.106) as

$$\begin{aligned} \xi_{\Phi IF12}^{1q} &= D_{AIF12}^{1q} + \xi_{PIF12}^{1q} \\ &= \lambda_c \overline{D_{A,FCB}^{1q}} + \lambda_{\Phi IF12}\left(17\Delta n_A^{1q} + 60\Delta n_{HMW12}^{1q}\right) + \xi_{PIF12}^{1q} \end{aligned} \tag{6.6.107}$$

Applying the three corrections to (6.6.97), the three user equations become

$$\begin{aligned} HMW12_u^{1q} - \lambda_{12}\overline{D_{HMW12,FCB}^{1q}} &= \lambda_{12}\widehat{N}_{u,12}^{q} + M_{HMW12} + \varepsilon_{HMW12} \\ \Phi IF12_u^{1q} - \lambda_c \overline{D_{A,FCB}^{1q}} - \xi_{PIF12}^{1q} &= \rho_u^{1q} + \lambda_{\Phi IF12}\left(17\widehat{N}_{u,1}^{q} + 60\widehat{N}_{u,12}^{q}\right) \\ &\quad + T_u^{1q} + M_{\Phi IF12} + \varepsilon_{\Phi IF12} \\ PIF12_u^{1q} - \xi_{PIF12}^{1q} &= \rho_u^{1q} + T_u^{1q} + M_{PIF12} + \varepsilon_{PIF12} \end{aligned} \tag{6.6.108}$$

with

$$\widehat{N}_{u,1}^{q} = N_{u,1}^{1q} + \Delta n_A^{1q} \qquad \widehat{N}_{u,12}^{q} = N_{u,12}^{1q} + \Delta n_{HMW12}^{1q} \tag{6.6.109}$$

The $HMW12$ and $\Phi IF12$ functions contain the same wide-lane ambiguity. It consists of the original wide-lane ambiguity plus an unknown number of wide-lane cycles in satellite hardware delay D_{HMW12}^{1q}. The ionospheric-free code bias ξ_{PIF12}^{1q} is needed for every epoch. The user can select any base satellite but must be able to identify the respective transmitted biases. The system includes $3(S-1)$ observations, three position coordinates, $2(S-1)$ ambiguities, and one tropospheric parameter. Since across-satellite differencing cancels the receiver clock errors and receiver hardware delays, the transmitted fractional cycle biases are more stable than those of PPP.

The fractional satellite hardware delays (6.6.102) and (6.6.106) for the $HMW12$ and $AIF12$ functions can be computed without knowledge of the network station coordinates. Both are geometry-free linear functions of carrier phases and pseudoranges. It follows that the observational noise and impact of the multipath are dominated by that of the pseudoranges, because the noise and multipath of the carrier phases are much smaller than those of the pseudoranges. Consequently, the observational noise and multipath impact of the computed satellite phase bias $\xi^{1q}_{\Phi IF12}$ in (6.6.107) is correspondingly large. However, the satellite code difference is still needed to complete this computation. Even though network station coordinates are not needed for the computations of the $D^{1q}_{HMW12,FCB}$ and $D^{1q}_{A,FCB}$ hardware delays, the coordinates are needed for computing ξ^{1q}_{PIF12}. If the latter biases can be obtained from the clock corrections of the IGS precise ephemeris, i.e., the equality $\xi^{1q}_{PIF12} = \xi^{1q}_{\text{IGS}}$ is valid, then the network station coordinates are not needed at all.

However, instead of computing $\xi^{1q}_{\Phi IF12}$ via (6.6.107), one can compute it more accurately using the function $\Phi IF12$. One could apply the procedure expressed in (6.6.80) to (6.6.87) to across-satellite differences. A possible drawback is that now the network station coordinates must be known. Also, one can apply the $AIF12$ and (6.6.107) approach to any dual-frequency observations, not just to L1 and L2 observations discussed above.

The recursive adjustment technique applies to all of the models because they contain a mix of epoch parameters and constant parameters. For example, in the one-step network case (6.6.61) and (6.6.67) the $\widehat{\xi}^{p}_{\Phi IF12}$, $\widehat{\xi}^{p}_{k,\Phi IF12}$, $\widehat{\xi}^{p}_{PIF12}$, and $\widehat{\xi}_{k,PIF12}$ are epoch parameters; the most active varying parameter is the receiver clock error. The $HMW12$ hardware delays $\widehat{D}^{2}_{HMW12}$ and $\widehat{d}_{2,HMW12}$ are fairly stable. Similarly, in the line-by-line case the fractional cycle bias $\xi^{p}_{a,FCB}$ varies rapidly while the hardware delay $D^{p}_{HMW12,FCB}$ varies more slowly. In the across-satellite single-differencing case, both hardware delays, $D^{1q}_{HMW12,FCB}$ and $D^{1q}_{A,FCB}$ vary slowly. In all cases the ambiguity parameters remain constant until cycle slips occur. In that case the ambiguity parameters must be reinitialized and some convergence time might be required depending on the number of slips. If the time between consecutive epochs is sufficiently small one might succeed in modeling the ionospheric change between the epochs and use across-time differences to determine the cycle slips and avoid or reduce re-convergence time.

6.7 TRIPLE-FREQUENCY SOLUTIONS

Special triple-frequency functions are considered that bring uniqueness to triple-frequency processing as opposed to classical dual-frequency methods. We basically discuss two types of solutions. The first one is the one-step batch solution in which all observations are combined and all parameters are estimated simultaneously. The second solution is TCAR (three-carrier ambiguity resolution), in which one attempts to resolve the ambiguities first and then computes the position coordinates of the station.

6.7.1 Single-Step Position Solution

Processing of triple- and dual-frequency observations does not conceptually differ much. In the triple-frequency case, the complete set of observations consists of the three pseudoranges and three carrier phases. As in the dual-frequency case, the original observables can be processed directly or first transformed into a set of linear independent functions, also called combinations, which may exhibit certain desirable characteristics. Consider the following example set:

$$\left.\begin{aligned}
P_1 &= \rho + I_{1,P} + T + M_{1,P} + \varepsilon_P \\
P_2 &= \rho + \beta_{(0,1,0)} I_{1,P} + T + M_{2,P} + \varepsilon_P \\
P_3 &= \rho + \beta_{(0,0,1)} I_{1,P} + T + M_{3,P} + \varepsilon_P \\
\Phi_1 &= \rho + \lambda_1 N_1 - I_{1,P} + T + M_{1,\Phi} + \varepsilon_\Phi \\
\Phi_{(1,-1,0)} &= \rho + \lambda_{(1,-1,0)} N_{(1,-1,0)} - \beta_{(1,-1,0)} I_{1,P} + T + M_{(1,-1,0),\Phi} + \varepsilon_{(1,-1,0),\Phi} \\
\Phi_{(1,0,-1)} &= \rho + \lambda_{(1,0,-1)} N_{(1,0,-1)} - \beta_{(1,0,-1)} I_{1,P} + T + M_{(1,0,-1),\Phi} + \varepsilon_{(1,0,-1),\Phi}
\end{aligned}\right\} \tag{6.7.1}$$

in which we have used a mixture of traditional notation and new triple-frequency subscript notation. We use the traditional subscript notation, which identifies the frequency by a single subscript when it is convenient and there is no concern of losing clarity. Examples of identity in notation are $\lambda_{(1,0,0)} = \lambda_1$ and $M_{(1,0,0),\Phi} = M_{1,\Phi}$. Checking the definition of auxiliary quantities given in (6.1.62), we readily see that the ionospheric scale factor $\beta_{(1,0,0)} = 1$ and any variance factor μ^2 with one nonzero index equals one.

Since this section exclusively deals with relative positioning between two stations using double differences, we have dropped the subscripts and superscripts that identify stations and satellites and also indicate the differencing operation. For example, we simply use P_1 instead of $P_{km,1}^{pq}$ to identify the pseudorange of the first frequency. In the simplified notation we, therefore, have the following double- differenced quantities: pseudorange P, scaled carrier phase Φ, topocentric satellite distance ρ, ionospheric delay $I_{1,P}$ at the first frequency, tropospheric delay T, integer ambiguity N, multipath M, and measurement noise ε.

In the model (6.7.1) we have chosen the original pseudoranges as observables. As to the carrier phase observation, we selected the extra-wide-lane $\Phi_{(0,1,-1)}$, the wide-lane $\Phi_{(1,-1,0)}$, and the original phase observation on the first frequency, Φ_1. Triple-frequency observations allow for additional combinations, many of which have desirable properties. Any of them can be used as long as the set is linearly independent. In all cases, it is assumed that variance-covariance propagation is fully applied to any functions of the original observables.

When estimating the positions in a network solution or even processing a single baseline, it might be advantageous to group the ambiguity parameters by narrow lane, wide lane, and extra wide lane and apply sequential estimation. With such a grouping of parameters, the required variance-covariance elements for estimating the extra-wide-lane integer ambiguities are conveniently located in the lower right submatrix or the top left submatrix of the full variance-covariance matrix. The ambiguity estimator could be used to identify the extra-wide-lane integer ambiguities and then constrain them. The smaller variance-covariance matrix resulting from implementing the extra wide-lane ambiguity constraints serves as a basis to estimate the wide-lane integer ambiguities. One can again take advantage of the grouping of the wide-lane ambiguity parameters. The number of remaining ambiguities, i.e., the wide lane and narrow lanes, is the same as in the case of traditional dual-frequency processing. After estimating and constraining the wide-lane integer ambiguities, the new variance-covariance matrix that is now even smaller in size, is the basis for estimating the narrow-lane ambiguities. Alternatively, of course, the search algorithm could operate on the full variance-covariance matrix and optimize the sequence of search itself.

The need to minimize the computation load during ambiguity resolution has resulted in a strong desire to estimate the extra-wide-lane ambiguities first. This can be done as described above, i.e., as part of the positioning solution estimate the extra wide lanes first, implement the integer constraints, and then apply the ambiguity estimator to the updated solutions containing less ambiguity parameters, and so on. Alternatively, one can estimate the extra-wide-lane ambiguities independently and prior to the positioning solution. The latter approach is the essence of the TCAR technique to be discussed below.

In the network or baseline solution with model (6.7.1), all correlations between the parameters are considered in the ambiguity resolution by way of utilizing the full variance-covariance matrix. Techniques like LAMBDA are optimal because they operate on the full variance-covariance matrix. Some correlations between parameters are ignored if integer ambiguities, such as the extra wide lanes, are resolved by TCAR techniques prior to the positioning solution. In that sense the one-step solution, which simultaneously searches on all integer ambiguities as part of the positioning solution, is optimal.

The tropospheric and ionospheric effects on the observations are as relevant to triple-frequency observations as they are to dual-frequency observations. These effects cancel in double differencing for a sort baseline per definition. For longer baselines, the tropospheric delay must be either estimated or corrected based on a tropospheric model or mitigated using available external network corrections. The same is true for the ionosphere in principle. However, triple-frequency observations provide the possibility of formulating ionospheric-reduced functions for longer baseline processing when the residual double difference ionosphere can become significant. In fact, observations from three or more frequencies make it possible to create functions of the original observables that to some degree balance noise, virtual wavelength, and ionospheric dependency. Generally speaking, for rapid and successful ambiguity fixing, it is beneficial to have functions that are affected by the ionosphere as little as possible, have a long wavelength relative to the remaining ionospheric delay, and yet exhibit minimal noise amplification.

Cocard et al. (2008) provides a thorough mathematical treatment to identify all phase combinations for GPS frequencies that exhibit the properties of low noise, reduced ionospheric dependency, and acceptable wavelengths. They group the functions (6.1.59) by the sum of the integer indices i, j, and k, i.e., $i+j+k=0$, $i+j+k=\pm 1$, etc. and demonstrate that two functions from the first group are needed, as well as one from another group. Among the many functions identified, only a small subset exhibits the desirable properties; this includes the triplet consisting of the two extra wide lanes (0,1,−1) and (1, −6,5), and the narrow lane (4,0, −3).

Feng (2008) also carries out an extensive investigation to identify the most suitable functions for the GPS, Galileo, and Beidou systems. He generalizes the search by minimizing a condition that not only considers the noise of the original observations but also includes noise factors for residual orbital errors, tropospheric errors, first- and second-order ionospheric errors, and multipath. This total noise is considered a function of the baseline length for a more realistic modeling of uncertainty. Because the GPS, Galileo, and Beidou satellite systems use in part different frequencies, the optimal set of combinations depends on the system. Additionally, the assumptions made for the modeling of the noise as a function of baseline length affects the outcome. He also identifies a number of combinations of interest, among them the three combinations given above for GPS.

Table 6.7.1 provides relevant values for the phase functions used in this section. For other relevant combinations, the reader is referred to the references. The numerical values listed are the wavelength λ, the ionospheric scale factor β, the variance factor μ^2, and a multipath factor ν. The definition of these quantities is given in (6.1.63). All values refer to GPS frequencies. The function (1, −6, 5) is indeed an extra wide lane because its wavelength is 3.258m, and (4,0, −3) is a narrow lane. The relative insensitivity of these two new functions regarding the ionosphere is evidenced from the small ionospheric scale factors of −0.074 and −0.0099. They should, therefore, be good candidates for the processing of longer baselines. However, their variance factors are high because of the close adjacency of the second and third frequencies.

Even though the new extra wide lane functions (1, −6, 5) show a very desirable low ionospheric dependency as compared to the other extra wide lane (0,1, −1), there is still a need for the traditional ionospheric-free function. In fact, with triple-frequency observations, we can formulate several dual-frequency ionospheric-free functions. Of special interest are the triple-frequency ionospheric-free functions that also minimize the variance. Consider the pseudorange and carrier phase functions

$$PC = aP_1 + bP_2 + cP_3 \tag{6.7.2}$$

$$\Phi C = a\Phi_1 + b\Phi_2 + c\Phi_3 \tag{6.7.3}$$

and the conditions of the factors

$$\left.\begin{aligned} a + b + c &= 1 \\ a + \frac{f_1^2}{f_2^2} b + \frac{f_1^2}{f_3^2} c &= 0 \\ a^2 + b^2 + c^2 &= \min \end{aligned}\right\} \tag{6.7.4}$$

TABLE 6.7.1 Selected Triple-Frequency Function Values for GPS Frequencies. The wavelength is in meters.

(i, j, k)	(4,0, −3)	(1,0, −1)	(1, −1,0)	(1, −6,5)	(0,1, −1)
$\lambda_{(i,j,k)}$	0.108	0.752	0.863	3.258	5.865
$\beta_{(i,j,k)}$	−0.0099	−1.339	−1.283	−0.074	−1.719
$\mu^2_{(i,j,k)}$	6.79	24	33	10775	1105
$\nu_{(i,j,k)}$	4	1	8	161	47

It follows from (6.1.59) to (6.1.66) that the first condition preserves the geometric terms, the second condition enforces the function to be ionospheric free, and the third condition minimizes the variance of the function. The third condition assumes that the standard deviations $\sigma_{\Phi_i} = \sigma_\Phi$ and $\sigma_{P_i} = \sigma_P$ are, respectively, the same for all frequencies. The general solution for the coefficients is

$$a = \frac{1 - F_a - F_b + 2F_a F_b}{2(1 - F_a + F_b^2)} \qquad b = F_a - aF_b \qquad c = 1 - a - b \tag{6.7.5}$$

$$F_a = \frac{f_1^2}{f_1^2 - f_3^2} \qquad F_b = \frac{f_1^2(f_2^2 - f_3^2)}{f_2^2(f_1^2 - f_3^2)} \tag{6.7.6}$$

For GPS frequencies, we have $a = 2.3269$, $b = -0.3596$, and $c = -0.9673$. These computed functions can be written in the standard form

$$PC = \rho + c\,dt - c\,d\bar{t} + T + \delta_{PC} + \varepsilon_{PC} \tag{6.7.7}$$

$$\Phi C = \rho + R + c\,dt - c\,d\bar{t} + T + \delta_{\Phi C} + \varepsilon_{\Phi C} \tag{6.7.8}$$

with $R = a\lambda_1 N_1 + b\lambda_2 N_2 + c\lambda_3 N_3$. The respective standard deviations can be computed as $\sigma_{PC} = 2.545\sigma_P$ and $\sigma_{\Phi C} = 2.545\sigma_\Phi$. Please note that the derivation (6.7.2) to (6.7.8) as presented refers to undifferenced observations. When viewed as double differences, the only changes are in (6.7.7) and (6.7.8), i.e., deletion of the clock terms, and the replacement of δ_{PC} with M_{PC} and $\delta_{\Phi C}$ with $M_{\Phi C}$. The ΦC function is presented in Hatch (2006), including the general form of the solution coefficients. Using only the first two conditions of (6.7.4) leads to geometry-free and ionospheric-free (GIF) solutions, which are popular in dual-frequency processing.

6.7.2 Geometry-Free TCAR

The idea behind the TCAR approach is to find three carrier phase linear combinations that allow integer ambiguity resolution in three consecutive steps. In a fourth step, the resolved integer ambiguities are considered known when estimating the receiver position in a geometry-based solution. One can either use the estimated integer combination or transform them to original ambiguities and use the latter in the position computation. This transformation must, of course, preserve the integer nature that imposes some restrictions on admissible combinations. Consider the following example of the transformations:

$$\begin{bmatrix} 0 & 1 & -1 \\ 1 & -6 & 5 \\ 4 & 0 & -3 \end{bmatrix} \begin{bmatrix} N_1 \\ N_2 \\ N_3 \end{bmatrix} = \begin{bmatrix} N_{(0,1,-1)} \\ N_{(1,-6,5)} \\ N_{(4,0,3)} \end{bmatrix} \tag{6.7.9}$$

$$\begin{bmatrix} N_1 \\ N_2 \\ N_3 \end{bmatrix} = \begin{bmatrix} -18 & -3 & 1 \\ -23 & -4 & 1 \\ -24 & -4 & 1 \end{bmatrix} \begin{bmatrix} N_{(0,1,-1)} \\ N_{(1,-6,5)} \\ N_{(4,0,-3)} \end{bmatrix} \tag{6.7.10}$$

For the original ambiguities to be integers, it is necessary that the elements of the matrix on the left side of (6.7.9) are integers and that the determinant is either plus or minus one. These conditions can readily be explained by computing the matrix inverse. Equation (A.3.4) shows a general way to compute the inverse. If the elements of the matrix are integers, then the cofactor matrix also contains integers, and if the determinant located in the denominator is plus or minus one, then the elements of the inverse matrix must be integers.

There are two approaches to TCAR. The first to be discussed is the geometry-free approach (GF-TCAR), in which the functions do not contain the topocentric satellite distance and the tropospheric delay. In the second approach, the geometry-based (GB-TCAR) approach, the topocentric satellite distance and tropospheric delay are present in the equations. However, the topocentric distance is not parameterized in terms of station coordinates. For each of these approaches, the double differences are processed separately and the respective ambiguity is determined by simple rounding in a sequential solution. Both approaches begin by resolving the extra-wide-lane ambiguity, proceed with estimating the wide-lane ambiguity, and then resolving the narrow-lane ambiguity. In deviation from the original idea of TCAR, which calls for consecutive estimation of these ambiguities, one can readily combine two or even all three steps into one solution.

We first review the geometry-free solutions in the context of dual-frequency observations. This type of a solution approach has been frequently used even during the time when only dual-frequency observations were available. For example, already Goad (1990) and Euler and Goad (1991) use the geometry-free model to study optimal filtering for the combined pseudorange and carrier phase observations for single and dual frequencies. We will discuss this model to demonstrate the reduction in correlation between estimated ambiguities due to wide-laning, and clarify the term extra-wide-laning as used traditionally during the dual-frequency era and its use today in connection with triple-frequency processing.

Taking the undifferenced pseudorange and carrier phase equations (6.1.28) and (6.1.32), carrying out the double differencing, and dropping the subscripts and superscripts that identify stations and satellites, the dual-frequency pseudoranges and carrier phases are written in the form

$$\begin{bmatrix} P_1 \\ P_2 \\ \Phi_1 \\ \Phi_2 \end{bmatrix} = \begin{bmatrix} 1 & 1 & 0 & 0 \\ 1 & \gamma_{12} & 0 & 0 \\ 1 & -1 & \lambda_1 & 0 \\ 1 & -\gamma_{12} & 0 & \lambda_2 \end{bmatrix} \begin{bmatrix} \rho + \Delta \\ I_{1,P} \\ N_1 \\ N_2 \end{bmatrix} + \begin{bmatrix} M_{1,P} \\ M_{2,P} \\ M_{1,\Phi} \\ M_{2,\Phi} \end{bmatrix} + \begin{bmatrix} \varepsilon_{1,P} \\ \varepsilon_{2,P} \\ \varepsilon_{1,\Phi} \\ \varepsilon_{2,\Phi} \end{bmatrix} \tag{6.7.11}$$

The auxiliary parameter Δ includes the tropospheric delay, and in case of undifferenced equations it includes also the clock corrections and hardware delays of receiver and satellite. Other parameters are the ionospheric delay $I_{1,P}$, and the ambiguities N_1 and N_2. The factor γ_{12} is given in (6.1.1). The parameters $\rho + \Delta$ and $I_{1,P}$ change with time, but the ambiguity parameters are constant unless there are cycle slips. Equation (6.7.11) is called the geometry-free model; it is valid for static or moving receivers and is readily applicable to estimations with recursive LSQ having a set of constants and a set of epoch parameters to be estimated.

Dropping the multipath term as usual, the matrix form of (6.7.11) is $\ell_b = \mathbf{Ax} + \varepsilon$. The $\mathbf{A}$ matrix contains constants that do not depend on the receiver-satellite geometry. Since the matrix has full rank, the parameters can be expressed as a function of observations, i.e., $\mathbf{x} = \mathbf{A}^{-1}\ell_b$. Applying variance-covariance propagation, one obtains $\Sigma_x = \mathbf{A}^{-1}\Sigma_{\ell_b}(\mathbf{A}^{-1})^T$. Next we consider the linear transformation $\mathbf{z} = \mathbf{Zx}$, with

$$\mathbf{Z} = \begin{bmatrix} 1 & 0 & 0 & 0 \\ 0 & 1 & 0 & 0 \\ 0 & 0 & 1 & -1 \\ 0 & 0 & 1 & 0 \end{bmatrix} \tag{6.7.12}$$

with variance-covariance matrix $\Sigma_z = \mathbf{Z}\Sigma_x\mathbf{Z}^T$. The new variables of $\mathbf{z}$ are $\rho + \Delta, I_{1,P}, N_{12}$, and N_1, with the wide-lane ambiguity being $N_{12} = N_1 - N_2$.

For numerical computations, we assume that the standard deviation of the carrier phases $\sigma_{1,\varphi}$ and $\sigma_{2,\varphi}$ are related as $\sigma_{2,\Phi} = \sigma_{1,\Phi}\sqrt{\gamma_{12}}$, and that the standard deviations of the pseudorange and carrier phases follow the relation $k = \sigma_P/\sigma_\Phi$ for both frequencies, where k is a constant. Assuming further that the observations are uncorrelated, the covariance matrix of the observations consists of diagonal elements $(k^2, \gamma_{12}k^2, 1, \gamma_{12})$, and is scaled by $\sigma^2_{1,\Phi}$. If we set k equal to 154, which corresponds to the ratio of the L1 GPS frequency and the P-code chipping rate and use $\sigma_{1,\Phi} = 0.002\,\text{m}$, then the standard deviations and the correlation matrix are, respectively,

$$\left(\sigma_{\rho+\Delta}, \sigma_I, \sigma_{1,N}, \sigma_{2,N}\right) = \left(0.99\ \text{m}, 0.77\ \text{m}, 9.22\ \text{cycL}_1, 9.22\ \text{cycL}_2\right)$$

$$\begin{bmatrix} \sigma_{\rho+\Delta} \\ \sigma_{I_{1,P}} \\ \sigma_{N_1} \\ \sigma_{N_2} \end{bmatrix} = \begin{bmatrix} 0.99 \\ 0.77 \\ 9.22 \\ 9.22 \end{bmatrix} \quad \mathbf{C}_x = \begin{bmatrix} 1 & -0.9697 & -0.9942 & -0.9904 \\ & 1 & 0.9904 & 0.9942 \\ & & 1 & 0.9995 \\ \text{sym} & & & 1 \end{bmatrix} \tag{6.7.13}$$

$$\left(\sigma_{\rho+\Delta}, \sigma_I, \sigma_w, \sigma_{1,N}\right) = \left(0.99\ \text{m}, 0.77\ \text{m}, 0.28\ \text{cycL}_\text{w}, 9.22\ \text{cycL}_1\right)$$

$$\begin{bmatrix} \sigma_{\rho+\Delta} \\ \sigma_{I_{1,P}} \\ \sigma_{N_{12}} \\ \sigma_{N_1} \end{bmatrix} = \begin{bmatrix} 0.99 \\ 0.77 \\ 0.28 \\ 9.22 \end{bmatrix} \quad \mathbf{C}_z = \begin{bmatrix} 1 & -0.9697 & -0.1230 & -0.9942 \\ & 1 & 0.1230 & 0.9904 \\ & & 1 & 0.0154 \\ \text{sym} & & & 1 \end{bmatrix} \tag{6.7.14}$$

Striking features of the epoch solution (6.7.13) are the equality of the standard deviation for both ambiguities with the number of digits given, and the high correlation between all parameters. Of particular interest is the shape and orientation of the ellipse of standard deviation for the ambiguities. The general expressions (2.7.79) to (2.7.83) can be applied to the third and fourth parameters. They could be drawn with respect to the perpendicular N_1 and N_2 axes, which carry the units L1 cycles and

L2 cycles. The computations show that the ellipse almost degenerates into a straight line with an azimuth of 45°, the semiminor and semimajor axes being 0.20 and 13.04, respectively.

The correlation matrix (6.7.14) shows a small correlation of 0.0154 between the wide-lane ambiguity and the L1 ambiguity. Furthermore, the correlations between the wide-lane ambiguity and both the topocentric distance and the ionospheric parameter have been reduced significantly. Considering the small standard deviation for the wide-lane ambiguity of 0.28 and the low correlations with other parameters, it seems feasible to estimate the wide-lane ambiguity from epoch solutions. The semiaxes of the ellipse of standard deviation for the ambiguities are 9.22 and 0.28, respectively. The azimuth of the semimajor axis with respect to the N_{12} axis is 89.97°, i.e., the ellipse is elongated along the N_1 direction. The correlation matrix still shows high correlations between N_1 and the ionosphere and topocentric distance, indicating that the estimation of the N_1 ambiguity is not that straightforward and will require a long observation set. If we consider the square root of the determinant of the covariance matrix to be a single number that measures correlation, then $(|\mathbf{C}_z|/|\mathbf{C}_x|)^{1/2} \approx 33$ implies a major decorrelation of the epoch parameters.

Assume that the double-difference wide-lane ambiguity has been fixed using the HMW function (6.1.48), which is implied in (6.7.11), then $AC2$ of (6.1.56) allows computation of the L1 double-difference ambiguity as

$$N_1 = \varphi_1 + \frac{f_1}{f_1 - f_2}(N_{12} - \varphi_{12}) + \frac{f_1 - f_2}{f_2} I_{1,\varphi} + M_{AC2} \approx \varphi_1 + 4.5[N_{12} - \varphi_{12}] + \cdots \tag{6.7.15}$$

Fortunately, this expression does not depend on the large pseudorange multipath terms, but only on the smaller carrier phase multipath. Given the GPS frequencies f_1 and f_2, and assuming that the wide-lane ambiguity has been incorrectly identified within 1 lane, then the computed L1 ambiguity changes by 4.5 cycles. The first decimal of the computed L1 ambiguity would be close to 5. However, since the L1 ambiguity is an integer, we can use that fact to decide between two candidate wide-lane ambiguities. This procedure is known as extra-wide-laning (Wübbena, 1990). It was an important tool in the dual-frequency era that helped to shorten the time of successful ambiguity fixing. It is important to note, however, that in triple-frequency processing, the terms extra-wide-laning or extra-wide-lane ambiguity refer to any dual- or triple-frequency frequency function whose corresponding wavelength is larger than the legacy dual-frequency wavelength $\lambda_{(1,-1,0)}$.

In the subsequent sections, we provide one or several algorithms for the resolution of the extra-wide-lane, wide-lane, and narrow-lane ambiguity and briefly discuss distinguishing properties regarding ionospheric dependency, formal standard deviation of the computed ambiguity, and multipath magnification. In order to provide numerical values to approximately judge the quality of the various solutions, we assume a standard deviation of 0.002 and 0.2 m for the carrier phase and pseudorange measurement, respectively.

6.7.2.1 *Resolving EWL Ambiguity* The extra-wide-lane (EWL) ambiguity $N_{(0,1,-1)}$ is easy to compute, possibly even in a single epoch. This is a direct result of the given pseudorange and carrier phase measurement accuracies, as well as the closeness of the GPS second and third frequencies. Two solutions are discussed. The first solution shows a reduced ionospheric dependency and the second solution is ionospheric free. Just to be sure, the expression "extra wide lane" as used here in connection with triple-frequency observations is not to be confused with extra-wide-laning as used in connection with (6.7.15).

Differencing $\boldsymbol{\Phi}_{(0,1,-1)}$ and $\boldsymbol{P}_2$: This solution differences the extra wide lane and the pseudorange. Differencing (6.1.65) and (6.1.66) gives the function

$$\Phi_{(0,1,-1)} - P_2 = \lambda_{(0,1,-1)} N_{(0,1,-1)} - (\beta_{(0,1,-1)} + \beta_{(0,1,0)}) I_{1,P} + M + \varepsilon \tag{6.7.16}$$

The hardware delay terms cancel as part of the double differencing. The symbol M, without any subscript or superscript, denotes the total double-differenced multipath of the function. The multipath of the pseudorange is the dominating part, i.e., $M \approx M_P$. Similarly, the symbol ε denotes the random noise of the function. Rearranging the equation to solve for the ambiguity gives

$$N_{(0,1,-1)} = \frac{\Phi_{(0,1,-1)} - P_2}{\lambda_{(0,1,-1)}} + \frac{\beta_{(0,1,-1)} + \beta_{(0,1,0)}}{\lambda_{(0,1,-1)}} I_{1,P} - \frac{M + \varepsilon}{\lambda_{(0,1,-1)}} \tag{6.7.17}$$

The EWL ambiguity solution still depends on the ionosphere because the factor of $I_{1,P}$ in (6.7.17), denoted henceforth as β_N, equals -0.012. A double-difference ionosphere of 1 m falsifies the ambiguity by merely one hundredth of an extra-wide-lane cycle. A similarly good ionospheric reduction is achieved if one were to use the third pseudorange P_3 instead of P_2.

Assuming that the carrier phases are stochastically independent and have the same variance, and assuming a similar property for the statistics of the pseudoranges (although in this particular case there is only one pseudorange used), and then applying variance propagation following (6.1.67), the variance of the EWL ambiguity is

$$\sigma_N^2 = \frac{\mu_{(0,1,-1)}^2 \sigma_\Phi^2 + \sigma_P^2}{\lambda_{(0,1,-1)}^2} = 32\sigma_\Phi^2 + 0.029\sigma_P^2 \tag{6.7.18}$$

Equation (6.7.18) results merely from propagation of stochastic independent random errors and does not reflect the multipath. The relatively large factor of 32 of the carrier phase variance is caused by the close location of the second and third frequency. As stated above, taking 0.002 and 0.2 m as standard deviations for the carrier phase and pseudorange, respectively, the formal standard deviation of the EWL ambiguity is $\sigma_N = 0.036$ (extra wide lanes).

The propagation of the multipath is more complicated. It is essentially unpredictable since it depends on each individual carrier phase and pseudorange as well as time (because the reflection geometry is a function of time). While the multipath is a perpetually worrisome unknown in precise positioning, its impact on the calculation of the ambiguity is significantly reduced in this particular case because of the EWL wavelength $\lambda_{(0,1,-1)}$ in the denominator of (6.7.17). The multipath effect on the ambiguity is $M_N \leq 0.17M$, where M is the multipath of $\Phi_{(0,1,-1)} - P_2$ and thus itself a function of M_Φ and M_P. Using factor $\nu_{(i,j,k)}$ of (6.1.63), one can compute the maximum value by adding the absolute values of phase and pseudorange combinations multipath. In this particular case, it actually is sufficient to approximate $M \approx M_P$ since $M_P \gg M_\Phi$, thus simply obtain $M_N \leq 0.17M_P$. There is no additional scaling since only one pseudorange is involved (and not a pseudorange combination).

We conclude that function (6.7.16) is a good candidate for estimating the EWL ambiguity because the ionospheric impact is small, the formal standard deviation of the ambiguity is low, and the multipath is significantly reduced.

Applying the HMW Function to Second and Third Frequency: An alternative way of computing the EWL ambiguity follows directly from (6.1.48) when applied to the second and third frequency:

$$\Phi_{(0,1,-1)} - P_{(0,1,1)} = \lambda_{(0,1,-1)} N_{(0,1,-1)} + M + \varepsilon \tag{6.7.19}$$

$$N_{(0,1,-1)} = \frac{\Phi_{(0,1,-1)} - P_{(0,1,1)}}{\lambda_{(0,1,-1)}} - \frac{M + \varepsilon}{\lambda_{(0,1,-1)}} \tag{6.7.20}$$

$$\sigma_N^2 = \frac{\mu^2_{(0,1,-1)}\sigma_\Phi^2 + \mu^2_{(0,1,1)}\sigma_P^2}{\lambda^2_{(0,1,-1)}} = 32\sigma_\Phi^2 + 0.015\sigma_P^2 \tag{6.7.21}$$

This solution is ionospheric free as to first-order ionospheric effects on the observation. The standard deviation is very close to the one determined for the previous solution, and the multipath is reduced by the same factor. Therefore, both solution approaches are essentially equivalent, although one might intuitively prefer the ionospheric free solution to alleviate any concerns about the ionosphere.

6.7.2.2 Resolving the WL Ambiguity Three solutions are discussed for resolving the wide-lane (WL) ambiguity. The significance of the ionospheric delay becomes more apparent as seen from the first solution presented. The second solution uses the HMW function applied to the first and second frequency, as has been traditionally done in the dual-frequency case. The third solution represents one of the modern approaches, which is ionospheric free and minimizes the variance.

Differencing $\overline{\Phi}_{(0,1,-1)}$ and $\Phi_{(1,-1,0)}$: Knowing the EWL integer ambiguity, we can readily write the ambiguity-corrected carrier phase extra wide lane as

$$\overline{\Phi}_{(0,1,-1)} = \Phi_{(0,1,-1)} - \lambda_{(0,1,-1)} N_{(0,1,-1)} \tag{6.7.22}$$

Subtracting the wide-lane carrier phase function from the ambiguity-corrected function gives

$$\overline{\Phi}_{(0,1,-1)} - \Phi_{(1,-1,0)} = -\lambda_{(1,-1,0)} N_{(1,-1,0)} + (-\beta_{(0,1,-1)} + \beta_{(1,-1,0)}) I_{1,P} + M + \varepsilon \tag{6.7.23}$$

$$N_{(1,-1,0)} = \frac{-\overline{\Phi}_{(0,1,-1)} + \Phi_{(1,-1,0)}}{\lambda_{(1,-1,0)}} + \frac{-\beta_{(0,1,-1)} + \beta_{(1,-1,0)}}{\lambda_{(1,-1,0)}} I_{1,P} + \frac{M + \varepsilon}{\lambda_{(1,-1,0)}} \tag{6.7.24}$$

$$\sigma_N^2 = \frac{\mu^2_{(0,1,-1)} + \mu^2_{(1,-1,0)}}{\lambda^2_{(1,-1,0)}} \sigma_\Phi^2 = (1485 + 44)\sigma_\Phi^2 = 1529\sigma_\Phi^2 \tag{6.7.25}$$

The ionospheric factor $\beta_N = 0.505$ is relatively large. An ionosphere of 1 m causes a change of one-half of the WL ambiguity. The standard deviation of the WL ambiguity is $\sigma_N = 39\sigma_\Phi$. Notice that the variance factor of the extra wide lane is much larger than that of the wide lane because of the relative closeness of the second and third frequencies. There is also a similarly unequal contribution to the total multipath which is $M_N \leq 64M_\Phi$, applying again the multipath factor of (6.1.68). There is no pseudorange multipath because the function does not include pseudoranges. The technique is best suited for short baselines, due to the residual impact of the ionosphere.

The above simple quality measures make clear that WL resolution should be expected to be more difficult than the EWL resolution. More observations will need to be taken over a longer period of time to reduce the noise to be able to identify the correct integer of the ambiguity. Unfortunately, when observations are taken over a longer period of time, the multipath variations can become a major concern.

Applying HMW Function to the First and Second Frequency: The HMW function provides an attractive alternative to the previous approach. We can readily write

$$N_{(1,-1,0)} = \frac{\Phi_{(1,-1,0)} - P_{(1,1,0)}}{\lambda_{(1,-1,0)}} - \frac{M + \varepsilon}{\lambda_{(1,-1,0)}} \tag{6.7.26}$$

$$\sigma_N^2 = \frac{\mu^2_{(1,-1,0)}\sigma_\Phi^2 + \mu^2_{(1,1,0)}\sigma_P^2}{\lambda^2_{(1,-1,0)}} = 44\sigma_\Phi^2 + 0.682\sigma_P^2 \tag{6.7.27}$$

This estimate of the wide-lane ambiguity is free of ionospheric effects and even has a good formal standard deviation of $\sigma_N = 0.17$, assuming the default values. As with any of the HMW functions, equation (6.7.26) contains a potentially large pseudorange multipath that is even slightly amplified because the wide-lane wavelength is less than 1 m, i.e., $M_N \leq 1.2M_\Phi$.

Ionospheric-Reduced and Minimum Variance: Zhao et al. (2014) propose to minimize the variance of the sum of the scaled pseudoranges and ambiguity-corrected extra wide lane. In addition, they introduce an adaptive factor that scales the ionospheric effect from zero (ionospheric free) to higher values that might relate to longer baselines. Consider

$$aP_1 + bP_2 + cP_3 + d\overline{\Phi}_{(0,1-1)} - \Phi_{(1-1,0)} = -\lambda_{(1-1,0)}N_{(1-1,0)} + \beta I_{1,P} + M + \varepsilon \tag{6.7.28}$$

$$\beta = a + b\beta_{(0,1,0)} + c\beta_{(0,0,1)} - d\beta_{(0,1,-1)} + (1+\kappa)\beta_{(1-1,0)} \tag{6.7.29}$$

$$N_{(1-10)} = \frac{-aP_1 - bP_2 - cP_3 - d\overline{\Phi}_{(0,1,-1)} + \Phi_{(1,-1,0)}}{\lambda_{(1,-1,0)}} + \frac{\beta}{\lambda_{(1,-1,0)}} I_{1,P} + \frac{M + \varepsilon}{\lambda_{(1,-1,0)}} \tag{6.7.30}$$

$$\left.\begin{aligned} a+b+c+d &= 1 \\ \beta &= 0 \\ (a^2+b^2+c^2)\sigma_P^2 + d^2\mu^2{}_{(0,1,-1)}\sigma_\Phi^2 &= \min \end{aligned}\right\} \tag{6.7.31}$$

The first condition in (6.7.31) assures the geometry-free part, i.e., the topocentric satellite distance and the tropospheric delay terms cancel. The second condition, which includes the adoptive factor κ, enforces the function (6.7.28) to be ionospheric free even if $\kappa = 0$. The idea is to increase κ with baseline length to allow a residual double-differenced ionosphere of $k\beta_{(1,-1,0)}\ I_{1,P}$. The third condition implies minimum variance, assuming (as is done throughout this section) that the variances of the three pseudoranges are the same. The conditions (6.7.31) together imply the need to compute a new set of coefficients (a, b, c, d) for every κ.

For the special case of $\kappa = 0$, one obtains $a = 0.5938$, $b = 0.0756$, $c = -0.0416$, and $d = 0.3721$, and the variance of the wide-lane ambiguity becomes

$$\sigma_N^2 = \frac{(a^2+b^2+c^2)\sigma_P^2 + (d^2\mu^2{}_{(0,1,-1)} + \mu^2_{(1,-1,0)})\sigma_\Phi^2}{\lambda^2_{(1,-1,0)}} = 0.484\sigma_P^2 + 250\sigma_\Phi^2 \tag{6.7.32}$$

Using again the default standard deviations for pseudoranges and carrier phases, we get $\sigma_N = 0.143$ for this ionospheric-free case. The maximum multipath $M_N \leq 0.82M_P + 29.6M_\Phi$ contains a significant phase contribution that comes from the EWL component. For a discussion on the cases of $\kappa \neq 0$, please see Zhao et al. (2014).

6.7.2.3 Resolving the NL Ambiguity Three solutions are presented for resolving the narrow-lane (NL) ambiguity N_3. All three solutions rely on ambiguity-corrected wide-lane carrier phase observations. The first one is also applicable to dual-frequency applications to resolve N_1 and shows a strong dependency in the ionospheric delay. The second and third solutions are of the ionospheric-free type and are characterized, as one would expect, by a very high standard deviation and multipath factor.

Differencing Ambiguity-Corrected WL and Original Phase: Since the wide-lane ambiguity is now known, we can compute

$$\overline{\Phi}_{(1,-1,0)} - \Phi_3 = -\lambda_3 N_3 + (-\beta_{(1,-1,0)} + \beta_{(0,0,1)})I_{1,P} + M + \varepsilon \tag{6.7.33}$$

$$N_3 = \frac{-\overline{\Phi}_{(1,-1,0)} + \Phi_3}{\lambda_3} + \frac{(-\beta_{(1,-1,0)} + \beta_{(0,0,1)})}{\lambda_3} I_{1,P} - \frac{M+\varepsilon}{\lambda_3} \tag{6.7.34}$$

$$\sigma_N^2 = \frac{\mu^2_{(1,-1,0)} + 1}{\lambda_3^2}\sigma_\Phi^2 = 522\sigma_\Phi^2 \tag{6.7.35}$$

The ionospheric factor of $\beta_N = 12.06$ causes a large amplification of the residual ionospheric carrier phase delays. The standard deviation of the ambiguity is $\sigma_N = 23\sigma_\Phi$. Even if (6.7.33) were differenced with respect to Φ_1 or Φ_2, the standard deviation would not change significantly because the wide lane is the largest contributor. The multipath is $M_N \leq 36M_\Phi$. Once the N_3 ambiguity is available, the other original ambiguities follow from

$$\begin{aligned} N_2 &= N_{(0,1,-1)} + N_3 \\ N_1 &= N_{(1,-1,0)} + N_2 \end{aligned} \tag{6.7.36}$$

Because of the high ionospheric dependency, this approach works best for short baselines.

In support of this approach, one might consider estimating the ionosphere. Differencing the ambiguity-corrected extra wide lane and wide lane gives

$$\overline{\Phi}_{(0,1,-1)} - \overline{\Phi}_{(1,-1,0)} = (-\beta_{(0,1,-1)} + \beta_{(1,-1,0)})I_{1,P} + M + \varepsilon \tag{6.7.37}$$

$$I_{1,P} = \frac{\overline{\Phi}_{(0,1,-1)} - \overline{\Phi}_{(1,-1,0)}}{-\beta_{(0,1,-1)} + \beta_{(1,-1,0)}} - \frac{M + \varepsilon}{-\beta_{(0,1,-1)} + \beta_{(1,-1,0)}} \tag{6.7.38}$$

$$\sigma^2_{I_{1,P}} = \frac{\mu^2_{(0.1,-1)} + \mu^2_{(1,-1,0)}}{(-\beta_{(0,1,-1)} + -\beta_{(1,-1,0)})^2}\sigma^2_\Phi = 6008\sigma^2_\Phi \tag{6.7.39}$$

The respective numerical quality values are $\sigma_1 = 78\sigma_\Phi$ and $M_1 \leq 126M_\Phi$. Applying the variance propagation to (6.7.34) for the carrier phases and the ionospheric delay to take both the observational noise and the uncertainty of the computed ionosphere into account, gives

$$\sigma^2_N = (522 + 12.06^2 \cdot 6008)\sigma^2_\Phi \tag{6.7.40}$$

with $\sigma_N = 935\sigma_\Phi$. This high standard deviation clearly indicates that computing the ionospheric delay first and then using it in (6.7.34) results in a high uncertainty for the ambiguity.

An alternative way of computing the ionosphere is equation (6.1.57), which uses the original observations explicitly. In the traditional notation the function is

$$\begin{aligned} AC3 &\equiv (\lambda_{12} - \lambda_{13})\varphi_1 - \lambda_{12}\varphi_2 + \lambda_{13}\varphi_3 - N_{12}\lambda_{12} + N_{13}\lambda_{13} \\ &= \left(\sqrt{\gamma_{12}} - \sqrt{\gamma_{13}}\right)I_{1,P} + M + \varepsilon \end{aligned} \tag{6.7.41}$$

However, this equation is identical to (6.7.37) after appropriate scaling, and therefore, does not offer a better way for computing the ionospheric delay. Given the significant impact of the residual ionosphere when computing the original ambiguity N_3, it is tempting to look again for an ionospheric-free solution.

WL Ambiguity-Corrected Triple-Frequency Phases: A possible candidate for computing the first ambiguity is equation (6.1.53), after all it is an ionospheric-free and geometry-free function. Solving the equation for N_1, the general form can be written as

$$N_1 = a\Phi_1 + b\Phi_2 + c\Phi_3 + dN_{12} + eN_{13} + M + \varepsilon \tag{6.7.42}$$

The ambiguity N_{13} is obtained from the previously resolved WL and EWL ambiguities as $N_{13} = N_{12} + N_{23}$. The numerical phase factors are $a = -143$, $b = -777$, and

$c = -634$. These values translate into a very large standard deviation of $\sigma_N = 1013\sigma_\Phi$ for the ambiguity and large multipath magnification of $M_N \leq 1554 M_\Phi$. These extraordinarily large values do not change even if (6.7.42) were to be formulated in terms of N_{23} instead of N_{13}.

Ionospheric-Free Function with Ambiguity-Corrected EWL and WL: Another approach to compute an ionospheric-free function is to utilize both ambiguity-corrected EWL and WL functions. Consider

$$a\overline{\Phi}_{(0,1,-1)} + b\overline{\Phi}_{(1,-1,0)} - \Phi_3 = -\lambda_3 N_3 + (-a\beta_{(0,1,-1)} - b\beta_{(1,-1,0)} + \beta_{(0,0,1)})I_{1,P} + M + \varepsilon \tag{6.7.43}$$

$$N_3 = \frac{-a\overline{\Phi}_{(0,1,-1)} - b\overline{\Phi}_{(1,-1,0)} + \Phi_3}{\lambda_3} + \beta_N I_{1,P} + \frac{M + \varepsilon}{\lambda_3} \tag{6.7.44}$$

$$\sigma^2_{N_3} = \frac{a^2\mu^2_{(0,1,-1)} + b^2\mu^2_{(1,-1,0)} + 1}{\lambda_3}\sigma^2_\Phi \tag{6.7.45}$$

$$\left.\begin{aligned} a + b &= 1 \\ a\beta_{(0,1,-1)} + b\beta_{(1,-1,0)} - \beta_{(0,0,1)} &= 0 \end{aligned}\right\} \tag{6.7.46}$$

$$a = \frac{\beta_{(0,0,1)} - \beta_{(1,-1,0)}}{\beta_{(0,1,-1)} - \beta_{(1,-1,0)}} \qquad b = 1 - a \tag{6.7.47}$$

The first condition in (6.7.46) assures a geometry-free solution, and the second condition makes the solution ionospheric free. The first phase factor is $a = -7.07$. The standard deviation and the multipath can be computed as $\sigma_N = 2818\sigma_\Phi$ and $M_N \leq 4686 M_\Phi$. These values are very high and render the solution to be of questionable value. Li et al. (2010) carried out the solution for the combination (0,1,−1), (1,−6,5), and (4,0,−3), resulting in $a = -0.039$, $\sigma_N = 997\sigma_\Phi$, and $M_N \leq 1596 M_\Phi$.

6.7.3 Geometry-Based TCAR

The geometric terms, such as the topocentric satellite distance and the tropospheric delay, are included in the solution. In order to make a solution possible, the mathematical model includes separate equations for pseudoranges and carrier phases. Because the tropospheric delay is explicitly included, it might be necessary to model or estimate this delay for long baselines. For short baselines, the tropospheric delay is lumped with the topocentric satellite distance. The goal of geometry-based TCAR (GB-TCAR) is still to estimate the integer ambiguities first in three separate steps and then estimate the position coordinates in a fourth step. The topocentric satellites distance is, therefore, not parameterized in terms of coordinates.

Even in the case of GB-TCAR, one also prefers to compute the extra wide-lane ambiguity $N_{(0,1,-1)}$ according to (6.7.19) using the HMW function. This is done because this function can be easily applied and works well. Instead of resolving the wide lanes $N_{(0,1,-1)}$ or $N_{(1,0,-1)}$ next, which certainly could be done, we resolve the extra wide lane $N_{(1,-6,5)}$, and then the narrow lane $N_{(4,0,-3)}$.

Resolving $\boldsymbol{N}_{(1,-6,5)}$: Consider the following model:

$$\left.\begin{aligned} PC &= \rho' + M_{PC} + \varepsilon_{PC} \\ \Phi_{(1,-6,5)} &= \rho' + \lambda_{(1,-6,5)} N_{(1,-6,5)} - \beta_{(1,-6,5)} I_{1,P} + M_{(1,-6,5)} + \varepsilon_{(1,-6,5),\Phi} \end{aligned}\right\} \tag{6.7.48}$$

where PC is the triple-frequency function (6.7.7) that is ionospheric free and minimizes the variances. The tropospheric delay is lumped with the topocentric satellite distance as $\rho' = \rho + T$. The standard deviation of the optimized pseudorange function is given above as $\sigma_{PC} = 2.545\sigma_P$, and the multipath is $M_{PC} = 3.6M_P$. Instead of PC, one could also use an ionospheric-free dual-frequency function because their standard deviations are not much larger. The ionospheric factor for the phase combination is $\beta_{(1,-6,5)} = -0.074$, making this function suitable for processing of long baselines. The standard deviation and the multipath are $\sigma_{(1,-6,5),\Phi} = 104\sigma_\Phi$ and $M_{(1,-6,5),\Phi} = 161M_\Phi$. The wavelength is $\lambda_{(1,-6,5)} = 3.258$ which, according to our adopted convention, actually is an EWL and not a WL function.

The model contains two types of parameters, the lumped parameter ρ' which is estimated for each epoch, and the ambiguity parameter which is constant as long as there is no cycle slip. Once the integer ambiguities have been determined, we can compute the traditional WL and EWL ambiguities as

$$\begin{bmatrix} N_{(1,-1,0)} \\ N_{(1,0,-1)} \end{bmatrix} = \begin{bmatrix} 1 & 5 \\ 1 & 6 \end{bmatrix} \begin{bmatrix} N_{(1,-6,5)} \\ N_{(0,1,-1)} \end{bmatrix} \tag{6.7.49}$$

which can then be used as known quantities.

Resolving the Narrow-Lane Ambiguity: Since $N_{(0,1,-1)}$, $N_{(1,-6,5)}$, and $N_{(1,0,-1)}$ are known, the narrow-lane ambiguity can be computed using one of the wide-lane functions as

$$\left.\begin{aligned} \overline{\Phi}_{(1,0,-1)} &= \rho' - \beta_{(1,0,-1)} I_{1,P} + M_{(1,0,-1)} + \varepsilon_{(1,0,-1),\Phi} \\ \Phi_{(4,0,-3)} &= \rho' + \lambda_{(4,0,-3)} N_{(4,0,-3)} - \beta_{(4,0,-3)} I_{1,P} + M_{(4,0,-3)} + \varepsilon_{(4,0,-3),\Phi} \end{aligned}\right\} \tag{6.7.50}$$

Several variations are possible. For example, instead of using the extra wide lane $\Phi_{(1,0,-1)}$, one might consider $\Phi_{(1,-6,5)}$. The latter provides more reduction in ionospheric impact at the expense of potentially a higher multipath.

6.7.4 Integrated TCAR

The various steps of GB-TCAR can, of course, be combined into one step which is referred to as an integrated TCAR (Vollath et.al., 1998). This model uses all observations simultaneously. In this example,

$$\left.\begin{aligned} P_1 &= (\rho + T) + I_{1,P} + M_{1,P} + \varepsilon_P \\ P_2 &= (\rho + T) + \beta_{(0,1,0)} I_{1,P} + M_{2,P} + \varepsilon_P \\ P_3 &= (\rho + T) + \beta_{(0,0,1)} I_{1,P} + M_{3,P} + \varepsilon_P \\ \Phi_1 &= (\rho + T) + \lambda_1 N_1 - I_{1,P} + M_{1,\Phi} + \varepsilon_\Phi \\ \Phi_{(4,0,-3)} &= (\rho + T) + \lambda_{(4,0,-3)} N_{(4.0.-3)} - \beta_{(4,0,-3)} I_{1,P} + M_{(4,0,-3),\Phi} + \varepsilon_{(4,0,-3),\Phi} \\ \Phi_{(1,-6,5)} &= (\rho + T) + \lambda_{(1,-6,5)} N_{(1.-6.5)} - \beta_{(1,-6,5)} I_{1,P} + M_{(1,-6,5),\Phi} + \varepsilon_{(1,0,-1),\Phi} \\ L_{1,P} &= I_{1,P} \end{aligned}\right\} \tag{6.7.51}$$

the EWL and NL carrier phase functions were chosen. For long baselines, the tropospheric delay might need to be modeled and parameterized separately. In epoch-by-epoch sequential processing, one would fix the extra wide lane first and continue processing epochs until the other ambiguities have been fixed. Since several ambiguities are estimated in one step, one can readily use search algorithms that take advantage of the full variance-covariance matrix and not neglect correlations between the parameters.

The system (6.7.51) contains an ionospheric observation. In the simplest case the initial value, conceptually identical to approximate values in adjustment terminology, could be zero and the ionospheric parameter would be allowed to adjust according to the assigned weight. If available, one could use an external ionospheric model to assign the initial value. In that case the residual ionospheric delays to be estimated would be small and the ambiguity estimation over longer baseline should be easier, depending on the accuracy of the external information.

The system (6.7.51) can readily be replaced with another set of functions, as long as they are independent. An interesting combination is the optimized pseudorange equation (6.7.2), the triple-frequency ionospheric-free phase function (6.1.58), and the dual-frequency ionospheric-free phase function (6.1.39). The ambiguities to be estimated would be N_1, N_{12}, and N_{13}.

6.7.5 Positioning with Resolved Wide Lanes

Following the TCAR philosophy, the station coordinates are estimated after all ambiguities have been resolved. Typically, one would prefer the resolved original integer ambiguities N_1, N_2, and N_3 for accurate positioning. However, when in need of rapid positioning with low accuracy, one can utilize the resolved EWL and WL ambiguities and avoid the additional difficulties of resolving the NL ambiguity. For example, consider the ambiguity-corrected EWL function

$$\overline{\Phi}_{(1,-6,5)} = \rho - \beta_{(1,-6,5)} I_{1,P} + T + M_{(1,-6,5),\Phi} + \varepsilon_{(1,-6,5),\Phi} \tag{6.7.52}$$

Since this function has a low ionospheric dependency, it is suitable for long baseline processing. Another candidate is function (6.1.58), given here in traditional notation:

$$\begin{aligned} AC4 &\equiv \lambda_{13}\left[\frac{\lambda_{12}}{\lambda_1}\varphi_1 - \left(\frac{\lambda_{12}}{\lambda_1} + \frac{\lambda_{13}}{\lambda_3}\right)\varphi_2 + \frac{\lambda_{23}}{\lambda_3}\varphi_3 - \frac{\lambda_{12}}{\lambda_1}N_{12} + \frac{\lambda_{23}}{\lambda_3}N_{23}\right] \\ &= \rho + T + M_{AC4} + \varepsilon_{AC4} \end{aligned} \tag{6.7.53}$$

The formal standard deviation of the function is $\sigma_{AC4} = 27\sigma_\varphi$. The multipath is $M \leq 41M_\varphi$.

For even coarser positioning, consider the special pseudorange function PC of (6.7.2), which was designed to minimize the variance. As a matter of interest, another pseudorange equation that includes all three pseudoranges can be readily derived from (6.7.53). Divide each φ_i by λ_i and replace the symbols φ_i with Φ_i. In a second and final step, replace the Φ_i by P_i and delete the ambiguity terms, giving

$$\frac{\lambda_{13}\lambda_{12}}{\lambda_1^2}P_1 - \frac{\lambda_{13}}{\lambda_2}\left(\frac{\lambda_{12}}{\lambda_1} + \frac{\lambda_{23}}{\lambda_3}\right)P_2 + \frac{\lambda_{13}\lambda_{23}}{\lambda_3^2}P_3 = \rho + T + M + \varepsilon \qquad (6.7.54)$$

The numerical values of the respective pseudorange factors are 17.88, −84.71, and 67.82. Such large factors cause a very high variance for the combination and a potentially large multipath magnification. Therefore, this function is not attractive for use.

As to GF-TCAR and short baselines, for which per definition the double-differenced ionosphere is negligible, only the formal standard deviation of the ambiguities and multipath magnification factor need to be examined. For the extra wide lane $N_{(0,1,-1)}$ and wide lane function $N_{(1,-1,0)}$, there are several acceptable choices. In both cases, the HMW function is among them. Approach (6.7.34) is best-suited for estimating the N_3 ambiguities since the other two candidates have a high standard deviation and high multipath magnification. The same functions also seem to be the preferred functions for long baselines. Clearly, in that case the ionospheric delay becomes noticeable, external information about the ionosphere should be considered.

In terms of GB-TCAR, the system (6.7.51) is the preferred one because all observational information is used together. Since the extra wide lanes $N_{(0,1,-1)}$ or $N_{(1,-6,5)}$ can generally be determined over a short period of time, possibly even with a single epoch of data, one might give preference to determining one of them separately and then constrain it.

6.8 SUMMARY

In this chapter, we addressed the basic GNSS positioning approaches. This chapter should be viewed together with Chapter 7, which provides all the details on RTK using recursive least-squares. In Section 6.1, we derived the basic pseudorange and carrier phase equation and then listed various undifferenced functions of these observables, including triple-frequency functions. The notation used in this chapter was explained, and efforts were made to keep the notation clear and systematic. We also referred to the special triple-frequency subscript notation, which has become popular in recent literature.

Section 6.2 referred to operational details of "things to know" for serious GNSS users. We emphasized the excellent "GNSS infrastructure" that is in place and ready to be used. Over the years, much effort has been made to establish various services of exemplary quality that make it easy for the user to get the best performance out of GNSS systems. Especially relevant are the services of the IGS and the various online computing services that accept original field observations.

Sections 6.3 and 6.4 referred to the well-established navigation solution using pseudoranges and the broadcast ephemeris for single-point positioning (nonlinear and linearized solutions), as well as relative positioning using carrier phases and pseudoranges with emphasis on static positioning. The dilution of precision factors was given. Although the ambiguity function technique does not seem to enjoy major popularity among users, it was presented to provide an alternative to the customary double-difference ambiguity fixing. Yet another alternative to double differencing was briefly presented, i.e., the equivalent undifferenced formulation.

Ambiguity fixing with LAMBDA was dealt with in Section 6.5. The popular ratio test was discussed, including one approach of discernibility that gives some guidance as to the best value to adopt for the ratio. It was mentioned that a lot of research has been done to improve the testing theory to assure that indeed the correct set of ambiguities is accepted. An example is the aperture theory developed by Teunissen. However, to keep the mathematics at a minimum, only respective references are cited for this research. Instead, a major subsection was provided to see what other disciplines are doing who have problems similar to ambiguity fixing in GNSS.

As to network-supported positioning in Section 6.6, the key parameters of PPP are the lumped parameter R_k^p and the ionospheric-free receiver and satellite code biases $\xi_{k,PIF12}$ and ξ_{PIF12}^P in (6.6.2). In the case of RTK, the differential corrections are $\Delta\Phi_k^p$ in (6.6.24) and ΔP_k^p in (6.6.33), which are transmitted to the user. There were three PPP-RTK solutions discussed—the single-frequency, the dual-frequency, and the across-satellite difference methods. For these three methods, the elements transmitted to the user are, respectively, $\{\xi_{\Phi,FCB}^p, \xi_P^p\}$ of (6.6.57), $\{\xi_{\Phi IF12}^1, D_{HMW12}^p, \xi_{PIF12}^p\}$ of (6.6.75), and $\{\overline{D_{HMW12,FCB}^{1q}}, \overline{D_{A,FCB}^{1q}}, \xi_{PIF12}^{1q}\}$ of (6.6.108).

In Section 6.7, the triple-frequency solutions were examined. The major difference between the single-step batch solution and TCAR is that the former uses all correlations between parameters when resolving the integer ambiguities. Of the TCAR solutions presented, the system that combines all observations is preferred because it allows all correlations to be utilized when fixing the integers. Both single-step solution and TCAR can be solved sequentially allowing the EWL ambiguities to be estimates first, followed by the WL and then the NL ambiguities. In terms of GF-TCAR and short baselines, for which per definition the double-differenced ionosphere is negligible, only the formal standard deviation of the ambiguities and multipath magnification factor need to be examined. As to EWL and WL functions, there are several acceptable choices. In both cases, the HMW function is among them. In terms of GB-TCAR, the system that uses all observational information is preferred. Since the extra wide lanes can generally be determined over a short period of time, possibly even with a single epoch of data, one might in this case give preference to determining these separately and then constrain them.

CHAPTER 7

REAL-TIME KINEMATICS RELATIVE POSITIONING

Real-time kinematics (RTK) is a high-precision positioning technique that uses carrier phase and pseudorange measurements in real time. The high-precision position calculations are performed at the rate of measurements at the rover station. The base station, which is located at a known position, transmits its raw data, appropriately formatted, through a data communication channel. Ultra high frequency (UHF), cellular Global System for Mobile Communications (GSM), Long Term Evolution (LTE), WiFi, or Internet channels can be used for data transmission. Usually the data is transmitted one way, from the base to the rover. One or several rovers can listen to a certain base station and difference their raw measurements with raw measurements from the base station to correct the position.

There are several formats in use for data transmission of full raw measurements or differential corrections. All formats compact the transmission of information that is necessary to cancel GNSS errors. Errors that do not depend on the position of a station, or are nearly independent of location, tend to vary slightly with position and have almost equal effects on the measurements of both stations. These include satellite clock errors, satellite ephemerides errors, and atmospheric delays. The rover, having available its own measurements and the measurements from the base station, is able to form across-receiver differences to calculate the high-precision position relative to the base. The rover computes across-receiver differences for all satellites observed simultaneously at the base and the rover. Across-receiver differencing is carried out in a uniform manner for all GNSS systems such as GPS, GLONASS, Galileo, QZSS, Beidou, and SBAS.

In this chapter the recursive least-squares estimation approach of Chapter 3 is applied exclusively. The notation of that chapter is in general also carried over.

The estimation uses across-receiver differences and not double differences. Two data sets of actual observations help in illustrating the numerical aspects.

We develop a unified scheme for processing multifrequency and multisystem observations in RTK mode. We present tables that allow a unique association of signals and satellite systems. The suitability of a linear model to express the frequency dependency of GLONASS receiver hardware is verified first, which then allows for ready incorporation of GLONASS observations into the processing scheme. Although the linearization of carrier phase and pseudorange observations has been addressed in the previous chapter, it is presented again in the context of across-receiver difference observables and for providing the linearized form of the light time iteration procedure discussed in Chapter 6.

We first apply the RTK algorithm to a short static baseline and generate figures to demonstrate the convergence of across-receiver fractional hardware carrier phase delays. The RTK kinematic processing solutions begin with a short line whose rover is allowed unconstrained motions. This is followed by the RTK dynamic processing of a short line, whose rover motion is described by a dynamic model, and by the processing of a long line in which the ionospheric delay is also described by a dynamic model. A separate section deals with the extension of the algorithm to allow the number of signals to vary, as is the case when satellites set or rise or loss of signal occurs due to blockage of the signal by physical objects along the line of sight to the satellites. New ambiguity parameters are introduced after cycle slips. In addition, a special section is dedicated to the detection and isolation of cycle slips. The approach selected borrows from procedures that are popular in compressive signal sensing theory. The slip history is considered a sparse event and is numerated by sparsely populated vectors or matrices. The second to the last section deals with ambiguity fixing. We identify one base satellite within a group, i.e., the group of GPS satellites, estimate the base satellite ambiguities as real-valued number and fix the double-difference ambiguities. The chapter closes with remarks on optimal software implementation.

7.1 MULTISYSTEM CONSIDERATIONS

When speaking about the GNSS navigation signal, we mean a combination of signals from different carrier frequencies and satellite systems such as GPS, GLONASS, Galileo, QZSS, SBAS, and Beidou. We note that not all combinations of systems and frequencies are possible. For example, Galileo does not transmit signals in the L2 frequency band and GPS does not transmit in the E6 band. Table 7.1.1 shows the currently available satellite systems and carrier frequencies expressed in MHz.

The GLONASS FDMA L1 and L2 signals are available on GLONASS M satellites. The integer number l taking values from the range

$$l \in [-7, 6] \tag{7.1.1}$$

is referred to as the frequency letter. Actually, the same letter has been allocated to two satellites having different satellite numbers. The respective satellites are located at opposite points in the same orbital plane and thus cannot be observed simultaneously by a station located near or on the earth surface. The expected GLONASS

TABLE 7.1.1 GNSS Name and Band Frequencies.

GNSS	Band	Frequency
GPS	L1	$154 \times 10.23 = 1575.42$
	L2	$120 \times 10.23 = 1227.6$
	L5	$115 \times 10.23 = 1176.45$
Galileo	L1	$154 \times 10.23 = 1575.42$
	E5a	$115 \times 10.23 = 1176.45$
	E5b	$118 \times 10.23 = 1207.14$
	E6	$125 \times 10.23 = 1278.75$
GLONASS FDMA	L1	$1602 + l \times 0.5625$
	L2	$1246 + l \times 0.4375$
Beidou (Compass)	B1	$152.5 \times 10.23 = 1561.098$
	B2 (E5b)	$118 \times 10.23 = 1207.14$
	B3	$124 \times 10.23 = 1268.52$
QZSS	L1	1575.42
	L2	1227.6
	L5	1176.45
	E6	1278.75
SBAS	L1	1575.42
	L5	1176.45

CDMA system (not shown in the table) will include the GLONASS CDMA L1, L2, L3, and L5. GPS, Galileo, SBAS, and QZSS signals are using the same time scale, while GLONASS and Beidou use their own time scale.

The basics of code and carrier phase processing are described in preceding chapters. Now we are starting to describe the numerical algorithms for recurrent processing of pseudorange and carrier phase measurements intended for both real-time kinematics (RTK) processing and postmission processing. The next section gives expressions for undifferenced measurements and across-receiver difference measurements for rover and base. These are the navigation equations since they connect measurements with the position of the rover station, which is the subject of the position determination.

7.2 UNDIFFERENCED AND ACROSS-RECEIVER DIFFERENCE OBSERVATIONS

Let S^* be the set of signals currently available and S_k be the set of signals available for processing at receiver k. For $s \in S_k$ we assume that the signal $s = (p, b)$ is represented by a pair of numbers consisting of an internal number p that uniquely identifies the navigation system, and the frequency band identifier b. Table 7.2.1 shows the internal number assignment we have chosen. Note that internal numbering is for internal use inside the receiver firmware, it is not standardized and certainly can differ by receiver

TABLE 7.2.1 Assignment of Internal Satellite Numbers.

Internal Satellite Number p	GNSS	Set of Available Frequency Bands F^p
$1, \ldots, 32$	GPS	L1, L2, L5
$33, \ldots, 56$	GLONASS FDMA	L1, L2
$57, \ldots, 86$	Galileo	L1, E5a, E5b, E6
$87, \ldots, 90$	QZSS	L1, L2, L5, E6
$91, \ldots, 120$	Beidou	B1, B2, B3
$121, \ldots, 143$	SBAS	L1, L5

TABLE 7.2.2 Current Assignment of GLONASS Satellite Number and Frequency Letters.

p	$l(p)$	p	$l(p)$	p	$l(p)$	p	$l(p)$
33	1	39	5	45	−2	51	3
34	−4	40	6	46	−7	52	2
35	5	41	−2	47	0	53	4
36	6	42	−7	48	−1	54	−3
37	1	43	0	49	4	55	3
38	−4	44	−1	50	−3	56	2

manufacturer. The range of the frequency band identifier depends on the satellite number. Let us denote the set of frequency bands available for the satellite p by F^p. The GLONASS satellite letters (7.1.1) are mapped to numbers in Table 7.2.1 (currently for $p = 33, \ldots, 56$) as shown in Table 7.2.2 according to the official site of the Russian Information Analytical Center (http://glonass-iac.ru/en/GLONASS).

Let Σ^p denote the satellite system corresponding to satellite p. The mapping $p \to \Sigma^p$ is defined in the Table 7.2.1. Note again, that Table 7.2.1 describes an exemplary mapping. Different manufacturers use different mappings.

We now present the fundamental set of navigation equations using notations introduced in Chapter 6 and generalized to the signal concept notation $s = (p, b)$ and $b \in F^p$, and indexing the station by k, and time of measurement by t. The pseudorange measurement equation becomes

$$P^p_{k,b}(t) = \rho^p_k(t) + c\,dt_k(t) - c\,dt^p(t) + \left(\frac{f^p_{L1}}{f^p_b}\right)^2 I^p_{k,L1}(t) + T^p_k(t)$$
$$+ d_{k,b,P} + M^p_{k,b,P} - D^p_{b,P} + \varepsilon^p_{k,b,P}(t) \tag{7.2.1}$$

The signal of the frequency band b emitted by the satellite p experiences a hardware delay at the receiver k by $d_{k,b,P}$. The corresponding satellite hardware delay is denoted by $D^p_{b,P}$. The code multipath delay of the signal emitted from satellite p at frequency band b and received by station k is denoted by $M^p_{k,b,P}$.

The carrier phase measurement has the form

$$\varphi^p_{k,b}(t) = \frac{f^p_b}{c}\rho^p_k(t) + f^p_b dt_k(t) - f^p_b dt^p(t) + N^p_{k,b}\left(t^p_{CS,k,b}\right) - \frac{1}{c}\frac{\left(f^p_{L1}\right)^2}{f^p_b} I^p_{k,L1}(t)$$
$$+ \frac{f^p_b}{c} T^p_k(t) + d^p_{k,b,\varphi} + M^p_{k,b,\varphi} - D^p_{b,\varphi} + \varepsilon^p_{k,b,\varphi} \tag{7.2.2}$$

where f^p_b is the carrier frequency of the signal. For example, according to Tables 7.1.1, 7.2.1, and 7.2.2 we have: $f^p_{L1} = 1575.42$ MHz, $f^p_{L2} = 1227.6$ MHz, $f^p_{L5} = 1176.45$ MHz for $p = 1, \ldots, 32$, $f^p_{L1} = 1602 + l(p) \times 0.5625$ MHz, $f^p_{L2} = 1246 + l(p) \times 0.4375$ MHz for $p = 33, \ldots, 56$, $f^p_{L1} = 1575.42$ MHz, $f^p_{E5a} = 1176.45$ MHz, $f^p_{E5b} = 1207.14$ MHz, $f^p_{E6} = 1278.75$ MHz for $p = 57, \ldots, 86$ and so on.

The symbol $t^p_{CS,k,b}$ in (7.2.2) is the exact time the last cycle slip happened. Note that the cycle slip leads to a jump, usually integer valued, of the carrier phase ambiguity. Half a cycle slip can also occur, lasting for several seconds until certain phase-locked loop (PLL) corrects its state to the nearest stable state. Other notation includes the carrier phase multipath delay $M^p_{k,b,\varphi}$ and the satellite hardware delay $D^p_{b,\varphi}$. The receiver hardware delay of a signal emitted at frequency band b by satellite p at station k usually depends on frequency. Assuming that

$$d^p_{k,b,\varphi} = d^0_{k,b,\varphi} + \frac{f^p_b}{c}\mu_{k,b,\varphi} \tag{7.2.3}$$

we are introducing a linear frequency dependence of the hardware delay as a reasonable first-order approximation. For all GPS L1 signals, having the same carrier frequency 1575.42 MHz (and therefore experiencing the same hardware delays), the second term in (7.2.3) can be ignored so one obtains $d^p_{k,L1,\varphi} \equiv d^0_{k,L1,\varphi}$ for $p = 1, \ldots, 32$. The same is true for GPS L2 signals, GPS L5 signals, and other signals except for GLONASS FDMA L1 and GLONASS FDMA L2, because each satellite has its own carrier frequency letter. In other words, considering dependency of hardware delay on the satellite number inside the system and on the frequency band makes sense only for the GLONASS system.

As discussed later, this first-order approximation is applicable in practice. The coefficients of the linear dependency $\mu_{k,b,\varphi}$ are available from a lookup table stored in the computer or receiver memory. Another option is to consider it a constant parameter that is to be estimated along with other parameters. Note that using the lookup table allows for a more precise representation and more thorough compensation of the hardware biases as compared to their linear representation. Additional details about the creation of a lookup table are given below.

All other notations used in (7.2.1) and (7.2.2) have been introduced in Chapter 6. Using the following notation for the carrier wavelengths

$$\lambda^p_b = \frac{c}{f^p_b} \tag{7.2.4}$$

we can present the carrier phase measurements equation (7.2.2) in themetric form:

$$\Phi^p_{k,b}(t) = \rho^p_k(t) + c\,dt_k(t) - c\,dt^p(t) + \lambda^p_b N^p_{k,b}\left(t^p_{CS,k,b}\right) - \left(\frac{\lambda^p_b}{\lambda^p_{L1}}\right)^2 I^p_{k,L1}(t) + T^p_k(t)$$
$$+ \lambda^p_b d^p_{k,b,\varphi} + M^p_{k,b,\Phi} - D^p_{b,\Phi} + \varepsilon_{b,\Phi}(t) \tag{7.2.5}$$

The terms $M^p_{k,b,\Phi}$, $D^p_{b,\Phi}$, and $\varepsilon_{b,\Phi}(t)$ denote carrier phase multipath, hardware-dependent biases of the satellite, and noise expressed in the metric form.

The error terms in equations (7.2.1), (7.2.2), and (7.2.5) can be divided into two groups—modeled errors and nonmodeled errors. The tropospheric delay can be modeled using one of the models described in Chapter 8. According to the tropospheric model, the delay term $T^p_k(t)$ is estimated using a rough approximation of the position and a priori atmospheric data such as temperature, pressure, and humidity. Improving the position solution iteratively also improves the tropospheric delay estimate, provided the iterations converge. Therefore, the tropospheric delay is considered as a correction to be compensated on the left side of (7.2.1) and (7.2.2), forming left-side terms

$$\overline{P}^p_{k,b}(t) = P^p_{k,b}(t) - T^p_k(t) \tag{7.2.6}$$

$$\overline{\varphi}^p_{k,b}(t) = \varphi^p_{k,b}(t) - \frac{1}{\lambda^p_b} T^p_k(t) \tag{7.2.7}$$

Note that we do not compensate for other terms that are specific to a certain satellite and common for different stations because these will vanish when calculating across-receiver differences.

The multipath error term is usually not modeled. We must accept its existence together with possibly other unmodeled errors. Not being able to directly compensate or estimate these unmodeled errors, we take into account their statistical properties, such as epoch-wise variance or across-epoch correlation. Combining all unmodeled terms with those on the right side of (7.2.1) and (7.2.2), we define the cumulative unmodeled errors $\overline{\varepsilon}^p_{k,b,P}(t)$ and $\overline{\varepsilon}^p_{k,b,\varphi}(t)$.

The navigation equations can be now presented in the form

$$\overline{P}^p_{k,b}(t) = \rho^p_k(t) + c\,dt_k(t) - c\,dt^p(t) + \left(\frac{f^p_{L1}}{f^p_b}\right)^2 I^p_{k,L1}(t)$$
$$+ d_{k,b,P} - D^p_{b,P} + \overline{\varepsilon}^p_{k,b,P}(t) \tag{7.2.8}$$

$$\overline{\varphi}^p_{k,b}(t) = \frac{1}{\lambda^p_b}\rho^p_k(t) + f^p_b dt_k(t) - f^p_b dt^p(t) + N^p_{k,b}\left(t^p_{CS,k,b}\right) - \frac{1}{\lambda^p_b}\left(\frac{f^p_{L1}}{f^p_b}\right)^2 I^p_{k,L1}(t)$$
$$+ d^p_{k,b,\varphi} - D^p_{b,\varphi} + \overline{\varepsilon}^p_{k,b,\varphi} \tag{7.2.9}$$

In equations (7.2.1), (7.2.2), (7.2.8), and (7.2.9), the symbol ρ_k^p denotes the distance that the signal travels from transmission at the satellite antenna to reception at the receiver antenna. The travel time in the vacuum is

$$\tau_k^p = \frac{\rho_k^p}{c} \tag{7.2.10}$$

Assuming that two receivers k and m observe the same satellite p, the across-receiver differences of pseudorange and carrier phase measurements that are introduced in Section 6.1.2 can be written as

$$\overline{P}_{km,b}^p(t) = \rho_k^p(t) - \rho_m^p(t) + c\,dt_{km}(t) + \left(\frac{f_{L1}^p}{f_b^p}\right)^2 I_{km,L1}^p(t)$$
$$+ d_{km,b,P} + \overline{\varepsilon}_{km,b,P}^p(t), \tag{7.2.11}$$

$$\overline{\varphi}_{km,b}^p(t) = \frac{1}{\lambda_b^p}\left(\rho_k^p(t) - \rho_m^p(t)\right) + f_b^p dt_{km}(t) + N_{km,b}^p\left(t_{CS,km,b}^p\right) - \frac{1}{\lambda_b^p}\left(\frac{f_{L1}^p}{f_b^p}\right)^2 I_{km,L1}^p(t)$$
$$+ d_{km,b,\varphi}^p + \overline{\varepsilon}_{km,b,\varphi}^p, \tag{7.2.12}$$

The symbol $t_{CS,km,b}^p = \max\left\{t_{CS,k,b}^p, t_{CS,m,b}^p\right\}$ denotes the time at which the last cycle slip occurred at either carrier phase $\varphi_{k,b}^p(t)$ or $\varphi_{m,b}^p(t)$, i.e., the time the latest cycle slip occurred in either receiver. As was mentioned in Section 6.1.2, errors or biases which are specific for satellite p vanish in (7.2.11) and (7.2.12). We also denote

$$S_{km} = S_k \cap S_m \tag{7.2.13}$$

as the set of signals common for both stations k and m. Across-receiver differences are available only for signals $s \in S_{km}$.

The across-receiver difference of the hardware delay $d_{km,b,\varphi}^p$ can be written as $d_{k,b,\varphi}^0$ for all signals except GLONASS FDMA L1 and GLONASS FDMA L2, as was previously mentioned. The first-order approximation of $d_{km,b,\varphi}^p$ for GLONASS signals has the form

$$d_{km,b,\varphi}^p = d_{km,b,\varphi}^0 + \frac{1}{\lambda_b^p}\mu_{km,b,\varphi} \tag{7.2.14}$$

which is similar to (7.2.3). Note that the constant term $d_{km,b,\varphi}^0$ always appears as a sum with the across-receiver carrier phase ambiguity $N_{km,b}^p$. Therefore, the across-receiver ambiguity is part of it and estimated along as a lumped parameter. For all signals except GLONASS FDMA, the hardware carrier phase delays disappear in the across-receiver difference. For GLONASS FDMA, the constant term $d_{km,b,\varphi}^0$ is combined with the across-receiver ambiguity while the linear term $\left(1/\lambda_b^p\right)\mu_{km,b,\varphi}$ is preserved. The coefficient $\mu_{km,b,\varphi}$ is the additional delay expressed in metric units. If receivers at stations k and m are absolutely identical, then we can assume that

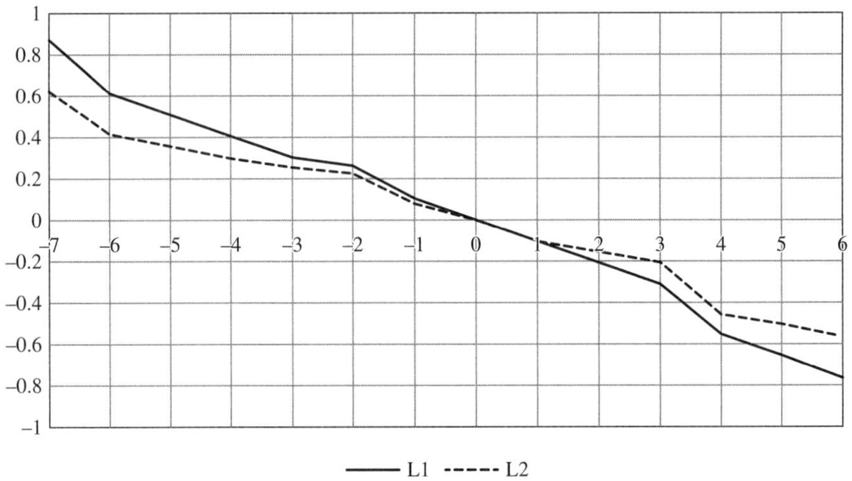

Figure 7.2.1 Across-receiver difference of GLONASS hardware carrier phase delay in cycles (vertical axis) as a function of the frequency letter (horizontal axis).

the across-receiver difference $\mu_{km,b,\varphi}$ vanishes, which is confirmed by real-world experience; however, if the pair of receivers are not identical (for example, if they are produced by different manufacturers), the problem of estimation of this value becomes critical. For example, for the pair Triumph-1 receiver by Javad GNSS and a Leica receiver, the values $d^p_{km,L1,\varphi}$ and $d^p_{km,L2,\varphi}$ expressed as a function of the carrier phase letter l are shown in Figure 7.2.1.

The figure confirms that first-order approximation (7.2.14) is reasonable because the first-order linear term dominates the variation. The constant terms $d^0_{km,L1,\varphi}$ and $d^0_{km,L2,\varphi}$ are chosen in such a way, that $d^p_{km,L1,\varphi} = 0$ and $d^p_{km,L2,\varphi} = 0$ for the zero letter or, in other words, for such p that $l(p) = 0$ in Table 7.2.2. They take large values and cannot be simply ignored because the ambiguity resolution would become impossible.

There are two ways to determine these constants. The first way requires long-term data collection of a zero baseline. Processing of the across-receiver and across-satellite differences allows for determination of the fractional parts of carrier phase ambiguity. The resulting averaged data is then stored in the software lookup tables. The second way considers the linear dependence (7.2.14) and estimates the "slope coefficients" $\mu_{km,L1,\varphi}$ and $\mu_{km,L2,\varphi}$ as additional constants along with other parameters during positioning. The linearization (7.2.14) is part of the linearized navigation equation scheme considered in the next section.

7.3 LINEARIZATION AND HARDWARE BIAS PARAMETERIZATION

Aiming to apply the linear estimation theory described in Chapter 3, let us linearize the navigation equations around a nominal location. Let $\left(x_{k,0}(t), y_{k,0}(t), z_{k,0}(t)\right)^T$ be the vector of approximate Cartesian coordinates of the station k at the time t. Note that location of antenna, location of receiver, and location of station have the same meaning throughout this chapter. Let the station m be located at the precisely known static position

$$\boldsymbol{x}_m = \begin{pmatrix} x_m \\ y_m \\ z_m \end{pmatrix} \tag{7.3.1}$$

Station k is considered unknown or only approximately known. The position of the station k can be expressed as

$$\boldsymbol{x}_k(t) = \begin{pmatrix} x_k(t) \\ y_k(t) \\ z_k(t) \end{pmatrix} = \begin{pmatrix} x_{k,0}(t) + dx_k(t) \\ y_{k,0}(t) + dy_k(t) \\ z_{k,0}(t) + dz_k(t) \end{pmatrix} \tag{7.3.2}$$

by adding corrections to the approximate coordinates. Stand-alone, or autonomous positioning using only pseudorange measurements as described below provides approximation accurate to within several meters or better. Therefore, the expected range of the corrections dx_k, dy_k, dz_k is a few meters. Let

$$\boldsymbol{x}^p(t - \tau_k^p) = \begin{pmatrix} x^p\left(t - \tau_k^p\right) \\ y^p(t - \tau_k^p) \\ z^p(t - \tau_k^p) \end{pmatrix} \tag{7.3.3}$$

be Cartesian coordinates of the satellite p at the time $\left(t - \tau_k^p\right)$. The signal travel distance ρ_k^p is defined in Section 6.2.1. In this section we expand this expression and represent it in the form

$$\rho_k^p\left(t, t - \tau_k^p\right) = \left\|\boldsymbol{x}_k(t) - \boldsymbol{x}^p\left(t - \tau_k^p\right)\right\| + d\rho_k^p \tag{7.3.4}$$

where $d\rho_k^p$ is the additional distance between satellite and receiver antennas that the signal travels due to the rotation of the earth. Let $\dot{\Omega}_e$ be the angular speed of the earth rotation, $\dot{\Omega}_e = 7.2921151467 \times 10^{-5}$ rad/ sec, and

$$\vec{\dot{\Omega}}_e = \begin{pmatrix} 0 \\ 0 \\ \dot{\Omega}_e \end{pmatrix} \tag{7.3.5}$$

be the angular rotation velocity vector expressed in ECEF, where the arrow means the vector. Since the earth rotation angle during the travel time is small, we can use the first-order approximation of the rotation matrix

$$\boldsymbol{R}_3(\theta) \approx \boldsymbol{I}_3 + \begin{bmatrix} 0 & \theta & 0 \\ -\theta & 0 & 0 \\ 0 & 0 & 0 \end{bmatrix} \tag{7.3.6}$$

where $\boldsymbol{I}_3$ is the 3×3 identity matrix, as well as

$$\boldsymbol{R}_3(\theta)\boldsymbol{x}^p(t-\tau_k^p) \approx \boldsymbol{x}^p\left(t-\tau_k^p\right) + \begin{bmatrix} 0 & \theta & 0 \\ -\theta & 0 & 0 \\ 0 & 0 & 0 \end{bmatrix} \boldsymbol{x}^p\left(t-\tau_k^p\right)$$

$$= \boldsymbol{x}^p\left(t-\tau_k^p\right) - \tau_k^p \vec{\dot{\Omega}}_e \times \boldsymbol{x}^p\left(t-\tau_k^p\right) \tag{7.3.7}$$

where the symbol $\times$ denotes the vector product. Preserving the first-order term in the Taylor series expansion of the expression (7.3.4), we have

$$\begin{aligned} &\rho_k^p\left(t, t-\tau_k^p\right) \\ &= \left\|\boldsymbol{x}_k(t) - \boldsymbol{R}_3(\theta)\boldsymbol{x}^p\left(t-\tau_k^p\right)\right\| \\ &\approx \left\|\boldsymbol{x}_k(t) - \boldsymbol{x}^p\left(t-\tau_k^p\right) - \tau_k^p \vec{\dot{\Omega}}_e \times \boldsymbol{x}^p\left(t-\tau_k^p\right)\right\| \\ &\approx \left\|\boldsymbol{x}_k(t) - \boldsymbol{x}^p\left(t-\tau_k^p\right)\right\| - \frac{\tau_k^p \left[\boldsymbol{x}_k(t) - \boldsymbol{x}^p\left(t-\tau_k^p\right)\right] \cdot \vec{\dot{\Omega}}_e \times \boldsymbol{x}^p\left(t-\tau_k^p\right)}{\left\|\boldsymbol{x}_k(t) - \boldsymbol{x}^p\left(t-\tau_k^p\right)\right\|} \end{aligned} \tag{7.3.8}$$

where the symbol $\cdot$ is used for the scalar product. Estimating τ_k^p as $\tau_k^p \approx \frac{\left\|\boldsymbol{x}_k(t)-\boldsymbol{x}^p\left(t-\tau_k^p\right)\right\|}{c}$ we rewrite the expression (7.3.8) in the form

$$\rho_k^p\left(t, t-\tau_k^p\right) \approx \left\|\boldsymbol{x}_k(t) - \boldsymbol{x}^p\left(t-\tau_k^p\right)\right\| - \frac{\left[\boldsymbol{x}_k(t) - \boldsymbol{x}^p\left(t-\tau_k^p\right)\right] \cdot \vec{\dot{\Omega}}_e \times \boldsymbol{x}^p\left(t-\tau_k^p\right)}{c} \tag{7.3.9}$$

The numerator in the second term of expression is a triple product of vectors, which is further expressed as

$$\begin{aligned} &\left[\boldsymbol{x}_k(t) - \boldsymbol{x}^p\left(t-\tau_k^p\right)\right] \cdot \vec{\dot{\Omega}}_e \times \boldsymbol{x}^p\left(t-\tau_k^p\right) \\ &= \det\left[\boldsymbol{x}_k(t) - \boldsymbol{x}^p\left(t-\tau_k^p\right) \;\middle|\; \vec{\dot{\Omega}}_e \;\middle|\; \boldsymbol{x}^p\left(t-\tau_k^p\right)\right] \\ &= \det\left[\boldsymbol{x}_k(t) \;\middle|\; \vec{\dot{\Omega}}_e \;\middle|\; \boldsymbol{x}^p\left(t-\tau_k^p\right)\right] - \det\left[\boldsymbol{x}^p\left(t-\tau_k^p\right) \;\middle|\; \vec{\dot{\Omega}}_e \;\middle|\; \boldsymbol{x}^p\left(t-\tau_k^p\right)\right] \\ &= -\det\left[\vec{\dot{\Omega}}_e \;\middle|\; \boldsymbol{x}_k(t) \;\middle|\; \boldsymbol{x}^p\left(t-\tau_k^p\right)\right] - 0 = -\det\begin{bmatrix} 0 & x_k & x^p \\ 0 & y_k & y^p \\ \dot{\Omega}_e & z_k & z^p \end{bmatrix} \\ &= \dot{\Omega}_e\left(x^p\left(t-\tau_k^p\right) y_k(t) - x_k(t) y^p\left(t-\tau_k^p\right)\right) \end{aligned} \tag{7.3.10}$$

where we took into account that a determinant with two equal columns vanishes and the permutation of columns changes the sign of a determinant.

Finally,

$$
\begin{aligned}
\rho_k^p\left(t, t-\tau_k^p\right) &= \left\|\boldsymbol{x}_k(t) - \boldsymbol{x}^p\left(t-\tau_k^p\right)\right\| + \frac{\dot{\Omega}_e}{c}\left(x^p\left(t-\tau_k^p\right) y_k(t) - x_k(t) y^p\left(t-\tau_k^p\right)\right) \\
&= \sqrt{\left(x_k(t) - x^p\left(t-\tau_k^p\right)\right)^2 + \left(y_k(t) - y^p\left(t-\tau_k^p\right)\right)^2 + \left(z_k(t) - z^p\left(t-\tau_k^p\right)\right)^2} \\
&\quad + \frac{\dot{\Omega}_e}{c}\left(x^p\left(t-\tau_k^p\right) y_k(t) - x_k(t) y^p\left(t-\tau_k^p\right)\right)
\end{aligned} \tag{7.3.11}
$$

This equation gives a first-order approximation to the iterative solution discussed at the end of Section 6.2.1. Linearization of (7.3.11) around the approximate point $(u_{k,0}(t), v_{k,0}(t), w_{k,0}(t))^T$ takes the following form:

$$
\rho_k^p\left(t, t-\tau_k^p\right) \approx \rho_{k,0}^p\left(t, t-\tilde{\tau}_k^p\right) + H_{k,1}^p\, dx_k(t) + H_{k,2}^p\, dy_k(t) + H_{k,3}^p\, dz_k(t) \tag{7.3.12}
$$

where

$$
\begin{aligned}
&\rho_{k,0}^p\left(t, t-\tilde{\tau}_k^p\right) \\
&= \sqrt{\left(x_{k,0}(t) - x^p\left(t-\tilde{\tau}_k^p\right)\right)^2 + \left(y_{k,0}(t) - y^p\left(t-\tilde{\tau}_k^p\right)\right)^2 + \left(z_{k,0}(t) - z^p\left(t-\tilde{\tau}_k^p\right)\right)^2} \\
&\quad + \frac{\dot{\Omega}_e}{c}\left[x^p\left(t-\tilde{\tau}_k^p\right) y_{k,0}(t) - x_{k,0}(t) y^p\left(t-\tilde{\tau}_k^p\right)\right]
\end{aligned} \tag{7.3.13}
$$

A first approximation $\tilde{\tau}_{k,b}^p$ for the travel time is obtained from the pseudorange observation as

$$
\tilde{\tau}_{k,b}^p = \frac{P_{k,b}^p(t)}{c} \tag{7.3.14}
$$

The partial derivatives, or directional cosines, are expressed as

$$
\begin{aligned}
H_{k,1}^p(t) &= \frac{x_{k,0}(t) - x^p\left(t-\tilde{\tau}_k^p\right)}{\sqrt{\left(x_{k,0}(t) - x^p\left(t-\tilde{\tau}_k^p\right)\right)^2 + \left(y_{k,0}(t) - y^p\left(t-\tilde{\tau}_k^p\right)\right)^2 + \left(z_{k,0}(t) - z^p\left(t-\tilde{\tau}_k^p\right)\right)^2}} \\
&\quad - \frac{\dot{\Omega}_e}{c} y^p\left(t-\tilde{\tau}_k^p\right) \\
H_{k,2}^p(t) &= \frac{y_{k,0}(t) - y^p\left(t-\tilde{\tau}_k^p\right)}{\sqrt{\left(x_{k,0}(t) - x^p\left(t-\tilde{\tau}_k^p\right)\right)^2 + \left(y_{k,0}(t) - y^p\left(t-\tilde{\tau}_k^p\right)\right)^2 + \left(z_{k,0}(t) - z^p\left(t-\tilde{\tau}_k^p\right)\right)^2}} \\
&\quad + \frac{\dot{\Omega}_e}{c} x^p\left(t-\tilde{\tau}_k^p\right) \\
H_{k,3}^p(t) &= \frac{z_{k,0} - z^p\left(t-\tilde{\tau}_k^p\right)}{\sqrt{\left(x_{k,0}(t) - x^p\left(t-\tilde{\tau}_k^p\right)\right)^2 + \left(y_{k,0}(t) - y^p\left(t-\tilde{\tau}_k^p\right)\right)^2 + \left(z_{k,0}(t) - z^p\left(t-\tilde{\tau}_k^p\right)\right)^2}}
\end{aligned} \tag{7.3.15}
$$

Finally, all necessary expressions are now available to express the across-receiver differences (7.2.11) and (7.2.12) in the linearized form,

$$\overline{P}^p_{km,b}(t) - \rho^p_{k,0}\left(t - \tilde{\tau}^p_k\right) + \rho^p_m\left(t - \tilde{\tau}^p_m\right) = H^p_{k,1}\, dx_k(t) + H^p_{k,2}\, dy_k(t) + H^p_{k,3}\, dz_k(t)$$
$$+ c\, dt_{km}(t) + \left(\frac{f^p_{L1}}{f^p_b}\right)^2 I^p_{km,L1}(t) + d_{km,b,P} + \overline{\varepsilon}^p_{km,b,P}(t) \tag{7.3.16}$$

$$\overline{\varphi}^p_{km,b}(t) - \frac{1}{\lambda^p_b}\left(\rho^p_{k,0}\left(t - \tilde{\tau}^p_k\right) - \rho^p_m\left(t - \tilde{\tau}^p_m\right)\right) = \frac{1}{\lambda^p_b}\left(H^p_{k,1}\, dx_k(t) + H^p_{k,2}\, dy_k(t) + H^p_{k,3}\, dz_k(t)\right)$$
$$+ f^p_b dt_{km}(t) + N^p_{km,b}\left(t^p_{CS,km,b}\right) - \frac{1}{\lambda^p_b}\left(\frac{f^p_{L1}}{f^p_b}\right)^2 I^p_{km,L1}(t) + d^p_{km,b,\varphi} + \overline{\varepsilon}^p_{km,b,\varphi} \tag{7.3.17}$$

Let n_{km} be the number of signals, counting pseudoranges and carrier phases, in set $s \in S_{km}$. The subscript km will be omitted whenever it does not lead to misunderstanding throughout this section, i.e., $S \equiv S_{km}$, $n \equiv n_{km}$, and so on. Assume that the across-receiver difference pseudorange and carrier phase observations are somehow ordered in set S,

$$S = \{s_1, \ldots, s_n\} \tag{7.3.18}$$

Let A be the $n \times 3$ matrix composed of the coefficients (7.3.15). More precisely, let

$$S = \{s_1, \ldots, s_n\} = \{(p_1, b_1), \ldots, (p_n, b_n)\} \tag{7.3.19}$$

then

$$\boldsymbol{A} = \begin{bmatrix} H^{p_1}_{k,1} & H^{p_1}_{k,2} & H^{p_1}_{k,3} \\ H^{p_2}_{k,1} & H^{p_2}_{k,2} & H^{p_3}_{k,3} \\ & \cdots & \\ H^{p_n}_{k,1} & H^{p_n}_{k,2} & H^{p_n}_{k,3} \end{bmatrix} \tag{7.3.20}$$

and the set of across-receiver difference equations (7.3.16) and (7.3.17) can be represented in the vector form.

Station m with a known position is called *the base*. The position of station k is to be determined. It can be either static, occupying a time-invariant position, or its position can vary in time. In the latter case, we have to solve the kinematic positioning problem. Station k is called *the rover* independently of whether it is static or roving. Real-time positioning is called t*he real-time kinematics* (RTK).

The position of the base station (7.3.1) and the approximate position of the rover station $X_{k,0}$ in (7.3.2) are known. The positions of the satellites $\boldsymbol{x}^p\left(t - \tilde{\tau}^p_m\right)$ and $\boldsymbol{x}^p\left(t - \tilde{\tau}^p_k\right)$ are calculated using the broadcast ephemeris or a precise ephemeris provided by, for example, the IGS. Given the known data, the quantities on the left of the linearized equations (7.3.16) can be calculated forming the n-dimensional vector

$$\boldsymbol{b}_P(t) = \begin{pmatrix} \overline{P}^{p_1}_{km,b_1}(t) - \rho^{p_1}_{k,0}\left(t - \tilde{\tau}^{p_1}_k\right) + \rho^{p_1}_m\left(t - \tilde{\tau}^{p_1}_m\right) \\ \overline{P}^{p_2}_{km,b_2}(t) - \rho^{p_2}_{k,0}\left(t - \tilde{\tau}^{p_2}_k\right) + \rho^{p_2}_m\left(t - \tilde{\tau}^{p_2}_m\right) \\ \vdots \\ \overline{P}^{p_n}_{km,b_n}(t) - \rho^{p_n}_{k,0}\left(t - \tilde{\tau}^{p_n}_k\right) + \rho^{p_n}_m\left(t - \tilde{\tau}^{p_n}_m\right) \end{pmatrix} \tag{7.3.21}$$

where $\overline{P}^{p_i}_{km,b_i}(t) = \overline{P}^{p_i}_{k,b_i}(t) - \overline{P}^{p_i}_{m,b_i}(t)$. The quantities $\overline{P}^{p_i}_{k,b_i}(t)$ and $\overline{P}^{p_i}_{m,b_i}(t)$ are defined in (7.2.6). The quantities on the left of the linearized equations (7.3.17) can be calculated the same way, forming the n-dimensional vector

$$\boldsymbol{b}_\varphi(t) = \begin{pmatrix} \overline{\varphi}^{p_1}_{km,b_1}(t) - \frac{1}{\lambda^{p_1}_{b_1}}\left(\rho^{p_1}_{k,0}\left(t - \tilde{\tau}^{p_1}_k\right) - \rho^{p_1}_m\left(t - \tilde{\tau}^{p_1}_m\right)\right) \\ \overline{\varphi}^{p_2}_{km,b_2}(t) - \frac{1}{\lambda^{p_2}_{b_2}}\left(\rho^{p_2}_{k,0}\left(t - \tilde{\tau}^{p_2}_k\right) - \rho^{p_2}_m\left(t - \tilde{\tau}^{p_2}_m\right)\right) \\ \vdots \\ \overline{\varphi}^{p_n}_{km,b_n}(t) - \frac{1}{\lambda^{p_n}_{b_n}}\left(\rho^{p_n}_{k,0}\left(t - \tilde{\tau}^{p_n}_k\right) - \rho^{p_n}_m\left(t - \tilde{\tau}^{p_n}_m\right)\right) \end{pmatrix} \tag{7.3.22}$$

where $\overline{\varphi}^{p_i}_{km,b_i}(t) = \overline{\varphi}^{p_i}_{k,b_i}(t) - \overline{\varphi}^{p_i}_{m,b_i}(t)$. The quantities $\overline{\varphi}^{p_i}_{k,b_i}(t)$ and $\overline{\varphi}^{p_i}_{m,b_i}(t)$ are defined in (7.2.7). Let us combine three independent variables of the linearized system (7.3.16) and (7.3.17) into the three-dimensional vector

$$d\boldsymbol{x}(t) = \left(dx_k(t), dy_k(t), dz_k(t)\right)^T \tag{7.3.23}$$

and denote

$$\xi(t) = c\,dt(t) \tag{7.3.24}$$

Let

$$\boldsymbol{e} = (1, 1, \cdots, 1)^T \tag{7.3.25}$$

be the vector having units at all positions, and let

$$\boldsymbol{i} = \left(I^{p_1}_{L1}, I^{p_2}_{L1}, \ldots, I^{p_n}_{L1}\right)^T \tag{7.3.26}$$

be the vector of across-receiver difference ionospheric delays related to all n satellites,

$$\boldsymbol{\Gamma} = \begin{bmatrix} \left(f^{p_1}_{L1}/f^{p_1}_{b_1}\right)^2 & 0 & \vdots & 0 \\ 0 & \left(f^{p_2}_{L1}/f^{p_2}_{b_2}\right)^2 & \vdots & 0 \\ \cdots & \cdots & \ddots & \vdots \\ 0 & 0 & \cdots & \left(f^{p_n}_{L1}/f^{p_n}_{b_n}\right)^2 \end{bmatrix} \tag{7.3.27}$$

And let

$$\boldsymbol{\Lambda} = \begin{bmatrix} \lambda_{b_1}^{p_1} & 0 & \vdots & 0 \\ 0 & \lambda_{b_2}^{p_2} & \vdots & 0 \\ \cdots & \cdots & \ddots & \vdots \\ 0 & 0 & \cdots & \lambda_{b_n}^{p_n} \end{bmatrix} \tag{7.3.28}$$

be diagonal matrices,

$$\boldsymbol{n} = \left(N_{b_1}^{p_1}\left(t_{CS,b_1}^{p_1}\right), N_{b_2}^{p_2}\left(t_{CS,b_2}^{p_2}\right), \ldots, N_{b_n}^{p_n}\left(t_{CS,b_n}^{p_n}\right)\right)^T \tag{7.3.29}$$

be the vector of carrier phase ambiguities, and, finally

$$\boldsymbol{d}_P = (d_{b_1,P}, d_{b_2,P}, \ldots, d_{b_n,P})^T \tag{7.3.30}$$

$$\boldsymbol{d}_\varphi = (d_{b_1,\varphi}^{p_1}, d_{b_2,\varphi}^{p_2}, \ldots, d_{b_n,\varphi}^{p_n})^T \tag{7.3.31}$$

be the vector of across-receiver pseudorange and phase receiver hardware biases. Then, ignoring noise terms, the systems (7.3.16) and (7.3.17) can be rewritten as

$$\boldsymbol{b}_P(t) = \boldsymbol{A}d\boldsymbol{x}(t) + \boldsymbol{e}\xi(t) + \boldsymbol{\Gamma}\boldsymbol{i}(t) + \boldsymbol{d}_P \tag{7.3.32}$$

$$\boldsymbol{b}_\varphi(t) = \boldsymbol{\Lambda}^{-1}\boldsymbol{A}d\boldsymbol{x}(t) + \boldsymbol{\Lambda}^{-1}\boldsymbol{e}\xi(t) + \boldsymbol{n} - \boldsymbol{\Lambda}^{-1}\boldsymbol{\Gamma}\boldsymbol{i}(t) + \boldsymbol{d}_\varphi \tag{7.3.33}$$

The linearized across-receiver pseudorange and carrier phase navigation equations (7.3.32) and (7.3.33) will be used in the rest of this chapter.

The hardware biases depend on the satellite system and the frequency band. They are constant or slowly varying in time due to variation of receiver temperature during operation, aging of electronic parts, and other physical reasons. Typically, receivers carry out one or several radio frequency conversions as the signal travels from the antenna to the digital processing component. Conversion from the carrier frequency to an intermediate frequency that is typically several tens of MHz, also referred to as down-conversion or frequency shifting, is performed independently for each radio frequency. Receivers have dedicated intermediate frequency channels, including an intermediate frequency oscillator for each frequency band. Therefore, a reasonable assumption is that each combination of satellite system and frequency corresponds to one bias specific to a certain intermediate frequency channel. The number of such combinations is generally less than the number of satellites. Each signal (p, b) corresponds to a pair (Σ^p, b) where $b \in F^p$.

In the following we will use symbols G for GPS, R for GLONASS, E for Galileo, B for Beidou, Q for QZSS, and S for the SBAS system. For example, for a dual-frequency and dual-system receiver supporting L1 and L2 bands for GPS and GLONASS, the pair (Σ^p, b) takes four values: (G, L1), (G, L2), (R, L1), and (R, L2). This means that there are four different variables $d_{L_1,G,P}$, $d_{L_2,G,P}$, $d_{L_1,R,P}$, and $d_{L_2,R,P}$ in vector (7.3.30). For carrier phase hardware biases, as already discussed in the previous section, there are four variables $d_{L_1,G,\varphi}$, $d_{L_2,G,\varphi}$, $d_{L_1,R,\varphi}$, and $d_{L_2,R,\varphi}$.

The entries of vector (7.3.31) related to GPS are $d_{L_1,G,\varphi}$ and $d_{L_2,G,\varphi}$. The entries of vector (7.3.31) related to GLONASS can be expressed according to (7.2.14) and Table 7.1.1 as

$$d^p_{L_1,\varphi} = d^0_{L_1,\varphi} + \frac{f^p_{L_1}}{c}\mu_{L_1,\varphi} = d^0_{L_1,\varphi} + \frac{1.602 \cdot 10^9 + 5.625 \cdot 10^5 l(p)}{c}\mu_{L_1,\varphi}$$
$$\equiv d_{L_1,R,\varphi} + l(p)\overline{\mu}_{L_1,R,\varphi} \tag{7.3.34}$$

$$d^p_{L_2,\varphi} = d^0_{L_2,\varphi} + \frac{f^p_{L_2}}{c}\mu_{L_2,\varphi} = d^0_{L_2,\varphi} + \frac{1.246 \cdot 10^9 + 4.375 \cdot 10^5 l(p)}{c}\mu_{L_2,\varphi}$$
$$\equiv d_{L_2,R,\varphi} + l(p)\overline{\mu}_{L_2,R,\varphi} \tag{7.3.35}$$

where $l(p)$ stands for the GLONASS letter number. In other words, variables $d_{L_1,G,\varphi}$, $d_{L_2,G,\varphi}$, $d_{L_1,R,\varphi}$, and $d_{L_2,R,\varphi}$ stand for the constant terms $d^0_{km,b,\varphi}$, whereas $\overline{\mu}_{L_1,R,\varphi}$ and $\overline{\mu}_{L_2,R,\varphi}$ denote "slope" values.

In the case of a multifrequency and multisystem receiver supporting

- L1, L2, and L5 bands for GPS
- L1 and L2 GLONASS
- L1, E5a, E5b, and E6 Galileo
- L1, L2, L5, and E6 QZSS
- L1 and L5 SBAS
- B1, B2, and B3 Beidou

the signals (L1 GPS, L1 Galileo, L1 SBAS, L1 QZSS), (L2 GPS, L2 QZSS), (L5 GPS, E5a Galileo, L5 SBAS, L5 QZSS), (E6 Galileo, E6 QZSS), respectively, can share the same channel. Therefore, there are ten combinations: (G/E/S/Q, L1), (G/Q, L2), (G/E/S/Q, L5), (E, E5b), (R, L1), (R, L2), (E/Q, E6), (B1, B), (B2, B), and (B3, B). It means that there are up to 10 different independent slowly varying or constant variables $d_{L_1,G/E/S/Q,P}$, $d_{L_2,G/Q,P}$, $d_{L_5,G/E/S/Q,P}$, $d_{E_{5b},E,P}$, $d_{L_1,R,P}$, $d_{L_2,R,P}$, $d_{E_6,E/Q,P}$, $d_{B_1,B,P}$, $d_{B_2,B,P}$, $d_{B_3,B,P}$ and up to 10 different independent variables $d_{L_1,G/E/S,Q,\varphi}$, $d_{L_2,G/Q,\varphi}$, $d_{L_5,G/E/S/Q,\varphi}$, $d_{E_{5b},E,\varphi}$, $d_{L_1,R,\varphi}$, $d_{L_2,R,\varphi}$, $d_{E_6,E/Q,\varphi}$, $d_{B_1,B,\varphi}$, $d_{B_2,B,\varphi}$, $d_{B_3,B,\varphi}$ among entries of the vectors (7.3.30) and (7.3.31), respectively, except for GLONASS. For GLONASS, the expressions (7.3.34) and (7.3.35) apply.

Note that the biases vector $\boldsymbol{d}_P$ and the single difference time variable $\xi(t)$ appear as a sum in equation (7.3.32). This means that one of the biases, say $d_{L_1,G/E/S/Q,P}$, can be combined with $\xi(t)$, while others can be differenced with $d_{L_1,G/E/S/Q,P}$. We therefore can formulate new bias variables,

$$\begin{aligned}
&\eta_1 = d_{L_2,G/Q,P} - d_{L_1,G/E/S/Q,P} && \eta_2 = d_{L_1,R,P} - d_{L_1,G/E/S/Q,P} && \eta_3 = d_{L_2,R,P} - d_{L_1,G/E/S/Q,P}\\
&\eta_4 = d_{L_5,G/E/S/Q,P} - d_{L_1,G/E/S/Q,P} && \eta_5 = d_{E_{5b},E,P} - d_{L_1,G/E/S/Q,P} && \eta_6 = d_{E_6,E/Q,P} - d_{L_1,G/E/S/Q,P}\\
&\eta_7 = d_{B_1,B,P} - d_{L_1,G/E/S/Q,P} && \eta_8 = d_{B_2,B,P} - d_{L_1,G/E/S/Q,P} && \eta_9 = d_{B_3,B,P} - d_{L_1,G/E/S/Q,P}
\end{aligned} \tag{7.3.36}$$

This reparameterization is sometimes referred to as establishing a bias datum. The linearized equations (7.3.32) can now be expressed in the form

$$\mathbf{b}_P(t) = \mathbf{A}\, d\mathbf{x}(t) + \mathbf{e}\xi(t) + \boldsymbol{\Gamma}\mathbf{i}(t) + \mathbf{W}_\eta \boldsymbol{\eta} \tag{7.3.37}$$

The bias vector $\boldsymbol{\eta}$ has the appropriate dimension m_η. It is three dimensional for dual-band and dual-system GPS/GLONASS receivers. In the case of the multiband, multisystem receiver considered above, the vector $\boldsymbol{\eta}$ is nine dimensional with variables defined in (7.3.36). It is one dimensional in the case of dual-band GPS-only receivers or single-band GPS/GLONASS receivers. There are three biases η_1, η_4, η_6 in the case of dual-system multiband (GPS L1, L2, L5) / (Galileo L1, E5a, E6) receivers.

The *bias allocation matrix* $\mathbf{W}_\eta$ has dimensions $n \times m_\eta$ and allocates a single bias, or none, to a certain signal. No bias is allocated to the signal $s_i = (p_i, b_i)$ if Σ^{p_i} is GPS, Galileo, SBAS, or QZSS, and $b_i = \text{L1}$. In this case, the row of $\mathbf{W}_\eta$ consists of zeroes. In the case of other signals $s_i = (p_i, b_i)$, the ith row $\mathbf{W}_{\eta,i}$ of the allocation matrix has one and only one unit entry while others are zero. The row is defined as follows:

$$\left.\begin{array}{lll}
\mathbf{W}_{\eta,i} = (0,0,0,0,0,0,0,0,0) & \text{if} & \Sigma^{p_i} = \text{GPS, Galileo, SBAS, or QZSS}, b_i = \text{L1} \\
\mathbf{W}_{\eta,i} = (1,0,0,0,0,0,0,0,0) & \text{if} & \Sigma^{p_i} = \text{GPS or QZSS}, b_i = \text{L2} \\
\mathbf{W}_{\eta,i} = (0,1,0,0,0,0,0,0,0) & \text{if} & \Sigma^{p_i} = \text{GLONASS}, b_i = \text{L1} \\
\mathbf{W}_{\eta,i} = (0,0,1,0,0,0,0,0,0) & \text{if} & \Sigma^{p_i} = \text{GLONASS}, b_i = \text{L2} \\
\mathbf{W}_{\eta,i} = (0,0,0,1,0,0,0,0,0) & \text{if} & \Sigma^{p_i} = \text{GPS, Galileo,SBAS, or QZSS}, b_i = \text{L5(E5a)} \\
\mathbf{W}_{\eta,i} = (0,0,0,0,1,0,0,0,0) & \text{if} & \Sigma^{p_i} = \text{Galileo}, b_i = \text{E5}b \\
\mathbf{W}_{\eta,i} = (0,0,0,0,0,1,0,0,0) & \text{if} & \Sigma^{p_i} = \text{QZSS or Galileo}, b_i = \text{E6} \\
\mathbf{W}_{\eta,i} = (0,0,0,0,0,0,1,0,0) & \text{if} & \Sigma^{p_i} = \text{Beidou}, b_i = \text{B1} \\
\mathbf{W}_{\eta,i} = (0,0,0,0,0,0,0,1,0) & \text{if} & \Sigma^{p_i} = \text{Beidou}, b_i = \text{B2} \\
\mathbf{W}_{\eta,i} = (0,0,0,0,0,0,0,0,1) & \text{if} & \Sigma^{p_i} = \text{Beidou}, b_i = \text{B3}
\end{array}\right\} \tag{7.3.38}$$

Consider a dual-band GPS/GLONASS receiver as an example. Suppose it tracks six GPS satellites and six GLONASS satellites. The total number of dual-band signals is 24. Let the signals be ordered in the following way: 6 GPS L1, 6 GPS L2, 6 GLONASS L1, and 6 GLONASS L2 signals. The biases allocation matrix presented in the linearized single-difference pseudorange equation (7.3.37) takes the following form:

$$\mathbf{W}_\eta^{\;T} = \begin{bmatrix} 0 & 0 & & 0 & 1 & 1 & & 1 & 0 & 0 & & 0 & 0 & 0 & & 0 \\ 0 & 0 & \vdots & 0 & 0 & 0 & \vdots & 0 & 1 & 1 & \vdots & 1 & 0 & 0 & \vdots & 0 \\ 0 & 0 & & 0 & 0 & 0 & & 0 & 0 & 0 & & 0 & 1 & 1 & & 1 \end{bmatrix} \tag{7.3.39}$$

Notice that the matrix is transposed.

Now consider hardware biases affecting the across-receiver carrier phase measurements in equation (7.3.33). For all signals except GLONASS, $d^p_{b,\varphi} \equiv d^0_{b,\varphi}$ is one of variables $d_{L_1,G/E/S/Q,\varphi}$, $d_{L_2,G/Q,\varphi}$, $d_{L_5,G/E/S/Q,\varphi}$, $d_{E_{5b},E,\varphi}$, $d_{E_6,E/Q,\varphi}$, $d_{B_1,B,\varphi}$, $d_{B_2,B,\varphi}$, and $d_{B_3,B,\varphi}$. For GLONASS signals we have expressions (7.3.34) and (7.3.35).

The vector $\boldsymbol{d}_\varphi$ appears as a sum with the ambiguity vector $\boldsymbol{n}$ in (7.3.33). The values $l(p)\overline{\mu}_{L_1,\varphi}$ and $l(p)\overline{\mu}_{L_2,\varphi}$ in (7.3.34) and (7.3.35) appear linearly dependent on the letter $l(p)$ of the GLONASS signal. Similarly as discussed above, each combination of satellite system and frequency corresponds to one certain carrier phase bias. In the case of multisystem and multifrequency receivers we have 10 different terms: $d_{L_1,G/E/S/Q,\varphi}$, $d_{L_2,G/Q,\varphi}$, $d_{L_5,G/E/S/Q,\varphi}$, $d_{E_{5b},E,\varphi}$, $d_{E_6,E/Q,\varphi}$, $d_{B_1,B,\varphi}$, $d_{B_2,B,\varphi}$, $d_{B_3,B,\varphi}$, $d_{L_1,R,\varphi}$, and $d_{L_2,R,\varphi}$. These terms are combined with ambiguities and thus destroy the integerness. The fractional part of the ambiguities will be common for all measurements inside the group of signals (p, b) having the same combination of system and frequency band (Σ^p, b). For example, ambiguities will have the same fractional part for signals among the (GPS/ Galileo/ SBAS/ QZSS, L1) group, or for signals among the (GLONASS, L2) group. Allowing different reference satellites for each group when forming double differences, we guarantee that the terms $d_{L_1,G/E/S/Q,\varphi}$, $d_{L_2,G/Q,\varphi}$, $d_{L_5,G/E/S/Q,\varphi}$, $d_{E_{5b},E,\varphi}$, $d_{E_6,E/Q,\varphi}$, $d_{B_1,B,\varphi}$, $d_{B_2,B,\varphi}$, $d_{B_3,B,\varphi}$, $d_{L_1,R,\varphi}$, and $d_{L_2,R,\varphi}$ vanish and thus allow integer fixing of double-difference ambiguities. Let the across-receiver ambiguity vector $\boldsymbol{n}$ be expressed in the form of concatenation of 10 groups:

$$\boldsymbol{n} = \begin{pmatrix} \boldsymbol{n}_{L_1,G/E/S/Q} \\ \boldsymbol{n}_{L_2,G/Q} \\ \boldsymbol{n}_{L_5,G/E/S/Q} \\ \boldsymbol{n}_{E_{5b},E} \\ \boldsymbol{n}_{L_1,R} \\ \boldsymbol{n}_{L_2,R} \\ \boldsymbol{n}_{E_6,E} \\ \boldsymbol{n}_{B_1,B} \\ \boldsymbol{n}_{B_2,B} \\ \boldsymbol{n}_{B_3,B} \end{pmatrix} \tag{7.3.40}$$

Let $\boldsymbol{n}_\alpha$ be the ambiguity vector of a certain group, $a = 1, \ldots, 10$. The actual number of groups depends on the receiver hardware. Each vector $\boldsymbol{n}_\alpha$ has its own dimension n_α. Also, let $\{v\}$ be the fractional part of the value v, $\{v\} = v - [v]$, with $[v]$ being the integer part of v. The ambiguities inside each group have common fractional parts

$$\{N_{a,i}\} = \{N_{a,r_a}\} \tag{7.3.41}$$

where $N_{a,i}$ denotes the ith entry of vector N_α. The reference ambiguity r_a in (7.3.41) is chosen independently for each group. The index i in (7.3.41) varies in the range $i = 1, \ldots, n_a$. An alternative form of the condition (7.3.41) is

$$\{n_a N_{a,i}\} = \left\{ \sum_{j=1}^{n_a} N_{a,j} \right\} \tag{7.3.42}$$

which does not depend on the choice of the reference ambiguity.

The variables $\overline{\mu}_{L_1,R,\varphi}$ and $\overline{\mu}_{L_2,R,\varphi}$ in (7.3.34) and (7.3.35) introduce the frequency dependency of the bias inside the L1 and L2 group for the GLONASS system. Therefore, we have only two groups of signals involving $\mu^p_{b,\varphi} \neq 0$. Let

$$\boldsymbol{\mu} = \begin{pmatrix} \overline{\mu}_{L_1,R,\varphi} \\ \overline{\mu}_{L_2,R,\varphi} \end{pmatrix} \tag{7.3.43}$$

and $\boldsymbol{W}_\mu$ be the $n \times 2$ *bias allocation matrix* corresponding to the bias vector $\boldsymbol{\mu}$ with rows defined as follows:

$$\left.\begin{array}{ll} \boldsymbol{W}_{\mu,i} = [l(p_i), 0] & \text{if } \Sigma^{p_i} = \text{GLONASS}, b_i = L_1 \\ \boldsymbol{W}_{\mu,i} = [0, l(p_i)] & \text{if } \Sigma^{p_i} = \text{GLONASS}, b_i = L_2 \\ \boldsymbol{W}_{\mu,i} = [0, 0] & \text{otherwise} \end{array}\right\} \tag{7.3.44}$$

For the example of the GPS/GLONASS/L1/L2 receiver introduced earlier, we have (the matrix is transposed for convenience),

$$\boldsymbol{W}^T_\mu = \left[\begin{array}{cccc|cccc|cccc|cccc} 0 & 0 & \vdots & 0 & 0 & 0 & \vdots & 0 & l_1 & l_2 & \vdots & l_{n_5} & 0 & 0 & \vdots & 0 \\ 0 & 0 & \vdots & 0 & 0 & 0 & \vdots & 0 & 0 & 0 & \vdots & 0 & l_1 & l_2 & \vdots & l_{n_6} \end{array}\right] \tag{7.3.45}$$

where n_5 is the number of ambiguities of the fifth group in (7.3.40) (GLONASS, L1). Similarly, n_6 denotes the number of (GLONASS, L2) ambiguities, which is not necessarily equal to n_5. The across-receiver differenced carrier phase linearized measurements in equation (7.3.33) can be expressed in the form

$$\boldsymbol{b}_\varphi(t) = \boldsymbol{\Lambda}^{-1}\boldsymbol{A}\, d\boldsymbol{x}(t) + \boldsymbol{\Lambda}^{-1}\boldsymbol{e}\xi(t) + \boldsymbol{n} - \boldsymbol{\Lambda}^{-1}\boldsymbol{\Gamma}\boldsymbol{i}(t) + \boldsymbol{W}_\mu\boldsymbol{\mu} \tag{7.3.46}$$

The variables $\boldsymbol{\mu}$ are absent if the hardware does not support GLONASS. The vector $\boldsymbol{\mu}$ becomes a scalar if only the L1 or L2 frequency band is available for GLONASS.

In summary, the linearized single-difference measurement equations are (7.3.37) and (7.3.46). The recursive estimation algorithms described in Chapter 3 can now be easily applied. In expression (7.3.46) the ambiguity vector is presented as a sum with $\boldsymbol{W}_\mu\boldsymbol{\mu}$. Therefore, we cannot distinguish between $\boldsymbol{n}$ and $\boldsymbol{W}_\mu\boldsymbol{\mu}$ until we impose the condition (7.3.41) or (7.3.42). The floating ambiguity estimation filter estimates the sum $\boldsymbol{n} + \boldsymbol{W}_\mu\boldsymbol{\mu}$. In the following sections we describe filtering algorithms. Real-valued ambiguities are commonly known as floating ambiguities. The determination of ambiguities subjected to the constraint (7.3.41), resulting in integer-valued ambiguities or fixed ambiguities, will be described in Section 7.10.

7.4 RTK ALGORITHM FOR STATIC AND SHORT BASELINES

Many surveying applications use RTK for static positioning. The base station is m and its actual coordinates (7.3.1) are known. The position of the rover station k is unknown and considered stationary. Its approximate position is

$$\boldsymbol{x}_{k,0} = \begin{pmatrix} x_{k,0} \\ y_{k,0} \\ z_{k,0} \end{pmatrix}$$

and its unknown position is expressed as

$$\boldsymbol{x}_k = \begin{pmatrix} x_k \\ y_k \\ z_k \end{pmatrix} = \begin{pmatrix} x_{k,0} + dx_k \\ y_{k,0} + dy_k \\ z_{k,0} + dz_k \end{pmatrix} \equiv \boldsymbol{x}_{k,0} + d\boldsymbol{x} \tag{7.4.1}$$

The across-receiver ionospheric delay partially cancels in equations (7.3.37) and (7.3.46). The residual delay has a magnitude that is approximately proportional to the baseline length. Assume that for each satellite the magnitude of the across-receiver ionospheric delay is constrained by the following expression:

$$|\boldsymbol{i}^p_{km}| \approx S \times 10^{-6} \times \|\boldsymbol{x}_k - \boldsymbol{x}_m\| \tag{7.4.2}$$

where the scaling factor S typically varies from 1 to 4 depending on the solar activity in the 11-year cycle, taking values ≈ 1 for years of low solar activity and ≈ 4 for high solar activity. This ionospheric modeling is true for typical lengths encountered in surveying, i.e., not greater than several tens of kilometers. The residual ionosphere, therefore, is about S millimeters for every kilometer of baseline length.

For short baselines of about 5 km or less, the across-receiver ionosphere delay can be neglected in navigation equations (7.3.37) and (7.3.46), which then take the form

$$\boldsymbol{b}_P(t) = \boldsymbol{A}\, d\boldsymbol{x} + \boldsymbol{e}\xi(t) + \boldsymbol{W}_\eta \eta \tag{7.4.3}$$

$$\boldsymbol{b}_\varphi(t) = \boldsymbol{\Lambda}^{-1}\boldsymbol{A}\, d\boldsymbol{x} + \boldsymbol{\Lambda}^{-1}\boldsymbol{e}\xi(t) + \boldsymbol{n} + \boldsymbol{\Lambda}^{-1}\boldsymbol{W}_\mu \boldsymbol{\mu} \tag{7.4.4}$$

The time-invariant parameters to be estimated in (7.4.3) and (7.4.4) are as follows: correction to the approximate rover position $d\boldsymbol{x}$, carrier phase ambiguity $\boldsymbol{n}$, pseudorange biases η defined in (7.3.36), and carrier phase biases $\boldsymbol{\mu}$ defined in (7.3.43). The last term is absent if there are no GLONASS measurements. The across-receiver carrier phase ambiguities are estimated as real-valued approximation, yielding the so-called floating ambiguity. The integer-valued (fixed) ambiguity solution will be considered later.

The ambiguity vector $\boldsymbol{n}$ includes the across-receiver windup terms (Section 6.2.4). Rotation of the antenna around the vertical axis generates one carrier phase cycle per each full rotation. However, this would not violate the assumption of a constant value of $\boldsymbol{n}$ since both the base and rover stations remain stationary. Later, when discussing the kinematic processing we will address this question in detail.

Estimation of the floating ambiguity is a necessary first step. The condition (7.3.41) is not taken into account and the term $\boldsymbol{n} + \boldsymbol{\Lambda}^{-1}\boldsymbol{W}_\mu \boldsymbol{\mu}$ in (7.4.4) can be estimated as constant real-valued floating ambiguity vector. So, at this stage we replace (7.4.4) with the equation

$$\boldsymbol{b}_{\varphi}(t) = \boldsymbol{\Lambda}^{-1}\boldsymbol{A}\ \ d\boldsymbol{x} + \boldsymbol{\Lambda}^{-1}\boldsymbol{e}\xi(t) + \boldsymbol{n} \tag{7.4.5}$$

The kinematic parameter to be estimated in (7.4.3) and (7.4.5) is $\xi(t)$ which is the across-receiver clock error. Therefore, we have a mixed set of variables, static and arbitrary varying. The numerical scheme for incremental least squares is described in Section 3.2.

Let $\boldsymbol{C}_P$ be a covariance matrix of the across-receiver pseudorange unmodeled errors including noise and multipath. Define it as a diagonal matrix, assuming that hardware noise and multipath errors are independent for different satellites, taking the form

$$\boldsymbol{C}_P = \mathrm{diag}\left((\sigma_{1,P})^2, (\sigma_{2,P})^2, \ldots, (\sigma_{n,P})^2\right) \tag{7.4.6}$$

Individual errors corresponding to signals (p, b) and (q, c) can be taken to be independent of each other, which suggests the diagonal form of the matrix (7.4.6). The variance $\sigma_{i,P}^2$, corresponding to the signal $s_i = (p_i, b_i)$, consists of two terms. The first term reflects a noise component that depends on the signal frequency band, and a second term reflects the variance of errors that are dependent on satellite elevation. The latter errors include a multipath generated by reflection from the ground and the residual ionosphere delay of signals that pass through the layer of the ionosphere at different elevation angles. A good practical assumption is

$$\sigma_{i,P}^2 = \sigma_{b_i,P}^2 + \left(\frac{\overline{\sigma}_P}{\varepsilon + \sin\alpha^{p_i}}\right)^2 \tag{7.4.7}$$

where $\sigma_{b_i,P}$ is a standard deviation of the noise component depending on the frequency band, α^{p_i} is the elevation of the satellite p_i, and ε and $\overline{\sigma}_P$ are constants. A usual choice that works in practice for most receivers is

$$\sigma_{b_i,P} \sim 0.25 - 2\,m \quad \varepsilon \sim 0.1 \quad \overline{\sigma}_P \sim 0.5 - 1\,m \tag{7.4.8}$$

Let $\boldsymbol{C}_{\varphi}$ be a covariance matrix of the across-receiver carrier phase noise. Using the same reasoning as above, we assume

$$\boldsymbol{C}_{\varphi} = \mathrm{diag}\left((\sigma_{1,\varphi})^2, (\sigma_{2,\varphi})^2, \ldots, (\sigma_{n,\varphi})^2\right) \tag{7.4.9}$$

$$(\sigma_{i,\varphi})^2 = (\sigma_{b_i,\varphi})^2 + \left(\frac{\overline{\sigma}_{\varphi}}{\varepsilon + \sin\alpha^{p_i}}\right)^2 \tag{7.4.10}$$

$$\sigma_{b_i,\varphi} \sim 0.01 - 0.05 \text{ cycle} \quad \varepsilon \sim 0.1 \quad \overline{\sigma}_{\varphi} \sim 0.01 - 0.05 \text{ cycle} \tag{7.4.11}$$

Covariance matrices (7.4.6) and (7.4.9) are used in the algorithm, described in Section 3.2.

Let us present the set of navigation equations (7.4.3) and (7.4.5) in the form (3.2.1). Recall the notation that $t+1$ does not necessarily mean time t incremented by one second. It means the discrete time instance immediately following the time instance t. The actual time step will be denoted by δt.

Consider the case of multisystem and multiple frequency band receivers. The structure of the navigation equations and the set of parameters were described in the previous section. We use $\boldsymbol{R}^n$ to denote the n-dimensional real-valued Euclidean space and $\boldsymbol{R}^{n\times m}$ for the space of real-valued $n \times m$ matrices. Let $\boldsymbol{y}$ be the vector of constant parameters (not including $\boldsymbol{\mu}$ in the floating solution for reasons discussed above),

$$\boldsymbol{y} = \begin{pmatrix} \eta \\ d\boldsymbol{x} \\ \boldsymbol{n} \end{pmatrix} \tag{7.4.12}$$

where $\boldsymbol{\eta} \in R^9$ is the vector of intersignal biases described in (7.3.36), $d\boldsymbol{x} \in \boldsymbol{R}^3$ is a vector of corrections to the rover position, and $\boldsymbol{n} \in \boldsymbol{R}^n$ is a vector of carrier phase ambiguities of signals, structured according to (7.3.40). The total number of signals is n.

The vector of time-varying parameters is one dimensional since only one variable $\xi(t)$ is time dependent. Rewrite measurement equations (7.4.3) and (7.4.5) in the form (3.2.1),

$$\boldsymbol{J}\xi(t) + \boldsymbol{W}(t)\boldsymbol{y} = \boldsymbol{b}(t) \tag{7.4.13}$$

where

$$\boldsymbol{J} = \begin{pmatrix} \boldsymbol{e} \\ \hdashline \boldsymbol{\Lambda}^{-1}\boldsymbol{e} \end{pmatrix} \in \boldsymbol{R}^{2n\times 1} \tag{7.4.14}$$

and $\boldsymbol{\Lambda} \in R^{n\times n}$ is a diagonal matrix of wavelengths, defined in (7.3.28),

$$\boldsymbol{W}(t) = \left[\begin{array}{c:c:c} \boldsymbol{W}_\eta & \boldsymbol{A}(t) & 0 \\ \hdashline 0 & \boldsymbol{\Lambda}^{-1}\boldsymbol{A}(t) & \boldsymbol{I}_n \end{array}\right] \tag{7.4.15}$$

The bias allocation matrix $\boldsymbol{W}_\eta \in \boldsymbol{R}^{n\times 9}$ is defined in (7.3.38). The zero matrices have appropriate size, and $\boldsymbol{I}_n$ denotes $n \times n$ identity matrix. The directional cosine matrix $\boldsymbol{A}(t) \in \boldsymbol{R}^{n\times 3}$ is defined in (7.3.20), and the time-varying entries of the matrix are defined in (7.3.15). The right-hand side vector $\boldsymbol{b}(t)$ has the form

$$\boldsymbol{b}(t) = \begin{pmatrix} \boldsymbol{b}_P(t) \\ \hdashline \boldsymbol{b}_\varphi(t) \end{pmatrix} \in \boldsymbol{R}^{2n} \tag{7.4.16}$$

The vectors $\boldsymbol{b}_P(t)$ and $\boldsymbol{b}_\varphi(t)$ are defined in (7.3.21) and (7.3.22), respectively.

In the following, we assume that the set of signals does not change during the operation of the algorithm, and that there are no cycle slips in the carrier phase measurements. Both assumptions will be relaxed later.

Let $\boldsymbol{C}$ be the diagonal matrix of measurements composed of blocks $\boldsymbol{C}_P$ and $\boldsymbol{C}_\varphi$ as defined above in (7.4.6) and(7.4.9),

$$\boldsymbol{C} = \left[\begin{array}{c|c} \boldsymbol{C}_P & 0 \\ \hline 0 & \boldsymbol{C}_\varphi \end{array}\right] \in \boldsymbol{R}^{2n\times 2n} \tag{7.4.17}$$

The Cholesky decomposition of the diagonal matrix $\boldsymbol{C}$ takes the simple form

$$\boldsymbol{C} = \Sigma^2 \quad \Sigma = \left[\begin{array}{c|c} \Sigma_P & 0 \\ \hline 0 & \Sigma_\varphi \end{array}\right] \tag{7.4.18}$$

with $\Sigma_P = \text{diag}(\sigma_{1,P}, \sigma_{2,P}, \dots, \sigma_{n,P})$ and $\Sigma_\varphi = \text{diag}(\sigma_{1,\varphi}, \sigma_{2,\varphi}, \dots, \sigma_{n,\varphi})$. The forward and backward solutions are simply

$$\boldsymbol{F}_\Sigma \boldsymbol{b} = \boldsymbol{B}_\Sigma \boldsymbol{b} = \Sigma^{-1}\boldsymbol{b} \tag{7.4.19}$$

Let $n_y = \dim(\boldsymbol{y})$ be the number of constant estimated parameters. Algorithm 2 (Table 3.2.1) of Chapter 3 takes the following form. Start with $t = t_0$, $\xi(t_0) = 0$, $\hat{\boldsymbol{D}}(t_0) \in R^{n_y \times n_y}$, $\hat{\boldsymbol{D}}(t_0) = 0$, and $\boldsymbol{y}(t_0) = 0$, and continue as described in Table 7.4.1. The updated estimate of the constant parameters vector is "disassembled" into the following parts:

$$\boldsymbol{y}^T(t+1) = \left(\eta^T(t+1), d\boldsymbol{x}^T(t+1), \boldsymbol{n}^T(t+1)\right) \tag{7.4.20}$$

The updated estimate of the time-varying scalar parameter $\xi(t+1)$ is the across-receiver single-difference clock error.

7.4.1 Illustrative Example

We processed 2660 epochs of raw data collected on February 15, 2013, by two Triumph-1 receivers by Javad GNSS. The data collection started at 07:47:58.00 and finished at 08:32:17.00. The raw data included dual-band GPS and GLONASS pseudorange and carrier phase data. The approximate ECEF position of the base in the WGS-84 frame was

$$\begin{aligned} \boldsymbol{x}_m &= \left(-2681608.127, -4307231.857, 3851912.054\right)^T \\ &= 37.390538°N, -121.905680°E, -11.06\,\text{m} \end{aligned} \tag{7.4.21}$$

The approximate ECEF position of the rover was

$$\begin{aligned} \boldsymbol{x}_{k,0} &= \left(-2681615.678, -4307211.353, 3851926.005\right)^T \\ &-37.390711°N, -121.905874°E, -13.25\,\text{m} \end{aligned} \tag{7.4.22}$$

TABLE 7.4.1 Algorithm for Static and Short Baseline RTK.

Compute the right-hand side vector $\boldsymbol{b}(t+1)$ according to (7.4.16), (7.3.21), and (7.3.22)	$\boldsymbol{b}(t+1) = \left(\begin{array}{c} \boldsymbol{b}_P(t+1) \\ \hline \boldsymbol{b}_\varphi(t+1) \end{array}\right) \in \boldsymbol{R}^{2n}$ $\boldsymbol{b}_P(t+1) = \begin{pmatrix} \overline{P}^{p_1}_{km,b_1}(t+1) - \rho^{p_1}_{k,0}(t+1-\tilde{\tau}^{p_1}_k) + \rho^{p_1}_m(t+1-\tilde{\tau}^{p_1}_m) \\ \overline{P}^{p_2}_{km,b_2}(t+1) - \rho^{p_2}_{k,0}(t+1-\tilde{\tau}^{p_2}_k) + \rho^{p_2}_m(t+1-\tilde{\tau}^{p_2}_m) \\ \vdots \\ \overline{P}^{p_n}_{km,b_n}(t+1) - \rho^{p_n}_{k,0}(t+1-\tilde{\tau}^{p_n}_k) + \rho^{p_n}_m(t+1-\tilde{\tau}^{p_n}_m) \end{pmatrix}$ $\boldsymbol{b}_\varphi(t+1) = \begin{pmatrix} \overline{\varphi}^{p_1}_{km,b_1}(t+1) - \frac{1}{\lambda^{p_1}_{b_1}}\left(\rho^{p_1}_{k,0}(t+1-\tilde{\tau}^{p_1}_k) - \rho^{p_1}_m(t+1-\tilde{\tau}^{p_1}_m)\right) \\ \overline{\varphi}^{p_2}_{km,b_2}(t+1) - \frac{1}{\lambda^{p_2}_{b_2}}\left(\rho^{p_2}_{k,0}(t+1-\tilde{\tau}^{p_2}_k) - \rho^{p_2}_m(t+1-\tilde{\tau}^{p_2}_m)\right) \\ \vdots \\ \overline{\varphi}^{p_n}_{km,b_n}(t+1) - \frac{1}{\lambda^{p_n}_{b_n}}\left(\rho^{p_n}_{k,0}(t+1-\tilde{\tau}^{p_n}_k) - \rho^{p_n}_m(t+1-\tilde{\tau}^{p_n}_m)\right) \end{pmatrix}$
Compute the linearized measurements matrix $\boldsymbol{W}(t+1)$ according to (7.4.15), (7.3.38), (7.3.20), and (7.3.15)	$\boldsymbol{W}(t+1) = \left[\begin{array}{ccc} \boldsymbol{W}_\eta & \boldsymbol{A}(t+1) & 0 \\ \hline 0 & \Lambda^{-1}\boldsymbol{A}(t+1) & \boldsymbol{I}_n \end{array}\right]$ $\boldsymbol{A}(t+1) = \begin{bmatrix} H^{p_1}_{k,1} & H^{p_1}_{k,2} & H^{p_1}_{k,3} \\ H^{p_2}_{k,1} & H^{p_2}_{k,2} & H^{p_3}_{k,3} \\ & \cdots & \\ H^{p_n}_{k,1} & H^{p_n}_{k,2} & H^{p_n}_{k,3} \end{bmatrix}$
Compute the vector $\boldsymbol{J}$	$\boldsymbol{J} = \left(\begin{array}{c} \boldsymbol{e} \\ \hline \Lambda^{-1}\boldsymbol{e} \end{array}\right) \in \boldsymbol{R}^{2n\times 1}$
Square root of diagonal covariance matrix according to (7.4.18)	$\Sigma = \left[\begin{array}{c\|c} \Sigma_P & 0 \\ \hline 0 & \Sigma_\varphi \end{array}\right]$
Weighing	$\overline{\boldsymbol{b}}(t+1) = \Sigma^{-1}\boldsymbol{b}(t+1)$ $\overline{\boldsymbol{J}} = \Sigma^{-1}\boldsymbol{J}$ $\overline{\boldsymbol{W}}(t+1) = \Sigma^{-1}\boldsymbol{W}(t+1)$
Compute the residual vector $\overline{\boldsymbol{r}}(t+1)$	$\overline{\boldsymbol{r}}(t+1) = \overline{\boldsymbol{b}}(t+1) - \overline{\boldsymbol{W}}(t+1)\boldsymbol{y}(t)$
Compute the projection matrix Π	$\gamma = \sqrt{\overline{\boldsymbol{J}}^T\overline{\boldsymbol{J}}}$ $\tilde{\boldsymbol{J}}^T = \gamma^{-1}\overline{\boldsymbol{J}}^T$ $\Pi = \boldsymbol{I}_n - \tilde{\boldsymbol{J}}\tilde{\boldsymbol{J}}^T$
Update the matrix $\hat{\boldsymbol{D}}(t)$	$\hat{\boldsymbol{D}}(t+1) = \hat{\boldsymbol{D}}(t) + \overline{\boldsymbol{W}}^T(t+1)\Pi\overline{\boldsymbol{W}}(t+1)$ $= \hat{\boldsymbol{D}}(t) + \overline{\boldsymbol{W}}^T(t+1)\overline{\boldsymbol{W}}(t+1) - \overline{\boldsymbol{W}}^T(t+1)\tilde{\boldsymbol{J}}\tilde{\boldsymbol{J}}^T\overline{\boldsymbol{W}}(t+1)$

(continued)

TABLE 7.4.1 (*Continued*)

Compute Cholesky decomposition of $\hat{\boldsymbol{D}}(t+1)$	$\hat{\boldsymbol{D}}(t+1) = \hat{\boldsymbol{L}}\hat{\boldsymbol{L}}^T$
Update the estimate of constant parameters $\boldsymbol{y}(t)$	$\boldsymbol{y}(t+1) = \boldsymbol{y}(t) + \boldsymbol{B}_{\hat{L}}\left(\boldsymbol{F}_{\hat{L}}\left(\overline{\boldsymbol{W}}^T(t+1)\,\boldsymbol{\Pi}\bar{\boldsymbol{r}}(t+1)\right)\right)$
Compute the second residual vector $\boldsymbol{r}'(t+1)$	$\boldsymbol{r}'(t+1) = \bar{\boldsymbol{b}}(t+1) - \overline{\boldsymbol{W}}(t+1)\boldsymbol{y}(t+1)$
Compute the estimate of across-receiver clock difference $\xi(t+1)$	$\xi(t+1) = \frac{1}{\gamma}\tilde{\boldsymbol{J}}^T\boldsymbol{r}'(t+1)$

The approximation of the baseline vector was

$$\boldsymbol{bl}_{km,0} = \boldsymbol{x}_{k,0} - \boldsymbol{x}_m = \left(-7.551, \quad 20.504, \quad 13.951\right)^T_{xyz} \tag{7.4.23}$$

whereas the known accurate vector was

$$\boldsymbol{bl}^*_{km} = \boldsymbol{x}_k - \boldsymbol{x}_m = \left(-9.960, \quad 20.634, \quad 15.402\right)^T_{xyz} \tag{7.4.24}$$

Transformation to the geodetic horizon (easting, northing, and up) gave us

$$\begin{aligned} \boldsymbol{bl}_{km,0} &= \left(-17.247, \quad 19.231, \quad -2.187\right)^T_{enu} \\ \boldsymbol{bl}^*_{km} &= \left(-19.361, \quad 19.677, \quad -0.382\right)^T_{enu} \end{aligned} \tag{7.4.25}$$

The algorithm was executed to calculate the correction vector $d\boldsymbol{x}$ in (7.4.1), along with the other parameters in (7.4.12) and the clock error $\xi(t)$. Ten satellites were chosen for the processing, such that the constellation would not change during processing as suggested in assumptions that we made. The set of GPS satellites is defined by their PRNs: 4, 9, 15, 17, 24, 28. The set of GLONASS satellites is defined by their letter numbers –7, 0, 2, and 4, which correspond to carrier frequencies seen in Table 7.4.2. The time of day in seconds varies from 28,078 to 30,737.

TABLE 7.4.2 GLONASS Letters and Frequencies.

Letter	L1 (MHz)	L2 (MHz)
–7	1598.0625	1242.9375
0	1602.0	1246.0
2	1603.125	1246.875
4	1604.25	1247.75

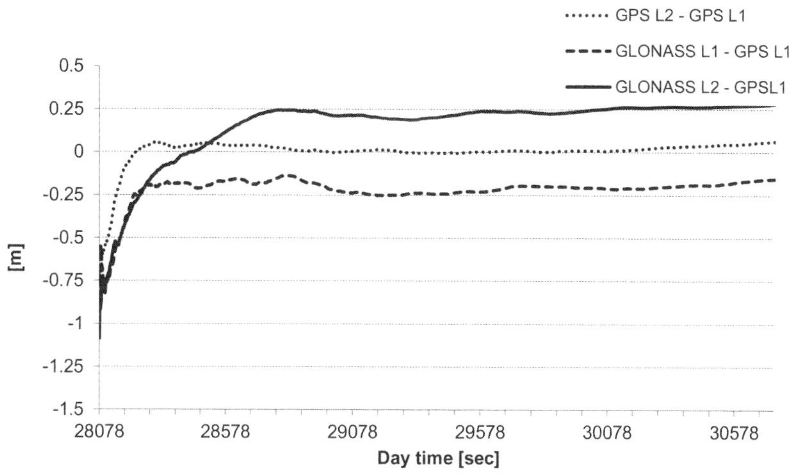

Figure 7.4.1 Across-receiver single-difference hardware bias estimates as a function of time.

Figure 7.4.1 shows the estimates of across-receiver hardware biases. The hardware biases' differences are small but nonzero. Their estimate converges as shown in the figure. Both receivers are of the same type and made with identical parts. Figure 7.4.2 illustrates the convergence of corrections $dx_k(t), dy_k(t), dz_k(t)$ transformed into the geodetic horizon format. The corrections converge to the vector

$$d\bar{\mathbf{x}} = \left(-2.122, \quad 0.451, \quad 1.797\right)^T_{enu} \tag{7.4.26}$$

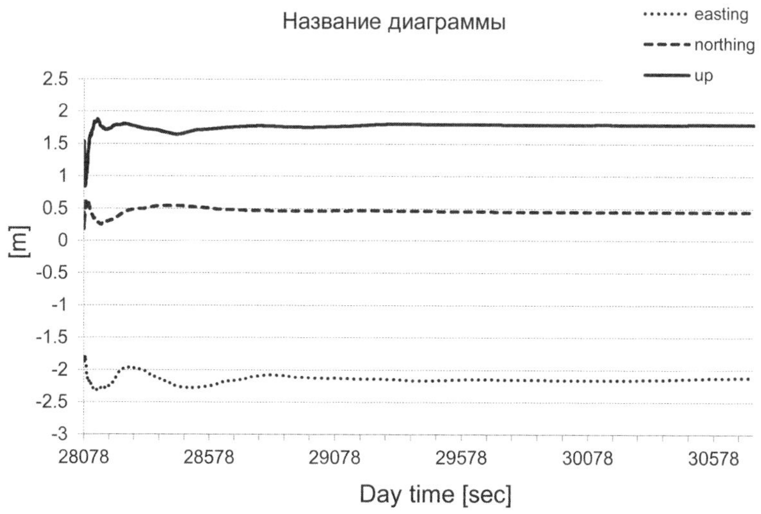

Figure 7.4.2 Corrections of the rover position expressed in easting, northing, and up [m].

which is in agreement with

$$\boldsymbol{bl}_{km}^{*} - \boldsymbol{bl}_{km,0} = \left(-2.114, \quad 0.446, \quad 1.805\right)_{enu}^{T} \tag{7.4.27}$$

which is the difference between correct baseline and initial approximation (7.4.25).

These results illustrate the convergence of the floating ambiguity solution to the correct solution. In order to investigate the convergence of the floating ambiguities satisfying condition (7.3.41), consider the floating estimates of the GPS L1 ambiguities for six satellites. First, wait until they converge, then calculate the integer parts of the settled values and subtract these integer values from the estimated float values. Figure 7.4.3 shows fractional parts converging to approximately equal values.

Figure 7.4.4 illustrates convergence of the GPS L2 ambiguity estimates. After compensation of the integer parts the same way as discussed above, they show convergence to a common fractional value which obviously differs from that corresponding to the L1 frequency band. This happens because different hardware channels have different heterodynes, which confirms that hardware delays are a function of frequency.

Figures 7.4.5 and 7.4.6 show the GLONASS ambiguity estimates for L1 and L2 frequency bands. These estimates have also been corrected by subtracting the integer parts to show how their fractional parts match. The ambiguities demonstrate convergence to real values that have approximately the same fractional values. As suggested earlier in this chapter, the coefficients of the frequency-dependent part of carrier phase biases introduced in (7.3.43) are included in the estimates of the floating ambiguities. This means that they must introduce different fractional parts for ambiguities that correspond to satellites having different frequency letters. The reason for not seeing this disagreement in the figures, or at least the difference are negligibly small, can be explained by the fact that the hardware of the base and rover receiver was identical, i.e., both used Triumph-1 receivers.

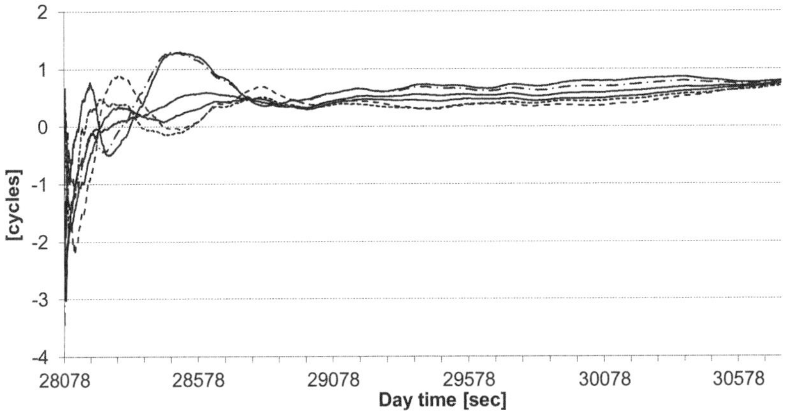

Figure 7.4.3 Convergence of the GPS L1 ambiguity estimates to a common fractional value.

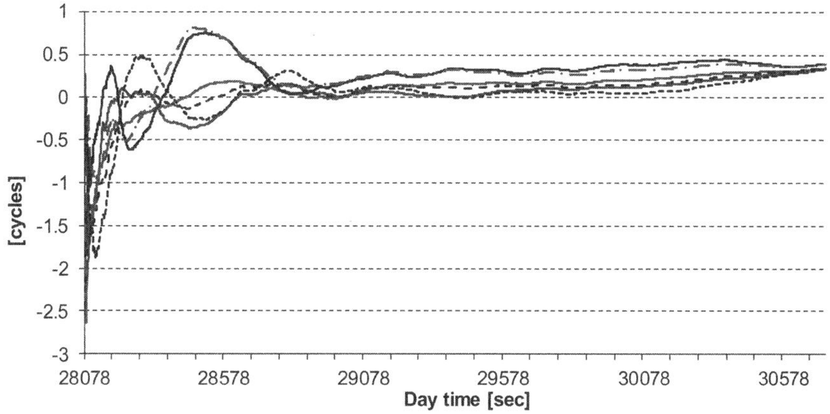

Figure 7.4.4 Convergence of the GPS L2 ambiguity estimates to common fractional value.

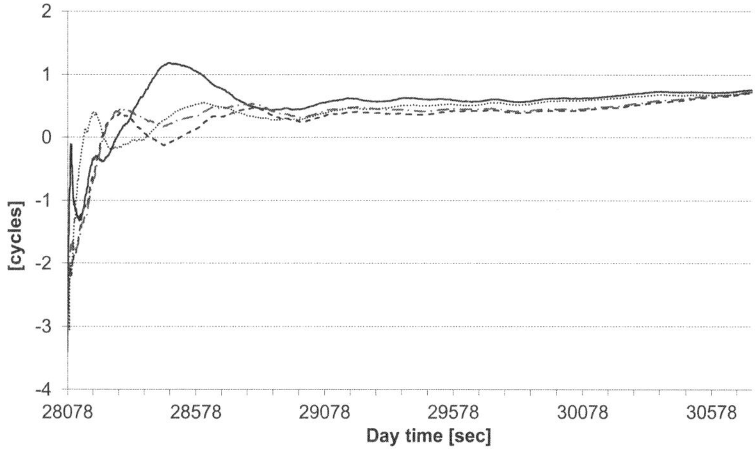

Figure 7.4.5 Convergence of the GLONASS L1 ambiguity estimates to a common fractional value.

Figures 7.4.3 to 7.4.6 illustrate convergence of across-receiver ambiguity estimates. The slow rate of convergence does not allow precise determination of the fractional parts, yet the figures allow us to make a qualitative conclusion about convergence. The ambiguities will jump to their correct integer value after they have been fixed. Fixing ambiguities for this example is considered in Section 7.10.

Figure 7.4.7 illustrates behavior of the time–varying parameter $\xi(t)$. The across-receivers' single-difference clock error varies in time because local oscillators vary independently. The figure shows the long-term instability of the local oscillators. The time drift with respect to the system time is periodically corrected in the receiver firmware. Normally, corrections happen when the difference exceeds

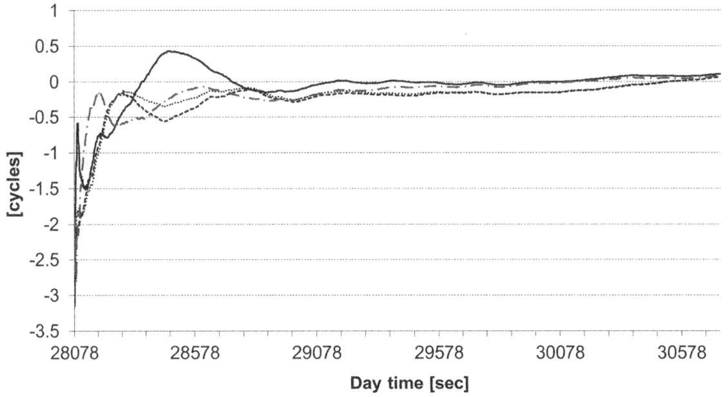

Figure 7.4.6 Convergence of the GLONASS L2 ambiguity estimates to a common fractional value.

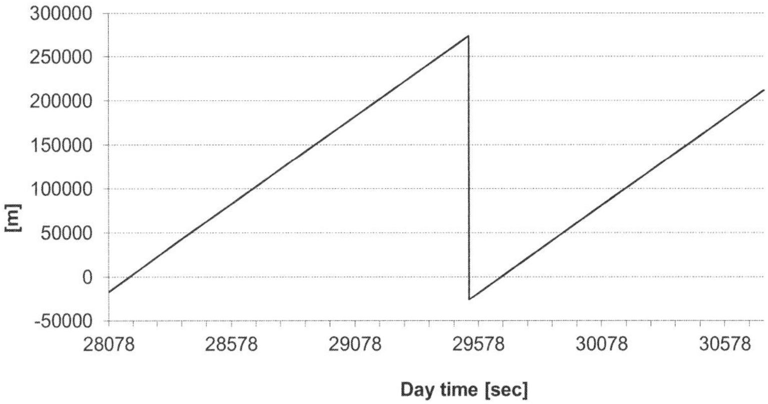

Figure 7.4.7 Across-receiver difference of the clock shift.

plus or minus half a millisecond. The clock is then corrected by one millisecond. Half a millisecond is approximately 149,896 m in the metric scale. Therefore, the single difference of the clock error can experience jumps of about 299,792 m, as can be seen from the figure.

In this section, we discussed the RTK algorithm for processing a short static baseline. The convergence of across-receiver hardware bias estimates, corrections to the rover position, and ambiguity estimates is illustrated. Note that the term ambiguity used in this and following sections differs of that used in Chapter 6, where it was understood to be a physical quantity having an integer-valued nature. Instead, in this chapter we allow across-receiver ambiguities to have a fractional part. Moreover, it includes hardware delays. In this chapter, when focusing on the computational aspects, the term ambiguity is a "lumped" parameter, aggregating all terms appearing as a sum with integer ambiguity in the navigation equations. The algorithm estimates ambiguity as real-valued vectors. The condition of having a common fractional part for ambiguities from each group of signals will be imposed later in Section 7.10, which is devoted to fixing across-receiver ambiguity. In the next section, we extend consideration of the numerical algorithms to the cases of a kinematic rover and different baseline lengths.

7.5 RTK ALGORITHM FOR KINEMATIC ROVERS AND SHORT BASELINES

The rover is allowed to change its position arbitrarily over time. The numerical scheme for parameters estimation using across-receiver differences of pseudorange and carrier phase measurements will be derived. This problem falls into the classification described in Section 3.2, where the static parameters are hardware biases and carrier phase ambiguities, and arbitrarily varying kinematic parameters are the across-receiver clock error and corrections to the approximate rover position.

Considering the navigation equations (7.4.3) and (7.4.5), we do not assume the corrections $\boldsymbol{x}$ to the rover position (7.4.1) to be a constant. Instead, the rover position, i.e., the approximate location, and the correction will be considered as time varying. This problem corresponds to the case of real-time kinematics. Equation (7.4.1) becomes

$$\boldsymbol{x}_k(t) = \begin{pmatrix} x_k(t) \\ y_k(t) \\ z_k(t) \end{pmatrix} = \begin{pmatrix} x_{k,0}(t) + dx_k(t) \\ y_{k,0}(t) + dy_k(t) \\ z_{k,0}(t) + dz_k(t) \end{pmatrix} \equiv \boldsymbol{x}_{k,0}(t) + d\boldsymbol{x}(t) \tag{7.5.1}$$

and the navigation equations (7.4.3) and (7.4.5) take the form

$$\boldsymbol{b}_P(t) = \boldsymbol{A}\boldsymbol{x}(t) + \boldsymbol{e}\xi(t) + \boldsymbol{W}_\eta \boldsymbol{\eta} \tag{7.5.2}$$

$$\boldsymbol{b}_\varphi(t) = \boldsymbol{\Lambda}^{-1}\boldsymbol{A}\boldsymbol{x}(t) + \boldsymbol{\Lambda}^{-1}\boldsymbol{e}\xi(t) + \boldsymbol{n} \tag{7.5.3}$$

The approximation $\boldsymbol{x}_{k,0}(t)$ of the rover position is not constant. An easy choice for the time-varying approximation is the stand-alone position of the rover computed from pseudoranges. Another possibility is using the previously estimated position $\boldsymbol{x}_k(t-1)$, assuming the dynamics are not too rapid. The vector of constant parameters now takes the form

$$\boldsymbol{y} = \begin{pmatrix} \boldsymbol{\eta} \\ \boldsymbol{n} \end{pmatrix} \tag{7.5.4}$$

while the vector of arbitrary varying parameters is

$$d\bar{\boldsymbol{x}}(t) = \begin{pmatrix} \xi(t) \\ d\boldsymbol{x}(t) \end{pmatrix} \tag{7.5.5}$$

We rewrite equations (3.2.1) in the form

$$\boldsymbol{J}(t)\, d\bar{\boldsymbol{x}}(t) + \boldsymbol{W}\boldsymbol{y} = \boldsymbol{b}(t) \tag{7.5.6}$$

where now

$$\boldsymbol{J}(t) = \left[\begin{array}{c|c} \boldsymbol{e} & \boldsymbol{A}(t) \\ \hline \Lambda^{-1}\boldsymbol{e} & \Lambda^{-1}\boldsymbol{A}(t) \end{array}\right] \in \boldsymbol{R}^{2n\times 4}, \qquad \boldsymbol{W} = \left[\begin{array}{c|c} \boldsymbol{W}_\eta & 0 \\ \hline 0 & \boldsymbol{I}_n \end{array}\right] \tag{7.5.7}$$

and apply the algorithm of Section 3.2 the same way as we did in the previous section. Additionally, all other notations used in the description of the algorithm have not changed.

The across-receiver carrier phase ambiguity vector $\boldsymbol{n}$ includes across-receiver windup terms. In the case of a kinematic rover, these terms may not be constant. This fact violates the assumption that $\boldsymbol{n}$ is a constant vector because the rover antenna can experience arbitrary rotations. However, for short baselines these rotations lead to the variation of all carrier phase observables by almost the same windup angle. This means that for short baselines, the windup angle almost does not violate the integer value of double-difference ambiguities; see Section 6.2.4. Working with across-receiver differences, the algorithm detects simultaneous variation of all carrier phase observables and compensates them in such a way that the ambiguity vector $\boldsymbol{n}$ remains constant. We call this procedure windup compensation.

Regarding implementation of the windup compensation we recognize that if all values $\overline{\varphi}^p_{km,b}(t+1)$ in the expression (7.3.22) experience the same variation w_{km}, then the corresponding components of the residual vector $\boldsymbol{r}_\varphi(t+1)$ defined as

$$\boldsymbol{r}(t+1) = \boldsymbol{b}(t+1) - \boldsymbol{W}\boldsymbol{y}(t), \boldsymbol{r}(t+1) = \left(\boldsymbol{r}_P^T(t+1), \boldsymbol{r}_\varphi^T(t+1)\right)^T \tag{7.5.8}$$

will experience the same variation according to equation (7.5.3) and expressions (7.3.22). Partitioning the vector $\boldsymbol{r}(t+1)$ in (7.5.8) into two parts corresponds to the partition (7.4.16). The variation is easily detectable if we compare the mean value

$$\overline{r}_\varphi = \frac{1}{n}\sum_{i=1}^{n} r_{\varphi,i}(t+1) \tag{7.5.9}$$

with the root mean square deviation

$$\overline{r}_{\varphi,0} = \sqrt{\frac{1}{n}\sum_{i=1}^{n} r^2_{\varphi,i}(t+1) - \overline{r}^2_\varphi} \tag{7.5.10}$$

By choosing a proper confidence level, we get the threshold value $\beta\overline{r}_{\varphi,0}$ for the detection criterion, where the scalar β is the multiple of the root mean square deviation. It is chosen as 3 or 4 for a 0.997 or 0.999 confidence level, respectively. We assume that the carrier phase windup happens if the following condition holds:

$$\overline{r}_\varphi > \beta\overline{r}_{\varphi,0} \tag{7.5.11}$$

If (7.5.11) holds, then all values $\overline{\varphi}^p_{km,b}(t+1)$ are updated by the same value $\overline{r}_\varphi$:

$$\overline{\varphi}^p_{km,b}(t+1) := \overline{\varphi}^p_{km,b}(t+1) - \overline{r}_\varphi \tag{7.5.12}$$

A full description of the short baseline kinematic rover RTK algorithm follows. Let $n_y = \dim(\boldsymbol{y}) = \dim(\boldsymbol{\eta}) + \dim(\boldsymbol{n})$. Starting with $t = t_0$, $d\bar{\boldsymbol{x}}(t_0) = 0$, $\hat{\boldsymbol{D}}(t_0) \in \boldsymbol{R}^{n_y \times n_y}$, $\hat{\boldsymbol{D}}(t_0) = 0$, and $y(t_0) = 0$, continue as described in Table 7.5.1.

The updated estimate of the constant parameters vector consists of the parts

$$\boldsymbol{y}^T(t+1) = \left(\eta^T(t+1), \boldsymbol{n}^T(t+1)\right) \tag{7.5.13}$$

The updated estimate of the time-varying parameter $d\bar{\boldsymbol{x}}(t+1)$ consists of the across-receiver clock shift and the correction vector of the rover position

$$d\bar{\boldsymbol{x}}^T(t+1) = \left(\xi(t+1), d\boldsymbol{x}^T(t+1)\right) \tag{7.5.14}$$

The updated estimate of the time-varying scalar parameter $\xi(t+1)$ is the across-receiver clock error. Vector $d\boldsymbol{x}(t+1)$ is the correction to the rover position.

7.5.1 Illustrative Example

The raw data of used in the Section 7.4.1 is now processed as if they were kinematics set. In other words, data collected in static mode will be processed by the algorithm described in the Table 7.5.1, which calls for an arbitrary varying position of the rover. The approximate ECEF position of the base is (7.4.21), the same as used in of Section 7.4.1.

Figure 7.5.1 shows the estimated hardware biases. The estimates and the time dependence pattern look much like those shown in Figure 7.4.1 for the static case. Although a larger number of time-varying parameters potentially leads to larger volatility of all parameter estimates, the hardware biases converge to the same values in both cases.

Figures 7.5.2 to 7.5.4 show components of the baseline $\boldsymbol{bl}_{km}(t) = \boldsymbol{x}_k(t) - \boldsymbol{x}_m$ calculated for the static and kinematic cases (Section 7.4.1 and the present section) i and kinematic (Example 7.5.2) cases in comparison with their precise values given in (7.4.24). In order to have a detailed look at the convergence properties for the two cases, only the last 600 epochs are presented. As can clearly be seen, the kinematic estimate shows volatility compared to the static estimate. The reasons for such volatility are noise and carrier phase multipath errors.

Finally, Figure 7.5.5 shows the scatter plot of easting and northing components of the baseline obtained for the kinematic and static cases, also shown for the last 600 epochs. The scatter plot of the static solution shows convergence of the recursive estimate. Differences between kinematic and static solutions can reach several centimeters for various reasons. First, the convergence from the initial data to the value settled around the accurate value takes time. Second, the multipath error varies with time, thus disturbing the instant corrections vector. Multipath is a result of signal reflection from a surface. For long occupation times it filters out in static processing.

The plots of the carrier phase ambiguity estimates look practically identical to those of the static case described in the Section 7.4.1 and are omitted for the sake of brevity. Across-receiver clock error estimates are also identical to those shown in Figure 7.4.7.

TABLE 7.5.1 Algorithm for Kinematic Rover and Short Baseline RTK.

Perform the windup compensation procedure as described in (7.5.9) and (7.5.12)	$\bar{r}_{\varphi} = \frac{1}{n}\sum_{i=1}^{n} r_{\varphi,i}(t+1)$ $\bar{r}_{\varphi,0} = \sqrt{\frac{1}{n}\sum_{i=1}^{n} r^2_{\varphi,i}(t+1) - \bar{r}^2_{\varphi}}$ if $\bar{r}_{\varphi} > \beta\bar{r}_{\varphi,0}$ then $\overline{\varphi}^{p}_{km,b}(t+1) := \overline{\varphi}^{p}_{km,b}(t+1) - \bar{r}_{\varphi}$
Compute the right-hand side vector $\boldsymbol{b}(t+1)$ according to (7.4.16), (7.3.21), and (7.3.22)	$\boldsymbol{b}(t+1) = \begin{pmatrix} \boldsymbol{b}_P(t+1) \\ \hdashline \boldsymbol{b}_{\varphi}(t+1) \end{pmatrix} \in \boldsymbol{R}^{2n}$ $\boldsymbol{b}_P(t+1) = \begin{pmatrix} \overline{P}^{p_1}_{km,b_1}(t+1) - \rho^{p_1}_{k,0}\left(t+1-\tilde{\tau}^{p_1}_k\right) + \rho^{p_1}_m\left(t+1-\tilde{\tau}^{p_1}_m\right) \\ \overline{P}^{p_2}_{km,b_2}(t+1) - \rho^{p_2}_{k,0}\left(t+1-\tilde{\tau}^{p_2}_k\right) + \rho^{p_2}_m\left(t+1-\tilde{\tau}^{p_2}_m\right) \\ \vdots \\ \overline{P}^{p_n}_{km,b_n}(t+1) - \rho^{p_n}_{k,0}\left(t+1-\tilde{\tau}^{p_n}_k\right) + \rho^{p_n}_m\left(t+1-\tilde{\tau}^{p_n}_m\right) \end{pmatrix}$ $\boldsymbol{b}_{\varphi}(t+1) = \begin{pmatrix} \overline{\varphi}^{p_1}_{km,b_1}(t+1) - \frac{1}{\lambda^{p_1}_{b_1}}\left(\rho^{p_1}_{k,0}\left(t+1-\tilde{\tau}^{p_1}_k\right) - \rho^{p_1}_m\left(t+1-\tilde{\tau}^{p_1}_m\right)\right) \\ \overline{\varphi}^{p_2}_{km,b_2}(t+1) - \frac{1}{\lambda^{p_2}_{b_2}}\left(\rho^{p_2}_{k,0}\left(t+1-\tilde{\tau}^{p_2}_k\right) - \rho^{p_2}_m\left(t+1-\tilde{\tau}^{p_2}_m\right)\right) \\ \vdots \\ \overline{\varphi}^{p_n}_{km,b_n}(t+1) - \frac{1}{\lambda^{p_n}_{b_n}}\left(\rho^{p_n}_{k,0}\left(t+1-\tilde{\tau}^{p_n}_k\right) - \rho^{p_n}_m\left(t+1-\tilde{\tau}^{p_n}_m\right)\right) \end{pmatrix}$
Compute the matrix $\boldsymbol{J}(t+1)$ according to (7.5.7), (7.3.20), and (7.3.15)	$\boldsymbol{J}(t+1) = \left[\begin{array}{c:c} \boldsymbol{e} & \boldsymbol{A}(t+1) \\ \hdashline \boldsymbol{\Lambda}^{-1}\boldsymbol{e} & \boldsymbol{\Lambda}^{-1}\boldsymbol{A}(t+1) \end{array}\right]$ $\boldsymbol{A}(t+1) = \begin{bmatrix} H^{p_1}_{k,1} & H^{p_1}_{k,2} & H^{p_1}_{k,3} \\ H^{p_2}_{k,1} & H^{p_2}_{k,2} & H^{p_3}_{k,3} \\ & \cdots & \\ H^{p_n}_{k,1} & H^{p_n}_{k,2} & H^{p_n}_{k,3} \end{bmatrix}$
Compute the matrix $\boldsymbol{W}$ according to (7.5.7) and (7.3.38)	$\boldsymbol{W} = \left[\begin{array}{c:c} \boldsymbol{W}_{\eta} & 0 \\ \hdashline 0 & \boldsymbol{I}_n \end{array}\right]$
Square root of diagonal covariance matrix according to (7.4.18)	$\Sigma = \left[\begin{array}{c:c} \Sigma_P & 0 \\ \hdashline 0 & \Sigma_{\varphi} \end{array}\right]$
Weighing	$\overline{\boldsymbol{b}}(t+1) = \Sigma^{-1}\boldsymbol{b}(t+1)$ $\overline{\boldsymbol{J}}(t+1) = \Sigma^{-1}\boldsymbol{J}(t+1)$ $\overline{\boldsymbol{W}} = \Sigma^{-1}\boldsymbol{W}$

TABLE 7.5.1 (*Continued*)

Compute the residual vector $\bar{\boldsymbol{r}}(t+1)$	$\bar{\boldsymbol{r}}(t+1) = \bar{\boldsymbol{b}}(t+1) - \overline{\boldsymbol{W}}\boldsymbol{y}(t)$
Compute Cholesky decomposition	$\boldsymbol{L}_J\boldsymbol{L}_J^T = \bar{\boldsymbol{J}}^T(t+1)\bar{\boldsymbol{J}}(t+1)$
Forward run substitution	$\tilde{\boldsymbol{J}}^T(t+1) = \boldsymbol{F}_{\boldsymbol{L}_J}(\bar{\boldsymbol{J}}^T(t+1))$
Compute the projection matrix	$\Pi = \boldsymbol{I}_{2n} - \tilde{\boldsymbol{J}}(t+1)\tilde{\boldsymbol{J}}^T(t+1)$
Update the matrix $\hat{\boldsymbol{D}}(t)$	$\hat{\boldsymbol{D}}(t+1) = \hat{\boldsymbol{D}}(t) + \overline{\boldsymbol{W}}^T\Pi\overline{\boldsymbol{W}}$ $= \hat{\boldsymbol{D}}(t) + \overline{\boldsymbol{W}}^T\overline{\boldsymbol{W}} - \overline{\boldsymbol{W}}^T\tilde{\boldsymbol{J}}(t+1)\tilde{\boldsymbol{J}}^T(t+1)\overline{\boldsymbol{W}}$
Compute Cholesky decomposition	$\hat{\boldsymbol{D}}(t+1) = \hat{\boldsymbol{L}}\hat{\boldsymbol{L}}^T$
Update the estimate of constant parameters $\boldsymbol{y}(t)$	$\boldsymbol{y}(t+1) = \boldsymbol{y}(t) + \boldsymbol{B}_{\hat{L}}\left(\boldsymbol{F}_{\hat{L}}\left(\overline{\boldsymbol{W}}^T(t+1)\,\Pi\bar{\boldsymbol{r}}(t+1)\right)\right)$
Compute the second residual vector $\boldsymbol{r}'(t+1)$	$\boldsymbol{r}'(t+1) = \bar{\boldsymbol{b}}(t+1) - \overline{\boldsymbol{W}}\boldsymbol{y}(t+1)$
Compute the updated estimate $d\bar{\boldsymbol{x}}(t+1)$	$d\bar{\boldsymbol{x}}(t+1) = \boldsymbol{B}_{L_{\bar{J}}}\tilde{\boldsymbol{J}}^T(t+1)\boldsymbol{r}'(t+1)$

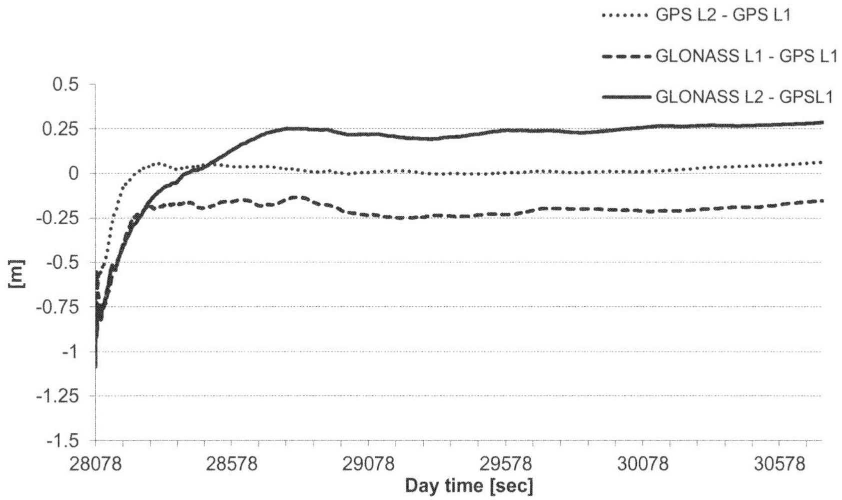

Figure 7.5.1 Hardware biases estimates for the kinematic case.

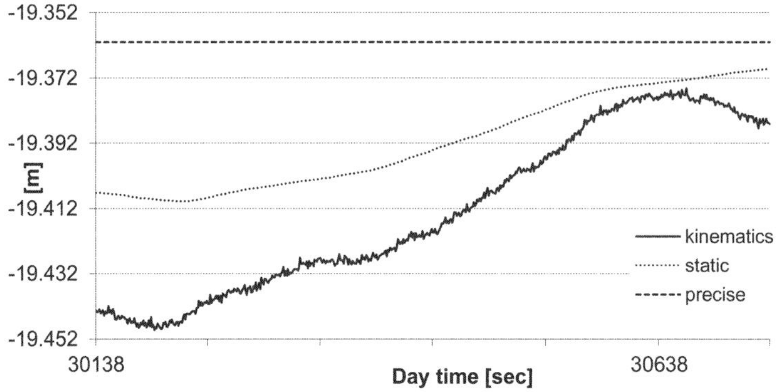

Figure 7.5.2 Comparison of the easting component of the baseline calculated for the kinematic and static cases with the precise value. The last 600 epochs are shown.

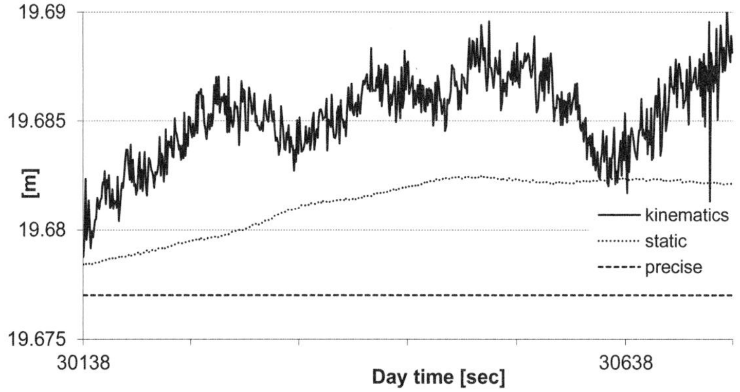

Figure 7.5.3 Comparison of the northing component of the baseline calculated for the kinematic and static cases with the precise value. Only the last 600 epochs are shown.

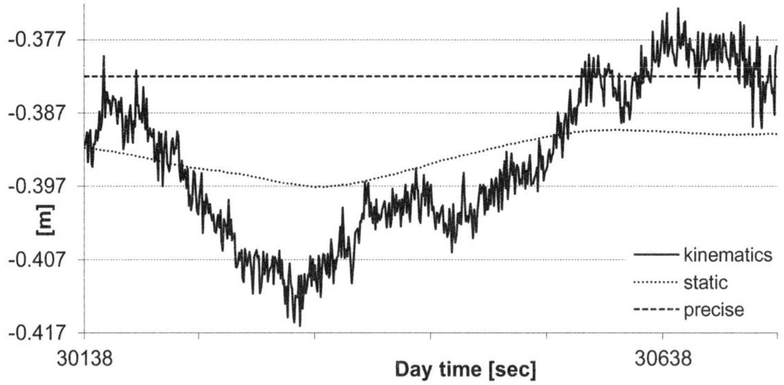

Figure 7.5.4 Comparison of the up component of the baseline calculated for the kinematic and static cases with the precise value. Only the last 600 epochs are shown.

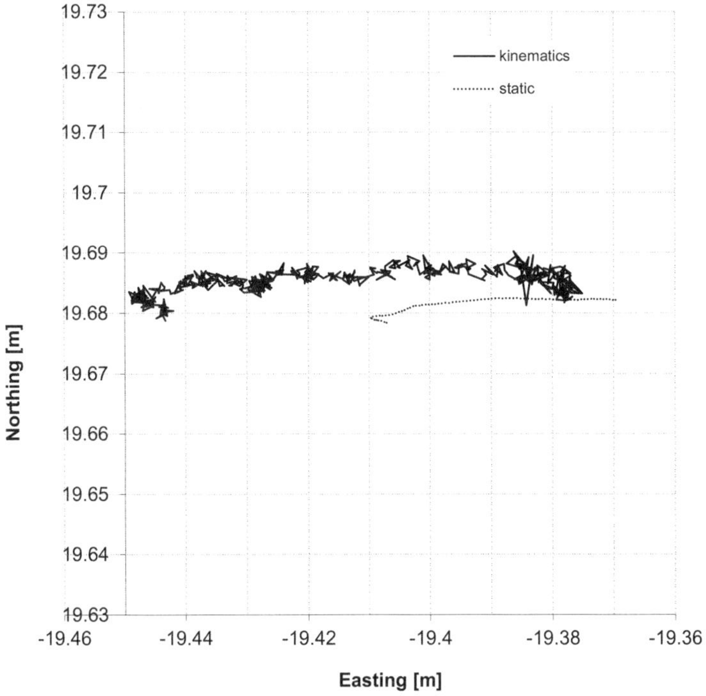

Figure 7.5.5 Scatter plot of easting and northing components for the baseline. The last 600 epochs are shown. Dashed line (static), solid (kinematic).

7.6 RTK ALGORITHM WITH DYNAMIC MODEL AND SHORT BASELINES

We derive the numerical scheme for estimating the time-varying position of the rover when it is subjected to dynamic constraints. This problem falls into the classification described in Section 3.5 where the static parameters are hardware biases and carrier phase ambiguities, the across-receiver clock error is the arbitrary varying kinematic parameter, and the kinematic parameters subjected to the dynamic model are corrections to the approximate rover position.

We use the same across-receiver linearized navigation equations (7.4.3) and (7.4.5) that were used in the previous section. The motion of the rover was constrained to static in Section 7.4, and it was not constrained at all in Section 7.5. Now we consider the intermediate problem of constraining the motion by a certain dynamic mode. Following this way, we will distinguish between kinematic and dynamic cases.

In some applications, the physical nature of the problem imposes the dynamic model. For example, dynamics of the wheeled robot are governed by certain nonholonomic differential equations. The dynamics of a solid body equipped by inertial sensors are also well known. In general, we can say that the rover does not move very aggressively, just like a vessel experiencing no significant roll and heave when in steady motion or making slow turns, or we can say that the rover is maneuvering rapidly. For such a general description of the dynamics, one often uses the model proposed in Singer (1970). In our notations the model looks like

$$\begin{aligned}
\boldsymbol{x}_k(t) &= \boldsymbol{x}_k(t-1) + \Delta t \boldsymbol{v}_k(t-1) + \frac{1}{2}\Delta t^2 \boldsymbol{a}_k(t-1) \\
\boldsymbol{v}_k(t) &= \boldsymbol{v}_k(t-1) + \Delta t \boldsymbol{a}_k(t-1) \\
\boldsymbol{a}_k(t) &= \gamma \boldsymbol{a}_k(t-1)
\end{aligned} \tag{7.6.1}$$

where $\boldsymbol{x}_k(t)$, $\boldsymbol{v}_k(t)$, and $\boldsymbol{a}_k(t)$ are position, velocity, and acceleration of the rover (kth station). The symbol Δt denotes the across-epoch time interval; note that $(t+1)$ refers to the epoch after epoch t and not a certain time increment. The factor γ is given by

$$\gamma = e^{-\alpha} \qquad \alpha = \frac{\Delta t}{T_{\text{corr}}} \tag{7.6.2}$$

The correlation time T_{corr} governs the volatility of acceleration. For example, if the rover experiences slow turns between long periods of steady movement, we can assume $T_{\text{corr}} = 60$ sec, while for atmospheric turbulence $T_{\text{corr}} = 1$ sec is more appropriate.

Assuming that at each epoch we have the position of the rover expressed as a sum of the approximation $\boldsymbol{x}_{k,0}(t)$ and correction $d\boldsymbol{x}(t)$ similar to (7.5.1), we can introduce the vector of state variables

$$\tilde{\boldsymbol{x}}(t) = \begin{pmatrix} d\boldsymbol{x}(t) \\ \boldsymbol{v}(t) \\ \boldsymbol{a}(t) \end{pmatrix} \tag{7.6.3}$$

and the matrix $\boldsymbol{F}$ (not depending on t)

$$\boldsymbol{F} = \left[\begin{array}{c:c:c} \boldsymbol{I}_3 & \Delta t \boldsymbol{I}_3 & \frac{1}{2}\Delta t^2 \boldsymbol{I}_3 \\ \hdashline 0 & \boldsymbol{I}_3 & \Delta t \boldsymbol{I}_3 \\ \hdashline 0 & 0 & \gamma \boldsymbol{I}_3 \end{array}\right] \in \boldsymbol{R}^{9\times 9} \tag{7.6.4}$$

Then the dynamic equations (3.3.1) take the form

$$\tilde{\boldsymbol{x}}(t) = \boldsymbol{F}\tilde{\boldsymbol{x}}(t-1) + \boldsymbol{f}(t) + \varepsilon(t) \tag{7.6.5}$$

where

$$\boldsymbol{f}(t) = \begin{pmatrix} \boldsymbol{x}_{k,0}(t-1) - \boldsymbol{x}_{k,0}(t) \\ 0 \\ 0 \end{pmatrix} \tag{7.6.6}$$

and

$$E\left(\varepsilon(t)\varepsilon^T(t)\right) = \boldsymbol{Q} = \sigma^2 \left[\begin{array}{c:c:c} \frac{\Delta t^5}{20}\boldsymbol{I}_3 & \frac{\Delta t^4}{8}\boldsymbol{I}_3 & \frac{\Delta t^3}{6}\boldsymbol{I}_3 \\ \hdashline \frac{\Delta t^4}{8}\boldsymbol{I}_3 & \frac{\Delta t^3}{3}\boldsymbol{I}_3 & \frac{\Delta t^2}{2}\boldsymbol{I}_3 \\ \hdashline \frac{\Delta t^3}{6}\boldsymbol{I}_3 & \frac{\Delta t^2}{2}\boldsymbol{I}_3 & \Delta t \boldsymbol{I}_3 \end{array}\right] \in \boldsymbol{R}^{9\times 9} \tag{7.6.7}$$

The quantity σ^2 is another parameter that depends on the expected dynamics (Singer, 1970). Note that the introduction of the vector $\boldsymbol{f}(t)$ in (7.6.5) extends the notation (3.3.1), while it only changes the "compute the projected estimate" step of the optimal estimation algorithm; see Table 3.5.1.

We now have everything prepared for application of the numerical scheme described in Section 3.5. The vector of constant parameters consists of hardware biases $\boldsymbol{\eta}$ and floating ambiguities $\boldsymbol{n}$, and takes the form

$$\boldsymbol{y} = \begin{pmatrix} \boldsymbol{\eta} \\ \boldsymbol{n} \end{pmatrix} \tag{7.6.8}$$

The single arbitrary varying parameter is $\xi(t)$, which is the across-receiver clock error. Rewrite the measurement equations (3.5.1) in the form

$$\boldsymbol{H}(t)\tilde{\boldsymbol{x}}(t) + \boldsymbol{J}\xi(t) + \boldsymbol{W}\boldsymbol{y} = \boldsymbol{b}(t) \tag{7.6.9}$$

where

$$\boldsymbol{H}(t) = \left[\begin{array}{c|c|c} \boldsymbol{A}(t) & 0 & 0 \\ \hline \boldsymbol{\Lambda}^{-1}\boldsymbol{A}(t) & 0 & 0 \end{array}\right] \in \boldsymbol{R}^{2n\times 9}, \boldsymbol{J} = \left(\begin{array}{c} \boldsymbol{e} \\ \hline \boldsymbol{\Lambda}^{-1}\boldsymbol{e} \end{array}\right) \in \boldsymbol{R}^{2n\times 1},$$

$$\boldsymbol{W} = \left[\begin{array}{c|c} \boldsymbol{W}_{\eta} & 0 \\ \hline 0 & \boldsymbol{I}_n \end{array}\right] \in \boldsymbol{R}^{2n\times(2n+9)} \tag{7.6.10}$$

and apply the algorithm given in Table 3.5.1. described in Section 3.5 the same way as we did in the previous section. The first three columns of the matrix $\boldsymbol{H}(t)$ correspond to the variables $d\boldsymbol{x}(t)$, while the next six columns correspond to $\boldsymbol{v}(t)$ and $\boldsymbol{a}(t)$. The dimensions of the matrix $\boldsymbol{W}$ in (7.6.10) are presented for the general case in (7.3.38). In the case of a GPS/GLONASS triple-band receiver, the matrix $\boldsymbol{W}$ has dimensions $2n \times (2n+3)$. All other notations used in the description of the algorithm have the same meaning as in the previous section. Let $n_y = \dim(\boldsymbol{y}) = \dim(\boldsymbol{\eta}) + \dim(\boldsymbol{n})$. Starting with $t = t_0$, $\tilde{\boldsymbol{x}}(t_0) = 0$, $\hat{\boldsymbol{D}}(t_0) \in \boldsymbol{R}^{(n_y+9)\times(n_y+9)}$, $\hat{\boldsymbol{D}}(t_0) = 0$, and $\boldsymbol{y}(t_0) = 0$, continue as described in Table 7.6.1.

7.6.1 Illustrative Example

The raw data used in Sections 7.4.1 and 7.5.1 will be processed taking into account a dynamic model. In other words, data collected in static mode will be processed by equation (7.6.1), allowing variation of the rover position in accordance to the dynamic model (7.6.4), (7.6.5), and (7.6.7), and allowing arbitrary variation of the across-receiver clock estimate.

Figure 7.6.1 compares the up component of the baseline vector calculated for the dynamic case (dashed line) and the kinematic case (solid line). Only the last 100 epochs are presented in order to easily see the difference. The figure shows a more conservative variation in the dynamic case, whereas the kinematic estimate demonstrates higher volatility. We used the following parameters of the dynamic model: $\alpha = 0.01\,\text{sec}^{-1}$ and $\sigma = 0.01\ \text{m}/\text{sec}^2$.

TABLE 7.6.1 Algorithm with Dynamic Model and Short Baselines.

Perform the windup compensation procedure as described in (7.5.9) and (7.5.12)	$\bar{r}_{\varphi} = \frac{1}{n}\sum_{i=1}^{n} r_{\varphi,i}(t+1)$ $\bar{r}_{\varphi,0} = \sqrt{\frac{1}{n}\sum_{i=1}^{n} r_{\varphi,i}^{2}(t+1) - \bar{r}_{\varphi}^{2}}$ if $\bar{r}_{\varphi} > \beta\bar{r}_{\varphi,0}$ then $\bar{\varphi}^{p}_{km,b}(t+1) := \bar{\varphi}^{p}_{km,b}(t+1) - \bar{r}_{\varphi}$
Compute the right-hand side vector $\boldsymbol{b}(t+1)$ according to (7.4.16), (7.3.21), and (7.3.22)	$\boldsymbol{b}(t+1) = \begin{pmatrix} \boldsymbol{b}_P(t+1) \\ \hdashline \boldsymbol{b}_{\varphi}(t+1) \end{pmatrix} \in \boldsymbol{R}^{2n}$ $\boldsymbol{b}_P(t+1) = \begin{pmatrix} \bar{P}^{p_1}_{km,b_1}(t+1) - \rho^{p_1}_{k,0}\left(t+1-\tilde{\tau}^{p_1}_{k}\right) + \rho^{p_1}_{m}\left(t+1-\tilde{\tau}^{p_1}_{m}\right) \\ \bar{P}^{p_2}_{km,b_2}(t+1) - \rho^{p_2}_{k,0}\left(t+1-\tilde{\tau}^{p_2}_{k}\right) + \rho^{p_2}_{m}\left(t+1-\tilde{\tau}^{p_2}_{m}\right) \\ \vdots \\ \bar{P}^{p_n}_{km,b_n}(t+1) - \rho^{p_n}_{k,0}\left(t+1-\tilde{\tau}^{p_n}_{k}\right) + \rho^{p_n}_{m}\left(t+1-\tilde{\tau}^{p_n}_{m}\right) \end{pmatrix}$ $\boldsymbol{b}_{\varphi}(t+1) = \begin{pmatrix} \bar{\varphi}^{p_1}_{km,b_1}(t+1) - \frac{1}{\lambda^{p_1}_{b_1}}\left(\rho^{p_1}_{k,0}\left(t+1-\tilde{\tau}^{p_1}_{k}\right) - \rho^{p_1}_{m}\left(t+1-\tilde{\tau}^{p_1}_{m}\right)\right) \\ \bar{\varphi}^{p_2}_{km,b_2}(t+1) - \frac{1}{\lambda^{p_2}_{b_2}}\left(\rho^{p_2}_{k,0}\left(t+1-\tilde{\tau}^{p_2}_{k}\right) - \rho^{p_2}_{m}\left(t+1-\tilde{\tau}^{p_2}_{m}\right)\right) \\ \vdots \\ \bar{\varphi}^{p_n}_{km,b_n}(t+1) - \frac{1}{\lambda^{p_n}_{b_n}}\left(\rho^{p_n}_{k,0}\left(t+1-\tilde{\tau}^{p_n}_{k}\right) - \rho^{p_n}_{m}\left(t+1-\tilde{\tau}^{p_n}_{m}\right)\right) \end{pmatrix}$
Compute the covariance matrix $\boldsymbol{Q}$ according to (7.6.7) and its Cholesky decomposition	$\boldsymbol{L_Q}\boldsymbol{L_Q}^{T} = \boldsymbol{Q}$
Compute forward and backward substitutions with the matrix $\boldsymbol{F}$ defined according to (7.6.4)	$\bar{\boldsymbol{F}} = \boldsymbol{F}_{\boldsymbol{L_Q}}(\boldsymbol{F})$ $\tilde{\boldsymbol{F}} = \boldsymbol{B}_{\boldsymbol{L_Q}}(\bar{\boldsymbol{F}})$
Compute the matrix $\boldsymbol{H}(t+1)$ according to (7.6.10), (7.3.20), and (7.3.15)	$\boldsymbol{H}(t) = \left[\begin{array}{c:c:c} \boldsymbol{A}(t) & 0 & 0 \\ \hdashline \Lambda^{-1}\boldsymbol{A}(t) & 0 & 0 \end{array}\right]$ $\boldsymbol{A}(t+1) = \begin{bmatrix} H^{p_1}_{k,1} & H^{p_1}_{k,2} & H^{p_1}_{k,3} \\ H^{p_2}_{k,1} & H^{p_2}_{k,2} & H^{p_3}_{k,3} \\ & \cdots & \\ H^{p_n}_{k,1} & H^{p_n}_{k,2} & H^{p_n}_{k,3} \end{bmatrix}$
Compute the vector $\boldsymbol{J}$ according to (7.6.10)	$\boldsymbol{J} = \begin{pmatrix} \boldsymbol{e} \\ \hdashline \Lambda^{-1}\boldsymbol{e} \end{pmatrix}$

TABLE 7.6.1 (*Continued*)

Compute the matrix $\boldsymbol{W}$ according to (7.6.10), (7.3.38)	$\boldsymbol{W} = \begin{bmatrix} \boldsymbol{W}_\eta & 0 \\ 0 & \boldsymbol{I}_n \end{bmatrix}$
Square root of diagonal covariance matrix according to (7.4.18)	$\Sigma = \begin{bmatrix} \Sigma_P & 0 \\ 0 & \Sigma_\varphi \end{bmatrix}$
Weighing	$\bar{\boldsymbol{b}}(t+1) = \Sigma^{-1}\boldsymbol{b}(t+1)$ $\bar{\boldsymbol{J}} = \Sigma^{-1}\boldsymbol{J}$ $\bar{\boldsymbol{W}} = \Sigma^{-1}\boldsymbol{W}$ $\bar{\boldsymbol{H}}(t+1) = \Sigma^{-1}\boldsymbol{H}(t+1)$
Compute the projection matrix Π	$\gamma = \sqrt{\bar{\boldsymbol{J}}^T\bar{\boldsymbol{J}}}$ $\tilde{\boldsymbol{J}} = \frac{1}{\gamma}\bar{\boldsymbol{J}}$ $\Pi = \boldsymbol{I}_{2n} - \tilde{\boldsymbol{J}}\tilde{\boldsymbol{J}}^T$
Compute the residual vector $\bar{\boldsymbol{r}}(t+1)$	$\bar{\boldsymbol{r}}(t+1) = \bar{\boldsymbol{b}}(t+1) - \bar{\boldsymbol{H}}(t+1)\tilde{\boldsymbol{x}}(t) - \bar{\boldsymbol{W}}\boldsymbol{y}(t)$
Compute the updating matrix $\boldsymbol{\Delta}(t+1)$	$\Delta(t+1) = \begin{bmatrix} \bar{\boldsymbol{F}}^T\bar{\boldsymbol{F}} & -\tilde{\boldsymbol{F}}^T & 0 \\ -\tilde{\boldsymbol{F}} & \boldsymbol{Q}^{-1} + \bar{\boldsymbol{H}}^T(t+1)\Pi\bar{\boldsymbol{H}}(t+1) & \bar{\boldsymbol{H}}^T(t+1)\Pi\bar{\boldsymbol{W}} \\ 0 & \bar{\boldsymbol{W}}^T\Pi\bar{\boldsymbol{H}}(t+1) & \bar{\boldsymbol{W}}^T\Pi\bar{\boldsymbol{W}} \end{bmatrix}$
Partitioning the matrix $\hat{\boldsymbol{D}}(t)$ in the block form and extend it adding zero blocks	$\hat{\boldsymbol{D}}(t) = \begin{bmatrix} \boldsymbol{D}^{xx} & \boldsymbol{D}^{xy} \\ \boldsymbol{D}^{xyT} & \boldsymbol{D}^{yy} \end{bmatrix}, \boldsymbol{D}^{xx} \in \boldsymbol{R}^{9\times 9}, \boldsymbol{D}^{xy} \in \boldsymbol{R}^{9\times n_y}, \boldsymbol{D}^{yy} \in \boldsymbol{R}^{n_y\times n_y}$ $\tilde{\boldsymbol{D}}(t) = \begin{bmatrix} \boldsymbol{D}^{xx} & 0 & \boldsymbol{D}^{xy} \\ 0 & 0 & 0 \\ \boldsymbol{D}^{xyT} & 0 & \boldsymbol{D}^{yy} \end{bmatrix} \in \boldsymbol{R}^{(n_y+18)\times(n_y+18)}$
Update the matrix $\tilde{\boldsymbol{D}}(t)$	$\boldsymbol{G}(t+1) = \tilde{\boldsymbol{D}}(t) + \Delta(t+1)$ $= \begin{bmatrix} \boldsymbol{D}^{xx} + \bar{\boldsymbol{F}}^T\bar{\boldsymbol{F}} & -\tilde{\boldsymbol{F}}^T & \boldsymbol{D}^{xy} \\ -\tilde{\boldsymbol{F}} & \boldsymbol{Q}^{-1} + \bar{\boldsymbol{H}}^T(t+1)\Pi\bar{\boldsymbol{H}}(t+1) & \bar{\boldsymbol{H}}^T(t+1)\Pi\bar{\boldsymbol{W}} \\ \boldsymbol{D}^{xyT} & \bar{\boldsymbol{W}}^T\Pi\bar{\boldsymbol{H}}(t+1) & \boldsymbol{D}^{yy} + \bar{\boldsymbol{W}}^T\Pi\bar{\boldsymbol{W}} \end{bmatrix}$
Compute Cholesky decomposition and the updated matrix $\hat{\boldsymbol{D}}(t+1)$	$\boldsymbol{G}(t+1) = \hat{\boldsymbol{L}}\hat{\boldsymbol{L}}^T = \begin{bmatrix} \boldsymbol{L} & 0 \\ \boldsymbol{K} & \boldsymbol{M} \end{bmatrix}\begin{bmatrix} \boldsymbol{L}^T & \boldsymbol{K}^T \\ 0 & \boldsymbol{M}^T \end{bmatrix}, \boldsymbol{L} \in \boldsymbol{R}^{9\times 9}, \boldsymbol{M} \in \boldsymbol{R}^{(n_y+9)\times(n_y+9)}$ $\hat{\boldsymbol{D}}(t+1) = \boldsymbol{M}\boldsymbol{M}^T$

(*continued*)

TABLE 7.6.1 (*Continued*)

Compute the updated estimate $\begin{pmatrix}\tilde{\boldsymbol{x}}(t+1)\\ \boldsymbol{y}(t+1)\end{pmatrix}$	$\begin{pmatrix}\tilde{\boldsymbol{x}}(t+1)\\ \boldsymbol{y}(t+1)\end{pmatrix} = \begin{pmatrix}\boldsymbol{F}\tilde{\boldsymbol{x}}(t)+\boldsymbol{x}_{k,0}(t)-\boldsymbol{x}_{k,0}(t+1)\\ \boldsymbol{y}(t)\end{pmatrix} + \boldsymbol{B}_M\left(\boldsymbol{F}_M\left(\left(\begin{matrix}\overline{\boldsymbol{H}}^T(t+1)\\ \overline{\boldsymbol{W}}^T\end{matrix}\right)\Pi\bar{\boldsymbol{r}}(t+1)\right)\right)$
Compute the second residual vector	$\boldsymbol{r}'(t+1)=\bar{\boldsymbol{b}}(t+1)-\overline{\boldsymbol{H}}^T(t+1)\boldsymbol{x}(t+1)-\overline{\boldsymbol{W}}\boldsymbol{y}(t+1)$
Compute the estimate of the across-receiver clock difference $\xi(t+1)$	$\xi(t+1)=\frac{1}{\gamma}\tilde{\boldsymbol{J}}^T\boldsymbol{r}'(t+1)$

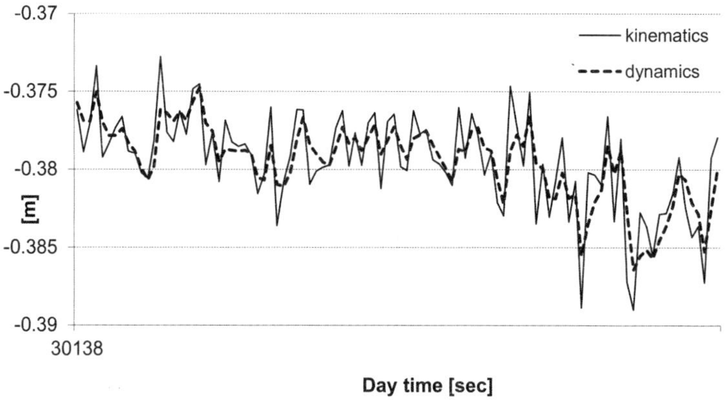

Figure 7.6.1 Comparison of up component of the baseline calculated for both kinematic and dynamic cases. The last 100 epochs are shown.

The other estimates, excluding northing and easting components of the baseline, are practically identical to those obtained in the kinematic case and are omitted for brevity.

In this section, we considered short kinematic baseline processing with the rover motion constrained by a dynamic model. In contrast with the previous section, no arbitrary variation of the rover position is acceptable. We considered the case of a very general dynamic model (7.6.1). In practical applications, such as machine control, the dynamic model should be chosen that best reflects the dynamics of the machine. For example, the road construction machines that have a body (probably rotating), a blade or boom and bucket, the physical quantities like masses, lengths of links, and inertia momentums can be used to describe the dynamic model.

If there is no specific information about the dynamics of the physical bodies, except knowing that the kinematics of the rover antenna cannot be absolutely arbitrary by allowing immediate and large changes of position, a reasonable practical recipe is to use model (7.6.1). Figure 7.6.1 illustrates how this model constrains the variation of position.

7.7 RTK ALGORITHM WITH DYNAMIC MODEL AND LONG BASELINES

If baselines are greater than about 5 km, the across-receiver ionospheric delays cannot be neglected in the navigation equations (7.3.37) and (7.3.46). The vector $\boldsymbol{i}(t) \in \boldsymbol{R}^n$ of the across-receiver ionospheric delay varies with time. Having additional n variables to estimate at each epoch, the redundancy of the system of navigation equations reduces because the difference in the number of equations and parameters reduces. There are two options to restrict variation of the ionospheric delay estimate:

1. Restrict variation of the ionospheric delay estimated at each epoch. The estimates related to different epochs are independent of each other.
2. Introduce the dynamic model that governs variation of the ionospheric delay from epoch to epoch.

If one considers the ionospheric delays as slowly varying parameters, one follows the second approach, which is the more general of the two. Following the second approach, include the vector $\boldsymbol{i}(t)$ in the set of estimated parameters that are subject to the dynamic model. Modify the measurement equations (7.6.9) as follows:

$$\boldsymbol{H}(t)\tilde{\boldsymbol{x}}(t) + \boldsymbol{H}_i\boldsymbol{i}(t) + \boldsymbol{J}\xi(t) + \boldsymbol{W}\boldsymbol{y} = \boldsymbol{b}(t) \tag{7.7.1}$$

where the matrix $\boldsymbol{H}_i(t)$ has the form

$$\boldsymbol{H}_i = \begin{bmatrix} \boldsymbol{\Gamma} \\ \hdashline -\boldsymbol{\Lambda}^{-1}\boldsymbol{\Gamma} \end{bmatrix} \tag{7.7.2}$$

according to equations (7.3.37) and (7.3.46), and the matrix $\boldsymbol{\Gamma}$ is defined by (7.3.27). Restrict the variation of the time-dependent vector $\boldsymbol{i}(t)$ by the dynamic equations

$$\boldsymbol{i}(t) = \gamma_i \boldsymbol{i}(t-1) + \boldsymbol{\varepsilon}_i(t) \tag{7.7.3}$$

where

$$\gamma_i = e^{-\Delta t/\tau_i} \tag{7.7.4}$$

with Δt being the across-epoch time difference, and τ_i is the correlation time reflecting the rate of variation of the ionospheric delay in time. A typical value for τ_i is 1200 sec. The white noise $\boldsymbol{\varepsilon}_i(t)$ has the covariance matrix $\sigma_i^2\boldsymbol{I}_n$, which provides the variance of the across-receiver ionosphere satisfying the condition (7.4.2). Let

$$\overline{\sigma}_i^2 = (S \times 10^{-6} \times \|\boldsymbol{x}_k - \boldsymbol{x}_m\|)^2 \tag{7.7.5}$$

where the factor S is defined in (7.4.2) to be an expected mean value of $\|\boldsymbol{i}(t)\|^2$. It follows from (7.7.3) that $\|\boldsymbol{i}(t)\|^2 = \|\gamma_i \boldsymbol{i}(t-1) + \epsilon_i(t)\|^2$. Then, assuming that the stochastic process $\boldsymbol{i}(t)$ is stationary and $\boldsymbol{i}(t)$ does not depend on $\boldsymbol{\epsilon}_i(t)$, we take a mean value of both sides of the last equality. We obtain

$$\sigma_i^2 = \left(1 - \gamma_i^2\right)\overline{\sigma}_i^2 \tag{7.7.6}$$

combine parameters that are subject to dynamic constraints into the new vector

$$\tilde{\tilde{\boldsymbol{x}}}(t) = \left[\begin{array}{c} \tilde{\boldsymbol{x}}(t) \\ \hdashline \boldsymbol{i}(t) \end{array}\right] \in \boldsymbol{R}^{9+n} \tag{7.7.7}$$

and introduce the matrices

$$\tilde{\tilde{\boldsymbol{H}}}(t) = \left[\boldsymbol{H}(t) \;\vdots\; \boldsymbol{H}_i\right] \in \boldsymbol{R}^{2n\times(n+9)} \tag{7.7.8}$$

$$\tilde{\tilde{\boldsymbol{F}}} = \left[\begin{array}{c:c} \boldsymbol{F} & 0 \\ \hdashline 0 & \gamma_i \boldsymbol{I}_n \end{array}\right] \in \boldsymbol{R}^{(n+9)\times(n+9)} \tag{7.7.9}$$

$$\tilde{\tilde{\boldsymbol{Q}}} = \left[\begin{array}{c:c} \boldsymbol{Q} & 0 \\ \hdashline 0 & \sigma_i^2 \boldsymbol{I}_n \end{array}\right] \in \boldsymbol{R}^{(9+n)\times(9+n)} \tag{7.7.10}$$

then apply the algorithm, described in the Table 7.6.1 with vector $\tilde{\boldsymbol{x}}(t)$ substituted by vector $\tilde{\tilde{\boldsymbol{x}}}(t)$, and matrices $\boldsymbol{H}(t)$, $\boldsymbol{F}$, and $\boldsymbol{Q}$ substituted by matrices (7.7.8), (7.7.9), and (7.7.10), respectively. For long baselines, the windup compensation procedure does not totally compensate the windup angle in double-difference carrier phase observables.

7.7.1 Illustrative Example

Consider a dual-band GPS and GLONASS data set collected for a 45 km baseline observed over 4075 sec. For $S = 2$, equation (7.7.5) gives $\sigma_i = 0.09$ m. The ionospheric correlation time is chosen as $\tau_i = 1200$ sec. Figure 7.7.1 illustrates the estimates of the baseline vector components using the algorithm described in Section 7.6 without ionospheric delay estimation (dashed line), and the algorithm described in the current section that includes the ionospheric delay estimation (solid black line). The dotted line shows the components of the known baseline. As can be seen, the ionosphere estimation improves the overall accuracy of the estimation. Moreover, the ambiguity resolution is faster because it allows compensation of the slowly varying ionospheric delay, which otherwise would appear as bias in the navigation equations if not compensated.

Figure 7.7.2 illustrates the ionospheric delay estimate for one GPS satellite (solid black line) and one GLONASS satellite (dashed line). In order to exclude the transient values, the figure shows the last 3000 epochs. For the sake of comparison, Figure 7.7.3 shows the estimate of the ionospheric delay for the same two satellites when using $\tau_i = 3600$ sec. A larger correlation time produces more stable estimates.

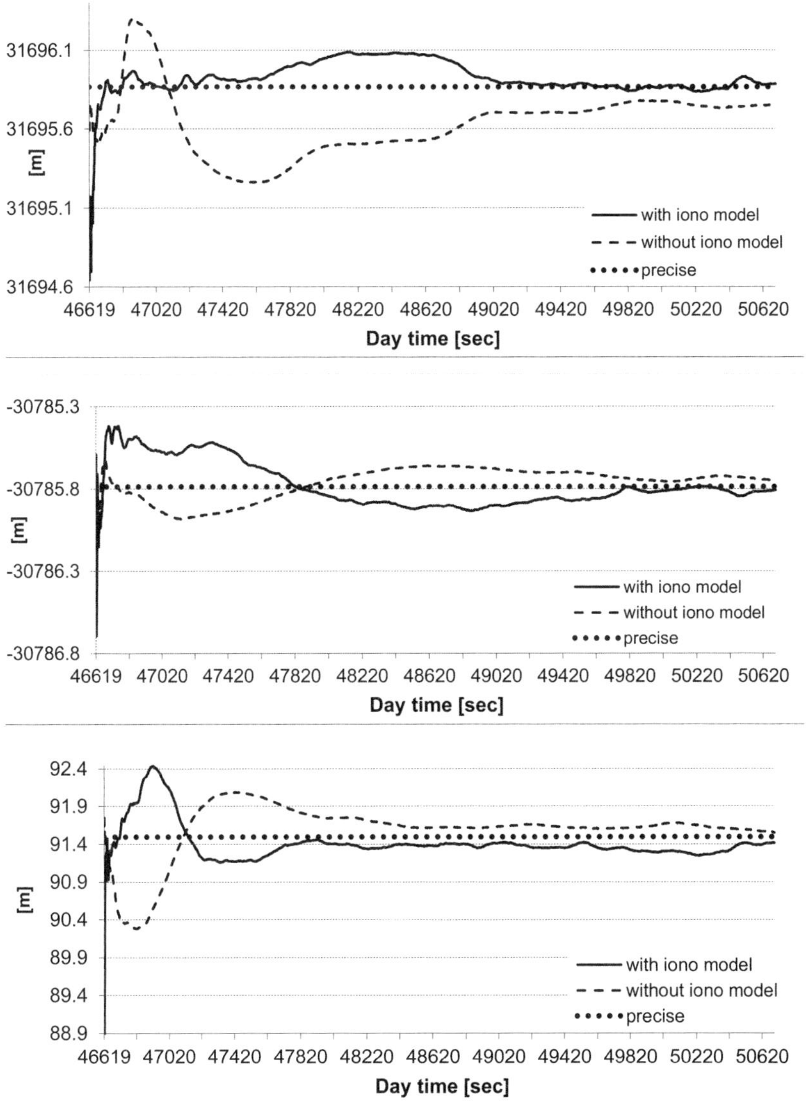

Figure 7.7.1 Easting, northing, and up components of the baseline calculated for the cases of estimating the residual ionospheric delay (solid line) and ignoring the ionospheric delay (dashed line). The dotted line shows the known values.

In this section we considered long baseline processing, in which case the across-receiver ionospheric difference cannot be neglected and the RTK processing engine must appropriately handle it. The most efficient approach is estimating it and thus compensating for it in the navigation equations. The ionospheric term appears in the navigation equations as an additional variable reducing redundancy. Effectively, we have n variables more to be estimated at each time epoch. On

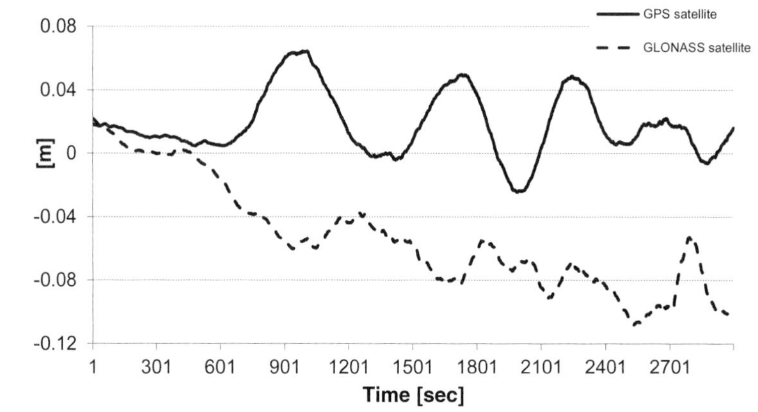

Figure 7.7.2 Across-receiver ionospheric delay estimate for one GPS and one GLONASS satellite obtained for the a priori ionospheric correlation time $\tau_i = 1200$ sec. The units are in meters.

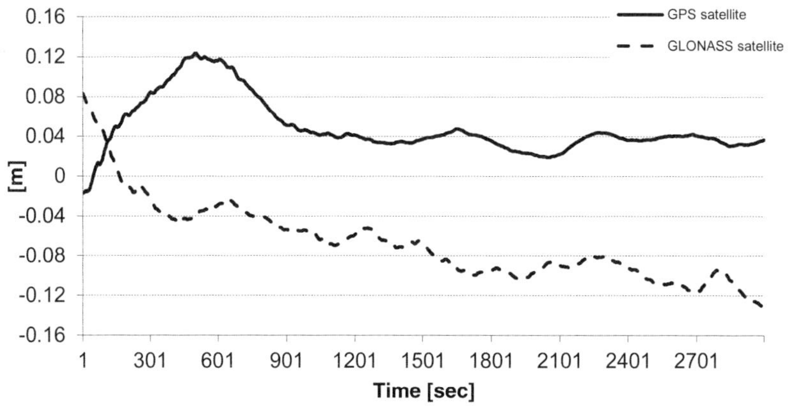

Figure 7.7.3 Ionosphere delay estimate for one GPS satellite and one GLONASS satellite obtained for the a priori ionosphere correlation time $\tau_i = 3600$ sec.

the other hand, it is obvious that the ionospheric delay at the adjacent epochs for each satellite cannot vary independently. Therefore, it is wise to apply a dynamic model constraining variation of the residual ionospheric delay estimate. We used a practically proven first-order dynamic model – the simplest one among all possible models. The above figures illustrate the behavior of the ionospheric delay estimates. Effectively, we can see from Figures 7.7.2 and 7.7.3 that the RTK processing engine picks up part of other real error terms, most likely the multipath, together with the ionospheric estimate. The positive effect of the ionospheric delay estimation on the position estimate is illustrated in Figure 7.7.1. The estimation of the ionospheric delay decreases a mean value of the position error and the dynamic model allows for better smoothing of the ionosphere estimate and, therefore, a smoothing of the position noise estimate.

7.8 RTK ALGORITHMS WITH CHANGING NUMBER OF SIGNALS

In the RTK algorithms in the previous sections we assumed the set of signals to be constant over time. The receiver was assumed to observe the same number of satellites and the receiver hardware and firmware were assumed to track a constant number of signals for each satellite. These assumptions greatly simplify the description of filtering schemes but naturally do not correspond to reality. Satellites rise and set while observation sessions are ongoing. The satellite visibility may change for a short period in kinematic applications when a vehicle passes a tree or other object. Even if one of the signals, say the C/A L1 GPS signal keeps tracking, another signal from the same satellite, say L2 GPS, may experience short-term interruption. The discontinuity may be as short as one or a few epochs, but nevertheless the number of signals is changing in time and the RTK algorithm must make accommodations to achieve optimal processing for such a discontinuous data flow.

Even if a tracked signal does not completely disappear, the continuity of the phase measurement may be affected. Discontinuity can occur when the carrier phase locked loop (PLL) loses continuity of tracking. In other words, the state of the PLL may jump into the vicinity of another stable point. This means that after some transient is settled, the carrier phase ambiguity will have changed by some unknown integer (cycles). The dimension of the state space is defined by the order of the PLL, which in turn depends on the number of integration operators in the closed loop. Usually the PLL state space is two or three dimensional, including the carrier phase, its rate of change, and probably second time derivative.

The ambiguities were considered constant in the description of the RTK algorithms. Now we accept the fact that they can vary. The detection of cycle slips was already mentioned in Section 6.2.3 when discussing geometry-free solutions. More numerical algorithms will be considered in the following section. The cycle slips leading to an unknown ambiguity jump will be treated as if the signal disappears at the current epoch and appears again at the next epoch, suggesting another value of $t^p_{CS,km,b}$ and another carrier phase ambiguity in the navigation equations (7.2.12) and (7.3.17).

First, consider the case when a new signal appears. Let the set of signals be $S = \{s_1, \ldots, s_n\}$ as suggested by equation (7.3.19) for epochs $t = t_0, \ldots, t_1$. Suppose the new signal s_{n+1} appears at the epoch $t_1 + 1$, so that starting with epoch $t_1 + 1$, we have

$$S_+ = \{s_1, \ldots, s_n, s_{n+1}\} \tag{7.8.1}$$

which has one more signal s_{n+1} compared to (7.3.19).

New signals might require a new carrier phase ambiguity and probably a new hardware bias if the new signal is the only one in the group having the same bias. These new constant parameters must be recursively estimated starting with epoch t_1 along with all other constant parameters. In Chapter 3, when discussing recurrent algorithms for various least-squares problems, we assumed that the vector of time-invariant parameters has a constant dimension that does not depend on time. For example, when discussing problem (3.1.1) we supposed that $\boldsymbol{y} \in \boldsymbol{R}^n$, while the number of observables $m(t)$ was varying with time. Let us now extend the problem setup, assuming that $\boldsymbol{y} \in \boldsymbol{R}^n$ for $t = 1, \ldots, t_1$ and $\boldsymbol{y} \in \boldsymbol{R}^{n+1}$ for $t = t_1 + 1, \ldots$.

We will apply Algorithm 1 given in Table 3.1.1 in Section 3.1 for problem (3.1.1), running it for sequential epochs $t = 1, \dots, t_1$. Starting with epoch $t_1 + 1$ we assume that the vector y has dimension $n + 1$. Moreover, after its optimal estimate $\boldsymbol{y}(t_1)$ is calculated, we can assume that it has dimension $n + 1$ for all time instants $t = 1, \dots, t_1 + 1, \dots$, with the $(n + 1)$th entry set to zero for $t = 1, \dots, t_1$. The measurement matrix is assumed to have the last column, i.e., the $n + 1$th column, containing zeroes

$$\boldsymbol{W}(t) = \begin{bmatrix} w_{11}(t) & \cdots & w_{1n}(t) & 0 \\ w_{21}(t) & \cdots & w_{2n}(t) & 0 \\ \cdots & \cdots & \cdots & \cdots \\ w_{m(t),1}(t) & \cdots & w_{m(t),n}(t) & 0 \end{bmatrix} \tag{7.8.2}$$

for $t = 1, \dots, t_1$. Starting with epoch $t_1 + 1$, the algorithm operates with dimension $n + 1$, but modifications are needed at the time instant t_1. The iteration of Algorithm 1 (see Table 3.1.1), corresponding to the epoch $t = t_1$, starts with the extension of the vector $\boldsymbol{y}(t_1)$ by adding $(n + 1)$th entry equal to zero

$$\boldsymbol{y}(t_1) \to \begin{bmatrix} \boldsymbol{y}(t_1) \\ \hline 0 \end{bmatrix} \tag{7.8.3}$$

Also, the calculation of matrix $\boldsymbol{D}(t_1 + 1)$ is split into two substeps:

a. Extend the $n \times n$ matrix $\boldsymbol{D}(t_1)$ by forming the $(n + 1) \times (n + 1)$ matrix

$$\boldsymbol{D}_{+}(t_1) = \left[\begin{array}{ccc|c} & & & 0 \\ & \boldsymbol{D}(t_1) & & \vdots \\ & & & 0 \\ \hline 0 & \cdots & 0 & 0 \end{array}\right] \in \boldsymbol{R}^{(n+1)\times(n+1)} \tag{7.8.4}$$

b. Calculate the matrix update

$$\boldsymbol{D}(t_1 + 1) = \boldsymbol{D}_{+}(t_1) + \overline{\boldsymbol{W}}^T(t_1 + 1)\overline{\boldsymbol{W}}(t_1 + 1) \tag{7.8.5}$$

All other steps are performed the same way as described in Algorithm 1 in Table 3.1.1. Obviously, the dimension can be extended by more than one at the time instant $t = t_1$. More than one zero will be added to the vector $\boldsymbol{y}(t_1)$ in (7.8.3) and more than one zero row and column will be added to the matrix $\boldsymbol{D}(t_1)$ in (7.8.4). The same reasoning holds for all other problem setups described in Sections 3.2 to 3.5.

The step "*Update the matrix* $\hat{\boldsymbol{D}}(t)$" of the algorithms in Tables 7.4.1 and 7.5.1, and the step "*Compute Cholesky decomposition and the updated matrix* $\hat{\boldsymbol{D}}(t + 1)$" of the algorithm in Table 7.6.1 must be modified in the same manner as (7.8.4) and (7.8.3), allowing estimation of the ambiguity of the new signal to start at the same epoch as it appears for the first time.

Now consider the case when the existing signal disappears. Let the set of signals be $S = \{s_1, \dots, s_n\}$ as suggested by (7.3.19) for epochs $t = t_0, \dots, t_1$, and suppose that one of the signals, say s_n without loss of generality, disappears at the epoch $t_1 + 1$ so that starting with the epoch $t_1 + 1$ we have

$$S_- = \{s_1, \dots, s_{n-1}\} \tag{7.8.6}$$

Again, going back to the problem in Section 3.1 of Chapter 3, and assume that the algorithm in Table 3.1.1 was running for sequential time instances $t = 1, \dots, t_1$, recursively estimating the vector of constant parameters $\boldsymbol{y} \in \boldsymbol{R}^n$ based on sequentially received measurements. Starting with time instant $t_1 + 1$, we assume that the vector y has dimension $n - 1$ and that the variable y_n is no longer represented in the observation model (3.1.1). How do we modify the numerical scheme described in Section 3.1 in such a way that it would generate an optimal estimate of vector $\boldsymbol{y}(t) \in \boldsymbol{R}^n$ for $t = 1, \dots, t_1$ and $\boldsymbol{z}(t) \in \boldsymbol{R}^{n-1}$ for $t = t_1 + 1, \dots$?

Suppose that the algorithm in Table 3.1.1 ran for sequential epochs $t = 1, \dots, t_1$. Let us look at expression (3.1.11) and present it in a more convenient form for further analysis. Expanding the parentheses in (3.1.11) and taking into account (3.1.13) and (3.1.15), we obtain

$$\begin{aligned}
\boldsymbol{I}(\boldsymbol{y}, t') &= \left(\overline{\boldsymbol{W}}(t')\boldsymbol{y} - \overline{\boldsymbol{b}}(t')\right)^T \left(\overline{\boldsymbol{W}}(t')\boldsymbol{y} - \overline{\boldsymbol{b}}(t')\right) \\
&= \boldsymbol{y}^T \boldsymbol{D}(t')\boldsymbol{y} - 2\boldsymbol{y}^T \overline{\boldsymbol{W}}^T(t')\overline{\boldsymbol{b}}(t') + \overline{\boldsymbol{b}}^T(t')\overline{\boldsymbol{b}}(t') \\
&= \boldsymbol{y}^T \boldsymbol{D}(t')\boldsymbol{y} - 2\boldsymbol{y}^T \boldsymbol{D}(t')\boldsymbol{D}^{-1}(t')\overline{\boldsymbol{W}}^T(t')\overline{\boldsymbol{b}}(t') + \overline{\boldsymbol{b}}^T(t')\overline{\boldsymbol{b}}(t') \\
&= \boldsymbol{y}^T \boldsymbol{D}(t')\boldsymbol{y} - 2\boldsymbol{y}^T \boldsymbol{D}(t')\boldsymbol{y}(t') + \overline{\boldsymbol{b}}^T(t')\overline{\boldsymbol{b}}(t')
\end{aligned} \tag{7.8.7}$$

Then adding and subtracting the term $\boldsymbol{y}^T(t')\boldsymbol{D}(t')\boldsymbol{y}(t')$ to the above expression, we arrive at the following expression:

$$\begin{aligned}
\boldsymbol{I}(\boldsymbol{y}, t') &= \boldsymbol{y}^T \boldsymbol{D}(t')\boldsymbol{y} - 2\boldsymbol{y}^T \boldsymbol{D}(t')\boldsymbol{y}(t') + \boldsymbol{y}(t')^T\boldsymbol{D}(t')\boldsymbol{y}(t') - \boldsymbol{y}(t')^T\boldsymbol{D}(t')\boldsymbol{y}(t') + \overline{\boldsymbol{b}}^T(t')\overline{\boldsymbol{b}}(t') \\
&= \left(\boldsymbol{y} - \boldsymbol{y}(t')\right)^T \boldsymbol{D}(t')\left(\boldsymbol{y} - \boldsymbol{y}(t')\right) - \overline{\boldsymbol{b}}^T\overline{\boldsymbol{W}}(t')\boldsymbol{D}^{-1}(t')\overline{\boldsymbol{W}}^T(t')\overline{\boldsymbol{b}}(t') + \overline{\boldsymbol{b}}^T(t')\overline{\boldsymbol{b}}(t') \\
&= \left(\boldsymbol{y} - \boldsymbol{y}(t')\right)^T \boldsymbol{D}(t')\left(\boldsymbol{y} - \boldsymbol{y}(t')\right) + \overline{\boldsymbol{b}}^T(t')(\boldsymbol{I} - \overline{\boldsymbol{W}}(t')\boldsymbol{D}^{-1}(t')\overline{\boldsymbol{W}}^T)\overline{\boldsymbol{b}}(t') \\
&= \left(\boldsymbol{y} - \boldsymbol{y}(t')\right)^T \boldsymbol{D}(t')\left(\boldsymbol{y} - \boldsymbol{y}(t')\right) + c
\end{aligned} \tag{7.8.8}$$

where the constant c does not depend on the variable $\boldsymbol{y}$. Obviously the quadratic function (7.8.8) takes its minimum at the point $\boldsymbol{y}(t')$. Then we can express the function $\boldsymbol{I}(\boldsymbol{y}, t' + 1)$, corresponding to the next time instant recursively using expressions (3.1.8) and (7.8.8),

$$\begin{aligned}
&\boldsymbol{I}(\boldsymbol{y}, t' + 1) \\
&= \boldsymbol{I}(\boldsymbol{y}, t') + \left(\overline{\boldsymbol{W}}(t' + 1)\boldsymbol{y} - \overline{\boldsymbol{b}}(t' + 1)\right)^T \left(\overline{\boldsymbol{W}}(t' + 1)\boldsymbol{y} - \overline{\boldsymbol{b}}(t' + 1)\right) \\
&= \left(\boldsymbol{y} - \boldsymbol{y}(t')\right)^T \boldsymbol{D}(t')\left(\boldsymbol{y} - \boldsymbol{y}(t')\right) + \left(\overline{\boldsymbol{W}}(t' + 1)\boldsymbol{y} - \overline{\boldsymbol{b}}(t' + 1)\right)^T \left(\overline{\boldsymbol{W}}(t' + 1)\boldsymbol{y} - \overline{\boldsymbol{b}}(t' + 1)\right)
\end{aligned} \tag{7.8.9}$$

The optimal estimate $\boldsymbol{y}(t'+1)$ minimizes the function (7.8.9). Equating the first derivatives of (7.8.9) to zero, we obtain the equation

$$\boldsymbol{D}(t')(\boldsymbol{y}-\boldsymbol{y}(t')) + \overline{\boldsymbol{W}}(t'+1)^T\left(\overline{\boldsymbol{W}}(t'+1)\boldsymbol{y} - \overline{\boldsymbol{b}}(t'+1)\right) = 0 \tag{7.8.10}$$

Solving it for $\boldsymbol{y}(t'+1)$, we obtain the same expressions as (3.1.16),

$$\boldsymbol{y}(t'+1) = \boldsymbol{y}(t') + \boldsymbol{D}^{-1}(t'+1)\overline{\boldsymbol{W}}^T(t'+1)\left(\overline{\boldsymbol{b}}(t'+1) - \overline{\boldsymbol{W}}(t'+1)\boldsymbol{y}(t')\right) \tag{7.8.11}$$

where $\boldsymbol{D}(t'+1) = \boldsymbol{D}(t') + \overline{\boldsymbol{W}}^T(t'+1)\overline{\boldsymbol{W}}(t'+1)$. See also expression (3.1.17).

Look at the least-squares problem (7.8.9) for $t' = t_1$. Let the vector of variables $\boldsymbol{y}$ and the vector of optimal estimate $\boldsymbol{y}(t_1)$ be split into two parts:

$$\boldsymbol{y} = \begin{bmatrix} \boldsymbol{z} \\ \hdashline \boldsymbol{y}_n \end{bmatrix} \quad \boldsymbol{z} \in \boldsymbol{R}^{n-1} \quad \boldsymbol{y}(t_1) = \begin{bmatrix} \boldsymbol{z}(t_1) \\ \hdashline \boldsymbol{y}_n(t_1) \end{bmatrix} \quad \boldsymbol{z}(t_1) \in \boldsymbol{R}^{n-1} \tag{7.8.12}$$

and the matrix $\boldsymbol{D}(t_1)$ be split accordingly:

$$\boldsymbol{D}(t_1) = \begin{bmatrix} \boldsymbol{D}_z(t_1) & \boldsymbol{d}_n(t_1) \\ \boldsymbol{d}^T{}_n(t_1) & d_{nn}(t_1) \end{bmatrix} \tag{7.8.13}$$

According to our assumption $\boldsymbol{z}(t) \in \boldsymbol{R}^{n-1}$ for $t = t_1 + 1, \ldots$ and, therefore, we can express the matrix $\boldsymbol{W}(t_1+1)$ in the form

$$\boldsymbol{W}(t_1+1) = \begin{bmatrix} w_{11}(t_1+1) & \cdots & w_{1,n-1}(t_1+1) & 0 \\ w_{21}(t_1+1) & \cdots & w_{2,n-1}(t_1+1) & 0 \\ \cdots & \cdots & \cdots & \cdots \\ w_{m(t_1+1),1}(t_1+1) & \cdots & w_{m(t_1+1),n-1}(t_1+1) & 0 \end{bmatrix} = \left[\boldsymbol{W}_z(t_1+1) \mid 0\right] \tag{7.8.14}$$

whose last column, the nth column, is zero. We can rewrite the problem (7.8.9) in the form

$$\begin{aligned}
\min_{\boldsymbol{y}\in\boldsymbol{R}^n} \boldsymbol{I}(\boldsymbol{y}, t_1+1) &= \min_{\boldsymbol{z}\in\boldsymbol{R}^n, \boldsymbol{y}_n\in\boldsymbol{R}^1} \boldsymbol{I}(\boldsymbol{z}, \boldsymbol{y}_n, t_1+1) = \min_{\boldsymbol{z}\in\boldsymbol{R}^n}\left(\min_{\boldsymbol{y}_n\in\boldsymbol{R}^1} \boldsymbol{I}(\boldsymbol{z}, \boldsymbol{y}_n, t_1+1)\right) \\
&= \min_{\boldsymbol{z}\in\boldsymbol{R}^n}\Big[(\boldsymbol{z}-\boldsymbol{z}(t_1))^T\boldsymbol{D}_z(t_1)(\boldsymbol{z}-\boldsymbol{z}(t_1)) \\
&\quad + \left(\overline{\boldsymbol{W}}_z(t_1+1)\boldsymbol{z} - \overline{\boldsymbol{b}}(t'+1)\right)^T\left(\overline{\boldsymbol{W}}_{\boldsymbol{z}}(t_1+1)\boldsymbol{z} - \overline{\boldsymbol{b}}(t'+1)\right) \\
&\quad + \min_{\boldsymbol{y}_n\in\boldsymbol{R}^1}\left(2(\boldsymbol{z}-\boldsymbol{z}(t_1))^T\boldsymbol{d}_n(t_1)(\boldsymbol{y}-\boldsymbol{y}(t_1)) + d_{nn}(\boldsymbol{y}-\boldsymbol{y}(t_1))^2\right)\Big]
\end{aligned} \tag{7.8.15}$$

Taking the internal minimum over the variable $\boldsymbol{y}_n \in \boldsymbol{R}^1$ in (7.8.15), we obtain

$$(\boldsymbol{y}-\boldsymbol{y}(t_1)) = -\frac{\boldsymbol{d}_n^T(t_1)(\boldsymbol{z}-\boldsymbol{z}(t_1))}{d_{nn}} \tag{7.8.16}$$

Substituting (7.8.16) back into (7.8.15) we present it in the form

$$\min_{\mathbf{z}\in\mathbf{R}^n} \bar{\mathbf{I}}(\mathbf{z}, t_1+1) = \min_{\mathbf{z}\in\mathbf{R}^n} \Big[\big(\mathbf{z} - \mathbf{z}(t_1)\big)^T \bar{\mathbf{D}}_z(t_1)\big(\mathbf{z} - \mathbf{z}(t_1)\big) + \big(\bar{\mathbf{W}}_z(t_1+1)\mathbf{z} - \bar{\mathbf{b}}(t'+1)\big)^T \big(\bar{\mathbf{W}}_z(t_1+1)\mathbf{z} - \bar{\mathbf{b}}(t'+1)\big)\Big] \tag{7.8.17}$$

where

$$\bar{\mathbf{D}}_z(t_1) = \mathbf{D}_z(t_1) - \frac{1}{d_{nn}}\mathbf{d}_n(t_1)\mathbf{d}_n^T(t_1) \tag{7.8.18}$$

Then, the solution to the problem (7.8.17) is given by the following expressions:

$$\mathbf{z}(t'+1) = \mathbf{z}(t') + \mathbf{D}_z^{-1}(t'+1)\bar{\mathbf{W}}_z^T(t'+1)\big(\bar{\mathbf{b}}(t'+1) - \bar{\mathbf{W}}_z(t'+1)\mathbf{z}(t')\big) \tag{7.8.19}$$

$$\mathbf{D}_z(t'+1) = \bar{\mathbf{D}}_z(t') + \bar{\mathbf{W}}_z^T(t'+1)\bar{\mathbf{W}}_z(t'+1) \tag{7.8.20}$$

which are obtained the same way that expression (7.8.11) was obtained above.

We note that the matrix $\mathbf{D}(t)$ now has a form similar to (7.8.13) for $t \geq t_1 + 1$. Since the matrix $W(t)$ has a form similar to (7.8.14) for all $t \geq t_1 + 1$, then columns $\mathbf{d}_n(t) \equiv \mathbf{d}_n(t_1)$ for all $t \geq t_1 + 1$, which follows from expression (3.1.15). It means that the algorithm described in Table 3.1.1 remains unchanged except for updating of the matrix $\mathbf{D}(t_1)$ and the optimal estimate. Updating of the matrix $\mathbf{D}(t_1)$ takes the form:

a. Present the $n \times n$ matrix $\mathbf{D}(t_1)$ in the form (7.8.13).
b. Calculate the $(n-1) \times (n-1)$ matrix update according to the expression

$$\mathbf{D}(t'+1) = \mathbf{D}_z(t_1) - \frac{1}{d_{nn}}\mathbf{d}_n(t_1)\mathbf{d}_n^T(t_1) + \bar{\mathbf{W}}_z^T(t'+1)\bar{\mathbf{W}}_z(t'+1) \tag{7.8.21}$$

following from expressions (7.8.20) and (7.8.18). Updating of the optimal estimate takes the form:

a. Present the n-dimensional estimate $\mathbf{y}(t_1)$ in the form of (7.8.12).
b. Calculate the estimate update:

$$\mathbf{y}(t+1) := \mathbf{z}(t) + \mathbf{B}_{L_{\mathbf{D}(t+1)}}\big(\mathbf{F}_{L_{\mathbf{D}(t+1)}}\bar{\mathbf{W}}^T(t+1)\bar{\mathbf{r}}(t+1)\big) \tag{7.8.22}$$

All remaining quantities are calculated according to the algorithm in Table 3.1.1. The same reasoning holds true for all other problems described in Sections 3.2, 3.4, and 3.5. The matrix $\hat{\mathbf{D}}(t)$ updating steps of the algorithms in Sections 7.4.1, 7.5.1, and 7.6.1 must be modified according to the procedure described above, allowing for "seamless" optimal estimation of the remaining ambiguities when one of the signals disappears.

Note again that when the cycle slip is detected in one of the carrier phase signals, the situation can be treated as a discontinuity of the carrier phase observation. One can proceed with a sequential application of two schemes, assuming that the signal disappears and reappears with another ambiguity.

7.9 CYCLE SLIP DETECTION AND ISOLATION

As was mentioned above, a discontinuity of phase measurement can occur on one or several signals. The individual PLL generating carrier phase measurement for a certain signal may lose the locked mode as a result of short-term shading or some other disturbance. Depending on the order of the PLL, its state vector can be two dimensional, three dimensional, or it can have even higher dimensions. In the first case, the PLL state vector consists of the carrier phase and its first derivative, also known as Doppler frequency. In the second case, the second derivative of the carrier phase, or rate of Doppler frequency, is added. Discontinuity of tracking results in a jump of the state into the vicinity of another stable point in the state space. This means that after some transient is settled, the carrier phase ambiguity has changed by some number of integer cycles. This means that the carrier phase ambiguity changes its integer value. One can deal with this by adding another ambiguity variable and changing the value of $t^{p}_{CS,km,b}$ in the navigation equation (7.3.17).

This section focuses on the detection of cycle slips using across-receiver across-epoch differences and making use of signal redundancy. We start with brief remarks on triple-difference cycle slip fixing and traditional dual-frequency geometry-free solutions, by providing some specifics for triple-frequency cycle slip detection, and briefly discussing the two-step method by Dai et al. (2008).

Triple-difference solution was considered in Section 6.1.4 as a tool for cycle slip detection. It is applicable to the static positioning case. The triple-difference method works independently for each carrier phase signal and each pair of satellites, using time redundancy and assuming that the measurements are sufficiently oversampled. The sampling period is equal to the across-epoch time interval. In the kinematic case, a sufficient sampling rate allows prediction of the term $\Delta\rho^{pq}_{km}$ in expression (6.1.91). Any of the curve fitting methods make it possible to predict the triple difference of the topocentric satellite distance. Other predictors known in the estimation theory can also be applied. A predicted triple difference of the topocentric satellite distances can be used to compensate the term $\Delta\rho^{pq}_{km}$ for the kinematic case when it cannot be neglected.

If more than one carrier frequency is available for each satellite, it is possible to use geometry-free combinations as mentioned in Section 6.2.3. Detection of cycle slips based on the geometry-free combination works if the cycle slip occurs on only one of the signals for a certain satellite. More generally, it works if cycle slips affecting different frequency signals do not cancel in the geometry-free combination. Even if a step change has been detected in the geometry-free combination, it is hard to judge upon which of the two signals, let us say corresponding to L1 or L2, the slip actually occurred. Therefore, to be on the safe side the processing engine must flag both signals as possibly having a cycle slip, even if it may result in a false alarm for one of them.

Cycle slip detection methods for triple-frequency observations are considered, for example, in Wu et al. (2010). The set of observations consists of the three pseudoranges and three carrier phases. These include L1, L2, L5 for GPS and QZSS; L1, E5b, E5a for Galileo; or B1, B2, B3 for the Beidou satellite system. Consider the across-receiver difference of the scaled carrier phase functions (6.1.26) for a certain satellite in the form (7.2.12):

$$\overline{\Phi}_1(t) \equiv \lambda_1\overline{\varphi}_1(t) = \rho(t) + \xi(t) + \lambda_1 N_1(t) - I_1(t) + \lambda_1 d_1 + \lambda_1\overline{\varepsilon}_1(t) \quad (7.9.1)$$

$$\overline{\Phi}_2(t) \equiv \lambda_2\overline{\varphi}_2(t) = \rho(t) + \xi(t) + \lambda_2 N_2(t) - (f_1/f_2)^2 I_1(t) + \lambda_2 d_2 + \lambda_2\overline{\varepsilon}_2(t) \quad (7.9.2)$$

$$\overline{\Phi}_3(t) \equiv \lambda_3\overline{\varphi}_3(t) = \rho(t) + \xi(t) + \lambda_3 N_3(t) - (f_1/f_3)^2 I_1(t) + \lambda_3 d_3 + \lambda_3\overline{\varepsilon}_3(t) \quad (7.9.3)$$

where the across-receiver difference symbol *km* is omitted for brevity. The across-receiver ambiguities $N_i(t)$ do not depend on t until a cycle slip occurs. The symbol Δ will be used to denote differencing across time between two adjacent epochs $t+1$ and t,

$$\begin{aligned}\lambda_i\,\Delta\overline{\varphi}_i(t+1,t) &= \Delta\rho(t+1,t) + c\,\Delta\xi(t+1,t) + \lambda_i\,\Delta N_i(t+1,t)\\ &\quad -\gamma_i\,\Delta I_i(t+1,t) + \lambda_i\,\Delta d_i(t+1,t) + \lambda\,\Delta\overline{\varepsilon}_i(t+1,t)\end{aligned} \quad (7.9.4)$$

where $\gamma_i = (f_1/f_i)^2$ and $\gamma_1 = 1$. Let us make the following simplifying assumptions:

a. Ionospheric delay does not significantly change between epochs, so the quantity $\Delta I_i(t+1,t)$ can be neglected in (7.9.4).
b. The hardware bias term is practically constant, so the cross-epoch bias $\lambda_i\,\Delta d_i(t+1,t)$ can be neglected.

The quantities $\Delta N_i(t+1,t)$, $i = 1,2,3$ represent exactly the cycle slips that occurred between epochs $t+1$ and t. Let us construct the multifrequency carrier phase combination:

$$\begin{aligned}\Delta\Phi_c &\equiv \sum_{i=1,2,3}\alpha_i\lambda_i\,\Delta\overline{\varphi}_i(t+1,t)\\ &= \left[\Delta\rho(t+1,t) + c\,\Delta\xi(t+1,t)\right]\sum_{i=1,2,3}\alpha_i + \sum_{i=1,2,3}\alpha_i\lambda_i\,\Delta N_i(t+1,t)\\ &\quad + \sum_{i=1,2,3}\alpha_i\lambda_i\,\Delta\overline{\varepsilon}_i(t+1,t)\end{aligned} \quad (7.9.5)$$

In order to preserve the geometric and clock terms in (7.9.5), we assume that coefficients α_i satisfy

$$\alpha_1 + \alpha_2 + \alpha_3 = 1 \quad (7.9.6)$$

Therefore (7.9.5) can be written as

$$\Delta\Phi_c = \Delta\rho(t+1,t) + c\,\Delta\xi(t+1,t) + \lambda_c\,\Delta N_c + \Delta\varepsilon_c \quad (7.9.7)$$

where $\lambda_c\,\Delta N_c \equiv \sum_{i=1,2,3}\alpha_i\lambda_i\,\Delta N_i(t+1,t)$ and $\Delta\varepsilon_c \equiv \sum_{i=1,2,3}\alpha_i\lambda_i\overline{\varepsilon}_i(t+1,t)$. The value ΔN_c will be called the *cycle slip of the carrier phase combination*. In order to keep the integer value of ΔN_c, assuming that cycle slips $\Delta N_i(t+1,t)$ are integers, the coefficients α_i satisfy additional conditions. Namely, the values

$$m_i = \frac{\lambda_i \alpha_i}{\lambda} \tag{7.9.8}$$

must be integer. Together with (7.9.6), one obtains the expression for the *wavelength of carrier phase combination*

$$\lambda_c = \frac{\lambda_1\lambda_2\lambda_3}{m_1\lambda_2\lambda_3 + m_2\lambda_1\lambda_3 + m_3\lambda_1\lambda_2} = \frac{1}{m_1/\lambda_1 + m_2/\lambda_2 + m_3/\lambda_3} \tag{7.9.9}$$

and corresponding frequency

$$f_c \equiv \frac{c}{\lambda_c} = m_1 f_1 + m_2 f_2 + m_3 f_3 \equiv f_{(m_1,m_2,m_3)} \tag{7.9.10}$$

using triple-frequency subscript notations as in Section 6.7.1. Taking in account expressions (7.9.7) to (7.9.9), the *across-receiver across-time carrier phase combination* $\Delta\varphi_c = \Delta\Phi_c/\lambda_c$ can be expressed as

$$\Delta\varphi_c = \Delta\overline{\varphi}_{(m_1,m_2,m_3)}(t+1,t) \tag{7.9.11}$$

The carrier phase noise is assumed to be Gaussian white noise with a standard deviation σ_φ for each of the three frequencies. Then the standard deviation of the combination noise $\Delta\varepsilon_c$ is expressed as

$$\sigma_c = \sqrt{2}\lambda_c\sigma_\varphi\sqrt{m_1^2 + m_2^2 + m_3^2} \tag{7.9.12}$$

with the $\sqrt{2}$ factor reflecting across-epoch differencing. The wavelength combination for selected integer parameters m_1, m_2, and m_3 is shown in the Table 7.9.1, with classic wide lane and extra wide lane as particular cases.

Let us compare the expression (7.9.7) with across-receiver across-epoch pseudorange observations with ionospheric and biases terms neglected according to assumptions (a) and (b). This pseudorange function is expressed as

$$\Delta P = \Delta\rho(t+1,t) + c\,\Delta\xi(t+1,t) + \Delta\varepsilon_P \tag{7.9.13}$$

with standard deviation of the noise term $\Delta\varepsilon_P$ estimated as $\sqrt{2}\sigma_P$. The factor $\sqrt{2}$ again reflects the across-epoch differencing, and σ_P is the standard deviation of the pseudorange noise. Then the cycle slip value of the carrier phase combination (7.9.11) can be estimated as

$$\Delta N_c = \frac{\Delta\Phi_c - \Delta P}{\lambda_c} + \frac{\Delta\varepsilon_c - \Delta\varepsilon_P}{\lambda_c} \tag{7.9.14}$$

with the second term being the cycle slip estimation error. Its standard deviation is estimated according to (7.9.12) as

$$\sigma_N = \sqrt{2}\cdot\sqrt{\sigma_\varphi^2(m_1^2 + m_2^2 + m_3^2) + \frac{\sigma_P^2}{\lambda_c^2}} \tag{7.9.15}$$

TABLE 7.9.1 Wavelength and Frequency of Selected GPS Carrier Phase Combinations.

m_1	m_2	m_3	f_c (MHz)	λ_c (m)
1	−1	0	347.82	0.862
1	0	−1	398.97	0.751
0	1	−1	51.15	5.861
1	−6	5	92.07	3.256
−9	2	10	40.92	7.326
−1	10	−9	112.53	2.664
3	0	−4	20.46	14.652
−1	8	−7	10.23	29.305

From (7.9.15) it can be seen that the precision of cycle slip determination depends on the noise of pseudorange and carrier phase observations and the wavelength of carrier phase combination. Under the same observation condition, the longer the carrier phase wavelength is and the smaller the absolute values m_i, the higher the precision of the cycle slip estimation. This means that the combination $\{3, 0, -4\}$ with $\lambda_c = 14.652$ is preferable to $\{-9, 2, 10\}$ with $\lambda_c = 7.326$ (see Table 7.9.1).

Another interesting approach to detection and correction of cycle slips in single GNSS receivers is proposed in Dai et al. (2008). They use the geometry-free combination for cycle slip detection. At the second stage, they combine the across-epoch increments of carrier phase and pseudorange measurements to correctly determine the cycle slip values using an integer search technique as applied in ambiguity resolution, because effectively the cycle slip is just an unknown increment of the carrier phase ambiguity. We introduce the generalization of the geometry-free combination if we replace the condition (7.9.6) by

$$\alpha_1 + \alpha_2 + \alpha_3 = 0 \tag{7.9.16}$$

Then (7.9.5) takes the form

$$\Delta\Phi_{(\alpha_1,\alpha_2,\alpha_3)}(t+1,t) \equiv \sum_{i=1,2,3} \alpha_i\lambda_i\,\Delta\overline{\varphi}_i(t+1,t) = \sum_{i=1,2,3} \alpha_i\lambda_i N_i(t+1,t) + \sum_{i=1,2,3} \alpha_i\lambda_i\Delta\overline{\varepsilon}_i(t+1,t) \tag{7.9.17}$$

The standard deviation of the noise term in (7.9.17) is expressed as

$$\sigma_c = \sqrt{2}\sigma_\varphi\sqrt{(\alpha_1^2\lambda_1^2 + \alpha_2^2\lambda_2^2 + \alpha_3^2\lambda_3^2)} \tag{7.9.18}$$

By choosing a proper confidence level, we get the critical value $\beta\sigma_c$ for the cycle slip detection criterion, where the factor β depends on the confidence level. It can be chosen, for example, as 3 for the confidence level 0.997. The following inequality allows checking whether cycle slips arose on one or more carrier phase signals of a specific satellite between two adjacent epochs:

$$|\Delta\Phi_{(\alpha_1,\alpha_2,\alpha_3)}(t+1,t)| > \beta\sigma_c \tag{7.9.19}$$

However, if (7.9.19) is not satisfied, one cannot state that the carrier phase data is not contaminated by cycle slips. There are special cycle slip groups which cannot be detected using the criterion (7.9.19). Consider, for example, the group of cycle slips $\{154, 120, 115\}$ for the GPS case. With obvious notations we have $\Delta N_{(\alpha_1,\alpha_2,\alpha_3)}\{154, 120, 115\} \equiv \Delta N_c(t+1,t) = 0$. Therefore $\Delta\Phi_{(\alpha_1,\alpha_2,\alpha_3)}(t+1,t) = \Delta\varepsilon_c$ and this cycle slip group is not detectable by criterion (7.9.19). Other groups of cycle slips can give nonzero but small values of $\lambda_c\ \Delta N_c(t+1,t)$, also precluding detection by criterion (7.9.19). If the cycle slip group satisfies the inequality

$$\left|\lambda_c\,\Delta N_c(t+1,t)\right| \le \sqrt{\beta^2-1}\sigma_c \tag{7.9.20}$$

then it cannot be detected by criterion (7.9.19). The cycle slip groups satisfying (7.9.20) are called in Dai et al. (2008) *insensitive groups*. For the Galileo case, an obvious example of the insensitive group is $\{154, 118, 115\}$ resulting in $\Delta N_{(\alpha_1,\alpha_2,\alpha_3)}\{154, 118, 115\} = 0$ for all scalars $(\alpha_1, \alpha_2, \alpha_3)$, thus satisfying (7.9.16).

Consider more examples. Let the scalars be chosen as $(-1, 5, -4)$. An example of an insensitive cycle slip group giving a nonzero but small value satisfying (7.9.20) is $\{51, 39, 38\}$. We have $|\Delta N_{(-1,5,-4)}\{51, 39, 38\}| = 0.0107$. If $\sigma_\varphi = 0.01$ and $\beta = 3$, then direct calculations with (7.9.18) give $\sqrt{\beta^2-1}\sigma_c = 0.0645$ which proves that the group $\{51, 39, 38\}$ is insensitive for the combination $(-1, 5, -4)$. Another group $\{31, 23, 23\}$ gives $|\Delta N_{(-1,5,-4)}\{31, 23, 23\}| = 0.783$, which does not satisfy (7.9.20) and can be detected by the criterion (7.9.19) for a given confidence level.

Let now the scalars be $(4, 1, -5)$. In this case, $\sqrt{\beta^2-1}\sigma_c = 0.0600$. The group $\{51, 39, 38\}$ [which was insensitive for combination $(-1, 5, -4)$] now gives $|\Delta N_{(4,1,-5)}\{51, 39, 38\}| = 0.0882$ which can be [detected by (7.9.19), while the group $\{31, 23, 23\}$ [detectable by (7.9.19) for previous combination] now shows insensitivity because $|\Delta N_{(4,1,-5)}\{31, 23, 23\}| = 0.0032$.

There might be different insensitive groups, depending on the values of $\alpha_1, \alpha_2, \alpha_3$. As was illustrated, using two proper geometry-free combinations can reduce the number of insensitive cycle slip groups.

Once the occurrence of cycle slips has been confirmed, one should quantify the value of the slips and remove them from the carrier phase observations by subtraction. Dai et al. (2008) give a cycle slip determination approach to fix the value of cycle slips for a single satellite. They use pseudoranges to determine the value of the cycle slips. By introducing the across-epoch pseudorange observables (7.9.13), the integer search method is applied as described in Sections 6.5.2 to 6.5.4.

Removal of the cycle slips from the carrier phase observations is not the only possible approach for handling cycle slips. Another option consists of resetting the ambiguity estimation channels affected by cycle slips in the recursive estimation algorithm. The method described in Section 7.8 can be used for this purpose.

The described methods have advantages such as individual analysis for each satellite or signal, but there are also disadvantages. A main disadvantage of individual analysis of triple differences is that for fast kinematics situations, the carrier phase measurements experience rapid and unpredictable variations, which can mask small cycle slips. Using geometry-free combinations in multiple-frequency receivers is not possible if some signals are temporarily unavailable; the technique is simply not applicable for use with single-frequency receivers. One can readily state that there is no one method for cycle slip detection and removal that is best for all situations.

7.9.1 Solutions Based on Signal Redundancy

In real-time kinematic processing we need approaches that work reliably under fast and aggressive dynamics, even when the number of available signals varies unpredictably. Here we discuss another group of methods based on the concept of signal redundancy. There must be at least five satellites available in two sequential epochs. Multiple-frequency signals are not necessary for each satellite. The method works for single-frequency cases as well, see Kozlov and Tkachenko (1998). Some satellites may have only L1 observables, while others could have L5 observables only, or L1 and L2 observables.

Consider multiple signals for two sequential epochs. We form across-receiver across-epoch differences and use the linearized form (7.3.46). Again, the symbol Δ will be used to denote differencing over time:

$$\begin{aligned}\Delta \boldsymbol{b}_{\varphi}(t+1,t) = {} & \boldsymbol{\Lambda}^{-1}\boldsymbol{A}\,\Delta d\boldsymbol{x}(t+1,t) + \boldsymbol{\Lambda}^{-1}\boldsymbol{e}\,\Delta\xi(t+1,t) + \Delta\boldsymbol{n}(t+1,t) \\ & -\boldsymbol{\Lambda}^{-1}\boldsymbol{\Gamma}\,\Delta\boldsymbol{i}(t+1,t) + \boldsymbol{\Lambda}^{-1}\boldsymbol{W}_{\mu}\,\Delta\boldsymbol{\mu}(t+1,t)\end{aligned} \tag{7.9.21}$$

Taking into account assumptions (a) and (b) formulated earlier in this section, we neglect the terms $\boldsymbol{\Lambda}^{-1}\boldsymbol{\Gamma}\Delta\boldsymbol{i}(t+1,t)$ and $\boldsymbol{\Lambda}^{-1}\boldsymbol{W}_{\mu}\Delta\boldsymbol{\mu}(t+1,t)$,

$$\Delta \boldsymbol{b}_{\varphi}(t+1,t) = \boldsymbol{\Lambda}^{-1}\boldsymbol{A}\,\Delta d\boldsymbol{x}(t+1,t) + \boldsymbol{\Lambda}^{-1}\boldsymbol{e}\,\Delta\xi(t+1,t) + \Delta\boldsymbol{n}(t+1,t) \tag{7.9.22}$$

When forming the right-side vector of (7.3.22) we had a nominal position $\left(x_{k,0}(t), y_{k,0}(t), z_{k,0}(t)\right)^T$ at each epoch as point of linearization, and $d\boldsymbol{x}(t)$ was the vector of corrections to this nominal position. Let us assume that the precise position $\boldsymbol{x}_k(t)$, or its estimate obtained at the epoch t, can serve as the nominal position for the next epoch $t+1$:

$$\boldsymbol{x}_{k,0}(t+1) = \boldsymbol{x}_k(t) \tag{7.9.23}$$

Therefore, the quantity $\Delta d\boldsymbol{x}(t+1,t)$ becomes an across-epoch position increment $\Delta\boldsymbol{x}(t+1,t)$ of the rover antenna. The expression for the residual vector $\Delta\boldsymbol{b}_{\varphi}(t+1,t)$ in (7.9.22) has the form

$$\Delta \boldsymbol{b}_{\varphi}(t+1,t) = \begin{pmatrix} \Delta\overline{\varphi}^{p_1}_{km,b_1}(t+1,t) - \frac{1}{\lambda^{p_1}_{b_1}}\left(\Delta\rho^{p_1}_k(t+1,t) - \Delta\rho^{p_1}_m(t+1,t)\right) \\ \Delta\overline{\varphi}^{p_2}_{km,b_2}(t+1,t) - \frac{1}{\lambda^{p_2}_{b_2}}\left(\Delta\rho^{p_2}_k(t+1,t) - \Delta\rho^{p_2}_m(t+1,t)\right) \\ \vdots \\ \Delta\overline{\varphi}^{p_n}_{km,b_n}(t+1,t) - \frac{1}{\lambda^{p_n}_{b_n}}\left(\Delta\rho^{p_n}_k(t+1,t) - \Delta\rho^{p_n}_m(t+1,t)\right) \end{pmatrix} \tag{7.9.24}$$

The across-time increment of topocentric geometric distance from the base station to the satellite is expressed as

$$\Delta\rho_m^{p_i}(t+1,t) = \rho_m^{p_i}\left(t+1-\tilde{\tau}_m^{p_i}(t+1)\right) - \rho_m^{p_1}\left(t-\tilde{\tau}_m^{p_i}(t)\right) \tag{7.9.25}$$

Since the position of the base is stationary, the value $\Delta\rho_m^{p_i}(t+1,t)$ reflects the change of distance caused by the motion of the satellite, though it also depends on the base station position. In the same way, since we assumed that $\boldsymbol{x}_{k,0}(t+1) = \boldsymbol{x}_k(t)$, the value

$$\Delta\rho_k^{p_i}(t+1,t) = \rho_{k,0}^{p_i}\left(t+1-\tilde{\tau}_m^{p_i}(t+1)\right) - \rho_k^{p_i}\left(t-\tilde{\tau}_m^{p_i}(t)\right) \tag{7.9.26}$$

will be defined by the motion of the satellite, though also being dependent on the position $\boldsymbol{x}_k(t)$. If the baseline is short, the quantities (7.9.25) and (7.9.26) compensate for each other in (7.9.24). This means that for short baselines, the across-receiver across-time carrier phase residuals are approximately equal to the across-receiver across-time carrier phase observations. For longer baselines this conclusion is not valid; however, the difference of $\Delta\boldsymbol{b}_\varphi(t+1,t)$ and $\Delta\boldsymbol{\varphi}(t+1,t)$ will show a slow varying function with time, depending on the motion of the satellites rather than on the motion of the rover. The shorter the baseline, the less it is time dependent.

The quantity $\Delta\boldsymbol{n}(t+1,t)$ in (7.9.22) is exactly the vector of cycle slips. Using notations

$$\boldsymbol{J} = \left[\boldsymbol{\Lambda}^{-1}\boldsymbol{e} \quad \boldsymbol{\Lambda}^{-1}\boldsymbol{A}\right] \in \boldsymbol{R}^{n\times 4} \tag{7.9.27}$$

$$\overline{\boldsymbol{x}}(t+1) = \begin{pmatrix}\Delta\xi(t+1,t)\\ \Delta\boldsymbol{x}(t+1,t)\end{pmatrix} = \boldsymbol{R}^4 \tag{7.9.28}$$

$$\boldsymbol{b}(t+1) = \Delta\boldsymbol{b}_\varphi(t+1,t) \tag{7.9.29}$$

$$\delta(t+1) = \Delta\boldsymbol{n}(t+1,t) \tag{7.9.30}$$

we can present equation (7.9.22) in the form

$$\boldsymbol{b}(t+1) = \boldsymbol{J}\overline{\boldsymbol{x}}(t+1) + \delta(t+1) + \varepsilon_\varphi(t+1) \tag{7.9.31}$$

where we have added the noise term $\varepsilon_\varphi(t+1)$. Suppose $\varepsilon_{\varphi,s} \sim n(0,\sigma_{\varepsilon_\varphi})$ and its covariance matrix is

$$\boldsymbol{C} = E(\varepsilon_\varphi\varepsilon_\varphi^T) = \sigma_{\varepsilon_\varphi}{}^2\boldsymbol{I}_n \tag{7.9.32}$$

where $\sigma_{\varepsilon_\varphi}$ is of the order of a hundredth of a cycle, e.g., 0.01 or 0.02, and $\boldsymbol{I}_n$ is the $n\times n$ identity matrix. The outlier vector $\delta(t+1)$ contains integer values. Another significant difference between noise $\varepsilon_\varphi(t+1)$ and outlier vector $\delta(t+1)$ is that the latter one is *sparse*.

The sparseness property of the cycle slip outlier vector $\delta(t+1)$ means that it consists mostly of zeroes. Probably all of its entries are zero if no cycle slips occurred between 2 consecutive epochs. Even if cycle slips occurred, they affect only a small number of signal measurements, for example, 1, 2, or 3 among a total number of 20 or so signals. Sparseness of the outlier vector is the key property that allows recovery and isolation of its nonzero entries. Here we will introduce an approach that explores the sparseness property.

The following terminology is borrowed from the literature on *compressive sensing*. The compressive sensing technique has received much attention in the literature devoted to the recovery of sparse signals. Sparseness occurs in different domains such as time domain, frequency domain, and space domain. We will be using some basic concepts adopted in this theory. More specifically, we will be dealing with outliers as sparse vectors since they affect only a few signals.

The set of nonzero entries of the vector $\boldsymbol{\delta}(t+1)$ is called "support set" and is denoted by

$$\text{supp}\left(\boldsymbol{\delta}(t+1)\right) = \left\{s : \delta_s(t+1) \neq 0\right\} \tag{7.9.33}$$

The lower index s indexes the entries of the vector $\boldsymbol{\delta}(t+1)$ connected to signals $s = 1, \ldots, n$. We use the "cardinality" notation to denote the number of elements in the set. Therefore,

$$\text{card}\left(\text{supp}\left(\boldsymbol{\delta}(t+1)\right)\right) \tag{7.9.34}$$

is exactly the number of nonzero entries in the vector $\boldsymbol{\delta}$. Another frequently used notation for the same quantity is l_0-norm.

Recall the definition of the l_q-norm of a vector. For an arbitrary vector δ, it is defined as

$$\|\boldsymbol{\delta}\|_{l_q} = \left(\sum_{s=1}^{n} |\delta_s|^q\right)^{1/q} \tag{7.9.35}$$

For example, if q takes values ∞, 2, 1, and 0, then the l_α-norm is expressed as

$$\|\boldsymbol{\delta}\|_{l_\infty} = \max_{s=1,\cdots,n} |\delta_s| \tag{7.9.36}$$

$$\|\boldsymbol{\delta}\|_{l_2} = \left(\sum_{s=1}^{n} |\delta_s|^2\right)^{1/2} \tag{7.9.37}$$

$$\|\boldsymbol{\delta}\|_{l_1} = \sum_{s=1}^{n} |\delta_s| \tag{7.9.38}$$

$$\|\boldsymbol{\delta}\|_{l_0} = \text{card}\{s : \delta_s \neq 0\} \tag{7.9.39}$$

where (7.9.36) gives the maximum modulus of the entries, expression (7.9.37) is an Euclidian norm, expression (7.9.38) is a sum of modulus of the entries, and (7.9.39) is a number of the nonzero entries.

Note that across-epoch carrier phase increments for each signal form the n-dimensional vector $\boldsymbol{b}(t+1)$, whereas the four-dimensional vector $\overline{\boldsymbol{x}}(t+1)$, which varies arbitrarily, can form only a four-dimensional subspace in the n-dimensional space, provided no cycle slips have occurred. This simply means that carrier phase variation cannot be arbitrary. The carrier phase measurements must vary consistently to each other, forming only a four-dimensional subspace when cycle slips are absent. In order to check if it is true, we must test the linear system

$$\boldsymbol{b}(t+1) = \boldsymbol{J}\overline{\boldsymbol{x}}(t+1) \tag{7.9.40}$$

for consistency. The zero hypothesis is that $\boldsymbol{\delta}(t+1) = 0$ in the system (7.9.31). Actually, the noise term $\boldsymbol{\varepsilon}_{\varphi}$ in (7.9.31) prevents consistency. However, we can check to see if the sum of squared residuals of the least-squares solution of (7.9.40) with covariance matrix (7.9.32) satisfies the statistical test χ^2_{n-4}. If it is satisfied, we can say that the system (7.9.40) is consistent with accuracy $\sigma_{\varepsilon_{\varphi}}$. If consistency of the system (7.9.40) is violated, then our assumption about absence of cycle slips is not true, thus $\boldsymbol{\delta}(t+1) \neq 0$. If that happens, we say that the presence of cycle slips has been detected. In order to find out which entry or entries of the sparse vector $\boldsymbol{\delta}(t+1)$ is nonzero, we apply a blunders detection technique.

Below we consider a modern approach recently developed in the digital signal processing literature dealing with *sparse signals detection* (Candez and Tao, 2005; Candez et al., 2005). Sparseness means that the vector to be recovered has a small number of nonzero entries, while zero entries dominate. Probably all entries are zero if no cycle slips occurred. Detection of the outliers can be referred to as recovery of the support set (7.9.33). We will compare different approaches to recovery of the vector of outliers. As we will see, each approach relates to the certain value of q in (7.9.35).

Let $\boldsymbol{x}^*$ be the least-squares solution to the system (7.9.40). Assuming that the covariance matrix of the noise is (7.9.32), we have

$$\boldsymbol{x}^* = (\boldsymbol{J}^T\boldsymbol{J})^{-1}\boldsymbol{J}^T\boldsymbol{b} \tag{7.9.41}$$

where the time symbol $t+1$ is omitted for brevity. Then the residual vector and weighted sum of squares can be calculated as

$$\boldsymbol{v} = \boldsymbol{b} - \boldsymbol{J}\boldsymbol{x}^* = \left(\boldsymbol{I}_n - \boldsymbol{J}(\boldsymbol{J}^T\boldsymbol{J})^{-1}\boldsymbol{J}^T\right)\boldsymbol{b} \tag{7.9.42}$$

$$V = \boldsymbol{v}^T\boldsymbol{C}^{-1}\boldsymbol{v} = \frac{1}{\sigma^2_{\varepsilon_{\varphi}}}\boldsymbol{v}^T\boldsymbol{v} \tag{7.9.43}$$

The expression (2.7.37) of Section 2.7.1 states that if $\varepsilon_{\varphi,s} \sim n(0, \sigma_{\varepsilon_{\varphi}})$, then the value (7.9.43) is distributed according to the χ^2 law with $n-4$ degrees of freedom. Let α denote the significance level. The inverse χ^2 distribution $T(\alpha, m)$ with m degrees of freedom is available from statistical tables and mathematical software. This means that if the value (7.9.43) does not exceed $T(\alpha, n-4)$ with probability $1-\alpha$ the zero hypothesis is accepted. If it exceeds, then the presence of cycle slips is detected.

Approach 1 (Exhaustive search of affected signals): Assuming that the redundancy is large enough for the removing of affected signals, we can isolate cycle slips. To remove signals affected by cycle slips, we perform an exhaustive search removing signals one by one, then pair by pair, then triple, by triple, and so on, until statistical tests comparing the value (7.9.43), calculated using the remaining signals with the threshold $T(\alpha, n-k-4)$ is satisfied. Here, k is the number of removed signals,

$k = 1, 2, 3, \ldots$, and $n - k$ is the number of remaining signals. The matrix $\boldsymbol{J}$ and the vector $\boldsymbol{b}$ now have dimensions $(n - k) \times 4$ and $(n - k) \times 1$. The search is performed until $n - k > 4$ or, in other words, as long as redundancy allows. The search is complete once either the statistical test is satisfied or redundancy does not allow removing any more signals.

Let us consider the data set used in Subsection 7.4.1 to illustrate how cycle slip detection and isolation works. It is a short baseline static data set. We assumed that the best available estimate of position $\boldsymbol{x}(t)$ is taken as a nominal position for the epoch $t + 1$ and, therefore, the quantity $\Delta d\boldsymbol{x}(t + 1, t)$ becomes the across-epoch position incremental of the rover antenna. For static conditions we can use the time-invariant nominal position of the rover antenna for all epochs, while the precise base position is used as the nominal position of the base antenna. The first 200 epochs are analyzed.

The data was carefully selected and analyzed beforehand to be sure that there were no cycle slips. After that, three signals were artificially affected by cycle slips:

- L1 signal of the satellite GPS 4
- L2 signal of the satellite GPS 9
- L1 signal of the satellite GLONASS with the frequency letter –7

The magnitude of the cycle slip is +1 for all three signals. All three cycle slips are inserted at epoch number 110. We know that small cycle slips are hard to detect and isolate. Also, "nested" cycle slips, or cycle slips that occur simultaneously, are hard to analyze. Therefore, the case we are analyzing is not trivial. Figure 7.9.1 shows the triple differences of residuals (7.9.24) for the L1 signal, calculated for satellites GPS 4 (which is affected by the cycle slip) and the satellite GPS 15 (which, as we know, is not affected because all three cycle slips were inserted manually).

As was stated earlier, if the condition (7.9.23) holds and the baseline is short (in our case it is only a few meters long) the triple difference of the topocentric geometric distance in (7.9.24) is close to the triple difference carrier phase. It is relatively small in the Figure 7.9.1 and the carrier phase cycle slip generates a single-epoch outlier

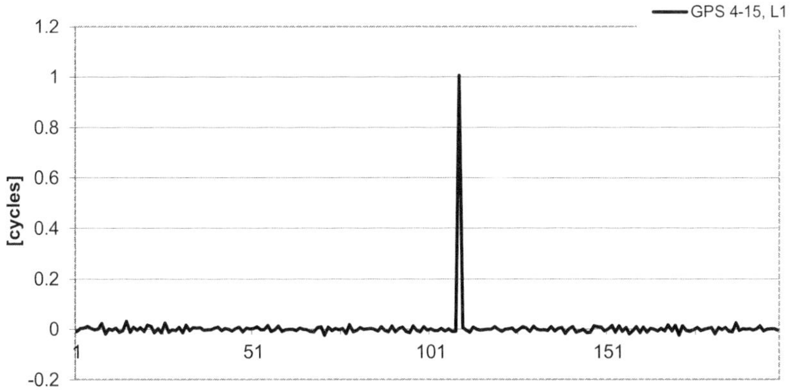

Figure 7.9.1 Triple-difference carrier phase residuals calculated for L1 signal with GPS satellites 4 and 15.

at the epoch 110, which is exactly the epoch when the artificial cycle slip occurred. Triple differences clearly show both the location and magnitude of the cycle slip. Disadvantages of such a straightforward use of triple differences are as follows:

a. It works properly if the reference satellite is not affected by the cycle slip.
b. The cycle slip is seen clearly because the rover was static. If the antenna experiences unpredictable motion between two epochs, the "legal" carrier phase variation caused by the motion can exceed tens or hundreds of cycles, while the "illegal" variation caused by the cycle slip can be as little as one cycle. Therefore, the motion component can totally mask the cycle slip.

Recall that we do not consider individual across-epoch variation of carrier phases. Instead, we analyze across-receiver across-time differences. Figure 7.9.2 shows plots of $\Delta b_{\varphi}^{p}(t+1,t)$ across-receiver across-time differences of residuals (7.9.24) for the same signals GPS 4 (L1) and GPS 15 (L1).

The single-epoch outlier in the dashed plot, caused by the cycle slip, is not easy to recognize because of the carrier phase variation due to across-receiver clock variation. On the other hand, this variation is almost identical for the two signals, GPS 4 (L1) and GPS 15 (L1), and that is the reason why it cancels in the triple differences of residuals shown in the previous figure. However, it must be emphasized again that the carrier phase variation caused by the rover antenna dynamics can be so large and so unpredictable that the cycle slip cannot be recognized when analyzing either triple differences or across-receiver across-time differences individually.

Let us now try the method described above, based on the statistical analysis of the residuals (7.9.42) and (7.9.43). First, run solutions of problem (7.9.40) and calculate the values (7.9.43) for each epoch. The results are presented in Figure 7.9.3. The threshold value $T(0.05, 16) = 26.29$ is shown in the figure as a dashed line. The value $\sigma_{\varepsilon_{\varphi}}$ is 0.01. The plot is clipped at the value 70, while V equals 21,279 at the epoch 110 where three outliers occurred.

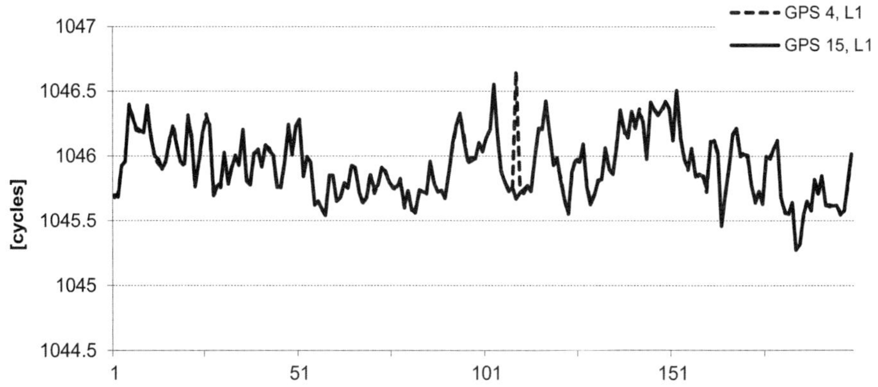

Figure 7.9.2 Across-receiver across-time differences of carrier phase residuals calculated for L1 signals of GPS 4 (dashed) and GPS 15 (solid).

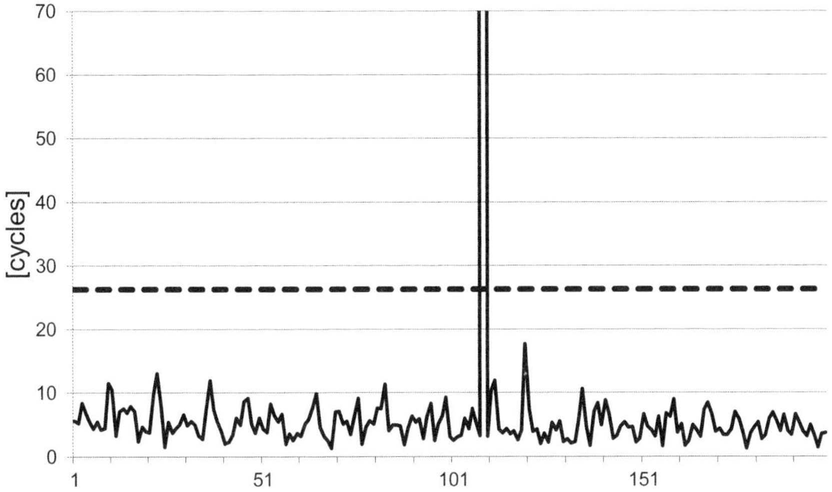

Figure 7.9.3 Weighted sum of residuals squared V (solid) compared with the χ^2 threshold value (dashed).

After performing the exhaustive search by removing signals one by one, then removing pairs, then removing triplets of signals, we finally found that the criterion

$$V \leq T(\alpha, n-k-4) \tag{7.9.44}$$

was satisfied with $k = 3$, $V = 4.586$, and $T(0.05, 13)=22.36$.

As a result of the search, exactly three signals affected by cycle slips (having numbers 1, 7, 12) were removed. Note that there is no need to calculate the solution to system (7.9.40) every time the set of signals changes during the exhaustive search. Suppose that the row $\boldsymbol{j}_s$ is added to the matrix $\boldsymbol{J}$ or removed from it. Then the matrix $\boldsymbol{J}^T\boldsymbol{J}$ involved in expression (7.9.41) is updated by the rank one matrix $\Delta = \boldsymbol{j}_s^T\boldsymbol{j}_s$. The rank one update method, described in Section A.3.9, allows updating of the Cholesky decomposition of matrix $\boldsymbol{J}^T\boldsymbol{J}$, resulting in the Cholesky decomposition of $\boldsymbol{J}^T\boldsymbol{J} \pm \Delta$. This method can be used for recalculating the solution of the linear system after one equation has been deleted or added (Golub et al.,1996).

The method works independently of the motion of the rover; it can be static or performing unpredictable maneuvers. The numerical scheme of the method works the same way because no assumptions about dynamics need to be made as the basis of the method.

Note again that we are using an exhaustive search, trying all combinations of measurements in order to find the vector $\boldsymbol{\delta}(t+1)$ with the least possible number of nonzero entries. We refer to this method as the l_0 optimization, as it minimizes the l_0-norm of the vector $\boldsymbol{\delta}(t+1)$ according to (7.9.34) and (7.9.39). We now describe another approach to the outlier detection, which promises less of a computational complexity because no exhaustive search will be involved.

Let us look at the problem (7.9.31) from another point of view, following the approach described in Candez et al. (2005). We want to correctly *recover* the vector $\bar{\mathbf{x}}(t+1)$ from the measurements $\mathbf{b}(t+1)$, which are corrupted by small noise $\varepsilon_{\varphi}(t+1)$ and sparse outliers $\boldsymbol{\delta}(t+1)$. The sparseness implies that only a small number of entries in the vector $\boldsymbol{\delta}(t+1)$ are nonzero. The magnitude of nonzero entries can be arbitrarily large. The key property of $\boldsymbol{\delta}(t+1)$ is the sparseness. For example, only 3 out of 20 entries being nonzero would be considered sparse. Is it possible to recover $\bar{\mathbf{x}}(t+1)$ exactly or estimate it almost exactly from data corrupted not only by noise but also by large outliers?

Approach 2 (Minimization of l_1-norm of residuals): Under reasonable conditions, the vector $\bar{\mathbf{x}}(t+1)$ can be recovered as a solution to the following l_1 minimization problem:

$$\mathbf{b}(t+1) = \mathbf{J}\bar{\mathbf{x}}(t+1) + \boldsymbol{\delta}(t+1)$$

$$\|\boldsymbol{\delta}(t+1)\|_{l_1} \to \min \tag{7.9.45}$$

provided that the vector of outliers is sufficiently sparse. The function $\|\boldsymbol{\delta}\|_{l_1} = \sum_{s=1}^{n} |\delta_s|$ was introduced in (7.9.38). The approach can be reformulated using the concept of the support set (7.9.33). Sufficient sparseness of the vector $\boldsymbol{\delta}(t+1)$ means that

$$\|\boldsymbol{\delta}\|_{l_0} \leq \rho n \quad \rho \ll 1 \tag{7.9.46}$$

Therefore, the true value of the vector $\bar{\mathbf{x}}(t+1)$ can be recovered almost exactly by solving the linear programming problem (7.9.45). "Almost" means that the recovered solution will be exact up to the precision induced by the magnitude $\sigma_{\varepsilon_{\varphi}}$ of the noise component of the error $\varepsilon_{\varphi}(t+1)$ in model (7.9.31). To see that (7.9.45) is a linear optimization problem, rewrite it in the following equivalent form:

$$\mathbf{b}(t+1) = \mathbf{J}\bar{\mathbf{x}}(t+1) + \mathbf{u} - \mathbf{v}$$

$$u_s \geq 0 \quad v_s \geq 0$$

$$\sum_{s=1}^{n} (u_s + v_s) \to \min \tag{7.9.47}$$

Almost exact recovery of the vector $\bar{\mathbf{x}}(t+1)$ means that the outlier vector $\boldsymbol{\delta}(t+1)$ will appear "almost" exactly (up to the magnitude of the noise) in the residuals

$$\mathbf{r} = \mathbf{b}(t+1) - \mathbf{J}\bar{\mathbf{x}}(t+1)$$

$$\mathbf{r} \approx \boldsymbol{\delta}(t+1) \tag{7.9.48}$$

Let us try this approach using the same example considered earlier. Direct application of the simplex method (Vanderbei, 2008) to the problem (7.9.47) results in the solution

$$\bar{\mathbf{x}}^*(t+1) \qquad \mathbf{u}^*, \mathbf{v}^* \tag{7.9.49}$$

and

$$\boldsymbol{\delta}^*(t+1) = \boldsymbol{u}^* - \boldsymbol{v}^* \tag{7.9.50}$$

Only entries of the vector $\delta^*(t+1)$ having numbers 1, 7, 12 (GPS L1 PRN4; GLONASS L1 letter -7, and GPS L2 PRN 9, respectively) are significantly nonzero

$$\begin{aligned} \delta_1^*(t+1) &= 0.9806 \\ \delta_7^*(t+1) &= 0.9951 \\ \delta_{12}^*(t+1) &= 0.9982 \end{aligned} \tag{7.9.51}$$

while others are either exactly zero or near zero with a magnitude of the order of $\sigma_{\varepsilon_\varphi}$. Below there are all 20 entries of the vector, and the 1th, 7th, and 12th entries are marked by a bar:

$$\begin{aligned} \delta^*(\mathrm{t}+1) = (&\overline{0.980}, 0, 0.0173, 0, 0, 0, \overline{0.995}, 0, 0, 0, 0, \overline{0.998}, 0, 0.003, 0, 0, 0.002, \\ &0, 0.002, 0, 0, 0, 0, 0, 0.006, 0, 0, 0.004, 0.002, 0.003, 0.002, 0, 0, 0, \\ &0, 0.013, 0, 0, 0, 0.00)^T \end{aligned} \tag{7.9.52}$$

To be precise, we are interested in the recovery of just the vector of outliers, not the vector $\overline{\boldsymbol{x}}(t+1)$. Correct recovery of $\boldsymbol{\delta}(t+1)$ means that the support set (7.9.33) has been correctly detected. Note again that we are interested in "almost" correct recovery of the support set, i.e., entries having near zero absolute values, comparable with $\sigma_{\varepsilon_\varphi}$, are identified as zero.

With the outlier vector support set having been identified, the problem of cycle slip isolation is completely solved. The signals affected by outliers can be removed, while the remaining signals are affected only by noise, which we have always accepted as unavoidable.

The simplex method is not the only known method for linear programming problems. There are many efficient methods, like interior point methods (Lustig et al., 1994), that show extremely efficient behavior that qualifies them for implementation in real-time software. The computational complexity of the linear optimization problem is polynomial dependent on the problem dimensions, while the computational load for exhaustive search grows exponentially as a function of the number of signals.

Simultaneous solution for the vector $\overline{\boldsymbol{x}}(t+1)$ and recovery of the outlier vector $\boldsymbol{\delta}(t+1)$ via solution to the problem (7.9.47) will be referred to as the l_1-optimization method. This method is faster but less precise then the l_0- optimization method. This means that the l_1-optimization method does not necessarily correctly recover the outlier support set, while the l_0 optimization always works correctly.

Approach 3 (OMP method): Now consider the third method, which is even less computationally involved than the l_1-optimization method; however, it is also less precise in recovery of the support set. Let $\boldsymbol{F}$ be the *annulator matrix* for the matrix $\boldsymbol{J}$,

$$\boldsymbol{FJ} = 0 \qquad \boldsymbol{J} \in \boldsymbol{R}^{n\times 4} \qquad \boldsymbol{F} \in \boldsymbol{R}^{(n-4)\times n} \tag{7.9.53}$$

This equation implies that the columns of matrix $\boldsymbol{J}$ belong to the kernel of the matrix $\boldsymbol{F}$, or that $\boldsymbol{J}$ spans the null space of $\boldsymbol{F}$. Let us multiply both sides of the equality (7.9.31) by the matrix $\boldsymbol{F}$:

$$\boldsymbol{F}\delta(t+1) = \boldsymbol{c}(t+1) + \xi_{\varphi}(t+1) \tag{7.9.54}$$

where $\boldsymbol{c}(t+1) = \boldsymbol{Fb}(t+1)$ and $\xi_{\varphi}(t+1) = -\boldsymbol{F}\varepsilon_{\varphi}(t+1)$. Ignoring noise, we are looking for solutions to the system (7.9.54). Note first that if cycle slips are absent, the system (7.9.40) is "almost" consistent and the vector $\boldsymbol{b}(t+1)$ belongs to the linear space spanned on the columns of matrix $\boldsymbol{J}$. This means that if cycle slips are absent, the vector $\boldsymbol{c}(t+1) = \boldsymbol{Fb}(t+1)$ is near zero. Otherwise, if cycle slips exist, the outlier vector satisfies the system (7.9.54) with the nonzero constant term $\boldsymbol{c}(t+1)$.

How can one find the annulator matrix? Consider the $\boldsymbol{QR}$ decomposition of the matrix $\boldsymbol{J}$ as described in section A.3.8,

$$\boldsymbol{QR} = \boldsymbol{J} \tag{7.9.55}$$

where the matrix $\boldsymbol{Q} \in \boldsymbol{R}^{n\times n}$ is orthonormal and the matrix $\boldsymbol{R} \in \boldsymbol{R}^{n\times 4}$ has the following structure:

$$\boldsymbol{R} = \begin{bmatrix} \boldsymbol{U} \\ \hline {}_{(n-4)}\boldsymbol{0}_4 \end{bmatrix} \qquad \boldsymbol{U} \in \boldsymbol{R}^{4\times 4} \quad {}_{(n-4)}\boldsymbol{0}_4 \in R^{(n-4)\times 4} \tag{7.9.56}$$

The matrix $\boldsymbol{U}$ is upper triangle. It directly follows from the previous two equations that the orthonormal matrix $\boldsymbol{Q}^T$ can be partitioned into two parts

$$\boldsymbol{Q}^T = \begin{bmatrix} \boldsymbol{G} \\ \hline \boldsymbol{F} \end{bmatrix} \tag{7.9.57}$$

with the matrix $\boldsymbol{F}$ obeying the property (7.9.53).

The linear system (7.9.54) is underdetermined because $\boldsymbol{F} \in \boldsymbol{R}^{(n-4)\times n}$. There are infinitely many solutions, but we are interested in a solution that has the least possible support set. In other words, we are looking for a solution with the least possible number of nonzero entries. In order to solve this problem, also arising in the context of a compressive sensing framework, we can apply the l_1 optimization

$$\begin{gathered} \boldsymbol{F}(\boldsymbol{u} - \boldsymbol{v}) = \boldsymbol{c}(t+1) \\ u_s \geq 0 \qquad v_s \geq 0 \\ \sum_{s=1}^{n} (u_s + v_s) \to \min \end{gathered} \tag{7.9.58}$$

which is equivalent to (7.9.47). Alternatively, we can apply the so-called *orthogonal matching pursuit (OMP)* algorithm (Cai and Wang, 2011). The OMP algorithm is described as follows:

1. Initialize the algorithm calculating residual and support set

$$\boldsymbol{y}^{(0)} = \boldsymbol{c}(t+1) \quad S^{(0)} = \varnothing \tag{7.9.59}$$

and set $k = 0$.

2. Check if the termination criterion

$$\frac{1}{\sigma^2_{\varepsilon_\varphi}} \|\boldsymbol{y}^{(k)}\|_{l_2} < T(\alpha, n-k-4) \tag{7.9.60}$$

is satisfied with some significance level α. The algorithm terminates if the criterion (7.9.60) is satisfied. Otherwise set $k := k+1$ and continue.

3. Find the index $s^{(k)}$ that solves the optimization problem

$$\max_{s=1,\ldots,n} \frac{\left|\boldsymbol{f}_s^T \boldsymbol{y}^{(k-1)}\right|}{\sqrt{\boldsymbol{f}_s^T \boldsymbol{f}_s}} \tag{7.9.61}$$

with $\boldsymbol{f}_s$ being the s-th column of the matrix $\boldsymbol{F}$.

4. Update the support set

$$S^{(k)} = S^{(k-1)} \cup \{s^{(k)}\} \tag{7.9.62}$$

5. Solve the least-squares problem

$$\delta^{(k)} = \left(\boldsymbol{F}^T\left(S^{(k)}\right)\boldsymbol{F}(S^{(k)})\right)^{-1}\boldsymbol{F}^T(S^{(k)})\boldsymbol{c}(t+1) \tag{7.9.63}$$

with the matrix $\boldsymbol{F}(S)$ being the matrix consisting of columns $\boldsymbol{f}_s$, $s \in S$, and calculate the residual vector

$$\boldsymbol{y}^{(k)} = \boldsymbol{c}(t+1) - \boldsymbol{F}(S^{(k)})\delta^{(k)} \tag{7.9.64}$$

6. Go to step 1.

Let us apply the OMP algorithm to the same example, GPS + GLONASS, L1 + L2, number of signals $n = 20$. Earlier we chose $\delta_\varphi = 0.01$. Let the significance level be the same as before, $\alpha = 0.95$. At epoch 110, we have the following results running the OMP algorithm:

$k = 0$ $\quad \frac{1}{\sigma^2_{\varepsilon_\varphi}} \|y^{(0)}\|_{l_2} = 21279$

$k = 1$ $\quad s_1 = 12 \quad S_1 = \{12\} \quad \frac{1}{\sigma^2_{\varepsilon_\varphi}} \|y^{(1)}\|_{l_2} = 13287$

$k = 2$ $\quad s_2 = 7 \quad S_2 = \{12, 7\} \quad \frac{1}{\sigma^2_{\varepsilon_\varphi}} \|y^{(2)}\|_{l_2} = 5281.6$

$k = 3$ $\quad s_3 = 1 \quad S_3 = \{12, 7, 1\} \quad \frac{1}{\sigma^2_{\varepsilon_\varphi}} \|y^{(3)}\|_{l_2} = 4.586$

and the algorithm terminates as $4.586 < T(0.95, 13) = 22.36$ and the criterion (7.9.60) is satisfied. So, the correct support set is found at the third iteration.

This method will be referred to as "the OMP method." It has complexity (computational load) comparable with $n^2 \times 4$ which is the lowest complexity among all methods considered. On the other hand, it has least precision in correctly recovering the support set, which as was already stated, is the set of signals affected by cycle slips or signals to be isolated.

Let us present the OMP method in a slightly modified form. Consider the residual (7.9.42) of the least-squares solution of problem (7.9.40). Taking into account the decomposition (7.9.55) and (7.9.56), we can write

$$\begin{aligned} \boldsymbol{r} &= \left(\boldsymbol{I}_n - \boldsymbol{J}(\boldsymbol{J}^T\boldsymbol{J})^{-1}\boldsymbol{J}^T\right)\boldsymbol{b}(t+1) \\ &= \boldsymbol{Q}\left(\boldsymbol{I}_n - \left[\begin{array}{c} \boldsymbol{U} \\ \hline {}_{(n-4)}\boldsymbol{0}_4 \end{array}\right](\boldsymbol{U}^T\boldsymbol{U})^{-1}\left[\boldsymbol{U}^T \;\middle|\; \boldsymbol{0}_{4,n-4}\right]\right)\boldsymbol{Q}^T\boldsymbol{b}(t+1) \\ &= \boldsymbol{Q}\left[\begin{array}{c|c} \boldsymbol{0}_4 & {}_4\boldsymbol{0}_{n-4} \\ \hline {}_{n-4}\boldsymbol{0}_4 & \boldsymbol{I}_{n-4} \end{array}\right]\boldsymbol{Q}^T\boldsymbol{b}(t+1) = \boldsymbol{F}^T\boldsymbol{F}\boldsymbol{b}(t+1) = \boldsymbol{F}^T\boldsymbol{c}(t+1) \end{aligned} \tag{7.9.65}$$

Consider the first iteration of the OMP method setting $k = 1$. The vector $\boldsymbol{y}^{(0)} = \boldsymbol{c}(t+1)$ and, therefore, taking into account (7.9.65), the step (7.9.61) chooses the maximum normalized entry of the residual

$$s^{(1)} = \underset{s=1,\cdots,n}{\mathrm{Arg\ max}} \frac{|\boldsymbol{f}_s^T\boldsymbol{c}(t+1)|}{\|\boldsymbol{f}_s\|_{l_2}} = \underset{s=1,\ldots,n}{\mathrm{Arg\ max}} \frac{|\boldsymbol{r}_s|}{\|\boldsymbol{f}_s\|_{l_2}} \tag{7.9.66}$$

where r_s denotes the sth entry of the residual vector $\boldsymbol{r}$. In other words, at the first iteration we have isolated the signal corresponding to the maximum value of the normalized residual (see data snooping in Section 2.9.2 for comparison).

At its first iteration, the OMP method isolates signals with the maximum value of the normalized residual entry. The isolated signal is removed from the linear system (7.9.40), which is solved again with the dimension reduced by one. Note that the sequential computation of solution to the linear system with the number of equations reduced by one can be efficiently performed using the "rank one update" technique.

7.10 ACROSS-RECEIVER AMBIGUITY FIXING

The across-receiver ambiguity vector is structured according to (7.3.40). Each individual vector $\boldsymbol{n}_a$ is subject to the constraint (7.3.41), which means that all ambiguities inside a certain group must have an identical fractional part. In equivalent formulation, this means that across-receiver across-satellite (double) differences must be integers. Let us denote the set of such ambiguity vectors by $\mathbb{N}$, which is exactly the set of vectors partitioned according to (7.3.40) satisfying the condition

$$\mathbb{N} = \left\{\boldsymbol{n} = (\boldsymbol{n}_1{}^T, \ldots, \boldsymbol{n}_{\bar{a}}{}^T)^T \;:\; \boldsymbol{n}_a \in \mathbb{N}_a, a = 1, \ldots, \bar{a}\right\} \tag{7.10.1}$$

$$\mathbb{N}_a = \left\{\boldsymbol{n}_a \;:\; \boldsymbol{n}_a = \hat{\boldsymbol{n}}_a + \alpha_a\boldsymbol{e}_{n_a}, \hat{\boldsymbol{n}}_a \in Z^{n_a}, a = 1, \ldots, \bar{a}\right\} \tag{7.10.2}$$

where $\bar{a} \leq 10$ is the number of ambiguity groups in the partitioning (7.3.40), $\boldsymbol{n}_a$ is the n_a-dimensional ambiguity vector of the certain group $a = 1, \ldots, \bar{a}$, $\hat{\boldsymbol{n}}_a$ is the integer-valued ambiguity vector (which can be considered a double-difference ambiguity vector), the real-valued common fractional part is denoted by α_a, and $\mathbf{e}_{n_a} = (1, \dots, 1)^T$ is the vector consisting of all units. The symbol Z^n denotes the set of integer-valued vectors (the lattice, generated by the standard basis).

Considering the value $\Delta\mathbf{v}^T\mathbf{P}\mathbf{v}$ in the expression (6.5.6), we first explain how it appears in the RTK filtering scheme. Then we explain how to use methods described in Sections 6.5.2 to 6.5.4 for ambiguity resolution. Sections 7.4.1, 7.5.1, and 7.6.1 deal with a vector of constant parameters, a time-varying vector of corrections to the approximate station positions, and optionally across-receiver ionospheric delays. The vector of constant parameters is denoted in (7.4.12), (7.5.4), and (7.6.8) by $\mathbf{y}$. It can be split into two parts similar to (6.5.1) using the following notation:

$$\mathbf{y} = \begin{bmatrix} \eta \\ \mathbf{n} \end{bmatrix} \quad \eta \in R^{n_\eta} \quad \mathbf{n} \in \mathbb{N} \tag{7.10.3}$$

The algorithms in Subsections 7.4.1, 7.5.1, and 7.6.1 recursively produce a real-valued estimate $\overline{\mathbf{y}}(t)$ of the parameter (7.10.3), along with time-varying parameter $\overline{\mathbf{x}}(t)$.

Omitting for brevity the symbol t of time dependence, we obtain the following expression of the change function $\Delta\mathbf{v}^T\mathbf{P}\mathbf{v}$:

$$q(\mathbf{y}) = (\mathbf{y} - \overline{\mathbf{y}})^T\mathbf{D}(\mathbf{y} - \overline{\mathbf{y}}) \tag{7.10.4}$$

with the matrix $\mathbf{D}$ recursively updated. The function (7.10.4) is subject to minimization over the vector variable $\mathbf{y}$, partitioned in two parts according to (7.10.3)

$$\min_{\mathbf{y}=\begin{bmatrix} \eta \\ \mathbf{n} \end{bmatrix},\ \mathbf{n}\in\mathbb{N}} q(\mathbf{y}) \tag{7.10.5}$$

The definition (7.10.2) is redundant. We can combine one of the entries of the vector $\overline{\mathbf{n}}_a \in Z^{n_a}$ with the real value α_a. Choosing this entry is equivalent to choosing a reference signal. For each group, one reference signal must be chosen. It is reasonable to choose the signal with the best signal-to-noise ratio (SNR). Using that signal to compute across-satellite differences introduces the least possible error to other signals. Let us denote the reference signal as r_a for the group a, $a = 1, \dots, \overline{a}$. Then

$$\mathbb{N}_a = \{\mathbf{n}_a : \ \mathbf{n}_a = \mathbf{E}_{a,r_a}\hat{\mathbf{n}}_{a,r_a} + \alpha_a\mathbf{e}_{n_a-1}, \hat{\mathbf{n}}_{a,r_a} \in Z^{n_a-1}, a = 1, \cdots, \overline{a}\} \tag{7.10.6}$$

where $\mathbf{E}_{a,r_a}$ is the $n_a \times (n_a - 1)$ matrix obtained by crossing out the r_ath column from the identity matrix $\mathbf{I}_{n_a}$ so that the row r_a consists of all zeroes,

$$\mathbf{E}_{a,r_a} = \begin{bmatrix} 1 & & & & & \\ & \ddots & & & & \\ & & 1 & & & \\ 0 & \cdots & 0 & 0 & \cdots & 0 \\ & & & 1 & & \\ & & & & \ddots & \\ & & & & & 1 \end{bmatrix} \leftarrow r_a \tag{7.10.7}$$

The real-valued parameters α_a can be combined with the vector $\boldsymbol{\eta}$ in the partition (7.10.3). We will do it in two steps. First, we present the matrix $\boldsymbol{D}$ and the vector $\bar{\boldsymbol{y}}$ in (7.10.4) in the block form according to the partition (7.10.3)

$$\boldsymbol{D} = \begin{bmatrix} \boldsymbol{D}_{\eta\eta} & \boldsymbol{D}_{\eta N} \\ \boldsymbol{D}_{\eta N}^T & \boldsymbol{D}_{NN} \end{bmatrix} \quad \bar{\boldsymbol{y}} = \begin{bmatrix} \bar{\eta} \\ \bar{\boldsymbol{n}} \end{bmatrix} \tag{7.10.8}$$

Application of the partial minimization [Appendix A.3.7, equation (A.3.88)] leads to the equivalent form of the problem (7.10.5),

$$\min_{\boldsymbol{n}\in\mathbb{N}} q_{\boldsymbol{n}}(\boldsymbol{n}) = \min_{N\in\mathbb{N}} (\boldsymbol{n}-\bar{\boldsymbol{n}})^T(\boldsymbol{D}_{\boldsymbol{nn}} - \boldsymbol{D}_{\eta\boldsymbol{n}}^T\boldsymbol{D}_{\eta\eta}^{-1}\boldsymbol{D}_{\eta\boldsymbol{n}})(\boldsymbol{n}-\bar{\boldsymbol{n}}) \tag{7.10.9}$$

where minimization is taken over the vector $\boldsymbol{n}$ from the set defined by conditions (7.10.1) and (7.10.6).

Let us denote

$$\boldsymbol{E}_r = \begin{bmatrix} \boldsymbol{E}_{1,r_1} & & \vdots & \\ & \boldsymbol{E}_{2,r_2} & \vdots & \\ \cdots & \cdots & \ddots & \vdots \\ & & \cdots & \boldsymbol{E}_{\bar{a},r_{\bar{a}}} \end{bmatrix} \in R^{n\times(n-\bar{a})} \tag{7.10.10}$$

$$\boldsymbol{G}_r = \begin{bmatrix} \boldsymbol{e}_{1,r_1} & & \vdots & \\ & \boldsymbol{e}_{2,r_2} & \vdots & \\ \cdots & \cdots & \ddots & \vdots \\ & & \cdots & \boldsymbol{e}_{\bar{a},r_{\bar{a}}} \end{bmatrix} \in R^{n\times\bar{a}} \tag{7.10.11}$$

where n is the number of signals, i.e., the dimension of the ambiguity vector. Then $\boldsymbol{n} \in \mathbb{N}$ can be presented as

$$\boldsymbol{n} = \boldsymbol{E}_r\hat{\boldsymbol{n}} + \boldsymbol{G}_r\alpha \tag{7.10.12}$$

where $\hat{\boldsymbol{n}} \in Z^{n-\bar{a}}$ is the integer-valued vector, and $\boldsymbol{\alpha}$ is the real-valued vector consisting of the real-valued reference ambiguities. The quadratic function $q_{\boldsymbol{n}}(\boldsymbol{n})$ defined in (7.10.9) can be rewritten as

$$q_{\boldsymbol{n}}(\hat{\boldsymbol{n}},\alpha) = (\boldsymbol{E}_r\hat{\boldsymbol{n}} + \boldsymbol{G}_r\alpha - \bar{\boldsymbol{n}})^T(\boldsymbol{D}_{\boldsymbol{nn}} - \boldsymbol{D}_{\eta\boldsymbol{n}}^T\boldsymbol{D}_{\eta\eta}^{-1}\boldsymbol{D}_{\eta\boldsymbol{n}})(\boldsymbol{E}_r\hat{\boldsymbol{n}} + \boldsymbol{G}_r\alpha - \bar{\boldsymbol{n}}) \tag{7.10.13}$$

in the form (A.3.89) of Appendix A. Application of the second step of the partial minimization leads to the integer quadratic minimization problem

$$\min_{\hat{\boldsymbol{n}}\in Z^{n-\bar{a}}}[\min_{\alpha\in R^{\bar{a}}} q_{\boldsymbol{n}}(\hat{\boldsymbol{n}},\alpha)] = \min_{\hat{\boldsymbol{n}}\in Z^{n-\bar{a}}} \hat{q}_{\boldsymbol{n}}(\hat{\boldsymbol{n}}) \tag{7.10.14}$$

with $\hat{q}_{\boldsymbol{n}}(\hat{\boldsymbol{n}})$ defined as

$$\hat{q}_{\boldsymbol{N}}(\hat{\boldsymbol{n}}) = (\boldsymbol{E}_r\hat{\boldsymbol{n}} - \bar{\boldsymbol{n}})^T \Pi_\alpha(\boldsymbol{E}_r\hat{\boldsymbol{n}} - \bar{\boldsymbol{n}}) \tag{7.10.15}$$

following expression (A.3.91) of Appendix A and $\boldsymbol{\Pi}_\alpha$ being defined according to (A.3.92) as

$$\boldsymbol{\Pi}_\alpha = \overline{\boldsymbol{D}} - \overline{\boldsymbol{D}}\boldsymbol{G}_r(\boldsymbol{G}_r^T\overline{\boldsymbol{D}}\boldsymbol{G}_r)^{-1}\boldsymbol{G}_r^T\overline{\boldsymbol{D}} \tag{7.10.16}$$

where

$$\overline{\boldsymbol{D}} = \boldsymbol{D}_{\boldsymbol{nn}} - \boldsymbol{D}_{\eta\boldsymbol{n}}^T\boldsymbol{D}_{\eta\eta}^{-1}\boldsymbol{D}_{\eta\boldsymbol{n}} \tag{7.10.17}$$

is matrix of the quadratic function (7.10.13). Finally, the function $\hat{q}_{\boldsymbol{n}}(\hat{\boldsymbol{n}})$ can be expressed in the form

$$\hat{q}_{\boldsymbol{n}}(\hat{\boldsymbol{n}}) = (\hat{\boldsymbol{n}} - \overline{\overline{\boldsymbol{n}}})^T\overline{\overline{\boldsymbol{D}}}(\hat{\boldsymbol{n}} - \overline{\overline{\boldsymbol{n}}}) \tag{7.10.18}$$

where

$$\overline{\overline{\boldsymbol{D}}} = \boldsymbol{E}_r^T\boldsymbol{\Pi}_\alpha\boldsymbol{E}_r \tag{7.10.19}$$

$$\overline{\overline{\boldsymbol{n}}} = (\boldsymbol{E}_r^T\boldsymbol{\Pi}_\alpha\boldsymbol{E}_r)^{-1}\boldsymbol{E}_r^T\boldsymbol{\Pi}_\alpha\overline{\boldsymbol{n}} \tag{7.10.20}$$

The integer search algorithms explained in Sections 6.5.2 and 6.5.4 can then be applied to the problem $\min_{\hat{\boldsymbol{n}}\in Z^{n-\bar{a}}} \hat{q}_N(\hat{\boldsymbol{n}})$.

Keep in mind that it is not necessary to explicitly calculate the matrix $\overline{\boldsymbol{D}}$ in (7.10.17). Instead, one calculates the Cholesky decomposition of matrix $\boldsymbol{D}$, partitioned into blocks as in (7.10.8), and makes use of expression (A.3.94) of Appendix A.

If GLONASS observables are involved in processing, then interchannel biases appear. As discussed at the end of Section 7.2, these biases can be either compensated or estimated. More precisely, the slope terms $\mu_{L1,R,\varphi}$ and $\mu_{L2,R,\varphi}$ can be estimated. Let the vector $\boldsymbol{\mu}$ be defined according to (7.3.43) and the matrix $\boldsymbol{W}_\mu$ be defined according to (7.3.45). Let us consider the real-valued vector $\boldsymbol{\mu}$ as the parameter to be estimated along with $\boldsymbol{\alpha}$ in (7.10.13),

$$\begin{aligned} &q_{\boldsymbol{n}}(\hat{\boldsymbol{n}}, \boldsymbol{\alpha}, \boldsymbol{\mu}) \\ &= (\boldsymbol{E}_r\hat{\boldsymbol{n}} + \boldsymbol{G}_r\alpha + \boldsymbol{W}_\mu\mu - \overline{\boldsymbol{n}})^T(\boldsymbol{D}_{\boldsymbol{nn}} - \boldsymbol{D}_{\eta\boldsymbol{n}}^T\boldsymbol{D}_{\eta\eta}^{-1}\boldsymbol{D}_{\eta\boldsymbol{n}})(\boldsymbol{E}_r\hat{\boldsymbol{n}} + \boldsymbol{G}_r\alpha + \boldsymbol{W}_\mu\mu - \overline{\boldsymbol{n}}) \\ &\quad + \rho\boldsymbol{\mu}^T\boldsymbol{\mu} \end{aligned} \tag{7.10.21}$$

Attempting to estimate both $\boldsymbol{\mu}$ and $\boldsymbol{\alpha}$ can make the problem (7.10.21) ill conditioned. A positive parameter ρ is added to avoid ill conditioning or near singularity of the matrix of the quadratic function

$$\hat{q}_{\boldsymbol{n}}(\hat{\boldsymbol{n}}) = \min_{\alpha\in R^{\bar{a}}, \mu\in R^2} q_{\boldsymbol{n}}(\hat{\boldsymbol{n}}, \boldsymbol{\alpha}, \boldsymbol{\mu}) \qquad \hat{\boldsymbol{n}} \in Z^{n-\bar{a}} \tag{7.10.22}$$

The greater ρ, the more constrained the variation of $\boldsymbol{\mu}$. The estimate of $\boldsymbol{\mu}$ will become near zero when ρ takes large values relative to $\max_i \overline{\boldsymbol{D}}_{ii}$. Recall that parameter $\boldsymbol{\mu}$

refers to the first-order linear approximation of the dependency of hardware carrier phase biases on frequency, as proposed in expressions (7.3.34) and (7.3.35). This linear approximation makes sense because in practice the biases behave as shown in Figure 7.2.1. The linear part of the across-receiver hardware carrier phase bias as shown in that figure is quite typical for two receivers from different manufacturers. Usually its value can be one or several cycles when the letter l runs over the range $[-7, 6]$. In other words, parameter μ takes values in the range of approximately $[-0.1, 0.1]$. A typical choice for the parameter ρ is 0.001 or 0.0001.

Figure 7.2.1 shows that across-receiver hardware biases violate the integer-valued nature of ambiguities and make fixing them impossible unless the biases are either compensated or estimated. In the first case, they are taken from the lookup table. The RTCM3 format allows for transmission of manufacturer and hardware version details from the base to the rover. Thus, the rover is able to read the appropriate lookup table. In the second case, if the rover does not know the hardware details of the base, or if it does not have the appropriate lookup table, it can try to estimate the coefficients of their linear approximation, or even estimate higher order coefficients.

7.10.1 Illustrative Example

Let us now continue with the example in Subsection 7.5.2. The ambiguities were fixed with a discernibility ratio of 4.78 at the first epoch. To see how the position changes when the ambiguities are fixed, we intentionally postponed the fixing to epoch 31, allowing the first 30 epochs to float. Figure 7.10.1 illustrates the variation of the up component of the baseline vector expressed in the geodetic horizon. Note that the RTK algorithm described in Section 7.5 estimates the position of the kinematic rover. The dashed black line shows the variation of the up component of the float solution for the first 30 epochs, followed by another 30 epochs of the fixed solution (the solid line).

Before considering the convergence of the estimated across-receiver ambiguity parameters, note that their values also aggregate hardware delays along with integer-valued number of carrier waves, giving them fractional parts. In Chapter 6 the term "across-receiver ambiguity" strictly referred to the integer-valued quantity. After the double-difference ambiguities have been fixed, the single-difference ambiguity estimates (with fraction) take their correct values as summarized in Tables 7.10.1 and 7.10.2. Ambiguities of four groups ($\overline{a} = 4$ for this example) have equal fractional parts inside each group.

When L1, L2, and L5 GPS signals are available along with L1 and L2 GLONASS; L1 and E5 Galileo; L1, L2, and L5 QZSS; L1 and L5 WAAS; and B1, B2, and B3 Compass, there can be more than 40 different ambiguities. Although fast integer search methods have been addressed in Sections 6.5.2 to 6.5.4, computation reduction issues remain significant. Moreover, when trying to fix all ambiguities together, there may be situations when a large error or an undetected cycle slip in one measurement channel prevents the whole set of ambiguities from being fixed successfully. There are two different ways to mitigate this problem:

- Partial fixing
- Fixing linear combinations of measurements

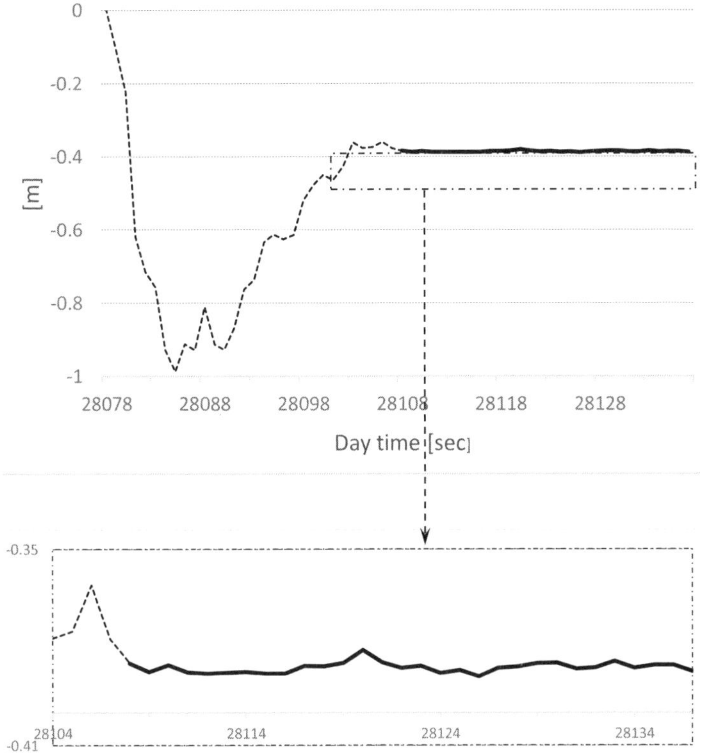

Figure 7.10.1 Up component of the baseline vector. The first 60 epochs are shown (top); first 30 are float, next 30 are fixed. The bottom panel shows an enlargement of same plot starting with epoch 28,104.

In the partial fixing approach, a set of the "best ambiguities" is identified and fixed first. The number of ambiguities in the best set should not exceed some predefined value, say 20 or 30, but should include ambiguities corresponding to signals with the highest SNR and a good geometry factor. Other criteria for selecting various subsets of ambiguities to be fixed are discussed in Cao et al. (2007). The criteria might be a satellite system, frequency, or a combination thereof.

While the set of best ambiguities is subjected to fixing, the other ambiguities not included in that set are kept floating. The partial minimization approach described in Appendix A.3.7 is used to decompose the least-squares problem (7.10.18) into two subproblems. Let vectors $\hat{\boldsymbol{n}} = (\hat{\boldsymbol{n}}_1^T, \hat{\boldsymbol{n}}_2^T)^T$ and $\bar{\bar{\boldsymbol{n}}} = (\bar{\bar{\boldsymbol{n}}}_1^T, \bar{\bar{\boldsymbol{n}}}_2^T)^T$ be partitioned into two parts. Then the problem (7.10.18) can be presented in the form (A.3.86). The internal minimum (A.3.88) is taken over the variable $\hat{\boldsymbol{n}}_2$ [the variable $\boldsymbol{y}$ in the formulation (A.3.86)]. It corresponds to the ambiguities that are kept floating, while the external minimum is taken over the variable $\hat{\boldsymbol{n}}_1$ [the variable $\boldsymbol{x}$ in (A.3.88)]. It corresponds to the ambiguities needing to be fixed. The minimization is performed over the integer-valued vectors $\hat{\boldsymbol{n}}_1$. The floating vector $\hat{\boldsymbol{n}}_2(\hat{\boldsymbol{n}}_1)$ (the argument of

TABLE 7.10.1 Fixed Values of the Estimated Across-Receiver GPS Ambiguities.

PRN	L1	L2
4	−29.2932	0.7662
9	11.7068	−8.2338
15	7.7068	15.7662
17	−22.2932	−8.2338
24	20.7068	11.7662
28	1.7068	1.7662

TABLE 7.10.2 Fixed Values of the Estimated Across-Receiver GLONASS Ambiguities.

Letter	L1	L2
−7	9.3922	6.0677
0	3.3922	2.0677
2	−22.6078	10.0677
4	−35.6078	−12.9323

conditional minimum) is then calculated according to expression (A.3.87). At this stage we have the integer-valued vector $\hat{\boldsymbol{n}}_1$ and the floating vector $\hat{\boldsymbol{n}}_2(\hat{\boldsymbol{n}}_1)$. Usually, if the ambiguities of the first primary set were fixed correctly with good discernibility ratios, the floating values $\hat{\boldsymbol{n}}_2(\hat{\boldsymbol{n}}_1)$ become close to the integer values. Straightforward rounding can complete the calculation. Alternatively, we can substitute the fixed vector $\hat{\boldsymbol{n}}_1$ into (7.10.18) and consider it as a quadratic function of the remaining variable $\hat{\boldsymbol{n}}_2$ subject to fixing.

Partial ambiguity fixing is discussed in Teunissen (1998), though it can be derived from the partial minimization scheme described in Appendix A.3.7. When keeping part of the ambiguities floating, they are combined with the parameter η in the partition (7.10.8) and (7.10.9), and are handled the same way as already discussed. After the primary set of ambiguities has been fixed, the set of real-valued parameters η, including the floating estimate of the secondary set of ambiguities, is updated accordingly. Then secondary ambiguities are fixed. If the number of ambiguities is very large, the fixing process can be divided into three or more stages. Consider only two subsets. When solving for the first subset (external problem), the matrix of the first problem is defined according to expression (A.3.88), taking into account the whole matrix (7.10.19). Therefore, the correlation between two subsets of estimated single-difference ambiguities will correctly be taken into account.

A drawback of sequential partial fixing is that the discernibility ratio estimated at each stage will in general be lower than the one estimated for the whole problem, assuming that there are no biases affecting the signals.

An advantage of sequential partial fixing is its ability to temporarily or permanently isolate the signal affected by the bias that prevents the fixing. Only one bad signal can decrease the discernibility ratio, thus not allowing the fixing of the whole set of ambiguities. A natural criterion for exclusion of a signal from the set to be fixed is a low SNR in combination with a relatively high satellite elevation. Usually it indicates either shading of the signal or a very large carrier phase multipath error. Sometimes the presence of large multipath error can be detected by comparing residuals of the carrier phase observables of the first and second (or third) frequencies for a certain satellite. If residuals have large values of a different sign, this can be an indication of the presence of significant carrier phase multipath. In this case, across-receiver single-difference ambiguities corresponding to observables supposedly affected by multipath, can be intentionally kept floating for several epochs. Thus, the partial fixing approach can be used not only to reduce the dimension of the integer search but also to isolate some measurements that could decrease the discernibility ratio if they are included in the fixing.

The algorithm of fixing across-receiver ambiguities has been considered in this section. The estimated across-receiver ambiguities include hardware delays as a sum with the integer-valued ambiguities. Hardware delays cause fractional parts of across-receiver ambiguities, which are common for each group of signals. The single-difference ambiguity approach allows for processing of arbitrary combinations of signals because no L1, L2, or L5 linear combinations are calculated. This allows processing of carrier phase and pseudorange observables sequentially, epoch by epoch, even if signals from some frequencies are temporarily or permanently not available at some epochs. If certain signals become unavailable at a certain epoch or become available again (either physically or due to a detected cycle slip), the methods described in Sections 7.8 and 7.9 allow for sequential optimal processing.

7.11 SOFTWARE IMPLEMENTATION

An RTK engine implementation can be divided into several computational processes that run in parallel:

1. Ambiguity filtering and resolution
2. Calculation of position, velocity, and time
3. Receiving and extrapolating the measurements from the base station, which can be either a real physical receiver or a virtual base station intentionally generated to serve the rover

The measurements are generated in the rover at a high update rate, usually 20 to 100 times per second, we say at 20 to 100 Hz, while measurements generated by the base are transmitted to the rover at a much lower rate, say once per second, i.e., at 1 Hz rate. The relatively low rate of transmission is explained by way of throughput and the reliability of the data link.

Ambiguity filtering and resolution are computationally intensive. The numerical schemes described in this chapter involve many matrix operations. Matrix dimensions depend on the number of ambiguities, which in turn depends on the number of available signals. On the other hand, the ambiguities are constant and do not vary until the cycle slip occurs. This means the ambiguities can be estimated at a low update rate, not necessarily at the rate the measurements are generated in the rover receiver. Because the data from the rover and base must be used when forming across-receiver differences, only measurements with matching time tags can be taken into the ambiguity estimation process. Therefore, the first process can be considered slow and of low priority. Note that real-time operating systems working in the receiver must share computational resources among parallel processes. This task is accomplished using a mechanism of priorities. The higher priority process interrupts the lower priority process when attempting to share common resources. Therefore, high priority is usually granted to processes operating with raw data acquisition and initial data handling. These processes are not intensive computationally and allow other lower priority processes to access processor resources after they complete their tasks. Low-priority processes perform computationally intensive tasks, processing relatively slow varying quantities.

The second process completes the computation of position, velocity, and clock shift. These calculations are based on the unbiased carrier phase measurements determined as a difference between carrier phase measurements and constant ambiguities estimated in the first process. Carrier phase measurements are normally generated at a high update rate, while ambiguities are estimated using a low update rate, i.e., the same rate as measurements come through the data link between the base and the rover. The second process contains relatively simple calculations. It has a higher priority compared with the first one.

The third computational process is responsible for receiving measurements from the base. It decodes measurements, formatted as RTCM 2 or 3. These formats are specifically designed for compact and reliable data transmission through the radio or GSM channels affected by irregular time delay. The RTCM messages carry either full carrier phase and pseudorange measurements or corrections to them. Corrections are usually residual values defined as a measurement with geometry and time terms compensated at the base station before transmission.

The measurements received from the base must be extrapolated ahead of time to compensate for the transmission delay and the difference between the high update rate of the rover measurements (e.g., 20 Hz) and low update rate of the base measurements (e.g., 1Hz). Extrapolation allows the rover software to have the base measurements match the same time instants of the rover measurements. Extrapolation can be performed for each signal using independent Kalman filters, based on the simple second- or third-order dynamic model. See Singer (1970) for examples.

CHAPTER 8

TROPOSPHERE AND IONOSPHERE

The impact of the atmosphere on GNSS observations is the focus of this chapter. In Section 8.1 we begin with a general overview of the troposphere and ionosphere as it relates to GPS satellite surveying and introduce the general form of the index of refraction. Section 8.2 addresses tropospheric refraction starting with the expression of refractivity as a function of partial pressure of dry air, partial water vapor pressure, and temperature. The equation for the zenith hydrostatic delay (ZHD) by Saastamoinen (1972), the expression for the zenith wet delay (ZWD) by Mendes and Langley (1999), and Niell (1996) to relate slant delays and zenith delays are given without derivation. We then establish the relationship between the zenith wet delay and precipitable water vapor (PWV).

Section 8.3 deals with tropospheric absorption and water vapor radiometers (WVR) that measure the tropospheric wet delay. We present and discuss the radiative transfer equation and the concept of brightness temperature. To demonstrate further the principles of the water vapor radiometer, we discuss the relevant absorption line profiles for water vapor, oxygen, and liquid water. This is followed by a brief discussion on retrieval techniques for computing wet delay and on radiometer calibration using tipping curves.

Section 8.4 concentrates on ionospheric refraction. We begin with the Appleton-Hartree formula for refraction index and derive expressions for the first-order, second-order, and third-order ionospheric delay of GNSS signals. We briefly address the masking of certain cycle slips in the GPS L1 and L2 phases, then focus on the popular single-layer ionospheric model to relate the vertical total electron content (VTEC) to the respective slant total electron content (STEC), and how VTEC is computed from ground GNSS observations. The chapter concludes with brief remarks on popular global ionospheric models.

8.1 OVERVIEW

The propagation media affect electromagnetic wave propagation at all frequencies, resulting in a bending of the signal path, time delays of arriving modulations, advances of carrier phases, scintillation, and other changes. In GNSS positioning one is primarily concerned with the arrival times of carrier modulations and carrier phases. Geometric bending of the signal path causes a small delay that is negligible for elevation angles above 5°. The propagation of electromagnetic waves through the various atmospheric regions varies with location and time in a complex manner. There are two major regions of the atmosphere of interest; these are the troposphere and ionosphere. Whereas positioning with GNSS requires careful consideration of the impacts of the atmosphere on the observations, GNSS, in turn, has become an important tool for studying the properties of the atmosphere. The propagation of electromagnetic signals in the GNSS frequency range, which is approximately the microwave region, is discussed in this chapter but only to the extent required for positioning.

Most of the mass of the atmosphere is located in the troposphere. We are concerned with the tropospheric delay of pseudoranges and carrier phases. For frequencies below 30 GHz, the troposphere behaves essentially like a nondispersive medium, i.e., the refraction is independent of the frequency of the signals passing through it. This tropospheric refraction includes the effect of the neutral, gaseous atmosphere. The effective height of the troposphere is about 40 km. The density in higher regions is too small to have a measurable effect. Mendes (1999) and Schüler (2001) are just two of many excellent references available to read about the details of tropospheric refractions. Typically, tropospheric refraction is treated in two parts. The first part is the hydrostatic component that follows the laws of ideal gases. It is responsible for a zenith delay of about 240 cm at sea level locations. It can be computed accurately from pressure measured at the receiver antenna. The more variable second part is the wet component, which is also called the nonhydrostatic wet component and is responsible for up to 40 cm of delay in the zenith direction. Computing the wet delay accurately is a difficult task because of the spatial and temporal variation of water vapor. As an example, Figure 8.1.1 shows the ZWD every 5 min for 11 consecutive

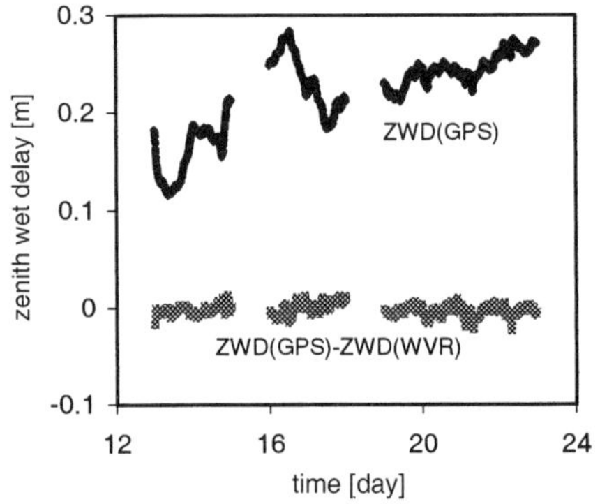

Figure 8.1.1 ZWD from GPS. (Data from Bar-Sever, JPL.)

days in July of 1999 at Lamont, Oklahoma, as determined by GPS. The figure also shows the difference in ZWD as determined by GPS and WVR. Both determinations agree within 1 cm. The gaps indicate times when suitable observations were not available.

The ionosphere covers the region between approximately 50 and 1500 km above the earth and is characterized by the presence of free (negatively charged) electrons and positively charged molecules called ions. The number of free elections varies with time and space and, consequently electron density profiles exhibit considerable variability. Often the spatial distribution of the ionosphere in height is divided into regions which themselves can be subdivided into layers. The lower region is the D region reaching up to about 90 km, followed by the E region up to about 150 km, and the F region up to about 600 km. The area above 600 km is called the topside of the ionosphere. These specific numerical height values are not universally agreed upon; therefore other values may be found in the literature. In GNSS applications we are concerned with the overall impact of the free electrons along the transmission line from satellite to receiver for the GNSS frequency range. The regional subdivisions and their respective characteristics, such as the phenomena of signal reflections for transmissions below 30 MHz, are not of interest to GNSS positioning although such properties are important for communication. Hargreaves (1992) and Davies (1990) are recommended texts for in-depth studies of the physics of the ionosphere.

Of concern in GNSS applications is the total electron content (TEC) which equals the number of free electrons in 1 m^2 column. When referring to the path along the receiver-satellite line of sight we talk about slant total electron content (STEC), and the TEC of a vertical column is called the vertical total electron content (VTEC). The TEC is often expressed in TEC units TECU, whereby $1\ \text{TECU} = 10^{16}\ \text{el}/\text{m}^2$. The free electrons delay pseudoranges and advance carrier phases by equal amounts as will be derived below. The size of this impact depends not only on the TEC value but also on the carrier frequency, i.e., the ionosphere is a dispersive medium. For the GNSS frequency range the delays or advances can amount to tens of meters.

Figure 8.1.2 shows a snapshot of the TEC. This GIM (global ionospheric map) shows the VTEC on March 7, 2000, at 03 UT. The map shows the typical global morphology of the ionosphere when the Appleton (equatorial) anomaly is well developed. There are two very strong peaks of ionization on either side of the geomagnetic equator. The peaks begin in the afternoon and stretch into the nighttime region. The peak values are relatively large since it is near solar maximum, although vertical TECs can be larger than the 140 to 150 TECU shown for this day. The peak of the ionosphere occurs typically in the equatorial region at 14:00 local time. Each dot is the location of a GPS receiver that contributed data to this GIM run. This particular GIM's time resolution is 15 min, i.e., a new map is available four times per hour. The GDGPS system (www.gdgps.net) now produces such global maps every 5 min.

The better the atmospheric parameters are known the more accurate can the respective corrections to GNSS observations be computed. One typically uses temperature, pressure, and humidity at the receiver antenna, as well as the TEC. GPS also contributes to the mapping of the spatial and temporal distribution of these atmospheric parameters. Figure 8.1.3 shows a schematic view of a low-earth orbiter (LEO)

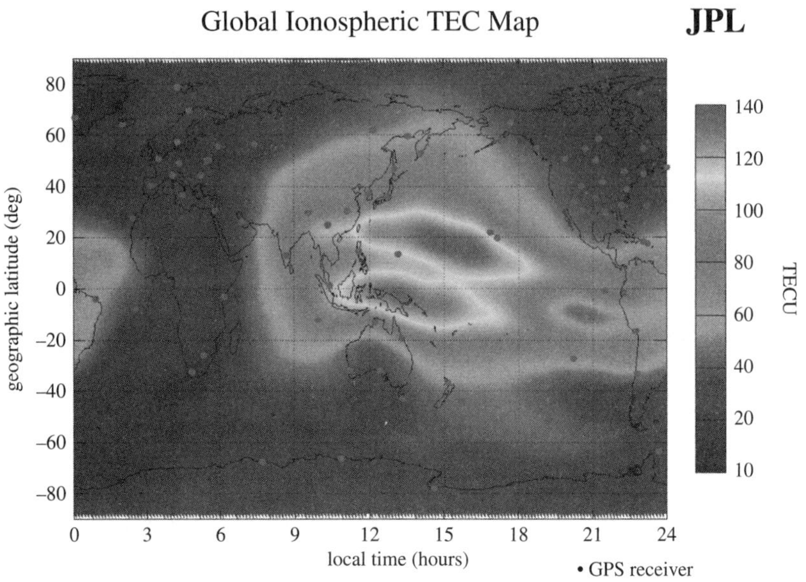

Figure 8.1.2 Snapshot of VTEC. (Courtesy of B. D. Wilson, JPL.)

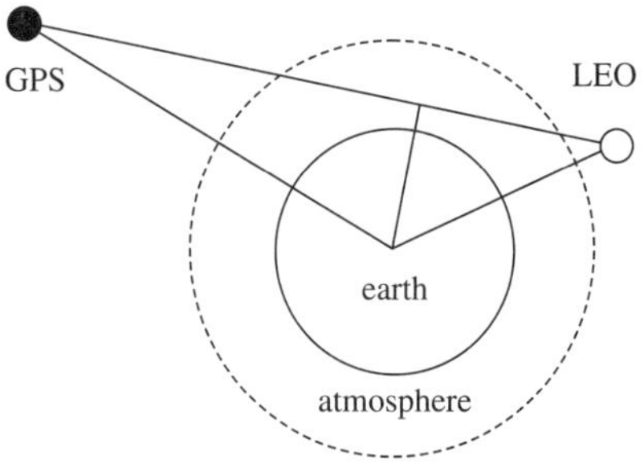

Figure 8.1.3 Schematic view of an LEO satellite and a GPS satellite configuration.

and a GPS satellite. As viewed from the LEO, an occultation takes place when the GPS satellite rises or sets behind the ionosphere and troposphere. When signals pass through the media they experience tropospheric delays, ionospheric code delays, and phase advances. Assuming the accurate position of the LEO is known and the LEO carries a GPS receiver, one can estimate atmospheric parameters and profiles of these parameters by comparing the travel time of the signal and the geometric distance between both satellites. One of the more extensive efforts for the spatial and temporal mapping of the atmosphere uses the FORMOSAT-3/COSMIC constellation which was launched in 2006. The constellation consists of six satellites equipped with GPS antennas. The system provides global coverage with almost uniform data distribution eliminating sparseness and lack of occultation data which hindered previous efforts.

Radio occultation will not be discussed further in this chapter because such details of modeling and processing respective observations are not within the scope of this book. Background information on occultation techniques and a thorough error analysis for retrieved profiles of refractivity, geopotential height of constant pressure levels, and temperature is available in Kursinski et al. (1997). Bust and Mitchel (2008) discuss the status of voxel-based 3-dimensional time-varying mapping of the ionosphere and the iterative pixel-based algebraic reconstructed technique (ART) and multiplicative algebraic reconstruction technique (MART) of ionospheric tomography. A comprehensive open access source on the subject is Liou et al. (2010). A special issue of *GPS Solutions* (vol. 14, number 1, 2010) has been published on the result of the COSMIC mission. A recent contributions to COSMIC radio occultations is Sokolovskiy et al. (2013) who studied the improvements provided by the new L2C GPS observations.

The general form of the index of refraction for electromagnetic waves can be written as a complex number

$$\overline{n} = \mu - i\chi \tag{8.1.1}$$

where μ and χ are related to refraction and absorption, respectively. Let A_0 denote the amplitude, we can write the equation of a wave as

$$A = A_0 e^{i(\omega t - \overline{n}\omega x/c)} = A_0 e^{i(\omega t - \mu\,\omega x/c)} e^{-\chi\,\omega x/c} \tag{8.1.2}$$

The wave propagates at speed c/μ, where c denotes the speed of light. The absorption in the medium is given by the exponential attenuation $e^{-\chi\omega x/c}$. The absorption coefficient is $\kappa = \omega\chi/c$. It is readily seen that the amplitude of the wave will reduce by factor e at distance $1/\kappa$.

For GPS frequencies and for frequencies in the microwave region, the index of refraction can be written as

$$\overline{n} = n + n'(f) + in''(f) \tag{8.1.3}$$

The medium is called dispersive if $\overline{n}$ is a function of the frequency. When applying (8.1.3) to the troposphere the real parts n and $n'(f)$ determine refraction that causes the delays in pseudoranges and carrier phases. The nondispersive part of the index of refraction is n. For frequencies in the microwave range the frequency-dependent real term $n'(f)$ causes delays around the millimeter level at 60 GHz and centimeter level at 300 GHz (Janssen, 1993, p. 218). In general, $n'(f)$ and $n''(f)$ are due to interactions with line resonances of molecules in the vicinity of the carrier frequency. The GPS frequencies are far from atmospheric resonance lines. The imaginary part $n''(f)$, however, quantifies absorption (emission) and is important to the WVR observable. When applying (8.1.3) to the ionosphere the term $n'(f)$ is important.

8.2 TROPOSPHERIC REFRACTION AND DELAY

The index of refraction is a function of the actual tropospheric path through which the ray passes, starting at the receiver antenna and continuing up to the end of the effective troposphere. Let s denote the distance; the delay due to refraction is

$$v = \int n(s)ds - \int ds = \int (n(s) - 1)ds \tag{8.2.1}$$

The first integral refers to the curved propagation path. The path is curved due to the decreasing index of refraction with height above the earth. The second integral is the geometric straight-line distance the wave would take if the atmosphere were a vacuum. The integration begins at the height of the receiver antenna.

Because the index of refraction $n(s)$ is numerically close to unity, it is convenient to introduce a separate symbol for the difference,

$$n(s) - 1 = N(s) \cdot 10^{-6} \tag{8.2.2}$$

$N(s)$ is called the refractivity. Great efforts have been made during the second part of the last century to determine the refractivity for microwaves. Examples of relevant literature are Thayer (1974) and Askne and Nordius (1987). The refractivity is usually given in the form

$$N = k_1 \frac{p_d}{T} Z_d^{-1} + k_2 \frac{p_{wv}}{T} Z_{wv}^{-1} + k_3 \frac{p_{wv}}{T_2} Z_{wv}^{-1} \tag{8.2.3}$$

where

p_d partial pressure of dry air (mbar). The dry gases of the atmosphere are, in decreasing percentage of the total volume: N_2, O_2, Ar, CO_2, Ne, He, Kr, Xe, CH_4, H_2, and N_2O. These gases represent 99.96% of the volume.

p_{wv} partial pressure of water vapor (mbar). Water vapor is highly variable but hardly exceeds 1% of the mass of the atmosphere. Most of the water in the air is from water vapor. Even inside clouds, precipitation and turbulence ensure that water droplet density remains low. This variability presents a challenge to accurate GNSS applications over long distances on one hand, but on the other hand opens up a new field of activity, i.e., remotely sensing the atmosphere for water vapor.

T absolute temperature in degrees Kelvin (K).

Z_d, Z_{wv} compressibility factors that take into account small departures in behavior of moist atmosphere and ideal gas. Spilker (1996, p. 528) lists the expressions. These factors are often set to unity.

k_1, k_2, k_3 physical constants that are based in part on theory and in part on experimental observations. Bevis et al. (1994) lists: $k_1 = 77.60$ K/mbar, $k_2 = 69.5$ K/mbar, $k_3 = 370100$ K^2/mbar.

Partial water vapor pressure and relative humidity R_h are related by the well-known expression, e.g., WMO (1961),

$$P_{wv[\text{mbar}]} = 0.01\, R_{h[\%]}\, e^{-37.2465+0.213166T-0.000256908T^2} \tag{8.2.4}$$

The two partial pressures are related to the total pressure p, which is measured directly, by

$$p = p_d + p_{wv} \tag{8.2.5}$$

The first term of (8.2.3) expresses the sum of distortions of electron charges of the dry gas molecules under the influence of an applied magnetic field. The second term refers to the same effect but for water vapor. The third term is caused by the permanent dipole moment of the water vapor molecule; it is a direct result of the geometry of the water vapor molecular structure. Within the GPS frequency range the third term is practically independent of frequency. This is not necessarily true for higher frequencies that are close to the major water vapor resonance lines. Equation (8.2.3) is further developed by splitting the first term into two terms, one that gives refractivity of an ideal gas in hydrostatic equilibrium and another term that is a function of the partial water vapor pressure. The large hydrostatic constituent can then be accurately computed from ground-based total pressure. The smaller and more variable water vapor contribution must be dealt with separately.

The modification of the first term (8.2.3) begins by applying the equation of state for the gas constituent $i, (i = d, i = wv)$,

$$p_i = Z_i \rho_i R_i T \tag{8.2.6}$$

where ρ_i is the mass density and R_i is the specific gas constant ($R_i = R/M_i$, where R is the universal gas constant and M_i is the molar mass). Substituting p_d in (8.2.6) into the first term in (8.2.3), replacing the ρ_d by the total density ρ and ρ_{wv}, and applying (8.2.6) for ρ_{wv} gives for the first term

$$k_1 \frac{p_d}{T} Z_d^{-1} = k_1 R_d \rho_d = k_1 R_d \rho - k_1 R_d \rho_{wv} = k_1 R_d \rho - k_1 \frac{R_d}{R_{wv}} \frac{p_{wv}}{T} Z_{wv}^{-1} \tag{8.2.7}$$

Substituting (8.2.7) in (8.2.3) and combining it with the second term of that equation gives

$$N = k_1 R_d \rho + k_2' \frac{p_{wv}}{T} Z_{wv}^{-1} + k_3 \frac{p_{wv}}{T^2} Z_{wv}^{-1} \tag{8.2.8}$$

The new constant k_2' is

$$k_2' = k_2 - k_1 \frac{R_d}{R_{wv}} = k_2 - k_1 \frac{M_{wv}}{M_d} \tag{8.2.9}$$

Bevis et al. (1994) gives $k_2' = 22.1$ K/mbar.

We can now define the hydrostatic and wet (nonhydrostatic) refractivity as

$$N_d = k_1 R_d \rho = k_1 \frac{p}{T} \tag{8.2.10}$$

$$N_{wv} = k_2' \frac{p_{wv}}{T} Z_{wv}^{-1} + k_3 \frac{p_{wv}}{T^2} Z_{wv}^{-1} \tag{8.2.11}$$

If we integrate (8.2.3) along the zenith direction using (8.2.10) and (8.2.11), we obtain the ZHD and ZWD, respectively,

$$ZHD = 10^{-6} \int N_d(h)\, dh \tag{8.2.12}$$

$$ZWD = 10^{-6} \int N_{wv}(h)\, dh \tag{8.2.13}$$

The hydrostatic refractivity N_d depends on total density ρ or the total pressure p. When integrating N_d along the ray path the hydrostatic equilibrium condition to ideal gases is applied. The integration of N_{wv} is complicated by the temporal and spatial variation of the partial water vapor pressure p_{wv} along the path.

8.2.1 Zenith Delay Functions

Even though the hydrostatic refractivity is based on the laws of ideal gases, the integration (8.2.12) still requires assumptions about the variation of temperature and gravity along the path. Examples of solutions for the ZHD are Hopfield (1969) and Saastamoinen (1972). Saastamoinen's solution is given in Davis et al. (1985) in the form

$$ZHD_{[\mathrm{m}]} = \frac{0.0022768 p_{0[\mathrm{mbar}]}}{1 - 0.00266 \cos 2\varphi - 0.00028 H_{[\mathrm{km}]}} \tag{8.2.14}$$

The symbol p_0 denotes the total pressure at the site whose orthometric height is H and latitude is φ. Note that 1 mbar equals 1 hPa.

The model assumptions regarding the wet refractivity are more problematic because of temporal and spatial variability of water vapor. Mendes and Langley (1999) analyzed radiosonde data and explored the correlation between the ZWD and the surface partial water vapor pressure $p_{wv,0}$. Their model is

$$ZWD_{[\mathrm{m}]} = 0.0122 + 0.00943 p_{wv,0[\mathrm{mbar}]} \tag{8.2.15}$$

Surface meteorological data should be used with caution in the estimation of the ZWD. Typical field observations can be influenced by "surface layer biases" introduced by micrometeorological effects. The measurements at the earth's surface are not necessarily representative of adjacent layers along the line of sight to the satellites. Temperature inversion can occur during nighttime when the air layers close to the ground are cooler than the higher air layers, due to ground surface radiation loss. Convection can occur during noontime when the sun heats the air layers near the ground.

Expressions exist that do not explicitly separate between ZHD and ZWD. In some cases, the models are independent of direct meteorological measurements. The latter typically derive their input from model atmospheres.

8.2.2 Mapping Functions

Tropospheric delay is shortest in zenith direction and increases with zenith angle ϑ as the air mass traversed by the signal increases. The exact functional relationship is again complicated by temporal and spatial variability of the troposphere. The mapping function models this dependency. We relate the slant hydrostatic and wet delays, SHD and SWD, to the respective zenith delays by

$$SHD = ZHD \cdot m_h(\vartheta) \tag{8.2.16}$$

$$SWD = ZWD \cdot m_{wv}(\vartheta) \tag{8.2.17}$$

The slant total delay (STD) is

$$STD = ZHD \cdot m_h(\vartheta) + ZWD \cdot m_{wv}(\vartheta) \tag{8.2.18}$$

The literature contains several models for the mapping functions m_h and m_{wv}. The one in common use is Niell's (1996) function,

$$m(\vartheta) = \frac{1 + \dfrac{a}{1 + \dfrac{b}{1 + c}}}{\cos\vartheta + \dfrac{a}{\cos\vartheta + \dfrac{b}{\cos\vartheta + c}}} + h_{[\mathrm{km}]}\left(\frac{1}{\cos\vartheta} - \frac{1 + \dfrac{a_h}{1 + \frac{b_h}{1+c_h}}}{\cos\vartheta + \dfrac{a_h}{\cos\vartheta + \dfrac{b_h}{\cos\vartheta + c_h}}}\right) \tag{8.2.19}$$

The coefficients for this expression are listed in Table 8.2.1 for m_h as a function of the latitude φ of the station. If $\varphi < 15°$ one should use the tabulated values for $\varphi = 15°$; if $\varphi > 75°$ use the values for $\varphi = 75°$; if $15° \le \varphi \le 75°$ linear interpolation applies. Before using the values in Table 8.2.1, however, the coefficients a, b, and c must be corrected for periodic terms following the general formula

$$a(\varphi, \mathrm{DOY}) = \tilde{a} - a_p \cos\left(2\pi \frac{\mathrm{DOY} - \mathrm{DOY}_0}{365.25}\right) \tag{8.2.20}$$

where DOY denotes the day of year and DOY_0 is 28 or 211 for stations in the southern or northern hemisphere, respectively. When computing the wet mapping

TABLE 8.2.1 Coefficients for Niell's Hydrostatic Mapping Function.

φ	$\tilde{a} \cdot 10^3$	$\tilde{b} \cdot 10^3$	$\tilde{c} \cdot 10^3$	$a_p \cdot 10^5$	$b_p \cdot 10^5$	$c_p \cdot 10^5$
15	1.2769934	2.9153695	62.610505	0	0	0
30	1.2683230	2.9152299	62.837393	1.2709626	2.1414979	9.0128400
45	1.2465397	2.9288445	63.721774	2.6523662	3.0160779	4.3497037
60	1.2196049	2.9022565	63.824265	3.4000452	7.2562722	84.795348
75	1.2045996	2.9024912	64.258455	4.1202191	11.723375	170.37206
	$a_h \cdot 10^5$	$b_h \cdot 10^3$	$c_h \cdot 10^3$			
	2.53	5.49	1.14			

TABLE 8.2.2 Coefficients for Niell's Wet Mapping Function.

φ	$a \cdot 10^4$	$b \cdot 10^3$	$c \cdot 10^2$
15	5.8021897	1.4275268	4.3472961
30	5.6794847	1.5138625	4.6729510
45	5.8118019	1.4572752	4.3908931
60	5.9727542	1.5007428	4.4626982
75	6.1641693	1.7599082	5.4736038

function, the height-dependent second term in (8.2.19) is dropped and the coefficients of Table 8.2.2 apply.

The Niell mapping function enjoys much popularity because it is accurate, independent of surface meteorology, and requires only station location and time as input. It is based on a standard atmosphere to derive average values for the coefficients and the seasonal amplitudes. Since compensation of tropospheric delay is of great importance in GNSS applications, in particular for long baselines where the tropospheric effects decorrelated or for accurate single-point positioning, active research has continued and various new solutions and approaches have become available. Soon after his innovative idea to generate a mapping function that does not require surface meteorology, Niell recognized certain shortcomings of using the standard atmosphere and began experimenting with a numerical weather model (NWM) to obtain an improved solution (Niell, 2000, 2001). He related the hydrostatic coefficients to the 200 mbar isobaric pressure level surface above the site. The result is referred to as the isobaric mapping function (IMF).

As the spatial resolution and accuracy of numerical weather models have improved, these models have become the backbone for developing modern tropospheric mapping functions. Boehm and Schuh (2004) introduced the Vienna mapping function (VMF) which exploited all the relevant data that can be extracted from an NWM. An updated version, called VMF1, is described in Boehm et al. (2006). These Vienna mapping functions are based on data from the European Center for Medium-Range Weather Forecasts (ECMWF) models. For extra demanding applications, VMF1 can generate the mapping function coefficients for individual stations, although the computational load is heavy for each such computation.

However, slightly less accurate but more accessible than the original VMF1, new implementations of the VMF1 are also available to users. These versions take as input a file that contains a global grid of points for which the needed information has been generated, and then determine the mapping function coefficients for the user site through spatial and temporal interpolation. The latest of these versions, called GPT2, has been introduced in Lagler et al. (2013). The input is an ASCII file generated by processing global monthly mean profiles for pressure, temperature, specific humidity, and geopotential (discretized at 37 pressure levels and 1° latitude and longitude), between the years 2001 and 2010 (10 years), available from the ERA-Interim, which is the latest global atmospheric reanalysis of the ECMWF (Dee et al., 2011). This file contains 120 monthly values for pressure, temperature, specific humidity, the a coefficients for the hydrostatic and wet mapping functions, and the temperature laps rate on a global grid of either 1° or 5°. For each entry the mean value and the annual and semiannual amplitudes are given. The current implementation requires the user to execute two simple Fortran subroutines, both of which are available at ftp://tai.bipm.org/iers/convupdt/chapter9/ for download. In addition to the file described above, the subroutine GPT2.F takes as input the ellipsoidal latitude, longitude, and height of the station and the time of observation. The output is pressure, temperature, temperature laps rate, water vapor pressure, and hydrostatic and wet mapping function coefficients a for the site. The second subroutine, VMF1_HT, takes as input the thus computed a coefficients, geographic location of the site, and time and zenith distance of observation, and outputs hydrostatic and wet mapping function values $m_h(\vartheta)$ and $m_{wv}(\vartheta)$. It uses b and c coefficients discussed in Boehm et al. (2006) and numerically corrected as explained in the comments section of subroutine VMF1_HT. For computing ZHD and ZWD the user has the option to use measured total surface pressure and partial water vapor pressure in (8.2.14) and (8.2.15) or similar expressions or use the output values from GPT2.F.

Whenever NWMs are used to determine the mapping function coefficients, there is always implicitly the technique of raytracing involved with the needed meteorological data derived from the NWM. Raytracing through the atmosphere can be used as the truth for comparisons if profiles of pressure, temperature, and relative humidity are available from radiosondes. Hobiger et al. (2008a,b) report on using NWM to compute the tropospheric slant delays directly and in real time.

8.2.3 Precipitable Water Vapor

The GPS observables directly depend on the STD. This quantity, therefore, can be estimated from GPS observations. One might envision the situation where widely spaced receivers are located at known stations and that the precise ephemeris is available. If all other errors are taken into consideration, then the residual misclosures of the observations are the STD. We could compute the ZHD from surface pressure measurements and a hydrostatic delay model. Using appropriate mapping functions, we could then compute the ZWD from (8.2.18) using the estimated STD. The ZWD can then be converted to precipitable water.

We begin with defining the integrated water vapor (IWV) along the vertical and the precipitable water vapor (PWV) as

$$IWV \equiv \int \rho_{wv}\, dh \tag{8.2.21}$$

$$PWV \equiv \frac{IWV}{\rho_w} \tag{8.2.22}$$

where ρ_w is the density of liquid water. In order to relate the ZWD to these measures, it is convenient to introduce the mean temperature T_m,

$$T_m \equiv \frac{\displaystyle\int \frac{p_{wv}}{T} Z_{wv}^{-1}\, dh}{\displaystyle\int \frac{p_{wv}}{T^2} Z_{wv}^{-1}\, dh} \tag{8.2.23}$$

The ZWD follows then from (8.2.13), (8.2.11), and (8.2.23) as

$$ZWD = 10^{-6} \left(k_2' + \frac{k_3}{T_m} \right) \int \frac{p_{wv}}{T} Z_{wv}^{-1}\, dh \tag{8.2.24}$$

To be precise let us recall that (8.2.24) represents the nonhydrostatic zenith delay. Using the state equation of water vapor gas,

$$\frac{p_{wv}}{T} Z_{wv}^{-1} = R_{wv} \rho_{wv} \tag{8.2.25}$$

in the integrand gives

$$ZWD = 10^{-6} \left(k_2' + \frac{k_3}{T_m} \right) R_{wv} \int \rho_{wv}\, dh \tag{8.2.26}$$

We replace the integrand in (8.2.26) by IWV according to (8.2.21) and then replace the specific gas constant R_{wv} by the universal gas constant R and the molar mass M_{wv}. The conversion factor Q that relates the zenith nonhydrostatic wet delay to the precipitable water then becomes

$$Q \equiv \frac{ZWD}{PWV} = \rho_w \frac{R}{M_{wv}} \left(k_2' + \frac{k_3}{T_m} \right) 10^{-6} \tag{8.2.27}$$

The constants needed in (8.2.27) are known with sufficient accuracy. The largest error contribution comes from T_m, which varies with location, height, season, and weather. The Q value varies between 5.9 and 6.5, depending on the air temperature. For warmer conditions, when the air can hold more water vapor, the ratio is toward the low end. Bevis et al. (1992) correlate T_m with the surface temperature T_0 and offer the model

$$T_{m[\mathrm{K}]} = 70.2 + 0.72 T_{0[\mathrm{K}]} \tag{8.2.28}$$

The following models for Q are based on radiosonde observations (Keihm, JPL, private communication),

$$Q = 6.135 - 0.01294(T_0 - 300) \tag{8.2.29}$$

$$Q = 6.517 - 0.1686\, PWV + 0.0181\, PWV^2 \tag{8.2.30}$$

$$Q = 6.524 - 0.02797\, ZWD + 0.00049\, ZWD^2 \tag{8.2.31}$$

If surface temperatures are not available, one can use (8.2.30) or (8.2.31), which take advantage of the fact that Q correlates with PWV (since higher PWV values are generally associated with higher tropospheric temperatures).

8.3 TROPOSPHERE ABSORPTION

This section deals briefly with some elements of remote sensing by microwaves. The interested reader may consult a general text on remote sensing. We recommend the book by Janssen (1993) because it is dedicated to atmospheric remote sensing by microwave radiometry. The material presented below very much depends on that source. Solheim's (1993) dissertation is also highly recommended for additional reading.

8.3.1 The Radiative Transfer Equation

The energy emission and absorption of molecules are due to transitions between allowed energy states. Several fundamental laws of physics relate to the emissions and absorptions of gaseous molecules. Bohr's frequency condition relates the frequency f of a photon emission or absorption to the energy levels E_a and E_b of the molecule and to Planck's constant h. Einstein's law of emission and absorption specifies that if $E_a > E_b$, the probability of stimulated emission of a photon by a transition from state a to state b is equal to the probability of absorption of a photon by a transition from b to a. These two probabilities are proportional to the incident energy at frequency f. Dirac's perturbation theory gives the conditions that must be fulfilled in order to enable the electromagnetic field to introduce transitions between states. For wavelengths that are very long compared to molecular dimensions, this operator is the dipole moment. This is the case in microwave radiometry. We typically observe the rotation spectra, corresponding to radiation emitted in transition between rotational states of a molecule having an electric dipole moment. The rotational motion of a diatomic molecule can be visualized as a rotation of a rigid body about its center of mass. Other types of transitions of molecular quantum states that emit at the ultraviolet, gamma, or infrared range are not relevant to sensing of water vapor. Although the atmosphere contains other polar gases, only water vapor and oxygen are present in enough quantity to emit significantly at microwave range.

Let $I(f)$ denote the instantaneous radiant power that flows at a point in a medium, over a unit area, per unit-frequency interval at a specified frequency f, and in a given direction per unit solid angle. As the signal travels along the path s, the power changes when it encounters sources and sinks of radiation. This change is described by the differential equation

$$\frac{dI(f)}{ds} = -I(f)\,\alpha + S \tag{8.3.1}$$

The symbol α denotes the absorption (describing the loss) and S is the source (describing the gain) into the given direction.

Scattering from other directions can lead to losses and gains to the intensity. In the following we will ignore scattering. We assume thermodynamic equilibrium, which means that at each point along the path s the source can be characterized by temperature T. The law of conservation of energy for absorbed and emitted energy relates the source and absorption as

$$S = \alpha\, B(f, T) \tag{8.3.2}$$

where

$$B(f, T) = \frac{2\pi h f^3}{c^2\,(e^{hf/(kT)} - 1)} \tag{8.3.3}$$

$B(f, T)$ is the Planck function, h is the Planck constant, k is the Boltzmann constant, T is the physical temperature, and c denotes the speed of light. Please consult the specialized literature for details on (8.3.3).

With stated assumptions, equation (8.3.1) becomes a standard differential equation with all terms depending only on the intensity along the path of propagation. The solution can be written as

$$I(f, 0) = I(f, s_0)e^{-\tau(s_0)} + \int_0^{s_0} B(f, T)\, e^{-\tau(s)}\, \alpha\, ds \tag{8.3.4}$$

$$\tau(s) = \int_0^{s} \alpha(s')\, ds' \tag{8.3.5}$$

Equation (8.3.4) is called the radiative transfer equation. $I(f, 0)$ is the intensity at the measurement location $s = 0$, and $I(f, s_0)$ is the intensity at some boundary location $s = s_0$. The symbol $\tau(s)$ denotes the optical depth or the opacity.

If $hf \ll kT$, as is the case for microwaves and longer waves, the denominator in (8.3.3) can be expanded in terms of $hf\,/\,kT$. After truncating the expansion, the Planck function becomes the Rayleigh-Jeans approximation

$$B(\lambda, T) \approx \frac{2f^2 kT}{c^2} = \frac{2kT}{\lambda^2} \tag{8.3.6}$$

The symbol λ denotes the wavelength. Expression (8.3.6) expresses a linear relationship between Planck function and temperature T. For a given opacity (8.3.5) the intensity (8.3.4) is proportional to the temperature of the field of view of the radiometer antenna, given (8.3.6).

The Rayleigh-Jeans brightness temperature $T_b(f\,)$ is defined by

$$T_b(f) \equiv \frac{\lambda^2}{2k} I(f) \tag{8.3.7}$$

$T_b(f)$ is measured in degrees Kelvin; it is a simple function of the intensity of the radiation at the measurement location. If we declare the space beyond the boundary s_0 as the background space, the Rayleigh-Jeans background brightness temperature can be written as

$$T_{b0}(f) \equiv \frac{\lambda^2}{2k} I(f, s_0) \tag{8.3.8}$$

Using definitions (8.3.7) and (8.3.8), the approximation (8.3.6), and $T = T_b$, the radiative transfer equation (8.3.4) becomes

$$T_b = T_{b0}\, e^{-\tau(s_0)} + \int_0^{s_0} T(s)\, \alpha\, e^{-\tau(s)}\, ds \tag{8.3.9}$$

This is Chandrasekhar's equation of radiative transfer as used in microwave remote sensing. For ground-based GPS applications, the sensor (radiometer) is on the ground ($s = 0$) and senses all the way to $s = \infty$. T_{b0} becomes the cosmic background temperature T_{cosmic}, which results from the residual cosmic radiation of outer space that is left from the Big Bang. Thus

$$T_b = T_{\text{cosmic}}\, e^{-\tau(\infty)} + \int_0^{\infty} T(s)\, \alpha\, e^{-\tau(s)}\, ds \tag{8.3.10}$$

$$\tau(\infty) = \int_0^{\infty} \alpha(s)\, ds \tag{8.3.11}$$

$$T_{\text{cosmic}} = 2.7 \text{ K} \tag{8.3.12}$$

The brightness temperature (8.3.10) depends on the atmospheric profiles of physical temperature T and absorption α. For the atmosphere the latter is a function of pressure, temperature, and humidity. Equation (8.3.10) represents the forward problem, i.e., given temperature and absorption profiles along the path, one can compute brightness temperature. The inverse solution of (8.3.10) is of much practical interest. It potentially allows the determination of atmospheric properties such as T and α, as well as their spatial distribution from brightness temperature measurements.

Consider the following special cases. Assume that the temperature T is constant. Neglecting the cosmic term, using $d\tau = \alpha\, ds$, the radiative transfer equation (8.3.10) becomes

$$T_b = T \int_0^{\tau(a)} e^{-\tau}\, d\tau = T\left(1 - e^{-\tau(a)}\right) \tag{8.3.13}$$

For a large optical depth $\tau(a) \gg 1$ we get $T_b = T$ and the radiometer acts like a thermometer. For a small optical path $\tau(a) \ll 1$ we get $T_b = T\, \tau(a)$. If the temperature is known, then $\tau(a)$ can be determined. If we also know the absorption properties of the constituencies, it might be possible to estimate the concentration of a particular constituent of the atmosphere.

For the sake of clarity, we reiterate that (8.3.7) defines the Rayleigh-Jeans brightness temperature. The thermodynamic brightness temperature is defined as the temperature of a blackbody radiator that produces the same intensity as the source being observed. The latter definition refers to the physical temperature, whereas the

Rayleigh-Jeans definition directly relates to the radiated intensity. The difference between both definitions can be traced back to the approximation implied in (8.3.6). A graphical representation of the differences is found in Janssen (1993, p. 10).

8.3.2 Absorption Line Profiles

Microwave radiometers measure the brightness temperature. In ground-based radiometry, the relevant molecules are water vapor, diatomic oxygen (O_2), and liquid water. Mathematical models have been developed for the absorption. For isolated molecules, the quantum mechanic transitions occur at well-defined resonance frequencies (line spectrum). Collision with other molecules broadens these spectral lines. When gas molecules interact, the potential energy changes due to changing relative position and orientation of the molecules. As a result, the gas is able to absorb photons at frequencies well removed from the resonance lines. Pressure broadening converts the line spectrum into a continuous absorption spectrum, called the line profile. The interactions and thus the broadening increase with pressure. Given the structure of molecules it is possible to derive mathematical functions for the line profiles. Because of the complexities of these computations and the presence of collisions, these functions typically require refinement with laboratory observations. The results are line profile models.

Figures 8.3.1 and 8.3.2 show line profiles for water vapor, oxygen, and liquid water computed with Fortran routines provided by Rosenkranz. (See also Rosenkranz, 1998). All computations refer to a temperature of 15°C. The top three lines in Figure 8.3.1 show the line profiles for water vapor for pressures of 700, 850, and 1013 mbar, and a water vapor density of 10 g/m^3. The maximum absorption occurs at the resonance frequency of 22.235 GHz. The effect of pressure broadening on the absorption curve is readily visible. Between about 20.4 and 23.8 GHz the absorption is less, the higher the pressure. The reverse is true in the wings of the line profile. In the vicinity of these two particular frequencies, the absorption is relatively independent of pressure. Most WVRs use at least one of these frequencies

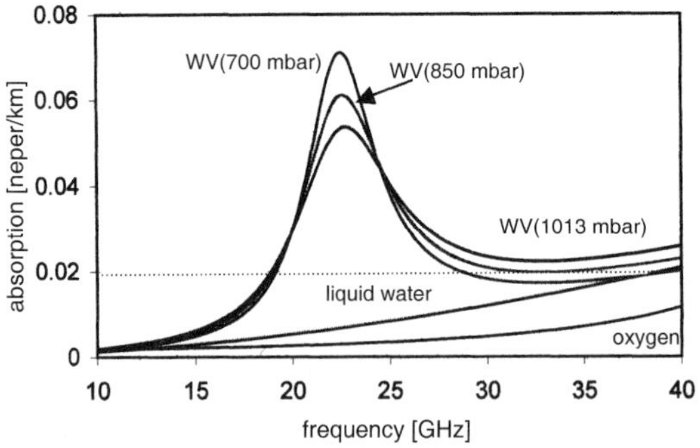

Figure 8.3.1 Absorption of water vapor, liquid water, and oxygen between 10 and 40 GHz.

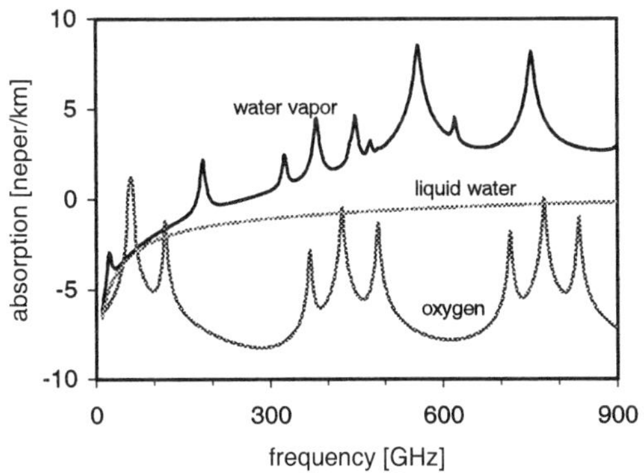

Figure 8.3.2 Absorption of water vapor, liquid water, and oxygen between 10 and 900 GHz.

to minimize the sensitivity of brightness temperature to the vertical distribution of water vapor. The water vapor absorption is fairly stable in regard to changes in frequency around 31.4 GHz. Dual-frequency WVRs for ground-based sensing of water vapor typically also use the 31.4 GHz frequency to separate the effects of water vapor from cloud liquid. The 31.4 GHz channel is approximately twice as sensitive to cloud liquid emissions as the channel near 20.4 GHz. The opposite is true for water vapor, allowing separate retrievals of the two most variable atmospheric constituents. The absorption line of oxygen in Figure 8.3.1 refers to a water vapor density of 10 g/m^3 and a pressure of 1013 mbar. The line of liquid water (suspended water droplets) is based on a water density of 0.1 g/m^3. The absorption used in the radiative transfer equation (8.3.10) is the sum of the absorption of the individual molecular constituencies, i.e.,

$$\alpha = \alpha_{wv}(f, T, p, \rho_{wv}) + \alpha_{lw}(f, T, \rho_{lw}) + \alpha_{ox}(f, T, p, \rho_{wv}) \tag{8.3.14}$$

The absorption units are typically referred to as neper per kilometer. The absorption unit refers to the fractional loss of intensity per unit distance (km) traveled in a logarithmic sense. That is, an absorption value of 1 neper/km would imply that the power would be attenuated by $1/e$ fractional amount over 1 km given that the absorption properties remained constant over that kilometer. A neper is the natural logarithm of a voltage ratio and is related to the dB unit as follows:

$$\text{dB} = \frac{20}{\ln(10)}\text{neper} \approx 8.686\ \text{neper} \tag{8.3.15}$$

The line profiles contain other maxima, as seen in Figure 8.3.2. A large maximum for water vapor at 183.310 GHz is relevant to water vapor sensing in airborne radiometry. The liquid water absorption increases monotonically with frequency in the microwave range. Oxygen has a band of resonance near 60 GHz. The oxygen absorption is well modeled with pressure and temperature measurements on the ground; the absorption is small compared to that of water vapor and nearly constant for a specific site because oxygen is mixed well in the air. The profiles of Figure 8.3.2 refer to a temperature of 15°C, a water vapor density of 10 g/m^3, a pressure of 1013 mbar, and a liquid water density 0.1 g/m^3.

Since the absorption of oxygen can be computed from the model and ground-based observations, it is possible to separate its known contribution in (8.3.10) and invert the radiative transfer equation to determine integrated water vapor and liquid water as a function of the observed brightness temperatures. Westwater (1978) provides a thorough error analysis for this standard dual-frequency case. The fact that at 23.8 GHz the absorption of water vapor is significantly higher than at 31.4 GHz (while the absorption of liquid water changes monotonically over that region), can be used to retrieve separately integrated water vapor and liquid water from the inversion of the radiative transfer equation. With more channels distributed appropriately over the frequency, one might be able to roughly infer the water vapor profiles as well as integrated water vapor and liquid water, or even temperature, vapor, and liquid profiles.

8.3.3 General Statistical Retrieval

Consider the following experiment. Use a radiosonde to measure the temperature and water vapor density profile along the vertical and use equations (8.2.21) and (8.2.22) to compute IWV and PWV. Compute the brightness temperature T_b from the radiative transfer equation (8.3.10) for each radiometer frequency using the frequency-dependent absorption model for water vapor $\alpha_{wv}(f, T, p, p_w)$ and oxygen absorption.

Figure 8.3.3 shows the result of such an experiment. The plot shows the observed T_b for WVR channels at 20.7 and 31.4 GHz. The data refer to a Bermuda radiosonde station and were collected over a 3-year period. The Bermuda site experiences nearly the full range of global humidity and cloud cover conditions. The scatter about the

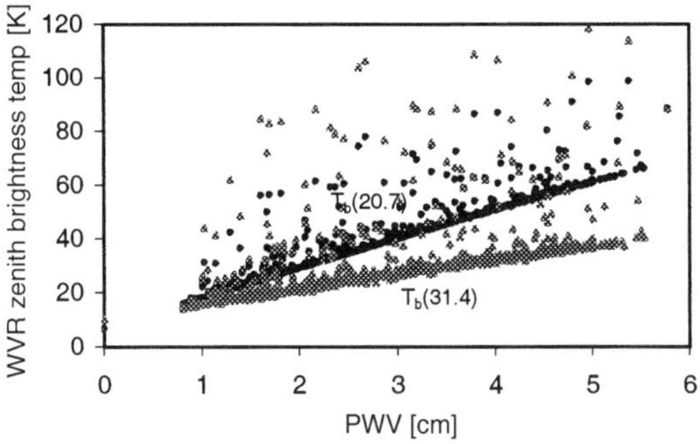

Figure 8.3.3 Brightness temperature versus precipitable water vapor. (Data source: Keihm, JPL.)

heavily populated "clear" lines is due to the occurrence of cloudy cases. The slopes of T_b (20.7) are approximately 2.2 times the slopes of the T_b (31.4). The scatter about the T_b (31.4) clear line is approximately twice as large as the scatter about the T_b (20.7) clear line. These results are indicative of the facts that (1) the sensitivity of T_b (20.7) to PWV is approximately 2.2 times greater than that of T_b (31.4) and (2) that the sensitivity of T_b (31.4) to liquid water is approximately 2 times greater than that of T_b (20.7). The sensitivity to liquid water is also illustrated in Figure 8.3.4, which shows T_b variations versus cloud liquid. Despite the large scatter (due to variable PWV), one can see that the slope of the T_b (31.4) data is approximately twice as large as the slope of the T_b (20.7) data.

Because of the relationships between ZWD, IWV, and PWV as seen by (8.2.26), (8.2.21), and (8.2.22), the strong correlation seen in Figure 8.3.3 between PWV and the brightness temperature makes a simple statistical retrieval procedure for the ZWD possible. Assume a radiosonde reference station is available to determine ZWD and that a WVR measures zenith $T_{20.7}$ and $T_{31.4}$. Using the model

$$ZWD = c_0 + c_{20.7} T_{20.7} + c_{31.4} T_{31.4} \tag{8.3.16}$$

we can estimate accurate retrieval coefficients $\hat{c}_0, \hat{c}_{20.7}$, and $\hat{c}_{31.4}$. When users operate a WVR in the same climatological region, they can then readily compute the ZWD at their location from the observed brightness temperature and the estimated regression coefficients. This statistical retrieval procedure can be generalized by using an expanded regression model in (8.3.16) and by incorporating brightness temperature measurements from several radiosonde references distributed over a region.

The opacity may also be used in this regression. In fact, opacity varies more linearly with PWV than does the brightness temperature T_b. At high levels of water vapor, or low elevation angles, the T_b measurements will eventually begin to saturate, i.e., the rate of T_b increase with increasing vapor will start to fall off. This is not true for opacity, which essentially remains linear with the in-path vapor abundance. Opacity is available from (8.3.11) but also can be conveniently related to the

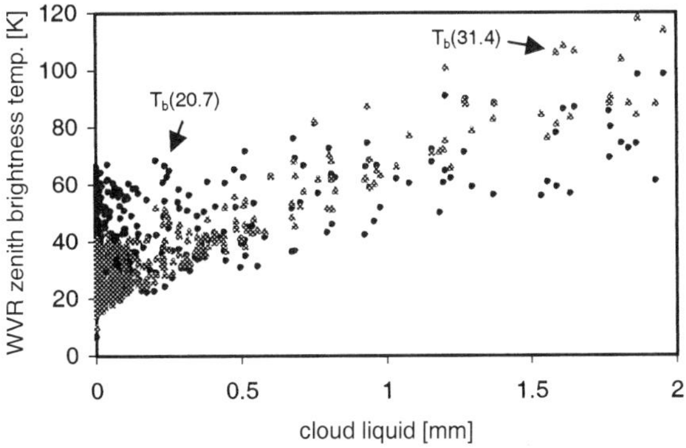

Figure 8.3.4 Brightness temperature versus cloud liquid. (Data source: Keihm, JPL.)

brightness temperature. Define mean radiation temperature T_{mr} as

$$T_{mr} \equiv \frac{\displaystyle\int_0^\infty T(s)\, \alpha(s)\, e^{-\tau(s)}\, ds}{\displaystyle\int_0^\infty \alpha(s)\, e^{-\tau(s)}\, ds} \tag{8.3.17}$$

This auxiliary quantity can be accurately estimated from climatologic data. Corrections with surface temperature permit T_{mr} estimates to be computed to a typical accuracy of $\sim$ 3 K. Using the relationship

$$\int_0^{\tau(\infty)} \alpha e^{-\tau}\, ds = 1 - e^{-\tau(\infty)} \tag{8.3.18}$$

where we used again $d\tau = \alpha\, ds$, the radiative transfer equation (8.3.10) can be written as

$$T_b = T_{cosmic}\, e^{-\tau(\infty)} + T_{mr}\left(1 - e^{-\tau(\infty)}\right) \tag{8.3.19}$$

which, in turn, can be rewritten as

$$\tau(\infty) = \ln\left(\frac{T_{mr} - T_{cosmic}}{T_{mr} - T_b}\right) \tag{8.3.20}$$

The opacities and brightness temperature show similarly high correlations with the wet delay. In fact, at low elevation angles the opacities correlate even better with the wet delay than do brightness temperatures.

If the user measures the brightness temperatures along the slant path rather than the zenith direction, the observed T_b must be converted to the vertical to estimate ZWD using (8.3.16). Given the slant T_b measurement at zenith angle ϑ, and an estimate of T_{mr}, the slant opacity can be computed and converted to the zenith opacity using the simple $1 / \cos(\vartheta)$ mapping function. The equivalent zenith T_b follows from (8.3.19). For elevation angles above 15° this conversion is very accurate.

For a specific site T_{mr} is computed from (8.3.17) using radiosonde data that typify the site. The variation of T_{mr} with slant angle is minimal for elevations down to about 20°. The value used for WVR calibration and water vapor retrievals can be a site average (standard deviation typically about 10 K), or can be adjusted for season to reduce the uncertainty. If surface temperatures T are available, then T_{mr} correlations with T can reduce the T_{mr} uncertainty to about 3 K.

8.3.4 Calibration of WVR

Because the intensity of the atmospheric microwave emission is very low, the WVR calibration is important. Microwave radiometers receive roughly a billionth of a watt in microwave energy from the atmosphere. The calibration establishes a relationship between the radiometer reading and the brightness temperature. Here we briefly discuss the calibration with tipping curves. This technique provides accurate brightness temperatures and the instrument gain without any prior knowledge of either.

Under the assumption that the atmosphere is horizontally homogeneous and that the sky is clear, the opacity is proportional to the thickness of the atmosphere. Clearly the amount of atmosphere sensed increases with the zenith angle. For zenith angles less than about 60° one might consider adopting the following model for the mapping function for the opacity:

$$m_\tau(\vartheta) \equiv \frac{\tau(\vartheta)}{\tau(\vartheta = 0)} = \frac{1}{\cos \vartheta} \tag{8.3.21}$$

Figure 8.3.5 shows an example of radiometer calibration using tipping. The opacity is plotted versus air mass. Looking straight up, the opacity of one air mass is observed. Looking at 30°, the opacity of two air masses is observed, etc. Since opacity is linear, we can extrapolate to zero air mass. At zero air mass, we have $m_\tau(\vartheta) = 0$ because there is no opacity for a zero atmosphere.

The calibration starts with a radiometer voltage (noise diode, labeled ND in Figure 8.3.5) reading N_{bb} of an internal reference object, which one might think of as a blackbody. The physical temperature of that object is T_{bb}. Let G denote the initial estimate of the gain factor (change in radiometer count reading over change in temperature). The observed brightness temperature at various zenith angles, measured by tipping the antenna, is then computed by

$$T(\vartheta) = T_{\text{bb}} - \frac{1}{G}\left(N_{\text{bb}} - N(\vartheta)\right) \tag{8.3.22}$$

The brightness temperatures are substituted into (8.3.20) to get the opacity. If the linear regression line through the computed opacities does not pass through the origin,

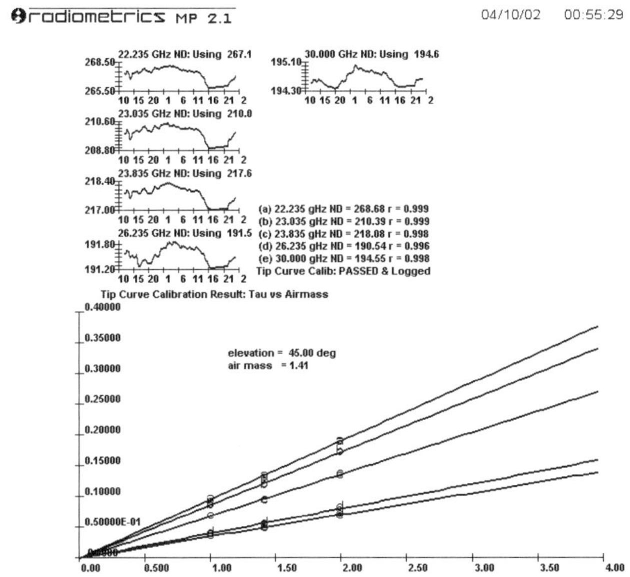

Figure 8.3.5 Tipping curve example. (Courtesy of R. Ware, Radiometrics Corporation, Boulder.)

the gain factor G is adjusted until it passes though the origin. If the regression coefficient of the linear fit is better than a threshold value, typically $r = 0.99$, the tip curve calibration is accepted. The time series in Figure 8.3.5 show the history of passed tip curve calibrations at various microwave frequencies.

The tipping curve calibration assumes that we know the microwave cosmic background brightness temperature $T_{\text{cosmic}} = 2.7$ K. Arno Penzias and Robert Wilson received the Nobel Prize for physics in 1978 for their discovery of the cosmic background radiation. Conducting their radio astronomy experiments, they realized a residual radiation that was characteristically independent of the orientation of the antenna.

8.4 IONOSPHERIC REFRACTION

Coronal mass ejections (CMEs) and extreme ultraviolet (EUV) solar radiation (solar flux) are the primary causes of the ionization (Webb and Howard, 1994). A CME is a major solar eruption. When passing the earth, it causes at times sudden and large geomagnetic storms, which generate convection motions within the ionosphere, as well as enhanced localized currents. The phenomena can produce large spatial and temporal variation in the TEC and increased scintillation in phase and amplitude. Complicating matters are coronal holes, which are pathways of low density through which high-speed solar wind can escape the sun. Coronal holes and CME are the two major drivers of magnetic activities on the earth. Larger magnetic storms are rare but may occur at any time.

Solar flux originates high in the sun's chromosphere and low in its corona. Even a quiet sun emits radio energy across a broad frequency spectrum, with slowly varying intensity. EUV radiation is absorbed by the neutral atmosphere and therefore cannot be measured accurately from ground-based instrumentation. Accurate determination of the EUV flux requires observations from space-based platforms above the ionosphere. A popular surrogate measure to the EUV radiation is the widely observed flux at 2800 MHz (10.7 cm). The 10.7 cm flux is useful for studying the ozone layer and global warming. However, Doherty et al. (2000) point out that predicting the TEC by using the daily values of solar 10.7 cm radio flux is not useful due to the irregular, and sometimes very poor, correlation between the TEC and the flux. The TEC at any given place and time is not a simple function of the amount of solar ionizing flux.

The transition from a gas to an ionized gas, i.e., plasma, occurs gradually. During the process, a molecular gas dissociates first into an atomic gas that, with increasing temperature, ionizes as the collisions between atoms break up the outermost orbital electrons. The resulting plasma consists of a mixture of neutral particles, positive ions (atoms or molecules that have lost one or more electrons), and negative electrons. Once produced, the free electron and the ions tend to recombine, and a balance is established between electron-ion production and loss. The net concentration of free electrons is what impacts electromagnetic waves passing through the ionosphere. In order for gases to be ionized, a certain amount of radiated energy must be absorbed. Hargreaves (1992, p. 223) gives maximum wavelengths for radiation needed to ionize various gases. The average wavelength is about 900 Å (1 Å equals

0.1 nm). The primary gases available at the upper atmosphere for ionization are oxygen, ozone, nitrogen, and nitrous oxide.

Because the ionosphere contains particles that are electrically charged and capable of creating and interacting with electromagnetic fields, there are many phenomena in the ionosphere that are not present in ordinary fluids and solids. For example, the degree of ionization does not uniformly increase with the distance from the earth's surface. As mentioned above, there are regions of ionization, historically labeled *D*, *E*, and *F*, that have special characteristics as a result of variation in the EUV absorption, the predominant type of ions present, or pathways generated by the electromagnetic field. The electron density is not constant within such a region and the transition to another region is continuous. Whereas the TEC determines the amount of pseudorange delays and carrier phase advances, it is the layering that is relevant to radio communication in terms of signal reflection and distance that can be bridged at a given time of the day. In the lowest *D* region, approximately 60 to 90 km above the earth, the atmosphere is still dense and atoms that have been broken up into ions recombine quickly. The level of ionization is directly related to radiation that begins at sunrise, disappears at sunset, and generally varies with the sun's elevation angle. There is still some residual ionization left at local midnight. The *E* region extends from about 90 to 150 km and peaks around 105 to 110 km. In the *F* region, the electrons and ions recombine slowly due to low pressure. The observable effect of the solar radiation develops more slowly and peaks after noon. During daytime this region separates into the *F*1 and *F*2 layers. The *F*2 layer (upper layer) is the region of highest electron density. The top part of the ionosphere reaches up to 1000 to 1500 km. There is no real boundary between the ionosphere and the outer magnetosphere.

Ionospheric convection is the main result of the coupling between the magnetosphere and ionosphere. While in low altitudes the ionospheric plasma co-rotates with the earth, at higher latitudes it is convecting under the influence of the large-scale magnetospheric electric field. Electrons and protons that speed along the magnetic field lines until they strike the atmosphere not only generate the spectacular lights of the aurora in higher latitudes, but they also cause additional ionization. Peaks of electron densities are also found at lower latitudes on both sides of the magnetic equator. The electric field and the horizontal magnetic field interact at the magnetic equator to raise ionization from the magnetic equator to greater heights, where it diffuses along magnetic field lines to latitudes approximately $\pm 15°$ to $20°$ on either side of the magnetic equator. The largest TEC values in the world typically occur at these so-called equatorial anomaly latitudes.

There are local disturbances of electron density in the ionosphere. On a small scale, irregularities of a few hundred meters in size can cause amplitude fading and phase scintillation of GPS signals. Larger disturbances of the size of a few kilometers can significantly impact the TEC. Amplitude fading and scintillation can cause receivers to lose lock, or receivers may not be able to maintain lock for a prolonged period of time. Scintillation on GPS frequencies is rare in the midlatitudes, and amplitude scintillation, even under geomagnetically disturbed conditions, is normally not large in the auroral regions. However, rapid phase scintillation can be a problem in both the equatorial and the auroral regions, especially for semicodeless L2 GPS receivers, as the bandwidth of such receivers might be too narrow to follow rapid phase scintillation effects. Strong scintillation in the equatorial region generally occurs in the postsunset to local midnight time period, or during geomagnetically quiet period, but, mostly during equinoctial months in years having high solar activity. Even during times of strong amplitude scintillation the likelihood of simultaneous deep amplitude fading to occur on more than one GPS satellite is small. Thus, a modern GPS receiver observing all satellites in view should be able to operate continuously through strong scintillation albeit with a continuously changing geometric dilution of precision (GDOP) due to the continually changing "mix" of GPS satellites in lock.

Sunspots are seen as dark areas in the solar disk. At the dark centers the temperature drops to about 3700 K from 5700 K for the surrounding photosphere. They are magnetic regions with field strengths thousands of times stronger than the earth's magnetic field. Sunspots often appear in groups with sets of two spots, one with positive (north) magnetic fields, and one with negative (south) magnetic fields. Sunspots have an approximate lifetime of a few days to a month. The systematic recording of these events began in 1849 when the Swiss astronomer Johann Wolf introduced the sunspot number. This number captures the total number of spots seen, the number of disturbed regions, and the sensitivity of the observing instrument. Wolf searched observatory records to tabulate past sunspot activities. He apparently traced the activities to 1610, the year Galileo Galilei first observed sunspots through his telescope (McKinnon, 1987). Sunspot activities follow a periodic variation, with a principal period of 11 years, as seen in Figure 8.4.1. The cycles are usually not symmetric. The time from minimum to maximum is shorter than the time from maximum to minimum.

Sunspots are good indicators of solar activities. Even though sunspots have a high correlation with CME and solar flux, there is no strict mathematical relationship between them. It can happen that GPS is adversely affected even when daily sunspot numbers are actually low. Kunches and Klobuchar (2000) point out that GPS operations are more problematic during certain years of the solar cycle and during certain months of those years. The years at or just after the solar maximum will be stormy, and the months near the equinoxes will contain the greatest number of storm days. Sunspots are good for long-term prediction of ionospheric states.

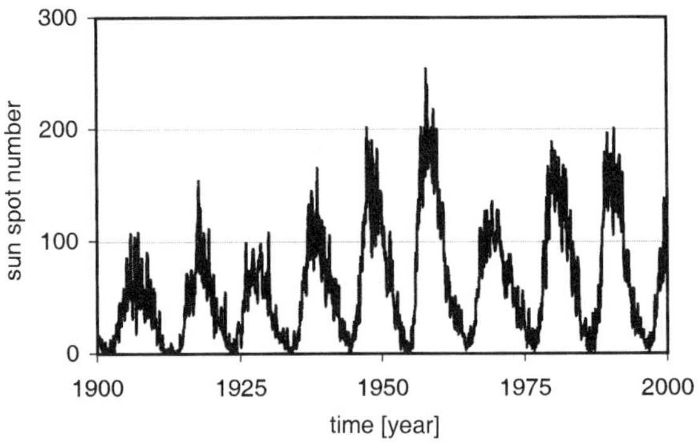

Figure 8.4.1 Sunspot numbers.

8.4.1 Index of Ionospheric Refraction

The Appleton-Hartree formula is usually taken as the start for developing the ionospheric index of refraction that is applicable to the range of GPS frequencies. The formula is valid for a homogeneous plasma that consists of electrons and heavy positive ions, a uniform magnetic field, and a given electron collision frequency. Following Davies (1990, p. 72), the Appleton-Hartree formula is

$$n^2 = 1 - \frac{X}{1 - iZ - \dfrac{Y_T^2}{2(1 - X - iZ)} \pm \sqrt{\dfrac{Y_T^4}{4(1 - X - iZ)^2} + Y_L^2}} \tag{8.4.1}$$

Since the goal is to find the ionospheric index of refraction that applies to the GNSS frequency f, several simplifications are permissible. The numerator X represents the squared ratio of f_p and f, where f_p is a natural gyrofrequency or plasma frequency by which free electrons cycle geomagnetic field lines. It is a basic constant of plasma (Davies 1990, pp. 21, 73). We can write

$$X = \frac{f_p^2}{f^2} \qquad f_p^2 = A_p N_e \qquad A_p = \frac{e^2}{4\pi^2 \varepsilon_0 m_e} = 80.6 \tag{8.4.2}$$

The electron density N_e is typically given in units of electrons per cubic meter (el/m^3). The symbol $e = 1.60218 \cdot 10^{-19}$ coulombs denotes the electron charge, $m_e = 9.10939 \cdot 10^{-31}$ kg is the electron mass, and $\varepsilon_0 = 8.854119 \cdot 10^{-12}$ faradays/m is the permittivity of free space. The element $Z = v/f$ is the ratio of the electron collision frequency v and the satellite frequency. This term quantifies the absorption. We simply set $Z = 0$ because it is negligible in the context of ionospheric refraction at GNSS frequencies. As a result of this simplification the expression of the refraction index is no longer complex. The symbols Y_T and Y_L denote the transversal and longitudinal component of the geomagnetic filed vector $\boldsymbol{B}$. For actual computations the International Geomagnetic Reference Field (Finlay et al., 2010) can be a source for data for accurate representation of the earth's magnetic field. Let B denote the modulus of this vector, then

$$Y_T = Y \sin\theta \qquad Y_L = Y \cos\theta \tag{8.4.3}$$

$$Y = \frac{f_g}{f} \qquad f_g = A_g B \qquad A_g = \frac{e}{2\pi m_e} \tag{8.4.4}$$

where θ is the angle between signal propagation direction and vector $\boldsymbol{B}$. Incorporating the specifications mentioned above, the Appleton-Hartree equation (8.4.1) takes the form

$$n^2 = 1 - \frac{X}{1 - \dfrac{Y^2 \sin^2\theta}{2(1 - X)} \pm \sqrt{\dfrac{Y^4 \sin^2\theta}{4(1 - X)^2} + Y^2 \cos^2\theta}} \tag{8.4.5}$$

The next step in the development is expanding (8.4.5) in a binomial series and retaining only relevant terms. Details of this expansion are given in Brunner and Gu (1991) or Petrie et al. (2011). The result is

$$n = 1 - \frac{A_p N_e}{2f^2} - \frac{A_p N_e A_g B|\cos\theta|}{2f^3} - \frac{A_p^2 N_e^2}{8f^4} \tag{8.4.6}$$

The second term in (8.4.6) is a first-order term, followed by the second- and third-order terms. Note that the power of the frequency in the denominators increases with order. The absolute value of the cosine term is required to handle RHCP signals over the full range of 2π. For approximate numerical evaluation one might assume average values such as $N_e \approx 10^{12}$ el/m^3 and $B = 5 \cdot 10^{-5}$ telsa. Within this approximation, only the second-order term depends on the geomagnetic vector.

Since the ionosphere is a dispersive medium, the phase velocity is a function of its frequency, i.e., the carrier phase and the phase of the modulation travel at different velocities. Consequently we need to introduce two indices of refraction. Henceforth, the phase index of refraction dealt with above is denoted by n_Φ and specifically refers to the GNSS carrier phase. Since the carrier is modulated by codes such as C/A-codes and P-codes, we also need a separate index of refraction for the propagation of these codes. It is called the group index of refraction and denoted by n_g. General physics provides the well-known relationship between these two indices of refraction as

$$n_g = n_\Phi + f\frac{dn_\Phi}{df} \tag{8.4.7}$$

Thus we only need to differentiate (8.4.6) with respect to the frequency, multiply by the frequency, and add it to the phase index of refraction, giving

$$n_g = 1 + \frac{A_p N_e}{2f^2} + \frac{A_p N_e A_g B|\cos\theta|}{f^3} + \frac{3A_p^2 N_e^2}{8f^4} \tag{8.4.8}$$

Comparing (8.4.6) and (8.4.8) we observe that the first-order terms have the same magnitude but opposite sign. The higher order terms also have opposite sign but differ by a factor 2 and 3, respectively. Since $N_e > 0$ we have $n_\Phi < 1$ and $n_g > 1$.

The impact of the ionosphere is given by

$$I_{impact} = \int (n(s) - 1)\, ds \tag{8.4.9}$$

whereby the integration takes place over the propagation path of the signal, with the index of refraction changing with distance. For the carrier phase and the pseudorange the impact is in units of meters,

$$I_{f,\Phi} = \int (n_\Phi - 1)\, ds = -\frac{q}{f^2} - \frac{s}{2f^3} - \frac{r}{3f^4} + \cdots \tag{8.4.10}$$

$$I_{f,P} = \int (n_g - 1)\, ds = \frac{q}{f^2} + \frac{s}{f^3} + \frac{r}{f^4} + \cdots \tag{8.4.11}$$

We readily see that the carrier phase experiences an advancement because $n_\Phi < 1$ and the pseudorange experiences a delay because $n_g > 1$ relative to vacuum travel for which $n_g = n_\Phi$ since the index of refraction would not depend on the frequency. The q, a, and r terms follow from (8.4.6) or (8.4.8) as

$$q = \frac{A_p}{2}\int N_e\,ds = 40.3\int N_e\,ds \tag{8.4.12}$$

$$s = A_p A_g \int N_e B|\cos\theta|\,ds = 2.256\cdot 10^{12}\int N_e B|\cos\theta|ds \tag{8.4.13}$$

$$r = \frac{3}{8}A_p^2\int N_e^2\,ds = 2437\int N_e^2\,ds \tag{8.4.14}$$

All expressions above are seen to be a function of the integral $\int N_e\,ds$ which represents the total number of free electron along a 1 m^2 column from the earth surface to the end of the ionosphere. It has become a common practice to call this integral the total electron content (TEC). Thus

$$TEC = \int N_e\,ds \tag{8.4.15}$$

Typically the TEC values range from 10^{16} to 10^{18}el/m^2. Often the total electron content is expressed in terms of TEC units (TECU), with one TECU being 10^{16} electrons per 1m^2 column.

The values q, s, and r determine the magnitude of the first-, second-, and third-order ionospheric delays for a given GNSS frequency. In order to evaluate these delays approximately, let us assume an ionosphere of 60 TECUs , giving a TEC of $60\cdot 10^{16}$ electrons in the 1 m^2 column. Using this value in (8.4.12) the first order ionospheric delay becomes $q/f^2 \approx 10$ m. Using a geomagnetic field of $B = 5\cdot 10^{-5}$ telsa in (8.4.13) the second-order ionospheric delay is $s/f^3 \approx 1$ cm. In order to evaluate r in (8.4.14) we assume that N_e is uniformly distributed over 100 km, giving $r/f^4 \approx 1$ mm for the third-order delay. Apparently the second-order ionospheric delay is at the level of 1 cm or less, and the third-order effect is at the level of the carrier phase measurement accuracy. Another way at looking at the impact of the large first-order ionospheric term is to realize that a change of 1.12% in TEC causes a single-cycle change in L1 assuming a maximum TEC of 10^{18} electrons.

In many applications dual-frequency observations are used to eliminate the ionospheric effects on the signal. Consider the basic dual-frequency ionospheric-free function (6.1.39),

$$\Phi IF12 \equiv \frac{f_1^2}{f_1^2 - f_2^2}\Phi_1 - \frac{f_2^2}{f_1^2 - f_2^2}\Phi_2 \tag{8.4.16}$$

It is readily verified that the large first-order ionospheric term q/f^2 cancels in this function. However, subjecting the second- and third-order terms to this function shows that neither of them cancels. Therefore, we have reached the important conclusion that only the first-order ionospheric effects cancel in the popular "ionospheric-free function" (8.4.16). The impact of the second-order term is

$$\frac{f_1^2}{f_1^2 - f_2^2} \cdot \frac{-s}{2f_1^3} - \frac{f_2^2}{f_1^2 - f_2^2} \cdot \frac{-s}{2f_2^3} = \frac{s}{2(f_1^2 + f_2^2)\, f_1 f_2} \tag{8.4.17}$$

Taking the same numerical values for TEC and B as used above, the second-order effect is approximately 0.0037 m. A similar computation for the third-order term yields ≈ 0.0008 m

If there are three frequency observations available, one can also compute the function $\Phi IF13$ [simply replace subscripts 2 by 3 in (8.4.16)]. It can then be verified that the function

$$IF_2 \equiv \Phi IF12 - \Phi IF13 \frac{(f_1^2 + f_3^2)\, f_3}{(f_1^2 + f_2^2)\, f_2} \neq f(q, s) \tag{8.4.18}$$

does not depend on the first and second order ionospheric effect since the s term cancels. The factor of $\Phi IF13$ is 0.724 in case of GPS. See Wang et al. (2005) for a method to compute the third-order ionospheric delay given three frequency observations.

Returning to the first-order ionospheric effect, equations (8.4.11) and (8.4.12) give the well-known expression

$$I_{f,P} = \frac{40.30}{f^2} TEC \tag{8.4.19}$$

As mentioned above, TEC of 10^{18} a change of about 1% or 1 TECU in electron content causes a change in range of about one L1 wavelength. The first-order ionospheric defects on two frequencies are related as

$$I_{1,P} = -I_{1,\Phi} = -\frac{c}{f_1} I_{1,\varphi} \tag{8.4.20}$$

$$I_{2,P} = -I_{2,\Phi} = -\frac{c}{f_2} I_{2,\varphi} \tag{8.4.21}$$

$$\frac{I_{1,P}}{I_{2,P}} = \frac{f_2^2}{f_1^2} \tag{8.4.22}$$

$$\frac{I_{1,\varphi}}{I_{2,\varphi}} = \frac{f_2}{f_1} \tag{8.4.23}$$

These relations are convenient when manipulating phase observation. The subscripts P and Φ refer to the pseudorange and scaled carrier phase, respectively, with the numerical values given in linear units of meters. The subscript φ indicates the unit of radians for the ionospheric phase advance. Figure 8.4.2 shows a log-log plot of the ionospheric delay as a function of frequency for various TEC values.

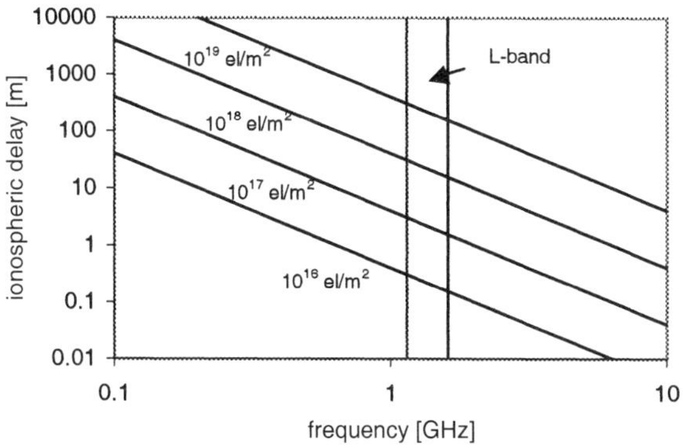

Figure 8.4.2 Ionospheric range correction.

The literature is rich with excellent contributions about ionospheric delay. Most of it can be found searching online resources. Reading this literature one notices that most publications refer to the seminal study by Hartmann and Leitinger (1984) on tropospheric and ionospheric effects for signal frequencies above 100 MHz. Details on higher-order ionospheric effects GPS signals can already be found in Bassiri and Hajj (1993) and Datta-Barua et al. (2006). An excellent review of the subject is provided by Petrie et al. (2011) which also includes an extensive list of references. The latter reference also reviews several approaches to quantify the geometric bending error, or the excess path length, which result from the difference of the geometric receiver-satellite range and the actual path the signal takes under the influence of ionospheric refraction. The bending effect has the same frequency dependency as the third-order ionospheric delay term and its size increases as the elevation angle decreases. Garcia-Fernandez et al. (2013) evaluate various approaches for computing second-order ionospheric corrections, such as ways to obtain TEC values and variations in ionospheric model assumption, and also examine deeper the relationship between network solutions and PPP in regards to these corrections. In general one must keep in mind that the ionospheric impacts on GNSS signals depend on the total electron content which varies with time and location.

Even though phase advancement and group delay are very important, they are not the only manifestations of the ionosphere on the signal propagation. Some of the phase variations are converted to amplitude variation by means of diffraction. The result can be an irregular but rapid variation in amplitude and phase, called scintillation. The signal can experience short-term fading by losing strength. Scintillations might occasionally cause phase-lock problems to occur in receivers. A receiver's bandwidth must be sufficiently wide not only to accommodate the normal rate of change of the geometric Doppler shift, (up to 1 Hz) but also the phase fluctuations due to strong amplitude and phase scintillation. These scintillation effects generally require a minimum receiver bandwidth of at least 3 Hz under severe fading and phase jitter conditions. If the receiver bandwidth is set to 1 Hz to deal with the rate of change of the geometric Doppler shift, and if the ionosphere causes an additional 1 Hz shift, the receiver might lose phase lock.

8.4.2 Ionospheric Function and Cycle Slips

Dual-frequency observations can eliminate the first-order ionospheric effects by forming the popular dual-frequency ionospheric-free functions. Consider the relevant terms of the ionospheric-free function (6.1.41)

$$\frac{f_1^2}{f_1^2 - f_2^2}\,\varphi_1 - \frac{f_1 f_2}{f_1^2 - f_2^2}\,\varphi_2 = \cdots \beta_{12} N_1 - \delta_{12} N_2 + \cdots \tag{8.4.24}$$

and the ionospheric function (6.1.44)

$$\varphi_1 - \frac{f_1}{f_2}\varphi_2 = \cdots + N_1 - \sqrt{\gamma_{12}} N_2 - (1 - \gamma_{12}) I_{1,\varphi} + \cdots \tag{8.4.25}$$

with $\beta_{12} = f_1^2/(f_1^2 - f_2^2)$, $\delta_{12} = f_1 f_2/(f_1^2 - f_2^2)$, and the squared ratio $\gamma_{12} = f_1^2/f_2^2$ as already defined in (6.1.1). Analyzing these dual-frequency carrier phase functions requires extra attention because certain cycle slip combinations on L1 and L2 generate almost identical effects. For example, consider the ionosphere-free phase observable (8.4.24). The ambiguities enter this function not as integers but in the combination of $\beta_{12}N_1 - \delta_{12}N_2$, causing noninteger change when cycle slips occur.

Table 8.4.1 lists in columns 1 and 2 small changes in the ambiguities and shows in columns 3 and 4 their effects on the ionospheric-free and the ionospheric phase

TABLE 8.4.1 Small Cycle Slips on Ionospheric-Free and Ionospheric Functions[a].

ΔN_1	ΔN_2	$\beta_{12}\,\Delta N_1 - \delta_{12}\,\Delta N_2$	$\Delta N_1 - \sqrt{\gamma_{12}}\,\Delta N_2$
±1	±1	±0.562	∓0.283
±2	±2	±1.124	∓0.567
±1	±2	∓1.422	∓1.567
±2	±3	±0.860	∓1.850
±3	±4	∓0.298	∓2.133
±4	±5	±0.264	∓2.417
±5	±6	±0.827	∓2.700
±6	±7	±1.389	∓2.983
±5	±7	±1.157	∓3.983
±6	±8	∓0.595	∓4.267
±7	±9	∓0.033	∓4.550
±8	±10	±0.529	∓4.833

[a]The numbers refer to GPS L1 and L2 frequencies.

functions. Certain combinations of both integers produce almost identical changes in the ionosphere-free phase function. For example, the change of (−7, −9) causes only a small change of 0.033 cycles in the ionospheric-free function. The changes (1, 1) and (8, 10) cause nearly indistinguishable changes of 0.562 and 0.529 cycles. If pseudorange positioning is accurate enough to resolve the ambiguities within three to four cycles, then some of these pairs can be identified. Also, analyzing the ionospheric function does not identify all pairs because several pairs generate the same changes within a couple of tenths of a cycle.

Table 8.4.2 shows an arrangement of integer pairs that have the same effect on the ionospheric function within a couple of hundredths of a cycle. For example, the impact of combinations (−2, −7) and (7, 0) differs by only 0.02 cycle. This amount is too small to be discovered in an observation series since it approaches the level of phase measurement accuracy.

8.4.3 Single-Layer Ionospheric Mapping Function

Although the ionosphere varies in thickness over time and location one often models it as an infinitesimal thin single layer at a certain height above the earth. Through simple geometric relations one derives an ionospheric mapping function which relates the slant TEC and the vertical electron content (VTEC) as a function of the satellites zenith angle. Figure 8.4.3 shows a spherical approximation of the respective geometry with earth radius R. The receiver is located at station k where the satellite appears at zenith angle z_k. The receiver-satellite line of sight intersects the single ionospheric layer, located at height h above the earth, at the ionospheric pierce point (IPP).

TABLE 8.4.2 Similar Effects of Selected Cycle Slips Pairs on the Ionospheric Function[a].

ΔN_1	ΔN_2	$\Delta N_1 - \sqrt{\gamma_{12}}\,\Delta N_2$	ΔN_1	ΔN_2	$\Delta N_1 - \sqrt{\gamma_{12}}\,\Delta N_2$
−2	−7	6.983	7	0	7.000
−2	−6	5.700	7	1	5.717
−2	−5	4.417	7	2	4.433
−2	−4	3.133	7	3	3.150
−2	−3	1.850	7	4	1.867
−2	−2	0.567	7	5	0.583
−2	−1	−0.718	7	6	−0.700
−2	0	−2.000	7	7	−1.983
2	0	2.000	−7	−7	1.983
2	1	0.717	−7	−6	0.700
2	2	−0.567	−7	−5	−0.583
2	3	−1.850	−7	−4	−1.867
2	4	−3.133	−7	−3	−3.150
2	5	−4.417	−7	−2	−4.433
2	6	−5.700	−7	−1	−5.717
2	7	−6.983	−7	0	−7.000

[a]The numbers refer to GPS L1 and L2 frequencies.

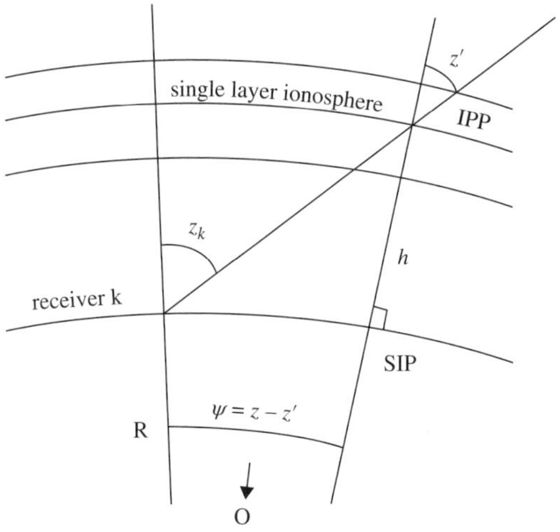

Figure 8.4.3 Spherical approximation of single-layer ionospheric model.

The satellite zenith angle at IPP is z'. Simple geometry shows that the geocentric angle of receiver and IPP is $\psi = z_k - z'$. The subionospheric point (SIP) is located at the intersection of the sphere and the geocentric line though the IPP.

For the sake of deriving the single-layer model let us now think of the ionosphere as a thin band centered at the infinitesimal thin single layer. We assume that the electrons are homogeneously distributed within this thin band. Since the ionospheric delay is proportional to the TEC it is also proportional to the distance traveled through the band. Therefore, we can relate the VTEC and the TEC as

$$VTEC = \cos z' \cdot TEC \tag{8.4.26}$$

Please note that such a simple model was also used to relate the slant and vertical opacity in the previous section. Applying plane trigonometry to the triangle O-k-IPP gives $\sin z' = R \sin z/(R + h)$, then the ionospheric mapping function $F(z_k)$ can be defined as

$$F(z_k) \equiv \frac{1}{\cos z'} = \left[1 - \left(\frac{R \sin z_k}{R + h}\right)^2\right]^{-1/2} \tag{8.4.27}$$

relating the slant TEC and VTEC at the zenith angle z_k as

$$TEC = F(z_k) \cdot VTEC \tag{8.4.28}$$

The location of the SIP can readily be obtained by applying spherical trigonometry to this spherical approximation. Let (φ_k, λ_k) denote the known geodetic latitude and longitude of the receiver. We can construct a spherical triangle whose sides are

$90 - \varphi_k$ and ψ, and enclose the azimuth α_k^p. The law of sine (A.1.1) and the law of cosine for sides (A.1.3) give

$$\sin(\lambda_{IPP} - \lambda_k) = \sin\psi_k \sin\alpha_k^p / \cos\varphi_{IPP} \tag{8.4.29}$$

$$\sin\varphi_{IPP} = \sin\varphi_k \cos\psi + \cos\varphi_k \sin\psi \cos\alpha_k^p \tag{8.4.30}$$

which determine the location of the subionospheric point. Alternative methods exist to compute this location.

8.4.4 VTEC from Ground Observations

The ionosphere can be estimated using the dual-frequency observations. Consider the ionospheric functions (6.1.45) and (6.1.46),

$$I4(t) = \lambda_1 N_1 - \lambda_2 N_2 - (1 - \gamma_{12}) I_{1,P} - d_{12,\Phi} + D_{12,\Phi} + \varepsilon_\Phi \tag{8.4.31}$$

$$I5(t) = (1 - \gamma_{12})\, I_{1,P} - d_{12,P} + D_{12,P} + \varepsilon_P \tag{8.4.32}$$

where $d_{12,\Phi}$ and $D_{12,\Phi}$ denote the interfrequency receiver hardware phase delay and satellite hardware phase delay, respectively, also called the across-frequency hardware delays. The $d_{12,P}$ and $D_{12,P}$ are the respective hardware code delays, called differential code bias (DCB) in Chapter 6. See also the discussion on interfrequency signal correction of Section 6.2.2.2. The pseudoranges should be corrected for these biases or corrections. The multipath term have been omitted from the equations.

Let us consider a continuous satellite arc of observations. During such a time span the receiver and satellite hardware delays can be considered constant. As a first step one needs to fix all cycle slips in (8.4.31). Next we compute the offset

$$\Delta_k^p = \frac{1}{n} \sum_{i=1}^{n} \left(I5_k^p + I4_k^p \right)_i \tag{8.4.33}$$

The summation goes over the n epochs of the arc. Perhaps one might adopt an elevation-dependent weighting scheme in (8.4.33) to take into account the change in measurement accuracy with elevation angle. The computed offset is subtracted from (8.4.31) which can then be modeled as

$$I4_k^p(t) - \Delta_k^p = -(1 - \gamma_{12})\, I_{k,1,P}^p(t) - d_k + D^p \tag{8.4.34}$$

per arc. For clarity we have added the subscript k to denote the receiver and the superscript p to identify the satellite. The terms d_k and D^p are the residual receiver and satellite hardware delays that can be taken as a constant over the time of the arc, but could bias the ionospheric estimates. Using the ionospheric mapping function (8.4.28) to relate the STEC at the receiver to the VTEC at the ionospheric pierce point, we can write

$$I_{k,1,P}^p(\lambda, \varphi, t) = F(z_k)\, I_{k,1,P}(\lambda_{IPP}, \varphi_{IPP}, t) \tag{8.4.35}$$

TABLE 8.4.3 Broadcast Ionospheric Model.

φ_k, λ_k geod. latitude and longitude of receiver [SC]	T = GPS time [sec]
α_k^p, β_k^p azimuth and altitude of satellite [SC]	α_n, γ_n broadcast coefficients

$$F_k^p = 1 + 16\left(0.53 - \beta_k^p\right)^3 \qquad \text{(a)}$$

$$\psi = \frac{0.0137}{\alpha_k^p + 0.11} - 0.022 \qquad \text{(b)}$$

$$\varphi_{IPP} = \begin{cases} \varphi_k + \psi \cos \alpha_k^p & \text{if } |\varphi_{IPP}| \le 0.416 \\ 0.416 & \text{if } \varphi_{IPP} > 0.416 \\ -0.416 & \text{if } \varphi_{IPP} < -0.416 \end{cases} \qquad \text{(c)}$$

$$\lambda_{IPP} = \lambda_k + \frac{\psi \sin \alpha_k^p}{\cos \varphi_{IPP}} \qquad \text{(d)}$$

$$\phi = \varphi_{IPP} + 0.064 \cos(\lambda_{IPP} - 1.617) \qquad \text{(e)}$$

$$t = \begin{cases} \lambda_{IPP} 4.32 \times 10^4 + T & \text{if } 0 \le t < 86400 \\ \lambda_{IPP} 4.32 \times 10^4 + T - 86400 & \text{if } t \ge 86400 \\ \lambda_{IPP} 4.32 \times 10^4 + T + 86400 & \text{if } t < 0 \end{cases} \qquad \text{(f)}$$

$$x = \frac{2\pi(t - 50400)}{P} \qquad \text{(g)}$$

$$P = \begin{cases} \sum_{n=0}^{3} \gamma_n \phi^n & \text{if } P \ge 72000 \\ 72000 & \text{if } P < 72000 \end{cases} \qquad \text{(h)}$$

$$A = \begin{cases} \sum_{n=0}^{3} \alpha_n \phi^n & \text{if } A \ge 0 \\ A = 0 & \text{if } A < 0 \end{cases} \qquad \text{(i)}$$

$$I_{k,1,P}^p = \begin{cases} c\, F_k^p \left[5 \times 10^{-9} + A\left(1 - \frac{x^2}{2} + \frac{x^4}{24}\right)\right] & \text{if } |x| < 1.57 \\ c\, F_k^p (5 \times 10^{-9}) & \text{if } |x| > 1.57 \end{cases} \qquad \text{(j)}$$

Conversion of SC unit: 1 SC = 180°

One could use the mapping function (8.4.27), or the one of Table 8.4.3, or one that is based on a realistic electron density profile model such as the extended slab density model by Coster et al. (1992).

Next, a function is needed that expresses the vertical ionospheric delay as a function of latitude and longitude. A simple representation could be a spherical harmonic expansion similar to (4.3.31),

$$VTEC(\varphi, \Delta\lambda) = \sum_{n=0}^{n_{\max}} \sum_{m=0}^{n} (C_{nm} \cos m\Delta\lambda + S_{nm} \sin m\Delta\lambda) \overline{P}_{nm}(\sin \varphi_k) \qquad (8.4.36)$$

where (φ, λ) is the point for which the VTEC is needed, $\Delta\lambda = \lambda - \lambda_0$, λ_0 is the longitude of the mean sun, and $\overline{P}_{nm}$ are the associated Legendre functions. The spherical harmonic coefficients (C_{nm}, S_{nm}) represent a parameterization of the global VTEC field. Since the TEC varies with time, even in a sun-referenced frame, such coefficients are only valid for a certain period of time before they would have to be updated. There are other parameterizations possible. For example, Mannucci et al. (1998) divides the surface of the earth into tiles (triangles) and estimates the vertical TEC at the vertices. Only observations that fall within the triangle are used to estimate the TEC at the vertices of that triangle, assuming that the TEC varies linearly within the triangle. The IGS GIM is available on a geographic latitude and longitude grid for certain epochs. Computing the VTEC at a certain location and time requires then a spatial and temporal interpolation.

Conceptually the spherical harmonic coefficients can be estimated as follows: Express the observations $I4_k^p$ as a function of spherical harmonic coefficients by substituting (8.4.35) into (8.4.34) , use (8.4.19) to convert vertical ionospheric delay to VTEC. In practice one prefers a parameterization in the geomagnetic sun-fixed frame because the TEC values depend the least on time in that frame.

Since the instrumental biases d_k and D^p are geometry independent, but the ionospheric delay depends on the azimuth and elevation of the satellite, the biases and the ionospheric parameters are estimable. However, the receiver and satellite hardware delays cannot be estimated separately unless either a receiver or satellite is introduced as a reference. Alternatively, one could impose the constraint $\Sigma D = 0$ (zero-mean reference), or follow Sardón et al. (1994) who do not combine the ionospheric pseudorange and carrier phase equations but lump together the ambiguities and the hardware phase delays and distribute the effect of hardware delays among other terms. Mannucci et al. (1998) avoid singularity by combing d_k and D^p and estimating only one constant per arc, called "phase-connected" arc of data.

8.4.5 Global Ionospheric Maps

The term global ionospheric map (GIM) refers to a mathematical expression or a set of data files to allow computing of the VTEC or the vertical ionospheric delay at any location on the earth at a specific instant of time. Equation (8.4.36) is an example.

8.4.5.1 IGS GIMs When thinking of a global ionospheric map the IGS products most likely come to mind first. These products were already introduced in Section 6.2.7.1. They generally serve as a standard for comparisons because of their high accuracy, and are a product of extensive international cooperation. Each of the IGS associate centers located around the world independently computes a global VTEC distribution from GNSS observations and shares its solution and related data with an associate combination center to compute the final IGS GIM. The data transfer is facilitated by an especially designed standard data structure called IONEX (ionospheric exchange format).

8.4.5.2 International Reference Ionosphere The International Reference Ionosphere (IRI) results from efforts of the Committee on Space and Research (COPSAR) and the International Union of Radio Science (URSI). The website http://iri.gsfc.nasa.gov/ provides background information on the IRI and allows users to enter input data and instantly receive results in digital or graphical form. It uses a variety of input data. In addition to monthly averages of electron density the IRI provides much information about the ionosphere that is of interest to the ionospheric specialist rather than the geodesist, such as electron temperature, ion temperature, and ion composition in the altitude range of 60 to1500 km. However, with assimilation of GNSS observation the IRI model becomes not only better and also more responsive to short-term phenomena. On the other side, the GNSS community uses the wealth of information that the IRI model provides to investigate the most optimal height of the single ionospheric shell for relating VTEC and STEC. The IRI model is data driven, i.e., it is an empirical model. The current version is IRI-2011. To learn more about the status of this model and its expected future development the reader is referred to Bilitza et al. (2011).

8.4.5.3 GPS Broadcast Ionospheric Model In order to support single-frequency positioning, the GPS broadcast message contains eight ionospheric model coefficients for computing the ionospheric group delay along the signal path. The respective algorithm was developed by Klobuchar (1987) and is listed in Table 8.4.3. See also IS-GPS-200G (2012, p. 123) or Klobuchar (1996). In addition to the broadcast coefficients, other input parameters are the geodetic latitude and longitude of the receiver, the azimuth and elevation angle of the satellite as viewed from the receiver, and the time. Note that several angular arguments are expressed in semicircles (SC). All auxiliary quantities in the middle portion of the table can be computed one at a time starting from the top. The function in the third part of the table has been multiplied with the velocity of light, in order to yield the slant group delay directly in meters. The algorithm presented here compensates about 50 to 60% of the actual group delay.

The Klobuchar algorithm is based on the single-layer model of the ionosphere. As discussed above, the assumption is that the TEC is concentrated in an infinitesimally thin spherical layer at a certain height, e.g., 350 km in this case. The model further assumes that the maximum ionospheric disturbance occurs at 14:00 local time. The mapping function F and other expressions are approximations to reduce computational complexity but are still of sufficient accuracy to meet the purpose of the algorithm. Most of the symbols have the same meaning as used above, e.g., the geodetic latitude and longitude of the ionospheric pierce point and the receiver are $(\varphi_{IPP}, \lambda_{IPP})$ and (φ_k, λ_k), respectively, and the geocentric receiver-IPP angle is ψ. The geomagnetic latitude of the ionospheric pierce point is ϕ, t is the local time, P is the period in seconds, x is the phase in radians, and A denotes the amplitude in seconds.

8.4.5.4 NeQuick Model NeQuick (Radicella, 2009) is a three-dimensional time-dependent global electron density model. NeQuick contains an analytical representation of the vertical profile of electron density, with continuous first derivative. It takes the characteristics of the various ionospheric layers such as location and thickness into account. The input of NeQuick is the positions of receiver and satellite, time, and ionization parameters such as the solar radio flux F10.7 or monthly smoothed sunspot numbers. It calculates the electron content at various places along the receiver-satellite path. The STEC follows from numerical integration. The NeQuick model has been adopted by Galileo for single-frequency users. There is especially a good motivation for adopting a high-quality ionospheric model for single-frequency Galileo users because the Galileo global navigation satellite system will feature the highly precise E5 AltBOC signal for precision code range measurements. According to OS-SIS-CD (2010) it is assumed that the model corrects 70% of the ionospheric delay in the Galileo frequency range. Each satellite transmits as part of the navigation message three ionization parameters for computing an effective ionization level parameter which replaces the solar flux input parameter F10.7. Schüler (2014) used the NeQuick model to attempt to derive an improved mapping function that would perform better at lower elevation angles than the standard single-layer mapping function.

8.4.5.5 Transmission to the User Various methods are in use to make the VTEC data available to the user. Users of IGS products can obtain the data via the Internet. In many applications it is sufficient to use the ionospheric information included with the satellite broadcast navigation message; examples are the streams based on the Klobuchar model in case of GPS or the NeQuick model in case of Galileo. Each of the GNSS satellite systems broadcasts such data. The situation is similar for Satellite Augmentation Systems (SBAS) such as the Wide Area Augmentation System (WAAS) in the United States, the European Geostationary Overlay Service (EGNOS), and the GPS aided GEO augmentation system (GAGAN). Each augmentation service draws observation from a network of reference stations, processes them at a center, and uploads the VTEC information typically to a geostationary satellite for rebroadcasting to the user.

Finally, for accurate relative positioning in surveying over short distances with carrier phase observations and ambiguity fixing, single-frequency users still depend on the elimination of ionospheric effects through across-receiver or double differencing, as mentioned in Sections 6.1.2.1 and 6.1.3.

CHAPTER 9

GNSS RECEIVER ANTENNAS

The receiver antenna is the first block in a chain of signal transformations to convert the signals emitted by the satellites into useful data. Some of the antenna features define the currently achievable accuracy of positioning. However, traditionally, the basics of applied electromagnetics are omitted in courses on satellite geodesy. To compensate for that, the first six sections address fundamentals of the electromagnetic field and antenna theory with focus on antennas useful for precise positioning.

In Section 9.1, plane and spherical electromagnetic waves of different types of polarization are discussed and widely used reference sources of radiation—a Hertzian and a half-wave dipole—are introduced. This allows treating the radiation of any practical antenna as interference of radiation of Hertzian dipoles, and in doing so, coming to a unified field representation that is valid for any arbitrary antenna. In addition, the widely used complex notation for time harmonic signals and dB scale are explained is this section.

Section 9.2 discusses antenna directivity and gain. The discussions on antenna pattern are supplemented with examples of a perfect antenna for satellite positioning, base station antennas, and rover antennas. It is shown that the cause of the largest error in regard to positioning is the conflict between antenna gain for low elevation satellites and the ability of an antenna to suppress multipath reflections associated with these satellites. The section ends with estimating the effective area of a typical GNSS receiving antenna.

Discussions of antenna phase center, phase center variations, and antenna calibrations are provided in Section 9.3. In general, an antenna phase center can be defined differently depending on the application. The expression "adopted in GNSS practice" refers to the averaging of deviations of position over a very long observation session; these deviations are caused by carrier phase delays and advances introduced by antenna phase pattern.

Section 9.4 is dedicated to multipath. What is commonly referred to as multipath is identified as a particular case of the broad area of diffraction phenomenon; spatial spreads of this phenomenon are estimated on the basis of Fresnel zones. Diffraction over a half-plane is used as an example since it allows for a complete analytical treatment. This example illustrates types of errors that occur in the transition from free line of sight to deep shadowing of satellite by obstacles. Multipath reflections from different kinds of soils are discussed in this section with a focus on generating a left-hand circular polarized signal. The antenna down-up ratio is introduced, and typical multipath-induced behavior of carrier phase residuals normally observed with positioning is examined.

Section 9.5 is a brief introduction to transmission line theory and practice. We explain why power transmission at GNSS frequencies requires different approaches as compared to low-frequencies transmission. The wave impedance of a line, antenna mismatch, and voltage standing-wave ratio (VSWR) terms are introduced.

Section 9.6 concludes the general overview of the GNSS antenna area. Electromagnetic noise reception from outer space, noise generation, and signal and noise propagation through electronic circuitry are analyzed. Additionally, the role of a low-noise amplifier is highlighted. The signal-to-noise ratio normally observed at the GNSS receiver output is estimated.

Finally, Section 9.7 is an overview of practical GNSS antenna designs. We begin with engineering formulas to estimate a common patch antenna performance. We continue with variants of patch antennas with artificial metal substrates that are useful for broadband GNSS applications. Then we turn to ground planes commonly used for multipath mitigation, including flat metal, impedance, and semitransparent ground planes. Special attention is given to antennas with a cutoff pattern required to achieve millimeter precision of positioning in real time. Antenna samples and relevant limitations regarding antenna size are discussed. The section concludes with array antennas and antenna manufacturing items. The majority of antenna samples presented in this section is based on the developments done by D. Tatarnikov and his colleagues of the Topcon Technology Center. Multiple references are provided for a broader view of GNSS antenna technology.

Several appendices contain supplementary material to support this chapter. The appendices mainly include concise mathematical formulations, often in regard to fundamental principles, that help one understand the roots of the material presented in this chapter. Appendix D deals with the basics of vector calculus. Although this material is found in advanced books on mathematics, the author feels that it is important to provide this summary for the sake of easy reference. Appendices E and F provide details about electromagnetic basics involved. Appendices G through I contain details about electromagnetic analysis for certain antenna types.

The notation used in this chapter differs in many cases from the notation used in other parts of the book. Vectors are indicated by an arrow placed atop a uppercase letter, such as $\vec{E}$, $\vec{H}$. This notation is typical with the antenna area and helps to differentiate physical vectors in three-dimensional space with mathematical multidimensional vectors widely used in other chapters. Unit vectors are indicated by lowercase letters and a subscript zero like in $\vec{x}_0, \vec{y}_0$. Coordinate frames are employed for certain electromagnetic situations. In most cases the coordinate frame is associated with the antenna under consideration. The frame in most cases is local to the electromagnetic problem and is not related to geodetic frames, as is typically discussed in other parts of the book.

9.1 ELEMENTS OF ELECTROMAGNETIC FIELDS AND ELECTROMAGNETIC WAVES

The principle quantities describing the electromagnetic field and the most important equations are discussed in this section. The goal is to touch upon basics which are required to characterize antennas used with GNSS receivers. For an in-depth treatment of engineering electromagnetics, the reader is referred to the classical reference Balanis (1989). A compact overview of antenna theory and antenna parameters could be found in the handbook by Lo and Lee (1993).

9.1.1 Electromagnetic Field

A scalar field in general is a scalar quantity distributed in space and time. It is convenient to consider a number of sensors located at different points in space measuring a physical quantity u. Introducing Cartesian coordinates (x, y, z) to mark the positions of sensors, one writes the scalar field u observed at a time instant t in the form

$$u = u(x, y, z, t) \tag{9.1.1}$$

A common example of a scalar field is temperature distribution in a room. Temperature could be different at different points and could vary over time.

A vector is a quantity that has both magnitude and direction. A typical example would be the wind speed, as it may vary from point to point in space and over time. The vector field as a function of space coordinates and time is introduced by writing

$$\vec{A} = \vec{A}(x, y, z, t) \tag{9.1.2}$$

In order to describe vector quantities, it is useful to project the vectors onto a coordinate system. For Cartesian coordinates one has

$$\vec{A}(x, y, z, t) = A_x(x, y, z, t)\vec{x}_0 + A_y(x, y, z, t)\vec{y}_0 + A_z(x, y, z, t)\vec{z}_0 \tag{9.1.3}$$

We mark unit vectors of Cartesian coordinates as $\vec{x}_0, \vec{y}_0, \vec{z}_0$. Other coordinate systems often used are spherical or cylindrical. Details are shown in Appendix D. Equation (9.1.3) represents a sum of projections multiplied by the corresponding unit vectors. Each projection generally is a function of space coordinates and time.

The examples mentioned above were related to some medium or "matter." The electromagnetic field does not require any "matter." In today's physics the electromagnetic field is considered another form of substance (compared to matter). It is capable of transporting energy across the space in the form of electromagnetic waves. In general, an electromagnetic field is a plurality of four vector fields: electric field intensity $\vec{E}$, electric flux density $\vec{D}$, magnetic field intensity $\vec{H}$, and magnetic flux density $\vec{B}$. The International System of Units (SI) is used throughout the chapter. The vector units in the SI system are: $\vec{E}$ (volts/meter), $\vec{D}$ (coulombs/ square meter), $\vec{H}$ (amperes/meter), and $\vec{B}$ (webers/square meter). For a majority of media that relate to the antenna area the vector $\vec{D}$ is strictly proportional to $\vec{E}$, and $\vec{B}$ is proportional to $\vec{H}$. Free space is the area where there is no matter. For vectors in free space one writes

$$\vec{D} = \varepsilon_0 \vec{E} \tag{9.1.4}$$

$$\vec{B} = \mu_0 \vec{H} \tag{9.1.5}$$

The parameters ε_0 and μ_0 are called absolute permittivity and permeability of free space. Using SI units

$$\varepsilon_0 = \frac{1}{36\pi} 10^{-9} \qquad \text{[farads/meter]} \tag{9.1.6}$$

$$\mu_0 = 4\pi 10^{-7} \qquad \text{[henries/meter\}} \tag{9.1.7}$$

To avoid mistakes one is to note that, as mentioned, with free space "there is no matter"; it would be incorrect to consider free space as having some kind of permittivity and permeability. The parameters ε_0 and μ_0 do not have specific physical meaning. Their appearance and numerical value comes from the use of SI units.

Electrical and magnetic properties of media are characterized by a relative (to that of free space) dielectric permittivity, ε, relative magnetic permeability, μ, and conductivity of the medium, σ. The terms dielectric and magnetic constants are also in use for the ε and μ, respectively. The parameters ε and μ are dimensionless, parameter σ is measured in units of 1/(ohms times meters). For a particular medium one writes

$$\vec{D} = \varepsilon_0 \varepsilon \vec{E} \tag{9.1.8}$$

$$\vec{B} = \mu_0 \mu \vec{H} \tag{9.1.9}$$

Conductivity σ establishes a relationship between electric field intensity $\vec{E}$ and conduction currents induced in the media. This is known as a differential form of Ohm's law. We will not use this material in explicit form in this chapter and refer the reader to Balanis (1989) for details. For the free space one takes $\varepsilon = \mu = 1$ and $\sigma = 0$. As to electromagnetics related to GNSS antenna, the majority of media is nonmagnetic and $\mu = 1$ could be accepted in most cases except for atmospheric plasma and special materials like ferrites. Typical values of ε and σ for selected media are listed in Table 9.1.1. The data is taken partly from Balanis (1989) and partly from Nikolsky (1978). One notes that dielectric permittivity ε ranges from unity up to several dozen. The conductivity σ covers 24 orders in magnitude starting from 10^{-17} for insulators and up to 10^7 for metals. The first and the last row in Table 9.1.1 indicate two useful limiting cases. The first one is called a perfect conductor with infinitely large

TABLE 9.1.1 Permittivity, Conductivity, and Loss Factor of Selected Media

Material	ε	$\sigma\ [\Omega\ \text{m}]^{-1}$	$\tan \Delta^e$
Perfect conductor	1	∞	∞
Copper	1	5.8×10^7	7×10^8
Gold	1	4×10^7	4.8×10^8
Aluminum	1	3.5×10^7	4.2×10^8
Iron	1	1×10^7	1.2×10^8
Seawater	80	$1 \ldots 4$	0.3
Natural freshwater	80	$10^{-3} \ldots 2.4 \times 10^{-2}$	7.5×10^{-4}
Wet soil	10–30	$3 \times 10^{-3} \ldots 3 \times 10^{-2}$	0.02
Dry soil	3–6	$1 \times 10^{-5} \ldots 2 \times 10^{-3}$	5×10^{-4}
Marble	8	$10^{-7} \ldots 10^{-9}$	1.5×10^{-10}
Quartz	4	2×10^{-17}	
Air	1.0005	0	0
Perfect insulator	ε	0	0

conductivity. This is a reasonable model for most metals at GNSS frequencies. The last one is a perfect insulator possessing some permittivity ε and zero conductivity. The last column in Table 9.1.1 shows a quantity referred to as dielectric loss factor. This quantity is described in Section 9.1.3.

Relationships (9.1.8) and (9.1.9) allow avoiding the use of vectors $\vec{D}$ and $\vec{B}$ for the majority of GNSS antenna-related considerations and allow characterizing the electromagnetic field by vectors $\vec{E}$ and $\vec{H}$. The relations of electric field $\vec{E}$ and magnetic field $\vec{H}$, while both vary in time and space, are the contents of the Maxwell equations, formulated by James C. Maxwell in the last third of the nineteenth century. These equations are one of the cornerstones of knowledge of the world around us. For the moment we are looking into an area where there are no sources of radiation, such as between the transmitting and receiving antennas. For the homogeneous nonconductive medium with parameters ε and μ the Maxwell equations read

$$\text{rot}\,\vec{H} = \varepsilon\varepsilon_0 \frac{\partial \vec{E}}{\partial t} \tag{9.1.10}$$

$$\text{rot}\,\vec{E} = -\mu\mu_0 \frac{\partial \vec{H}}{\partial t} \tag{9.1.11}$$

The details of partial differentiation of vectors and of the use of the rot operator are provided in Appendix D.

9.1.2 Plane Electromagnetic Wave

The plane electromagnetic wave is one of the most basic and simplest solutions to equations (9.1.10) and (9.1.11). Let both vectors $\vec{E}$ and $\vec{H}$ vary along a certain direction in space. Align the z axis of the coordinate frame with this direction. It is assumed there are no variations of the fields in the x and y directions. The medium is taken as unbounded free space with $\varepsilon = \mu = 1$. A general representation of a linear polarized plane electromagnetic wave propagating in free space in z direction is

$$\vec{E} = E_0 u(ct - z)\vec{x}_0 \tag{9.1.12}$$

$$\vec{H} = \frac{E_0}{\eta_0} u(ct - z)\vec{y}_0 \tag{9.1.13}$$

Here

$$\eta_0 = \sqrt{\frac{\mu_0}{\varepsilon_0}} = 120\pi \qquad [\text{ohm}] \tag{9.1.14}$$

is referred to as an intrinsic impedance of free space and

$$c = \frac{1}{\sqrt{\varepsilon_0 \mu_0}} \approx 3 \cdot 10^8 \qquad [\text{m/ sec}] \tag{9.1.15}$$

is velocity of light in free space. With (9.1.12) and (9.1.13) E_0 is constant and $u(s)$ is a wave profile. Both could be arbitrary and set up by the source. In order to prove (9.1.12) and (9.1.13) one is to substitute these expressions into (9.1.10) and (9.1.11), make use of the expression (D.12) of Appendix D taking $\partial/\partial x = \partial/\partial y = 0$, and employ the relationship $\partial u(ct - z)/\partial t = -c\partial u(ct - z)/\partial z$.

The main features of the wave are illustrated in Figure 9.1.1. Let the initial distribution of fields along z axes at a time instant $t = 0$ be shown as solid lines at the left panel. Let vector fields at a point z_a be shown by thick arrows. Within the time increment Δt the distribution will move along the z axes by the distance $c\,\Delta t$. Fields indicated by thick arrows will now be observed at point $z_a' = z_a + c\,\Delta t$. This is shown

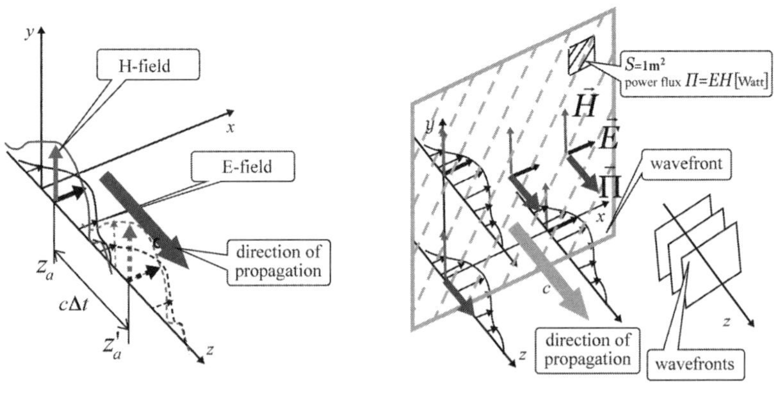

Figure 9.1.1 Fields of a plane wave propagating along z axis.

by dotted lines and arrows. The parameter c of (9.1.15) is the velocity of this motion. In short, one says that it is a vector field distribution that moves in space.

Furthermore, with expressions (9.1.12) and (9.1.13) the vectors $\vec{E}$ and $\vec{H}$ are perpendicular to each other and are arranged in such a way that if looking along the direction of travel the rotation from $\vec{E}$ to $\vec{H}$ is seen as clockwise. One says that these vectors constitute a right-hand triad with the direction of wave travel. Vectors $\vec{E}$ and $\vec{H}$ do not have projections onto the direction of travel. Such a wave is called a transverse wave. The vectors $\vec{E}$ and $\vec{H}$ are directly proportional to each other in terms of magnitude with the intrinsic impedance η_0 in (9.1.14) being a proportionality coefficient. In SI units this coefficient is measured in ohms. In this regard we note that it is incorrect to take the intrinsic impedance as a kind of "free space resistivity" to the wave travel. Here again, similar to the discussion of ε_0 and μ_0, in free space there is no matter to provide resistance. The value and dimensions of the free space intrinsic impedance comes from the SI units.

Expressions (9.1.12) and (9.1.13) indicate that there is no dependence of fields on x and y coordinates as was previously mentioned. However, these expressions are not to be considered as an optical ray focused around the z axis. Instead, an electromagnetic plane wave defined by expressions (9.1.12) and (9.1.13) is a distribution of fields occupying the entire space in such a way that along any line parallel to the z axes these distributions are identical. In other words, each of two vectors $\vec{E}$ and $\vec{H}$ has the same magnitude and direction at any point in a plane perpendicular to the z axis. This is illustrated further in the right panel in Figure 9.1.1. In general, a surface of all the points for which vectors $\vec{E}$ (and $\vec{H}$) are identical is called a wavefront. Wavefronts associated with the wave of (9.1.12) and (9.1.13) are planes, which is why such a wave is called a plane one. One may view the process of wave travel as a plurality of wavefronts moving across space with the velocity of light. This is illustrated at the bottom right corner of the figure. Each front has a certain value of vectors $\vec{E}$ and $\vec{H}$ associated with it.

An electromagnetic wave transports energy. There is a certain flux of power through each 1m^2 of an imaginary plane perpendicular to the direction of wave travel The power flux density is characterized by the Poynting vector

$$\vec{\Pi} = [\vec{E}, \vec{H}] \quad [\text{watts/square meter}] \tag{9.1.16}$$

Here and further on in this chapter a cross-product of two vectors is indicated by brackets like $[\cdot]$. The magnitude of the Poynting vector equals the power per 1 m^2 of a front. The direction of this vector points in the power flux direction. For a plane wave it coincides with the direction of the wave travel. The Poynting vector is the same at any point of a plane wavefront.

Finally, the orientation of vectors $\vec{E}$ and $\vec{H}$ with respect to x-y coordinates in general could be arbitrary but in such a way that they constitute a right-hand triad with the direction of wave travel. Thus, the relation

$$\vec{E} = \eta_0[\vec{H}, \vec{z}_0] \tag{9.1.17}$$

holds. Here $\vec{z}_0$ points in the direction of wave motion.

A particular but important case of a plane wave occurs when the fields exhibit time harmonic alternation with fixed frequency f measured in units of [hertz]. The frequency is related to the period of alternation in time T in seconds as

$$f = \frac{1}{T} \tag{9.1.18}$$

Along with frequency f an angular frequency ω (also referred to as circular frequency) is in use. The angular frequency is measured in radians per second and is related to f and T as

$$\omega = 2\pi f = \frac{2\pi}{T} \tag{9.1.19}$$

A time harmonic plane wave in free space satisfies the general expressions (9.1.12) and (9.1.13), but the profile u takes the form

$$u = \cos(\omega t - kz + \psi_0) \tag{9.1.20}$$

Here

$$k = \omega\sqrt{\varepsilon_0 \mu_0} \tag{9.1.21}$$

is called a wavenumber or a propagation constant, and the argument of the cosine function in (9.1.20) is called an instantaneous phase,

$$\psi(z, t) = \omega t - kz + \psi_0 \tag{9.1.22}$$

with ψ_0 being an initial phase observed at a point $z = 0$ at time instant $t = 0$. The instantaneous phase (9.1.22) is constant across any plane $z = \text{const}$, that is, across any wavefront.

The process of wave motion in the z direction is illustrated at the left panel in Figure 9.1.2 for the vector $\vec{E}$.

The distribution of vector $\vec{H}$ is similar and perpendicular to $\vec{E}$. Let ψ_0 in (9.1.22) be zero. Then at a time instant $t = 0$ the instantaneous phase (9.1.22) at a point $z = 0$

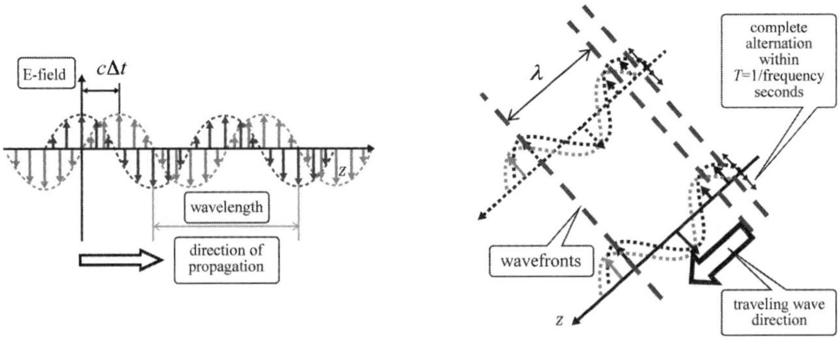

Figure 9.1.2 Time harmonic plane wave propagating along z axis.

would be zero. With time increment Δt this value would move to the point Δz such that

$$\omega \Delta t - k \Delta z = 0 \tag{9.1.23}$$

The velocity of this motion,

$$v_p = \frac{\Delta z}{\Delta t} = \frac{\omega}{k} = c = \frac{1}{\sqrt{\varepsilon_0 \mu_0}} \tag{9.1.24}$$

is called a phase velocity. For free space it equals the velocity of light (9.1.15). Combining (9.1.21) and (9.1.24) one may rewrite (9.1.20) in the form

$$u = \cos\left(\omega\left(t - \frac{z}{v_p}\right) + \psi_0\right) \tag{9.1.25}$$

This expression shows that with z advancing, the wave field exhibits a phase delay equal to the product of ω and time interval z/v_p. The latter is required to cover distance z when traveling with the velocity v_p.

As seen with expression (9.1.20) and Figure 9.1.2 the fields of a time harmonic plane wave exhibit periodicity not only in time but also in space. The spatial period λ is defined as a distance along the direction of wave motion which causes 2π phase delay to the field intensities. Thus

$$(\omega t - k(z + \lambda) + \psi_0) - (\omega t - kz + \psi_0) = -2\pi \tag{9.1.26}$$

yielding

$$\lambda = \frac{2\pi}{k} \tag{9.1.27}$$

The distance λ is called a wavelength. Making use of (9.1.21), (9.1.24) and (9.1.19), one may write

$$\lambda = v_p T \tag{9.1.28}$$

This gives another definition of wavelength. Namely, a wavelength is such a distance that a wavefront, associated with some certain value of the fields, covers within a time interval equal to the period T. Another useful expression for the wavelength would be

$$\lambda = \frac{v_p}{f} \tag{9.1.29}$$

For free space this comes to

$$\lambda = \frac{c}{f} \tag{9.1.30}$$

or

$$\lambda \quad [\text{cm}] = \frac{30}{f\,[\text{GHz}]} \tag{9.1.31}$$

This last expression gives a convenient rule for calculating a free space wavelength.

With the wavelength given, it is convenient to invert (9.1.27) and write

$$k = \frac{2\pi}{\lambda} \tag{9.1.32}$$

Using (9.1.32), it is instructive to rewrite (9.1.20) in the form

$$u = \cos\left(2\pi\left(\frac{t}{T} - \frac{z}{\lambda}\right) + \psi_0\right) \tag{9.1.33}$$

This clearly indicates the periodicity in time and space. The right panel of Figure 9.1.2 illustrates a spatial distribution of the E field. Wavefronts perpendicular to the drawing plane are shown as dashed lines. It is worth mentioning that the H field exhibits same alternation and is in-phase with the E field. The $\vec{H}$ vector is aligned parallel to the y axes perpendicular to the drawing.

The expression for the Poynting vector (9.1.16) also holds true with the time harmonic wave. It provides instantaneous value of power flux density at any point in space and any instant of time. Using (9.1.16) with (9.1.12), (9.1.13), and (9.1.20) one has

$$\vec{\Pi} = \frac{|E_0|^2}{\eta_0}\cos^2(\omega t - kz + \psi_0)\vec{z}_0 = \frac{1}{2}\frac{|E_0|^2}{\eta_0}(1 + \cos(2(\omega t - kz + \psi_0)))\vec{z}_0 \tag{9.1.34}$$

A power flux averaged over the period of an alternation is of interest in most cases. Averaging over the period T yields

$$\tilde{\vec{\Pi}} = \frac{1}{T}\int_0^T \vec{\Pi}\, dt = \frac{1}{2}\frac{|E_0|^2}{\eta_0}\vec{z}_0 \tag{9.1.35}$$

This shows that time-averaged power flux density associated with a time harmonic plane wave is the same at any point in space and at any instant in time.

If the wave propagates not in free space but rather in some medium with parameters ε, μ, then expressions (9.1.12), (9.1.13), and (9.1.20) hold true, but instead of k and η_0 one is to use

$$k_m = \omega\sqrt{\varepsilon\mu\varepsilon_0\mu_0} \tag{9.1.36}$$

$$\eta_m = \sqrt{\frac{\mu}{\varepsilon}}\eta_0 \tag{9.1.37}$$

We use subscript m (medium) to distinguish these parameters with free space. Following the derivations of (9.1.24), one concludes that the phase velocity takes the form

$$v_{p;m} = \frac{c}{\sqrt{\varepsilon\mu}} \tag{9.1.38}$$

It is one $\sqrt{\varepsilon\mu}$th of the velocity of light in free space. Hence the wavelength (9.1.28) will also be less by the factor $\sqrt{\varepsilon\mu}$ as compared to that of free space,

$$\lambda_m = \frac{\lambda}{\sqrt{\varepsilon\mu}} \tag{9.1.39}$$

The quantity η_m in (9.1.37) is called an intrinsic impedance of the medium. It differs with that of free space by the factor $\sqrt{\mu/\varepsilon}$. Regarding this term a note should be made similar to the one made for free space impedance. At the moment we are considering a nonconductive lossless medium, it would be incorrect to think that such a medium provides a kind of "resistance" to the wave travel. Intrinsic impedance (9.1.37) is just a proportionality coefficient between electric and magnetic field intensities. In Section 9.1.3 we will discuss the wave traveling across a lossy medium.

The just-discussed time harmonic plane wave is an idealization to the real electromagnetic process. This wave occupies the entire space and lasts for unlimited time. First, using the plane wave model would be justified in Section 9.1.4 while discussing spherical waves radiated by an actual source. In regards to the latter, one is to consider a group velocity term.

Any time-limited signal could be represented by a plurality of time harmonic alternations. This is done by Fourier transform (Poularikas, 2000). Let us take two harmonics with close frequencies ω_1, ω_2 such that the difference $\omega_2 - \omega_1 = 2\Delta\omega << \omega_{1,2}$. We assume that the amplitudes of the harmonics are the same at both frequencies. Let $k_{1,2}$ be the wavenumbers associated with frequencies ω_1, ω_2 and let $\psi_{1,2}$ be the initial phases. For the total electric field distribution in space and time one has

$$\begin{aligned} E &= E_0(\cos(\omega_1 t - k_1 z + \psi_1) + \cos(\omega_2 t - k_2 z + \psi_2)) \\ &= 2E_0 \cos(\Delta\omega t - \Delta k z + \Delta\psi)\cos(\omega t - kz + \psi) \end{aligned} \tag{9.1.40}$$

Here $\omega_{1,2} = \omega \pm \Delta\omega$, $k_{1,2} = k \pm \Delta k$, and $\psi_{1,2} = \psi \pm \Delta\psi$. Expression (9.1.40) shows that the total field could be viewed as a plane wave with angular frequency ω and wavenumber k. This wave is called a carrier. The amplitude of this wave exhibits slow variations in time and space following the envelope $\cos(\Delta\omega t - \Delta k z - \Delta\psi)$. This envelope moves in a positive z direction with a velocity equal to

$$v_g = \frac{\Delta\omega}{\Delta k} \tag{9.1.41}$$

This velocity is recognized as an envelope or a signal velocity, and is called a group velocity. It is proven in Balanis (1989) that the group velocity is also the velocity of the power. Taking the limit $\Delta\omega \to 0$, one rewrites (9.1.41) in the form

$$v_g = \frac{d\omega}{dk} \tag{9.1.42}$$

For the free space case, making use of expressions (9.1.21) and (9.1.24), one has $v_g = v_p = c$ and both group and phase velocities equal to that of light. Further, a medium with ε, μ being constant over a frequency range of interest is referred to as nondispersive. From (9.1.36) and (9.1.38) one has $v_g = v_p = c/\sqrt{\varepsilon\mu}$ for such a medium. The nondispersive medium just decreases the velocities by a factor of $\sqrt{\varepsilon\mu}$. Finally, if a medium exhibits a permittivity or permeability variations as a function of frequency, it is called a dispersive medium. For a dispersive medium the phase and group velocities are different. A typical example is ionospheric plasma (this was discussed in Chapter 8).

We conclude this section with an important remark. The GNSS signals spectrum is shown schematically in Figure 5.4.1. From the point of view of antenna techniques the details on signal power distribution over the spectrum are not of much interest, instead, a frequency band defined by the lower and upper frequency of each signal spectrum is. By absolute bandwidth Δf one refers to the difference between upper frequency of the band f_2 and lower frequency f_1,

$$\Delta f = f_2 - f_1 \tag{9.1.43}$$

However, with electromagnetics related to antenna area and signal propagation, not an absolute but rather a relative bandwidth δf is important. This is the bandwidth Δf related to the center frequency of a band f_0. Relative bandwidth is normally expressed as a percentage,

$$\delta f = \frac{\Delta f}{f_0} 100\% \tag{9.1.44}$$

with

$$f_0 = \frac{f_2 + f_1}{2} \tag{9.1.45}$$

For instance, using Figure 5.4.1., the relative bandwidth of the GPS L2 signal is about 2%. One may check that the relative bandwidths of all the other GNSS signals are also a low percentage. With electromagnetics, signals with relative bandwidth of a few percent are referred to as narrowband. This is not to be confused with communications terminology where GNSS signals are referred to as broadband due to their pseudorandom noise structure.

The electromagnetic properties of the majority of media related to propagation of GNSS signals do not exhibit vast variations over frequency. Thus, the reason to point out narrowband signals is that topics like signal propagation and reflection can be discussed assuming the signal to be a perfect time harmonic with fixed frequency equal to (9.1.45) or, equivalently, the carrier frequency. This simplifies the analysis greatly and is widely used with this chapter.

Furthermore, it is currently not practical to consider antenna designs having as many frequency channels as there are signals. Instead, the signals are grouped into two subbands: the lower GNSS band ranging from 1160 to 1300 MHz (left panel of Figure 5.4.1) and the upper GNSS band, including augmentation systems like Omnistar, ranging from 1545 to 1610 MHz (right panel of Figure 5.4.1). The relative bandwidth of the lower GNSS band is 12% and that of the upper GNSS band is

4%. The relative bandwidth of the entire GNSS band is 32.5%. Later, in Section 9.7, we will see that these values affect antenna design to a large extent.

Based on this understanding, a free space wavelength of GNSS signals is 25.9 cm at the lowest frequency of the lower GNSS band and shortens to 18.6 cm at the highest frequency of the upper GNSS band. As we will see, a wavelength constitutes a natural scale for antenna-related considerations. For quick estimation purposes, it is convenient to recall that the wavelength of GNSS signals in free space is about 20 cm.

9.1.3 Complex Notations and Plane Wave in Lossy Media

Expressions (9.1.12), (9.1.13), and (9.1.20) give just one example of what is referred to as a time harmonic field with fixed frequency ω. Generally, with time harmonic alternation, the directions, amplitudes, and phases of $\vec{E}$ and $\vec{H}$ vectors vary from point to point. In Cartesian coordinates the most general representation for the time harmonic field is

$$\begin{aligned}\vec{E}(x,y,z,t) &= E_{0x}(x,y,z)\cos(\omega t+\psi_x(x,y,z))\vec{x}_0\\ &+ E_{0y}(x,y,z)\cos(\omega t+\psi_y(x,y,z))\vec{y}_0\\ &+ E_{0z}(x,y,z)\cos(\omega t+\psi_z(x,y,z))\vec{z}_0\end{aligned} \tag{9.1.46}$$

Here $E_{0x,y,z}(x,y,z)$ and $\psi_{x,y,z}(x,y,z)$ are amplitudes and initial phases of the corresponding projections. They are functions of position.

With trigonometric representations like (9.1.46), linear transformations such as summation, differentiation, and integration change the function. For instance, consider $d\sin(\omega t)/dt = \omega\cos(\omega t)$. For convenience, it is common to make use of the exponential form $e^{i\omega t}$. Here i is the imaginary unit $i=\sqrt{-1}$. For the exponential form the differentiation and integration are equivalent to multiplication, i.e., $de^{i\omega t}/dt = i\omega e^{i\omega t}$. Let $u(t)$ be a time harmonic quantity with amplitude U_0 and initial phase ψ_0 such that

$$u(t) = U_0\cos(\omega t+\psi_0) \tag{9.1.47}$$

Instead of (9.1.47) one writes

$$\widetilde{u}(t) = U_0 e^{i(\omega t+\psi_0)} = \left(U_0 e^{i\psi_0}\right)e^{i\omega t} \tag{9.1.48}$$

The tilde symbol is temporarily used to denote a complex time harmonic quantity. The term in parenthesis in (9.1.48) is the complex amplitude. It contains amplitude and initial phase and does not vary with time. We temporarily denote the complex amplitude also referred to as phasor by a dot placed atop the quantity and write

$$\dot{U} = U_0 e^{i\psi_0} \tag{9.1.49}$$

Thus

$$\widetilde{u}(t) = \dot{U}e^{i\omega t} \tag{9.1.50}$$

It is instructive to apply complex notations to the sum of time harmonic quantities with the same angular frequency ω. Let us have a number of time harmonic signals with amplitudes U_q and initial phases ψ_q. Here the index $q = 1, 2, \ldots, Q$ counts signals, and Q is the total number of signals. For the sum of these signals one has

$$\widetilde{u}_\Sigma(t) = \sum_{q=1}^{Q} U_q e^{i(\omega t+\psi_q)} = e^{i\omega t}\left(\sum_{q=1}^{Q} U_q e^{i\psi_q}\right) \tag{9.1.51}$$

According to the rules of complex algebra, the term in parenthesis is a complex quantity with amplitude (module) $U_{0\Sigma}$ and phase ψ_Σ such that

$$\dot{U}_\Sigma = U_{0\Sigma} e^{i\psi_\Sigma} = \sum_{q=1}^{Q} U_q e^{i\psi_q} \tag{9.1.52}$$

and

$$\widetilde{u}_\Sigma(t) = \sum_{q=1}^{Q} U_q e^{i(\omega t+\psi_q)} = \dot{U}_\Sigma e^{i\omega t} \tag{9.1.53}$$

Thus, the sum is also time harmonic with the same frequency.

Derivations similar to the one just done are valid with vectors. The complex form of (9.1.46) is

$$\begin{aligned}
&\tilde{\vec{E}}(x, y, z, t)\\
&\quad = E_{0x}(x, y, z) e^{i(\omega t+\psi_x(x,y,z))}\vec{x}_0 + E_{0y}(x, y, z) e^{i(\omega t+\psi_y(x,y,z))}\vec{y}_0\\
&\qquad + E_{0z}(x, y, z) e^{i(\omega t+\psi_z(x,y,z))}\vec{z}_0\\
&\quad = \left[E_{0x}(x, y, z)\, e^{i\psi_x(x,y,z)}\vec{x}_0 + E_{0y}(x, y, z) e^{i\psi_y(x,y,z)}\vec{y}_0 + E_{0z}(x, y, z) e^{i\psi_z(x,y,z)}\vec{z}_0\right] e^{i\omega t}
\end{aligned} \tag{9.1.54}$$

The expression in brackets in (9.1.54) is the complex amplitude of the vector

$$\dot{\vec{E}}(x, y, z) = \begin{bmatrix} E_{0x}(x, y, z)\, e^{i\psi_x(x,y,z)}\vec{x}_0 + E_{0y}(x, y, z) e^{i\psi_y(x,y,z)}\vec{y}_0 \\ +E_{0z}(x, y, z) e^{i\psi_z(x,y,z)}\vec{z}_0 \end{bmatrix} \tag{9.1.55}$$

Thus one has

$$\tilde{\vec{E}}(x, y, z, t) = \dot{\vec{E}}(x, y, z) e^{i\omega t} \tag{9.1.56}$$

As seen, dependency of time and spatial coordinates is separated. The complex amplitude of the vector contains position-dependent quantities only.

If complex amplitudes are known, one may always reconstruct a real time-dependent quantity by multiplying with time-dependent factor $e^{i\omega t}$ and taking a real part. For instance, applying

$$u(t) = \mathrm{Re}\left\{\dot{U} e^{i\omega t}\right\} \tag{9.1.57}$$

yields (9.1.47) and applying

$$\vec{E}(x, y, z, t) = \text{Re}\left\{ \dot{\vec{E}}(x, y, z)\, e^{i\omega t} \right\} \tag{9.1.58}$$

yields (9.1.46).

In the treatment of electromagnetic fields the most important advantage of complex notations is the separation of dependency of time and spatial coordinates as mentioned (please check (9.1.54)). Furthermore, the factor $e^{i\omega t}$ is common with all the equations and drops out. In this way an essential simplification is achieved. Writing both vectors $\vec{E}$ and $\vec{H}$ in the form of (9.1.56) and substituting into (9.1.10) yields

$$e^{i\omega t}\text{rot}\, \dot{\vec{H}} = i\omega\varepsilon\varepsilon_0 \dot{\vec{E}} e^{i\omega t} \tag{9.1.59}$$

Hence

$$\text{rot}\, \dot{\vec{H}} = i\omega\varepsilon\varepsilon_0 \dot{\vec{E}} \tag{9.1.60}$$

One can proceed similarly with equation (9.1.11).

An additional advantage of complex notations is that permittivity and conductivity of the medium can be accounted for in a unified way. This is done by introducing a complex permittivity

$$\dot{\varepsilon} = \left(\varepsilon - i\frac{\sigma}{\omega\varepsilon_0} \right) \tag{9.1.61}$$

We refer to Balanis (1989) for details of derivation. This complex permittivity can be rewritten in the form

$$\dot{\varepsilon} = \varepsilon(1 - i \tan \Delta^e) \tag{9.1.62}$$

Here $\tan \Delta^e$ is called the electric loss tangent,

$$\tan \Delta^e = \frac{\sigma}{\omega\varepsilon\varepsilon_0} \tag{9.1.63}$$

Sometimes with material specifications an electric loss tangent is referred to as a "dissipation factor." The reason is that the imaginary portion of the dielectric permittivity is related to dissipation of the electromagnetic energy in the medium. We show $\tan \Delta^e$ for 1.5-GHz frequency in the third column of Table 9.1.1.

To be able to account for magnetic losses in the medium one writes magnetic permeability in a complex form similar to (9.1.62)

$$\dot{\mu} = \mu(1 - i \tan \Delta^m) \tag{9.1.64}$$

Here $\tan \Delta^m$ is a magnetic loss tangent. There is no need to go into the details of the topic of magnetic loss tangent since it will not be used explicitly in this chapter.

For all subsequent discussions, we omit the dots with all complex quantities for the sake of simplicity of writing. Throughout this chapter it is assumed that time harmonic quantities are represented by complex amplitudes unless the opposite is stated. Maxwell equations for complex amplitudes of the vectors $\vec{E}$ and $\vec{H}$ take the form

$$\text{rot } \vec{H} = i\omega\varepsilon\varepsilon_0\vec{E} \tag{9.1.65}$$

$$\text{rot } \vec{E} = -i\omega\mu\mu_0\vec{H} \tag{9.1.66}$$

These equations do not contain time as an independent variable. However, one should always recall the rule in (9.1.58) for reconstructing the actual time-dependent process.

Now we look into power flux density. Assuming that the phases of the $\vec{E}$ and $\vec{H}$ vectors could differ from each other, performing averaging over period T similar to (9.1.35) yields

$$\vec{\Pi} = {}^1\!/\!_2[\vec{E}, \vec{H}^*] \tag{9.1.67}$$

The superscript * indicates a complex conjugate throughout this chapter. In what follows we omit the tilde like in expression (9.1.35) and always use the time-averaged value defined by this expression. The Poynting vector (9.1.67) is in general complex. Its real part is known as an active power flux density related to the power carried by an electromagnetic wave. The imaginary part is known as a reactive power flux density. It is related to reactive electromagnetic power which continuously bounces back and forth between space and a source of radiation and is primarily located in close proximity of the source. More on this power will be provided in Section 9.1.4.

The expressions for a time harmonic plane wave in complex notation take the form

$$\vec{E} = E_0 e^{-ik_m z}\vec{x}_0 \tag{9.1.68}$$

$$\vec{H} = \frac{E_0}{\eta_m} e^{-ik_m z}\vec{y}_0 \tag{9.1.69}$$

They could be proven by substituting them into (9.1.65) and (9.1.66), and making use of (9.1.36) and (9.1.37). For complex permittivity and permeability, the wavenumber (9.1.36) and the intrinsic impedance (9.1.37) become complex. Substituting (9.1.62) and (9.1.64) into (9.1.36) yields

$$k_m = \omega\sqrt{\varepsilon_0\mu_0\varepsilon\mu(1 - i\tan\Delta^e)(1 - i\tan\Delta^m)} = \beta - i\alpha \tag{9.1.70}$$

Here $\beta > 0$ is a real part of the wavenumber and $\alpha > 0$ is an imaginary part. Using the rule (9.1.58) one arrives at a real time-dependent form of (9.1.68) as

$$\vec{E}(z, t) = |E_0|e^{-\alpha z}\cos(\omega t - \beta z + \psi_0)\vec{x}_0 \tag{9.1.71}$$

and proceeds similarly with (9.1.69).

The real part β of the wavenumber is referred to as phase constant or propagation constant. In comparison to (9.1.20) it is seen that β defines the phase velocity and the wavelength. The imaginary part α is known as an attenuation constant. Expression (9.1.71) shows that the wave field now decays along the direction of travel. This is what one would assume because the imaginary part α comes from electromagnetic energy losses in the medium. By definition, a skin depth δ is a distance associated with the amplitude decay by a factor of e. Using this definition, from (9.1.71) it follows that

$$\delta = 1/\alpha \tag{9.1.72}$$

For lossy media, a plane wave shows distinctly different behavior for insulators and conductors. Good dielectrics, also referred to as insulators, are characterized by

$$\tan \Delta^e << 1 \tag{9.1.73}$$

From (9.1.70) assuming $\tan \Delta^m = 0$ one has

$$\beta \approx \omega\sqrt{\varepsilon_0\mu_0\varepsilon\mu} \tag{9.1.74}$$

$$\alpha \approx \beta\frac{\tan \Delta^e}{2} \tag{9.1.75}$$

Comparing (9.1.71) to (9.1.36), one recognizes that the phase velocity and wavelength coincide with the lossless case. Because of (9.1.73) the inequality $\beta >> \alpha$ holds. Thus, we have $\delta >> \lambda_m$. In short, one may say that the wave propagates through good dielectric almost like through a lossless medium, with the skin depth largely exceeding the wavelength in the medium. This is what happens with the wave propagating through the troposphere, for instance.

With good conductors

$$\tan \Delta^e >> 1 \tag{9.1.76}$$

and from (9.1.70) one has

$$\beta \approx \alpha \approx \omega\sqrt{\varepsilon_0\mu_0\varepsilon\mu}\sqrt{\frac{\tan \Delta^e}{2}} \tag{9.1.77}$$

Thus the wavelength in conductors decreases compared to that in free space by orders of magnitude due to the $\sqrt{\tan \Delta^e/2}$ factor in (9.1.77). The skin depth in conductors is

$$\delta = \frac{\lambda_m}{2\pi} \tag{9.1.78}$$

This means that the wave traveling through conductors almost disappears at the distances of fractions of wavelength in the medium. In other words, the electromagnetic field and electric currents are concentrated within a thin layer near the surface of the conductor.

We take two examples. For gold in Table 9.1.1, expressions (9.1.77) and (9.1.78) give a skin depth for GNSS frequencies of $2 \cdot 10^{-3}$mm. That is the reason that most electronic components like printed circuit boards and parts of antennas are plated with a thin layer of gold. This is to decrease loss of signal energy and to provide environmental protection at the same time. Another example is seawater. Here from Table 9.1.1 one obtains $\tan \Delta^e \approx 0.3$. Materials with $\tan \Delta^e$ of the order of unity are called semiconductors. They exhibit some of the properties of conductors and insulators. In this case, one is to use the general expression (9.1.70) to find out that skin depth (9.1.72) for seawater at GNSS frequencies is about 2 cm. Thus at a depth of several centimeters counting from the sea surface the GNSS signal is completely absorbed by the water.

9.1.4 Radiation and Spherical Waves

The simplest and widely used model of a source of radiation is the Hertzian dipole. By such one means an elementary source in the form of time harmonic electric current filament with the length $L << \lambda$. The dipole is an appropriate model for one of the first antennas constructed by Heinrich Hertz in the late nineteenth century.

We assume that the medium is free space. We use the spherical coordinate system (Figure 9.1.3, left panel; also see Appendix D for details of vector representation in spherical coordinates). A dipole is placed at the origin and is parallel to the zenith axes. We assume time harmonic radiation with fixed frequency ω. The complex amplitudes of fields radiated by the dipole and observed at a point with coordinates (r, θ, ϕ) are

$$\vec{H} = \vec{\phi}_0 IL \frac{1}{4\pi}\left(\frac{1}{r^2} + \frac{ik}{r}\right) e^{-ikr} \sin\theta \tag{9.1.79}$$

$$\vec{E} = -IL \frac{1}{4\pi}\frac{\eta_0}{k}\left(\vec{r}_0 \frac{2}{r^2}\left(\frac{i}{r} - k\right)\cos\theta + \vec{\theta}_0 \frac{1}{r}\left(\frac{i}{r^2} - \frac{k}{r} - ik^2\right)\sin\theta\right) e^{-ikr} \tag{9.1.80}$$

Here I is the electric current in units of amperes flowing through the dipole. Details of the derivations for these expressions are shown in Appendix E.

It is common to subdivide the space around the dipole into three regions. The first one is referred to as a reactive near-field region. It is located at the distances $r << \lambda$ immediately surrounding the dipole. The third one is a far-field region with $r >> \lambda$. Finally, the second region is sometimes referred to as a radiating near-field region. It is located between the two. We begin with the field characterization in the reactive near-field region.

Because of the inequality $r << \lambda$, one keeps only the highest terms of $1/r$ in (9.1.79) and (9.1.80). These terms provide the main contribution to the fields.

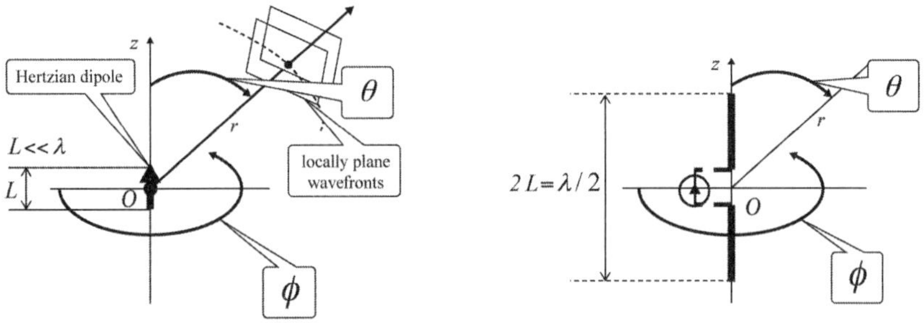

Figure 9.1.3 Coordinate frames for Hertzian and half-wave dipoles.

Also here $kr << 1$ and the exponential term could be omitted. Thus one has

$$\vec{H} = \vec{\phi}_0 IL \frac{1}{4\pi} \frac{1}{r^2} \sin\theta \tag{9.1.81}$$

$$\vec{E} = -iIL \frac{1}{4\pi} \frac{\eta_0}{k} \frac{1}{r^3} (\vec{r}_0 2\cos\theta + \vec{\theta}_0 \sin\theta) \tag{9.1.82}$$

We see that the electric and magnetic field intensities are 90° shifted in phase from each other. The imaginary unit coefficient of electric field points to that. One concludes that with fields (9.1.81) and (9.1.82) the Poynting vector (9.1.67) is purely imaginary. One says that in the near-field region a reactive power is being accumulated. This power is not radiated into space but instead is concentrated in close proximity of the source. This is true not only with dipole but rather with any antenna. The reactive power concentration is an undesirable but in general unavoidable property. This, generally speaking, is what makes bandwidth of any antenna limited. More on reactive power contribution to the antenna characteristics will be discussed in Section 9.5.2.

At the far-field region one omits all the higher order terms of $1/r$ in expressions (9.1.79) and (9.1.80) and arrives at

$$\vec{H} = \vec{\phi}_0 IL \frac{ik}{4\pi} \frac{e^{-ikr}}{r} \sin\theta \tag{9.1.83}$$

$$\vec{E} = \vec{\theta}_0 IL\eta_0 \frac{ik}{4\pi} \frac{e^{-ikr}}{r} \sin\theta \tag{9.1.84}$$

The electric and magnetic fields are now in-phase. Substituting these last expressions into (9.1.67) yields

$$\vec{\Pi} = \vec{r}_0 \frac{1}{2} |I|^2 \eta_0 \frac{(kL)^2}{(4\pi r)^2} \sin^2\theta \tag{9.1.85}$$

The Poynting vector is now real. It has only one projection and it is pointed outward toward the larger distances r. This indicates power flux toward larger distances.

Next, with expressions (9.1.83) and (9.1.84) the electric and magnetic field intensities decay as r^{-1} with r increase, and power flux density (9.1.85) decays as r^{-2}. We are to calculate the total radiated power P_Σ as a power flux through some imaginary sphere of radius r centered at the dipole. Recall that the surface of a sphere grows as $4\pi r^2$ with r increase. Thus the total power P_Σ should be independent of r. Indeed, using (9.1.85) and (D.26) from Appendix D, one has

$$P_\Sigma = \int_0^\pi \int_0^{2\pi} \frac{1}{2} |I|^2 \eta_0 \frac{(kL)^2}{(4\pi r)^2} \sin^3\theta \, r^2 d\phi \, d\theta = \text{const}(r) \tag{9.1.86}$$

In other words, through each imaginary sphere centered at the dipole the total flux is the same. This is in agreement with the energy conservation low since with free space there is no matter to absorb the radiated energy. In this regard, in the literature the power flux density decay as r^{-2} is occasionally referred to as radiation loss.

This should not be confusing, the total power (9.1.86) remains always the same and is just being spread over the imaginary spheres with larger radii.

The power distribution is not homogeneous over the sphere. The field amplitudes are proportional to $\sin\theta$ and power flux density (9.1.85) is proportional to $\sin^2\theta$. One says that the Hertzian dipole possesses some directivity. The fields and power flux get maximum values in the direction perpendicular to that of the current ($\theta = \pi/2$) and are zero in the directions along the current ($\theta = 0; \pi$). The term $\sin\theta$ represents what is known as radiation pattern or antenna pattern. More on antenna patterns will be discussed in Section 9.2.1.

Next, we see that the phase of fields (9.1.83) and (9.1.84) exhibits a delay $-kr$ with distance growth. Employing (9.1.32), (9.1.28), and (9.1.24) one notes that

$$kr = 2\pi\frac{r}{cT} = 2\pi\frac{\tau_r}{T} \tag{9.1.87}$$

Here τ_r is the time interval which is required to cover the distance r while traveling with velocity of light c. The quantity $-kr$ is referred to as path delay. This delay is constant over a sphere with radius r. Recalling the definition of the wavefront from Section 9.1.2, one says that the front of the wave represented by expressions (9.1.83) and (9.1.84) is a sphere. This is referred to as a spherical wave. The common sense illustration to a spherical wave could be a well-known circular wavefront caused by a float at the water surface.

Finally, the fields (9.1.83) and (9.1.84) do not have projections onto the direction of the wave travel $\vec{r}_0$, they constitute a right-hand triad with said direction and are proportional to each other with η_0 being the proportionality coefficient. This means that locally, within some small area around an observation point, a spherical wave is constructed exactly the same way as the plane wave, as discussed in previous sections. The wavefront of said plane wave could be taken as a plane tangential to said sphere at the observation point. An example of such fronts is shown schematically in the left panel of Figure 9.1.3. This property will remain common for all antennas. This was one of the reasons we discussed a plane wave in previous sections.

It should be mentioned that GNSS satellites are about 20,000 km from the earth's surface. Any local area at the earth's surface, like a building or even an entire city, is negligibly small compared to that distance. This means that within such areas the fields associated with GNSS signals could be viewed as plane waves arriving from the satellites. Finally, with radiating near-field region a transition from reactive near fields to the far fields is observed.

Now we turn to what is called an interference of waves radiated by a plurality of dipoles. This will give us a method to calculate the field radiated by any arbitrary antenna. First, we take a half-wave dipole as an example.

A half-wave dipole is probably the simplest antenna that could actually be constructed. This antenna consists of just two pieces of straight thin wires of about a quarter-wavelength each. This is shown in the right panel in Figure 9.1.3. The signal is transported to the narrow gap between the wires by a transmission line shown in the figure as dashed lines. Transmission line basics will be discussed further in

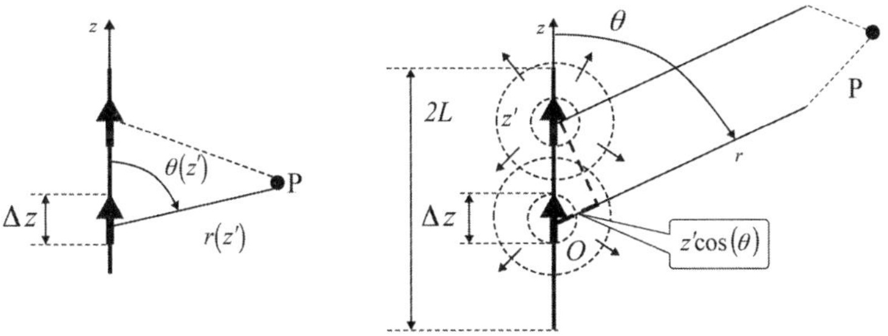

Figure 9.1.4 Calculations of fields radiated by a half-wave dipole.

Section 9.5. The electric current distribution along the dipole has a form of half the cosine-like wave, namely, using the coordinate frame of the figure,

$$I(z) = I_0 \cos\left(\frac{\pi}{2L} z\right) \tag{9.1.88}$$

Here $I(z)$ is the electric current at a point with coordinate z at the dipole and I_0 is the current at the center of the dipole. This value is set by the source. The symbol L is a length of one arm of the dipole which is equal to a quarter of the wavelength.

We calculate the field radiated by a half-wave dipole as observed at a point P in space. For that purpose the dipole is subdivided into a set of Hertzian dipoles each with an infinitesimal length Δz. Two such Hertzian dipoles are shown in the left panel in Figure 9.1.4.

The field $\Delta\vec{E}$ radiated by a Hertzian dipole located at coordinate z' is

$$\Delta\vec{E} = -I(z')\frac{1}{4\pi}\frac{\eta_0}{k}\left(\begin{array}{c} \vec{r}_0\frac{2}{(r(z'))^2}\left(\frac{i}{r(z')} - k\right)\cos(\theta(z')) \\ +\vec{\theta}_0\frac{1}{r(z')}\left(\frac{i}{(r(z'))^2} - \frac{k}{r(z')} - ik^2\right)\sin(\theta(z')) \end{array}\right)e^{-ikr(z')}\Delta z' \tag{9.1.89}$$

Here $r(z')$ is the distance counted from the Hertzian dipole to observation point P, and $\theta(z')$ is the corresponding angle counted from the z axis. Both quantities are functions of coordinate z'. The total field is a sum (integral) of the fields (9.1.89) radiated by all the Hertzian dipoles along the half-wave dipole. Thus one has

$$\vec{E} = -I_0\frac{1}{4\pi}\frac{\eta_0}{k}\int_{-L}^{L}\cos\left(\frac{\pi}{2L}z'\right)\left(\begin{array}{c} \vec{r}_0\frac{2}{(r(z'))^2}\left(\frac{i}{r(z')} - k\right)\cos(\theta(z')) \\ +\vec{\theta}_0\frac{1}{r(z')}\left(\frac{i}{(r(z'))^2} - \frac{k}{r(z')} - ik^2\right)\sin(\theta(z')) \end{array}\right)e^{-ikr(z')}dz' \tag{9.1.90}$$

In the general case this expression cannot be simplified. This shows that in the reactive near-field region and the radiating near-field region the fields are quite complex. But for the far-field region the situation is different.

To calculate field intensities in the far-field region one omits all the higher order terms of $1/r$ similar to the far-field derivation of Hertzian dipole. The right panel of Figure 9.1.4. shows an observation point P at the far-field region of the half-wave dipole. Please note that the distance to P is assumed to be very large compared to the dipole length $2L$. Two radius vectors go to P originating from two points on a dipole. One is the dipole center located at the origin and another one is a point with coordinate z'. The radius vectors are essentially parallel because of the very large distance to P. In other words one may say that with (9.1.90) the angle θ and the distance r would be the same for all the points on a dipole coinciding with those at the origin. This holds true except for the exponent. From the right triangle shown as thick dashed lines, it follows that

$$r(z') \approx r - z' \cos\theta \tag{9.1.91}$$

The reason for using (9.1.91) with the exponent is that the path difference $z' \cos\theta$ is not negligible compared to the wavelength. Thus the product of this path difference with the wavenumber k of (9.1.32) in the exponent may provide contributions comparable to 2π. Factoring terms independent on z' out of the integral (9.1.90) yields

$$\vec{E} = \vec{\theta}_0 I_0 \eta_0 \frac{ik}{4\pi} \frac{e^{-ikr}}{r} \sin\theta \int_{-L}^{L} \cos\left(\frac{\pi}{2L} z'\right) e^{ikz' \cos\theta} dz' \tag{9.1.92}$$

This expression shows that for the far-field region one obtains the total field in the form of a spherical wave: the amplitude decays as $1/r$ with distance and the phase delay equals $-kr$. The integral in (9.1.92) tells us that the field is a sum (integral) of partial spherical waves radiated by Hertzian dipoles. These are shown as dashed circles in the right panel in Figure 9.1.4. Each partial spherical wave is taken with the amplitude of the corresponding Hertzian dipole equal to $\cos(\pi z'/2L)$. The partial waves arrive at the observation point P with an extra phase delay or phase advance due to path difference. These delays and advances are a function of the direction set up by angle θ.

We can now consider a generalization. Currents flowing over parts of any antenna could be subdivided into a set of elementary Hertzian dipoles. These dipoles would radiate partial spherical waves. The total wave in the far-field region of an antenna is an interference of such partial waves. For any arbitrary antenna the total wave is expressed as follows:

$$\vec{E}(r,\theta,\phi) = \frac{e^{-ikr}}{r} \left(\vec{\theta}_0 U_\theta F_\theta(\theta,\phi) e^{i\Psi_\theta(\theta,\phi)} + \vec{\phi}_0 U_\phi F_\phi(\theta,\phi) e^{i\Psi_\phi(\theta,\phi)} \right) \tag{9.1.93}$$

$$\vec{H}(r,\theta,\phi) = \frac{1}{\eta_0} [\vec{r}_0, \vec{E}(r,\theta,\phi)] \tag{9.1.94}$$

The derivations of these expressions are given in Appendix E. The fields (9.1.93) and (9.1.94) constitute a spherical wave traveling in all directions from the radiator. The wave has the same main features as has been discussed in regard to the Hertzian and half-wave dipoles. Namely, the wave amplitude decays as r^{-1} with distance and its phase exhibits progressive delay as $-kr$. The wave is a locally plane one since the $\vec{E}$ and $\vec{H}$ fields are orthogonal to each other, are proportional to each other with intrinsic impedance being a constant of proportionality, do not have projections onto the direction of the wave travel $\vec{r}_0$, and constitute a right-hand triad with said direction.

The difference between expressions (9.1.83), (9.1.84), (9.1.93), and (9.1.94) is that now, in general, the vectors $\vec{E}$ and $\vec{H}$ have projections onto both basis vectors $\vec{\theta}_0$ and $\vec{\phi}_0$. If one travels along the imaginary sphere of some radius r, then these projections are not constant over the sphere. The amplitudes of these projections are functions of angles θ and ϕ. The angles define the direction toward the observation point. The functions $F_\theta(\theta, \phi)$ and $F_\phi(\theta, \phi)$ are real functions with peak values equal to unity. One says that functions $F_\theta(\theta, \phi)$ and $F_\phi(\theta, \phi)$, representing field intensity as a function of directions in space, constitute the radiation pattern of the source. In regard to antennas, this is also referred to as antenna pattern. In general, the patterns for θth and ϕth projections differ. Also, the directions for which $F_\theta(\theta, \phi)$ and $F_\phi(\theta, \phi)$ reach peak values are in general different. The antenna pattern is one of the most important characteristics of an antenna. Antenna patterns of GNSS user antennas will be discussed in details in Section 9.2

Further, functions $\Psi_\theta(\theta, \phi)$ and $\Psi_\phi(\theta, \phi)$ of (9.1.93) indicate that, in general, phase delays of vector projections are not constant over the imaginary sphere in the far-field region. This means that the wavefronts are not exactly spherical (see more discussion in Section 9.3). However, it is customary to refer to the antenna far field as to a spherical wave. Phases $\Psi_\theta(\theta, \phi)$ and $\Psi_\phi(\theta, \phi)$ are functions of direction indicated by angles θ and ϕ. These functions are called antenna phase patterns. By definition, an antenna phase pattern shows radiated field phase as a function of directions in space. In general, phase patterns for θth and ϕth projections differ. Antenna phase pattern is another important characteristic for GNSS applications. It defines what is known as antenna phase center and phase center variations (PCV).

The constants U_θ and U_ϕ in (9.1.93) are defined by antenna structure and signal source voltage applied to the antenna input. Sometimes they are referred to as normalization constants. These constants are, in general, complex.

One is to note that for a given antenna the modules $|U_\phi|$ and $|U_\phi|$ are well defined by the requirement that the peak values of $F_\theta(\theta, \phi)$ and $F_\phi(\theta, \phi)$ are equal to unity. Indeed, these modules are actual field intensities of the corresponding projections at the distance r in the directions where the radiation patterns are maximal. In contrast, only the phases of products $e^{-ikr} U_\theta e^{i\Psi_\theta(\theta,\phi)}$ and $e^{-ikr} U_\phi e^{i\Psi_\phi(\theta,\phi)}$ are actual field phases with (r, θ, ϕ) given. This means that the phase patterns $\Psi_\theta(\theta, \phi)$ and $\Psi_\phi(\theta, \phi)$ are defined up to constant terms: the constants could always be added to $\Psi_\theta(\theta, \phi)$ and $\Psi_\phi(\theta, \phi)$ with proper changes of arguments (phases) of U_ϕ and U_ϕ. It will be shown in Section 9.3 that one must account for this uncertainty of phase patterns.

To summarize, in spherical coordinates the point location is characterized by three coordinates: radial distance r and two angles θ and ϕ. In the far-field region the field intensity dependency on the radial coordinate is the same for all the antennas. What distinguishes one antenna from the other is field intensity dependencies on angles θ and ϕ. These are manifested by antenna radiation pattern and phase pattern.

9.1.5 Receiving Mode

Thus far we have discussed the radiation of electromagnetic waves. However, the GNSS user antennas are essentially of the receiving type. The bridge between transmitting and receiving modes of antenna operation is established by the reciprocity theorem (Balanis, 1989).

Applied to the antenna pattern case, the reciprocity theorem states that the patterns for the transmitting and receiving modes are identical. As an illustration consider two antennas located in free space (Figure 9.1.5)

On the left of the figure a signal source is connected to antenna *A* which radiates power into space. The electromagnetic waves reach antenna *B* and induce currents on the elements of antenna *B*. The currents flow through the input of the receiver connected to antenna *B*. On the right the inverted case is shown. Here a signal generator is connected to antenna *B* and the receiver is connected to antenna *A*. The reciprocity theorem for the case could be phrased as follows: if the signal generators are the same in both cases then the signals at the receiver's inputs are the same. Starting with this point we consider the situation shown in Figure 9.1.6.

At the left panel, position and orientation of antenna *A* is fixed. Antenna *B* moves along the sphere centered at *A* in such a way that the orientation of antenna *B* with respect to the line *A* to *B* is fixed. Both antennas are in the far-field region of each other. It follows from the reciprocity theorem that whichever antenna *A* or *B* is transmitting or receiving, the signal at the receiver input will be proportional to the antenna pattern

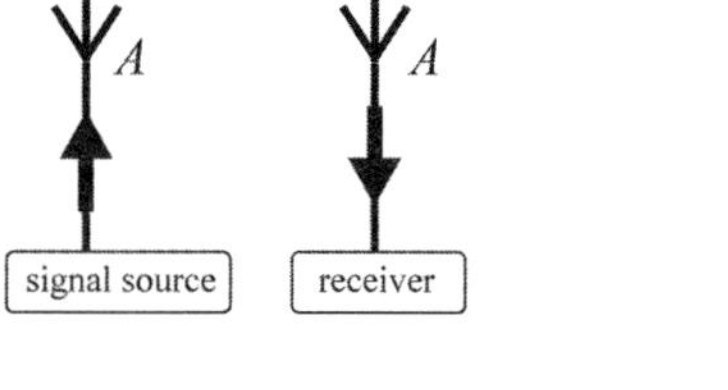

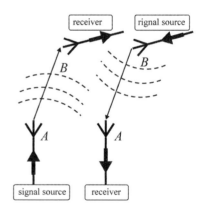

Figure 9.1.5 Reciprocity theorem.

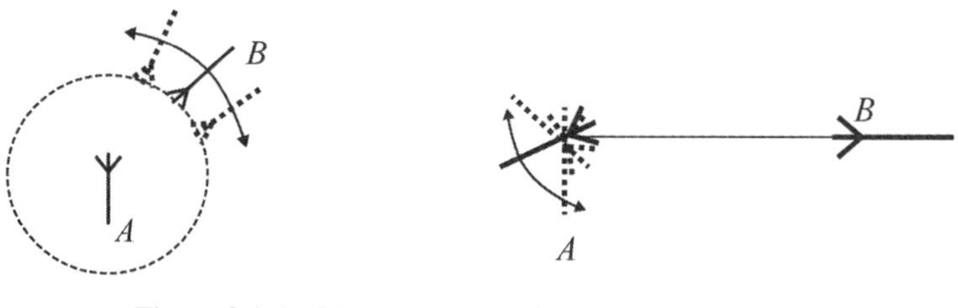

Figure 9.1.6 Measurements of antenna pattern of *A*.

of A. At the right panel another case is shown. Here antenna B is fixed and A is rotated with respect to the line A to B. The signal at the receiver input is proportional to the antenna pattern of A regardless of whichever antenna is transmitting or receiving.

The identity of antenna patterns for receiving and transmitting modes simplifies antenna-related considerations to a large extent. For many cases it is more convenient to treat the transmitting mode of the antenna rather than the receiving mode. This will be used throughout the chapter. Figure 9.1.6 illustrates the practical ways for antenna pattern measurements.

9.1.6 Polarization of Electromagnetic Waves

We consider a time harmonic electromagnetic plane wave and write it in real time-dependent form:

$$\vec{E}_1 = E_0 \cos(\omega t - kz + \psi_0)\vec{x}_0 \tag{9.1.95}$$

$$\vec{H}_1 = \frac{E_0}{\eta_0} \cos(\omega t - kz + \psi_0)\vec{y}_0 \tag{9.1.96}$$

The E-field distributions along the z coordinate have been discussed already in regard to Figure 9.1.2. Now we take some fixed coordinate, $z = 0$ for instance, and plot the E-field magnitude versus time. This is shown in Figure 9.1.7. We conclude that the E vector alternates in space and time being always parallel to some fixed direction. In our case it is x axis. The H field possesses the same properties while being always parallel to the y axis. Such a wave with vector $\vec{E}$ (and $\vec{H}$) being always parallel to a certain direction is said to be linear polarized.

Now along with the plane wave defined by expressions (9.1.95) and (9.1.96) we take another wave. This second wave has the same amplitude E_0. The vectors $\vec{E}$ and $\vec{H}$ of this second wave are 90° rotated in space and 90° delayed in phase with respect to corresponding vectors of the first wave. We write

$$\vec{E}_2 = E_0 \cos\left(\omega t - kz + \psi_0 - \frac{\pi}{2}\right)\vec{y}_0 \tag{9.1.97}$$

$$\vec{H}_2 = -\frac{E_0}{\eta_0} \cos\left(\omega t - kz + \psi_0 - \frac{\pi}{2}\right)\vec{x}_0 \tag{9.1.98}$$

One notes the "minus" sign in (9.1.98), which is needed for the vectors $\vec{E}$ and $\vec{H}$ to constitute a right-hand triad with the direction of travel (z axis).

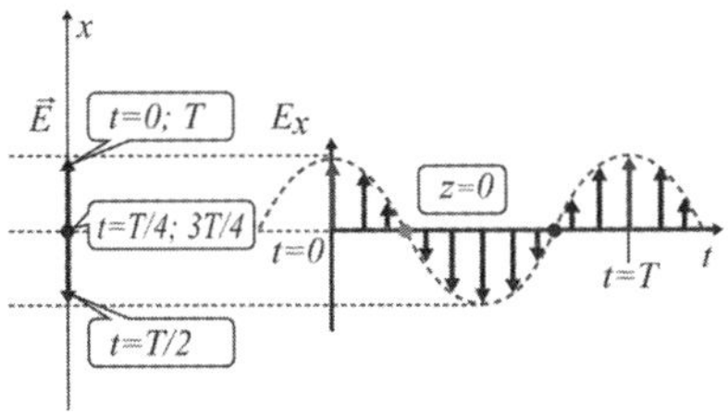

Figure 9.1.7 ***E*-field intensity of a linear polarized wave versus time.**

Next, we focus on a wave that is the sum of waves 1 and 2. The E-field intensity of this wave is denoted as $\vec{E}_\Sigma$ such that

$$\vec{E}_\Sigma = \vec{E}_1 + \vec{E}_2 \tag{9.1.99}$$

A similar expression holds for the $\vec{H}$ field. For any specific point in space, for instance, point $z = 0$, one has (omitting the initial phase ψ_0 for simplicity of writing)

$$\vec{E}_1(z = 0) = E_0 \cos(\omega t)\vec{x}_0 \tag{9.1.100}$$

$$\vec{E}_2(z = 0) = E_0 \sin(\omega t)\vec{y}_0 \tag{9.1.101}$$

$$\vec{E}_\Sigma(z = 0) = E_0 \cos(\omega t)\vec{x}_0 + E_0 \sin(\omega t)\vec{y}_0 \tag{9.1.102}$$

We check the behavior of $\vec{E}_\Sigma$ with time. Snapshots for three time instants, namely $t = 0$, $t = T/8$, and $t = T/4$ are shown in Figure 9.1.8.

First we note that as seen with (9.1.102) for any time instant the absolute value (module) of $\vec{E}_\Sigma$ remains constant, i.e., $|\vec{E}_\Sigma| = \sqrt{E_x^2 + E_y^2} = |E_0|$. Next, for $t = 0$ (left panel) one has $\vec{E}_1(z = 0) = E_0\vec{x}_0$, $\vec{E}_2(z = 0) = 0$ and $\vec{E}_\Sigma(z = 0) = E_0\vec{x}_0$. For $t = T/8$ (middle panel) the corresponding values are $\vec{E}_1(z = 0) = E_0 \cos(\pi/4)\vec{x}_0$, $\vec{E}_2(z = 0) = E_0 \sin(\pi/4)\vec{y}_0$, and $\vec{E}_\Sigma(z = 0) = E_0 \cos(\pi/4)\vec{x}_0 + E_0 \sin(\pi/4)\vec{y}_0$. Finally, for $t = T/4$ (right panel) we have $\vec{E}_1(z = 0) = 0$, $\vec{E}_2(z = 0) = E_0\vec{y}_0$, and $\vec{E}_\Sigma(z = 0) = E_0\vec{y}_0$. It is suggested the reader verify that for $t = T$ the field intensity coincides with that for $t = 0$. If one has taken another point in space with a certain z coordinate, then the same analysis will hold; all the vectors will receive a delay in phase equal to $-kz$. One concludes that at any point in space the vector $\vec{E}_\Sigma$ rotates as time advances while remaining always constant in magnitude. This vector makes a complete circle within the time interval of one period T. In Figure 9.1.8 the z axis is directed toward the observer. This is marked by a thick dot at the origin. If one watches the rotation of vector $\vec{E}_\Sigma$ from the top of $\vec{z}_0$, the rotation is seen in a counterclockwise direction.

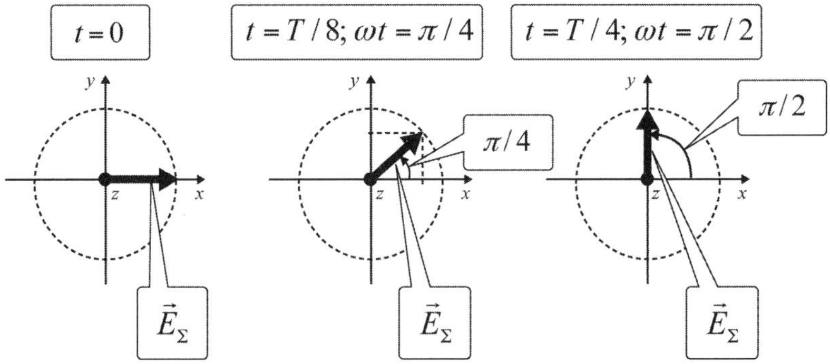

Figure 9.1.8 Snapshots of a circular polarized field versus time.

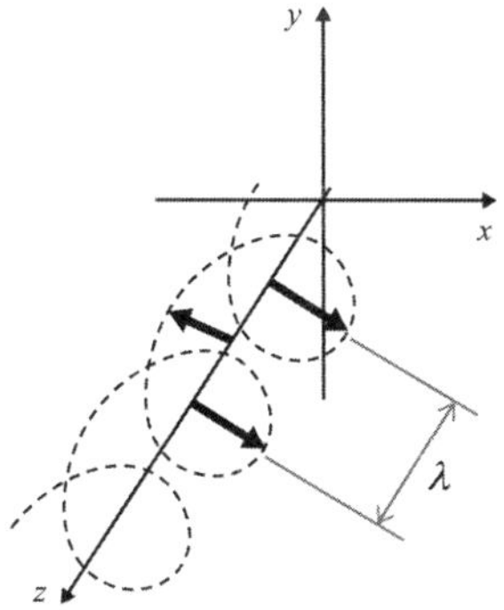

Figure 9.1.9 Circular polarized field versus space coordinate.

In the same way one may take any instant of time, for example, $t = 0$, and analyze the behavior of the vectors versus space coordinate z. One writes

$$\vec{E}_1(t = 0) = E_0 \cos(-kz)\vec{x}_0 \tag{9.1.103}$$

$$\vec{E}_2(t = 0) = E_0 \sin(-kz)\vec{y}_0 \tag{9.1.104}$$

$$\vec{E}_\Sigma(t = 0) = E_0 \cos(-kz)\vec{x}_0 + E_0 \sin(-kz)\vec{y}_0 \tag{9.1.105}$$

It is suggested that the reader check a set of points along the z axes, for instance, $z = 0$, $z = \lambda/8$, and $z = \lambda/4$ to find out that the rotation is similar as described above. We conclude that at any time instant the vector $\vec{E}_\Sigma$ is distributed in space in such a way that it exhibits a rotation in counterclockwise direction if one watches from the top of the direction of the wave propagation (z axis). The vector $\vec{E}_\Sigma$ is always constant in magnitude and makes a complete circle within the distance of one wavelength. This is schematically shown in Figure 9.1.9. The vector $\vec{H}_\Sigma$ of the wave exhibits the same behavior, while being always perpendicular to $\vec{E}_\Sigma$.

A wave with vectors $\vec{E}$ and $\vec{H}$ exhibiting rotations in time and space in a counterclockwise direction when watched from the top of the direction of wave propagation, being always constant in magnitude and making a complete circle within the time interval equal to the period of alternation and the space increment equaling the wavelength, is called right-hand circular polarized (RHCP). We note that if one took the $\vec{E}_2$ and $\vec{H}_2$ vectors not delayed but rather advanced 90° in phase with respect to $\vec{E}_1$ and $\vec{H}_1$ such that

$$\vec{E}_2 = E_0 \cos\left(\omega t - kz + \psi_0 + \frac{\pi}{2}\right)\vec{y}_0 \tag{9.1.106}$$

$$\vec{H}_2 = -\frac{E_0}{\eta_0} \cos\left(\omega t - kz + \psi_0 + \frac{\pi}{2}\right)\vec{x}_0 \tag{9.1.107}$$

then the same rotations would be observed in a clockwise direction. Such a wave is called left-hand circular polarized (LHCP).

What we have just said proves *statement A*: A circular polarized wave is a sum of two linear polarized waves. Vectors $\vec{E}$ (and $\vec{H}$) of the waves have the same amplitudes, are 90° rotated in space with respect to each other, and are 90° shifted in phase with respect to each other. The circular polarized wave is said to be RHCP if the rotation is observed in a counterclockwise direction when watched from the top of the direction of wave propagation and it is LHCP if the rotation is in the opposite direction.

For further derivations it is more convenient to use complex amplitudes. For RHCP and LHCP waves traveling in positive direction of the z axis one writes

$$\vec{E}_{RHCP} = E_0 e^{-ikz} \frac{1}{\sqrt{2}} (\vec{x}_0 - i\vec{y}_0) \tag{9.1.108}$$

$$\vec{E}_{LHCP} = E_0 e^{-ikz} \frac{1}{\sqrt{2}} (\vec{x}_0 + i\vec{y}_0) \tag{9.1.109}$$

$$\vec{H}_{RHCP(LHCP)} = \frac{1}{\eta_0} \left[\vec{E}_{RHCP(LHCP)}, \vec{z}_0 \right] \tag{9.1.110}$$

The constant $1/\sqrt{2}$ in (9.1.108) and (9.1.109) is introduced for purposes of normalization as will be seen below. Now it is straightforward to prove the inverse statement, *statement B*: A linear polarized wave is a sum of two circular polarized waves of equal magnitudes but opposite directions of rotation. Indeed, using (9.1.108) and (9.1.109) and setting up

$$\vec{E}_1 = \frac{1}{\sqrt{2}} \left(\vec{E}_{LHCP} + \vec{E}_{RHCP} \right) = E_0 e^{-ikz} \vec{x}_0 \tag{9.1.111}$$

$$\vec{E}_2 = \frac{1}{\sqrt{2}} \frac{1}{i} \left(\vec{E}_{LHCP} - \vec{E}_{RHCP} \right) = E_0 e^{-ikz} \vec{y}_0 \tag{9.1.112}$$

one has two linear polarized waves. Snapshots for the vector (9.1.112) at time instants $t = 0$, $t = T/8$, and $t = T/4$ at a point $z = 0$ are illustrated in Figure 9.1.10. To obtain these figures one is to transfer the complex amplitudes in (9.1.112) to real time-dependent quantities using the rule (9.1.58).

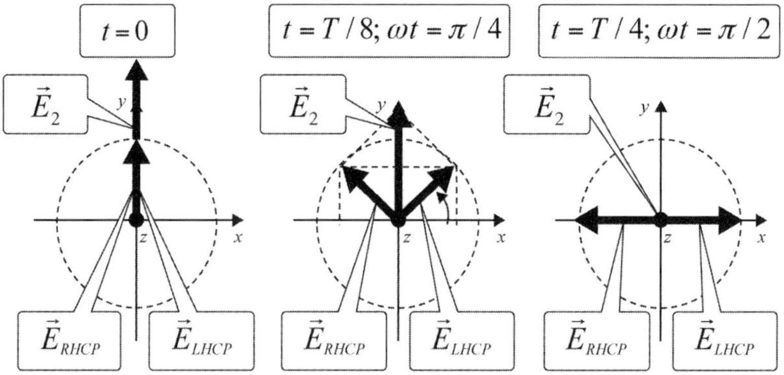

Figure 9.1.10 Linear polarized wave as a sum of two circular polarized waves.

As was discussed in Section 9.1.4, a plane wave could be viewed as a local representation of an actual spherical wave radiated by a source. One may say that expression (9.1.93) represents the far field of a source as a sum of two linear polarized spherical waves: with vectors $\vec{E}$ having the θth and the ϕth projections. However, one can write an equivalent representation of the same field in terms of RHCP and LHCP components. For this purpose we introduce circular polarized basis vectors:

$$\vec{p}_0^{RHCP} = \frac{1}{\sqrt{2}}(\vec{\theta}_0 - i\vec{\phi}_0) \tag{9.1.113}$$

$$\vec{p}_0^{LHCP} = \frac{1}{\sqrt{2}}(\vec{\theta}_0 + i\vec{\phi}_0) \tag{9.1.114}$$

These vectors have a unit length (module)

$$\left|\vec{p}_0^{RHCP}\right| = \left|\vec{p}_0^{LHCP}\right| = 1 \tag{9.1.115}$$

and are orthogonal to each other such that the dot product

$$\vec{p}_0^{RHCP}\,\vec{p}_0^{LHCP}\,* = 0 \tag{9.1.116}$$

Transformation of (9.1.113) and (9.1.114) to a real time-dependent form makes it clear that the vector (9.1.113) rotates in a counterclockwise direction if watched from the top of the $\vec{r}_0$ vector of the spherical coordinate frame. The vector (9.1.114) rotates in a clockwise direction. Vectors (9.1.113) and (9.1.114) could be viewed as another basis in the plane perpendicular to $\vec{r}_0$. These vectors along with the pair $(\vec{\theta}_0, \vec{\phi}_0)$ are particular cases of what is known as polarization basis. Vectors (9.1.113) and (9.1.114) are circular polarized while $\vec{\theta}_0$ and $\vec{\phi}_0$ are of orthogonal linear polarizations.

The field transformation from linear polarized representation (9.1.93) to circular polarization is performed by the regular routine of basis transformation. According to (9.1.113) and (9.1.114) the transformation matrix is

$$\overline{\overline{A}} = \frac{1}{\sqrt{2}}\begin{bmatrix} 1 & -i \\ 1 & i \end{bmatrix} \tag{9.1.117}$$

Thus one has

$$\vec{E}(r,\theta,\phi) = \frac{e^{-ikr}}{r}\begin{pmatrix} \vec{p}_0^{RHCP} U_{RHCP} F_{RHCP}(\theta,\phi)\, e^{i\Psi_{RHCP}(\theta,\phi)} \\ +\vec{p}_0^{LHCP} U_{LHCP} F_{LHCP}(\theta,\phi) e^{i\Psi_{LHCP}(\theta,\phi)} \end{pmatrix} \tag{9.1.118}$$

with

$$U_{RHCP}F_{RHCP}(\theta,\phi)e^{i\Psi_{RHCP}(\theta,\phi)} = \frac{1}{\sqrt{2}}\begin{pmatrix} U_\theta F_\theta(\theta,\phi)\, e^{i\Psi_\theta(\theta,\phi)} \\ +iU_\phi F_\phi(\theta,\phi) e^{i\Psi_\phi(\theta,\phi)} \end{pmatrix} \tag{9.1.119}$$

$$U_{LHCP}F_{LHCP}(\theta,\phi)e^{i\Psi_{LHCP}(\theta,\phi)} = \frac{1}{\sqrt{2}}\begin{pmatrix} U_\theta F_\theta(\theta,\phi)\, e^{i\Psi_\theta(\theta,\phi)} \\ -iU_\phi F_\phi(\theta,\phi) e^{i\Psi_\phi(\theta,\phi)} \end{pmatrix} \tag{9.1.120}$$

Here the functions $F_{RHCP(LHCP)}(\theta,\phi)$ are real with peak values equal to unity. These functions are antenna patterns in terms of RHCP (LHCP) components, $\Psi_{RHCP(LHCP)}(\theta,\phi)$ are corresponding phase patterns, and $U_{RHCP(LHCP)}$ are complex normalization constants. Since the expression (9.1.94) for the H-field intensity holds true with the E-field representation (9.1.118), one is able to write the Poynting vector (9.1.67) in the form

$$\vec{\Pi} = \frac{1}{2}[\vec{E},\vec{H}^*] = \frac{1}{2\eta_0}\frac{1}{r^2}(|U_{RHCP}|^2(F_{RHCP}(\theta,\phi))^2 + |U_{LHCP}|^2(F_{LHCP}(\theta,\phi))^2)\vec{r}_0 \tag{9.1.121}$$

This expression shows a total power flux as a sum of RHCP and LHCP fluxes.

A simple rule in the antenna field says that transmitting and receiving antennas should be matched in terms of polarization. For instance, if one antenna radiates a linear polarized wave, with the vector $\vec{E}$ being aligned in a certain direction, and another antenna receiving the radiation and being linear polarized but perpendicular to that of the transmitting antenna, then the received signal will be zero. The same is true for the case when a transmitting antenna is RHCP and the receiving one is LHCP. This provides one of the reasons why GNSS signals are chosen to be circular polarized.

Let us just assume for a moment that linear polarization is chosen for GNSS positioning. Let us further assume that the antenna of one satellite transmits a certain linear polarized signal, say with the vector $\vec{E}$ parallel to the north-south line of the local horizon. If the user antenna is also linear polarized and aligned in the west-east direction, there will be a complete loss of signal. Therefore, to make such a system functional, it would be necessary that not only all satellite antennas be parallel to each other but also that all the user antennas would have to be parallel to the same line or direction. This is obviously impractical. With a circular polarized signal all these problems are avoided.

However, expressions (9.1.93) and (9.1.118) show that in general two linear or two circular polarized components are radiated. By principal or co-polarization one means a desired type behavior of the vector orientation. An undesired type is designated as cross polarization. For instance, with expression (9.1.93) if a linear polarization with the θth component of electric field intensity is desirable then this component will be referred to as co-polarized while the ϕth is called cross polarized. Closer to the GNSS case, if with expression (9.1.118) a RHCP component is desirable, it will be referred to as co-polarized and LHCP as cross polarized. For the latter case we rewrite (9.1.118) in the form

$$\vec{E}(r,\theta,\phi) = \frac{e^{-ikr}}{r}U_{RHCP}\begin{bmatrix} F_{RHCP}(\theta,\phi)\, e^{i\Psi_{RHCP}(\theta,\phi)}\vec{p}_0^{\,RHCP} \\ +\alpha^{cross}F_{LHCP}(\theta,\phi)e^{i\Psi_{LHCP}(\theta,\phi)}\vec{p}_0^{\,LHCP}\end{bmatrix} \tag{9.1.122}$$

Here

$$\alpha^{cross} = \frac{U_{LHCP}}{U_{RHCP}} \tag{9.1.123}$$

is a coefficient which relates cross-polarization pattern (LHCP) to the principal one (RHCP). This coefficient could be easily obtained with an appropriate antenna measurements technique by measuring the LHCP component intensity versus RHCP component intensity for some fixed direction, for instance, the direction where $F_{RHCP}(\theta, \phi)$ reaches peak value. We will see later that a representation in the form of (9.1.122) is adopted with the GNSS user antenna data. Coefficient α^{cross} is angular independent and is sometimes referred to as a normalization constant for the cross-polarized component. Using (9.1.122) the total power flux (9.1.121) will read

$$\vec{\Pi} = \frac{1}{2}[\vec{E}, \vec{H}^*] = \frac{1}{2\eta_0}\frac{1}{r^2}|U_{RHCP}|^2\left[F^2_{RHCP}(\theta, \phi) + (\alpha^{cross})^2 F^2_{LHCP}(\theta, \phi)\right]\vec{r}_0 \tag{9.1.124}$$

Sometimes the total power pattern is important. This is the (normalized) total power flux density as function of direction in space:

$$F^2(\theta, \phi) = \frac{1}{F^2_{max}(\theta, \phi)}\left(F^2_{RHCP}(\theta, \phi) + (\alpha^{cross})^2 F^2_{LHCP}(\theta, \phi)\right) \tag{9.1.125}$$

Here $F^2_{max}(\theta, \phi)$ is the maximal value. One note is to follow—as mentioned at the end of Section 9.1.4—the antenna phase pattern defined up to a constant term. With expression (9.1.122) it is convenient to take the term α^{cross} as a positive real number by relating the constant phase term to the phase pattern of cross-polarized component (to Ψ_{LHCP}).

We close this section with the most general type of polarization, i.e., elliptical polarization. This is illustrated in Figure 9.1.11.

In the left panel the vector $\vec{E}$ observed at some point in space is shown. The vector rotates and makes a complete cycle within the period T with the end of the vector tracing an ellipse. Right-hand and left-hand elliptical polarizations are defined the same way as with circular polarizations. It is common to characterize elliptical polarization by an axial ratio. By definition the axial ratio α_{ar} is the ratio of the semiaxes of an ellipse, namely,

$$\alpha_{ar} = \frac{b}{a} \tag{9.1.126}$$

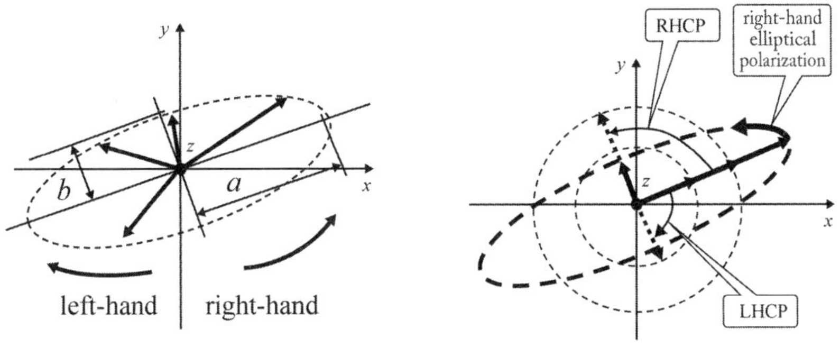

Figure 9.1.11 Elliptical polarization.

Linear polarization and circular polarization are particular cases of elliptical polarization—for linear polarization $\alpha_{ar} = 0$ and for circular polarization $\alpha_{ar} = 1$.

In regard to elliptical polarization statements C and D hold. *Statement C*: An elliptically polarized wave is a sum of two linear polarized waves if at least one of the conditions of statement A is violated. *Statement D*: An elliptically polarized wave is a sum of RHCP and LHCP waves. It has right-hand or left-hand rotation depending on which component, RHCP or LHCP, is dominating in magnitude. We omit proofs of statements C and D but illustrate statement D with the right panel in Figure 9.1.11. Here a major semiaxes of an ellipse occurs at the time instant when the vectors $\vec{E}$ of two circular polarized waves coincide and a minor semiaxes occurs when said vectors are opposite to each other. Thus one writes

$$\begin{cases} |\vec{E}_{RHCP}| + |\vec{E}_{LHCP}| = a \\ |\vec{E}_{RHCP}| - |\vec{E}_{LHCP}| = b \end{cases} \tag{9.1.127}$$

The axial ratio of an ellipse α_{ar} could be readily obtained by antenna measurement techniques. Once α_{ar} is known, then by inverting (9.1.127) one has

$$\frac{|\vec{E}_{LHCP}|}{|\vec{E}_{RHCP}|} = \frac{1 - \alpha_{ar}}{1 + \alpha_{ar}} \tag{9.1.128}$$

This relationship allows the estimation of α^{cross} once the axial ratio is known. Polarization properties of the GNSS user antenna are important in many aspects. This is discussed further in Sections 9.2 and 9.4.

9.1.7 The dB Scale

The decibel (dB) is a convenient and commonly adopted unit to compare quantities exhibiting a large range in magnitude. This is in particular true if the system response is proportional not to the quantity but rather to the logarithm of the quantity. The most common example is probably the human ear. Human ear responds to air pressure coming from a sound. This pressure varies within an extremely large scale. If the ear response was linearly proportional to air pressure and tuned for the best reception of regular human speech volume, the humans would be deaf to the rustling grass and might not survive the noise of an aircraft. Instead, the response of the ear is proportional to the logarithm of air pressure. Figure 9.1.12 shows a plot of

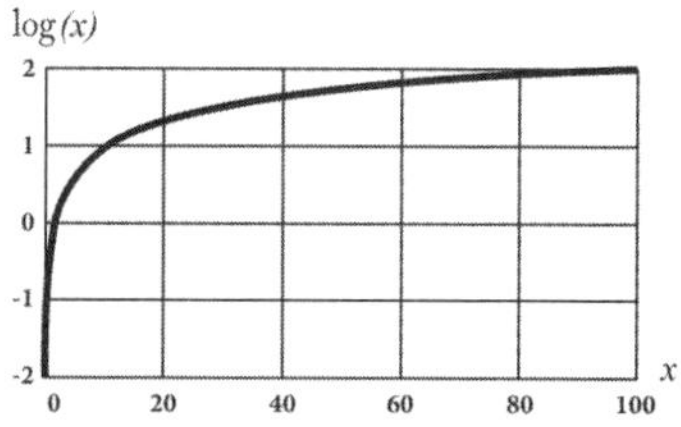

Figure 9.1.12 Logarithm function.

the logarithm function. This function emphasizes small values and suppresses large ones. For instance, if a quantity changes by two orders of magnitude, then logarithm changes by a factor of log(100) = 2 times.

Prior to discussing the dB scale one is to make an observation. As seen from (9.1.35) the power flux is proportional to the square of the electric field intensity. Another example of a squared functional relation is known from elementary physics. Namely, the power P of a direct electric current is a product of current I and voltage U:

$$P = IU \tag{9.1.129}$$

By applying Ohm's law this can be rewritten as

$$P = I^2 R = U^2 \frac{1}{R} \tag{9.1.130}$$

where R stands for resistance. Thus the power is proportional to current or voltage squared. This rule could be generalized. Each time one speaks about some quantity like current, voltage, or electric and magnetic field intensities, the power related to the quantity will be proportional to the quantity squared. Let Q be such a quantity, P_Q be corresponding power, and Q_0 be some fixed reference value. Then Q/Q_0 is said to be a relative value of Q (related to Q_0), or $P_Q/P_{Q0} = Q^2/{Q_0}^2$ is said to be a relative power of Q.

By definition a related quantity in decibel units is

$$Q_{[\mathrm{dB}]} = 20 \log\left(\frac{Q}{Q_0}\right) = 10 \log\left(\frac{P_Q}{P_{Q_0}}\right) \tag{9.1.131}$$

Here dB units are marked by subscript dB in brackets. Back conversion from dB to magnitudes and powers is

$$Q = Q_0 10^{Q_{[\mathrm{dB}]}/20} \tag{9.1.132}$$

$$P_Q = P_{Q_0} 10^{Q_{[\mathrm{dB}]}/10} \tag{9.1.133}$$

One is to note that the value of Q in dB is the same for relative magnitudes and powers (9.1.131). Hence while using dB one does not have to state if magnitudes or powers are meant. Typical figures to memorize are: +3 dB means twice the power or $\sqrt{2} \approx 1.4$ of magnitude, –3 dB is half-power or $1/\sqrt{2} \approx 0.7$ magnitude, +6 dB is two times magnitude or 4 times power, and –6 dB is half-magnitude or quarter power.

A good example of the use of the dB scale relevant to GNSS antennas is low-noise amplifier (LNA) gain. As will be discussed in Section 9.6 the LNA goes directly after the antenna and is responsible for setting up the GNSS receiver system noise figure. The LNA gain normally is specified in technical documents, e.g., a 30 ± 2 dB specification is typical. This means that ± 2 dB gain change would not affect the system performance. One may note that 30 dB means 1000 times in terms of power while 32 dB means 1585 times power amplification and 28 dB means 631 times. Therefore, in terms of output signal power the difference is significant. But again, for system performance ± 2 dB of LNA gain is normally not that important. Thus, in this case decibels provide the scale that is more suitable for the problem.

Finally, another reason to use dB scale is just practical convenience. Many formulas look like multiplications of terms, in which case one is just to add numbers in the dB scale. The dB scale is commonly adopted in electronic engineering and in GNSS antenna documentation in particular and will be frequently used throughout the chapter.

9.2 ANTENNA PATTERN AND GAIN

An antenna pattern defines the response of an antenna as a function of direction the signal is radiated to or arrives from. The antenna pattern properties determine currently achievable precision of positioning in typical environments. In this section the main features of antenna pattern are discussed in sequential order. We follow the commonly adopted method, namely analyzing the pattern in transmitting mode and making the "bridge" to receiving mode using the reciprocity theorem. The section ends with the satellite signal power estimates at the antenna output.

9.2.1 Receiving GNSS Antenna Pattern and Reference Station and Rover Antennas

It has been stated is Section 9.1.4 that an antenna pattern represents the distribution of radiated field intensity or power as a function of directions in space. Practical ways to measure the antenna pattern have been discussed in connection with Figure 9.1.6. The method illustrated in the left panel of the figure is actually in use with very large antennas like paraboloid reflector antennas or antenna arrays for radioastronomy or radars. The sensor, antenna *B*, could be carried by a helicopter. Typical GNSS user antennas are of the order of meters at most. With such antennas indoor testing can be performed in anechoic chambers. An anechoic chamber is a room with walls, ceiling,

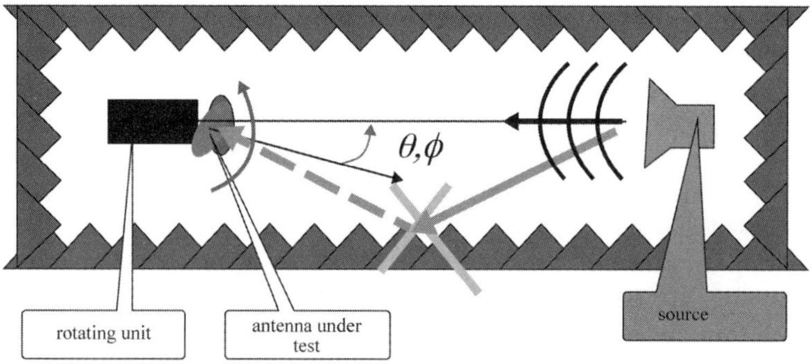

Figure 9.2.1 Antenna pattern measurement with anechoic chamber.

and a floor covered with special electromagnetic absorbing material. This is shown schematically in Figure 9.2.1.

The material absorbs the electromagnetic wave impinging upon it and thus prevents the wave from being reflected. The free space conditions are simulated. With anechoic chamber tests, an arrangement shown in the right panel of the Figure 9.1.6 is normally used. The antenna under test (antenna A) is rotated. The position and orientation of antenna B is fixed. Whichever antenna is radiating or receiving, the received signal is a function of the rotation angles and is proportional to the pattern of antenna A.

In this section we omit details related to polarization, leaving this discussion for Section 9.2.3. To begin with, we assume that only the principal polarization component is present in the far field. We denote an antenna pattern as $F(\theta, \phi)$. This is a real function of angles θ, ϕ normalized in such a way that the peak value equals to unity. This can always be done with calculated or measured data. Since the power flux density (Poynting vector) is proportional to the squared field intensity, the antenna pattern in terms of power is $F^2(\theta, \phi)$.

We begin with some examples first. As seen from (9.1.84) and (9.1.85) a Hertzian dipole pattern is

$$F(\theta, \phi) = \sin\theta \tag{9.2.1}$$

It is a function of θ but not ϕ. This might be clear because of the ideal rotational symmetry of the dipole with respect to azimuth (Figure 9.1.3).

Also of interest is that $F(\theta, \phi) = 0$ for $\theta = 0$ or π. This means that the dipole does not radiate along its axis. Next, $F(\theta, \phi) = 1$ for $\theta = \pi/2$ (90°). This means that the dipole current mostly radiates in the direction perpendicular to its axis. Now we turn to the half-wave dipole. Performing integration with (9.1.92) and normalizing the pattern to a peak value of unity yields

$$F(\theta, \phi) = \frac{\cos\left(\dfrac{\pi}{2}\cos(\theta)\right)}{\sin\theta} \tag{9.2.2}$$

The main features of this pattern are similar to (9.2.1). The pattern does not depend on ϕ because of rotational symmetry. The half-wave dipole does not radiate along its axis but at a maximum in the direction perpendicular to its axis.

Now we look at a common way to plot an antenna pattern. Sometimes polar plots are used. The top-left panel of Figure 9.2.2 shows polar plots for two patterns, the Hertzian dipole (9.2.1) and the half-wave dipole (9.2.2).

In general, an antenna pattern plot represents a surface F as a function of two variables, θ and ϕ. The patterns (9.2.1) and (9.2.2) are independent of ϕ. The corresponding surfaces would be tori which are homogeneous in azimuth. However, it is customary to use plain figures assuming either θ or ϕ to be constant. One may note that with spherical coordinates the angle θ varies within the range $0 \le \theta \le \pi$ (Appendix D). However, for illustrative purposes it is customary to allow θ to vary within the range $-\pi \le \theta \le \pi$, or from –180° to +180°, thus creating a complete great circle section of a pattern (IEEE Standard, 2004). Since patterns like (9.2.1) and (9.2.2) are independent of ϕ, the plots of Figure 9.2.2 are valid for any ϕ. The plots

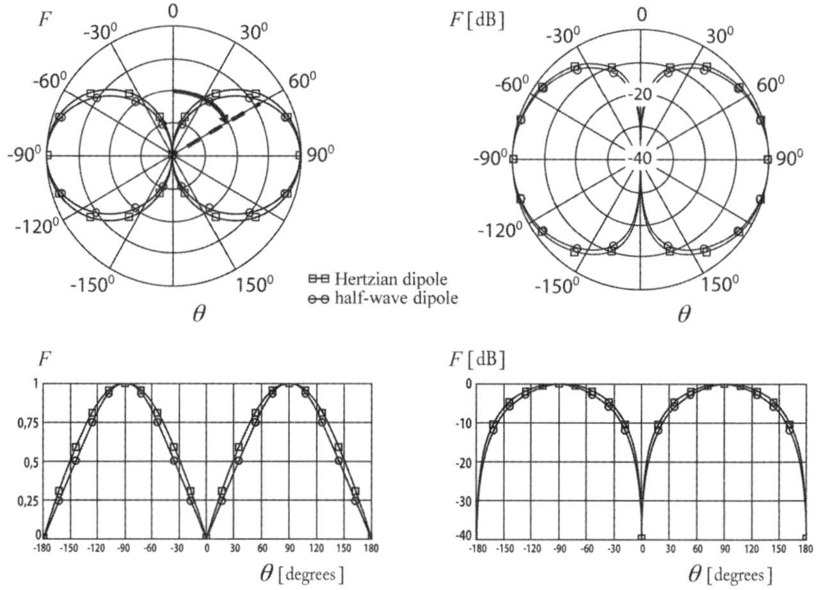

Figure 9.2.2 Radiation patterns of Hertzian and half-wave dipoles.

are generated using common rules of polar plots, namely angle θ is counted from the dipole axis and the length of the segment (shown as dashed line) is counted from the origin. This length is proportional to the pattern reading.

As seen in Figure 9.2.2 the Hertzian and half-wave dipoles have almost the same pattern shape. This illustrates that current segments have about the same radiation patterns if the length of the segment varies from being negligibly small compared to the wavelength up to about half of a wavelength [actually, up to about one wavelength (Lo and Lee, 1993)]. Half-wavelength size is chosen not because of antenna pattern advantages but for purposes of matching the antenna with the feeding cable. Such matching conditions are discussed further in Section 9.5. The top-right panel of Figure 9.2.2 illustrates the same two patterns in dB units. The peak value equal to unity is 0 dB. Zero reading is minus infinity in dB. Another way to illustrate the pattern is to use a Cartesian plot. An example is shown for the same patterns in relative units (bottom-left panel) and in dB (bottom-right panel).

Now we turn to the GNSS user antenna. Figure 9.2.3 illustrates a typical situation. The user antenna is installed at an open site about 2 m above the ground surface. We place the coordinate origin somewhere at the antenna. The vertical axis points to the zenith. The angle θ is referred to as the zenith angle such that $\theta = 0$ stands for zenith direction, $\theta = \pi/2$ is for horizon, and $\theta = \pi$ is for nadir. Sometimes the elevation angle θ^e is also of use. This angle is measured relative to the local horizon. Angle ϕ is the local azimuth.

First, one considers a perfect GNSS receiving antenna. Such an antenna has an ideal rotational symmetry with respect to zenith axes. Thus the antenna pattern is

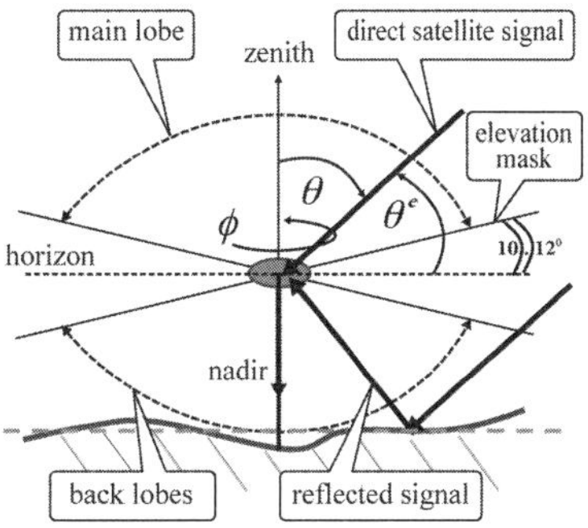

Figure 9.2.3 Coordinate frame for receiving GNSS antenna over undelaying terrain.

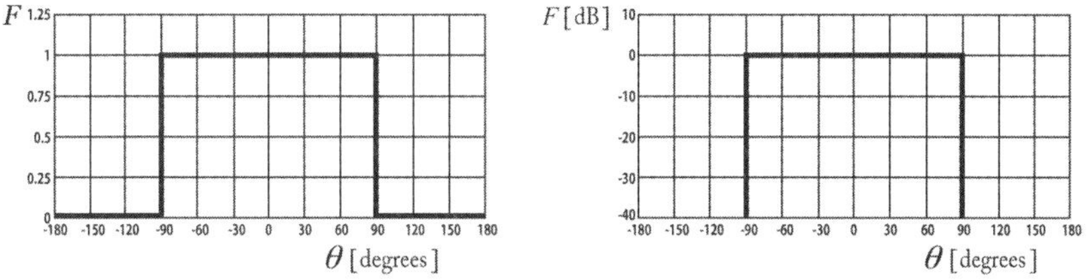

Figure 9.2.4 Perfect GNSS user antenna pattern.

independent of ϕ. Such an ideal antenna would provide an unbiased position in the horizontal plane. Despite such an ideal antenna one still has to deal with multipath, which is considered an essential error source in today's high-precision GNSS positioning. For good GNSS sites selection the only source of multipath is reflections from terrain below the antenna. Therefore, for the complete rejection of multipath signals, a perfect antenna would have to have a zero antenna pattern in directions below the horizon. In order to have equal reception capabilities for all satellites in view the antenna would have to have a constant pattern for directions above the horizon. In normalized form this constant is unity. Hence a perfect GNSS user antenna would have a step-like pattern equal to unity for directions from zenith down to horizon and equal to zero from horizon to nadir (left panel in Figure 9.2.4). Right panel of the Figure 9.2.4 illustrates the same pattern in dB scale. The pattern is to have the RHCP component only to match with the type of polarization radiated by the satellites.

It is well established in antenna theory that the speed by which the pattern changes with angles is generally proportional to the antenna size in wavelength scale. The step-like pattern of Figure 9.2.4 would require an antenna of infinitely large size [the reader might wish to check Lopez (1979) for a detailed discussion of the vertical array antenna size needed to approach the step-like pattern of Figure 9.2.4, more discussions are also provided in Sections 9.4.4, 9.7.4, and 9.7.7).

We now review the pattern of a GNSS user antenna that is of practical size. We use the pattern in the form of normalized power flux (9.1.125), express this pattern in dB, and plot it schematically in Figure 9.2.5. A high degree of rotational symmetry with respect to the vertical axis is assumed and only the region $0 \leq \theta \leq \pi$ is shown.

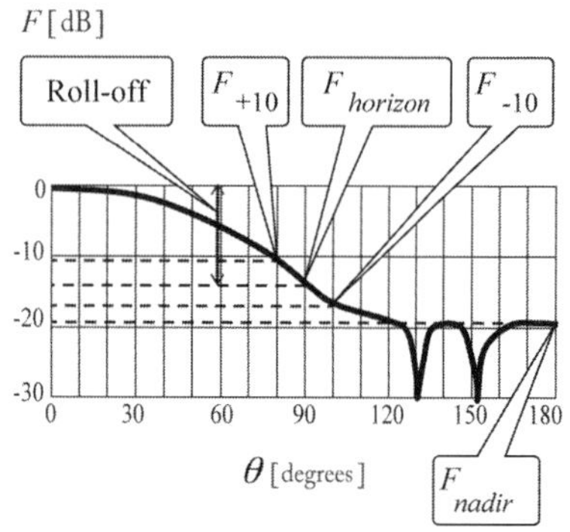

Figure 9.2.5 Typical pattern of a GNSS user antenna.

A typical receiver antenna pattern has a maximum value equal to unity (or 0 dB) normally in the direction of zenith. Then the pattern readings typically decrease as θ increases. Normally an elevation mask is used in processing. Typically the elevation mask is 10° to 12°. The reason for the mask angle is that satellite signals below those elevations are likely to be obstructed by natural and human-made obstacles. At the same time, the reliable tracking of low elevated satellites is absolutely necessary for high-precision positioning to keep the DOP factor at minimal levels. This means that the pattern readings for low elevations should be as high as possible (as small as possible in absolute value in the dB scale with a minus sign). We mark the pattern reading for 10° elevation as F_{+10}. This reading determines the system's ability to track low elevation satellites. In the literature and in special antenna documentation another pattern reading is sometimes mentioned. This is the reading for horizon direction $F_{horizon}$. The antenna pattern value in the direction of the horizon versus the zenith is often referred to as a roll-off. With typical designs the F_{+10} reading is slightly less than roll-off (in absolute value in dB). The angular region within the top semisphere above 10° to12° elevation is referred to as a major lobe of the GNSS user antenna pattern.

Now we discuss directions below the horizon. The pattern reading for 10° below the horizon is marked as F_{-10}. The reason is that given the specular reflections model, see Section 9.4, for satellite elevations of 10° and higher the reflections from underlying terrain will come from 10° and lower below the horizon. Thus the F_{-10} reading indicates a kind of weakest case from a multipath suppression standpoint. This reading sets up the largest multipath error contribution. Thus the reading is desired to be as small as possible (as large as possible in absolute value in dB with a minus sign). So far for typical GNSS user antennas F_{-10} is 5 to 6 dB below F_{+10}. With θ increasing below the horizon, the pattern values tend to decrease. Thus the multipath error associated with high elevated satellites normally is smaller as compared to low elevation multipath. This relation is analyzed in more detail in Section 9.4. There could be some sharp drops and local maximums of the pattern (these are referred to as back lobes). Finally we mark the F_{nadir} reading for $\theta = 180°$. This reading defines suppression of multipath associated with high elevation satellites.

Therefore, one encounters conflicting requirements that actually govern large portions of antenna design in high precision GNSS. With the antenna size generally given it is not possible to achieve both high F_{+10} and low F_{-10}. The collision between low elevated satellites tracking and multipath protection, while keeping the receiving GNSS antennas to practical size, is the major reason for achieving centimeter accuracy instead of millimeter positioning accuracy in real time!

Currently, there are two distinctly different types of the GNSS receiving antennas adopted in practice. The first is referred to as a reference station antenna. Currently, such antennas are normally about 40 cm (two wavelengths) in size and are installed at adequate open-sky sites. These antennas are to have the best multipath protection possible to ensure the highest quality reference data. The standard is the choke ring ground plane antenna.

The choke ring antenna was initially developed by the Jet Propulsion Laboratory (JPL) of NASA. This antenna has been serving the geodetic community for more than 20 years. Some of the design considerations of this antenna are discussed in Section 9.7.4 along with newly suggested developments. Figure 9.2.6 illustrates the CR4 antenna of Topcon Corporation. This antenna is a version of the original JPL choke ring antenna with Dorne and Margolin antenna element. In Figure 9.2.7 the antenna pattern for L1 and L2 signals of GPS are plotted in typical polar format.

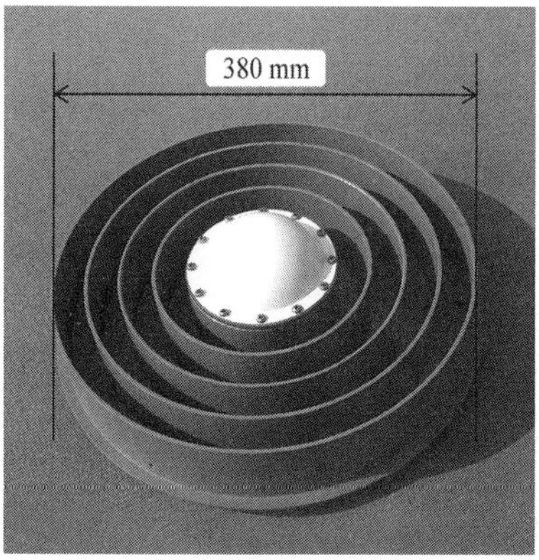

Figure 9.2.6 CR4 reference station antenna of Topcon Corp. JPL—original choke ring ground plane with Dorne & Margolin antenna element.

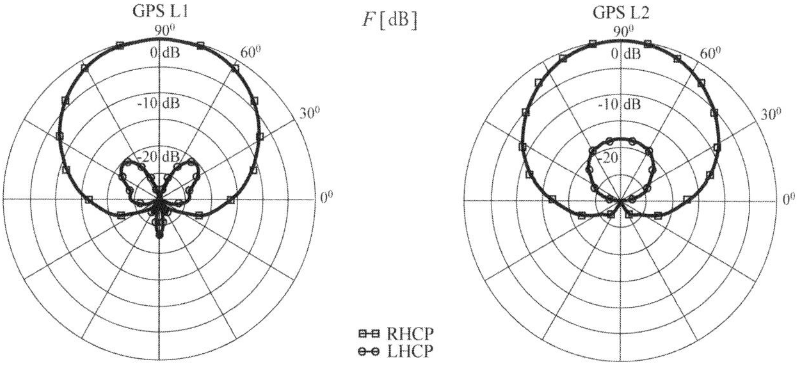

Figure 9.2.7 Antenna patterns of CR4 antenna for 1575 MHz (left panel) and 1227 MHz (right panel) frequencies.

Patterns for RHCP and LHCP components are shown separately as adopted in practice. The LHCP pattern is related to the RHCP pattern using expression (9.1.122). For directions above the horizon the RHCP component is of interest. The F_{+10} readings here are –14 to –15 dB for L1 and L2 signals. Approaching nadir the LHCP component is dominating. For this antenna F_{nadir} is –25 to –30 dB.

The second antenna type is referred to as a rover antenna, which is for practical use in the field. The art of design here is to provide compact and light-weight designs and multipath protection that is sufficient for standard accuracy of positioning. With the rover antennas, the F_{+10} reading normally is somewhat higher in order for the receiver to keep track of the low elevated satellites even under hard practical conditions. As an example, Figure 9.2.8 shows the rover antenna MGA8 of Topcon Corporation. One may notice the difference in size compared to the choke ring (see more about this antenna in Section 9.7.5). The RHCP and LHCP patterns of this antenna are shown in Figure 9.2.9 for L1 and L2 signals, respectively. The F_{+10} readings are −7 to −8 dB while F_{nadir} is about –20 dB.

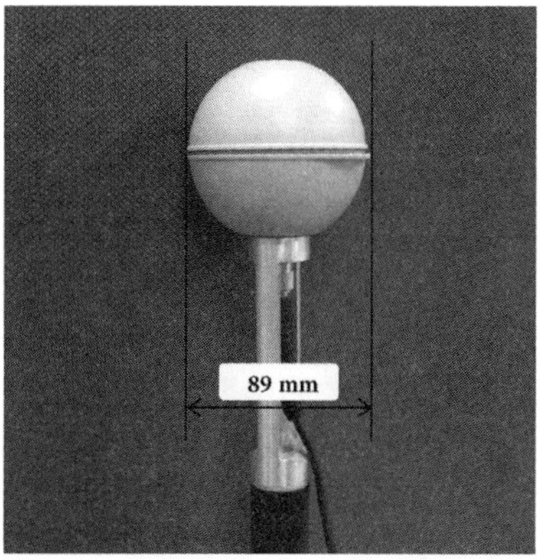

Figure 9.2.8 Dual-frequency rover antenna MGA8 of Topcon Corp.

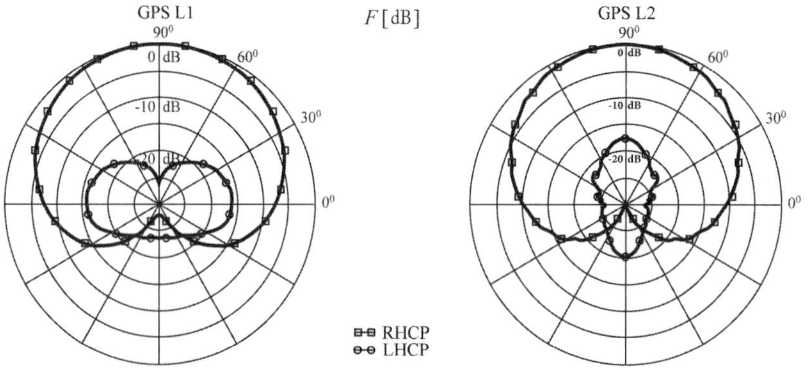

Figure 9.2.9 Antenna patterns of MGA8 antenna for 1575 MHz (left panel) and 1227 MHz (right panel) frequencies.

9.2.2 Directivity

Intuitively and from practical experience one knows that if one is to replace an antenna by a more "directional" antenna the signal strength might get "better." We will look into this phenomenon and discuss what we mean by saying better in the case of GNSS applications. Once again, we consider the transmitting mode of the antenna and refer to the equality of the antenna pattern in the receiving mode (Section 9.1.5).

We begin with the ideal isotropic radiator, which is an antenna having equal radiation intensity in all directions. Sometimes such radiators are called hypothetical because it can be proven that it is not possible to achieve such properties in practice. However, this hypothetical radiator is a convenient reference for directional properties of actual antennas, thus it is widely used.

Let P_Σ stand for the total power radiated by the isotropic radiator. One observes the radiation on an imaginary sphere in the far-field region. If the power is distributed uniformly in all directions, then the flux through each unit element of the sphere will be the total power divided by the total area of the sphere. The latter is $4\pi r^2$ where the radius of the sphere is r. Hence the power flux per unit square (module of the Poynting vector) is

$$\Pi_{isotropic} = \frac{P_\Sigma}{4\pi r^2} \tag{9.2.3}$$

It is the same in any direction fixed by angles θ and ϕ. Now we replace the isotropic radiator with the actual antenna that has a pattern of $F(\theta, \phi)$. The pattern is normalized such that its peak value equals unity. We assume for the moment that only the principal polarization component is radiated. We further assume that the total radiated power P_Σ is the same as that of the ideal isotropic radiator. The power flux through different unit elements of the sphere, however, will no longer be the same. The flux will depend on angles θ and ϕ and will be proportional to the antenna power pattern,

$$\Pi_{actual}(\theta, \phi) = \frac{P_\Sigma}{4\pi r^2} D_0 F^2(\theta, \phi) \tag{9.2.4}$$

The constant of proportionality is denoted by D_0. This constant is called antenna directivity. It shows the gain in power flux versus an ideal isotropic antenna for the direction where $F(\theta, \phi) = 1$. In other words,

$$\Pi_{max\ actual}(\theta, \phi) = \frac{P_\Sigma}{4\pi r^2} D_0 \tag{9.2.5}$$

Thus, the exact definition of directivity is as follows: We consider the direction in space in which the actual antenna radiates the peak signal. The antenna directivity shows how many times the power flux density in this direction would grow if one were to replace an ideal isotropic radiator with an actual antenna, assuming the same total radiated power by both antennas.

The meaning of directivity comes from general considerations of energy conservation law. An ideal isotropic radiator radiates with equal intensity in all directions. Since an actual antenna has some directivity, it will radiate in some directions less

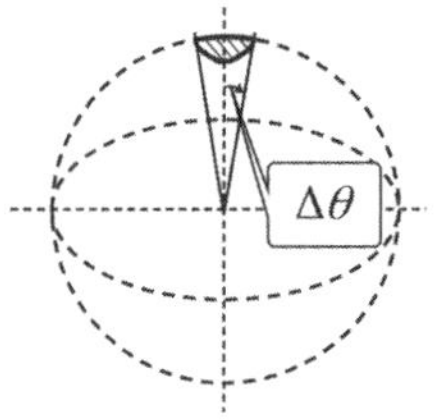

Figure 9.2.10 Directivity calculation of a narrow-beam antenna.

than the ideal isotropic antenna. Due to conservation of total power it must radiate more in other directions. The coefficient showing the power flux density growth in the direction in which the peak power flux is observed is the directivity.

We want to stress that the antenna itself is what is called a passive component. It does not contain any amplifiers of the signal at the antenna input. One is not to be confused with a low-noise amplifier (LNA), which is included into the user GNSS antenna housing but it actually goes after the antenna. What is called antenna directivity comes just from antenna pattern properties, the antenna is able to radiate more power in some directions just because it radiates less in other directions when compared to an isotropic source.

Let's look at some examples. First, we consider an antenna radiating all the power within a very narrow beam having an angular width of $\Delta\theta$ (Figure 9.2.10).

Narrow-beam antennas are used for radar purposes. For an imaginary sphere of radius r, the spherical area within the beam is

$$S = r^2 \pi \, \Delta\theta^2 \tag{9.2.6}$$

The power flux density through this area will be total radiated power divided by S, namely

$$\Pi = \frac{P_\Sigma}{r^2 \pi \, \Delta\theta^2} \tag{9.2.7}$$

The power flux density for an isotropic radiator is given in (9.2.2). Using the definition of the directivity D_0, one has

$$D_0 = \frac{\dfrac{P_\Sigma}{r^2 \pi \Delta\theta^2}}{\dfrac{P_\Sigma}{4\pi r^2}} = \frac{4\pi}{\pi \, \Delta\theta^2} \tag{9.2.8}$$

Thus for this case, the directivity is the full solid angle of 4π steradian divided by that of a beam. By making antennas with a narrow beam one can greatly increase the power flux. Radar antennas have a directivity of 30 dB (or 1000 times) and more.

The next example refers to the perfect GNSS receiver antenna of the previous section. In transmitting mode, such an antenna radiates uniformly in all directions within the semisphere above the horizon but will not radiate at all within the semisphere below the horizon. The area of the imaginary semisphere above the horizon is one-half of that of the full sphere. Hence,

$$D_0 = \frac{\dfrac{P_\Sigma}{2\pi r^2}}{\dfrac{P_\Sigma}{4\pi r^2}} = 2 \rightarrow +3 \text{ dB} \tag{9.2.9}$$

The directivity is equal to 2 or, equivalently, +3 dB. That is what one would achieve with a perfect GNSS antenna for both the satellite tracking and the multipath rejection point of view.

In order to derive the general formulas for the directivity, one must note that because of the power conservation law, the total radiated power results from integrating the power flux density over the imaginary sphere, thus

$$P_\Sigma = \int_0^{2\pi} \int_0^{\pi} \Pi(r, \theta, \phi) r^2 \sin(\theta) d\theta d\phi \tag{9.2.10}$$

Here, r is the radius of the sphere. Substituting (9.2.4) yields

$$P_\Sigma = \int_0^{2\pi} \int_0^{\pi} \frac{P_\Sigma}{4\pi r^2} D_0 F^2(\theta, \phi) r^2 \sin(\theta) d\theta d\phi \tag{9.2.11}$$

and

$$D_0 = \frac{4\pi}{\displaystyle\int_0^{2\pi} \int_0^{\pi} F^2(\theta, \phi) \sin(\theta) d\theta d\phi} \tag{9.2.12}$$

For a good GNSS user antenna with rotational symmetry with respect to the vertical axis, taking $F(\theta, \phi) = F(\theta)$ and performing integration over ϕ simplifies this expression to

$$D_0 = \frac{2}{\displaystyle\int_0^{\pi} F^2(\theta) \sin(\theta)\, d\theta} \tag{9.2.13}$$

Finally, the directivity pattern is

$$D(\theta, \phi) = D_0 F^2(\theta, \phi) \tag{9.2.14}$$

This pattern shows gain or loss in power flux density versus an isotropic radiator as a function of direction in space. The same quantity in dB scale reads

$$D(\theta, \phi)_{[\text{dB}]} = D_{0[\text{dB}]} + F(\theta, \phi)_{[\text{dB}]} \tag{9.2.15}$$

Since the antenna pattern is less than or equal to unity, $F(\theta, \phi)_{[\text{dB}]} \leq 0$ holds. Therefore (9.2.15) shows the power flux decrease for a particular direction related to maximum.

We look at more directivity examples. For the Hertzian dipole, substituting the antenna pattern (9.2.1) into (9.2.13), yields

$$D_0 = \frac{2}{\displaystyle\int_0^{\pi} \sin^3(\theta)\ d\theta} = 1.5 \rightarrow +1.8 \text{ dB} \tag{9.2.16}$$

For half-wave dipole employing (9.2.2),

$$D_0 = \frac{2}{\displaystyle\int_0^{\pi} \left(\frac{\cos\left(\frac{\pi}{2}\cos(\theta)\right)}{\sin(\theta)} \right)^2 \sin(\theta)\ d\theta} = 1.64 \rightarrow +2.15 \text{ dB} \tag{9.2.17}$$

It has been mentioned that the antenna pattern for the Hertzian dipole is a bit wider compared to the half-wave dipole (see Figure 9.2.2). This results in a slight decrease in the directivity of the Hertzian dipole.

Finally we turn to directivity estimates that resemble a typical GNSS user antenna. We start with replotting Figure 9.2.5 in relative units in terms of power (Figure 9.2.11, left panel). An ideal GNSS user antenna power pattern is shown as a dashed line for comparison. We note that a 12- to 18-dB roll-off for the direction of the horizon in the relative power units scale means about 1/10 or less compared to zenith. Levels of about –20dB, which are typical for the back lobes area of the pattern, are 0.01 compared to zenith, and are shown schematically in the left panel in Figure 9.2.11.

Now we take note of the denominator in the directivity formula (9.2.13). We use an azimuth-independent version. Note that the "tail" of the pattern after 90° does not contribute any significant value to the integral because $F^2(\theta)$ is small. Therefore, for the user antenna directivity calculations it is permissible to integrate only over the top semisphere,

$$D_0 \approx \frac{2}{\displaystyle\int_0^{\pi/2} F^2(\theta)\sin(\theta)\, d\theta} \tag{9.2.18}$$

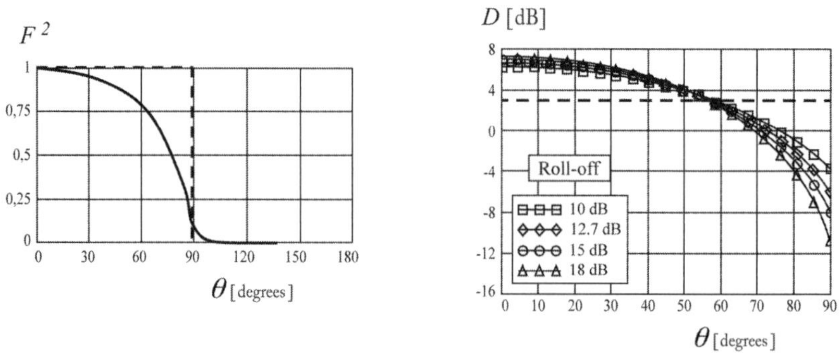

Figure 9.2.11 Directivity estimates of a typical GNSS user antenna.

But for directions in the top semisphere a typical user antenna pattern is smooth and could be approximated by a simple function.

We choose a cosine function with the only degree of freedom being the $F_{horizon}$ value. We approximate the pattern as

$$F(\theta) = \frac{\Delta + \cos(\theta)}{\Delta + 1} \tag{9.2.19}$$

This gives $F(0) = 1$ for the zenith direction, and for horizon direction we get

$$F_{horizon} = F\left(\frac{\pi}{2}\right) = \frac{\Delta}{\Delta + 1} \tag{9.2.20}$$

By varying Δ one can approximate a practical pattern. For instance, with $\Delta = 0.3$ one has a 12.7-dB roll-off which is a typical practical figure. Substituting the approximation (9.2.19) into (9.2.18) and using $\Delta = 0.3$, yields +6.7-dB directivity. This is a typical figure for a GNSS user antenna.

Next, we look into signal power versus zenith angle θ. For this purpose we consider the antenna directivity pattern (9.2.14). For a given roll-off value in dB one calculates Δ by inverting (9.2.20),

$$\Delta = \frac{10^{F_{horizon}[\mathrm{dB}]/20}}{1 - 10^{F_{horizon}[\mathrm{dB}]/20}} \tag{9.2.21}$$

This expression is substituted into (9.2.19) and then into (9.2.18) and (9.2.14). This way one has the directivity pattern (9.2.14) with the roll-off values $F_{horizon}$ as a parameter. This is plotted in the right panel in Figure 9.2.11 in dB scale. The perfect GNSS antenna directivity pattern is also shown as the dashed line for comparison. This antenna has a constant directivity pattern equal to +3 dB.

We see that if a typical user antenna has 12.7-dB roll-off, then the directivity for the horizon is about 4 dB below a perfect antenna. This means that the typical user antenna will receive 4 dB less signal power from low elevation satellites compared to a perfect one. For 18-dB roll-off the loss is about 7 dB. It has been pointed out previously that signal power for low elevation satellites is important from a reliability of signal tracking standpoint. In actuality, power losses against a perfect antenna will be even slightly larger due to a number of loss factors discussed in Section 9.2.4. Additionally, with a roll-off increase in dB scale the directivity for zenith slightly increases. For 12.7-dB roll-off one has 6.7-dB zenith directivity while for 18-dB roll-off the directivity is 7.2 dB. This stays in agreement with the main directivity features: with a roll-off increase, the pattern becomes somewhat narrower and, thus, directivity for zenith D_0 should grow.

Given estimates are important for two reasons. First, one is to note that slight variations of zenith directivity D_0 are associated with larger variations in the directivity toward low elevated satellites. From this standpoint the zenith directivity, sometimes shown in the user antenna documentation, is not that informative. Second, the growth of the zenith directivity D_0 is associated with a decrease in the low elevation satellites' signal reception capabilities. Thus, larger zenith directivity generally means worse GNSS antenna performance! It is worth mentioning that this kind of result, which may sound paradoxical, is based on the general energy conservation law and no engineering effort can be applied to overcome this. We conclude that the better GNSS user antenna would have a wider pattern for the top semisphere with less zenith directivity. However, this is not to affect the multipath rejection capabilities. For more detailed technical information regarding GNSS antenna directivity, the reader is referred to Rao et al. (2013).

9.2.3 Polarization Properties of the Receiving GNSS Antenna

We begin with an example of a practical GNSS antenna. We take a circular metal ground plane and place two crossed half-wave dipoles over it (Figure 9.2.12, left panel). By means of a feed network, not shown in the figure, a high-frequency voltage is provided to both dipoles in such a way that the two voltages have the same amplitude and a 90° phase shift with respect to one another. The voltages induce currents over the dipoles. The currents are shown schematically as dashed arrows. The amplitudes of the currents are equal and denoted by I_0. The y-dipole current is 90° delayed in phase with respect to the x-dipole current. This is indicated by the $-i$ term with the y-dipole current.

An observation is in order. Specifics and flavor of engineering electromagnetics manifests itself by two governing equations (9.1.10) and (9.1.11) and libraries of books solving them for cases of particular interest. This resembles both the power of fundamental physics to catch the main features of a large area in a compact format, and sophistication of engineering to derive practically relevant conclusions from these "simple" principles. With engineering electromagnetics, similar to many other engineering branches, two methods are available for most cases: either more or less approximate or heuristic analytical estimates or exact computer simulations with special software packages. We will come across this many times in this chapter. In regard to the antenna shown in the left panel in Figure 9.2.12 no "simple" closed-form solution for the fields is available so far because the dipoles induce a

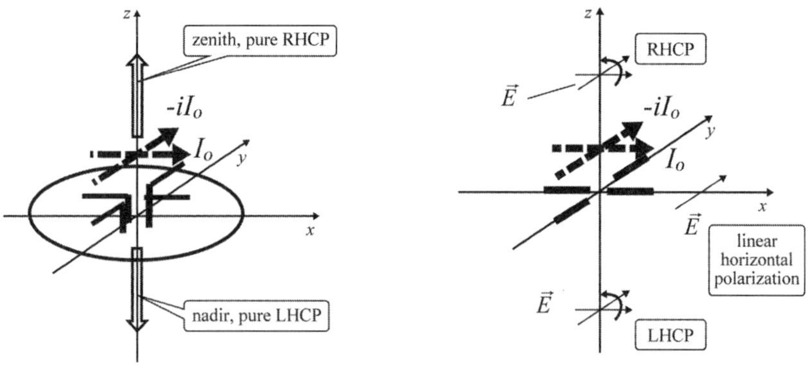

Figure 9.2.12 Cross-dipole antenna.

complex current distribution onto the ground plane. Some estimates and ground plane design guidelines will be discussed further in Section 9.7. As for this case, useful closed-form expressions for the far fields could be derived for a limited case of no ground plane. This design is not practical but it provides a general overview of polarization properties of the circular polarized antennas. Let us turn to a new case.

Consider two half-wave dipoles radiating in free space (right panel in Figure 9.2.12) and excited by voltages of above. As discussed in Section 9.1.6, the electromagnetic field could be equivalently represented as a sum of linear polarized waves or as a sum of circular polarized waves (far-field derivations for a cross-dipole system are shown in Appendix E). On the basis of the linear polarized wave, the far field is

$$\vec{E} = -iI_0\eta_0\frac{e^{-ikr}}{2\pi r}\left(F_\theta(\theta,\phi)e^{i\Psi_\theta(\theta,\phi)}\vec{\theta}_0 + F_\phi(\theta,\phi)e^{i\Psi_\phi(\theta,\phi)}\vec{\phi}_0\right) \tag{9.2.22}$$

Here

$$F_\theta(\theta,\phi)e^{i\Psi_\theta(\theta,\phi)} = \cos\theta(f(n_x)\cos\phi - if(n_y)\sin\phi) \tag{9.2.23}$$

$$F_\phi(\theta,\phi)e^{i\Psi_\phi(\theta,\phi)} = -i(f(n_y)\cos\phi - if(n_x)\sin\phi) \tag{9.2.24}$$

with

$$f(u) = \frac{\cos\left(\frac{\pi}{2}u\right)}{1-(u)^2} \tag{9.2.25}$$

For the circular polarized basis, the same field is

$$\vec{E} = -iI_0\eta_0\frac{e^{-ikr}}{\sqrt{2}\pi r}\left(F_{RHCP}(\theta,\phi)e^{i\Psi_{RHCP}(\theta,\phi)}\vec{p}_0^{\,RHCP} + F_{LHCP}(\theta,\phi)e^{i\Psi_{LHCP}(\theta,\phi)}\vec{p}_0^{\,LHCP}\right) \tag{9.2.26}$$

where

$$F_{RHCP}(\theta,\phi)e^{i\Psi_{RHCP}(\theta,\phi)} = \frac{1}{2}\begin{bmatrix}\cos\theta(f(n_x)\cos\phi - if(n_y)\sin\phi)\\ +(f(n_y)\cos\phi - if(n_x)\sin\phi)\end{bmatrix} \tag{9.2.27}$$

$$F_{LHCP}(\theta,\phi)e^{i\Psi_{LHCP}(\theta,\phi)} = \frac{1}{2}\begin{bmatrix}\cos\theta(f(n_x)\cos\phi - if(n_y)\sin\phi)\\ -(f(n_y)\cos\phi - if(n_x)\sin\phi)\end{bmatrix} \tag{9.2.28}$$

Directional cosines n_x and n_y are given by (E.14) and (E.15) of Appendix E and are illustrated in Figure E.2.

We take the case $\theta = 0$ (zenith) first. From (9.2.26), (9.2.27), and (9.2.28) one has $F_{RHCP} = 1$, $\Psi_{RHCP} = -\phi$, and $F_{LHCP} = 0$. This means that for the zenith direction such a cross-dipole antenna radiates a pure RHCP field. The importance of the phase pattern Ψ_{RHCP} being linear progressive with respect to azimuth will be discussed further in Section 9.3. Now we consider the horizon direction $\theta = \pi/2$. It is more convenient to use the linear polarized basis for the moment with expressions (9.2.22), (9.2.23), and (9.2.24). We see that the vertical (the θ th) component vanishes in this direction and the only remaining one is the horizontal (the ϕ th) component.

In short, one says that in the horizon direction a no-ground plane cross-dipole antenna radiates a horizontal linear polarized field. Finally, we take the nadir direction $\theta = \pi$. From (9.2.26), (9.2.27), and (9.2.28) one immediately recognizes that $F_{RHCP} = 0$ and $F_{LHCP} = 1$. Hence, a cross-dipole antenna radiates a pure LHCP field.

This basic result could be understood merely on the basis of symmetry. As was discussed in Section 9.1.4, in the direction perpendicular to a dipole the dipole radiates an electric field parallel to its axis. The electric field of a cross-dipole system follows dipole current relationships in amplitude and phase for the zenith direction. In other words, if one watches the electric field from the top of the z axis, it will rotate in a counterclockwise direction, manifesting RHCP. Due to the symmetry of the system, the field behavior in nadir direction will be the same. But now if one watches the rotation from the negative z axis, it will be seen in a clockwise direction, thus manifesting LHCP.

Next, there should be a direction in space where the rotation changes from right hand to left hand. Due to symmetry this must be the local horizon. Let us fix the x-axis direction with the local horizon plane. The dipole parallel to the x axis does not radiate in the x axis direction (see Section 9.1.4), instead, the y dipole radiates its maximum. Thus, within the plane of the local horizon, the polarization is linear horizontal.

The ground plane shown in the left panel in Figure 9.2.12 changes these results twofold. The ground plane will act like a mirror, making the radiation in an upward direction much more as compared to the downward direction. Thus the RHCP component at zenith direction will be much larger in magnitude compared to the LHCP component at nadir. That is what is required to suppress multipath reflections coming from underneath the antenna. Then, in general, an antenna with a ground plane will not have a pure linear polarization in horizon direction. Instead it will have an elliptical polarization with the RHCP component dominating.

A more general situation is seen in Figure 9.2.13, which shows a schematic illustration of the GNSS user antenna polarization properties. One can say that if the antenna radiates a pure RHCP field in the direction of the zenith, it will provide a pure LHCP radiation in antizenith or nadir. This follows from basic symmetry as discussed above. However, with these rotational properties, there should be some transition area where

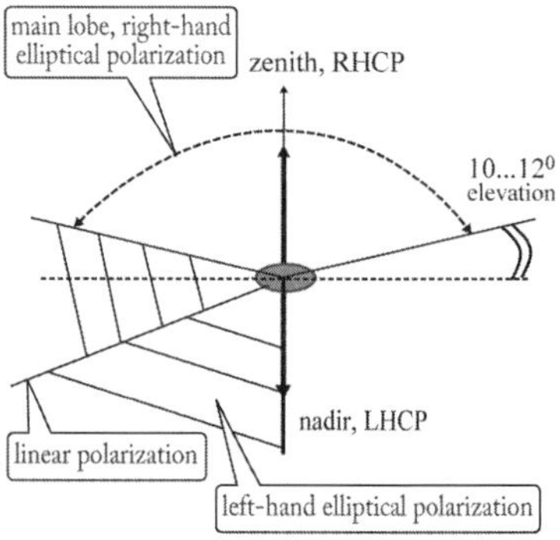

Figure 9.2.13 Polarization properties of a receiving GNSS antenna (schematically).

the rotation changes its direction. As mentioned in Section.9.2.1, the angular region from the zenith down to about 10° to 12° elevation could be referred to as the main lobe of the GNSS antenna pattern. Within this angular region, the polarization normally is almost perfectly right-hand circular for most of the GNSS antennas. Then the angular region of transition starts. In this region the polarization is getting more and more elliptical, but still has the right-hand direction of rotation. At angles of about 40° to 60° below horizon the polarization becomes linear. After that the LHCP component dominates (see Figures 9.2.7 and 9.2.9 illustrating RHCP and LHCP patterns for the choke ring and MGA8 antennas).

We complete this discussion by addressing the polarization loss term. We have established that in most directions a typical GNSS user antenna will have a somewhat elliptical polarization. This could be represented as a sum of pure RHCP and pure LHCP components. In Section 9.2.2 we discussed directivity assuming that only the principal polarization component is radiated. Now we take the cross-polarized component into consideration. We use power flux density in the form of (9.1.121). The total radiated power is

$$P_{\Sigma} = P_{RHCP} + P_{LHCP} \tag{9.2.29}$$

Here the power associated with co-(RHCP) and cross-(LHCP) polarized components is

$$P_{RHCP(LHCP)} = \frac{1}{2\eta_0} |U_{RHCP(LHCP)}|^2 \int_0^{\pi} \int_0^{2\pi} (F_{RHCP(LHCP)}(\theta, \phi))^2 \sin\theta \, d\phi \, d\theta \tag{9.2.30}$$

The power associated with the cross-polarized component is useless. We define polarization efficiency for the GNSS user antenna as

$$\chi_{pol} = \frac{P_{RHCP}}{P_{RHCP} + P_{LHCP}} \tag{9.2.31}$$

The term χ_{pol} is one of the loss factors to be considered with antenna gain (see next Section 9.2.4). Let us estimate polarization efficiency for the choke ring antenna using patterns shown in Figure 9.2.7. First, as we have done in Section 9.2.2, we neglect radiation in the bottom half of the sphere. This is possible because both RHCP and LHCP components provide only small contributions to directivity in this area. Now one notes that for most of the directions in the top semisphere the LHCP component relative to RHCP is about –15 dB or less. For estimation purposes we take –10 dB or 0.1 in related power units as an overestimate. This gives $\chi_{pol} = 0.9$ or -0.4 dB.

Finally, with the patterns seen in Figures 9.2.7 and 9.2.9 one may estimate the axial ratio term inverting the relationship (9.1.128). This axial ratio will be a function of θ. For instance, for L2 frequency of a choke ring antenna (Figure 9.2.7, right panel) at zenith, we take LHCP pattern reading as –18 dB relative to RHCP. This gives 0.126 in relative units. Expression (9.1.128) shows then that $\alpha_{ar} = 0.78$. For the horizon ($\theta = \pi/2$) taking –10 dB LHCP relative to RHCP, one has $\alpha_{ar} = 0.51$. We see that the axial ratio tends to decrease from zenith to horizon. This is common for all GNSS user antennas as discussed above. Normally α_{ar} is specified for the zenith direction in user antenna documentation. A typical requirement is for α_{ar} to not be less than 0.7 or −3 dB. However, the reason to limit the α_{ar} is not just polarization loss. Normally the decrease of α_{ar} (the growth in dB scale with minus sign) for the zenith indicates the loss of rotational symmetry of the antenna with respect to azimuth. As will be discussed in Section 9.3, this in turn results in the antenna phase center offset from the vertical axis, providing a biased position in the horizontal plane.

9.2.4 Antenna Gain

Antenna gain shows the actual gain or loss of signal power radiated in a certain direction against an ideal isotropic radiator. In Section 9.2.2, we discussed the gain that comes from directivity. Now we take loss factors into consideration. These losses are generally unavoidable. Once again, we keep discussing a transmitting mode of antenna functionality.

Consider a signal generator connected to the antenna by a transmission line (Figure 9.2.14). We will discuss some of transmission line basics in Section 9.5. For now, we just note that a part of the power P_g provided by the generator will be unavoidably reflected back from the antenna input. The physical phenomenon causing such a reflection is known as a mismatch. The reflected power will be lost. By χ_{ret} we denote a parameter known as return loss. This is the part of the power P_g lost due to reflections. One writes

$$P_a = (1 - \chi_{ret})P_g \tag{9.2.32}$$

Here, P_a is the useful power actually taken from the generator. Next, some of the power P_a will be absorbed by the antenna body. An antenna, like any other real-world body, absorbs electromagnetic field energy. We introduce χ_a as antenna efficiency and write

$$P_\Sigma = \chi_a P_a \tag{9.2.33}$$

where P_Σ is actually the power radiated into space. Finally, as discussed in the previous section, some of the radiated power is associated with the cross-polarized component of the field. This power is lost. Thus the final useful power is

$$P_{useful} = \chi_{pol}\chi_a(1 - \chi_{ret})P_g \tag{9.2.34}$$

This is the power that will be distributed in space due to the directivity properties of the antenna.

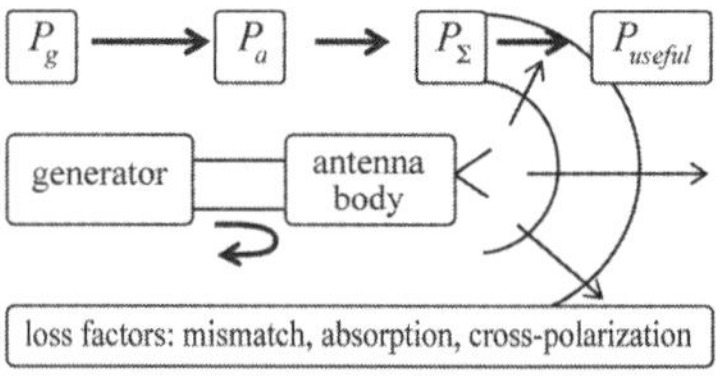

Figure 9.2.14 Antenna gain calculation.

By definition, the antenna gain G is the ratio of the radiation intensity in a given direction to the radiation intensity that would be obtained if the power accepted by the antenna were radiated isotropically (IEEE Standard, 2004). The reason for introducing antenna gain is its ability to show actual advantages or disadvantages of one antenna against another in terms of radiated power flux. The gain could be actually obtained by replacing the antenna under test with another one whose gain is known from precise calculations or measurements. By comparison to derivations in Section 9.2.2, one writes

$$G_0 = \chi_{pol}\chi_a(1 - \chi_{ret})D_0 \tag{9.2.35}$$

This is referred to as maximal gain or just gain. An antenna gain pattern for the receiving GNSS antenna is

$$G(\theta, \phi) = G_0 F^2_{RHCP}(\theta, \phi) \tag{9.2.36}$$

where $F^2_{RHCP}(\theta, \phi)$ is a power pattern of the RHCP component.

It is worth mentioning that for the transmitting mode of an antenna, the loss factors introduced are somewhat less important compared to the receiving mode. With the transmitting mode, the losses could be potentially compensated by an increase in power of a signal generator. On the contrary, with the receiving mode an antenna is the first sensor with the chain of transformation from the arriving signal to obtaining useful data. Each of the loss factors provides a certain signal damage that cannot be compensated for. This will be discussed in more detail in Section 9.6 in relation to the signal-to-noise ratio observed at the user GNSS receiver output. The natural requirement for high-precision GNSS user antenna designs is to keep all loss factors as small as possible.

We are to stress again that an antenna is a passive component, meaning no additional power is consumed from other sources for purposes of signal amplification. As seen with (9.2.35), the antenna gain comes only from the directivity property. All the loss factors contribute to a decrease in gain.

We look at some relevant numbers. The polarization efficiency χ_{pol} has been characterized already in the previous section. The antenna efficiency χ_a is kept at the highest levels by means of careful design and using appropriate materials. A typical estimate is for χ_a not to exceed –1 dB. We will see later that return loss χ_{ret} typically does not exceed –10 dB.

We conclude this section with a topic the user might come across while dealing with GNSS antennas—the gain transformation from one standard to another. Antenna directivity with respect to an isotropic radiator is quite easy to evaluate and discuss. However, as was mentioned already, an isotropic radiator cannot be constructed in practice. That is why another antenna is sometimes used as a standard to evaluate gain. The goal is to have the gain related to an antenna that could actually be built. Directivity related to the new standard $D_{0;new\ standard}$ is

$$D_{0;new\ standard} = \frac{\Pi_{actual}}{\Pi_{new\ standard}} = \frac{\Pi_{actual}}{\Pi_{isotropic}} \Big/ \frac{\Pi_{new\ standard}}{\Pi_{isotropic}} \tag{9.2.37}$$

Here Π stands for power flux density (module of the Poynting vector) as before. The half-wave dipole antenna is often used as a standard because half-wave dipole antenna can easily be constructed and tested. At the same time, the directivity of a half-wave dipole is known exactly; see (9.2.17). In dB units one has

$$G_{[\mathrm{dBd}]} = G_{[\mathrm{dBic}]} - 2.15 \tag{9.2.38}$$

Here, special notations commonly used are shown. dBd units denote the gain related to a dipole, while dBic is the gain related to the isotropic radiator.

9.2.5 Antenna Effective Area

Our goal is to make use of all the above derivations to estimate the signal power delivered to the output of the user GNSS antenna. We begin with the antenna effective area term.

In Figure 9.2.15, we show schematically the power flux from the signal source arriving at the antenna. For GNSS applications the signal source is the satellite. Let Π_{sat} be the power flux density from the satellite. The antenna collects this power and makes it available to the antenna output. The power flux density is measured in watts per square meter. The received power, $P_{received}$, is measured in watts. The received power is directly proportional to the arriving flux. Thus

$$P_{received} = \Pi_{sat} S_{eff} \tag{9.2.39}$$

The constant of proportionality S_{eff} should be measured in square meters. This term is referred to as the antenna effective area in the direction of the signal.

One is to note that, in general, the effective area is not related to an actual surface on the antenna elements. For instance, the effective area of a dipole antenna should not be treated as a sum of surfaces of wires comprising the dipole. Instead, one may think that a receiving antenna collects all the power flux that comes through a certain equivalent effective area. This effective area is not a constant for the given antenna, but rather it is proportional to the antenna power pattern. This is the case because of the reciprocity theorem discussed in Section 9.1.5.

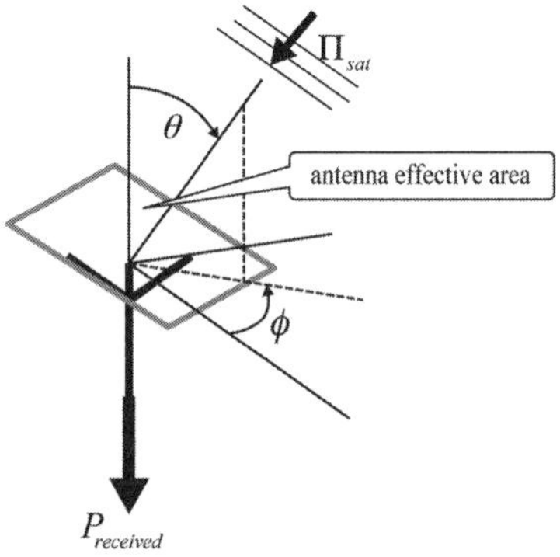

Figure 9.2.15 Definition of antenna effective area.

The key expression that relates the effective area with the antenna gain is

$$S_{eff}(\theta, \phi) = \frac{\lambda^2}{4\pi} G(\theta, \phi) \tag{9.2.40}$$

Here, $G(\theta, \phi)$ is the antenna gain pattern in transmitting mode. This relation constitutes the "bridge" between the transmitting and receiving mode antenna functionality. To estimate the receiving properties of the antenna, one is to employ this expression and make use of the gain derivations for the transmitting mode.

By substituting (9.2.35) and (9.2.26) into (9.2.40) one accounts for directivity and loss factors,

$$S_{eff}(\theta, \phi) = \frac{\lambda^2}{4\pi} \chi_{pol} \chi_a (1 - \chi_{ret}) D_0 F^2_{RHCP}(\theta, \phi) \tag{9.2.41}$$

Then the received power (9.2.39) in explicit form is

$$P_{received} = P_{0sat} \chi_{pol} \chi_a (1 - \chi_{ret}) D_0 F^2_{RHCP}(\theta, \phi) \tag{9.2.42}$$

Here, P_{0sat} is the standard power received by the user antenna with a unit gain. This value is specified by satellite developers. We take it as 10^{-16} W.

It is common to use (9.2.42) in the dB scale rather than in power units. For this purpose it is customary to use not 1 W but rather a 1-mW power as a reference. The power related to 1 mW has a special designation "dBm"—reads as "dB to mW." The received power (9.2.42) in dBm is

$$P_{received[\mathrm{dBm}]} = -130 + D_{0[\mathrm{dB}]} + F_{RHCP[\mathrm{dB}]}(\theta, \phi) + \chi_{pol[\mathrm{dB}]} + \chi_{a[\mathrm{dB}]} + (1 - \chi_{ret})_{[\mathrm{dB}]} \tag{9.2.43}$$

We note that all the terms except for $D_{0[\mathrm{dB}]}$ are negative, thus decreasing the power. From this standpoint the power received from low elevated satellites is of special interest. As discussed in Section 9.2.1, this power is F_{+10} dB below zenith. In actuality, this power is even smaller due to a number of loss factors such as directivity of a satellite antenna and increase in path propagation loss. We list all the other loss factors for convenience: $\chi_{pol[\mathrm{dB}]} = -0.4\,\mathrm{dB}$, $\chi_{a[\mathrm{dB}]} = -1\,\mathrm{dB}$, $(1 - \chi_{ret})_{[\mathrm{dB}]} = -0.45\,\mathrm{dB}$.

It is instructive to mention that if one takes $D_0 = +6\,\mathrm{dB}$, the loss estimates just given above, and assuming a GNSS wavelength of 20 cm, the expression (9.2.41) gives an antenna effective area of $S_{eff} \approx 80\ \mathrm{cm}^2$ for the direction where $F^2_{RHCP}(\theta, \phi)$ reaches unity (normally in zenith). This is independent to antenna type.

9.3 PHASE CENTER

The antenna phase center and phase center variations are quantities that explicitly characterize a receiving GNSS antenna as a geodetic instrument. The physical nature of these quantities originates from the antenna phase pattern. This section starts with an overview of phase pattern details.

9.3.1 Antenna Phase Pattern

Antenna phase pattern has been introduced in Section 9.1.4. In the receiving mode, the phase pattern shows carrier phase delays or advances caused by the antenna as a function of the direction from which the signal arrives. Assuming an RHCP signal is radiated by a satellite, we consider only the principal polarization component of the receiving antenna pattern which is RHCP. We denote the phase pattern as $\Psi(\theta, \phi)$ and omit the RHCP designation for simplicity of writing. The phase pattern is usually measured in radians or degrees. The direct way of measuring $\Psi(\theta, \phi)$ would be the same as with antenna pattern measurements shown in Figure 9.2.1. The antenna under test and the source are located in the anechoic chamber at the far-field region of each other. The antenna under test is rotated against some fixed center of rotation. Signal phase delays or advances as a function of angular rotation generate the phase pattern.

One immediately recognizes two important points. First, the antenna phase pattern is accurate up to a constant term, because the distance between the source and antenna under test can be arbitrary as long as they are at the far-field region of each other. The reader is referred to discussions at the end of Section 9.1.4. We will see in the next section that the uncertainty of this constant term should be properly accounted for. Second, the thus defined antenna phase pattern is related to an adopted center of rotation. We will call this center of rotation the antenna reference point (ARP). For the discussion about antenna patterns in Section 9.2.1 the center of rotation was not that important because the focus was on magnitudes of fields. Let us displace the center of rotation against ARP. If the displacement is small compared to the distance between the two antennas, the antenna pattern readings will not be affected. But if the displacement is not small in wavelength scale, then the associated path delay change will be noticeable compared to 2π.

First, we look into the transformation of phase pattern if the center of rotation is changed. One is to consider Figure 9.3.1.

We prefer to think in terms of the situation shown in the left panel Figure 9.1.6. Namely, let the antenna under test be fixed and the source be rotated. Let the initial

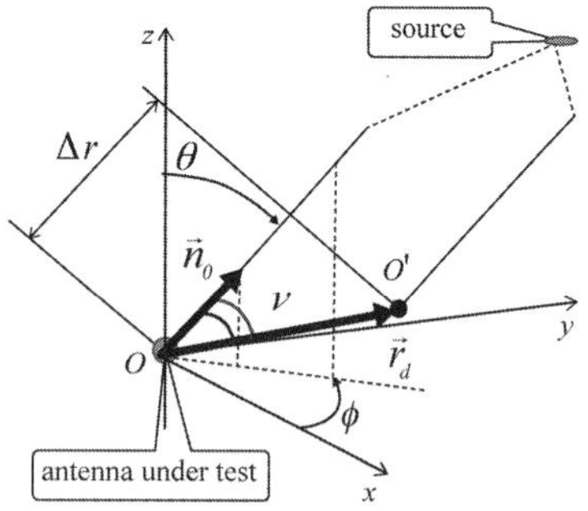

Figure 9.3.1 Antenna phase pattern transformation with the center of rotation change.

center of rotation be O and the new one be O'. The displacement vector $\vec{r}_d$ connects the old and new centers. The angles θ and ϕ give the direction of the source in the coordinate frame associated with the antenna under test, $\vec{n}_0$ is a unit vector pointing toward the source. We use a logic that has already been applied in Section 9.1.4: as long as the displacement $|\vec{r}_d|$ is small compared to the distance to the source, the directions to the source from O and O' are essentially parallel. Let the "old" phase pattern be $\Psi(\theta, \phi)$. This "old pattern" is measured with the source rotating with respect to the center O. When the source rotates with respect to the new center O' the signal exhibits extra delays or advances due to propagation path change by a factor Δr. From the right triangle with Figure 9.3.1

$$\Delta r = |\vec{r}_d| \cos \nu = \vec{r}_d \vec{n}_0 \tag{9.3.1}$$

Here ν is the angle between $\vec{r}_d$ and $\vec{n}_0$. Introducing Cartesian projections of $\vec{r}_d$

$$\vec{r}_d = x_d \vec{x}_0 + y_d \vec{y}_0 + z_d \vec{z}_0 \tag{9.3.2}$$

and $\vec{n}_0$ (directional cosines)

$$n_x = \sin\theta \cos\varphi \tag{9.3.3}$$

$$n_y = \sin\theta \sin\varphi \tag{9.3.4}$$

$$n_z = \cos\theta \tag{9.3.5}$$

one writes the dot product at the right hand side of (9.3.1) as

$$\Delta r = \vec{r}_d \vec{n}_0 = x_d \sin(\theta)\cos(\phi) + y_d \sin(\theta)\sin(\phi) + z_d \cos(\theta) \tag{9.3.6}$$

Thus the carrier phase delays or advances will be

$$\Delta\Psi = k\Delta r \tag{9.3.7}$$

Here k is a wavenumber (9.1.32), in which λ denotes the wavelength. The total phase of the field will be recognized as the antenna phase pattern related to the new origin $\Psi'(\theta, \phi)$ such that

$$\Psi'(\theta, \phi) = \Psi(\theta, \phi) - \Delta\Psi \tag{9.3.8}$$

Making use of (9.3.6), (9.3.7) yields

$$\Psi'(\theta, \phi) = \Psi(\theta, \phi) - \frac{2\pi}{\lambda}(x_d \sin(\theta)\cos(\phi) + y_d \sin(\theta)\sin(\phi) + z_d \cos(\theta)) \tag{9.3.9}$$

Therefore the change in phase pattern due to the center of rotation displacement equals a change in path delay for each direction (θ, ϕ), expressed in angular units.

An important property of circular polarized antennas and of receiving GNSS antennas, in particular, has been mentioned already in Section 9.2.3. Namely, the phase pattern of an RHCP antenna can be written in the form

$$\Psi(\theta, \phi) = -\phi + \Psi_1(\theta, \phi) \tag{9.3.10}$$

The first term on the right-hand side is a liner progressive phase delay, i.e., the azimuth. This term is common for all the RHCP antennas. The second term is the remaining phase pattern. The linear progressive term gives rise to the so-called windup correction (see Section 6.2.4)

In order to finalize the overview of antenna phase pattern features, one is to note that an antenna phase pattern is obviously a periodic function of azimuth ϕ with period 2π. Such functions could be expanded in Fourier series with respect to azimuth. For a function of the zenith angle θ and azimuth ϕ an expansion into spherical harmonics is possible. Generally, spherical harmonics are widely used in many areas. For fundamental treatment in regards to fields theory see Morse and Feshbach (1953). A spherical harmonic GPS antenna phase pattern expansion was presented in Rothacher et al. (1995). Spherical harmonics are orthogonal in the space of functions defined at $0 \le \theta \le \pi$ and $0 \le \phi \le 2\pi$, thus constituting a basis in the space. The relevant expansion of an antenna phase pattern is

$$\Psi_1(\theta, \phi) = \sum_{n=0}^{\infty} \sum_{m=0}^{n} (A_{mn} \cos(m\phi) + B_{mn} \sin(m\phi)) P_n^m(\cos\theta) \tag{9.3.11}$$

Here $P_n^m(\cos\theta)$ are Legendre functions. Orthogonality conditions for these functions and other useful formulas can be found in Abramowitz and Stegun (1972) or Gradshteyn and Ryzhik (1994). The coefficients A_{mn} and B_{mn} can be determined using measured data for $\Psi_1(\theta, \phi)$ employing orthogonality.

9.3.2 Phase Center Offset and Variations

We begin with a definition. An antenna is said to have an ideal phase center if its phase pattern $\Psi_1(\theta, \phi)$ in (9.3.10) is constant or can be made constant by the transformation (9.3.9). For the first case, the phase center location is at the center of rotation to which the pattern is referred to, thus coinciding with the ARP. In the second case the phase center is offset from the center of rotation by the vector $\vec{r}_d$.

This definition is qualitatively clear. If the antenna has an ideal phase center it introduces an extra carrier phase delay or advance that is the same in all directions. Since the phase pattern is defined up to constant term, this extra delay or advance could be set to zero. In transmitting mode, in the far-field region such an antenna will be recognized as a point that radiates ideal spherical wavefronts. Unfortunately, no real-world antennas have an ideal phase center. Moreover, in general, the definition of what one takes as phase center depends on the application.

Let us look at a qualitative illustration first by observing a float oscillating on the water surface. If the float possesses perfect rotation symmetry, say like a cylinder

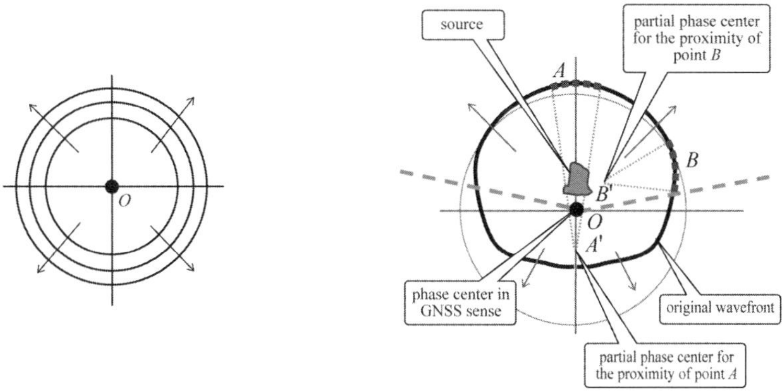

Figure 9.3.2 Ideal circular and real wavefronts and associated phase centers.

perpendicular to the surface, then the wavefronts generated by the float would be perfect circles as shown in the left panel in Figure 9.3.2. However, with the float having arbitrary shape the wavefronts get disturbed (right panel of the figure). Then, in the proximity of an observation point A the wavefront can be approximated by a circular arc shown by the thick dotted line. Let A' be the center of curvature of this arc. An observer located in the proximity of A will recognize the wavefronts as circles originating from A'. Similarly, an observer at point B will recognize wavefronts as originating from B'. Points A' and B' are referred to as partial phase centers, but these will not be of interest in regards to GNSS positioning.

In order to define the phase center term relevant to receiving GNSS antennas, we consider a typical situation of differential satellite positioning. In Figure 9.3.3 a base station and rover antenna are shown. Each antenna has its ARP. In practice the ARP is usually fixed at the antenna axis at the base of the threads plane. We assume that phase patterns of base and rover antennas $\Psi^{base}(\theta, \phi)$ and $\Psi^{rover}(\theta, \phi)$ are known with respect to the corresponding ARPs. The baseline vector $\vec{r}$ with projections (x, y, z) connects the ARPs of the base and rover. The goal of satellite positioning is to measure $\vec{r}$. We look closely into this procedure.

We use the logic of the previous section. We assume that the baseline length is much smaller compared to the distance to the satellite. Thus, one may take the satellite

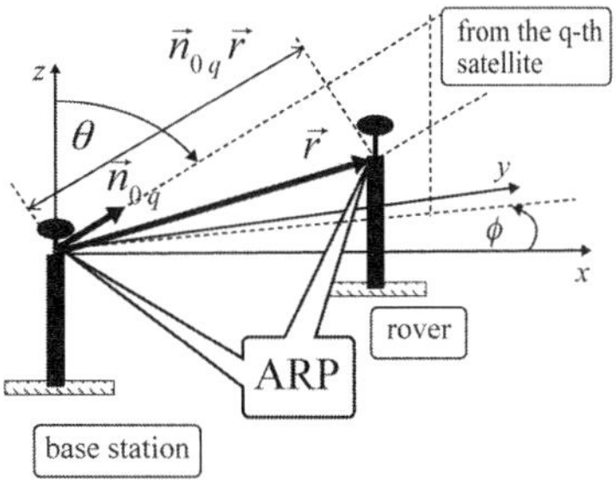

Figure 9.3.3 Calculation of phase center of GNSS user antenna.

directions from the base and rover as being essentially parallel. Let us assume that at some instant of time there is a satellite q seen at angles θ_q, ϕ_q. Let $\vec{n}_{0q}$ be a unit vector pointing toward satellite q. The Cartesian projections of $\vec{n}_{0q}$ are directional cosines:

$$
\begin{aligned}
n_{xq} &= \sin(\theta_q)\cos(\phi_q) \\
n_{yq} &= \sin(\theta_q)\sin(\phi_q) \\
n_{zq} &= \cos(\theta_q)
\end{aligned}
\tag{9.3.12}
$$

Let $\psi_q^{base(rover)}$ be the propagation path delay in radians for the signal traveling to the base (rover). We are interested in ψ_q, which is the difference in path delays between rover and base:

$$\psi_q = \psi_q^{rover} - \psi_q^{base} \tag{9.3.13}$$

Let the distance from satellite q to the base (rover) be $r_q^{base(rover)}$. Similar to the discussion in regards to Figure 9.3.1 one writes

$$\psi_q = -kr_q^{rover} - (-kr_q^{base}) = k\vec{n}_{0q}\vec{r} \tag{9.3.14}$$

Here k is a wavenumber (9.1.32). Note, that we omit all the tropospheric and ionospheric details because they cancel when differencing (9.3.14) for short baselines.

Now we turn to the equation that characterizes a practical observation case. For this purpose we introduce $\hat{\psi}_q$. This is the GNSS observable (the output of the respective receivers). One may write

$$\hat{\psi}_q = \psi_q + \Delta\psi_q + \tau \tag{9.3.15}$$

Here ψ_q is the exact value (9.3.13), $\Delta\psi_q$ is the direction-dependent error term, and τ is the constant error term. This constant term is related to hardware delays and initial clock offset. Similarly let $\hat{\vec{r}}$ be the GNSS estimate of the exact baseline $\vec{r}$. The estimate differs from the exact value by the error $\Delta\vec{r}$:

$$\hat{\vec{r}} = \vec{r} + \Delta\vec{r} \tag{9.3.16}$$

Then, the GNSS observation equation is

$$\hat{\psi}_q = k\vec{n}_{0q}\hat{\vec{r}} \tag{9.3.17}$$

which states that the GNSS observable is related to the baseline estimate in the same way as the exact carrier phase difference is related to the exact baseline.

By subtracting (9.3.14) in (9.3.17) and using (9.3.16) one has the error equation

$$k\vec{n}_{0q}\Delta\vec{r} = \Delta\psi_q + \tau \tag{9.3.18}$$

We proceed with analyzing this equation by making certain assumptions. The first assumption will be that there are no other error sources contributing to the extra phase delays or advances except for the base and rover antennas. These delays or advances are represented by the phase patterns of these antennas.

One immediately recognizes that for the carrier phase difference between base and rover the linear azimuth-dependent term in (9.3.10) will be common to both phase patterns and will cancel. From now on we will drop the subscript 1 with the remaining phase pattern at the right-hand side of (9.3.10) and always consider the phase pattern with the linear azimuth-dependent term subtracted. Thus

$$\Delta\psi_q = \Psi^{rover}(\theta_q, \phi_q) - \Psi^{base}(\theta_q, \phi_q) \tag{9.3.19}$$

This equation is valid if the distance r between the base and the rover antennas is small compared to the distance to satellite. Otherwise the windup correction arises.

Next, for the moment we assume that the base antenna has an ideal phase center at its reference point and define its phase pattern to zero. We drop the identification "rover" with the rover antenna phase pattern for the sake of simplicity of writing and have

$$\Delta\psi_q = \Psi(\theta_q, \phi_q) \tag{9.3.20}$$

Thus, we write (9.3.18) as

$$k\vec{n}_{0q}\,\Delta\vec{r} = \Psi(\theta_q, \phi_q) + \tau \tag{9.3.21}$$

Within a given observation session there are a total number of Q angles available from simultaneous observations at both stations. Therefore, one has Q equations like (9.3.21) which together make a system of linear algebraic equations

$$k\vec{n}_{0q}\,\Delta\vec{r} - \tau = \Psi(\theta_q, \phi_q); \quad q = 1, 2 \cdots Q \tag{9.3.22}$$

of four unknowns—the three components of the $\Delta\vec{r}$ vector and the angular independent delay term τ. The number of equations in (9.3.22) is much larger than four. Note that the goal here is to define the phase center location via the phase pattern with the assumption there are no other error sources in the observation session except for the phase pattern of the rover antenna. This means a common clock would be used in a practical implementation.

By definition one says that the least-squares solution of the system (9.3.22) will be the estimate of the rover antenna phase center offset $\vec{r}_{pc}$ relative to its ARP:

$$\vec{r}_{pc} = \Delta\vec{r} \tag{9.3.23}$$

This definition is quite reasonable. This situation is shown by the vector diagram in Figure 9.3.4 where we can see that the offset $\vec{r}_{pc}$ is the difference between the baseline estimate and actual baseline under the assumptions made.

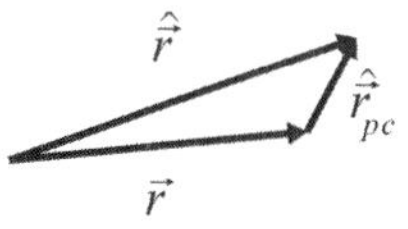

Figure 9.3.4 Phase center offset definition.

The least-squares solution to (9.3.22) is

$$S = \sum_{q=1}^{Q}(-k\vec{n}_{0q}\vec{r}_{pc} + \Psi(\theta_q, \phi_q) + \tau)^2 \rightarrow \min \tag{9.3.24}$$

Please note that up to this point a somewhat practical procedure has been discussed. Certainly there are no base station antennas having an ideal phase center and there are always additional error sources such as multipath. We will make certain corrections later and return to this procedure when we discuss antenna calibrations in the next section.

Finally we realize that the estimate of the rover antenna phase center offset as described above has an obvious drawback—it depends on the geometry of the observation session. Namely it depends on the total number of observations Q and on the respective angles θ_q, ϕ_q. To avoid this uncertainty one makes an additional assumption that during the observation session the satellites move in such a way that their paths cover the whole top semisphere continuously and homogeneously. Then we switch from the summation in (9.3.24) to integration. The exact formula is

$$S = \int_0^{2\pi}\int_0^{(\pi/2)-\alpha} \{-(k\sin(\theta)\cos(\phi)x_{pc} + k\sin(\theta)\sin(\phi)y_{pc} + k\cos(\theta)z_{pc}) + \Psi(\theta, \phi) + \tau\}^2 \sin(\theta)d\theta d\phi \rightarrow \min \tag{9.3.25}$$

Here x_{pc}, y_{pc}, z_{pc} are the exact rover antenna phase center offsets with respect to its reference point. The angle α is the elevation mask used with the observation session as there were no satellites considered below zenith angle $(\pi/2 - \alpha)$. Some comments on this formula are to follow.

First, as has been discussed previously, the phase pattern is defined up to a constant term. This constant is part of the constant term τ in (9.3.25). Thus τ includes the uncertainty of the phase pattern, hardware delays like length of cable connecting the antenna with the receiver, and initial clock offset. All of these constants do not affect the phase center location. Because of this, the antenna phase patterns can always be standardized by subtracting a constant term, for instance, by setting up the readings for zenith direction to zero. Next, the antenna phase center location is obviously a function of elevation mask. This should always be accounted for with positioning algorithms.

Finally, let the antenna have an ideal phase center located at (x_{pc}, y_{pc}, z_{pc}). Then, as follows from (9.3.9), its phase pattern with respect to the ARP is

$$\Psi(\theta, \phi) = k\sin(\theta)\cos(\phi)x_{pc} + k\sin(\theta)\sin(\phi)y_{pc} + k\cos(\theta)z_{pc} \tag{9.3.26}$$

Thus the formula (9.3.25) provides the solution to the problem of where to put a hypothetical antenna with an ideal phase center in such a way that its phase pattern would best fit the actual rover antenna phase pattern in the least-squares sense. This definition of antenna phase center is quite certain. By IEEE Standard (2004) it is referred to as an average phase center over the coverage region above the elevation mask. Getting back to the right panel in Figure 9.3.2 one may say that the phase center in the GNSS sense would be at point O. This point is the center of a circle best fitting the actual wavefront for directions within the angular sector of interest shown as thick dashed lines.

Now we turn to the solution of (9.3.25). Following a regular least-squares routine, the unknowns $(x_{pc}, y_{pc}, z_{pc}, \tau)$ are the solution to the system of linear algebraic equations

$$\frac{\partial S}{\partial x_{pc}} = 0 \qquad \frac{\partial S}{\partial y_{pc}} = 0 \qquad \frac{\partial S}{\partial z_{pc}} = 0 \qquad \frac{\partial S}{\partial \tau} = 0 \tag{9.3.27}$$

Taking partial derivations of the integrand, performing integration, and solving (9.3.27) yields

$$x_{pc} = \frac{\lambda}{2\pi^2} \frac{\int_0^{2\pi} \int_0^{(\pi/2)-\alpha} \Psi(\theta, \phi) \sin(\theta)^2 \cos(\phi) d\theta d\phi}{\int_0^{(\pi/2)-\alpha} \sin^3(\theta)\, d\theta} \tag{9.3.28}$$

$$y_{pc} = \frac{\lambda}{2\pi^2} \frac{\int_0^{2\pi} \int_0^{(\pi/2)-\alpha} \Psi(\theta, \phi) \sin(\theta)^2 \sin(\phi) d\theta d\phi}{\int_0^{(\pi/2)-\alpha} \sin^3(\theta)\, d\theta} \tag{9.3.29}$$

$$z_{pc} = \frac{\lambda}{4\pi^2} \frac{\dfrac{\int_0^{2\pi} \int_0^{(\pi/2)-\alpha} \psi(\theta, \phi) \cos(\theta) \sin(\theta) d\theta d\phi}{\int_0^{(\pi/2)-\alpha} \cos(\theta) \sin(\theta)\, d\theta} - \dfrac{\int_0^{2\pi} \int_0^{(\pi/2)-\alpha} \psi(\theta, \phi) \sin(\theta) d\theta d\phi}{\int_0^{(\pi/2)-\alpha} \sin(\theta)\, d\theta}}{\dfrac{\int_0^{(\pi/2)-\alpha} \cos(\theta)^2 \sin(\theta)\, d\theta}{\int_0^{(\pi/2)-\alpha} \cos(\theta) \sin(\theta)\, d\theta} - \dfrac{\int_0^{(\pi/2)-\alpha} \cos(\theta) \sin(\theta)\, d\theta}{\int_0^{(\pi/2)-\alpha} \sin(\theta)\, d\theta}} \tag{9.3.30}$$

We see, that in general the phase center is offset from the ARP in all three coordinates (x, y, z). Thus, generally, the user antenna should be oriented with respect to the local horizon.

An important particular case exists if the antenna possesses rotational symmetry with respect to the vertical axis. In this case its phase pattern is a function of the elevation angle θ only,

$$\Psi(\theta, \phi) = \Psi(\theta) \tag{9.3.31}$$

and from (9.3.28) and (9.3.29) it follows that

$$x_{pc} = y_{pc} = 0 \tag{9.3.32}$$

In practice such an antenna is sometimes referred to as zero centered. GNSS positioning with a zero-centered antenna would be independent of antenna rotation with respect to the vertical axis.

What remains is to note that if the base antenna does not possess an ideal phase center at its ARP then the carrier phase error would be like (9.3.19) and the error to positioning $\Delta\vec{r}$ in (9.3.18) would just be

$$\Delta\vec{r} = \vec{r}_{pc}^{\,rover} - \vec{r}_{pc}^{\,base} \tag{9.3.33}$$

Here $\vec{r}_{pc}^{\,base(rover)}$ is the phase center offset of the base (rover) antenna, respectively.

We next discuss the term phase center variation (PCV) which is used in GNSS practice. Once the phase center is defined let us transfer the antenna phase pattern from ARP to the phase center. This new phase pattern has the special designation PCV. Employing (9.3.9), one has

$$PCV(\theta, \phi) = \Psi(\theta, \phi) - \frac{2\pi}{\lambda}(x_{pc}\sin(\theta)\cos(\phi) + y_{pc}\sin(\theta)\sin(\phi) + z_{pc}\cos(\theta)) \tag{9.3.34}$$

In short, a PCV refers to the antenna phase pattern that is related to the phase center. Normally the PCV is expressed in length units rather than in angular units. The transformation coefficient with (9.3.34) is $\lambda/2\pi$; thus, the nature of the PCV term is clear. After the phase center has been defined, what remains of the phase pattern after transformation (9.3.34) looks like a slight phase center variation as expressed as a function of angles θ and ϕ.

We conclude with a remark. By substituting (9.3.11) into (9.3.28) and (9.3.29), the phase center offset in the horizontal plane is represented as a sum of contributions of the terms of (9.3.11). One may notice that all terms in the expansion (9.3.11) do not contribute to the phase center offset in the horizontal plane except for the term $m = 1$. Thus, claims of zero offset in the horizontal plane imply an antenna phase pattern (and PCV) that is not strictly constant in azimuth ϕ but rather possesses a certain degree of rotational symmetry with respect to azimuth—namely, with the term $m = 1$ vanishing. But the just-defined phase center refers to the mean value of the position within a very long observation session (strictly—with satellite tracks homogeneously covering the entire top the semisphere). Real-time positioning with about 10 satellites may exhibit noticeable deviations from the phase center if the PCV is large.

9.3.3 Antenna Calibrations

The practical procedures to determine the antenna phase center and the PCV are known as antenna calibrations. There are three procedures in use: anechoic chamber calibrations, relative calibrations and absolute calibrations.

Anechoic chamber calibration has already been discussed. Once the phase pattern is known by anechoic chamber measurements, one applies (9.3.28) to (9.3.30) and (9.3.34) to obtain phase center offset (PCO) and then the PCV. Using an anechoic chamber as an antenna-specific instrument potentially allows for detailed antenna characterization; phase center motion versus frequency, for instance, is of prime interest (Schupler and Clark, 2000). The difficulties of employing anechoic chamber calibrations are of a practical nature.

The typical accuracy requirement for the GNSS user antenna phase pattern measurements is 1 mm. Multiplying by $2\pi/\lambda$, with λ being the wavelength and taking $\lambda =$ 20 cm, one has an acceptable error of 0.031 rad. We will see later in Section 9.4 that the phase error in radians approximately equals the magnitude of the multipath-reflected signal in relative units. So the reflected signal magnitude should be less than 0.031 in relative units or –30 dB. But there is not just one multipath signal, say reflection from the floor as shown with Figure 9.2.1; there is also multipath from the walls and the ceiling. There are also implications as to disturbances from the equipment and setup arrangements. Taking these additional error sources into account, one arrives at a requirement for the anechoic properties of the chamber to be below –40 dB. These are very stringent requirements.

Relative antenna calibration has been the practical solution to the problem for more than two decades. The relative calibration technique was developed and has been used at the antenna calibration center operated by National Geodetic Survey (NGS) of the United States (Mader, 1999).

In order to understand relative calibrations we revisit the procedure of the previous section again. We drop the assumption that the base antenna has an ideal phase center and use the carrier phase delay error in the form of (9.3.19) and the positioning error as (9.3.33). However, now we use these equations in the reverse order. Let us assume that a baseline between the ARPs of base and rover is known a priori from direct measurements, e.g., as obtained from optical instruments. In Figure 9.3.5 points *A* and *B* represent the base and rover ARPs, respectively. Let the base antenna phase center offset *CA* be known a priori by some other measurements, e.g., from anechoic chamber calibrations. Then we take actual field GNSS carrier phase observations and determine a baseline *CD* which is the baseline between the two phase centers. After that,

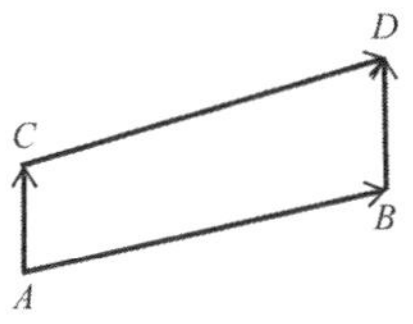

Figure 9.3.5 Relative antenna calibrations procedure.

the rover phase center offset *DB* is obtained from the vector quadrangle. Now the difference between phase center offsets (9.3.33) is substituted into equation (9.3.22) and the difference between phase patterns (9.3.19) is calculated for the set of angles (θ_q, ϕ_q) for that observation session. This is a practical relative calibration procedure.

Obviously this procedure does not allow the determination of the rover antenna phase pattern if the pattern of the base is not known. However, the power of this approach is that, if the same base antenna is always used as the standard, there is no actual need for its phase pattern. It could be set to zero. This is because in any high-precision GNSS signal processing algorithms that use single or double differencing, it is the difference between the phase patterns of the two antennas at the ends of a baseline that is relevant. If each of the two patterns has the same error equal to the antenna phase pattern of the standard, then this error would cancel in the difference as a common term. A database of calibrated antennas can be found at the NGS website http://www.ngs.noaa.gov/ANTCAL.

Finally, we discuss the so-called absolute antenna calibrations that were developed at the University of Hannover and the GEO++ company in Germany (Wübbena et al., 1996, 2000). Absolute calibrations give the antenna phase pattern directly, similar to anechoic chamber calibrations. However, for absolute calibrations there is no need for the chamber. GNSS satellites are used as signal sources instead.

The schematic of this situation is shown in Figure 9.3.6. Let the antenna under test—the rover—be rotated by a special robotic device similar to the rotations in the anechoic chamber. The base station antenna is being kept at a fixed position. Let satellite q be at the direction expressed by the angles θ_q, ϕ_q. At the time instant t_1 let the rover be inclined and rotated in such a way that, in the rover local coordinate system, the satellite q corresponds to the angles $\theta_{1;q}, \phi_{1;q}$. For this situation, the difference in carrier phase delays between rover and base will be as follows:

$$\Delta\psi_{1;q} = \Psi^{rover}(\theta_{1;q}, \phi_{1;q}) - \Psi^{base}(\theta_q, \phi_q) + ARP\,path\,delay_1 \qquad (9.3.35)$$

In this equation the first term is the rover phase pattern reading for angles $\theta_{1;q}, \phi_{1;q}$, the second term is the base antenna phase pattern reading for the angles θ_q, ϕ_q, and the

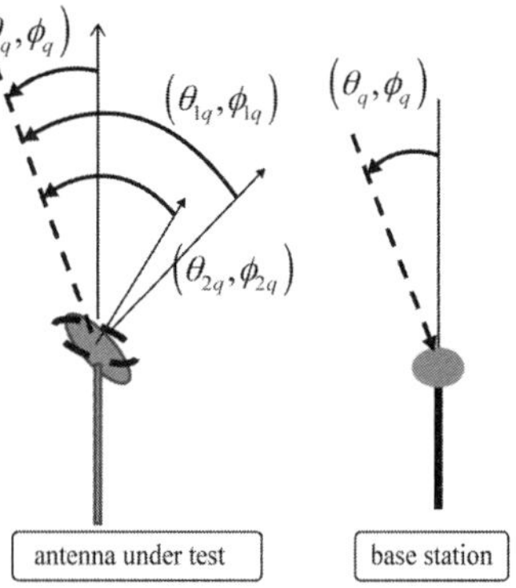

Figure 9.3.6 Absolute antenna calibrations (schematically).

third term is the path delay of the baseline between the two antenna reference points. The latter is known precisely by calibrating the robot. The corresponding equation at time instant t_2 is

$$\Delta\psi_{2;q} = \Psi^{rover}(\theta_{2;q}, \phi_{2;q}) - \Psi^{base}(\theta_q, \phi_q) + ARP\,path\ delay_2 \tag{9.3.36}$$

Please note that for the latter time instant the rover antenna is additionally rotated and inclined such that it "faces" satellite q by the angles $\theta_{2;q}, \phi_{2;q}$. If the robot can rotate the rover antenna fast enough, then, within the time t_1 to t_2, the satellite q has not moved significantly. The base antenna still "sees" satellite q under the same angles θ_q, ϕ_q. Then by differencing $\Delta\psi_{2;q}$ and $\Delta\psi_{1;q}$ the base antenna phase pattern cancel. Thus one has

$$\Delta\Delta\psi_{21;q} = \Delta\psi_{2;q} - \Delta\psi_{1;q} = \Psi^{rover}(\theta_{2;q}, \phi_{2;q}) - \Psi^{rover}(\theta_{1;q}, \phi_{1;q}) \tag{9.3.37}$$

The term on the left-hand side is referred to as time difference, i.e., the difference in carrier phase delays for two time instants. Note: known path delays are supposed to be subtracted from both the right- and left-hand side of (9.3.37).

Following this procedure, one gets the rover antenna phase pattern incrementally as angle-by-angle difference. If we define the phase delay in the zenith direction to be zero, then one may use the observed incremental changes to build a complete rover phase pattern in the sequence as given below:

$$\begin{aligned} \Psi^{rover}(0,0) &= 0 \\ \Psi^{rover}(\theta_{1;q}, \phi_{1;q}) &= \Delta\Delta\psi_{10;q} \\ \Psi^{rover}(\theta_{2;q}, \phi_{2;q}) &= \Delta\Delta\psi_{21;q} + \Delta\Delta\psi_{10;q} \end{aligned} \tag{9.3.38}$$

There certainly is a multipath error with any other GNSS observations. At any open site there will be multipath reflections from surrounding terrain arriving at the rover antenna. To eliminate these reflections from the pattern, the above procedure is to be performed with all the visible satellites in parallel, followed by averaging. Amendments to the just-described schematic accounting for satellites motion across the sky and other details of the procedure can be found in the references cited above or at the website of GEO + + company http://www.geopp.de/. For details about the absolute antenna calibrations at the U.S. NGS the reader is referred to Bilich and Mader (2010).

9.3.4 Group Delay Pattern

The antenna introduces delays not only to carrier phase but also to GNSS signal codes. In general, a delay in "signal" propagation is referred to as a group delay measured in units of time. This quantity is a function of the direction the signal arrives from; it is designated as a group delay pattern $\tau_g(\theta, \phi)$.

We follow the derivations of expression (9.1.40). For the signal $u(t)$ comprising two harmonics at close frequencies ω_1 and ω_2, we rewrite (9.1.40) in the form

$$u(t) = 2u_0 \cos(\Delta\omega(t - \tau_g) - \Delta kz) \cos(\omega t - kz - \psi) \tag{9.3.39}$$

Here, the group delay is

$$\tau_g = \frac{\Delta\psi}{\Delta\omega} \tag{9.3.40}$$

Taking the limit $\Delta\omega \to 0$, one has

$$\tau_g = \frac{d\psi}{d\omega} \tag{9.3.41}$$

Thus a group delay pattern is

$$\tau_g(\theta, \phi) = \frac{d\Psi(\theta, \phi)}{d\omega} \tag{9.3.42}$$

It equals the derivative of the phase pattern with respect to the angular frequency of the carrier.

In principle, one may introduce group delay center and variations for pseudoranges instead of carrier phases following exactly the derivations of the previous sections. However, in practice this approach is not commonly adopted. The reason is that normally with correct antenna designs that are capable of covering the GNSS signal bands (Figure 5.4.1, also see the remark at the end of Section 9.1.2), the pattern (9.3.42) is smooth enough such that the pattern contribution to the total error of code-differential techniques is negligible. For more material on antenna group delay see Lopez (2010) and Rao et al. (2013).

9.4 DIFFRACTION AND MULTIPATH

Multipath is one of the most frequently mentioned error sources in connection with high-precision positioning. Multipath is a particular case of diffraction phenomena; diffraction over a half-plane is an appropriate example to illustrate the semishadowing case. We begin this section with a discussion about diffraction. Multipath reflections from terrain underlying the receiving antenna and the role of antenna down-up ratio to mitigate the error are discussed after that.

9.4.1 Diffraction Phenomena

On the way from the satellite to the receiver the radiated signal encounters obstacles. These could be natural obstacles like trees or human-made obstacles like buildings. The term "diffraction" in general refers to a phenomenon that occurs when the wave interacts with an obstacle. Figure 9.4.1 illustrates the case. Here the electromagnetic field of an incident wave is illustrated by a plurality of wavefronts shown in solid lines.

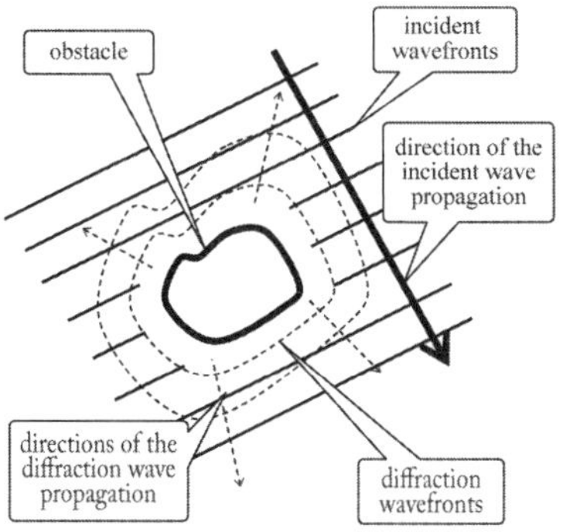

Figure 9.4.1 Diffraction phenomenon.

As a result of this interaction, a so-called diffraction field is generated. Diffraction wavefronts are shown as dashed lines. The diffraction field affects the amplitude and phase distribution of the incident field. Diffraction phenomena occur with waves of any kind, such as acoustic, electromagnetic, or say water surface waves. A common-sense example could be ocean waves interacting with an isolated rock in a bay. The branch of science treating diffraction phenomena is known as wave diffraction theory. A complete list of references on this subject would be endless. In regard to electromagnetics, the foundations are provided by Morse and Feshbach (1953), Felsen and Marcuvitz (2003), Fock (1965), Keller (1962), Ufimtsev (2003), Kouyoumjian and Pathak (1974), Balanis (1989), and references therein. One is to note that so-called strict analytical or closed-form solutions are available for a very limited number of obstacle models such as spheres, cylinders, or wedges. For cases where obstacles are much smaller or much larger compared to the wavelength, a group of asymptotic methods is available. In more general cases, thorough numerical simulations with special software packages are applied.

Back to GNSS, we note that the diffraction of satellite signals depends on the obstacle configuration and on what the obstacle is made of. We will see later in this section that with specular reflections the numbers are different for, let's say, dry or wet soil if assumed as a reflective surface. Also the GNSS diffraction scenery changes with time. For example, cars move, trees bend under the force of the wind, which makes exact simulations unrealistic. Thus, in general, it is not practical to estimate the potential diffraction-related errors to GNSS positioning with the accuracy high enough to compensate for them. Our focus will be on the main features of the phenomena and what can be done antenna-wise to mitigate the errors. The natural scale for diffraction problems is the wavelength λ. Similar to the previous sections we take $\lambda = 20$ cm for estimation purposes.

It has been noted already that the direct satellite signal is not like an optical ray having a negligibly small cross section. Instead in proximity of the user antenna a direct signal is a plane wave distributed in space. In general, if one disturbs the electromagnetic field of the wave at any point in space then the signal at the receiving antenna will also get disturbed. Our goal for the moment is to estimate a spatial area that must be free of obstacles so that the direct path can be considered nondisturbed.

An area in space that is relevant for the wave propagation from the transmitting (satellite) antenna to the receiving one (the user antenna) is referred to as a Fresnel zone. In general, the Fresnel zone consists of an infinite number of concentric ellipsoids of rotation with the transmitting and receiving antennas being the focal points. The ellipsoids are shown as dashed lines at the top-left panel in Figure 9.4.2. The outer radius ρ of the nth ellipsoid at the distance r_1 and r_2 from the transmitter and the receiver, respectively, is

$$\rho = \sqrt{n}\sqrt{\lambda \frac{r_1 r_2}{r_1 + r_2}} \tag{9.4.1}$$

The main contribution comes from the fields within the first few ellipsoids. For estimation purposes we take three of them. The third ellipsoid is referred to as the third Fresnel zone. We will call them simply Fresnel zone for short. For our case the satellite is the transmitter, $r_1 >> r_2$ and for $n = 3$

$$\rho \approx \sqrt{3\lambda r_2} \tag{9.4.2}$$

For instance, at the distance $r_2 = 10\,\text{m}$ from the receiving antenna we have $\rho = 2.45\,\text{m}$.

Now we briefly overview cases that may occur depending on where the obstacle is located within a Fresnel zone. If the obstacle covers part of the Fresnel zone (top-left panel in Figure 9.4.2), the satellite signal at the antenna output may still happen to be strong enough for the user receiver to track it, but the carrier phase would be affected. This is referred to as partial shadowing.

Another case is deep shadowing. This case occurs when the Fresnel zone is completely blocked by a large obstacle like a tall building (Figure 9.4.2, top-right panel). Deep shadowing normally leads to loss of tracking of the satellite signal by the user receiver. If signals of many satellites are blocked, then the user receiver is unable to provide a position. This is what happens in natural or urban canyons.

The third case is when the obstacles are located outside the Fresnel zone of the direct signal. Diffraction wavefronts caused by the obstacles are then referred to as reflections. Such reflected fields could happen to be strong at the user antenna proximity. When reflections from one or more sources arrive at the antenna along with the direct signal, this is referred to as multipath (Figure 9.4.2, bottom-left panel). Multipath reflections from terrain that is undelaying the user antenna are generally unavoidable. We will focus on multipath in subsequent sections.

Finally, the fourth case is referred to as near-field effects. One notes that an interaction of an incident wave and an obstacle could be viewed as generating a diffracted wave if the distance from the antenna to the obstacle is large compared to the antenna dimensions and the wavelength. Otherwise, an obstacle and an antenna are said to be in the near-field region of each other (see Section 9.1.4 for near-field region discussion). A body located in the near-field region of an antenna, strictly speaking, should be viewed as a part of the antenna. This body would affect the antenna pattern in terms of magnitude, phase, and antenna frequency response. In GNSS positioning, however, a receiving antenna could happen to be installed in close proximity to such

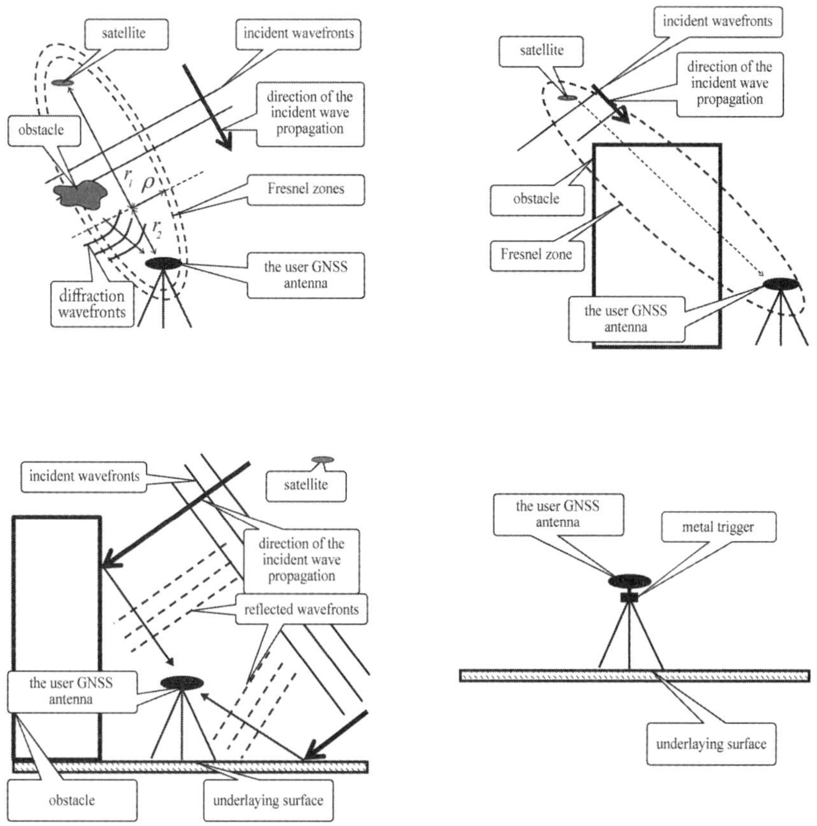

Figure 9.4.2 Diffraction cases relevant to GNSS positioning.

a body in a way that the disturbances are relatively small; the antenna performance would appear to be "almost" normal. For instance, this is the case if a body is located underneath the antenna. As was discussed in Section 9.2.1, the antenna gain in the directions below the local horizon is small compared to directions in the top semi-sphere. In antenna transmitting mode, a body located underneath is illuminated by a relatively weak field, thus having a small impact on the antenna characteristics. The same is true for the receiving mode due to reciprocity. However, errors in positioning introduced by such a body may happen to be noticeable (Dilssner et al., 2008). In this case, we adopt terminology from the GNSS literature and call such disturbances to GNSS observables a near-field multipath. A typical example would be a metal trigger normally used with an antenna on a tripod (Figure 9.4.2, bottom-right panel). If an antenna is too close to the trigger (say less than a wavelength of 20 cm), an extra error in the vertical coordinate produced by the trigger may reach noticeable values up to about half a centimeter. As discussed by Wübbena et al. (2011), near-field effects manifest themselves differently as compared to "regular" multipath reflections and, thus, could be identified in signal processing.

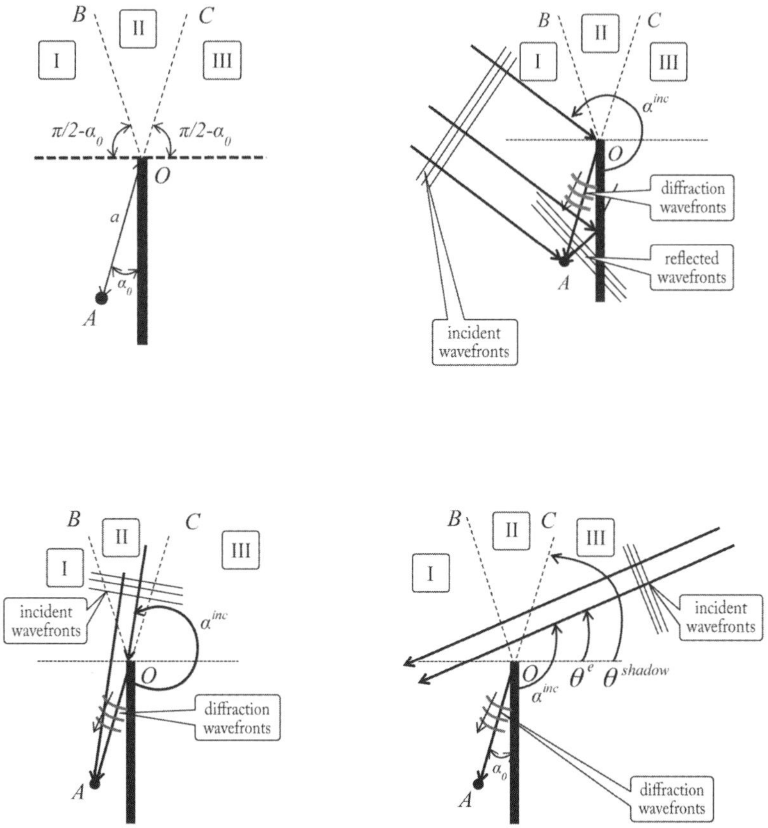

Figure 9.4.3 Diffraction over half-plane.

The above analysis based on Fresnel zones is in all respects an approximate one. The purpose of the analysis was to estimate the spatial spread of the processes. Now we are to illustrate the wave diffraction effects in more detail. In particular, we look into phenomena which occur in transition from partial to deep shadowing. For this purpose, we consider a case that allows complete analytical treatment.

Please look at the top-left panel in Figure 9.4.3. Here, a half-plane is shown as a thick solid line. The half-plane is the model of a tall building for instance. The half-plane is vertical and infinite in the downward direction and in the direction perpendicular to the drawing plane. We assume the half-plane to be a perfect conductor. The half-plane edge is located at the origin (point O). The local horizon is shown as a thick dashed line. We assume that the receiving antenna is located at point A (shown by a thick dot). Our goal is to calculate the field at point A when it is excited by a plane wave arriving from a satellite. The solution to the diffraction problem is described in Appendix F.

We assume that the distance a from point A to the half-plane edge is large compared to the wavelength. Let the angle α_0 fix the direction toward A with respect to the half-plane (top-left panel, Figure 9.4.3). We subdivide the top semisphere into three angular sectors. Let α^{inc} be the angle from which the incident wave (emitted by a satellite) arrives. If the wave arrives from the directions within sector I (top-right panel, Figure 9.4.3) such that $\pi + \alpha_0 \leq \alpha^{inc} \leq 3\pi/2$, then the receiving antenna is illuminated (using optical analogy) by three types of waves. The first one is an incident wave from the satellite, and the second is the wave reflected by the half-plane. The third one is a diffraction wave originating from the half-plane edge. If an incident wave arrives from the direction within sector II (bottom-left panel, Figure 9.4.3) such that $\pi - \alpha_0 \leq \alpha^{inc} \leq \pi + \alpha_0$, then the receiving antenna is illuminated by an incident wave and a diffraction wave. Finally, with an incident wave arriving within sector III (bottom-right panel Figure 9.4.3) such that $\pi/2 \leq \alpha^{inc} \leq \pi - \alpha_0$, the antenna at a point A is within the shadow region of an incident wave and there is no reflected wave in this region. The antenna is illuminated by diffraction wave only.

The incident and reflected waves mentioned above are referred to as geometrical optics fields. The waves are of the plane wave type discussed in Sections 9.1.2 and 9.1.3. The reflected wave manifests what is known as multipath arriving from the top semisphere. One might wish to check the derivations of Section 9.4.3 to estimate the area of the half-plane responsible for forming the reflected wave field. We need to mention that, staying with a perfectly conductive model of a half-plane, if an incident wave is RHCP then the reflected wave is LHCP.

A diffraction wave is of a so-called cylindrical type. This wave originates from the half-plane edge. The wavefronts of this wave are cylinders with the half-plane edge being an axis. At distances far from the origin, the cylindrical wave is locally a plane wave, meaning that in close proximity of an observation point, the vector field is constructed the same way as for a plane wave (compare to spherical waves discussion in Section 9.1.4). The difference with a spherical wave is that the field intensities decay as $1/\sqrt{r}$, with r increasing. Here, r is the distance from the source (half-plane edge).

The just described representation of the fields is valid if the direction of propagation of an incident wave is far from the shadow boundaries. In Figure 9.4.3 the lines OB and OC are shadow boundaries for reflected and incident waves, respectively. If an incident wave propagation direction is within narrow angular sectors immediately surrounding shadow boundaries, then the representation does not hold. One is to use a more general representation in the form of Fresnel integral (expression F.5 of Appendix F).

For angular sectors I and II, the cylindrical wave provides a minor contribution to the total field compared to an incident (and reflected) wave. However, for sector III the cylindrical wave is the only term. We look into this case.

We introduce the angle θ^{shadow} (see the bottom-right panel in Figure 9.4.3) such that

$$2\pi - \alpha_0 = \theta^{shadow} + 3\pi/2 \tag{9.4.3}$$

and satellite elevation angle θ^e such that

$$\theta^e = \alpha^{inc} - \pi/2 \tag{9.4.4}$$

Then, at an observation point A, a satellite with elevation angle $\theta^e < \theta^{shadow}$ is shadowed by the half-plane. In the proximity of point A the diffraction field is a local plane wave propagating from the origin O. An antenna response to the diffraction field will be as to a plane wave arriving from the direction θ^{shadow}.

Thus, the signal magnitude S^d at the antenna output is the product of two terms

$$S^d = F(\theta^{shadow})D \tag{9.4.5}$$

Here, $F(\theta^{shadow})$ is the antenna pattern reading for the elevation angle θ^{shadow}, and D is the diffraction term (see Appendix F)

$$D = \frac{1}{2\sqrt{\pi}\sqrt{2ka}\sin\left(\dfrac{\theta^{shadow} - \theta^e}{2}\right)} \tag{9.4.6}$$

This expression is valid when $\sqrt{2ka}\sin((\theta^{shadow} - \theta^e)/2) > 1$. At the same time, the carrier phase error of the diffracted signal as a function of the direct signal (which would occur if there was no half-plane), is

$$\Delta\psi^d = -ka(1 - \cos(\theta^{shadow} - \theta^e)) - \frac{\pi}{4} \tag{9.4.7}$$

Next, we look into data illustrating what happens if satellite elevation is slightly below θ^{shadow}.

In Figure 9.4.4 we plot D in dB and the carrier phase error in cycles (in fractions of 2π) versus satellite elevation angle θ^e. The values $a = 50\lambda$ (equal to 10 m assuming wavelength of 20 cm) and $\theta^{shadow} = 60°$ are adopted. As seen from the plots if $\theta^e = 50°$ (that is, 10° below θ^{shadow}), the error (9.4.7) approaches a cycle. With this θ^e the diffraction coefficient (9.4.6) provides an extra 18-dB signal attenuation to $F(\theta^{shadow})$. With today's sensitive receivers such a signal is potentially strong enough to be tracked. Thus, an erroneous ambiguities resolution for such a satellite is likely to occur. With smaller θ^e, the user antenna appears in the deeper shadow and the satellite signal is likely to be lost. These are the main features of phenomena which occur in transition from partial to deep shadowing.

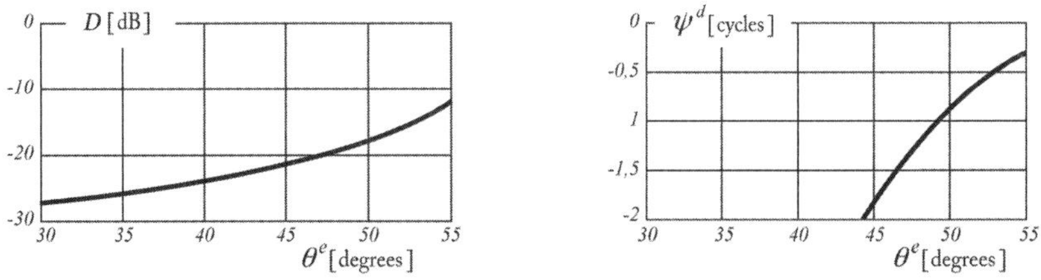

Figure 9.4.4 Amplitude and carrier phase errors with direct satellite signal shadowed by half-plane.

We conclude this discussion by mentioning that although real-time corrections to diffraction-related effects are so far impractical, increasing computer power allows for the modeling of these effects even for complex sites. Such simulations are of value for receiver design and test purposes. See discussion in Chen et al. (2009), Rigden and Elliott (2006), and Weiss et al. (2008)

9.4.2 General Characterization of Carrier Phase Multipath

Let a direct signal with amplitude U_0^{direct} and phase ψ^{direct} and a number of reflected signals with amplitudes $U_{0;q}$ and phases ψ_q be observed at the antenna output. The index $q = 1, 2, \ldots, Q$ identifies reflected signals. Using the derivations of Section 9.1.3, expression (9.1.52), one replaces all the reflected signals by one total reflected signal with amplitude U_0^{refl} and phase ψ^{refl} by writing

$$U_0^{refl} e^{i\psi^{refl}} = \sum_{q=1}^{Q} U_{0;q} e^{i\psi_q} \tag{9.4.8}$$

For short, we call this signal just the reflected signal. We denote the ratio of the reflected and direct signal amplitude as α^{refl}, such that

$$\alpha^{refl} = \frac{U_0^{refl}}{U_0^{direct}} \tag{9.4.9}$$

and denote the difference in the phase of reflected and direct signal as

$$\Delta\psi^{refl} = \psi^{refl} - \psi^{direct} \tag{9.4.10}$$

We consider the case where the reflected signal is weaker than the direct signal such that $\alpha^{refl} < 1$. For the total signal at the antenna output, one has

$$U_0^{\Sigma} e^{i\psi^{\Sigma}} = U_0^{direct} e^{i\psi^{direct}} + U_0^{refl} e^{i\psi^{refl}} = U_0^{direct} e^{i\psi^{direct}} \left(1 + \alpha^{refl} e^{i\Delta\psi^{refl}}\right) \tag{9.4.11}$$

Denote the expression in parenthesis as

$$\left(1 + \alpha^{refl} e^{i\Delta\psi^{refl}}\right) = \alpha^{mult} e^{i\psi^{mult}} \tag{9.4.12}$$

Thus, the total signal becomes

$$U_0^{\Sigma} e^{i\psi^{\Sigma}} = \left(U_0^{direct} \alpha^{mult}\right) e^{i(\psi^{direct} + \psi^{mult})} \tag{9.4.13}$$

This equation shows that the presence of the reflected signal leads to a multipath amplitude error α^{mult} and multipath carrier phase error ψ^{mult}.

The expression in parenthesis on the left-hand side of (9.4.12) can be analyzed on the complex plane (Figure 9.4.5). Here we are interested in the sum of two vectors. The first vector is equal to unity. The second vector has magnitude α^{refl} and angle

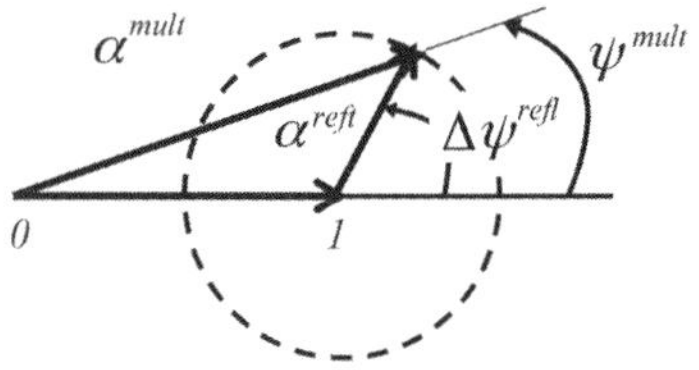

Figure 9.4.5 Representation of direct and reflected signals on a complex plane.

$\Delta\psi^{refl}$ relative to the first one. The magnitude (module) of the sum of two vectors gives the amplitude error

$$\alpha^{mult} = |1 + \alpha^{refl} e^{i\Delta\psi^{refl}}| = \sqrt{1 + 2\alpha^{refl}\cos(\Delta\psi^{refl}) + (\alpha^{refl})^2} \tag{9.4.14}$$

The argument of the sum gives carrier phase error

$$\psi^{mult} = \arctan\frac{\alpha^{refl}\sin(\Delta\psi^{refl})}{1 + \alpha^{refl}\cos(\Delta\psi^{refl})} \tag{9.4.15}$$

Now we are interested in the behavior of these errors as a function of α^{refl} and $\Delta\psi^{refl}$. We start with the amplitude error.

As seen from (9.4.14) and Figure 9.4.5 the amplitude error α^{mult} achieves a maximum of $(1 + \alpha^{refl})$ when $\Delta\psi^{refl} = 0$ and a minimum $(1 - \alpha^{refl})$ when $\Delta\psi^{refl} = \pm\pi$. For any other $\Delta\psi^{refl}$ the inequality holds.

$$1 - \alpha^{refl} < \alpha^{mult} < 1 + \alpha^{refl} \tag{9.4.16}$$

In the left panel in Figure 9.4.6 we illustrate α^{mult} in dB units as a function of $\Delta\psi^{refl}$ for different relative amplitudes of the reflected signal α^{refl}. As seen, unless α^{refl} approaches unity, the amplitude error α^{mult} stays within the range of a few dB. This may not constitute a problem to the receiver except for low elevated satellites. As was discussed in Section 9.2 the signal power at the antenna output for low elevation satellites is less than for high elevation satellites. A further decrease in signal power due

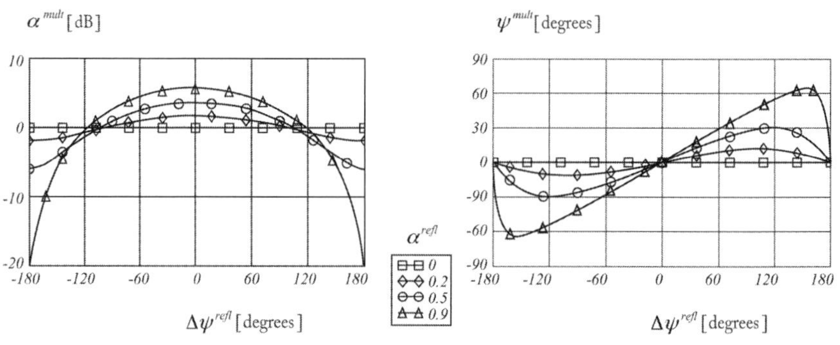

Figure 9.4.6 Amplitude and carrier phase errors caused by multipath.

to multipath may lead to the malfunctioning of the phase lock loops of the receiver. This is recognized as a cycle slip. When α^{refl} approaches unity, large signal drops occur when $\Delta\psi^{refl} \approx \pm\pi$. In communications theory this is referred to as multipath induced fading.

Now we turn to the carrier phase error (9.4.15). For a weak multipath signal with $\alpha^{refl} << 1$, using the approximation $\tan x \approx x$ for $|x| << 1$, yields

$$\psi^{mult} \approx \alpha^{refl} \sin(\Delta\psi^{refl}) \tag{9.4.17}$$

Thus the phase error ψ^{mult} oscillates with $\Delta\psi^{refl}$, and α^{refl} is the amplitude of these oscillations. In short, one may say that the multipath phase error in radians would be within $\pm\alpha^{refl}$. In the case of a strong multipath when $\alpha^{refl} = 1$, one obtains from (9.4.15):

$$\psi^{mult} = \arctan \frac{\sin(\Delta\psi^{refl})}{1 + \cos(\Delta\psi^{refl})} = \arctan \frac{\sin\left(\frac{\Delta\psi^{refl}}{2}\right)}{\cos\left(\frac{\Delta\psi^{refl}}{2}\right)} = \frac{\Delta\psi^{refl}}{2} \tag{9.4.18}$$

Assuming $-\pi < \Delta\psi^{refl} < \pi$, one has a maximal multipath carrier phase error $|\psi^{mult}|_{max} = \pi/2$. We plot (9.4.15) as a function of $\Delta\psi^{refl}$ and different α^{refl} in the right panel in Figure 9.4.6.

Summarizing, if the ratio of the total multipath signal amplitude and direct signal amplitude is $\alpha^{refl} \leq 0.6, \ldots, 0.8$, then the amplitude error can be about several dB and the carrier phase error about 60° at most. Under such circumstances loss of signal tracking does not occur and normally the receiver is able to provide correct ambiguities resolution. A multipath error is less destructive compared to shadowing from this point of view. However, multipath is in general unavoidable because reflections from the ground surface below the user antenna always exist.

9.4.3 Specular Reflections

There is an important case of multipath that is referred to as specular reflections. Figure 9.4.7 illustrates a direct wave from a satellite impinging upon a large plane surface. For the moment, we consider the surface to be an ideal unbounded plane and

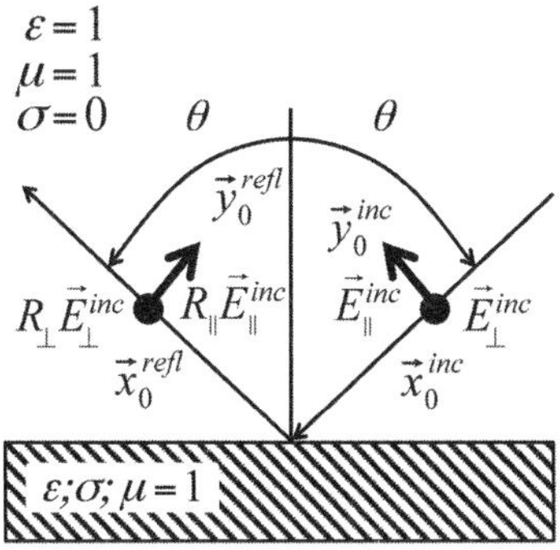

Figure 9.4.7 Definition of Fresnel coefficients with specular reflections.

the medium below the surface to be homogeneous. This may not look like a practically relevant model, but we already mentioned that our goal is not to derive exact formulas to account for multipath and multipath errors but to discuss the main features of the phenomena. Later in this section we will make some estimates on how large the surface should be for the unbounded plane model to be valid. Under the assumptions made, the wave diffraction problem allows for a closed-form solution. We will focus on the case when the ground surface undelaying the user antenna produces the reflections. But the same derivations are valid for any large plane surface, such as a wall of a tall building or a canyon.

Figure 9.4.7 also shows a reflected wave. This wave travels away from the surface in such a way that the angle of travel direction with respect to the normal of the surface is θ, which is also the angle of the incident wave direction. Such reflections are called specular. The amplitude of the reflected wave differs from the amplitude of the incident wave by a reflection coefficient known as the Fresnel coefficient. It is a function of the incident wave frequency, angle θ, and the parameters of the media. We are assuming the incident wave to be propagating in air. As was discussed in Section 9.1.1, the free space approximation is relevant for this case. The medium below the surface is characterized by permittivity ε and conductivity σ (see Table 9.1.1 in Section 9.1.1). We assume that this medium is nonmagnetic, thus its permeability is $\mu = 1$. There will also be a wave propagating inside the medium. This wave is called transmitted (we are not interested in this wave).

In order to calculate the reflection coefficient one must distinguish between two types of incident wave polarization. The first type is called perpendicular polarization. This is a linear polarization with the electric field vector of the incident wave being perpendicular to the drawing plane. This vector is designated by $\vec{E}_{\perp}^{inc}$ and is marked by a dot in Figure 9.4.7. Another type is called parallel polarization. Here, vector $\vec{E}_{||}^{inc}$ is in the drawing plane. It is worth mentioning that the electric field vector should always be perpendicular to the direction of wave propagation (Section 9.1.2). The two types of polarization make a vector basis in the plane perpendicular to the direction of wave travel. The reflected wave amplitudes are (Balanis, 1989)

$$E_{\perp}^{refl} = R_{\perp} E_{\perp}^{inc} \tag{9.4.19}$$

$$E_{||}^{refl} = R_{||} E_{||}^{inc} \tag{9.4.20}$$

with the reflection coefficients for perpendicular and parallel polarization, respectively,

$$R_{\perp} = \frac{\eta' \cos\theta - \xi}{\eta' \cos\theta + \xi} \tag{9.4.21}$$

$$R_{||} = -\frac{\eta' \xi - \cos\theta}{\eta' \xi + \cos\theta} \tag{9.4.22}$$

Here,

$$\eta' = \frac{\eta_m}{\eta_0} = \frac{1}{\sqrt{\varepsilon(1 - i \tan \Delta^e)}} \tag{9.4.23}$$

is the intrinsic impedance of the medium (9.1.37) related to that of free space, $\tan\Delta^e$ is the electric loss tangent (9.1.63) and the cosine of the complex refraction angle is

$$\xi = \sqrt{1 - \frac{\sin^2\theta}{\varepsilon(1 - i\tan\Delta^e)}} \tag{9.4.24}$$

Please note that expressions (9.4.21) and (9.4.22) give the reflection coefficients in complex form. Therefore, these expressions contain both magnitude (module) and phase. One is to fix the branch of the square root of the complex numbers in (9.4.23), (9.4.24) with negative imaginary portion similar to (9.1.70). We turn to computational results.

Figure 9.4.8 shows modules and phases of reflection coefficients for several media as a function of the angle θ. Modules of $R_\perp$ and $R_{||}$ are shown at the left-hand side of the figure. First, please note that copper is always a perfect conductor. Copper surfaces

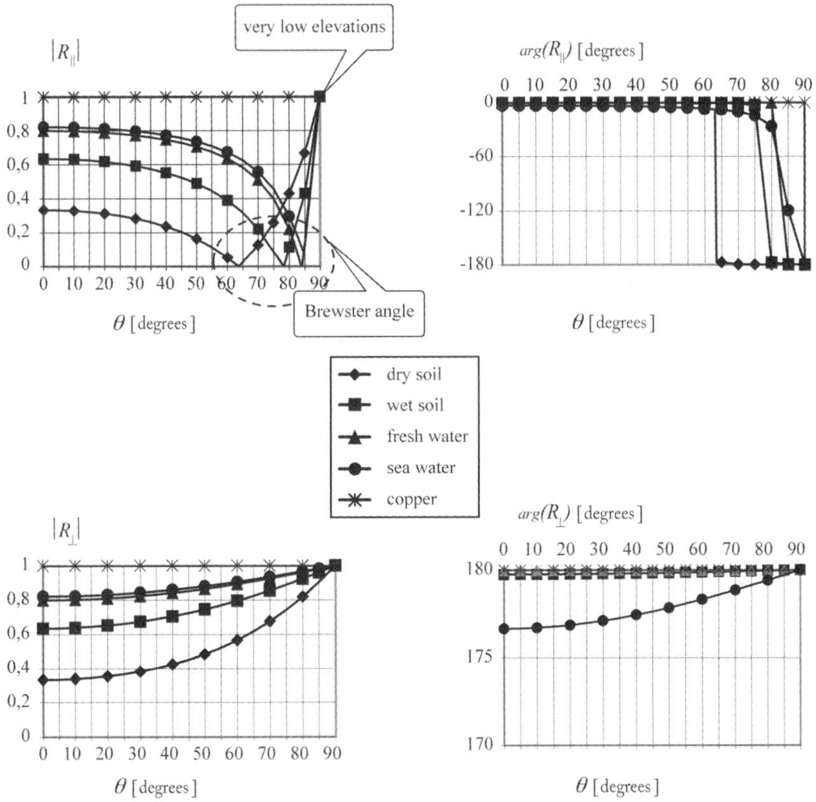

Figure 9.4.8 Modules and phases of reflection coefficients for a linear polarized incident wave.

reflect the incident wave like a mirror with modules of $R_{\perp}$ and $R_{||}$ reaching unity. Now let us recognize an important property. All of the media exhibit perfect mirror properties for grazing directions with respect to the surface. For ground reflections the angle θ is counted from the zenith, and low grazing angles close to $\theta \approx 90\,°$ are low elevations with respect to the local horizon. In this angle range the reflection coefficients modules are about unity for all media.

We are focused on parallel polarization (top plots in Figure 9.4.8). Here, for all media, except for perfect conductors, there exists a so-called Brewster angle. For this angle the reflection coefficient reaches a distinct minimum that is almost zero (top-left panel). For the incident wave arriving at the Brewster angle, all the power will go inside the media without reflection. Finally for angles close to zenith with $\theta \approx 0$, we see a variety of values. Most of them are around 0.5 which means half-magnitude reflection. For low elevations the phases of $R_{||}$ (top-right panel) are all about 180° which implies antiphase reflection. For angles close to zenith the phases are about zero which implies in-phase reflection. Rapid change occurs around the Brewster angle.

Next, in the bottom plots we show the relevant results for perpendicular polarization. There is no Brewster angle for this polarization. All the media behave like a perfect mirror for low elevation angles and show reflection coefficient of about one-half for incident wave arrival directions close to the zenith (bottom-left panel). The phases of $R_{\perp}$ (bottom-right panel) are all about 180° which means antiphase reflection for all media.

Now we look into how to treat an RHCP incident wave that is relevant for GNSS applications. First we introduce a set of unit vectors in the planes perpendicular to the directions that the waves travel (see Figure 9.4.7). Here are basis vectors $(\vec{x}_0^{inc}, \vec{y}_0^{inc})$ associated with the incident wave, and $(\vec{x}_0^{refl}, \vec{y}_0^{refl})$ associated with the reflected wave. For the incident wave to be RHCP, one writes

$$\vec{E}^{inc} = E_0{}^{inc} \frac{1}{\sqrt{2}} \left(\vec{x}_0^{inc} - i\vec{y}_0^{inc} \right) \tag{9.4.25}$$

where $E_0{}^{inc}$ is amplitude. The reflected wave takes the form

$$\vec{E}^{refl} = E_0{}^{inc} \frac{1}{\sqrt{2}} \left(R_{\perp}\vec{x}_0^{refl} - iR_{||}\vec{y}_0^{refl} \right) \tag{9.4.26}$$

Now we introduce circular polarized basis vectors associated with the reflected wave similar to Section 9.1.6,

$$\vec{p}_0^{RHCP} = \frac{1}{\sqrt{2}} \left(\vec{x}_0^{refl} - i\vec{y}_0^{refl} \right) \tag{9.4.27}$$

$$\vec{p}_0^{LHCP} = \frac{1}{\sqrt{2}} \left(\vec{x}_0^{refl} + i\vec{y}_0^{refl} \right) \tag{9.4.28}$$

and perform a basis transformation similar to that described in regard to expression (9.1.118). Thus, one has

$$\vec{E}^{refl} = E_0^{inc}\left\{\frac{R_\perp + R_{||}}{2}\vec{p}_0^{RHCP} + \frac{R_\perp - R_{||}}{2}\vec{p}_0^{LHCP}\right\} \tag{9.4.29}$$

We denote

$$R_{RHCP} = \frac{R_\perp + R_{||}}{2} \tag{9.4.30}$$

$$R_{LHCP} = \frac{R_\perp - R_{||}}{2} \tag{9.4.31}$$

and arrive at the following conclusion: If an RHCP wave is incident onto a flat surface, then it will be reflected with an RHCP coefficient (9.4.30); additionally, an LHCP component will be generated. Its coefficient is shown in (9.4.31). We write the reflection coefficients in exponential form:

$$R_{RHCP(LHCP)} = |R_{RHCP(LHCP)}|e^{i\psi_{R;RHCP(LHCP)}} \tag{9.4.32}$$

and plot the modules and phases in Figure 9.4.9. The module $|R_{RHCP}|$ (top-left panel) equals unity for low elevations of about $\theta \approx 90$. So the reflected wave is almost totally RHCP. But for zenith directions it decreases to zero, which means that the reflected wave will be totally LHCP. Module $|R_{LHCP}|$ (bottom-left panel) is about one-half for most of the directions, which means that the reflected LHCP field magnitude is about one-half the incident RHCP wave magnitude. Only for low elevations will it decrease to zero. This should be the case because for those elevations the RHCP component has about unity relative amplitude. Phases of the RHCP and LHCP components (shown at the right-hand side of the figure) are close to 180°. Thus, antiphase reflection occurs for most angles.

Now one needs to check the diagram in Figure 9.2.13 again. As was stated in Section 9.2.3, a GNSS user antenna is left-hand circular polarized for directions close to nadir. Then polarization becomes elliptical, which means the antenna is sensitive to both RHCP and LHCP components, and finally for directions close to the horizon, the antenna is close to RHCP. As seen in Figure 9.4.9, the reflected signal has exactly the same properties. So, any GNSS user antenna is almost perfectly matched with ground reflected signals by polarization. The antenna does not filter out the signals reflected by the ground due to polarization properties.

To finalize the discussion we are to estimate the area on the surface that defines the reflected field intensity in the proximity of the user antenna. We should mention again that figures like 9.4.7 are to be viewed as schematic. Incident and reflected fields are spatial processes. In Figure 9.4.10 we show the reflective surface with the user antenna installed at the height h over it. For a given θ the distance s from the antenna nadir to the point of reflection C (left panel) is

$$s = h \tan\theta \tag{9.4.33}$$

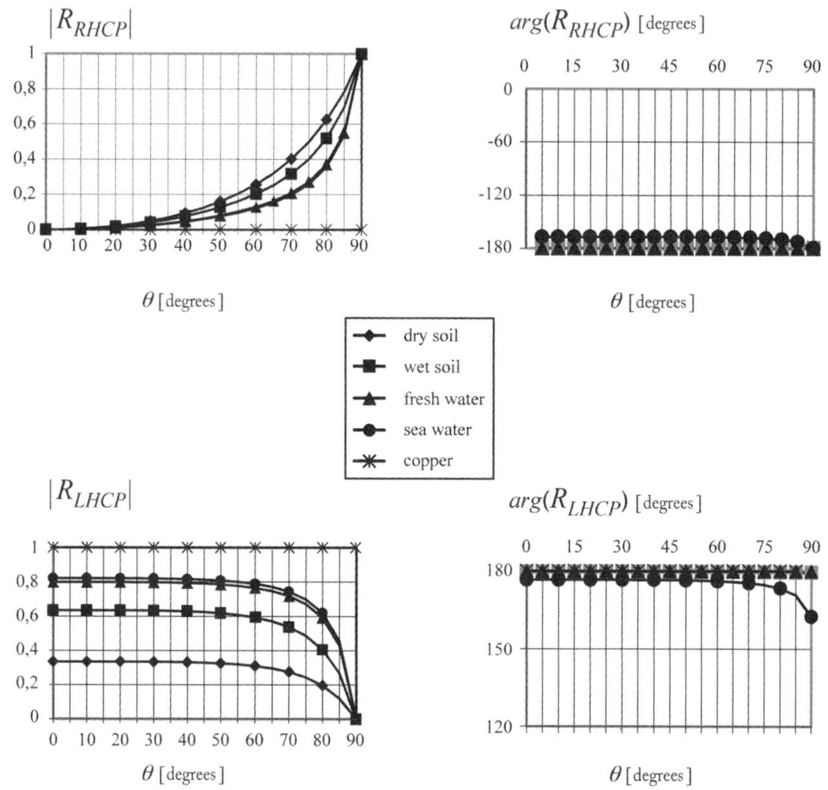

Figure 9.4.9 Modules and phases of reflection coefficients for RHCP incident wave.

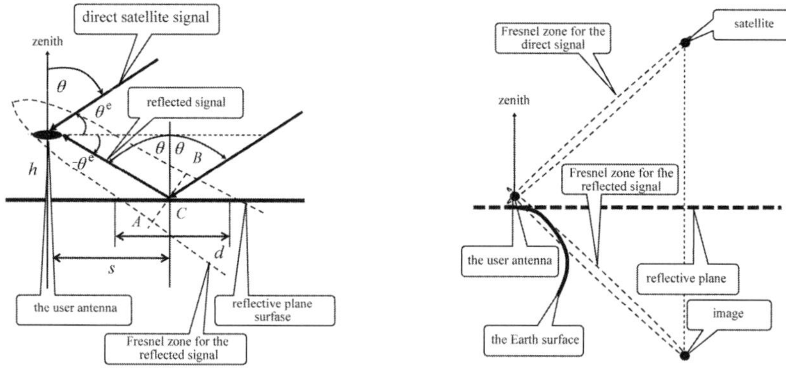

Figure 9.4.10 Reflective area for the specular reflection from the earth surface.

By image approach (Appendix E) one may view the reflection as signal generation by an image of a real source. This is illustrated in the right panel. Shown is the reflective plane which is tangential to the earth surface at the user antenna location. The actual source is the satellite and the image is located opposite the source under the reflective plane at the same distance from the plane. Two Fresnel zones are shown as dashed lines; one is for the direct satellite signal and another one is for the signal generated by the image. Now we are back to the left panel. Here the Fresnel zone for the reflected signal in the user antenna proximity is shown as dashed lines. The wave propagation path distance from the point C to the antenna is $h/\cos\theta$, and from (9.4.2) the cross section of the Fresnel zone is a circle with diameter

$$|AB| = 2\sqrt{3(\lambda h)/\cos\theta} \tag{9.4.34}$$

Thus the footprint of the Fresnel zone at the surface is an ellipse with the semiminor axis (perpendicular to the drawing) equal to $|AB|/2$ and semimajor axis equal to $d/2$, where

$$d = |AB|/\cos\theta = 2\sqrt{3(\lambda h)}/(\cos\theta)^{3/2} \tag{9.4.35}$$

We call this area a reflective area for short. For instance with $h = 2$ m and $\lambda = 20$ cm the reflective area for the zenith direction $\theta = 0$ is a circle with a radius of about 1 m centered at antenna nadir, for a zenith angle $\theta = 45^\circ$ it is an ellipse with semimajor axis $d/2 \approx 1.2$ m centered at point C which is 2 m away from the nadir, and with the lowest elevation angle of 10° ($\theta = 80^\circ$) the semimajor axis is about 15 m with the center being 11 m away from the nadir.

In conclusion, the surface roughness and inconsistency of the parameters of the reflective medium will certainly change the values of the reflection coefficients but may not change the main features of the phenomena. This is in particular true for low elevations, when the reflective area is large compared to the wavelength. This justifies the wide range of multipath estimates based on a specular reflections assumption. For more accurate estimates accounting for surface roughness, the reader is referred to Beckman and Spizzichino (1987).

9.4.4 Antenna Down-Up Ratio

As shown above, the relative amplitudes of the circular polarized components of the wave reflected by a plane surface are about 0.5. The way to reduce the multipath error for positioning is to make the antenna less sensitive to signals coming from below the horizon. This can be done by means of shaping the antenna pattern. We are now turning to that discussion. Our goal with the remaining section and the following sections is to estimate the carrier phase multipath error as a function of antenna pattern characteristics.

We consider the situation of Figure 9.4.10 and take h to be about 2 m. For now it will be more convenient to use the satellite elevation angle θ^e instead of the zenith angle θ (see the left panel of the figure for angles designations). The elevation angle θ^e is counted from the local horizon plane. With the specular reflections model the reflected signal would be coming from the $-\theta^e$ direction. We use the approach of Section 9.2.5 and characterize the signal at the antenna output by means of the antenna effective area. Now we are interested not in powers but rather in complex amplitudes. We use the antenna pattern in the form of (9.1.122) which relates the LHCP component to RHCP by the normalization coefficient α^{cross}. For the direct satellite signal at the antenna output U^{direct} using (9.2.42) one writes

$$U^{direct} = \sqrt{P_{0sat} D_0 \chi_{pol} \chi_a (1 - \chi_{ret})} F_{RHCP}(\theta^e) e^{i\Psi_{RHCP}(\theta^e)} e^{-i\psi^{direct}} \tag{9.4.36}$$

With this expression the term under the square root is the total received power in the direction where the RHCP pattern reaches unity. For a GNSS user antenna this direction normally is the zenith. Then we show that the direct signal is proportional to the RHCP antenna pattern component. This is so because the satellite signal is RHCP. We assume a high degree of rotational symmetry of the user antenna pattern with respect to azimuth. For this reason, the antenna pattern $F_{RHCP}(\theta^e)$ and phase pattern $\Psi_{RHCP}(\theta^e)$ are only functions of elevation angle and not azimuth. Finally, with (9.4.36) the term ψ^{direct} is the carrier phase path delay for the signal arriving from the satellite.

A similar expression for the reflected signal reads

$$\begin{aligned} U^{refl} &= \sqrt{P_{0sat} D_0 \chi_{pol} \chi_a (1 - \chi_{ret})} \\ &\times \begin{bmatrix} |R_{RHCP}|\, e^{i\psi_{R;RHCP}} F_{RHCP}(-\theta^e) e^{i\Psi_{RHCP}(-\theta^e)} \\ + |R_{LHCP}| e^{i\psi_{R;LHCP}} \alpha^{cross} F_{LHCP}(-\theta^e) e^{i\Psi_{LHCP}(-\theta^e)} \end{bmatrix} e^{-i(\psi^{direct} + \Delta\psi^{path})} \end{aligned} \tag{9.4.37}$$

Here we have the same square root of the total received power for zenith as with expression (9.4.36). The sum of contributions from RHCP and LHCP reflected signals is shown in brackets. Each contribution is proportional to the corresponding antenna pattern reading for directions below horizon. The modules and phases of the reflection coefficients $R_{RHCP(LHCP)}$ are shown explicitly. Finally we have added an extra path delay, $\Delta\psi^{path}$, to the exponent of the reflected signal.

Now we turn to expression (9.4.15). It shows that the multipath carrier phase error is a function of the relative amplitude of the reflected signal and extra phase delay relative to the direct signal. Using (9.4.36) and (9.4.37), one writes

$$\alpha^{refl} = \frac{|U^{refl}|}{|U^{direct}|} \tag{9.4.38}$$

$$\Delta\psi^{refl} = \arg(U^{refl}) - \arg(U^{direct}) \tag{9.4.39}$$

One notes that even with the assumption made, there are too many unknowns in expressions (9.4.38) and (9.4.39). These are the terms depending on soil reflection coefficients. Also, a path delay $-\Delta\psi^{path}$ can be specified for the infinite flat surface model. But in practice, a large or small flat surface would have some irregularities like minor hills and gaps. These irregularities could easily be of the order of centimeters, which makes them sufficiently large compared to the wavelength. Therefore, strictly speaking, the antenna height above the surface becomes uncertain in the wavelength scale. This illustrates once again that the direct way of estimating multipath and eliminating the carrier phase error is so far impractical. The best one can do is find a way to reduce the error by means of proper antenna design.

For estimation purposes, one may consider worst cases regarding multipath signal strength. The first way would be to take the surface as a perfect conductor. Applying $\tan \Delta^e \to \infty$, with the results of the previous section one has $R_{RHCP} = 0$, $|R_{LHCP}| = 1$, $\psi_{R;LHCP} = \pi$, and expressions (9.4.38) and (9.4.39) take the form

$$\alpha^{refl} = \frac{|U^{refl}|}{|U^{direct}|} = \frac{\alpha^{cross} F_{LHCP}(-\theta^e)}{F_{RHCP}(\theta^e)} \tag{9.4.40}$$

$$\Delta\psi^{refl} = -\Delta\psi^{path} + \pi + \Psi_{LHCP}(-\theta^e) - \Psi_{RHCP}(\theta^e) \tag{9.4.41}$$

Expression (9.4.40) has a simple meaning. It is the ratio of the antenna pattern reading for a specific angle below the horizon to the reading for the same angle above horizon. The LHCP component is taken for the first one and RHCP for the second. This ratio sometimes is referred to as an antenna down- up ratio. In short, one may say that the reflected signal magnitude at the antenna output related to the direct signal equals the antenna down-up ratio. However, the expression (9.4.40) has certain disadvantages. In particular as seen from Figures 9.2.7 and 9.2.9 the RHCP component of the antenna pattern exceeds the LHCP component for the majority of directions below the horizon except for the angular area close to nadir. Thus the expression (9.4.40) would be an underestimate of the multipath error since only LHCP is taken into account for directions below the horizon.

Another way to estimate α^{refl} and $\Delta\psi^{refl}$ would be to use the antenna pattern $F(\theta, \phi)$ in the form of a square root of total power pattern (9.1.125). This would mean that the antenna is considered to be perfectly matched with the direct and reflected signals in terms of polarization and that the total relative reflected power equals unity. We will follow this way. We introduce the antenna down-up ratio $DU(\theta^e)$ as

$$DU(\theta^e) = \frac{F(-\theta^e)}{F(\theta^e)} \tag{9.4.42}$$

Thus, (9.4.38) will read

$$\alpha^{refl} = DU(\theta^e) \tag{9.4.43}$$

We will see later that $\Delta\psi^{path}$ is rapidly changing as a function of elevation while the phase patterns are somewhat smooth. Within the approximations made we omit the antenna phase patterns and (9.4.41) reduces to

$$\Delta\psi^{refl} = -\Delta\psi^{path} + \pi \tag{9.4.44}$$

These last two expressions along with (9.4.15) define the carrier phase multipath error at the antenna output. We leave detailed discussion for the next section and proceed with some estimates first.

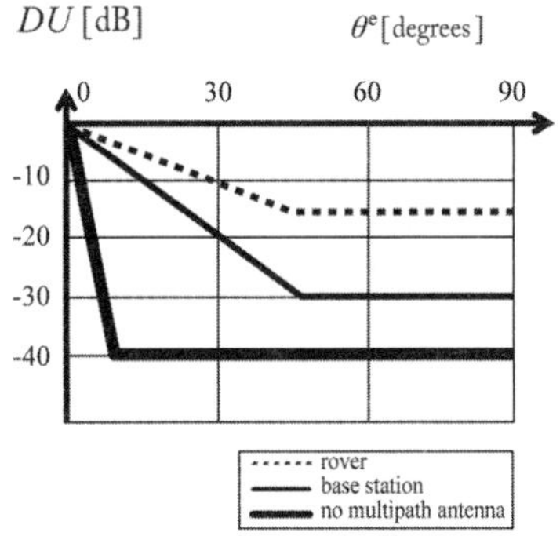

Figure 9.4.11 Approximation of down-up ratio versus satellite elevation for typical receiving GNSS antennas.

An antenna pattern as a whole and the down-up ratio in particular depend on many features of the antenna design. We will not go into detail here, leaving some considerations for Section 9.7. Instead we will look at typical data valid for most of the receiving antennas for high-precision GNSS positioning. The diagram in Figure 9.4.11 shows three curves. The first one (dashed line) is for a typical rover antenna. The second one (solid line) is for a base station antenna with a choke ring ground plane (see Section 9.2.1). The down-up ratios are plotted in dB units as a function of elevation angle. We see that for the horizon direction the down-up ratio equals 0 dB. This means that neither antenna can provide suppression for a reflected signal coming from the horizon direction along with the direct signal. Then, for a reflected signal coming from a small angle below the horizon, the direct signal should come from the same small angle above the horizon. An antenna of the size adopted in GNSS practice cannot distinguish between two signals coming from almost the same direction. Thus down-up values for very small elevations are small. As the satellite elevation angle increases the angular difference between a direct and reflected signal increases and the down-up ratio increases in absolute value providing more suppression of the reflected signal. The way the down-up ratio changes with elevation could be rather complex depending on the particular antenna type. We just show trends by linear functions. As was mentioned in Section 9.2.1, normally an elevation mask of 10° is used with the positioning algorithms. The down-up reading for 10° elevation is the ratio of F_{-10}/F_{+10} readings introduced in Section 9.2.1. With typical antennas, the down-up reading is about several dB with minus sign. Finally, for high elevation angles the down-up reaches some typical values, which are about –15 dB for the rover antenna and about –30 dB for the choke ring antenna. We will see in the next section that this difference defines the actual accuracy of positioning with these two antenna types. This is the practical reason to distinguish between rover and base station antennas as discussed in Section 9.2.1

For the moment we estimate the multipath carrier phase errors using data from Figure 9.4.11. For small relative amplitudes α^{refl} the multipath carrier phase error expressed in radians does not exceed α^{refl} (Section 9.4.2). The α^{refl} in turn equals the down-up ratio. For rover antennas we take -15 dB as a good estimate for high elevations. This amounts to 0.178 in relative units; therefore, the carrier phase error will not exceed 0.178 radians or 10.2° for high elevated satellites. Taking a down-up value of –30 dB for the choke ring antenna gives 0.032 in relative units, which makes 1.8° multipath carrier phase error.

Now the question is: What kind of down-up ratio do we actually need? Let us take a 1-mm positioning error in real time as the goal. Assuming that the DOP factor of the satellite constellation geometry is 3, we will allow a multipath error of 0.33 mm. Converting from millimeters to radians, we multiply by $2\pi/\lambda$ with $\lambda = 20\,\text{cm}$. This gives an error of 0.01 radians, which corresponds to –40 dB down-up. Thus, for 1-mm accuracy with RTK positioning, the down-up curve should have a quick drop from 0 up to −40 dB within the 10° angular sector. This is illustrated in Figure 9.4.11 by the curve named "no multipath antenna" (thick solid line). See more considerations on "no multipath antennas" in Counselman (1999). Such behavior means that the antenna pattern rapidly increases for the top semisphere starting at the horizon and rapidly decreases in value for angles within the bottom semisphere. As was mentioned already in Section 9.2.1, to achieve this kind of performance the antenna size in wavelength scale should be noticeable. One may wish to check Lopez (2008), Lopez (2010), and Thornberg et al. (2003) to learn of vertical array antennas with total lengths exceeding 2 m that actually realize this kind of behavior. A large ground plane antenna has been presented by Tatarnikov and Astakhov (2013, 2014). See additional discussion is Sections 9.7.4 and 9.7.7.

9.4.5 PCV and PCO Errors Due to Ground Multipath

We continue the discussion on ground reflections. For the moment our focus is carrier phase path delay $\Delta\psi^{path}$ of the reflected signal. To estimate $\Delta\psi^{path}$, it is convenient to view the reflected signal as arriving at the antenna image rather than bouncing off the reflective surface. From the right triangle in Figure 9.4.12, the path delay ΔL is

$$\Delta L = 2h \sin \theta^e \tag{9.4.45}$$

The carrier phase delay $\Delta\psi^{path}$ is

$$\Delta\psi^{path} = k\Delta L \tag{9.4.46}$$

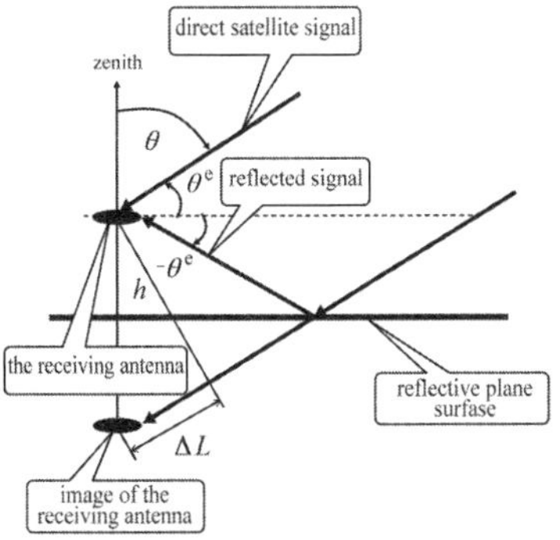

Figure 9.4.12 Carrier phase path delay difference for reflected signal versus direct signal.

and from (9.4.44) we obtain

$$\Delta\psi^{refl} = -2kh\sin\theta^e + \pi \tag{9.4.47}$$

Here k is a wavenumber (9.1.32). Using (9.4.43) one writes the carrier phase error (9.4.15) as

$$\psi^{mult} = \arctan\frac{DU(\theta^e)\sin(\Delta\psi^{refl})}{1 + DU(\theta^e)\cos(\Delta\psi^{refl})} \tag{9.4.48}$$

Here $\Delta\psi^{refl}$ is defined by (9.4.47) (Elòsegui et al., 1995).

We estimate the behavior of ψ^{mult} as a function of θ^e. We consider the practical case with the antenna being 2 m above the ground. With our assumed wavelength of 20 cm one has $2kh = 40\pi$. In expression (9.4.47), a slight change in $\sin\theta^e$ is multiplied by a large value of $2kh$ and results in a large change in the trigonometric functions argument in (9.4.48). Thus the multipath error ψ^{mult} oscillates as a function of the elevation angle θ^e. This error completes a period with an elevation angle increment $\Delta\theta^e$, which provides 2π increment to the delay (9.4.47). Thus, one may write

$$|\Delta\psi^{refl}| \approx \left|\frac{d\psi^{refl}}{d\theta^e}\right|\Delta\theta^e = 2kh\cos\theta^e\Delta\theta^e \tag{9.4.49}$$

If $\Delta\psi^{refl} = 2\pi$, then

$$\Delta\theta^e \approx \frac{\lambda}{2h\cos\theta^e} \tag{9.4.50}$$

Here, λ is a wavelength. We see that the period of the multipath error (9.4.48) is inversely proportional to the antenna height and increases with satellite elevation. For an antenna height of 2 m above the ground and a wavelength of 20 cm one has a period of carrier phase error oscillations of 3° for elevation around 10°, 4° for 45° elevation and 32° for a satellite elevation of 85°.

Regarding the magnitude of carrier phase error, we recall that for small elevation angles the antenna down-up ratio is about unity, and from (9.4.48) it follows that the carrier phase error reaches its maximum value in such a case. For high elevation angles, the antenna down-up is small and the carrier phase error is simply proportional to the down-up value. The multipath carrier phase error (9.4.48) is plotted in Figure 9.4.13 as a function of the elevation angle for the down-up curves of Figure 9.4.11, assuming an antenna height of $h = 2$ m. In regards to the curves named "rover" and "base station" one may note that this type of behavior is normally observed for the carrier phase residuals of positioning algorithms. The curve for "no multipath antenna" shows almost zero values for $\theta^e > 10°$ as predictable.

The features discussed above can be viewed from another angle. Expression (9.4.44) was derived by neglecting the antenna phase pattern. This is equivalent to the assumption that the antenna had an ideal phase center. Expression (9.4.48) then can be viewed as a phase pattern of a system comprised of an antenna and a reflective surface. Indeed, in transmitting mode one is to invert the arriving rays to departing ones in Figure 9.4.12. Then, in the far-field region the radiated field would be the sum (interference) of direct and reflected waves resulting in the phase pattern

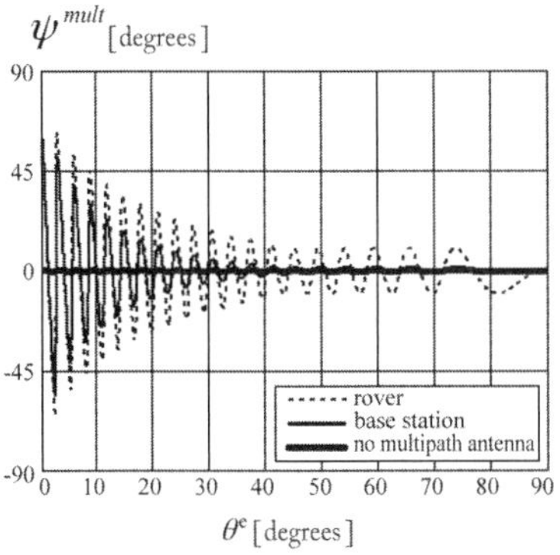

Figure 9.4.13 Carrier phase multipath error versus satellite elevation for typical receiving GNSS antennas.

of (9.4.48). The same phase pattern will hold true for the receiving mode due to reciprocity.

It is of interest to estimate an extra offset of the phase center due to thus defined phase pattern. One writes

$$\Psi(\theta, \phi) = \psi^{mult}(\theta) \tag{9.4.51}$$

and substitutes this into (9.3.28) to (9.3.30). Please note that we use zenith angle θ in (9.4.51) instead of elevation angle $\theta^e = \pi/2 - \theta$. One recognizes that the horizontal offsets (9.3.28) and (9.3.29) are zero. This is due to an assumption of an ideal rotational symmetry of the reflective surface with respect to the azimuth. The vertical offset is plotted in Figure 9.4.14 as a function of height h of the antenna above ground. Calculations are done using the three down-up curves of Figure 9.4.11. One notes that within a practical range h of about 2 m, a rover antenna provides a vertical phase center offset instability of about $\pm$ 2 mm, and the choke ring base station antenna about $\pm$1 mm. "No multipath antenna" would show instability of much less than 1 mm. As the height h increases the phase pattern (9.4.51) would exhibit more rapid variations versus θ (see (9.4.47)). These variations would be averaged out with phase center offset calculations. Thus as height increases, the instabilities mentioned above are less noticeable. However, one should recognize that this data characterizes

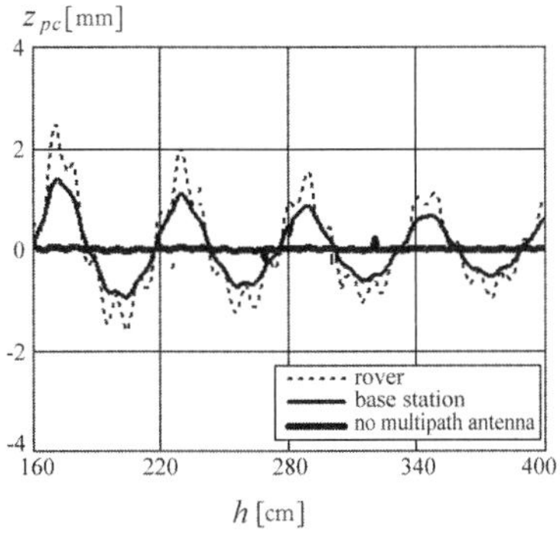

Figure 9.4.14 Phase center offset in vertical coordinate caused by reflections from undelaying terrain.

just the mean phase center offset. Real-time variations are much larger since they are proportional to the peak values in Figure 9.4.13 and the DOP factor.

9.5 TRANSMISSION LINES

Normally, an antenna is connected to the receiver by coaxial cable. Mismatch and signal losses are phenomena to be accounted for. We discuss these phenomena in the broader frame of transmission lines topics.

9.5.1 Transmission Line Basics

We begin by illustrating the difference between low-frequency connection, say two wires connecting the computer with the wall outlet, and radio frequency connection. A pair-wire line is commonly used for AC power supply. For distances inside a room or a building one normally does not have to be concerned with the consistency of the geometrical properties of the line, such as diameter of wires, distance between wires, radius of wire bends, and the like. The only requirement is for the line to transfer AC power of the desired amount. However, with GNSS antennas a specially designed cable is used. The goal here is different, namely to obtain the match with an antenna output and a receiver input.

In order to understand the difference, we look into the processes taking place inside a line. Let us consider a segment of a transmission line. Many different types of lines are in use. The derivations below are applicable to most of them. Since it is the goal to look at what is going on between a GNSS user antenna and a receiver, a coaxial cable is of primary interest. We assume that the cross section of the line is always the same. A coaxial cable consists of an outer conductor, the shield, and an inner conductor, the wire (Figure 9.5.1). These two conductors are separated by a dielectric filling. We introduce the z coordinate along the line.

Let $U(z,t)$ be the signal voltage across the line that is observed at the cross section with coordinate z and at time instant t. For a coaxial cable $U(z,t)$ is the voltage between the shield and the wire. Note that if one connects the computer to the wall outlet one has no doubts that the voltage at the computer input is the same as at the outlet. However, in general this is not true; there is a change in voltage for

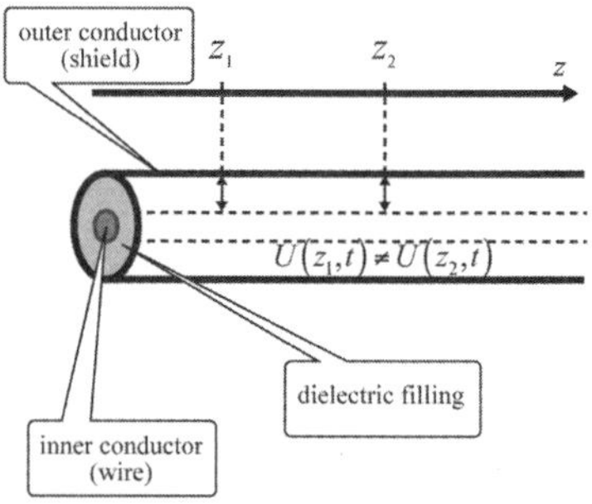

Figure 9.5.1 Coaxial cable with z coordinate along the line.

different positions along the line. At some particular time instant t the voltage U at cross-section z_1 is not the same as the voltage at cross-section z_2. We want to stress that the difference comes not from the conductor resistivity or from any other losses like radiation from the line. For the moment, we will discuss a perfect loss-less line. Signal losses in a line are discussed further with Section 9.5.3.

The signal $U(z,t)$ propagates along the line in the form of an electromagnetic wave. We write this wave in a real time-harmonic form (Section 9.1.2) as

$$U(z,t) = U_0 \cos\left(2\pi\left(\frac{t}{T} - \frac{z}{\Lambda}\right)\right) \tag{9.5.1}$$

Here, U_0 is the amplitude in volts, and T is the period of the voltage alternation in seconds. We mark the wavelength in the line by Λ to distinguish it with the free space wavelength λ. For coaxial cables and pair-wire lines, Λ coincides with (9.1.39), where ε is the permittivity of the medium filling the line, and $\mu = 1$. For coaxial cable the typical values are $\varepsilon = 2...4$. Expression (9.5.1) in particular illustrates why transmission lines work so differently in cases of an AC power supply and a GNSS antenna connection. For instance, the frequency of AC power in Europe is 50 HZ. Assuming a pair-wire line is filled with air, one has $\Lambda = 6 \cdot 10^6$ m. The distance the voltage travels from the outlet to a computer is about say 2 to 3 m. If we take z equal to zero or to 3 m, and a wavelength of 6000 km, then indeed according to (9.5.1) the voltages at the source and the recipient is the same at any instant in time. In the case of GNSS signals at frequency $f = 1.5 \cdot 10^9$ Hz one has $\Lambda = 10...15$ cm. Thus, the cable length is normally much greater than the wavelength in the line. Therefore, at any time instant one may find completely different voltage values along the line.

We introduce a parameter

$$\beta = \frac{2\pi}{\Lambda} \tag{9.5.2}$$

called propagation constant in the line. We rewrite (9.5.1) in a canonical form [see (9.1.20)] as

$$U(z,t) = U_0 \cos(\omega t - \beta z) \tag{9.5.3}$$

Figure 9.5.2 represents a general view of a signal transmission via a line. There is a signal source (or a generator), and a signal recipient (or a load), connected by a line. Note that we point the z axis from the load toward the generator and we assume

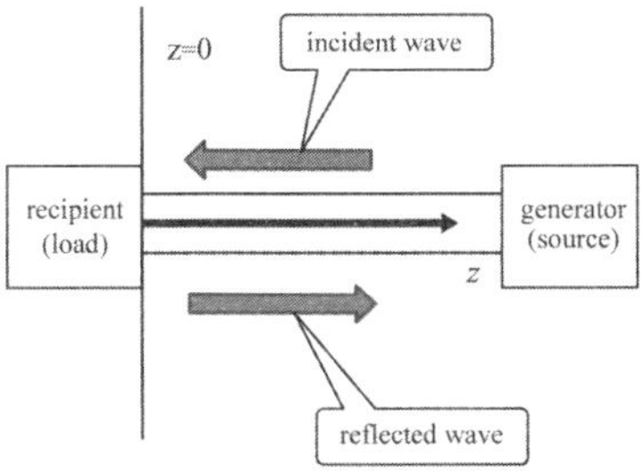

Figure 9.5.2 General schematic for a transmission via a line.

the load to be connected to the line at cross-section $z = 0$. As is normal with time harmonic processes, we are mostly interested in amplitudes and phases of the voltage alternation at different points of the line. For that purpose, we turn to complex notations. A wave traveling toward the load is called an incident. We write it in complex form as

$$U^{inc} = U_0^{inc} e^{i\beta z} \tag{9.5.4}$$

Here, U_0^{inc} is a complex amplitude containing amplitude and initial phase. Please note that this wave is traveling in the opposite direction with respect to the z axis. That is why the sign in the exponential term is positive. In general, it is not possible to make an incident wave power that is totally absorbed by a load. Some portion of an incident wave power would be reflected by a load and would propagate back in the form of a reflected wave. This wave takes the form

$$U^{refl} = U_0^{refl} e^{-i\beta z} \tag{9.5.5}$$

Here, U_0^{refl} is a complex amplitude. The minus sign in the exponential term indicates the wave traveling in positive z direction. A basic quantity

$$\Gamma = \frac{U_0^{refl}}{U_0^{inc}} \tag{9.5.6}$$

is called the reflection coefficient from the load. The value of this coefficient is a function of the properties of the particular line and the particular load. The reflection coefficient is the relative complex amplitude of a reflected wave versus an incident wave. From this definition it follows that $0 < |\Gamma| < 1$. Please note that this quantity is complex. In general, the reflected wave has somewhat different amplitude and different initial phase as compared to the incident wave.

We look at what kind of amplitude distribution occurs along the line when both the incident and the reflected wave propagate. We introduce the total voltage U_Σ as a sum of an incident and reflected wave voltages and write

$$U_\Sigma = U_0^{inc} e^{i\beta z} + U_0^{refl} e^{-i\beta z} = U_0^{inc} e^{i\beta z} \left(1 + |\Gamma| \, e^{i\psi_\Gamma} e^{-i2\beta z}\right) \tag{9.5.7}$$

Here, the module and phase of reflection coefficient are shown explicitly. The module of the total voltage U_Σ shows amplitudes of the voltage at different cross sections of the line,

$$|U_\Sigma| = |U_0^{inc}||1 + |\Gamma| e^{i(\psi_\Gamma - 2\beta z)}| \tag{9.5.8}$$

One notes that the amplitudes are no longer the same as was the case when only the incident wave propagated. The second factor on the right-hand side of (9.5.8) shows that there is a time-invariant distribution of amplitudes along the line. This second factor is analyzed the same way it was in expression (9.4.12).

Figure 9.5.3 shows two vectors in a complex plane. The first vector is unity and the second has the magnitude $|\Gamma|$ and the angle $(\psi_\Gamma - 2\beta z)$ with respect to the first one.

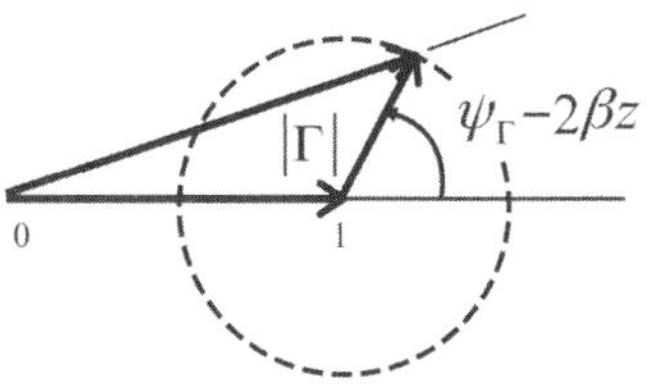

Figure 9.5.3 Incident, reflected, and total voltage vectors on a complex plane.

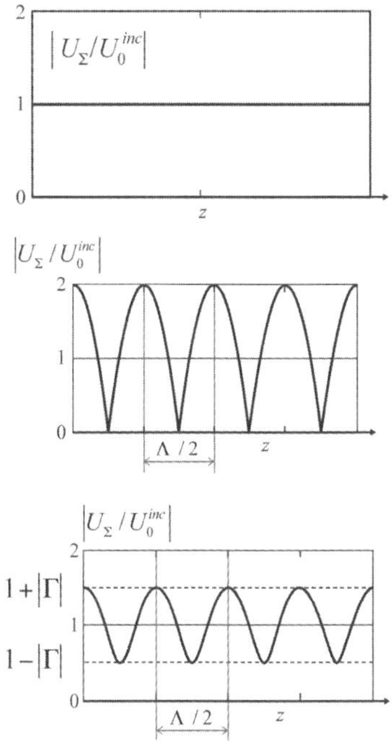

Figure 9.5.4 Traveling wave, standing wave, and mixed wave voltages.

With a change in coordinate z, the second vector rotates making a complete circle within the distance equal to half the wavelength Λ. We consider two special cases first.

If $\Gamma = 0$, one is back to just an incident wave traveling. Here $|U_\Sigma| = |U_0^{inc}|$ and the amplitudes of the voltage alternation are the same along the line and equal to that of the incident wave (top panel in Figure 9.5.4). This is referred to as the traveling wave mode of a transmission line. It is the most desirable mode of operation. It should be clear that with pure traveling wave mode, all the incident wave power is being absorbed by the load.

Now we take $|\Gamma| = 1$. This would mean a complete reflection of the incident wave back toward the generator. According to (9.5.8) the voltage achieves zero amplitudes at cross sections which are called nodes that are located at z_n coordinates such that

$$\psi_\Gamma - 2\beta z_n = (2n + 1)\pi \tag{9.5.9}$$

with n being an integer. There are cross sections where the amplitude of voltage oscillation reaches its maximum value equal to $2|U_0^{inc}|$. These cross sections are called loops or antinodes. They are located at coordinates z_n where

$$\psi_\Gamma - 2\beta z_n = 2n\pi \tag{9.5.10}$$

At all other points the amplitudes are between zero and $2|U_0^{inc}|$ (see the middle panel in Figure 9.5.4). This mode is referred to as a pure standing wave. No power is absorbed by the load. As seen with expressions (9.5.9) and (9.5.10), the distance between two successive loops or two nodes is one-half of the wavelength in the line. Similarly, the distance between a loop and neighboring node is a quarter of the wavelength. It is worth mentioning that the system of nodes and loops does not move along the line as time advances. That is the reason this wave is called a standing wave.

In general cases, one has $0 < |\Gamma| < 1$ and a system of minimum and maximum amplitudes occurs. As seen in the bottom panel of the Figure 9.5.4 the voltage amplitudes at minimums $|U_{min}|$ is

$$|U_{min}| = |U_0^{inc}|(1 - |\Gamma|) \tag{9.5.11}$$

The minimums occur at z, thus satisfying (9.5.9). The voltage amplitudes at the maximums are

$$|U_{max}| = |U_0^{inc}|(1 + |\Gamma|) \tag{9.5.12}$$

The maximums are located where (9.5.10) holds. This described mode of a transmission line operation is sometimes referred to as mixed waves mode. The power is partly absorbed by the load and partly reflected back to the generator.

By definition, the voltage standing-wave ratio (VSWR) is a ratio of the voltage alternation amplitudes at maximums and minimums. Employing (9.5.11) and (9.5.12), one has

$$VSWR = \frac{1 + |\Gamma|}{1 - |\Gamma|} \tag{9.5.13}$$

For the traveling wave mode $VSWR = 1$; for standing wave $VSWR = +\infty$. VSWR is the parameter often shown in GNSS antenna specifications.

The traveling wave mode is often referred to as a perfect match of line and load, while the pure standing wave mode is referred to as a complete mismatch between line and load. The reason for looking for the lowest possible VSWR is not merely to reduce the waste of power that will go back to the generator. In actuality, the generator could be mismatched with the line as well. In that case, the wave would be going back and forth inside the cable like between two partially reflecting mirrors. This situation is known as a resonator. Vast variations of the line response as a function of signal frequency may happen in this case. With the GNSS user antenna case the antenna works as generator and the receiver is the signal recipient. For proper functionality, the VSWR should be limited with both the antenna and the receiver sides. Normally $VSWR \leq 2$ is considered acceptable.

Finally we look at what is required to achieve a perfect match of a load with the transmission line. The transmission line is characterized by a parameter known as characteristic impedance. The term wave impedance is also in use. This impedance is defined by the dimensions of the parts constituting the line and the materials of which these parts are made. For coaxial cable the wave impedance is

$$W_{coax} = \frac{60}{\sqrt{\varepsilon}} \ln \frac{D}{d} \qquad \text{(ohm)} \tag{9.5.14}$$

where D is the diameter of the shield, d is the diameter of the wire, and ε is the permittivity of the filling. With GNSS user equipment a standard value of wave impedance is 50 ohms (Ω) for the cable connecting the antenna with the receiver. Here a note should be made similar to that made in Section 9.1.2 while discussing the intrinsic impedance of a medium. It would be erroneous to think of wave impedance as resistivity the line provides to signal propagation. So far, we have been discussing a loss-less cable. A characteristic or wave impedance of the line is a proportionality coefficient between the traveling wave voltage and the electric current flowing on the conductors. Similar to the derivation of (9.1.35), one may say that if the traveling wave voltage amplitude is U_0 then the power P the wave is carrying is

$$P = {}^1\!/\!_2 |U_0|^2 W_{coax} \tag{9.5.15}$$

We turn to the characterization of load. For each fixed frequency, the load could be characterized by a parameter that is referred to as an input impedance Z_{load}. This impedance is measured in ohms. Normally it is a sum of two parts:

$$Z_{load} = R_{load} + iX_{load} \tag{9.5.16}$$

The real part of the impedance R_{load} characterizes the ability of the load to absorb electric power and use that power for purposes for which the load was originally intended. This part is sometimes called active impedance. The imaginary part of the impedance X_{load} characterizes the ability to store electric power inside the load without utilizing it. It is called reactive impedance. It might be known from elementary physics that an element that absorbs electric power is a resistor. Elements storing the power could be capacitors or inductances. Whatever complicated load there is, within some narrow frequency band it is equivalent to a mix of resistors, capacitors, and inductances when viewed from the standpoint of input impedance.

The key equation that illustrates the relationship between the input impedance of the load, the properties of the line, and the reflection coefficient is

$$\Gamma = \frac{Z_{load} - W_{line}}{Z_{load} + W_{line}} \tag{9.5.17}$$

This expression is valid not only for coaxial cable but for lines of any kind. That is why the designation W_{line} is used for the wave impedance instead of W_{coax}. Expression (9.5.17) follows from Ohm's law.

From expression (9.5.17), the so-called matching conditions can be seen. We see that $\Gamma = 0$ holds if and only if two conditions are true:

$$R_{load} = W_{line} \tag{9.5.18}$$

$$X_{load} = 0 \tag{9.5.19}$$

In short, one says that for the load to be matched with the line, it should have purely active impedance equal to the characteristic impedance of the line. In any other case some mismatch would happen, which would result in a reflected wave, as mentioned above.

Now we mention that, as seen from (9.5.17), reflections will occur if Z_{load} changes or W_{line} changes. This justifies the requirement that high-quality coaxial cable be used for GNSS applications, as W_{line} should always remain constant throughout the line. Regarding the initial example of AC power transmission, it can be shown that if the length of the line is negligible compared to the wavelength in the line, one does not have to be concerned with W_{line}. That is the reason why normally the diameters of the two wires used for AC power transmission or the distance between them is not that important.

In regard to coaxial cable, from (9.5.14) we see that if we multiply D and d by the same factor, then the characteristic impedance would not change. So from the standpoint of matching conditions, one can say that thick cables with a large outer diameter and very thin cables work the same. However, due to practical cable manufacturing circumstances, in general, thicker cables would have less loss of signal. Cable signal loss is discussed further in Section 9.5.3. Note, both D and d are to be much smaller compared to the wavelength, otherwise, the higher order modes will propagate inside the cable (Balanis, 1989).

9.5.2 Antenna Frequency Response

Now we will look at an antenna as a transmission line load in the transmitting mode. The receiving mode will be treated using the reciprocity theorem.

The transmitting antenna generates an electromagnetic field in the surrounding space. As discussed in Section 9.1.4, there are three distinctly different regions of space from the standpoint of electromagnetic field properties. The region directly surrounding the antenna is the near field. The near-field region is known as an area where electromagnetic power is predominantly being stored. Then a radiating near-field region follows. Here the electromagnetic field transforms from storing the energy to radiating it. The far-field region is an area where mostly electromagnetic waves exist. Electromagnetic waves transport the power from the antenna to outer space. Just a short note regarding the AC power supply example of the previous section, with a wavelength of 6000 km, the entire city is within the near-field region of the AC power supply network. The radiated power in this region is negligible compared to the power stored or consumed in the network. This is true especially since the distances between wire conductors comprising a network are negligible compared to the wavelength. These are reasons why one normally does not have to account for the network antenna properties.

In regard to antennas with dimensions comparable to or exceeding a wavelength, the amount of stored power in the near-field region completely depends on the antenna design and the frequency of a signal. This stored power contributes to what is called the imaginary part of the antenna input impedance, or input reactance $X_A(\omega)$. We show this as a function of frequency to emphasize that $X_A(\omega)$ normally strongly depends on frequency. Next, the power radiated into the far field is actually taken from the signal source. So from that point of view, the antenna works like a resistor that absorbs power from the input. This contributes to what is called the real part of the input impedance $R_A(\omega)$. This quantity also depends on frequency. We write the total input impedance as

$$Z_A(\omega) = R_A(\omega) + iX_A(\omega) \tag{9.5.20}$$

and apply the derivations of the previous section.

One has a transmission line loaded by an antenna with input impedance as in (9.5.20). The incident wave traveling toward the antenna would be partially reflected due to mismatch between the antenna and the line. The reflection coefficient Γ is as given by (9.5.17) with $Z_A(\omega)$ introduced instead of Z_{load}. The matching conditions (9.5.18) and (9.5.19) take the form of

$$R_A(\omega) = W_{line} \tag{9.5.21}$$

$$X_A(\omega) = 0 \tag{9.5.22}$$

Obviously, these conditions hold only for a fixed frequency. But now one may recall that each GNSS signal occupies a frequency band (see discussion at the end of Section 9.1.2). It is not possible to make $R_A(\omega)$ and $X_A(\omega)$ strict constants within some frequency range, such as GPS L1 or L2, for example. Moreover, the second condition (9.5.22) is known as the so-called resonant condition. It could be implemented by means of precise antenna tuning and would stay true only over a negligibly small frequency range. That is why normally the antenna would not be strictly matched with the line for the complete signal frequency band.

To overcome this difficulty one usually applies some practical requirements for antenna mismatch within the desired band. Normally, the VSWR never reaches unity. Therefore, a strict match with the line never happens. Instead, the VSWR is less than some prerequired level, say less than 2 within the antenna frequency band. There is rapid growth with frequency, i.e., mismatch, beyond that band. See actual data regarding TA-5 antenna in Figure 9.7.20.

We turn to some numbers. For the given refection coefficient $|\Gamma|$, the amount of the reflected power is $|\Gamma|^2$. This power was designated as return loss χ_{ret} in Section 9.2.4, thus

$$\chi_{ret} = |\Gamma|^2 \tag{9.5.23}$$

For the typical requirements of VSWR to be less than 2, the inversion of (9.5.13) gives $|\Gamma| < 0.33$ and $\chi_{ret} \approx -10\,\text{dB}$ as stated in Section 9.2.4. The same signal losses would be observed in the receiving mode by reciprocity. The useful received power is then proportional to $(1 - \chi_{ret})$ and the loss of useful power is $10\log(1 - \chi_{ret}) = -0.45$ dB as was stated previously.

We finalize this discussion with a note. If there are no limitations in terms of space available for an antenna, one may achieve potentially any low VSWR within the desired frequency band. However, with today's technology due to the successes of microelectronics, the receiver antenna is one of the bulkiest components of the equipment. Thus the natural trend is antenna down-sizing. In this regard there exists the fundamental Chu limit (Chu, 1948) which states that antenna dimensions cannot be made less than a certain value, assuming a given VSWR and antenna efficiency (Section 9.2.4) within the desired frequency band. This limit cannot be overcome. From this perspective the art of antenna design is to develop more compact GNSS antennas approaching the Chu limit. One may mention that with the common microstrip patch antennas discussed in Section 9.7.1, there is still a potential for down-sizing.

9.5.3 Cable Losses

While propagating along the cable the traveling wave always loses some of its energy since a real-world body applies some resistance to electric currents. Losses in the media filling the cable also apply. To account for all these losses, we rewrite Equation (9.5.3) in the same manner as for the plane wave propagating through a lossy medium (9.1.71):

$$U(z,t) = U_0 e^{-\alpha z}\cos(\omega t - \beta z) \tag{9.5.24}$$

This equation shows that the traveling wave voltage exponentially decays with distance as z increases. The parameter α is called attenuation constant (Section 9.1.3). The power would decay as squared voltage. So, for power one has

$$P(z) = P_0 e^{-2\alpha z} \tag{9.5.25}$$

Here, P_0 is the traveling wave power at the source. The dB equivalent of equation (9.5.25) is

$$10\log\frac{P(L)}{P_0}\ [\text{dB}] = -20\alpha L\log(e) \tag{9.5.26}$$

We see that the power losses in dB at the output of the cable with length L are linearly proportional to the length. For that reason, the attenuation constant is often used in units of dB per meter. The conversion rule is

$$\alpha\,[\text{dB/m}] = 20\log(e)\alpha \tag{9.5.27}$$

Thus, cable losses $\chi_{cable[\text{dB}]}$ in dB within the cable of length L are

$$\chi_{cable[\text{dB}]} = \alpha[\text{dB/m}]L[\text{m}] \tag{9.5.28}$$

The quantity $\alpha[\text{dB/m}]$ is one of the most important parameters specified by cable manufacturers.

As a practical example, a medium quality cable connecting the GNSS antenna with the receiver normally has about 0.5 dB/m losses. Typically, a 10-dB signal drop is allowed between the antenna output and the receiver input, as such a drop would not affect the signal-to noise ratio (which we will discuss in the next section). A good estimate is 2 dB for losses in connectors. Actually this is an overestimate but it is always reasonable to have some extra power available. The remaining 8 dB for cable losses would give a permissible cable length of 16 m, which is a practical figure.

9.6 SIGNAL-TO-NOISE RATIO

We tend to think of the real-world data as being somewhat "noisy." By "noise," one means a stochastic signal which appears "by itself," thus damaging the useful signal. By commonsense experience one knows that if the noise is too large compared to the useful signal, no useful information would be extracted. With radio-receiving systems like GNSS user equipment the received signal power itself is low. If there were no noise, then the lack of signal power would be readily compensated by signal amplification. Instead what is of prime importance is a proportion between signal power and noise power. In short, this proportion is commonly referred to as signal-to-noise ratio (SNR). Indeed, any signal amplification will increase the signal power and the "initial" noise power equally, keeping the SNR unchanged at its best. In actuality, an unavoidable property of real-world bodies is that they absorb some signal power impinging on them and generate extra noise. Consequently, special actions need to be taken to prevent initial SNR from being decreased by signal processing. From this standpoint, it is clear that the main objective must be to obtain the highest possible SNR at the very first step, which is the antenna. In this section we are discussing noise generation and signal and noise propagation through the user antenna. We will see what needs to be done for no noticeable SNR degradation to occur after the antenna. Practically relevant examples in regards to the cable running from antenna to receiver will be included in the discussion. We finalize the section with estimates of SNR observed at the receiver output.

9.6.1 Noise Temperature

We begin with Figure 9.6.1. As shown here, an antenna receives the useful signal from satellites along with noise coming from space. The antenna is a real-world body, thus some signal loss and noise increase is observed. So at the very first step, a certain SNR degradation is obtained. Then there may be a short piece of transmission line (cable). Cables provide signal attenuation (Section 9.5.3) and more noise generation. Thus, a natural desire is to exclude this cable and to move the first stage of amplification right next to the antenna. This first stage has a special property. It is designed in such a way that no further SNR degradation will occur. This first stage is known as a low-noise amplifier (LNA). What is surrounded with the circle at the diagram is sometimes referred to as an active GNSS antenna. Normally, the LNA is incorporated into the

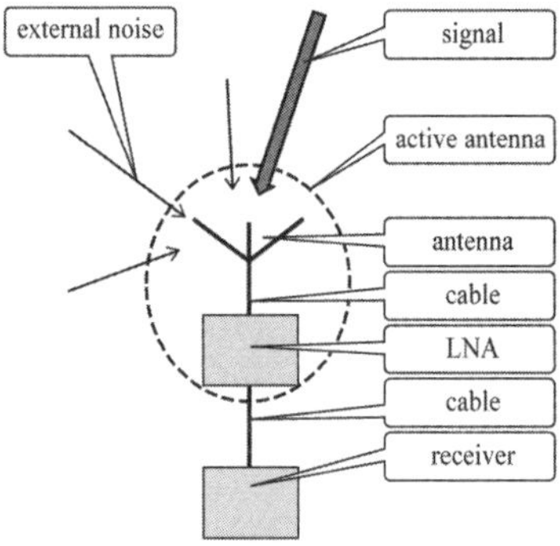

Figure 9.6.1 Block diagram for signal and noise analysis with the receiving GNSS system.

same housing as the antenna itself. After that there is a cable that connects the antenna with the receiver and, finally, the receiver circuitry.

We turn to electromagnetic noise characterization. According to the laws of fundamental physics, each body emits stochastic electromagnetic energy, also known as noise. The power of noise is distributed over the entire radio frequency band starting from the very low frequencies to infrared emissions. Over the range of GNSS frequencies one may consider the power of the noise to be distributed homogeneously over the spectrum. Let S_n stand for noise power per unit frequency band. This is called noise power spectral density. The noise power P_n within the frequency range Δf will be

$$P_n = S_n \Delta f \tag{9.6.1}$$

Due to homogeneous noise power spectral density, the larger the Δf the more noise power there is. That is one of the reasons that the electronic receiving system should be sensitive only over the desired frequency range and reject all other frequencies. This desired frequency range is called a passband. This passband should be as close as possible to the desired signal bandwidth. Another reason to limit the passband is to suppress all the other signals arriving to the antenna except for the useful ones. In the GNSS case, such are cell phone signals, satellite communication links, and the like.

The value of S_n in (9.6.1) depends on the material of the body, the processes inside the body, the size and shape of the body, and the temperature. We consider the temperature measured in degrees Kelvin. A body at absolute zero degrees Kelvin will not emit any power. At any temperature greater than absolute zero, a body will emit some electromagnetic noise. But the dependence of S_n on the material and size of a body makes the situation uncertain and inconvenient to use. The practical approach to overcome this difficulty is as follows. A body is called an ideal blackbody if it absorbs all the electromagnetic power arriving on it. This absorbed power generally would cause the temperature of the body to increase. If we want the blackbody to be in thermal balance with the environment, due to the fundamental energy conservation law, the blackbody should emit the same electromagnetic power it absorbs. Hence, one comes to the conclusion that the ability of the blackbody to emit electromagnetic power should be a function of the actual temperature of the environment only and not the details of the blackbody construction. At radiofrequencies, the following expression holds:

$$S_{n,blackbody} = k_B T_{actual} \tag{9.6.2}$$

Here, T_{actual} is the actual temperature of the environment in the Kelvin scale. The proportionality coefficient $k_B = 1.38 \cdot 10^{-23}[\mathrm{W}/^{\mathrm{o}}\ \mathrm{K}]$ is the universal Boltzman constant. The noise power generated by a blackbody within a frequency band Δf is

$$P_{n,blackbody} = k_B T_{actual} \Delta f \tag{9.6.3}$$

Other bodies would emit a different noise power than an ideal blackbody. We turn to noise emitted by a real body and make use of the blackbody approach.

The noise power spectral density S_n as emitted by a real body is commonly written in the form

$$S_n = k_B T_n \tag{9.6.4}$$

As opposed to (9.6.2), here T_n is referred to as the equivalent noise temperature of a body. It is not a measure of actual temperature. Instead, T_n is a temperature that an ideal blackbody should have in order to generate the same noise power as a real body under consideration; thus the blackbody serves as a standard. The real body might generate more or less noise as compared to the blackbody at a given temperature. Therefore, equivalent noise temperature T_n generally does not coincide with the actual temperature of the environment. This noise is referred to as a thermal noise because the noise power spectral density of an ideal blackbody is a function of only the temperature. There could be other kinds of noise associated with electronic circuitry, but these are less significant to our considerations and we are omitting them.

Making use of (9.6.4) the noise power is

$$P_n = k_B T_n \Delta f \tag{9.6.5}$$

An important note is to come at this point. The noise of interest is called white noise. By definition, white noise is such a noise for which the following two statements hold: (a) the power spectral density is homogeneous within the frequency range, and, (b) the noise components are completely uncorrelated. An important property of white noise is that the total noise power from several different sources is the sum of the noise power of each source:

$$P_{n,\Sigma} = \sum_q P_{n,;q} \tag{9.6.6}$$

Here $P_{n,\Sigma}$ is total noise power and $P_{n,q}$ is the noise power of the qth source. Because of the additivity property of (9.6.6) one may analyze noise sources separately.

9.6.2 Characterization of Noise Sources

We begin with external noise characterization. The main noise contributors are sky, stars, sun, ground, and human activities. By sky noise we mean the noise whose sources are distributed in a continuous manner over the top semisphere with respect to the user antenna. These sources are gases comprising the atmosphere and galactic noises. By stars and sun noise, we mean the set of discrete sources which are seen under specific angles above the local horizon. Certain stars are known to radiate large amounts of noise power, but the sun is the largest noise radiator. By ground noise we refer to all the noise power coming from underneath the user antenna. The materials comprising the underlying surface contribute greatly to that noise. And finally there is the human-made noise. This noise comes from human activity, such as industrial noise, transportation noise, etc.

Following considerations of the antenna effective area (Section 9.2.5), one concludes that the antenna response to different external noise sources generally depends on their angular location with respect to the antenna pattern. The angular location of the sources varies with time. Also, the physical temperature of the sky and the ground surface vary with time of day and season. All this makes exact external noise calculations complex, but we will see later that the external noise contribution is significant but not the largest. Thus, some averaged estimate would be of practical use. Using materials of Lo and Lee (1993), the average external noise temperature T_{ext} received by the user antenna could be adopted as

$$T_{ext} = 100\,\mathrm{K} \tag{9.6.7}$$

Next, we turn to antenna noise. The antenna is treated as a passive circuit or passive unit. The term "passive" is used when referring to circuits or units that do not require any external power source for their operation. A passive circuit or unit transforms the signal by ideally keeping the signal power constant. In our case, an antenna and the coaxial cable connecting the antenna with the receiver would be examples of passive circuits. Due to resistance for electric currents, a passive circuit introduces power losses to the signal and generates some noise. Let χ stand for the efficiency of the circuit. The antenna efficiency has already been introduced in expression (9.2.33). In general, the efficiency χ relates the signal power at the circuit input P_{input} to that at the output $P_{output,signal}$,

$$P_{output,signal} = \chi P_{input} \tag{9.6.8}$$

The efficiency χ equals to unity for the ideal case and is always less then unity for a real-world unit.

As was mentioned already, the signal power absorption is always related to noise generation. The relationship between the efficiency χ and noise power P_n is given by the fundamental Nyquist theorem (Lo and Lee, 1993). For our case it could be stated as follows: If the unit with efficiency χ is in thermodynamic balance with the environment with temperature T_0, then the noise power generated within frequency band Δf is

$$P_n = k_B T_0 (1 - \chi)\Delta f \tag{9.6.9}$$

For noise estimation purposes T_0 always equals some standard value. This value is adopted as 20° C or approximately 290 K. Thus we take

$$T_0 = 290\,\text{K} \tag{9.6.10}$$

Comparing (9.6.9) to (9.6.4), one writes the equivalent noise temperature of a passive unit as

$$T_\chi = T_0(1 - \chi) \tag{9.6.11}$$

We mark this temperature as T_χ to highlight that the noise is related to the efficiency of the unit. The limiting case with $\chi = 1$ has already been referred to as an ideal or loss-less circuit. Noise power is zero. The other limiting case is $\chi = 0$. Sometimes, this is called a perfectly matched load because such a circuit totally absorbs all the input power. The outcome of this circuit will be just noise with the equivalent noise temperature of (9.6.10). Thus, the noise power generated by a perfectly matched load equals that of a blackbody at standard temperature. For the antenna we take χ_a as –1 dB or 0.8 in relative units as stated in Section 9.2.4. Equation (9.6.11) then yields the antenna noise temperature T_a as

$$T_a = 58\,\text{K} \tag{9.6.12}$$

Next, we consider losses in the cable which connects the antenna to the receiver. With (9.5.28) for a cable having $\alpha = 0.5\,\text{dB/m}$ and $L = 20$ m one obtains $\chi_{cable} = -10\,\text{dB}$ or 0.1 in relative units. It may sound surprising that we treat as typical a cable that provides only 10% of the signal power to the output. But this is actually a typical example, as discussed in Section 9.5.3. Later we will see why such functionality does not lead to loss of signal quality. Using (9.6.11), one has an equivalent noise temperature of the cable as

$$T_{cable} = 261\,\text{K} \tag{9.6.13}$$

Now we turn to an active circuit. By "active" one commonly means a circuit that requires an external power source for its operation. We take an amplifier as an example. An amplifier increases the signal power by a gain factor G_{amp}, with $G_{amp} > 1$. The signal power amplification is obtained at the expense of power consumption from an external source. In most cases, this is a DC source. Signal power transformation for the amplifier will look like

$$P_{output} = G_{amp} P_{input} \tag{9.6.14}$$

Here P_{input} and P_{output} are input and output power, respectively. An amplifier is treated differently than a passive circuit. There are four main processes taking place in parallel inside an amplifier: the DC source power consumption, input power amplification, some unavoidable input power absorption, and noise generation. Let us now discuss the commonly adopted way to treat the amplifier.

Let the amplifier noise power be generated right at the amplifier input and then passed through a noise-less amplifier. Thus, instead of a noise power P_n at an amplifier output, an equivalent noise power at an amplifier input P'_n is used. This P'_n is related to P_n as

$$P'_n = \frac{P_n}{G_{amp}} \tag{9.6.15}$$

Then, one has three power terms passing together through a noise-less amplifier: the signal power P_{signal}, the incoming noise power $P_{n,inc}$, and the equivalent noise power of an amplifier P'_n. At the amplifier output, all three are amplified by a gain factor.

To characterize an amplifier one has to know G_{amp} and P'_n. To measure these quantities, two experiments are performed. The first could be called "signal measurement," where a strong signal source is applied to the amplifier input. Its power $P_{signal;1}$ is considered to be known and much larger than the noise. By measuring the output power $P_{output;1}$, one has

$$G_{amp} = \frac{P_{output;1}}{P_{signal;1}} \tag{9.6.16}$$

The second experiment is "noise measurement", where one applies some known noise power from a noise standard $P_{n,standard}$. The output power $P_{output;2}$ is measured. This output power is

$$P_{output;2} = G_{amp}\left(P_{n,standard} + P'_n\right) \tag{9.6.17}$$

Thus

$$P'_n = \frac{P_{output;2}}{G_{amp}} - P_{n,standard} \tag{9.6.18}$$

Instead of power P'_n, it is common to characterize the amplifier with an amplifier noise factor N. This parameter is often shown in GNSS user antenna documentation. By definition,

$$N = \frac{P'_n + P_{n,standard}}{P_{n,standard}} \tag{9.6.19}$$

Thus, with the above noise measurement experiment,

$$N = \frac{P_{output;2}}{G_{amp}P_{n,standard}} \tag{9.6.20}$$

In practice, a perfectly matched load is used as a noise standard. Its noise power equals to that of a blackbody, thus

$$P_{n;standard} = k_B T_0 \,\Delta f \tag{9.6.21}$$

Once N is known, and involving (9.6.18), (9.6.19), and (9.6.21), one has

$$P'_n = k_B T_0 (N - 1)\Delta f \tag{9.6.22}$$

Comparing this expression to (9.6.5), one has equivalent noise temperature at the amplifier input

$$T_{amp} = T_0(N - 1) \tag{9.6.23}$$

Today's good LNAs of receiving GNSS antennas provide $G_{amp} = 30\,\text{dB}$ (1000 times in related units) and $N = 1.5$ dB, or 1.41 in related units. Using this figure, one obtains from expression (9.6.23) the noise temperature of LNA

$$T_{LNA} = 119\,\text{K} \tag{9.6.24}$$

Thus, all the noise sources are known and we turn to signal and noise propagation through the chain of the units of interest. Our focus will be on the role of LNA.

9.6.3 Signal and Noise Propagation through a Chain of Circuits

First, let us consider an amplifier with gain G_1 and noise factor N_1. Let P_s be incoming signal power and $P_{n;inc}$ be incoming noise power. The SNR of the incoming signal is

$$SNR_{inc} = \frac{P_s}{P_{n;inc}} \tag{9.6.25}$$

At point 0 (Figure 9.6.2, left panel) we add the equivalent noise power of the amplifier $P'_{n;1}$ related to noise factor N_1 as shown by (9.6.22). Then we allow both signal and the total noise to propagate through a noise-less amplifier in accordance with above. At the output (point 1) one has an SNR in the form

$$SNR_1 = \frac{P_s G_1}{\left(P_{n;inc} + P'_{n;1}\right) G_1} = \frac{P_s}{P_{n;inc} + P'_{n;1}} \tag{9.6.26}$$

This SNR is degraded compared to that of the incoming signal (9.6.25) due to the noise generated by the amplifier.

Next, consider a chain of two amplifiers (Figure 9.6.2, middle panel). Let the gains of the amplifiers be $G_{1,2}$, respectively. The input of the chain is point 0, the input of the second amplifier is point 1, and the output of the chain is point 2. Total noise power at point 2 is

$$P_{n;2} = G_2 \left[G_1 \left(P_{n;inc} + P'_{n;1} \right) + P'_{n;2} \right] \tag{9.6.27}$$

This expression is constructed as follows. We take the incoming noise power and add the equivalent noise power of the first amplifier $P'_{n;1}$. This gives the total noise power

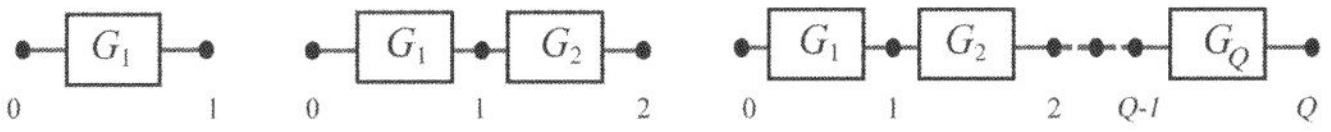

Figure 9.6.2 Signal and noise propagation through a chain of amplifiers.

at point 0. Then we allow this noise to propagate through the noise-less amplifier with gain G_1. This simply means multiplying the sum of two powers in parenthesis by the gain. Thus we have the noise power at point 1 that is coming from the first amplifier. Next we add the equivalent noise power $P'_{n;2}$ of the second amplifier. This gives a total noise power at point 1(see the term in brackets in (9.6.27)) and then we multiply this by a gain G_2 and get finally the noise power at point 2. In a straightforward manner one has the signal power at point 2, which is just the product of the incoming signal power and the two gains

$$P_{s;2} = G_2 G_1 P_s \tag{9.6.28}$$

Now we construct signal-to-noise ratio at the output (point 2),

$$SNR_2 = \frac{P_{s;2}}{P_{n;2}} = \frac{G_2 G_1 P_{s;0}}{G_2 \left[G_1 \left(P_{n;0} + P'_{n;1} \right) + P'_{n;2} \right]} = \frac{P_{s;0}}{\left(P_{n;0} + P'_{n;1} \right) + \frac{1}{G_1} P'_{n;2}} \tag{9.6.29}$$

We see that if the gain of the first amplifier is large enough such that

$$\frac{1}{G_1} P'_{n;2} << P_{n;0} + P'_{n;1} \tag{9.6.30}$$

then the SNR at the chain output does not depend on the second amplifier,

$$SNR_2 \approx \frac{P_s}{P_{n;inc} + P'_{n;1}} = SNR_1 \tag{9.6.31}$$

This SNR is defined by the incoming signal and noise powers and by the noise power of the first amplifier only.

Thus, for the incoming SNR not to degrade significantly, the first amplifier is to have the large gain and the smallest possible noise factor. That is what the LNA is all about. Special techniques are known to meet both of these requirements. One may verify that for LNA with about 30 dB gain (or 1000 times in relative units) the inequality (9.6.30) really holds unless the noise of the second amplifier is extremely large.

Now we perform the same analysis for the chain of Q amplifiers (see the right panel in Figure 9.6.2). Similarly as above, one has the total noise power at the output

$$P_{n;Q} = G_Q \left[G_{Q-1} \left[\cdots G_2 \left[G_1 \left[P_{n;inc} + P'_{n;1} \right] + P'_{n;2} \right] + \cdots P'_{n;Q-1} \right] + P'_{n;Q} \right] \tag{9.6.32}$$

Here, G_q and $P'_{n;q}$ are a gain and an equivalent noise power of the qth amplifier, respectively. Please note that the only noise power that is multiplied by all the gains is the sum of the incoming noise power and the first amplifier. The total signal power at the output is

$$P_{s;Q} = G_Q G_{Q-1} \ldots G_2 G_1 P_s \tag{9.6.33}$$

For the SNR at the output, one writes

$$(SNR_Q)^{-1} = \frac{G_Q G_{Q-1} \dots G_2 G_1 [P_{n;inc} + P'_{n;1}] + G_Q G_{Q-1} \dots G_2 P'_{n;2} + \dots G_Q P'_{n;Q}}{G_Q G_{Q-1} \dots G_2 G_1 P_s}$$

$$= \frac{P_{n;inc} + P'_{n;1}}{P_s} + \frac{P'_{n;2}}{G_1 P_s} + \frac{P'_{n;3}}{G_2 G_1 P_s} + \dots \frac{P'_{n;Q}}{G_{Q-1} \dots G_2 G_1 P_s} \tag{9.6.34}$$

If $G_1 >> 1$, then

$$SNR_Q \approx \frac{P_s}{P_{n;inc} + P'_{n;1}} SNR_1 \tag{9.6.35}$$

is equal to the SNR observed after the first amplifier. The noise temperature T_{chain} of a chain will look like

$$T_{chain} = \left[(N_1 - 1) + \frac{N_2 - 1}{G_1} + \frac{N_3 - 1}{G_2 G_1} + \dots \frac{N_Q - 1}{G_{Q-1} \dots G_2 G_1} \right] T_0 \tag{9.6.36}$$

Here, N_q, with $q = 1, \dots, Q$, is a noise factor of the qth amplifier. Expression (9.6.36) is known as Friis' formula. Under conditions already mentioned, we neglect all the terms in brackets in (9.6.36) except for the first one and obtain

$$T_{chain} = T_1 \tag{9.6.37}$$

Thus, the noise temperature of a chain equals that of the first amplifier.

Let us turn to an example relevant to practical GNSS antenna use. Let the LNA be connected to the cable with efficiency χ (Figure 9.6.3, left panel). Our goal is to find conditions for the SNR at the output of a chain to coincide with that at the output of LNA. We write the total noise power at the output of the chain as

$$P_{n;2} = \chi G_1 (P_{n;inc} + P'_{n;1}) + P_{n;\chi} \tag{9.6.38}$$

Here, one has the sum of incoming noise power and the equivalent noise power of the first amplifier that is amplified by the gain factor G_1 and then absorbed by the efficiency factor χ. Also, a noise of a circuit with efficiency χ is generated. The signal power at the chain output is

$$P_{s;2} = \chi G_1 P_s \tag{9.6.39}$$

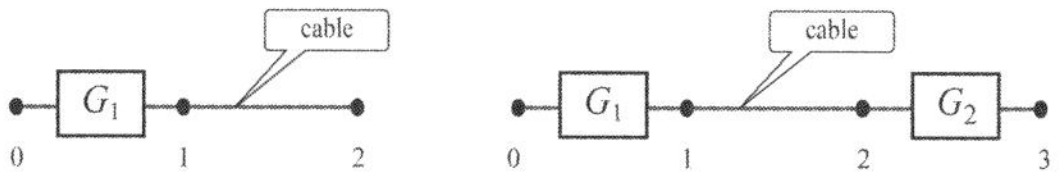

Figure 9.6.3 Signal and noise propagation through a cable run with amplifier insert.

It is just the incoming signal power that is amplified first and partially absorbed afterward. The SNR at the output is

$$SNR_2 = \frac{P_{s;2}}{P_{n;2}} = \frac{\chi G_1 P_s}{\chi G_1 \left(P_{n;inc} + P'_{n;1}\right) + P_{n;\chi}} = \frac{\chi P_s}{\chi \left(P_{n;inc} + P'_{n;1}\right) + \frac{1}{G_1} P_{n;\chi}} \tag{9.6.40}$$

Then, similar to the derivations of above, if the gain G_1 is sufficiently large, one may neglect the second term in the denominator, and the common factor χ cancels. Thus, the cable "disappears," and SNR_2 will take the form of (9.6.31) and will not depend on the properties of the cable.

However, one notes that the actual condition to neglect the second term in the denominator in (9.6.40) reads

$$\chi G_1 \left(P_{n;inc} + P'_{n;1}\right) >> k_B T_0 (1 - \chi) \Delta f \tag{9.6.41}$$

Here, we have introduced the noise power of a cable from (9.6.9). Thus, if the cable run is too long, then $\chi \to 0$ and the left-hand side in (9.6.41) diminishes, however large the gain G_1 is. The cable will work as a perfectly matched load with SNR at the output tending to be zero. This situation may occur with long cables that are typical with GNSS reference network antenna installations.

To overcome this difficulty one is to consider a chain shown in the right panel in Figure 9.6.3. Here the first amplifier, the LNA, is connected via cable to the second one. At the output of a chain (point 3) one has

$$\begin{aligned} SNR_3 = \frac{P_{s;3}}{P_{n;3}} &= \frac{G_2 \chi G_1 P_{s;0}}{G_2 \chi G_1 \left(P_{n;0} + P'_{n;1}\right) + G_2 \left(P_{n;\chi} + P'_{n;2}\right)} \\ &= \frac{P_{s;0}}{\left(P_{n;0} + P'_{n;1}\right) + \frac{1}{\chi G_1} \left(P_{n;\chi} + P'_{n;2}\right)} \end{aligned} \tag{9.6.42}$$

This SNR equals to (9.6.31) if

$$P_{n;0} + P'_{n;1} >> \frac{1}{\chi G_1} \left(P_{n;\chi} + P'_{n;2}\right) \tag{9.6.43}$$

The second amplifier "disappears" along with the cable, from SNR standpoint. The noise power at the output (point 3) will be

$$P_{n;3} \approx G_2 \chi G_1 (P_{n;0} + P'_{n;1}) \tag{9.6.44}$$

It is now proportional to the gain G_2 (along with the signal!). Thus, one may continue the cable run with a line amplifier insert, provided that initial noise amplified by G_2 is much larger than the noise contribution of the next step. One is to note, of course, that the larger the losses to be compensated, the larger the gains and the smaller the noise figures of the amplifiers must be.

To conclude this discussion we need to mention that instead of efficiency χ of a passive circuit, one may operate with an equivalent gain G such that

$$G = \chi < 1 \tag{9.6.45}$$

and with noise factor

$$N = \frac{1}{\chi} \tag{9.6.46}$$

This way is adopted in literature such that a chain of active and passive circuits are analyzed in a unified format employing Friis' formula (9.6.36) and SNR of (9.6.34). We prefer not do that, however, and keep the efficiency χ for passive circuits.

9.6.4 SNR of the GNSS Receiving System

We combine all derivations to estimate the SNR that is normally observed at the receiver input. In order to obtain the respective function we look again at Figure 9.6.1, assume there is no cable between the antenna and LNA, take expression (9.2.42) for the signal power at the antenna output, and neglect polarization losses and mismatch. The expression for the $SNR_{r;input}$ at the receiver input will look like

$$SNR_{r.input} = \frac{10^{-16} D_0 F^2(\theta, \phi) \chi_a G_{LNA} \chi_{cable}}{\Delta f k_B [(T_{ext}\chi_a + T_0(1-\chi_a) + T_0(N_{LNA}-1)) G_{LNA}\chi_{cable} + T_0(1-\chi_{cable})]} \tag{9.6.47}$$

In the numerator we have the power received by the antenna (9.2.42), amplified by the factor of LNA gain, G_{LNA}, and then partially lost due to attenuation in the cable. In the denominator, we express the noise powers via equivalent noise temperatures. The common k_B and frequency bandwidth Δf terms were factored out. What remains in the brackets is actually the overall noise temperature of the chain: the first term is an external noise with the temperature T_{ext} attenuated by a factor of antenna efficiency χ_a, the second term is the noise generated by the antenna due to efficiency χ_a, and the third term is LNA noise. These three terms are amplified by the LNA via the factor G_{LNA} and then attenuated by the cable. The last term is the cable noise, taking the efficiency χ_{cable} into account.

We introduce T_Σ as the noise temperature of an active antenna, which is comprised of external noise, antenna noise, and LNA noise. One has

$$T_\Sigma = T_{ext}\chi_a + T_0(1-\chi_a) + T_0(N_{LNA}-1) \tag{9.6.48}$$

All terms of this expression have been discussed already. Now we summarize numerical values. We take (9.6.7), (9.6.12), and (9.6.24) and get

$$T_\Sigma = 257\,\mathrm{K} \tag{9.6.49}$$

Assuming an LNA gain of 30 dB or 1000 times in relative units, we get $T_\Sigma G_{LNA}\chi_{cable} = 25{,}700\,\mathrm{K}$. Compared to this figure the noise generated by a cable as expressed by (9.6.13) is about 100 times less. This result is expected. It is the large LNA gain that plays the major role here. Once we neglect the cable noise in the denominator of (9.6.47), we immediately arrive at

$$SNR_{R.input} = \frac{10^{-16} D_0 F^2(\theta, \phi) \chi_a}{\Delta f k_B [T_{ext} \chi_a + T_0(1 - \chi_a) + T_0(N_{LNA} - 1)]} = SNR_{A.output} \quad (9.6.50)$$

The $SNR_{A;input}$ stands for the SNR at the output of the active antenna.

First we conclude what has been mentioned already: Even under rather hard conditions of 0.1 cable efficiency (a 90% power loss), a big LNA gain keeps the SNR unchanged. Then we see that the SNR does not actually depend on the LNA gain G_{LNA}; the gain just has to be large. Recall the initial discussions in Section 9.1.7: a ± 2 dB, which looks like a relatively large variation in terms of absolute power, is not important for SNR.

We now arrive at the final point. Instead of using the SNR in terms of power, some another quantity is often used. The term

$$S_\Sigma = k_B T_\Sigma \quad (9.6.51)$$

represents the power spectral density of an active antenna noise. SNR in the form of

$$SNR = 10 \log \left(\frac{10^{-16} D_0 F^2 (\theta, \phi) \chi_a}{S_\Sigma} \right) \quad (9.6.52)$$

is considered to have dimensions [dB · Hz]. This ratio is adopted as the signal quality indicator at the antenna output. By way of derivations from the previous section, one may state that, with a properly designed system, this value should stay unchanged throughout signal transformations. Using 6 dB directivity for the antenna (4 in relative units) and collecting all the numbers above, the formulae (9.6.52) yields $SNR \approx 50$[dB · Hz] for zenith direction. This is a typical figure normally reported by a GNSS receiver.

9.7 ANTENNA TYPES

This section provides an overview of antenna types adopted by current GNSS receiving technology. Some details about derivations can be found in the Appendices. For basics about the electromagnetic instrumentation involved, the reader is referred again to Balanis (1989).

9.7.1 Patch Antennas

Microstrip patch antennas remain one of the most common types for GNSS surveying due to their compact profile and manufacturing simplicity. These antennas have been the focus of antenna developers for more than three decades. For guidance on theory and design of these antennas the reader is referred to Waterhouse (2003), Pozar and Schaubert (1995), Garg et al. (2001), and Kumar and Ray (2003). Exact computer simulations using electromagnetics software packages are in broad use with designers. However, the main features of patch antenna performance relevant for GNSS applications can be illustrated with a simplified approach known as single cavity mode approximation. This approach is chosen below. Details of derivations are shown in Appendix G. Within the scope of the discussion the main disadvantage of the single-mode approximation is that an antenna bandwidth appears to be about two times larger than actually achievable. More accurate formulas to estimate the antenna Q factor can be found in Garg et al. (2001).

We start with a linear polarized antenna. It consists of the dielectric substrate 1, the metal patch 2, and ground plane 3 (the left panel in Figure 9.7.1). The patch is connected to the probe 4, which in turn is connected to the inner conductor of the coaxial feed 5 via a hole in the ground plane. Another option is that the probe 4 is connected to the microstrip line 6 (right panel in the figure). For circular polarized antennas, it is common to choose a patch of rectangular or circular shape. We will stay with the rectangular shape because it is easier to analyze and implement. The main features of circular patch functionality are similar to those of rectangular ones. Designations of antenna constitutive parameters are seen in the figure. Here, a_x is the patch size along the symmetry axis where the probe is located (x axis), a_y is the patch size along the y axis perpendicular to the x axis, ε is permittivity of substrate, h is substrate thickness, x_{pr} is the probe offset from the center of symmetry, and r_{pr} is the probe radius. The substrate thickness h normally is several hundredths of the free space wavelength, and substrate permittivity ε is normally about 3 to 4, reaching about 30 with compact designs.

A microstrip patch antenna is of the resonant type. In its classical implementation it works within a narrow frequency band around the resonant frequency. Within this frequency band, the $\vec{E}$ - and $\vec{H}$-field distribution in space immediately surrounding the patch takes a certain special form referred to as a TM_{10} cavity mode. Spatial distribution of the $\vec{E}$ field is schematically illustrated in Figure 9.7.2 (left panel). In

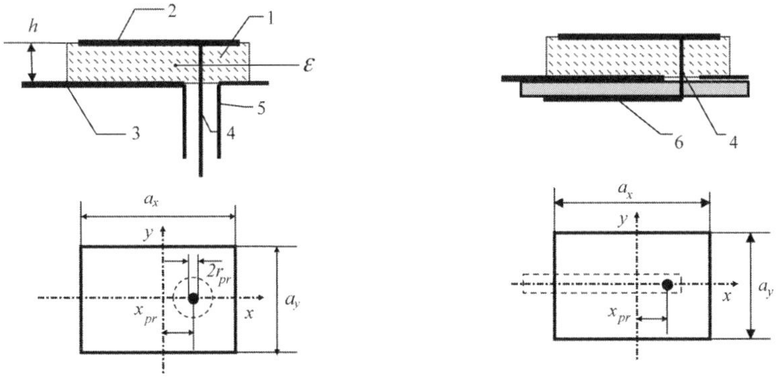

Figure 9.7.1 Linear polarized microstrip patch antenna.

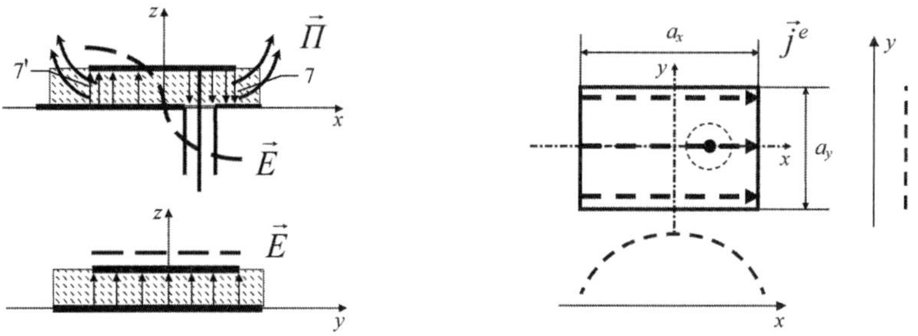

Figure 9.7.2 TM$_{10}$ cavity mode of patch antenna.

the space between a patch and a ground plane the vector $\vec{E}$ is perpendicular to the two surfaces. Resonance occurs when

$$a_x \approx \frac{\lambda_0}{2\sqrt{\varepsilon}} \tag{9.7.1}$$

The symbol λ_0 is the free space wavelength at resonant frequency. Similarly, as in previous sections, we consider the transmitting mode of the antenna. In transmitting mode, the antenna radiates power into free space through the slots parallel to the y axis. These slots are marked as 7, 7′ in the top-left panel in Figure 9.7.2 and are formed by the patch edges and the ground plane. The E-field distribution along said slots is homogeneous (bottom-left panel. The E-field distribution along the slots formed by patch edges parallel to the x axis and the ground plane is antiphase (dashed line in top-left panel). Power radiation via these slots is neglected. The electric current $\vec{j}$ associated with the resonant mode and flowing at the patch is illustrated in the right panel in Figure 9.7.2. This current is homogeneous with respect to the y axis and takes the form of half a cosine-like wave with respect to the x axis. The radiated field has the dominant component of the vector $\vec{E}$ parallel to the x axis.

The input impedance of the antenna is (see Appendix G)

$$Z_{inp} = \frac{1}{G + iB} + iX_L \tag{9.7.2}$$

with

$$G = \frac{2G_\Sigma}{\sin^2\left(\frac{\pi}{a_x} x_{pr}\right)} \tag{9.7.3}$$

$$B = \frac{2\left\{B_\Sigma + \frac{a_x a_y}{\eta_0 4kh}\left[k^2\varepsilon - \left(\frac{\pi}{a_x}\right)^2\right]\right\}}{\sin^2\left(\frac{\pi}{a_x} x_{pr}\right)} \tag{9.7.4}$$

In these expressions, k is a free space wavenumber (9.1.32), η_0 is free space intrinsic impedance (9.1.14), and

$$Y_\Sigma = G_\Sigma + iB_\Sigma \tag{9.7.5}$$

is radiating admittance of a slot (Garg et al., 2001) with

$$G_\Sigma = \frac{1}{2}\frac{ka_y}{\eta_0}\left(1 - \frac{(kh)^2}{24}\right) \tag{9.7.6}$$

$$B_\Sigma = \frac{1}{2\pi}\frac{ka_y}{\eta_0}(3.135 - 2\log(kh)) \tag{9.7.7}$$

The last term in (9.7.2) is the inductance of a probe

$$X_L = \frac{1}{2\pi}\eta_0 kh(0.1159 - \ln(k\sqrt{\varepsilon}r_{pr})) \tag{9.7.8}$$

Expression (9.7.2) shows that input impedance exhibits a frequency response similar to that of a parallel resonant circuit. Equating B of (9.7.4) to zero at resonant frequency f_0 and solving with respect to a_x yields

$$a_x = \frac{\lambda_0}{2\sqrt{\varepsilon}}\left[\sqrt{1 + \left(\frac{2}{\pi\sqrt{\varepsilon}}\frac{h}{\lambda_0}\left(3.135 - 2\log\left(\frac{2\pi h}{\lambda_0}\right)\right)\right)^2} - \frac{2}{\pi\sqrt{\varepsilon}}\frac{h}{\lambda_0}\left(3.135 - 2\log\left(\frac{2\pi h}{\lambda_0}\right)\right)\right] \tag{9.7.9}$$

This expression shows that the resonant size is slightly less compared to (9.7.1) due to capacitance (9.7.7). With substrate thickness $h << \lambda_0$, the maximal active impedance reached at resonant frequency is

$$R_{inp\ max\,[\text{ohm}]} = G_\Sigma^{-1} \approx 60\frac{\lambda_0}{a_y}\sin^2\left(\frac{\pi}{a_x}x_{pr}\right) \tag{9.7.10}$$

This expression defines the probe displacement x_{pr}, which is required to match the antenna with the feeder. The antenna Q factor is defined by frequency bandwidth Δf where the active impedance exceeds $1/2R_{inp\ max}$. This bandwidth is given by

$$\frac{\Delta f}{f_0}_{[\%]} = \frac{1}{Q}100 = \frac{4h}{\lambda_0\sqrt{\varepsilon}}100 \tag{9.7.11}$$

Thus the bandwidth is set up by substrate parameters. For typical GNSS applications, assuming $\lambda_0 = 20\,\text{cm}$, $h = 5\,\text{mm}$, and $\varepsilon = 4$, expression (9.7.11) gives 5% bandwidth. In actuality an antenna with such parameters possesses slightly less than 3% bandwidth.

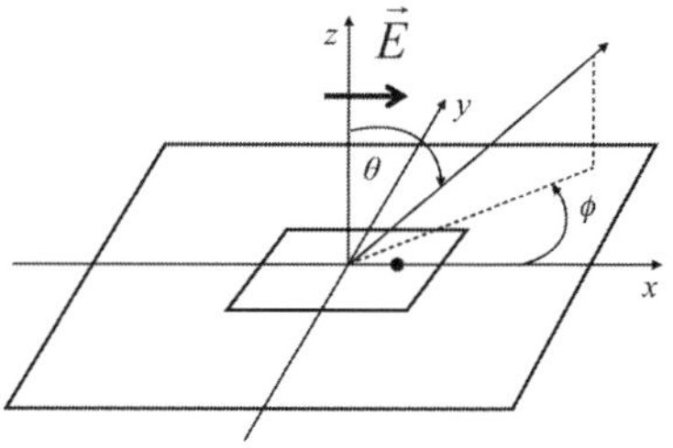

Figure 9.7.3 Coordinate frame for calculation of patch antenna far field.

The nonnormalized antenna pattern in terms of the θth and the ϕth components is

$$F_\theta = N_x \cos\theta \cos\phi + N_y \cos\theta \sin\phi - N_z \sin\theta \tag{9.7.12}$$

$$F_\phi = N_x \sin\phi - N_y \cos\phi \tag{9.7.13}$$

Here θ is the zenith angle and ϕ is the azimuth counted from the x axis as shown in Figure 9.7.3. With these expressions,

$$N_x = -I\frac{4a_y}{kh\eta_0 \sin\left(\frac{\pi}{a_x}x_{pr}\right)(G+iB)}\frac{\cos u}{1-\left(\frac{2u}{\pi}\right)^2}\frac{\sin v}{v}\sin w \tag{9.7.14}$$

$$N_y = 0 \tag{9.7.15}$$

$$N_z = I\left[2he^{ikx_{pr}\sin\theta\cos\phi} + \frac{8(\varepsilon-1)ka_xa_y}{\eta_0\pi^2 \sin\left(\frac{\pi}{a_x}x_{pr}\right)(G+iB)}\frac{u\cos u}{1-\left(\frac{2u}{\pi}\right)^2}\frac{\sin v}{v}\right]\frac{\sin w}{w} \tag{9.7.16}$$

The symbol I denotes a probe current, and

$$u = \frac{ka_x}{2}\sin\theta\cos\phi \tag{9.7.17}$$

$$v = \frac{ka_y}{2}\sin\theta\sin\phi \tag{9.7.18}$$

$$w = kh\cos\theta \tag{9.7.19}$$

As seen from (9.7.12) to (9.7.19), in the plane $\phi = 0$ or π, the component F_ϕ equals zero and the only remaining component is F_θ. This component belongs to the plane, and in the direction $\theta = 0$, this component is parallel to the x axis where the probe is located. This is the principal polarization component. There is no cross-polarized component F_ϕ in this plane. In general, it is common with linear polarized antennas to distinguish E and H planes of symmetry. The E plane is a plane that contains the principal polarization component of vector $\vec{E}$. The same is true for H plane in regard

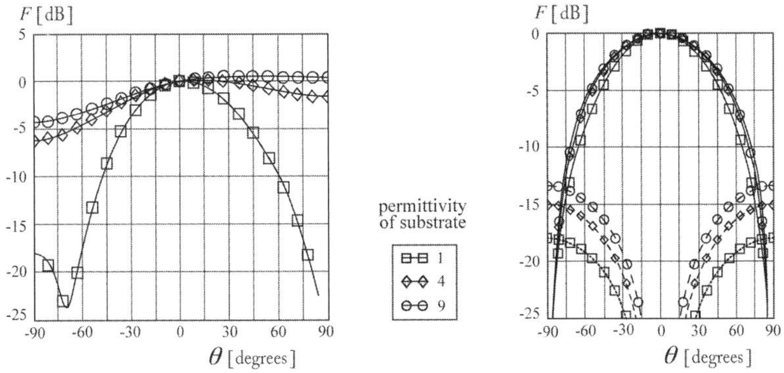

Figure 9.7.4 Patch antenna pattern.

to vector $\vec{H}$. For the current case, the plane defined by $\phi = 0$ and $\phi = \pi$ is the E plane of the antenna and the plane $\phi = \pi/2$ and $\phi = 3\pi/2$ is the H plane. The principal polarization component of the E field in the latter plane is F_ϕ. This component is parallel to the x axis; however, within that plane we generally have $F_\theta \neq 0$. This is a cross-polarized component. In other planes with respect to azimuth ϕ, both principal and cross-polarized components are also observed.

The E-plane patterns calculated using the above expressions are illustrated in Figure 9.7.4 for three antennas. These antennas differ by substrates permittivity. The substrates thickness h is $0.025\lambda_0$ or 5 mm if one takes $\lambda_0 = 20$ cm. The patch is assumed to be square with dimensions $a_x = a_y$ given by (9.7.9). The values are 0.464, 0.24, and $0.163\lambda_0$, respectively. In each case the probe location is chosen by (9.7.10) to match with a 50-Ω feeder. As seen with the plots, the E-plane patterns (left panel) are not symmetric with respect to the zenith direction $\theta = 0$. This is due to radiation of the probe which is offset from the center of symmetry. An antenna with air substrate $\varepsilon = 1$ possesses a narrower pattern compared to others; the pattern of this antenna is significantly nonsymmetrical and has a deep drop. This property, along with the large size of this antenna, limits its use in GNSS applications in spite of having the largest bandwidth according to (9.7.11). As permittivity ε growths, the patch size decreases, the probe approaches the center of symmetry, and the pattern becomes smoother and wider. One is to note that the changes just mentioned are accompanied by an antenna bandwidth (9.7.11) decrease.

Right panel in Figure 9.7.4 illustrates the same patterns for the H plane. The principal polarization components are shown as solid lines and cross-polarized components are shown as dashed lines. The cross-polarized component equals zero for the zenith direction as was already mentioned. However, this component is growing in the directions closer to the horizon ($\theta = 90°$). With substrate permittivity growth, the cross-polarized component intensity also increases.

For correct interpretation of the antenna patterns just described, one is to note that formulas (9.7.14) to (9.7.16) are derived using the antenna ground plane approximation as an unbounded plane. With a ground plane of finite size, the pattern

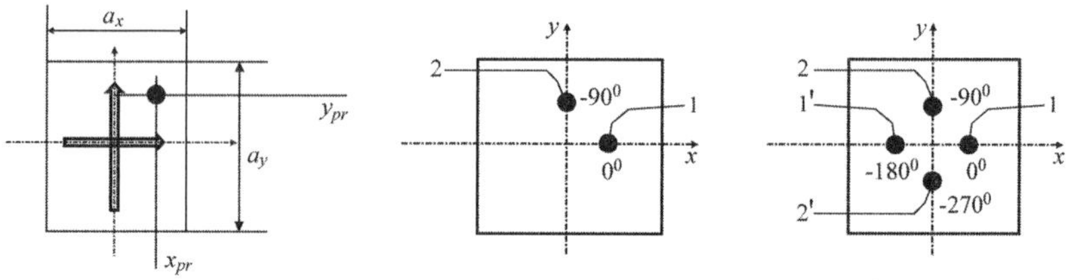

Figure 9.7.5 Circular polarized patch antenna with one, two, and four probes excitation.

reading in the E plane decreases by an extra 6 dB in directions close to horizon ($\theta = 90°$). This is explained further in Section 9.7.3.

To achieve a circular polarization one must excite two similar cavity modes (TM_{10} and TM_{01}) with respect to the x and y axes of symmetry. These two modes are said to be orthogonal as the power balances [expressions (G.4) and (G.6) of Appendix G] of these modes are independent of each other. One possible way to achieve the excitation of the two modes is to use one probe located near a diagonal of the rectangular patch (Figure 9.7.5, left panel). The patch dimensions a_x and a_y are then to be chosen as

$$a_{x,y} = a(1 \mp \Delta f/(2f_0)) \tag{9.7.20}$$

Here, a is resonant size (9.7.9) for given λ_0 and $\Delta f/f_0$ is a relative bandwidth (9.7.11). The patch currents associated with these two resonant modes are shown schematically in the left panel in Figure 9.7.5 as block arrows. Using expression (G.17), and as was done for the TM_{01} mode, one may make sure that under conditions (9.7.20), the patch current parallel to the y axis is 90° delayed in phase compared to the one parallel to the x axis. By choosing proper displacements of probe x_{pr} and y_{pr}, the magnitudes of the E-field components associated with these currents are made equal to each other for the zenith direction. Thus, an RHCP field is achieved in zenith. However, the bandwidth with such an antenna is limited not by mismatch but rather by polarization properties. The frequency bandwidth within which the axial ratio is greater than 0.7 is about half compared to (9.7.11). Also, such an antenna possesses a phase center offset in the horizontal plane. This is due to the probe displacement from the vertical axis of symmetry.

Another way to achieve circular polarization is to use a square patch with the size defined by (9.7.9) with either two (middle panel, Figure 9.7.5) or four (right panel, Figure 9.7.5) excitation probes located at symmetry axes. Probes are equally offset from the center. Probes 1 and 2 in the middle panel excite two orthogonal resonant modes with respect to x and y axes. The same holds with pairs of probes 1, 1′ and 2, 2′ in the right panel. The electric current amplitudes of the probes are to be equal to each other. Probe 2 in the middle panel is to be 90° phase delayed with respect to probe 1. This is achieved by employing a feed network connected to the probes. With four probes excitation, the probe currents are to have 90° progressive phase delay. The drawback of the two-probe version is that the pattern is not exactly symmetrical with respect to the vertical axis as has been discussed already. This is associated with

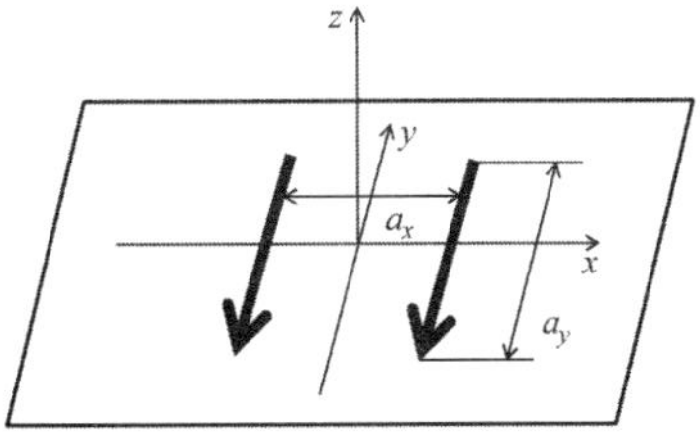

Figure 9.7.6 Two magnetic current segments as a model of patch antenna radiation.

the phase center offset in the horizontal plane, whereas with the four-probe excitation the antenna possesses an exact rotational symmetry.

If one neglects the antenna pattern symmetry consideration, then a useful simplified estimate for the pattern holds. As mentioned, with linear polarized mode the main radiation comes from slots 7, 7′ in Figure 9.7.2. These slots could be modeled, see Lier and Jacobsen (1983), as equivalent magnetic current segments placed onto the ground plane (Figure 9.7.6). For details of the magnetic current definition the reader is referred to Appendix E. In the E plane the radiation pattern of these currents is

$$F(\theta) = \cos\left(k\frac{a_x}{2}\sin\theta\right) \tag{9.7.21}$$

Here, the ground plane approximation is assumed as an unbounded plane and thus, this expression is valid for $-\pi/2 \leq \theta \leq \pi/2$. As an example, let the desired pattern roll-off from the zenith to the horizon be 10 dB. Then, as was mentioned already for the unbounded ground plane model, the roll-off is 6 dB less making it equal to 4 dB. Expression (9.7.21) then gives the antenna patch size $a_x \approx \lambda/4$ and from (9.7.1) the permittivity of the substrate is $\varepsilon \approx 4$. These are typical figures for a GNSS user antenna.

As was mentioned previously, patch antennas are narrowband. In their canonical implementation they are not suitable for dual-frequency L1/L2 functionality, which is required for real-time positioning. One way to overcome this difficulty is to arrange L1 and L2 antenna elements in planar concentric format. As an example, Figure 9.7.7

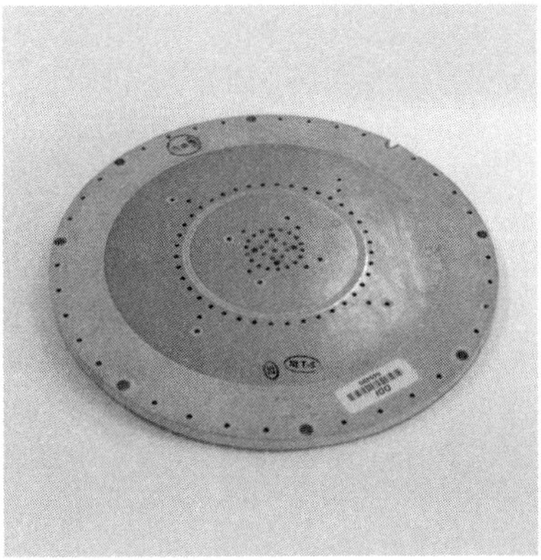

Figure 9.7.7 Dual frequency concentric patch antenna board of Legant antenna.

illustrates the dual-frequency Legant antenna board. This antenna was designed in the mid-90s at Javad Positioning Systems (JPS) under the leadership of V. Filippov, in which D. Tatarnikov participated as a senior scientist. The diameter of the antenna is 150 mm, the inner shorted circular patch serves for L1 GPS/GLONASS signals, and the outer ring with the inner shorted wall is for L2 signals. This antenna has been registered as Legant by JPS and later by Topcon Corp. The board was also used with Regant and CR3 choke ring antennas. Other considerations for dual-frequency concentric patch antennas for applications to GNSS positioning can be found in Boccia et al. (2007) and Basilio et al. (2007).

With an increasing demand to reduce the size of compact integrated units intended for field applications, the stacked multifrequency patch antennas have become of interest. Various design approaches are known and could be potentially employed. The related discussion can be found in Kumar and Ray (2003) and Rao et al. (2013), Gao et al. (2014) and Chen et al. (2012). The Topcon PGA1 dual-frequency antenna stack is shown in Figure 9.7.8 as a practical example. Top and bottom patch antennas are for GPS/GLONASS L1 and L2 frequencies, respectively. The patch of the bottom antenna serves at the same time as a ground plane for the top antenna. Both patch antennas are fabricated with ceramic substrates of 5 mm thickness. The LNA board is located directly under the stack. This board also contains the microstrip feed network. The footprint of the bottom portion is 90 × 90 mm. This stack has been manufactured for more than a decade.

With the GNSS frequency bandwidth extension for Galileo and Compass systems and L5 and L3 signals of GPS and GLONASS, the substrate thickness h of the patch antenna is to be increased. As seen with (9.7.11), for a 10% relative bandwidth assuming $\varepsilon = 4$ and a wavelength λ of 20 cm, one has $h = 1$ cm. In actuality, this is an underestimate as mentioned previously. To cover 12% of the lower GNSS band (see discussion at the end of Section 9.1.2) the substrate thickness increases up to 2 cm. Such thick substrates would contribute to antenna weight and cause manufacturing complications. As an alternative, light-weight metal structures simulating dielectric properties can be considered. Such structures are referred to as artificial dielectrics. Designs relevant to GNSS applications have been patented by Tatarnikov

Figure 9.7.8 PGA1 antenna stack of Topcon Corp.

Figure 9.7.9 Full-wave GNSS Fence antenna of Topcon Corp.

et al. (2008a, 2013b). Details of treatment presented in Tatarnikov (2008b, 2009) are summarized in Appendix H.

As an example, Figure 9.7.9 illustrates the Topcon Fence antenna covering the entire GNSS band. Such antennas are sometimes referred to as a full-wave antenna. The antenna is a stack of two patch antennas. The stack has a total height of 22 mm, and the equivalent dielectric constant of artificial substrates is about 4. The weight of the stack is 150 g.

9.7.2 Other Types of Antennas

A variety of antennas have been developed for different applications of GNSS positioning. The reader can find a detailed discussion of these antennas in Rao et al. (2013), Chen et al. (2012) and references therein. For satellite surveying, however, only a limited number of antenna types have been adopted so far, as there are strict limitations in terms of size and weight. At the same time, as was mentioned in previous sections, given the frequency bandwidth the loss factor generally appears too severe when downsizing the antenna. The SNR degradation associated with increased antenna noise temperature is undesirable for RTK algorithms, as it leads to cycle slips and problems with ambiguity fixing. This is particularly true for low elevated satellites. Thus, one may say that for survey applications the art of antenna designers is to create compact low-loss antennas. Last but not least, cost efficiency is also a factor. One is to note a remarkable modification of the spiral antenna known as Pinwheel. The reader is referred to Kunysz (2000) and Rao et al. (2013) for details.

9.7.3 Flat Metal Ground Planes

Generally, the purpose of the ground plane is to decrease the antenna gain for directions below the horizon, thus suppressing multipath coming from underneath. Different types of ground planes adopted in receiving GNSS antenna designs are discussed in this and two subsequent sections. In order to avoid confusion, the following terminology is used: the antenna installed over a ground plane is referred to as an antenna element, and a combination of the antenna element and a ground plane is referred to as an antenna system.

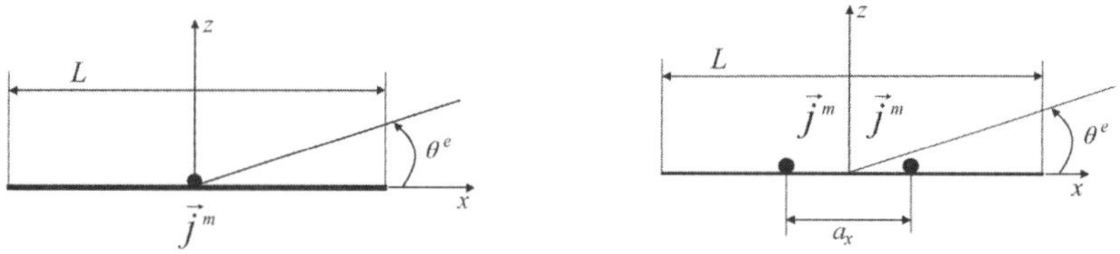

Figure 9.7.10 Two-dimensional models of a flat metal ground plane.

A flat metal ground plane is the most common arrangement with GNSS user antenna designs. The details of ground plane functionality can differ considerably, depending on the antenna element type. We begin with low profile antenna elements, such as patch elements.

Useful estimates of the ground plane performance can be derived from the two-dimensional model seen in Figure 9.7.10. It shows a ground plane in the form of a strip of size (width) L. The strip is infinite in the direction perpendicular to the drawing. The ground plane is excited by the magnetic line current placed at the center (left panel). This current is referred to as a source. Without a ground plane, the source has an omnidirectional pattern in the x-z plane. The ground plane is assumed to be perfectly conducting.

The thus-defined problem is canonical with electromagnetic wave diffraction theory. We will employ the edge waves approach of Ufimtsev (1962, 2003) in a simplified format. Following his approach, one considers the electric currents induced at the ground plane surface by the source. The electric current is a sum of two terms: physical optics (PO) current and edge waves current. The PO current is the portion of a current that occurs if the ground plane is an entire unbounded plane, and the edge waves current is excited by the ground plane edges. However, estimates in Tatarnikov (2008c) have shown that the edge waves current provides comparatively small correction for the case when the ground plane is of the order of half a wavelength or larger. For this model, we neglect the edge wave's current contribution. The PO current $\vec{j}^e$ has an explicit form (Appendix E):

$$\vec{j}^e = U^m \frac{k}{4\eta_0} 2H_0^{(2)}(k|x|)\vec{x}_0 \tag{9.7.22}$$

Here, U^m is the source amplitude and $H_0^{(2)}(s)$ is the zeroth-order Hankel function of the second kind. Using the first term of the Hankel function asymptotic form for large argument (E.43), it is seen that the current (9.7.22) decays as $(k|x|)^{-1/2}$ for large distances from the source. This is an important point to remember. The total radiated field is a sum of the radiation of a source in free space and that of the current (9.7.22). This latter radiation can be represented as radiation of the currents (9.7.22) flowing through an unbounded plane minus the radiation of tails, located at $|x| > L$.

For these tails, the asymptotic form of (E.43) holds. Employing the asymptotic form, one obtains the expression for the radiation pattern

$$F(\theta^e) = \left\{ \begin{pmatrix} 2 \\ 0 \end{pmatrix} - \frac{2}{\sqrt{\pi}} e^{i(\pi/4)} \left(\sin \frac{\theta^e}{2} \int_{\sqrt{k(L/2)(1+\cos\theta^e)}}^{\infty} e^{-it^2} dt \right.\right.$$
$$\left.\left. + \begin{pmatrix} +1 \\ -1 \end{pmatrix} \cos \frac{\theta^e}{2} \int_{\sqrt{k(L/2)(1-\cos\theta^e)}}^{\infty} e^{-it^2} dt \right) \right\}; \begin{pmatrix} 0 < \theta^e \leq \pi \\ -\pi \leq \theta^e < 0 \end{pmatrix} \quad (9.7.23)$$

This pattern is normalized to that of the source radiating in free space.

For directions not too close to grazing the ground plane ($\theta^e = 0, \pi$), such as when $k(L/2)(1 \mp \cos\theta^e)$ is not too small, keeping in (9.7.23) the first term of the asymptotic expansion (F.7) of the Fresnel integral, one comes to the expression

$$F(\theta^e) = \left\{ \begin{pmatrix} 2 \\ 0 \end{pmatrix} - \frac{1}{\sqrt{\pi}} \frac{e^{-i[k(L/2)+(\pi/4)]}}{\sqrt{kL}} \right.$$
$$\left. \left(e^{-ik(L/2)\cos\theta^e} \tan\frac{\theta^e}{2} + e^{ik(L/2)\cos\theta^e} \cot\frac{\theta^e}{2} \right) \right\}; \begin{pmatrix} 0 < \theta^e \leq \pi \\ -\pi \leq \theta^e < 0 \end{pmatrix} \quad (9.7.24)$$

This expression says that for directions of top semisphere ($\theta^e > 0$), the radiation is almost doubled with respect to the free space radiation of the source. Thus, for high enough elevation angles the ground plane works almost like a perfect mirror. Next, there are two diffraction waves originating from the ground plane edges. The amplitudes of these waves are of the order of $(kL)^{-1/2}$. For directions above the ground plane ($\theta^e > 0$), these diffraction waves contribute to pattern "waving." For directions below the ground plane ($\theta^e < 0$), the radiation pattern is defined by these two waves.

For $\theta^e = 0$, one is to use the equation (F.6) with expression (9.7.23) to find out that $F(0) = 1$ holds. In other words, the radiation in the direction grazing the ground plane is ½ magnitude (–6 dB) related to what would be observed if the ground plane was unbounded. This was mentioned already in Section 9.7.1.

Now we turn to the model of the patch antenna element in the form of two magnetic line currents (Figure 9.7.6). We now refer to these currents as the source. The corresponding two-dimensional model is shown in the right panel in Figure 9.7.10. We rewrite the expression (9.7.21) for the radiation pattern of the source using elevation angle θ^e for convenience:

$$F^{\infty}(\theta^e) = \cos\left(k\frac{a_x}{2}\cos\theta^e\right) \quad (9.7.25)$$

We denote this pattern now as $F^{\infty}(\theta^e)$ to indicate that this pattern holds with the ground plane model in the form of an unbounded plane. For a ground plane of finite size L, assuming that the magnetic line currents are not too close to the ground plane edges, derivations similar to those used with (9.7.23) show that

$$F(\theta^e) = \left\{ \begin{pmatrix} 2 \\ 0 \end{pmatrix} F^{\infty}(\theta^e) - F^{\infty}(0)\frac{2}{\sqrt{\pi}}e^{i(\pi/4)} \right.$$
$$\left. \left(\sin\frac{\theta^e}{2}\int_{\sqrt{k(L/2)(1+\cos\theta^e)}}^{\infty} e^{-it^2}dt + \begin{pmatrix} +1 \\ -1 \end{pmatrix}\cos\frac{\theta^e}{2}\int_{\sqrt{k(L/2)(1-\cos\theta^e)}}^{\infty} e^{-it^2}dt \right) \right\} \tag{9.7.26}$$

with $0 < \theta^e \leq \pi/2$ and $-\pi/2 \leq \theta^e < 0$. Thus the radiation in the bottom half-space $-\pi \leq \theta^e < 0$ is proportional to the source pattern in the direction grazing the ground plane [$F^{\infty}(0)$ term in 9.7.26]. This illustrates the collision mentioned in Section 9.2.1, i.e., to decrease the antenna gain for directions below the horizon with the given ground plane size, one is to decrease $F^{\infty}(0)$ which is not permissible from the standpoint of low elevated satellites tracking.

The pattern of (9.7.26) and the down-up ratio is illustrated in Figure 9.7.11 for a number of the ground planes with different L. The down-up ratio is defined as

$$DU(\theta^e) = \frac{F(-\theta^e)}{F(\theta^e)} \tag{9.7.27}$$

A patch length a_x equal to 0.25 wavelengths is used in (9.7.25). One may mention that the down-up ratios for the ground plane sizes of 0.5 and 1 wavelengths are almost same. As such, an increase in the ground plane size from about 0.5 to 1 wavelength would not provide a significant advantage in terms of multipath protection. As the size L grows, the down-up ratio grows in absolute value, thus it improves. However, the field in the bottom half-space decreases as $(kL)^{-1/2}$ versus L. Thus, such down-up improvement versus L is not that fast. From this standpoint, a significant increase in the ground plane size up to 1m and more as sometimes suggested, may not be recognized as a way to noticeably improve multipath protection. On the other hand, more

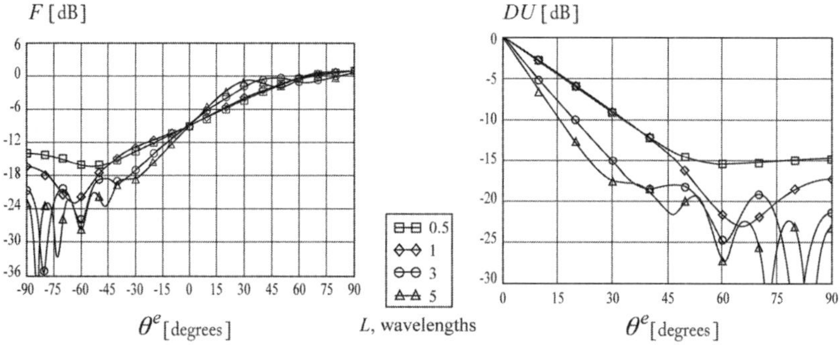

Figure 9.7.11 Radiation pattern and down-up ratio of patch antenna element over a flat metal ground plane.

accurate estimates of the edge waves contribution show that with a ground plane size L less than ½ wavelength, the down-up ratio rapidly degrades as L decreases (Tatarnikov 2008c). Antenna systems with such small flat metal ground planes generally do not meet the regular requirements of GNSS positioning accuracy.

It is to be noted that, staying within the PO approximation to the currents, the expression (9.7.26) gives a correct approximation to the pattern of a real 3-dimentional ground plane (disk). This gives value to the considerations given above and holds true except for narrow angular sector close to nadir direction. The reason for the difference is that with the 2-dimensional model, the excitation of the ground plane edges by the current (9.7.22) decays as $(kL)^{-1/2}$ with L increases. This results in the field in nadir direction having the same order of magnitude as a function of L. For a disk the current would decay as ρ^{-1} following the field intensities decrease in the far-field region. Here, ρ is the distance from the source, but the perimeter of the ground plane would grow proportionally to the ground plane radius. Thus, for sufficiently large ground planes, the contribution from the field diffracted over the edges in nadir direction would stay constant within the PO approximation. This contribution would be proportional to the illumination of the ground plane edges by the antenna element, or, in other words, to the antenna element pattern reading in the direction grazing the ground plane. This, once again, illustrates that a significant increase, within practically reasonable margins, of the flat metal ground plane size might not be a practical solution for improving multipath mitigation.

All of this suggests that a practically reasonable ground plane size is within 0.5 to 1 wavelength range. As seen in Figure 9.7.11, to achieve the down-up ratio of –15 dB for medium and high elevated satellites, a ground plane of about half-wave size is sufficient. With the lowest GNSS frequency this is about 13 to 14 cm. Commercially available rover antenna designs for field applications confirm this.

One is to note that in order to mitigate the conflict between low elevation gain and multipath protection, different kinds of ground plane "endings" have been suggested. Such endings are a kind of frame around the ground plane edges. The purpose of the frame is to decrease the amount of power diffracted over the edges in the direction underneath the antenna See Popugaev and Wansch (2014), Li et al. (2005), Timoshin and Soloviev. (2000), Westfall and Stephenson (1999), and Maqsood et al. (2013) for more detail.

Now we turn to another antenna element type that is referred to as cross dipoles. First, consider the Hertzian dipole raised some distance h over an infinite metal plane (Figure 9.7.12). Introducing the image (Appendix E) and using expression (9.1.79), one writes the total magnetic field intensity at the ground plane surface in the form

$$H_y = 2IL\frac{1}{4\pi}\left(\frac{1}{r^2} + \frac{ik}{r}\right)\frac{h}{r}e^{-ikr} \tag{9.7.28}$$

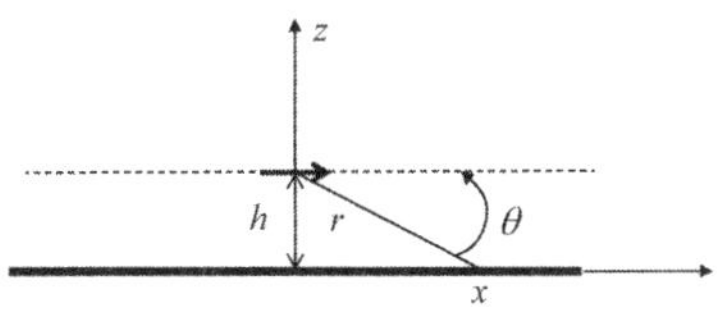

Figure 9.7.12 Hertzian dipole over flat metal ground plane.

where $r = \sqrt{h^2 + x^2}$. It follows that for $x >> h$, $H_y \sim x^{-2}$ holds. Employing the boundary condition (E.44) of Appendix E, one concludes that for a large enough distance x the electric current induced at the ground plane surface decays as x^{-2}. The same derivations hold for the dipole placed perpendicular to the drawing plane. The cross-dipole antenna element normally is installed at a height $h = \lambda/4$ over the ground plane. Thus, the currents induced at the ground plane will decay as ρ^{-2}, with distance ρ counted from the cross-dipole footprint. This is faster than in the case of low profile antennas. This illustrates that potential advantages for multipath protection could be achieved using a cross-dipole antenna element over a sufficiently large flat metal ground plane. See Counselman (1999) and Counselman and Steinbrecher (1987) for a discussion of one of the first antenna systems developed for high-precision applications.

9.7.4 Impedance Ground Planes

We consider the two-dimensional problem of Figure 9.7.10 but now assume that the strip of width L supports the impedance boundary condition (E.48). We say the strip forms an impedance surface. It is assumed that there is some structure underneath the strip that creates the surface impedance Z_S. More on these types of structures is given below.

For the moment we take the limit $L \to \infty$, thus turning to an infinite impedance plane. Then, the excitation problem of the left panel in Figure 9.7.10 is solved using the expansion of an incident field into a spectrum of plane waves. The derivation technique has been known since the first work of the German physicist A. Sommerfeld, and it is named after him. A complete treatment of the technique can be found in Felsen and Marcuvitz (2003). We are looking for the case when Z_S is pure imaginary with a negative (capacitive) imaginary portion. The result is

$$E_\tau \approx -\frac{U^m}{\sqrt{2\pi}} k e^{i(3\pi/4)} \frac{\eta_0}{Z_S} \frac{e^{-ikx}}{(kx)^{3/2}} \tag{9.7.29}$$

$$F^\infty(\theta^e) = \left(1 + \frac{Z_s}{\eta_0}\right) \frac{\sin\theta^e}{\sin\theta^e + \frac{Z_s}{\eta_0}} \tag{9.7.30}$$

The symbol E_τ denotes a component of the electrical field tangential to the ground plane at distances x from the source such that $kx \gg 1$, and $F^\infty(\theta^e)$ is the normalized radiation pattern. Details of derivation for the case and related discussion can be found in Tatarnikov et al. (2005). We note that the field intensity decays as $(kx)^{-3/2}$, which is faster than in the case where the surface is a perfect conductor. One may say that an impedance surface "forces" the radiation to leave faster with distance from the source than it would if the surface was a flat conductive one. Thus, with a ground plane of finite size L, less power will reach the ground plane edges and diffract over them in the directions underneath. This suggests that antenna systems with capacitive impedance ground planes have better multipath protection compared to flat metal ones. Next, as seen in (9.7.29), in order to decrease the fields further one is to make the impedance Z_S high in absolute value. One says that the ground plane is to realize a high capacitive impedance surface (HCIS). Finally, as seen in (9.7.30), the natural drawback of this approach is that the HCIS generally contributes to the narrowing of the radiation pattern. In the limit of an infinite plane, the initially omnidirectional source will have a null in the direction grazing the ground plane ($\theta^e = 0$). This narrowing is undesirable, as it may result in difficulties with low elevation satellite tracking.

These facts are known in the antenna community. HCIS is used in many antenna applications. In the current literature, artificial magnetic conductor (AMC) and perfect magnetic conductor (PMC) terms are used as a substitute for HCIS. However, a somewhat old-fashioned HCIS seems preferable at the moment to indicate that we are talking capacitive impedance and to differentiate with PMC, which is strictly valid at the limit $Z_S \to \infty$.

Similar to flat conductive ground planes, the art of antenna design is to establish the best proportion between multipath suppression capabilities and antenna gain for low elevation angles. Commonly known and adopted is a choke ring ground plane. Details of this remarkable antenna design can be found in Tranquilla et al. (1994), and see Figures 9.2.6 and 9.2.7 regarding image and radiation patterns. With the choke ring ground plane, the impedance surface passes through the choke groove openings (dashed line in the left panel in Figure 9.7.13). In order to form capacitive surface impedance, the groove depth is slightly larger than a quarter of the wavelength. The frequency response of the imaginary portion of Z_S is schematically shown in the right panel of the figure. The frequency f_C is known as a cut-off frequency. At frequencies below f_C, a so-called surface wave propagates along the impedance surface, destroying the ground plane functionality. A large variation of antenna performance characteristics is observed at frequencies approaching f_C. In actual designs f_C is slightly below the GPS L5 frequency. On the other hand, the impedance Z_S has to be large in absolute value. The operational frequency band of the antenna is between 1160 and 1610 MHz. Approaching 1610 MHz (the upper limit of the upper GNSS band, see Section 9.1.2), the impedance generally degrades. In order to provide more consistent surface impedance over the GNSS band, a dual-depth choke ring has been

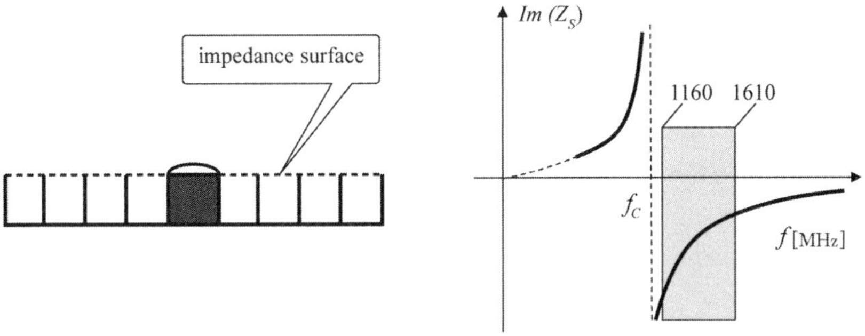

Figure 9.7.13 Impedance structure of choke grooves and frequency response (schematically).

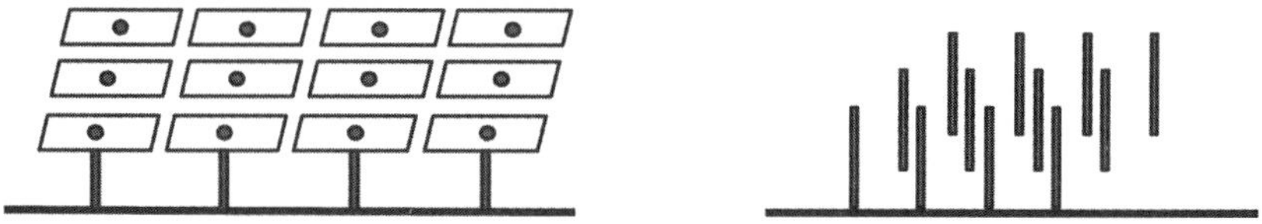

Figure 9.7.14 Impedance structures of mushrooms and straight pins.

developed by JPS company (Ashjaee et al., 2001). In that design, a diaphragm is inserted into the grooves such that it is transparent for the signals of the lower GNSS band and shorted for the upper band.

Another structure that provides capacitive impedance is sometimes referred to as mushrooms (Sievenpiper et al., 1999). This structure, seen in the left panel in Figure 9.7.14, is a dense grid of metal plates (patches) shorted to the ground plate by vertical standoffs. Normally, standoffs would be formed by plated vias with printed board technology. The height of the plates over the ground plate is of the order of 1/10 of the wavelength. GNSS receiving antennas with the mushrooms ground planes are presented by McKinzie et al., (2002), Baracco et al., (2008), Bao et al., (2006), Baggen et al., (2008), and Rao and Rosario (2011).

An easy-to-make structure with broadband response is composed of straight pins as discussed in King et al., (1983) and Tatarnikov et al. (2011a). The pin length is to slightly exceed a quarter of the free space wavelength λ. The pins are arranged in a regular grid with 0.1 to 0.2λ spacing (right panel in Figure 9.7.14).

With the intention of keeping the advantages of the HCIS ground plane in terms of multipath protection and at the same time increasing the antenna system gain for directions close to the horizon, a nonplanar HCIS ground plane is considered. Initial developments were published in Kunysz (2003). These developments resulted in the conical pyramidal choke ring antenna of NovAtel and Leica Geosystems. The details can be found in Rao et al. (2013). In the laboratory of Topcon Moscow Center, convex impedance structures have been analyzed, resulting in the design described in Tatarnikov et al. (2011,a,b, 2013a). The main features could be expressed as follows. Consider the left panel in Figure 9.7.15. Intuitively, it is clear that if the radius R of the HCIS curvature is small enough, then the antenna system pattern would tend to be that of an antenna element without the ground plane with broad coverage of directions in the top semisphere but with weak multipath protection. On the other hand, with R

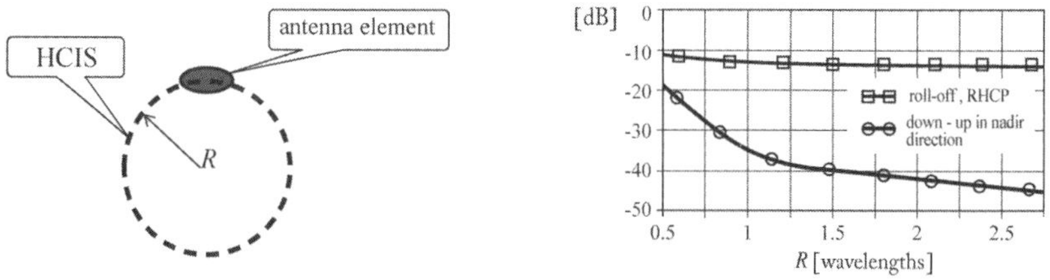

Figure 9.7.15 Schematic view and performance characteristics of convex impedance ground plane.

increase the pattern would tend to be that of a planar HCIS with potentially good multipath protection but with decreased gain for low elevation satellites. However, as R increases the antenna system gain toward low elevation angles degrades slowly, staying almost constant within a practically reasonable range of R. At the same time, the down-up ratio improves rapidly. This is illustrated in the right panel in Figure 9.7.15. The plots are calculated assuming the HCIS is in the form of a complete sphere to simplify the estimates. The bottom portion of the sphere does not provide a significant contribution since only a small portion of radiated power reaches this area due to the "forcing" effect of HCIS as mentioned above.

Thus, a variety of the ground plane designs could be suggested that possess both improved gain toward low elevation satellites and improved multipath protection when compared to commonly adopted values. As an example, Figure 9.7.16 shows the PN-A5 antenna of Topcon Corporation. The pin structure mentioned above is used to form the HCIS. The ground plane has a shape of a half-sphere with the external diameter of the ground plane chosen to fit commonly adopted radomes recommended by the International GNSS Service (IGS). In Figure 9.7.17 the antenna patterns of the PN-A5 antenna are shown. One might compare the patterns with those of traditional

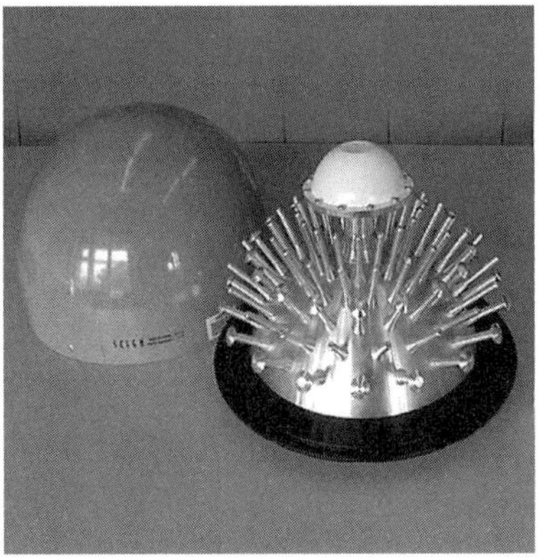

Figure 9.7.16 PN-A5 antenna of Topcon Corp.

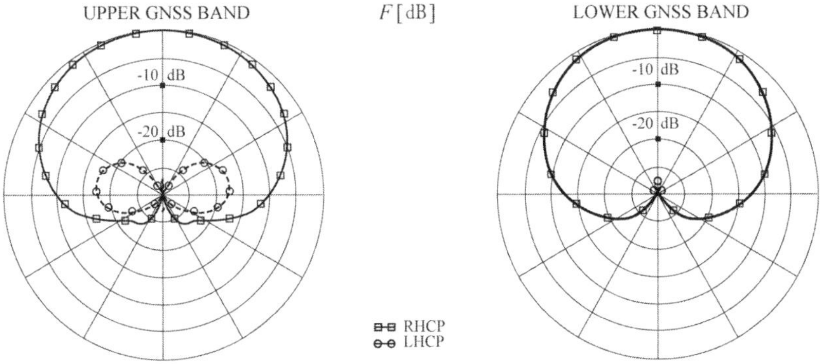

Figure 9.7.17 Radiation patterns of PN-A5 antenna.

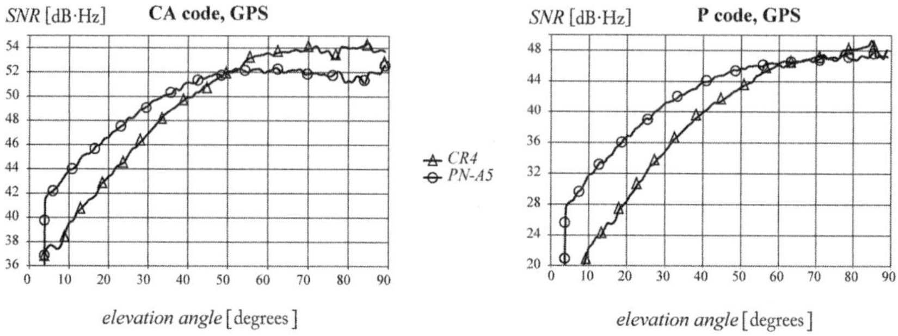

Figure 9.7.18 SNR versus satellite elevation for PN-A5 antenna compared to standard choke ring antenna.

choke ring seen in Figure 9.2.7. The type of improvements in signal strength offered by the convex design is illustrated in Figures 9.7.18. The figures show SNR values versus satellite elevation as reported by a geodetic receiver. For the GPS C/A-code (left panel), the increase in SNR for low elevations is about 4dB and for P2-code (right panel) it is up to 10 dB when compared to the CR4 choke ring antenna. One may note a decrease in SNR for high elevations in the case of the PN-A5 antenna. This is in agreement with the main antenna gain features discussed in Section 9.2: With antenna pattern broadening, the zenith gain decreases, provided that all the loss factors stay unchanged. Such a decrease in the zenith gain does not affect the satellite tracking capabilities as the signal power is high in this direction.

Another feature to be mentioned regarding this design is phase center stability in the vertical coordinate as a function of frequency. The measured data is illustrated in Figure 9.7.19 in comparison to the CR4 choke ring antenna of Topcon, which is a version of the original JPL design. It is seen that the cut-off frequency of the common choke ring antenna is slightly below the lowest frequency of the GNSS band, which is 1160 MHz. See related discussion in regards to Figure 9.7.13 above. Approaching the cut-off frequency the phase center displacement of the choke ring antenna is of the order of centimeters. On the other hand, the phase center displacement of PN-A5 is below 5 mm within the entire GNSS range. One should mention that the

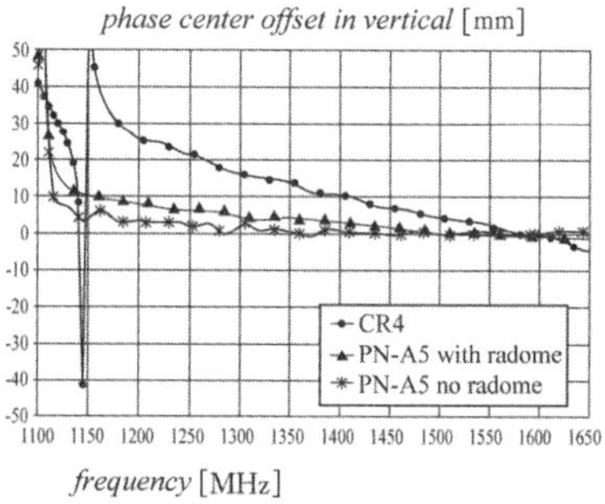

Figure 9.7.19 Phase center offset in vertical coordinate for PN-A5 antenna versus frequency compared to that of CR4 choke ring.

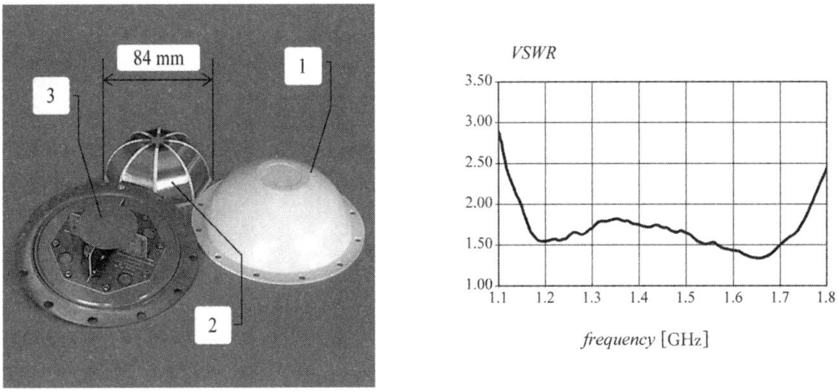

Figure 9.7.20 TA-5 antenna element of Topcon. Corp.

IGS-type radome contributes to an extra displacement versus frequency of the order of a half-centimeter. This feature of the radome will most likely be addressed with further developments.

A specially designed antenna element is used with PN-A5. This antenna element is TA-5 of Topcon Corp. The element continues the line of volumetric low-loss geodetic antenna designs initially represented by the Dorne and Margolin antenna. The TA-5 element is a cup comprising an array of convex patches (Figure 9.7.20; Tatarnikov et al., 2013c). The main features of the electromagnetic background as provided in Tatarnikov et al. (2009a) are summarized in Appendix I. The excitation unit 3 in Figure 9.7.20 is electromagnetically coupled with the array of patches 2. The VSWR observed at the input of one of two linear polarized channels of the excitation unit is shown in the right panel of the figure. Here, $VSWR \leq 2$ within the frequency range that exceeds 40%.

Pins HCIS, being a light-weight broadband structure, potentially allows consideration of large impedance ground plane antennas. With such antennas, the down-up ratio approaches the curve named "no multipath antenna" in Section 9.4.4. in regards to Figure 9.4.11. Thus, millimeter accuracy of real-time positioning can be addressed.

As mentioned above, placing an antenna element straight onto a plane HCIS of sufficiently large size might not be recognized as a solution because of antenna pattern narrowing. On the other hand, from a geometrical optics perspective it might be clear that by raising an antenna element above the ground plane (Figure 9.7.21), one would generally increase the illumination of the space below the horizon. By doing so,

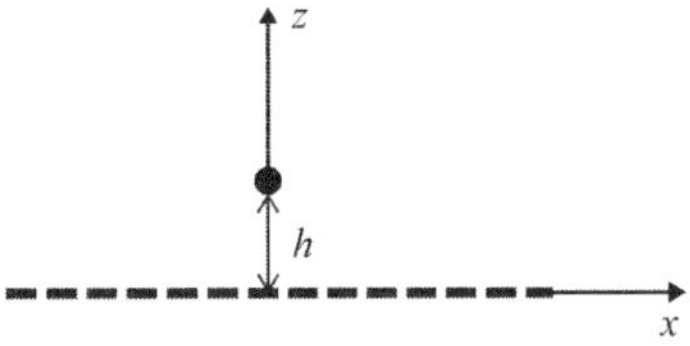

Figure 9.7.21 Antenna element raised against HCIS ground plane (schematically).

the multipath rejection capabilities of the ground plane would degrade. However, as was discussed in Tatarnikov and Astakhov (2013), for a ground plane of sufficiently large size with height h increasing, the antenna system gain toward low elevations grows (improves) rapidly while multipath rejection capabilities (down-up ratio) remains almost unchanged. This has served as justification for the consideration of large ground planes. An increase of height h might be accompanied by antenna pattern disturbances in the top semisphere. It has been shown in the referenced paper that the disturbances are small if the antenna element possesses its own down-up ratio in nadir direction of −12 to −15 dB or better. The image in the left panel in Figure 9.7.22 shows a scaled model operating at 5700 MHz. Here the HCIS is formed by pins. The diameter of the ground plane is about 71 cm (13.5λ) and a microstrip patch antenna with a local flat ground plane serves as the antenna element. Such an antenna element has a down-up ratio of −15 dB in the direction toward the impedance surface. The right panel in the figure shows the measured radiation pattern of this system. As seen, the antenna pattern reading for 12° elevation is about −9 dB, down-up at 12° elevation is better than −20 dB, and the antenna pattern reading in nadir direction is less than −40 dB relative to the zenith. A receiving GNSS antenna system realizing this principle is shown in Figure 9.7.23. The antenna system comprises a straight pin ground plane with a total diameter of 3 m and an antenna element mounted at a height

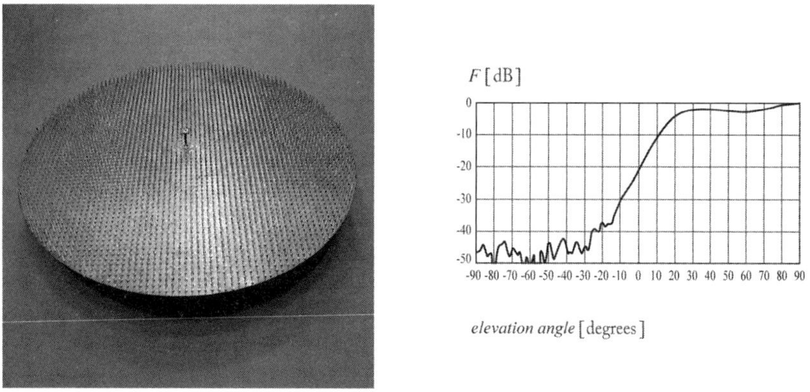

Figure 9.7.22 Scaled model and radiation pattern of large impedance ground plane antenna system operating at 5700MHz.

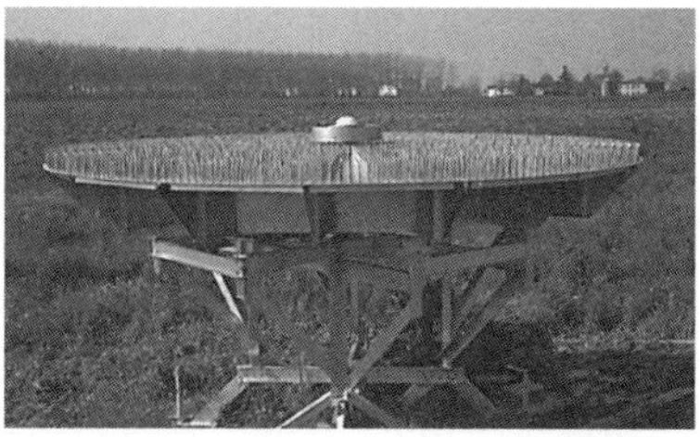

Figure 9.7.23 GNSS receiving antenna with large impedance ground plane.

of about 7 to 8 cm. One can say that the requirements for the antenna element mentioned above are met by the majority of commercially available GNSS antennas for geodesy and surveying. Different types of such antennas have been tested as antenna elements, yielding practically the same results. Shown in the image is the choke ring antenna element.

The actual precision of GNSS positioning in real time is illustrated by tests conducted at an open-sky test range. Three short baselines of about 30 m length were involved: (a) between two standard choke ring antennas as base and rover, (b) between two large impedance ground plane antennas seen in Figure 9.7.23, and c) zero baseline. In the latter case, two geodetic-grade GNSS receivers are connected to one antenna via a splitter. A zero baseline is free from multipath errors and illustrates system noise level. Real-time errors in the vertical coordinate for these three cases are shown in Figure 9.7.24. In each plot, the results of system noise smoothing by moving window of 10 samples are illustrated by a thick line. As seen, for large impedance ground plane antennas the remaining multipath error falls below system noise and is

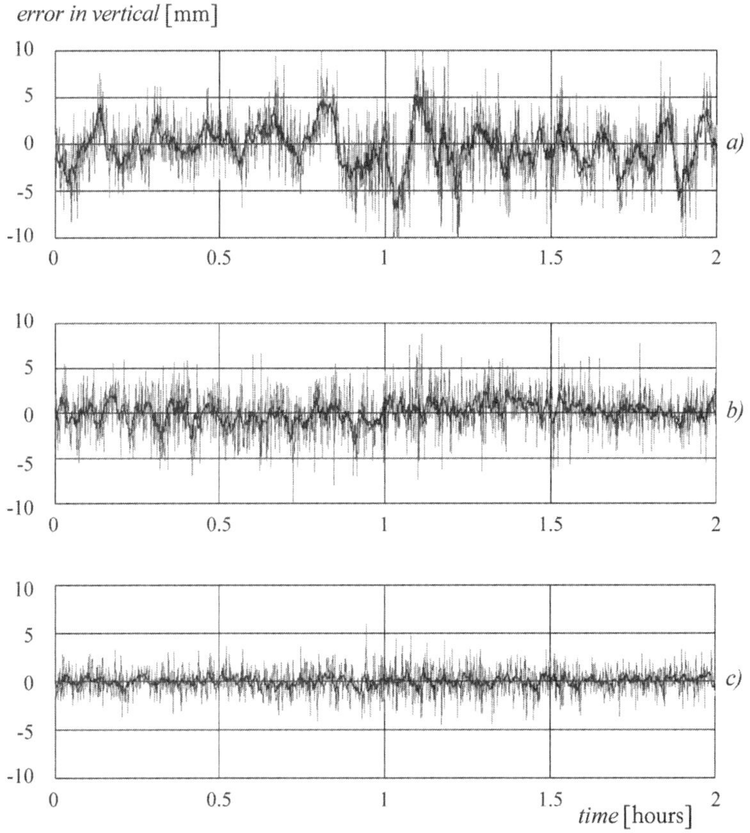

Figure 9.7.24 Real-time error in vertical coordinate for the antenna with large impedance ground plane (*b*) compared to standard choke ring (*a*) and zero baseline (*c*).

estimated to be ±1.5 to 2 mm. More details in regard to this antenna system performance can be found in Tatarnikov and Astakhov (2014), Mader et al. (2014).

The above designs employed surfaces and structures that formed a capacitive impedance. For the case of inductive impedance, a surface wave is excited as has been mentioned. However, a ground plane composed of a structure supporting the surface wave propagation could also be considered. This has been presented by Sciré-Scapppuzo and Makarov (2009).

9.7.5 Vertical Choke Rings and Compact Rover Antenna

Rather than arranging choke grooves in a planar format as discussed above, one may consider grooves arranged in a vertical stack. This is the vertical choke ring antenna discussed by Lee et al. (2004). Similar considerations have been applied by Soutiaguine et al. (2004), resulting in another design solution.

The insert in Figure 9.7.25 shows an array of two magnetic line currents perpendicular to the drawing plane. If the distance d between currents is small compared to the wavelength and the currents amplitudes $U_{1,2}$ are related as

$$U_2 = U_1 e^{i(\pi - kd)} \tag{9.7.31}$$

then the radiation pattern is a classical cardioid

$$f(\theta) = \frac{1 + \cos(\theta)}{2} \tag{9.7.32}$$

shown in Figure 9.7.25. As seen from the plot, the radiation in the nadir direction ($\theta \approx \pi$) is suppressed. The left panel in Figure 9.7.26 shows a stack of two patch antennas. Patch 1 is active and connected to the LNA in the receiving GNSS case. A choke groove underneath patch 1 is filled with dielectric. In order to form capacitive impedance, the groove depth slightly exceeds a quarter of the wavelength into the media filling. Thus, this groove is implemented in the form of patch antenna 2 with the total size being near resonance (9.7.1). This arrangement was referred to as antiantenna in Soutiaguine et al. (2004) and Tatarnikov et al. (2005). The patch

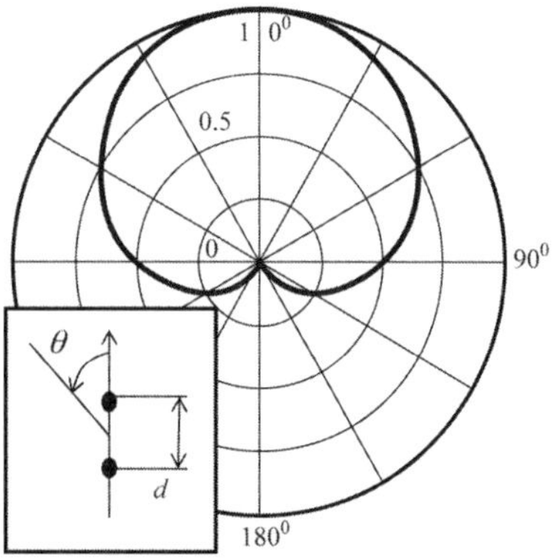

Figure 9.7.25 Cardioid pattern of two omnidirectional sources arranged in a vertical stack.

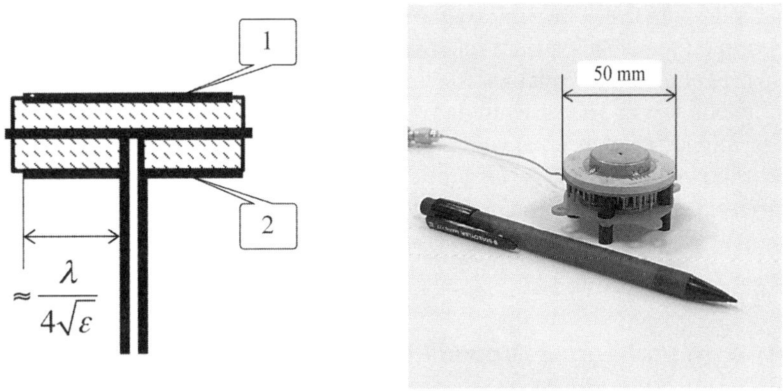

Figure 9.7.26 Single-frequency antenna stack employing antiantenna approach.

antenna 2 is passive. It is excited by patch 1 by means of electromagnetic coupling. The system is tuned in such a way that the cardioid pattern of (9.7.32) is approximated.

This approach allows one to overcome the limitation mentioned in Section 9.7.3. Namely, the minimally allowed ground plane size is about half the wavelength. The antiantenna replacing the ground plane is smaller by a factor of $\sqrt{\varepsilon}$. Here ε is the permittivity of the medium filling patch antenna 2. Thus, a compact multipath-protected rover antenna could be considered. As an example, the right panel in Figure 9.7.26 illustrates a single-frequency L1 antenna by Topcon Corp. The antenna was developed in the early 2000s and was intended for code-differential techniques for geo-information systems (GIS). To provide a bandwidth sufficient for GPS/GLONASS functionality, the antiantenna substrate is formed by metal comb boards in accordance with Tatarnikov et al. (2008a) (see discussion in Section 9.7.1 and Appendix H). The antenna does not require any ground plane.

In order to provide dual-frequency functionality, a four-patch structure has been developed. In the left panel in Figure 9.7.27 the patches 1 and 2 are active and patches

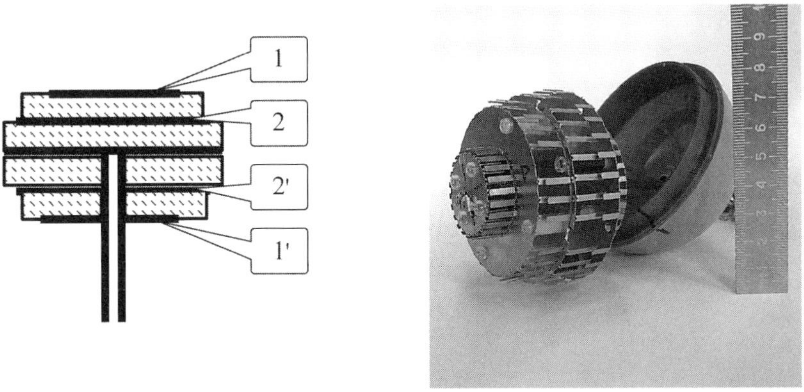

Figure 9.7.27 MGA8 dual-frequency antenna stack of Topcon Corp. employing antiantenna approach.

$1'$ and $2'$ are passive. Practical example of this approach is illustrated in the right panel of the figure. This quadri-patch system is utilized in the MGA8 antenna presented in Section 9.2.1. With this structure, capacitive substrates discussed in Section 9.7.1 are employed. The structure inscribes into a sphere of 8.5 cm diameter. It has been mentioned in connection with Figure 9.2.9 that this antenna provides multipath suppression of about 20 dB in the directions close to nadir. The antenna is about 40% less in size compared to typical patch antenna implementations with flat metal ground planes. The permittivity of the artificial substrates employed with the last two designs is about 4 to 6.

9.7.6 Semitransparent Ground Planes

As discussed above, GNSS antennas with planar ground planes are subject to a conflict—the necessity to keep antenna gain for low elevations at acceptable level leads to illuminating the ground plane edges by comparatively strong fields, which in turn increases radiation in the directions below the horizon. In the receiving mode this manifests itself as a multipath error. A potential way to solve the conflict relates to the use of a thin sheet of resisting materials with the ground plane. Intuitively, if the material possesses some resistivity, then the wave traveling toward the edges would be partly absorbed. Such thin resistive sheets are known as R-cards. Electromagnetics related to using R-cards in ground plane design is presented in Rojas et al. (1995). An optimized structure of a GNSS antenna ground plane is patented by Westfall (1997). The Zephyr antenna of Trimble Navigation has been presented by Krantz et al. (2001).

In other developments, one considers a grid of conductors with impedances Z imbedded into the grid (Figure 9.7.28). If the grid is dense, as is the case when a unit cell is small compared to the wavelength, the grid performs like a homogeneous thin sheet. In general, Z could be arbitrary complex. With pure resistive Z a resistive sheet is realized.

A dense grid can be analyzed employing averaged boundary conditions of the kind presented in Kontorovich et al. (1987), leading to the following:

$$\vec{E}_\tau^+ = \vec{E}_\tau^- = \vec{E}_\tau \tag{9.7.33}$$

$$\hat{Z}_g[\vec{n}_0, \vec{H}_\tau^+ - \vec{H}_\tau^-] = \hat{Z}_g \vec{j}_S^e = \vec{E}_\tau \tag{9.7.34}$$

The superscripts +/− denote the field intensities at the top and bottom sides of the thin sheet approximating the grid, τ marks the field components tangential

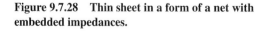

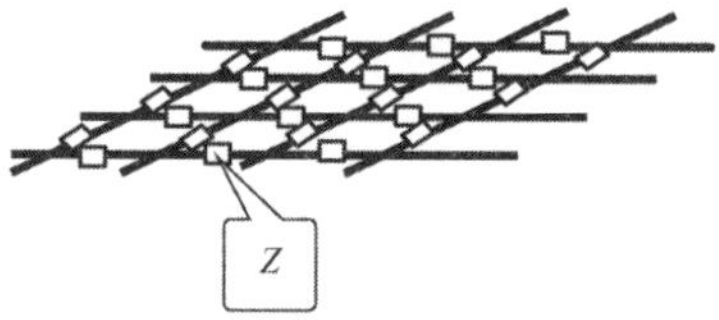

Figure 9.7.28 Thin sheet in a form of a net with embedded impedances.

to the sheet, $\vec{j}_S^e$ is the electrical current induced at the sheet, and $\hat{Z}_g$ is called grid impedance, a tensor operator in the general case (see Tretyakov, 2003, for details). The mechanism of the wave interacting with such a material may be referred to as semitransparency. Namely, the wave radiated by the source is partly reflected by the sheet and partly penetrates it. The potential performance of semitransparent ground planes with pure imaginary or complex imbedded impedance Z has been initially discussed in Tatarnikov (2008d) and is further summarized in Tatarnikov (2012). For the two-dimensional model seen in the left panel in Figure 9.7.10, with semitransparent ground plane, the PO approximation to the averaged electric current reads as

$$\vec{j}_S^e \approx \vec{x}_0 U^m \frac{k}{4\eta_0} 2H_0^{(2)}(kx)Q(q) \tag{9.7.35}$$

where

$$q = \sqrt{2}e^{-i(\pi/4)}\frac{Z_g}{W_0}\sqrt{kx} \tag{9.7.36}$$

$$Q(q) = 1 - iqe^{-q^2}\sqrt{\pi}(1 \mp \Phi(\pm iq)); \quad \mathrm{Im}(q)^{>}/_{<}0 \tag{9.7.37}$$

and

$$\Phi(q) = \frac{2}{\sqrt{\pi}}\int_0^q e^{-t^2}\,dt \tag{9.7.38}$$

is the probability integral. These expressions show that for $\mathrm{Im}Z_g \leq 0$ the current decays as $(kx)^{-3/2}$, which is faster than for a perfectly conductive case. For $\mathrm{Im}Z_g > 0$, the current grows due to initial surface wave formation and then rapidly vanishes. Thus, a complex Z_g potentially offers more flexibility in terms of current control. More details and an implementation example related to GNSS antenna design can be found in Tatarnikov (2012).

9.7.7 Array Antennas

An array antenna consists of a group of antennas, referred to as antenna elements that operate simultaneously. Figure 9.7.29 shows $2N+1$ antenna elements, indicated by

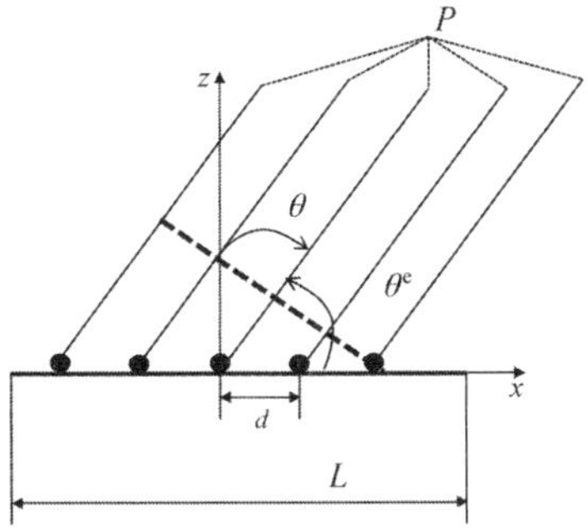

Figure 9.7.29 Linear array antenna.

thick dots and placed at increment d along the x axis. We take the total number of elements as odd for simplicity. The case using an even number of elements is treated similarly. Let all the antenna elements be of the same type, let $f(\theta)$ be the radiation pattern of one antenna element, and let all the antenna elements be excited by complex voltages U_q, with $q = -N, \ldots, N$, applied to their inputs. Consider a direction in space identified by the angle θ. We now employ the approach already discussed in Section 9.1.4. Consider an observation point P in the far-field region of the antenna. In that case the directions of the waves propagating to the antenna elements are essentially parallel. The signal from the qth antenna element will arrive with a phase delay or advance equal to $kqd \sin\theta$, where k is a wavenumber. Thus, the total signal from the array is

$$F(\theta) = f(\theta) \sum_{q=-N}^{N} U_q e^{ikqd \sin(\theta)} \tag{9.7.39}$$

This expression is known as pattern multiplication theorem. It shows that the array pattern is a product of two terms: the radiation pattern of an antenna element and the array factor. We designate the array factor as $S(\theta)$ and write

$$S(\theta) = \sum_{q=N}^{N} U_q e^{ikqd \sin(\theta)} \tag{9.7.40}$$

In order to illustrate the array factor, let all the antenna elements be excited with the same amplitudes and with linear progressive phase shift as a function of the location of the antenna element within the array, such that

$$U_q = e^{-iq\psi} \qquad \psi = \text{const} \tag{9.7.41}$$

Substituting this expression into (9.7.40) and summing the geometrical progression yields

$$S(\theta) = \frac{\sin\left((2N+1)\dfrac{kd\sin\theta - \psi}{2}\right)}{\sin\left(\dfrac{kd\sin\theta - \psi}{2}\right)} \tag{9.7.42}$$

We take the direction in space for which the path delays of different antenna elements are compensated by the phase shift ψ, such that

$$\psi = kd \sin\theta_0 \tag{9.7.43}$$

In that direction, the array factor (9.7.42) reaches a maximal value equal to the number of elements $2N + 1$. This is referred to as the main beam of the array. The phase shift ψ is controlled by an array feeding system. By changing ψ, the main beam moves as a function of θ according to (9.7.43). This is referred to as beam steering.

The beam steering capability of an array is illustrated in Figure 9.7.29 for three values of θ_0. The array factor (9.7.42) is plotted by thick dashed lines. We have adopted $2N + 1 = 11$ elements and an interelements spacing of $d = 0.5\lambda$. The elevation angle θ^e is used instead of zenith angle θ. One is to note that if an array forms a beam in broadside direction $\theta_0 = 0$ (top panel), the width of the beam $\Delta\theta_0$ is inversely proportional to the total length of the array. From (9.7.42) it follows that [see Mailloux (2005) and Van Trees (2002) for details]

$$\Delta\theta_0 \approx 51^\circ \frac{\lambda}{(2N+1)d} \tag{9.7.44}$$

While steering the beam off the broadside direction, generally the beam width grows. This is driven by the size of the projection of the total length of an array onto the main beam direction (see Figure 9.7.29 and the middle panel in Figure 9.7.30) such that the beam width obeys

$$\Delta\theta \approx \frac{\Delta\theta_0}{\cos\theta} \tag{9.7.45}$$

The main beam is surrounded by local maximums referred to as side lobes. With the array factor (9.7.42), the largest side lobe level is about–13 dB.

However, with beam steering another phenomenon occurs in addition to just the widening of the beam. If for a given orientation of the main beam θ_0 an integer p

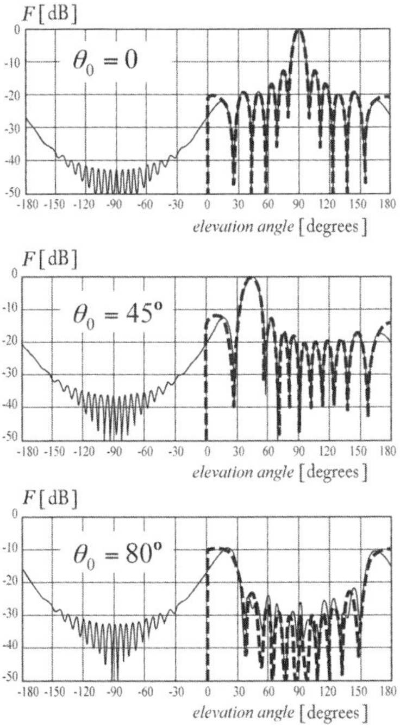

Figure 9.7.30 Radiation patterns of the 11-elements linear array.

could be found such that $|\sin\theta_0 + p(\lambda/d)| \le 1$, then another beam would be formed in the direction θ_p such that

$$\sin\theta_p = \sin\theta_0 + p\frac{\lambda}{d} \tag{9.7.46}$$

This beam is referred to as a grating lobe. With $d = 0.5\lambda$, if the main beam is steered to $\theta = 90°$, the grating lobe is observed at $\theta = -90°$. The bottom panel in Figure 9.7.30 shows the case where the main beam is steered to 10° elevation with respect to the array. Grating lobe formation in the direction of $\theta^e = 180°$ is seen.

The array above is referred to as linear since all the elements are aligned in one direction. In order to provide beam steering within the semisphere, planar arrays are used. Array factor derivations for planar arrays is straightforward. The reader is referred to Mailloux (2005) and Van Trees (2002) for details.

However, it is to be noted that with the pattern multiplication theorem the electromagnetic coupling of antenna elements is totally neglected. This coupling manifests itself as antenna element mismatch with the feeding lines, the mismatch being a function of the main beam direction and the antenna element location within the array. Due to this coupling, the radiation pattern of antenna element varies as a function of the element location. Moreover, normally a vast mismatch of the array elements with the feeding lines is observed when the main beam, or a grating lobe, is oriented in the direction close to grazing the array. This is known as scan blindness. Fundamental guidance accounting for all these effects is available. The reader is referred to Hansen (2009), Amitay et al. (1972), and Mailloux (2005). The multiplication theorem is in broad use just as a first-order estimate.

Regarding GNSS receiving antennas, it is of interest to check the capability of an array to suppress multipath signals coming from underneath the ground plane. We return to Figure 9.7.29 but now view the array elements as magnetic line currents arranged on a ground plane of size L. Following the derivations of Section 9.7.3 one comes to the expression for the radiation pattern of an element displaced by a factor of qd from the origin

$$\begin{aligned} F_q(\theta^e) &= e^{ikqd\cos\theta^e} \\ &\times \left\{ \begin{pmatrix} 2 \\ 0 \end{pmatrix} - \frac{2}{\sqrt{\pi}} e^{i(\pi/4)} \left(\sin\frac{\theta^e}{2} \int_{\sqrt{k[(L/2)+qd](1+\cos\theta^e)}}^{\infty} e^{-it^2}\,dt \right.\right. \\ &\left.\left. + \begin{pmatrix} +1 \\ -1 \end{pmatrix} \cos\frac{\theta^e}{2} \int_{\sqrt{k[(L/2)-qd](1-\cos\theta^e)}}^{\infty} e^{-it^2}\,dt \right) \right\} \end{aligned} \tag{9.7.47}$$

with $0 < \theta^e \le \pi$ and $-\pi \le \theta^e < 0$, respectively. We see that this pattern is a function of the distances of the elements to the ground plane edges. Thus, strictly speaking, the pattern multiplication theorem (9.7.39) does not hold for this case. In order to calculate the array pattern, one is to multiply the pattern (9.7.47) by the excitation voltage U_q (9.7.41), (9.7.43) and sum up all the patterns. That way, the solid thin curves in Figure 9.7.30 are obtained. It is assumed that the ground plane of size L equals six wavelengths. One may say that the ground plane edges provide a minor contribution to the pattern in the top semisphere. However, the excitation of the ground plane edges defines the pattern readings in the directions underneath the array. In this regard, when the main beam is steered toward low elevations, the illumination of the array edges grows. As seen in the bottom panel in Figure 9.7.30, the down-up ratio for 10° is of the order of several dB. Thus, the array does not offer significant advantages in multipath protection compared to a single element.

Beam steering is just a particular case of a more broad area of array techniques known as beamforming. The reader is referred to Van Trees (2002) for general guidance. By proper choice of excitation voltages in (9.7.39), an array may form several independent beams or perform a null steering rather than beam steering. The latter is of special interest for GNSS applications.

By null steering one means a situation where an array pattern exhibits a deep drop, a null, in the direction of an undesired signal. Thus, an intentional or unintentional interference could be mitigated along with multipath arrivals from directions in the top hemisphere. In general, a planar array is capable of forming several independent nulls. If the direction of undesired signal arrival was known, then the corresponding pattern could be synthesized employing one of the deterministic procedures known as conventional beamformers. However, this is normally not the case.

For unknown directions of arrival, a correlation of time series of desired and actual signals is estimated and employed. This potentially has an advantage of not only estimating the direction of arrival but also helping with the resolution of close arrivals. The corresponding techniques are referred to as adaptive processing, optimum processing, or space-time adaptive processing. A complete treatment of the approach goes beyond the scope of this discussion. The reader is referred to Van Trees (2002), Gupta and Moore (2001) and Rao et al. (2013) for additional detail.

Getting back to mitigation of ground reflections as of the largest error source for high-precision positioning, one is to note that significant advantages have been achieved using vertical arrays instead of horizontal ones. The insert in Figure 9.7.31 shows a vertical array of five elements. The elements are supposed to be omnidirectional. The interelements spacing is ¼ of the wavelength. Let the excitation voltages be

$$U_q = e^{-ikqd} cos\left(\pi \frac{q}{5}\right); \qquad q = -2, \ldots, 2 \qquad (9.7.48)$$

This means that the elements are excited by a wave traveling upward. The excitation of the first and last element is suppressed. The corresponding radiation pattern shown

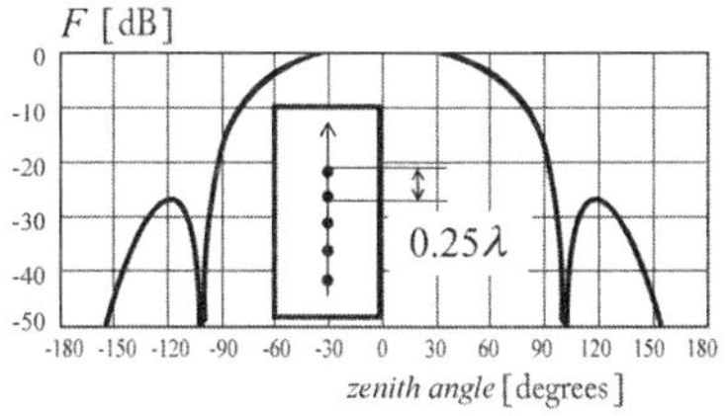

Figure 9.7.31 Radiation pattern of the vertical 5-element array.

in Figure 9.7.31 covers the entire top semisphere with a sharp drop, i.e., cut-off, while crossing the horizon. This is what is required to approach a "no-multipath" antenna, as discussed in previous sections. More accurate synthesis of such an array is presented in Counselman (1999).

However, as mentioned, the array factor estimates suffer from a "lack of electromagnetic." Developments and optimizations performed with electromagnetic computer simulation software have resulted into the remarkable vertical array antenna of Lopez (2010). The array is slightly over 2 m high, which is more than 10 wavelengths, and about 10 cm in diameter. It provides about 40 dB of ground multipath suppression starting at 5° below the horizon.

As a further development, one may consider a spherical array rather than vertical. A spherical array potentially offers an advantage of having both a good degree of suppression of multipath signals coming from underneath and a beamforming or null-steering capability with respect to the azimuth. Thus, multipath coming from the top semisphere may also be mitigated. A related discussion is presented in Tatarnikov and Astakhov (2012). As an example, Figure 9.7.32 shows an image of one meridional ring of such an array antenna. Experimentally, the measured pattern is shown in the right panel in the figure. Multipath suppression is 20 dB starting at 10°. The diameter of the ring is 65 cm.

9.7.8 Antenna Manufacturing Issues

In the manufacturing of antennas for high-precision positioning the main focus is on consistency and reliability of antenna calibration data. This is particularly true when the antenna comprises an ensemble of small parts, such as capacitive substrates or pins structures, as seen in Figure 9.7.9 and 9.7.16. Practical experience has shown that one of the most sensitive indicators is stability of the antenna phase center in the horizontal plane.

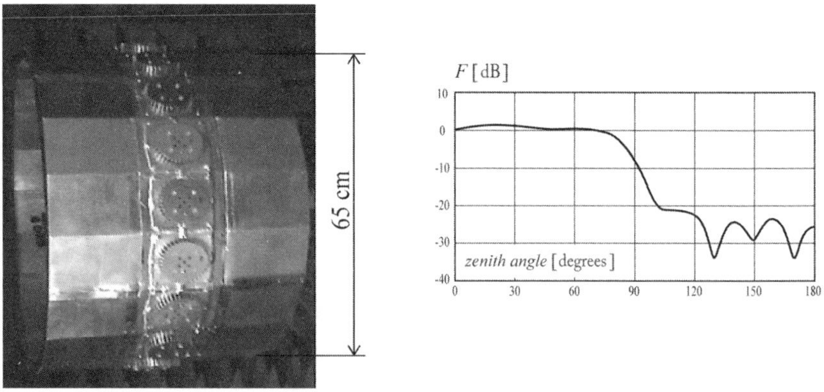

Figure 9.7.32 Experimental sample and radiation pattern of a meridional ring of spherical array.

Figure 9.7.33 PN-A5 antenna installed at rotational robot.

A convenient tool to ensure consistency of manufacturing is the absolute antenna calibration procedure described in Wübbena et al. (1996, 2000). In order to measure the phase center offset in the horizontal plane, a robot rotating the antenna under test with respect to the vertical axis has been used. Figure 9.7.33 shows such a robot installed on the roof top of the antenna manufacturing facility. Applying a one-axis limitation to the procedure just referenced allows achieving an accuracy of horizontal phase center offset measurements in the order of 0.3 to 0.4 mm within a few rotations. The corresponding software has been developed by I. Soutiaguine of Topcon Center.

APPENDIX A

GENERAL BACKGROUND

This appendix provides mathematical and statistical material that is handy to have available in a classroom situation to support key derivations or conclusions given in the main chapters. The appendix begins with a listing of expressions from spherical trigonometry. The rotation matrices are given along with brief definitions of positive and negative rotations. The sections on eigenvalues, matrix partitioning, Cholesky decomposition, partial minimization, rank one update, linearization, and statistics contain primary reference material for the least-squares Chapters 2 and 3. The subsection on the distribution of simple functions of random variables primarily supports these chapters also.

A.1 SPHERICAL TRIGONOMETRY

The sides of a spherical triangle are defined by great circles. A great circle is an intersection of the sphere with a plane that passes through the center of the sphere. It follows from geometric consideration of the special properties of the sphere that great circles are normal sections and geodesic lines. Figure A.1.1 shows a spherical triangle with corners (A, B, C), sides (a, b, c), and angles (α, β, γ). Notice that the sequence of the elements in the respective triplets is consistent, counterclockwise in this case. The sides of the spherical triangle are given in angular units. In many applications, one of the vertices of the spherical triangle represents the North or South Pole. Documentation of the expressions listed below is readily available from the mathematical literature. Complete derivations can be found in Sigl (1977).

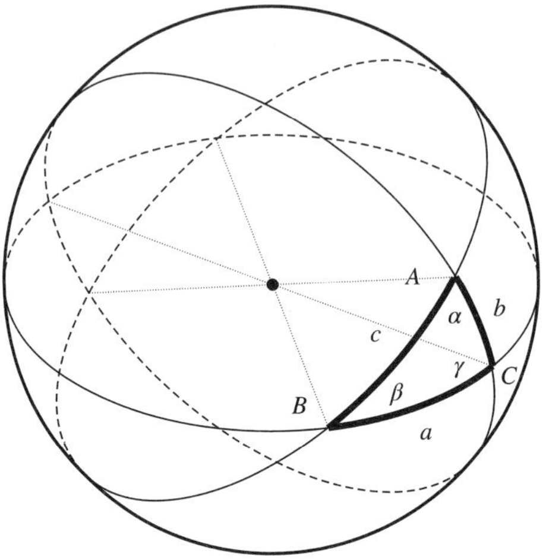

Figure A.1.1 Spherical triangle.

Law of Sine

$$\frac{\sin\ a}{\sin\ \alpha} = \frac{\sin\ b}{\sin\ \beta} \tag{A.1.1}$$

$$\frac{\sin a}{\sin \alpha} = \frac{\sin c}{\sin \gamma} \tag{A.1.2}$$

Law of Cosine for Sides

$$\begin{aligned}
\cos\ a &= \cos\ b\ \cos\ c + \sin\ b\ \sin\ c\ \cos\ \alpha \\
\cos\ b &= \cos\ c\ \cos\ a + \sin\ c\ \sin\ a\ \cos\ \beta \\
\cos\ c &= \cos\ a\ \cos\ b + \sin\ a\ \sin\ b\ \cos\ \gamma
\end{aligned} \tag{A.1.3}$$

Law of Cosine for Angles

$$\begin{aligned}
\cos\ \alpha &= -\cos\ \beta\ \cos\ \gamma + \sin\ \beta\ \sin\ \gamma\ \cos\ \ a \\
\cos\ \beta &= -\cos\ \gamma\ \cos\ \alpha + \sin\ \gamma\ \sin\ \alpha\ \cos\ b \\
\cos\ \gamma &= -\cos\ \alpha\ \cos\ \beta + \sin\ \alpha\ \sin\ \beta\ \cos\ c
\end{aligned} \tag{A.1.4}$$

Five Argument Formulas

$$\begin{aligned}
\sin\ a\ \cos\ \beta &= \cos\ b\ \sin\ c - \sin\ b\ \cos\ c\ \cos\ \alpha \\
\sin\ b\ \cos\ \gamma &= \cos\ c\ \sin\ a - \sin\ c\ \cos\ a\ \cos\ \beta \\
\sin\ c\ \cos\ \alpha &= \cos\ a\ \sin\ b - \sin\ a\ \cos\ b\ \cos\ \gamma
\end{aligned} \tag{A.1.5}$$

$$\begin{aligned}
\sin a \cos \gamma &= \cos c \sin b - \sin c \cos b \cos \alpha \\
\sin b \cos \alpha &= \cos a \sin c - \sin a \cos c \cos \beta \\
\sin c \cos \beta &= \cos b \sin a - \sin b \cos a \cos \gamma
\end{aligned} \tag{A.1.6}$$

$$\begin{aligned}
\sin \alpha \cos b &= \cos \beta \sin \gamma + \sin \beta \cos \gamma \cos a \\
\sin \beta \cos c &= \cos \gamma \sin \alpha + \sin \gamma \cos \alpha \cos b \\
\sin \gamma \cos a &= \cos \alpha \sin \beta + \sin \alpha \cos \beta \cos c
\end{aligned} \tag{A.1.7}$$

$$\begin{aligned}
\sin \alpha \cos c &= \cos \gamma \sin \beta + \sin \gamma \cos \beta \cos a \\
\sin \beta \cos a &= \cos \alpha \sin \gamma + \sin \alpha \cos \gamma \cos b \\
\sin \gamma \cos b &= \cos \beta \sin \alpha + \sin \beta \cos \alpha \cos c
\end{aligned} \tag{A.1.8}$$

Four Argument Formulas

$$\begin{aligned}
\sin \alpha \cot \beta &= \cot b \sin c - \cos c \cos \alpha \\
\sin \alpha \cot \gamma &= \cot c \sin b - \cos b \cos \alpha \\
\sin \beta \cot \gamma &= \cot c \sin a - \cos a \cos \beta \\
\sin \beta \cot \alpha &= \cot a \sin c - \cos c \cos \beta \\
\sin \gamma \cot \alpha &= \cot a \sin b - \cos b \cos \gamma \\
\sin \gamma \cot \beta &= \cot b \sin a - \cos a \cos \gamma
\end{aligned} \tag{A.1.9}$$

Gauss (Delambre, Mollweide) Formulas—not all permutations listed

$$\sin \frac{\alpha}{2} \sin \frac{b+c}{2} = \sin \frac{a}{2} \cos \frac{\beta - \gamma}{2} \tag{A.1.10}$$

$$\sin \frac{\alpha}{2} \cos \frac{b+c}{2} = \cos \frac{a}{2} \cos \frac{\beta + \gamma}{2} \tag{A.1.11}$$

$$\cos \frac{\alpha}{2} \sin \frac{b-c}{2} = \sin \frac{a}{2} \sin \frac{\beta - \gamma}{2} \tag{A.1.12}$$

$$\cos \frac{\alpha}{2} \cos \frac{b-c}{2} = \cos \frac{a}{2} \sin \frac{\beta + \gamma}{2} \tag{A.1.13}$$

Napier Analogies—not all permutations listed

$$\begin{aligned}
\tan \frac{a+b}{2} &= \tan \frac{c}{2} \frac{\cos \dfrac{\alpha - \beta}{2}}{\cos \dfrac{\alpha + \beta}{2}} \\
\tan \frac{a-b}{2} &= \tan \frac{c}{2} \frac{\sin \dfrac{\alpha - \beta}{2}}{\sin \dfrac{\alpha + \beta}{2}}
\end{aligned} \tag{A.1.14}$$

$$\begin{aligned}\tan\frac{\alpha+\beta}{2} &= \cot\frac{\gamma}{2}\frac{\cos\dfrac{a-b}{2}}{\cos\dfrac{a+b}{2}}\\ \tan\frac{\alpha-\beta}{2} &= \cot\frac{\gamma}{2}\frac{\sin\dfrac{a-b}{2}}{\sin\dfrac{a+b}{2}}\end{aligned} \tag{A.1.15}$$

Half Angle Formulas

$$s = (a+b+c)/2 \tag{A.1.16}$$

$$k = \sqrt{\frac{\sin(s-a)\ \sin(s-b)\ \sin(s-c)}{\sin\ s}} \tag{A.1.17}$$

$$\begin{aligned}\tan\frac{\alpha}{2} &= \frac{k}{\sin(s-a)}\\ \tan\frac{\beta}{2} &= \frac{k}{\sin(s-b)}\\ \tan\frac{\gamma}{2} &= \frac{k}{\sin(s-c)}\end{aligned} \tag{A.1.18}$$

Half Side Formulas

$$\sigma = (\alpha+\beta+\gamma)/2 \tag{A.1.19}$$

$$k' = \sqrt{\frac{\cos(\sigma-\alpha)\cos(\sigma-\beta)\ \cos(\sigma-\gamma)}{-\cos\ \sigma}} \tag{A.1.20}$$

$$\begin{aligned}\tan\frac{a}{2} &= \frac{\cos(\sigma-\alpha)}{k'}\\ \tan\frac{b}{2} &= \frac{\cos(\sigma-\beta)}{k'}\\ \tan\frac{c}{2} &= \frac{\cos(\sigma-\gamma)}{k'}\end{aligned} \tag{A.1.21}$$

L'Huilier-Serret Formulas

$$M = \sqrt{\frac{\tan\dfrac{s-a}{2}\cdot\tan\dfrac{s-b}{2}\cdot\tan\dfrac{s-c}{2}}{\tan\dfrac{s}{2}}} \tag{A.1.22}$$

$$\tan\frac{\varepsilon}{4} = M\cdot\tan\frac{s}{2} \tag{A.1.23}$$

$$\tan\left(\frac{\alpha}{2}-\frac{\varepsilon}{4}\right)=M\cdot\cot\frac{s-a}{2}$$
$$\tan\left(\frac{\beta}{2}-\frac{\varepsilon}{4}\right)=M\cdot\cot\frac{s-b}{2} \qquad \text{(A.1.24)}$$
$$\tan\left(\frac{\gamma}{2}-\frac{\varepsilon}{4}\right)=M\cdot\cot\frac{s-c}{2}$$

The symbol ε denotes the spherical angular excess. The area of spherical triangle can be expressed as

$$\Delta=\varepsilon\, r^2 \qquad \text{(A.1.25)}$$

where r denotes the radius of the sphere.

A.2 ROTATION MATRICES

Rotations between coordinate systems are very conveniently expressed in terms of rotation matrices. The rotation matrices

$$\boldsymbol{R}_1(\theta)=\begin{bmatrix}1 & 0 & 0\\ 0 & \cos\theta & \sin\theta\\ 0 & -\sin\theta & \cos\theta\end{bmatrix} \qquad \text{(A.2.1)}$$

$$\boldsymbol{R}_2(\theta)=\begin{bmatrix}\cos\theta & 0 & -\sin\theta\\ 0 & 1 & 0\\ \sin\theta & 0 & \cos\theta\end{bmatrix} \qquad \text{(A.2.2)}$$

$$\boldsymbol{R}_3(\theta)=\begin{bmatrix}\cos\theta & \sin\theta & 0\\ -\sin\theta & \cos\theta & 0\\ 0 & 0 & 1\end{bmatrix} \qquad \text{(A.2.3)}$$

describe rotations by the angle θ of a right-handed coordinate system around the first, second, and third axes, respectively. The rotation angle is positive for a counterclockwise rotation, as viewed from the positive end of the axis about which the rotation takes place. The result of successive rotations depends on the specific sequence of the individual rotations. An exception to this rule is differentially small rotations for which the sequence of rotations does not matter.

A.3 LINEAR ALGEBRA

In general, in science and engineering we find nonlinear mathematical relationships between the observations and other quantities such as coordinates, height, area, and volume. One may only think about the common occurrence of expressing a distance observation as a function of Cartesian coordinates. Seldom is there a natural linear relation between observations as there is in spirit leveling. Least-squares adjustment and statistical treatment require that nonlinear mathematical relations be linearized. The result is a set of linear equations that is the starting point for further development which involves such elements as eigenvalues and eigenvectors, matrix partitioning, and Cholesky decomposition. Fortunately, using linear algebra in this context does not require memorization of proofs and theorems.

A.3.1 Determinants and Matrix Inverse

Let the elements of a matrix $\boldsymbol{A}$ be denoted by a_{ij}, where the subscript i denotes the row and j the column. A $u \times u$ square matrix $\boldsymbol{A}$ has a uniquely defined determinant, denoted by $|\boldsymbol{A}|$, and said to be of order u. The determinant of a 1×1 matrix equals the matrix element. The determinant of $\boldsymbol{A}$ is expressed as a function of determinants of submatrices of size $(u - 1) \times (u - 1)$, $(u - 2) \times (u - 2)$, etc. until the size 2 or 1 is reached. The determinant is conveniently expressed in terms of minors and cofactors.

The minor can be computed for each element of the matrix. It is equal to the determinant after the respective row and column have been deleted. For example, the minor for $i = 1$ and $j = 2$ is

$$m_{12} = \begin{vmatrix} a_{21} & a_{23} & \cdots & a_{2u} \\ a_{31} & a_{33} & \cdots & a_{3u} \\ \vdots & \vdots & \cdots & \vdots \\ a_{u1} & a_{u3} & \cdots & a_{uu} \end{vmatrix} \tag{A.3.1}$$

The cofactor c_{ij} is equal to plus or minus the minor, depending on the subscripts i and j,

$$c_{ij} = (-1)^{i+j} m_{ij} \tag{A.3.2}$$

The determinant of $\boldsymbol{A}$ can now be expressed as

$$|\boldsymbol{A}| = \sum_{j=1}^{u} a_{kj}\, c_{kj} \tag{A.3.3}$$

The subscript k is fixed in (A.3.3) but can be any value between 1and u, i.e., the determinant can be computed based on the minors for any one of the u rows or columns. Of course, the determinant (A.3.1) can be expressed as a function of determinants of matrixes of size $(u - 2) \times (u - 2)$, etc.

Determinants have many useful properties. For example, the rank of a matrix equals the order of the largest nonsingular square submatrix, i.e., the largest order for a nonzero determinant that can be found. The determinant is zero and the matrix is singular if the columns or rows of $\boldsymbol{A}$ are linearly dependent. The inverse of the square matrix can be expressed as

$$\boldsymbol{A}^{-1} = \frac{1}{|\boldsymbol{A}|} \boldsymbol{C}^T \tag{A.3.4}$$

where $\boldsymbol{C}$ is the cofactor matrix consisting of the elements c_{ij} given in (A.3.2). The product of the matrix and its inverse equals the identity matrix, i.e., $\boldsymbol{AA}^{-1} = \boldsymbol{I}$ and $\boldsymbol{A}^{-1}\boldsymbol{A} = \boldsymbol{I}$. These simple relations do not hold for generalized matrix inverses that can be computed for singular or even rectangular matrices. Information on generalized inverses is available in the standard mathematical literature. Given that $\boldsymbol{A}$, $\boldsymbol{B}$, and $\boldsymbol{C}$ are nonsingular square matrices, the inverse of a product

$$(\boldsymbol{ABC})^{-1} = \boldsymbol{C}^{-1}\boldsymbol{B}^{-1}\boldsymbol{A}^{-1} \tag{A.3.5}$$

equals the product of the switched factors and their inverses.

Computation techniques for inverting nonsingular square matrices abound in linear algebra textbooks. In many cases the matrices to be inverted show a definite pattern and are often sparsely populated. When solving large systems of equations, it might be necessary to take advantage of these patterns in order to reduce the computation load. Some applications might produce ill-conditioned and numerically near-singular matrices that require special attention. Very useful subroutines for dealing with large and patterned matrices for geodetic applications are given in Milbert (1984).

A.3.2 Eigenvalues and Eigenvectors

Let $\boldsymbol{A}$ denote a $u \times u$ matrix and $\boldsymbol{x}$ be a $u \times 1$ vector. If $\boldsymbol{x}$ fulfills the equation

$$\boldsymbol{Ax} = \lambda \boldsymbol{x} \tag{A.3.6}$$

it is called an eigenvector, and the scalar λ is the corresponding eigenvalue. Equation (A.3.6) can be rewritten as

$$(\boldsymbol{A} - \lambda \boldsymbol{I})\boldsymbol{x} = \boldsymbol{0} \tag{A.3.7}$$

If $\boldsymbol{x}_0$ denotes a solution of (A.3.7) and α is a scalar, then $\alpha\boldsymbol{x}_0$ is also a solution. It follows that (A.3.7) provides only the direction of the eigenvector. There exists a nontrivial solution for $\boldsymbol{x}$ if the determinant is zero, i.e.,

$$|\boldsymbol{A} - \lambda \boldsymbol{I}| = 0 \tag{A.3.8}$$

This is the characteristic equation. It is a polynomial of the uth order in λ, providing u solutions λ_i, with $i = 1, \ldots, u$. Some of the eigenvalues can be zero, real, or even complex. Equation (A.3.7) provides an eigenvector $\boldsymbol{x}_i$ for each eigenvalue λ_i.

For a symmetric matrix, all eigenvalues are real. Although the characteristic equation might have multiple solutions, the number of zero eigenvalues is equal to the rank defect of the matrix. The eigenvectors are mutually orthogonal,

$$\boldsymbol{x}_i^T \boldsymbol{x}_j = 0 \tag{A.3.9}$$

For positive definite matrices all eigenvalues are real and positive. Let the normalized eigenvectors $\boldsymbol{x}_i/\|\boldsymbol{x}_i\|$ be denoted by $\boldsymbol{e}_i$; we can then arrange the normalized eigenvectors into the matrix

$$\boldsymbol{E} = [\boldsymbol{e}_1 \quad \boldsymbol{e}_2 \; \cdots \; \boldsymbol{e}_u] \tag{A.3.10}$$

The matrix $\boldsymbol{E}$ is an orthonormal matrix with the property

$$\boldsymbol{E}^T = \boldsymbol{E}^{-1} \tag{A.3.11}$$

Positive definite matrices play an important role in least-squares estimation.

A.3.3 Eigenvalue Decomposition

Consider again a $u \times u$ matrix $\boldsymbol{A}$ and the respective matrix $\boldsymbol{E}$ that consists of linearly independent normalized eigenvectors. The product of these two matrices can be written as

$$\begin{aligned} \boldsymbol{AE} &= [\boldsymbol{Ae}_1 \; \boldsymbol{Ae}_2 \quad \cdots \quad \boldsymbol{Ae}_u] \\ &= [\lambda_1\boldsymbol{e}_1 \; \lambda_2\boldsymbol{e}_2 \quad \cdots \; \lambda_u\boldsymbol{e}_u] \\ &= \boldsymbol{E\Lambda} \end{aligned} \tag{A.3.12}$$

where $\boldsymbol{\Lambda}$ is a diagonal matrix with eigenvalues λ_i at the diagonal. Multiplying this equation by $\boldsymbol{E}^T$ from the right and making use of (A.3.11), one gets the well-known eigenvalue decomposition

$$\boldsymbol{A} = \boldsymbol{E\Lambda E}^T \tag{A.3.13}$$

Multiplying (A.3.12) from the left with $\boldsymbol{E}^T$ gives

$$\boldsymbol{E}^T\boldsymbol{AE} = \boldsymbol{\Lambda} \tag{A.3.14}$$

This form expresses the diagonalization theorem, i.e., multiplying the matrix $\boldsymbol{A}$ from the right by $\boldsymbol{E}$ and from the left by $\boldsymbol{E}^T$ gives a diagonal matrix with nonzero diagonal elements. Multiplying this equation from the left and the right by $\boldsymbol{\Lambda}^{-1/2}$ and setting $\boldsymbol{D} = \boldsymbol{E\Lambda}^{-1/2}$, we obtain

$$\boldsymbol{D}^T\boldsymbol{AD} = \boldsymbol{I} \tag{A.3.15}$$

Taking the inverse of (A.3.14) and applying rules (A.3.5) and (A.3.11) gives

$$\boldsymbol{E}^T\boldsymbol{A}^{-1}\boldsymbol{E} = \boldsymbol{\Lambda}^{-1} \tag{A.3.16}$$

If the $u \times u$ matrix $\boldsymbol{A}$ is positive semidefinite with rank $R(\boldsymbol{A}) = r < u$, an equation similar to (A.3.14) can be found. Consider the matrix

$$\boldsymbol{E} = [{}_u\boldsymbol{F}_{r\,u}\boldsymbol{G}_{u-r}] \tag{A.3.17}$$

where the column of $\boldsymbol{F}$ consists of the normalized eigenvectors that pertain to the r nonzero eigenvalues. The submatrix $\boldsymbol{G}$ consists of $u - r$ eigenvectors that pertain to the $u - r$ zero eigenvalues. The columns of $\boldsymbol{F}$ and $\boldsymbol{G}$ span the column and null space of the matrix $\boldsymbol{A}$, respectively. Because of (A.3.6) and zero eigenvalues it follows that $\boldsymbol{A}\,\boldsymbol{G} = \boldsymbol{0}$. Therefore,

$$\boldsymbol{E}^T\boldsymbol{A}\boldsymbol{E} = \begin{bmatrix} \boldsymbol{E}^T \\ \boldsymbol{G}^T \end{bmatrix} \boldsymbol{A}[\boldsymbol{F}\,\boldsymbol{G}] = \begin{bmatrix} \boldsymbol{F}^T\boldsymbol{A}\boldsymbol{F} & \boldsymbol{0} \\ \boldsymbol{0} & \boldsymbol{0} \end{bmatrix} = \begin{bmatrix} \boldsymbol{\Lambda} & \boldsymbol{0} \\ \boldsymbol{0} & \boldsymbol{0} \end{bmatrix} \tag{A.3.18}$$

where the diagonal submatrix $\boldsymbol{\Lambda}$ contains r nonzero eigenvalues. Setting $\boldsymbol{D} = (\boldsymbol{F}\boldsymbol{\Lambda}^{-1/2} \vdots \boldsymbol{G})$ it follows that

$$\boldsymbol{D}^T\boldsymbol{A}\boldsymbol{D} = \begin{bmatrix} \boldsymbol{I} & \boldsymbol{0} \\ \boldsymbol{0} & \boldsymbol{0} \end{bmatrix} \tag{A.3.19}$$

in which case $\boldsymbol{I}$ is the $r \times r$ identity matrix.

A.3.4 Quadratic Forms

Let $\boldsymbol{A}$ denote a $u \times u$ matrix and $\boldsymbol{x}$ a vector of size u; then $\boldsymbol{x}^T\boldsymbol{A}\boldsymbol{x} = v$ is a quadratic form. The matrix $\boldsymbol{A}$ is called positive semidefinite if $\boldsymbol{x}^T\boldsymbol{A}\boldsymbol{x} \geq 0$ for all $\boldsymbol{x}$. In the special case that

$$\boldsymbol{x}^T\boldsymbol{A}\boldsymbol{x} > 0 \tag{A.3.20}$$

is valid for all $\boldsymbol{x}$, the matrix $\boldsymbol{A}$ is positive definite. The properties of such a matrix include:

1. $R(\boldsymbol{A}) = u$ (full rank).
2. $a_{ij} > 0$ for all i.
3. The inverse $\boldsymbol{A}^{-1}$ is positive definite.
4. If $\boldsymbol{B}$ is an $n \times u$ matrix with rank $u < n$, then the matrix $\boldsymbol{B}^T\boldsymbol{A}\boldsymbol{B}$ is positive definite. This property can readily be used to quickly generate a positive definite matrix. Assume a $\boldsymbol{B}$ matrix with full column rank, take $\boldsymbol{A} = \boldsymbol{I}$ and compute the product $\boldsymbol{B}^T\boldsymbol{A}\boldsymbol{B}$. If $R(\boldsymbol{B}) = r < u$, then $\boldsymbol{B}^T\boldsymbol{A}\boldsymbol{B}$ is positive semidefinite.
5. Let $\boldsymbol{D}$ be a $q \times q$ matrix formed by deleting $u - p$ rows and the corresponding $u - p$ columns of $\boldsymbol{A}$. Then $\boldsymbol{D}$ is positive definite.

A necessary and sufficient condition for a symmetric matrix to be positive definite is that the principal minor determinants be positive; i.e.,

$$a_{11} > 0, \qquad \begin{vmatrix} a_{11} & a_{12} \\ a_{21} & a_{22} \end{vmatrix} > 0 \qquad \ldots, \qquad |\boldsymbol{A}| > 0 \tag{A.3.21}$$

or that all eigenvalues are real and positive. This latter condition is often useful in verifying computationally that a particular matrix is positive definite. By computing the eigenvalues one can ascertain whether or not the matrix is singular. The number of zero eigenvalues equals the rank deficiency. Extremely small eigenvalues might indicate a near-singularity or ill-conditioning of the matrix. In such numerical computations the impact of rounding errors and other numerical considerations must be taken into account when making judgments about singularity or near-singularity of a matrix.

If $\mathbf{A}$ is positive definite, then $\mathbf{x}^T\mathbf{A}\mathbf{x} = v$ represents the equation of a u-dimensional ellipsoid expressed in a Cartesian coordinate system (x). The center of the ellipsoid is at $\mathbf{x} = \mathbf{0}$. The rotation transformation

$$\mathbf{x} = \mathbf{E}\mathbf{y} \tag{A.3.22}$$

expresses the quadratic form in the (y) coordinate system,

$$\mathbf{y}^T\mathbf{E}^T\mathbf{A}\mathbf{E}\mathbf{y} = v \tag{A.3.23}$$

Since the matrix $\mathbf{E}$ consists of normalized eigenvectors we can use (A.3.14) to obtain the simple expressions

$$\mathbf{y}^T \boldsymbol{\Lambda}\, \mathbf{y} = y_1^2\, \lambda_1 + y_2^2\, \lambda_2 + \cdots + y_u^2\, \lambda_u = v \tag{A.3.24}$$

This expression can be written as

$$\frac{y_1^2}{v/\lambda_1} + \frac{y_2^2}{v/\lambda_2} + \cdots + \frac{y_u^2}{v/\lambda_u} = 1 \tag{A.3.25}$$

This is the equation for the u-dimensional ellipsoid in the principal axes form, i.e., the coordinate system (y) coincides with the principal axes of the hyperellipsoid, and the lengths of the principal axes are proportional to the reciprocal of the square root of the eigenvalues. All eigenvalues are positive because the matrix $\mathbf{A}$ is positive definite. Equation (A.3.22) also determines the orientation between the (x) and (y) coordinate systems. If $\mathbf{A}$ has a rank defect, the dimension of the hyperellipsoid is $R(\mathbf{A}) = r < u$.

Differentiation of a quadratic form is a common method for determining the minimum in least squares. Let the vectors $\mathbf{x}$ and $\mathbf{y}$ be of dimension u and let the $u \times u$ matrix $\mathbf{A}$ contain constants. Consider the quadratic form

$$w = \mathbf{x}^T\mathbf{A}\,\mathbf{y} = \mathbf{y}^T\mathbf{A}^T\mathbf{x} \tag{A.3.26}$$

Because (A.3.26) is a 1×1 matrix, the expression can be transposed. This fact is used frequently to simplify expressions when deriving least-squares solutions. The total differential dw is

$$dw = \frac{\partial w}{\partial \mathbf{x}} d\mathbf{x} + \frac{\partial w}{\partial \mathbf{y}} d\mathbf{y} \tag{A.3.27}$$

The vectors $d\mathbf{x}$ and $d\mathbf{y}$ contain the differentials of the components of $\mathbf{x}$ and $\mathbf{y}$, respectively. From (A.3.27) and (A.3.26) it follows that

$$dw = \mathbf{y}^T\mathbf{A}^T d\mathbf{x} + \mathbf{x}^T\mathbf{A}\, d\mathbf{y} \tag{A.3.28}$$

If the matrix $\mathbf{A}$ is symmetric, then the total differential of

$$\phi = \mathbf{x}^T\mathbf{A}\mathbf{x} \tag{A.3.29}$$

is of interest. The partial derivatives are written in form of u equations as

$$\frac{\partial \phi}{\partial \boldsymbol{x}} \equiv \left[\frac{\partial \phi}{\partial x_1} \quad \cdots \quad \frac{\partial \phi}{\partial x_u}\right]^T = 2\,\boldsymbol{A}\boldsymbol{x} \tag{A.3.30}$$

The partial derivatives $\partial\phi/\partial x_k$ at the kth component can readily be verified as

$$\begin{aligned}\frac{\partial \boldsymbol{x}^T\boldsymbol{A}\boldsymbol{x}}{\partial x_k} &= \frac{\partial}{\partial x_k}\left(\sum_{j=1}^{u}\sum_{i=1}^{u} x_i x_j a_{ij}\right)\\ &= \sum_{j=1}^{u} x_j a_{kj} + \sum_{i=1}^{u} x_i a_{ik} = 2\sum_{j=1}^{u} a_{kj}x_j\\ &= (2\boldsymbol{A}\boldsymbol{x})_k \end{aligned} \tag{A.3.31}$$

because $\boldsymbol{A}$ is symmetric.

Equation (A.3.30) is used for deriving least-squares solutions, which requires locating the stationary point or the minimum for a quadratic function. The procedure is to take the partial derivatives with respect to all variables and equate them to zero. While the details of the least-squares derivations are given in Chapter 2, the following example serves to demonstrate the principle of minimization using matrix notation.

Let $\boldsymbol{B}$ denote an $n \times u$ rectangular matrix with $n > u$, $\boldsymbol{\ell}$ is a vector of size n, and $\boldsymbol{P}$ an $n \times n$ symmetric weight matrix that can include the special case $\boldsymbol{P} = \boldsymbol{I}$. The elements of $\boldsymbol{B}$, $\boldsymbol{\ell}$, and $\boldsymbol{P}$ are constants. The least-squares solution of

$$\boldsymbol{v} = \boldsymbol{B}\boldsymbol{x} + \boldsymbol{\ell} \tag{A.3.32}$$

requires determining $\boldsymbol{x}$ such that $\phi(\boldsymbol{x}) \equiv \boldsymbol{v}^T\boldsymbol{P}\boldsymbol{v} = \min$. Actually, equation (A.3.32) can be identified as the observation equation adjustment, but this association is not relevant for the current exercise. First, we compute the vector of partial derivatives,

$$\begin{aligned}\frac{\partial \boldsymbol{v}^T\boldsymbol{P}\boldsymbol{v}}{\partial \boldsymbol{x}} &= \frac{\partial}{\partial \boldsymbol{x}}\left[(\boldsymbol{B}\boldsymbol{x} + \boldsymbol{\ell})^T\boldsymbol{P}(\boldsymbol{B}\boldsymbol{x} + \boldsymbol{\ell})\right]\\ &= \frac{\partial}{\partial \boldsymbol{x}}(\boldsymbol{x}^T\boldsymbol{B}^T\boldsymbol{P}\boldsymbol{B}\boldsymbol{x} + 2\boldsymbol{\ell}^T\boldsymbol{P}\boldsymbol{B}\boldsymbol{x} + \boldsymbol{\ell}^T\boldsymbol{P}\boldsymbol{\ell})\\ &= 2\boldsymbol{B}^T\boldsymbol{P}\boldsymbol{B}\boldsymbol{x} + 2\boldsymbol{B}^T\boldsymbol{P}\boldsymbol{\ell}\end{aligned} \tag{A.3.33}$$

One can readily see how (A.3.30) was used to differentiate the first term. Equation (A.3.28) applies to the differentiation of the second term. The derivative of the third term is zero because it is a constant. Next, as is done in all minimization problems in calculus, we equate the partial derivatives to zero,

$$\frac{\partial \boldsymbol{v}^T\boldsymbol{P}\boldsymbol{v}}{\partial \boldsymbol{x}} = 2\boldsymbol{B}^T\boldsymbol{P}\boldsymbol{B}\boldsymbol{x} + 2\boldsymbol{B}^T\boldsymbol{P}\boldsymbol{\ell} = \boldsymbol{0} \tag{A.3.34}$$

The matrix equation (A.3.34) represents u equations to be solved for the u unknowns $\boldsymbol{x}$. As is customary, we denote the solution by $\hat{\boldsymbol{x}}$,

$$\hat{\boldsymbol{x}} = -(\boldsymbol{B}^T \boldsymbol{P} \boldsymbol{B})^{-1} \boldsymbol{B}^T \boldsymbol{P} \boldsymbol{\ell} \tag{A.3.35}$$

to emphasize that it is the least-squares solution that minimizes $\boldsymbol{v}^T \boldsymbol{P} \boldsymbol{v}$. As is further known from calculus, the condition (A.3.34) strictly speaking assures only that a stationary point has been found and that higher order derivatives would be needed to verify that a minimum has been identified. Using the properties of positive definite matrices one can readily verify that indeed a minimum has been achieved instead of computing and checking second-order derivatives. Finally, let us note that if $\boldsymbol{B}$ has full column rank then $\boldsymbol{B}^T \boldsymbol{P} \boldsymbol{B}$ is a positive definite matrix.

Let us now consider a more general construction called a quadratic function. It is a sum of the quadratic form and a linear form

$$\phi(\boldsymbol{x}) = \boldsymbol{x}^T \boldsymbol{A} \boldsymbol{x} + 2\boldsymbol{g}^T \boldsymbol{x} + c \tag{A.3.36}$$

where the matrix $\boldsymbol{A}$ is positive definite, $\boldsymbol{g}$ is a u-dimensional vector, and c is a constant. The least-squares problem can be obviously expressed as (A.3.36). Note that a function is called convex if the segment connecting any two points on the graph of the function lies above the graph. The function (A.3.36) is convex in its variables, because the matrix $\boldsymbol{A}$ is positive definite. It achieves a unique minimum at the point $\hat{\boldsymbol{x}}$ obtained by equating the partial derivatives to zero, similar to (A.3.34), i.e., $2\boldsymbol{A}\boldsymbol{x} + 2\boldsymbol{g} = \boldsymbol{0}$ which results in the expression

$$\hat{\boldsymbol{x}} = -\boldsymbol{A}^{-1} \boldsymbol{g} \tag{A.3.37}$$

Let us substitute the expression $\boldsymbol{g} = -\boldsymbol{A}\hat{\boldsymbol{x}}$ into (A.3.36); then add and subtract the expression $\hat{\boldsymbol{x}}^T \boldsymbol{A} \hat{\boldsymbol{x}}$, we obtain the following expression:

$$\begin{aligned} \phi(\boldsymbol{x}) &= (\boldsymbol{x} - \hat{\boldsymbol{x}})^T \boldsymbol{A} (\boldsymbol{x} - \hat{\boldsymbol{x}}) + c - \hat{\boldsymbol{x}}^T \boldsymbol{A} \hat{\boldsymbol{x}} \\ &\equiv (\boldsymbol{x} - \hat{\boldsymbol{x}})^T \boldsymbol{A} (\boldsymbol{x} - \hat{\boldsymbol{x}}) + c' \end{aligned} \tag{A.3.38}$$

It proves the following statement. The quadratic function is equal to the quadratic form over the variables $(\boldsymbol{x} - \hat{\boldsymbol{x}})$ up to the constant c', not affecting the vector (A.3.37), minimizing it.

A.3.5 Matrix Partitioning

Consider the following partitioning of the nonsingular square matrix $\boldsymbol{N}$,

$$\boldsymbol{N} = \begin{bmatrix} \boldsymbol{N}_{11} & \boldsymbol{N}_{12} \\ \boldsymbol{N}_{21} & \boldsymbol{N}_{22} \end{bmatrix} \tag{A.3.39}$$

where $\boldsymbol{N}_{11}$ and $\boldsymbol{N}_{22}$ are square matrices, although not necessarily of the same size. Let $\boldsymbol{Q}$ denote the partitioned inverse matrix

$$\mathbf{Q} = \mathbf{N}^{-1} = \begin{bmatrix} \mathbf{Q}_{11} & \mathbf{Q}_{12} \\ \mathbf{Q}_{21} & \mathbf{Q}_{22} \end{bmatrix} \tag{A.3.40}$$

such that the sizes of the $\mathbf{N}_{ij}$ and $\mathbf{Q}_{ij}$, respectively, are the same. Equations (A.3.39) and (A.3.40) imply the four relations

$$\mathbf{N}_{11}\mathbf{Q}_{11} + \mathbf{N}_{12}\mathbf{Q}_{21} = \mathbf{I} \tag{A.3.41}$$

$$\mathbf{N}_{11}\mathbf{Q}_{12} + \mathbf{N}_{12}\mathbf{Q}_{22} = \mathbf{0} \tag{A.3.42}$$

$$\mathbf{N}_{21}\mathbf{Q}_{11} + \mathbf{N}_{22}\mathbf{Q}_{21} = \mathbf{0} \tag{A.3.43}$$

$$\mathbf{N}_{21}\mathbf{Q}_{12} + \mathbf{N}_{22}\mathbf{Q}_{22} = \mathbf{I} \tag{A.3.44}$$

The solutions for the submatrices $\mathbf{Q}_{ij}$ are carried out according to the standard rules for solving a system of linear equations, with the restriction that the inverse is defined only for square submatrices. Multiplying (A.3.41) from the left by $\mathbf{N}_{21}\,\mathbf{N}_{11}^{-1}$ and subtracting the product from (A.3.43) gives

$$\mathbf{Q}_{21} = -\left(\mathbf{N}_{22} - \mathbf{N}_{21}\mathbf{N}_{11}^{-1}\mathbf{N}_{12}\right)^{-1}\mathbf{N}_{21}\mathbf{N}_{11}^{-1} \tag{A.3.45}$$

Multiplying (A.3.42) from the left by $\mathbf{N}_{21}\mathbf{N}_{11}^{-1}$ and subtracting the product from (A.3.44) gives

$$\mathbf{Q}_{22} = \left(\mathbf{N}_{22} - \mathbf{N}_{21}\mathbf{N}_{11}^{-1}\mathbf{N}_{12}\right)^{-1} \tag{A.3.46}$$

Substituting (A.3.46) in (A.3.42) gives

$$\mathbf{Q}_{12} = -\mathbf{N}_{11}^{-1}\mathbf{N}_{12}\left(\mathbf{N}_{22} - \mathbf{N}_{21}\mathbf{N}_{11}^{-1}\mathbf{N}_{12}\right)^{-1} \tag{A.3.47}$$

Substituting (A.3.45) in (A.3.41) gives

$$\mathbf{Q}_{11} = \mathbf{N}_{11}^{-1} + \mathbf{N}_{11}^{-1}\mathbf{N}_{12}\left(\mathbf{N}_{22} - \mathbf{N}_{21}\mathbf{N}_{11}^{-1}\mathbf{N}_{12}\right)^{-1}\mathbf{N}_{21}\mathbf{N}_{11}^{-1} \tag{A.3.48}$$

An alternative solution for $\mathbf{Q}_{11}$, $\mathbf{Q}_{12}$, $\mathbf{Q}_{21}$, and $\mathbf{Q}_{22}$ is readily obtained. Multiplying (A.3.43)from the left by $\mathbf{N}_{12}\mathbf{N}_{22}^{-1}$ and subtracting the product from (A.3.41) gives

$$\mathbf{Q}_{11} = \left(\mathbf{N}_{11} - \mathbf{N}_{12}\mathbf{N}_{22}^{-1}\mathbf{N}_{21}\right)^{-1} \tag{A.3.49}$$

Substituting (A.3.49) in (A.3.43) gives

$$\mathbf{Q}_{21} = -\mathbf{N}_{22}^{-1}\mathbf{N}_{21}\left(\mathbf{N}_{11} - \mathbf{N}_{12}\mathbf{N}_{22}^{-1}\mathbf{N}_{21}\right)^{-1} \tag{A.3.50}$$

Premultiplying (A.3.44) by $\mathbf{N}_{12}\mathbf{N}_{22}^{-1}$ and subtracting (A.3.42) gives

$$\mathbf{Q}_{12} = -\left(\mathbf{N}_{11} - \mathbf{N}_{12}\mathbf{N}_{22}^{-1}\mathbf{N}_{21}\right)^{-1}\mathbf{N}_{12}\mathbf{N}_{22}^{-1} \tag{A.3.51}$$

Substituting (A.3.51) in (A.3.44) gives

$$\mathbf{Q}_{22} = \mathbf{N}_{22}^{-1} + \mathbf{N}_{22}^{-1}\mathbf{N}_{21}\left(\mathbf{N}_{11} - \mathbf{N}_{12}\mathbf{N}_{22}^{-1}\mathbf{N}_{21}\right)^{-1}\mathbf{N}_{12}\mathbf{N}_{22}^{-1} \tag{A.3.52}$$

Usually the matrix partitioning technique is used to reduce the size of large matrices that must be inverted or to derive alternative expressions. Because these matrix identities are frequently used, and because they look somewhat puzzling unless one is aware of the simple solutions given above, they are summarized here again to be viewed at a glance,

$$\begin{bmatrix} \boldsymbol{N}_{11}^{-1}+\boldsymbol{N}_{11}^{-1}\boldsymbol{N}_{12}\left(\boldsymbol{N}_{22}-\boldsymbol{N}_{21}\boldsymbol{N}_{11}^{-1}\boldsymbol{N}_{12}\right)^{-1}\boldsymbol{N}_{21}\boldsymbol{N}_{11}^{-1} & -\boldsymbol{N}_{11}^{-1}\boldsymbol{N}_{12}\left(\boldsymbol{N}_{22}-\boldsymbol{N}_{21}\boldsymbol{N}_{11}^{-1}\boldsymbol{N}_{12}\right)^{-1} \\ -\left(\boldsymbol{N}_{22}-\boldsymbol{N}_{21}\boldsymbol{N}_{11}^{-1}\boldsymbol{N}_{12}\right)^{-1}\boldsymbol{N}_{21}\boldsymbol{N}_{11}^{-1} & \left(\boldsymbol{N}_{22}-\boldsymbol{N}_{21}\boldsymbol{N}_{11}^{-1}\boldsymbol{N}_{12}\right)^{-1} \end{bmatrix}$$
$$= \begin{bmatrix} \left(\boldsymbol{N}_{11}-\boldsymbol{N}_{12}\boldsymbol{N}_{22}^{-1}\boldsymbol{N}_{21}\right)^{-1} & -\left(\boldsymbol{N}_{11}-\boldsymbol{N}_{12}\boldsymbol{N}_{22}^{-1}\boldsymbol{N}_{21}\right)^{-1}\boldsymbol{N}_{12}\boldsymbol{N}_{22}^{-1} \\ -\boldsymbol{N}_{22}^{-1}\boldsymbol{N}_{21}\left(\boldsymbol{N}_{11}-\boldsymbol{N}_{12}\boldsymbol{N}_{22}^{-1}\boldsymbol{N}_{21}\right)^{-1} & \boldsymbol{N}_{22}^{-1}+\boldsymbol{N}_{22}^{-1}\boldsymbol{N}_{21}\left(\boldsymbol{N}_{11}-\boldsymbol{N}_{12}\boldsymbol{N}_{22}^{-1}\boldsymbol{N}_{21}\right)^{-1}\boldsymbol{N}_{12}\boldsymbol{N}_{22}^{-1} \end{bmatrix} \tag{A.3.53}$$

A.3.6 Cholesky Decomposition

For positive definite matrices the Cholesky decomposition, also known as the square root method, is an efficient way to solve systems of equations and to invert a positive definite matrix. A $u \times u$ positive definite matrix $\boldsymbol{N}$ can be decomposed as the product of a lower triangular matrix $\boldsymbol{L}$ and an upper triangular matrix $\boldsymbol{L}^T$,

$$\boldsymbol{N} = \boldsymbol{L}\boldsymbol{L}^T \tag{A.3.54}$$

with $\boldsymbol{L}$ being called the Cholesky factor. If $\boldsymbol{E}$ is an orthonormal matrix with property (A.3.11) then $\boldsymbol{B} = \boldsymbol{L}\boldsymbol{E}$ is also a Cholesky factor because $\boldsymbol{L}\boldsymbol{E}\boldsymbol{E}^T\boldsymbol{L}^T = \boldsymbol{L}\boldsymbol{L}^T$.

The lower and upper triangular matrices have several useful properties. For example, the eigenvalues of the triangular matrix equal the diagonal elements. The determinant of a triangular matrix equals the product of the diagonal elements. Because the determinant of a matrix product is equal to the product of the determinants of the factors, it follows that $\boldsymbol{N}$ is singular if any one of the diagonal elements of $\boldsymbol{L}$ is zero. This fact can be used conveniently during the computation of $\boldsymbol{L}$ to eliminate parameters that cause a singularity or observations that are linearly dependent.

The Cholesky algorithm provides the instruction for computing the lower triangular matrix $\boldsymbol{L}$. The elements of $\boldsymbol{L}$ are

$$l_{jk} = \begin{cases} \sqrt{n_{jj} - \sum\limits_{m=1}^{j-1} l_{jm}^2} & \text{for} \quad k = j \\ \dfrac{1}{l_{kk}}\left(n_{jk} - \sum\limits_{m=1}^{k-1} l_{jm} l_{km}\right) & \text{for} \quad k < j \\ 0 & \text{for} \quad k > j \end{cases} \tag{A.3.55}$$

where $1 \le j \le u$, $1 \le k \le u$, and u is the size of $\boldsymbol{N}$. The Cholesky algorithm preserves the pattern of leading zeros in the rows and columns of $\boldsymbol{N}$, as can be readily verified.

For example, if the first x elements in row y of $\boldsymbol{N}$ are zero, then the first x elements in row y of $\boldsymbol{L}$ are also zero. Taking advantage of this fact speeds up the computation of $\boldsymbol{L}$ for a large system that exhibits significant patterns of leading zeros. The algorithm (A.3.55) begins with the element l_{11}. Subsequently, the columns (or rows) can be computed sequentially from 1 to u, whereby previously computed columns (or rows) remain unchanged while the next one is computed.

The triangular matrix $\boldsymbol{L}$ can readily be partitioned according to partition (A.3.39) as

$$\boldsymbol{L} = \begin{bmatrix} \boldsymbol{L}_{11} & \boldsymbol{0} \\ \boldsymbol{L}_{21} & \boldsymbol{L}_{22} \end{bmatrix} \tag{A.3.56}$$

and then $\boldsymbol{N}$ of (A.3.54) can be written in terms of submatrices as

$$\boldsymbol{N} = \begin{bmatrix} \boldsymbol{L}_{11}\boldsymbol{L}_{11}^T & \boldsymbol{L}_{11}\boldsymbol{L}_{21}^T \\ \boldsymbol{L}_{21}\boldsymbol{L}_{11}^T & \boldsymbol{L}_{21}\boldsymbol{L}_{21}^T + \boldsymbol{L}_{22}\boldsymbol{L}_{22}^T \end{bmatrix} \tag{A.3.57}$$

Applying (A.3.46) to (A.3.57), the inverse of $\boldsymbol{Q}_{22}$ in (A.3.40) becomes

$$\boldsymbol{Q}_{22}^{-1} = \boldsymbol{L}_{22}\boldsymbol{L}_{22}^T \tag{A.3.58}$$

Depending on the application one might group the parameters such that (A.3.58) can be used directly in subsequent computations, i.e., the needed inverse $\boldsymbol{Q}_{22}^{-1}$ is a simple function of the Cholesky factors that had been computed previously.

Straightforward calculations using (A.3.39) and (A.3.57) give

$$\boldsymbol{N}_{11} = \boldsymbol{L}_{11}\boldsymbol{L}_{11}^T \tag{A.3.59}$$

$$\boldsymbol{L}_{21} = \left(\boldsymbol{L}_{11}^{-1}\boldsymbol{N}_{21}^T\right)^T \tag{A.3.60}$$

$$\boldsymbol{Q}_{22}^{-1} = \boldsymbol{L}_{22}\boldsymbol{L}_{22}^T = \boldsymbol{N}_{22} - \boldsymbol{L}_{21}\boldsymbol{L}_{21}^T = \boldsymbol{N}_{22} - \boldsymbol{N}_{21}\boldsymbol{N}_{11}{}^{-1}\boldsymbol{N}_{21}^T \tag{A.3.61}$$

The diagonal elements of $\boldsymbol{L}$ are not necessarily unity. Consider a unitriangular matrix $\boldsymbol{G}$ with elements taken from $\boldsymbol{L}$ such that $g_{jk} = l_{jk}/l_{kk}$ and a new diagonal matrix $\boldsymbol{D}$ such that $d_{jj} = l_{jj}^2$, then

$$\boldsymbol{L} = \boldsymbol{G}\sqrt{\boldsymbol{D}} \tag{A.3.62}$$

$$\boldsymbol{N} = \boldsymbol{L}\boldsymbol{L}^T = \boldsymbol{G}\boldsymbol{D}\boldsymbol{G}^T \tag{A.3.63}$$

Because $\boldsymbol{N}$ is a positive definite matrix, the diagonal elements of $\boldsymbol{G}$ are $+1$.

Assume that we wish to solve the system of equations

$$\boldsymbol{N}\boldsymbol{x} = \boldsymbol{u} \tag{A.3.64}$$

An initial thought might be to compute the inverse of $\boldsymbol{N}$ and compute $\boldsymbol{x} = \boldsymbol{N}^{-1}\boldsymbol{u}$. However, the unknown $\boldsymbol{x}$ can be solved without explicitly inverting the matrix and thus reduce the computational load. Instead, the first step in solving $\boldsymbol{x}$ is to substitute

(A.3.54) for $\boldsymbol{N}$ and premultiply the resulting equation with $\boldsymbol{L}^{-1}$, obtaining the triangular equations

$$\boldsymbol{L}^T\boldsymbol{x} = \boldsymbol{L}^{-1}\boldsymbol{u} \tag{A.3.65}$$

Denoting the right-hand side of (A.3.65) by $\boldsymbol{c}_u$, then multiplying from the left by $\boldsymbol{L}$ and replacing $\boldsymbol{L}\boldsymbol{L}^T\boldsymbol{x}$ by $\boldsymbol{u}$, we obtain the two equations

$$\boldsymbol{L}\boldsymbol{c}_u = \boldsymbol{u} \tag{A.3.66}$$

$$\boldsymbol{L}^T\boldsymbol{x} = \boldsymbol{c}_u \tag{A.3.67}$$

We first solve $\boldsymbol{c}_u$ from (A.3.66), called the forward solution, starting with the first element. Using the thus computed $\boldsymbol{c}_u$, the back solution (A.3.67) yields the parameters $\boldsymbol{x}$, starting with the last element.

In least squares, the auxiliary quantity

$$l = -\boldsymbol{c}_u^T\boldsymbol{c}_u = -(\boldsymbol{L}^{-1}\boldsymbol{u})^T\,(\boldsymbol{L}^{-1}\boldsymbol{u}) = -\boldsymbol{u}^T\boldsymbol{N}^{-1}\boldsymbol{u} \tag{A.3.68}$$

is needed to compute $\boldsymbol{v}^T\boldsymbol{P}\boldsymbol{v}$ (see Table 2.5.1). This term is always needed because it relates to the chi-square test that assesses the quality of the adjustment. The Cholesky algorithm provides l from $\boldsymbol{c}_u$ without explicitly using the inverses of $\boldsymbol{N}$ and $\boldsymbol{L}$.

Computing the inverse requires a much bigger computational effort than merely solving the system of equations. In computing the inverse, the first step is to make u solutions of the type (A.3.66) to obtain the columns of $\boldsymbol{C}$,

$$\boldsymbol{L}\boldsymbol{C} = \boldsymbol{I} \tag{A.3.69}$$

where $\boldsymbol{I}$ is the $u \times u$ identity matrix. This is followed by u solutions of the type (A.3.67), using the columns of $\boldsymbol{C}$ for $\boldsymbol{c}_u$, to obtain the respective u columns of the inverse of $\boldsymbol{N}$.

The Cholesky factor $\boldsymbol{L}$ can be used directly to compute uncorrelated observations. From (A.3.54) it follows that premultiplying $\boldsymbol{N}$ with $\boldsymbol{L}^{-1}$ and postmultiplying it with the transpose gives the identity matrix. Therefore, the Cholesky factor $\boldsymbol{L}$ can be used in ways similar to the matrix $\boldsymbol{D}$ in (A.3.15). Let $\boldsymbol{L}$ now denote the Cholesky factor of the covariance matrix of the observations Σ_{ℓ_b}, then the transformation (2.7.10) can be written as

$$\boldsymbol{L}^{-1}\boldsymbol{v} = \boldsymbol{L}^{-1}\boldsymbol{A}\boldsymbol{x} + \boldsymbol{L}^{-1}\boldsymbol{\ell} \tag{A.3.70}$$

Denoting the transformed observations by a bar, we get

$$\bar{\boldsymbol{v}} = \bar{\boldsymbol{A}}\boldsymbol{x} + \bar{\boldsymbol{\ell}} \tag{A.3.71}$$

$$\boldsymbol{L}\,\bar{\boldsymbol{\ell}} = \boldsymbol{\ell} \tag{A.3.72}$$

$$\boldsymbol{L}\bar{\boldsymbol{A}}_\alpha = \boldsymbol{A}_\alpha \tag{A.3.73}$$

The subscript α in (A.3.73) indicates the column. The matrix $\overline{\boldsymbol{A}}$ and the vector $\overline{\ell}$ can be computed directly from the forward solutions (A.3.73) and (A.3.72) using $\boldsymbol{L}$. The inverse $\boldsymbol{L}^{-1}$ is not required explicitly. Upon completion of the adjustment the residuals follow from

$$\boldsymbol{L}\overline{\boldsymbol{v}} = \boldsymbol{v} \tag{A.3.74}$$

It is at times advantageous to work with decorrelated observations. Examples are horizontal angle observations or even GPS vectors. Decorrelated observations can be added one at a time to the adjustment, whereas correlated observations should be added by sets. See also Section 2.8.6 for a discussion of decorrelated redundancy numbers.

A.3.7 Partial Minimization of Quadratic Functions

Let $w(\boldsymbol{x}, \boldsymbol{y})$ be the quadratic function of two vector variables, n-dimensional vector $\boldsymbol{x}$ and m-dimensional vector $\boldsymbol{y}$

$$w(\boldsymbol{x}, \boldsymbol{y}) = \boldsymbol{x}^T\boldsymbol{A}\boldsymbol{x} + 2\boldsymbol{y}^T\boldsymbol{B}\boldsymbol{x} + \boldsymbol{y}^T\boldsymbol{D}\boldsymbol{y} + 2\boldsymbol{x}^T\boldsymbol{g} + 2\boldsymbol{y}^T\boldsymbol{h} + c \tag{A.3.75}$$

where all matrices and vectors have appropriate dimensions, c is a scalar, and the matrix

$$\boldsymbol{N} = \begin{bmatrix} \boldsymbol{A} & \boldsymbol{B}^T \\ \boldsymbol{B} & \boldsymbol{D} \end{bmatrix} \tag{A.3.76}$$

is positive definite. Then the solution to the minimization problem

$$\min_{\boldsymbol{x},\boldsymbol{y}} w(\boldsymbol{x}, \boldsymbol{y}) \tag{A.3.77}$$

is expressed as

$$\boldsymbol{z}^* = \begin{bmatrix} \boldsymbol{x}^* \\ \boldsymbol{y}^* \end{bmatrix} = -\boldsymbol{N}^{-1}\boldsymbol{k} \qquad \boldsymbol{k} = \begin{bmatrix} \boldsymbol{g} \\ \boldsymbol{h} \end{bmatrix} \tag{A.3.78}$$

The solution (A.3.78) does not depend on the scalar value c. We can solve the problem using matrix partitioning expressions (A.3.53). Another way, sometimes more convenient, yet mathematically equivalent consists in dividing of the minimization operation into two sequentially applied minimizations, over the variable $\boldsymbol{y}$ and over the variable $\boldsymbol{x}$

$$\min_{\boldsymbol{x},\boldsymbol{y}} w(\boldsymbol{x}, \boldsymbol{y}) = \min_{\boldsymbol{x}} \left[\min_{\boldsymbol{y}} w(\boldsymbol{x}, \boldsymbol{y}) \right] \tag{A.3.79}$$

Denoting the expression in the brackets by $v(\boldsymbol{x})$,

$$v(\boldsymbol{x}) = \min_{\boldsymbol{y}} w(\boldsymbol{x}, \boldsymbol{y}) \tag{A.3.80}$$

we come to the expression

$$\min_{\boldsymbol{x},\boldsymbol{y}} w(\boldsymbol{x}, \boldsymbol{y}) = \min_{\boldsymbol{x}} v(\boldsymbol{x}) \tag{A.3.81}$$

Let $\boldsymbol{y}(\boldsymbol{x})$ be the value of $\boldsymbol{y}$ minimizing the function (A.3.80) given fixed value $\boldsymbol{x}$. It is called the argument of the conditional minimum. Equating to zero the vector of the first partial derivatives $\partial w(\boldsymbol{x},\boldsymbol{y})/\partial \boldsymbol{y} = 2\boldsymbol{D}\boldsymbol{y} + 2\boldsymbol{B}\boldsymbol{x} + 2\boldsymbol{h} = \boldsymbol{0}$ and get

$$\boldsymbol{y}(\boldsymbol{x}) = -\boldsymbol{D}^{-1}(\boldsymbol{B}\boldsymbol{x} + \boldsymbol{h}) \tag{A.3.82}$$

Now taking into account expressions (A.3.80) and (A.3.82) and obtain

$$\begin{aligned} v(\boldsymbol{x}) = w\big(\boldsymbol{x},\boldsymbol{y}(\boldsymbol{x})\big) &= \boldsymbol{x}^T\boldsymbol{A}\boldsymbol{x} - 2\big(\boldsymbol{D}^{-1}(\boldsymbol{B}\boldsymbol{x}+\boldsymbol{h})\big)^T\boldsymbol{B}\boldsymbol{x} \\ &\quad + \big(\boldsymbol{D}^{-1}(\boldsymbol{B}\boldsymbol{x}+\boldsymbol{h})\big)^T\boldsymbol{D}\boldsymbol{D}^{-1}(\boldsymbol{B}\boldsymbol{x}+\boldsymbol{h}) + 2\boldsymbol{x}^T\boldsymbol{g} - 2\big(\boldsymbol{D}^{-1}(\boldsymbol{B}\boldsymbol{x}+\boldsymbol{h})\big)^T\boldsymbol{h} + c \\ &= \boldsymbol{x}^T\boldsymbol{A}\boldsymbol{x} - (\boldsymbol{B}\boldsymbol{x}+\boldsymbol{h})^T\boldsymbol{D}^{-1}(\boldsymbol{B}\boldsymbol{x}+\boldsymbol{h}) + 2\boldsymbol{x}^T\boldsymbol{g} + c \\ &= \boldsymbol{x}^T(\boldsymbol{A} - \boldsymbol{B}^T\boldsymbol{D}^{-1}\boldsymbol{B})\boldsymbol{x} + 2\boldsymbol{x}^T(\boldsymbol{g} - \boldsymbol{B}^T\boldsymbol{D}^{-1}\boldsymbol{h}) + \bar{c} \end{aligned} \tag{A.3.83}$$

with $\bar{c} = c + \boldsymbol{h}^T\boldsymbol{D}^{-1}\boldsymbol{h}$. Therefore, the function $v(\boldsymbol{x})$ is quadratic with the matrix part $\boldsymbol{A} - \boldsymbol{B}^T\boldsymbol{D}^{-1}\boldsymbol{B}$ and the vector $2(\boldsymbol{g} - \boldsymbol{B}^T\boldsymbol{D}^{-1}\boldsymbol{h})$ as linear part. It takes its minimum at the point

$$\boldsymbol{x}^* = -(\boldsymbol{A} - \boldsymbol{B}^T\boldsymbol{D}^{-1}\boldsymbol{B})^{-1}(\boldsymbol{g} - \boldsymbol{B}^T\boldsymbol{D}^{-1}\boldsymbol{h}) \tag{A.3.84}$$

Finally, substituting (A.3.84) into the expression for the argument of the conditional minimum (A.3.82), we obtain

$$\begin{aligned} \boldsymbol{y}^* &= \boldsymbol{y}(\boldsymbol{x}^*) = -\boldsymbol{D}^{-1}(\boldsymbol{B}\boldsymbol{x}^* + \boldsymbol{h}) \\ &= \boldsymbol{D}^{-1}\boldsymbol{B}(\boldsymbol{A} - \boldsymbol{B}^T\boldsymbol{D}^{-1}\boldsymbol{B})^{-1}\boldsymbol{g} - (\boldsymbol{D}^{-1} + \boldsymbol{D}^{-1}\boldsymbol{B}(\boldsymbol{A} - \boldsymbol{B}^T\boldsymbol{D}^{-1}\boldsymbol{B})^{-1}\boldsymbol{B}^T\boldsymbol{D}^{-1})\boldsymbol{h} \end{aligned} \tag{A.3.85}$$

Note again that expressions (A.3.84) and (A.3.85) can be obtained as result of block-wise inversion of the matrix $\boldsymbol{N}$ in expression (A.3.78).

If the quadratic function is expressed in the form

$$\begin{aligned} w(\boldsymbol{x},\boldsymbol{y}) &= (\boldsymbol{z} - \bar{\boldsymbol{z}})^T\boldsymbol{M}(\boldsymbol{z} - \bar{\boldsymbol{z}}) \\ &= (\boldsymbol{x} - \bar{\boldsymbol{x}})^T\boldsymbol{A}(\boldsymbol{x} - \bar{\boldsymbol{x}}) + 2(\boldsymbol{y} - \bar{\boldsymbol{y}})^T\boldsymbol{B}(\boldsymbol{x} - \bar{\boldsymbol{x}}) + (\boldsymbol{y} - \bar{\boldsymbol{y}})^T\boldsymbol{D}(\boldsymbol{y} - \bar{\boldsymbol{y}}) \end{aligned} \tag{A.3.86}$$

then the argument of the conditional minimum takes the form

$$\boldsymbol{y}(\boldsymbol{x}) = \bar{\boldsymbol{y}} - \boldsymbol{D}^{-1}\boldsymbol{B}(\boldsymbol{x} - \bar{\boldsymbol{x}}) \tag{A.3.87}$$

and the quadratic function $v(\boldsymbol{x})$ is expressed as

$$v(\boldsymbol{x}) = \min_{\boldsymbol{y}} w(\boldsymbol{x},\boldsymbol{y}) = w\big(\boldsymbol{x},\boldsymbol{y}(\boldsymbol{x})\big) = (\boldsymbol{x} - \bar{\boldsymbol{x}})^T(\boldsymbol{A} - \boldsymbol{B}^T\boldsymbol{D}^{-1}\boldsymbol{B})(\boldsymbol{x} - \bar{\boldsymbol{x}}) \tag{A.3.88}$$

As it follows from (A.3.88), the function $v(\boldsymbol{x})$ takes its real-valued minimum at the point $\bar{\boldsymbol{x}}$ and, after substitution into (A.3.87), the argument of the conditional minimum becomes $\boldsymbol{y}(\bar{\boldsymbol{x}}) = \bar{\boldsymbol{y}} - \boldsymbol{D}^{-1}\boldsymbol{B}(\bar{\boldsymbol{x}} - \bar{\boldsymbol{x}}) = \bar{\boldsymbol{y}}$. The last result is obviously predictable since the vector $\bar{\boldsymbol{z}}$ minimizes the quadratic function (A.3.86). But, if the variable $\boldsymbol{x}$ takes values from the integer-valued space $\boldsymbol{Z}^n$, not from n-dimensional real-valued space $\boldsymbol{R}^n$, the minimization of the function $v(\boldsymbol{x})$ is impossible in the explicit closed form. Instead, the integer-valued search must be applied. After the integer search is performed and integer-valued minimizer $\hat{\boldsymbol{x}}$ is found, the expression (A.3.87) completes computations resulting in the real-valued vector $\boldsymbol{y}(\hat{\boldsymbol{x}}) = \bar{\boldsymbol{y}} - \boldsymbol{D}^{-1}\boldsymbol{B}(\hat{\boldsymbol{x}} - \bar{\boldsymbol{x}})$.

Consider another case of the quadratic function, used in ambiguity fixing algorithms. Let it be expressed in the form

$$w(\boldsymbol{x},\boldsymbol{y}) = (\boldsymbol{Ex} + \boldsymbol{Gy} - \bar{\boldsymbol{a}})^T\boldsymbol{D}(\boldsymbol{Ex} + \boldsymbol{Gy} - \bar{\boldsymbol{a}}) \tag{A.3.89}$$

with the positive definite matrix $\boldsymbol{D}$. Then the argument of conditional minimum takes the form

$$\boldsymbol{y}(\boldsymbol{x}) = -(\boldsymbol{G}^T\boldsymbol{DG})^{-1}\boldsymbol{G}^T\boldsymbol{D}(\boldsymbol{Ex} - \bar{\boldsymbol{a}}) \tag{A.3.90}$$

and after substituting it into (A.3.89) we have the function $v(\boldsymbol{x})$ taking the form

$$\begin{aligned}
v(\boldsymbol{x}) &= \min_{\boldsymbol{y}} w(\boldsymbol{x},\boldsymbol{y}) = w\big(\boldsymbol{x},\boldsymbol{y}(\boldsymbol{x})\big) \\
&= \big(\boldsymbol{Ex} + \boldsymbol{Gy}(\boldsymbol{x}) - \bar{\boldsymbol{a}}\big)^T\boldsymbol{D}\big(\boldsymbol{Ex} + \boldsymbol{Gy}(\boldsymbol{x}) - \bar{\boldsymbol{a}}\big) \\
&= \big(\boldsymbol{Ex} - \bar{\boldsymbol{a}} - \boldsymbol{G}(\boldsymbol{G}^T\boldsymbol{DG})^{-1}\boldsymbol{G}^T\boldsymbol{D}(\boldsymbol{Ex} - \bar{\boldsymbol{a}})\big)^T\boldsymbol{D} \\
&\qquad \times \big(\boldsymbol{Ex} - \bar{\boldsymbol{a}} - \boldsymbol{G}(\boldsymbol{G}^T\boldsymbol{DE})^{-1}\boldsymbol{G}^T\boldsymbol{D}(\boldsymbol{Ex} - \bar{\boldsymbol{a}})\big) \\
&= \big(\boldsymbol{Ex} - \bar{\boldsymbol{a}}\big)^T\big(\boldsymbol{D} - \boldsymbol{DG}(\boldsymbol{G}^T\boldsymbol{DG})^{-1}\boldsymbol{G}^T\boldsymbol{D}\big)(\boldsymbol{Ex} - \bar{\boldsymbol{a}}) \\
&= (\boldsymbol{Ex} - \bar{\boldsymbol{a}})^T\Pi_{\boldsymbol{y}}(\boldsymbol{Ex} - \bar{\boldsymbol{a}}) \\
&= \boldsymbol{x}^T\boldsymbol{E}^T\Pi_{\boldsymbol{y}}\boldsymbol{Ex} - 2\boldsymbol{x}^T\boldsymbol{E}^T\Pi_{\boldsymbol{y}}\bar{\boldsymbol{a}}
\end{aligned} \tag{A.3.91}$$

with

$$\Pi_{\boldsymbol{y}} = \boldsymbol{D} - \boldsymbol{DG}(\boldsymbol{G}^T\boldsymbol{DG})^{-1}\boldsymbol{G}^T\boldsymbol{D} \tag{A.3.92}$$

Note again, that if the variable $\boldsymbol{x}$ takes values from the integer valued space $\boldsymbol{Z}^n$, minimization of the function $v(\boldsymbol{x})$ is performed using the integer-valued search explained in the Section 6.5. After the integer search is performed and integer-valued minimizer $\hat{\boldsymbol{x}}$ is found, the expression (A.3.90) completes the computations resulting in the real-valued vector $\boldsymbol{y}(\hat{\boldsymbol{x}})$.

It is assumed that the matrix $(\boldsymbol{A} - \boldsymbol{B}^T\boldsymbol{D}^{-1}\boldsymbol{B})$ in expressions (A.3.83) and (A.3.88) is positive definite. The matrix $\boldsymbol{E}^T\Pi_{\boldsymbol{y}}\boldsymbol{E}$ in the expression (A.3.91) is also assumed to be positive definite.

Let the matrix $\boldsymbol{N}$ allow Cholesky decomposition expressed in the block form as

$$\boldsymbol{N} = \begin{bmatrix} \boldsymbol{L} & \boldsymbol{0} \\ \boldsymbol{K} & \boldsymbol{M} \end{bmatrix} \begin{bmatrix} \boldsymbol{L}^T & \boldsymbol{K}^T \\ \boldsymbol{0} & \boldsymbol{M}^T \end{bmatrix} \tag{A.3.93}$$

Using (A.3.61) we readily have

$$(\boldsymbol{A} - \boldsymbol{B}^T\boldsymbol{D}^{-1}\boldsymbol{B}) = \boldsymbol{M}\boldsymbol{M}^T \tag{A.3.94}$$

which means that the matrix inversion $(\boldsymbol{A} - \boldsymbol{B}^T\boldsymbol{D}^{-1}\boldsymbol{B})^{-1}$ in the expression (A.3.84) can be replaced with the forward and backward solutions with matrices $\boldsymbol{M}$ and $\boldsymbol{M}^T$, respectively, as explained in expressions (A.3.66) and (A.3.67) of Section A.3.6.

Now consider the structured least-squares problem

$$\boldsymbol{A}_i\boldsymbol{y}_i + \boldsymbol{B}_i\boldsymbol{x} = \boldsymbol{b}_i \quad i = 1, \ldots, N \tag{A.3.95}$$

The vector $\boldsymbol{x}$ is n-dimensional, each vector $\boldsymbol{y}_i$ has the dimension m_i, and each vector $\boldsymbol{b}_i$ has dimension l_i . The matrices have appropriate size. The block-diagonal system (A.3.95) with coordinating blocks is schematically shown below:

$$\begin{bmatrix} \boldsymbol{A}_1 & & & & \boldsymbol{B}_1 \\ & \boldsymbol{A}_2 & & & \boldsymbol{B}_2 \\ & & \ddots & & \vdots \\ & & & \boldsymbol{A}_N & \boldsymbol{B}_N \end{bmatrix} \times \begin{bmatrix} \boldsymbol{y}_1 \\ \boldsymbol{y}_2 \\ \vdots \\ \boldsymbol{y}_N \\ \boldsymbol{x} \end{bmatrix} = \begin{bmatrix} \boldsymbol{b}_1 \\ \boldsymbol{b}_2 \\ \vdots \\ \boldsymbol{b}_N \end{bmatrix} \tag{A.3.96}$$

The least-squares problem for the system (A.3.96) reads as

$$\min_{\boldsymbol{x},\boldsymbol{y}_1,\ldots,\boldsymbol{y}_N} w(\boldsymbol{x}, \boldsymbol{y}_1, \ldots, \boldsymbol{y}_N) = \min_{\boldsymbol{x},\boldsymbol{y}_1,\ldots,\boldsymbol{y}_N} \sum_{i=1}^{N} \|\boldsymbol{A}_i\boldsymbol{y}_i + \boldsymbol{B}_i\boldsymbol{x} - \boldsymbol{b}_i\|^2 \tag{A.3.97}$$

We show how the partial minimization approach leads to a decomposition of the structured least squares into the series of separate minimization problems and one "coordinating" minimization problem. The same way as it is done in (A.3.79), we can rewrite (A.3.97) in the form

$$\min_{\boldsymbol{x},\boldsymbol{y}_1,\ldots,\boldsymbol{y}_N} \sum_{i=1}^{N} \|\boldsymbol{A}_i\boldsymbol{y}_i + \boldsymbol{B}_i\boldsymbol{x} - \boldsymbol{b}_i\|^2 = \min_{\boldsymbol{x}} \left[\min_{\boldsymbol{y}_1,\ldots,\boldsymbol{y}_N} \sum_{i=1}^{N} \|\boldsymbol{A}_i\boldsymbol{y}_i + \boldsymbol{B}_i\boldsymbol{x} - \boldsymbol{b}_i\|^2 \right] \tag{A.3.98}$$

The internal minimum in (A.3.98) as a function of $\boldsymbol{x}$ is denoted by $v(\boldsymbol{x})$. It is split into N-independent minimizations

$$v(\boldsymbol{x}) = \min_{\boldsymbol{y}_1,\ldots,\boldsymbol{y}_N} \sum_{i=1}^{N} \|\boldsymbol{A}_i\boldsymbol{y}_i + \boldsymbol{B}_i\boldsymbol{x} - \boldsymbol{b}_i\|^2 = \sum_{i=1}^{N} \left[\min_{\boldsymbol{y}_i} \|\boldsymbol{A}_i\boldsymbol{y}_i + \boldsymbol{B}_i\boldsymbol{x} - \boldsymbol{b}_i\|^2 \right] \tag{A.3.99}$$

Let

$$\boldsymbol{y}_i(\boldsymbol{x}) = -(\boldsymbol{A}_i^T\boldsymbol{A}_i)^{-1}\boldsymbol{A}_i^T(\boldsymbol{b}_i - \boldsymbol{B}_i\boldsymbol{x}) \tag{A.3.100}$$

be the argument of the conditional minimum. Then

$$\begin{aligned} v(\boldsymbol{x}) &= \sum_{i=1}^{N} \|\boldsymbol{A}_i\boldsymbol{y}_i(\boldsymbol{x}) + \boldsymbol{B}_i\boldsymbol{x} - \boldsymbol{b}_i\|^2 \\ &= \sum_{i=1}^{N} \| -\boldsymbol{A}_i\left(\boldsymbol{A}_i^T\boldsymbol{A}_i\right)^{-1}\boldsymbol{A}_i^T(\boldsymbol{b}_i - \boldsymbol{B}\boldsymbol{x})\boldsymbol{y}_i(\boldsymbol{x}) + \boldsymbol{B}_i\boldsymbol{x} - \boldsymbol{b}_i\|^2 \\ &= \sum_{i=1}^{N} \|\Pi_i\left(\boldsymbol{b}_i - \boldsymbol{B}_i\boldsymbol{x}\right)\|^2 \end{aligned} \tag{A.3.101}$$

is the quadratic function of $\boldsymbol{x}$ with

$$\Pi_i = I_{l_i} - \boldsymbol{A}_i(\boldsymbol{A}_i^T\boldsymbol{A}_i)^{-1}\boldsymbol{A}_i^T \quad (\Pi_i)^2 = \Pi_i \tag{A.3.102}$$

being the projection matrices. Then the vector $\boldsymbol{x}^*$ minimizing the function (A.3.101) is calculated as

$$\boldsymbol{x}^* = \left(\sum_{i=1}^{N} \boldsymbol{B}_i^T\Pi_i\boldsymbol{B}_i\right)^{-1} \sum_{i=1}^{N} \boldsymbol{B}_i^T\Pi_i\boldsymbol{b}_i \tag{A.3.103}$$

Therefore, a sufficient condition for solvability of problem (A.3.97) is non-singularity (actually, positive definiteness) of the matrices $\boldsymbol{A}_i^T\boldsymbol{A}_i$ and $\sum_{i=1}^{N}\boldsymbol{B}_i^T\Pi_i\boldsymbol{B}_i$. The solution consists of two stages. First, the coordinating problem is solved resulting in (A.3.103). Then the series of solutions $\boldsymbol{y}_i^* = \boldsymbol{y}_i(\boldsymbol{x}^*)$ is calculated according to (A.3.100).

The partial minimization approach can be used for decomposition of structured least-squares problems of a form more general than (A.3.96). It can be applied in cases when the pattern of the matrix is organized in such a way that fixing of certain group of variables (coordinating variables) splits the minimization problem into a series of independent minimization subproblems. The coordinating problem

$$\min_{\boldsymbol{x}} v(\boldsymbol{x}) \tag{A.3.104}$$

is solved first, followed by series of independent minimization subproblems. The technique of partial minimization is referred to in the geodetic literature as Helmert blocking or Helmert-Wolf blocking (Wolf, 1978). During the readjustment of the NAD 1983 the technique was implemented to optimize the adjustment methodology and execution (Schwarz and Wade, 1990).

A.3.8 *QR* Decomposition

This subsection introduces one more matrix decomposition. Let $\boldsymbol{J}$ be the $n \times m$ matrix with full column rank, which means its columns $\boldsymbol{j}_1, \boldsymbol{j}_2, \ldots, \boldsymbol{j}_m$ are linearly independent.

The $\boldsymbol{QR}$ decomposition of the matrix $\boldsymbol{J}$ is defined as follows:

$$\boldsymbol{J} = \boldsymbol{QR} \qquad \boldsymbol{R} = \left[\begin{array}{c} \boldsymbol{U} \\ \hdashline {}_{n-m}\boldsymbol{0}_m \end{array}\right] \tag{A.3.105}$$

where the $n \times n$ matrix $\boldsymbol{Q}$ is orthonormal and the $m \times m$ matrix $\boldsymbol{U}$ is upper triangular. Two main applications of this decomposition [also called factorization because (A.3.105) expresses $\boldsymbol{J}$ as a product of two matrix factors] are

1. Calculation of the Cholesky decomposition of matrix $\boldsymbol{J}^T\boldsymbol{J}$ without its explicit calculation. From (A.3.105) it follows

$$\boldsymbol{J}^T\boldsymbol{J} = \boldsymbol{R}^T\boldsymbol{Q}^T\boldsymbol{QR} = \boldsymbol{R}^T\boldsymbol{R} = \boldsymbol{U}^T\boldsymbol{U} = \boldsymbol{LL}^T \tag{A.3.106}$$

with $\boldsymbol{L}$ being defined as $\boldsymbol{L} = \boldsymbol{U}^T$.

2. Calculation of the annulator matrix. Partitioning of the matrix $\boldsymbol{Q}^T$ into two parts,

$$\boldsymbol{Q}^T = \left[\begin{array}{c} \boldsymbol{G} \\ \hdashline \boldsymbol{F} \end{array}\right] \tag{A.3.107}$$

gives

$$\left[\begin{array}{c} \boldsymbol{GJ} \\ \hdashline \boldsymbol{FJ} \end{array}\right] = \boldsymbol{Q}^T\boldsymbol{J} = \boldsymbol{R} = \left[\begin{array}{c} \boldsymbol{U} \\ \hdashline {}_{(n-m)}\boldsymbol{0}_m \end{array}\right] \tag{A.3.108}$$

which proves that $\boldsymbol{FJ} = {}_{(n-m)}\boldsymbol{0}_m$ and the $(n-m) \times n$ matrix $\boldsymbol{F}$ annulates the matrix $\boldsymbol{J}$.

The matrix $\boldsymbol{J}$ will now be processed column by column. A sequence of orthonormal matrices $\boldsymbol{Q}^{(k)}$ will be constructed to operate on $\boldsymbol{J}$ to reduce the number of nonzero entries. The superscript (k) denotes the step of the algorithm, while the subscript k enumerates the columns of matrix $\boldsymbol{J}$. Let us start with the first column $\boldsymbol{j}_1$ and construct the orthonormal matrix $\boldsymbol{Q}^{(1)}$, called *the Hausholder matrix*, as follows:

$$\boldsymbol{Q}^{(1)} = \boldsymbol{I}_n - 2\tau_1\boldsymbol{h}_1\boldsymbol{h}_1^T \tag{A.3.109}$$

where the column vector $\boldsymbol{h}_1$ is defined as

$$\boldsymbol{h}_1 = \boldsymbol{j}_1 + \alpha_1\boldsymbol{e}_1 \tag{A.3.110}$$

$$\boldsymbol{e}_1 = (1, 0, \ldots, 0)^T \tag{A.3.111}$$

and scalar parameters α_1 and τ_1 are chosen to satisfy the following conditions:

$$\boldsymbol{Q}^{(1)}\boldsymbol{j}_1 = \|\boldsymbol{j}_1\|\boldsymbol{e}_1 \tag{A.3.112}$$

$$\frac{1}{\boldsymbol{h}_1^T\boldsymbol{h}_1} = \tau_1 \tag{A.3.113}$$

The last condition guaranties that the matrix (A.3.109) is orthonormal. We can readily verify that

$$\boldsymbol{Q}^{(1)^T}\boldsymbol{Q}^{(1)} = \boldsymbol{I}_n - 4\tau_1\boldsymbol{h}_1\boldsymbol{h}_1^T + 4\tau_1^2\boldsymbol{h}_1^T\boldsymbol{h}_1\boldsymbol{h}_1\boldsymbol{h}_1^T = \boldsymbol{I}_n \tag{A.3.114}$$

The symbol $\boldsymbol{I}_n$ is used to denote the $n \times n$ identity matrix. Moreover, we have $\boldsymbol{Q}^{(1)}\boldsymbol{h}_1 = \boldsymbol{h}_1 - 2\tau_1\boldsymbol{h}_1\boldsymbol{h}_1^T\boldsymbol{h}_1 = -\boldsymbol{h}_1$ and $\boldsymbol{Q}^{(1)}\boldsymbol{x} = \boldsymbol{x}$ for each vector $\boldsymbol{x}$ satisfying the condition $\boldsymbol{h}_1^T\boldsymbol{x} = 0$. The last two conditions mean that $\boldsymbol{Q}^{(1)}$ defines the mirror reflection relative to the plane $\boldsymbol{h}_1^T\boldsymbol{x} = 0$. Also, note that the orthonormal matrix $\boldsymbol{Q}^{(1)}$ is symmetric which means that it is involutory: $\boldsymbol{Q}^{(1)}\boldsymbol{Q}^{(1)} = \boldsymbol{I}_n$.

Now it follows from (A.3.109) and (A.3.110) that

$$\begin{aligned}\boldsymbol{Q}^{(1)}\boldsymbol{j}_1 &= \boldsymbol{j}_1 - 2\tau_1(\boldsymbol{j}_1 + \alpha_1\boldsymbol{e}_1)(\boldsymbol{j}_1 + \alpha_1\boldsymbol{e}_1)^T\boldsymbol{j}_1 \\ &= \boldsymbol{j}_1 - 2\tau_1\boldsymbol{j}_1(\|\boldsymbol{j}_1\|^2 + \alpha_1 j_{11}) - 2\tau_1\alpha_1\boldsymbol{e}_1(\|\boldsymbol{j}_1\|^2 + \alpha_1 j_{11}) \\ &= [1 - 2\tau_1(\|\boldsymbol{j}_1\|^2 + \alpha_1 j_{11})]\boldsymbol{j}_1 - 2\tau_1\alpha_1\boldsymbol{e}_1(\|\boldsymbol{j}_1\|^2 + \alpha_1 j_{11})\end{aligned} \tag{A.3.115}$$

Let us choose the parameter α_1 such that the term in brackets in the last expression becomes zero,

$$1 - 2\tau_1\left(\|\boldsymbol{j}_1\|^2 + \alpha_1 j_{11}\right) = 0 \tag{A.3.116}$$

Then it follows from (A.3.115) that

$$\boldsymbol{Q}^{(1)}\boldsymbol{j}_1 = -2\tau_1\alpha_1\boldsymbol{e}_1\left(\|\boldsymbol{j}_1\|^2 + \alpha_1 j_{11}\right) \tag{A.3.117}$$

and the combination of conditions (A.3.113) and (A.3.116) gives

$$\begin{aligned}\boldsymbol{h}_1^T\boldsymbol{h}_1 &= \|\boldsymbol{j}_1\|^2 + 2\alpha_1 j_{11} + \alpha_1^2 = \frac{1}{\tau_1} \\ &= 2\left(\|\boldsymbol{j}_1\|^2 + \alpha_1 j_{11}\right)\end{aligned} \tag{A.3.118}$$

It follows from the last equality that $\alpha_1^2 = \|\boldsymbol{j}_1\|^2$ or

$$\alpha_1 = \pm\|\boldsymbol{j}_1\| \tag{A.3.119}$$

Finally, the combination of (A.3.116), (A.3.117), and (A.3.119) gives

$$\boldsymbol{Q}^{(1)}\boldsymbol{j}_1 = \mp\|\boldsymbol{j}_1\|\boldsymbol{e}_1 \tag{A.3.120}$$

which means that the orthonormal matrix $\boldsymbol{Q}^{(1)}$ transforms the vector $\boldsymbol{j}_1$ into $\|\boldsymbol{j}_1\|\boldsymbol{e}_1$ if we choose $\alpha_1 = -\|\boldsymbol{j}_1\|$, and it transforms $\boldsymbol{j}_1$ into $-\|\boldsymbol{j}_1\|\boldsymbol{e}_1$ if $\alpha_1 = \|\boldsymbol{j}_1\|$. In both cases only the first entry of the transformed vector is nonzero. Thus, we have

$$\boldsymbol{Q}^{(1)}\boldsymbol{J} = \left[\begin{array}{c|ccc} \mp\|\boldsymbol{j}_1\| & * & \cdots & * \\ \hline 0 & & & \\ \vdots & & \boldsymbol{J}^{(2)} & \\ 0 & & & \end{array}\right] \tag{A.3.121}$$

where the symbol $*$ denotes some nonzero scalars, and $\boldsymbol{J}^{(2)}$ is some $(n-1)\times(m-1)$ matrix. Both sign options are possible in (A.3.119), but for the sake of numerical stability it is worthwhile to choose the sign of α_1 coinciding with the sign of the scalar j_{11} in order to reduce the probability of having small values $\|\boldsymbol{j}_1\|^2 + \alpha_1 j_{11}$ in (A.3.118).

Then we construct the matrix

$$\boldsymbol{Q}^{(2)} = \left[\begin{array}{c|c} 1 & 0 \quad \cdots \quad 0 \\ \hline 0 & \\ \vdots & \boldsymbol{I}_{n-1} - 2\tau_2 \boldsymbol{h}_2 \boldsymbol{h}_2^T \\ 0 & \end{array}\right] \tag{A.3.122}$$

which, when multiplied from left with matrix (A.3.121), leaves its first row and the first column unchanged and thus reducing the first column of matrix $\boldsymbol{J}^{(2)}$ into a form similar to (A.3.120). Applying m transformations gives

$$\boldsymbol{Q}^{(m)} \cdots \boldsymbol{Q}^{(2)} \boldsymbol{Q}^{(1)} \boldsymbol{J} = \left[\begin{array}{c} \boldsymbol{U} \\ \hline {}_{n-m}\boldsymbol{0}_m \end{array}\right] \tag{A.3.123}$$

The last condition defines the matrix $\boldsymbol{Q}$ in (A.3.105) as

$$\boldsymbol{Q} = \boldsymbol{Q}^{(1)} \boldsymbol{Q}^{(2)} \cdots \boldsymbol{Q}^{(m)} \tag{A.3.124}$$

In the last equality we took into account that $\boldsymbol{Q}^{(k)} = \boldsymbol{Q}^{(k)^T}$.

This completes the description of the $\boldsymbol{QR}$ decomposition as a series of mirror orthonormal transformation converting the $n \times m$ matrix $\boldsymbol{J}$ into upper triangular form (A.3.105).

A.3.9 Rank One Update of Cholesky Decomposition

For the overdetermined system

$$\boldsymbol{Jx} = \boldsymbol{b} \tag{A.3.125}$$

consider the least-squares problem

$$(\boldsymbol{Jx} - \boldsymbol{b})^T (\boldsymbol{Jx} - \boldsymbol{b}) \to \min \tag{A.3.126}$$

resulting in the solution

$$\boldsymbol{x}^* = (\boldsymbol{J}^T \boldsymbol{J})^{-1} \boldsymbol{J}^T \boldsymbol{b} \tag{A.3.127}$$

where the $n \times m$ $(n > m)$ matrix $\boldsymbol{J}$ is supposed to have a full rank. When looking for outliers affecting the right-hand side vector $\boldsymbol{b}$, the overdetermined system (A.3.125) can be solved repeatedly with a different number of equations. Namely, we may want to strike one of the equations from the system (A.3.125) or add one equation, changing the dimension n by one to either $n-1$ or $n+1$. Let

$$\boldsymbol{j}^T \boldsymbol{x} = b \tag{A.3.128}$$

be an equation to be stricken or added to the system (A.3.125). In the first case the term $\boldsymbol{J}^T\boldsymbol{J}$ in (A.3.127) will be changed to $\boldsymbol{J}^T\boldsymbol{J}-\boldsymbol{jj}^T$, and in the second case it will be changed to $\boldsymbol{J}^T\boldsymbol{J}+\boldsymbol{jj}^T$. Also, the term $\boldsymbol{J}^T\boldsymbol{b}$ will be changed to $\boldsymbol{J}^T\boldsymbol{b}\mp\boldsymbol{j}b$. Aiming to apply the Cholesky decomposition to the solution (A.3.127) we consider the following problem: given a positive definite matrix $\boldsymbol{J}^T\boldsymbol{J}$ in the form of Cholesky decomposition

$$\boldsymbol{J}^T\boldsymbol{J}=\boldsymbol{L}\boldsymbol{L}^T \tag{A.3.129}$$

find the Cholesky decomposition of the new matrix $\boldsymbol{J}^T\boldsymbol{J}+\alpha\boldsymbol{jj}^T$,

$$\boldsymbol{J}^T\boldsymbol{J}+\alpha\boldsymbol{jj}^T=\overline{\boldsymbol{L}}\,\overline{\boldsymbol{L}}^T \tag{A.3.130}$$

where $\alpha=\pm1$. Obvious straightforward calculation of (A.3.130) takes the number of calculations proportional to m^3. We are looking for a less computationally intensive way, assuming that decomposition (A.3.129) is already done. The matrix update $\alpha\boldsymbol{jj}^T$ has a rank of one. Calculating the decomposition (A.3.130), given the decomposition (A.3.129), is called the *rank one update of Cholesky decomposition* (Yang, 1977; Golub and Van Loan, 1996). The algorithm described in this subsection has a computational complexity proportional to m^2.

Let us express (A.3.130) in the equivalent form

$$\left[\boldsymbol{j}\;\vdots\;\boldsymbol{L}\right]\boldsymbol{D}\begin{bmatrix}\boldsymbol{j}^T\\ \hdashline \boldsymbol{L}^T\end{bmatrix}=\overline{\boldsymbol{L}}\,\overline{\boldsymbol{L}}^T \tag{A.3.131}$$

where the $(m+1)\times(m+1)$ diagonal matrix $\boldsymbol{D}$ has the form

$$\boldsymbol{D}=\begin{bmatrix}\alpha & \boldsymbol{0}_m^T\\ \boldsymbol{0}_m & I_m\end{bmatrix} \tag{A.3.132}$$

and the $m\times(m+1)$ matrix $\left[\boldsymbol{j}\;\vdots\;\boldsymbol{L}\right]$ has the pattern

$$\left[\boldsymbol{j}\;\vdots\;\boldsymbol{L}\right]=\begin{bmatrix}* & * & 0 & 0 & \cdots & 0\\ * & * & * & 0 & \cdots & 0\\ \vdots & \cdots & \cdots & & & \\ * & * & * & * & * & *\end{bmatrix} \tag{A.3.133}$$

The symbol $*$ denotes an arbitrary nonzero value. The idea of the algorithm consists of sequential application of simple orthonormal transformations $\boldsymbol{Q}^{(k)}$ and upper triangular transformations $\boldsymbol{U}^{(k)}$ to reduce (A.3.133) into the form

$$\left[\boldsymbol{j}\;\vdots\;\boldsymbol{L}\right]\boldsymbol{Q}^{(1)}\boldsymbol{U}^{(1)}\boldsymbol{Q}^{(2)}\boldsymbol{U}^{(2)}\cdots\boldsymbol{Q}^{(m)}\boldsymbol{U}^{(m)}=\left[\boldsymbol{0}\;\vdots\;\boldsymbol{L}^{(m)}\right]=\begin{bmatrix}0 & * & 0 & 0 & \cdots & 0\\ 0 & * & * & 0 & \cdots & 0\\ \vdots & \cdots & \cdots & & & \\ 0 & * & * & * & * & *\end{bmatrix} \tag{A.3.134}$$

such that

$$\boldsymbol{J}^T\boldsymbol{J} + \alpha\boldsymbol{j}\boldsymbol{j}^T = \left[\boldsymbol{0} \mid \boldsymbol{L}^{(m)}\right] \boldsymbol{D}^{(m)} \left[\begin{array}{c} \boldsymbol{0}^T \\ \hline \boldsymbol{L}^{(m)T} \end{array}\right] \tag{A.3.135}$$

where the matrix $\boldsymbol{D}^{(m)}$ is diagonal.

Let $\boldsymbol{Q}^{(1)}$ be the orthonormal matrix of the form

$$\boldsymbol{Q}^{(1)} = \left[\begin{array}{c|c|c|c|c} c_1 & -s_1 & 0 & & 0 \\ \hline s_1 & c_1 & 0 & & 0 \\ \hline 0 & 0 & 1 & & \\ \hline & & & \ddots & \\ \hline 0 & 0 & & & 1 \end{array}\right] \tag{A.3.136}$$

where the numbers c_1 and s_1, $c_1^2 + s_1^2 = 1$ are chosen in such a way as to reduce the first row of the matrix (A.3.133) into the following form,

$$\left(j_1 \mid l_{11} \mid 0 \mid \cdots \mid 0\right) \boldsymbol{Q}^{(1)} = \left(0 \mid \bar{l}_{11} \mid 0 \mid \cdots \mid 0\right) \tag{A.3.137}$$

In the last expression j_1 and l_{11} are the only two nonzero entries marked by $*$ in the first row of the matrix pattern (A.3.133). Straightforward calculations give

$$\bar{l}_{11} = \sqrt{j_1^2 + l_{11}^2}, \; c_1 = \frac{l_{11}}{\bar{l}_{11}}, \; s_1 = -\frac{j_1}{\bar{l}_{11}} \tag{A.3.138}$$

The matrix (A.3.136) is called *the Givens matrix*, or *the planar rotation matrix* because only two coordinates are involved in the orthonormal transformation. Let us denote

$$\left[\boldsymbol{j} \mid \boldsymbol{L}\right] \boldsymbol{Q}^{(1)} = \left[\begin{array}{c|ccccc} 0 & \bar{l}_{11} & 0 & 0 & \cdots & 0 \\ * & * & * & 0 & \cdots & 0 \\ \vdots & \cdots & \cdots & & & \\ * & * & * & * & * & * \end{array}\right] = \left[\boldsymbol{j}^{(1)} \mid \bar{\boldsymbol{L}}^{(1)}\right] \tag{A.3.139}$$

Since $\boldsymbol{Q}^{(1)}$ is orthonormal, we can rewrite (A.3.131) in the equivalent form

$$\begin{aligned} \left[\boldsymbol{j} \mid \boldsymbol{L}\right] \boldsymbol{D} \left[\begin{array}{c} \boldsymbol{j}^T \\ \hline \boldsymbol{L}^T \end{array}\right] &= \left[\boldsymbol{j} \mid \boldsymbol{L}\right] \boldsymbol{Q}^{(1)}\boldsymbol{Q}^{(1)^T}\boldsymbol{D}\boldsymbol{Q}^{(1)}\boldsymbol{Q}^{(1)^T} \left[\begin{array}{c} \boldsymbol{j}^T \\ \hline \boldsymbol{L}^T \end{array}\right] \\ &= \left[\boldsymbol{j}^{(1)} \mid \bar{\boldsymbol{L}}^{(1)}\right] \boldsymbol{Q}^{(1)^T}\boldsymbol{D}\boldsymbol{Q}^{(1)^T} \left[\begin{array}{c} \boldsymbol{j}^{(1)^T} \\ \hline \bar{\boldsymbol{L}}^{(1)^T} \end{array}\right] \end{aligned} \tag{A.3.140}$$

where the matrix $\boldsymbol{Q}^{(1)^T}\boldsymbol{D}\boldsymbol{Q}^{(1)}$ is no longer diagonal. Instead, it has the form

$$\boldsymbol{Q}^{(1)^T}\boldsymbol{D}\boldsymbol{Q}^{(1)} = \left[\begin{array}{c|c|c|c|c} g_1 & e_1 & 0 & & 0 \\ \hline e_1 & h_1 & 0 & & 0 \\ \hline 0 & 0 & 1 & & \\ \hline & & & \ddots & \\ \hline 0 & 0 & & & 1 \end{array}\right] \tag{A.3.141}$$

where straightforward calculations give

$$g_1 = \alpha c_1^2 + s_1^2,\ h_1 = \alpha s_1^2 + c_1^2,\ e_1 = (1-\alpha)c_1 s_1 \tag{A.3.142}$$

The matrix (A.3.141) is then expressed in the form of $\boldsymbol{UDU}^T$ decomposition which is similar to Cholesky decomposition, but has an upper triangular matrix $\boldsymbol{U}$ and a diagonal matrix $\boldsymbol{D}$. Thus, we have

$$\boldsymbol{Q}^{(1)^T}\boldsymbol{D}\boldsymbol{Q}^{(1)} = \boldsymbol{U}^{(1)}\boldsymbol{D}^{(1)}\boldsymbol{U}^{(1)^T} \tag{A.3.143}$$

where

$$\boldsymbol{U}^{(1)} = \begin{bmatrix} 1 & \gamma_1 & 0 & & 0 \\ 0 & 1 & 0 & & 0 \\ 0 & 0 & 1 & & \\ & & & \ddots & \\ 0 & 0 & & & 1 \end{bmatrix} \qquad \boldsymbol{D}^{(1)} = \begin{bmatrix} \alpha_1 & 0 & 0 & & 0 \\ 0 & \beta_1 & 0 & & 0 \\ 0 & 0 & 1 & & \\ & & & \ddots & \\ 0 & 0 & & & 1 \end{bmatrix} \tag{A.3.144}$$

with coefficients α_1, β_1, and γ_1 defined as

$$\beta_1 = \alpha s_1^2 + c_1^2,\ \gamma_1 = \frac{(1-\alpha)c_1 s_1}{\beta_1} \qquad \alpha_1 = \alpha c_1^2 + s_1^2 - \gamma_1^2\beta_1 \tag{A.3.145}$$

which can be checked by a direct substitution using expressions (A.3.141) and (A.3.142). To complete the first step, denote

$$\left[\boldsymbol{j}^{(1)} \mid \overline{\boldsymbol{L}}^{(1)}\right]\boldsymbol{U}^{(1)} = \begin{bmatrix} 0 & \bar{l}_{11} & 0 & 0 & \cdots & 0 \\ * & * & * & 0 & \cdots & 0 \\ \vdots & \cdots & \cdots & & & \\ * & * & * & * & * & * \end{bmatrix} = \left[\boldsymbol{j}^{(1)} \mid \boldsymbol{L}^{(1)}\right] \tag{A.3.146}$$

where multiplication of the matrix $\left[\boldsymbol{j}^{(1)} \mid \boldsymbol{L}^{(1)}\right]$ by the matrix $\boldsymbol{U}^{(1)}$ having the form (A.3.144) does not change the column $\boldsymbol{j}^{(1)}$, changing only the first column of matrix $\overline{\boldsymbol{L}}^{(1)}$, however, leaving its first entry $\bar{l}_{11}$ unchanged. Thus, the first step of the transformations is completed with the representation

$$\boldsymbol{J}^T\boldsymbol{J} + \alpha\boldsymbol{j}\boldsymbol{j}^T = \left[\boldsymbol{j}^{(1)} \mid \boldsymbol{L}^{(1)}\right]\boldsymbol{D}^{(1)}\begin{bmatrix} \boldsymbol{j}^{(1)} \\ \boldsymbol{L}^{(1)} \end{bmatrix} \tag{A.3.147}$$

where the matrix $\boldsymbol{D}^{(1)}$ is diagonal and the matrix $\boldsymbol{L}^{(1)}$ has a pattern shown in the expression (A.3.146). The computational complexity of this step is proportional to m.

The second step is aimed to zero the second entry of the column $\boldsymbol{j}^{(1)}$. To accomplish this goal we construct the orthonormal matrix

$$\boldsymbol{Q}^{(2)} = \begin{bmatrix} c_2 & 0 & -s_2 & & 0 \\ 0 & 1 & 0 & & 0 \\ s_2 & 0 & c_2 & & \\ & & & \ddots & \\ 0 & 0 & & & 1 \end{bmatrix} \tag{A.3.148}$$

and the upper triangular and diagonal matrices

$$\boldsymbol{U}^{(2)} = \begin{bmatrix} 1 & 0 & \gamma_2 & & 0 \\ 0 & 1 & 0 & & 0 \\ 0 & 0 & 1 & & \\ & & & \ddots & \\ 0 & 0 & & & 1 \end{bmatrix} \qquad \boldsymbol{D}^{(2)} = \begin{bmatrix} \alpha_2 & 0 & 0 & & 0 \\ 0 & \beta_1 & 0 & & 0 \\ 0 & 0 & \beta_2 & & \\ & & & \ddots & \\ 0 & 0 & & & 1 \end{bmatrix} \tag{A.3.149}$$

resulting in a representation similar to (A.3.147),

$$\boldsymbol{J}^T\boldsymbol{J} + \alpha \boldsymbol{j}\boldsymbol{j}^T = \left[\boldsymbol{j}^{(2)} \;\middle|\; \boldsymbol{L}^{(2)}\right] \boldsymbol{D}^{(2)} \begin{bmatrix} \boldsymbol{j}^{(2)} \\ \hline \boldsymbol{L}^{(2)} \end{bmatrix} \tag{A.3.150}$$

where

$$\left[\boldsymbol{j}^{(2)} \;\middle|\; \boldsymbol{L}^{(2)}\right] = \left[\boldsymbol{j}^{(1)} \;\middle|\; \boldsymbol{L}^{(1)}\right] \boldsymbol{Q}^{(2)}\boldsymbol{U}^{(2)} = \begin{bmatrix} 0 & \bar{l}_{11} & 0 & 0 & \cdots & 0 \\ 0 & * & \bar{l}_{22} & 0 & \cdots & 0 \\ \vdots & \cdots & \cdots & & & \\ * & * & * & * & * & * \end{bmatrix} \tag{A.3.151}$$

Multiplication of the matrix $\left[\boldsymbol{j}^{(1)} \;\middle|\; \boldsymbol{L}^{(1)}\right]$ by the matrix $\boldsymbol{Q}^{(2)}$ does not spoil the first zero entry in vector $\boldsymbol{j}^{(1)}$, while setting zero in the second entry of $\boldsymbol{j}^{(2)}$. Also, multiplication by the matrix $\boldsymbol{U}^{(2)}$ does not spoil the two zero entries of column $\boldsymbol{j}^{(2)}$. This follows from the structure of matrices (A.3.148) and (A.3.149).

The updating process is repeated resulting in the representation

$$\boldsymbol{J}^T\boldsymbol{J} + \alpha \boldsymbol{j}\boldsymbol{j}^T = \left[\boldsymbol{0}_m \;\middle|\; \boldsymbol{L}^{(m)}\right] \boldsymbol{D}^{(m)} \begin{bmatrix} \boldsymbol{0}_m^T \\ \hline \boldsymbol{L}^{(m)} \end{bmatrix} \tag{A.3.152}$$

at the end of the mth step. Each step takes the number of calculations proportional to m. Thus, the total computational complexity of the rank one update is estimated as m^2. The matrix $\boldsymbol{D}^{(m)}$ is $(m+1) \times (m+1)$ diagonal,

$$\boldsymbol{D}^{(m)} = \begin{bmatrix} \alpha_m & \boldsymbol{0}_m^T \\ \boldsymbol{0}_m & \overline{\boldsymbol{D}} \end{bmatrix} \qquad \overline{\boldsymbol{D}} = \begin{bmatrix} \beta_1 & & \\ & \ddots & \\ & & \beta_m \end{bmatrix} \tag{A.3.153}$$

where the values $\beta_i > 0$ if and only if the matrix $\boldsymbol{J}^T\boldsymbol{J} + \alpha\boldsymbol{j}\boldsymbol{j}^T$ is positive definite. Obviously this is true if $\alpha = 1$. However, it can lose positive definiteness if $\alpha = -1$. A positive diagonal $\overline{\boldsymbol{D}}$ allows us to judge whether $\boldsymbol{J}^T\boldsymbol{J} + \alpha\boldsymbol{j}\boldsymbol{j}^T$ is positive definite. Finally, if it is true, the algorithm completes with the answer

$$\overline{\boldsymbol{L}} = \boldsymbol{L}^{(m)}\boldsymbol{D}^{1/2} \tag{A.3.154}$$

Otherwise, if one of β_i is zero (negative values are impossible) we stop with the conclusion that the system (A.3.125) became underdetermined after exclusion of one equation.

To complete calculation of the new solution (A.3.127) note that

$$\overline{\boldsymbol{x}}^* = (\overline{\boldsymbol{LL}}^T)^{-1}(\boldsymbol{J}^T\boldsymbol{b} + \alpha \boldsymbol{j}b) \tag{A.3.155}$$

The explicit matrix inversion in (A.3.155) is not computed in practice, instead it is replaced by the forward and backward substitution solutions described in equations (A.3.66) and (A.3.67). The straightforward calculations of (A.3.127) would take nm^2 arithmetic operations to compute the matrix $\boldsymbol{J}^T\boldsymbol{J}$ and another m^3 operations to compute the Cholesky decomposition. The rank one update takes m^2 operations. Forward and backward solutions take another m^2 operations. Therefore, the rank one update saves computations if the number m is large.

A.4 LINEARIZATION

Observations are often related by nonlinear functions of unknown parameters. The adjustment algorithm uses a linear functional relationship between the observations and the parameters and uses iterations to account for the nonlinearity. To perform an adjustment, one must therefore linearize these relationships. Expanding the functions in a Taylor series and retaining only the linear terms accomplishes this. Consider the nonlinear function

$$y = f(x) \tag{A.4.1}$$

which has one variable x. The Taylor series expansion of this function is

$$y = f(x_o) + \left.\frac{\partial y}{\partial x}\right|_{x_0} dx + \frac{1}{2!}\left.\frac{\partial^2 y}{\partial x^2}\right|_{x_0} dx^2 + \cdots \tag{A.4.2}$$

The linear portion is given by the first two terms

$$\overline{y} = f(x_o) + \left.\frac{\partial y}{\partial x}\right|_{x_0} dx \tag{A.4.3}$$

The derivative is evaluated at the point of expansion x_0. At the point of expansion, the nonlinear function is tangent to the linearized function. The functions separate by

$$\varepsilon = y - \overline{y} \tag{A.4.4}$$

as x departs from the expansion point x_0. The linear form (A.4.3) is a sufficiently accurate approximation of the nonlinear relation (A.4.1) only in the vicinity of the point of expansion.

Typically, an application contains more than one parameter. The above concept of linearization, therefore, is generalized to higher dimensions. The expansion of a two-variable function

$$z = f(x, y) \tag{A.4.5}$$

is

$$z = f(x_0, y_0) + \left.\frac{\partial z}{\partial x}\right|_{x_0, y_0} dx + \left.\frac{\partial z}{\partial y}\right|_{x_0, y_0} dy + \cdots \tag{A.4.6}$$

The point of expansion is $P(x = x_0,\ y = y_0)$. The linearized form

$$\bar{z} = f(x_0, y_0) + \left.\frac{\partial z}{\partial x}\right|_{x_0, y_0} dx + \left.\frac{\partial z}{\partial y}\right|_{x_0, y_0} dy \tag{A.4.7}$$

represents the tangent plane on the surface (A.4.5) at the expansion point. A generalization for the expansion of multivariable functions is readily seen. If n functions are related to u variables as in

$$\mathbf{y} = \mathbf{f}(\mathbf{x}) = \begin{bmatrix} f_1(\mathbf{x}) \\ f_2(\mathbf{x}) \\ \vdots \\ f_n(\mathbf{x}) \end{bmatrix} = \begin{bmatrix} f_1(x_1, x_2, \cdots, x_u) \\ f_2(x_1, x_2, \cdots, x_u) \\ \vdots \\ f_n(x_1, x_2, \cdots, x_u) \end{bmatrix} \tag{A.4.8}$$

the linearized form is

$$\bar{\mathbf{y}} = \mathbf{f}(\mathbf{x}_0) + \left.\frac{\partial \mathbf{f}}{\partial \mathbf{x}}\right|_{x_0} d\mathbf{x} \tag{A.4.9}$$

where

$$\frac{\partial \mathbf{f}}{\partial \mathbf{x}} = {}_n\mathbf{G}_u = \begin{bmatrix} \frac{\partial f_1}{\partial x_1} & \frac{\partial f_1}{\partial x_2} & \cdots & \frac{\partial f_1}{\partial x_u} \\ \frac{\partial f_2}{\partial x_1} & \frac{\partial f_2}{\partial x_2} & \cdots & \frac{\partial f_2}{\partial x_u} \\ \vdots & \vdots & \cdots & \vdots \\ \frac{\partial f_n}{\partial x_1} & \frac{\partial f_n}{\partial x_2} & \cdots & \frac{\partial f_n}{\partial x_u} \end{bmatrix} \tag{A.4.10}$$

The point of expansion is $P(\mathbf{x} = \mathbf{x}_0)$. Every component of $\mathbf{y}$ is differentiated with respect to every variable. Thus, the matrix $\mathbf{G}$ has as many columns as there are parameters, and as many rows as there are components in $\mathbf{y}$. The components of $\mathbf{f}(\mathbf{x}_0)$ are equal to the respective functions evaluated at $\mathbf{x}_0$.

Although linearization is a general mathematical concept, it is needed in least-squares adjustment in surveying, geodesy, and navigation because the mathematical models are typically nonlinear. Apart from linear regression, the mathematical relationships that relate observations and parameters are typically nonlinear in engineering and science. Well-known exceptions are leveling networks and GNSS vector network as discussed in other parts of this book. In these rare cases the linearization and the concept of point of expansion are not needed. In nonlinear mathematical models, however, the point of expansion is defined by assumed values for the parameters. During the iterations the point of expansion moves closer to the true value of the parameters. The adjustment is said to have converged if additional iterations do not change the results. It follows that under normal circumstances the vector $d\mathbf{x}$ in (A.4.9), representing the difference between the adjusted and approximate parameters, is very small during the last iteration. Therefore, any higher order terms truncated in (A.4.9) are negligible, indeed. In terms of notation $d\boldsymbol{x}$ is identical to the $\boldsymbol{x}$ in (A.3.32). The details of adjustments are discussed in Chapters 2 and 3.

A.5 STATISTICS

We briefly summarize statistical aspects in this section that are sufficient for understanding routine applications of statistics in adjustments, in particular the derivation of the distribution of $\boldsymbol{v}^T\boldsymbol{P}\boldsymbol{v}$ and the basic test for accessing the validity of an adjustment. One-dimensional distributions, hypothesis testing, distributions of simple functions of random variables, multivariate normal distribution, and variance-covariance propagation are addressed. The reader is referred for in-depth treatment of statistics to the standard statistical literature.

To start with, let as review some basic terminology: An *observation*, or a statistical event, is the outcome of a statistical experiment such as throwing a dice or measuring an angle or a distance. A *random variable* is the outcome of an event. The random variable is denoted by a tilde. Thus, $\tilde{x}$ is a random variable and $\tilde{\boldsymbol{x}}$ is a vector of random variables. However, we will often not use the tilde to simplify the notation when it is unambiguous as to which symbol represents the random variable. The *population* is the totality of all events. It includes all possible values that the random variable can have. The population is described by a finite set of parameters, called the population parameters. The normal distribution, e.g., describes such a population and is completely specified by the mean and the variance. A *sample* is a subset of the population. For example, if the same distance is measured 10 times, then these 10 measurements are a sample of all the possible measurements. A *statistics* represents an estimate of the population parameters or functions of these parameters. It is computed from a sample. For example, the 10 measurements of the same distance can be used to estimate the mean and the variance of the normal distribution. *Probability* is related to the frequency of occurrence of a specific event. Each value of the random variable has an associated probability. The *probability density function* relates the probability to the possible values of the random variable.

A.5.1 One-Dimensional Distributions

The chi-square, normal, t, and F distributions are listed. These are, of course, the most basic distribution and are merely listed for the convenience of the student. In addition, we introduce the mean and the variance.

Probability Density and Accumulative Probability: If $f(x)$ denotes the probability density function, then

$$P(a \le \tilde{x} \le b) = \int_a^b f(x)\,dx \tag{A.5.1}$$

is the probability that the random variable $\tilde{x}$ assumes a value in the interval $[a, b]$. For $f(x)$ to be a probability function of the random variable $\tilde{x}$, it has to fulfill certain conditions. First, $f(x)$ must be a nonnegative function, because there is always an outcome of an experiment, i.e., the observation can be positive, negative, or even zero. Second, the probability that a sample (observation) is one of all possible outcomes should be 1. Thus the density function $f(x)$ must fulfill the following conditions:

$$f(x) \geq 0 \tag{A.5.2}$$

$$\int_{-\infty}^{\infty} f(x)\, dx = 1 \tag{A.5.3}$$

The integration is taken over the whole range (population) of the random variable. Conditions (A.5.2)and (A.5.3) imply that the density function is zero at minus infinity and plus infinity. The probability

$$P(\tilde{x} \leq x) = F(x) = \int_{-\infty}^{x} f(t)\, dt \tag{A.5.4}$$

is called the cumulative distribution function. It is a nondecreasing function because of condition (A.5.2).

Mean: The mean, also called the expected value of a continuously distributed random variable, is defined as

$$\mu_x = E(\tilde{x}) = \int_{-\infty}^{\infty} x f(x)\, dx \tag{A.5.5}$$

The mean is a function of the density function of the random variable. The integration is extended over the whole population. Equation (A.5.5) is the analogy to the weighted mean in the case of discrete distributions.

Variance: The variance is defined by

$$\sigma_{\tilde{x}}^2 = E(\tilde{x} - \mu_x)^2 = \int_{-\infty}^{\infty} (x - \mu_x)^2 f(x)\, dx \tag{A.5.6}$$

The variance measures the spread of the probability density in the sense that it gives the expected value of the squared deviations from the mean. A small variance therefore indicates that most of the probability density is located around the mean.

Chi-Square Distribution: The chi-square density function is given by

$$f(x) = \begin{cases} \dfrac{1}{2^{r/2}\,\Gamma(r/2)} x^{(r/2)-1}\, e^{-x/2} & x > 0 \\ 0 & \text{elsewhere} \end{cases} \tag{A.5.7}$$

The symbol r denotes a positive integer and is called the degree of freedom. The mean, also called the expected value, equals r, and the variance equals $2r$. The degree of freedom is sufficient to describe completely the chi-square distribution. The symbol Γ denotes the well-known gamma function, which is dealt with in books on advanced calculus and can be written as

$$\Gamma(g) = (g - 1)! \tag{A.5.8}$$

$$\Gamma\left(g + \frac{1}{2}\right) = \frac{\sqrt{\pi}\, \Gamma(2g)}{2^{2g}\, \Gamma(g)} \tag{A.5.9}$$

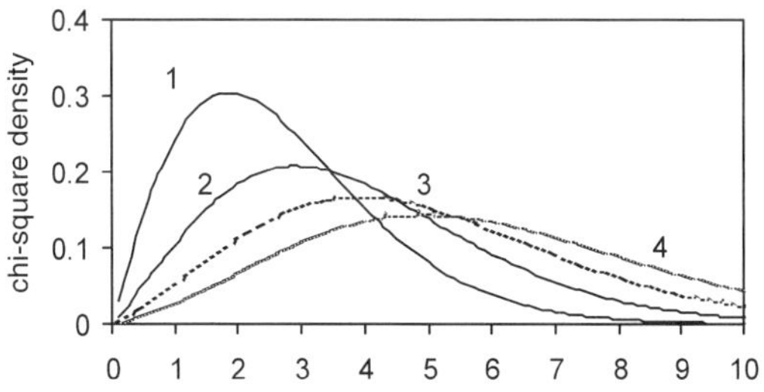

Figure A.5.1 Chi-square distribution of various degrees of freedom.

for positive integer g. Examples of the chi-square distribution for small degrees of freedom are given in Figure A.5.1. The probability that the random variable $\tilde{x}$ is less than w_α is

$$P(\tilde{x} < w_\alpha) = \int_0^{w_\alpha} f(x)\,dx = 1 - \alpha \tag{A.5.10}$$

Regarding notation, equation (A.5.10) implies that to the right of w_α there is the probability α; the integration from w_α to infinity equals α. If the random variable $\tilde{x}$ has a chi-square distribution with r degrees of freedom, then we use the notation $\tilde{x} \sim \chi_r^2$.

The distribution (A.5.7) is more precisely called the central chi-square distribution. The noncentral chi-square is a generalization of this distribution. The density function does not have a simple closed form; it consists of an infinite sum of terms. If $\tilde{x}$ has a noncentral chi-square distribution, this is expressed by $\tilde{x} \sim \chi_{r,\lambda}^2$ where λ denotes the noncentrality parameter. The mean is

$$E(\tilde{x}) = r + \lambda \tag{A.5.11}$$

as opposed to just r for the central chi-square distribution.

Normal Distribution: The density function of the normal distribution is

$$f(x) = \frac{1}{\sigma\sqrt{2\pi}} e^{-(x-\mu)^2/2\sigma^2} \qquad -\infty < x < \infty \tag{A.5.12}$$

where μ and σ^2 denote the mean and the variance. The notation $\tilde{x} \sim n(\mu, \sigma^2)$ is usually used. The two parameters μ and σ completely describe the normal distribution. See Figure A.5.2. The normal distribution has the following characteristics:

1. The distribution is symmetric about the mean.
2. The maximum density is at the mean.
3. For small variances, the maximum density is larger and the slopes are steeper than in the case of large variances.
4. The inflection points are at $x = \mu \pm \sigma$.

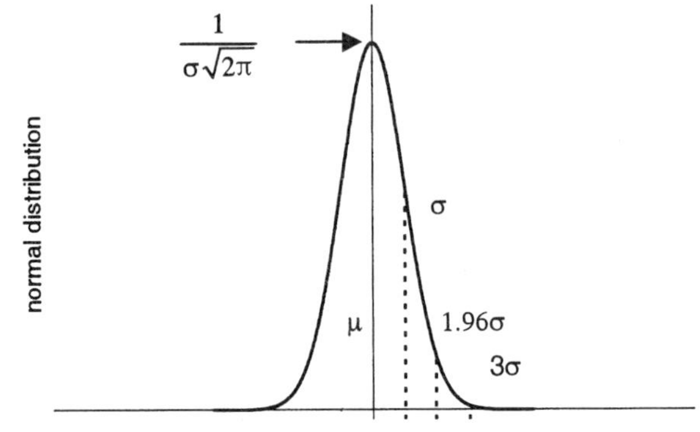

Figure A.5.2 Normal density function.

If $\tilde{x} \sim n(\mu, \sigma^2)$, then the transformed variable

$$\tilde{w} = \frac{\tilde{x} - \mu}{\sigma} \sim n(0, 1) \tag{A.5.13}$$

has a normal distribution with zero mean and unit variance. The random variable $\tilde{w}$ is said to have a standardized normal distribution. The density function for $\tilde{w}$ is

$$f(w) = \frac{1}{\sqrt{2\pi}} e^{-w^2/2} \qquad -\infty < w < \infty \tag{A.5.14}$$

The probability that the random variable $\tilde{x}$ is less than w_α is

$$P(\tilde{x} < w_\alpha) = \int_0^{w_\alpha} f(w)\, dw \tag{A.5.15}$$

Table A.5.1 lists selected values that are frequently quoted. For a normal distribution, in about 68% of all cases the observations fall within one standard deviation from the mean, and only every 370th observation deviates from the mean by more than 3σ. Therefore, the 3σ value is sometimes taken as the limit to what is regarded as random error. Any larger deviation from the mean is usually considered a blunder. Statistically, large errors cannot be avoided, but their occurrence is unlikely. The 3σ criterion is not necessarily applicable in least-squares adjustments because the adjusted random variables are multivariate distributed and are correlated.

TABLE A.5.1 Selected Values from the Normal Distribution

x	σ	2σ	3σ	0.674σ	1.645σ	1.960σ
$N(x) - N(-x)$	0.6827	0.9544	0.9973	0.5	0.90	0.95

***t* Distribution**: Assume that $\tilde{w} \sim n(0,\ 1)$ and $\tilde{v} \sim \chi_r^2$ are two stochastically independent random variables with unit normal and chi-square distribution, respectively. The random variable

$$\tilde{t} = \frac{\tilde{w}}{\sqrt{\tilde{v}/r}} \tag{A.5.16}$$

has a t distribution with r degrees of freedom. The distribution function is

$$f(t_r) = \frac{\Gamma\ [(r+1)/2]}{\sqrt{\pi r}\ \Gamma(r/2)} \left[1 + \frac{t^2}{r}\right]^{-(r+1)/2} \qquad -\infty < t < \infty \tag{A.5.17}$$

The density function (A.5.17) is symmetric with respect to $t = 0$. See Figure A.5.3. Furthermore, if $r = \infty$ then the t distribution is identical to the standardized normal distribution

$$t_\infty = n(0,\ 1) \tag{A.5.18}$$

The density in the vicinity of the mean (zero) is smaller than for the unit normal distribution, whereas the reverse is true at the extremities of the distribution. The t distribution converges rapidly toward the normal distribution. If the random variable $\tilde{w} \sim n(\delta,\ 1)$ is normal distributed with unit variance but with a nonzero mean, then the function (A.5.16) has a noncentral t distribution with r degrees of freedom and a noncentrality parameter δ.

***F* Distribution:** Consider two stochastically independent random variables, $\tilde{u} \sim \chi_{r_1}^2$ and $\tilde{v} \sim \chi_{r_2}^2$, distributed as chi-square with r_1 and r_2 degrees of freedom, respectively. The random variable

$$\tilde{F} = \frac{\tilde{u}/r_1}{\tilde{v}/r_2} \tag{A.5.19}$$

has the density function

$$f(F_{r_1,r_2}) = \frac{\Gamma\ [(r_1+r_2)/2]\ (r_1/r_2)^{r_1/2}}{\Gamma\ (r_1/2)\ \Gamma(r_2/2)} \frac{F^{(r_1/2)-1}}{(1+r_1F/r_2)^{(r_1+r_2)/2}} \qquad 0 < F < \infty \tag{A.5.20}$$

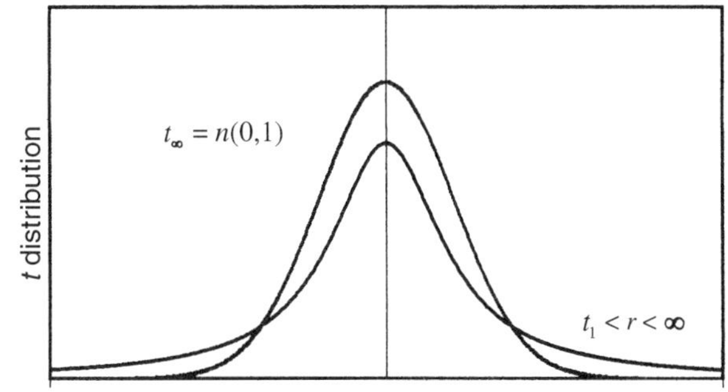

Figure A.5.3 Probability density function of the *t* distribution.

This is the F distribution with r_1 and r_2 degrees of freedom. The mean, or the expected value, is

$$E(F_{r_1,r_2}) = \frac{r_2}{r_2 - 2} \tag{A.5.21}$$

for $r_2 > 2$. Care should always be taken to identify the sequence of degrees of freedom properly since the density function is not symmetric in these variables. See Figure A.5.4. The following relationship

$$F_{r_1,r_2,\alpha} = \frac{1}{F_{r_2,r_1,1-\alpha}} \tag{A.5.22}$$

holds. The F distribution is related to the chi-square and the t distributions as follows:

$$\frac{\chi_r^2}{r} \sim F_{r,\infty} \tag{A.5.23}$$

$$t_r^2 \sim F_{1,r} \tag{A.5.24}$$

If $\tilde{u} \sim \chi^2_{r_1,\lambda}$ has a noncentral chi-square distribution with r_1 degrees of freedom and a noncentrality parameter λ, then the function F in (A.5.19) has a noncentral F distribution with r_1 and r_2 degrees of freedom and noncentrality parameter λ. The mean for the noncentral distribution is

$$E(F_{r_1,r_2,\lambda}) = \frac{r_2}{r_2 - 2}\left(1 + \frac{\lambda}{r_1}\right) \tag{A.5.25}$$

A.5.2 Distribution of Simple Functions

There are several functions of random variables that are useful in least-squares estimation. Assume that $(\tilde{x}_1, \tilde{x}_2, \cdots, \tilde{x}_n)$ are n stochastically independent variables, each having a normal distribution, with different means μ_i and variances σ_i^2. Then the linear function

$$\tilde{y} = k_1\tilde{x}_1 + k_2\tilde{x}_2 + \cdots + k_n\tilde{x}_n \tag{A.5.26}$$

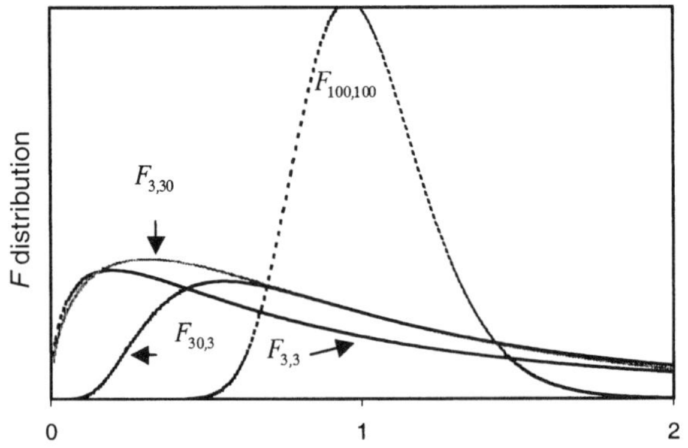

Figure A.5.4 ***F* distribution.**

is distributed as

$$\tilde{y} \sim n\left(\sum_i^n k_i\mu_i, \sum_i^n k_i^2\sigma_i^2\right) \tag{A.5.27}$$

If the random variable $\tilde{w}$ has a standardized normal distribution $\tilde{w} \sim (0, 1)$, then the square of the standardized normal distribution

$$\tilde{v} = \tilde{w}^2 \sim \chi_1^2 \tag{A.5.28}$$

has a chi-square distribution with one degree of freedom.

Assume that $(\tilde{x}_1, \tilde{x}_2, \ldots, \tilde{x}_n)$ are n stochastically independent random variables, each having a chi-square distribution with differing r_i degrees of freedom. Then the random variable

$$\tilde{y} = \tilde{x}_1 + \tilde{x}_2 + \ldots + \tilde{x}_n \tag{A.5.29}$$

is distributed

$$\tilde{y} \sim \chi^2_{\sum r_i} \tag{A.5.30}$$

The degree of freedom equals the sum of the individual degrees of freedom.

Assume $(\tilde{x}_1, \tilde{x}_2, \ldots, \tilde{x}_n)$ are n stochastically independent random variables, each having a normal distribution. The means are nonzero. Then

$$\tilde{y} \sim \sum \tilde{w}^2 = \sum^n \left(\frac{\tilde{x}_i - \mu_i}{\sigma_i}\right)^2 \sim \chi_n^2 \tag{A.5.31}$$

Assume that $(\tilde{x}_1, \tilde{x}_2, \ldots, \tilde{x}_n)$ are n stochastically independent normal random variables with different means μ_i and variances σ_i^2. Then the sum of squares

$$\tilde{y} = \sum \tilde{x}_i^2 \sim \chi^2_{n,\lambda} \tag{A.5.32}$$

has a noncentral chi-square distribution. The degree of freedom is n and the noncentrality parameter is

$$\lambda = \sum \frac{\mu_i^2}{\sigma_i^2} \tag{A.5.33}$$

A.5.3 Hypothesis Tests

A hypothesis is a statement about the parameters of a distribution. A test of a hypothesis is a rule that, based on the sample values, leads to a decision to accept or reject the null hypothesis. A test statistic is computed from the sample values (the observations) and from the specifications of the null hypothesis. If the test statistic falls within a critical region, the null hypothesis is rejected. For example, $\tilde{\boldsymbol{v}}^T\boldsymbol{P}\tilde{\boldsymbol{v}}$ is a test statistic having a chi-square distribution. The computed test statistic is $\hat{\boldsymbol{v}}^T\boldsymbol{P}\hat{\boldsymbol{v}}$. The specification of the zero hypothesis could be that the a posteriori and a priori variance of unit weight are the same.

If the null hypothesis H_0 is true, the computed value may fall inside the critical region because the sample statistic is computed from sample values (observations). There is a probability α that this can happen. One speaks of a type-I error if the hypothesis H_0 is rejected although it is true; the probability of a type-I error is α, which, incidentally, is also the significance level of the test. However, there is a probability that the sample statistics falls in the critical region when H_0 is false (and hence H_1 is true). That probability is denoted by $1 - \beta$ and represents the area under the density function $f(t|H_1)$ from t_α to ∞ in Figure A.5.5. If the alternative hypothesis H_1 is true and the sample statistic does not fall in the critical region, one would mistakenly accept H_0 and commit a type-II error. The probability of committing a type-II error is β.

Figure A.5.5 displays the probability density functions of the test statistics under the specifications of the null hypothesis H_0 and the alternative hypothesis H_1. The figure also shows the critical region for which the null hypothesis is rejected, and the alternative hypothesis is accepted if the computed sample statistics t falls in that region. Thus, reject H_0 if

$$t > t_\alpha \tag{A.5.34}$$

The shape and location of the density function of the test statistics under the alternative hypothesis depend on the specifications of the alternative hypothesis. Thus, the probability of a type-II error, β, depends on the specifications of H_1. A desirable approach in statistical testing would be to minimize the probability of both types of errors. However, this is not practical, because all distributions of the alternative hypotheses, which, in general, are of the noncentral type, would have to be computed. Figure A.5.5 shows that the probability β increases as α decreases. A common procedure is to fix the probability of a type-I error to, say, $\alpha = 0.05$, and not compute β.

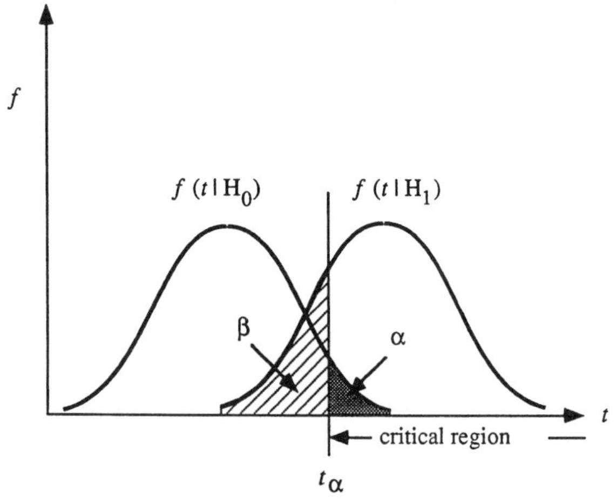

Figure A.5.5 Example of probability distributions of test statistics and critical region.

The rule (A.5.34) is a one-tail test in the upper end of the distribution. Depending on the situation, it might be desirable to employ a two-tail test. In that case the null hypothesis is rejected if

$$|t| > t_{\alpha/2} \tag{A.5.35}$$

and the distribution under H_0 is symmetric. It is rejected if

$$t > t_{\alpha/2} \tag{A.5.36}$$

$$t < t_{1-\alpha/2} \tag{A.5.37}$$

and the distribution is not symmetric. The critical regions are at both tails of the distribution, with each tail covering a probability area of $\alpha/2$.

However, much effort has gone into research as to how the magnitude of β can be controlled (Baarda, 1968). After all, committing a type-II error implies accepting the null hypothesis even though the alternative hypothesis is true. For example, it could mean that it has been concluded that no deformation took place even though actual deformations occurred. Such an error could be costly in many respects. In Section 2.8.2 some consideration is given to the type-II error in regards to blunder detection and internal and external reliability, again based on Baarda's work. Section 6.5.3 considers type-II errors in regard to ambiguity fixing.

The goodness-of-fit test is a simple and useful example of statistical testing. Assume we wish to test a series of observations to determine whether they come from a certain population with a specified distribution. We subdivide the observation series into n bins. Let n_i denote the number of observations in bin i. The subdivision should be such that $n_i \geq 5$. Compute for each bin the expected number d_i of observations based on the hypothetical distribution. It can be shown that

$$\chi^2 = \sum_{i=1}^{n} \frac{(n_i - d_i)^2}{d_i} \tag{A.5.38}$$

is distributed approximately as χ^2_{n-1}. The zero hypothesis states that the sample is from the specified distribution. Reject H_0 at a $100\alpha\%$ significance level if

$$\chi^2 > \chi^2_{n-1,\,\alpha} \tag{A.5.39}$$

This test could be used to verify that normalized residuals belong to $n(0,\ 1)$.

A.5.4 Multivariate Distributions

Any function $f(x_1,\ x_2, \ldots, x_n)$ of n continuous variables $\tilde{x}_i$ can be a joint multivariate density function provided that

$$f(x_1,\ x_2, \ldots,\ x_n) \geq 0 \tag{A.5.40}$$

$$\int_{-\infty}^{\infty} \cdots \int_{-\infty}^{\infty} f(x_1,\ x_2, \ldots,\ x_n)\, dx_1\ \cdots\ dx_n = 1 \tag{A.5.41}$$

It follows as a natural extension from (A.5.4) that

$$P(\tilde{x}_1 < a_1, \ldots, \tilde{x}_n < a_n) = \int_{-\infty}^{a_1} \cdots \int_{-\infty}^{a_n} f(x_1, x_2, \ldots, x_n)\, dx_1 \cdots dx_n \quad \text{(A.5.42)}$$

The marginal density of a subset of random variables $(x_1, x_2, \ldots, x_p)$ is

$$g(x_1, x_2, \ldots, x_p) = \int_{-\infty}^{\infty} \cdots \int_{-\infty}^{\infty} f(x_1, x_2, \ldots, x_n)\, dx_{p+1}\, dx_{p+2} \cdots dx_n \quad \text{(A.5.43)}$$

Stochastic Independence: The concept of stochastic independence is required when dealing with multivariate distributions. Two sets of random variables, $(\tilde{x}_1, \ldots, \tilde{x}_p)$ and $(\tilde{x}_{p+1}, \ldots, \tilde{x}_n)$, are stochastically independent if the joint density function can be written as a product of the two respective marginal density functions, e.g.,

$$f(x_1, x_2, \ldots, x_n) = g_1(x_1, x_2, \ldots, x_p)\, g_2(x_{p+1}, x_{p+2}, \ldots, x_n) \quad \text{(A.5.44)}$$

Vector of Means: The expected value for the individual parameter x_i is

$$\mu_{x_i} = E(\tilde{x}_i) = \int_{-\infty}^{\infty} \cdots \int_{-\infty}^{\infty} x_i f(x_1, x_2, \ldots, x_n)\, dx_1\, dx_2 \cdots dx_n \quad \text{(A.5.45)}$$

In vector notation the expected values of all parameters are

$$E(\tilde{\mathbf{x}}) = \left[E(\tilde{x}_1) \quad \cdots \quad E(\tilde{x}_n)\right]^T \quad \text{(A.5.46)}$$

Variance: The variance of an individual parameter is given by

$$\sigma_{x_i}^2 = E(\tilde{x}_i - \mu_{x_i})^2 = \int_{-\infty}^{\infty} \cdots \int_{-\infty}^{\infty} (x_i - \mu_{x_i})^2 f(x_1, x_2, \ldots, x_n)\, dx_1 \cdots dx_n \quad \text{(A.5.47)}$$

Covariance: For multivariate distributions, another quantity called the covariance becomes important. The covariance describes the statistical relationship between two random variables. The covariance is

$$\begin{aligned} \sigma_{x_i, x_j} &= E[(x_i - \mu_{x_i})(x_j - \mu_{x_j})] \\ &= \int_{-\infty}^{\infty} \cdots \int_{-\infty}^{\infty} (x_i - \mu_{x_i})(x_j - \mu_{x_j}) f(x_1, x_2, \ldots, x_n)\, dx_1 \cdots dx_n \end{aligned} \quad \text{(A.5.48)}$$

Whereas the variance is always larger than or equal to zero, the covariance can be negative, positive, or even zero.

Correlation Coefficients: The correlation coefficient of two random variables is defined as

$$\rho_{x_i, x_j} = \frac{E[(\tilde{x}_i - \mu_{x_i})(\tilde{x}_j - \mu_{x_j})]}{\sigma_{x_i}\sigma_{x_j}} = \frac{\sigma_{x_i, x_j}}{\sigma_{x_i}\sigma_{x_j}} \quad \text{(A.5.49)}$$

Therefore, the correlation coefficient equals the covariance divided by the respective standard deviations. An important property of the correlation coefficient is that

$$-1 \leq \rho_{x_i,x_j} \leq 1 \tag{A.5.50}$$

If two random variables are stochastically independent, then the covariance (and thus the correlation coefficient) is zero. By making use of (A.5.44) for the density function of stochastically independent random variables, we can write (A.5.48) as

$$\begin{aligned}\sigma_{x_i,x_j} &= \int_{-\infty}^{\infty}\int_{-\infty}^{\infty}(x_i-\mu_{x_i})\,(x_j-\mu_{x_j})\,g_i(x_i)g_j(x_j)\,dx_i\,dx_j \\ &= \int_{-\infty}^{\infty}(x_i-\mu_{x_i})g_i(x_i)\,dx_i\int_{-\infty}^{\infty}(x_j-\mu_{x_j})\,g_j(x_j)\,dx_j\end{aligned} \tag{A.5.51}$$

These integrals are zero because of the definition of the mean. The converse, i.e., zero correlation, implies stochastic independence is valid only for the multivariate normal distribution.

Variance-Covariance Matrix: Equations (A.5.45), (A.5.48), and (A.5.49) can be used to express the variances, covariances, and correlations for all components in the random vector $\tilde{\boldsymbol{x}}$. Consider the random vector

$$\tilde{\boldsymbol{x}} - \boldsymbol{\mu}_x = \begin{bmatrix}\tilde{x}_1 - \mu_{x_1} & \cdots & \tilde{x}_n - \mu_{x_n}\end{bmatrix}^T \tag{A.5.52}$$

then the $(n \times n)$ variance-covariance matrix $\boldsymbol{\Sigma}_x$ and correlation matrix $\mathbf{C}$ are

$$\boldsymbol{\Sigma}_x = E\left[(\tilde{\boldsymbol{x}}-\boldsymbol{\mu}_x)\,(\tilde{\boldsymbol{x}}-\boldsymbol{\mu}_x)^T\right] = E\left[\tilde{\boldsymbol{x}}\tilde{\boldsymbol{x}}^T - \boldsymbol{\mu}_x\boldsymbol{\mu}_x^T\right] \tag{A.5.53}$$

$$\boldsymbol{\Sigma}_x = \begin{bmatrix}\sigma_{x_1}^2 & \sigma_{x_1,x_2} & \cdots & \sigma_{x_1,x_n} \\ & & \cdots & \sigma_{x_2,x_n} \\ & & \ddots & \vdots \\ \text{sym} & & & \sigma_{x_n}^2\end{bmatrix} \quad \mathbf{C} = \begin{bmatrix}1 & \rho_{x_1,x_2} & \cdots & \rho_{x_1,x_n} \\ & & \cdots & \rho_{x_2,x_n} \\ & & \ddots & \vdots \\ \text{sym} & & & 1\end{bmatrix} \tag{A.5.54}$$

The variance-covariance matrix is symmetric because switching the subscripts in (A.5.48) only switches factors. The expectation operator E is applied to each matrix element. The variance-covariance matrix is often referred to simply as covariance matrix for the sake of brevity. The correlation matrix is also symmetric, the diagonal elements equal 1, and the off-diagonal elements range from -1 to $+1$.

A.5.5 Variance-Covariance Propagation

The purpose of variance-covariance propagation is to compute the variances and covariances of linear functions of random variables. Nonlinear functions must first be linearized. Variance-covariance propagation is applicable to single random variables or to vectors of random variables.

Propagation: Usually we are more interested in a linear function of the random variables than in the random variables themselves. Typical examples are the adjusted coordinates used to compute distances and angles. From the definition of the mean, it follows that for a constant c

$$E(c) = c\int_{-\infty}^{\infty} f(x)\,dx = c \tag{A.5.55}$$

and

$$E\,(c\tilde{x}) = cE(\tilde{x}) \tag{A.5.56}$$

The expected value (mean) of a constant equals the constant. Because the mean is a constant, it follows that

$$E\,[E(\tilde{x})] = \mu_x \tag{A.5.57}$$

Relations (A.5.55) and (A.5.56) also hold for multivariate density functions, as can be seen from (A.5.45). Let $\tilde{y} = \tilde{x}_1 + \tilde{x}_2$ be a linear function of random variables, then

$$\begin{aligned} E(\tilde{x}_1 + \tilde{x}_2) &= \int_{-\infty}^{\infty}\int_{-\infty}^{\infty} (x_1 + x_2) f(x_1, x_2)\,dx_1\,dx_2 \\ &= \int_{-\infty}^{\infty}\int_{-\infty}^{\infty} x_1 f(x_1, x_2)\,dx_1\,dx_2 + \int_{-\infty}^{\infty}\int_{-\infty}^{\infty} x_2 f(x_1, x_2)\,dx_1\,dx_2 \\ &= E\,(\tilde{x}_1) + E\,(\tilde{x}_2) \end{aligned} \tag{A.5.58}$$

Thus, the expected value of the sum of two random variables equals the sum of the individual expected values. By combining (A.5.55) and (A.5.58), we can compute the expected value of a general linear function of random variables. Thus, if the elements of the $n \times u$ matrix $\boldsymbol{A}$ and the vector $\boldsymbol{a}_0$ are constants and

$$\tilde{\boldsymbol{y}} = \boldsymbol{a}_0 + \boldsymbol{A}\tilde{\boldsymbol{x}} \tag{A.5.59}$$

then the expected value is

$$E\,(\tilde{\boldsymbol{y}}) = \boldsymbol{a}_0 + \boldsymbol{A}\,E(\tilde{\boldsymbol{x}}) \tag{A.5.60}$$

This is the law for propagating the mean. The law of variance-covariance propagation is as follows:

$$\begin{aligned} \boldsymbol{\Sigma}_y &\equiv E\left[(\tilde{\boldsymbol{y}} - \boldsymbol{\mu}_y)\,(\tilde{\boldsymbol{y}} - \boldsymbol{\mu}_y)^T\right] \\ &= E\left\{[\tilde{\boldsymbol{y}} - E(\tilde{\boldsymbol{y}})]\,[\tilde{\boldsymbol{y}} - E(\tilde{\boldsymbol{y}})]^T\right\} \\ &= E\left\{[\tilde{\boldsymbol{y}} - \boldsymbol{a}_0 - \boldsymbol{A}E(\tilde{\boldsymbol{x}})]\,[\tilde{\boldsymbol{y}} - \boldsymbol{a}_0 - \boldsymbol{A}E(\tilde{\boldsymbol{x}})]^T\right\} \\ &= E\left\{[\boldsymbol{A}\tilde{\boldsymbol{x}} - \boldsymbol{A}E(\tilde{\boldsymbol{x}})]\,[\boldsymbol{A}\tilde{\boldsymbol{x}} - \boldsymbol{A}E(\tilde{\boldsymbol{x}})]^T\right\} \\ &= \boldsymbol{A}E\left\{[\tilde{\boldsymbol{x}} - E(\tilde{\boldsymbol{x}})]\,[\tilde{\boldsymbol{x}} - E(\tilde{\boldsymbol{x}})]^T\right\}\boldsymbol{A}^T \\ &= \boldsymbol{A}\boldsymbol{\Sigma}_x\boldsymbol{A}^T \end{aligned} \tag{A.5.61}$$

The first line in expression (A.5.61) is the general expression for the variance-covariance matrix of the random variable $\tilde{\boldsymbol{y}}$ according to definition (A.5.53); $\boldsymbol{\mu}_y$ is the expected value of $\tilde{\boldsymbol{y}}$. The third line follows by substituting (A.5.60) for the expected value of $\tilde{\boldsymbol{y}}$. Equation (A.5.59) has been substituted in the third line for $\tilde{\boldsymbol{y}}$, and, finally, the $\boldsymbol{A}$ matrix has been factored out. Thus the variance-covariance matrix of the random variable $\tilde{\boldsymbol{y}}$ is obtained by pre- and postmultiplying the variance-covariance matrix of the original random variable $\tilde{\boldsymbol{x}}$ by the coefficient matrix $\boldsymbol{A}$ and its transpose. The constant term $\boldsymbol{a}_0$ cancels. This is the law of variance-covariance propagation for linear functions of random variables. The covariance matrix Σ_y is a full matrix in general.

A.5.6 Multivariate Normal Distribution

This section considers some details specifically for the multivariate normal distribution. The multivariate normal distribution is especially appealing because the marginal distributions derived from multivariate normal distributions are also normally distributed. An extensive treatment of this distribution is, once again, found in the standard statistical literature. In order to simplify notation, the tilde is not used to identify random variables. The random nature of variables can be readily deduced from the context.

Let $\boldsymbol{x}$ be a vector with n random components with a mean of

$$E(\boldsymbol{x}) = \boldsymbol{\mu} \tag{A.5.62}$$

and a covariance matrix of

$$E\,[(\boldsymbol{x} - \boldsymbol{\mu})(\boldsymbol{x} - \boldsymbol{\mu})^T] = {}_n\Sigma_n \tag{A.5.63}$$

If $\boldsymbol{x}$ has a multivariate normal distribution, then the multivariate density function is

$$f(x_1, \ldots, x_n) = \frac{1}{(2\pi)^{n/2}|\Sigma|^{1/2}} e^{-(\boldsymbol{x}-\boldsymbol{\mu})^{\mathrm{T}}\Sigma^{-1}(\boldsymbol{x}-\boldsymbol{\mu})/2} \tag{A.5.64}$$

The mean and the covariance matrix completely describes the multivariate normal distribution. The notation

$${}_n\boldsymbol{x}_1 \sim N_n\,({}_n\boldsymbol{\mu}_1,\ {}_n\Sigma_n) \tag{A.5.65}$$

is used. The dimension of the distribution is n.

In the following, some theorems on multivariate normal distributions are given without proofs. These theorems are useful in deriving the distribution of $\boldsymbol{v}^T\boldsymbol{P}\boldsymbol{v}$ and some of the basic statistical tests in least-squares adjustments. If $\boldsymbol{x}$ is multivariate normal

$$\boldsymbol{x} \sim N(\boldsymbol{\mu},\ \Sigma) \tag{A.5.66}$$

and

$$\boldsymbol{z} = {}_m\boldsymbol{D}_n\boldsymbol{x} \tag{A.5.67}$$

is a linear function of the random variable, where $\boldsymbol{D}$ is a $m \times n$ matrix of rank $m \le n$, then Theorem 1 states that

$$\boldsymbol{z} \sim N_m\,(\boldsymbol{D}\,\boldsymbol{\mu},\ \boldsymbol{D}\Sigma\boldsymbol{D}^T) \tag{A.5.68}$$

is a multivariate normal distribution of dimension m. The mean and variance of the random variable $\boldsymbol{z}$ follow from the laws for propagating the mean (A.5.60) and variance covariances (A.5.61).

If $\boldsymbol{x}$ is multivariate normal $\boldsymbol{x} \sim N(\boldsymbol{\mu}, \ \boldsymbol{\Sigma})$, then Theorem 2 states that the marginal distribution of any set of components of $\boldsymbol{x}$ is multivariate normal with means, ***variances***, and covariances obtained by taking the proper component of $\boldsymbol{\mu}$ and $\boldsymbol{\Sigma}$. For example, if

$$\boldsymbol{x} = \begin{bmatrix} \boldsymbol{x}_1 \\ \boldsymbol{x}_2 \end{bmatrix} \sim N\left(\begin{bmatrix} \boldsymbol{\mu}_1 \\ \boldsymbol{\mu}_2 \end{bmatrix}, \begin{bmatrix} \boldsymbol{\Sigma}_{11} & \boldsymbol{\Sigma}_{12} \\ \boldsymbol{\Sigma}_{21} & \boldsymbol{\Sigma}_{22} \end{bmatrix} \right) \tag{A.5.69}$$

then the marginal distribution of $\boldsymbol{x}_2$ is

$$\boldsymbol{x}_2 \sim N(\boldsymbol{\mu}_2, \ \boldsymbol{\Sigma}_{22}) \tag{A.5.70}$$

The same law holds, of course, if the set contains only one component, say x_i. The marginal distribution of x_i is then

$$\boldsymbol{x}_i \sim n\ (\mu_i, \ \Sigma_i^2) \tag{A.5.71}$$

If $\boldsymbol{x}$ is multivariate normal, Theorem 3 states that a necessary and sufficient condition that two subsets of the random variables are stochastically independent is that the covariances are zero. For example, if

$$\begin{bmatrix} \boldsymbol{x}_1 \\ \boldsymbol{x}_2 \end{bmatrix} \sim N\left(\begin{bmatrix} \boldsymbol{\mu}_1 \\ \boldsymbol{\mu}_2 \end{bmatrix}, \begin{bmatrix} \boldsymbol{\Sigma}_{11} & \boldsymbol{0} \\ \boldsymbol{0} & \boldsymbol{\Sigma}_{22} \end{bmatrix} \right) \tag{A.5.72}$$

then $\boldsymbol{x}_1$ and $\boldsymbol{x}_2$ are stochastically independent. If one set of normally distributed random variables is uncorrelated with the remaining variables, the two sets are independent. The proof of the above theorem follows from the fact that the density function can be written as a product of $f_1(\boldsymbol{x}_1)$ and $f_2(\boldsymbol{x}_2)$ because of the special form of the density function (A.5.64)

APPENDIX B

THE ELLIPSOID

The ellipsoid of rotation is a geometric structure for mathematical formulations and computations. For example, the observables of the 3D geodetic model refer to the ellipsoidal normal and the geodetic horizon, whereas the observables of the ellipsoidal model are the angle between geodesics and the length of the geodesic on the ellipsoidal surface. In the case of the conformal mapping model, the ellipsoidal surface is mapped conformally. Details of these mathematical models are given in Sections 4.5 and 4.6. Because the ellipsoid is important as a computational reference and as a means to express position coordinates, the ellipsoid and the related geometry are summarized here. Since only ellipsoids of rotation have been adopted in practical geodesy and triaxial ellipsoids have been limited to theoretical studies, we will use the term *ellipsoid* for reasons of brevity to mean ellipsoid of rotation. Such an ellipsoid is generated when rotating an ellipse around the semiminor axis.

The expressions for computing on the ellipsoidal surface and on the conformal mapping plane are deeply rooted in differential geometry. Working expressions typically utilize series expansions that are simplified by truncating insignificant terms (having specific applications in terms of position accuracy and area in mind). The algebraic work necessary to arrive at working expressions is considerable and not at all obvious to the novice. Prior to the introduction of electronic computers, there was a strong interest in producing computationally efficient expressions. The expressions have been extensively documented in the geodetic literature, although some of this literature is now old and is even out of print. Many of the derivations are documented in Leick (2002).

The mathematical literature offers plenty of excellent texts on differential geometry. Differential geometry of course deals in general terms with surfaces. While this section focuses on the ellipsoid, the universality of expressions will occasionally be emphasized. The reader is advised to consult the mathematical literature if a more precise and comprehensive exposition of differential geometry is desired than is offered in the "tailored" approach of this appendix.

B.1 GEODETIC LATITUDE, LONGITUDE, AND HEIGHT

A popular way to give positions in 3D space is by means of geodetic latitude, geodetic longitude, and geodetic height. To be sure, these quantities are often referred to as ellipsoidal latitude, ellipsoidal longitude, and ellipsoidal height. Regardless of what one calls them, it is important to realize that they refer to an ellipsoid and not to a sphere, and thus are conceptually and numerically different from spherical latitude, longitude, and height. Another popular way of giving positions in space is Cartesian coordinates. It follows that the geodetic and Cartesian coordinate triplets are mathematically related.

Figure B.1.1 shows an ellipse with semimajor axis a and semiminor axis b. In the (ξ, η) coordinate system, the equation of the ellipse has the familiar form

$$\frac{\xi^2}{a^2} + \frac{\eta^2}{b^2} = 1 \tag{B.1.1}$$

Two parameters are sufficient to define the ellipse. Often the semimajor axis a and the flattening f, or a and the eccentricity e, are used to define an ellipse. These auxiliary quantities are related by

$$f = \frac{a-b}{a} \tag{B.1.2}$$

$$e^2 = 2f - f^2 \tag{B.1.3}$$

The figure also shows the tangent to the ellipse at some point A. The normal to this tangent intersects the semiminor axis at point C. The symbol N is used to denote the segment $\overline{AC}$. The angle φ is the geodetic latitude and equals the angle between the normal and the semimajor axis. It follows readily that

$$\xi = N \cos \varphi \tag{B.1.4}$$

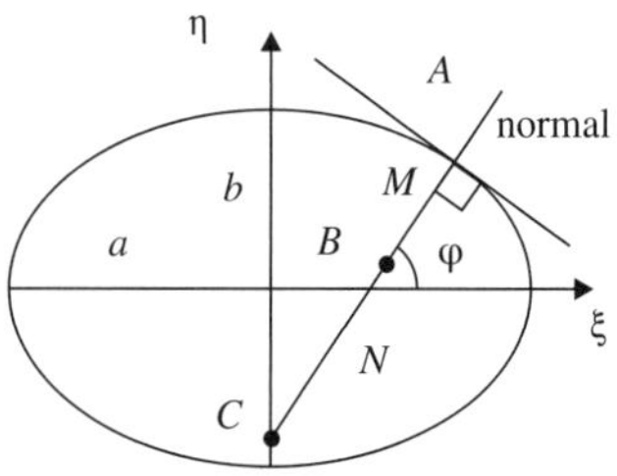

Figure B.1.1 Elements of the ellipse.

Upon stepping deeper into the geometry of the ellipse, it is found that

$$\eta = N(1 - e^2)\sin\varphi \tag{B.1.5}$$

and

$$N = \frac{a}{(1 - e^2\ \sin^2\varphi)^{1/2}} \tag{B.1.6}$$

Additional interpretation of N will be given below. The symbol M in Figure B.1.1 denotes the segment $\overline{AB}$ taken along the normal, i.e., the perpendicular of the tangent. M equals the radius of curvature of the ellipse at point A. Stepping again deeper into the geometry of the ellipse, we find that the radius of curvature can be expressed as

$$M = \frac{a\ (1 - e^2)}{(1 - e^2\ \sin^2\varphi)^{3/2}} \tag{B.1.7}$$

Note that the variable in expressions (B.1.4) to (B.1.7) is the geodetic latitude.

Rotating the ellipse of Figure B.1.1 around the η axis generates the ellipsoid of rotation, or simply the ellipsoid. Figure B.1.2 shows such an ellipsoid and the associated Cartesian and geodetic coordinates. The Cartesian coordinate system (x) = (x, y, z) has its origin at the center of the ellipsoid, the z axis coincides with the semiminor axis, and the x and y axes are located in the equatorial plane of the ellipsoid. The directions of the x and z axes and the center of the ellipsoid are typically fixed by conventions. The ellipsoidal normal through a space point P, i.e., a point on the physical earth surface, intersects with the z axis because of the rotational symmetry of the ellipsoid; however, it does not pass through the origin of the Cartesian coordinate system because of the flattening of the ellipsoid. The length of the ellipsoidal normal from P to the ellipsoid is the geodetic height h. The angle between the ellipsoidal normal

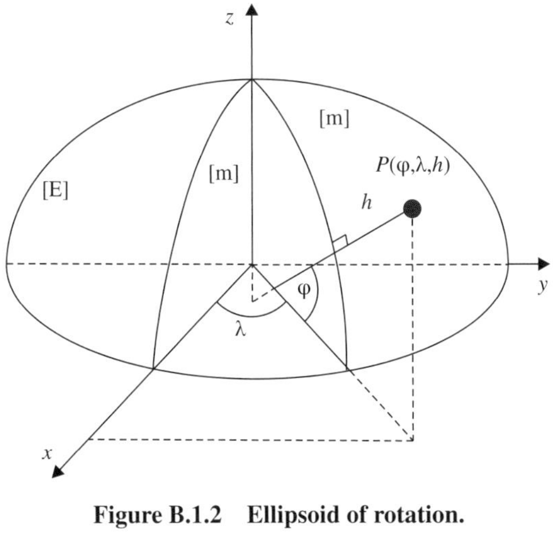

Figure B.1.2 Ellipsoid of rotation.

and the equatorial plane is the geodetic latitude φ in accordance with the definition given earlier.

According to the construction of the ellipsoid, any intersection of the ellipsoid [E] with a plane that contains the z axis generates an ellipse that is called the *geodetic* meridian [m]. The geodetic longitude λ is then defined as the angle between two geodetic meridian planes and counted positive eastward starting at the x axis. Therefore, the triplet of geodetic coordinates (φ, λ, h) completely describes the position of a point in space.

The plane at spatial point P, which is perpendicular to the point's ellipsoidal normal, defines the local geodetic horizon. This is the primary horizontal reference plane in the 3D geodetic model. Notice the distinction between the local geodetic horizon and the local astronomic horizon introduced elsewhere in this book (the latter is perpendicular to the plumb line at P).

Constant geodetic latitude and longitude lines trace the familiar lattice of meridians and parallels on the surface [E]. In differential geometry, such lines are called curvilinear lines [φ] and [λ], and (φ, λ) are called curvilinear coordinates. It is to be understood that the term *curvilinear* refers to a general surface and not just to the ellipsoid. A plane that contains the surface normal, in this case the ellipsoidal surface normal, is called a normal plane. The intersection of a normal plane with the surface (the ellipsoid) is a normal section.

With this terminology, the geodetic meridians [λ] are simply normal sections generated by a normal plane that contains the z axis. Consider the special case of a normal plane at P that is rotated with respect to the plane of the meridian by 90°. This is called the prime vertical normal plane. It also intersects the ellipsoid along a normal section, denoted by [pv]. The value N in (B.1.6) is the radius of curvature of that normal section [pv]. In fact, the student of differential geometry will recognize (B.1.4) as an application of the famous Meusnier theorem, which relates the radius of curvature of a general surface curve to the radius of curvature of the normal section when both curves have a common tangent. In this case, the general surface curve is the parallel [φ].

Given the geometric interpretation of the radius of curvatures of the meridian and the prime vertical normal sections, the curious student probably suspects the existence of another important relationship. It is Euler's equation which relates the radius of curvature R of a normal section in general direction α to the radius of curvature of the meridian and primer vertical normal sections as

$$\frac{1}{R} = \frac{\cos^2 \alpha}{M} + \frac{\sin^2 \alpha}{N} \tag{B.1.8}$$

The symbol α denotes the geodetic azimuth, i.e., the angle between two normal planes having the ellipsoidal normal at P in common. This is precisely the azimuth used in the 3D geodetic model. Equations (B.1.6), (B.1.7), and (B.1.8) imply $M \leq R \leq N$. Deeper study of differential geometry would reveal that the directions of the meridian and the prime vertical belong to the special group of directions that are perpendicular to each other and for which the curvatures (reciprocal of radius of curvatures) take on maximum and minimum values. These are the principal directions.

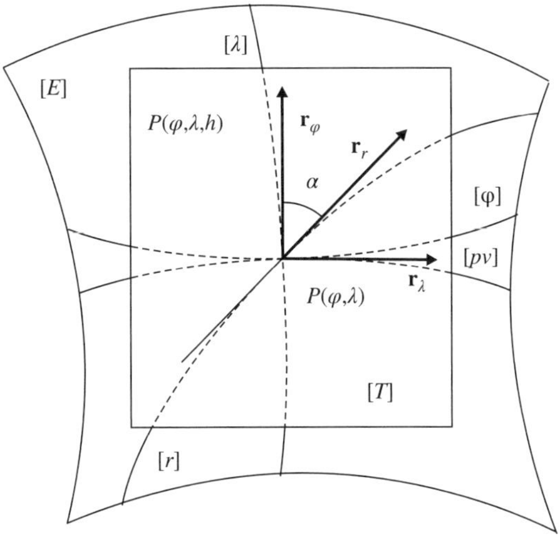

Figure B.1.3 Sections on the ellipsoid.

Figure B.1.3 shows various intersections. The tangent plane $[T]$ of the ellipsoidal surface $[E]$ at $P(\varphi, \lambda, h = 0)$ is spanned by the tangent vectors $\boldsymbol{r}_\varphi$ and $\boldsymbol{r}_\lambda$ of the meridian $[\lambda]$ and parallel $[\varphi]$. The nonnormal section $[\varphi]$ and the normal section $[pv]$ have the tangent $\boldsymbol{r}_\lambda$ in common. The azimuth α of the general normal section $[r]$ is the angle between the respective normal planes or, equivalently, the angle in the tangent plane between $\boldsymbol{r}_\varphi$ and $\boldsymbol{r}_r$. The angle between $\boldsymbol{r}_\varphi$ and $\boldsymbol{r}_\lambda$ is 90° because they represent the principal directions.

The Cartesian coordinates $(\mathrm{x}) = (x, y, z)$ can be expressed as a function of the geodetic coordinates (φ, λ, h) using (B.1.4) and (B.1.5), and the geometry shown in Figure B.1.2, as follows:

$$x = (N + h) \cos \varphi \cos \lambda \tag{B.1.9}$$

$$y = (N + h) \cos \varphi \sin \lambda \tag{B.1.10}$$

$$z = \left[N(1 - e^2) + h\right] \sin \varphi \tag{B.1.11}$$

The inverse solution, i.e., expressing the triplet (φ, λ, h) as a function of (x, y, z) involves a nonlinear mathematical relationship. The longitude follows straightforwardly from (B.1.9) and (B.1.10) as

$$\tan \lambda = \frac{y}{x} \tag{B.1.12}$$

One needs to pay attention to the quadrant of the longitude λ. In geodesy the longitude is typically positive eastward starting from the x axis and counting from 0° to 360°, i.e., $0° \leq \lambda < 360°$. Others give east $0° \leq \lambda(\mathrm{E}) \leq 180°$ or west

$0° < \lambda(\mathrm{W}) < 180°$ longitudes counting from 0° to 180°, respectively, or give negative values in the region $-180° < \lambda < 0°$. The geodetic latitude follows from the nonlinear equation (B.1.11) using some iterative technique. For this purpose it is convenient to rewrite (B.1.11) as

$$\tan\varphi = \frac{z}{\sqrt{x^2+y^2}}\left(1+\frac{e^2 N \sin\varphi}{z}\right) \tag{B.1.13}$$

and use

$$\varphi_{\text{initial}} = \tan^{-1}\left[\frac{z}{\left(1-e^2\right)\sqrt{x^2+y^2}}\right] \tag{B.1.14}$$

on the right-hand side of (B.1.13) to start the iteration. The iteration stops after successive solutions yield negligible changes in the geodetic latitude. After convergence, the geodetic height follows from

$$h = \frac{\sqrt{x^2+y^2}}{\cos\varphi} - N \tag{B.1.15}$$

as can be readily verified.

The differential relations between the Cartesian and geodetic coordinates are

$$\begin{bmatrix} dx \\ dy \\ dz \end{bmatrix} = \boldsymbol{J}(\varphi,\lambda,h)\begin{bmatrix} d\varphi \\ d\lambda \\ dh \end{bmatrix} \tag{B.1.16}$$

with transformation matrix $\boldsymbol{J}(\varphi,\lambda,h)$ being

$$\boldsymbol{J}(\varphi,\lambda,h) = \begin{bmatrix} -(M+h)\cos\lambda\,\sin\varphi & -(N+h)\cos\varphi\sin\lambda & \cos\varphi\cos\lambda \\ -(M+h)\sin\lambda\sin\varphi & (N+h)\cos\varphi\cos\lambda & \cos\varphi\sin\lambda \\ (M+h)\cos\varphi & 0 & \sin\varphi \end{bmatrix} \tag{B.1.17}$$

Obtaining the partial derivatives in such compact forms requires some algebraic work. For these and other compact forms to be developed later, it is helpful to take note of the following partial derivatives:

$$\frac{\partial(N\cos\varphi)}{\partial\varphi} = -M\sin\varphi \tag{B.1.18}$$

$$\frac{\partial(N\sin\varphi)}{\partial\varphi} = \frac{M\cos\varphi}{1-e^2} \tag{B.1.19}$$

$$\frac{\partial(M\sin\varphi)}{\partial\varphi} = \frac{M}{N\cos\varphi}\left[(2N-3M)\sin^2\varphi + N\right] \tag{B.1.20}$$

$$\frac{\partial(M\cos\varphi)}{\partial\varphi} = \frac{M}{N}(2N-3M)\sin\varphi \tag{B.1.21}$$

TABLE B.1.1 Dimensions of Important Ellipsoids

Datum	Ellipsoid	a [m]	$1/f$
NAD27	Clarke 1866	6378206.4	294.9786982
WGS72	WGS72	6378135.0	298.26
NAD83	GRS80	6378137.0	298.257222101
WGS84	WGS84	6378137.0	298.257223563

It might be comforting to know that the formulations given above are all that is needed to deal with the 3D geodetic model. Curvature is the only element that has thus far been taken from the realm of differential geometry. The elements of the geodesic or even conformal mapping have not yet been required. These facts account for the relative mathematical simplicity of the 3D geodetic model.

Table B.1.1 lists the defining values of a sample of ellipsoids that are in use today or have some historical relevancy. The size of the ellipsoid is usually identified with a name. One speaks of a *datum* if the size of the ellipsoid and its location with respect to the earth is defined. The semiaxes of a typical earth ellipsoid differ by about $a - b \approx 21$ km. If the ellipsoid is scaled to 1 m, this difference is just 3 mm.

B.2 COMPUTATION OF THE ELLIPSOIDAL SURFACE

The two-dimensional ellipsoidal and conformal mapping models require the geodesic line and the solution of geodesic triangles (triangles whose sides are geodesic lines) on the ellipsoidal surface. Because the respective expressions are based on differential geometry, this section offers a brief summary of the relevant material. Several expressions are given in general form and are valid for any smooth surface whose second derivatives exist and are continuous. While (φ, λ) continue to represent the geodetic latitude and longitude of the ellipsoid, they could easily be more generally interpreted as curvilinear coordinates on other surfaces.

B.2.1 Fundamental Coefficients

Equations (B.1.9) to (B.1.11) for the ellipsoid can be written in a compact and general form as

$$\boldsymbol{r}(\varphi, \lambda) = \begin{bmatrix} x(\varphi, \lambda) & y(\varphi, \lambda) & z(\varphi, \lambda) \end{bmatrix}^T \tag{B.2.1}$$

In fact, we can view (B.2.1) as the equation of general surface whose second derivatives exist and are continuous. The tangent vector to the λ curvilinear line is given by

$$\boldsymbol{r}_\varphi = \frac{\partial \boldsymbol{r}(\varphi, \lambda)}{\partial \varphi} = \begin{bmatrix} \dfrac{\partial x(\varphi, \lambda)}{\partial \varphi} & \dfrac{\partial y(\varphi, \lambda)}{\partial \varphi} & \dfrac{\partial z(\varphi, \lambda)}{\partial \varphi} \end{bmatrix}^T \tag{B.2.2}$$

Similarly, the tangent vector to the φ curvilinear line is

$$\boldsymbol{r}_\lambda = \frac{\partial \boldsymbol{r}(\varphi, \lambda)}{\partial \lambda} \tag{B.2.3}$$

The surface expression (B.2.1) can formally be expanded in a Taylor series. Let the point of expansion be at $\boldsymbol{r}(\varphi, \lambda)$ and let the differential increments be denoted as $d\varphi$ and $d\lambda$. Limiting the expansion to second-order terms gives

$$\boldsymbol{r}(\varphi + d\varphi, \lambda + d\lambda) = \boldsymbol{r}(\varphi, \lambda) + \boldsymbol{r}_{\varphi} d\varphi + \boldsymbol{r}_{\lambda} d\lambda + \frac{1}{2}\{\boldsymbol{r}_{\varphi\varphi}\, d\varphi^2 + 2\boldsymbol{r}_{\varphi\lambda}\, d\varphi\, d\lambda + \boldsymbol{r}_{\lambda\lambda} d\lambda^2\} + \cdots \tag{B.2.4}$$

It can be readily visualized that the first part of this expression,

$$\boldsymbol{t}(\varphi, \lambda) = \boldsymbol{r}(\varphi, \lambda) + \boldsymbol{r}_{\varphi}\, d\varphi + \boldsymbol{r}_{\lambda}\, d\lambda \tag{B.2.5}$$

represents the tangent plane $[T]$ which is located at $\boldsymbol{r}(\varphi, \lambda)$ and spanned by the vectors $\boldsymbol{r}_{\varphi}$ and $\boldsymbol{r}_{\lambda}$. The total differential

$$d\boldsymbol{r} = \boldsymbol{r}_{\varphi} d\varphi + \boldsymbol{r}_{\lambda} d\lambda \tag{B.2.6}$$

is a vector in the tangent plane and represents the linearized surface distance on $[E]$ from $p(\varphi, \lambda)$ to $P(\varphi + d\varphi, \lambda + d\lambda)$. See Figure B.2.1. The square of the length of the total differential is

$$\begin{aligned} ds^2 &= d\boldsymbol{r} \cdot \mathrm{d}\boldsymbol{r} \\ &= \boldsymbol{r}_{\varphi} \cdot \boldsymbol{r}_{\varphi}\, d\varphi^2 + 2\boldsymbol{r}_{\varphi} \cdot \boldsymbol{r}_{\lambda}\, d\varphi\, d\lambda + \boldsymbol{r}_{\lambda} \cdot \boldsymbol{r}_{\lambda}\, d\lambda^2 \\ &= E\, d\varphi^2 + 2F\, d\varphi\, d\lambda + G\, d\lambda^2 \end{aligned} \tag{B.2.7}$$

This is the first fundamental form. The quantities E, F, G are called, since Gauss, the first fundamental coefficients. Properties of the surface that can be expressed as a function of the first fundamental coefficients are called intrinsic properties. The totality of intrinsic properties of the surface is called the intrinsic geometry of the surface. Using vector identities, one can verify that

$$\begin{aligned} EG - F^2 &= (\boldsymbol{r}_{\varphi} \cdot \boldsymbol{r}_{\varphi})(\boldsymbol{r}_{\lambda} \cdot \boldsymbol{r}_{\lambda}) - (\boldsymbol{r}_{\varphi} \cdot \boldsymbol{r}_{\lambda})^2 = (\boldsymbol{r}_{\varphi} \times \boldsymbol{r}_{\lambda}) \cdot (\boldsymbol{r}_{\varphi} \times \boldsymbol{r}_{\lambda}) \\ &= \|\boldsymbol{r}_{\varphi} \times \boldsymbol{r}_{\lambda}\| > 0 \end{aligned} \tag{B.2.8}$$

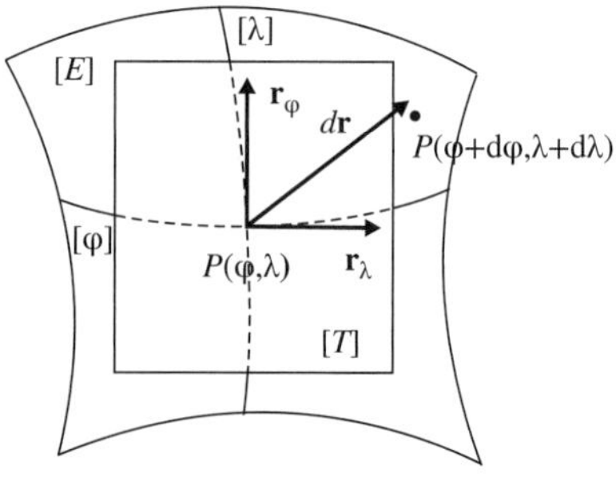

Figure B.2.1 The total differential.

and that $E > 0$ and $G > 0$. For orthogonal curvilinear lines, we have $F = 0$ because $\boldsymbol{r}_\varphi \cdot \boldsymbol{r}_\lambda = 0$. Evaluating the fundamental coefficients for the ellipsoidal surface [E] gives

$$E = M^2 \tag{B.2.9}$$

$$F = 0 \tag{B.2.10}$$

$$G = N^2 \cos^2\varphi \tag{B.2.11}$$

$$ds^2 = M^2 d\varphi^2 + N^2 \cos^2\varphi \, d\lambda^2 \tag{B.2.12}$$

The last term in (B.2.4),

$$\boldsymbol{p} = \frac{1}{2}\{\boldsymbol{r}_{\varphi\varphi} \, d\varphi^2 + 2\boldsymbol{r}_{\varphi\lambda} \, d\varphi \, d\lambda + \boldsymbol{r}_{\lambda\lambda} \, d\lambda^2\} \tag{B.2.13}$$

represents the deviation of a second-order surface approximation from the tangent plane. The vectors $\boldsymbol{r}_{\varphi\varphi}$ and $\boldsymbol{r}_{\lambda\lambda}$ contain the respective second partial derivative with respect to λ and φ, and $\boldsymbol{r}_{\varphi\lambda}$ contains the mixed derivatives. Introducing the surface normal $\boldsymbol{e}$ as

$$\boldsymbol{e} = \frac{\boldsymbol{r}_\varphi \times \boldsymbol{r}_\lambda}{\|\boldsymbol{r}_\varphi \times \boldsymbol{r}_\lambda\|} = \frac{\boldsymbol{r}_\varphi \times \boldsymbol{r}_\lambda}{\sqrt{EG - F^2}} \tag{B.2.14}$$

then the orthogonal distance of the second-order approximation to the tangent plane is

$$\begin{aligned} d &= -\boldsymbol{e} \cdot \boldsymbol{p} \\ &= \frac{1}{2}\{-\boldsymbol{e} \cdot \boldsymbol{r}_{\varphi\varphi} \, d\varphi^2 - 2\boldsymbol{e} \cdot \boldsymbol{r}_{\varphi\lambda} \, d\varphi \, d\lambda - \boldsymbol{e} \cdot \boldsymbol{r}_{\lambda\lambda} \, d\lambda^2\} \\ &= \frac{1}{2} \{D \, d\varphi^2 + 2D' \, d\varphi \, d\lambda + D'' d\lambda^2 \end{aligned} \tag{B.2.15}$$

Expression (B.2.15) is the second fundamental form and the elements (D, D', D'') are called, since Gauss, the second fundamental coefficients. For the ellipsoid these coefficients have the simple form

$$D = N \cos^2\varphi \tag{B.2.16}$$

$$D' = 0 \tag{B.2.17}$$

$$D'' = M \tag{B.2.18}$$

The partial derivatives (B.1.18) to (B.1.21) are very helpful in verifying this simple form.

B.2.2 Gauss Curvature

At every point of a smooth surface there are two perpendicular directions along which the curvature attains a maximum and a minimum value. These are the principal directions. Denoting the respective principal radius of curvatures by R_1 and R_2, a deeper study of differential geometry reveals

$$K \equiv \frac{1}{R_1 R_2} = \frac{DD'' - D'^2}{EG - F^2} = \frac{1}{MN} \tag{B.2.19}$$

where K is called the Gauss curvature. The latter part of (B.2.19) expresses the value of K for the ellipsoid. In general, if the curvilinear lines also happen to coincide with the principal directions, then $D' = 0$. It can be shown that the numerator $DD'' - D'^2$ can be expressed as a function of the first fundamental coefficients and their partial derivatives.

Since the denominator in (B.2.19) is always positive, the numerator determines the sign of K. A point is called *elliptic* if $K > 0$. In the neighborhood of an elliptic point, the surface lies on one side of the tangent plane. A point is called *hyperbolic* if $K < 0$. In the neighborhood of a hyperbolic point, the surface lies on both sides of the tangent plane. A point is *parabolic* if $K = 0$, in which case the surface may lie on either side of the tangent plane.

For $K = 0$ one of the values R_1 and R_2 must be infinite as follows from (B.2.19). If this occurs at every point of the surface one family of the principal directions must be straight lines. Examples are cylinders or cones. Such surfaces are called developable surfaces and can be reshaped into a plane without stretching and tearing.

B.2.3 Elliptic Arc

If s denotes the length of the arc of the ellipse from the equator, or simply the ecliptic arc, then

$$s = \int_0^{\varphi} \sqrt{E}\, d\varphi = \int_0^{\varphi} M\, d\varphi \tag{B.2.20}$$

There is no closed expression for the integral in (B.2.20). The following series expansion (Snyder, 1979) is frequently used

$$s = a\left[\left(1 - \frac{e^2}{4} - \frac{3e^4}{64} - \frac{5e^6}{256}\right)\varphi - \left(\frac{3e^2}{8} + \frac{3e^4}{32} + \frac{45e^6}{1024}\right)\sin 2\varphi \right.$$
$$\left. + \left(\frac{15e^4}{256} + \frac{45e^6}{1024}\right)\sin 4\varphi\right] \tag{B.2.21}$$

The inverse solution, i.e., given the elliptic arc with respect to the equator and computing the geodetic latitude, is available iteratively starting with the initial value

$$\varphi_{\text{initial}} = \frac{s}{a} \tag{B.2.22}$$

B.2.4 Angle

An angle on a surface is defined as the angle between two tangents. The angle, therefore, is a measure in the tangent plane. Figure B.2.2 shows two curves, f_1 and f_2,

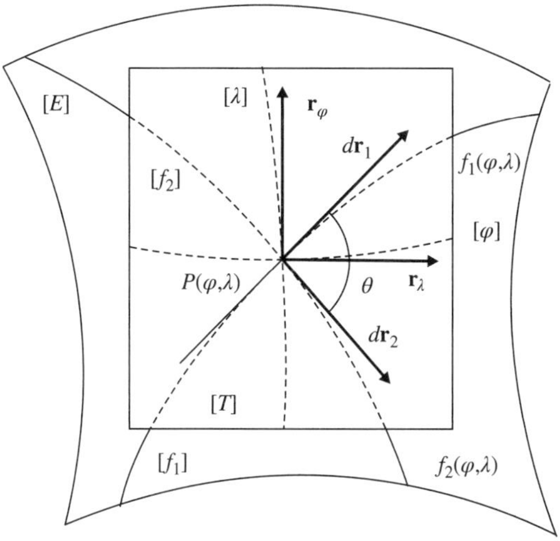

Figure B.2.2 Definition of surface angle.

on the surface that could be implicitly defined as $f_1(\varphi, \lambda) = 0$ and $f_2(\varphi, \lambda) = 0$. The differentials $(d\varphi_1, d\lambda_1)$ and $(d\varphi_2, d\lambda_2)$, which follow from differentiating these two functions, determine the tangent vectors as

$$d\boldsymbol{r}_1 = \boldsymbol{r}_\varphi \, d\varphi_1 + \boldsymbol{r}_\lambda \, d\lambda_1 \tag{B.2.23}$$

$$d\boldsymbol{r}_2 = \boldsymbol{r}_\varphi \, d\varphi_2 + \boldsymbol{r}_\lambda d\lambda_2 \tag{B.2.24}$$

Thus the expression for the angle becomes

$$\begin{aligned} \cos\theta &= \frac{dr_1 \cdot dr_2}{\|dr_1\| \, \|dr_2\|} \\ &= \frac{E\, d\varphi_1 \, d\varphi_2 + F(d\varphi_1 \, d\lambda_2 + d\varphi_2 \, d\lambda_1) + G d\lambda_1 \, d\lambda_2}{\sqrt{E d\varphi_1^2 + 2F \, d\varphi_1 \, d\lambda_1 + G \, d\lambda_1^2} \sqrt{E \, d\varphi_2^2 + 2F \, d\varphi_2 \, d\lambda_2 + G \, d\lambda_2^2}} \end{aligned} \tag{B.2.25}$$

Equation (B.2.25) is useful in verifying the conformal property in mapping.

B.2.5 Isometric Latitude

The first fundamental form (B.2.12) relates a differential change of curvilinear coordinates to the corresponding surface distance within first-order approximation. One can readily visualize that on the equator a respective change in φ or λ by one arc second traces about the same distance. This is not the case close to the pole because of the convergence of the meridians. Consider a new curvilinear parameter q, which is defined by the differential relation

$$dq \equiv \frac{M}{N \cos\varphi} d\varphi \tag{B.2.26}$$

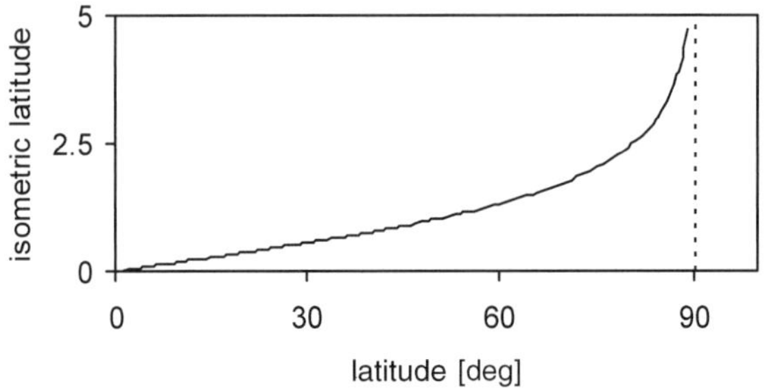

Figure B.2.3 Isometric latitude.

Substituting (B.2.26) in first fundamental form (B.2.12) gives

$$ds^2 = N^2 \cos^2\varphi (dq^2 + d\lambda^2) \tag{B.2.27}$$

Equation (B.2.27) clearly shows that the same changes in dq and $d\lambda$ cause the same change ds at a given point. Integrating (B.2.26)gives

$$q = \ln\left[\tan\left(45^\circ + \frac{\varphi}{2}\right)\left(\frac{1 - e\sin\varphi}{1 + e\sin\varphi}\right)^{e/2}\right] \tag{B.2.28}$$

The new parameter q is called the isometric latitude. It is a function of the geodetic latitude and reaches infinity at the pole. See Figure B.2.3. Because q is constant when φ is constant, the lines $q =$ constant are parallels on the ellipsoid. Equal incremented q parallels are spaced increasingly closer as one approaches the pole. The pair q and λ are called isometric curvilinear coordinates which trace, respectively, a lattice of isometric *curvilinear* lines $[q]$ and $[\lambda]$ on the ellipsoid.

The inverse solution, i.e., given the isometric latitude q and computing the geodetic latitude φ, is solved though iterations. Equation (B.2.28) can be written as

$$\tan\left(45^\circ + \frac{\varphi}{2}\right) = \varepsilon^q\left(\frac{1 + e\sin\varphi}{1 - e\sin\varphi}\right)^{e/2} \tag{B.2.29}$$

The symbol ε denotes the base of the natural system of logarithms ($\varepsilon = 2.71828\cdots$). It must not be confused with the eccentricity of the ellipsoid, which is assigned the symbol e in this book. The iteration begins by taking $e = 0$ on the right side of (B.2.29), giving

$$\varphi_{\text{initial}} = 2\tan^{-1}(\varepsilon^q) - \frac{\pi}{2} \tag{B.2.30}$$

B.2.6 Differential Equation of the Geodesic

Probably the best-known property of the geodesic line (or simply the geodesic) is that it is the shortest surface line between two points on the surface. This property determines the differential equations of the geodesic. Differential geometry offers

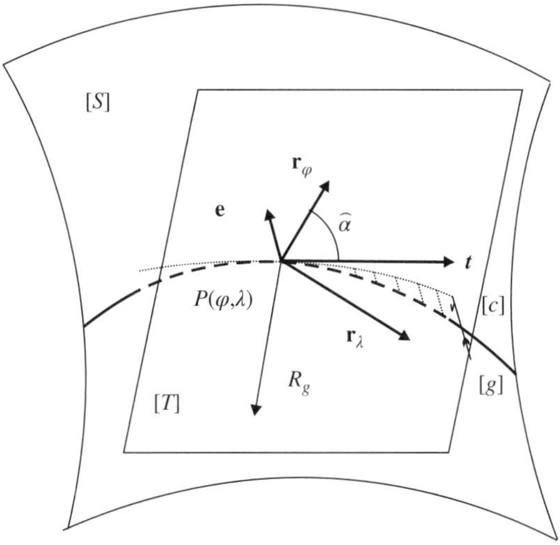

Figure B.2.4 Geodesic curvature.

other equivalent definitions of the geodesic. Consider Figure B.2.4, which shows a general surface $[S]$, the tangent plane $[T]$, and surface normal $\boldsymbol{e}$ at a point $P(\varphi, \lambda)$. Let there be a curve $[g]$ on $[S]$ that passes through $P(\varphi, \lambda)$. The tangent on this space curve is denoted by $\boldsymbol{t}$. This tangent is located in the tangent plane spanned by $\boldsymbol{r}_\varphi$ and $\boldsymbol{r}_\lambda$. Next, project curve $[g]$ orthogonally on the tangent plane in the differential neighborhood of $P(\varphi, \lambda)$. This generates a curve $[c]$ that is located in the tangent plane and has the tangent $\boldsymbol{t}$ in common with $[g]$. Like any plane curve, the curve $[c]$ has a curvature at the point $P(\varphi, \lambda)$, which is denoted here by κ_g. This is the geodesic curvature. It is related to the geodesic radius of curvature R_g by

$$\kappa_g = \frac{1}{R_g} \tag{B.2.31}$$

It can be shown that the geodesic curvature κ_g is a function of the first fundamental coefficients and their derivatives.

The situation described above and depicted in Figure B.2.4 for $P(\varphi, \lambda)$ can be conceptually repeated for every point of the curve $[g]$, i.e., for every point one can visualize the tangent plane and the orthogonal projection of $[g]$ in the differential neighborhood of the point of tangency. The curve $[g]$ is a geodesic if the geodesic curvature is zero at all these points, or, equivalently, the radius of the geodesic curvature is infinite. Because the geodesic radius of curvature is infinite, the projection of the geodesic on the tangent plane is a straight line in the differential neighborhood of $P(\varphi, \lambda)$. This geometric definition of the geodesic is also sufficient to determine the differential equations of $[g]$.

Differential geometry offers yet another definition of the geodesic that is frequently stated. Assume that expressions for the three Cartesian coordinates of $[g]$ are given as a function of some free parameter s. Differentiating each component once with respect to s gives the tangent vector $\boldsymbol{t}$; differentiating twice gives another vector called the principal normal of the curve $[g]$. It can be shown that the tangent vector and the principal normal of the curve are perpendicular. Next, the curves $[g]$ and $[c]$ can be viewed as curves on a general cylinder that is perpendicular to the tangent plane. Viewed like that, the curves $[c]$ and $[g]$ represent a normal section and a general section on the cylinder that have the tangent $\boldsymbol{t}$ in common. The respective radii of curvature are related by Meusnier's theorem. In this view the radius of curvature of the normal section $[c]$ is R_g. If R_g is to go to infinity, then Meusnier's theorem implies that the principal normal of $[g]$ and the surface normal $\boldsymbol{e}$ coincide.

The definition of the geodesic does not restrict the geodesics to plane curves. In fact, the geodesic will have, in general, curvature and torsion. However, the definition lends itself to some interpretation of "straightness." Consider a virtual surveyor who operates a virtual theodolite on the ellipsoidal surface. A first step in operating an actual theodolite is to level it, i.e., to align the vertical axis with the plumb line. In this example, the virtual surveyor will align the vertical axis with the surface normal. He is then tasked to stake out a straight line using differentially short sightings. He would begin setting up the instrument at the initial (first) point and stake out the second point using the azimuth $\widehat{\alpha}$. Next he would set up at the second point, backsight to the first point, and turn an angle of 180° to stake out the third point, and so on. In the mind of the virtual surveyor, he is staking out a straight line whereas he actually stakes out a geodesic using differentially short sightings.

Let $\widehat{\alpha}$ denote the azimuth of the geodesic, i.e., the angle between the tangent on the λ curvilinear line and the tangent on the geodesic as shown in Figure B.2.7, and let $\widehat{s}$ denote the length of the geodesic on [S]. Using the definition of the geodesic given above, the differential equations for the geodesic on a general surface can be developed as

$$\frac{d\varphi}{d\widehat{s}} = \frac{\sin\widehat{\alpha}}{\sqrt{E}} \tag{B.2.32}$$

$$\frac{d\lambda}{d\widehat{s}} = \frac{\cos\widehat{\alpha}}{\sqrt{G}} \tag{B.2.33}$$

$$\frac{d\widehat{\alpha}}{d\widehat{s}} = \frac{1}{\sqrt{EG}}\left(\frac{\partial\sqrt{G}}{\partial\varphi}\cos\widehat{\alpha} - \frac{\partial\sqrt{E}}{\partial\lambda}\sin\widehat{\alpha}\right) \tag{B.2.34}$$

In case of the ellipsoid $[E]$ the respective equations are

$$\frac{d\varphi}{d\widehat{s}} = \frac{\cos\widehat{\alpha}}{M} \tag{B.2.35}$$

$$\frac{d\lambda}{d\widehat{s}} = \frac{\sin\widehat{\alpha}}{N\cos\varphi} \tag{B.2.36}$$

$$\frac{d\widehat{\alpha}}{d\widehat{s}} = \frac{1}{N}\tan\varphi\sin\widehat{\alpha} \tag{B.2.37}$$

Figure B.2.5 shows a geodesic triangle whose corners consist of the pole $P(\varphi = 90°)$ and the points $P_1(\varphi_1, \lambda_1)$ and $P_2(\varphi_2, \lambda_2)$. The sides of this triangle are

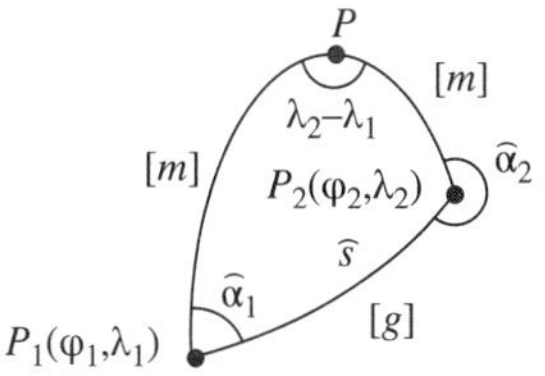

Figure B.2.5 Geodesic triangle.

the meridians, which can be readily identified as geodesic lines, and the geodesic line from P_1 to P_2. At the heart of the ellipsoidal computations are the so-called direct and inverse problems. In the case of the direct problem, the geodetic latitude and longitude at one station, say, $P_1(\varphi_1, \lambda_1)$, and the geodesic azimuth and distance $(\widehat{\alpha}_1, \widehat{s})$ to another point, are known; the geodetic latitude φ_2, longitude λ_2, and back azimuth $\widehat{\alpha}_2$ are required. Formerly, the direct solution is written as

$$\begin{bmatrix} \varphi_2 \\ \lambda_2 \\ \widehat{\alpha}_2 \end{bmatrix} = \begin{bmatrix} d_1\left(\varphi_1, \lambda_1, \widehat{\alpha}_1, \widehat{s}\right) \\ d_2(\varphi_1, \lambda_1, \widehat{\alpha}_1, \widehat{s}) \\ d_3(\varphi_1, \lambda_1, \widehat{\alpha}_1, \widehat{s}) \end{bmatrix} \tag{B.2.38}$$

For the inverse problem, the geodetic latitudes and longitudes of $P_1(\varphi_1, \lambda_1)$ and $P_2(\varphi_2, \lambda_2)$ are given, and the forward and back azimuths and the length of the geodesic are required, i.e.,

$$\begin{bmatrix} \widehat{s} \\ \widehat{\alpha}_1 \\ \widehat{\alpha}_2 \end{bmatrix} = \begin{bmatrix} i_1\left(\varphi_1, \lambda_1, \varphi_2, \lambda_2\right) \\ i_2(\varphi_1, \lambda_1, \varphi_2, \lambda_2) \\ i_3(\varphi_1, \lambda_1, \varphi_2, \lambda_2) \end{bmatrix} \tag{B.2.39}$$

Most solutions of (B.2.35) to (B.2.37) rely on extensive series expansions with intermittent truncation of small terms. Various innovative approaches have been implemented to keep the number of significant terms small and yet achieve accurate solutions. Some solutions are valid only for short lines, while others apply to intermediary long lines, or even to lines that go all around the ellipsoid.

B.2.7 The Gauss Midlatitude Solution

Table B.2.1 summarizes the Gauss midlatitude solution (Grossman, 1976, pp. 101–106). The term *midlatitude* indicates that the point of expansion in the series developments is mean latitude and/or longitude between $P_1(\varphi_1, \lambda_1)$ and $P_2(\varphi_2, \lambda_2)$. The inverse solution begins by first evaluating the auxiliary expressions shown in the first section of the table, followed by the expressions in the second section. The first step for the direct solution requires the computation of approximate geodetic latitude and longitude for station $P_2(\varphi_2, \lambda_2)$ as indicated in the third section. These initial coordinates are used to evaluate the auxiliary quantities of the first section, which, in turn, are used to compute improved coordinates for station P_2 from the remaining expressions of the third section. The direct solution is iterated until convergence is achieved.

The linearized form of the inverse solution is important when computing (adjusting) networks on the ellipsoid. The truncated expressions of the partial

TABLE B.2.1 The Gauss Midlatitude Solution

Auxiliary Terms: $\Delta\varphi = \varphi_2 - \varphi_1$; $\Delta\lambda = \lambda_2 - \lambda_1$

$$\varphi = \frac{\varphi_1 + \varphi_2}{2}; \quad t = \tan\varphi; \ \eta^2 = \frac{e^2}{1-e^2}\cos^2\varphi; \quad V^2 = 1 + \eta^2; \ f_1 = 1/M; \ f_2 = 1/N$$

$$f_3 = \frac{1}{24}; \ f_4 = \frac{1+\eta^2 - 9\eta^2 t^2}{24V^4}; \ f_5 = \frac{1-2\eta^2}{24}; \ f_6 = \frac{\eta^2(1-t^2)}{8V^4}; \ f_7 = \frac{1+\eta^2}{12};$$

$$f_8 = \frac{3+8\eta^2}{24V^4}$$

Inverse Solution Given $(\varphi_1, \lambda_1, \varphi_2, \lambda_2)$, compute $(\widehat{s}, \widehat{\alpha}_1, \widehat{\alpha}_2)$

$$\widehat{s}\sin\widehat{\alpha} = \frac{1}{f_2}\Delta\lambda\cos\varphi\left[1 - f_3(\Delta\lambda\sin\varphi)^2 + f_4\Delta\varphi^2\right] \tag{a}$$

$$\widehat{s}\cos\widehat{\alpha} = \frac{1}{f_1}\Delta\varphi\cos\frac{\Delta\lambda}{2}\left[1 + f_5(\Delta\lambda\cos\varphi)^2 + f_6\Delta\varphi^2\right] \tag{b}$$

$$\Delta\widehat{\alpha} = \Delta\lambda\sin\varphi\left[1 + f_7(\Delta\lambda\cos\varphi)^2 + f_8\Delta\varphi^2\right] \tag{c}$$

$$\widehat{s} = \sqrt{(\widehat{s}\sin\widehat{\alpha})^2 + (\widehat{s}\cos\widehat{\alpha})^2} \tag{d}$$

$$\widehat{\alpha} = \tan^{-1}\left(\frac{\widehat{s}\sin\widehat{\alpha}}{\widehat{s}\cos\widehat{\alpha}}\right) \tag{e}$$

$$\widehat{\alpha}_1 = \widehat{\alpha} - \frac{\Delta\widehat{\alpha}}{2} \tag{f}$$

$$\widehat{\alpha}_2 = \widehat{\alpha} + \frac{\Delta\widehat{\alpha}}{2} \pm \pi \tag{g}$$

Direct Solution Given $(\varphi_1, \lambda_1, \widehat{s}, \widehat{\alpha}_1)$, compute $(\varphi_2, \lambda_2, \widehat{\alpha}_2)$

$$\lambda_2 \approx \lambda_1 + \frac{\widehat{s}\sin\widehat{\alpha}_1}{N_1\cos\varphi_1} \tag{h}$$

$$\varphi_2 \approx \varphi_1 + \frac{\widehat{s}\cos\widehat{\alpha}_1}{M_1} \tag{i}$$

Iteration $(\varphi_1, \lambda_1, \varphi_2, \lambda_2)$: reevaluate auxiliary terms

$$\Delta\widehat{\alpha} = \Delta\lambda\sin\varphi\left[1 + f_7(\Delta\lambda\cos\varphi)^2 + f_8\Delta\varphi^2\right] \tag{j}$$

$$\widehat{\alpha} = \widehat{\alpha}_1 + \frac{\Delta\widehat{\alpha}}{2} \tag{k}$$

$$\widehat{\alpha}_2 = \widehat{\alpha} + \frac{\Delta\widehat{\alpha}}{2} \pm \pi \tag{l}$$

$$\lambda_2 = \lambda_1 + f_2\frac{\widehat{s}\sin\widehat{\alpha}}{\cos\varphi}\left[1 + f_3(\Delta\lambda\sin\varphi)^2 - f_4\Delta\varphi^2\right] \tag{m}$$

$$\varphi_2 = \varphi_1 + f_1\frac{\widehat{s}\cos\widehat{\alpha}}{\cos(\Delta\lambda/2)}\left[1 - f_5(\Delta\lambda\cos\varphi)^2 - f_6\Delta\varphi^2\right] \tag{n}$$

derivatives in

$$d\widehat{s} = \frac{\partial i_1}{\partial\varphi_1}d\varphi_1 + \frac{\partial i_1}{\partial\lambda_1}d\lambda_1 + \frac{\partial i_1}{\partial\varphi_2}d\varphi_2 + \frac{\partial i_1}{\partial\lambda_2}d\lambda_2 \tag{B.2.40}$$

$$d\widehat{\alpha}_1 = \frac{\partial i_2}{\partial\varphi_1}d\varphi_1 + \frac{\partial i_2}{\partial\lambda_1}d\lambda_1 + \frac{\partial i_2}{\partial\varphi_2}d\varphi_2 + \frac{\partial i_2}{\partial\lambda_2}d\lambda_2 \tag{B.2.41}$$

are listed in Table B.2.2.

TABLE B.2.2 Partial Derivatives of the Geodesic on the Ellipsoid

	$d\varphi_1$	$d\lambda_1$	$d\varphi_2$	$d\lambda_2$
$d\widehat{s}$	$-M_1 \cos\widehat{\alpha}_1$	$N_2 \cos\varphi_2 \sin\widehat{\alpha}_2$	$-M_2 \cos\widehat{\alpha}_2$	$-N_2 \cos\varphi_2 \sin\widehat{\alpha}_2$
$d\widehat{\alpha}_1$	$\dfrac{M_1 \sin\widehat{\alpha}_1}{\widehat{s}}$	$\dfrac{N_2 \cos\varphi_2 \cos\widehat{\alpha}_2}{\widehat{s}}$	$\dfrac{M_2 \sin\widehat{\alpha}_2}{\widehat{s}}$	$-\dfrac{N_2 \cos\varphi_2 \cos\widehat{\alpha}_2}{\widehat{s}}$

B.2.8 Angular Excess

The Gauss-Bonnet theorem of differential geometry provides an expression for the sum of interior angles $\widehat{\delta}_i$ of a general polygon (continuous curvature) on a surface

$$\sum_{i=1}^{v} \widehat{\delta}_i = (v-2)\cdot\pi + \int_C \kappa_g \, ds + \iint_{\text{area}} K \, dA \tag{B.2.42}$$

For the sum of the interior angles of a geodesic triangle one readily obtains

$$\widehat{\delta}_1 + \widehat{\delta}_2 + \widehat{\delta}_3 = \pi + \varepsilon \tag{B.2.43}$$

with

$$\varepsilon = \iint_{\text{area}} K \, dA \tag{B.2.44}$$

because $\kappa_g = 0$. The sum of the angles of the geodesic triangle differs from π by the double integral of the Gauss curvature taken over the area of the triangle. The sum of the interior angles of a geodesic triangle is greater than, less than, or equal to π, depending on whether the Gauss curvature is positive, negative, or zero. There is angular excess for the geodesic triangle on the ellipsoid because $K > 0$. On the unit sphere, the excess in angular measurement is called the spherical excess. It equals the area of the triangle, i.e., $\varepsilon = A$, because $k = 1$ on the unit sphere.

B.2.9 Transformation in a Small Region

The following is an example of what might be called a "similarity transformation" on the ellipsoid. Consider a cluster of stations, $i = 1, \ldots, m$, each having two sets of coordinates $(\varphi_{o,i}, \lambda_{o,i})$ and $(\varphi_{n,i}, \lambda_{n,i})$ on the same ellipsoid. The subscripts o and n may be interpreted as "old" and "new." The goal is to establish a simple transformation between the coordinates.

The two-dimensional transformation is done with the tools developed in this appendix. First, we compute the center of figure (φ_c, λ_c) of the stations in the n set by simply averaging latitudes and longitudes, respectively. Next, consider the geodesics that connect the center of figure (φ_c, λ_c) with the positions $(\varphi_{n,i}, \lambda_{n,i})$. The discrepancies $(\varphi_{o,i} - \varphi_{n,i})$ and $(\lambda_{o,i} - \lambda_{n,i})$ take on the role of observation to be used to compute the transformation parameters by least squares. We define four transformation parameters as follows: the translation of the center of figure $(d\varphi_c, d\lambda_c)$, the common azimuth rotation $d\widehat{\alpha}_c$ at the center of figure, and common scale factor $1 - \Delta$ for all geodesics going from the center of figure to the individual points. Thus,

$$\boldsymbol{x} = [d\lambda_c \quad d\varphi_c \quad \Delta \quad d\widehat{\alpha}_c]^T \tag{B.2.45}$$

Since the discrepancies $(\varphi_{o,i} - \varphi_{n,i})$ and $(\lambda_{o,i} - \lambda_{n,i})$ are small quantities, the coefficients listed in Table B.2.2 represent the linear mathematical model of the adjustment. The observation equations for the mixed adjustment model are

$$\begin{aligned}\Delta\widehat{s}_{ci} &= -M_i \cos\widehat{\alpha}_{ic}\,(\varphi_{o,i} - \varphi_{n,i}) - M_c \cos\widehat{\alpha}_{ci}\, d\varphi_c \\ &\quad - N_i \cos\varphi_i \sin\widehat{\alpha}_{ic}\,(\lambda_{o,i} - \lambda_{n,i}) + N_i \cos\varphi_i \sin\widehat{\alpha}_{ic}\, d\lambda_c\end{aligned} \tag{B.2.46}$$

$$\begin{aligned}d\widehat{\alpha}_c &= \frac{M_c}{\widehat{s}_{ci}} \sin\widehat{\alpha}_{ci}\, d\varphi_c + \frac{M_i}{\widehat{s}_{ci}} \sin\widehat{\alpha}_{ic}\,(\varphi_{o,i} - \varphi_{n,i}) \\ &\quad - \frac{N_i}{\widehat{s}_{ci}} \cos\varphi_i \cos\widehat{\alpha}_{ic}\,(\lambda_{o,i} - \lambda_{n,i}) + \frac{N_i}{\widehat{s}_{ci}} \cos\varphi_i \cos\widehat{\alpha}_{ic}\, d\lambda_c\end{aligned} \tag{B.2.47}$$

The respective submatrices of $\boldsymbol{B}, \boldsymbol{A}$, and $\boldsymbol{w}$ for station i are

$$\begin{matrix} \varphi_{n,i} & \lambda_{n,i} & \varphi_{o,i} & \lambda_{o,i} \end{matrix}$$

$$\boldsymbol{B} = \begin{bmatrix} M_i \cos\widehat{\alpha}_{ic} & N_i \cos\varphi_i \sin\widehat{\alpha}_{ic} & -M_i \cos\widehat{\alpha}_{ic} & -N_i \cos\varphi_i \sin\widehat{\alpha}_{ic} \\ -\dfrac{M_i}{\widehat{s}_{ci}} \sin\widehat{\alpha}_{ic} & \dfrac{N_i}{\widehat{s}_{ic}} \cos\varphi_i \cos\widehat{\alpha}_{ci} & \dfrac{M_i}{\widehat{s}_{ci}} \sin\widehat{\alpha}_{ic} & -\dfrac{N_i}{\widehat{s}_{ci}} \cos\varphi_i \cos\widehat{\alpha}_{ic} \end{bmatrix} \tag{B.2.48}$$

$$\begin{matrix} d\varphi_c & d\lambda_c & \Delta & d\widehat{\alpha}_c \end{matrix}$$

$$\boldsymbol{A} = \begin{bmatrix} -M_c \cos\widehat{\alpha}_{ci} & N_i \cos\varphi_i \sin\widehat{\alpha}_{ic} & -\widehat{s}_{ci} & 0 \\ \frac{M_c}{\widehat{s}_{ci}} \sin\widehat{\alpha}_{ci} & \frac{N_i}{\widehat{s}_{ci}} \cos\varphi_i \cos\widehat{\alpha}_{ic} & 0 & -1 \end{bmatrix} \tag{B.2.49}$$

$$\boldsymbol{w} = \begin{bmatrix} -M_i \cos\widehat{\alpha}_{ic}(\varphi_{o,i} - \varphi_{n,i}) & -N_i \cos\varphi_i \sin\widehat{\alpha}_{ic}(\lambda_{o,i} - \lambda_{n,i}) \\ \frac{M_i}{\widehat{s}_{ci}} \sin\widehat{\alpha}_{ic}(\varphi_{o,i} - \varphi_{n,i}) & -\frac{N_i}{\widehat{s}_{ci}} \cos\varphi_i \cos\widehat{\alpha}_{ic}(\lambda_{o,i} - \lambda_{n,i}) \end{bmatrix} \tag{B.2.50}$$

Once the adjusted transformation parameters are available, we can compute the position of the adjusted center of figure and the length and azimuth for any geodesics as follows:

$$\varphi_{o,c} = \varphi_{n,c} + d\varphi_c \tag{B.2.51}$$

$$\lambda_{o,c} = \lambda_{n,c} + d\lambda_c \tag{B.2.52}$$

$$\widehat{s}_{o,ci} = \widehat{s}_{n,ci} + \Delta\widehat{s}_{ci} \tag{B.2.53}$$

$$\widehat{\alpha}_{o,ci} = \widehat{\alpha}_{n,ci} + d\widehat{\alpha}_c \tag{B.2.54}$$

With (B.2.51) through (B.2.54) the positions of stations in the o system can be computed by using the direct solution given in Table B.2.1.

APPENDIX C

CONFORMAL MAPPING

The conformal property means that an angle between lines on the original equals the angle of their images. One must keep in mind that an angle is defined as the angle between tangents.

The first section begins with conformal mapping of planes using complex functions. It serves two purposes. First, it demonstrates in a rather simple manner the difference between conformality and similarity transformation. Second, it gives the technique for transforming the isometric plane into one of the desired standard conformal mappings, such as the ones by Mercator or Lambert. The next section gives the general formulation of conformality between general surfaces, making use of the first fundamental coefficients. Section C.3 gives the details about the isometric plane, and Section C.4 deals with those conformal mappings that are generally used in surveying. The most important ones are the transverse Mercator mapping and the Lambert conformal mapping. For example, all but one of the U.S. state plane coordinate systems are based on these mappings. An exception is a system in Alaska that uses the oblique Mercator mapping. The latter is not discussed here.

Clearly, conformal mapping has a long history with many individuals having made significant contributions. The historically inclined reader may consult the specialized literature for a full exposition of this interesting aspect. It might not be easy to delineate individual contributions in all cases. This is in part true because concepts were sometimes formulated before the appropriate mathematical tools became available.

C.1 CONFORMAL MAPPING OF PLANES

The complex number z can be given in one of the following three well-known equivalent forms:

$$z = \lambda + iq = r(\cos\theta + i\sin\theta) = re^{i\theta} \tag{C.1.1}$$

The symbols λ and q denote the real and imaginary parts, respectively, and are typically graphed as Cartesian coordinates. The polar form, the middle part of (C.1.1), is specified by the magnitude r and the argument θ. The third part of (C.1.1) is the Euler form. The reader is referred to the mathematical literature to brush up on the algebra with complex numbers, if necessary. A function of complex numbers such as

$$w = f(z) \tag{C.1.2}$$

is called complex mapping. The variable $z = \lambda + iq$ represents points on the original which are to be mapped, and $w = x + iy$ represents the respective images or the map.

$$x + iy = f(\lambda + iq) \tag{C.1.3}$$

Separating the real and imaginary parts, we can write

$$x = x(\lambda, q) \tag{C.1.4}$$

$$y = y(\lambda, q) \tag{C.1.5}$$

The derivative of the complex function (C.1.2) plays a key role in assuring that the complex mapping is conformal. The image of the increment Δz is

$$\Delta w = f(z + \Delta z) - f(z) \tag{C.1.6}$$

Analogous to computing the derivative for real functions, the derivative of a complex function follows from the limit

$$\frac{dw}{dz} \equiv f'(z) = \lim_{\Delta z\to 0} \frac{f(z + \Delta z) - f(z)}{\Delta z} = \lim_{\Delta z\to 0} \frac{\Delta w}{\Delta z} \tag{C.1.7}$$

In contrast to the case of real functions, the increment Δz has a direction; one has virtually an infinite number of possibilities of letting Δz go to zero. If the limit exists and is independent of the manner in which Δz approaches zero, then the function $f(z)$ is called differentiable. It is proven in the mathematical literature that the Cauchy-Riemann equations

$$\frac{\partial x}{\partial \lambda} = \frac{\partial y}{\partial q} \tag{C.1.8}$$

$$\frac{\partial x}{\partial q} = -\frac{\partial y}{\partial \lambda} \tag{C.1.9}$$

represent necessary and sufficient conditions for the derivative to exist. In that case, the actual derivative is given by

$$f'(z) = \frac{\partial x}{\partial \lambda} + i\frac{\partial y}{\partial \lambda} = \frac{\partial y}{\partial q} - i\frac{\partial x}{\partial q} \tag{C.1.10}$$

In terms of interpreting (C.1.2) as conformal mapping, it is advantageous to rewrite (C.1.7) using Euler's form of complex numbers, i.e.,

$$\Delta z = |\Delta z| e^{i\theta} \tag{C.1.11}$$

$$\Delta w = |\Delta w|\, e^{i\varphi} \tag{C.1.12}$$

$$f'(z) = \lim_{\Delta z \to 0} \frac{|\Delta w|\, e^{i\varphi}}{|\Delta z|\, e^{i\theta}} = \lim_{\Delta z \to 0} \frac{|\Delta w|}{|\Delta z|} e^{i(\varphi - \theta)} = |f'(z)| e^{i\gamma} \tag{C.1.13}$$

The symbols θ and φ denote here the arguments of the respective differential numbers Δz and Δw. Since the derivative exists (we will consider only functions that fulfill the Cauchy-Riemann conditions), both the magnitude $|f'(z)|$ and the argument of the derivative

$$\gamma = \varphi - \theta \tag{C.1.14}$$

are independent of the manner in which Δz approaches zero. The mapping (C.1.2) in the differential neighborhood of z is

$$|\Delta w| = |f'(z)||\Delta z| \tag{C.1.15}$$

According to (C.1.14), the argument of the image is

$$\arg \Delta w = \arg \Delta z + \arg f'(z) \tag{C.1.16}$$

Equations (C.1.15) and (C.1.16) allow the following interpretation: for complex mapping $w = f(z)$, assuming that the derivative exists, the length of an infinitesimal distance $|\Delta z|$ on the original is scaled by the factor $|f'(z)|$. This factor is solely a function of z and is independent of the direction of Δz. Similarly, the difference in the direction of the original Δz and its image Δw, $\arg \Delta w - \arg \Delta z$, is independent of the direction of the original Δz because the argument $\arg f'(z)$ is independent of Δz. Consequently, two infinitesimal segments at z will be mapped into two images that enclose the same angle, $\varphi_1 - \varphi_2 = \theta_1 - \theta_2$, or

$$\varphi_1 - \theta_1 = \gamma \tag{C.1.17}$$

$$\varphi_2 - \theta_2 = \gamma \tag{C.1.18}$$

Figure C.1.1 shows the mapping of two points in the differential neighborhood of z. The differential figures $(z_2 - z - z_1)$ and $(w_2 - w - w_1)$ differ by translation, rotation, and scale. The conformal mapping $f(z)$ does not change the angles between differentially located points; consequently, infinitesimally small figures are similar.

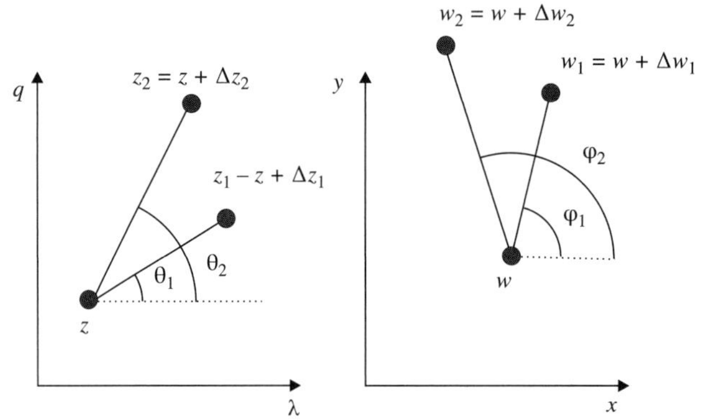

Figure C.1.1 Conformal mapping in differential neighborhood.

The scale factor of the mapping follows from (C.1.10):

$$k = |f'(z)| = \sqrt{\left(\frac{\partial x}{\partial \lambda}\right)^2 + \left(\frac{\partial y}{\partial \lambda}\right)^2} = \sqrt{\left(\frac{\partial x}{\partial q}\right)^2 + \left(\frac{\partial y}{\partial q}\right)^2} \tag{C.1.19}$$

The rotation angle γ, which will later be identified as the meridian convergence, follows from

$$\tan \gamma = \frac{\partial y/\partial \lambda}{\partial x/\partial \lambda} = -\frac{\partial x/\partial q}{\partial y/\partial q} \tag{C.1.20}$$

The following example should demonstrate the idea of conformal mapping. Using $z = \lambda + iq$ and $w = x + iy$ the simple mapping function

$$w = z^2 \tag{C.1.21}$$

gives

$$x = \lambda^2 - q^2 \tag{C.1.22}$$

$$y = 2\lambda q \tag{C.1.23}$$

Thus, the coordinates are $x = \lambda^2 - q^2$ and $y = 2\lambda q$. The partial derivatives

$$\frac{\partial x}{\partial \lambda} = \frac{\partial y}{\partial q} = 2\lambda \tag{C.1.24}$$

$$\frac{\partial x}{\partial q} = -\frac{\partial y}{\partial \lambda} = -2q \tag{C.1.25}$$

satisfy the Cauchy-Riemann equations and are continuous over the (λ, q) plane. The derivative is

$$f'(z) = \frac{\partial x}{\partial \lambda} + i\frac{\partial y}{\partial \lambda} = \frac{\partial y}{\partial q} - i\frac{\partial x}{\partial q} = 2\lambda + i2q \tag{C.1.26}$$

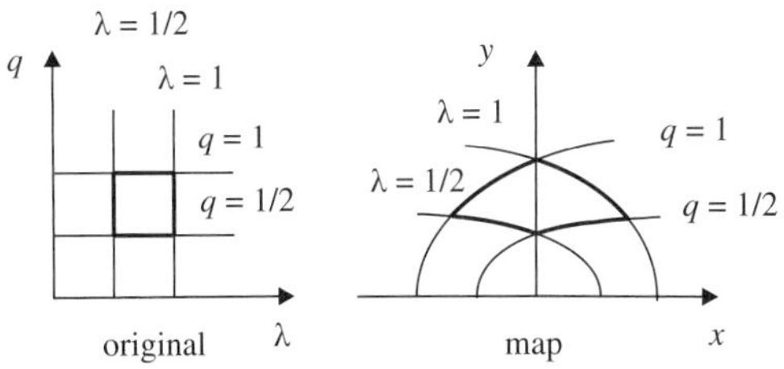

Figure C.1.2 Simple conformal mapping between planes.

Images of the lines λ = constant = c_1 follow from the mapping equations (C.1.22) and (C.1.23) upon setting $\lambda = c_1$ and eliminating q,

$$y = \pm\sqrt{4c_1^4 - 4c_1^2 x} \tag{C.1.27}$$

Similarly we obtain for the images of the lines q = constant = c_2 as

$$y = \pm\sqrt{4c_2^4 - 4c_2^2 x} \tag{C.1.28}$$

The scale in the differential neighborhood of z follows from (C.1.19) and (C.1.26) as

$$k = |f'(z)| = \sqrt{4\lambda^2 + 4q^2} \tag{C.1.29}$$

The rotation in the same differential neighborhood is, according to (C.1.16) and (C.1.26),

$$\arg f'(z) = \tan^{-1}\frac{q}{\lambda} \tag{C.1.30}$$

Figure C.1.2 shows this mapping. Mathematically, any lines parallel to the q or λ axes map into parabolas. Using differential calculus it can be readily verified that the images of the parametric curves map into a family of orthogonal curves. The same tools can be used to verify that the angle between the general lines $f_1(\lambda, q) = 0$ and $f_2(\lambda, q) = 0$ will be the same on the map. Note that the scale and the rotation angle vary continuously with location. The square and its image cannot be related by a similarity transformation.

C.2 CONFORMAL MAPPING OF GENERAL SURFACES

The approach is to find conditions for the first fundamental coefficients to assure that conformality is achieved. This general formulation is valid for conformal mapping of any surface, e.g., mapping the ellipsoid on the sphere, sphere on a plane, ellipsoid on a plane, etc.

Let the surfaces $[S]$ be expressed in terms of curvilinear coordinates (u, v),

$$\left.\begin{aligned} x &= x(u, v) \\ y &= y(u, v) \\ z &= z(u, v) \end{aligned}\right\} \tag{C.2.1}$$

This surface is to be mapped conformally on the surface $[\overline{S}]$,

$$\left.\begin{aligned} \overline{x} &= x'(\overline{u}, \overline{v}) \\ \overline{y} &= y'(\overline{u}, \overline{v}) \\ \overline{z} &= z'(\overline{u}, \overline{v}) \end{aligned}\right\} \tag{C.2.2}$$

whose curvilinear coordinates are denoted by $(\overline{u}, \overline{v})$. The mapping equations

$$\overline{u} = \overline{u}(u, v) \tag{C.2.3}$$

$$\overline{v} = \overline{v}(u, v) \tag{C.2.4}$$

relate both sets of curvilinear coordinates. These mapping equations of course are not arbitrary but must eventually be derived such that the mapping is conformal. Substituting these equations in the surface representation (C.2.2) gives

$$\left.\begin{aligned} \overline{x} &= \overline{x}(u, v) \\ \overline{y} &= \overline{y}(u, v) \\ \overline{z} &= \overline{z}(u, v) \end{aligned}\right\} \tag{C.2.5}$$

Equations (C.2.5) express the image surface $[\overline{S}]$ as a function of the curvilinear coordinates of the original surface $[S]$. The first fundamental forms (B.2.7) for both surfaces are

$$ds^2 = E\,du^2 + 2F\,du\,dv + G\,dv^2 \tag{C.2.6}$$

$$d\overline{s}^2 = \overline{E}\,du^2 + 2\overline{F}\,du\,dv + \overline{G}\,dv^2 \tag{C.2.7}$$

The conformal property is given in terms of the condition on the first fundamental coefficients

$$k^2(u, v) \equiv \frac{\overline{E}}{E} = \frac{\overline{F}}{F} = \frac{\overline{G}}{G} \tag{C.2.8}$$

That conditions (C.2.8) indeed assure conformality as can be verified by computing the angle between the two curves $f_1(u, v) = 0$ and $f_2(u, v) = 0$ on $[S]$ and between the respective images on $[\overline{S}]$. Equation (B.2.25) gives the angle on the original as

$$\cos(ds_1, ds_2) = \frac{E\,du_1\,du_2 + F\,(du_1\,dv_2 + du_2\,dv_1) + G\,dv_1\,dv_2}{\sqrt{E\,du_1^2 + 2F\,du_1\,dv_1 + G\,dv_1^2}\sqrt{E\,du_2^2 + 2F\,du_2\,dv_2 + G\,dv_2^2}} \tag{C.2.9}$$

Since the image surface has been expressed in terms of curvilinear coordinates (u, v) of the original, and since the functions $f_1(u, v) = 0$ and $f_2(u, v) = 0$ apply to the mapped lines as well, it follows that the angle on the image is given by

$$\cos(d\overline{s}_1, d\overline{s}_2) = \frac{\overline{E}\, du_1\, du_2 + \overline{F}(du_1\, dv_2 + du_2\, dv_1) + \overline{G}\, dv_1\, dv_2}{\sqrt{\overline{E}\, du_1^2 + 2\overline{F}\, du_1\, dv_1 + \overline{G}\, dv_1^2}\sqrt{\overline{E}\, du_2^2 + 2\overline{F}\, du_2\, dv_2 + \overline{G}\, dv_2^2}} \tag{C.2.10}$$

Replacing $\overline{E}$, $\overline{F}$, and $\overline{G}$ with k^2E, k^2G, and k^2G, respectively, following (C.2.8), one readily sees that

$$\cos(d\overline{s}_1, d\overline{s}_2) = \cos(ds_1, ds_2) \tag{C.2.11}$$

and therefore the angle enclosed by the tangents on f_1 and f_2 is preserved. The point scale factor for the mapping is

$$k(u, v) = \frac{d\overline{s}}{ds} \tag{C.2.12}$$

As an example, one might verify the general condition (C.2.8) for the simple conformal mapping (C.1.21) between two planes. Following the general notation, the equations for the original (C.2.1) have the simple form $y = q$ and $x = \lambda$. The respective first fundamental coefficients are $E = G = 1$ and $F = 0$. The expressions for the image surface (C.2.2) are $\overline{x} = x$ and $\overline{y} = y$. Substituting the mapping equations (C.1.22) and (C.1.23) into the image surface expressions gives $\overline{x} = \lambda^2 - q^2$ and $\overline{y} = 2\lambda q$. The first fundamental coefficients are $\overline{E} = \overline{G} = 4\lambda^2 + 4q^2$ and $\overline{F} = 0$. It follows that the condition (C.2.8) is indeed fulfilled for this simple mapping.

C.3 ISOMETRIC PLANE

An especially simple situation arises if the curvilinear coordinates (u, v) on the original are isometric and orthogonal. The curvilinear coordinates (q, λ), where q denotes the isometric latitude given in (B.2.28), form such an isometric net on the ellipsoid. The first fundamental form becomes, according to (B.2.27),

$$ds^2 = N^2 \cos^2 \varphi\, (dq^2 + d\lambda^2) \tag{C.3.1}$$

which implies that $E = G = N^2 \cos^2 \varphi$ and $F = 0$. The first step in utilizing the isometric curvilinear coordinates (q, λ) for conformal mapping is to consider the mapping equations

$$x = \lambda \tag{C.3.2}$$

$$y = q \tag{C.3.3}$$

and interpret (x, y) as Cartesian coordinates, i.e., the expressions for the image surface simply are

$$\overline{x} = \lambda \tag{C.3.4}$$

$$\overline{y} = q \tag{C.3.5}$$

and $\overline{E} = \overline{G} = 1$ and $\overline{F} = 0$. The first fundamental coefficients meet the condition (C.2.8). The point scale factor for this mapping is

$$k^2 = \frac{dq^2 + d\lambda^2}{E(dq^2 + d\lambda^2)} = \frac{1}{N^2 \cos^2 \varphi} \tag{C.3.6}$$

We may, therefore, conclude that one way of creating a conformal mapping of a general surface to a plane is to establish an isometric net on the original and then interpret the isometric coordinates as Cartesian coordinates and call the result the isometric mapping plane.

In a subsequent step, the isometric plane can be mapped conformally onto another mapping plane by the analytic function

$$x + iy = f(\lambda + iq) \tag{C.3.7}$$

The implied mapping equations are

$$x = x(q, \lambda) \tag{C.3.8}$$

$$y = y(q, \lambda) \tag{C.3.9}$$

where (x, y) denote the coordinates in the final map. The point scale factor of such a sequential conformal mapping equals the product of that of the individual mappings. According to (C.2.12) and (C.1.19), we have

$$\begin{aligned} k &= \frac{ds_{IP}}{ds}\frac{d\overline{s}}{ds_{IP}} = k_{IP} \cdot k_{IP \to Map} \\ &= \frac{\sqrt{(\partial x/\partial \lambda)^2 + (\partial y/\partial \lambda)^2}}{N \cos \varphi} = \frac{\sqrt{(\partial x/\partial q)^2 + (\partial y/\partial q)^2}}{N \cos \varphi} \end{aligned} \tag{C.3.10}$$

Additional specifications that the complex function must fulfill will assure that a conformal map with the desired properties will be obtained.

C.4 POPULAR CONFORMAL MAPPINGS

The transverse Mercator and Lambert conformal mappings are the most popular mappings used in geodetic computations. Not only do they serve as the basis for the

U.S. state plane coordinate systems but they are also widely used by other countries as the national mapping system. Since the respective mapping equations can be easily programmed, these mappings are suitable for local mapping as well. The equatorial Mercator mapping is presented first because it follows in such a straightforward manner from the isometric plane. The transverse Mercator and the Lambert conformal mapping will then be discussed. Finally, the polar conformal mapping is specified. Most of the derivations related to this appendix are compiled in Leick (2002).

C.4.1 Equatorial Mercator

The equatorial Mercator mapping (EM) is a linear mapping of the isometric plane such that the equator of the ellipsoid and its image on the map are of the same length. See Figure C.4.1. This is accomplished by

$$x + iy = a(\lambda + iq) \tag{C.4.1}$$

where the symbol a denotes the semimajor axis of the ellipsoid. The mapping equations become

$$x = a\lambda \tag{C.4.2}$$

$$y = aq \tag{C.4.3}$$

This map is simply a magnification of the isometric plane. The meridians map into straight lines that are parallel to the y axis; y is zero at the equator. The mapped parallels are also straight lines and are parallel to the x axis, but spacing increases toward the poles for the same latitude increment. The equator is mapped equidistantly.

The meridian convergence is zero because mapped meridians are parallel to the y axis. Zero meridian convergence can be readily verified by applying expression (C.1.20) to the mapping equations (C.4.2) and (C.4.3). The point scale factor is, according to (C.3.10),

$$k = \frac{a}{N \cos \varphi} \tag{C.4.4}$$

This point scale factor does not depend on the longitude; $k = 1$ on the equator and the value increases with latitude. This makes this mapping attractive for use in regions close to the equator.

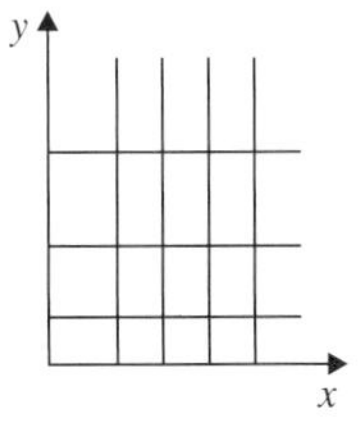

Figure C.4.1 Equatorial Mercator map.

Any meridian can serve as central meridian or zero meridian, with which the y axis coincides. For example, the meridian, which passes through the middle of the mapping area, can be the zero meridian at which $x = 0$. Furthermore, the point scale factor must not be confused with the scale of a conventional map, which is the ratio of a plotted distance over the mapped distance. The point scale factor is a characteristic of the mapping and changes with location, whereas the scale of a map is dictated by the size of the plotting paper and the area to be plotted.

The loxodrome is a curve that intersects consecutive meridians at the same azimuth. It can be readily visualized that the loxodrome maps into a straight line for the equatorial Mercator mapping.

C.4.2 Transverse Mercator

The specifications for the transverse Mercator mapping (TM) are as follows:

1. Apply conformal mapping conditions.
2. Adopt a central meridian λ_0 that passes more or less through the middle of the area to be mapped. For reasons of convenience, relabel the longitudes starting with $\lambda = 0$ at the central meridian.
3. Let the mapped central meridian coincide with the y axis of the map. Assign $x = 0$ to the image of the central meridian.
4. The length of the mapped central meridian should be k_0 times the length of the corresponding elliptic arc, i.e., at the central meridian $y = k_0\, s_\varphi$.

The derivation of the transverse Mercator mapping begins with the isometric plane and a suitable complex function f for (C.3.7). Condition 4 specifies the image of the central meridian and implies

$$0 + ik_0 s_\varphi = f(0 + iq) \tag{C.4.5}$$

This function can be expanded in a Taylor series, which provides an opportunity to impose the Cauchy-Riemann conditions on the partial derivatives. The general picture of the transverse Mercator map is shown in Figure C.4.2. The image of the central

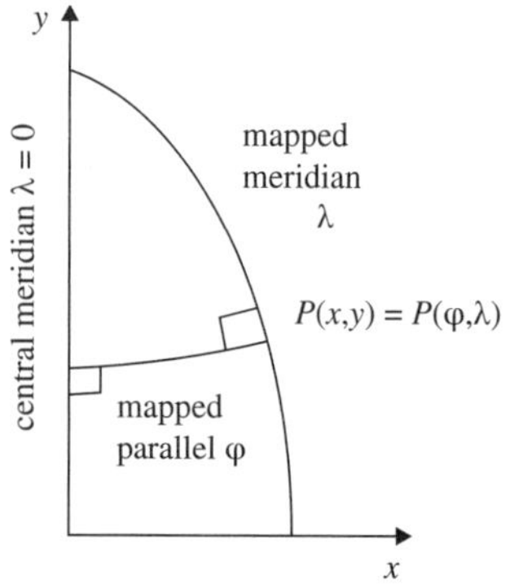

Figure C.4.2 Transverse Mercator map.

TABLE C.4.1 Transverse Mercator Direct Mapping

$$\frac{x}{k_0 N} = \lambda \cos\varphi + \frac{\lambda^3 \cos^3\varphi}{6}(1 - t^2 + \eta^2) + \frac{\lambda^5 \cos^5\varphi}{120}(5 - 18t^2 + t^4 + 14\eta^2 - 58t^2\eta^2) \qquad \text{(a)}$$

$$\frac{y}{k_0 N} = \frac{s}{N} + \frac{\lambda^2}{2}\sin\varphi\cos\varphi + \frac{\lambda^4}{24}\sin\varphi\cos^3\varphi\,(5 - t^2 + 9\eta^2 + 4\eta^4) + \frac{\lambda^6}{720}\sin\varphi\cos^5\varphi(61 - 58t^2 + t^4 + 270\eta^2 - 330t^2\eta^2) \qquad \text{(b)}$$

$$\gamma = \lambda \sin\varphi\left[1 + \frac{\lambda^2\cos^2\varphi}{3}\left(1 + 3\eta^2 + 2\eta^4\right) + \frac{\lambda^4\cos^4\varphi}{15}(2 - t^2)\right] \qquad \text{(c)}$$

meridian is a straight line; all other meridians are curved lines coming together at the pole and being perpendicular to the image of the equator. The latter coincides with the x axis. The mapped parallels are, of course, perpendicular to the mapped meridians; however, they are not circles but mathematically complex curves.

The mapping equations for the direct mapping from $P(\varphi, \lambda)$ to $P(x, y)$ are listed in Table C.4.1. In these expressions, the longitude λ is counted positive to the east, starting at the central meridian. All quantities that depend on the latitude must be evaluated at φ. Equation (B.1.6) gives the expression for the radius of curvature N of the prime vertical section. The symbol s denotes the length of the elliptic arc from the equator to φ as given by (B.2.21). The symbols t and η are used for brevity and mean:

$$t = \tan\varphi \qquad \text{(C.4.6)}$$

$$\eta^2 = \frac{e^2}{1 - e^2}\cos^2\varphi \qquad \text{(C.4.7)}$$

The inverse solution for mapping $P(x, y)$ to $P(\varphi, \lambda)$ is given in Table C.4.2. All latitude-dependent terms in this table must be evaluated for the so-called footpoint latitude φ_f. The footpoint is a point on the central meridian obtained by drawing a parallel to the x axis through the point $P(x, y)$. Given the y coordinate, the foot-point latitude can be computed iteratively from (B.2.21). Because of condition 4, the following relation holds:

$$s_f = \frac{y}{k_0} \qquad \text{(C.4.8)}$$

where s_f is the length of the central meridian from the equator to the footpoint. Substitute (C.4.8) in (B.2.22) and solve φ_f iteratively.

The expression for the point scale factor is

$$\frac{k}{k_0} = 1 + \frac{\lambda^2}{2}\cos^2\varphi\,(1 + \eta^2) + \frac{\lambda^4}{24}\cos^4\varphi(5 - 4t^2 + 14\eta^2 + 13\eta^4 - 28t^2\eta^2 + 4\eta^6 - 48t^2\eta^4 - 24t^2\eta^6) + \frac{\lambda^6}{720}\cos^6\varphi(61 - 148t^2 + 16t^4) \qquad \text{(C.4.9)}$$

TABLE C.4.2 Transverse Mercator Inverse Mapping

$$\varphi = \varphi_f - \frac{t}{2}(1+\eta^2)\left(\frac{x}{k_0 N}\right)^2 + \frac{t}{24}(5+3t^2+6\eta^2-6\eta^2 t^2-3\eta^4-9t^2\eta^4)\left(\frac{x}{k_0 N}\right)^4$$
$$-\frac{t}{720}(61+90t^2+45\eta^4+107\eta^2-162\eta^2 t^2-45t^4\eta^2)\left(\frac{x}{k_0 N}\right)^6 \qquad \text{(a)}$$

$$\lambda \cos\varphi_f = \frac{x}{k_0 N} - \frac{1}{6}\left(\frac{x}{k_0 N}\right)^3(1+2t^2+\eta^2)$$
$$+\frac{1}{120}\left(\frac{x}{k_0 N}\right)^5(5+28t^2+24t^2+6\eta^2+8t^2\eta^2) \qquad \text{(b)}$$

This equation shows that the scale factor k increases primarily with longitude. In fact, isoscale lines run more or less parallel to the image of the central meridian. Since the mapping distortions increase as k departs from 1, the factor k_0 is an important element of design. By selecting $k_0 < 1$, one allows some distortion at the central meridian for the benefit of having less distortion away from the central meridian. In this way, the longitudinal coverage of the area of a map can be extended given a level of acceptable distortion.

The appearance of the TM mapping expressions reveals the fact that they have been obtained from series expansions. Consequently, the expressions are accurate only as long as the truncation errors are negligible. Note the symmetries with respect to the central meridian, $x(-\lambda) = -x(\lambda)$ and $y(-\lambda) = y(\lambda)$, and the equator, $y(-\varphi) = -y(\varphi)$ and $x(-\varphi) = x(\varphi)$. These TM expressions are also given in Thomas (1952, pp. 96–103), who lists some additional higher-order terms.

The transverse Mercator mapping of the ellipsoidal as given above is attributed to Gauss, who used his extensive developments in differential geometry to study conformal mapping of general surfaces. Other scientists further refined Gauss's basic developments in order to produce expressions suitable for calculation, which was a necessity before computers became available. Most notable are contributions by L. Krüger. Lee (1976) presents closed or exact formulas for the transverse Mercator mapping with respect to the ellipsoid; these elliptical expressions were programmed by Dozier (1980). Lee further discusses other variations of the transverse Mercator mapping, in addition to one with constant scale factor along the mapped central meridian presented here. Finally, it should be emphasized that Lambert (1772) already gave expressions for the transverse Mercator mapping with respect to the sphere.

C.4.3 Lambert Conformal

The specifications for the Lambert conformal (LC) mapping are:

1. Apply conformal mapping conditions.
2. Adopt a central meridian λ_0 that passes more or less through the middle of the area to be mapped. For reasons of convenience, relabel the longitudes, starting with $\lambda = 0$ at the central meridian.

3. Let the mapped central meridian coincide with the y axis of the map. Assign $x = 0$ for the image of the central meridian.
4. Map the meridians into straight lines passing through the image of the pole; map the parallels into concentric circles around the image of the pole. Select a standard parallel φ_0 that passes more or less through the middle of the area to be mapped. The length of the mapped standard parallel is k_0 times the length of the corresponding ellipsoidal parallel. The point scale factor along any mapped parallel is constant. Start counting $y = 0$ at the image of the standard parallel.

The general picture of the Lambert conformal map is shown in Figure C.4.3. The mapping is singular at the pole, which is the reason why the angle of the mapped meridian is λ', and not λ, at the pole. Denoting the distance from the mapped parallel to the pole by r, the pair (λ', r) are polar coordinates that form a set of orthogonal curvilinear lines on the map. The first fundamental form for this choice of coordinates is

$$d\overline{s}^2 = dr^2 + r^2 d\lambda'^2 = r^2 \left(\frac{dr^2}{r^2} + d\lambda'^2 \right) \tag{C.4.10}$$

We observe that (λ', r) is not an isometric net. The same increments in dr and $d\lambda'$ result in different changes of $d\overline{s}$. If we define the auxiliary coordinate

$$dq' \equiv -\frac{dr}{r} \tag{C.4.11}$$

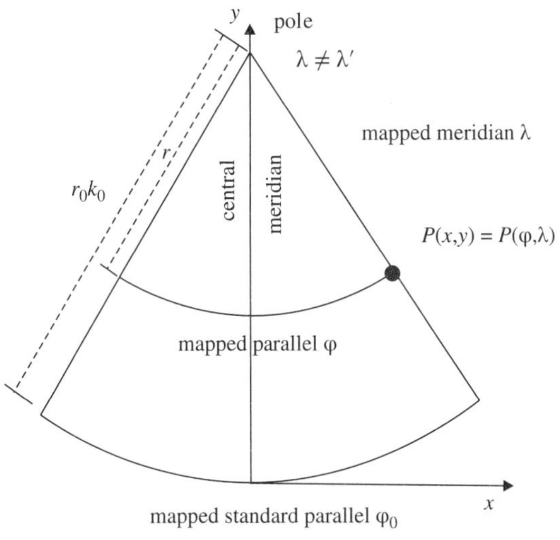

Figure C.4.3 Lambert conformal mapping.

then (λ', q') indeed constitutes an isometric net in the mapping plane. The integration of (C.4.11) gives

$$q' = -\int_{k_0 r}^{r} \frac{dr}{r} = -(\ln r - \ln k_0 r_0) = -\ln \frac{r}{k_0 r_0} \tag{C.4.12}$$

At the standard parallel φ_0 we have $r = k_0 r_0$ and $q' = 0$. The negative sign in (C.4.11) takes care of the fact that q' increases toward the pole, whereas r decreases. Equal incremented quadrilaterals of (q', λ') decrease as the pole is approached. The Lambert conformal mapping is now specified by

$$\lambda' + iq' = \alpha[\lambda + i(q - q_0)] \tag{C.4.13}$$

where $\alpha = \sin \varphi_0$ and q_0 is the isometric latitude of the standard parallel. The value of the constant α is derived on the basis of condition 4.

The expressions for the direct and inverse mapping are listed in Tables C.4.3 and C.4.4. See Thomas (1952, p. 117) or Leick (2002) for a complete derivation. The symbol $\varepsilon = 2.71828\ldots$ denotes the base of the natural system of logarithm and should not be confused with the eccentricity e of the ellipsoid. The inverse solution gives the isometric latitude first, which can then be readily converted to the geodetic latitude.

There is no series expansion involved. However, attention must be given to numerical accuracy when converting q to φ. The point scale factor is

$$k = \frac{k_0 N_0 \cos \varphi_0}{N \cos \varphi} \varepsilon^{-(q-q_0)\sin \varphi_0} \tag{C.4.14}$$

TABLE C.4.3 Lambert Conformal Direct Mapping

$x = k_0 N_0 \cot \varphi_0 \; \varepsilon^{-\Delta q \sin \varphi_0} \sin(\lambda \sin \varphi_0)$	(a)
$y = k_0 N_0 \cot \varphi_0 \; [1 - \varepsilon^{-\Delta q \sin \varphi_0} \cos(\lambda \sin \varphi_0)]$	(b)
$\gamma \equiv \lambda' = \lambda \sin \varphi_0$	(c)

TABLE C.4.4 Lambert Conformal Inverse Mapping

$\tan \lambda' = \dfrac{x}{k_0 N_0 \cot \varphi_0 - y}$	(a)
$r = \dfrac{k_0 N_0 \cot \varphi_0 - y}{\cos \lambda'}$	(b)
$\lambda = \dfrac{\lambda'}{\sin \varphi_0}$	(c)
$\Delta q = -\dfrac{1}{\sin \varphi_0} \ln \left(\dfrac{r}{k_0 N_0 \cot \varphi_0} \right)$	(d)
$q = q_0 + \Delta q$	(e)

TABLE C.4.5 Conversion from Two Standard Parallels to One Standard Parallel

$\varphi_0 = \sin^{-1}\left[\dfrac{\ln(N_1 \cos\varphi_1) - \ln(N_2 \cos\varphi_2)}{q_2 - q_1}\right]$	(a)
$k_0 = \dfrac{N_1 \cos\varphi_1}{N_0 \cos\varphi_0}\varepsilon^{(q_1-q_0)\sin\varphi_0} = \dfrac{N_2 \cos\varphi_2}{N_0 \cos\varphi_0}\varepsilon^{(q_2-q_0)\sin\varphi_0}$	(b)

Note that (k_0, φ_0) or, equivalently, (k_0, q_0) specifies the expressions for the Lambert conformal mapping. The area of smallest distortion is along the image of the standard parallel in the east-west direction; as one departs from the standard parallel, the distortions increase in the north-south direction. By selecting $k_0 < 1$ it is possible to reduce the distortions at the northern and southern extremities of the mapping area by allowing some distortions in the vicinity of the standard parallel. Whenever $k_0 < 1$ there are two parallels, one south and one north of the standard parallel, along which the point scale factor k equals 1, i.e., these two parallels are mapped without distortion in length.

The designer of the map has the choice of either specifying k_0 and φ_0 or the two parallels for which $k = 1$. In the latter case, one speaks of a two-standard-parallel Lambert conformal mapping. If the Lambert conformal mapping is specified by two standard parallels φ_1 and φ_2 with $k_1 = k_2 = 1$, then k_0 and φ_0 follow from the expression of Table C.4.5.

In the special case of $\varphi_0 = 90°$, the Lambert conformal mapping becomes the polar conformal mapping. The expressions are obtained by noting the following mathematical limit:

$$F \equiv \lim_{\varphi_0 \to 90°} N_0(\cos\varphi_0)\varepsilon^{q_0} = \frac{2a^2}{b}\left(\frac{1-e}{1+e}\right)^{e/2} \tag{C.4.15}$$

where the symbols a and b denote the semiaxis of the ellipsoid and the general relation $b/a = \sqrt{1 - e^2}$ has been used. Using (C.4.15), the equations for the polar conformal mapping become

$$x = k_0 F\varepsilon^{-q} \sin\lambda \tag{C.4.16}$$

$$y = k_0 F\varepsilon^{-q} \cos\lambda \tag{C.4.17}$$

$$\gamma = \lambda \tag{C.4.18}$$

$$k = \frac{k_0 F\varepsilon^{-q}}{N \cos\varphi} \tag{C.4.19}$$

As is the case with the Lambert conformal mapping, the meridians are straight lines radiating from the image of the pole and the parallels are concentric circles around the pole. The y axis coincides with the 180° meridian. See Figure C.4.4 for details. There is no particular advantage in selecting the central meridian of the area to be mapped

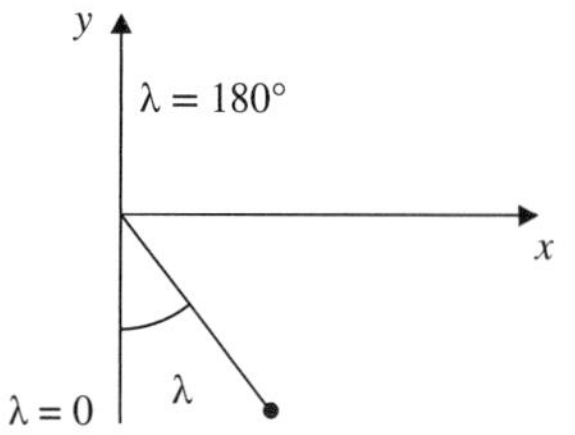

Figure C.4.4 Polar conformal mapping.

as the zero meridian. One, therefore, will usually select the Greenwich meridian. The point scale factor is k_0 at the pole.

The polar conformal mapping is not a stereographic projection of the ellipsoid. Only in the special case of $e = 0$ do the expressions given above transition to those of the stereographic projection of the sphere (polar aspect) with the point of perspective being at the South Pole. The mapping is not only stereographic in the case of the sphere (because it is projected from a single point of perspective), but it is also azimuthal because the sphere is projected on the tangent plane. In the case of the oblique aspect of the stereographic projection, the sphere is projected on the tangent plane other than at the pole, with the center of projection being located on the sphere diametrically opposite the point of tangency. The ellipsoidal form of the oblique aspect is not perspective (no one single center of projection) in order to meet

TABLE C.4.6 Legend for U.S. State Plane Coordinate System Defining Constants

	Mapping	
T	Transverse Mercator	
L	Lambert conformal	
O	Oblique Mercator	
UTM	Universal transverse Mercator	
1:M	Scale reduction at central meridian	
	Conversion factors	
Meters	U.S. survey feet	International feet
152,400.3048	500,000.0	
213,360.0		700,000.0
304,800.6096	1,000,000.0	
609,600.0		2,000,000.0
609,601.2192	2,000.000.0	
914,401.8289	3,000,000.0	
1.	3.28083333333	
1.		3.28083989501
0.3048		1.
1200/3937	1.	
0.30480060960	1.	

TABLE C.4.7 U.S. State Plane Coordinate Systems. West longitudes are listed 1983 defining constants.

State Zone	SPCS Zone	Type T/L/O	Latitude Origin [DD MM]	Longitude Origin [DD MM]	False North [m]	False East [m]	$\frac{1}{1-k_0}$	Standard Parallels South [DD MM]	Standard Parallels North [DD MM]
Alabama									
East	0101	T	30 30	85 50	0.0	200000.0	25000		
West	0102	T	30 00	87 30	0.0	600000.0	15000		
Alaska									
Zone1	5001	0	57 00	133 40	-5000000.0	5000000.0	10000	Axis Az	$\tan^{-1}(-3/4)$
Zone 2	5002	T	54 00	142 00	0.0	500000.0	10000		
Zone 3	5003	T	54 00	146 00	0.0	500000.0	10000		
Zone 4	5004	T	54 00	150 00	0.0	500000.0	10000		
Zone 5	5005	T	54 00	154 00	0.0	500000.0	10000		
Zone 6	5006	T	54 00	158 00	0.0	500000.0	10000		
Zone 7	5007	T	54 00	162 00	0.0	500000.0	10000		
Zone 8	5008	T	54 00	166 00	0.0	500000.0	10000		
Zone 9	5009	T	54 00	170 00	0.0	500000.0	10000		
Zone 10	5010	L	51 00	176 00	0.0	1000000.0		51 50	53 50
Arizona									
East	0201	T	31 00	110 10	0.0	213360.0	10000		
Central	0202	T	31 00	111 55	0.0	213360.0	10000		
West	0203	T	31 00	113 45	0.0	213360.0	15000		
Arkansas									
North	0301	L	34 20	92 00	0.0	400000.0		34 56	36 14
South	0302	L	32 40	92 00	400000.0	400000.0		33 18	34 46

(*continued*)

TABLE C.4.7 (*Continued*)

State Zone	SPCS Zone	Type T/L/O	Latitude Origin [DD MM]	Longitude Origin [DD MM]	False North [m]	False East [m]	$\frac{1}{1-k_0}$	Standard Parallels South [DD MM]	Standard Parallels North [DD MM]
California									
Zone 1	0401	L	39 20	122 00	500000.0	2000000.0		40 00	41 40
Zone 2	0402	L	37 40	122 00	500000.0	2000000.0		38 20	39 50
Zone 3	0403	L	36 30	120 30	500000.0	2000000.0		37 04	38 26
Zone 4	0404	L	35 20	119 00	500000.0	2000000.0		36 00	37 15
Zone 5	0405	L	33 30	118 00	500000.0	2000000.0		34 02	35 28
Zone 6	0406	L	32 10	116 15	500000.0	2000000.0		32 47	33 53
Colorado									
North	0501	L	39 20	105 30	304800.6096	914401.8289		39 43	40 47
Central	0502	L	37 50	105 30	304800.6096	914401.8289		38 27	39 45
South	0503	L	36 40	105 30	304800.6096	914401.8289		37 14	38 26
Connecticut	0600	L	40 50	72 45	152400.3048	304800.6096		41 12	41 52
Delaware	0700	T	38 00	75 25	0.0	200000.0	200000		
Florida									
East	0901	T	24 20	81 00	0.0	200000.0	17000		
West	0902	T	24 20	82 00	0.0	200000.0	17000		
North	0903	L	29 00	84 30	0.0	600000.0		29 35	30 45
Georgia									
East	1001	T	30 00	82 10	0.0	200000.0	10000		
West	1002	T	30 00	84 10	0.0	700000.0	10000		

Hawaii									
Zone 1	5101	T	18 50	155 30	0.0	500000.0	30000		
Zone 2	5102	T	20 20	156 40	0.0	500000.0	30000		
Zone 3	5103	T	21 10	158 00	0.0	500000.0	100000		
Zone 4	5104	T	21 50	159 30	0.0	500000.0	100000		
Zone 5	5105	T	21 40	160 10	0.0	500000.0	∞		
Idaho									
East	1101	T	41 40	112 10	0.0	200000.0	19000		
Central	1102	T	41 40	114 00	0.0	500000.0	19000		
West	1103	T	41 40	115 45	0.0	800000.0	15000		
Illinois									
East	1201	T	36 40	88 20	0.0	300000.0	40000		
West	1202	T	36 40	90 10	0.0	700000.0	17000		
Indiana									
East	1301	T	37 30	85 40	250000.0	100000.0	30000		
West	1302	T	37 30	87 05	250000.0	900000.0	30000		
Iowa									
North	1401	L	41 30	93 30	1000000.0	1500000.0		42 04	43 16
South	1402	L	40 00	93 30	0.0	500000.0		40 37	41 47
Kansas									
North	1501	L	38 20	98 00	0.0	400000.0		38 43	39 47
South	1502	L	36 40	98 30	400000.0	400000.0		37 16	38 34

(*continued*)

TABLE C.4.7 ***(Continued)***

State Zone	SPCS Zone	Type T/L/O	Latitude Origin [DD MM]	Longitude Origin [DD MM]	False North [m]	False East [m]	$\frac{1}{1-k_0}$	Standard Parallels South [DD MM]	Standard Parallels North [DD MM]
Kentucky									
North	1601	L	37 30	84 15	0.0	500000.0		37 58	38 58
South	1602	L	36 20	85 45	500000.0	500000.0		36 44	37 56
Louisiana									
North	1701	L	30 30	92 30	0.0	1000000.0		31 10	32 40
South	1702	L	28 30	91 20	0.0	1000000.0		29 18	30 42
Offshore	1703	L	25 30	91 20	0.0	1000000.0		26 10	27 50
Maine									
East	1801	T	43 40	68 30	0.0	300000.0	10000		
West	1802	T	42 50	70 10	0.0	900000.0	30000		
Maryland	1900	L	37 40	77 00	0.0	400000.0		38 18	39 27
Massachusetts									
Mainland	2001	L	41 00	71 30	750000.0	200000.0		41 43	42 41
Island	2002	L	41 00	70 30	0.0	500000.0		41 17	41 29
Michigan									
North	2111	L	44 47	87 00	0.0	8000000.0		45 29	47 05
Central	2112	L	43 19	84 22	0.0	6000000.0		44 11	45 42
South	2113	L	41 30	84 22	0.0	4000000.0		42 06	43 40
Minnesota									
North	2201	L	46 30	93 06	100000.0	800000.0		47 02	48 38
Central	2202	L	45 00	94 15	100000.0	800000.0		45 37	47 03
South	2203	L	43 00	94 00	100000.0	800000.0		43 47	45 13

Mississippi									
East	2301	T	29 30	88 50	0.0	300000.0	20000		
West	2302	T	29 30	90 20	0.0	700000.0	20000		
Missouri									
East	2401	T	35 50	90 30	0.0	250000.0	15000		
Central	2402	T	35 50	92 30	0.0	500000.0	15000		
West	2403	T	36 10	94 30	0.0	850000.0	17000		
Montana	2500	L	44 15	109 30	0.0	600000.0		45 00	49 00
Nebraska	2600	L	39 50	100 00	0.0	500000.0		40 00	43 00
Nevada									
East	2701	T	34 45	115 35	8000000.0	200000.0	10000		
Central	2702	T	34 45	116 40	6000000.0	500000.0	10000		
West	2703	T	34 45	118 35	4000000.0	800000.0	10000		
New Hampshire	2800	T	42 30	71 40	0.0	300000.0	30000		
New Jersey	2900	T	38 50	74 30	0.0	150000.0	10000		
New Mexico									
East	3001	T	31 00	104 20	0.0	165000.0	11000		
Central	3002	T	31 00	106 15	0.0	500000.0	10000		
West	3003	T	31 00	107 50	0.0	830000.0	12000		
New York									
East	3101	T	38 50	74 30	0.0	150000.0	10000		
Central	3102	T	40 00	76 35	0.0	250000.0	16000		
West	3103	T	40 00	78 35	0.0	350000.0	16000		

(continued)

TABLE C.4.7 (*Continued*)

State Zone	SPCS Zone	Type T/L/O	Latitude Origin [DD MM]	Longitude Origin [DD MM]	False North [m]	False East [m]	$\frac{1}{1-k_0}$	Standard Parallels South [DD MM]	Standard Parallels North [DD MM]
Long Island	3104	L	40 10	74 00	0.0	300000.0		40 40	41 02
North Carolina	3200	L	33 45	79 00	0.0	609601.22		34 20	36 10
North Dakota									
North	3301	L	47 00	100 30	0.0	600000.0		47 26	48 44
South	3302	L	45 40	100 30	0.0	600000.0		46 11	47 29
Ohio									
North	3401	L	39 40	82 30	0.0	600000.0		40 26	41 42
South	3402	L	38 00	82 30	0.0	600000.0		38 44	40 02
Oklahoma									
North	3501	L	35 00	98 00	0.0	600000.0		35 34	36 46
South	3502	L	33 20	98 00	0.0	600000.0		33 56	35 14
Oregon									
North	3601	L	43 40	120 30	0.0	2500000.0		44 20	4600
South	3602	L	41 40	120 30	0.0	1500000.0		42 20	44 00
Pennsylvania									
North	3701	L	40 10	77 45	0.0	600000.0		40 53	41 57
South	3702	L	39 20	77 45	0.0	600000.0		39 56	40 58
Rhode Island	3800	T	41 05	71 30	0.0	100000.0	160000		
South Carolina	3900	L	31 50	81 00	0.0	609600.0		32 30	34 50
South Dakota									
North	4001	L	43 50	100 00	0.0	600000.0		44 25	45 41
South	4002	L	42 20	100 20	0.0	600000.0		42 50	44 24
Tennessee	4100	L	34 20	86 00	0.0	600000.0		35 15	36 25

Texas									
North	4201	L	34 00	101 30	1000000.0	200000.0		34 39	36 11
North Central	4202	L	31 40	98 30	2000000.0	600000.0		32 08	33 58
Central	4203	L	29 40	100 20	3000000.0	700000.0		30 07	31 53
South Central	4204	L	27 50	99 00	4000000.0	600000.0		28 23	30 17
South	4205	L	25 40	98 30	5000000.0	300000.0		26 10	27 50
Utah									
North	4301	L	40 20	111 30	10000000.0	500000.0		40 43	41 47
Central	4302	L	38 20	111 30	20000000.0	500000.0		39 01	40 39
South	4303	L	36 40	111 30	30000000.0	500000.0		37 13	38 21
Vermont	4400	T	42 30	72 30	0.0	500000.0	28000		
Virginia									
North	4501	L	37 40	78 30	20000000.0	3500000.0		38 02	39 12
South	4502	L	36 20	78 30	10000000.0	3500000.0		36 46	37 58
Washington									
North	4601	L	47 00	120 50	0.0	500000.0		47 30	48 44
South	4602	L	45 20	120 30	0.0	500000.0		45 50	47 20
West Virginia									
North	4701	L	38 30	79 30	0.0	600000.0		39 00	40 15
South	4702	L	37 00	81 00	0.0	600000.0		37 29	38 53
Wisconsin									
North	4801	L	45 10	90 00	0.0	600000.0		45 34	46 46
Central	4802	L	43 50	90 00	0.0	600000.0		44 15	45 30
South	4803	L	42 00	90 00	0.0	600000.0		42 44	44 04
Wyoming									
East	4901	T	40 30	105 10	0.0	200000.0	16000		
East Central	4902	T	40 30	107 20	100000.0	400000.0	16000		
West Central	4903	T	40 30	108 45	0.0	600000.0	16000		
West	4904	T	40 30	110 05	100000.0	800000.0	16000		
Puerto Rico	5200	L	17 50	66 26	200000.0	200000.0		18 02	18 26
Virgin Islands	5200	L	17 50	66 26	200000.0	200000.0		18 02	18 26

the conformal property. Readers desiring information on oblique aspect mappings are referred to the specialized literature.

In order to emphasize that the mappings discussed in this section are derived from the conformality condition, the term *mapping* has been used consistently. For example, we prefer to speak of transverse Mercator or Lambert conformal mapping instead of the transverse Mercator or Lambert conformal projection. There is not one single point of perspective for these mappings of the ellipsoid.

C.4.4 SPC and UTM

Each state and U.S. possession has a state plane coordinate system (SPC) defined for use in surveying and mapping (Stem, 1989). Many of the state plane coordinate systems use the transverse Mercator mapping. States with large east-west extent use the Lambert conformal mapping. Many states divide their region into zones and use both the transverse Mercator and the Lambert conformal mapping for individual zones. The exception to this scheme is the state plane coordinate system for the panhandle of Alaska, for which the oblique Mercator mapping is used.

The defining constants for the U.S. state plane coordinate system of 1983 are given in Tables C.4.6 and C.4.7. Table C.4.7 also contains the adopted values of false north and false east and a four-digit code to identify the projection. The "origin" as given in Table C.4.7 is not identical with the origin of the coordinate system in Figures C.4.2 and C.4.3. The state plane coordinates refer to the NAD83 ellipsoid. The specifications of the UTM mapping are given in Table C.4.8. These specifications must be taken into account when using the mapping equations given in this appendix. The National Geodetic Survey and other mapping agencies make software available on the Internet for computing coordinates for the officially adopted mappings.

TABLE C.4.8 UTM Mapping System Specifications

UTM zones	6° in longitude (exceptions exist)
Limits in latitude	$-80° < \text{latitude} < 80°$
Longitude of origin	Central meridian of each zone
Latitude of origin	0° (equator)
Units	Meter
False northing	0 m at equator, increases northward for northern hemisphere 10, 000, 000 m at equator, decreases southward for southern hemisphere
False easting	500, 000 m at the central meridian, decreasing westward
Central meridian scale	0.9996 (exceptions exist)
Zone numbers	Starting with 1 centered 177° W and increasing eastward to zone 60 entered at 177° E (exceptions exist)
Limits of zones and overlap	Zones are bounded by meridians that are multiples of 6° W and 6° E of Greenwich
Reference ellipsoid	Depending on region and datum, e.g., GRS80 in United State for NAD83

It is the surveyor's choice to use a state plane coordinate system or to generate a local mapping by merely specifying k_0 (usually 1) and the central meridian and/or the standard parallel. In the latter case the mapping reductions are small. If, in addition, a local ellipsoid is specified, then most of the reductions can be neglected for small surveys. While these specifications might lead to a reduction in the computational load, which, in view of modern computer power is not critical any longer, it increases the probability that reductions are inadvertently neglected when they should not be.

APPENDIX D

VECTOR CALCULUS AND DELTA FUNCTION

We provide an overview of vector calculus statements used throughout Chapter 9. For a detailed treatment of the subject the reader is referred to specialized texts in electromagnetic field theory such as Balanis (1989) or mathematics handbooks like Korn and Korn (1968).

In order to characterize scalar and vector fields, we introduce coordinate systems. A general treatment of orthogonal coordinates can be found in Morse and Feshbach (1953). The three coordinate systems used most often are Cartesian, cylindrical, and spherical. Cartesian coordinates are illustrated in Figure D.1.

Here one has three orthogonal axes — x, y, and z — with origin O. A standard basis of three unit vectors corresponding to those axes is designated as $(\vec{x}_0, \vec{y}_0, \vec{z}_0)$. Let vector $\vec{A}$ be measured by a sensor located at point M with coordinates (x, y, z). To expand the vector $\vec{A}$ into a sum of Cartesian projections one first moves the basis unit vectors $(\vec{x}_0, \vec{y}_0, \vec{z}_0)$ to the point M. This is done by a parallel translation and is shown as dashed arrows. Next, one projects the vector $\vec{A}$ onto those dashed axes. We denote these projections as $A_x(x, y, z, t), A_y(x, y, z, t)$, and $A_z(x, y, z, t)$. These projections in general are functions of the coordinates (x, y, z) and time t. Thus one writes

$$\vec{A}(x, y, z, t) = A_x(x, y, z, t)\vec{x}_0 + A_y(x, y, z, t)\vec{y}_0 + A_z(x, y, z, t)\vec{z}_0 \qquad \text{(D.1)}$$

The cylindrical coordinate system is shown in Figure D.2. One of the Cartesian axis is designated as z. In the plane perpendicular to this axis are the polar coordinates r and ϕ, where r is the distance from the z axis and ϕ is the azimuth. At point M with coordinates (r, ϕ, z) the three basis vectors are $\vec{r}_0$, $\vec{\phi}_0$, and $\vec{z}_0$. Vector $\vec{r}_0$ is

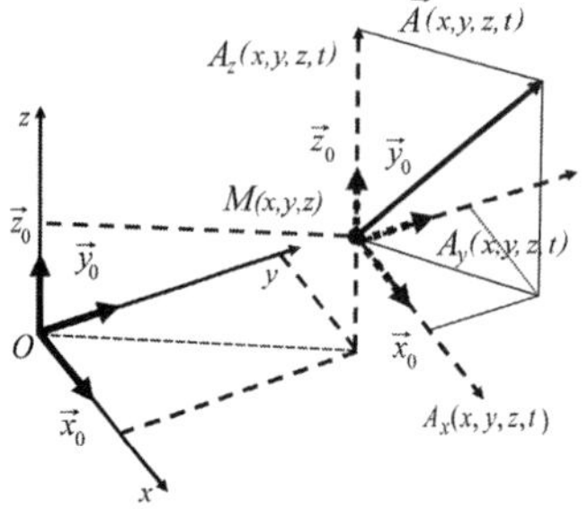

Figure D.1 Vectors representation in Cartesian coordinates.

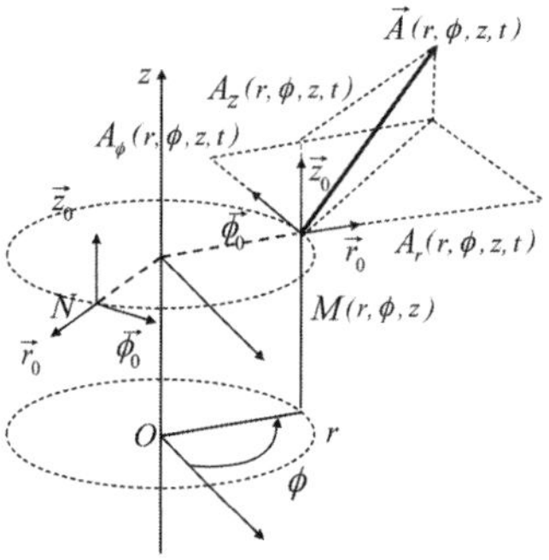

Figure D.2 Vectors representation in cylindrical coordinates.

a continuation of the radius vector in the plane perpendicular to the z axis. It points outward, i.e., away from the z axis. The vector $\vec{\phi}_0$ is tangential to the circle for which $z = \text{const}$ and $r = \text{const}$, and it points toward the increase of azimuth φ. The vector $\vec{z}_0$ is a unit vector of the Cartesian axis z. The vector $\vec{A}$ as measured at point $M(r, \phi, z)$ can be written as

$$\vec{A}(r, \phi, z, t) = A_r(r, \phi, z, t)\vec{r}_0 + A_\phi(r, \phi, z, t)\vec{\phi}_0 + A_z(r, \phi, z, t)\vec{z}_0 \qquad \text{(D.2)}$$

At another point N, the vectors $\vec{r}_0$ and $\vec{\phi}_0$ are related to those at point M by rotation.

Finally, the third coordinate system is the spherical one. It is shown in Figure D.3. The coordinates of the observation point M are the radial distance r from the origin, the polar angle θ (also referred to as zenith or inclination angle), and the azimuthal angle ϕ. The angle θ is counted from a fixed direction. In most GNSS antenna applications this direction points toward the local zenith. Thus θ varies within the range $0 \le \theta \le \pi$. The azimuth angle ϕ is counted from another fixed direction within the plane perpendicular to the zenith direction, called the local horizon. This angle varies within the range $0 \le \phi \le 2\pi$. In geodetic and surveying applications the azimuth is counted from the north direction and is typically denoted by α. At a point M with coordinates (r, θ, ϕ) the three basis unit vectors are $\vec{r}_0$, a continuation of radius vector

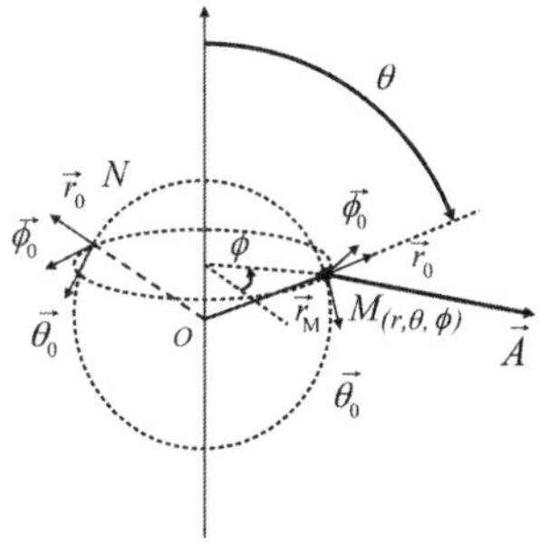

Figure D.3 Vectors representation in spherical coordinates.

$\vec{r}_M$ pointing outward toward larger distances, $\vec{\theta}_0$ is tangential to the sphere r = const and to the arc ϕ = const on that sphere. This vector points toward the increase of θ. Finally, vector $\vec{\phi}_0$ is tangential to the circle θ = const, r = const and it points toward the increase of azimuth ϕ. The vector $\vec{A}$ as measured at point $M(r, \theta, \phi)$ takes the form

$$\vec{A}(r,\theta,\phi,t) = A_r(r,\theta,\phi,t)\vec{r}_0 + A_\theta(r,\theta,\phi,t)\vec{\theta}_0 + A_\phi(r,\theta,\phi,t)\vec{\phi}_0 \tag{D.3}$$

At any other point N the directions of all three basis unit vectors will change.

All the coordinate systems are right handed. This means that an observer located on top of the third basis vector and looking at the origin will see the rotation from the first vector toward the second one in a counterclockwise direction. For coordinates and vector projection conversions between coordinate systems, the reader is referred to Korn and Korn (1968) and Lo and Lee (1993).

In the complex amplitudes method, the vector projections do not depend on time t but rather are complex values containing amplitudes and initial phases. We briefly state some computation rules for such vectors. The absolute value, or the module, is

$$|\vec{A}| = \sqrt{|A_x|^2 + |A_y|^2 + |A_z|^2} \tag{D.4}$$

The absolute value of the components $|A_{x,y,z}|$ is called the module of projections as defined by the common rules of complex algebra. The multiplication by a complex number α is

$$\alpha\vec{A} = \alpha A_x\vec{x}_0 + \alpha A_y\vec{y}_0 + \alpha A_z\vec{z}_0 \tag{D.5}$$

If two vectors are defined (measured) at the same spatial point, then the dot product is

$$\vec{A}\vec{B}^* = A_xB_x^* + A_yB_y^* + A_zB_z^* \tag{D.6}$$

Here superscript * means complex conjugate. In particular

$$\vec{A}\vec{A}^* = |\vec{A}|^2 \tag{D.7}$$

A cross product is

$$[\vec{A}, \vec{B}^*] = \begin{vmatrix} \vec{x}_0 & \vec{y}_0 & \vec{z}_0 \\ A_x & A_y & A_z \\ B_x^* & B_y^* & B_z^* \end{vmatrix} \tag{D.8}$$

Here, || stands for determinant. For vectors with real values all said rules reduce to common vector algebra.

Differential and integral vector operations are of much use. Particularly important are the grad, div, rot, and ∇^2 operators. The latter is Laplacian. The designation curl is also in use the in literature instead of rot. First, one is to note partial differentiation of vector $\vec{A}$ with respect to parameters (either spatial coordinate or time). For instance,

$$\frac{\partial}{\partial t}\vec{A} = \frac{\partial}{\partial t}A_x(x, y, z, t)\vec{x}_0 + \frac{\partial}{\partial t}A_y(x, y, z, t)\vec{y}_0 + \frac{\partial}{\partial t}A_z(x, y, z, t)\vec{z}_0 \tag{D.9}$$

With Cartesian coordinates, we obtain

$$\text{grad } u = \frac{\partial u}{\partial x}\vec{x}_0 + \frac{\partial u}{\partial y}\vec{y}_0 + \frac{\partial u}{\partial z}\vec{z}_0 \tag{D.10}$$

$$\text{div}\,\vec{A} = \frac{\partial A_x}{\partial x} + \frac{\partial A_y}{\partial y} + \frac{\partial A_z}{\partial z} \tag{D.11}$$

$$\text{rot}\,\vec{A} = \left(\frac{\partial}{\partial y}A_z - \frac{\partial}{\partial z}A_y\right)\vec{x}_0 + \left(\frac{\partial}{\partial z}A_x - \frac{\partial}{\partial x}A_z\right)\vec{y}_0 + \left(\frac{\partial}{\partial x}A_y - \frac{\partial}{\partial y}A_x\right)\vec{z}_0 \tag{D.12}$$

$$\nabla^2 = \frac{\partial^2}{\partial x^2} + \frac{\partial^2}{\partial y^2} + \frac{\partial^2}{\partial z^2} \tag{D.13}$$

In cylindrical coordinates,

$$\text{grad } u = \frac{\partial u}{\partial r}\vec{r}_0 + \frac{1}{r}\frac{\partial u}{\partial \phi}\vec{\phi} + \frac{\partial u}{\partial z}\vec{z}_0 \tag{D.14}$$

$$\text{div}\,\vec{A} = \frac{1}{r}\frac{\partial}{\partial r}(rA_r) + \frac{1}{r}\frac{\partial A_\phi}{\partial \phi} + \frac{\partial A_z}{\partial z} \tag{D.15}$$

$$\text{rot}\,\vec{A} = \frac{1}{r}\left(\frac{\partial A_z}{\partial \phi} - \frac{\partial A_\phi}{\partial z}\right)\vec{r}_0 + \left(\frac{\partial A_r}{\partial z} - \frac{\partial A_z}{\partial r}\right)\vec{\phi}_0 + \left(\frac{1}{r}\frac{\partial}{\partial r}(rA_\phi) - \frac{1}{r}\frac{\partial A_r}{\partial \phi}\right)\vec{z}_0 \tag{D.16}$$

$$\nabla^2 = \frac{1}{r}\frac{\partial}{\partial r}\left(r\frac{\partial}{\partial r}\right) + \frac{1}{r^2}\frac{\partial^2}{\partial \phi^2} + \frac{\partial^2}{\partial z^2} \tag{D.17}$$

and in spherical coordinates,

$$\text{grad } u = \frac{\partial u}{\partial r}\vec{r}_0 + \frac{1}{r}\frac{\partial u}{\partial \theta}\vec{\theta}_0 + \frac{1}{r\sin\theta}\frac{\partial u}{\partial \phi}\vec{\phi}_0 \tag{D.18}$$

$$\operatorname{div}\vec{A} = \frac{1}{r^2}\frac{\partial}{\partial r}(r^2 A_r) + \frac{1}{r\sin\theta}\frac{\partial}{\partial\theta}(A_\theta\sin\theta) + \frac{1}{r\sin\theta}\frac{\partial A_\phi}{\partial\phi} \tag{D.19}$$

$$\operatorname{rot}\vec{A} = \frac{1}{r\sin\theta}\left(\frac{\partial}{\partial\theta}\left(A_\phi\sin\theta\right) - \frac{\partial A_\theta}{\partial\phi}\right)\vec{r}_0 + \frac{1}{r}\left(\frac{1}{\sin\theta}\frac{\partial A_r}{\partial\phi} - \frac{\partial}{\partial r}\left(rA_\phi\right)\right)\vec{\theta}_0 + \frac{1}{r}\left(\frac{\partial}{\partial r}\left(rA_\theta\right) - \frac{\partial A_r}{\partial\theta}\right)\vec{\phi}_0 \tag{D.20}$$

$$\nabla^2 = \frac{1}{r^2}\frac{\partial}{\partial r}\left(r^2\frac{\partial}{\partial r}\right) + \frac{1}{r^2\sin\theta}\frac{\partial}{\partial\theta}\left(\sin\theta\frac{\partial}{\partial\theta}\right) + \frac{1}{r^2\sin^2\theta}\frac{\partial^2}{\partial\phi^2} \tag{D.21}$$

With these expressions u is an arbitrary scalar field and $\vec{A}$ is a vector field.

For vector flux calculation, one considers an imaginary closed surface S. The surface is subdivided into a set of small segments (Figure D.4).

We introduce a differential vector $\vec{dS}$. This vector is perpendicular to S and points outward of S. Its module dS equals the area of the segment. We expand the vector $\vec{A}$ into the component $\vec{A}_\tau$ tangential to S and into a normal component $\vec{A}_n$. The component $\vec{A}_n$ contributes to the flux through S. By definition, a flux through a segment is

$$d\Phi = A_n\,dS = \vec{A}\,\vec{dS} \tag{D.22}$$

The flux through the surface S is a sum (integral) over all segments

$$\Phi = \oint_S \vec{A}\,\vec{dS} \tag{D.23}$$

The flux through the surface of the sphere with radius r centered at the origin is calculated using spherical coordinates. One has

$$\vec{dS} = \vec{r}_0 r^2 \sin\theta\,d\theta\,d\phi \tag{D.24}$$

$$\vec{A}\,\vec{dS} = A_r r^2 \sin\theta\,d\theta\,d\phi \tag{D.25}$$

thus

$$\Phi = \int_0^\pi \int_0^{2\pi} A_r(r,\theta,\phi) r^2 \sin\theta\,d\phi\,d\theta \tag{D.26}$$

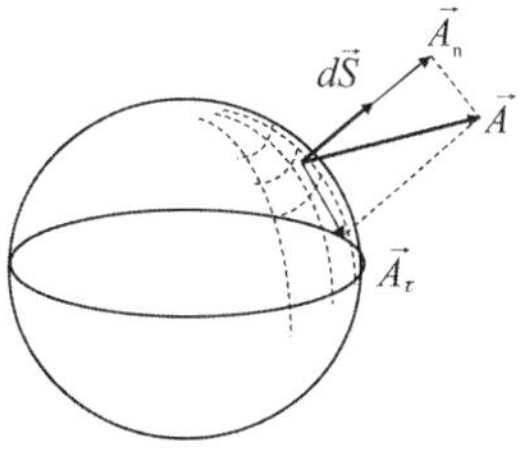

Figure D.4 Calculation of a vector flux.

With integral transformations, the Green's, divergence and the Stokes' theorems hold. These are not used explicitly in Chapter 9. The reader is referred to Balanis (1989) or Korn and Korn (1968) for details.

Dirac delta function $\delta(x)$ is widely used to model sources of radiation. By definition, for an arbitrary function $f(x)$

$$\int_a^b f(x')\delta(x-x')\,dx' = \begin{cases} f(x) & \text{if} \quad a < x < b \\ 0 & \text{otherwise} \end{cases} \tag{D.27}$$

In particular

$$\int_a^b \delta(x-x')\,dx' = \begin{cases} 1 & \text{if} \quad a < x < b \\ 0 & \text{otherwise} \end{cases} \tag{D.28}$$

The delta function $\delta(x-a)$ could be viewed as a limit $\Delta x \to 0$ of a stepwise function defined over the x axis, and being equal $1/\Delta x$ if x belongs to the interval Δx centered at a and otherwise equal to zero.

APPENDIX E

ELECTROMAGNETIC FIELD GENERATED BY ARBITRARY SOURCES, MAGNETIC CURRENTS, BOUNDARY CONDITIONS, AND IMAGES

Consider a source represented by electric current density $\vec{j}^e$ measured in *amperes/square meter*. This vector characterizes the electric current distributed in space and flowing through an imaginary unit surface area (Balanis (1989)). The Maxwell equations with the source given read as follows:

$$\operatorname{rot}\vec{H} = i\omega\varepsilon_0\vec{E} + \vec{j}^e \tag{E.1}$$

$$\operatorname{rot}\vec{E} = -i\omega\mu_0\vec{H} \tag{E.2}$$

We assume time harmonic fields with fixed angular frequency ω. Medium is assumed to be free space.

In order to solve these equations, the electric vector potential $\vec{A}^e$ is introduced by the relationships

$$\vec{H} = \operatorname{rot}\vec{A}^e \tag{E.3}$$

$$\vec{E} = -i\omega\mu_0\vec{A}^e + \frac{1}{i\omega\varepsilon_0}\operatorname{grad}\operatorname{div}\vec{A}^e \tag{E.4}$$

The thus introduced vector potential satisfies the Helmholtz equation

$$\nabla^2\vec{A}^e + k^2\vec{A}^e = -\vec{j}^e \tag{E.5}$$

The Laplacian ∇^2 and grad and div operators are given in Appendix D, and k is the wavenumber (9.1.32). Once (E.5) is solved, the magnetic field intensity is calculated employing (E.3). For a spatial area outside the source, instead of (E.4), one may use

$$\vec{E} = \frac{1}{i\omega\varepsilon_0}\operatorname{rot}\vec{H} \tag{E.6}$$

A solution to (E.5) is

$$\vec{A}^e = \frac{1}{4\pi}\int_V \vec{j}^e \frac{e^{-ik|\vec{r}-\vec{r}'|}}{|\vec{r}-\vec{r}'|}dV' \tag{E.7}$$

Here, both vectors $\vec{j}^e$ and $\vec{A}^e$ are represented by Cartesian projections, and V is the spatial area occupied by the source. The radius vector $\vec{r}$ points toward the observation point M and radius vector $\vec{r}'$ points to the variable point of integration (Figure E.1, left panel).

For the Hertzian dipole placed at the origin (Figure E.1, right panel) one writes (E.7) in the form

$$\vec{A}^e = \frac{1}{4\pi}\vec{j}^e\frac{e^{-ikr}}{r}\Delta V = \frac{1}{4\pi}j^e SL\frac{e^{-ikr}}{r}\vec{z}_0 = \frac{IL}{4\pi}\frac{e^{-ikr}}{r}\vec{z}_0 \tag{E.8}$$

Here, ΔV is the volume occupied by the dipole. This volume is written as a product of cross-section S and length L. In order to transform to spherical coordinates at an observation point M (see the right panel in Figure E.1) one has

$$\vec{z}_0 = \vec{r}_0\cos\theta - \vec{\theta}_0\sin\theta \tag{E.9}$$

Substituting (E.9) into (E.8) and then into (E.3) and (E.6), using expression (D.20) for the rot operator yields (9.1.79) and (9.1.80).

For the far-field region, expressions (E.7), (E.3), and (E.6) simplify for an arbitrary source. If r is much larger compared to wavelength λ and dimensions of the source, then for the denominator of the integrand of (E.7), one can write

$$|\vec{r}-\vec{r}'| \approx r \tag{E.10}$$

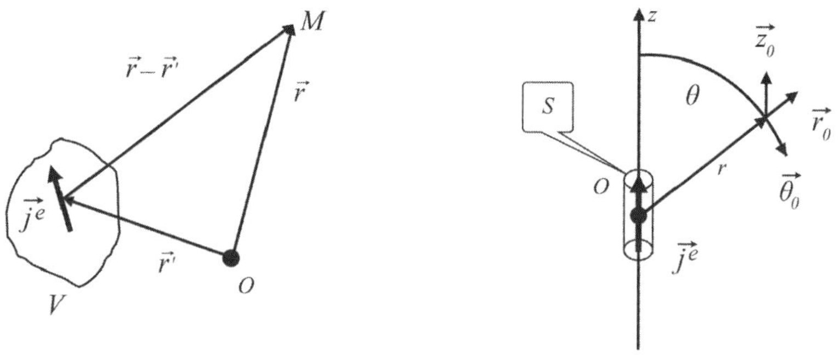

Figure E.1 Calculation of fields radiated by an arbitrary source and Hertzian dipole.

The exponent in (E.7) is treated as follows. First, Cartesian coordinates of an observation point are introduced by writing

$$x = rn_x \tag{E.11}$$

$$y = rn_y \tag{E.12}$$

$$z = rn_z \tag{E.13}$$

Here,

$$n_x = \sin\theta\cos\phi \tag{E.14}$$

$$n_y = \sin\theta\sin\phi \tag{E.15}$$

$$n_z = \cos\theta \tag{E.16}$$

are directional cosines (projections) of the unit vector $\vec{n}_0$ directed toward an observation point (Figure E.2).

Let the Cartesian coordinates of the variable point of integration be (x', y', z'). Keeping the first term of the Tailor expansion with respect to r'/r, one writes

$$|\vec{r} - \vec{r}'| = \sqrt{(x-x')^2 + (y-y')^2 + (z-z')^2} = \sqrt{r^2 - 2(xx' + yy' + zz') + r'^2}$$
$$\approx r\left(1 - \frac{xx' + yy' + zz'}{r^2}\right) = r - x'n_x - y'n_y - z'n_z \tag{E.17}$$

Hence, (E.7) could be rewritten in the form

$$\vec{A}^e = \frac{1}{4\pi}\frac{e^{-ikr}}{r}\vec{N}^e \tag{E.18}$$

Here,

$$N^e_{x,y,z} = \int_V j^e_{x,y,z}(x', y', z')e^{ik(n_x x' + n_y y' + n_z z')}dV' \tag{E.19}$$

are Cartesian coordinates of the auxiliary vector $\vec{N}^e$, sometimes referred to as an electric vector of radiation. We transfer from Cartesian to spherical projections of $\vec{N}^e$

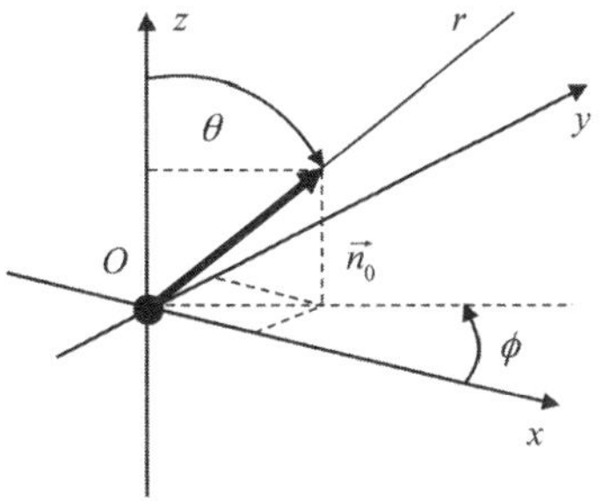

Figure E.2 Cartesian projections of a unit vector directed toward observation point.

using the transformation formulas of Lo and Lee (1993):

$$N_r^e = N_x^e \sin\theta\cos\phi + N_y^e \sin\theta\sin\phi + N_z^e \cos\theta \tag{E.20}$$

$$N_\theta^e = N_x^e \cos\theta\cos\phi + N_y^e \cos\theta\sin\phi - N_z^e \sin\theta \tag{E.21}$$

$$N_\phi^e = -N_x^e \sin\phi + N_y^e \cos\phi \tag{E.22}$$

Finally, the expression for the vector $\vec{A}^e$ in the far-field region takes the form

$$\vec{A}^e = \frac{1}{4\pi}\frac{e^{-ikr}}{r}\left(\vec{r}_0 N_r^e + \vec{\theta}_0 N_\theta^e + \vec{\phi}_0 N_\phi^e\right) \tag{E.23}$$

This expression is substituted into (E.3) and further into (E.6). While calculating the rot operators (D.20), one omits all the higher order terms with regard to $1/r$ except for the major one. This way one obtains the expressions

$$E_\theta = -i\frac{e^{-ikr}}{2\lambda r}\eta_0 N_\theta^e \tag{E.24}$$

$$E_\phi = -i\frac{e^{-ikr}}{2\lambda r}\eta_0 N_\phi^e \tag{E.25}$$

$$E_r = H_r = 0 \tag{E.26}$$

$$\vec{H} = \frac{1}{\eta_0}\left[\vec{r}_0, \vec{E}\right] \tag{E.27}$$

Here η_0 is given by (9.1.14). The Poynting vector (9.1.67) is

$$\vec{\Pi} = \eta_0 \frac{1}{8\lambda^2 r^2}\left(|N_\theta^e|^2 + |N_\phi^e|^2\right)\vec{r}_0 \tag{E.28}$$

In general, N_θ^e and N_ϕ^e are complex quantities. We write them in exponential form

$$N_{\theta,\phi}^e(\theta,\phi) = C_{\theta,\phi} F_{\theta,\phi}(\theta,\phi) e^{i\Psi_{\theta,\phi}(\theta,\phi)} \tag{E.29}$$

Here, $C_{\theta,\phi}$ are angular-independent constants, $F_{\theta,\phi}(\theta,\phi)$ are real functions with peak values equal to unity, and $\Psi_{\theta,\phi}(\theta,\phi)$ are angular-dependent phase functions. Collecting all the constants into $U_{\theta,\phi}$, one obtains (9.1.93) and (9.1.94).

As an example, we calculate the far field of the cross-dipole antenna radiating in free space (Figure 9.2.12, right panel). For the dipole parallel to the x axis, the electric current density is

$$j_x^e = I_{0x}\cos\left(\frac{\pi}{2L}x\right)\delta(y)\delta(z) \tag{E.30}$$

The symbol I_{0x} denotes the current in the middle of the dipole, and L is the length of one arm of the dipole. As to the use of delta functions, see Appendix D. Substituting (E.30) into (E.19) and setting $L = \lambda/4$ yields

$$N_x^e = I_{0x}\frac{\lambda}{\pi} f\left(n_x\right) \tag{E.31}$$

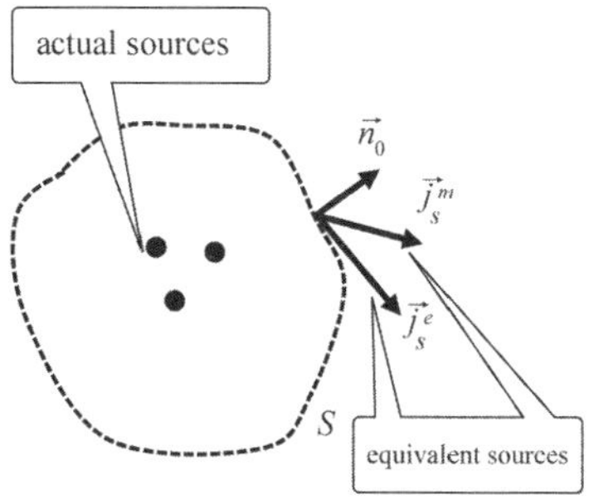

Figure E.3 Equivalence theorem.

The function $f(u)$ is defined by (9.2.25). Similarly, for the dipole oriented along the y axis, one has

$$N_y^e = I_{0y}\frac{\lambda}{\pi}f\left(n_y\right) \tag{E.32}$$

Substituting (E.31) and (E.32) into (E.21) and (E.22) and further into (E.24) and (E.25), and setting $I_{0y} = -iI_{0x} = -iI_0$, one arrives at (9.2.22).

Electric current source $\vec{j}^e$ in (E.1) was assumed to be the actual conduction current associated with electric charges motion. However, in many cases it is convenient to replace the actual sources with equivalent ones. Equivalence is understood in such a way that the $\vec{E}$ and $\vec{H}$ fields generated by actual and equivalent sources coincide within the spatial area of interest. The approach originates from the Huygens principle. The exact statement is known as an equivalence theorem (Balanis, (1989)). Let S be an imaginary closed surface containing all the sources of radiation and $\vec{E}_\tau$ and $\vec{H}_\tau$ be components of the original fields tangential to S. Then, fields outside S could be viewed as radiation of equivalent electric and magnetic surface currents $\vec{j}_S^e$ and $\vec{j}_S^m$, distributed on S and related to the original fields intensities as

$$\left[\vec{n}_0, \vec{H}_\tau\right] = \vec{j}_S^e \tag{E.33}$$

$$-\left[\vec{n}_0, \vec{E}_\tau\right] = \vec{j}_S^m \tag{E.34}$$

Here $\vec{n}_0$ is the unit normal vector to S pointing outside S (Figure E.3). This way, an equivalent magnetic current is introduced.

Magnetic currents do not exist in a physical sense as there are no realizable magnetic charges. Equivalent magnetic currents are useful each time the electric field intensity distribution is known or could be established from physical considerations (e.g., magnetic line currents modeling the patch antenna radiation in Section 9.7.1). With magnetic currents as sources, the Maxwell equations for free space read

$$\text{rot}\,\vec{H} = i\omega\varepsilon_0\vec{E} \tag{E.35}$$

$$\text{rot}\,\vec{E} = -i\omega\mu_0\vec{H} - \vec{j}^m \tag{E.36}$$

A solution to these equations could be accomplished by introducing the magnetic vector potential

$$\vec{E} = \text{rot}\,\vec{A}^m \tag{E.37}$$

and following the derivations carried out for electric currents. However, Equations (E.1) and (E.2), (E.35) and (E.36) obey the duality principle such that by changing $\vec{j}^e \to -\vec{j}^m$, $\vec{E} \to \vec{H}, \varepsilon_0 \to -\mu_0$, a solution to an "electrical" problem could be transferred into a related "magnetic" one.

Sometimes the main features of the electromagnetic phenomena can be captured employing two-dimensional approximations, which simplifies the related problems to a large extent. This approach is used in Section 9.7 in regard to ground planes. Consider a homogeneous electric line current, infinitely long in the z direction and placed at the origin of the coordinate frame such that

$$\vec{j}^e = I_0 \vec{z}_0 \delta(x)\delta(y) \tag{E.38}$$

Here, I_0 is the amplitude. Using cylindrical coordinates (Appendix D), and assuming $\partial/\partial z = 0$, the solution to (E.1) and (E.2) is

$$\vec{H} = \frac{I_0}{4i} k H_1^{(2)}(kr)\vec{\phi}_0 \tag{E.39}$$

$$\vec{E} = \frac{I_0}{4i}\frac{k^2}{i\omega\varepsilon_0} H_0^{(2)}(kr)\vec{z}_0 \tag{E.40}$$

By duality, fields excited by magnetic line current with amplitude U_0 are

$$\vec{E} = -\frac{U_0}{4i} k H_1^{(2)}(kr)\vec{\phi}_0 \tag{E.41}$$

$$\vec{H} = \frac{U_0}{4i}\frac{k^2}{i\omega\mu_0} H_0^{(2)}(kr)\vec{z}_0 \tag{E.42}$$

The medium was free space in both cases. With these expressions the $H_{0,1}^{(2)}(s)$ are Hankel functions of zero and first order and second kind (Gradshteyn and Ryshik, 1994; Abramowitz and Stegun, 1972).

We analyze expressions (E.39) to (E.42) in the far-field region. Employing asymptotic expansion for the large argument,

$$H_n^{(2)}(s) \approx \sqrt{\frac{2}{\pi s}} \cdot e^{-i\left(s - n\frac{\pi}{2} - \frac{\pi}{4}\right)} \tag{E.43}$$

one makes sure that the expressions describe cylindrical waves, with wavefronts being concentric cylinders with respect to the z axis. These waves are locally plane ones (see discussion on spherical waves in Section 9.1.4). The field intensities decay as $1/\sqrt{kr}$ with distance r. Thus the total power flux through an imaginary cylinder with unit length along the z axis stays independent of r.

Real-world bodies consist of media with different ε, μ, and σ. In most cases, a transition area between the media is a thin layer with a cross section of the order of atomic dimensions. By neglecting such a layer the parameters ε, μ, and σ exhibit discontinuities. Fields on both sides of the discontinuity are related by boundary conditions. Taking metals as perfect electric conductors (PEC), the following two conditions hold at the boundary between air (free space) and PEC:

$$\vec{j}_S^e = [\vec{n}_0, \vec{H}_\tau] \tag{E.44}$$

$$\vec{E}_\tau = 0 \tag{E.45}$$

Here, $\vec{j}_S^e$ is the density of the electric current induced at the PEC surface, subscript τ marks vector components tangential to the surface, and $\vec{n}_0$ is the unit normal vector directed from the PEC toward the air.

Perfect magnetic conductors do not exist in a physical sense due to the absence of physical magnetic currents and charges. However, there are artificial media with surfaces behaving like perfect magnetic conductors (PMC). At such surfaces, the following boundary conditions hold:

$$\vec{j}_S^m = -[\vec{n}_0, \vec{E}_\tau] \tag{E.46}$$

$$\vec{H}_\tau = 0 \tag{E.47}$$

Here, $\vec{j}_S^m$ is the density of equivalent magnetic current induced at the PMC surface. In a more general sense, the impedance boundary conditions

$$\vec{E}_\tau = Z_S[\vec{n}_0, \vec{H}_\tau] \tag{E.48}$$

hold. Here, Z_S is the surface impedance. It is assumed that there is a structure that produces Z_S (for examples see Sections 9.7.1, 9.7.4, and Appendix H). The boundary conditions at the surface of PEC and PMC are particular cases of (E.48) with $Z_S = 0$ and $Z_S \to \infty$, respectively. There are other types of boundary conditions, in particular the ones used in Section 9.7.6.

Fields radiated by a source over a boundary can be calculated by replacing the boundary by an image of a source. This approach simplifies to a large extent if the

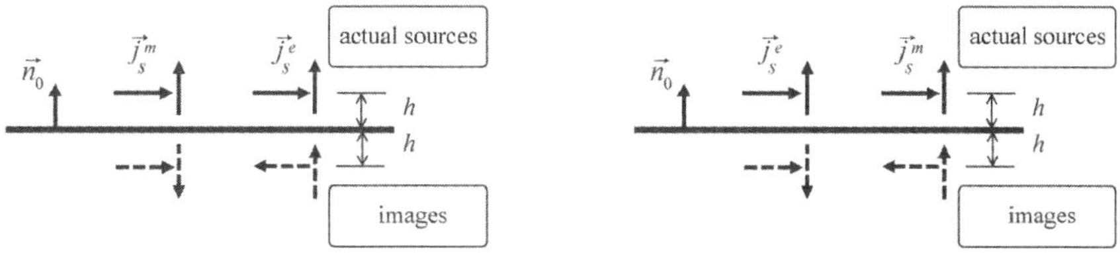

Figure E.4 Images of the sources over PEC (left) and PMC (right) boundaries.

boundary is an ideal infinite plane. In particular, if electric and magnetic dipoles are placed at some distance h over the PEC or PMC boundary, then the images are located at the same distance below the boundary. The phase relationships between actual sources and the images are illustrated in Figure E.4 for PEC (left panel) and PMC (right panel) boundaries. Using above derivations, one can make sure that the boundary conditions (E.45) and (E.47) hold in respective cases. This explains why the radiation of the magnetic line current gets doubled when the current is put onto the PEC plane; see expression (9.7.24). As another example, substituting (E.42) into (E.44) and accounting for the image leads to (9.7.22).

APPENDIX F

DIFFRACTION OVER HALF-PLANE

The cylindrical coordinate frame relevant for the problem is shown in Figure F.1. The z axis coincides with the half-plane edge at the origin. An arbitrary point P is characterized by a distance r and angle α. The values $\alpha = \pi/2$ and $\alpha = 3\pi/2$ span the local horizon, and $\alpha = \pi$ is the local zenith. The half-plane corresponds to angles $\alpha = 0$ and $\alpha = 2\pi$. The incident plane wave is arriving from angle α^{inc} is such that $\pi/2 < \alpha^{inc} < 3\pi/2$. We will be looking for the fields observed at point A located at a distance from the edge $a >> \lambda$ and at the angle $\alpha = 2\pi - \alpha_0$.

The solution to the wave diffraction problem is known for two cases of incident wave polarization. The E polarization is when the vector $\vec{E}$ has only an E_z component. The H polarization occurs when the same holds for the vector $\vec{H}$. The total field observed at any point is (Ufimtsev, 1962)

$$E_z = E_{0z}\left[u\left(r, \alpha - \alpha^{inc}\right) - u\left(r, \alpha + \alpha^{inc}\right)\right] \tag{F.1}$$

with E polarization and

$$H_z = H_{0z}\left[u\left(r, \alpha - \alpha^{inc}\right) + u\left(r, \alpha + \alpha^{inc}\right)\right] \tag{F.2}$$

with H polarization. Here, E_{0z} and H_{0z} are constants. For the angular region of interest, i.e., $2\pi \leq \alpha \leq 3\pi/2$, the function $u(r, \alpha \mp \alpha^{inc})$ is

$$u(r, \alpha - \alpha^{inc}) = v(r, \alpha - \alpha^{inc}) + \begin{cases} e^{ikr\cos\left(\alpha - \alpha^{inc}\right)}; 3\pi/2 \leq \alpha^{inc} < \pi - \alpha_0 \\ 0; \pi - \alpha_0 \leq \alpha^{inc} \leq \pi/2 \end{cases} \tag{F.3}$$

$$u(r, \alpha + \alpha^{inc}) = v(r, \alpha + \alpha^{inc}) + \begin{cases} e^{ikr\cos(\alpha+\alpha^{inc})}; 3\pi/2 \leq \alpha^{inc} < \pi + \alpha_0 \\ 0; \pi + \alpha_0 \leq \alpha^{inc} \leq \pi/2 \end{cases} \tag{F.4}$$

with function $v(r, \xi)$ being

$$v(r, \xi) = e^{ikr\cos\xi} \frac{e^{i(\pi/4)}}{\sqrt{\pi}} \int_{\infty\cos(\xi/2)}^{\sqrt{2kr}\cos(\xi/2)} e^{-iq^2} dq \tag{F.5}$$

Here, the lower limit of integration is always infinity with a sign equal to that of $\cos(\xi/2)$. The integral at the right-hand side of (F.5) is known as the Fresnel integral (Abramowitz and Stegun, 1972). For a zero argument the value is

$$\int_0^{\infty} e^{-iq^2} dq = \frac{\sqrt{\pi}}{2} e^{-i(\pi/4)} \tag{F.6}$$

If the argument is large enough, then the first term of the asymptotic expansion is

$$\int_s^{\infty} e^{-iq^2} dq \approx \frac{e^{-is^2}}{2is} \tag{F.7}$$

For $s > 1$, the error in (F.7) is less than 20% in module and less than 15° in phase. By writing

$$e^{ikr\cos(\alpha-\alpha^{inc})} = e^{-ik(x(-\cos\alpha^{inc})+y(-\sin\alpha^{inc}))} \tag{F.8}$$

one recognizes the exponential term in (F.3) as a plane wave arriving from the direction α^{inc}. Indeed, with expression (F.8), the Cartesian coordinates x, y are introduced (Figure F.1). The wavefront in (F.8) is a plane,

$$(-\cos\alpha^{inc})x + (-\sin\alpha^{inc})y = \text{const} \tag{F.9}$$

with a unit normal vector pointing in the direction of propagation as

$$\vec{n}_0 = (-\cos\alpha^{inc})\vec{x}_0 + (-\sin\alpha^{inc})\vec{y}_0 \tag{F.10}$$

Similarly, the exponential term in (F.4) is a plane wave reflected by the half-plane.

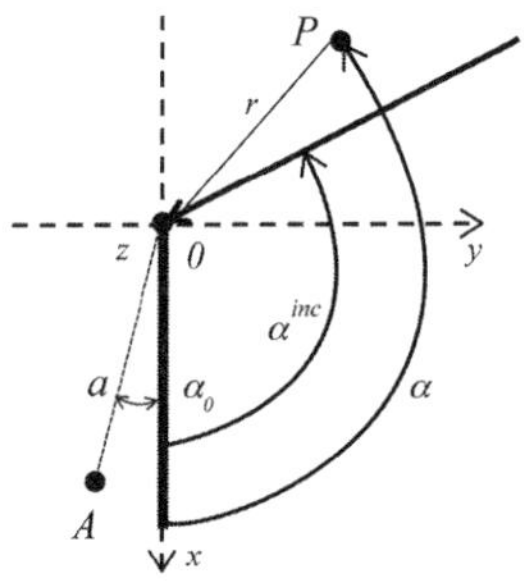

Figure F.1 Coordinate frame for diffraction over half-plane.

To construct an incident RHCP wave out of E and H polarized waves, one is to do as follows: for very large distances from the half-plane, one neglects the contribution of Fresnel integral in (F.4). For the H polarization, one employs (9.1.65) along with (F.2), (F.8) and (D.12) to find the corresponding E field in the form

$$\vec{E} = H_{0z}\eta_0 \left(\sin\alpha^{inc}\vec{x}_0 - \cos\alpha^{inc}\vec{y}_0\right) e^{-ik(x(-\cos\alpha^{inc})+y(-\sin\alpha^{inc}))} \tag{F.11}$$

This vector is perpendicular to (F.10), as it should be for the plane wave. Thus, for the sum of the E- and H-polarized waves to constitute an RHCP, one is to set up

$$E_{0z} = \frac{1}{\sqrt{2}} E_0 \tag{F.12}$$

$$H_{0z} = i\frac{1}{\eta_0\sqrt{2}} E_0 \tag{F.13}$$

Here E_0 is amplitude. Doing the same with (F.4) and employing (F.12) and (F.13), one recognizes a reflected wave as LHCP.

At a point A by setting up $\alpha = 2\pi - \alpha_0$ in (F.3) and (F.4) one notes that if an incident wave crosses the line OB of Figure 9.4.3 while passing from angular sector I to angular sector II (compare top-right and bottom-left panels), then the plane wave contribution to the reflected field jumps to zero [see (F.4)]. However, due to (F.6) the Fresnel integral compensates for this discontinuity and makes the total reflected field continuous. The same is true with respect to the incident field and line OC (bottom-left and bottom-right panels).

The contributions of the Fresnel integrals in (F.3) and (F.4) are called diffraction waves. We are looking into these waves in the proximity of point A. Let an incident wave direction of arrival be outside of the vicinity of lines OB and OC such that $\sqrt{2kr}\left|\cos\left(\frac{\alpha\mp\alpha^{inc}}{2}\right)\right| \geq 1$ holds. By using (F.7) with (F.5) and also (F.1) and (F.2), and then calculating the vector $\vec{E}$ for H polarization and vector $\vec{H}$ for E polarization using (9.1.65) and (9.1.66) and the rot operator in cylindrical coordinates given by (D.16), one finds that the diffraction field is a collection of cylindrical waves originating from the half-plane edge. These waves are locally plane ones. We do that in explicit form for the case when the incident wave arrives within the angular sector III (bottom-right panel in Figure 9.4.3). Accounting for (F.12) and (F.13) one finds that the second terms in brackets in (F.1) and (F.2) constitute an LHCP field. Assuming the RHCP antenna located at point A, this field is neglected. For the RHCP field the result is

$$\vec{E}^{d,RHCP} = E_0 \frac{e^{-ikr}}{\sqrt{kr}} \frac{e^{-i(\pi/4)}}{2\sqrt{2}\sqrt{\pi}} \frac{1}{\sin\left(\frac{\theta^{shadow} - \theta^e}{2}\right)} \frac{1}{\sqrt{2}}(\vec{z}_0 + i\vec{\varphi}_0) \tag{F.14}$$

Here angles θ^{shadow} and θ^{e} are introduced according to (9.4.3) and (9.4.4). Note that if there were no half-plane, then the incident wave would take the form

$$\vec{E}^{inc} = E_0 e^{-ikr\cos(\theta^{shadow}-\theta^{e})}\frac{1}{\sqrt{2}}\left(\vec{z}_0 + i\vec{\varphi}_0\right) \tag{F.15}$$

Setting up the distance $r = a$ in these last two expressions and comparing them in magnitudes (modules), one gets (9.4.6), while comparison in phases yields (9.4.7).

APPENDIX G

SINGLE CAVITY MODE APPROXIMATION WITH PATCH ANTENNA ANALYSIS

Consider the rectangular linear polarized patch antenna (Figure 9.7.1). The TM_{10} cavity mode is defined by expressions

$$\vec{E} = E_0 \sin\left(\frac{\pi}{a_x}x\right)\vec{z}_0 \tag{G.1}$$

$$\vec{H} = \frac{1}{-i\omega\mu_0}\text{rot}\,\vec{E} = -E_0\, i\, \frac{\pi}{a_x}\,\frac{1}{k\eta_0}\,\cos\left(\frac{\pi}{a_x}x\right)\vec{y}_0 \tag{G.2}$$

Here, E_0 is the amplitude defined by the probe current, ω is the angular frequency, and k is a free space wavenumber. The coordinate frame and antenna dimensions are shown in Figure 9.7.1. Expressions (G.1) and(G.2) are valid for the area between the patch and the ground plane, namely

$$\begin{Bmatrix} -a_x/2 \le x \le a_x/2 \\ -a_y/2 \le y \le a_y/2 \\ 0 \le z \le h \end{Bmatrix} \tag{G.3}$$

In order to define E_0 as excited by the probe current, the power balance is considered,

$$P_{pr} = P_v + P_\Sigma \tag{G.4}$$

Here,

$$P_{pr} = -\frac{1}{2}\int_V \vec{j}^{\,*}_{pr}\,\vec{E}\,dV \tag{G.5}$$

is power radiated by the probe,

$$P_v = i\omega \int_V \left(\frac{\mu_0 \left|\vec{H}\right|^2}{2} - \frac{\varepsilon_0 \varepsilon \left|\vec{E}\right|^2}{2} \right) dV \tag{G.6}$$

is power stored within the area (G.3) and

$$P_\Sigma = \int_{S_\Sigma} \vec{\Pi} d\vec{s} \tag{G.7}$$

is power radiated in free space. In these expressions V is the area (G.3), and S_Σ is the surface of the slots 7, 7′ in Figure 9.7.2. Substituting (G.1) and (G.2) into (G.6) yields

$$P_v = -\frac{i}{4} \frac{a_x a_y}{k\eta_0} \left(k^2 \varepsilon - \left(\frac{\pi}{a_x} \right)^2 \right) |E_0|^2 h \tag{G.8}$$

Since the probe radius r_{pr} and substrate thickness h are much smaller compared to the free space wavelength λ, with (G.5), one takes the probe as a current filament and writes

$$\vec{j}_{pr} = I\delta(x - x_{pr})\delta(y)\vec{z}_0 \tag{G.9}$$

Here, I is the electric current of the probe. The delta functions were introduced in Appendix D. Using (G.1) and (G.9) with (G.5) yields

$$P_{pr} = -\frac{1}{2} I^* E_0 h \sin\left(\frac{\pi}{a_x} x_{pr} \right) \tag{G.10}$$

The radiated power (G.7) is expressed as

$$P_\Sigma = 2\frac{|U|^2 Y_\Sigma^*}{2} \tag{G.11}$$

Here, the voltage U at the patch edge located at $x = \pm a_x/2$ is

$$U = E_0 h \tag{G.12}$$

The symbol Y_Σ denotes the radiating admittance (9.7.5). Substituting (G.8), (G.10), and (G.11) into (G.4) and solving with respect to E_0 yields

$$E_0 = -\frac{I \sin\left[(\pi/a_x)x_{pr}\right]}{2\left\{Y_\Sigma + (ia_x a_y)/(4\eta_0 kh)\left[k^2\varepsilon - (\pi/a_x)^2\right]\right\} h} \tag{G.13}$$

The total input power is expressed in the form

$$P_{inp} = P_{pr} + P_L \tag{G.14}$$

Here, P_{pr} is defined by (G.10) with E_0 defined by (G.13), and

$$P_L = i\frac{|I|^2 X_L}{2} \tag{G.15}$$

is the power related to the probe inductance X_L defined by (9.7.8). Writing the input power (G.14) in the form

$$P_{inp} = 1/2\,|I|^2 Z_{inp} \tag{G.16}$$

one has the input impedance Z_{inp} in the form of (9.7.2).

In order to calculate the antenna radiation pattern, one needs to account for the following sources of radiation: probe current, patch current, and polarization currents of the substrate. The probe current is given in (G.9). The patch current is related to the magnetic field tangential to the patch by boundary conditions (E.44), yielding

$$\vec{j}_{patch} = i\frac{1}{k\eta_0}\frac{\pi}{a_x}\frac{I}{\sin\left[(\pi/a_x)x_{pr}\right](G+iB)h}\cos\left(\frac{\pi}{a_x}x\right)\delta(z-h)\vec{x}_0 \tag{G.17}$$

The polarization currents in the substrate are defined by [Balanis (1989)]

$$\vec{j}_{pol} = i\omega\varepsilon_0(\varepsilon - 1)\vec{E} \tag{G.18}$$

The term "polarization" in this case stands for the process of orientational polarization in dielectrics. This is not to be confused with plane wave polarization as discussed in Section 9.1.6. We assume that with compact designs, the substrate does not exceed the limits of the patch. Using (G.13) with (G.1) and employing (9.7.3) and (9.7.4), yields

$$\vec{j}_{pol} = -\; i\omega\varepsilon_0(\varepsilon - 1)\frac{I}{\sin\left[(\pi/a_x)x_{pr}\right](G+iB)h}\sin\left(\frac{\pi}{a_x}x\right)\vec{z}_0 \tag{G.19}$$

Vectors (G.9), (G.17), and (G.19) are substituted into (E.19) yielding (9.7.12) to (9.7.16). An unbounded, perfectly conductive ground plane is taken into consideration employing images approach (Appendix E).

To analyze the TM_{01} mode instead of (G.1), one is to write

$$\vec{E} = E_0 \sin\left(\frac{\pi}{a_y}y\right)\vec{z}_0 \tag{G.20}$$

and perform the same derivations with the probe displaced along the y axis.

APPENDIX H

PATCH ANTENNAS WITH ARTIFICIAL DIELECTRIC SUBSTRATES

Consider the left panel of the Figure H.1, showing a structure comprising top and bottom metal planes with the dense periodic system of metal ribs attached to the bottom plane (Silin and Sazonov, 1971). The characteristic equation for the wave traveling in the x direction in the area $0 \leq z \leq d$ is

$$(k_z d)\tanh(k_z d) = -i\frac{Z_s}{\eta_0}kd \tag{H.1}$$

Here, Z_s is a surface impedance formed by the ribs,

$$Z_s = -i\eta_0 \tan(kb) \tag{H.2}$$

Assuming b and $d << \lambda$, the solution to (H.1) is

$$k_z{}^2 = -k^2\frac{b}{d} \tag{H.3}$$

Thus,

$$k_x{}^2 = k^2\left(1 + \frac{b}{d}\right) \tag{H.4}$$

and the slowdown factor β, related to equivalent effective dielectric constant ε_{eff} of an artificial medium, is

$$\beta = \sqrt{\varepsilon_{eff}} = \sqrt{1 + \frac{b}{d}} \tag{H.5}$$

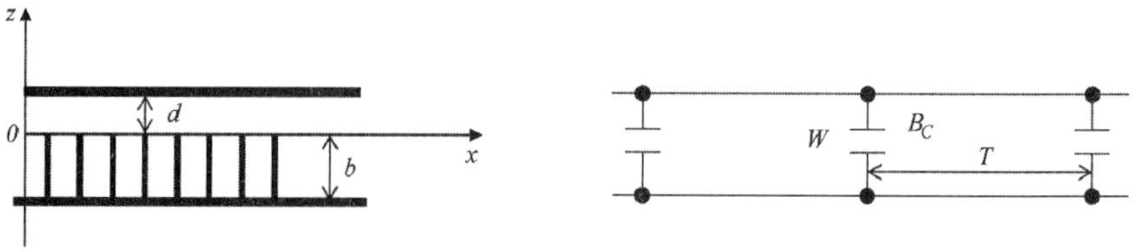

Figure H.1 Parallel plate waveguide partially filled with metal ribs and equivalent chain circuit.

Thus, for a narrow gap d such that $d \ll b$, a potentially high ε_{eff} can be achieved.

The same result can also be viewed differently. Let B_C be the susceptance of a capacitor formed by a rib and a top plane, and T be the spacing between the ribs. One arrives at an equivalent chain circuit as seen in the right panel of the figure. The characteristic equation for the chain is (Tretyakov, 2003)

$$\cos(k\beta T) = \cos(kT) - B_C W \sin(kT)/2 \tag{H.6}$$

where W is the characteristic impedance of a line. For a patch antenna with width a, the line can be viewed as a cut of a parallel plate waveguide with said width and filled with air, thus

$$W = \eta_0 \frac{h}{a} \tag{H.7}$$

Here $h = b + d$ is the thickness of the substrate. For $k\beta T << 1$, the solution to (H.6) is

$$\beta = \sqrt{1 + \frac{B_C W}{kT}} \tag{H.8}$$

Thus, the chain is nondispersive and equivalent to a homogeneous medium. Calculations with the help of (H.6) show that for $T \sim 10^{-2}\lambda$ undesirable dispersion becomes noticeable if $\beta \approx 7...10$ or larger. Such high β values are probably beyond practical interest for patch antennas in survey applications.

For circular polarized antennas, one might be interested in a two-dimensional lattice of pins rather than ribs. Accurate simulations with method of moments (Peterson et al., 1998) code have shown that, within practical limitations on pins radii and spacing between pins, $\varepsilon_{eff} \sim 4$ can be achieved. As an example, Figure H.2 shows a patch antenna stack covering the entire GNSS band. The height of the stack is 35 mm, and pin substrates with $\varepsilon_{eff} \approx 4$ are utilized.

Further, with a TM_{10} cavity mode of a patch antenna, the pins located at the central area of a patch are excited by comparatively weak fields [see expression (G.1)] and thus can be removed.

Compare the Q factors of two equivalent circuits in Figure H.3. Here the left panel corresponds to the "regular" patch antenna with dielectric substrate and slowdown factor of $\beta = \sqrt{\varepsilon}$; and ε is the permittivity of the substrate. Due to the symmetry of the TM_{10} cavity mode with respect to the y axis of Figure 9.7.1, one is able to treat a

Figure H.2 Full wave GNSS patch antenna stack formed with pins substrates.

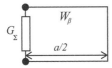

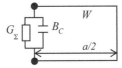

Figure H.3 Equivalent circuits for the patch antenna Q-factor estimates.

quarter-wavelength circuit shorted at one end. The radiating admittance G_Σ is given by (9.7.6), and susceptance (9.7.7) is neglected. For a square patch that is required for circular polarized antenna, one writes $a_x = a_y = a$ where a is defined by (9.7.1) and uses this with (9.7.6). The length of the stub at the left panel in Figure H.3 is $a/2$. The characteristic impedance W_β is taken similar to (H.7), but accounting for the medium with a slowdown factor of β, thus

$$W_\beta = W\frac{1}{\beta} \tag{H.9}$$

The Q factor of this circuit is given by (9.7.11). With the right panel in Figure H.3, one assumes that the line is filled with air and W is given by (H.7). However, let the same equivalent factor β be achieved by introducing a capacitor with susceptance B_C, thus

$$a = \frac{\lambda_0}{2\beta} \tag{H.10}$$

Here, λ_0 is the resonant wavelength. The resonant condition for this circuit reads

$$B_C W|_{\omega_0} = \cot\left(k\frac{a}{2}\right)|_{\omega_0} = \cot\left(\frac{\pi}{2\beta}\right) \tag{H.11}$$

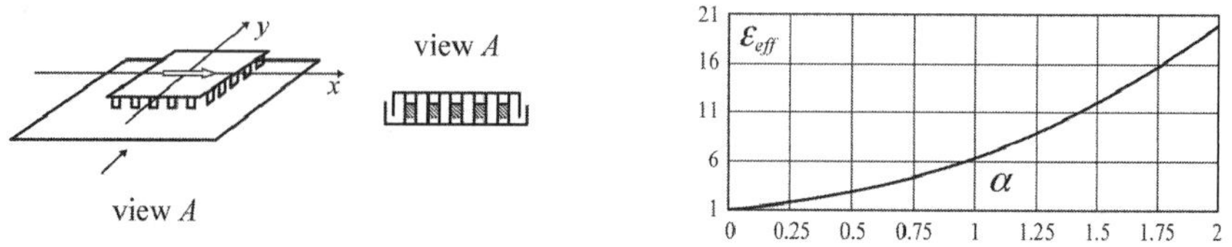

Figure H.4 General view and effective dielectric constant of a substrate versus constitutive parameter of a patch antenna with capacitive frame.

All the quantities are taken at resonant frequency ω_0. For the Q factor, one has

$$Q = \frac{1}{2G_\Sigma}\omega_0 \frac{d}{d\omega}\left[B_C - \frac{1}{W}\cot\left(k\frac{a}{2}\right)\right]_{\omega_0} = \frac{\lambda_0}{4h}\left[\frac{2}{\pi}\cot\left(\frac{\pi}{2\beta}\right) + \frac{1}{\beta \sin^2\left(\frac{\pi}{2\beta}\right)}\right]$$
$$\approx \frac{\lambda_0}{4h}\beta\frac{8}{\pi^2} \tag{H.12}$$

When compared to (9.7.11), this estimate shows that the capacitive load provides a slightly better Q factor than the dielectric substrate.

Now one considers a patch antenna with a substrate in the form of a capacitive frame around a perimeter (Figure H.4, left panel). Each capacitor is formed by a pair of metal legs coming from the patch and the ground plane. The homogenization technique to estimate the effective dielectric permittivity ε_{eff} of such a substrate is now described.

For linear polarization parallel to the x axis, we view the patch as a piece of microstrip line aligned in the x direction. Similar to (H.8), one writes the equivalent slowdown factor β_1 in the line as

$$\beta_1 = \sqrt{1 + \frac{2B_C W}{kT}} \tag{H.13}$$

Here, it accounts for two rows of capacitors with susceptance B_C. Next, consider the system of capacitors parallel to the y axis. In this case, an equivalent circuit as seen in the right panel in Figure H.3 holds. The difference with the above is that now the line is filled with media having slowdown factor (H.13), and the wave impedance of the line W_1 is

$$W_1 = W\frac{1}{\beta_1} \tag{H.14}$$

For $T \ll a$ the total number of capacitors aligned parallel to the y axis is $\approx a/T$, and instead of (H.11), the resonant condition will read

$$B_C\frac{a}{T}W_1 = \cot\left(k\beta_1\frac{a}{2}\right) \tag{H.15}$$

Now we introduce the total equivalent slowdown factor $\beta = \sqrt{\varepsilon_{eff}}$, such that

$$\beta = \beta_1 \beta_2 \tag{H.16}$$

and the patch size a is given by (H.10). Expressions (H.13), (H.15), and (H.16) define β as a function of B_C, h, and T. By introducing the constitutive parameter

$$\alpha = B_C \eta_0 \frac{h}{T} \tag{H.17}$$

and employing (H.7) and (H.10) with (H.13), one has

$$\beta_1 = \sqrt{1 + 2\beta_1 \beta_2 \alpha \frac{1}{\pi}} \tag{H.18}$$

Employing (H.14) with (H.15), one has

$$\alpha = \beta_1 \cot\left(\pi \frac{1}{2\beta_2}\right) \tag{H.19}$$

Substituting (H.19) in (H.18) and solving for β_1 yields

$$\beta_1 = \frac{1}{\sqrt{1 - (2\beta_2/\pi)\cot(\pi/2\beta_2)}} \tag{H.20}$$

Substituting (H.20) in (H.19) and further using (H.16) with (H.20), one arrives at

$$\begin{aligned} \beta &= \frac{\beta_2}{\sqrt{1 - (2\beta_2/\pi)\cot(\pi/2\beta_2)}} \\ \alpha &= \frac{\cot(\pi/2\beta_2)}{\sqrt{1 - (2\beta_2/\pi)\cot(\pi/2\beta_2)}} \end{aligned} \tag{H.21}$$

These expressions define β as a function of α in parametric form with β_2 as the parameter. We plot this as a function $\varepsilon_{eff}(\alpha)$ in the right panel of Figure H.4. Assuming (H.17), this curve gives effective dielectric permittivity ε_{eff} as a function of the constitutive parameters of the structure.

APPENDIX I

CONVEX PATCH ARRAY GEODETIC ANTENNA

Consider a two-dimensional patch (strip) of size L placed at a distance h over the shorted end of a parallel plate waveguide of width D (Figure I.1).

Let the current distribution over the patch coincide with that of a TM_{10} cavity mode (Appendix G), namely

$$\vec{j} = j_0 \cos\left(\frac{\pi}{L}\left(x - \frac{L}{2}\right)\right)\delta(z-h)\vec{x}_0 \tag{I.1}$$

The impedance Z is introduced by the relationship

$$P = 1/2|j_0|^2 Z \tag{I.2}$$

Here, P is radiated power. Expanding the fields into a set of eigenmodes of a parallel plate waveguide yields

$$Z = R + iX_L - iX_C \tag{I.3}$$

where

$$R = \eta_0 \frac{(C_0)^2}{D}\sin^2(kh) \tag{I.4}$$

$$X_L = \eta_0 \frac{(C_0)^2}{D}\frac{1}{2}\sin(2kh) \tag{I.5}$$

$$X_C = \eta_0 \frac{1}{D}\sum_{n=2}^{\infty}\frac{|\Gamma_n|}{k}(C_n)^2\left(1 - e^{-2|\Gamma_n|h}\right) \tag{I.6}$$

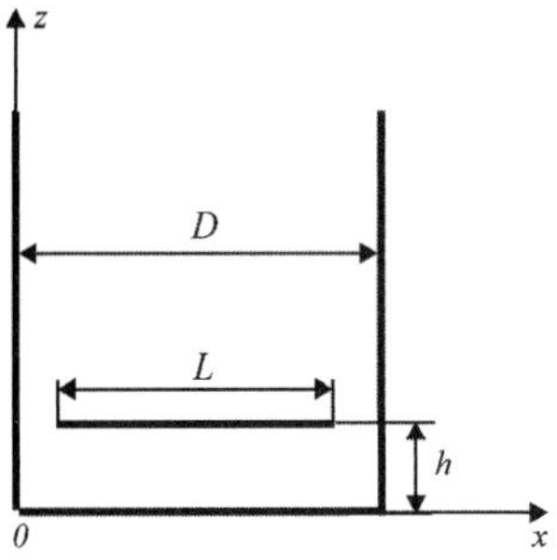

Figure I.1 Two-dimensional model of a patch (strip) in a parallel plate waveguide.

$$C_n = \int_{D/2-L/2}^{D/2+L/2} \cos\left(\frac{\pi}{L}\left(x - \frac{D}{2}\right)\right) \cos(\chi_n x) dx \tag{I.7}$$

$$\chi_n = n\frac{\pi}{D} \tag{I.8}$$

$$\Gamma_n = \sqrt{k^2 - \chi_n{}^2} \tag{I.9}$$

In these expressions it is assumed that $D < \lambda/2$ holds. Decreasing the patch size at resonance is achieved by quasi-static capacitance (I.6). For h being of the order of

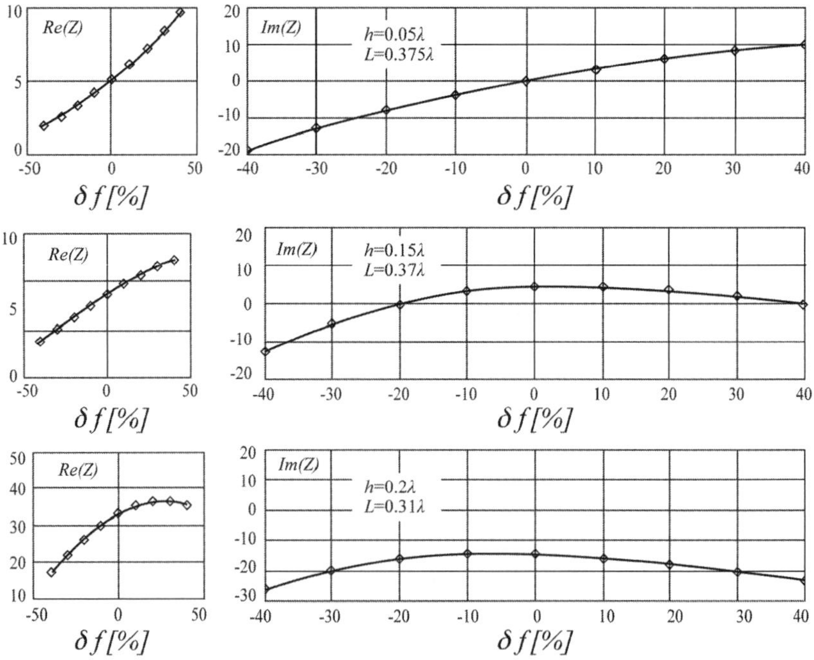

Figure I.2 Frequency response of the impedance *Z* associated with the patch.

hundredths of the wavelength, the impedance (I.3) exhibits series resonance typical of a patch at a small height over the ground plane. This is illustrated in the top panel in Figure I.2 for $h = 0.05\lambda$. However, for $h \approx \lambda/8$ the X_L behavior versus frequency in (I.5) becomes convex (see middle panel in Figure I.2). Thus, a substantial frequency bandwidth expansion is achieved. With further h increases, the capacitance (I.6) dominates (bottom panel).

The patch in the parallel plate waveguide is equivalent to an infinite array of patches with narrow gaps between them. The above analysis suggests that there exists an optimal height over the ground plane from the frequency bandwidth standpoint. A three-dimensional analogy is a semispherical cup containing an array of coupled convex patches as shown in Figure 9.7.20. The number of patches is chosen so as to provide azimuthal variations of the radiation pattern below 0.1 dB. Optimization techniques have resulted in a cup diameter of 84 mm, which is about one-third of the wavelength at the lowest frequency of the GNSS band.

REFERENCES

Abramowitz, M., and I. A. Stegun (Eds.) (1972) Handbook of Mathematical Functions with Formulas, Graphs, and Mathematical Tables. Dover Publications, New York.

Agrell, E., E. Eriksson, A. Vardy, and K. Zeger (2002) Closest Point Search in Lattices. IEEE Trans. Inform. Theory, 48(8):2201–2214.

Amitay, N., V. Galindo, and C. P. Wu Theory and Analysis of Phased Array Antenna. Wiley-Interscience, New York (1972).

Argus, D. F., and R. G. Gordon (1991) No-Net-Rotation Model of Current Plate Velocities Incorporating Plate Rotation Model NUVEL-1. Geophys. Res. Lett., 18(11):2039–2042.

Ashjaee, J., V. Filippov, D. Tatarnikov, A. Astakhov, and I. Soutiaguine (2001) Dual-Frequency Choke-Ring Ground Planes. U.S. Patent 6278407.

Askne, J., and H. Nordius (1987) Estimation of Tropospheric Delay for Microwaves from Surface Weather Data. Radio Sci., 22(3):379–386.

Awange, J., and E. Grafarend (2002a) Algebraic Solutions of GPS Pseudo-Ranging Equations. GPS Solutions, 5(4):20–32.

Awange, J., and E. Grafarend (2002b) Nonlinear Adjustment of GPS Observations of Type Pseudorange. GPS Solutions, 5(4):80–96.

Baarda, W. (1967) Statistical Concepts in Geodesy. Publication on Geodesy, New Series, 2(4). Netherlands Geodetic Commission Delft, The Netherlands.

Baarda, W. (1968) A Testing Procedure for Use in Geodetic Networks. Publication on Geodesy, New Series, 2(5). Netherlands Geodetic Commission Delft, The Netherlands.

Babai, L. (1986) On Lovász' Lattice Reduction and the Nearest Lattice Point Problem. Combinatorica, 6(1):1–13.

Badekas, J. (1969) Investigations Related to the Establishment of a World Geodetic System. Department of Geodetic Science, Ohio State University, OSUR 124.

Baggen, R., M. Martínez-Vázquez, J. Leiss, S. Holzwarth, L. Salghetti Drioli, and P. De Maagt (2008) Low Profile GALILEO Antenna Using EBG Technology. IEEE Trans. AP, 56(3):667–674.

Balanis, C. A. (1989) Advanced Engineering Electromagnetics. Wiley, New York.

Bancroft, S. (1985) An Algebraic Solution of the GPS Equations. IEEE Trans. Aerospace Electron. Syst., 21(7):56–59.

Bao, X. L., G. Ruvio, M. J. Ammann, and M. John (2006) A Novel GPS Patch Antenna on a Fractal Hi-Impedance Surface Substrate. IEEE Antennas Wireless Propag. Lett., 5, 323–326.

Baracco, J. M., L. Salghetti-Drioli, and P. De Maagt (2008) AMC Low Profile Wideband Reference Antenna for GPS and GALILEO Systems. IEEE Trans. AP, 56(8):2540–2547.

Barker, B. C., J. W. Betz, J. E. Clark, J. T. Correia, J. T. Gillis, S. Lazar, K. A. Rehborn, and J. R. Straton (2000) Overview of the GPS M Code Signal. Proc. ION NTM 2000. Institute of Navigation, Anaheim, CA, pp. 542–549.

Bar-Sever, Y. (1996) A New Model for GPS Yaw Attitude. J. Geodesy, 70(11):714–723.

Basilio, L. I., R. L. Chen, J. T. Williams, and D. R. Jackson (2007) A New Planar Dual-Band GPS Antenna Designed for Reduced Susceptibility to Low-Angle Multipath, IEEE Trans. AP, 55(8), 2358–2366.

Bassiri, S., and G. A. Hajj (1993) Higher-Order Ionospheric Effects on the Global Positioning System Observables and Means of Modeling Them. Manuscripta Geodaetica, 18(5), 280–289.

Beckman, P., and A. Spizzichino (1987) The Scattering of Electromagnetic Waves from Rough Surfaces. Artech House Boston, USA.

Beidou (2013) Beidou Navigation Satellite System Signal In Space Interface Control Document. Open Service Signal (Version 2.0), December 2013, China Satellite Navigation Office.

Bellman, R., and R. Kalaba (1966) Dynamic Programming and Modern Control Theory. Academic Press New York, NY.

Betz, J. W. (2002) Binary Offset Carrier Modulations for Radionavigation. Navigation, 48(4):227–246.

Bevis, M., S. Businger, T. A. Herring, C. Rocken, R. A. Anthes, and R. H. Ware (1992) GPS Meteorology: Remote Sensing of Water Vapor Using the Global Positioning System. J. Geophys. Res., 97(D14):15787–15801.

Bevis, M., S. Businger, S. Chriswell, T. Herring, R. Anthes, C. Rocken, and R. Ware (1994) GPS Meteorology: Mapping Zenith Wet Delay onto Perceivable Water. J. Appl. Meteorol., 33(3):379–386.

Bilich, A., and G. L. Mader (2010) GNSS Absolute Antenna Calibration at the National Geodetic Survey. Proc. ION-GNSS-2010. Institute of Navigation, Portland, OR, pp. 1369–1377.

Bilitza, D., L. A. McKinnell, B. Reinisch, and T. Fuller-Rowell (2011) The International Reference Ionosphere Today and in the Future. J. Geod., 85(12):909–920.

Blewitt, G. (1989) Carrier-Phase Ambiguity Resolution for the Global Positioning System Applied to Geodetic Baselines up to 2000 km. J. Geophys. Res., 94(B8):10187–10203.

Blomenhofer, H., G. Hein, and D. Walsh (1993) On-the-Fly Phase Ambiguity Resolution for Precise Aircraft Landing. Proc. ION GPS 1993. Institute of Navigation, 2:821–830.

Boccia, L., G. Amendola, and G. Di Massa (2007) Performance Evaluation of Shorted Annular Patch Antennas for High-Precision GPS Systems. IET Microw. Antennas Propag., 1(2):465–471.

Bock, Y., R. I. Abbot, C. C. Counselman, S. A. Gourevitch, and R. W. King (1985) Establishment of three-dimensional geodetic control by interferometry with the Global Positioning System. JGR, 90(B9):7689–7703.

Boehm, J., and H. Schuh (2004) Vienna Mapping Functions in VLBI analyses. Geophys. Res. Lett., 31, L01603.

Boehm, J., B. Werl, and H. Schuh (2006) Troposphere Mapping Functions for GPS and Very Long Baseline Interferometry from European Center for Medium-Range Weather Forecasts Operational Analysis Data. Geophys. Res. Lett., 111, B02406.

Borre, K., D. M. Akos, N. Bertelsen, P. Rinder, and S. H. Jensen (2007) A Software-Defined GPS and Galileo Receiver: A single-Frequency Approach. Birkhäuser.

Brunner, F. K., and M. Gu (1991) An Improved Model for the Dual Frequency Ionospheric Correction of GPS Observations. Manuscripta Geodaetica, 16:205–214.

Brown, A. (1989) Extended differential GPS. Navigation, 36(3):265–285.

Buchheim, C., A. Caprara, and A. Lodi (2010) An Effective Branch-and-Bound Algorithm for Convex Quadratic Integer Programming. Eisenbrand, F. and Shepherd, B. (Eds.), IPCO 2010: The 14th Conference on Integer Programming and Combinatorial Optimization. pp. 285–298. http://dl.acm.org.

Bust, G. S., and C. N. Mitchell (2008) History, Current State, and Future Directions of Ionospheric Imaging. Rev. Geophys., 46, RG1003.

Byun, S., G. A. Hajj, and L. E. Young (2002) Development and Application of GPS Signal Multipath Simulator. Radio Sci., 37(6):1–23.

Cai, T., and L. Wang (2011) Orthogonal Matching Pursuit for Sparse Signal Recovery with Noise. IEEE Trans. Inform. Theory, 57(7):4680–4688.

Candez, E., and T. Tao (2005) Decoding by Linear Programming. IEEE Trans. Inform. Theory, 57(12):4203–4215.

Candez, E., M. Rudelson, and T. Tao (2005) Error Correction via Linear Programming. Proc. 46th Annual IEEE Symposium on Foundations of Computer Science. FOCS 2005 Pittsburg, PA, pp. 668–681. http://dl.acm.org.

Cao, W., K. O'Keefe, and M. E. Cannon (2007) Partial Ambiguity Fixing within Multiple Frequencies and Systems. Proc. ION-GNSS-2006. Institute of Navigation, Fort Worth, TX, pp. 312–323.

Chaffee, J., and J. Abel (1994) On the Exact Solutions of Pseudorange Equations. IEEE Trans. Aerospace Electron. Syst., 30(4):1021–1030.

Chen, A., A. Chabory, A. C. Escher, C. Macabiau (2009) Development of a GPS Deterministic Multipath Simulator for an Efficient Computation of the Positioning Errors. Proc. ION GNSS 2009, September. Institute of Navigation, Savannah, GA, pp. 2378–2390.

Chen X., C. G. Parini, B. Collins, Y. Yao, M. U. Rehman (2012) Antennas for Global Navigation Satellite Systems, Wiley.

Chen, D., and G. Lachapelle (1995) A Comparison of the FASF and Least Squares Search Algorithms for Ambiguity Resolution on the Fly. Navigation, 42(2):371–390.

Chu, L. J. (1948) Physical Limitations of Omni-Directional Antennas. J. Appl. Phys., 19(12):1163–1175.

Cocard, M., S. Bourgon, O. Kamali, and P. Collins (2008) A Systematic Investigation of Optimal Carrier Phase Combinations for Modernized Triple-Frequency GPS. J. Geod., 82:555–564.

Collins, J., and A. Leick (1985) Analysis of Macrometer Networks with Emphasis on the Montgomery (PA) County Survey. Proc. Positioning with GPS-1985. NGS, Rockville, MD, pp. 667–693.

Collins, P. (2008) Isolating and Estimating Undifferenced GPS Integer Ambiguities. Proc. ION NTM 2008. Institute of Navigation, San Diego, CA, pp. 720–732.

Coster, A. J., E. M. Gaposchkin, and L. E. Thornton (1992) Real-Time Ionospheric Monitoring System Using the GPS. Navigation, 39(2):191–204.

Counselman, C. C. (1999) Multipath-Rejecting GPS Antenna. Proc. IEEE, 87(1):86–91.

Counselman, C. C., and S. A. Gourevitch (1981) Miniature Interferometer Terminals for Earth Surveying: Ambiguity and Multipath with Global Positioning System. IEEE Trans. Geosci. Remote Sens., GE-19(4):244–252.

Counselman C. C. and D. H. Steinbrecher (1987) Circularly Polarized Antenna for Satellite Positioning Systems Patent US 4,647,942.

Coyne, G. V., M. A. Hoskin, and O. Pedersen (Eds.) (1983) Gregorian Reform of the Calendar. Proc. Vatican Conference to Commemorate Its 400th Anniversary 1582–1982. Pontifica Academia Scientiarum, Vatican City.

Dai, Z., S. Knedlik, and O. Loffeld (2008) Real-Time Cycle-Slip Detection and Determination for Multiple Frequency GNSS. Proc. 5th Workshop on Positioning, Navigation and Communication. IEEE Xplore Digital Library, pp. 37–43.

Datta-Barua, S., W. T. Todd, J. Blanch and P. Enge (2006) Bounding Higher Order Ionosphere Errors for the Dual Frequency GPS User. Proc. ION GNSS 2006, September. Institute of Navigation, Fort Worth, TX, 1377–1392.

Davies, K. (1990) Ionospheric Radio. IEE Electromagnetic Waves Series 31. Peter Peregrinus, London.

Davis, J. L., T. A. Herring, I. I. Shapiro, A. E. E. Roger, and G. Elgered (1985) Geodesy by Radio Interferometry: Effects of Atmospheric Modeling Errors on Estimates of Baseline Length. Radio Sci., 20(6):1593–1607.

Dee, D. P., et al. (2011) The ERA-Interim Reanalysis: Configuration and Performance of the Data Assimilation System. Q. J. R. Meteorol. Soc., 137:553–597.

DeMets, C., R. G. Gordon, D. F. Argus, and S. Stein (1990) Current Plate Motions. Geophys. J. Int., 101, 425–478.

DeMets, C., R. G. Gordon, D. F. Argus, and S. Stein (1994) Effect of Recent Revisions to the Geomagnetic Reversal Time Scale on Estimates of Current Plate Motions. Geophys. Res. Lett., 21(20):2191–2194.

DeMets, C., R. G. Gordon, and D. F. Argus (2010) Geologically Current Plate Motions. Geophys. J. Int., 181:1–80.

van Dierendonck, A. J., P. C. Fenton, and T. J. Ford (1992) Theory and Performance of Narrow Correlator Spacing in a GPS Receiver. Navigation, 39(3):265–283.

Dilssner F., G. Seeber, G. Wübbena, and M. Schmitz (2008) Impact of Near-Field Effects on the GNSS Position Solution. Proc. ION GNSS 2008, September. Institute of Navigation, Savannah, GA, pp. 612–624.

Doherty, P. H., J. A. Klobuchar, and J. M. Kunches (2000) The Correlation between Solar 10.7 cm Radio Flux and Ionospheric Range Delay. GPS Solutions, 3(4): 75–79.

Dong, D., and Y. Bock (1989) Global Positioning System Network Analysis with Phase Ambiguity Resolution Applied to Crustal Deformation Studies in California. J. Geophys. Res., 94(B4):3949–3966.

Dozier, J. (1980) Improved Algorithm for Calculation of UTM and Geodetic Coordinates. NOAA TR NESSS 81. NGS, Rockville, MD.

Du, L., Z. Zhang, J. Zhang, L. Liu, R. Guo, and F. He (2014) An 18-Element GEO Broadcast Ephemeris Based on Non-Singular Elements. GPS Solutions. (Available online).

Elósegui P., J. L. Davis, R. T. K. Jaldehag, J. M. Johansson, A. E. Niell, and I. I. Shapiro (1995) "Geodesy using the global positioning system: The effects of signal scattering on estimates of site position," J. Geophys. Res., vol. 100, no. B7, pp. 9921–9934, June 10, 1995.

Engelis, T., R. Rapp, and Y. Bock (1985) Measuring Orthometric Height Differences with GPS and Gravity Data. Manuscripta Geodaetica, 10(3):187–194.

Escobal, P. R. (1965) Methods of Orbit Determination. Wiley, New York.

Euler, H. J., and C. C. Goad (1991) On Optimal Filtering of GPS Dual-Frequency Observations without Orbit Information. Bull. Géodés., 65(2):130–143.

Euler, H. J., and H. Landau (1992) Fast Ambiguity Resolution On-the-Fly for Real-Time Applications. Proc. 6th International Geodetic Symposium on Satellite Positioning. DMA and OSU, Columbus, OH, pp. 650–658.

Euler, H. J., and B. Schaffrin (1990) On a Measure for the Discernibility between Different Ambiguity Solutions in the Static-Kinematic GPS Mode. Proc. Kinematic Systems in Geodesy, Surveying, and Remote Sensing. Calgary, Alberta, Canada. Springer Verlag, Heidelberg, Germany pp. 285–295.

Euler, H. J., C. R. Keenan, B. E. Zebhauser, and G. Wübbena (2001) Study of a Simplified Approach in Utilizing Information from Permanent Reference Station Arrays. Proc. ION GPS 2001. Institute of Navigation, Salt Lake City, UT, pp. 379–391.

Feess, W., J. Cox, E. Howard, and K. Kovach (2013) GPS Inter-Signal Corrections (ISCs) Study. Proc. ION GNSS 2013. Nashville, TN, pp. 951–958.

Felsen, L., and N. Marcuvitz (2003) Radiation and Scattering of Waves. Wiley, Hoboken, NJ.

Feng, Y. (2008) GNSS Three Carrier Ambiguity Resolution Using Ionosphere-Reduced Virtual Signals. J. Geod., 82:847–862.

Fenton, P. C., W. H. Falkenberg, T. J. Ford, K. K. Ng, and A. J. van Dierendonck (1991) NovAtel's GPS Receiver: The High Performance OEM Sensor of the Future. Proc. ION GPS 1991. Institute of Navigation, Albuquerque, NM, pp. 49–58.

Fincke, U., and M. Pohst (1985) Improved Methods for Calculating Vectors of Shortest Length in a Lattice, Including a Complexity Analysis. Math. Computat., 44(April):463–471.

Finlay, C. C., et al. (2010) International Geomagnetic Reference Field: The Eleventh Generation. Geophys. J. Int., 183:1216–1230.

van Flandern, T. C., and K. F. Pulkkinen (1979) Low Precision Formulae for Planetary Positions. Astron. J. Suppl. Ser., 41:391–411.

Fliegel, H. F., and T. E. Gallini (1989) Radiation Pressure Models for Block II GPS Satellites. Proc. 5th International Geodetic Symposium on Satellite Positioning. DMA, Las Cruses, NM, pp. 789–798.

Fliegel, H. F., W. A. Fees, W. C. Layton, and N. W. Rhodus (1985) The GPS Radiation Force Model. Proc. Positioning with GPS-1985. NGS, Rockville, MD, pp. 113–119.

Fliegel, H. F., T. E. Gallini, and E. R. Swift (1992) Global Positioning System Radiation Force Model for Geodetic Applications. J. Geophys. Res., 97(B1):559–568.

Fock, V. A. (1965) Electromagnetic Diffraction and Propagation Problems. International Series of Monographs on Electromagnetic Waves, Volume 1, Pergamon Press.

Fontana, R. D., W. Cheung, P. M. Novak, and T. A. Stansell (2001a) The New L2 Civil Signal. Proc. ION GPS 2001. Salt Lake City, UT, pp. 617–631.

Fontana, R. D., W. Cheung, and T. A. Stansell (2001b) The Modernized L2 Civil Signal. GPS World, 12(9):28–34.

Frei, E., and G. Beutler (1990) Rapid Static Positioning Based on the Fast Ambiguity Resolution Approach "FARA." Theory and First Results. Manuscripta Geodaetica, 15(6):325–356.

Fu, Z., A. Hornbostel, J. Hammesfahr, and A. Konovaltsev (2003) Suppression of Multipath and Jamming Signals by Digital Beamforming for GNSS/Galileo Applications. GPS Solutions, 6(4):257–264.

Gabor, M. J., and R. S. Nerem (1999) GPS Carrier Phase Ambiguity Resolution Using Satellite-Satellite Single Difference. Proc. ION GPS 1999. Institute of Navigation, Nashville, TN, pp. 1569–1578.

Galileo (2010) European GNSS (Galileo) Open Service Signal In Space Interface Control Document (OS SIS ICD, Issue 1.1). Available online.

Gallager, R. G. (1963) Low Density Parity Check Codes. MIT Press, Cambridge, MA.

Gao, S., Q Luo, F. Zhu (2014) Circularly Polarized Antennas, Wiley Online Library.

Garcia-Fernandez, M., S. D. Desai, M. D. Butala, and A. Komjathy (2013) Evaluation of Different Approaches to Modelling the Second-Order Ionospheric Delay on GPS Measurements. JGR: Space Phys., 118, 7864–7873.

Garg, R., P. Bhartia, I. Bahl, and A. Ittipiboon (2001) Microstrip Antenna Design Handbook. Artech House Norwood, MA.

Ge, M., G. Gendt, M. Rothacher, C. Shi, and J. Liu (2008) Resolution of GPS carrier-phase ambiguities in precise point positioning (PPP) with daily observations. J Geod 82(7):389–399.

Gelb, A. (1974) Applied Optimal Estimation. Cambridge, MA: The MIT Press.

Geng J., X. Meng, A. H. Dodson, and F. N. Teferle (2010) Integer ambiguity resolution in precise point positioning: method comparison, J Geod 84:569–581.

Georgiadou, Y. and A. Kleusberg (1988) On carrier signal multipath effects in relative GPS positioning. Manuscripta Geodaetica, 13(3):172–179.

Gill, P., W. Murrey, and M. Wright (1982) Practical Optimization. Emerald Group Publishing Emerald Group Publishing Limited, Howard House, Wagon Lane, Bingley, UK.

GLONASS (2008) GLONASS Inteface Control Document, Edition 5.1. Russia Institute of Space Device Engineering. Available: ftp://ftp.kiam1.rssi.ru/pub/gps/lib/icd/IKD-redakcia%205.1%20ENG.pdf.

Goad, C. C. (1985) Precise Relative Position Determination Using Global Positioning System Carrier Phase Measurements in a Nondifference Mode. Proc. Positioning with GPS-1985. NGS, Rockville, MD, pp. 347–356.

Goad, C. C. (1998) Single-Site GPS Models. P. J. G. Teunissen and A. Kleusberg., (Eds.), GPS for Geodesy. Springer Verlag, Wien, pp. 446–449.

Goad, C. C. (1990) Optimal Filtering of Pseudoranges and Phases from Single-frequency GPS Receivers. Navigation, 37(3):191–204.

Goad, C. C., and A. Mueller (1988) An Automated Procedure for Generating an Optimum Set of Independent Double Difference Observables Using Global Positioning System Carrier Phase Measurements, Manuscripta Geodaetica, 13(6):365–369.

Golub, G. H., and C. F. Van Loan (1996) Matrix Computations Johns, 3rd ed. John Hopkins University Press, Baltimore, MD.

Gourevitch, S. A., S. Sila-Novitsky, and F. van Diggelen (1996) The GG24 Combined GPS+GLONASS Receiver. Proc. ION GPS 1996. Institute of Navigation, Kansas City, MO, pp. 141–145.

Gradshteyn, I. S., and I. M. Ryzhik (1994) Table of Integrals, Series and Products. Academic Press, New York.

Grafarend, E. (2000) Mixed Integer-Real Valued Adjustment (IRA) Problems. GPS Solutions, 4(1):31–45.

Grafarend, E. (2006) Linear and Nonlinear Models: Fixed Effects, Random Effects, and Mixed Models. Berlin, Germany.

Grafarend, E., and J. Shan (2002) GPS Solutions: Closed Forms, Critical and Special Configurations of P4P. GPS Solutions, 5(3):29–41.

Greenwalt, C. R., and M. E. Shultz (1962) Principles of Error Theory and Cartographic Applications. ACIC Technical Report No. 96. Aeronautical Chart and Information Center, St. Louis, MO.

Grossman, W. (1976) Geodätische Rechnungen und Abbildungen in der Landesvermessung. Witter Verlag, Stuttgart, Germany.

Gupta, I. J. and T. D. Moore (2001) Space-Frequency Adaptive Processing (SFAP) for Interference Suppression in GPS Receivers Proc. ION NTM 2001. Institute of Navigation, Long Beach, CA, pp. 377–385.

Hansen, R. B. (2009) Phased Array Antennas, 2-nd Ed. Wiley.

Hargreaves, J. K. (1992) The Solar-Terrestrial Environment. Cambridge University Press, Cambridge.

Hartmann, G. K., and R. Leitinger (1984) Range Errors Due to Ionospheric and Tropospheric Effects for Signal Frequencies above 100 MHz. Bull. Geodes., 58:109–136.

Hatch, R. R. (1982) The Synergism of GPS Code and Carrier Measurements. Proc. Third International Geodetic Symposium on Satellite Doppler Positioning, Las Cruces, New Mexico. Defense Mapping Agency (now NIA - National Intelligence Agency), Springfield, VA pp. 1213–1232.

Hatch, R. R. (1990) Instantaneous Ambiguity Resolution. Proc. Kinematic Systems in Geodesy, Surveying and Remote Sensing. IAG Symposium 107. Springer Verlag, Heidelberg, Germany pp. 299–308.

Hatch, R. R. (2006) A New Three-Frequency, Geometry-Free, Technique for Ambiguity Resolution. Proc. ION GNSS 2006. Institute of Navigation, Fort Worth, TX, pp. 309–316.

Hatch, R. R., R. Keegan, and T. A. Stansell (1992) Kinematic receiver technology from Magnavox. *Proc. 6th Intern. Geodetic Symp. on Sat. Pos.*, 174–181, DMA (now NIA), Springfield, VA.

Heck, B. (1987) Rechenverfahren und Auswertemodelle der Landesvermessung. Herbert Wichmann Verlag, Karlsruhe.

Hegarty, C. J., E. D. Powers, and B. Fonvile (2005) Accounting for Timing Biases between GPS, Modernized GPS, and Galileo Systems. Proc. ION GNSS 2005, Institude of Navigation, Long Beach, CA, September, 2401–2407.

Heiskanen, W. A., and H. Moritz (1967) Physical Geodesy. Freeman, San Francisco.

Hilla, S., and M. Jackson (2000) GPS Toolbox: Spanning Trees. GPS Solutions, 3(3):65–68.

Hobiger, T., R. Ichikawa, T. Takasu, Y. Koyama, and T. Kondo (2008a) Ray-Traced Troposphere Slant Delay for Precision Point Positioning. Earth Science Space, 60, e1–e4.

Hobiger, T., R. Ichikawa, Y. Koyama, and T. Kondo (2008b) Fast and Accurate Ray-Tracing Algorithms for Real-Time Space Geodetic Applications using Numerical Weather Models. J. Geophys. Res., 113, D20302.

Hopfield, H. S. (1969) Two-Quartic Tropospheric Refractivity Profile for Correcting Satellite Data. J. Geophys. R., 74(18):4487–4499.

Hristow, W. K. (1955) Die Gausschen und Geographischen Koordinaten auf dem Ellipsoid von Krassowsky. VEB Verlag Technik, Berlin.

Institute of Electrical and, Electronics Engineers (IEEE) (2004) Definitions of Terms for Antennas. IEEE Std. 145–1993.

International Earth Rotation Service (IERS) (2002) www.iers.org.

International GNSS Service (IGS) (2014) http://igs.org.

IS-GPS-200G (2012) Navstar GPS Space Segment/Navigation User Interface. Available online. www.gps.gov.

IS-GPS-705C (2012) Navstar GPS Space Segment/User Segment L5 Interfaces. Available online. www.gps.gov.

IS-GPS-800C (2012) Navstar GPS Space Segment/User Segment L1C Interface. Available online. www.gps.gov.

Janssen, M. A. (Ed.) (1993) Atmospheric Remote Sensing by Microwave Radiometry. Wiley, New York.

de Jonge, P. J., and C. C. J. M. Tiberius (1996) The LAMBDA Method for Integer Ambiguity Estimation: Implementation Aspects. Delft Geodetic Computing Center LGR Series, No. 12. Available: www.geo.tudelft.nl/mgp.

Jorgensen, P. S. (1986) Relativity Correction in GPS User Equipment. Proc. Position Location and Navigation System 1986 (PLANS). IEEE, New York, pp. 177–183.

Kannan, R. (1983) Improved Algorithms for Integer Programming and Related Lattice Problems. Proc. 15th ACM Symposium on Theory of Computing April. Boston, MA, pp. 193–206. www.amazon.com.

Kannan, R. (1987) Minkovski's Convex Body Theorem and Integer Programming. Math. Oper. Res., 12(April.):415–440.

Kaplan, E. D. (Ed.) (1996) Understanding GPS Principles and Applications. Artech House, Norwood, MA.

Kaplan, G. H. (2005) The IAU Resolutions on Astronomical Reference Systems, Time Scales and Earth Rotation Models. Circular No. 179. U.S. Naval Observatory, Washington, DC.

Kaula, W. M. (1962) Development of the Lunar and Solar Disturbing Functions for a Close Satellite. Astron. J., 67(5):300–303.

Kaula, W. M. (1966) Theory of Satellite Geodesy. Blaisdell Publishing Company, Waltham, MA.

Keller, J. B. (1962) Geometrical Theory of Diffraction, J. Opt. Soc., 52(2) pages 116–130.

King, R. J., D. V. Thiel, and K. S. Park (1983) The synthesis of surface reactance using an artificial dielectric. IEEE Trans. Antennas and Propagation., 31:471–476.

Klobuchar, J. A. (1987) Ionospheric Time-Delay Algorithm for Single-Frequency GPS Users. IEEE Trans. Aerospace Electron. Syst., AES-23(3):325–331.

Klobuchar, J. A. (1996) Ionospheric Effects on GPS. In Global Positioning System: Theory and Applications, Vol. I, Parkinson, B. W., J. J. Spilker, P. Axelrad and P. Enge (Eds.), American Institute of Aeronautics and Astronautics, Inc., Washington DC, pp. 513–514.

Kneissl, M. (1959) Mathematische Geodäsie. Handbuch der Vermessungskunde (Jordan–Eggert–Kneissl). Metzler, Stuttgart, Germany.

Koch, K. R. (1988) Parameter Estimation and Hypothesis Testing in Linear Models. Springer Verlag, New York.

Kok, J. (1984) On Data Snooping and Multiple Outlier Testing. NOAA Technical Report NOS NGS 30. National Geodetic Information Center, NOAA, Silver Spring, MD.

Kontorovich, M. I., M. I. Astrakhan, V. P. Akimov, and G. A. Fersman (1987) Electromagnetics of Mesh Structures (Electrodinamica setchatuh struktur). Raio I Sviaz, Moscow, (in Russian).

Korkine, A., and G. Zolotareff (1873) Sur les Formes Quadratiques, Mathematische Annalen, 6:366–389.

Korn, G. A., and T. A. Korn (1968) Mathematical Handbook for Scientists and Engineers: Definitions, Theorems and Formulas for Reference and Review. McGraw Hill, New York.

Kouyoumjian, R. G., and P. H. Pathak (1974) A Uniform Geometrical Theory of Diffraction for an Edge in a Perfectly Conducting Surface, Proc. IEEE, 62(10):1448–1461.

Kozlov, D., and M. Tkachenko (1998) Instant RTK cm with Low Cost GPS and GLONASS C/A Receivers. Navigation, 45(2):137–147.

Krabill, W. B. and C. F. Martin (1987) Aircraft positioning using Global Positioning System carrier phase data. Navigation, 34(1):1–21.

Krantz E., S. Riley, and P. Large (2001) The Design and Performance of the Zephyr Geodetic Antenna. Proc. ION GPS 2001. Institute of Navigation, Salt Lake City, UT, pp. 1942–1951.

Kumar, G., and K. P. Ray (2003) Broadband Microstrip Antennas. Artech House, Norwood, MA.

Kunches, J. M., and J. A. Klobuchar (2000) Some Aspects of the Variability of Geomagnetic Storms. GPS Solutions, 4(1):77–78.

Kunysz, W. (2000) High Performance GPS Pinwheel Antenna. Proc. ION GNSS 2000, Institute of Navigation, September, Salt Lake City, UT, pp. 2506–2511.

Kunysz, W. (2003) A Three Dimensional Choke Ring Ground Plane Antenna. Proc. ION GNSS 2003. Institute of Navigation, Portland, OR, pp. 1883–1888.

Kursinski, E. R., G. A. Hajj, J. T. Schofield, R. P. Linfield, and K. R. Hardy (1997) Observing the Earth's Atmosphere with Radio Occultation Measurements Using the Global Positioning System. J. Geophys. Res., 102(D19):23429–23465.

Lachapelle, G., H. Sun, M. E. Cannon, and G. Lu (1994) Precise Aircraft-to-Aircraft Positioning Using a Multiple Receiver Configuration. Proc. ION NTM 1994. Institute of Navigation, San Diego, CA, pp. 793–799.

Ladd, L. W., C. C. Counselman, and S. A. Gourevitch (1985) The Macrometer II Dual-Band Interferometric Surveyor. Proc. Positioning with GPS-1985. NGS, Rockville, MD, pp. 175–180.

Lagler, K., M. Schindelegger, J. Boehm, H. Krásná, and T. Nilsson (2013) GPT2: Empirical Slant Delay Model for Radio Space Geodetic Techniques. Geophys. Res. Lett., 40:1069–1073.

Lambert, J. H. (1772) Notes and Comments on the Composition of Terrestrial and Celestial Maps. Michigan Geographical Publication No. 8. Translated by W. R. Tobler, 1972. University of Michigan, Ann Arbor, MI.

Land, A., and A. Doig (1960) An Automatic Method of Solving Discrete Programming Problems. Econometrica, 28(3):497–520.

Lannes, A. (2013) On the Theoretical Link between LLL-Reduction and LAMBDA-Decorrelation. J. Geodesy, 87:323–335.

Laurichesse, D., and F. Mercier (2007) Integer Ambiguity Resolution on Undifferenced GPS Phase Measurements and Its Application to PPP. Proc. ION GNSS 2007. Institute of Navigation, Fort Worth, TX, pp. 839–848.

Lee, L. P. (1976) Conformal Projections Based on Elliptic Function. Monograph No. 16. Supplement No. 1 to Canadian Cartographer, Vol. 13. Department of Geography, York University. University of Toronto Press, Toronto, Canada.

Lee,Y., M. Kirchner, S. Ganguly, and S. Suman (2004) Multiband L5 Capable GPS Antenna with Reduced Backlobes. Proc. ION GNSS 2004. Long Beach, CA, Institute of Navigation pp. 1523–1530.

Leick, A. (2002) Lectures in GPS, Geodesy, and Adjustments—A Supplement. Unpublished notes. Department of Spatial Information Science and Engineering, University of Maine, Orono, ME.

Leick, A., and M. Emmons (1994) Quality Control with Reliability for Large GPS Networks. Surv. Eng., 120(1):26–41.

Leick, A., and B. H. W. van Gelder (1975) On Similarity Transformations and Geodetic Network Distortions Based on Doppler Satellite Observations. OSUR 235. Department of Geodetic Science, Ohio State University, Columbus, OH.

Leick, A., J. Li, J. Beser, and G. Mader (1995) Processing GLONASS Carrier Phase Observations—Theory and First Experience. Proc. of ION GPS 1995. Institute of Navigation, Palm Springs, CA, pp. 1041–1047.

Leick, A., J. Beser, P. Rosenboom, and B. Wiley (1998) Assessing GLONASS Observation. Proc. ION GPS 1998. Institute of Navigation, Nashville, TN, pp. 1605–1612.

Lemoine, F. G., et al. (1998) The Development of the Joint NASA GSFC and the National Imagery and Mapping Agency (NIMA) Geopotential Model EGM96. NASA/TP-1998–206861. NASA Goddard Space Flight Center, Greenbelt, MD.

Lenstra, A. K., H. W. Lenstra, Jr. and L. Lovász (1982) Factoring Polynomials with Rational Coefficients. Math. Ann., 261:515–534.

Li, B., Y. Feng, and Y. Shen (2010) Three Carrier Ambiguity Resolution: Distance-Independent Performance Demonstrated Using Semi-Generated Triple Frequency GPS Signals. GPS Solutions, 14(2):177–184.

Li, R. L., G. DeJean, M. M. Tentzenis, J. Papapolymerou, and J. Laskar (2005) Radiation-Pattern Improvement of Patch Antennas on a Large-Size Substrate Using a Compact Soft-Surface Structure and Its Realization on LTCC Multilayer Technology. IEEE Trans. AP, 53(1):200–208.

Lichten, S. M. and J. S. Border (1987) Strategies for high precision GPS orbit determination. JGR, 92, (B12):12751–12762.

Lier, E., and K. Jakobsen (1983) Rectangular Microstrip Patch Antennas with Infinite and Finite Ground Plane Dimensions. IEEE Trans. AP, 31(6):978–984.

Liou, Y. A., A. G. Pavelyev, S. Matugov, O. I. Yakovlev, and J. Wickert (2010) Radio Occultation Method for Remote Sensing of the Atmosphere and Ionosphere. InTech.

Lo, Y. T., and S. W. Lee (1993) (Eds.) Antenna Handbook, Vol. 1, Fundaments and Mathematical Techniques. Chapman & Hall, New York.

Lopez, A. R. (1979) Sharp Cutoff Radiation Patterns. IEEE Trans. AP, 27(6):820–824.

Lopez, A. R. (2008) LAAS/GBAS Ground Reference Antenna with Enhanced Mitigation of Ground Multipath. Proc. ION NTM 2008. Institute of Navigation, San Diego, CA, pp. 389–393.

Lopez, A. R. (2010) GPS Landing System Reference Antenna. IEEE Antennas Propagat. Mag., 52(1):104–113.

Loyer, S., F. Perosanz, F. Mercier, H. Capdeville, and J. C. Marty (2012) Zero-difference GPS ambiguity resolution at CNES-CLS IGS Analysis Center. J. Geod. 86(11):991–1003.

Lustig, I. J., R. E. Marsten, and D. F. Shanno (1994) Interior Point Methods for Linear Programming: Computational State of the Art. ORSA J. Comput., 6(2):1–14.

Macchi-Gernot, F., M. Petovello, and G. Lachapelle (2010) Combined Acquisition and Tracking Methods for GPS L1 C/A and L1C Signals. Int. J. Navig.Observ. Volume 2010, 19 pages.

Mader, G. L. (1986) Dynamic Positioning Using GPS Carrier Phase Measurements. Manuscripta Geodaetica, 11(4):272–277.

Mader, G. L. (1999) GPS Antenna Calibration at the National Geodetic Survey. GPS Solutions, 3(1):50–58.

Mader, G. L., and F. Czopek (2001) Calibrating the L1 and L2 Phase Centers of a Block IIA Antenna. Proc. ION-GPS-2001. Institute of Navigation, Salt Lake City, UT, pp. 1979–1984.

Mader G. L., A. Bilich, and D. Tatarnikov (2014) BigAnt Engineering and Experimental Results, IGS Workshop, Pasadena, CA, available online, www.ngs.noaa.gov.

Mailloux, R. J. (2005) Phased Array Antenna Handbook. Artech House, Norwood, MA.

Majithiya, P., K. Khatri and J. K. Hota (2011) Indian Regional Navigation Satellite System. Inside GNSS, January/February:40–46.

Mannucci, A. J., B. D. Wilson, D. N. Yuan, C. H. Ho, U. J. Lindqwister, and T. F. Runge (1998) A Global Mapping Technique for GPS-Derived Ionospheric Total Electron Content Measurements. Radio Science, 33(3):565–582.

Maqsood M., S. Gao, T. W. C. Brown, M. Unwin, R. Van Steenwijk, and J. D. Xu (2013) A Compact Multipath Mitigating Ground Plane for Multiband GNSS Antennas. IEEE Trans. AP, 61(5), 2775–2782.

McCarthy, D. D. (Ed.) (1996) IERS Conventions 1996. IERS Technical Note 21. Paris Observatory.

McKinnon, J. A. (1987) Sunspot Numbers: 1610–1985 Based on the Sunspot Activity in the Years 1620–1960. Report UAG-95. National Academy of Sciences, Washington, DC.

McKinzie, III W. E., R. Hurtado, and W. Klimczak (2002) Artificial Magnetic Conductor Technology Reduces Size and Weight for Precision GPS Antennas. Proc. ION NTM 2002. Institute of Navigation, San Diego, CA, pp. 448–459.

Meehan, T. K., and L. E. Young (1992) On-Receiver Signal Processing for GPS Multipath Reduction. Proc. 6th International Geodetic Symposium on Satellite Positioning. DMA and OSU, Columbus, OH, pp. 200–208.

Melbourne, W. G. (1985) The Case for Ranging in GPS-Based Geodetic Systems. Proc. Positioning with GPS-1985. NGS, Rockville, MD, pp. 373–386.

Mendes, V. B. (1999) Modeling the Neutral-Atmosphere Propagation Delay in Radiometric Space Techniques. Ph.D. Dissertation. Department of Geodesy and Geomatics Engineering Technical Report No. 199. University of New Brunswick, Fredericton, Canada.

Mendes, V. B., and R. B. Langley (1999) Tropospheric Zenith Delay Prediction Accuracy for High-Precision GPS Positioning and Navigation. Navigation, 46(1):25–34.

Milbert, D. G. (1984) Heard of Gold: Computer Routines for Large Space, Least Squares Computations. NOAA TM No. NGS-39. NOAA, Siver Spring, MD.

Misra, P., and P. Enge (2006) Global Positioning System Signals, Measurements, and Performance. Ganga-Jamuna Press, Lincoln, MA.

Molodenskii, M. S., V. F. Eremeev, and M. I. Yurkina (1962) Methods for Study of the External Gravitational Field and Figure of the Earth. Translation from Russian. National Technical Information Services, Springfield, VA.

Montenbruck, O., A. Hauschild, P. Steigenberger, U. Hugentobler, P. Teunissen, and S. Nakamura (2013) Initial Assessment of the COMPASS/Beidou-2 Regional Navigation Satellite System. GPS Solutions, 17(2):211–222.

Moritz, H. (1984) Geodetic Reference System 1980. Bull. Géodés., 58(3):388–398.

Morse, P., and H. Feshbach (1953) Methods of Theoretical Physics, Part I. McGraw Hill, New York.

Mueller, I. I. (1964) Introduction to Satellite Geodesy. F. Ungar, New York.

Mueller, I. I. (1969) Spherical and Practical Astronomy as Applied to Geodesy. F. Ungar, New York.

Niell, A. E. (1996) Global Mapping Functions for the Atmospheric Delay at Radio Wavelengths. J. Geophys. Res., 101(B2):3227–3246.

Niell, A. E. (2000) Improved Atmospheric Mapping Functions for VLBI and GPS. Earth Planets Space, 52(10):699–702.

Niell, A. E. (2001) Preliminary Evaluation of Atmospheric Mapping Functions on Numerical Weather Models. Phys. Chem. Earth (A), 26(6–8):475–480.

Nikolsky, V. V. (1978) Electromagnetics and Radio Waves Propagation (Electrodinamika I Rasprostranenie Radiovoln). Nauka, Moscow (in Russian).

OS-SIS-CD (2010) Open Service Signal In Space Interface Control Document. European Union.

Øvstedal, O. (2002) Absolute Positioning with Single-frequency GPS Receivers. GPS Solutions, 5(4):33–44.

Parkinson, B. W., J. J. Spilker, P. Axelrad, and P. Enge (Eds.) (1996) Global Positioning System: Theory and Applications. Progress in Aeronautics and Astronautics, American Institute of Aeronautics and Astronautics, Inc., Washington DC Vols. 163 and 164.

Pavlis, N. K., A. Holmes Simon, S. C. Kenyon, and J. K. Factor (2012) The Development and Evaluation of the Earth Gravitational Model 2008 (EGM2008). J. Geophys. Res., 117, B04406.

Pearson, C., and R. Snay (2013) Introducing HTDP 3.1 to Transform Coordinates across Time and Spatial Reference Frames. GPS Solutions, 17(1):1–15.

Peterson, A. F., S. L. Ray, and R. Mittra (1998) Computational Methods for Electromagnetics. IEEE Press, New York.

Petit, G., and B. Luzum (Eds.) (2010) IERS Conventions 2010. IERS Technical Note 36. Verlag des Bundesamts für Kartographie und Geodäsie. Available online. www.iers.org.

Petrie, E. J., M. Hernández-Pajares, P. Spalla, P. Moore, and M. A. King (2011) A Review of Higher Order Ionospheric Refraction Effects on Dual Frequency GPS. Surv. Geophys., 32:197–253.

Pohst, M. (1981) On the Computation of Lattice Vectors of Minimal Length, Successive Minima and Reduced Basis with Applications. ACM SIGSAM Bull., 15:37–44.

Pope, A. J. (1971) Transformation of Covariance Matrices Due to Changes in Minimal Control. Paper presented at the American Geophysical Union Fall Meeting, San Francisco.

Pope, A. J. (1976) The Statistics of Residuals and the Detection of Outliers. NOAA TR NOS. NGS 1. NOAA, Silver Spring, MD.

Popugaev, A., and R. Wansch (2014) Antenna Device for Transmitting and Receiving Electromagnetic Signals. Patent. U.S. 8624792 B2.

Poularikas, A. D. (2000) (Ed.) The Transforms and Applications Handbook, 2nd ed. CRC Press, Boca Raton, FL.

Povalyaev, A. (1997) Using Single Differences for Relative Positioning in GLONASS. Proc. ION GPS 1997. Institute of Navigation, Kansas City, MO, pp. 929–934.

Pozar, D. M., and D. H. Schaubert (Eds.) (1995) Microstip Antennas. IEEE, Wiley, New York.

Pozzobon, O., C. Wullems, M. Detratti, T. Stansell, K. Hudnut, and R. Keegan (2011) Future Wave. GPS World, April:30–41.

Pratt, M., B. Burke, and P. Misra (1997) Single Epoch Integer Ambiguity Resolution with GPS-GLONASS L1 Data. Proc. 53rd Annual Meeting of the Institute of Navigation. Albuquerque, NM, pp. 691–699.

Qin, X., S. Gourevitch, and M. Kuhl (1992) Very precise GPS - development status and results. *Proc. ION-GPS-92*, Institute of Navigation, 615–624.

QZSS (2013) Japan Aerospace Exploration Agency. Quasi-Zenith Satellite System Navigation Service. Interface Specification for QZSS (IS-QZSS), V1.5, March. Available online. www.qzss.jaxa.jp.

Raby, P., and P. Daly (1993) Using the GLONASS System for Geodetic Surveys. Proc. ION GPS 1993. Salt Lake City, UT, pp. 1129–1138.

Radicella, S. M. (2009) The NeQuick Model Genesis, Uses and Evolution. Ann. Geophys., 52(3):417–422.

Rao, B. R., and E. N. Rosario (2011) Electro-Textile Ground Planes for Multipath and Interference Mitigation in GNSS Antennas Covering 1.1 to 1.6 GHz. Proc. ION GNSS 2011. Institute of Navigation, Portland, OR, pp. 732–745.

Rao, B. R., W. Kunysz, P. Fante, and K. McDonald (2013) GPS/GNSS Antennas, Artech House, Norwood, MA.

Rapoport, L. (1997) General-Purpose Kinematic/Static GPS/GLONASS Post-Processing Engine. Proc. ION GPS 1997. Kansas City, MO, pp. 1757–1772.

Remondi, B. W. (1984) Using the Global Positioning System (GPS) Phase Observable for Relative Geodesy: Modeling, Processing, and Results. NOAA, reprint of doctoral dissertation. Center for Space Research, University of Texas at Austin, Austin, TX.

Remondi, B. W. (1985) Performing Centimeter-Level Surveys in Seconds with GPS Carrier Phase: Initial Results. Navigation, 32(4):386–400.

Rigden, G. J., and J. R. Elliott (2006) 3dM—A GPS Receiver Antenna Site Evaluation Tool. Proc. ION NTM 2006. Institute of Navigation, Monterey, CA, pp. 554–563.

Rojas, R. G., D. Colak, M. F. Otero, and W. D. Burnside (1995) Synthesis of Tapered Resistive Ground Plane for a Microstrip Antenna. Antennas and Propagation Society International Symposium, AP-S. Digest, pp. 1224–1227.

Rosenkranz, P. W. (1998) Water Vapor Microwave Continuum Absorption: A Comparison of Measurement and Models. Radio Science, 33(4):919–928.

Roßbach, U. (2001) Positioning and Navigation Using the Russian Satellite System GLONASS. Publication 70. University FAF Munich, Section Geodesy and Geo Information, Munich.

Rothacher, M., S. Schaer, L. Mervart, G. Beutler (1995) Determination of Antenna Phase Center Variations Using GPS Data. IGS Workshop, May 15–17, Potsdam, Germany. Available online. http://igscb.jpl.nasa.gov.

Ruland, R., and A. Leick (1985) Application of GPS to a High Precision Engineering Survey Network. Proc. Positioning with GPS-1985. NGS, Rockville, MD, pp. 483–493.

Saastamoinen, J. (1972) Atmospheric Correction for the Troposphere and Stratosphere in Radio Ranging of Satellites. Geophysical Monograph 15. Use of Artificial Satellites for Geodesy. American Geophysical Union, Washington, DC, pp. 247–251.

Sardón, E., A. Rius, and N. Zarraoa (1994) Estimation of the Transmitter and Receiver Differential Biases and the Ionospheric Total Electron Content from Global Positioning System Observations. Radio Sci., 29(3):577–586.

Schmitz, M., G. Wübbena, and G. Boettcher. (2002) Test of Phase Center Variations of Various GPS Antennas and Some Results. GPS Solutions, 6(1,2):18–27.

Schnorr, C. (1987) A Hierarchy of Polynomial Time Lattice Reduction Algorithms. Theor. Comput. Sci. 53(2–3):201–224.

Schnorr, C., and M. Euchner (1994) Lattice Basis Reduction: Improved Practical Algorithms and Solving Subset Sum Problems. Math. Program, 66:181–191.

Schomaker, M. C., and R. M. Berry (1981) Geodetic Leveling. NOAA Manual 3. NGS, Rockville, MD.

Schrijver, A. (1986) Theory of Linear and Integer Programming. Wiley, New York.

Schüler, T. (2001) On Ground-Based GPS Tropospheric Delay Estimation. Schriftenreihe, Vol. 73. Studiengang Geodäsie und Geoinformation, Univeristät der Bundeswehr München.

Schüler, T. (2014) Single-Frequency Single-Site VTEC Retrieval Using the NeQuick Ray Tracer of Obliquity Factor Determination. GPS Solutions,18(1):115–122.

Schupler, B. R., and T. A. Clark (2000) High Accuracy Characterization of Geodetic GPS Antennas Using Anechoic Chamber and Field Tests. Proc. ION GPS 2000. Institute of Navigation, Salt Lake City, UT, pp. 2499–2505.

Schwarz, C. R., and E. B. Wade (1990) North American Datum of 1983: Project Methodology and Execution. Bull. Geod. 64:28–62.

Sciré-Scappuzzo, F., and S. N. Makarov (2009) A Low-Multipath Wideband GPS Antenna with Cutoff or Non-Cutoff Corrugated Ground Plane. IEEE Trans. AP, 57(1): 33–46.

SDCM (2012) SDCM Interface Control Document, Radiosignals and Digital Data Structure of GLONASS Wide Area Augmentation System, System of Differential Correction and Monitoring, Edition 1. Available: www.sdcm.ru.

Seeber, G., and G. Wübbena (1989) Kinematic Positioning with Carrier Phases and "On-the-Way" Ambiguity Resolution. Proc. 5th International Geodetic Symposium on Satellite Positioning. DMA, Las Cruces, NM, pp. 600–609.

Shi, C., Q. Zhao, Z. Hu, and J. Liu (2013) Precise Relative Positioning Using Real Tracking Data from COMPASS GEO and IGSO Satellites. GPS Solutions, 17(1):103–119.

Sigl, R. (1977) Ebene und Sphärische Trigonometrie mit Anwendungen auf Kartographie, Geodäsie und Astronomie. Herbert Wichmann Verlag, Karlsruhe, Germany.

Sievenpiper, D., L. Zhang, R. F. J. Broas, N. G. Alexopoulos, and E. Yablonovitch (1999) High-Impedance Electromagnetic Surfaces with a Forbidden Frequency Band. IEEE Trans. MTT, 47(11):2059–2074.

Silin, R. A., and V. P. Sazonov (1971) Slow-Wave Structures. National Lending Library for Science and Technology, Boston Spa Eng.

Simsky, A. (2006) Three's the Charm. Triple-Frequency Combinations in Future GNSS. Inside GNSS, July/August:38–41.

Singer, R. A. (1970) Estimating Optimal Tracking Filter Performance for Manned Maneuvering Targets. IEEE Trans. Aerospace Electron. Syst., AES-6(4):473–483.

Snyder, J. P. (1979) Calculating Map Projections for the Ellipsoid. Amer. Cartogr., 6(1):67–76.

Snyder, J. P. (1982) Map Projections Used by the U.S. Geological Survey. Geological Survey Bulletin 1532. United States Printing Office, Washington, DC.

Sokolovskiy, S. V., W. S. Schreiner, Z. Zeng, D. C. Hunt, Y.-H. Kuo, T. K. Meehan, T. W. Stecheson, A. J. Mannucci, C. O. Ao (2013) Use of the L2C signal for Inversions of GPS Radio Occultation Data in the Neutral Atmosphere. GPS Solutions, doi:10.1007/s10291-013-0340-x.

Soler, T., and J. Marshall (2002) Rigorous Transformation of Variance-Covariance Matrices of GPS Derived Coordinates and Velocities. GPS Solutions, 6(2):76–90.

Soler, T., and B. H. W. van Gelder (1987) On Differential Scale Changes and the Satellite Doppler System Z-Shift. Geophys. J. R. Astr. Soc., 91(3):639–656.

Soler, T., J. Han, and N. Weston (2014) On Deflection of the Vertical Components and Their Transformations. J. Surv. Eng., 140(2), 04014005.

Solheim, F. S. (1993) Use of Pointed Water Vapor Radiometer Observations to Improve Vertical GPS Surveying Accuracy. Doctoral thesis, University of Colorado.

Soutiaguine, I., D. Tatarnikov, A. Astakhov, V. Filippov, A. Stepanenko (2004) Antenna Structures for Reducing the Effects of Multipath Radio Signals. Patent U.S. 6,836, 247 B2.

Spilker, J. J. (1996) Tropospheric Effects on GPS. GPS Theory Appl. 1:517–546.

Spliker, J. J. (1996) Tropospheric Effects on GPS. In Global Positioning System: Theory and Applications, Vol. I, Parkinson, B. W., J. J. Spilker, P. Axelrad and P. Enge (Eds.), American Institute of Aeronautics and Astronautics, Inc., Washington DC, pp. 517–546.

Springer, T. A., G. Beutler, and M. Rothacher (1999) A New Solar Radiation Pressure Model for GPS Satellites. GPS Solutions, 2(3):50–62.

SPS (2008) Global Positioning System Standard Positioning Service Performance Standard, 4th ed. Available www.gps.gov online.

Stem, J. E. (1989) State Plane Coordinate Systems of 1983. NOAA Manual No. NGS 5. NOAA, Silver Spring, MD.

Talbot, N. C. (1993) Centimeters in the Field, a Users Perspective of Real-Time Kinematic Positioning in a Production Environment. *Proc. ION-GPS-93*, Institute of Navigation, 589–598.

Tang, T., C. Deng, C. Shi, and J. Liu (2014) Triple Frequency Carrier Ambiguity Resolution for Beidou Navigation Satellite System. GPS Solutions. (Available online.) Accepted for publication; www.SpringerLink.com.

Tatarnikov, D., I. Soutiaguine, V. Filippov, A. Astakhov, A. Stepanenko, and P. Shamatulsky (2005) Multipath Mitigation by Conventional Antennas with Ground Planes and Passive Vertical Structures, GPS Solutions, 9(3):194–201.

Tatarnikov, D., A. Astakhov, P. Shamatulsky, I. Soutiaguine, and A. Stepanenko (2008a) Patch Antenna with Comb Substrate. EP Patent 1684381 B1. U.S. Patent 7,710,324 B2.

Tatarnikov, D. (2008b) Patch Antennas with Artificial Dielectric Substrates, Antennas, 1(128):35–45, Radiotechnika, Moscow, (in Russian).

Tatarnikov, D. (2008c) Ground Planes of Antennas for High Precision Satellite Positioning. Part 1. Flat Conductive and Impedance Ground Planes. Antennas, 4 (131):6–19, Radiotechnika, Moscow, (in Russian).

Tatarnikov, D. (2008d). Ground Planes of Antennas for High Precision Satellite Positioning. Part 2. Semi-Transparent Ground Planes. Antennas, 6(131):3–13, Radiotechnika, Moscow, (in Russian).

Tatarnikov, D. (2009) Enhanced Bandwidth Patch Antennas with Artificial Dielectrics Substrates for High Precision Satellite Positioning. IWAT IEEE International Workshop on Antenna Technology Small Antennas and Novel Metamaterials, March, Santa Monica, CA.

Tatarnikov, D., A. Astakhov, and A. Stepanenko (2009a) Broadband Volumetric Patch Antennas. Antennas, 3:52–58 Radiotechnika, Moscow, (in Russian).

Tatarnikov, D., A. Astakhov, A. Stepanenko, P. Shamatulsky, S. Yemelianov, and I. Soutiaguine (2009b) Topcon Full Wave RTK Antennas Based on Artificial Dielectric Technology. Proc. ION GNSS 2009. Institute of Navigation, Savannah, GA, pp. 420–424.

Tatarnikov, D., A. Astakhov, and A. Stepanenko (2011a) Broadband Convex Impedance Ground Planes for Multi-System GNSS Reference Station Antennas. GPS Solutions, 15(2):101–108.

Tatarnikov, D., A. Astakhov, and A. Stepanenko (2011b) GNSS Reference Station Antenna with Convex Impedance Ground Plane: Basics of Design and Performance Characterization Proc. ION ITM 2011. Institute of Navigation, San Diego, CA, pp. 1240–1245.

Tatarnikov, D. (2012) Semi-Transparent Ground Planes Excited by Magnetic Line Current. IEEE Trans. AP, 60(6):2843–2852.

Tatarnikov, D., and A. Astakhov (2012) Spherical Array Antenna with Π-shaped Pattern and Reduced Fields Intensities in Shadow Region. Proc. VI Russian National Conf. "Radars and Communication Systems." IRE RAN, Moscow, November 3–6 (in Russian).

Tatarnikov, D., and A. Astakhov (2013) Large Impedance Ground Plane Antennas for mm-Accuracy of GNSS Positioning in Real Time. Progress in Electromagnetics Research Symposium, August, Stockholm, Sweden, Proc. PIERS, pp. 1825–1829.

Tatarnikov, D., and A. Astakhov (2014) Approaching Millimeter Accuracy of GNSS Positioning in Real Time with Large Impedance Ground Plane Antennas. Proc. ION ITM 2014. Institute of Navigation, San Diego, CA, pp. 844–848.

Tatarnikov, D., A. Astakhov, and A. Stepanenko (2013a) Broadband Convex Ground Planes for Multipath Rejection. U.S. Patent 8,441,409 B2.

Tatarnikov, D., A. Astakhov, A. Stepanenko, and P. Shamatulsky (2013b) Patch Antenna with Capacitive Elements. U.S. Patent 8,446,322 B2.

Tatarnikov, D., A. Stepanenko, A. Astakhov, and V. Filippov (2013c) Compact Circular Polarized Antenna with Expanded Frequency Bandwidth. EP Patent 2335316 B1.

Tetewsky, A. K., and F. E. Mullen (1997) Carrier Phase Wrap-up Induced by Rotating GPS Antennas. GPS World, 8(2):51–57.

Tetewsky, A., J. Ross, A. Soltz, N. Vaughn, J. Anszperger, C. O'Brian, D. Graham D. Craig, and J. Lozow (2009) Making Sense of Inter-Signal Corrections. InsideGNSS, July/August:37–47.

Teunissen, P. J. G. (1993) Least Squares Estimation of Integer GPS Ambiguities. Invited Lecture, Section IV Theory and Methodology. IAG General Meeting, Beijing, China, August 1993. Available online: http://pages.citg.tudelft.nl/fileadmin/Faculteit/CiTG/Over_de_faculteit/Afdelingen/Afdeling_Geoscience_and_Remote_Sensing/pubs/PT_BEIJING93.PDF.

Teunissen, P. J. G. (1994) A New Method for Fast Carrier Phase Ambiguity Resolution. Proc. IEEE Position, Location and Navigation Symposium PLANS'94, pp. 662–673. available online at http://www.academia.edu/661699/_1994_A_new_method_for_fast_carrier_phase_ambiguity_estimation a general URL might be www.IEEE.org, but I could not find anything there.

Teunissen, P. J. G. (1997) On the Widelane and Its Decorrelating Property. J. Geodesy, 71(9):577–587.

Teunissen, P. J. G. (1998) Success Probability of Integer GPS Ambiguity Rounding and Bootstrapping. J. Geodesy, 72(10):606–612.

Teunissen, P. J. G. (1999) An Optimality Property of the Integer Least-Squares Estimator. J. Geodesy, 73(5):587–593.

Teunissen, P. J. G. (2003) Integer Aperture GNSS Ambiguity Resolution. Artif. Satellites, 38(3):79–88.

Teunissen, P. J. G., and S. Verhagen (2007) On GNSS Ambiguity Acceptance Tests. Proc. IGNSS Symposium 2, University of New South Wales, Sydney, Australia, December 4–6.

Teunissen, P. J. G, O. Odijk, and B. Zhang (2010) PPP-RTK: Results of CORS Network-Based PPP with Integer Ambiguity Resolution. Journal of Aeronautics, Astronautics and Aviation. Series A, 42(4):223–230.

Thayer, G. D. (1974) An Improved Equation for Radio Refractive Index of Air. Radio Sci., 9(10):803–807.

Thoelert, S., Montenbruck O, and M. Meurer (2014) IRNSS-1A — Signal and Clock Characterization of the Indian Regional Navigation System. GPS Solutions, 18(1): 147–152.

Thomas, P. D. (1952) Conformal Projections in Geodesy and Cartography. Special Publication No. 251. U.S. Department of Commerce, washington, DC.

Thornberg, D. B., D. S. Thornberg, M. F. Dibenedetto, M. S. Braasch, F. van Graas, and C. Bartone (2003) LAAS Integrated Multipath-Limiting Antenna, Navigation, 50(2): 117–130.

Timoshin, V. G., and A. M. Soloviev (2000) Microstrip Antenna with an Edge Ground Structure. U.S. Patent 6049309.

Tranquilla, J. M., J. P. Carr, and H. M. Al-Rizzo (1994) Analysis of a Choke Ring Ground Plane for Multipath Control in Global Positioning System (GPS) Applications, IEEE Proc. AP, 42(7):905–911.

Tretyakov, S. (2003) Analytical Modeling in Applied Electromagnetics. Artech House, Norwood, MA.

van Trees, H. L. (2002) Optimum Array Processing: Part IV of Detection, Estimation, and Modulation Theory. Wiley, Hoboken, NJ.

Tsui, J. B. Y. (2005) Fundamentals of Global Positioning System Receivers: A Software Approach. Wiley, Hoboken, NJ.

Ufimtsev, P. Ya. (2003) Theory of Edge Diffraction in Electromagnetics. Tech. Science Press, Encino, CA.

Ufimtsev, P. Ya. (1962) Method of Edge Waves in the Physical Theory of Diffraction. Foreign Technology Division, Wright-Patterson AFB, OH.

Vanderbei, R. J. (2008) Linear Programming: Foundations and Extensions, 3rd ed. International Series in Operations Research & Management Science, Vol. 114. Springer Verlag. Heidelberg, Germany.

Vaniček, P., and D. E. Wells (1974) Positioning of Horizontal Geodetic Datums. Can. Surv., 28(5):531–538.

Veis, G. (1960) Geodetic Use of Artificial Satellites. Smithsonian Contrib. Astrophys., 3(9):95–161.

Veitsel, V. A., A. V. Zhdanov, and M. I. Zhodzishsky (1998) The Mitigation of Multipath Errors by Strobe Correlators in GPS/GLONASS Receivers. GPS Solutions, 2(2):38–45.

Verhagen, S. (2004) Integer Ambiguity Validation: An Open Problem? GPS Solutions, 8(1):36–43.

Verhagen, S., and P. J. G. Teunissen (2006a) On the Probablility Density Function of the GNSS Ambiguity Residuals. GPS Solutions, 10(1):21–28.

Verhagen, S., and P. J. G. Teunissen (2006b) New Global Navigation Satellite System Ambiguity Resolution Method Compared to Existing Approaches. J. Guidance Control Dyn., 29(4):981–991.

Vincenty, T. (1979) The HAVAGO Three-dimensional Adjustment Program. NOAA TM NOS-NGS 17. NOAA, Silver Spring, MD.

Vollath, U., S. Birnbach, H. Landau, J. M. Fraile-Ordonez, and M. Martin-Neira (1998) Analysis of Three-Carrier Ambiguity Resolution (TCAR) Technique for Precise Relative Positioning in GNSS-2. Proc. ION GPS 1998. Nashville, TN, pp. 417–426.

Vollath, U., A. Buecherl, H. Landau, C. Pagels, and B. Wagner (2000) Multi-Base RTK Positioning Using Virtual Reference Stations. Proc. ION GPS 2000. Salt Lake City, UT, pp. 123–131.

Wang, J., Y. Feng, and C. Wang (2010) A Modified Inverse Integer Cholesky Decorrelation Method and Performance on Ambiguity Resolution. J. Global Position. Syst., 9(2):156–165.

Wang, J., C. Rizos, M. P. Stewart, and A. Leick (2001) GPS and GLONASS Integration: Modeling and Ambiguity Resolution Issues. GPS Solutions, 5(1):55–64.

Wang, J., M. P. Stewart, and M. Tsakiri (1998) A Discrimination Test Procedure for Ambiguity Resolution On-the-Fly. J. Geodesy, 72(11):644–653.

Wang, Z., Y. Wu, K. Zhang, and Y. Meng (2005) Triple-Frequency Method for High-Order Ionospheric Refractive Error Modelling in GPS Modernization. J. Global Position. Syst. 4(1):291–295.

Wanninger, L. (1997) Real-Time Differential GPS Error Modeling in Regional Reference Station Networks. Proc. IAG Scientific Assembly, Rio de Janeiro, September. IAG Symposia 118, Springer Verlag, 86–92.

Waterhouse, R. B. (2003) Microstrip Patch Antennas: A Designers Guide. Springer Science. Kluwer Academic Publishers, Norwell, MA.

Webb, D. F., and R. A. Howard (1994) The Solar Cycle Variation of Coronal Mass Ejections and the Solar Wind Mass Flux. J. Geophys. Res., 99(A4):4201–4220.

Weiss, J. P., P. Axelrad, and S. Anderson (2008) A GNSS Code Multipath Model for Semi-Urban, Aircraft, and Ship Environments. Navigation, 54(4):293–307.

Westfall, B. G. (1997) Antenna with R-Card Ground Plane. U.S. Patent 5,694,136.

Westfall, B. G., and K. B. Stephenson (1999) Antenna with Ground Plane Having Cutouts, U.S. Patent 5,986,615.

Westwater, E. R. (1978) The Accuracy of Water Vapor and Cloud Liquid Determination by Dual-Frequency Ground-Based Microwave Radiometry. Radio Sci., 13(4):677–685.

Whitehead, M. L., G. Penno, W. J. Feller, I. C. Messinger, W. L. Bertiger, R. J. Muellerschoen, B. A. Iijima, and G. Piesinger (1998) A Close Look at Satloc's Real-Time WADGPS System. GPS Solutions, 2(2):45–63.

World Meteorological Organization (WMO) (1961) Guide to Meteorological Instrument and Observing Practices, 2nd ed., No. 8. WHO, Geneva.

Wolf, H. (1963) Die Grundgleichungen der dreidimensionalen Geodäsie in elementarer Darstellung. Zeitschrift für Vermessungswesen, 88(6):225–233.

Wolf, H. (1978) The Helmert Block Method—Its Origin and Development. Proc. Second International Symposium on Problems Related to the Redefinition of the North American Geodetic Networks, April 24–28. NOAA, Rockville, MD, pp. 319–326.

Wu, J. T., S. C. Wu, G. A. Hajj, W. I. Bertiger, and S. M. Lichen (1993) Effects of Antenna Orientation on GPS Carrier Phase. Manuscripta Geodaetica, 18(2):91–98.

Wu, Y., S. G. Jin, Z. M. Wang, and J. B. Liu (2010) Cycle Slip Detection Using Multi-Frequency GPS Carrier Phase Observations: A Simulation Study. Adv. Space Res., 46:144–149.

Wübbena, G. (1990) Zur Modellierung von GPS-Beobachtungen für die hochgenaue Positionsbestimmung. Dissertation, Universität Hannover.

Wübben, W., D. Seethaler, J. Jaldén, and G. Matz (2011) Lattice Reduction: A Survey with Applications in Wireless Communications. IEEE Signal Process. Maga., 28(3):70–91.

Wübbena, G. (1985) Software Developments for Geodetic Positioning with GPS Using TI-4100 Code and Carrier Measurements. Proc. Precise Positioning with GPS. Rockville, MD, National Geodetic Survey (NGS) pp. 402–412.

Wübbena, G., F. Menge, M. Schmitz, G. Seeber, and C. Völksen (1996) A New Approach for Field Calibration of Absolute Antenna Phase Center Variations. Proc. ION GPS 1996. Institute of Navigation, Kansas City, MO, pp. 1205–1214.

Wübbena, G., M. Schmitz, F. Menge, V. Böder, and G. Seeber (2000) Automated Absolute Field Calibration of GPS Antennas in Real Time. Proc. ION GPS 2000. Institute of Navigation, Salt Lake City, UT, pp. 2512–2522.

Wübbena, G., M. Schmitz, and A. Prüllage (2011) On GNSS Station Calibration of Near-Field Multipath in RTK Networks Int.Symp. on GNSS, Space-Based and Ground-Based Augmentation Systems and Applications, October, Berlin, Germany http://www.geopp.de/media/docs/pdf/gpp_gnss11_conf_i.pdf.

Yang, W. H. (1977) A Method for Updating Cholesky Factorization of a Band Matrix. Comput. Methods Appl. Mech. Engi., 12:281–288.

Zhao, Q., Z. Dai, Z. Hu, B. Sun, C. Shi and J. Liu (2014) Three-Carrier Ambiguity Resolution Using the Modified TCAR Method. GPS Solutions. Available online. www .SpringerLink.com.

Zebhauser, B. E., H. J. Euler, and C. R. Keenan (2002) A Novel Approach for the Use of Information from Reference Station Networks Conforming to RTCM V2.3 and Future V3.0. Proc. ION NTM 2002. Institute of Navigation, San Diego, CA, pp. 863–876.

Zhdanov, A. V., M. I. Zhodzishsky, V. A. Veitsel, and J. Ashjaee (2001) Evolution of Multipath Error Reduction with GPS Signal Processing. GPS Solutions, 5(1):19–28.

Zhou, Y., and Z. He (2013) Variance Reduction of GNSS Ambiguity in (Inverse) Paired Cholesky Decorrelation Transformation. GPS Solutions. Available online. 18(4):509–517.

Ziebart, M., P. Cross, and S. Adhya (2002) Modeling Photon Pressure: The Key to High-Precision GPS Satellite Orbits. GPS World, 13(1):43–50.

Zinoviev, A. E. (2005) Using GLONASS in Combined GNSS Receivers: Current Status. Proc. ION GNSS 2005. Long Beach, CA, pp. 1046–1057.

Zumberge, J. F. (1998) Automated GPS Data Analysis Service. GPS Solutions, 2(3):76–78.

Zumberge, J. F., M. B. Heflin, D. C. Jefferson, M. M. Watkins, and F. H. Webb (1998a) Precise Point Processing for the Efficient and Robust Analysis of GPS Data from Large Networks. J. Geophys. Res., 102(B3):5005–5017.

Zumberge, J. F., M. M. Watkins, and F. H. Webb (1998b) Characteristics and Application of Precise GPS Clock Solutions Every 30 Seconds. Navigation, 44(4):449–456.

AUTHOR INDEX

SUBJECT INDEX